MASS TRANSFER IN MULTIPHASE SYSTEMS AND ITS APPLICATIONS

Edited by **Mohamed El-Amin**

INTECHOPEN.COM

Mass Transfer in Multiphase Systems and its Applications
Edited by Mohamed El-Amin

CBS, Edition **2017**

Published by InTech
Janeza Trdine 9, 51000 Rijeka, Croatia

Publishing Process Manager Sandra Bakic

Technical Editor InTech DTP team

Cover Designer InTech Design Team

Additional hard copies can be obtained from orders@intechopen.com

Mass Transfer in Multiphase Systems and its Applications, Edited by Mohamed El-Amin
p. cm.
ISBN 978-953-307-215-9

INTECH

open science | open minds

Contents

Contents

Contents

Preface

This book covers a number of developing topics in mass transfer processes in multiphase systems for a variety of applications. The book effectively blends theoretical, numerical, modeling, and experimental aspects of mass transfer in multiphase systems that are usually encountered in many research areas such as chemical, reactor, environmental and petroleum engineering. From biological and chemical reactors to paper and wood industry and all the way to thin film, the 31 chapters of this book serve as an important reference for any researcher or engineer working in the field of mass transfer and related topics.

The first chapter focuses on the description and modeling of mass transfer processes occurring between two fluid phases in a porous medium, while the second chapter is concerned with the basic principles underlying transport phenomena and chemical reaction in single- and multi-phase systems in porous media. Chapter 3 introduces the multiphase modeling of thermomechanical behavior of early-age silicate composites. The surfactant transfer in multiphase liquid systems under conditions of weak gravitational convection is presented in Chapter 4.

In the fifth chapter the volumetric mass transfer coefficient for multiphase mechanically agitated gas–liquid and gas–solid–liquid systems is obtained experimentally. Further, gas-liquid mass transfer analysis in an unbaffled vessel agitated by unsteadily forward-reverse rotating multiple impellers is provided in Chapter 6. Chapter 7 discuses the kinetic model of flotation based on the theory of mass transfer in gas-liquid bubbly flows. The eighth chapter deals with experimental investigation of mass transfer and wall shear stress, and their interaction at the concurrent gas-liquid flow in a vertical tube, in a channel with flow turn, and in a channel with abrupt expansion. The laminar mixed convection with mass transfer and phase change of flow reversal in channels is studied in the ninth chapter, and the tenth chapter exemplifies the theoretical aspects of the liquid-liquid extraction with and without a chemical reaction and the dimensioning of the extractors with original experimental work and interpretations.

The eleventh chapter introduces analysis of the existing theories and concepts of solid-state diffusion mass transfer in metals during mechanical alloying. In Chapter 12 the mass transfer coefficient is given for different situations (liquid-liquid, liquid-gas and liquid-solid) of two-phase mass transfer of steelmaking processes. Chapter 13 discusses the effect of Marangoni Instability on thin liquid film, thinker liquid layer and mass transfer devices.

Chapter 14 provides a review of model permitting the determination of wood drying rate represented by an overall mass transfer coefficient and a driving force. Moreoever, transport phenomena in paper and wood-based panels' production are discussed in Chapter 15.

In the sixteenth chapter the effect of lutetium doping on oxygen permeability in polycrystalline alumina wafers exposed to steep oxygen potential gradients was evaluated at high temperatures to investigate the mass-transfer phenomena. Mass transfer investigation of organic acid extraction with trioctylamine and aliquat336 dissolved in various solvents is introduced in Chapter 17.

Chapter 18 is focused on the development of semi-theoretical methods for calculation of gas holdup, interfacial area and liquid-phase mass transfer coefficients in gas-liquid and slurry bubble column reactors. Chapter 19 investigates the influence of mass transfer and kinetics on biodiesel (fatty acid alkyl) production process.

The physical-mathematical model of heat and mass transfer and condensation capture of fine dust on fluid droplets dispersed in jet scrubbers is suggested and analyzed in Chapter 20, while Chapter 21 is devoted to investigate mass transfer in filtration combustion processes.

The twenty-second chapter describes the merits of using hollow fiber supported liquid membranes (HFSLM), one of liquid membranes in supported (not clear) structures, and how mass transfer involves step-by-step in removing arsenic (As) and mercury (Hg). The twenty-third chapter presents an overview of the various mechanisms contributing to particulate drying in a bubbling fluidized bed and the mass transfer coefficient corresponding to each mechanism. A mathematical model and numerical simulation for hydriding/dehydriding process in a tubular type MH reactor packed with LaNi5 were provided in the twenty-fourth chapter. Chapter 25 is dedicated to the mass transfer coefficient around freely moving active particles in the dense phase of a fluidized bed. Chapter 26 is aimed at reviewing transport across biological membranes, with an emphasis on intestinal absorption, its model analysis and permeability prediction. The objective of the twenty-seventh chapter is to provide comprehensive information on theoretical-experimental analysis of coupled heat and mass transfer in packed bed drying of shrinking particles. The twenty-eighth chapter focuses on vapour-liquid mass transfer influence on the prediction of RD column behaviour neglecting the liquid-solid and intraparticle mass transfer. Mass transfer through catalytic membrane layer is studied in Chapter 29. In chapter 30 three types of bioreactors and stirred tank applied to biological systems are introduced and a mathematical model is developed. Finally in Chapter 31 analytical solutions of mass transfer around a prolate or an oblate spheroid immersed in a packed bed are obtained.

Mohamed Fathy El-Amin
Physical Sciences and Engineering Division
King Abdullah University of Science and Technology (KAUST)

1

Mass and Heat Transfer During Two-Phase Flow in Porous Media - Theory and Modeling

Jennifer Niessner[1] and S. Majid Hassanizadeh[2]
[1]*Institute of Hydraulic Engineering, University of Stuttgart, Stuttgart*
[2]*Department of Earth Sciences, Faculty of Geosciences, Utrecht University, Utrecht*
[1]*Germany*
[2]*The Netherlands*

1. Introduction

1.1 Motivation

This chapter focusses on the description and modeling of mass transfer processes occurring between two fluid phases in a porous medium. The principle underlying physical process comprises a transport of particles from one phase to the other phase. This process takes place across fluid–fluid interfaces (see Fig. 1) and may constitute evaporation, dissolution, or condensation, for example.

Such mass transfer processes are crucial in many applications involving flow and transport in porous media. Major examples are found in soil science (where the evaporation from soils is of interest), soil and groundwater remediation (like thermally-enhanced soil vapor extraction where dissolution, evaporation, and condensation play a role), storage of carbon dioxide in the subsurface (where the dissolution of carbon dioxide in the surrounding groundwater is a crucial storage mechanism), CO_2-enhanced oil recovery (where after primary and secondary recovery, carbon dioxide is injected into the reservoir in order to mobilize an additional 8-20 per cent of oil), and various industrial porous systems (such as certain types of fuel cells).

Let us have a closer look at a few of these applications and identify where interphase mass transfer is relevant. Four specific examples are shown in Fig. 2 and briefly described.

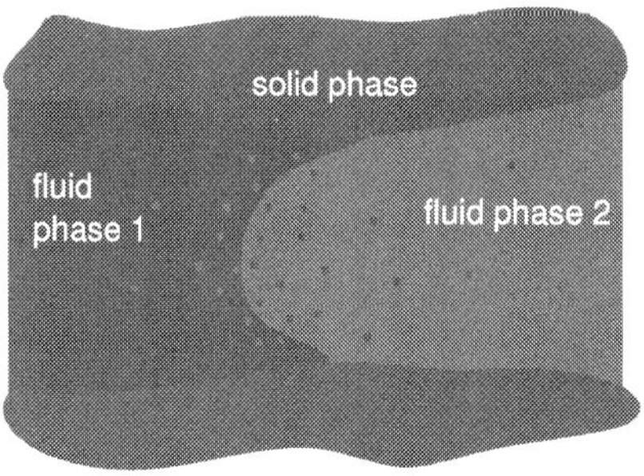

Fig. 1. Mass transfer processes (evaporation, dissolution, condensation) imply a transfer of particles across fluid–fluid interfaces.

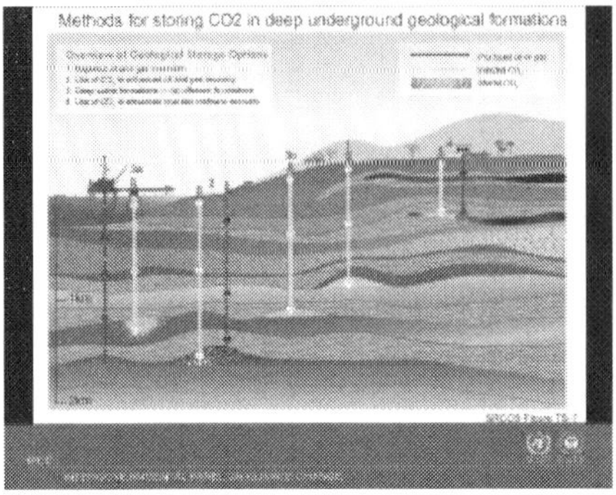

(a) Carbon dioxide storage in the subsurface (figure from IPCC (2005))

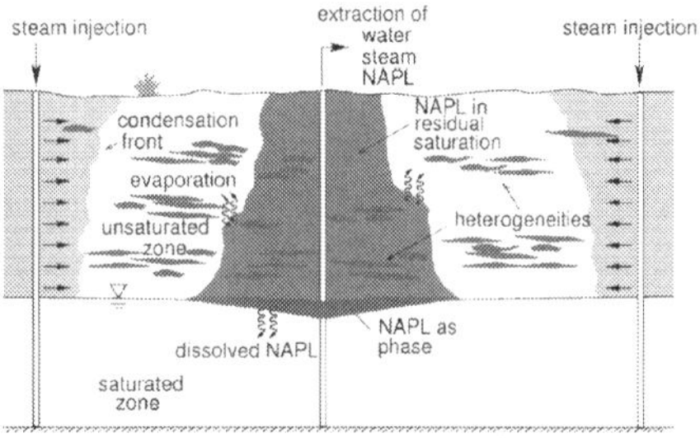

(b) Soil contamination and remediation

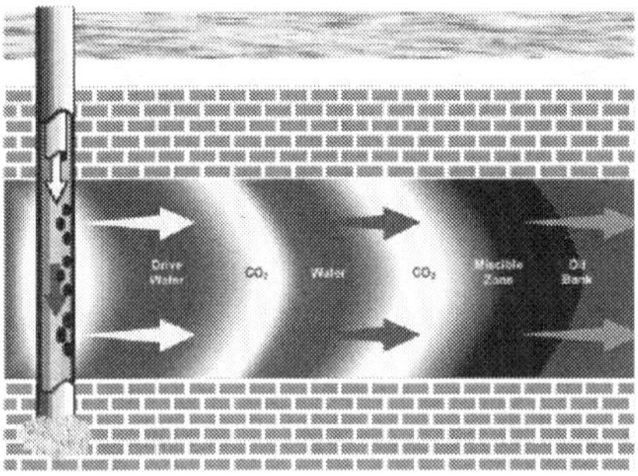

(c) Enhanced oil recovery (figure from `www.oxy.com`)

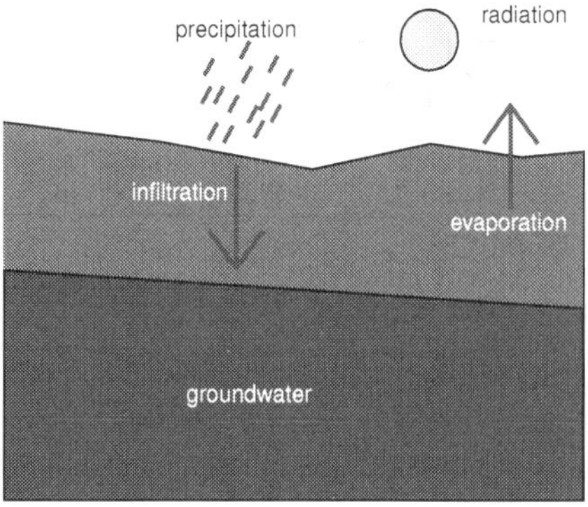

(d) Evaporation from soil

Fig. 2. Four applications of flow and transport in porous media where interphase mass transfer is important

(a) **Carbon capture and storage** (Fig. 2 (a)) is a recent strategy to mitigate the greenhouse effect by capturing the greenhouse gas carbon dioxide that is emitted e.g. by coal power plants and inject it directly into the subsurface below an impermeable caprock. Here, three different storage mechanisms are relevant on different time scales: 1) The capillary barrier mechanism of the caprock. This geologic layer is meant to keep the carbon dioxide in the storage reservoir as a separate phase. 2) Dissolution of the carbon dioxide in the surrounding brine (salty groundwater). This is a longterm storage mechanism and involves a mass transfer process as carbon dioxide molecules are "transferred" from the gaseous phase to the brine phase. 3) Geochemical reactions which immobilize the carbon dioxide through incorporation into the rock matrix.

(b) Shown in Fig. 2 (b) is a cartoon of a light **non-aqueous phase liquid (LNAPL) soil contamination** and its clean up by injection of steam at wells located around the contaminated soil. The idea behind this strategy is to mobilize the initially immobile (residual) LNAPL by evaporation of LNAPL component at large rates into the gaseous phase. The soil gas is then extracted by a centrally located extraction well. It means that the remediation mechanism relies on the evaporation of LNAPL component which represents a mass transfer from the liquid LNAPL phase into the gaseous phase.

(c) In order to produce an additional 8-20% of oil after primary and secondary recovery, carbon dioxide may be injected into an oil reservoir, e.g. alternatingly with water, see Fig. 2 (c). This is called **enhanced oil recovery**. The advantage of injecting carbon dioxide lies in the fact that it dissolves in the oil which in turn reduces the oil viscosity, and thus, increases its mobility. The improved mobility of the oil allows for an extraction of the otherwise trapped oil. Here, an interphase mass transfer process (dissolution) is responsible for an improved recovery.

(d) The last example (Fig. 2 (d)) shows the upper part of the soil. The water balance of this part of the subsurface is extremely important for agriculture or plant growth in general. Plants do not grow well under too wet or too dry conditions. One of the very important factors influencing this water balance (besides surface runoff and infiltration) is the **evaporation of water from the soil**, which is again an interphase mass transfer process.

1.2 Purpose of this work

Mass transfer processes are essential in a large variety of applications—the presented examples only show a small selection of systems. A common feature of all these applications is the fact that the relevant processes occur in relatively large domains such that it is not possible to resolve the pore structure and the fluid distribution in detail (left hand side of Fig. 3). Instead, a macro-scale approach is needed where properties and processes are averaged over a so-called representative elementary volume (right hand side of Fig. 3). This means that the common challenge in all of the above-mentioned applications is how to describe mass transfer processes on a macro scale. This transition from the pore scale to the macro scale is illustrated in Fig. 3 where on the left side, the pore-scale situation is shown (which is impossible to be resolved in detail) while on the right hand side, the macro-scale situation is shown.

2. Overview of classical mass transfer descriptions

2.1 Pore-scale considerations

In order to better understand the physics of interphase mass transfer, which is essential to provide a physically-based description of this process, we start our considerations on the pore

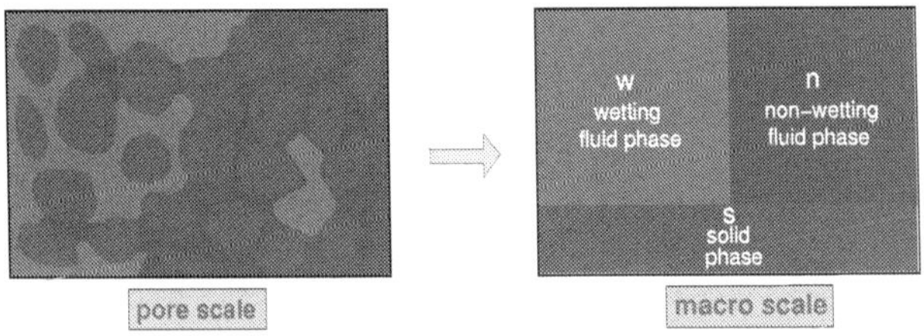

Fig. 3. Pore-scale versus macro-scale description of flow and transport in a porous medium.

scale. From there, we try to get a better understanding of the macro-scale physics of mass transfer, which is our scale of interest.

In Fig. 1 we have seen that interphase mass transfer is inherently a pore-scale process as it—naturally—takes place across fluid–fluid interfaces. Let us imagine a situation where two fluid phases, a wetting phase and a non-wetting phase, are brought in contact as shown in Fig. 4. Commonly, when the two phases are brought in contact (time $t = t_0$), equilibrium is quickly established directly at the interface. With respect to mass transfer, this means that the concentration of non-wetting phase particles in the wetting phase at the interface as well as the concentration of wetting-phase particles in the non-wettting phase at the interface are both at their equilibrium values, $C^2_{1,eq}$ and $C^1_{2,eq}$. At $t = t_0$, away from the interface, there is still no presence of α-phase particles in the β-phase. At a later time $t = t_1$, concentration profiles develop within the phases. However, within the bulk phases, the concentrations are still different from the respective equilibrium concentration at the interface. Considering a still later point of time, $t = t_2$, the equilibrium concentration is finally reached everywhere in the bulk phases.

These considerations show that mass transfer on the pore scale is inherently a kinetic process that is very much related to phase-interfaces. But how is this process represented on the macro scale, i.e. on a volume-averaged scale? This is what we will focus on in the next section.

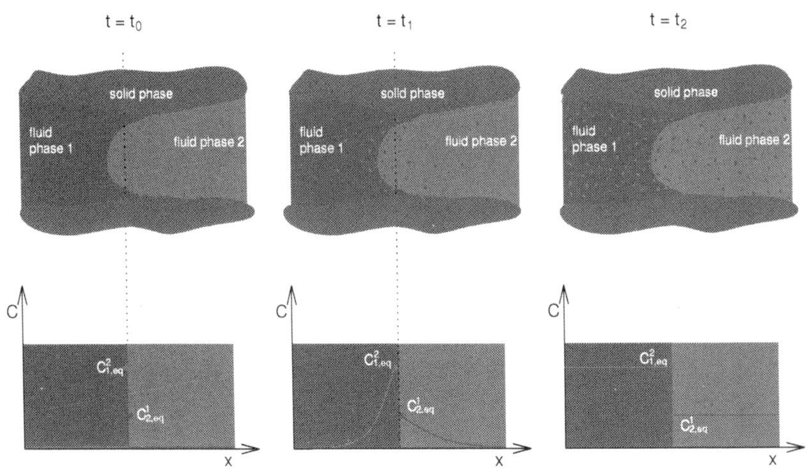

Fig. 4. Pore-scale picture of interphase mass transfer.

2.2 Current macro-scale descriptions

In Fig. 3, we illustrated the fact that when going from the pore scale to the macro scale, information about phase-interfaces is lost. The only information present on the macro scale is related to volume ratios, such as porosity and fluid saturations. But, as mentioned earlier, mass transfer is strongly linked to the presence of interfaces and interfacial areas and all the information about phase-interfaces disappears on the macro scale. This means that the description of mass transfer on the macro scale is not straight forward. Classical approaches for describing mass transfer generally rely on one of the following two principles:

1. assumption of local chemical equilibrium within an averaging volume or
2. kinetic description based on a fitted (linear) relationship.

These two classical approaches will be discussed in more detail in the following. An alternative approach which accounts for the phase-interfacial area will be presented later in Sec. 3.

2.2.1 Local equilibrium assumption

The assumption of local chemical equilibrium within an averaging volume means that the equilibrium concentrations are reached instantaneously everywhere within an averaging volume (in both phases). That means it is assumed that everywhere within the averaging volume, the situation at $t = t_2$ in Fig. 4 is reached from the beginning ($t = t_0$). Thus, at each point in the wetting phase and at each point in the non-wetting phase within the averaging volume, the equilibrium concentration of the components of the other phase is reached. This is an assumption which may be good in case of fast mass transfer, but bad in case of slow mass transfer processes. To be more precise, the assumption that the composition of a phase is at or close to equilibrium may be good if the characteristic time of mass transfer is small compared to that of flow. However, if large flow velocities occur as e.g. during air sparging, the local equilibrium assumption gives completely wrong results, see Falta (2000; 2003) and van Antwerp et al. (2008). We will investigate and quantify this issue later in Sec. 3.3

.

Local equilibrium models for multi-phase systems have been introduced and developed e.g. by Miller et al. (1990); Powers et al. (1992; 1994); Imhoff et al. (1994); Zhang & Schwartz (2000) and have been used and advanced ever since. Let us consider a system with a liquid phase (denoted by subscript l) and a gaseous phase (denoted by subscript g) composed of air and water components. Then, Henry's Law is employed to determine the mole fraction of air in the liquid phase, while the mole fraction of water in the gas phase is determined by assuming that the vapor pressure in the gas phase is equal to the saturation vapor pressure. Denoting the water component by superscript w and the air component by superscript a, this yields

$$x_l^a = p_g^a \cdot H_{l-g}^a \tag{1}$$

$$x_g^w = \frac{p_{sat}^w}{p_g}, \tag{2}$$

where x_l^a $[-]$ is the mole fraction of air in the liquid phase, x_g^w $[-]$ is the mole fraction of water in the gaseous phase, H_{l-g}^a $\left[\frac{1}{Pa}\right]$ is the Henry constant for the dissolution of air in the liquid phase, p_{sat}^w $[Pa]$ is the saturation vapor pressure of water, p_g^a $[Pa]$ is the partial pressure of air in the gas phase while p_g $[Pa]$ is the gas pressure. The remaining mole fractions result simply from the condition that mole fractions in each phase have to sum up to one,

$$x_l^w = 1 - x_l^a \tag{3}$$

$$x_g^a = 1 - x_g^w . \tag{4}$$

Note that while for a number of applications the equilibrium mole fractions are constants or merely a function of temperature, in our case, they will be functions of space and time as pressure and the composition of the phases changes.

2.2.2 Classical kinetic approach

Kinetic mass transfer approaches are traditionally applied to the dissolution of contaminants in the subsurface which form a separate phase from water, the so-called non-aqueous phase liquids (NAPLs). If such a non-aqueous phase liquid is heavier than water, it is called "dense non-aqueous phase liquid" or DNAPL. When an immobile lense of DNAPL is present at residual saturation (i.e. at a saturation which is so low that the phase is immobile) and dissolves into the surrounding groundwater, the kinetics of this mass transfer process usually plays an important role: the dissolution of DNAPL is a rate-limited process. This is also the case when a pool of DNAPL is formed on an impermeable layer. In these relatively simple cases, only the mass transfer of a DNAPL component from the DNAPL phase into the water phase has to be considered. For these cases, classical models acknowledge the fact that the rate of mass transfer is highly dependent (proportional to) interfacial area and assume a first-order rate of kinetic mass transfer between fluid phases in a porous medium on a macroscopic (i.e. volume-averaged) scale which can be expressed as (see e.g. Mayer & Hassanizadeh (2005)):

$$Q^{\kappa}_{\alpha\to\beta} = k^{\kappa}_{\alpha\to\beta} a_{\alpha\beta} (C^{\kappa}_{\beta,s} - C^{\kappa}_{\beta}), \tag{5}$$

where $Q^{\kappa}_{\alpha\to\beta}$ $\left[\frac{kg}{m^3 s}\right]$ is the interphase mass transfer rate of component κ from phase α to phase β, $k^{\kappa}_{\alpha\to\beta}$ $\left[\frac{m}{s}\right]$ is the mass transfer rate coefficient, $a_{\alpha\beta}$ $\left[\frac{1}{m}\right]$ is the specific interfacial area separating phases α and β, $C^{\kappa}_{\beta,s}$ $\left[\frac{kg}{m^3}\right]$ is the solubility limit of component κ in phase β, and finally, C^{κ}_{β} $\left[\frac{kg}{m^3}\right]$ is the actual concentration of component κ in phase β. The actual concentration is not larger than the solubility limit, $C^{\kappa}_{\beta} \leq C^{\kappa}_{\beta,s}$. The case $C^{\kappa}_{\beta} = C^{\kappa}_{\beta,s}$ corresponds to the local equilibrium case.

In the absence of a physically-based estimate of interfacial area in classical kinetic models, the mass transfer coefficient $k^{\kappa}_{\alpha\to\beta}$ and the specific interfacial area $a_{\alpha\beta}$ are often lumped into one single parameter (Miller et al. (1990); Powers et al. (1992; 1994); Imhoff et al. (1994); Zhang & Schwartz (2000)). This yields, in a simplified notation,

$$Q = k(C_s - C). \tag{6}$$

Here, C_s is the solubility limit of the DNAPL component in water and C is its actual concentration. The lumped mass transfer coefficient k $\left[\frac{1}{s}\right]$ is commonly related to a modified Sherwood number Sh by

$$k = Sh \frac{D_m}{d_{50}^2}, \tag{7}$$

where D_m $\left[\frac{m^2}{s}\right]$ is the aqueous phase molecular diffusion coefficient, and d_{50} $[m]$ is the mean size of the grains. The Sherwood number is then related to Reynold's number Re and DNAPL saturation S_n $[-]$ by

$$Sh = pRe^q S_n^r, \tag{8}$$

where p, q, and r are dimensionless fitting parameters. This is a purely empirical relationship. Although interphase mass transfer is proportional to specific interfacial area in the original Eq. (5), this dependence cannot explicitly be accounted for as the magnitude of specific interfacial area is not known.

An alternative classical approach for DNAPL pool dissolution has been proposed by Falta (2003) who modeled the dissolution of DNAPL component by a dual domain approach for a case with simple geometry. For this purpose, they divided the contaminated porous medium into two parts: one that contains DNAPL pools and one without DNAPL. For their simple case, the dual domain approach combined with an analytical solution for steady-state advection and dispersion provided a means for modeling rate-limited interphase mass transfer. While this approach provided good results for the case of simplified geometry, it might be oversimplified for the modeling of realistic situations.

3. Interfacial-area-based approach for mass transfer description

3.1 Theoretical background

Due to a number of deficiencies of the classical model for two-phase flow in porous media (one of which is the problem in describing kinetic interphase mass transfer on the macro scale), several approaches have been developed to describe two-phase flow in an alternative and thermodynamically-based way. Among these are a rational thermodynamics approach by Hassanizadeh & Gray (1980; 1990; 1993b;a), a thermodynamically constrained averaging theory approach by Gray and Miller (e.g. Gray & Miller (2005); Jackson et al. (2009)), mixture theory (Bowen (1982)) and an approach based on averaging and non-equilibrium thermodynamics by Marle (1981) and Kalaydjian (1987). While Marle (1981) and Kalaydjian (1987) developed their set of constitutive relationships phenomenologically, Hassanizadeh & Gray (1990; 1993b); Jackson et al. (2009), and Bowen (1982) exploited the entropy inequality to obtain constitutive relationships. To the best of our knowledge, the two-phase flow models of Marle (1981); Kalaydjian (1987); Hassanizadeh & Gray (1990; 1993b); Jackson et al. (2009) are the only ones to include interfaces explicitly in their formulation allowing to describe hysteresis as well as kinetic interphase mass and energy transfer in a physically-based way. In the following, we follow the approach of Hassanizadeh & Gray (1990; 1993b) as it includes the spatial and temporal evolution of phase-interfacial areas as parameters which allows us to model kinetic interphase mass transfer in a much more physically-based way than is classically done.

It has been conjectured by Hassanizadeh & Gray (1990; 1993b) that problems of the classical two-phase flow model, like the hysteretic behavior of the constitutive relationship between capillary pressure and saturation, are due to the absence of interfacial areas in the theory. Hassanizadeh and Gray showed (Hassanizadeh & Gray (1990; 1993b)) that by formulating the conservation equations not only for the bulk phases, but additionally for interfaces, and by exploiting the residual entropy inequality, a relationship between capillary pressure, saturation, and specific interfacial areas (interfacial area per volume of REV) can be derived. This relationship has been determined in various experimental works (Brusseau et al. (1997); Chen & Kibbey (2006); Culligan et al. (2004); Schaefer et al. (2000); Wildenschild et al. (2002);

Chen et al. (2007)) and computational studies (pore-network models and CFD simulations on the pore scale, see Reeves & Celia (1996); Held & Celia (2001); Joekar-Niasar et al. (2008; 2009); Porter et al. (2009)). The numerical work of Porter et al. (2009) using Lattice Boltzman simulations in a glass bead porous medium and experiments of Chen et al. (2007) show that the relationship between capillary pressure, the specific fluid-fluid interfacial area, and saturation is the same for drainage and imbibition to within the measurement error. This allows for the conclusion that the inclusion of fluid–fluid interfacial area into the capillary pressure–saturation relationship makes hysteresis disappear or, at least, reduces it down to a very small value. Niessner & Hassanizadeh (2008; 2009a;b) have modeled two-phase flow—using the thermodynamically-based set of equations developed by Hassanizadeh & Gray (1990)—and showed that this interfacial-area-based model is indeed able to model hysteresis as well as kinetic interphase mass and also energy transfer in a physically-based way.

3.2 Simplified model

After having presented the general background of our interfacial-area-based model, we will now proceed by discussing the mathematical model. The complete set of balance equations based on the approach of Hassanizadeh & Gray (1990) is too large to be handled numerically. In order to do numerical modeling, simplifying assumptions need to be made. In the following, we present such a simplified equation system as was derived in Niessner & Hassanizadeh (2009a).

This set of balance equations can be described by six mass and three momentum balance equations. These numbers result from the fact that mass balances for each component of each phase and the fluid–fluid interface (that is 2×3) while momentum balances are given for the bulk phases and the interface. Governing equations were derived by Hassanizadeh & Gray (1979) and Gray & Hassanizadeh (1989; 1998) for the case of flow of two pure fluid phases with no mass transfer. Extending these equations to the case of two fluid phases, each made of two components, we obtain the following equations.

mass balance for phase components ($\kappa = w, a$):

$$\begin{aligned} &\frac{\partial \left(\phi S_l \bar{\rho}_l \bar{X}_l^\kappa\right)}{\partial t} + \nabla \cdot \left(\phi S_l \bar{\rho}_l \bar{X}_l^\kappa \underline{\bar{v}}_l\right) - \nabla \cdot \left(\phi S_l \underline{\bar{j}}_l^\kappa\right) \\ = \; & \bar{\rho}_l Q_l^\kappa + \frac{1}{V} \int_{A_{lg}} \left[\rho_l X_l^\kappa \left(\underline{v}_{lg} - \underline{v}_l\right) + \underline{j}_l^\kappa\right] \cdot \underline{n}_{lg} \, dA \end{aligned} \tag{9}$$

$$\begin{aligned} &\frac{\partial \left(\phi S_g \bar{\rho}_g \bar{X}_g^\kappa\right)}{\partial t} + \nabla \cdot \left(\phi S_g \bar{\rho}_g \bar{X}_g^\kappa \underline{\bar{v}}_g\right) - \nabla \cdot \left(\phi S_g \underline{\bar{j}}_g^\kappa\right) \\ = \; & \bar{\rho}_g Q_g^\kappa + \frac{1}{V} \int_{A_{lg}} \left[\rho_g X_g^\kappa \left(\underline{v}_g - \underline{v}_{lg}\right) - \underline{j}_g^\kappa\right] \cdot \underline{n}_{lg} \, dA \end{aligned} \tag{10}$$

mass balance for lg-interface components ($\kappa = w, a$):

$$\begin{aligned} &\frac{\partial \left(\bar{\Gamma}_{lg} \bar{X}_{lg}^\kappa a_{lg}\right)}{\partial t} + \nabla \cdot \left(\bar{\Gamma}_{lg} \bar{X}_{lg}^\kappa a_{lg} \underline{\bar{v}}_{lg}\right) - \nabla \cdot \left(\underline{\bar{j}}_{lg}^\kappa a_{lg}\right) - \\ = \; & \frac{1}{V} \int_{A_{lg}} \left[\rho_l X_l^\kappa \left(\underline{v}_l - \underline{v}_{lg}\right) - \underline{j}_l^\kappa - \rho_g X_g^\kappa \left(\underline{v}_g - \underline{v}_{lg}\right) + \underline{j}_g^\kappa\right] \cdot \underline{n}_{lg} \, dA \end{aligned} \tag{11}$$

momentum balance for phases:

$$\begin{aligned} &\frac{\partial\left(\phi S_l \bar{\rho}_l \underline{\bar{v}}_l\right)}{\partial t} + \nabla\cdot\left(\phi S_l \bar{\rho}_l \underline{\bar{v}}_l \underline{\bar{v}}_l\right) - \nabla\cdot\left(\phi S_l \underline{T}_l\right) \\ = &\frac{1}{V}\int_{A_{lg}} \left[\rho_l \underline{v}_l \left(\underline{v}_{lg} - \underline{v}_l\right) + \underline{t}_l\right]\cdot \underline{n}_{lg}\, dA \end{aligned} \tag{12}$$

$$\begin{aligned} &\frac{\partial\left(\phi S_g \bar{\rho}_g \underline{\bar{v}}_g\right)}{\partial t} + \nabla\cdot\left(\phi S_g \bar{\rho}_g \underline{\bar{v}}_g \underline{\bar{v}}_g\right) - \nabla\cdot\left(\phi S_g \underline{T}_g\right) \\ = &\frac{1}{V}\int_{A_{lg}} \left[\rho_g \underline{v}_g \left(\underline{v}_g - \underline{v}_{lg}\right) - \underline{t}_g\right]\cdot \underline{n}_{lg}\, dA \end{aligned} \tag{13}$$

momentum balance for lg-interface:

$$\begin{aligned} &\frac{\partial\left(\bar{\Gamma}_{lg} \underline{\bar{v}}_{lg} a_{lg}\right)}{\partial t} + \nabla\cdot\left(\bar{\Gamma}_{lg} \underline{\bar{v}}_{lg} a_{lg} \underline{\bar{v}}_{lg}\right) - \nabla\cdot\left(\underline{T}_{lg} a_{lg}\right) \\ = &\frac{1}{V}\int_{A_{lg}} \left[\rho_l \underline{v}_l \left(\underline{v}_l - \underline{v}_{lg}\right) - \underline{t}_l - \rho_g \underline{v}_g \left(\underline{v}_g - \underline{v}_{lg}\right) + t_g\right]\cdot \underline{n}_{lg}\, dA, \end{aligned} \tag{14}$$

where the overbars denote volume-averaged (macro-scale) quantities. Here, $X_\alpha^\kappa\ [-]$ is the mass fraction of component κ in phase α, t is time, $Q_\alpha^\kappa\ \left[\frac{m^3}{s}\right]$ is an external source of component κ in phase α, $\underline{j}_\alpha^\kappa\ \left[\frac{kg\cdot m^4}{s}\right]$ is the diffusive flux of component κ in phase α, V is the magnitude of the averaging volume, A_{lg} denotes the interfaces separating the l-phase and the g-phase in an averaging volume, $\underline{v}_{lg}\ \left[\frac{m}{s}\right]$ is the velocity of the lg-interface, and $\underline{n}_{lg}$ is the unit vector normal to A_{lg} and pointing into the g-phase. Furthermore, $X_{lg}^\kappa\ [-]$ is the mass fraction of component κ in the lg-interface, $\underline{j}_{lg}^\kappa\ \left[\frac{kg\cdot m^4}{s}\right]$ is the diffusive flux of component κ in the lg-interface, $\underline{t}_\alpha\ \left[\frac{kg}{m\cdot s^2}\right]$ is the α-phase micro-scale stress tensor, $\underline{T}_\alpha\ \left[\frac{kg}{s^2}\right]$ is the α-phase macro-scale stress tensor, and $\underline{T}_{lg}\ \left[\frac{kg}{s^2}\right]$ is the macro-scale lg-interfacial stress tensor.

In the following, we provide a simplified version of Eq. (9) through (14). First, we assume that the composition of the interface does not change. This is a reasonable assumption as long as no surfactants are involved. This reduces the number of balance equations to eight, as we can sum up the mass balance equations for interface components. Furthermore, we assume that momentum balances can be simplified so far that we end up with Darcy-like equations for both bulk phases and interface. We further proceed by applying Fick's law to relate the micro-scale diffusive fluxes $\underline{j}_\alpha^\kappa$ to the local concentration gradient resulting in the following approximation:

$$\underline{j}_\alpha^\kappa \cdot \underline{n}_{lg} = \pm\frac{\rho_\alpha D^\kappa}{d^\kappa} a_{lg}\left(X_{\alpha,s}^\kappa - X_\alpha^\kappa\right), \tag{15}$$

where $D^\kappa\ \left[\frac{m^2}{s}\right]$ is the micro-scale Fickian diffusion coefficient for component κ, $d^\kappa\ [m]$ is the diffusion length of component κ, $X_{\alpha,s}^\kappa\ [-]$ is the solubility limit of component κ in phase α (i.e. the mass fraction corresponding to local equilibrium), and $X_\alpha^\kappa\ [-]$ is the micro-scale mass fraction of component κ in phase α at a distance d^κ away from the interface. Niessner &

Hassanizadeh (2009a) obtained the following determinate set of macro-scale equations:

$$\frac{\partial\left(\phi S_l \bar{\rho}_l \bar{X}_l^w\right)}{\partial t} + \nabla\cdot\left(\bar{\rho}_l \bar{X}_l^w \underline{\bar{v}}_l\right) - \nabla\cdot\left(\underline{\bar{j}}_l^w\right) = \rho_l Q_l^w - \frac{D^w \bar{\rho}_g}{d^w} a_{lg}\left(X_{g,s}^w - X_g^w\right) \quad (16)$$

$$\frac{\partial\left(\phi S_l \bar{\rho}_l \bar{X}_l^a\right)}{\partial t} + \nabla\cdot\left(\bar{\rho}_l \bar{X}_l^a \underline{\bar{v}}_l\right) - \nabla\cdot\left(\underline{\bar{j}}_l^a\right) = \rho_l Q_l^a + \frac{D^a \bar{\rho}_l}{d^a} a_{lg}\left(X_{l,s}^a - X_l^a\right) \quad (17)$$

$$\frac{\partial\left(\phi S_g \bar{\rho}_g \bar{X}_g^w\right)}{\partial t} + \nabla\cdot\left(\bar{\rho}_g \bar{X}_g^w \underline{\bar{v}}_g\right) - \nabla\cdot\left(\underline{\bar{j}}_g^w\right) = \rho_g Q_g^w + \frac{D^w \bar{\rho}_g}{d^w} a_{lg}\left(X_{g,s}^w - X_g^w\right) \quad (18)$$

$$\frac{\partial\left(\phi S_g \bar{\rho}_g \bar{X}_g^a\right)}{\partial t} + \nabla\cdot\left(\bar{\rho}_g \bar{X}_g^a \underline{\bar{v}}_g\right) - \nabla\cdot\left(\underline{\bar{j}}_g^a\right) = \rho_g Q_g^a - \frac{D^a \bar{\rho}_l}{d^a} a_{lg}\left(X_{l,s}^a - X_l^a\right) \quad (19)$$

$$\frac{\partial a_{lg}}{\partial t} + \nabla\cdot\left(a_{lg}\underline{v}_{lg}\right) = E_{lg} \quad (20)$$

$$\underline{\bar{v}}_l = -K\frac{S_l^2}{\mu_l}\left(\nabla p_l - \bar{\rho}_l \underline{g}\right) \quad (21)$$

$$\underline{\bar{v}}_g = -K\frac{S_g^2}{\mu_g}\left(\nabla p_g - \bar{\rho}_g \underline{g}\right) \quad (22)$$

$$\underline{\bar{v}}_{lg} = -K_{lg}\nabla a_{lg} \quad (23)$$

$$p_c = p_g - p_l \quad (24)$$

$$S_l + S_g = 1 \quad (25)$$

$$X_l^w + X_l^a = 1 \quad (26)$$

$$X_g^w + X_g^a = 1 \quad (27)$$

$$a_{lg} = a_{lg}(S_l, p_c), \quad (28)$$

The macro-scale mass fluxes $\underline{\bar{j}}_\alpha^\kappa$ are given by a Fickian dispersion equation,

$$\underline{\bar{j}}_\alpha^\kappa = -\rho_\alpha \bar{D}_\alpha^\kappa \nabla\left(\bar{X}_\alpha^\kappa\right), \quad (29)$$

where $\bar{D}_\alpha^\kappa$ is the macro-scale dispersion coefficient. Note that in Eq. (16) through (19), we have acknowledged the fact that an internal source of a component in one of the phases must correspond to a sink of that component of equal magnitude in the other phase. The solubility limits $X_{g,s}^w$ and $X_{l,s}^a$ are obtained from a local equilibrium assumption at the fluid–fluid interface. If, for example, we consider a two-phase–two-component air–water system solubility limits with respect to mole fractions are given by Eqs. (1) and (2).

3.3 Is a kinetic approach necessary?

Depending on the parameters, initial conditions, and boundary conditions of the system, kinetics might be important for mass transfer. If so, then it may not be sufficient to use a classical local equilibrium model instead of the more complex interfacial-area-based model. To allow for a decision, Niessner & Hassanizadeh (2009a) make the system of equations (16) through (28) dimensionless and study the dependence of kinetics on Damköhler number and Peclet number.

To do so, they define dimensionless variables:

$$t^* = \frac{t v_R}{\phi L}, \qquad \nabla^* = L\nabla, \qquad \bar{v}^*_\alpha = \frac{\bar{v}_\alpha}{v_R}, \qquad Q^{\kappa *}_\alpha = \frac{Q^\kappa_\alpha L}{X^\kappa_{\alpha,s} v_R}, \tag{30}$$

$$a^*_{lg} = \frac{a_{lg}}{a_{R,lg}}, \qquad \bar{D}^{\kappa *}_\alpha = \frac{\bar{D}^\kappa_\alpha}{D_{R,\alpha}}, \qquad \underline{v}^*_{lg} = \frac{\phi}{v_R}\underline{v}_{lg}, \tag{31}$$

$$E^*_{lg} = \frac{a_{R,lg}\phi L}{v_R} E_{lg}, \qquad \underline{g}^*_\alpha = \frac{\bar{\rho}_\alpha g L}{p_R}, \qquad K^*_{lg} = \frac{\phi K_{lg}}{L a_{R,lg} v_R}, \tag{32}$$

$$p^*_\alpha = \frac{p_\alpha}{p_R}, \qquad p^*_c = \frac{p_c}{p_R}, \qquad \rho^*_g = \frac{\bar{\rho}_g}{\bar{\rho}_l} \qquad \mu^*_l = \frac{\mu_l}{\mu_g}. \tag{33}$$

Here, ρ^*_g is density ratio, μ^*_l is viscosity ratio, v_R is a reference velocity, L is a characteristic length, $a_{R,lg}$ is a reference specific interfacial area, $D_{R,\alpha}$ is a reference dispersion coefficient, and p_R is a reference pressure. We assume that p_R and v_R can be chosen such that $\frac{K p_R}{\mu_l v_R L} = 1$. Also, Peclet number Pe_α and Damköhler number Da^κ are defined by

$$\mathrm{Pe}_\alpha = \frac{v_R L}{D_{R,\alpha}}, \qquad \mathrm{Da}^\kappa = \frac{D^\kappa L a_{R,lg}}{d v_R}. \tag{34}$$

These definitions lead to the following dimensionless form of Eq. (16) through (28):

$$\frac{\partial}{\partial t^*}\left(S_l \bar{X}^w_l\right) + \nabla^* \cdot \left(\bar{X}^w_l \underline{v}^*_l\right) - \nabla^* \cdot \left(\frac{D^{w*}_l}{\mathrm{Pe}_l}\nabla^* \bar{X}^w_l\right) = Q^{w*}_l - \mathrm{Da}^w a^*_{lg}\rho^*_g\left(X^w_{g,s} - \bar{X}^w_g\right) \tag{35}$$

$$\frac{\partial}{\partial t^*}\left(S_l \bar{X}^a_l\right) + \nabla^* \cdot \left(\bar{X}^a_l \underline{v}^*_l\right) - \nabla^* \cdot \left(\frac{D^{a*}_l}{\mathrm{Pe}_l}\nabla^* \bar{X}^a_l\right) = Q^{a*}_l + \mathrm{Da}^a a^*_{lg}\left(X^a_{l,s} - \bar{X}^a_l\right) \tag{36}$$

$$\frac{\partial}{\partial t^*}\left(S_g \bar{X}^w_g\right) + \nabla^* \cdot \left(\bar{X}^w_g \underline{v}^*_g\right) - \nabla^* \cdot \left(\frac{D^{w*}_g}{\mathrm{Pe}_g}\nabla^* \bar{X}^w_g\right) = Q^{w*}_g + \mathrm{Da}^w a^*_{lg}\left(X^w_{g,s} - \bar{X}^w_g\right) \tag{37}$$

$$\frac{\partial}{\partial t^*}\left(S_g \bar{X}^a_g\right) + \nabla^* \cdot \left(\bar{X}^a_g \underline{v}^*_g\right) - \nabla^* \cdot \left(\frac{D^{a*}_g}{\mathrm{Pe}_g}\nabla^* \bar{X}^a_g\right) = Q^{a*}_g - \mathrm{Da}^a a^*_{lg}\frac{1}{\rho^*_g}\left(X^a_{l,s} - \bar{X}^a_l\right) \tag{38}$$

$$\frac{\partial a^*_{lg}}{\partial t^*} + \nabla^* \cdot \left(a^*_{lg}\underline{v}^*_{lg}\right) = E^*_{lg} \tag{39}$$

$$\underline{v}^*_l = -S^2_l\left(\nabla^* p^*_l - \underline{g}^*_l\right) \tag{40}$$

$$\underline{v}^*_g = -S^2_g \mu^*_l\left(\nabla^* p^*_g - \underline{g}^*_g\right) \tag{41}$$

$$\underline{v}^*_{lg} = -K^*_{lg} \nabla^* a^*_{lg} \quad (42)$$
$$p^*_c = p^*_g - p^*_l \quad (43)$$
$$S_l + S_g = 1 \quad (44)$$
$$\bar{X}^w_l + \bar{X}^a_l = 1 \quad (45)$$
$$\bar{X}^w_g + \bar{X}^a_g = 1 \quad (46)$$
$$a^*_{lg} = a^*_{lg}(S_l, p^*_c). \quad (47)$$

In order to investigate the importance of kinetics, we define $\mathrm{Pe} := \mathrm{Pe}_l = \mathrm{Pe}_g$ and $\mathrm{Da} := \mathrm{Da}^w = \mathrm{Da}^a$ and vary Pe and Da independently over five orders of magnitude. Therefore, we consider a numerical example where dry air is injected into a horizontal (two-dimensional) porous medium of size 0.7 m $\times$ 0.5 m that is almost saturated with water (initial and boundary water saturation of 0.9).

Fig. 5 shows a comparison of actual mass fractions $\bar{X}^a_l$ and $\bar{X}^w_g$ to solubility mass fractions $X^a_{l,s}$ and $X^w_{g,s}$ for five different Damköhler numbers. Therefore, a cut through the domain is shown and two different time steps are compared.

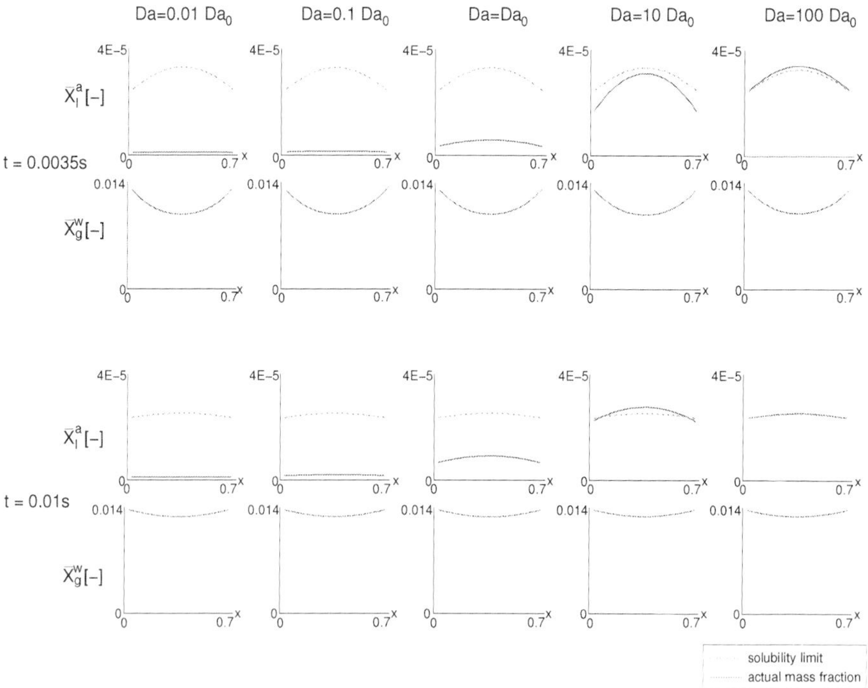

Fig. 5. Solubility limits $X^a_{l,s}$ and $X^w_{g,s}$ and actual mass fractions $\bar{X}^a_l$ and $\bar{X}^w_g$ for two different time steps (0.0035 s and 0.01 s) and 5 different Damköhler numbers.

It can be seen that the system is practically instantaneously in equilibrium with respect to the mass fraction $\bar{X}_g^w$ (water mass fraction in the gas phase) for the whole range of considered Damköhler numbers (see the second and forth row of graphs). With respect to the mass fraction $\bar{X}_l^a$ (air mass fraction in the water phase), for low Damköhler numbers and early times, the system is far from equilibrium (see the first and third row of graphs). With increasing time and with increasing Damköhler number, the system approaches equilibrium. As for high Damköhler numbers mass transfer is very fast, an "overshoot" occurs and the system becomes oversaturated before it reaches equilibrium.
One might argue that the considered time steps are extremely small and not relevant for the time scale relevant for the whole domain. However, what happens at this very early time has a large influence on the state of the system at all subsequent times.
It turned out that for different Peclet numbers, there is no difference in results. That means that kinetic interphase mass transfer is independent of Peclet number, at least within the four orders of magnitude considered here.

4. Extension to heat transfer

The concept of describing mass transfer based on modeling the evolution of interfacial areas using the thermodynamically-based approach of Hassanizadeh & Gray (1990; 1993b) can be extended to describing interphase heat transfer as well. The main difference between interphase mass and heat transfer is that, in addition to fluid–fluid interfaces, heat can also be transferred across fluid–solid interfaces, see Fig. 6.
Similarly to mass transfer, classical two-phase flow models describe heat transfer on the macro scale by either assuming local thermal equilibrium within an averaging volume or by formulating empirical models to describe the transfer rates. The latter is necessary as classically, both fluid–fluid and fluid–solid interfacial areas are unknown on the macro scale. And similiarly to mass transfer, we can use the thermodynamically-based approach of Hassanizadeh & Gray (1990; 1993b) which includes both fluid–fluid and fluid–solid interfacial areas in order to describe mass transfer in a physically-based way. We can

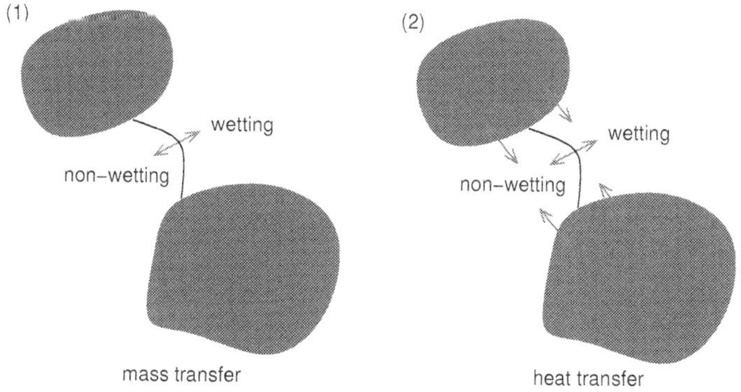

Fig. 6. Mass transfer takes place across fluid–fluid interfaces (left hand side) and heat transfer across fluid–fluid as well as fluid–solid interfaces (right hand side)

also perform a dimensional analysis and derive dimensionless numbers that help to decide whether kinetics of heat transfer needs to be accounted for or whether a local equilibrium model is sufficiently accurate on the macro scale. For more details on these issues, we refer to Niessner & Hassanizadeh (2009b).

5. Macro-scale example simulations

For the numerical solution of the system of Eq.s (16) through (28), we use a fully-coupled vertex-centered finite element method (an in-house code) which not only conserves mass locally, but is also applicable to unstructured grids. For time discretization, a fully implicit Eulerian approach is used, see e.g. Bastian et al. (1997); Bastian & Helmig (1999). The nonlinear system is linearized using a damped inexact Newton-Raphson solver, and the linear system is subsequently solved using a Bi-Conjugate Gradient Stabilized method (known as BiCGStab method). Full upwinding is applied to the flux terms of the bulk phase equations, but in the interfacial area flux term, central weighting is used.

5.1 Evaporator

As a simulation example, we consider a setup which is relevant in many industrial processes where a product needs to be concentrated (e.g. foods, chemicals, and salvage solvents) or dried through evaporation of water. The aqueous solution containing the desired product is fed into the evaporator mostly consisting of micro-channels and then passes a heat source. Heat converts the water in the solution into vapor and the vapor is subsequently removed from the solution. We model the heating and evaporation process through a setup as shown in Fig. 7.

This means we consider a horizontal domain that is closed along the sides (top and bottom in the figure) and that is subjected to a gradient in wetting phase pressure from left to right in the undisturbed situation. This system is assumed to be at the following initial conditions: a temperature of 293 K, gas phase at atmospheric pressure, water saturation of 0.9, a corresponding capillary pressure based on the primary drainage curve, and mass fractions that correspond to the local chemical equilibrium conditions as prescribed by Henry's Law and Raoult's Law (Lüdecke & Lüdecke (2000)). The porous medium is heated with a rate of Q_S in a square-shaped part of the domain, causing the evaporation of water (see Fig. 7). Note that the heat source heats up only the walls of the microchannels (the solid phase) and the heat is then transferred from the solid phase to the water phase.

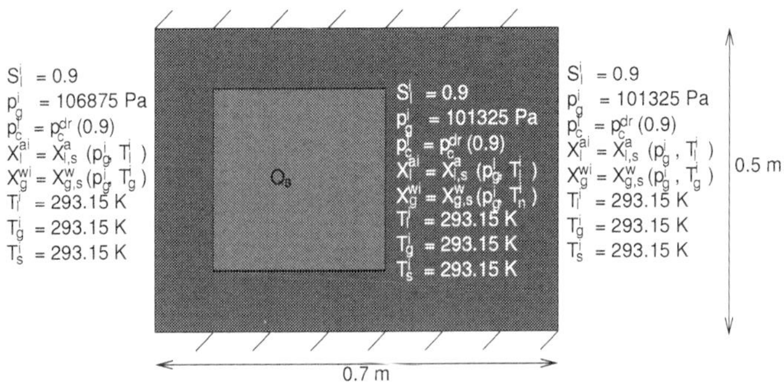

Fig. 7. Setup of the numerical example: water passes a heat source and is evaporated.

For comparison, the same setup is used for simulations using a classical two-phase flow model, where—in the absence of interfacial areas as parameters—local chemical and thermal equilibrium is assumed. This means that the heat source cannot be defined for the solid phase only; instead, the applied heat will instantaneously lead to a heating of all three phases.
The distribution of water saturation and water–gas specific interfacial area are shown in Fig. 8 at the time of 17 seconds after the heat source is switched on. The saturation distribution given by the interfacial area-based model is compared to that of the classical model. Obviously, water saturation decreases in the heated region due to the evaporation of water. Downstream of the heated zone, water saturation increases indicating that in the colder regions, water condenses again. Due to the decrease in water saturation, gas–water interfacial areas are created. The classical model predicts a much lower decrease in water saturation in the heated region.
Fig. 9 shows the mass fraction of air dissolved in the liquid phase after 17 seconds. There, also the equilibrium value (solubility limit) predicted by the interfacial area-based model is shown. Clearly, chemical non-equilibrium effects occur, but the classical model predicts approximately the same result as the equilibrium values in the interfacial area-based model. This is due to the fact that the classical two-phase flow approach always assumes local equilibrium and can only represent mass fractions corresponding to the equilibrium values.
The analogous comparison is shown in Fig. 10 with respect to the mass fractions of vapor in the gas phase.
Here, the mass fractions in the interfacial area-based model are very close to the equilibrium values, but larger differences to the classical model can be detected.

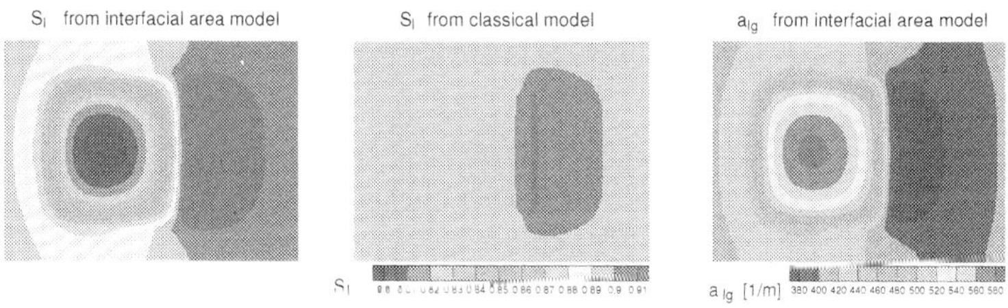

Fig. 8. Water saturation and water–gas specific interfacial area after 17s.

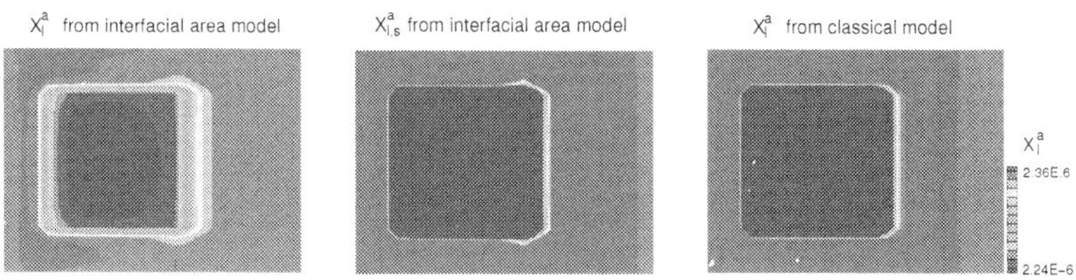

Fig. 9. Mass fraction of air in the liquid phase after 17s. Left: results using the interfacial area-based model, middle: equilibrium values given by the interfacial area-based model, right: results given by the classical model.

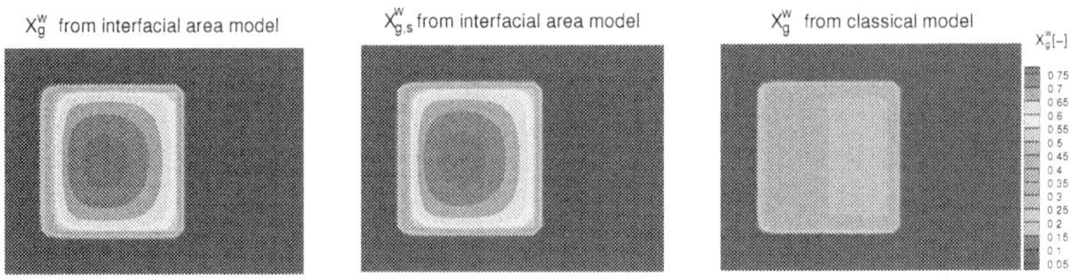

Fig. 10. Mass fractions of vapor in the gas phase after 17s. Left: results using the interfacial area-based model, middle: equilibrium values given by the interfacial area-based model, right: results given by the classical model.

Fig. 11 shows the temperatures of the three phases (liquid l, gas g, solid s) after 17s using the interfacial area-based model and the classical model. It can be seen that a lower temperature rise is predicted by the classical model than by the interfacial area-based model.

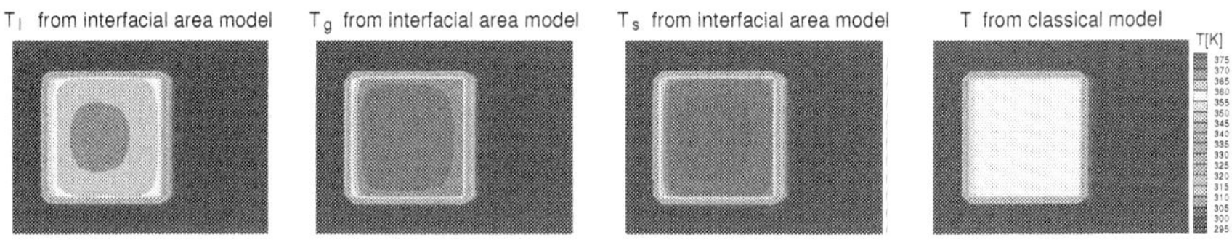

Fig. 11. Temperatures of the phases after 17s using the interfacial area-based model (three pictures on left) and the classical model (right picture).

Fig. 12 shows the temperature differences between the phases using the interfacial area-based model. We find that there are differences of up to 8 K between the phases. The classical two-phase model assumes local thermal equilibrium, i.e. $T_w = T_n = T_s$. In reality, however, the heat exchange between the phases is restricted by the respective interfacial areas and the heat equalization does not take place instantaneously.

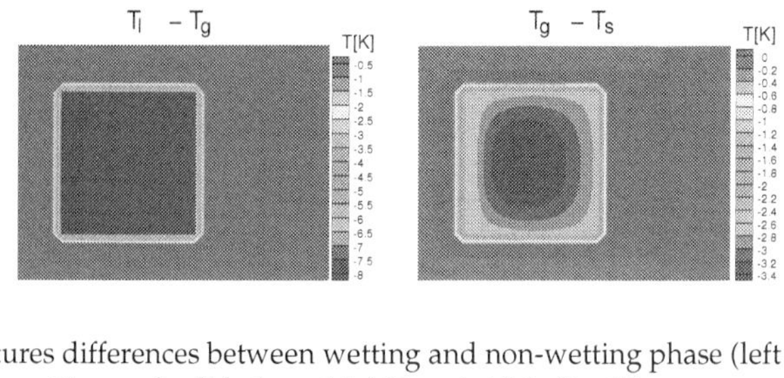

Fig. 12. Temperatures differences between wetting and non-wetting phase (left hand side) and between non-wetting and solid phase (right hand side) after 17s.

5.2 Drying of a porous medium

As a second example, we consider the drying of an initially almost water-saturated porous medium through injection of hot dry air (50 °C). This process is relevant in the textile, construction, and paper industries as well as in medical applications. The setup is shown in Fig. 13. For comparison, this setup is also simulated using the classical model. Note that in the classical model, due to the assumption of local chemical and thermal equilibrium, it is impossible to apply a source of hot dry air. Instead, the air supplied to the system instantaneously redistributes among the phases in order to yield equilibrium composition. The temperature of the air source will instantaneously heat up all three phases.

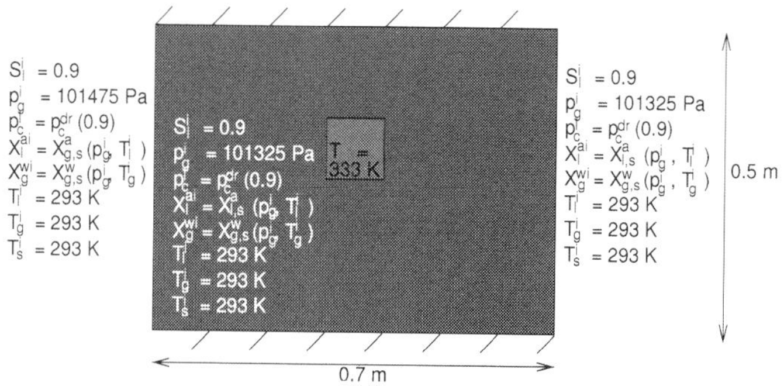

Fig. 13. Setup of the numerical example: drying of a porous medium through injection of warm dry air.

Fig. 14 shows the water saturation and the specific gas–water interfacial area after 12 seconds. The water saturation decreases due to the gas injection and interfaces are produced. The saturation distribution is different if the classical model is used. This may be due to the fact that the source term needs to be specified differently for the two models (see the above comments on the problem description).

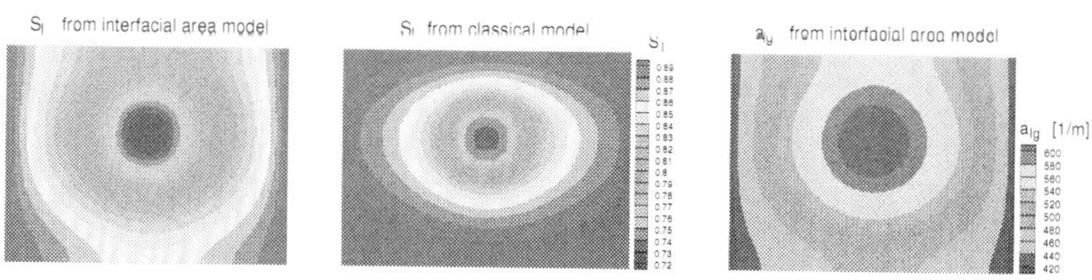

Fig. 14. Water saturation (using both interfacial area-based and classical model) and specific fluid–fluid interfacial area using the interfacial area-based model after 12s.

In Fig. 15, the mass fraction of air disolved in the water phase is shown after 12s. The actual mass fraction distribution using the interfacial-area-based approach is shown in the left picture. The mass fraction corresponding to local chemical equilibrium is shown in the middle figure. For comparison, the mass fraction distribution resulting from the classical model is shown in the right figure. Obviously, significant deviations from equilibrium can be detected. This implies that chemical non-equilibrium is very important in this example. The

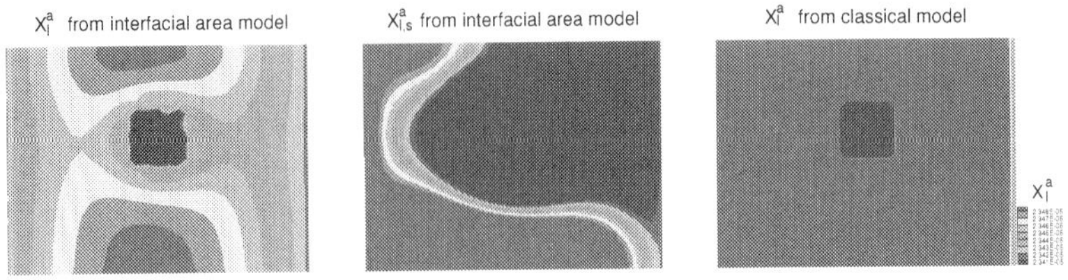

Fig. 15. Mass fractions of air in water. Left: interfacial-area-based model, actual mass fractions; middle: equilibrium mass fractions. Right: classical model.

equilibrium mass fractions of the interfacial area-based model are also very different from that predicted by the classical model, probably due to the differently specified source.
Fig. 16 shows the same for water vapor in the gas phase. Here, the deviation from chemical equilibrium is also significant, but the equilibrium mass fractions using the interfacial area-based model and the classical model are very similar.

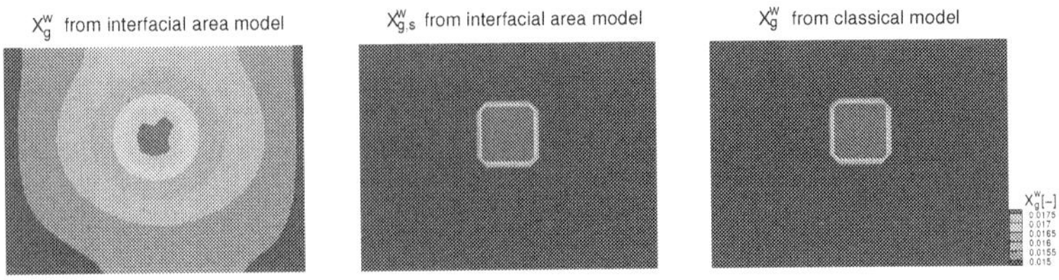

Fig. 16. Mass fractions of vapor in the gas phase. Left: interfacial-area-based model, actual mass fractions; middle: equilibrium mass fractions. Right: classical model.

In Fig. 17, the temperatures of the three phases (liquid l, gas g, and solid phase s) resulting from the interfacial area-based model are shown and compared to the temperature given by the classical model. Unlike the mass fractions, the temperatures are not very far from equilibrium (maximum difference in phase temperatures is approximately 0.17 K, see also Fig. 18). An interesting aspect is that the temperature difference is lower in the middle of the injection zone than in the surrounding area. This is due to the fact that specific interfacial area is at a maximum in this middle part leading to higher heat transfer rates in this region and phase temperature closer to each other and thus, closer to thermal equilibrium.

6. Summary and conclusions

In this chapter, the issue of interphase mass transfer during two-phase flow in a porous medium has been discussed. Starting from pore-scale considerations, the classical approaches for describing mass transfer have been presented which—due to the absence of interfacial area as a parameter—either assume local equilibrium within an averaging volume or use empirical approaches to describe the kinetics.

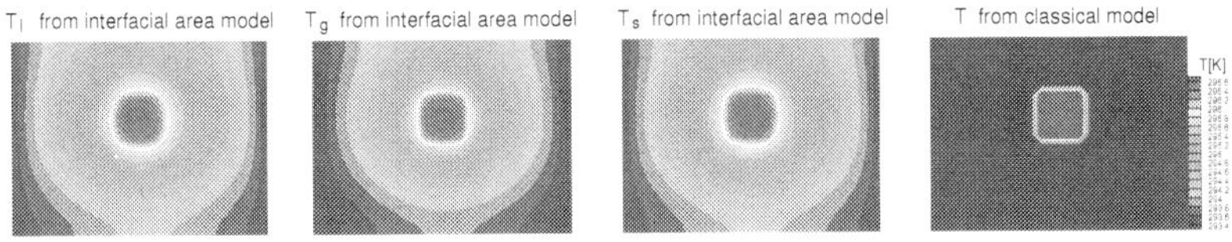

Fig. 17. Temperatures of the phases using the interfacial area-based model (three pictures on left hand side) and the classical model (right hand side picture).

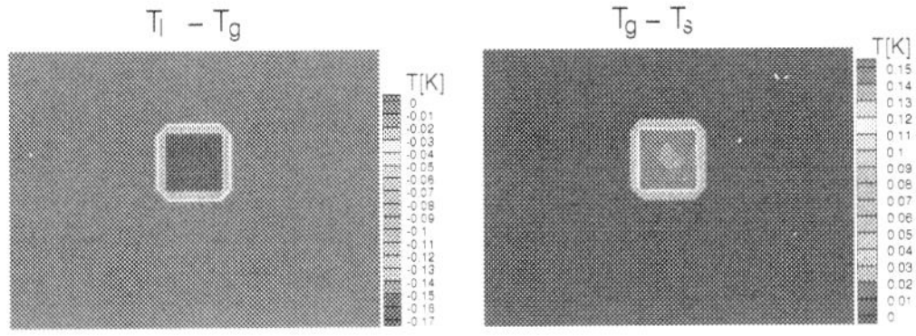

Fig. 18. Temperature differences between the phases.

As an alternative, a thermodynamically-based model was presented which explicitly accounts for the presence of interfaces and describes their evolution in space and time. Due to the knowledge of interfacial area, kinetic interface mass transfer can be modeled in a physically-based way. In order to decide whether kinetics of mass transfer needs to be accounted for or whether an equilibrium model would give sufficiently good results, a dimensional analysis was carried through. Also, the concept was extended to kinetic interphase heat transfer where in addition to fluid–fluid interfaces, fluid–solid interfaces are important. Two examples have illustrated the issues presented: in our results, we could observe that in the drying example, chemical non-equilibrium is significant. In the evaporator example, contrarily, thermal non-equilibrium is very pronounced.
In order to advance the description of real life systems, such as those described in Sec. 1.1, an important future step would be to apply the presented model concept to practical applications and to verify the results by comparison to experimental data.

7. Acknowledgements

We thank Benjamin Ahrenholz for providing the interfacial - area - capillary pressure - saturation relationships from Lattice - Boltzman simulations that entered the numerical simulations shown in this chapter.

8. References

Bastian, P., Birken, K., Johannsen, K., Lang, S., Eckstein, K., Neuss, N., Rentz-Reichert, H. & Wieners, C. (1997). UG - A Flexible Software Toolbox for Solving Partial Differential Equations, Computing and Visualization in Science, 1(1):27–40.

Bastian, P. & Helmig, R. (1999). Efficient Fully-Coupled Solution Techniques for Two Phase Flow in Porous Media. Parallel Multigrid Solution and Large Scale Computations.,

Advances in Water Resources 23: 199–216.

Bowen, R. (1982). Compressible Porous Media Models by Use of the Theory of Mixtures, *International Journal of Engineering Science* 20(6): 697–735.

Brusseau, M., Popovicova, J. & Silva, J. (1997). Characterizing gas–water interfacial and bulk-water partitioning for gas phase transport of organic contaminants in unsaturated porous media, *Environmental Sciences Technology* 31: 1645–1649.

Chen, D., Pyrak-Nolte, L., Griffin, J. & Giordano, N. (2007). Measurement of interfacial area per volume for drainage and imbibition, *Water Resources Research* 43(12).

Chen, L. & Kibbey, T. (2006). Measurement of air–water interfacial area for multiple hysteretic drainage curves in an unsaturated fine sand, *Langmuir* 22: 6674–6880.

Culligan, K., Wildenschild, D., Christensen, B., Gray, W., Rivers, M. & Tompson, A. (2004). Interfacial area measurements for unsaturated flow through a porous medium, *Water Resources Research* 40: 1–12.

Falta, R. W. (2000). Numerical modeling of kinetic interphase mass transfer during air sparging using a dual-media approach, *Water Resources Research* 36(12): 3391–3400.

Falta, R. W. (2003). Modeling sub-grid-block-scale dense nonaqueous phase liquid (DNAPL) pool dissolution using a dual-domain approach, *Water Resources Research* 39(12).

Gray, W. & Hassanizadeh, S. (1989). Averaging theorems and averaged equations for transport of interface properties in multiphase systems, *International Journal of Multi-Phase Flow* 15: 81–95.

Gray, W. & Hassanizadeh, S. (1998). Macroscale continuum mechanics for multiphase porous-media flow including phases, interfaces, common lines, and common points, *Advances in Water Resources* 21: 261–281.

Gray, W. & Miller, C. (2005). Thermodynamically Constrained Averaging Theory Approach for Modeling of Flow in Porous Media: 1. Motivation and Overview, *Advances in Water Resources* 28(2): 161–180.

Hassanizadeh, S. M. & Gray, W. G. (1979). General Conservation Equations for Multi-Phase Systems: 2. Mass, Momenta, Energy, and Entropy Equations, *Advances in Water Resources* 2: 191–203.

Hassanizadeh, S. M. & Gray, W. G. (1980). General Conservation Equations for Multi-Phase Systems: 3. Constitutive Theory for Porous Media Flow, *Advances in Water Resources* 3: 25–40.

Hassanizadeh, S. M. & Gray, W. G. (1990). Mechanics and Thermodynamics of Multiphase Flow in Porous Media Including Interphase Boundaries, *Advances in Water Resources* 13(4): 169–186.

Hassanizadeh, S. M. & Gray, W. G. (1993a). Thermodynamic Basis of Capillary Pressure in Porous Media, *Water Resources Research* 29(10): 3389 – 3405.

Hassanizadeh, S. M. & Gray, W. G. (1993b). Toward an improved description of the physics of two-phase flow, *Advances in Water Resources* 16(1): 53–67.

Held, R. & Celia, M. (2001). Modeling support of functional relationships between capillary pressure, saturation, interfacial area and common lines, *Advances in Water Resources* 24: 325–343.

Imhoff, P., Jaffe, P. & Pinder, G. (1994). An experimental study of complete dissolution of a nonaqueous phase liquid in saturated porous media, *Water Resources Research* 30(2): 307–320.

IPCC (2005). *Carbon Dioxide Capture and Storage. Special Report of the Intergovernmental Panel on Climate Change*, Cambridge University Press.

Jackson, A., Miller, C. & Gray, W. (2009). Thermodynamically constrained averaging theory approach for modeling flow and transport phenomena in porous medium systems: 6. two-fluid-phase flow, *Advances in Water Resources* 32(6): 779–795.

Joekar-Niasar, V., Hassanizadeh, S. M. & Leijnse, A. (2008). Insights into the relationship among capillary pressure, saturation, interfacial area and relative permeability using pore-scale network modeling, *Transport in Porous Media* 74: 201–219.

Joekar-Niasar, V., Hassanizadeh, S. M., Pyrak-Nolte, L. J. & Berentsen, C. (2009). Simulating drainage and imbibition experiments in a high-porosity micro-model using an unstructured pore-network model, *Water Resources Research* 45(W02430, doi:10.1029/2007WR006641).

Kalaydjian, F. (1987). A Macroscopic Description of Multiphase Flow in Porous Media Involving Spacetime Evolution of Fluid/Fluid Interface, *Transport in Porous Media* 2: 537 – 552.

Lüdecke, C. & Lüdecke, D. (2000). *Thermodynamik*, Springer, Berlin.

Marle, C.-M. (1981). From the pore scale to the macroscopic scale: Equations governing multiphase fluid flow through porous media, *Proceedings of Euromech 143*, pp. 57–61. Delft, Verruijt, A. and Barends, F. B. J. (eds.).

Mayer, A. & Hassanizadeh, S. (2005). *Soil and Groundwater Contamination: Nonaqueous Phase Liquids*, American Geophysical Union.

Miller, C., Poirier-McNeill, M. & Mayer, A. (1990). Dissolution of trapped nonaqueous phase liquids: Mass transfer characteristics, *Water Resources Research* 21(2): 77–120.

Niessner, J. & Hassanizadeh, S. (2008). A Model for Two-Phase Flow in Porous Media Including Fluid–Fluid Interfacial Area, *Water Resources Research* . 44, W08439, doi:10.1029/2007WR006721.

Niessner, J. & Hassanizadeh, S. (2009a). Modeling kinetic interphase mass transfer for two-phase flow in porous media including fluid–fluid interfacial area, *Transport in Porous Media* . doi:10.1007/s11242-009-9358-5.

Niessner, J. & Hassanizadeh, S. (2009b). Non-equilibrium interphase heat and mass transfer during two-phase flow in porous media—theoretical considerations and modeling, *Advances in Water Resources* 32: 1756–1766.

Porter, M., Schaap, M. & Wildenschild, D. (2009). Lattice-boltzmann simulations of the capillary pressure-saturation-interfacial area relationship for porous media, *Advances in Water Resources* 32(11): 1632–1640.

Powers, S., Abriola, L. & Weber, W. (1992). An experimental investigation of nonaqueous phase liquid dissolution in saturated subsurface systems: steady state mass transfer rates, *Water Resources Research* 28: 2691–2706.

Powers, S., Abriola, L. & Weber, W. (1994). An experimental investigation of nonaqueous phase liquid dissolution in saturated subsurface systems: transient mass transfer rates, *Water Resources Research* 30(2): 321–332.

Reeves, P. & Celia, M. (1996). A functional relationship between capillary pressure, saturation, and interfacial area as revealed by a pore-scale network model, *Water Resources Research* 32(8): 2345–2358.

Schaefer, C., DiCarlo, D. & Blunt, M. (2000). Experimental measurements of air–water interfacial area during gravity drainage and secondary imbibition in porous media, *Water Resources Research* 36: 885–890.

van Antwerp, D., Falta, R. & Gierke, J. (2008). Numerical Simulation of Field-Scale Contaminant Mass Transfer during Air Sparging, *Vadose Zone Journal* 7: 294–304.

Wildenschild, D., Hopmans, J., Vaz, C., Rivers, M. & Rikard, D. (2002). Using X-ray computed tomography in hydrology. Systems, resolutions, and limitations, *Journal of Hydrology* 267: 285–297.

Zhang, H. & Schwartz, F. (2000). Simulating the *in situ* oxidative treatment of chlorinated compounds by potassium permanganate, *Water Resources Research* 36(10): 3031–3042.

2

Solute Transport With Chemical Reaction in Single- and Multi-Phase Flow in Porous Media

M.F. El-Amin[1],
Amgad Salama[2] and Shuyu Sun[1]
[1]*King Abdullah University of Science and Technology (KAUST)*
[2]*Konkuk University*
[1]*Kingdom of Saudi Arabia*
[2]*South Korea*

1. Introduction

Transport phenomena in porous media describe the motion of fluids in media of porous structure which may be accompanied by heat/mass transfer and/or chemical reactions. While transport phenomena in fluid continua have been, to a large extent, very much comprehended, the subject matters in porous media are still under careful investigation and extensive research. Several reasons may be invoked to explain the difficulties associated with the study of transport phenomena in porous media. Probably the most obvious one is the fact that fluids move in porous media in complex, tortuous, and random passages that are even unknown a priori. Consequently, the governing laws may not be solved in any sense for the apparent difficulties in defining flow boundaries. Further complexities may be added should there exists heat transfer mechanisms associated with the flow and the interactions of heat transfer between the moving fluid and the solid matrix. Moreover, chemical reactions describe essential feature of transport in porous media. It is hardly to find transport processes in porous media without chemical reaction of some sort or another. Chemical reactions in porous media can occur naturally as a result of the interactions between the moving fluid and the surface of the solid matrix. These kinds of chemical reactions, which are usually slow, are pertinent to groundwater geochemistry, or it can be made to occur by utilizing the porous media surfaces to catalyze chemical reactions between reacting fluids. The study of these complex processes in porous media necessitate complete information about the internal structure of the porous media, which is far beyond the reach of our nowadays capacities. A fundamental question, thus, arises, in what framework do we need to cast the study of transport in porous media? In other words, do we really need to get such complete, comprehensive information about a given porous medium in order to gain useful information that could help us in our engineering applications? Do we really need to know the field variables distribution at each single point in the porous medium in order to be able to predict the evolution of this system with time, for example? Is it possible to make precise measurements within the porous media for field variables? And, even if we might be able to gain such detailed information, are we going to use them in their primitive forms for further analysis and development? The answer to these kind of questions may be that, for the sake of engineering applications, we do not need such a complete, comprehensive details, neither will we be able to obtain them nor will they be useful in their primitive form. In other words, field variables distribution will be randomly distributed and we would, in general, need to average them in order to gain statistically useful integral information. These ideas, in fact, enriched researchers' minds on their search for an appropriate framework to study phenomena in porous media. That is, if we need to average the pointwise field variables to get useful information, should not we might, as well, look for doing such kind of averaging on our way to investigating porous media. That is, is it possible to upscale our view to porous media such that we get smooth variables that represent integral information about the behavior of field variables not at a single point but within a volume of the porous medium surrounding this point? It turns out that researchers have appealed to this strategy several times on their way to explore the behavior of systems composed of innumerable building blocks. Thermodynamics, solid mechanics, fluid mechanics etc. are examples of sciences that adopted this approach by assuming the medium as continuum. Now, is it feasible to, also, treat phenomena occurring in porous media as continua? The answer to this question turns out to be yes as will be explained in the next section.

2. Framework

Salama & van Geel (2008a) provided an interesting analogy that sheds light on the possibility to adopt the continuum approach to phenomena occurring in porous media. They stated that an observer closer to a given porous medium will be able to see details of the porous medium (at least at the surface) that an observer from far distant would, generally, ignore. To the distant observer, the medium looks smooth and homogeneous like a continuum, Fig.1. It is exactly this point of view that we seek and it remains interesting to estimate that minimum distance that our observer would have to stay to get the continuum feeling of the medium. Of course moving beyond this distance would result in no significant improvement in the continuum picture. However, moving too far without having established the continuum feeling such that the extents of the domain enter the scene implies that it would not be possible to establish the continuum picture. This analogy, in fact, gives us an idea on how to properly define upscaling to porous media and hence establish continua. That is we need an upscaling scale (volume) that is much larger than small scale heterogeneity (e.g., pore diameter) and small enough such that it does not encompass the domain boundaries.

The mathematical machinery that provide such an upscaled description to phenomena occurring in porous media includes theory of mixtures, the method of volume averaging, method of homogenization, etc. In the framework of theory of mixtures, global balance equations are written based on the assumption of the existence of macroscale field variables, which are employed in the global balance equations. A localized version of these equations may then be obtained through mathematical manipulations to get what is called the macroscopic point equations. In the frame work of the method of volume averaging, on the other hand, the microscale conservation laws adapted to fluid continua filling the interstitial space are subjected to some integral operators over representative volume which size is understood within certain set of length scale constraints. These length scale constraints were introduced for proper upscaling based on the pioneering work of several researchers including (Whitaker, Gray, Hassanizadeh, Bear, Bachmat, Quintard, Slattery, Cushman, Marle and many others). Recently, (Salama & van Geel, 2008a) have postulated the conditions required for proper upscaling such that one may get the correct set of equations subject to the constraints pertinent to adhering to these conditions, as will be explained later. This set of macroscopic point equations, in which macroscopic field variables are defined at yet a larger scale than their microscopic counterpart, represents the governing conservation laws at the new macroscale. In this chapter we will be concerned mainly with the method of volume averaging.

3. Requirements for proper averaging

The upscaling process implies a one to one mapping between two domains one of which represents the actual porous medium and the second represents a fictitious continuous domain. Any point in the actual porous medium domain may lay on either the solid phase or the fluid phase and thus one can define a phase function which equals one if the point is in the fluid phase and is zero if it lays elsewhere. Over the actual porous medium domain, field variables are defined only over fluid phases (i.e., they only have values in the corresponding fluid-phase and zero elsewhere). These variables and/or their derivatives may not be continuous, particularly at the interfaces. The corresponding point over the fictitious medium, on the other hand, lays over a continuum where it is immaterial to talk about particular phase. Moreover, the macroscopic field variables defined over the fictitious medium are continuous over the whole domain, except possibly at the external boundaries. These macroscopic field variables represent the behavior of the fluid continuum contained within a certain volume (called representative elementary volume, REV) and are assigned to a single point in the fictitious domain. In order to succeed in establishing correctly this mapping process, extensive amount of research work have been conducted since the second half of the last century by several research groups including the pioneers mentioned earlier. On reviewing this work, (?) and recently Salama & van Geel (2008a) proposed a set of requirements such that proper upscaling may be achieved. They require that,

1. The smoothed macroscopic variables are free from pore-related heterogeneity.
2. The smoothed macroscopic variables do not depend on the size of the averaging volume.

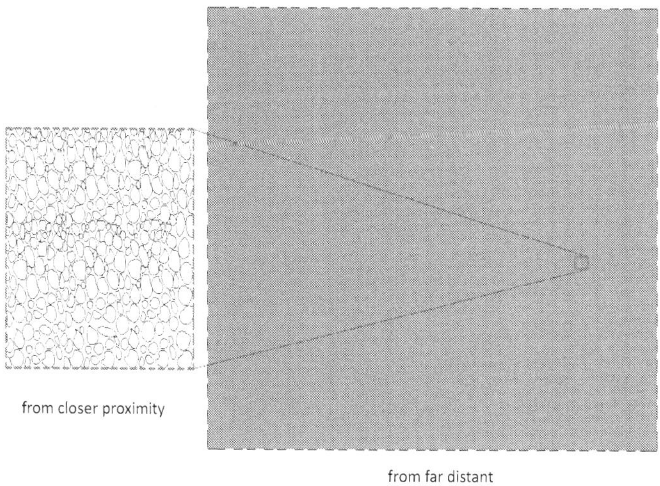

Fig. 1. View of two different observers at different proximity from a given porous medium domain.

3. The extent of the domain under study is large enough compared with the size of the averaging volume.
4. The amount of any conservative quantity (mass, momentum, energy, etc.) within any given volume (of the size of the REV or larger) is the same if evaluated over the actual porous medium domain or the corresponding fictitious domain. And, similarly, the flux of any conservative quantity across any surface (of the size of REA or larger) is also the same in both the actual and the fictitious domains.

These requirements necessitate the followings:

- An averaging volume exists for every macroscopic field variable.
- A common range of averaging volume may be found for all the field variables.

Adhering to the first three requirements, (Whitaker, 1967) indicated that if ℓ_β represents a length scale pertinent to the internal microscopic structure of the porous medium (typical pore or grain diameter) and if L represents a length scale associated with the extent of the domain of interest, then the length scale of the averaging volume should be such that it satisfies the following constraints

I. $\ell >> \ell_\beta$

II. $\ell << L$

Adhering to the fourth requirement, on the other hand, Salama & van Geel (2008a) were able to establish the proper averaging operator. They indicated that if we consider any conservative, intensive quantity, ψ_β, which may be scalar (e.g., mass of certain species per unit volume, energy per unit mass, etc.), or vector (e.g., linear momentum per unit mass). The total amount of ψ_β should equal to that evaluated over the same volume in the fictitious porous medium, Fig.2, which may be evaluated as:

$$\psi_{total} = \int_{REV} \rho_\beta(\mathbf{r},t)\psi_\beta(\mathbf{r},t)\gamma_\beta(\mathbf{r})dv \tag{1}$$

where ρ_β is the density of the β-phase, ψ_β is an intensive quantity, γ_β is the phase function, and $\mathbf{r}$ represents the position vector spanning the REV. The time, t, in the argument of the indicator function, γ_β represents the scenario of moving interfaces (e.g. immiscible multiphase system). In our case, however, the time may be omitted due to the fact that for our solid-fluid systems, the interface boundaries are assumed fixed in space. Refer to Fig.2 for a geometrical illustration.

The total amount of ψ_β should equal to that evaluated over the same volume in the fictitious porous medium, Fig.3, which may be evaluated as:

$$\psi_{total} = \int_{REV} \left\langle \rho_\beta \right\rangle^\beta (\mathbf{r},t) \left\langle \psi_\beta \right\rangle^\beta (\mathbf{r},t)\epsilon_\beta(\mathbf{r})dv \tag{2}$$

where $\left\langle \rho_\beta \right\rangle^\beta$ and $\left\langle \psi_\beta \right\rangle^\beta$ represent the intrinsic phase average of the β-phase density and intensive quantity as described earlier in Eq. (2), and ϵ_β is the porosity function over the REV. That is,

$$\int_{REV} \rho_\beta(\mathbf{r},t)\psi_\beta(\mathbf{r},t)\gamma_\beta(\mathbf{r},t)dv = \int_{REV} \left\langle \rho_\beta \right\rangle^\beta (\mathbf{r},t) \left\langle \psi_\beta \right\rangle^\beta (\mathbf{r},t)\epsilon_\beta(\mathbf{r})dv \tag{3}$$

From the condition that the averaged quantities are assigned to the centroid of the averaging volume, (Salama & van Geel, 2008a) postulated that,

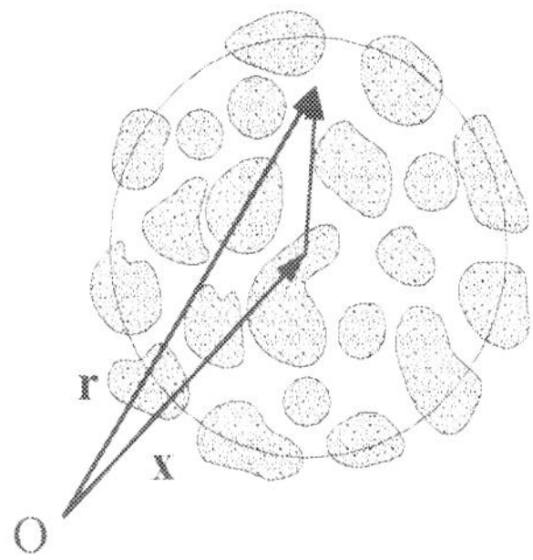

Fig. 2. An REV.

i. $\int_{REV} \gamma_\beta(\mathbf{r},t)dv = \epsilon_\beta(\mathbf{x},t)v = \int_{REV} \epsilon_\beta(\mathbf{r},t)dv$

ii. $\int_{REV} \rho_\beta(\mathbf{r},t)\gamma_\beta(\mathbf{r})dv = \left\langle \rho_\beta \right\rangle^\beta (\mathbf{x},t) \left\langle \psi_\beta \right\rangle^\beta (\mathbf{x},t)v = \int_{REV} \left\langle \rho_\beta \right\rangle^\beta (\mathbf{r},t)\epsilon_\beta(\mathbf{r},t)dv$

iii. $\int_{REV} \rho_\beta(\mathbf{r},t)\psi_\beta(\mathbf{r},t)\gamma_\beta(\mathbf{r})dv = \left\langle \rho_\beta \right\rangle^\beta (\mathbf{x},t) \left\langle \psi_\beta \right\rangle^\beta (\mathbf{x},t)\epsilon_\beta(\mathbf{x},t)v =$
$\int_{REV} \left\langle \rho_\beta \right\rangle^\beta (\mathbf{r},t) \left\langle \psi_\beta \right\rangle^\beta (\mathbf{r},t)\epsilon_\beta(\mathbf{r},t)dv$

where $\left\langle \rho_\beta \right\rangle^\beta (\mathbf{x},t) \left\langle \psi_\beta \right\rangle^\beta (\mathbf{x},t)$ and $\epsilon_\beta(\mathbf{x})$ are the intrinsic-phase average of ρ_β, ψ_β and the porosity at the centroid of the REV, respectively (Fig. 2).
This indicates that sampling the same volume (REV) in both the actual porous medium and the fictitious one should yield the same conservative quantity. In fact this equation may be taken as the definition for the averaging processes. That is,

$$\left\langle \psi_\beta \right\rangle^\beta (\mathbf{x},t) = \frac{1}{\left\langle \rho_\beta \right\rangle^\beta \epsilon_\beta v} \int_v \rho_\beta(\mathbf{r},t)\psi_\beta(\mathbf{r},t)\gamma_\beta(\mathbf{r})dv \qquad (4)$$

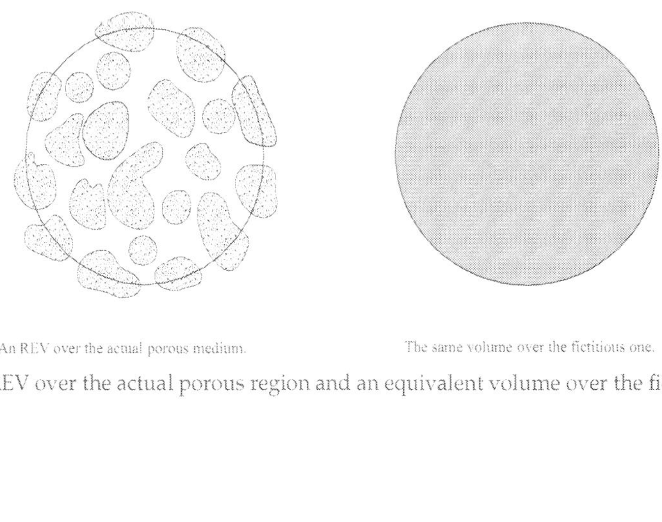

Fig. 3. An REV over the actual porous region and an equivalent volume over the fictitious one.

which was introduced by (Hassanizadeh & Gray, 1979b;a; 1980) and was called the mass average of ψ_β. In situations where β-phase is incompressible, we have $\left\langle \rho_\beta \right\rangle^\beta = \rho_\beta$ and the above equation reduces to Eq. (2) and also, if ψ_β represents a per unit volume quantity, for example the concentration of the β-phase, the density may be omitted. In other words the above mentioned three postulates may be used to define the averaging operators. Moreover, these postulates also suggest that: $\left\langle \left\langle \psi_\beta \right\rangle \right\rangle^\beta = \left\langle \psi_\beta \right\rangle$ which will show to simplify mathematical manipulation. Now for incompressible fluids Eq. (7) implies that the product $\left\langle \psi_\beta \right\rangle^\beta (\mathbf{r}, t)\epsilon_\beta(\mathbf{r})$ changes linearly over the REV. (Salama & van Geel, 2008a) further indicated that, there are at least three possibilities for the product $\left\langle \psi_\beta \right\rangle^\beta (\mathbf{r}, t)\epsilon_\beta(\mathbf{r})$ to change linearly within the REV:

1. both $\langle \psi \rangle^\beta$ and ϵ_β remain constant within the REV.
2. $\langle \psi \rangle^\beta$ changes linearly and ϵ_β remains constant.
3. $\langle \psi \rangle^\beta$ remains constant and ϵ_β changes linearly.

These requirements are rather restrictive, that is, if we allow $\langle \psi \rangle^\beta$ and ϵ_β to vary within the averaging volume such that their product, which is apparently nonlinear, they required that the nonlinearity within the averaging volume is relatively small. They defined the criteria for this case as: if ℓ_ϵ represents the length scale over which significant variation in porosity occur within the REV and ℓ_ψ represents that length scale over which significant deviation from the straight line variation of the intrinsic phase average of the conservative quantity $\left\langle \psi_\beta \right\rangle^\beta$ may occur, then they introduced their celebrated inequality

III. $\ell << min(\ell_\epsilon, \ell_\psi)$

Now adhering to these length scale constraints, it may be possible, in principle, to establish the continuum view to transport phenomena in porous media.

4. Consequences

As with our experience upon adopting the continuum hypothesis to the science of fluid mechanics, rather than the primitive Newtonian mechanics at the molecular scale an apparently different formulation needed to be adopted and several things arise. Probably the most obvious one is the fact that the state variables are modified. That is velocity vectors, for example, are, now, no longer associated with particles, rather they represent an integral behavior of a collection of several many particles contained within an averaging volume. The fluctuations of the actual particles velocity around continuum velocity suggest that two different mechanisms for heat and momentum transfer be hypothesized. One is associated with the transport of momentum and energy along with the continuum velocity and the other associated with the fluctuating components that appear as a surface flux in the continuum conservation laws, (Leal, 2007). We expect that these mechanisms become even more pronounced when adopting the continuum hypothesis to porous media. As an example, the dispersion of passive solute in pure liquids is very much influenced by the diffusion coefficient which is a macroscopic property of the medium. In porous media, on the other hand, dispersion becomes more pronounced and is no longer a property of the different fluids; it also depends on the internal geometrical structure of the porous medium. A second consequence of the continuum hypothesis is an uncertainty in the boundary conditions to be used in conjunction with the resulting macroscopic equations for motion and heat and mass transfer (Salama & van Geel, 2008b). A third consequence is the fact that the derived macroscopic point equations contain terms at the lower scale. These terms makes the macroscopic equations unclosed. Therefore, they need to be represented in terms of macroscopic field variables though parameters that me be identified and measured.

5. Single-phase flow modeling

5.1 Conservation laws

Following the constraints introduced earlier to properly upscale equations of motion of fluid continuum to be adapted to the upscaled continuum of porous medium, researchers and scientists were able to suggest the governing laws at the new continuum. They may be written for incompressible fluids as:

Continuity

$$\nabla \cdot \left\langle \mathbf{v}_\beta \right\rangle = 0 \tag{5}$$

Momentum

$$\begin{aligned} &\rho_\beta \frac{\partial \langle \mathbf{v}_\beta \rangle}{\partial t} + \rho_\beta \left\langle \mathbf{v}_\beta \right\rangle^\beta \cdot \nabla \left\langle \mathbf{v}_\beta \right\rangle^\beta = -\nabla \left\langle p_\beta \right\rangle^\beta + \rho_\beta g + \\ &\mu_\beta \nabla^2 \left\langle \mathbf{v}_\beta \right\rangle^\beta - \frac{\mu_\beta}{K} \left\langle \mathbf{v}_\beta \right\rangle - \frac{\rho_\beta F \epsilon_\beta}{\sqrt{K}} \left| \left\langle \mathbf{v}_\beta \right\rangle \right| \left\langle \mathbf{v}_\beta \right\rangle \end{aligned} \tag{6}$$

Energy

$$\sigma \frac{\partial \left\langle T_\beta \right\rangle^\beta}{\partial t} + \left\langle \mathbf{v}_\beta \right\rangle \cdot \nabla \left\langle T_\beta \right\rangle^\beta = k \nabla^2 \left\langle T_\beta \right\rangle^\beta \pm Q \tag{7}$$

Solute transport

$$\epsilon \frac{\partial \left\langle c_\beta \right\rangle^\beta}{\partial t} + \left\langle \mathbf{v}_\beta \right\rangle \cdot \nabla \left\langle c_\beta \right\rangle^\beta = \nabla \cdot \left(D \cdot \nabla \left\langle c_\beta \right\rangle^\beta \right) \pm S \tag{8}$$

where $\left\langle \mathbf{v}_\beta \right\rangle^\beta$ and $\left\langle p_\beta \right\rangle^\beta$ epresent the intrinsic average velocity and pressure, respectively and $\left\langle \mathbf{v}_\beta \right\rangle$ is the superficial average velocity, $\mathbf{v} = \sqrt{u^2 + v^2}$, $\sigma = (\rho C_p)_M / (\rho C_p)_f$, $k = (k_M / (\rho C_p)_f$, is the thermal diffusivity. From now on we will drop the averaging operator, $\langle \rangle$, to simplify notations. The energy equation is written assuming thermal equilibrium between the solid matrix and the moving fluid. The generic terms, Q and S, in the energy and solute equations represent energy added or taken from the system per unit volume of the fluid per unit time and the mass of solute added or depleted per unit volume of the fluid per unit time due to some source (e.g., chemical reaction which depends on the chemistry, the surface properties of the fluid/solid interfaces, etc.). Dissolution of the solid phase, for example, adds solute to the fluid and hence $S > 0$, while precipitation depletes it, i.e., $S < 0$. Organic decomposition or oxidation or reduction reactions may provide both sources and sinks. Chemical reactions in porous media are usually complex that even in apparently simple processes (e.g., dissolution), sequence of steps are usually involved. This implies that the time scale of the slowest step essentially determines the time required to progress through the sequence of steps. Among the different internal steps, it seems that the rate-limiting step is determined by reaction kinetics. Therefore, the chemical reaction source term in the solute transport equation may be represented in terms of rate constant, k, which lumps several factors multiplied by the concentration, i.e.,

$$S = kf(s) \tag{9}$$

where k has dimension time^{-1} and the form of the function f may be determined experimentally, (e.g., in the form of a power law). Apparently, the above set of equations is nonlinear and hence requires, generally, numerical techniques to provide solution (finite difference, finite element, boundary element, etc.). However, in some simplified situations, one may find similarity transformations to transform the governing set of partial differential equations to a set of ordinary differential equations which greatly simplify solutions. As an example, in the following subsection we show the results of using such similarity transformations in investigating the problem of natural convection and double dispersion past a vertical flat plate immersed in a homogeneous porous medium in connection with boundary layer approximation.

5.2 Examble: Chemical reaction in natural convection

The present investigation describes the combined effect of chemical reaction, solutal, and thermal dispersions on non-Darcian natural convection heat and mass transfer over a vertical flat plate in a fluid saturated porous medium (El-Amin et al., 2008). It can be described as follows: A fluid saturating a porous medium is induced to flow steadily by the action of buoyancy forces originated by the combined effect of both heat and solute concentration on the density of the saturating fluid. A heated, impermeable, semi-infinite vertical wall with both temperature and concentration kept constant is immersed in the porous medium. As heat and species disperse across the fluid, its density changes in space and time and the fluid is induced to flow in the upward direction adjacent to the vertical plate. Steady state is reached when both temperature and concentration profiles no longer change with time. In this study, the inclusion of an n-order chemical reaction is considered in the solute transport equation. On the other hand, the non-Darcy (Forchheimer) term is assumed in the flow equations. This term accounts for the non-linear effect of pore resistance and was first introduced by Forchheimer. It incorporates an additional empirical (dimensionless) constant, which is a property of the solid matrix, (Herwig & Koch, 1991). Thermal and mass diffusivities are defined in terms of the molecular thermal and solutal diffusivities, respectively. The Darcy and non-Darcy flow, temperature and concentration fields in porous media are observed to be governed by complex interactions among the diffusion and convection mechanisms as will be discussed later. It is assumed that the medium is isotropic with neither radiative heat transfer nor viscous dissipation effects. Moreover, thermal local equilibrium is also assumed. Physical model and coordinate system is shown in Fig.4.

The x-axis is taken along the plate and the y-axis is normal to it. The wall is maintained at constant temperature and concentration, T_w and C_w, respectively. The governing equations for the steady state scenario [as given by (Mulolani & Rahman, 2000; El-Amin, 2004) may be presented as:

Continuity:

$$\frac{\partial u}{\partial x} + \frac{\partial v}{\partial y} = 0 \tag{10}$$

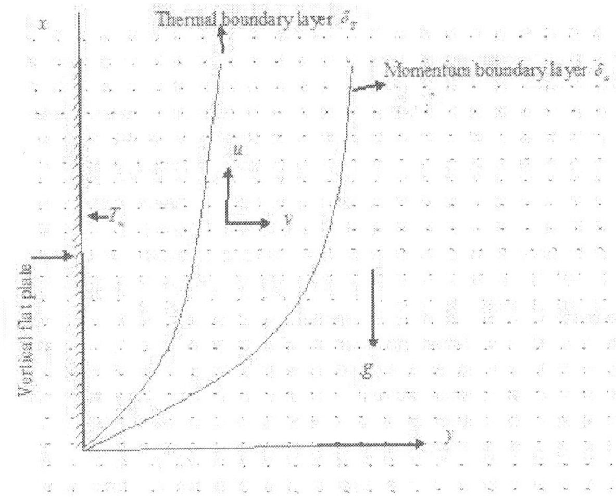

Fig. 4. Physical model and coordinate system.

Momentum:

$$u\frac{c\sqrt{K}}{\nu}u\left|\mathbf{v}\right| = -\frac{K}{\mu}\left(\frac{\partial p}{\partial x}+\rho g\right) \tag{11}$$

$$v\frac{c\sqrt{K}}{\nu}v\left|\mathbf{v}\right| = -\frac{K}{\mu}\left(\frac{\partial p}{\partial y}\right) \tag{12}$$

Energy:

$$u\frac{\partial T}{\partial x}+v\frac{\partial T}{\partial y} = \frac{\partial}{\partial x}\left(\alpha_x\frac{\partial T}{\partial x}\right)+\frac{\partial}{\partial y}\left(\alpha_y\frac{\partial T}{\partial y}\right) \tag{13}$$

Solute transport:

$$u\frac{\partial C}{\partial x}+v\frac{\partial C}{\partial y} = \frac{\partial}{\partial x}\left(D_x\frac{\partial C}{\partial x}\right)+\frac{\partial}{\partial y}\left(D_y\frac{\partial C}{\partial y}\right)-K_0\left(C-C_\infty\right)^n \tag{14}$$

Density

$$\rho = \rho_\infty\left[1-\beta^*\left(T-T_\infty\right)-\beta^{**}\left(C-C_\infty\right)\right] \tag{15}$$

Along with the boundary conditions:

$$\begin{aligned} &y=0: v=0, T_w=const., C_w=const.;\\ &y\to\infty: u=0, T\to T_\infty, C\to C_\infty \end{aligned} \tag{16}$$

where β^* is the thermal expansion coefficient β^{**} is the solutal expansion coefficient. It should be noted that u and v refers to components of the volume averaged (superficial) velocity of the fluid. The chemical reaction effect is acted by the last term in the right hand side of Eq. (14), where, the power n is the order of reaction and K_0 is the chemical reaction constant. It is assumed that the normal component of the velocity near the boundary is small compared with the other component of the velocity and the derivatives of any quantity in the normal direction are large compared with derivatives of the quantity in direction of the wall. Under these assumptions, Eq. (10) remains the same, while Eqs. (11)- (15) become:

$$u+\frac{c\sqrt{K}}{v}u^{2}=-\frac{K}{\mu}\left(\frac{\partial p}{\partial x}+\rho g\right) \tag{17}$$

$$\frac{\partial p}{\partial y}=0 \tag{18}$$

$$u\frac{\partial T}{\partial x}+v\frac{\partial T}{\partial y}=\frac{\partial}{\partial y}\left(\alpha_{y}\frac{\partial T}{\partial y}\right) \tag{19}$$

$$u\frac{\partial C}{\partial x}+v\frac{\partial C}{\partial y}=\frac{\partial}{\partial y}\left(D_{y}\frac{\partial C}{\partial y}\right)-K_{0}\left(C-C_{\infty}\right)^{n} \tag{20}$$

Following (Telles & V.Trevisan, 1993), the quantities of α_y and D_y are variables defined as $\alpha_y=\alpha+\gamma d|v|$ and $D_y=D+\zeta d|v|$ where, α and D are the molecular thermal and solutal diffusivities, respectively, whereas $\gamma d|v|$ and $\zeta d|v|$ represent dispersion thermal and solutal diffusivities, respectively. This model for thermal dispersion has been used extensively (e.g., (Cheng, 1981; Plumb, 1983; Hong & Tien, 1987; Lai & Kulacki, 1989; Murthy & Singh, 1997) in studies of non-Darcy convective heat transfer in porous media. Invoking the Boussinesq approximations, and defining the velocity components u and v in terms of stream function ψ as: $u=\partial\psi/\partial y$ and $v=-\partial\psi/\partial x$, the pressure term may be eliminated between Eqs. (17) and (18) and one obtains:

$$\frac{\partial^{2}\psi}{\partial y^{2}}+\frac{c\sqrt{K}}{v}\frac{\partial}{\partial y}\left(\frac{\partial\psi}{\partial y}\right)^{2}=\left(\frac{Kg\beta^{*}}{\mu}\frac{\partial T}{\partial y}+\frac{Kg\beta^{**}}{\mu}\frac{\partial C}{\partial y}\right)\rho_{\infty} \tag{21}$$

$$\frac{\partial\psi}{\partial y}\frac{\partial T}{\partial x}-\frac{\partial\psi}{\partial x}\frac{\partial T}{\partial y}=\frac{\partial}{\partial y}\left[\left(\alpha+\gamma d\frac{\partial\psi}{\partial y}\right)\frac{\partial T}{\partial y}\right] \tag{22}$$

$$\frac{\partial\psi}{\partial y}\frac{\partial C}{\partial x}-\frac{\partial\psi}{\partial x}\frac{\partial C}{\partial y}=\frac{\partial}{\partial y}\left[\left(D+\zeta d\frac{\partial\psi}{\partial y}\right)\frac{\partial C}{\partial y}\right]-K_{0}\left(C-C_{\infty}\right)^{n} \tag{23}$$

Introducing the similarity variable and similarity profiles (El-Amin, 2004):

$$\eta=Ra_{x}^{1/2}\frac{y}{x},f(\eta)=\frac{\psi}{\alpha Ra_{x}^{1/2}},\theta(\eta)=\frac{T-T_{\infty}}{T_{w}-T_{\infty}},\phi(\eta)=\frac{C-C_{\infty}}{C_{w}-C_{\infty}} \tag{24}$$

The problem statement is reduced to:

$$f''+2F_{0}Ra_{d}f'f''=\theta'+N\phi' \tag{25}$$

$$\theta''+\frac{1}{2}f\theta'+\gamma Ra_{d}\left(f'\theta''+f''\theta'\right)=0 \tag{26}$$

$$\phi''+\frac{1}{2}Lef\phi'+\zeta LeRa_{d}\left(f'\phi''+f''\phi'\right)-Sc\lambda\frac{Gc}{Re_{x}^{2}}\phi^{n=0} \tag{27}$$

As mentioned in (El-Amin, 2004), the parameter $F_0=c\sqrt{K}\alpha/vd$ collects a set of parameters that depend on the structure of the porous medium and the thermo physical properties of the fluid saturating it, $Ra_d=Kg\beta^{*}(T_w-T_\infty)d/\alpha v$ is the modified, pore-diameter-dependent

Rayleigh number, and $N = \beta^{**}(C_w - C_\infty)/\beta^* v$ is the buoyancy ratio parameter. With analogy to (Mulolani & Rahman, 2000; Aissa & Mohammadein, 2006), we define Gc to be the modified Grashof number, Re_x is local Reynolds number, Sc and λ are Schmidt number and non-dimensional chemical reaction parameter defined as $Gc = \beta^{**}g(C_w - C_\infty)^2 x^3/\nu^2$, $Re_x = u_r x/\nu$, $Sc = \nu/D$ and $\lambda = K_0 \alpha d(C_w - C_\infty)^{n-3}/Kg\beta^{**}$, where the diffusivity ratio Le (Lewis number) is the ratio of Schmidt number and Prandtl number, and $u_r = \sqrt{g\beta^* d(T_w - T_\infty)}$ is the reference velocity as defined by (Elbashbeshy, 1997).
Eq. (27) can be rewritten in the following form:

$$\phi'' + \frac{1}{2}Lef\phi' + \zeta LeRa_d\left(f'\phi'' + f''\phi'\right) - \chi\phi^n = 0 \tag{28}$$

With analogy to (Prasad et al., 2003; Aissa & Mohammadein, 2006), the non-dimensional chemical reaction parameter χ is defined as $\chi = Sc\lambda Gc/Re_x^2$. The boundary conditions then become:

$$f(0) = 0, \theta(0) = \phi(0) = 1, f'(\infty) = \theta(\infty) = \phi(\infty) = 0 \tag{29}$$

It is noteworthy to state that $F_0 = 0$ corresponds to the Darcian free convection regime, $\gamma = 0$ represents the case where the thermal dispersion effect is neglected and $\zeta = 0$ represents the case where the solutal dispersion effect is neglected. In Eq. (16), $N > 0$ indicates the aiding buoyancy and $N < 0$ indicates the opposing buoyancy. On the other hand, from the definition of the stream function, the velocity components become $u = (\alpha Ra_x/x)f'$ and $v = -(\alpha Ra_x^{1/2}/2x)[f - \eta f']$. The local heat transfer rate which is one of the primary interest of the study is given by $q_w = -k_e(\partial T/\partial y)|_{y=0}$, where, $k_e = k + k_d$ is the effective thermal conductivity of the porous medium which is the sum of the molecular thermal conductivity k and the dispersion thermal conductivity k_d. The local Nusselt number Nu_x is defined as $Nu_x = q_w x/(T_w - T_\infty)k_e$. Now the set of primary variables which describes the problem may be replaced with another set of dimensionless variables. This include: a dimension less variable that is related to the process of heat transfer in the given system which may be expressed as $Nu_x/\sqrt{Ra_x} = -[1 + \gamma Ra_d F'(0)]\theta'(0)$. Also, the local mass flux at the vertical wall that is given by $j_w = -D_y(\partial C/\partial y)|_{y=0}$ defines another dimensionless variable that is the local Sherwood number is given by, $Sh_x = j_w x/(C_w - C_\infty)D$. This, analogously, may also define another dimensionless variable as $Sh_x/\sqrt{Ra_x} = -[1 + \zeta Ra_d F'(0)]\phi'(0)$.
The details of the effects of all these parameters are presented in (El-Amin et al., 2008). We, however, highlight the role of the chemical reaction on this system. The effect of chemical reaction parameter χ on the concentration as a function of the boundary layer thickness η and with respect to the following parameters: $Le = 0.5$, $F_0 = 0.3$, $Ra_d = 0.7$, $\gamma = \zeta = 0.0$, $N = 0.1$ are plotted in Fig.5. This figure indicates that increasing the chemical reaction parameter decreases the concentration distributions, for this particular system. That is, chemical reaction in this system results in the consumption of the chemical of interest and hence results in concentration profile to decrease. Moreover, this particular system also shows the increase in chemical reaction parameter χ to enhance mass transfer rates (defined in terms of Sherwood number) as shown in Fig.6. It is worth mentioning that the effects of chemical reaction on velocity and temperature profiles as well as heat transfer rate may be negligible. Figs. 7 and 8 illustrate, respectively, the effect of Lewis number Le on Nusselt number and Sherwood number for various with the following parameters set as $\chi = 0.02$, $Ra_d = 0.7$, $F_0 = 0.3$, $N = -0.1$, $\gamma = 0.0$. The parameter ζ seems to reduce the heat transfer rates especially with higher Le number as shown in Fig. 7. In the case of mass transfer rates

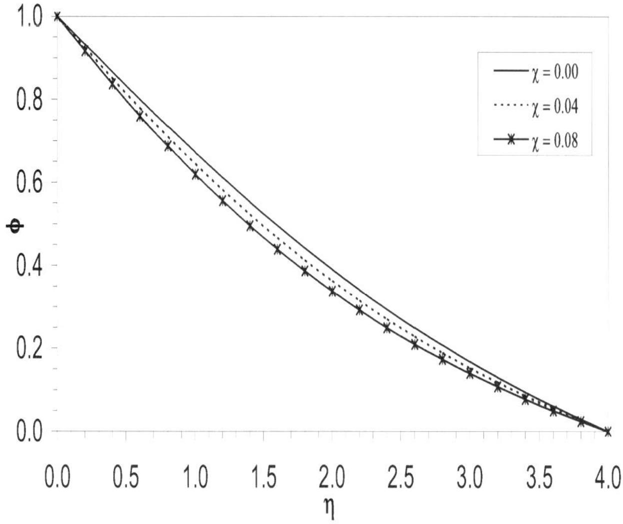

Fig. 5. Variation of dimensionless concentration with similarity space variable η for different χ ($Le = 0.5$, $F_0 = 0.3$, $Ra_d = 0.7$, $\gamma = \zeta = 0.0$, $N = -0.1$).

(defined in terms of Sherwood number), Fig. 8 illustrates that the parameter ζ enhances the mass transfer rate with small values of $Le \tilde{<} 1.55$ and the opposite is true for high values of $Le \tilde{>} 1.55$. This may be explained as follows: for small values of Le number, which indicates

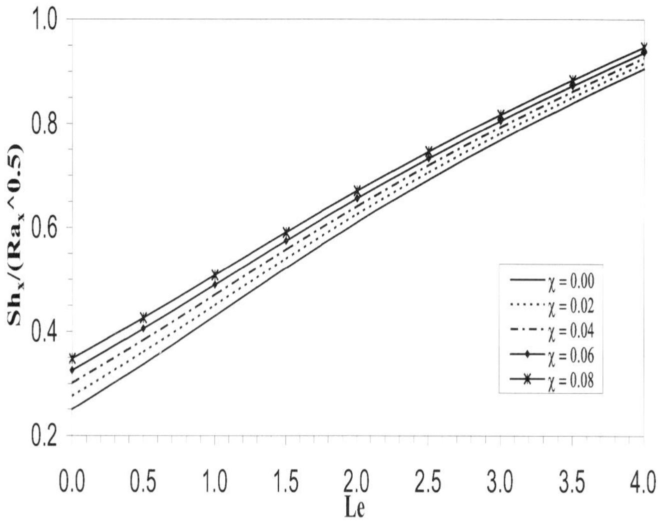

Fig. 6. Effect of Lewis number on Sherwood number for various χ ($F_0 = 0.3$, $Ra_d = 0.7$, $\gamma = \zeta = 0.0$, $N = -0.1$).

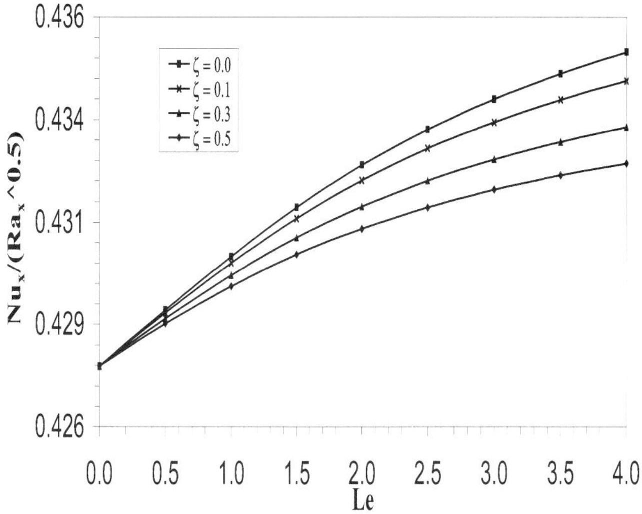

Fig. 7. Variation of Nusselt number with Lewis number for various ζ ($\chi = 0.02$, $F_0 = 0.3$, $Ra_d = 0.7$, $\gamma = 0.0$, $N = -0.1$).

that mass dispersion outweighs heat dispersion, the increase in the parameter ζ causes mass dispersion mechanism to be higher and since the concentration at the wall is kept constant this increases concentration gradient near the wall and hence increases Sherwood number. As

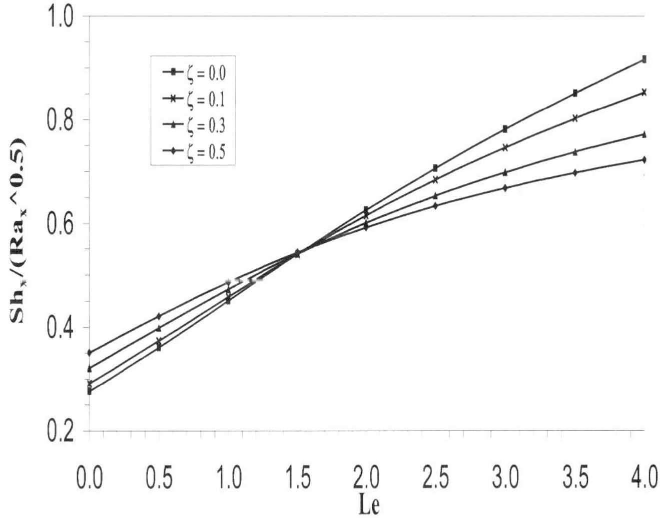

Fig. 8. Effect of Lewis number on Sherwood number for various ζ ($\chi = 0.02$, $F_0 = 0.3$, $Ra_d = 0.7$, $\gamma = 0.0$, $N = -0.1$).

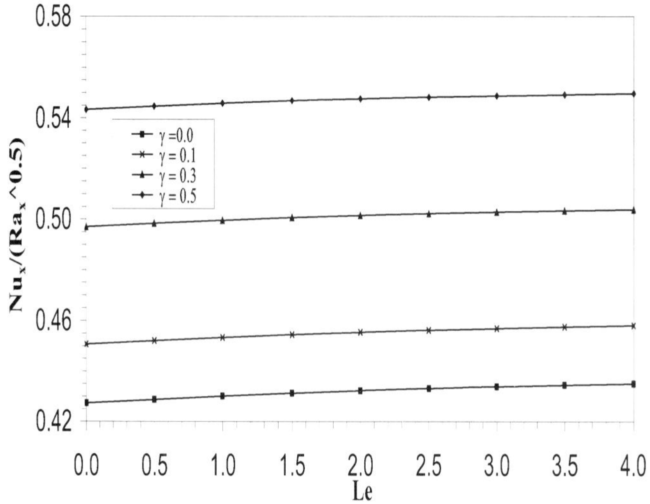

Fig. 9. Variation of Nusselt number with Lewis number for various γ ($\chi = 0.02$, $F_0 = 0.3$, $Ra_d = 0.7$, $\zeta = 0.0$, $N = -0.1$).

Le increases ($Le > 1$), heat dispersion outweighs mass dispersion and with the increase in ζ concentration gradient near the wall becomes smaller and this results in decreasing Sherwood number. Fig. 9 indicates that the increase in thermal dispersion parameter enhances the heat transfer rates.

6. Multi-phase flow modeling

Multi-phase systems in porous media are ubiquitous either naturally in connection with, for example, vadose zone hydrology, which involves the complex interaction between three phases (air, groundwater and soil) and also in many industrial applications such as enhanced oil recovery (e.g., chemical flooding and CO_2 injection), Nuclear waste disposal, transport of groundwater contaminated with hydrocarbon (NAPL, DNAPL), etc. Modeling of Multi-phase flows in porous media is, obviously, more difficult than in single-phase systems. Here we have to account for the complex interfacial interactions between phases as well as the time dependent deformation they undergo. Modeling of compositional flows in porous media is, therefore, necessary to understand a number of problems related to the environment (e.g., CO_2 sequestration) and industry (e.g., enhanced oil recovery). For example, CO_2 injection in hydrocarbon reservoirs has a double benefit, on the one side it is a profitable method due to issues related to global warming, and on the other hand it represents an effective mechanism in hydrocarbon recovery. Modeling of these processes is difficult because the several mechanisms involved. For example, this injection methodology associates, in addition to species transfer between phases, some substantial changes in density and viscosity of the phases. The number of phases and compositions of each phase depend on the thermodynamic conditions and the concentration of each species. Also, multi-phase compositional flows have varies applications in different areas such as nuclear reactor safety analysis (Dhir, 1994), high-level radioactive waste repositories (Doughty & Pruess, 1988), drying of porous solids and soils (Whitaker, 1977), porous heat pipes (Udell, 1985), geothermal energy production (Cheng, 1978), etc. The mathematical formulation of the transport phenomena are governed by conservation principles for each phase separately and by appropriate interfacial conditions between various phases. Firstly we give the general governing equations of multi-phase, multicomponent transport in porous media. Then, we provide them in details with analysis for two- and three-phase flows. The incompressible multi-phase compositional flow of immiscible fluids are described by the mass conservation in a phase (continuity equation), momentum conservation in a phase (generalized Darcy's equation) and mass conservation of component in phase (spices transport equation). The transport of N-components of multi-phase flow in porous media are described by the molar balance equations. Mass conservation in phase α :

$$\frac{\partial(\phi\rho_\alpha S_\alpha)}{\partial t} = -\nabla\cdot(\rho_\alpha u_\alpha) + q_\alpha \tag{30}$$

Momentum conservation in phase α:

$$u_\alpha = -\frac{Kk_{r\alpha}}{\mu_\alpha}(\nabla p_\alpha + \rho_\alpha g\nabla z) \tag{31}$$

Energy conservation in phase α:

$$\frac{\partial}{\partial t}(\epsilon\rho_\alpha S_\alpha h_\alpha) + \nabla\cdot(\rho_\alpha u_\alpha h_\alpha) = \nabla\cdot(\epsilon S_\alpha k_\alpha \nabla T) + \bar{q}_\alpha \tag{32}$$

Mass conservation of component i in phase α:

$$\frac{\partial(\phi c z_i)}{\partial t} + \nabla\cdot\sum_\alpha c_\alpha x_{\alpha i} u_\alpha = \nabla\cdot\left(\phi D^i_\alpha \nabla(c z_i)\right) + F_i, \quad i = 1,\cdots,N \tag{33}$$

where the index α denotes to the phase. S, p, q, u, k_r, ρ and μ are the phase saturation, pressure, mass flow rate, Darcy velocity, relative permeability, density and viscosity, respectively. c is the overall molar density; z_i is the total mole fraction of ith component; c_α is the phase molar densities; $x_{\alpha i}$ is the phase molar fractions; and F_i is the source/sink term of the ith component which can be considered as the phase change at the interface between the phase α and other phases; and/or the rate of interface transfer of the component i caused by chemical reaction (chemical non-equilibrium). D^i_α is a macroscopic second-order tensor incorporating diffusive and dispersive effects. The local thermal equilibrium among phases has been assumed, ($T_\alpha = T, \forall\alpha$), and k_α and $\bar{q}_\alpha$ represent the effective thermal conductivity of the phase α and the interphase heat transfer rate associated with phase α, respectively. Hence, $\sum_\alpha \bar{q}_\alpha = q$, q is an external volumetric heat source/sink (Starikovicius, 2003). The phase enthalpy k_α is related to the temperature T by, $h_\alpha = \int_0^T c_{p\alpha} dT + h^0_\alpha$. The saturation S_α of the phases are constrained by, $c_{p\alpha}$ and h^0_α are the specific heat and the reference enthalpy oh phase α, respectively.

$$\sum_\alpha S_\alpha = 1 \tag{34}$$

One may defined the phase saturation as the fraction of the void volume of a porous medium filled by this fluid phase. The mass flow rate q_α, describe sources or sinks and can be defined by the following relation (Chen, 2007),

$$q = \sum_j \rho^j q^j \delta(x - x^j) \tag{35}$$

$$q = -\sum_j \rho^j q^j \delta(x - x^j) \tag{36}$$

The index j represents the points of sources or sinks. Eq. (35) represents sources and q^j represents volume of the fluid (with density ρ^j) injected per unit time at the points locations x^j, while, Eq. (36) represents sinks and q^j represents volume of the fluid produced per unit time at x^j.
On the other hand, the molar density of wetting and nonwetting phases is given by,

$$c_\alpha = \sum_{i=1}^{N} c_{\alpha i} \tag{37}$$

where $c_{\alpha i}$ is the molar densities of the component i in the phase α. Therefore, the mole fraction of the component i in the respective phase is given as,

$$x_{\alpha i} = \frac{c_{\alpha i}}{c_\alpha}, \quad i = 1, \cdots, N \tag{38}$$

The mole fraction balance implies that,

$$\sum_{i=1}^{N} x_{\alpha i} = 1 \tag{39}$$

Also, for the total mole fraction of ith component,

$$\sum_{i=1}^{N} z_i = 1 \tag{40}$$

Alternatively, Eq. (32) can be rewritten in the following form,

$$\frac{\partial}{\partial t}\left(\phi \sum_\alpha c_\alpha x_{\alpha i} S_\alpha\right) + \nabla \cdot \sum_\alpha c_\alpha x_{\alpha i} u_\alpha = \nabla \cdot \left(\phi c_\alpha S_\alpha D_\alpha^i \nabla x_{\alpha i}\right) + F_i, \quad i = 1, \cdots, N \tag{41}$$

F_i may be written as,

$$F_i = \sum_\alpha x_{\alpha i} q_\alpha, \quad i = 1, \cdots, N \tag{42}$$

where q_α is the phase flow rate given by Eqs. (35), (36). From Eqs. (32) and (41), one may deduce,

$$cz_i = \sum_\alpha c_\alpha x_{\alpha i} S_\alpha = \sum_\alpha c_{\alpha i} S_\alpha, \quad i = 1, \cdots, N \tag{43}$$

If one uses the total mass variable X of the system (Nolen 1973; Young and Stephenson 1983),

$$X = \sum_\alpha c_\alpha S_\alpha \tag{44}$$

Therefore,

$$1 = \sum_{\alpha} \frac{c_\alpha S_\alpha}{X} = \sum_{\alpha} C_\alpha \tag{45}$$

where C_α is the mass fraction phase α, respectively.
The quantity,

$$k_\alpha = K k_{r\alpha} \tag{46}$$

is known as effective permeability of the phase α. The relative permeability of a phase is a dimensionless measure of the effective permeability of that phase. It is the ratio of the effective permeability of that phase to the absolute permeability. Also, it is interesting to define the quantities m_α which is known as mobility ratios of phases α, respectively are given by,

$$m_\alpha = \frac{k_\alpha}{\mu_\alpha} \tag{47}$$

The capillary pressure is the the difference between the pressures for two adjacent phases α_1 and α_2, given as,

$$p_{c\alpha_1\alpha_2} = p_{\alpha_1} - p_{\alpha_2} \tag{48}$$

The capillary pressure function is dependent on the pore geometry, fluid physical properties and phase saturations. The two phase capillary pressure can be expressed by Leverett dimensionless function $J(S)$, which is a function of the normalized saturation S,

$$p_c = \gamma \left(\frac{\phi}{K} \right)^{\frac{1}{2}} J(S) \tag{49}$$

The $J(S)$ function typically lies between two limiting (drainage and imbibition) curves which can be obtained experimentally.

6.1 Two-phase compositional flow

The governing equations of two-phase compositional flow of immiscible fluids are given by,
Mass conservation in phase α:

$$\frac{\partial(\phi \rho_\alpha S_\alpha)}{\partial t} = -\nabla \cdot (\rho_\alpha u_\alpha) + q_\alpha \quad \alpha = w, n \tag{50}$$

Momentum conservation in phase α:

$$u_\alpha = -\frac{K k_{r\alpha}}{\mu_\alpha} (\nabla p_\alpha + \rho_\alpha g \nabla z) \quad \alpha = w, n \tag{51}$$

Mass conservation of component i in phase α:

$$\frac{\partial(\phi c z_i)}{\partial t} + \nabla \cdot (c_w x_{wi} u_w + c_n x_{ni} u_n) = F_i, \quad i = 1, \cdots, N \tag{52}$$

where the index α denotes to the wetting (w) and non-wetting (n), respectively. S, p, q, u, k_r, ρ and μ are the phase saturation, pressure, mass flow rate, Darcy velocity, relative permeability, density and viscosity, respectively. c is the overall molar density; z_i is the total mole fraction of ith component; c_w, c_n are the wetting- and nonwetting-phase molar densities; x_{wi}, x_{ni} are the wetting- and nonwetting-phase molar fractions; and F_i is the source/sink term of the ith component. The saturation S_α of the phases are constrained by,

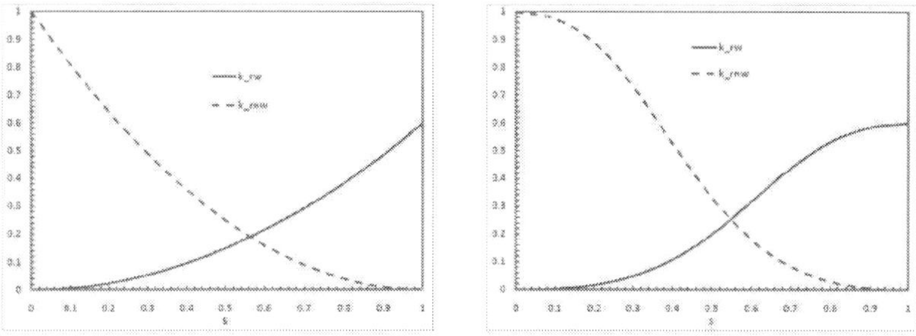

(a) Corey approximation (b) LET approximation

Fig. 10. Relative permeabilities.

$$S_w + S_n = 1 \tag{53}$$

The normalized wetting phase saturation S is given by,

$$S = \frac{S_w - S_{0w}}{1 - S_{nr} - S_{0w}} \quad 0 \leq S \leq 1 \tag{54}$$

where S_{0w} is the irreducible (minimal) wetting phase saturation and S_{nr} is the residual (minimal) non-wetting phase saturation. The expression of relation between the relative permeabilities and the normalized wetting phase saturation S, given as,

$$k_{rw} = k_{rw}^0 S^a \tag{55}$$

$$k_{rn} = k_{rn}^0 (1 - S)^b \tag{56}$$

The empirical parameters a and b can be obtained from measured data either by optimizing to analytical interpretation of measured data, or by optimizing using a core flow numerical simulator to match the experiment. $k_{rw}^0 = k_{rw}(S = 1)$ is the endpoint relative permeability to water, and $k_{rn}^0 = k_{rn}(S = 0)$ is the endpoint relative permeability to the non-wetting phase. For example, for the Corey power-law correlation, $a = b = 2$, $k_{rn}^0 = 1$, $k_{rw}^0 = 0.6$, for water-oil system see Fig.10a. Another example of relative permeabilities correlations is LET model which is more accurate than Corey model. The LET-type approximation is described by three empirical parameters L,E and T. The relative permeability correlation for water-oil system has the form,

$$k_{rw} = \frac{k_{rw}^0 S^{L_w}}{S^{L_w} + E_w (1 - S)^{T_w}} \tag{57}$$

and

$$k_{rn} = \frac{(1 - S)^{L_n}}{(1 - S)^{L_n} + E_n S^{T_n}} \tag{58}$$

The parameter E describes the position of the slope (or the elevation) of the curve. Fig. 10b shows LET relative permeabilities with $L = E = T = 2$ and $k_{rw}^0 = 0.6$ for water-oil system.

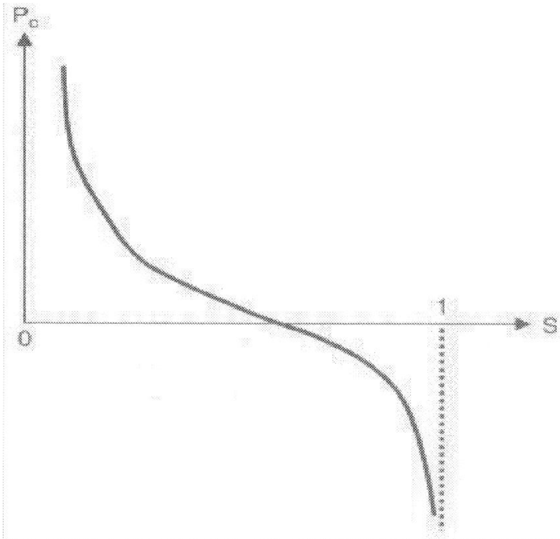

Fig. 11. Capillary pressure as a function of normalized wetting phase saturation.

Also, there are Corey- and LET-correlations for gas-water and gas-oil systems similar to the oil-water system. Correlation of the imbibition capillary pressure data depends on the type of application. For example, for water-oil system, see for example, (Pooladi-Darvish & Firoozabadi, 2000), the capillary pressure and the normalized wetting phase saturation are correlated as,

$$p_c = -B \ln S \tag{59}$$

where B is the capillary pressure parameter, which is equivalent to $\gamma\left(\frac{\phi}{K}\right)^{\frac{1}{2}}$, in the general form of the capillary pressure, Eq. (49), thus, $B \equiv -\gamma\left(\frac{\phi}{K}\right)^{\frac{1}{2}}$ and $J(S) \equiv \ln S$. Note that $J(S)$ is a scalar non-negative function. Capillary pressure as a function of normalized wetting phase (e.g. water) saturation is shown in Fig. 11. Also, the well known (van Genuchten, 1980; Brooks & Corey, 1964) capillary pressure formulae which can be written as,

$$p_c = p_0(S^{-1/m} - 1)^{1-m}, \quad 0 < m < 1 \tag{60}$$

$$p_c = p_d S^{-1/\lambda}, \quad 0.2 < \lambda < 3 \tag{61}$$

where p_0 is characteristic capillary pressure and p_d is called entry pressure.
The capillary pressure p_c is defined as a difference between the non-wetting and wetting phase pressures,

$$p_c = p_n - p_w \tag{62}$$

the total velocity defined as,

$$u = u_w + u_n \tag{63}$$

the total mobility is given by,

$$m(S) = m_w(S) + m_n(S) \tag{64}$$

the fractional flow functions are,

$$f_w(S) = \frac{m_w(S)}{m(S)}, \quad f_n(S) = \frac{m_n(S)}{m(S)} \tag{65}$$

and the density difference is,

$$\Delta\rho = \rho_n - \rho_w \tag{66}$$

On the other hand, the molar density of wetting and nonwetting phases is given by,

$$c_w = \sum_{i=1}^{N} c_{wi}, \quad c_n = \sum_{i=1}^{N} c_{ni} \tag{67}$$

where c_{wi} and c_{ni} are the molar densities of component i in the wetting phase and nonwetting phase phases, respectively. Therefore, the mole fraction of component i in the respective phase is given as,

$$x_{wi} = \frac{c_{wi}}{c_w}, \quad x_{ni} = \frac{c_{ni}}{c_n}, \quad i = 1, \cdots, N \tag{68}$$

The mole fraction balance implies that,

$$\sum_{i=1}^{N} x_{wi} = 1, \quad \sum_{i=1}^{N} x_{ni} = 1 \tag{69}$$

Also, for the total mole fraction of ith component,

$$\sum_{i=1}^{N} z_i = 1 \tag{70}$$

Alternatively, Eq. (52) can be rewritten in the following form,

$$\frac{\partial}{\partial t}\left[\phi\left(c_w x_{wi} S_w + c_n x_{ni} S_n\right)\right] + \nabla \cdot \left(c_w x_{wi} u_w + c_n x_{ni} u_n\right) = F_i, \quad i = 1, \cdots, N \tag{71}$$

F_i may be written as,

$$F_i = x_{wi} q_w + x_{ni} q_n, \quad i = 1, \cdots, N \tag{72}$$

where q_w and q_n are wetting phase and nonwetting phase phase flow rate, respectively. From Eqs. (32) and (72), one may deduce,

$$c z_i = c_w x_{wi} S_w + c_n x_{ni} S_n = c_{wi} S_w + c_{ni} S_n, \quad i = 1, \cdots, N \tag{73}$$

If one uses the total mass variable X of the system (Nolen, 1973; Young & Stephenson, 1983),

$$X = c_w S_w + c_n S_n \tag{74}$$

Therefore,

$$1 = \frac{c_w S_w}{X} + \frac{c_w S_n}{X} = C_w + C_n \tag{75}$$

where C_w and C_n are mass fractions of wetting- and nonwetting-phase of the system, respectively. It is noted that,

$$C_w = 1 - C_n \tag{76}$$

The total mole fraction of ith component, z_i, in terms of one phase (wetting phase) mass fraction and the wetting- and nonwetting-phase molar fractions, is given by,

$$z_i = C_w x_{wi} + (1 - C_w) x_{ni}, \quad i = 1, \cdots, N \tag{77}$$

The pressure equation can be obtained, using the concept of volume-balance, as follows,

$$\phi C_f \frac{\partial p}{\partial t} + \sum_{i=1}^{N} \bar{V}_i \nabla \cdot (c_w x_{wi} u_w + c_n x_{ni} u_n) = \sum_{i=1}^{N} \bar{V}_i F_i, \quad i = 1, \cdots, N \tag{78}$$

where C_f is the total fluid compressibility and V_i is the total partial molar volume of the ith component. The distribution of the each component inside the two phases is restricted to the stable thermodynamic equilibrium in terms of phases' fugacities, f_{wi} and f_{ni} of the ith component. The stable thermodynamic equilibrium is given by minimizing the Gibbs free energy of the system Bear (1972); Chen (2007),

$$f_{wi}(p_w, x_{w1}, x_{w2}, \cdots, x_{wN}) = f_{ni}(p_n, x_{n1}, x_{n2}, \cdots, x_{nN}), \quad i = 1, \cdots, N \tag{79}$$

The fugacity of the ith component is defined by,

$$f_{\alpha i} = p_\alpha x_{oi} \phi_{\alpha i}, \quad \alpha = w, n, \quad i = 1, \cdots, N \tag{80}$$

$\phi_{\alpha i}, \alpha = w, n$ is the fugacity coefficient of the ith component which will be defined below. The phase and volumetric behaviors, including the calculations of the fugacities, are modeled using the Peng-Robinson equation of state (Peng & Robinson, 1976). Introducing the pressure of the phase, $p_/alpha$, which is given by Peng-Robinson two-parameter equation of state as,

$$p_\alpha = \frac{RT}{V_\alpha - b_\alpha} - \frac{a_\alpha(T)}{V_\alpha(V_\alpha + b_\alpha) + b_\alpha(V_\alpha - b_\alpha)}, \quad \alpha = w, n \tag{81}$$

$$a_\alpha = \sum_{i=1}^{N} \sum_{j=1}^{N} x_{i\alpha} x_{j\alpha} (1 - \kappa_{ij}) \sqrt{a_i a_j}, \quad b_\alpha = \sum_{j=1}^{N} x_{i\alpha} b_i, \quad \alpha = w, n \tag{82}$$

where R is the universal gas of constant, T is the temperature, V_α is the molar volume of the phase α, κ_{ij} is a binary interaction parameter between the components i and j, a_i and a_i are empirical factor for the pure component i given by,

$$a_i = \Pi_{ia} \alpha_i \frac{R^2 T_{ic}^2}{p_{ic}}, \quad b_i = \Pi_{ib} \frac{R T_{ic}}{p_{ic}}, \quad i = 1, \cdots, N \tag{83}$$

T_{ic} and p_{ic} are the critical temperature and pressure,

$$\Pi_{ia} = 0.45724, \quad \Pi_{ib} = 0.077796, \quad \alpha_i = \left(1 - \lambda_i \left[1 - \sqrt{\frac{T}{T_{ic}}}\right]\right)^2, \quad \lambda_i = 0.37464 + 1.5432\omega_i - 0.26992\omega_i^2, \quad i = 1, \cdots, N \tag{84}$$

Eq. (72) can be rewritten in the following cubic form,

$$Z_\alpha^3 - (1 - B_\alpha)Z_\alpha^2 + (A_\alpha - 2B_\alpha - 3B_\alpha^2)Z_\alpha - (A_\alpha B_\alpha - B_\alpha^2 - B_\alpha^2) = 0, \quad \alpha = w, n \tag{85}$$

where Z_α is the compressibility factor given by,

$$Z_\alpha = \frac{b_\alpha V_\alpha}{RT}, \quad \alpha = w, n \tag{86}$$

(Chen, 2007) explained how to solve the cubic algebraic equation, Eq. (64). The fugacity coefficient $\phi_{\alpha i}$ of the ith component is defined in terms of the compressibility factor Z_α as,

$$\ln \phi_{\alpha i} = \frac{b_i}{b_\alpha}(Z_\alpha - 1) - \ln(Z_\alpha - B_\alpha) - \frac{A_\alpha}{2\sqrt{2}B_\alpha}\left(\frac{2}{a_\alpha}\sum_{j=1}^{N} x_{j\alpha}(1 - \kappa_{ij})\sqrt{a_i a_j} - \frac{b_i}{b_\alpha}\right) \cdot \ln\left(\frac{Z_\alpha + (1+\sqrt{2})B_\alpha}{Z_\alpha - (1-\sqrt{2})B_\alpha}\right) \tag{87}$$

Deriving of this equation can be found in details in (Chen, 2007).
Using Eqs. (51)- (53) and (62)- (65) with some mathematical manipulation one can find,

$$\frac{\partial(\phi\rho_w S_w)}{\partial t} = -\nabla \cdot \rho_w f_w(S)\left\{Km_n(S)\left(\frac{dp_c}{dS_w}\nabla S_w - \Delta\rho g \nabla z\right) + u\right\} + q_w \tag{88}$$

Alternatively, in terms of pressure the flow equations may be rewritten in the form,

$$\frac{\partial(\phi\rho_w S_w)}{dp_c}\left(\frac{\partial p_n}{\partial t} - \frac{\partial p_w}{\partial t}\right) = \nabla \cdot \rho_w\left\{\frac{Kk_{rw}}{\mu_w}(\nabla p_w - \rho_w g \nabla z)\right\} + q_w \tag{89}$$

$$\frac{\partial(\phi\rho_n(1 - S_w))}{dp_c}\left(\frac{\partial p_n}{\partial t} - \frac{\partial p_w}{\partial t}\right) = \nabla \cdot \rho_n\left\{\frac{Kk_{rn}}{\mu_n}(\nabla p_n - \rho_n g \nabla z)\right\} + q_n \tag{90}$$

Both models, Eq. (88) and Eqs. (89)- (90) are used intensively especially in the field of oil reservoir simulations.

6.2 Three-phase compositional flow

In three-phase compositional flow the governing equations will not has a big difference from the two-phase case. In this section we introduce the main points which distinguish the three-phase flow. On the other hand, we consider the black oil model as an example of the three-phase compositional flow instead of considering the general case to investigate such kind of complex flow. The black oil model is water-oil-gas system such that water represents the aqueous phase and oil represents oleic phase. The hydrocarbon in a reservoir is almost consists of oil and gas. Water is being naturally in the reservoir or injected in the secondary stage of oil recovery. Also, gas may be found naturally or/and injected as CO_2 injection for the enhanced oil recovery stage. The governing equations may be extended to the three-phase flow. The generalized Darcy's law with mass transfer equations will remain the same as in Eqs. (30) and (31) with considering $\alpha = w, o, g$, thus each phase is represented by two equations, continuity and momentum. The index α denotes to the water (w), oil (o) and gas (g), respectively. The solute transport equations is modified to suite the three-phase compositional flow as follow,

$$\frac{\partial(\phi c z_i)}{\partial t} + \nabla \cdot \left(c_w x_{wi} u_w + c_o x_{oi} u_o + c_g x_{gi} u_g \right) = F_i, \quad i = 1, \cdots, N \tag{91}$$

or

$$\frac{\partial}{\partial t} \left[\phi \left(c_w x_{wi} S_w + c_o x_{oi} S_o + c_g x_{gi} S_g \right) \right] + \nabla \cdot \left(c_w x_{wi} u_w + c_o x_{oi} u_o + c_g x_{gi} u_g \right) = x_{wi} q_w + x_{oi} q_o + x_{gi} q_g, \quad i = 1, \cdots, N \tag{92}$$

Following (Stone, 1970; 1973) we assume that the water-oil and oil-gas relative permeabilities are given as the two-phase case,

$$k_{rw}(S_w) = k_{rw}^0 \left(\frac{S_w - S_{wc}}{1 - S_{wc} - S_{orw}} \right)^{n_w} \tag{93}$$

$$k_{rg}(S_g) = k_{rg}^0 \left(\frac{S_g - S_{gr}}{1 - S_{wc} - S_{org} - S_{gr}} \right)^{n_g} \tag{94}$$

where S_{wc} is the connate water saturation, S_{org} is the residual oil saturation to gas, S_{orw} is the residual oil saturation to water, S_{gr} is the residual gas saturation to water. $S_w = 1 - S_{orw}$. The intermediate-wetting phase (oil phase) relative permeabilities are given by,

$$k_{row}(S_w) = k_{row}^0 \left(\frac{1 - S_w - S_{orw}}{1 - S_{wc} - S_{orw}} \right)^{n_{ow}} \tag{95}$$

$$k_{rog}(S_g) = k_{rog}^0 \left(\frac{1 - S_{wc} - S_{org} - S_g}{1 - S_{wc} - S_{org} - S_{gr}} \right)^{n_{og}} \tag{96}$$

The intermediate-wetting phase relative permeability is given by,

$$k_{ro}(S_w, S_g) = \frac{k_{row} k_{rog}}{k_{norm}} \tag{97}$$

k_{norm} may be setting as one or given by another formula as in the literature which will not mention here for breif.

6.3 Numerical methods for multi-phase flow

Much progress in the last three decades in numerical simulation of multi-phase flow with compositional and chemical effect. Both first-order finite difference and finite volume methods are used. First-order finite difference schemes has numerical dispersion issue, while the first-order finite volume has powerful features when used for two-phase flow simulation (Leveque, 2002). However, the later one has some limitations when applied to fractured media (Monteagudo & Firoozabadi, 2007). Also, higher-order methods have less numerical dispersion and more accurate flow field calculations than the first-order methods. The combined mixed-hybrid finite element (MHFE) and discontinuous Galerkin (DG) methods have been used to simulate two-phase flow by (Hoteit & Firoozabadi, 2005; 2006; Mikyska & Firoozabadi, 2010). In the combined MHFE-DG methods, MHFE is used to solve the pressure equation with total velocity, and DG method is used to solve explicitly the species transport equations. Therefore, the parts are coupled using scheme such as the iterative IMplicit Pressure and Explicit Concentration (IMPEC) scheme. Also, (Sun et al., 2002) have used combined MHFE-DG methods to miscible displacement problems in porous media.

The DG method (Wheeler, 1987; Sun & Wheeler, 2005a;b; 2006) is derived from variational principles by integration over local cells, thus it is locally mass conservative by construction. In addition, the DG method has low numerical diffusion because higher-order approximations are used within cells and the cells interfaces are weakly enforced through the bilinear form. DG method is efficiently implementable on unstructured and nonconforming meshes.
The MHFE methods are based on a variational principle expressing an equilibrium or saddle point condition that can be satisfied locally on each element (Brezzi & Fortin, 1991). It has an indefinite linear system of equations for pressure (scalar) and the total velocity (vector) but they definitized by appending as extra degrees of freedom the average pressures at the element edges.

7. References

Aissa, W. A. & Mohammadein, A. A. (2006). Chemical reaction effects on combined forced and free convection flow of water at 4 c past a semi-infinite vertical plate, *J. Eng. Sci. Assiut Univ.* 34: 1225–1237.

Bear, J. (1972). *Dynamics of Fluids in Porous Media*, Elsevier, New York.

Brezzi, F. & Fortin, M. (1991). *Mixed and hybrid finite element methods*, Springer–Verlag, New York.

Brooks, R. H. & Corey, A. T. (1964). Hydraulic properties of porous media, *Hydrology Papers* 3.

Chen, Z. (2007). *Reservoir simulation: mathematical techniques in oil recovery*, SIAM, USA.

Cheng, P. (1978). Heat transfer in geothermal systems, *Advances in Heat Transfer* 14: 1–105.

Cheng, P. (1981). Thermal dispersion effects on non-darcy convection flows in a saturated porous medium, *Lett. Heat Mass Transfer* 8: 267–270.

Dhir, V. K. (1994). Boiling and two-phase flow in porous media, *Annu. Rev. Heat Transfer* 5: 303–350.

Doughty, C. & Pruess, K. (1988). A semianalytical solution for heat-pipe effects near high-level nuclear waste packages buried in partially saturated geological media, *Int. J. Heat Mass Transfer* 31: 79–90.

El-Amin, M. F. (2004). Double dispersion effects on natural convection heat and mass transfer in non-darcy porous medium, *Appl. Math. Comp.* 156: 1–17.

El-Amin, M. F., Aissa, W. A. & Salama, A. (2008). Effects of chemical reaction and double dispersion on non-darcy free convection heat and mass transfer, *Transport in Porous Media* 75: 93–109.

Elbashbeshy, E. M. A. (1997). Heat and mass transfer along a vertical plate with variable surface tension and concentration in the presence of magnetic field, *Int. J. Eng. Sci. Math. Sci.* 4: 515–522.

Hassanizadeh, S. M. & Gray, W. (1979a). General conservation equations for multi-phase systems 2. mass, momenta, energy and entropy equations, *Advances in Water Resources* 2: 191–208.

Hassanizadeh, S. M. & Gray, W. (1980). General conservation equations for multi-phase systems 3. constitutive theory for porous media, *Advances in Water Resources* 3: 25–40.

Hassanizadeh, S. M. & Gray, W. G. (1979b). General conservation equations for multi-phase systems 1. averaging procedure, *Advances in Water Resources* 2: 131–144.

Herwig, H. & Koch, M. (1991). Natural convection momentum and heat transfer in saturated highly porous mediaan asymptotic approach, *Heat Mass Transfer* 26: 169–174.

Hong, J. T. & Tien, C. L. (1987). Analysis of thermal dispersion effect on vertical plate natural convection in porous media, *Int. J. Heat Mass Transfer* 30: 143–150.

Hoteit, H. & Firoozabadi, A. (2005). Multicomponent fluid flow by discontinuous galerkin mand mixed methods in unfractured and fractured media, *Water Resour. Res.* 41: W11412.

Hoteit, H. & Firoozabadi, A. (2006). Compositional modeling by the combined discontinuous galerkin mand mixed methods, *SPE Journal* .

Lai, F. C. & Kulacki, F. A. (1989). Thermal dispersion effect on non-darcy convection from horizontal surface in saturated porous media, *Int. J. Heat Mass Transfer* 32: 971–976.

Leal, L. G. (2007). *Advanced transport phenomena*, Cambridge University Press.

Leveque, R. J. (2002). *Finite volume methods for hyperbolic problems*, Cambridge Texts in Applied Mathematics, Cambridge University Press.

Mikyska, J. & Firoozabadi, A. (2010). Implementation of higher-order methods for robust and efficient compositional simulation, *J. Comput. Physics* 229: 2898–2913.

Monteagudo, J. E. P. & Firoozabadi, A. (2007). Control-volume model for simulation of water injection in fractured media: incorporating matrix heterogeneity and reservoir wettability effects, *SPE Journal* .

Mulolani, I. & Rahman, M. (2000). Similarity analysis for natural convection from a vertical plate with distributed wall concentration, *Int. J. Math. Math. Sci.* 23: 319–334.

Murthy, P. V. S. N. & Singh, P. (1997). Thermal dispersion effects on non-darcy natural convection with lateral mass flux, *Heat Mass Transfer* 33: 1–5.

Nolen, J. S. (1973). Numerical simulation of compositional phenomena in petroleum reservoirs, *Reprint Series, SPE, Dallas, 11, 268284.*

Peng, D.-Y. & Robinson, D. B. (1976). A new two-constant equation of state, *Industrial and Engineering Chemistry Fundamentals* 15: 59–64.

Plumb, O. A. (1983). The effect of thermal dispersion on heat transfer in packed bed boundary layers, *In: Proc. 1st ASME/JSME Thermal Engng. Joint Conf.*, Vol. 2, pp. 17–21.

Pooladi-Darvish, M. & Firoozabadi, A. (2000). Co-current and counter-current imbibition in a water-wet matrix block, *SPE Journal* 5: 3–11.

Prasad, K. V., Abel, S. & Datti, P. S. (2003). Diffusion of chemically reactive species of a non-newtonian fluid immersed in a porous medium over a stretching sheet, *Int. J. Non-Linear Mech.* 38: 651–657.

Salama, A. & van Geel, P. J. (2008a). Flow and solute transport in saturated porous media: 1 the continuum hypothesis, *J Porous Media* 11: 403–413.

Salama, A. & van Geel, P. J. (2008b). Flow and solute transport in saturated porous media: 2 violating the continuum hypothesis, *J Porous Media* 11: 421–441.

Starikovicius, V. (2003). The multiphase fl ow and heat transfer in porous media, *Berichte des Fraunhofer ITWM* 55: 1–30.

Stone, H. (1970). Probability model for estimating three-phase relative permeability, *JPT* 214.

Stone, H. (1973). Estimation of three-phase relative permeability and residual oil data, *J. Cdn. Pet. Tech.* 13: 53.

Sun, S., Riviere, B. & Wheeler, M. F. (2002). A combined mixed finite element and discontinuous galerkin method for miscible displacement problems in porous media, *Proceedings of International Symposium on Computational and Applied PDEs, Zhangjiajie National Park of China, 321-341.*

Sun, S. & Wheeler, M. F. (2005a). Discontinuous galerkin methods for coupled flow and reactive transport problems, *Appl. Num. Math.* 52: 273–298.

Sun, S. & Wheeler, M. F. (2005b). Symmetric and nonsymetric discontinuous galerkin methods for reactive transport in porous media, *SIAM J. Numer. Anal.* 43: 195–219.

Sun, S. & Wheeler, M. F. (2006). Analysis of discontinuous galerkin methods for multi-components reactive transport problem, *Comput. Math. Appl.* 52: 637–650.

Telles, R. S. & V.Trevisan, O. (1993). Dispersion in heat and mass transfer natural convection along vertical boundaries in porous media, *Int. J. Heat Mass Transfer* 36: 1357–1365.

Udell, K. S. (1985). Heat transfer in porous media considering phase change and capillarity – the heat pipe effect, *Int. J. Heat Mass Transfer* 28: 485–495.

van Genuchten, M. (1980). A closed-form equation for predicting the hydraulic conductivity of unsaturated soils, *Soil Sci. Soc. Am. J.* 44: 892–898.

Wheeler, M. F. (1987). An elliptic collocation finite element method with interior penalties, *SIAM J. Numer. Anal.* 15: 152–161.

Whitaker, S. (1967). Diffusion and dispersion in porous media, *AIChE* 13: 420–427.

Whitaker, S. (1977). Simultaneous heta, mass, and momentum transfer in porous media: A theory of drying, *Advances in Heat Transfer* 13: 119–203.

Young, L. C. & Stephenson, R. E. (1983). A generalized compositional approach for reservoir simulation, *SPE Journal* 23: 727–742.

3

Multiphase Modelling of Thermomechanical Behaviour of Early-Age Silicate Composites

Jiří Vala
Brno University of Technology, Faculty of Civil Engineering
Czech Republic

1. Introduction

The reliable prediction of thermomechanical behaviour of early-age silicate composites is a complicated multiphysical and multiscale problem, containing a lot of open questions. However, silicate mixtures, namely fresh concrete, are the most commonly used materials in building constructions throughout the world, thus such prediction is of great practical significance. The most important modelling outputs are the macroscopic effective strain, stress, temperature, moisture etc. time evolutions, driven by chemical reactions of particular clinker minerals with water. Every realistic model is then expected to include thermo-, chemo- and hygromechanical processes and phase changes, involving all available microstructural information related to the real porous medium.

The deformation of a material sample or a building construction made from silicate composites has to be analyzed at least as the superposition of

- reversible elastic deformation,
- viscous material flow,
- volume changes, unlike remaing contributions independent of external loads.

The crucial external and internal influences are:

- internal hydration heat, generated by the hydration hydraulic processes,
- ambient temperature variation, connected with ambient humidity variation (natural or artificial ones),
- external mechanical loads.

The significant physical (and chemical) processes are:

a) thermal deformation,
b) autogenous shrinkage,
c) carbonation,
d) elastic and creep deformation,
e) additional thermal deformation,
f) drying shrinkage and swelling.

In the first period of intense hydration a), accompanied by b), is dominant. In the later period the role of a) decreases, but the effect of c) has to be taken into account. The external mechanical loads cause d) (creep especially in the earliest age), the external temperature changes simultaneously force e), modified by f).

The traditional approach to the modelling of such complex physical and technical problems is the phenomenological one, as discussed in (Bažant, 2001): the effect of changes of density, porosity, permeability, compressive strength, etc. on material behaviour is lumped together to some model parameters, which must be identified by long-lasting tests in the whole range of model applicability. On the contrary, the so-called CCBM ("Computational Cement-Based Material") approach, suggested in (Maruyama et al., 2001), develops the original idea of (Tomosawa, 1997): the slight generalization of its (seemingly simple) form

$$\dot{\epsilon} = \Phi(\epsilon, \epsilon_*), \qquad \dot{\epsilon}_* = \Psi(\epsilon, \epsilon_*)$$

where ϵ is the radius of an unhydrated cement particle, ϵ_* its total radius including hydrate, dot symbols refer (everywhere in this chapter) to derivatives with respect to the time $t \geq 0$ and Φ, Ψ are (in general rather complicated) material characteristics with hidden ϵ, ϵ_* (but not with their time derivatives) again. The analysis of (Maruyama et al., 2001) assumes ideal spherical particles, hydration products adherent to such particles (whose size distribution is approximated by a special Rosin-Ramler function), water diffusing through the hydrate layer and chemically reacting with cement, up to interparticle contact effects; the amount of water is controlled by the pore structure, modified by hydration reactions of cement constituents and corresponding heat generation. Particular cement constituents, namely alite (C_3S, typically 65 % of the total mass in the Portland cement), belite (C_2S, 15 %), aluminate phase (C_3A, 7 %), ferrite phase (C_4AF, 8 %), etc., have their own densities and hydration reactions, generating hydration heat.

For various types of cement we have different hydration degree Γ, introduced as

$$\Gamma := \frac{\mu^h}{\mu^h_\infty}$$

where μ^h denotes the (usually increasing) mass of skeleton (and corresponding sink of liquid water mass) ans μ^h_∞ the final mass of hydrated (chemically combined) water in a volume unit; alternatively (cf. (Gawin et al., 2006a), p. 309)

$$\Gamma := \frac{Q^h}{Q^h_\infty}$$

in terms of the heat Q^h released during hydration and of its final value Q^h_∞. However, it is difficult to guarantee above sketched model assumption in building practice, applying also (not single-sized) additional aggregate; thus Γ is usually quantified from macroscopic experiments (as adiabatic calorimetric or isothermal strength evolution tests) not from such microstructural considerations. Consequently Γ can be evaluated by (Gawin et al., 2006a), p. 309, from an auxiliary evolution problem of type

$$\dot{\Gamma} = A(\Gamma, \phi, T)$$

with an a priori known real function A.

During the same hydration process the non-negligible vapour mass source μ^e, caused by the liquid water evaporation or desorption, occurs, too. Unfortunately, unlike μ^h, no reasonable constitutive relation is available for the direct evaluation of μ^e.
Clearly, the reliable prediction of material behaviour applicable to real building objects during hydration needs some multiscale analysis. The mechanistic approach (Pichler et al., 2007) makes it possible to consider above sketched effects explicitly because they appear directly in the model equations, distinguishing between 4 length scales, characterized as

I) anhydrous-cement scale (typical length of a representative volume element from 10-8 to 10-6 m), in more details decomposed into 3 subscales, where the qualitative estimate of activity of four main clinker phases, water and air requires the detailed micromechanical evaluation of corresponding chemical reactions,

II) cement-paste scale (from 10^{-6} to 10^{-4} m),

III) mortar scale (about 10^{-2} m),

IV) macroscale (about 10^{-1} m).

The analysis of capillary depression at scale I) (considering membrane forces on solid/liquid, solid/gas and liquid/gas interfaces), of ettringite formation at scale II), of autogenous deformation at scales II) and III), referring to the Hill homogenization lemma (see (Dormieux et al., 2006), p. 105), must be completed by the interpretation of such multiscale results at scale IV). However, different physical and chemical processes studied at particular scales do not admit proper and physically transparent mathematical analysis of two- and more-scale convergence, as discussed in (Cioranescu & Donato, 1999), (Vala, 2006) or (Efendiev et al., 2009), including its non-periodic (formally complicated) generalization, introduced in (Nguentseng, 2003-4).
The approach (Gawin et al., 2006a) applies certain mechanistic-type method to obtain the governing equations only, using the averaging hybrid mixture theory: the developments starts at the micro-scale and balance equations for particular phases and interfaces are introduced at this level and then averaged for obtaining macroscopic balance equations. Four phases are distinguished: solid skeleton, liquid water, vapour and dry air, whose densities are considered (under the passive air assumption) as constants; the whole hygro-thermo-chemo-mechanical process is then studied as the time evolution of capillary pressure, gas pressure, temperature and displacement of points related to the reference (initial) configuration, driven by balance equations of classical thermodynamics and conditioned by corresponding constitutive laws. The detailed geometrical analysis (Sanavia et al., 2002) (without phase changes) offers the possibility to extend such considerations beyond the assumption of small deformations and involve some elements of fracture mechanics.
The development, laboratory testing and computational simulations of new materials, namely those for the application of advanced engineering structures, belong to the research priorities of the Faculty of Civil Engineering of Brno University of Technology. Moreover, these activities should be intensified thanks to the proposed complex research institution AdMaS ("Advanced Materials and Structures") in the near future. The long-time behaviour of massive structures, especially its bearing value, durability and user properties, is typically conditioned by the early-age heat, moisture, etc. treatment, modified by foundations, subgrades, reinforcement and connecting members; thus the aim is to design the whole building process to minimize the development of significant tensile stresses to avoid the the danger of cracking, or even to force volume changes and corresponding final stresses appropriate for the future use of a structure. The deeper understanding of decissive

processes in early-age materials that effects volume changes is therefore needed, although no closed physical and mathematical models are available and all simplified calculations contain empirical parameters and functions, whose identification, supported by laboratory measurements or in situ observations, generates separate non-trivial problems, not discussed in details here. However, we shall demonstrate how the thermomechanical analysis of balance of mass, (linear and angular) momentum and energy for computational HAM ("heat, air and moisture") models in civil engineering is able to be extended to a complex computational model, including the mass source or solid skeleton related to the hydration process (and corresponding sink of liquid water mass), as well as the vapour mass source caused by the liquid water evaporation or desorption, using some micromechanical arguments from the theory of porous media.

2. Mixture components

To analyze the phase changes in an early-age silicate composite, we shall consider four material phases:

- solid material, identified by an index s,
- liquid water, identified by an index w,
- water vapour, identified by an index v,
- dry air, identified by an index a.

In addition to partial derivatives of scalar quantities ψ with respect to time, i. e.

$$\dot{\psi} := \partial\psi/\partial t,$$

we shall introduce also the partial derivatives of such quantities with respect to x_i, $i \in \{1,2,3\}$, $x = (x_1, x_2, x_3)$ being a Cartesian coordinate system in the three-dimensional Euclidean space R^3,

$$\psi_{,i} := \partial\psi/\partial x_j.$$

In the case of real vector variables with values in $\mathcal{R}^3$ we shall write ψ briefly instead of (ψ_1, ψ_2, ψ_3). Even in the case of matrix variables in the with values in $\mathcal{R}^{3\times 3}$, the space of real matrices of the third order, we shall write ψ only instead of ψ_{ij}, $i,j \in \{1,2,3\}$. Consequently, due to the preceding notation, unlike the matrix elements ψ_{ij} from $\psi \in \mathcal{R}^{3\times 3}$ with $i,j \in \{1,2,3\}$, we have e. g. for $\psi \in \mathcal{R}^3$

$$\psi_{i,j} := \partial\psi_i/\partial x_j.$$

We shall assume that the scale bridging between particular scales (if relevant scale material data are available) can be done by means of the averaging of model variables. The basic (averaged) variables in our model, related to a representative volume element, are:

- 4 intrinsic phase densities $R = (\rho^s, \rho^w, \rho^v, \rho^a)$,
- 12 components of phase velocities $V = (v_i^s, v_i^w, v_i^v, v_i^a)$ with $i \in \{1,2,3\}$,
- 3 fluid pressures $P = (p^w, p^v, p^a)$,
- 1 (absolute) temperature ϑ.

Intrinsic phase densities evidently do not reflect the amount of particular phases in a volume unit; thus it is useful to define (real) phase densities

$$\rho_\varepsilon := \eta^\varepsilon \rho^\varepsilon$$

formally for any phase index $\varepsilon \in \{s, w, v, a\}$ where

$$\begin{aligned} \eta^s &= (1-n), \\ \eta^w &= ns, \\ \eta^v &= n(1-s), \\ \eta^a &= n(1-s)(1-\phi). \end{aligned}$$

Unfortunately it is not easy to evaluate the porosity n, the saturation degree s and the relative humidity (the volume fraction occupied by water vapour in the total gaseous phase) ϕ. Nevertheless, similarly to ρ_ε, $\varepsilon \in \{s, v, w, a\}$, being motivated by the Dalton law by (Bermúdez de Castro, 2005), p. 111, we can also introduce pressures

$$p_\varepsilon := \eta^\varepsilon p^\varepsilon .$$

We can evaluate also the "macroscopic" mixture density

$$\rho := \rho_s + \rho_w + \rho_v + \rho_a .$$

Let us also remark that, from he point of view of solid phase, fluid pressures P are accompanied by a (partial) Cauchy stress tensor compound from components τ_{ij} with $i, j \in \{1,2,3\}$, whose (indirect) relation to V will be discussed later.

The porosity can be evaluated from the finite strain analysis by (Sanavia et al., 2002), p. 139. The multiphase medium at the macroscopic level can be described as the superposition of all phases ε, whose material point with coordinates $x_i^{\varepsilon 0}$ in the reference configuration Ω in R^3 occupies a point with coordinates $x_i^\varepsilon(t)$; zero indices are related to the reference configuration, here in the initial time $t = 0$. In the Lagrangian description of the motion the position of each material point can be expressed as

$$x_i^\varepsilon(t) = x_i^{\varepsilon 0} + u_i^\varepsilon(x_i^{\varepsilon 0}, t)$$

where $u_i^\varepsilon(x_i^{\varepsilon 0}, t)$ denotes the displacement at chosen time and zero indices are related to a reference configuration, here in time $t = 0$. Thus

$$F_{ij}^\varepsilon(x^{\varepsilon 0}, t) := \frac{\partial x_i^\varepsilon(x^{\varepsilon 0})}{\partial x_j^{\varepsilon 0}} = \delta_{ij} + \frac{\partial u_i^\varepsilon(x^{\varepsilon 0}, t)}{\partial x_j^{\varepsilon 0}}$$

can be taken as a deformation characteristic, δ being a Kronecker symbol, consequently

$$n = 1 - (1 - n_0)(\det F^s)^{-1}$$

where $n_0(x^{s0})$ denotes the porosity in the reference configuration. In the linearized geometry this access in not available: e. g. (Gawin et al., 2006a), p. 310 presents an empirical (nearly linear) function $n(\Gamma)$, justified by the correlation analysis, exploiting the experimental database of (De Schutter, 2002) where hardening cement pastes with 5 different values of water/cement ratio are studied.

The saturation degree s comes from the microporomechanical considerations (cf. (Dormieux et al., 2006), p. 247); their usual result (verified by laboratory measurements) is the so-called sorption isotherm, a real function $s(p^c)$ where p^c is the capillary pressure, determined (in general) as

$$p^c = \Xi(p^w, p^v, p^a)$$

for certain real function Ξ. For temperatures normally encountered in building structures and for capillary saturation range (Gawin et al., 2006a), pp. 305 and 311, recommends the simplified relation

$$p^c = p^v + p^a - p^w,$$

justified by an exploitation of the entropy inequality by means of the Coleman-Noll method; for more details and substantial generalizations see (Gray, 2000), p. 482.
To formulate the basic balance laws of classical thermodynamics, consequently the announced system of partial differential or integral equations of evolution, let us introduce the simple notation of (Vala, 2006), p. 33, applied to particular phases identified by indices $\varepsilon \in \{s, w, v, a\}$. If ω^ε is a source corresponding to a scalar quantity ψ^ε then the conservation of a scalar quantity ψ^ε reads

$$\dot{\psi}^\varepsilon + (\psi^\varepsilon v_i^\varepsilon)_{,i} = \omega^\varepsilon; \qquad (1)$$

this will be exploited in the following three sections.

3. Mass balance

For an arbitrary $\varepsilon \in \{s, w, v, a\}$ let us insert a scalar quantity

$$\psi^\varepsilon = \rho_\varepsilon$$

and a corresponding source term ω^ε into (1) where

$$\begin{aligned} \omega^s &= -\dot{\mu}^h, \\ \omega^w &= \dot{\mu}^h - \dot{\mu}^e, \\ \omega^v &= \dot{\mu}^e, \\ \omega^a &= 0. \end{aligned}$$

To simplify our notations, we shall apply the Einstein summation convention for indices from the set $\{1,2,3\}$ without additional explanations, e. g. in the following equation the sum over $j \in \{1,2,3\}$ is omitted formally. The resulting system of partial differential equations of evolution is

$$\begin{aligned} \dot{\rho}_s + (\rho_s v_j^s)_{,j} &= -\dot{\mu}^h, \\ \dot{\rho}_w + (\rho_w v_j^w)_{,j} &= \dot{\mu}^h - \dot{\mu}^e, \\ \dot{\rho}_v + (\rho_v v_j^v)_{,j} &= \dot{\mu}^e, \\ \dot{\rho}_a + (\rho_a v_j^a)_{,j} &= 0. \end{aligned} \qquad (2)$$

Let us notice (here and similarly for other balance laws, too, without explicit comments) that the system (2) contains the standard left-hand-side additive terms, well-known e. g. from (Bermúdez de Castro, 2005), p. 4 (for one phase), and from (Dormieux et al., 2006), p. 9 (for

more phases), supplied by the additional right-hand-side additive terms, containing variables μ^h, determined from the hydration degree Γ, and μ^e, a priori unknown.
For the effective numerical calculations, finite element techniques (cf. (Efendiev et al., 2009), p. 47) or finite volume ones (cf. (Efendiev et al., 2009), p. 52) should be applied. Therefore we need to rewrite (2) (and analogous equations in the following sections) into the integral form, able to include boundary conditions in a simple way. Let $\partial\Omega$ be the boundary of a domain Ω in R^3. Let us introduce the additional notation

$$\begin{aligned} (\varsigma,\psi) &:= \int_\Omega \varsigma(x)\psi(x)\,\mathrm{d}x, \\ \langle\varsigma,\psi\rangle &:= \int_{\partial\Omega} \varsigma(x)\psi(x)\,\mathrm{d}s(x) \end{aligned}$$

for any functions ς and ψ (integrable in the needed sense) where $\mathrm{d}x$ refers to the standard Lebesgue measure and $\mathrm{d}s(x)$ to the surface Hausdorff measure. The integration by parts (based on the Green-Ostrogradskii theorem) then gives

$$\begin{aligned} (\varsigma,\dot\rho_s) + \langle\varsigma,\rho_s v_j^s \nu_j\rangle - (\varsigma_{,j},\rho_s v_j^s) &= -(\varsigma,\dot\mu^h), \\ (\varsigma,\dot\rho_w) + \langle\varsigma,\rho_w v_j^w \nu_j\rangle - (\varsigma_{,j},\rho_w v_j^w) &= (\varsigma,\dot\mu^h - \dot\mu^e), \\ (\varsigma,\dot\rho_v) + \langle\varsigma,\rho_v v_j^v \nu_j\rangle - (\varsigma_{,j},\rho_v v_j^v) &= (\varsigma,\dot\mu^e), \\ (\varsigma,\dot\rho_a) + \langle\varsigma,\rho_a v_j^a \nu_j\rangle - (\varsigma_{,j},\rho_a v_j^a) &= 0 \end{aligned} \quad (3)$$

for any test function ς (more precisely: sufficiently smooth, otherwise only in sense of distributions, from an appropriate function space which may be problem-specific);

$$\nu(x) = (\nu_1(x),\nu_2(x),\nu_3(x))$$

means the local unit outward normal vector on (sufficiently smooth) $\partial\Omega$.

4. Momentum balance

The momentum balance includes the linear balance and the angular balance. Since we consider a non-polar continuum, the second one forces only the symmetry of any Cauchy stress tensor τ. It remains to apply the first one. For an arbitrary $\varepsilon \in \{s,w,v,a\}$ and (step-by-step) for particular $i \in \{1,2,3\}$ let us insert a scalar quantity

$$\psi^\varepsilon = \rho_\varepsilon v_i^\varepsilon$$

and a corresponding source term

$$\omega^\varepsilon = \sigma_{ij,j}^\varepsilon + \rho_\varepsilon(g_i - a_i^\varepsilon + \theta_i^\varepsilon)$$

into (1) where

$$\sigma_{ij}^s = \tau_{ij}$$

for the solid phase and

$$\sigma_{ij}^\varepsilon = -\delta_{ij} p_\varepsilon$$

for all other phases with $\varepsilon \in \{w,v,a\}$, the acceleration $g = (g_1,g_2,g_3)$ generates $\rho_\varepsilon g$, the volume density of a gravitational force, the acceleration $a^\varepsilon = (a_1^\varepsilon,a_2^\varepsilon,a_3^\varepsilon)$, corresponding to the velocity

$v^\varepsilon = (v_1^\varepsilon, v_2^\varepsilon, v_3^\varepsilon)$, generates the volume density of an inertia force, and the acceleration $\theta_i^\varepsilon = (\theta_1^\varepsilon, \theta_2^\varepsilon, \theta_3^\varepsilon)$ generates the volume density due to mechanical interaction with other phases. The total Cauchy stress, introduced (unlike the partial Cauchy stress τ) as

$$\sigma := \sigma^s + \sigma^w + \sigma^v + \sigma^a,$$

then has the components

$$\sigma_{ij} = \tau_{ij} - \delta_{ij}(p_w + p_v + p_a).$$

The Clapeyron law (see (Gawin et al., 2006a), p. 305) enables us to evaluate both gas pressures p^v and p^a (consequently p_v and p_a, too) as certain functions $p^v(\rho^v, \vartheta^v)$ and $p^a(\rho^a, \vartheta^a)$. Unfortunately, for the remaining pressure p^w (or p_w) we do not know such constitutive relation.

In the linear momentum balance equations we need to calculate all velocities v^ε and accelerations a^ε from corresponding displacements u^ε, $\varepsilon \in \{s, w, v, a\}$. By the chain rule we have

$$\begin{aligned} v_i^\varepsilon(x^{\varepsilon 0}, t) &= \dot{u}_i^\varepsilon(x^{\varepsilon 0}, t) + u_{i,j}^\varepsilon(x^{\varepsilon 0}, t)\, v_j^\varepsilon(x^{\varepsilon 0}, t), \\ a_i^\varepsilon(x^{\varepsilon 0}, t) &= \dot{v}_i^\varepsilon(x^{\varepsilon 0}, t) + v_{i,j}^\varepsilon(x^{\varepsilon 0}, t)\, v_j^\varepsilon(x^{\varepsilon 0}, t). \end{aligned}$$

The terms $u_{i,j}^\varepsilon v_j^\varepsilon$ and $v_{i,j}^\varepsilon v_j^\varepsilon$, can be understood in the sense of scalar products $\operatorname{grad} u^\varepsilon \cdot v^\varepsilon$, $\operatorname{grad} v^\varepsilon \cdot v^\varepsilon$ by (Sanavia, 2001), p. 139 (though the construction of such general grad-operator is not quite trivial). More precisely, the geometry of structured continua is described using fiber bundles and Riemanian manifold in (Yavari & Marsden, 2009), p. 8. Another approach, explained in (Bermúdez de Castro, 2005), p. 195, results in the ALE ("Arbitrary Lagrangian-Eulerian") formulations by (Bermúdez de Castro, 2005), p. 195. Nevertheless, both such generalizations lead to complicated, reader-unfriendly expressions.

The classical constitutive relation for the solid phase between τ, u^s, v^s, etc., considers a linearized sufficiently small strain tensor and its additive decomposition into several parts, typically to the linear elastic and the power-law viscoelastic (creep) ones, containing facultative corrections due to microcracking, as in (Gawin et al., 2006a), p. 343, and (in more details) in (Gawin et al., 2006b), p. 519, with help of special mechanical and chemical damage parameters. The finite deformation theory needs some multiplicative decomposition of F^s in the form

$$F^s = F^{s1} F^{s2} \ldots F^{sM}$$

into a finite number $M > 1$ of matrix components; the constitutive relation is then characterized by a function

$$\tau(F^{s1}, \ldots, F^{sM}, \dot{F}^{s1}, \ldots, \dot{F}^{sM}, \ldots).$$

Especially in the case $M = 2$ (Sanavia et al., 2002), p. 143 combines the first standard reversible elastic component with the second irreversible one from the Drucker-Prager plasticity model with linear isotropic hardening, Neff (2008) combines elasticity with nonlocal linear kinematic hardening due to dislocation interaction, etc. Recently Majorana (2010) suggests even $M = 9$, taking into account i) elastic deformation, ii) plastic deformation, iii) damage, iv) cracking, v) creep, vi) shrinkage, vii) lits, viii) thermal strain, ix) autogenous strain, covering most items a)-e) from *Introduction*. However, for all simulation of real processes the theoretical extent of such decomposition must be supported by the competence in the laboratory identification of needed material characteristics and in the estimate of their uncertainty, imported into constitutive relations.

The evaluation of $\theta_i^s, \theta_i^w, \theta_i^v$ and θ_i^a for all $i \in \{1,2,3\}$ comes from the Darcy law, whose general (nonlinear) form, following (Dormieux et al., 2006), p. 50, based on the knowledge of certain material functions F_1, F_2 and F_3, valid for each $\varepsilon \in \{w,v,a\}$, is

$$v_i^\varepsilon - v_i^s = F_i^\varepsilon(\theta_1^\varepsilon, \theta_2^\varepsilon, \theta_3^\varepsilon, p^\varepsilon, \eta^\varepsilon).$$

The special (linear) form of these relations, suggested in (Sanavia, 2001), p. 9 (this part of (Sanavia, 2001) contains much more technical details than its later revision (Sanavia et al., 2002)), is

$$K_{ij}^\varepsilon(\rho_\varepsilon \theta_j^\varepsilon - p_{,j}^\varepsilon) + \kappa^\varepsilon \eta^\varepsilon (v_i^\varepsilon - v_i^s) = 0$$

where, for a fixed $\varepsilon \in \{w,v,a\}$, K_{ij}^ε are components of a symmetric permeability matrix and κ^ε is a (positive) dynamic viscosity. Such constitutive relations have to be supplied by the constraint by (Sanavia, 2001), p. 8,

$$\theta_i^s + \theta_i^w + \theta_i^v + \theta_i^a = 0.$$

The resulting system of partial differential equations of evolution, unified (for brevity) for all $\varepsilon \in \{w,v,a\}$ and $i \in \{1,2,3\}$ is

$$(\rho_\varepsilon v_i^\varepsilon)^{\cdot} + (\rho_\varepsilon v_i^\varepsilon v_j^\varepsilon)_{,j} = -p_{,i}^\varepsilon + \rho_\varepsilon(g_i - a_i^\varepsilon + \theta_i^\varepsilon). \tag{4}$$

For the solid phase we receive similarly for all $i \in \{1,2,3\}$

$$(\rho_s v_i^s)^{\cdot} + (\rho_s v_i^s v_j^s)_{,j} = \tau_{ij,j} + \rho_s(g_i - a_i^s + \theta_i^w + \theta_i^v + \theta_i^a). \tag{5}$$

We can see that for negligible values of all components of v^s and θ^s (5) this result degenerates to classical Cauchy equilibrium conditions, well-know in building statics (with zero a^s) and dynamics (with nonzero a^s),

$$\rho_s a_i^s + \tau_{ij,j} = -\rho_s g_i.$$

Let us repeat the approach with a test function ς from the previous section and apply it to (4) and (5). For any $i \in \{1,2,3\}$ the integration by parts with $\varepsilon \in \{w,v,a\}$ gives

$$\begin{aligned}&(\varsigma, (\rho_\varepsilon v_i^\varepsilon)^{\cdot}) + \langle \varsigma, \rho_\varepsilon v_i^\varepsilon v_j^\varepsilon \nu_j \rangle - (\varsigma_{,j}, \rho_\varepsilon v_i^\varepsilon v_j^\varepsilon) \\ &= -\langle \varsigma, p^\varepsilon \nu_i \rangle + (\varsigma_{,i}, p^\varepsilon) + (\varsigma, \rho_\varepsilon(g_i - a_i^\varepsilon + \theta_i^\varepsilon)),\end{aligned} \tag{6}$$

whereas for the solid phase the analogous result is

$$\begin{aligned}&(\varsigma, (\rho_s v_i^s)^{\cdot}) + \langle \varsigma, \rho_s v_i^s v_j^s \nu_j \rangle - (\varsigma_{,j}, \rho_s v_i^s v_j^s) \\ &= \langle \varsigma, \tau_{ij} \nu_j \rangle - (\varsigma_{,j}, \tau_{ij}) + (\varsigma, \rho_s(g_i - a_i^s - \theta_i^w - \theta_i^v - \theta_i^a))\end{aligned} \tag{7}$$

for any test function ς.

5. Energy balance

Let us introduce four (in general temperature-variable) heat capacities $c^\varepsilon(\vartheta)$ as prescribed material characteristics and four heat fluxes $q^\varepsilon = (q_1^\varepsilon, q_2^\varepsilon, q_3^\varepsilon)$ as additional variables. Similarly to the mass balance, for an arbitrary $\varepsilon \in \{s,w,v,a\}$, let us insert a scalar quantity

$$\psi^\varepsilon = \frac{1}{2}\rho_\varepsilon(v_i^\varepsilon)^2 + \rho_\varepsilon c^\varepsilon \vartheta$$

and a corresponding source term

$$\omega^\varepsilon = (\sigma_{ij,j}^\varepsilon + q_i^\varepsilon)_{,i} + (\rho_\varepsilon(g_i - a_i^\varepsilon + \theta_i^\varepsilon))_{,i} + \varpi^\varepsilon$$

into (1) where the exchange of energy between phases is accounted using the additive terms

$$\begin{aligned}
\varpi^s &= -\dot{\mu}^h \zeta^h ,\\
\varpi^w &= \dot{\mu}^h \zeta^h - \dot{\mu}^e \zeta^e ,\\
\varpi^v &= \dot{\mu}^e \zeta^e ,\\
\varpi^a &= 0
\end{aligned}$$

with specific enthalpies ζ^h and ζ^e (introduced in (Bermúdez de Castro, 2005), p. 110): ζ^h that of hydration (related to the mass unit of chemically bound water) and ζ^e that of evaporation, following (Gawin et al., 2006a), p. 303. The resulting system of partial differential equations of evolution is

$$\begin{aligned}
&(\rho_s v_i^s v_i^s + 2\rho_s c^s \vartheta)^{\cdot} + ((\rho_s v_i^s v_i^s + 2\rho_s c^s \vartheta) v_j^s)_{,j}\\
&= 2\tau_{ij,ij} - 2q_{i,i}^s + 2(\rho_s(g_i - a_i^s - \theta_i^w - \theta_i^v - \theta_i^a))_{,i} - 2\dot{\mu}^h \zeta^h ,\\
&(\rho_w v_i^w v_i^w + 2\rho_w c^w \vartheta)^{\cdot} + ((\rho_w v_i^w v_i^w + 2\rho_w c^w \vartheta) v_j^w)_{,j}\\
&= -2p_{i,i}^w - 2q_{i,i}^w + 2(\rho_w(g_i - a_i^w + \theta_i^w))_{,i} + 2\dot{\mu}^h \zeta^h - 2\dot{\mu}^e \zeta^e ,\\
&(\rho_v v_i^v v_i^v + 2\rho_v c^v \vartheta)^{\cdot} + ((\rho_v v_i^v v_i^v + 2\rho_s c^w \vartheta) v_j^v)_{,j}\\
&= -2p_{i,i}^v - 2q_{i,i}^v + 2(\rho_v(g_i - a_i^v + \theta_i^v))_{,i} + 2\dot{\mu}^e \zeta^e ,\\
&(\rho_a v_i^a v_i^a + 2\rho_s c^s \vartheta)^{\cdot} + ((\rho_a v_i^a v_i^a + 2\rho_s c^s \vartheta) v_j^a)_{,j}\\
&= -2p_{i,i}^a - 2q_{i,i}^a + 2(\rho_v(g_i - a_i^a + \theta_i^a))_{,i} .
\end{aligned} \tag{8}$$

All heat fluxes can be computed using the Fourier law by (Bermúdez de Castro, 2005), p. 42: in general we have

$$q_i^\varepsilon + \lambda_{ij}^\varepsilon \vartheta_{,j} = 0$$

for every $\varepsilon \in \{s,w,v,a\}$ and $i \in \{1,2,3\}$ and prescribed material characteristics

$$\lambda_{ij}^\varepsilon(\vartheta, \vartheta_{,1}, \vartheta_{,2}, \vartheta_{,3}) .$$

Supposing that all these characteristics are independent of derivatives of ϑ, in the linearized case the elements λ_{ij} generate certain symmetrical matrix of real factors, allowed to be functions of ϑ. Under the assumption of isotropy material we obtain moreover $\lambda_{ij} = \lambda_* \delta_{ij}$ for some real factor $\lambda_*^\varepsilon(\vartheta)$; this is just the crucial simplification of (Gawin et al., 2006a), p. 313,

$$q_i^\varepsilon = -\lambda_*^\varepsilon \vartheta_{,i} .$$

Repeating the approach with some test function ς again and applying it to (8), from the integration by parts we obtain

$$
\begin{aligned}
& (\varsigma,(\rho_s v_i^s v_i^s + 2\rho_s c^s \vartheta)^\cdot) \\
& + \langle \varsigma,(\rho_s v_i^s v_i^s + 2\rho_s c^s \vartheta) v_j^s \nu_j \rangle - (\varsigma_{,j},(\rho_s v_i^s v_i^s + 2\rho_s c^s \vartheta) v_j^s) \\
= \; & 2\langle \varsigma,(\tau_{ij,j} - 2q_i^s)\nu_i \rangle - 2(\varsigma_{,i},\tau_{ij,j} - 2q_i^s) \\
& + 2\langle \varsigma,\rho_s(g_i - a_i^s - \theta_i^w - \theta_i^v - \theta_i^a)\nu_i \rangle - 2(\varsigma_{,i},\rho_s(g_i - a_i^s - \theta_i^w - \theta_i^v - \theta_i^a)) - 2(\varsigma,\dot{\mu}^h \xi^h),
\end{aligned}
$$

$$
\begin{aligned}
& (\varsigma,(\rho_w v_i^w v_i^w + 2\rho_w c^w \vartheta)^\cdot) \\
& + \langle \varsigma,(\rho_w v_i^w v_i ws + 2\rho_w c^w \vartheta) v_j^w nu_j \rangle - (\varsigma_{,j},(\rho_w v_i^w v_i^w + 2\rho_w c^w \vartheta) v_j^w) \\
= \; & -2\langle \varsigma,(p_i^w + q_i^w)\nu_i \rangle + 2(\varsigma_{,}i, p_i^w + q_i^w) \\
& + 2\langle \varsigma,(\rho_w(g_i - a_i^w + \theta_i^w))\nu_i \rangle - 2(\varsigma_{,i},\rho_w(g_i - a_i^w + \theta_i^w)) + 2(\varsigma,\dot{\mu}^h \xi^h - \dot{\mu}^e \xi^e), \qquad (9) \\
& (\varsigma,(\rho_v v_i^v v_i^v + 2\rho_v c^v \vartheta)^\cdot) \\
& + \langle \varsigma,(\rho_v v_i^v v_i^v + 2\rho_v c^v \vartheta) v_j^v \nu_j \rangle - (\varsigma_{,j},\rho_v v_i^v v_i^v + 2\rho_v c^v \vartheta) v_j^v) \\
= \; & -2\langle \varsigma,(p_i^v + q_i^v)\nu_i \rangle + 2(\varsigma_{,i}, p_i^v + q_i^v) \\
& + 2\langle \varsigma,\rho_v(g_i - a_i^v + \theta_i^v)\nu_i \rangle - 2(\varsigma_{,i},\rho_v(g_i - a_i^v + \theta_i^v)) + 2(\varsigma,\dot{\mu}^e \xi^e), \\
& (\varsigma,(\rho_a v_i^a v_i^a + 2\rho_a c^a \vartheta)^\cdot) \\
& + \langle \varsigma,(\rho_a v_i^a v_i^a + 2\rho_a c^a \vartheta) v_j^a)\nu_j \rangle - (\varsigma_{,j},\rho_a v_i^a v_i^a + 2\rho_a c^a \vartheta) v_j^a) \\
= \; & -2\langle \varsigma,(p_i^a + q_i^a)\nu_i \rangle + 2(\varsigma_{,i}, p_i^a + q_i^a) \\
& + 2\langle \varsigma,\rho_a(g_i - a_i^a + \theta_i^a)\nu_i \rangle - 2(\varsigma_{,i},\rho_a(g_i - a_i^a + \theta_i^a)).
\end{aligned}
$$

for any test function ς.

6. Initial and boundary conditions

Since the class of admissible problems (moreover, not defined properly yet) is rather large, we shall not try to create a list of all physically reasonable types boundary conditions. However, the following considerations demonstrate the methods of implementation of boundary conditions into above derived systems of evolution.

Let us remind the boundary integrals occurring in the integral forms of evolution equations. Those from (3) contain functions

$$\rho_s v_j^s \nu_j, \quad \rho_w v_j^w \nu_j, \quad \rho_v v_j^v \nu_j, \quad \rho_a v_j^a \nu_j,$$

those from (6) and (7), in addition to functions

$$\rho_s v_i^s v_j^s \nu_j, \quad \rho_w v_i^w v_j^w \nu_j, \quad \rho_v v_i^v v_j^v \nu_j, \quad \rho_a v_i^a v_j^a \nu_j$$

also functions

$$\tau_{ij}\nu_j, \quad p^w \nu_i, \quad p^v \nu_i, \quad p^a \nu_i,$$

finally those from (9) functions in addition to functions

$$\rho_s v_i^s v_i^s v_j^s \nu_j, \quad \rho_s v_i^w v_i^w v_j^w \nu_j, \quad \rho_s v_i^v v_i^v v_j^v \nu_j, \quad \rho_s v_i^a v_i^a v_j^a \nu_j$$

still other functions

$$\rho_s c^s \vartheta v_j^s \nu_j, \qquad \rho_w c^w \vartheta v_j^w \nu_j, \qquad \rho_v c^v \vartheta v_j^v \nu_j, \qquad \rho_a c^a \vartheta v_j^a \nu_j,$$

$$\tau_{ij,j}\nu_i, \qquad p_i^w \nu_i, \qquad p_i^v \nu_i, \qquad p_i^a \nu_i,$$

$$q_i^s \nu_i, \qquad q_i^w \nu_i, \qquad q_i^v \nu_i, \qquad q_i^a \nu_i,$$

$$\rho_s g_i \nu_i, \qquad \rho_w g_i \nu_i, \qquad \rho_v g_i \nu_i, \qquad \rho_a g_i \nu_i,$$

$$\rho_s a_i^s \nu_i, \qquad \rho_w a_i^w \nu_i, \qquad \rho_v a_i^v \nu_i, \qquad \rho_a a_i^a \nu_i,$$

$$(\rho_w - \rho_s)\theta_i^w \nu_i, \qquad (\rho_v - \rho_s)\theta_i^v \nu_i, \qquad (\rho_a - \rho_s)\theta_i^a \nu_i .$$

Solutions of the resulting system of equations (3), (6), (7) and (9), supplied by needed constitutive equations, should be included in certain spaces of abstract functions, mapping each time $t \geq 0$ to some Sobolev, Lebesgue, etc. function space $\mathcal{S}$. Initial values of all independent variables (whose appropriate choice will be discussed later) are supposed to be prescribed in time $t = 0$. Boundary conditions on $\partial\Omega$ or its part for the resulting system of equations (3), (6), (7) and (9), supplied by needed constitutive equations, can be then divided into three groups:

a) conditions built in the definition of spaces V, usually prescribed values of abstract functions from V on $\partial\Omega$ (in sense of traces) at any admissible time,

b) conditions exploiting the knowledge of above listed functions or their linear combinations,

c) other conditions.

All subsequent examples work with any x from $\partial\Omega$ or its certain part and with arbitrary time $t \geq 0$. The typical example of a) is

$$\vartheta = \vartheta_*$$

for the a priori known temperature $\vartheta_*(x,t)$. Another (technically more complicated, but homogeneous) example of a) can be

$$\rho v_i \nu_i = 0$$

with the carefully defined "effective" velocity

$$v_i := (\rho_s v_i^s + \rho_w v_i^w \nu_i + \rho_a v_i^a \nu_i)/\rho .$$

The classical example of b), known even from elementary statics, is

$$\sigma_{ij}\nu_j = f_i$$

with $i \in \{1,2,3\}$ for the prescribed surface load

$$f(x,t) = (f_1(x,t), f_2(x,t), f_3(x,t)) .$$

Indirectly this is a boundary condition for u (or v, etc.), due to stress-strain constitutive relations. Its analogy for heat transfer

$$q_i \nu_i = q_*$$

works with the total heat flux vector

$$q := \rho_s q^s + \rho_w q^w + \rho_v q^v + \rho_a q^a$$

for the prescribed heat flux $q_*(x,t)$.

One could expect a similar result for the total diffusive flux vector

$$r := \rho_w(v_i^w - v_i^s) + \rho_v(v_i^v - v_i^s) + \rho_a(v_i^a - v_i^s),$$

as applied in (Gawin et al., 2006a), p. 316, but this is limited by the facultative nonlinearity of the Darcy law. If the form of F_i^ε with $i \in \{1,2,3\}$ and $\varepsilon \in \{w,v,a\}$. does not admit the explicit evaluation of all corresponding vectors θ^ε as linear combinations of $v^\varepsilon - v^s$, we must refer to c). In general, to satisfy such boundary conditions properly, additional techniques, as the application of Lagrange multipliers, Kuhn-Tucker conditions, etc., are necessary.

7. Exact solutions and iterative algorithms

The mathematical solvability of engineering problems, whose formulation comes from physically reasonable and transparent considerations, depends critically on the choice:

a) of the set of primary variables,

b) of the choice spaces $\mathcal{S}$ of functions, abstract functions, etc.,

c) of the proper formulation of boundary and initial conditions.

Nevertheless, our problem of early-age time-dependent behaviour of silicate composites is rather complicated, thus most authors bring new versions of a), applying different mathematical, physical and technical simplifications. From the most simple algebraic relations, avoiding differential and integral calculus completely, referring to empirical relations and selected statistical techniques, as presented in (Moon et al., 2005) or in (Bentz, 2008), through one evolution equation, usually of heat conduction, supplied by a lot of empirical dependencies, justified by extensive experimental works, it is possible to trace the development to such advanced physical and mathematical models, as to the system of 10 differential equations of evolution with 10 primary variables and numerous constitutive relations in (Gawin et al., 2006b), p. 519.

To reduce the number of empirical relations, typically motivated by some micromechanical considerations, but not fully compatible with quantitative lower-scale computations, we are ready to work with 20 primary variables. In *Mixture components* we have introduced 20 variables $\mathcal{R}$, $\mathcal{V}$, $\mathcal{P}$ and ϑ. Only 2 of these variables, p^v and p^a, can be evaluated outside the system of (differential or integral) equations of evolution. On the other hand, regardless of the fact that the proper evaluation of Γ forces solution of an additional ordinary differential equation, no constitutive relations are available just to μ^e and φ. This motivates us to reorder our set of variables slightly, introducing

$$\mathcal{T} = (\theta, p^w, \mu^e, \varphi).$$

Therefore we can take $\mathcal{R}$, $\mathcal{V}$ and $\mathcal{T}$ as primary variables; finally we have 20 primary variables and 20 equations in evolution: (2), (4) with (5) and (8), or (3), (6) with (7) and (9), alternatively. We have noticed that the second version is much more frequently used for numerical computations.

Because of the complexity of the problem, the mathematical verification of existence, uniquiness, regularity etc. of solutions of initial and boundary problems, corresponding in the case of weak solutions to (3), (6) with (7) and (9), or in the case of strong (classical) solutions to (2), (4) with (5) and (8), due to admissible classes of initial and boundary conditions, contains a lot of open questions. The crucial problem seems to be just in b), i. e. in the definition of some spaces of (generalized) functions with the good properties, expected for the solutions.

Even a very special case with a liquid phase only, whose some important additive terms are neglected, derived directly from (2), (4) and (8), namely

$$\begin{aligned}
\dot{\rho}_w + (\rho_w v_j^w)_{,j} &= 0, \\
(\rho_w v_1^w)^{\cdot} + (\rho_w v_1^w v_j^w)_{,j} &= -p_{,1}^w + \rho_w (g_1 - a_1^w), \\
(\rho_w v_2^w)^{\cdot} + (\rho_w v_2^w v_j^w)_{,j} &= -p_{,2}^w + \rho_w (g_2 - a_2^w), \\
(\rho_w v_3^w)^{\cdot} + (\rho_w v_3^w v_j^w)_{,j} &= -p_{,3}^w + \rho_w (g_3 - a_3^w), \\
(\rho_w v_i^w v_i^w)^{\cdot} + ((\rho_w v_i^w v_i^w) v_j^w)_{,j} &= -2p_{i,i}^w + 2(\rho_w (g_i - a_i^w))_{,i},
\end{aligned}$$

generates (with standard constitutive relations) the so-called Navier-Stokes existence and smoothness problem, one of the *Millenium Prize Problems*, formulated by the Clay Mathematics Institute; for its complete definition see *www.claymath.org/millennium/Navier-Stokes_Equations/Official_Problem_Description.pdf*.

Although the formal solvability of Navier-Stokes equations is not clear, various methods have been developed successfully to analyze their approximate solutions. This can motivated us to the design of an iterative algorithm for numerical simulation of our much more complex physical (and chemical) process, clearly with expected difficulties in any convergence analysis. The precise form of such iterative algorithm, constructing, step-by-step in time, a finite-dimensional approximation of solution of (3), (6) with (7) and (9), depends substantially on c), discussed briefly in the previous section. However, the main idea, coming from some finite element or finite volume technique and from the Rothe method of discretization in time, at least at a finite time interval, can be:

1. set $\mathcal{R}$, $\mathcal{V}$ and $\mathcal{T}$ by the initial conditions at $t = 0$,
2. add a time step length to t, preserving $\mathcal{R}$, $\mathcal{V}$ and $\mathcal{T}$,
3. solve $\mathcal{R}$ from some linearized version of (3), evaluate the correction $\varepsilon_{\mathcal{R}}$ of $\mathcal{R}$,
4. solve $\mathcal{V}$ from the linearized version of (6 and 7), evaluate the correction $\varepsilon_{\mathcal{V}}$ of $\mathcal{V}$,
5. solve $\mathcal{T}$ from the linearized version of (9), evaluate the correction $\varepsilon_{\mathcal{T}}$ of $\mathcal{T}$,
6. if $\varepsilon_{\mathcal{R}}$, $\varepsilon_{\mathcal{V}}$ and $\varepsilon_{\mathcal{T}}$ are sufficiently small, return to 3,
7. if the final time is reached, stop the computation, otherwise return to 2.

Evidently, preconditioning, mesh adaptivity, stabilization techniques and other technical manipulations are useful to be implemented into such algorithm to force its robustness and effectivity.

8. Computational simulations

Most non-trivial physical and technical studies of the problems of early-age behaviour of silicate composites, or (in particular) of concrete, of mortar pastes, etc., include or refer to some numerical experiment, modelling or simulation, applying finite element or finite (control) volume (rarely finite difference or meshless) techniques together with methods for time discretization, as that of lines, of characteristics or of discretization in time, including some microstructural data. However, no specialized commercial or research software code concentrated to such problems is available; one could expect that the complete development of such a code (in such languages for scientific computing as Fortran or C++) would be expensive and time-consuming and would require a team of specialists in many fields of knowledge, including those outside information technologies. This results in the following rough classification of typically applied software packages:

- commercial software for HAM analysis as WUFI (Fraunhofer Institut Holzkirchen) or DELPHIN (Technische Universität Dresden),
- large commercial software systems as ANSYS, ABAQUS, etc. (not very flexible, offering only weak support for specific properties of silicate composites)
- specialized software for computations in civil engineering, namely ATHENA (Červenka Consulting Prague) for calculations of strain and stress distributions in concrete structures at various stages of their existence, including the concrete/reinforcement cooperation, the prediction of fracture and the behaviour under extreme loads, or CESAR-LCPC (Ram Caddsys Chennai, India) for mechanical, structural and geotechnical calculations involving even certain analysis of phenomena associated with young hardening concrete,
- user-friendly environment MATLAB/COMSOL for the support of development of original software.

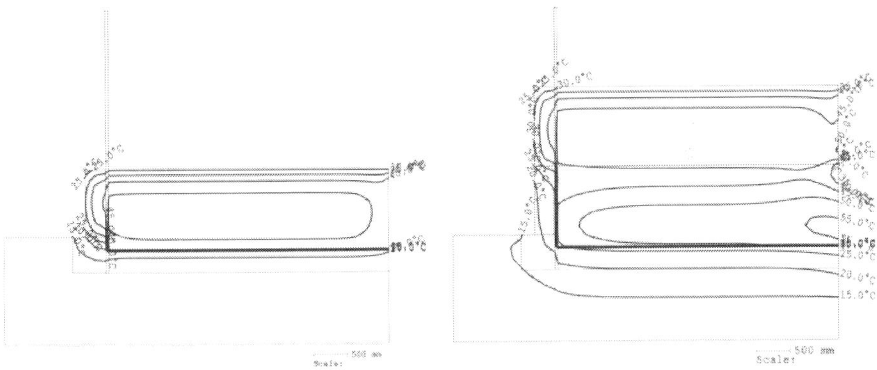

(a) (b)

Fig. 1. Distribution of: (a) $\vartheta(x_1,x_2)$ for fixed x_3 at $t = 24$s, (b) $\vartheta(x_1,x_2)$ for fixed x_3 at $t = 96$s.

The attempts to predict long-time properties of concrete and other silicate mixtures from the proper analysis of physical and chemical processes during their hardening at the Faculty of Civil Engineering of Brno University of Technology have its own history, involving all above sketched approaches. The original software is in progress, being still far from covering all above discussed micro- and macrostructural aspects. For illustration see (Vala et al., 2009) with numerical simulation of the time-variable thermomechanical behaviour of a massive concrete bridge structure in the Czech Republic (2.5 m thick slab has been formed in 5 layers in 5 time periods) during first 45 hours of its existence; the thermochemical evaluation of hydration heats corresponds to 9 dominant minerals included in the Portland cement.
Another practical example (not published yet) refers to a massive concrete foundation, prepared for the vibrational compactor. Fig. 1-3 document the temperature development in time, using the isotherms for $t \in \{24, 96, 144, 312, 408\}$ s (where their redistribution seems to be most interesting) in one half of a typical cut through such structure, built in successive time steps.

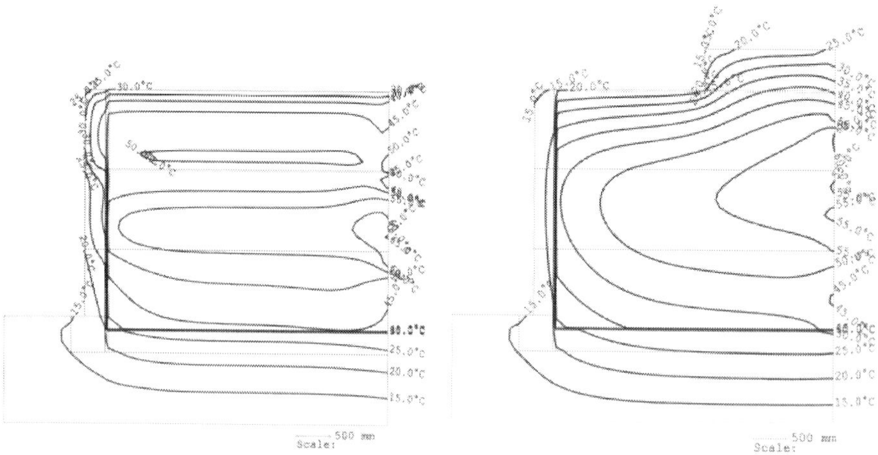

(a) (b)

Fig. 2. Distribution of: (a) $\vartheta(x_1, x_2)$ for fixed x_3 at $t = 144\,\mathrm{s}$, (b) $\vartheta(x_1, x_2)$ for fixed x_3 at $t = 312\,\mathrm{s}$.

9. Conclusion

We have derived a rather general (but physically transparent) model of thermomechanical behaviour of early-age silicate composites, including four phases, their interactions and phase changes. The related software experiments, motivated from building practice, exploit, up to now, only a minor part of this model.

We have mentioned some mathematical and numerical difficulties, reflected in the intricacy of software codes. From the practical point of view, even more unpleasant complication are connected with the (often unstable and ill-conditioned) inverse problems of identification of material characteristics, typically as nonlinear functions of several variables, in their reliability and correlation with (often only qualitative) microstructural information – cf. (Aly & Sanjayan, 2009) and Chap. 15 of (Kosmatka et al., 2002), called *Volume Changes of Concrete*. Simultaneously still other generalization, not built in our theory explicitly, are needed in practically motivated technical calculation, as thermal radiation of buildings, discussed in (Šťastník, 2007), or nonlocal modelling of micro- and macrocracking and damage by (Kozák & Vlček, 2005). Continued research seems to require more expensive both laboratory equipments and computer hardware and software, thus its extent depends on the grant support of some complex research project, as that mentioned in *Introduction*, in the near future.

10. Acknowledgement

This research has been partially supported by the Ministry of Education, Youth and Sports of Czech Republic No. MSM 0021630511, and by the project of specific research of Brno University of Technology No. FAST-S-10-17.

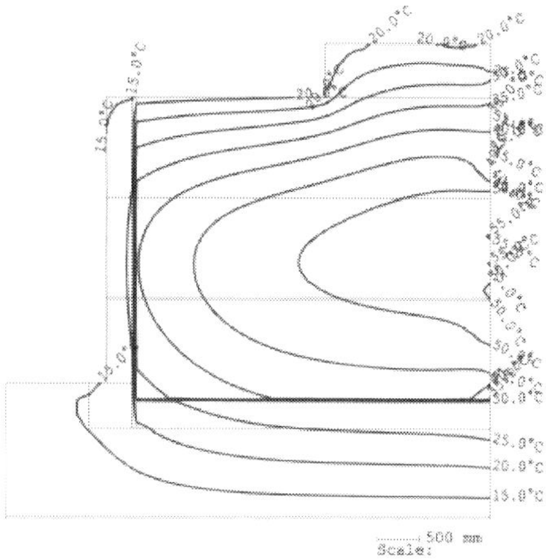

Fig. 3. Distribution of $\vartheta(x_1, x_2)$ for fixed x_3 at $t = 408\,\mathrm{s}$

11. References

Aly, T. & Sanjayan, J. G. Mechanism of early age shrinkage of concretes. *Materials and Structures*, Vol. 42, 431-468.

Bažant, Z. P. (2001). Prediction of concrete creep and shrinkage: past, present and future. *Nuclear Engineering and Design*, Vol. 203, 27-38.

Bentz, D. P. (2008). A review of early-age properties of cement-based materials. *Cement and Concrete Research*, Vol. 38, 196-204.

Bermúdez de Castro, A. (2005). *Continuum Thermomechanics*. Birkhäuser, Boston.

Cioranecu, D. & Donato, P. *An Introduction to Homogenization*. Oxford University Press.

De Schutter, G. Influence of hydration reaction on engineering properties of hardening concrete, *Materials and Structures*, Vol. 35, 453-461.

Dormieux, L., Kondo, D. & Ulm, F.-J. (2006). *Microporomechanics*. John Wiley & Sons, Hoboken (New Jersey).

Efendiev, Y. & Hou, T. Y. (2009). *Multiscale Finite Element Method – Theory and Applications*. Springer, Berlin.

Gawin, D., Pesavento, F. & Schrefler, B. A. (2006). Hygro-thermo-chemo/mechanical modelling of concrete at early ages and beyond. *International Journal for Numerical Methods in Engineering*, Vol. 67, 297-331 (Part I) and 332-363 (Part II).

Gawin, D., Pesavento, F. & Schrefler, B. A. (2006). Modelling damage processes of concrete at high temperature with thermodynamics of multi-phase porous media. *Journal of Theoretical and Applied Mechanics*, Vol. 44, 505-532.

Gray, W. G.(2000) Macroscale equilibrium conditions for two-phase flow in porous media. *International Journal of Multiphase Flow*, Vol. 26, 467-501.

Kosmatka, S. H., Kerkhoff, B. & Paranese, W. C. (2002). *Design and Control of Concrete Mixtures*. Portland Cement Association, Skokie (Illinois).

Kozák, V. & Vlček, L. (2005). The Weibull stress parameters calibration upon the toughness scaling model between cracks having different constraint. *Nuclear Engineering and Design*, Vol. 235, 1889–1896.

Majorana, C. E. (2010). An approach for modelling concrete spalling in finite strain. *Mathematics and Computers in Simulation*, Vol. 80, 1694-1712.

Moon, J.-H., Rajabipour, F., Pease, B. & Weiss, J. (2005). Autogenous shrinkage, residual stress and cracking in cementitious composites: the influence of internal and external restraint. *Proceedings of 4th International Research Seminar Self-Desiccation and Its Importance in Concrete Technology* in Gaithersburg (Maryland), 1–20.

Maruyama, I., Matsushita, T. & Noguchi, T. (2007). Numerical modeling of Portland cement hydration based on particle kinetic model and multi-component concept. *Proceedings of 12th International Congress on the Chemistry of Cement* in Montreal, TH1-08.3, 12 pp.

Neff, P. (2008). Remarks on invariant modelling in finite strain gradient plasticity. *Technische Mechanik*, Band 28, Heft 1, 13-21.

Nguetseng, G. (2003-4). Homogenization structures ans applications. *Zeitschrift für Analysis und ihre Anwendungen*, Vol. 22, 73-107 (Part I), and Vol. 23, 482-508 (Part II).

Pichler, Ch., Lackner, R. & Mang, H. A. (2007). A multiscale micromechanics model for the autogenous-shrinkage deformation of early-age cement-based materials. *Engineering Fracture Mechanics*, Vol. 74, 34-58.

Sanavia, L. (2001). A formulation for an unsaturated porous medium undergoing large inelastic strains. In: *Erwin Stein Award '2001*, International Center for Computational Engineering Sciences (Universität Hannover), 42 pp.

Sanavia, L., Schrefler, B. A. & Steinmann, P. (2002). A formulation for an unsaturated porous medium undergoing large inelastic strains, *Computational Mechanics*, Vol. 28, 137-151.

Steinhauser, M. O. (2008). *Computational Multiscale Modelling of Fluids and Solids – Theory and Applications*. Springer, Berlin.

Šťastník, S. (2007). On the surface degradation of building claddings caused by thermal radiation. *WTA Journal*, Vol. 2007, 223-238.

Tomosawa, F. Development of a kinetic model for hydration of cement. *Proceedings of 10th International Congress on the Chemistry of Cement* in Gothenburg, 2-51.

Vala, J. (2006). Two-scale finite element techniques in engineering mechanics. *Journal of Mechanical Engineering*, Vol. 57, 127-140.

Vala, J. & Šťastník, S. (2004). On the thermal stability in dwelling structures. *Building Research Journal*, Vol. 52, 31-56.

Vala, J., Šťastník, S. & Kozák, V. (2009). Computational thermomechanical modelling of early-age silicate composites. *Proceedings of the International Conference on Numerical Analysis and Applied Mathematics* in Rethymno, Part I, 232-235.

Yavari, A. & Marsden, J. E. (2009). Covariant balance laws in continua with microstructure. *Mathematical Physics*, Vol. 63, 1-42.

4

Surfactant Transfer in Multiphase Liquid Systems under Conditions of Weak Gravitational Convection

Konstantin Kostarev[1], Andrew Shmyrov[1], Andrew Zuev[1] and Antonio Viviani[2]
[1]*Institute of Continuous Media Mechanics,*
[2]*Seconda Università di Napoli*
[1]*Russia,*
[2]*Italy*

1. Introduction

To day, investigation of mass transfer between a drop and a surrounding medium is one of the most promising and at the same time the most complicated fields of research. The relevance of these investigations should be attributed to the fact that drops participate in many technological processes including extraction, mixing, dissolution, etc. The complexity of studies is caused by small dimensions and spherical shape of the drops which severely restricts the possibilities of experimental investigation (Henschke & Pfennig , 1999; Agble & Mendes-Tatsis, 2000; Kosvintsev & Reshetnikov, 2001; Amar *et al.*, 2005; Waheed *et al*, 2002). As a result the majority of studies into the process of mass transfer between a drop and an ambient fluid cover theoretical aspects of the problem and are based on different simplifications and assumptions on the character and the level of interaction between mass transfer mechanisms (Myshkis. *et al*, 1987; Subramanian *et al.*, 2001; Bratukhin *et al.*, 2001). Therefore the results of these studies require experimental verification especially in the case of surfactant diffusion, which can locally change the surface tension at the interface and initiate the solutocapillary motion even in initially homogeneous solutions (Wegener *et al.*, 2007, 2009a, 2009b; Burghoff & Kenig, 2006).
The character of interaction between buoyancy and capillary convections generally strongly depends on the gravity level. At most times it is suggested that the buoyancy plays the main role under normal gravity and that the capillary forces dominate in microgravity conditions. However at normal gravity conditions the free convection motion provides conditions for recovering the percentage composition of the liquid mixtures near the interface. With reduction in the gravity level the outflow of both the diffusing component and depleted mixture decreases. Lowering of the concentration gradient normal to the interface reduces the probability of formation of the concentration inhomogeneities along this boundary. Therefore the possibility for the occurrence of solutocapillary motion during mass transfer processes under microgravity conditions is still an open question.
In the autumn of 2007 the space experiment "Diffusion of a surfactant from a drop" was carried out during the orbital flight of the spacecraft Foton-M3 (Kostarev *et al.*, 2010). The performed experiment was aimed at studying the development of the capillary convection

during surfactant diffusion. In the framework of terrestrial simulation of the next space experiment we carried out a series of laboratory tests to provide insight into the nature of surfactant transfer from the liquid to the drop in a thin horizontal layer. The main results of our space and terrestrial investigations of mass transfer between a drop and a surrounding medium are presented below.

2. Experimental apparatus and technique

In terrestrial simulation of microgravity conditions the use of the horizontal liquid layer of small thickness gave us twofold benefits. First, it allowed a maximum reduction of intensity of the gravitational flow and second it allowed us to use a specific experimental technique – generation of a drop in the form of a short and wide vertical cylinder. The drop is bounded at the top and the bottom by the solid surfaces and has a free lateral surface which offers new possibilities for visualization of the surfactant distribution inside and outside the drop by means of optical methods. During laboratory experiments we investigated diffusion of isopropyl alcohol (used as a surfactant) from the drop of its mixture with chlorobenzene to the surrounding water (section 3, dissolving drop) or contrariwise from its ambient aqueous solution to the drop of chlorobenzene (section 5, absorbing drop). The isopropyl alcohol is lighter than water, which, in turn, has a lower density than chlorobenzene. All fluids and their solutions used in tests were transparent. The isopropyl alcohol is well soluble both in water and chlorobenzene, although the magnitudes of its solubility in these fluids differ essentially. In particular, an equilibrium distribution of alcohol between the solution and the drop is reached when its concentration in water C_0 and chlorobenzene C_d is 10% and 1.2%, respectively (Fig. 1). The limiting solubility of pure chlorobenzene in water is as low as 0.05% at 30°C and that of water in chlorobenzene is even lower. However, the reciprocal solubility of these two fluids increases with a growth of alcohol concentration in both fluids. Therefore, at low surfactant concentrations the absorbing drop during mass transfer adds only one component of the mixture – alcohol, whereas at high concentrations we might expect generation of appreciable counter-flows of the base fluids.

Visualization of the surfactant distribution was made with the help of the Fizeau interferometer (Gustafson *et al.*, 1968; Kostarev & Pshenichnikov, 2004). In accordance with the goals of experiment the interferometer was subjected to modernization, which consisted in creation of an additional unit including mirror *9* tilted at an angle of 45° to the laser beam axis, cuvette with a horizontal fluid layer *6* and video camera *8* to record the transmitted-light image of the drop (Fig. 2). The axis of rotation of the unit coincided with the optical axis of the parallel beam formed by collimator *5* of the interferometer. The center of the mirror and the beam axis also coincided in such a way that the laser beam was perpendicular to the plane layer containing the drop at any orientation of the cuvette with respect to the vector of the gravitational force. The selected scheme of the experimental setup enabled us to control the intensity of the gravitational action by way of changing the angle of inclination of the plane layer. However it also required an additional adjustment for horizontal positioning of the layer achieved with the error of less than 30″.

The interferometer cell in the form of rectangular parallelepiped 1.2 mm thick, 90 mm length and 50 mm width was used to create an element of an infinite thin horizontal layer. The surfactant distribution in the liquid layer inside the cell was visualized on its wide side. In the isothermal case the results of visualization of the concentration inhomogeneities in the fluid solutions were obtained as a system of interference bands, representing isolines of equal optical path length. In case of the mixture composition varies only across the transmitted light beam the interference bands could be identified with certain values of the surfactant concentration. Thus, for the layer 1.2 mm thick a transition from one band to another corresponded to a variation in the alcohol content in water from 0.27% at C_0 = 5% to 0.81% at C_0 = 45% (Zuev & Kostarev, 2006). For the chlorobenzene mixture a similar transition occurred due to a change in the alcohol concentration by 0.10%. The initial diameter D_0 of the cylindrical drops injected into the liquid layer with a medical syringe varied from 3 to 9 mm. The ambient temperature of experiments was (23±1)°C.

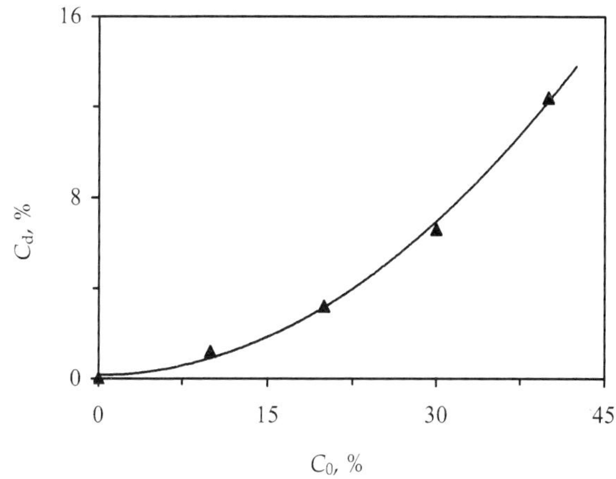

Fig. 1. Equilibrium concentration of isopropyl alcohol in the drop versus its concentration in the surrounding solution

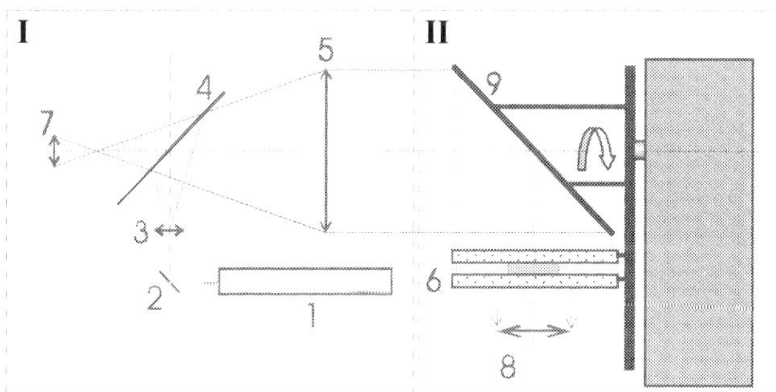

Fig. 2. Schematics of the experimental setup: *1* – laser; *2* – rotating mirror; *3* – micro-lens; *4* – semi-transparent mirror; *5* – lens-collimator; 6 – plane layer with a drop; *7* and *8* – video cameras, *9* – tilting mirror

3. Surfactant diffusion from drop (terrestrial simulation)

For investigation of surfactant dissolution process we used the chlorobenzene drops with he initial mass concentration of the isopropyl alcohol C_d ranged from 1% to 20%. A typical series of interferograms of the concentration field generated around a dissolving drop of the mixture containing surfactant is shown in Fig. 3. It is seen that the process of surfactant diffusion begins concurrently with the formation of the drop (Fig. 3,*a*) and even earlier and has three-dimensional character despite a small thickness of the layer and its horizontal position. The alcohol, escaping from the drop, did not have time to mix with water due to a low diffusion and therefore it floated up forming a thin layer along the upper boundary of the cell. A similar situation could be observed inside the drop – chlorobenzene, which had already got rid of the alcohol, sank along the lateral walls of the drop and moved along the bottom towards its center. As a result, vertical gradients of the surfactant concentration were formed both in the drop and in the layer. Capillary convection occurred practically immediately after formation of the drop. It developed in the form of three-dimensional nonstationary cells symmetrically formed at both sides of the interface. In a short time the size of the cells became comparable with the drop radius, which provided conditions for a rapid decrease of the surfactant concentration due to a continuing transfer of the mixture from the central region of the drop to the interface. Note that the capillary flow had a rather weak effect on the gravitational flow responsible for the propagation of the concentration front in a direction away from the drop boundary (such level of the interaction manifests itself in a dramatic distinction between two types of the convective motion shown in Fig. 3,*b*). At the same time, the boundary of the concentration front had still the traces of the originating cell flow (Figs. 3,*b*–3,*d*).

A decrease of the surfactant content in the drop smoothed down the concentration differences at the interface and the capillary flow decayed. After this the evolution of the concentration field was governed solely by the buoyancy force, which essentially simplified its structure (Fig. 3,*c*). As long as the amount of surfactant in the drop remained rather high, regeneration of the vertical solutal stratification at the interface led again to the development of the intensive capillary convection (Fig. 3,*d*). However, the arising cell motion continued for no more than a few seconds and was followed by the gravitational flow with essentially lower characteristic velocities. Depending on the initial surfactant concentration the number of such "outbursts" of the capillary convection could vary in the range from one or two at $C_d = 5\%$ to eight at $C_d = 20\%$ (it is to be noted that the number of outbursts markedly varied from test to test at the same value of C_d). The period of the alternation of different convective flow patterns was rather short (it lasted approximately two minutes at $C_d = 20\%$ for a drop with $D_0 = 5.6$ mm). Completion of the surfactant dissolution from the drop proceeded under conditions of quasi-diffusion. Depending on the values of C_d and D_0 the time of full dissolution varied from 7 to 10 minutes, which was tens times shorter than the characteristic times of diffusion.

At described series of interferograms the transmitted beam passed through the optical medium, whose properties were changing along direction of the light propagation. This rendered the interpretation of the current interference pattern impossible (because in the considered situation it was the main source of information concerning the two unknown quantities – the amount and extent of the concentration inhomogeneity). On the other hand, the visualized distribution of the isolines of equal optical path length was formed by the field of surfactant concentration which enabled us to watch its propagation throughout the cell volume, to estimate the intensity of its variation using the rate of change of the interference bands at the selected points (e.g. in drop center) and also to define the characteristic times of the main stages of the dissolution process (Kostarev *et al.*, 2007).

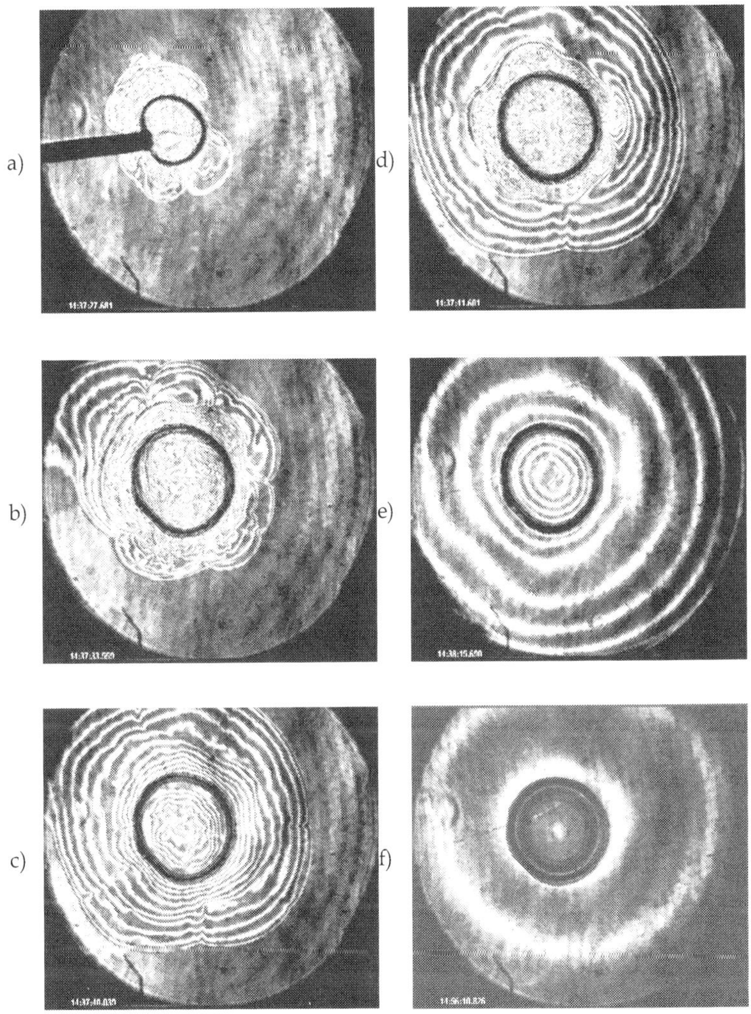

Fig. 3. Distribution of concentration during dissolution of the alcohol from the drop. D_0 = 4.7 mm with C_d = 15% in a horizontal layer 1.2 mm thick. *t*, sec: 1 (*a*), 7 (*b*), 13 (*c*), 15 (*d*), 49 (*e*), 580 (*f*)

In view of the fact that on the interferogram a transition from one interference band to another cannot be correlated with a certain variation of the concentration value, the relationships describing evolution of the concentration field will be presented without revaluation, i.e. as time variation of the number of interference bands *N* at the selected points. However we propose to retain the term "distribution of concentration" for discussion of the visualization results bearing yet in mind that the field structure is three-dimensional.

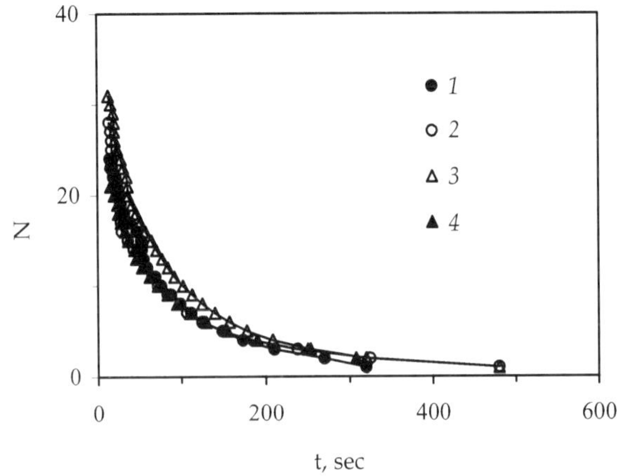

Fig. 4. Time variation of surfactant concentration in center of a drops with diameter of D_0 = 5.0 mm at different initial surfactant concentrations: C_d, %: 10 (*1*); 15 (2); 20 (3)

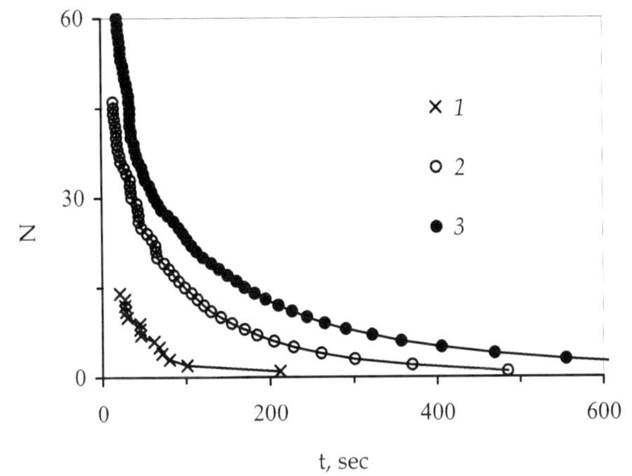

Fig. 5. Time variation of concentration of surfactant for drops at C_d = 15% with different initial diameters D_0, mm: 4.2 (*1*); 6.2 (2); 8.8 (3)

As it is evident from the presented interferograms, the process of surfactant dissolution from the drop proceeds in a quite symmetrical manner, suggesting that at the center of the drop concentration of the surfactant is kept maximal. Fig. 4 shows its variation with time for drops having close diameters but different initial surfactant concentration. It is readily seen that over the selected range of C_d the obtained curves coincide. Since observation of the drop has been made up to the moment of complete surfactant dissolution, such behavior of the curves means that by the time of first measurements, coinciding with the time of cessation of the intensive Marangoni convection (see Fig. 3,*c*), the content of the surfactant at the center of the drops with different C_d reaches the same value. From this observation follows the conclusion that a reduction of the difference in the initial surfactant content between different drops (even by two times) occurs at the stage of their formation and development of intensive capillary motion, i.e. during the first 10–12 seconds elapsed after the start of surfactant dissolution.

As we know now, in a horizontal layer, over a rather wide range of C_d variation of surfactant concentration in the drops is described by the same relationship, no matter how many "outbursts" of capillary convection interrupting the gravitational mode of dissolution have occurred. Therefore it is of interest to us to investigate variation of surfactant concentration at the center for drops with different initial diameters (Fig. 5). As might be expected the time of complete withdrawal of the surfactant increases with the size of the drop.

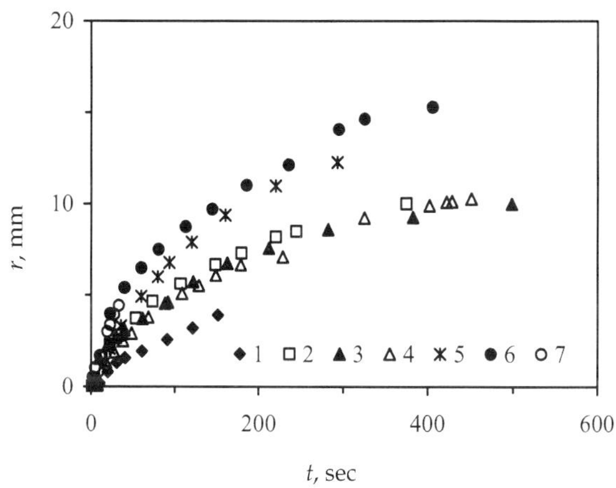

Fig. 6. Time variation of concentration front position for drops with different initial surfactant concentrations: C_0, %: 1 (*1*), 3 (*2*), 5 (*3*), 10 (*4*), 15 (*5*), 20 (*6*), 30 (*7*)

The experiments made have revealed also some specific features of the dissolution process for drops with a low content of the surfactant. Among these is the non-linear dependence of the diffusion front velocity on the surfactant concentration (curves *2* and *3* in Fig. 6). It has been found that intensification of the dissolution is caused by the oscillations of the free surface of a quiescent drop, which occurs due to a periodic onset of the local capillary convection at low alcohol concentrations (C_d ~ 3-5%). Further increase in the initial surfactant concentration results in cessation of the surface oscillations (by this time the Marangoni convection spreads over the whole surface of the drop), which reduces the rate of dissolution (at C_d from 7 to 11%, curve *4*). It should be noted that similar periodic ejections of the surfactant from the interface at small values of C_d can be observed in the case of free (spherical) drops of a binary mixture which dissolves in hydroweightlessness conditions (Plateau technique). They occur as weak vertical oscillations of the drop or as periodic up-and-down motions of the emulsion over the drop surface (at C_d ~1%) (Kostarev, 2005).

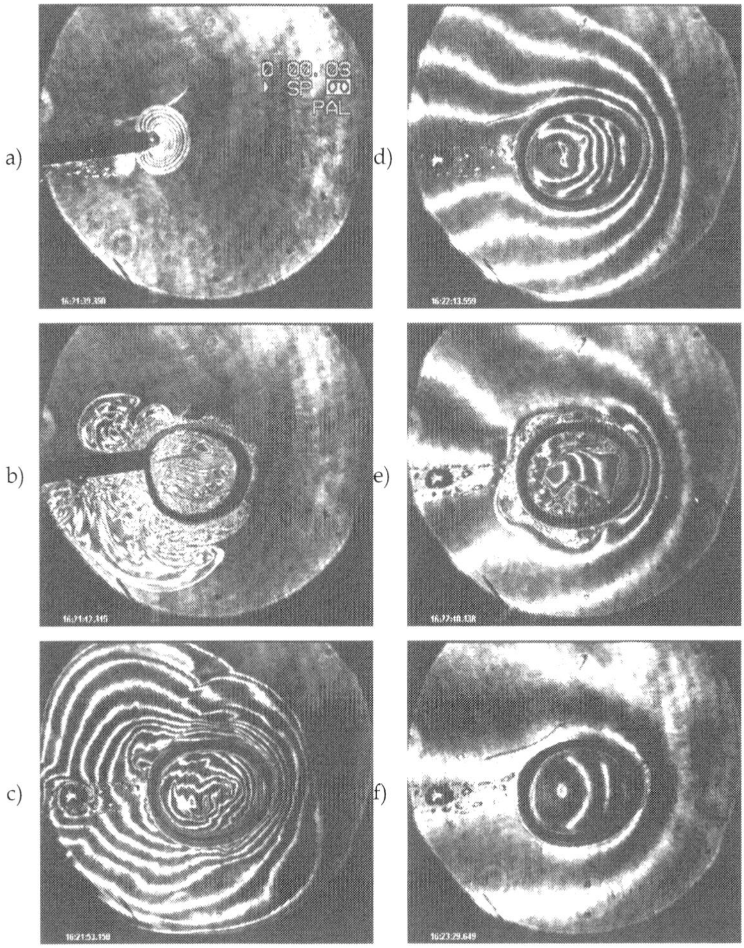

Fig. 7. Distribution of concentration during dissolution of the alcohol from the drop D_0 = 5.1 mm with C_d = 10% in the inclined layer, 1.2mm thick. Angle of inclination $\alpha = 9°$. *t*, sec: 1 (*a*), 4 (*b*), 15 (*c*), 35 (*d*), 62 (*e*), 111 (*f*)

Since our interest in dissolution of a drop in a thin horizontal layer is primarily dictated by the prospects for simulation of the diffusion processes in liquid systems with non-uniform distribution of surfactants in microgravity, it seems reasonable to give special attention to a change in the structure of the concentration field in a slightly inclined layer. Shown in Fig. 7 are the series of interferograms of the concentration field generated during surfactant dissolution from the drop in the layer inclined at an angle $\alpha = 9°$.

As in the case of a horizontal layer, the diffusion of the surfactant began already in the course of drop formation (Figs. 7,*a*–7,*b*) and the alcohol escaping from the drop spread chiefly in the direction of layer rising (Fig. 7,*c*). The concentration field inside the drop also underwent rearrangement – the zone of maximum surfactant concentration shifted and ceased to coincide with the geometrical center of the drop (Fig. 7,*d*). Although the capillary convection is formally independent of the direction and the magnitude of gravitational force, nonetheless the latter has indirect effect on the process. An "outburst" of the Marangoni convection began at the upper part of the drop, accumulating alcohol, which had already permeated through the drop surface but had not left it yet (Fig. 7,*e*). Thereafter the wave of the capillary motion began to spread downward along the interface. As in the previous case, the process of dissolution ceased in a quasi-diffusion mode, during which the center of the concentration field remained shifted relative the drop center (Fig 7,*f*).

The tests demonstrated that despite the rearrangement of the concentration field the intensity of mass transfer from the drop does not change with increasing α (at least up to α ~ 12°). Evidently this is because the change in the concentration field is nothing but its displacement as a whole with respect to the drop center at a distance proportional to the angle of inclination (Fig. 8). In this case, a decrease in the total flux of the surfactant from the lower part of the drop into the surrounding medium is compensated by an increase of the flux from the upper part.

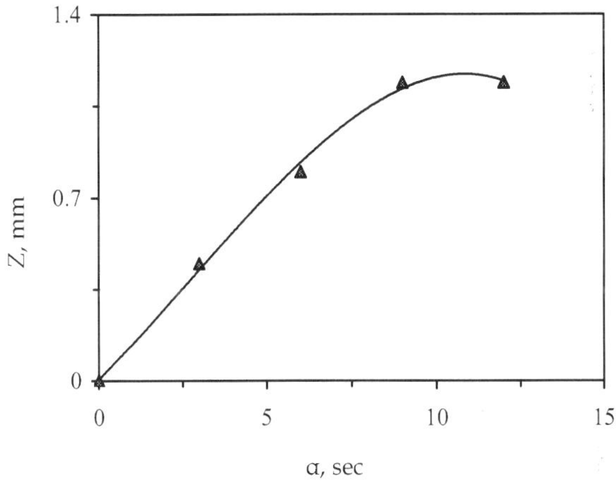

Fig. 8. Displacement of the concentration field center in the drop relative its center during surfactant dissolution in the inclined layer. D_0 = 5.0 mm, C_d – 10%

4. Surfactant diffusion from drop (space experiment)

4.1 Experimental setup

To obtain the drop in microgravity conditions we used a mixture containing 85% (by mass) of chlorobenzene and 15% of isopropyl alcohol. The distilled water was used as a fluid surrounding the drop. To prevent air bubble formation during long-time storage in microgravity conditions the water and the binary mixture before pouring into the cuvette were degassed by heating them for a long time up to the boiling point. For our experiment we designed and manufactured a small-scale Fizeau interferometer (Fig. 9) (Kostarev *et al.*, 2007).

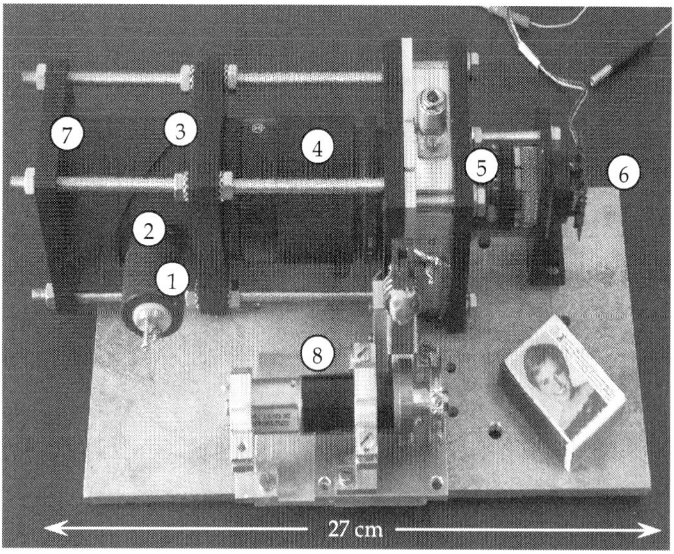

Fig. 9. Scheme of the setup for studying heat/mass transfer processes in microgravity conditions: *1* — laser; *2* —micro-lens; *3* — semi-transparent mirror; *4* — lens-collimator; *5*—Hele-Show cell with a drop; *6* and *7* — video cameras, *8* - device for drop formation

The collimator block of the interferometer consisting of microlens *2*, semitransparent mirror *3*, and lens-collimator *4* generated a plane-parallel coherent light beam of diameter 38 mm, emerging from a semiconductor laser *1*. The interferometer was equipped with two analog video cameras *6* and *7*, operating respectively with the reflected beam and the beam transmitted through the cuvette *5*. Camera *6* registered the interferograms of the temperature and concentration fields in the whole volume of the cuvette and camera *7* was intended to make more detailed records of the process evolution in the central part of the cuvette. The frequency of both cameras was 25 frames a second and the resolving power was 540×720 lines.

For experimental cuvette we chose the Hele-Shaw cell, which was a thin gap 1.2 mm thick between two plane-parallel glass plates *l* with semitransparent mirror coating (Fig. 10). The cell was encased in a metal frame *2* and formed a working cell of the interferometer adjusted to a band of the infinite width. The insert *3* placed in the gap had a hollow, which was used as a cuvette working cavity 35 mm in diameter. The cavity was filled with a base fluid through the opening *4* which was then used to locate a thermal expansion compensator *5*. A drop of a binary mixture was formed with the help of a needle of medical syringe which was placed along the cavity diameter. The needle was connected to the binary mixture supply system, which included a bellows for a liquid mixture, multi-step engine with a cam gear, and a device for needle decompression. The backbone of this device is a movable bar connected by means of inextensible thread to a cam mechanism. Prior to experiment the bar was inserted in the needle and the gap between them was sealed. After voltage was applied to the engine the cam began to rotate and first removed the bar from the needle and then bore against the wall of the bellows. As a result, the channel of the needle turned to be open to a flow of the binary mixture from the bellows to the cuvette cavity, where it wetted the side glass walls and formed a cylindrical drop of prescribed volume at the center of the cuvette. After this the engine was automatically stopped.

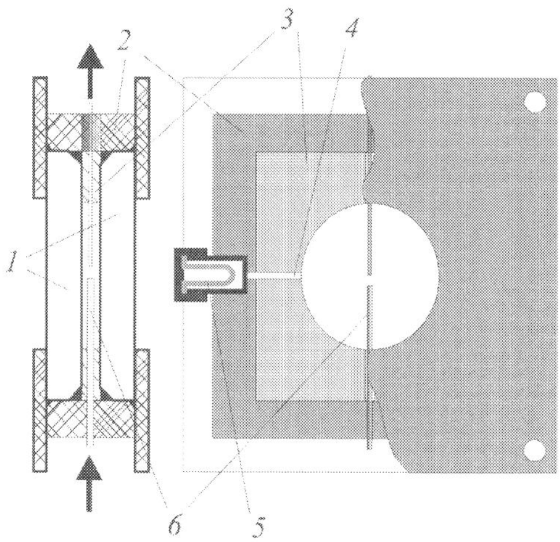

Fig. 10. Scheme of Hele-Show cell for studying mass transfer processes:
1 – interferometric glasses, *2* – metal frame, *3* – plastic insert, *4* – opening, *5* – thermal expansion compensator, *6* – tube for drop formation

To determine the flow structure formed during experiment we used the ability of the composed liquid system to form a non-transparent emulsion of water in chlorobenzene and emulsion of chlorobenzene in water while the alcohol diffused through the interface. Focusing the camera *7* (see Fig. 9) on the mid-plane of the cuvette turned the drops of emulsion into analogues of small light - scattering particles which provided flow visualization inside and outside the drop on the background of interferogram of the concentration field.

The total time of the test was 52 minutes. The experiment was carried out at ambient temperature (20±1)°C.

4.2 Results

The analysis of video records showed perfect consistency between the performed experiment and its cyclogram. As the experimental program envisaged during first five minutes records of the interefernce patterns were made by video camera shooting the central part of thc cuvette. The absence of the alcohol distribution near the end of the syringe needle suggested that its hermetic sealing was kept up to the beginning of the experiment. There were no air bubbles on the video records made by the camera, and neither there were the intereference bands on the periphery of the images which could be indicative of non-uniform heating of the cuvette.

After switching on the multi-step engine the needle is unsealed so that the binary mixture can be readily supplied to water filling the cuvette. It is to be noted that at first a rather large volume of aqueous solution of the alcohol (up to 6.5 µl) is ejected from the needle (Fig. 11,*a*) and only after this the needle forms a drop of a binary mixture with a clear-cut interphase boundary (Fig. 11,*b*). The maximal concentration of the alcohol in this drop is about 5.5%. The reason for appearance of the aqueous solution of the alcohol in the needle is penetration of water from the cavity into the channel of the needle after removal of the sealing bar.

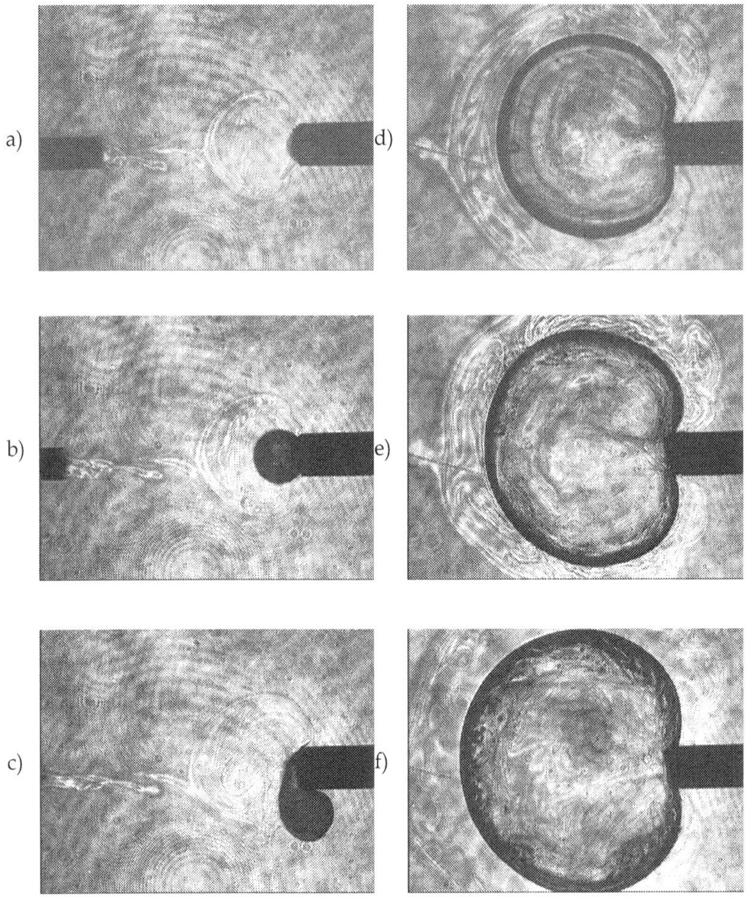

Fig. 11. Evolution of the concentration field and flow structure during the drop formation. External diameter of needle is 1.0 mm. Time from the test outset *t*, sec: 4.5 (*a*), 4.9 (*b*), 5.1 (*c*), 8.2 (*d*), 8.6 (*e*), 10.2 (*f*)

Since ejection of the drop is precede by appearance of the alcohol solution the drop during first seconds of its existence is in a quiescent state being surrounded by a similar media. Then as the drop grows it comes into contact with pure water which leads to initiation of intensive solutocapillary motion of the drop surface due to generation of the surface tension difference. As a result, the drop executes several oscillatory motions with large amplitude (Fig. 11,*c*) like a drop in the known Lewis and Pratt experiment (Lewis & Pratt, 1953). The oscillations of the drop extend the boundaries of the region containing liquid enriched with the alcohol. This leads to a decrease of the concentration gradient at the interface and the growing drop again quiets down. Approximately at the same time the drop reaches the lateral boundaries of the cavity gradually changing its shape from a spherical to a cylindrical. As the drop turns into a cylindrical drop its internal structure becomes more and more observable.

Further increase of the drop size is accompanied by formation of several internal solutal zones in the vicinity of its boundary. In these zones transfer of the surfactant occurs with quasi- diffusional velocities (the flow of the mixture caused by injection of the solution into the drop is concentrated in its central part, Fig. 11,*d*). A jump-wise increase of the feeding velocity by 2 times (at t = 8.3 sec) causes an abrupt intensification of the motion inside the drop, which breaks down the established concentration field. Again the vortex flow spreads over the whole drop while pure water reaches the surface of the drop in the zone of its contact with needle. This leads to a repeated outburst of capillary convection (Fig. 11,*e*) Development of the Marangoni convection is accompanied by deformation of the drop itself and the surrounding front of surfactant concentration, which gains a quasi-periodic structure. Then, as in the first case, a flow of the diffused surfactant into the zone, where pure water comes into contact with the solution of the drop, causes retardation of the capillary motion, which coincides in time with cessation of mixture supply to the drop (Fig. 11,*f*). At this moment the maximal diameter d_0 of drop reaches 6.2 mm.

When the forced motion in the drop ceases, the water emulsion, captured during surfactant diffusion at the time of intensive convection and kept near the drop surface, begins to penetrate deep into the drop. The penetrating emulsion forms a clearly delineated layer (Fig. 12,*a*). In the gap between this layer and the drop surface one can watch the formation of a light layer of the mixture, which has lost most of the surfactant due to diffusion (a decrease in the surfactant concentration leads to an abrupt change in the index of light refraction, Fig. 12,*b*). Propagation of emulsion with velocities of about 0.2 mm/sec is supported by a slow large-scale motion of the mixture in the drop caused by capillary eddies formed near the needle (Fig. 12,*c*).

The source of eddy formation is the alcohol, which diffuses from the needle after cessation of mixture supply. It creates a surfactant concentration difference at the free surfaces of the air bubbles, generated near the needle due to a decrease of air solubility in the fluid of the drop, in which alcohol concentration decreases. Most of the bubbles are formed at the drop boundary and then migrate deeper and deeper into the drop under the action of capillary forces. The average diameter of the bubble is of order 0.1 mm, and the maximum diameter is ~ 0.3 mm. Apart from the eddy flow there is a slow capillary motion of the mixture over the drop surface toward its far pole (opposite to the tip of the needle). This flow also contributes to the emulsion motion.

The motion of emulsion is accompanied by coalescence of part of its droplets, which settle down on the walls of the cavity. (The arising droplets can be seen due to a trace of emulsion kept in the stagnation zone behind them, Fig. 12,*d*). As the emulsion layer moves away from the boundary, one can observe development of a small-cell motion near the bubbles, which are in the diffusion gradient of the surfactant at the drop boundary. What is interesting, detachment of these bubbles from the drop interface as well as the local displacement of the drop boundary due to reduction of its area during diffusion of the surfactant causes a specific stochastic fluttering of the drop surface.

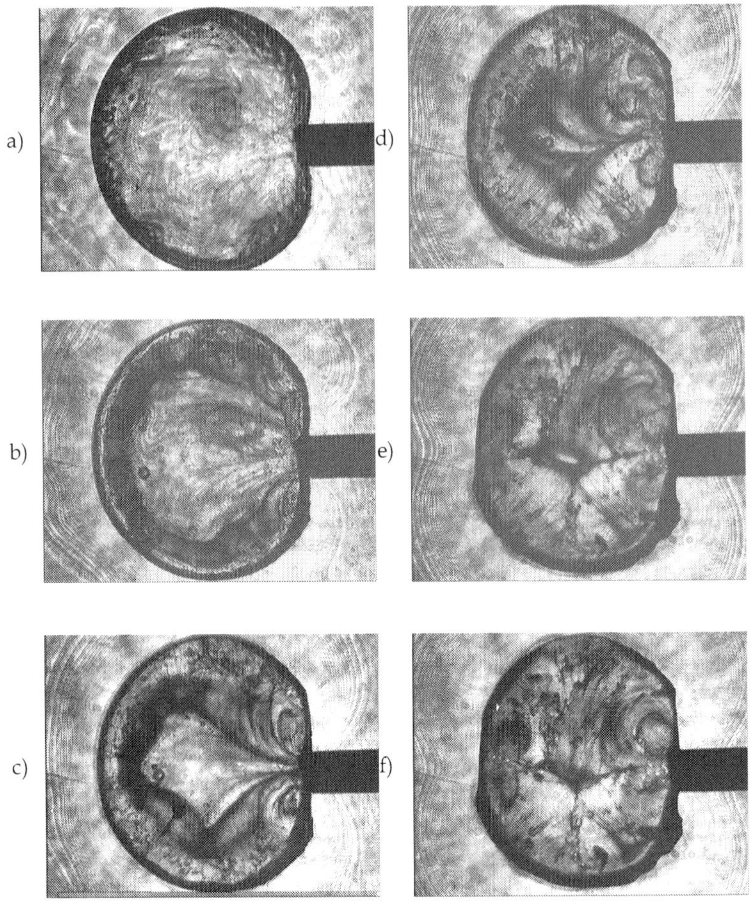

Fig. 12. Evolution of the concentration field and flow structure during surfactant dissolution process. Time from the test outset *t*, sec: 15.5 (*a*), 38.7 (*b*,) 88.4 (*c*), 163 (*d*), 251 (*e*), 273 (*f*)

The cell motion reaches maximum intensity at the drop pole opposite to the tip of the needle (evidently due to the absence of flows generated earlier in this region). The air bubbles formed in the immediate neighborhood, accelerates the motion of mixture from the center of the drop and also favors the development of the cell motion. The arising flow is of three-dimensional pattern and as the times goes it occupies an increasingly growing space and even deforms the solutal front, which moves away from the drop (Fig. 12,*e*). Simultaneously there occurs another interesting phenomenon - settling of gas bubbles at the cavity walls. During sedimentation the bubbles take the form of irregular semi-spheres. This effect can be observed at the time when the bubbles are in the emulsion cloud.
After five minutes from the beginning of drop formation the velocity of the fluid flow in the drop decreases practically to zero (Fig. 12,*f*) and the volume of the drop reduces to 0.9 of its maximum (Fig. 13).

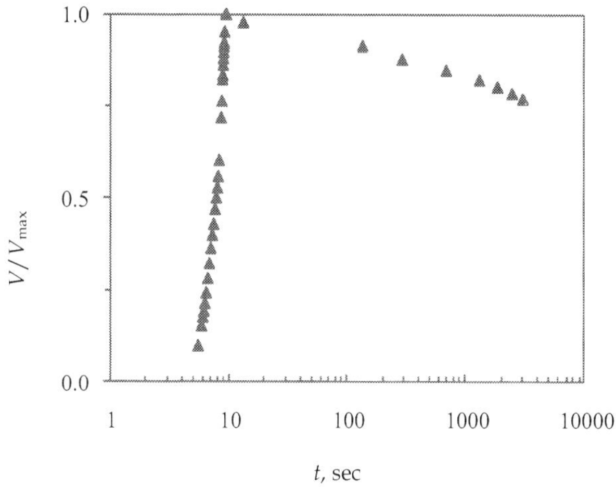

Fig. 13. Variation of the relative volume of dissolving drop with time

Fig. 14 shows a series of interferograms of the concentration field in the vicinity and inside the drop at different moments of time. As it is seen from the figure, in the following, after first five minutes, the only observable events are gradual dissolution of the emulsion and variation of the surfactant concentration gradient in the radial direction. Apart from this a decrease of the alcohol concentration in the region near the drop boundary leads to a growth of the surface tension and, as a result, to local jump-wise displacement of the drop boundary while the area of the latter decreases.

Hence, based on the analysis of the obtained video records we can draw a conclusion that during diffusion of the surfactant from the drop at least three times there were favorable conditions for the development of the intensive Marangoni convection. However, all the observed capillary flows rapidly decayed. In the first two cases - during drop formation and a change in the mixture feeding regime - the Marangoni convection decay was provoked by a flow of a surrounding fluid with higher surfactant concentration into the zone of the capillary motion, which eliminates the concentration difference along the interface. On the other hand, generation of such surfactant distribution in the vicinity of the drop was made possible only in the absence of the Archimedean force with the result that molecular diffusion became a governing factor in the process of mass transfer outside the drop. In the third case, when the Marangoni convection was initiated near the gas bubbles inside the drop, the reason of its decay was a decrease of surfactant percentage in the mixture filling the needle channel and rapid recovery of the surfactant distribution homogeneity in the region near the bubbles due to convective mixing.

Video recording of the central part of the cuvette has made it possible to trace the evolution of initiated flows, whereas records of the general view of the drop (Fig. 15,*a*) allow us to investigate propagation of the concentration front from the dissolving drop of the binary mixture and to establish the relationship between the position of its boundary and time (Fig. 15,*b*). It is readily seen that propagation of the concentration front is described fairly well by the power law with the exponent 1/3. Over the whole period of the experiment position of the concentration front changed by 5 mm suggesting that propagation of the surfactant occurred with diffusion velocities. According to measurements made in the second half of the experiment, the concentration filed has no preferential direction of propagation, which allows us to suppose that the value of the residual accelerations on the satellite "Foton-M 3" during the experiment was no higher than 10^{-4} g_0.

a)

b)

c)

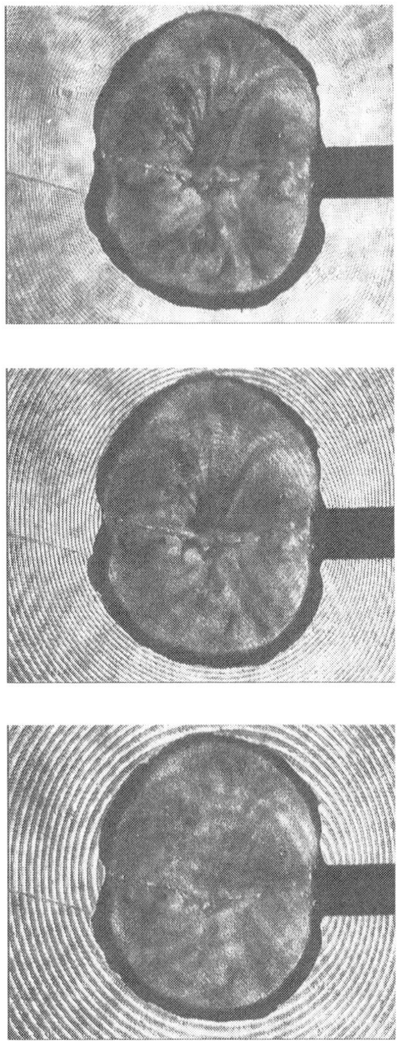

Fig. 14. Evolution of the concentration field and flow structure during surfactant dissolution process. Time from the test outset *t*, min: 15.2 (*a*), 19.3 (*b*), 28.8 (*c*)

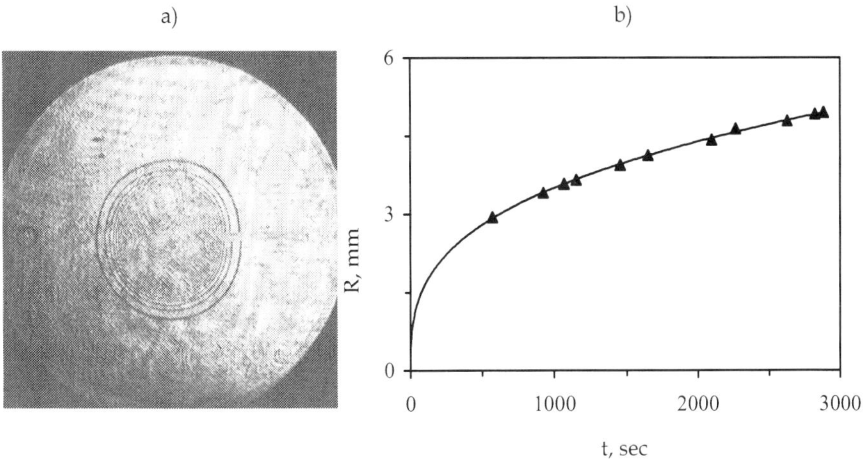

Fig. 15. View of distribution of surfactant concentration around drop (*a*) and variation of the boundary position (*b*) of concentration field with time

A detailed structure of the concentration field generated by surfactant diffusion (Fig. 16) was defined by making use of the computer-aided overlapping of images obtained from both video cameras (such an approach was used to solve the problem with relatively low value of their resolving power). Since only one video recorder was used for recording, we chose for overlapping the images shot by different camera with the shortest time span between them. It should be emphasized that distribution of the surfactant concentration outside the drop was obtained by way of direct measurements, which were made based on the interferograms starting from the field (front) boundary. It was impossible to apply this approach inside the drop because of the unknown value of the surfactant concentration at the drop boundary. This value remained unknown due its jump-wise change provoked by a difference in the chemical potentials. We determined it from the curves, which were plotted based on the coefficient of the equilibrium distribution of the isopropyl alcohol in the chlorobenze-water system under the assumption that the dissolution process was quasi-equilibrium due to predominance of diffusion. For the selected system of fluids the value of this coefficient was 0.12 in 15% water solution of the alcohol and as the alcohol concentration decreased, this coefficient also decreased to 0.10.

As it follows from Fig. 17, alcohol concentration at the drop surface at the time when the motion of the mixture in the drop ceases, is ~ 14%, which actually coincides with the initial concentration of the alcohol in the binary mixture of the drop. Note that at the end of the test the surfactant concentration at the drop boundary decreased approximately by two times. This reduction can be approximated with the least error by the exponential relationship. For the sake of comparison, on the Earth under conditions of maximum suppression of the gravitational convection the drop of the same volume completely lost the alcohol in a matter of ten minutes.

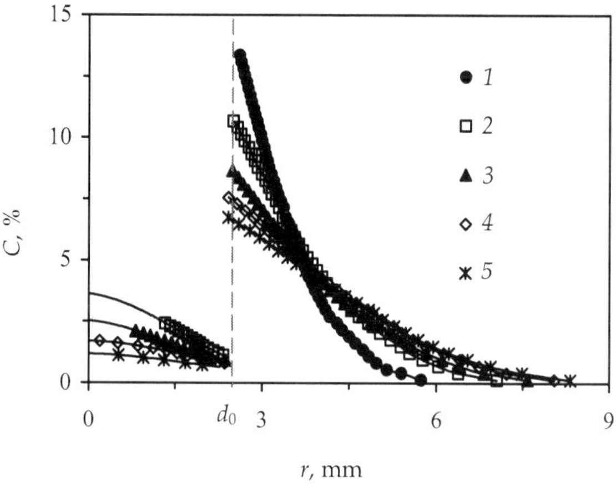

Fig. 16. Radial distributions of surfactant concentration in the vicinity and inside the drop at various time moments. t, min: 4.9 (*1*), 21.3 (2), 31.0 (3), 42.5 (4), 50.7 (5)

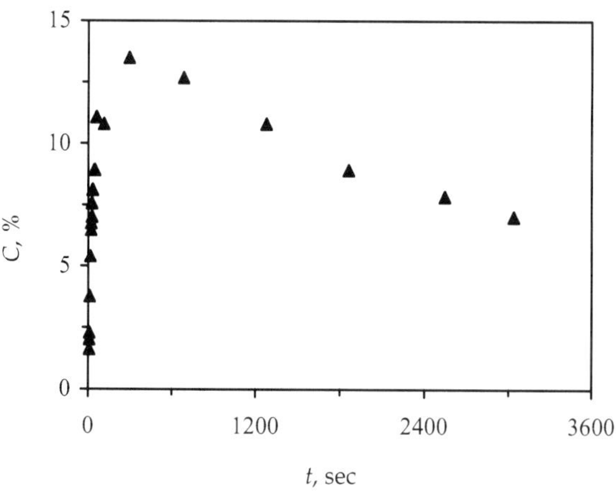

Fig. 17. Variation of surfactant concentration in water near the drop surface with time

5. Drop saturation by surfactant from homogeneous solution

In this series of test the initial mass concentration C_0 of the alcohol in the solution ranged from 1% to 50%. Fig. 18 shows two series of the interferograms reflecting the evolution of the alcohol distribution in the drop at different initial concentrations of alcohol in the solution. It is seen that at $C_0 \leq 10\%$ absorption of alcohol occurs without the development of the capillary convection in spite of the interference of the gravity force, which generates the vertical concentration gradient and, accordingly, the gradient of the surface tension. The latter seemed to be not large enough to produce shear stresses capable of deforming the adsorbed layer, which was formed at the interface from the impurities found in water (Birikh *et al.*, 2009). In the absence of the Marangoni convection, alcohol was slowly accumulated at the upper part of the drop. As soon as an increasingly growing thickness of the alcohol layer caused an additional optical beam path difference of half of the light wave length, the color of the drop on the interferogram changed (Figs. 18,*b*-18,*c*).

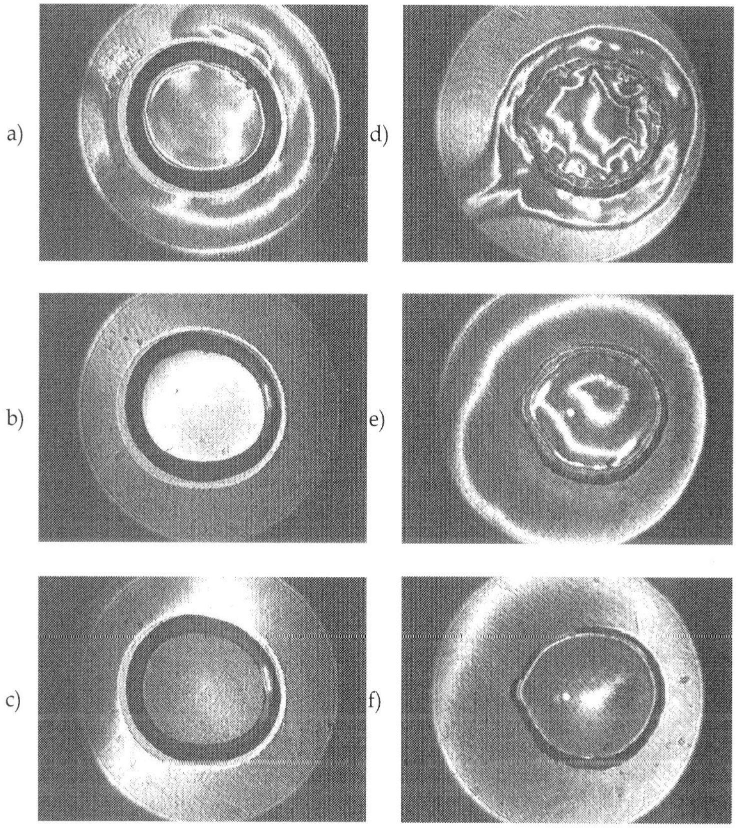

Fig. 18. Saturation of the chlorobenzene drop with isopropyl alcohol from its solution filling a horizontal Hele-Shaw cell. C_0 = 10%, D_0 = 5.2 mm; *t*, min: 0.5 (*a*), 7.0 (*b*), 10.0 (*c*); C_0 = 20%, D_0 = 5.4 mm; *t*, min: 0.1 (*d*), 1.0 (*e*), 6.0 (*f*)

At $C_0 \sim 20\%$ the gradients of surfactant concentration become higher than the threshold values which results in the development of a fine-cell capillary motion at the drop surface. With the growth of surfactant concentration the velocity of the surfactant flow into the drop increases and the gravitational force has not managed to smooth the layer of alcohol along the drop diameter. Due to the fact that the layer is formed from alcohol, rising along the lateral surface of the drop, its thickness is found to be larger at the layer edges. On the interferograms the radial variation in the layer thickness is represented by a system of concentric isolines, which merge with the passage of time at the center of the drop (Figs. 18,*d* -18,*e*). On the drop periphery one can readily see a thin layer of rising alcohol, which has diffused through the interface (Fig. 18,*f*).

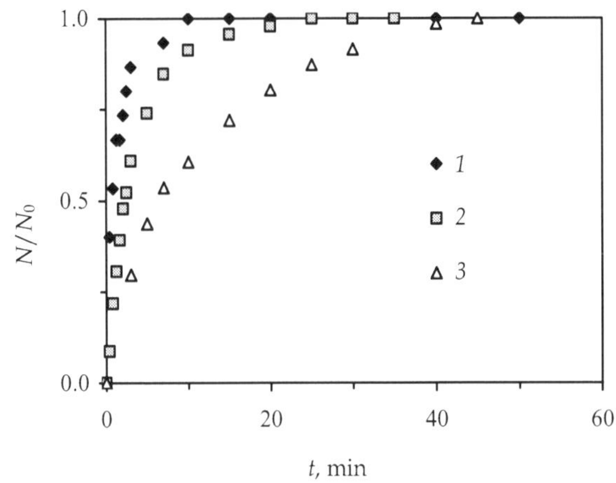

Fig. 19. The relative number of isolines inside the drop as a function of time for different concentrations of solution C_0, %: 10 (*1*), 20 (*2*), 30 (*3*). The initial drop area $S_0 \sim 35$ mm^2

In Fig. 19, a relative number of the interference bands vanishing at the center of the drop is plotted versus the time elapsed from the moment of drop formation for solutions of different concentrations (for normalization we used the maximum number of the bands obtained in each cases). The analysis of the curve behavior shows that duration of the phase, in which a convective mass transfer dominates over a pure diffusion transfer, rapidly increases with the growth of surfactant concentration in the solution.

Absorption of alcohol from its solution should cause a growth of the drop volume, which in the case of the cylindrical drop with the fixed thickness is manifested as an increase of its area. Indeed, saturation of the drop with alcohol was accompanied by a change of its area. However, this variation was of a complicated nature, which was specified by the initial concentration of the surfactant in the solution (Fig. 20). Such a behavior of the absorbing drop can be explained by an increase in the reciprocal solubility of water and chlorobenzene with a growth of concentration of their common solvent, viz., the content of alcohol in their solutions. The maximum increase (~20%) in the drop area was observed at the initial stage of its saturation at C_0 = 40% (curve *3* in Fig. 20). Then the drop began to dissolve due to increasing diffusion of chlorobenzene in the surrounding solution. A delay in dissolution was caused by the insufficient content of the surfactant inside the drop. The desired concentration was achieved only some time after injection of the drop into the solution. In the solution containing 45% of alcohol (curve *4* in Fig. 20) concentration of the absorbed surfactant at the drop boundary immediately reached the value, at which chlorobenzene began to dissolve in the surrounding solution. The process of the surfactant extraction turned into the reverse - dissolution of the extragent.

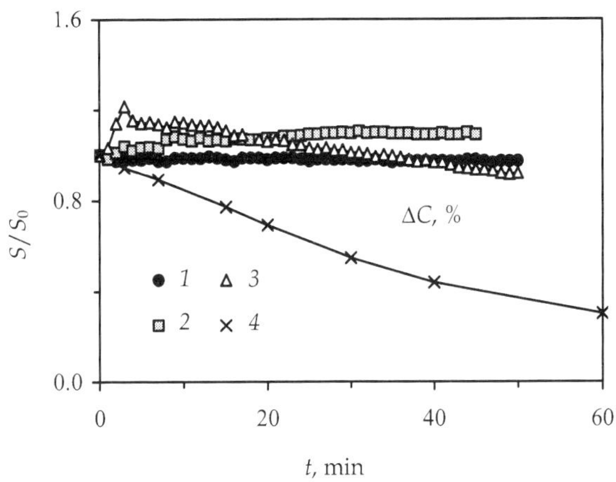

Fig. 20. The relative area of the chlorobenzene drop as a function of time for isopropyl solutions of various concentrations. The initial drop area $S_0 \sim 35$ mm^2. C_0, %: 10 (*1*), 30 (*2*), 40 (*3*), 45 (*4*)

Dissolution of chlorobenzene gave rise to an intensive gravitational flow in the surrounding solution. This process every so often was accompanied by generation of both the vertical and longitudinal (lying in the layer plane) difference of the surfactant concentration at the drop surface provoking the development of a large-scale capillary flow.

In our case, such a flow occurred in the form of two quasi-stationary vortices lying in the horizontal plane (Fig. 21). Note that the flow was sustained by a small difference (~3%) in the surfactant concentration between the "western" and "eastern" poles of the drop. The initial solution entrained by the flow reached the drop surface in the form of the concentration "tongue", which then split into the streams flowing round the drop. As they spread along the drop surface they lost part of alcohol due to its diffusion. After that they again merged into a single flux, which drifted away from the drop carrying part of alcohol-chlorobenzene mixture.

In turn, alcohol penetrated into the drop and formed two fluxes, which first moved along the interface and then after collision at the western pole of the drop spread deep into the drop generating a two-vortex flow. The flows initiated inside and near the drop closely resemble in structure the convective flow near the drop absorbing the surfactant from its stratified solution in the vertical Hele-Shaw cell (Kostarev *et al.*, 2007). However, under conditions of maximum suppression of gravitational convection the considered flow remains quasi-stationary - in contrast to the flow in a vertical cell, which at the similar surfactant concentration differences is of the pronounced oscillatory nature. At C_0 = 50% the phase boundary between the drop and the solution disappears leading to the formation of a three-component liquid mixture. Phase separation of the fluid system vanishes within the first two minutes after placing the drop into the surfactant solution.

a)

b)

c)

Fig. 21. Distribution of concentration of isopropyl alcohol during its absorption by the drop of chlorobenzene from the solution with initial concentration C_0 = 45%. D_0 = 6.5 mm. t, min: 0 (*a*); 40 (*b*); 190 (*c*, general view of the cell flow with the drop in center)

a)
b)
c)

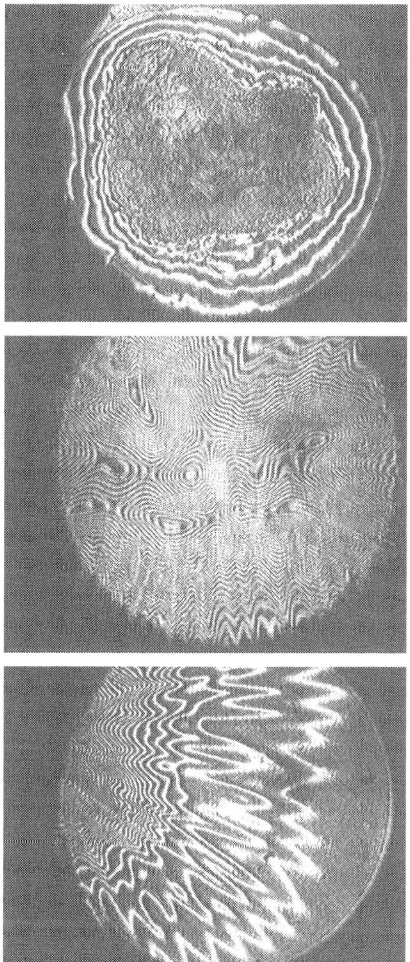

Fig. 22. Dissolution of a drop of chlorobenzene in 50% solution of alcohol. *t*, min: 1.25 (*a*); 9.0 (*b*); 11.2 (*c*)

Fig. 22 illustrates the main stages of drop dissolution in the surrounding solution at C_0 = 50%. Absorption of alcohol at the moment of drop formation diminishes the interface surface tension to an extent that it turns to be impossible to form a cylindrical drop even in one-millimeter clearance. Nevertheless, the original drop has an interface with sufficient curvature, owing to which the probing light beam scatters. As a result the drop interface is seen on the interferogram as a grey spot (Fig. 22,*a*). Then the drop surface increases because the drop spreads out over the cavity bottom as the surface tension continuously decreases. Further dissolution is accompanied by deformation of both the drop interface and the boundary between the drop and a solid substrate. Disappearance of the interface in the fluid system makes the mixture optically transparent because the jumps of the refraction coefficient have vanished. Fig. 22,*b* presents the interferogram of the optical inhomogeneity caused by inhomogeneous concentration of chlorobenzene at the place of the former drop. It is seen that a transition zone between the solution and the mixture with higher content of chlorobenzene has a well-defined relief, which is similar to the relief of the free surface of the drop. In Fig. 22,*c* we can see part of the concentration inhomogeneity picture with typical distribution of chlorobenzene in alcohol in the form of "fingers", whose origination can be attributed to a flow of chlorobenzene down the spatial inhomogeneities of the transition zone.

The period of spatial perturbations of the drop-solution interface, repeated further by the relief of the transition zone of the concentration inhomogeneity, is defined by the physico-chemical properties of the interacting fluids. As for the solution, these properties are strongly dependent on surfactant concentration. Indeed, according to the results of measurement the angular distance between the "fingers" increases with a growth of the surfactant solution concentration (Fig. 23).

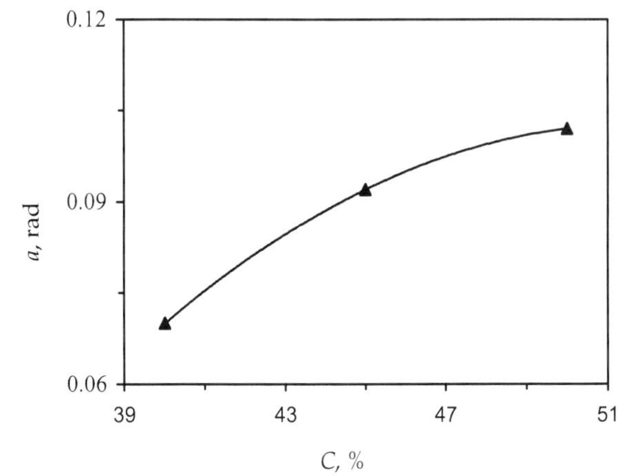

Fig. 23. Variation of the angle between "fingers" with the growth of initial concentration of the surfactant solution

6. Conclusion

As seen from a comparison of space and terrestrial experiments, the data obtained in microgravity differ markedly from the results of the laboratory study of surfactant diffusion in a thin horizontal layer. In the later case a small coefficient of surfactant diffusion facilitated generation of the density difference in the originally homogeneous surrounding fluid. This difference was large enough for development of the gravitational motion inside and outside the drop. Nevertheless a relative contribution of the Marangoni convection to the process considerably increased – the action of the capillary forces mainly determined the rate of mass transfer at the beginning of surfactant diffusion. Only a decrease in the surfactant concentration below some critical value led to regeneration of the gravitational-diffusion mechanism of mass transfer, yet its action was repeatedly interrupted by the

"outbursts" of intensive capillary convection. Because of a three-dimensional structure of the flows in the laboratory conditions, the application of the interference method posed many difficulties; only the qualitative characteristics were used for description of the evolution of concentration fields. However, the laboratory experiments allowed us to make an optimal choice of the parameters for the space experiment and to prepare its preliminary cyclogram.
The space experiment "Diffusion of the surfactant from a drop" was successful. The sensitivity of the new setup proved to be an order of magnitude higher than that of the shadow device "Pion-M" used on the "Mir" space station. The weight of the setup was reduced by more than an order of magnitude, and its external dimensions were also down sized. The interferometer can operate in the automatic regime allowing performance of experiments onboard the unmanned space vehicles. The obtained data supported the view that in microgravity conditions the mass transfer processes involving surfactant diffusion through the interface can proceed without excitation of intensive capillary convection.
Because of diffusion in the absence of capillary convection, the surfactant concentration reaches rather high values at both sides of the drop surface. This leads to much greater extent of mutual dissolution of the base fluids compared to the terrestrial conditions. Therefore, the interface failure in such a system can occur at much lower concentration of the surfactant which plays the role of a common solvent for both the base fluids. On the one hand, this allowed a more exact computation of the value of this concentration, which is an essential prerequisite for development of the theoretical grounds of phase transitions in three-component liquid systems. On the other hand, reduction of the limiting concentration opens the way to intensive experimental studies of the hydrodynamic effects involving the interface failure.
At the same time, based on terrestrial simulation data, we can expect the development of a large-scale Marangoni flow around the drop absorbing the surfactant from the homogeneous surfactant solution in microgravity conditions. Such a different evolution of mass transfer processes in the surfactant solution emphasizes again the importance of experimental investigations in real microgravity conditions.

7. Acknowledgements

The work was supported by Russian Foundation of Fundamental Research under project No 09-01-00484, joint project of SB, UB and F-EB of RAS (116/09-C-1-1005) and the program of the Department of Power Engineering, Mechanical Engineering, Mechanics and Control Processes of RAS No 09-T-1-1005.

8. References

Agble, D. & Mendes-Tatsis, M.A. (2000). The effect of surfactant on interfacial mass transfer in binary liquid-liquid systems. *J. Heat and Mass Transfer*, Vol. 43, p. 1025.

Amar, A.; Groß-Hardt, E.; Khrapitchev, A.A.; Stapf, S.; Pfennig, A. & Blümich, B. (2005). Visualizing flow vortices inside a single levitated drop. *J. of Magnetic Resonance*, Vol. 177, No. 1, pp. 74–85.

Bratukhin, Yu.K.; Kosvintsev, S.R. & Makarov, S.O. (2001). Droplet motion induced by diffusion of soluble surfactant to the external medium: theory. *Colloid J.*, Vol. 63, No. 3, pp. 326–332.

Burghoff, S. & Kenig, E.Y. (2006). A CFD model for mass transfer and interfacial phenomena on single droplets. *AIChE J.*, Vol. 52, No. 12, pp. 4071-4078.

Gustafson, S.E.; Kjellander, R.A.E. & Rolf, A.E. (1968). An interferometer for direct recording of refractive index distributions. *Z. Naturforch.*, Vol. 23a, No. 2, pp. 242-246.

Henschke, M. & Pfennig, A. (1999). Mass transfer enhancement in single drop extraction experiments. *AIChE J.*, Vol. 45, No. 10, pp. 2079-2084.

Kostarev, K.G. & Pshenichnikov, A.F. (2004). Gravitational convection of a liquid mixture in a horizontal cylindrical gap at moderate Grasshoff numbers. *Cosmic research*, Vol. 42, No. 2, pp. 115-122.

Kostarev, K.G. (2005). The study of the extraction of surface-active component of a binary liquid from model ("cylindrical") droplets. *Colloid J.*, Vol. 67, No. 3, pp. 357-362.

Kostarev, K.G.; Levtov, V.L.; Romanov, V.V.; Shmyrov, A.V.; Zuev, A.L. & Viviani, A. (2007). Experimental investigation of mass transfer in fluid systems in microgravity conditions. *Proceedings of 58th Int. Astronautical Congress*, Hyderabad, India, 2007. Paper № IAC-2007-A2.4.06. 10 p.

Kostarev, K.G.; Levtov, V.L.; Romanov, V.V.; Shmyrov, A.V. & Viviani, A. (2010). Experimental study of surfactant transfer in fluid systems in microgravity conditions. *Acta Astronautica*, Vol. 66, No. 3-4, pp. 427-433.

Kosvintsev, S.R. & Reshetnikov, D.G. (2001). Droplet motion induced by diffusion of soluble surfactant to the external medium: experiment. *Colloid J.*, Vol. 63, No. 3, pp. 318-325.

Lewis, J.B. & Pratt, H.R.C. (1953). Instabilities induced by mass transfer and deformable fluid interfaces. *Nature*, Vol. 171, pp. 1155—1161.

Myshkis, A.D.; Babskii, V.G.; Kopachevskii, N.D.; Slobozhanin, L.A. & Tyuptsov, A.D. (1987). *Low-Gravity Fluid Mechanics*, Springer, the Netherlands.

Subramanian, R.S. & Balasubramaniam, R. (2001). *The motion of bubbles and drops in reduced gravity*, Cambridge University Press, Cambridge, UK.

Waheed, M.; Henschke, M. & Pfennig, A. (2002). Mass transfer by free and forced convection from single spherical liquid drops. *J. of Heat and Mass Transfer*, Vol. 45, No. 22, pp. 4507-4514.

Wegener, M.; Grunig, J.; Stuber, J. & Paschedag, A.R. (2007). Transient rise velocity and mass transfer of a single drop with interfacial instabilities - experimental investigations. *Chemical Engineering Science*, Vol. 62, No. 11, pp. 2967-2978.

Wegener, M.; Fevre, M.; Paschedag, A.R. & Kraume, M. (2009a). Impact of Marangoni instabilities on the fluid dynamic behaviour of organic droplets. *J. of Heat and Mass Transfer*, Vol. 52, No. 11-12, pp. 2543-2551.

Wegener, M.; Paschedag, A.R. & Kraume, M. (2009b). Mass transfer enhancement through Marangoni instabilities during single drop formation. *J. of Heat and Mass Transfer*, Vol. 52, No. 11-12, pp. 2673-2677.

Zuev, A.L. & Kostarev, K.G. (2006). Oscillation of a convective flow round the air bubble in a vertically stratified solution of a surfactant. *J. Experimental and Theoretical Physics*, Vol. 130, No. 2, pp. 363-370.

5

Mass Transfer in Multiphase Mechanically Agitated Systems

Anna Kiełbus-Rąpała and Joanna Karcz
West Pomeranian University of Technology, Szczecin
Poland

1. Introduction

In this chapter, the results of the experimental studies concerning the volumetric mass transfer coefficient k_La obtained for mechanically agitated gas - liquid and gas - solid - liquid systems are discussed.

Mechanically agitated gas - liquid and gas -solid - liquid systems are widely used in many processes, for example oxidation, fermentation or wastewater treatment. In such cases, oxygen mass transfer between gas and liquid phases in the presence of solid particles can be described and analyzed by means of the volumetric mass transfer coefficient k_La. In the gas - liquid and gas - solid - liquid systems, the k_La coefficient value is affected by many factors such as geometrical parameters of the vessel, type of the impeller, operating parameters of the process (impeller speed, aeration rate), properties of the continuous phase (density, viscosity, surface tension, etc.) and also by the type, size and loading of solid particles.

To improve the efficiency of the processes conducted in gas - liquid and gas - solid - liquid three - phase systems two or more impellers on the common shaft are often used (Kiełbus—Rąpała & Karcz, 2009). Multiple - impeller stirred vessels due to the advantages such as increased gas hold - up, higher residence time of gas bubbles, superior liquid flow characteristics and lower power consumption per impeller are becoming more important comparing with single - impeller systems (Gogate et al., 2000). As the number of energy dissipation points increased with an increase in the impellers number on the same shaft, there is likely to be an enhancement in the gas hold - up due to gas redistribution, which results into higher values of volumetric gas - liquid mass transfer coefficient (Gogate et al., 2000).

Correct design of the vessel equipped with several agitators, therefore, the choice of the adequate configuration of the impellers for a given process depends on many parameters. That, which of the parameters will be the most important depends on the kind of process, which will be realized in the system. For the less oxygen demanding processes the designer attention is focused on the mixing intensity much more than on the volumetric mass transfer coefficient. When the most important thing is to achieve high mass transfer efficiency of the process, the agitated vessel should be such designed that the configuration of the agitators used ensure to get high both mixing intensity and the mass transfer coefficient values (Kiełbus-Rąpała & Karcz, 2010).

1.1 Survey of the results for gas-liquid system

Good mass transfer performance requires large interface area between gas and liquid and a high mass transfer coefficient The obtaining the good dispersion of the gas bubbles in the liquid agitated depends on the respectively high agitator speed, therefore the characteristics of the gas-liquid flow are intensively studied (Paul et al., 2004); Harnby et al., 1997).

In literature, there are a large number of correlations for the prediction of volumetric mass transfer coefficients k_La in mechanically stirred gas-liquid systems (Linek et al., 1982, 1987; Nocentini et al., 1993; Gogate & Pandit, 1999; Vasconcelos et al., 2000; Markopoulos et al., 2007). Markopoulos et al. (2007) compared the results which were obtained by other authors for agitated Newtonian and non-Newtonian aerated liquids. The comparative analysis carried out by Markopoulos et al. (2007) shows that significant disagreement is observed taking into account the form of the correlations for k_La. Therefore, no single equation exists representing all of the mass transfer data given in literature. The differences can mainly be ascribed to the differences in the geometry of the system, the range of operational conditions, capability to gas bubbles coalescence and the measurement method used. For a given geometry of the agitated vessel and liquid properties, volumetric mass transfer coefficient k_La for gas-liquid system is often described in literature by means of the following dependences

$$k_La = f(P/V_L, w_{og}) \tag{1}$$

or

$$k_La = f(n, w_{og}) \tag{2}$$

where: P/V_L - specific power consumption, w_{og} – superficial gas velocity, n – agitator speed. Most often, empirical correlations have the form of the following equations

$$k_La = C_1(P_{G-L}/V_L)^{a1}(w_{og})^{a2} \tag{1a}$$

or

$$k_La = C_2(n)^{a3}(w_{og})^{a4} \tag{2a}$$

Detailed values of the exponents a1 and a2 in Eq. (1a) obtained by different authors are given in paper by Markopoulos et al. (2007).

Recently, Pinelli (2007) has studied the role of small bubbles in gas-liquid mass transfer in agitated vessels and analyzed two-fraction model for non-coalescent or moderately viscous liquids. Martin et al. (2008) analyzed the effect of the micromixing and macromixing on the k_La values. This study shows that each mixing scale has a particular effect on the mass transfer rate.

It is worth to notice that values of the k_La coefficient can be depended on the measurement method used (Kiełbus-Rąpała & Karcz, 2009). Although a number of techniques were developed to measure the k_La values, the unsteady-state gassing-out dynamic methods (Van't Riet, 1979; Linek et al., 1982; Linek et al., 1987; Linek et al., 1996; Ozkan et al., 2000; Lu et al, 2002, Garcia-Ochoa & Gomez, 2009) are preferably used as they are fast, experimentally simple and applicable in various systems. These methods are used in many variants. For example, Machon et al. (1988) used a variation of the dynamic method in which the gas from liquid was firstly removed by vacuum and the system was then aerated for some time. When aeration was stopped and the bubbles escaped from the liquid, the steady state concentration of dissolved oxygen was measured.

Apart from experimental studies, new methods are recently used to prediction the values of the volumetric mass transfer coefficient. Lemoine & Morsi (2005) applied artificial neural network to analyze mass transfer process in a gas-liquid system. Using CFD technique, Moilanen et al. (2008) modeled mass transfer in an aerated vessel of working volume 0.2 m^3 which was equipped with Rushton, Phasejet or Combijet impeller.

1.2 Survey of the results for gas-solid-liquid system

The critical impeller speed for the solid suspension in liquid can be defined as the impeller speed at which no particles resting on the bottom of the vessel longer then 1- 3s (Zwietering, 1958). In multiple -impeller system introduction of gas into solid - liquid system implicate an increase in the impeller speed required for particles suspension just like in a single - impeller system. For solid suspension multiple impeller systems would prove to be disadvantageous if the distance between impellers is greater than the diameter of the impeller as there is an increase in the critical impeller speed for solid suspension (Gogate et al., 2000).

The data on the multiple - impeller systems working in the three - phase gas - solid - liquid systems are limited (Kiełbus-Rąpała & Karcz, 2009). Various aspects of the three - phase system stirring in the vessel equipped with more than one impeller on the common shaft were the aim of the studies in papers (Dutta & Pangarkar, 1995; Dohi et al., 1999, 2004; Roman & Tudose, 1997; Majirowa et al, 2002; Jahoda et al., 2000; Jin & Lant, 2004). The impellers used were usually Rushton turbines or pitched blade turbine. Dutta & Pangarkar (1995) and Dohi et al. (1999) analyzed experimentally the critical impeller speed needed to gas dispersion simultaneously with solid suspension in the systems, with four and three impellers, respectively. Critical impeller speed in multiple - impeller system can be modified easily by choosing optimal impeller combination. Power consumption in multiple - impeller systems was studied by Dohi et al., 1999, 2004; Majirowa et al., 2002; Roman & Tudose, 1997. Investigations in such systems concerned also gas hold - up (Dutta & Pangarkar, 1995; Dohi et al., 1999, 2004; Majirowa et al., 2002; Jahoda et al., 2000) and mixing time (Dutta & Pangarkar, 1995; Dohi et al., 1999; Jahoda et al., 2000). Hydrodynamics problems in the system stirred by means of triple impellers were analysed by Jin & Lant (2004). Authors compared hydrodynamic conditions in the stirred vessel with the air - lift and bubble column.

Although k_La coefficients were widely studied in the two - phase gas - liquid multiple - impeller systems (Arjunwadkar et. al, 1998; Fujasowa et al., 2007; Puthli et al., 2005; Moucha et al., 2003; Linek et al., 1996; Wu, 1995; Yoshida et al., 1996; Yawalkar et al., 2002) and many papers (Alper et al., 1980; Brehm et al., 1985; Lu et al., 1993; Kralj & Sinic, 1984; Chapman et al., 1983; Ruthiya et al., 2003) regard k_La measurements in the three - phase system stirred in the vessel equipped with single - impeller, the information in literature on this parameter in multi impeller gas - solid - liquid system is substantial.

On the basis of the literature data, the effect of solid particles on gas - liquid oxygen mass transfer rate in single - impeller systems at low solid concentrations analyzed Kiełbus–Rąpała & Karcz (2009) and Karcz & Kiełbus–Rąpała (2006). This effect was investigated by many authors (Alper et al., 1980; Brehm et al., 1985; Oguz et al., 1987; Lu et al., 1993; Kralj &

No	Ref.	Type of solids	Size of particles	Solids concentration	k_La coefficient Value
1.	Chandrasekaran & Sharma, 1977	activated carbon 0–2 mass %	<100 μm	0–0.2 mass % 0.2–2 mass %	increases 1.5 times Constant
2.	Joosten et al., 1977	Polypropylene, sugar, glass beads, 0–40% vol.	53-105, 250 μm 74-105 μm 53 μm, 88 μm	0–20% vol. 20–40% vol.	Constant rapidly decreases
3.	Alper et al., 1980	activated carbon, quartz sand 0–2 mass %	< 5 μm < 5 μm	0–0.2 mass % 0.2–2 mass %	increases 1.5 times constant
4.	Lee et al., 1982	glass beads 0–50 mass %	56 μm	0–30 mass % 30–50 mass %	decreases rapidly decreases slowly
5.	Chapmann et al., 1983	glass ballotini 3 and 20 mass %		0–3 mass % 3–20 mass %	slightly decreases significantly decreases
6.	Kralj & Sincic, 1984	activated carbon 0.25 and 0.5 mass %		the whole range	Constant
7.	Brehm et al., 1985	Al_2O_3 0–10% vol.	50 μm, 300 μm	0–5% vol. 5–10% vol.	increases 1.5 times Decreases
8.	Greaves & Loh, 1985	ion exchange resin 0–40 mass % glass ballotini 0–50 mass %	780 ± 70 μm 665 μm, 1.300 μm	0–15 mass % 15–50 mass %	constant decreases rapidly
9.	Bartos & Satterfield, 1986	glass beads 0–30% vol.	60 μm	the whole range	decreases with solid concentration increasing
10.	Mills et al., 1987	glass beads 0–40% vol.	66 μm	the whole range	decreases with solid concentration increasing
11.	Lu & Liang, 1990; cit. in Lu et al., 1993	kaolin 0–4 mass %	5.7 μm	the whole range	decreases with solid concentration increasing
12.	Lu et al., 1993	kaolin powder 0–4.5% vol. yeast 0–8% vol.	5.5 μm 3.7 μm	the whole range	increases to maximum (c.a. 2%) and then decreases
13.	Özbek & Gayik, 2001	Schotchbrite™ pads pieces (biomass): 5-25% vol.	0.1625 cm³, 0.65 cm³, 1.4625 cm³	the whole range	decreases with solid concentration increasing
14.	Kiełbus & Karcz, 2006	Sea sand 0.5–5 mass %	335 μm	0.5 mass % 2.5–5mass %	Increases Decreases
15.	Littlejohns & Daugulis, 2007	nylon 6.6, glass beads; Silicone rubber, styrene-butadiene copolymer	2.59 mm 6 mm 2.5 mm 3.59 mm	167g/l	Increases up to 268% Reducing by up to 63%

Table 1. The effect of presence and concentration of solid particles on k_La values determined by different authors

Sinic, 1984; Chapman et al., 1983, Ozkan et al., 2000; Kordac & Linek, 2010). The results of their studies were not always similar. The effect of presence and concentration of solid particles on volumetric mass transfer coefficient k_La determined by different authors is compared in Table 1.

Volumetric mass transfer coefficient value increased (Alper et al., 1980; Brehm et al., 1985; Lu et al., 1993), was constant (Kralj & Sinic, 1984) or slightly decreased (Chapman et al., 1983) with adding the particles in the system. Such differences can suggest that very important are properties of the particles, containing particle size (Ozbek & Gayik, 2001), density, hydrophobicity, oxygen diffusivity (Zhang et al., 2006; Littlejohns et al., 2007) and concentration. Various mechanisms were proposed for describing an enhancement of k_La caused by the particles. Shuttling effect, where absorptive particles enter the liquid boundary layer, absorbing dissolved gas and then desorbing this gas when back in the bulk phase. This effect is used especially for small particles with a size equal or smaller than the gas - liquid boundary layer (Ruthiya et al., 2003). For inert particles, without absorptive properties, the possible mechanisms which explain an enhancement of k_la include boundary layer mixing and gas - liquid interface changing. Boundary layer mixing, involving an increase in k_L due to turbulence at gas - liquid interface (Ruthiya et al., 2003; Kluytmants et al., 2003; Zhang et al., 2006), which causes a larger refreshment rate of liquid boundary layer by mixing with the bulk fluid. Changes in the gas - liquid interfacial area can be result of the particles presence at the gas - liquid interface, which can collide and interact with the gas - liquid interface or may induce turbulence near or at it, leading to a smaller effective diffusion layer or causing coalescence inhibition and an increase in a interfacial area (Ruthiya et al., 2003). However, a simple mechanism describing all cases does not exist. Comparing the data shown in Table 1 for a high concentration of the particles in the agitated three phase system it is worth to notice that values of the k_La coefficient rather decrease when the solid concentration increases (for example data given by Joosten et al., 1977; Lee et al., 1982; Brehm et al., 1985).

The results of the volumetric mass transfer coefficient k_La for the agitated three phase gas-solid-liquid system are usually correlated by different authors in the form of the following dependence

$$k_La = f(P/V_L, w_{og}, \ldots) \tag{3}$$

Some correlations presented in literature are shown in Table 2.

Galaction et al. (2004) conducted the k_La coefficient measurements in the three - phase system containing various type of microorganisms. Authors studied the effect of concentration and morphology of biomass, specific power input and superficial air velocity on volumetric gas - liquid mass transfer coefficient value in the stirred vessel equipped with two turbine stirrers. They proposed the following mathematical correlation

$$k_La = aX^{\alpha}(P_{G\text{-}L\text{-}S}/V_L)^{\beta}w_{og}^{\gamma} \tag{4}$$

describing the influence of considered factors on k_La. The values of the parameters a, α, β and γ in Eq. (4) are given in Table 3. On the basis of the measurements authors have found that, in the bioreactor, the value of k_La depends not only on operating and geometrical parameters but also on the type and morphology of the microorganisms in the system.

Author	Equation	System	Impeller	D [m]	Operating parameters	Comments
Kralj & Sincic 1984	$k_L a = 0.109\left(\frac{P_{G-L-S}}{V_L}\right)^{0.213} w_{og}^{0.34}$	Aqueous solution of NaOH - air - active carbon	Rushton turbine; $d/D = 0.33$	0.1	n = 12.5-50 1/s; $w_{og} = 10^{-3}$-10^{-2} m/s; X = 0.25, 0.5 mass%	required the presence of the antifoams (0.03-0.1% mass.)
Brehm et al. 1985	$k_L a = 0.5612\left(\frac{P_{G-L-S}}{V_L}\right)^{0.65}\left(\frac{\eta_m}{\eta}\right)^{0.47} w_{og}^{0.4}$	water or aqueous solution of polyetylen glycol – air – different solid particles	Turbine with 4 flat blades; $d/D = 0.5$	0.145	n = 8.33-13.33 1/s; Vg = 0.139-2.78 x10^{-4} m^3/s ; $X \le 10$ % obj.	solids: Al_2O_3, Fe_2O_3 kiselguhr, glass beads (d_p = 50-300 μm); sand (d_p = 300 μm and powder); 185 exp. points; Δ±11.5%
Brehm & Oguz 1988	$k_L a = 3.07 \cdot 10^{-3} \frac{\left(\frac{P_{G-L-S}}{V_L}\right)^{0.75} w_{og}^{0.5\alpha} D_L^{0.5}}{\eta_m^{0.34}\sigma_L^3}$	water (or various organic liquids) – air – various solid particles	Turbine with 4 flat blades; $d/D = 0.5$	0.145	η_m =1.3-40.7 x10^3 Pas; P_{G-L-S}/V_L = 0.75-6.3 kW/m^3; w_{og} = 0.84-4.2 m/s; σ_L = 24.3-71.8 x10^{-3} N/m; D_L = 0.68-2.41 m^2/s;	Particles: $CaCO_3$, Fe_2O_3, $BaSO_4$, n [1/s] = const = 10.83; $\alpha = (\sigma_{water}/\sigma_L)^{1/2}$ Δ±11%

Table 2. The correlations of the volumetric mass transfer coefficient $k_L a$ for the agitated gas-solid-liquid systems and the applicability of the equations

Type of biomass	A	α	β	γ	±Δ, %
Propionibacterium shermanii	6.586	-0.282	-0.0286	0.429	7.8
Saccharomyces cerevisiae	52.44	-0.702	-0.0762	0.514	6.8
Penicillium chrysogenum (aggregates)	0.193	-0.269	0.0288	0.257	8.4
Penicillium chrysogenum (free structure)	33.59	-1.012	-0.0463	0.94	7.6

Table 3. The values of the coefficient and exponents in Eq. (4)

2. Experimental

In this part of the chapter, our experimental results of the mass transfer coefficient k_La obtained for the gas - liquid or gas - solid - liquid systems agitated in a baffled vessel equipped with single high-speed impeller or the system of the two impellers are presented.
Experimental studies were conducted in cylindrical transparent vessel of the inner diameter D = 0.288 m. Geometrical parameters of the vessel used are shown in Fig. 1. The vessel, equipped with a flat bottom and four baffles of the width $B = 0.1D$, was filled with liquid up to the height $H = D$ (Fig. 1a) and H = 2D (Fig. 1b). The working liquid volumes were V_L = 0.02 m^3 and V_L = 0.04 m^3, respectively. In the vessel with single impeller the impeller was placed at a hight $h = 0.33D$. In the tall vessel two high-speed impellers were located on the common shaft. The distance of the impeller from the bottom of the vessel was $h_1 = 0.167H$ for the lower and $h_2 = 0.67H$ for the upper impeller, respectively. All impellers have diameter $d = 0.33D$. Gas was introduced into liquid through a ring shaped sparger, of the diameter $d_d = 0.7d$. The sparger with 44 symmetrically drilled holes of 2 mm in diameter was located at the distance $e = 0.5h_1$ from the bottom of the vessel. To avoid surface aeration of the liquid which could disturb the measurements, the vessel was covered at the top with a transparent lid.

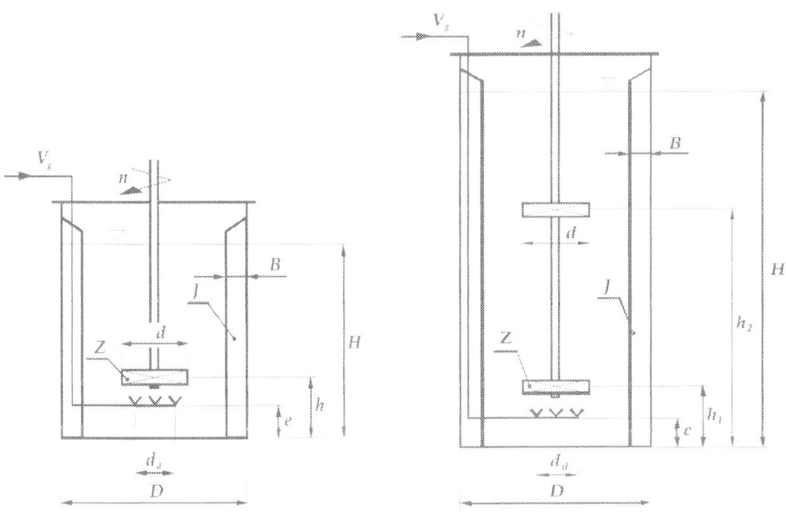

Fig. 1. Geometrical parameters of the stirred vessel; D = 0.288m; $d = 0.33D$; a) $H = D$; b) $H = 2D$

Four types of high-speed agitators (Fig. 2) differing in generated fluid pumping mode were tested: Rushton and Smith turbines (radial flow impellers), A 315 (mixed flow) and HE 3 (axial flow impeller). As a single agitator were used: RT, CD 6 and A 315.
The measurements were carried out for gas-liquid and gas-solid liquid systems. Distilled water were used as a continuous phase, whilst air as gas phase. The second dispersed phase in a three-phase system was fraction of sea sand particles with mean diameter d_p = 335 μm

and density ρ_p = 2600 kg/m³. The measurements were conducted under various aeration rates, impeller speeds, and solid particles concentration. The measurements were performed for three different solid concentrations, X = 0.5, 2.5 and 5 mass %. The experiments were conducted at five different values of gas flow rate from the following range: V_g (m³/s) ∈ ⟨0; 5.56 × 10⁻³⟩ (superficial gas velocity w_{og} (m/s) ∈ ⟨0; 8.53 × 10⁻³⟩). Power consumption was measured using the strain gauge method.

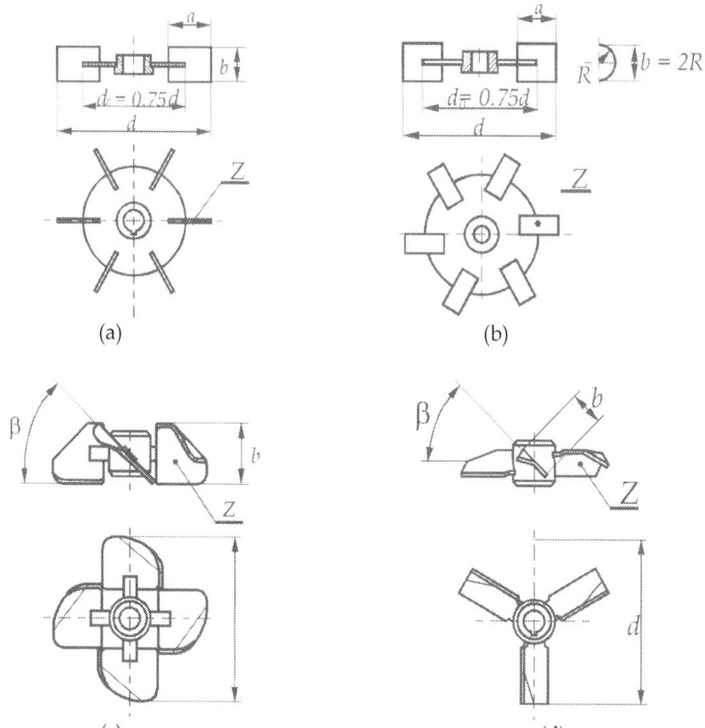

Fig. 2. Agitators used in the study; a) Rushton disc turbine (RT), b) Smith disc turbine (CD 6), c) A 315, d) HE 3

All measurements were carried out within the turbulent regime of the fluid flow in the agitated vessel. During the measurements conducted in the gas-solid-liquid system, impeller speeds higher than the critical impeller speeds n_{JSG} for the three-phase system were maintained. Critical agitator speeds in the gas–solid–liquid system were evaluated on basis of the Zwietering criterion (Zwietering, 1958).

To determine the volumetric gas–liquid mass transfer coefficients k_La the unsteady-state gassing-out dynamic method (Van't Riet, 1979; Ozkan et al., 2000; Lu et al., 2002) was used. The dynamic methods are used in many variants. In this study, the variant in which, at constant values of the impeller speed and gas flow rates, the kind of gas introduced into the vessel is changed was used. The gases were air and nitrogen. At the beginning of the experiment, in order to remove oxygen, nitrogen was passed through the system. When the value of dissolved oxygen was lowered, air replaced nitrogen. After interchanging $N_2 \rightarrow$ air, absorption of oxygen followed. The change of the oxygen concentration dissolved in liquid was measured by means of fast oxygen probe (galvanic type) coupled with an oxygen meter (CO-551, ELMETRON). The time delay of the measuring probe was ca. 3 s. This sensor was immersed in the liquid and placed at half distance between the baffles.

In calculation of k_La, the response time of the probe should be taken into account (Kiełbus-Rąpała & Karcz, 2009). Response time is defined as the time required by the probe used in the study to measure 63 % of the overall value of the oxygen concentration change (Van't Riet, 1979) and it is related to the oxygen diffusion through the membrane of the probe. According to this fact, the measured values are sufficiently reliable (error below 6 %) only for the response time of the probe shorter or equal to $(1/k_La)$. Therefore, in the experiments, evaluation of the oxygen probe response time was made by measuring the oxygen concentration changes after moving the probe from a liquid saturated by nitrogen to a saturated oxygen solution. The results indicate that using this probe, the k_La values of less than 0.1 s^{-1} are reliable. All values obtained in the measurements were lower than 0.1 s^{-1}, so the error resulting from the response time τ could be neglected.

The k_La coefficient values were determined from the slope of the plot $\ln[(C^*-C_0)/(C^*-C)] = f(t)$. In this dependence, t denotes time whilst C^*, C_0, and C describe saturation concentration of gas in liquid, and concentrations of the gas at time $t = 0$ and t, respectively. The calculation was performed using the least-squares method.

3. The results

3.1 The results for gas-liquid system

Experimental investigations of the mass transfer process in the two-phase gas-liquid systems enable to determine the changes of the volumetric mass transfer coefficient values with operating parameters (impeller speed n and superficial gas velocity w_{og}). The results for the vessel with single impeller are represented by the data obtained for CD 6 impeller (Fig. 3). In this fig. the dependencies of k_La = f(n) for various values of w_{og} are compared. It follows from the results that for all agitator tested the k_La coefficient value significantly increased with the increase both impeller speed and superficial gas velocity.

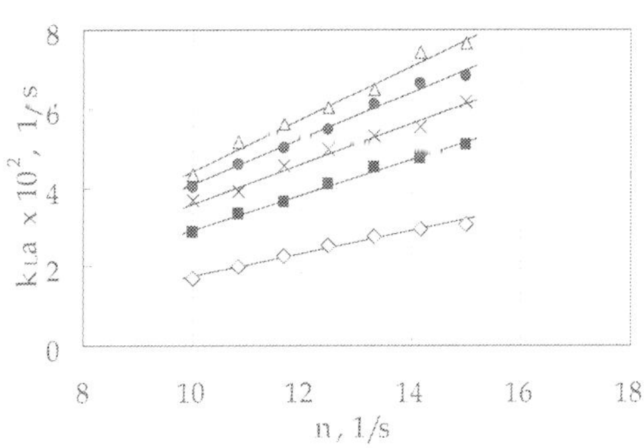

Fig. 3. The dependence of k_La = f(n) for single Smith turbine (CD 6) working in a gas-liquid system; varied values of $w_{og} \times 10^3$ m/s: (◊) 1.71; (▪) 3.41; (x) 5.12; (•) 6.82; (Δ) 8.53

The data, concerning double impeller systems, obtained for various values of superficial gas velocity w_{og}, are presented in Fig. 4a for the vessel with A 315 - Rushton turbine system and in Fig. 4b for the vessel with Smith turbine–Rushton turbine system. It can be observed that similarly to the vessel with single impeller, volumetric mass transfer coefficient is strongly affected by both impeller speed and superficial gas velocity. The differences in the k_La values between particular values of superficial gas velocity w_{og} are greater in the case of Smith turbine - Rushton turbine configuration.

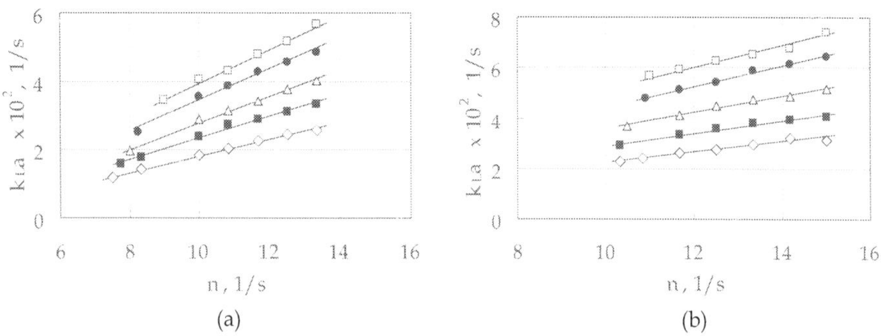

Fig. 4. The dependence of k_La = f(n) for two configurations of the agitators: a) A 315 (lower)–RT (upper), b) CD 6 (lower) –RT (upper); gas–liquid system; varied values of $w_{og} \times 10^3$ m/s: (◊) 1.71; (■) 2.56; (Δ) 3.41; (•) 5.12; (□) 6.82

According to the literature data our results confirmed that in a gas–liquid system the volumetric mass transfer coefficient is strongly affected by the type of the agitator used for mixing of the system. The comparison of the k_La coefficient values obtained for three single agitators: RT, CD 6 and A 315 are presented in Fig. 5. In the vessel with radial flow both Rushton and Smith turbines the values of k_La coefficient were significantly higher compared to the values characterized the vessel equipped with mixed flow A 315 impeller. This tendency was observed in the whole range of the measurements.

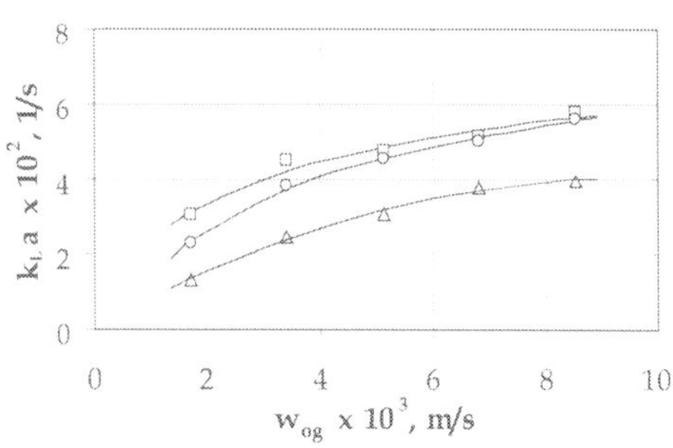

Fig. 5. The effect of a single impeller type on the k_La coefficient value in a gas-liquid system; n = const = 11.67 1/s; (Δ) A 315; (○) CD 6; (□) RT

The effect of the impeller configuration on the volumetric mass transfer coefficient k_La value is presented in Figs. 6 and 7. The results concerning mass transfer processes conducting in the vessel equipped with two agitators on a common shaft showed that the type of both lower and upper impeller strongly affects k_La coefficient values. In Fig. 6 the dependencies k_La = f(n) for different impeller configurations working in a gas-liquid system were compared. Fig. 6a presents the effect of the type of an upper agitator, whilst Fig. 6b the effect of the type of a lower agitator.

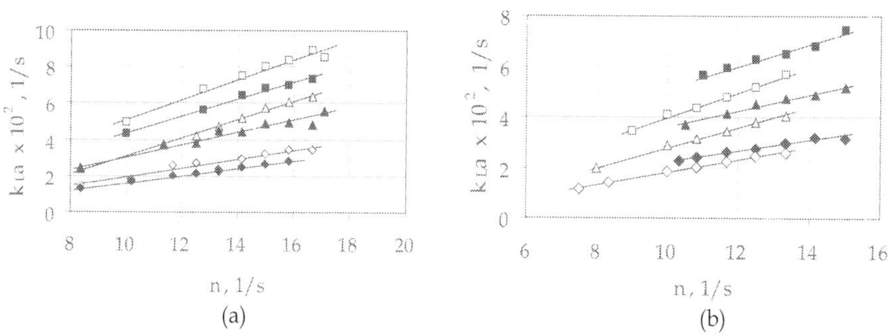

Fig. 6. The effect of the configuration of agitators on the k_La coefficient value; a) the effect of an upper agitator; empty points: RT–A 315 (upper), filled points: RT–HE 3 (upper); b) the effect of lower agitator; empty points: A 315 (lower)–RT; filled points: CD 6 (lower)–RT (upper); gas–liquid system; varied values of $w_{og} \times 10^3$ m/s: (◊♦) 1.71; (Δ▲) 3.41; (□■) 6.82

In the two-phase distilled water–air system significantly higher values of the volumetric mass transfer coefficient were obtained using the set of the impellers with A 315 as an upper one. Comparison of the k_La coefficient values for three chosen values of superficial gas velocity w_{og} and both configurations differ in an upper impeller is presented in Fig. 6a, 7a. The A 315 impeller (data represent by empty points in Fig. 6a), which generates radial-axial fluid flow in the vessel, let to ensure more advantageous conditions for conducting of mass transfer process, comparing with generating typical axial liquid circulation HE 3 impeller (data represent by filled points in Fig. 6a). At the constant values of agitator speeds, the differences in k_La coefficient values increase with the increase of quantity of gas phase introduced into the liquid volume. At the gas velocity w_{og} = 6,28×10^{-3} m/s, k_La values in the system agitated by RT–A 315 configuration were over 20 % higher, than those obtained analogously using RT–HE 3 set of agitators.

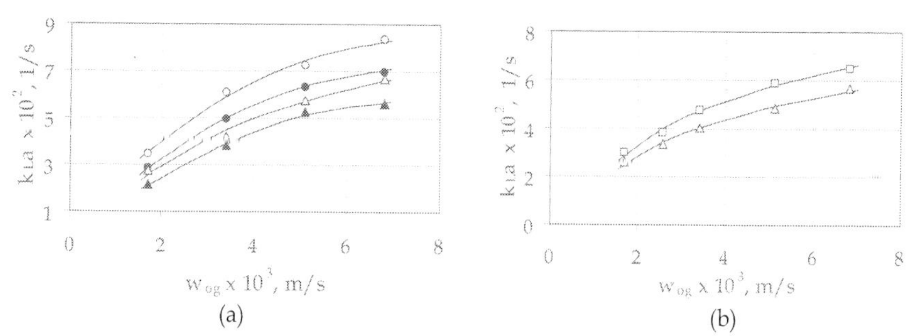

Fig. 7. Dependence $k_La = f(w_{og})$ for different impeller configuration; a) various upper impeller: filled points: RT-HE 3, empty points: RT-A 315, n = 12.5 1/s (triangles), n = 15.83 1/s (circles); b) various lower impellers: (Δ) A 315-RT, (□) CD 6-RT; n = 13.33 1/s;

Comparison of the results obtained for double-impeller system differ in a lower impeller (Fig. 6b) shows, that at constant value of the impeller speed, the highest values of the $k_L a$ coefficient were obtained using Smith turbine–Rushton turbine configuration (data represent by filled points in Fig. 6b). At higher values of superficial gas velocity the differences between volumetric mass transfer coefficient values adequate to each configuration increased. The A 315 impeller, which differs in the generated profile of the fluid circulation from the turbine, is less effective as a lower impeller for the mass transfer processes in the agitated vessel comparing with CD 6 impeller (Fig. 7b).
The results of the studies of the volumetric mass transfer coefficient in the gas-liquid system were described mathematically, by means of the dependency, which connect $k_L a$ coefficient with unit power consumption ($P_{G\text{-}L}/V_L$, W/m^3) and superficial gas velocity (w_{og}, m/s):

$$k_L a = A\left(\frac{P_{G-L}}{V_L}\right)^B w_{og}{}^C \tag{5}$$

The values of the coefficient A and exponents B, C in this Eq. are collected in Table 4 for single impeller system and in Table 5 for all tested configurations of double impeller. In these tables mean relative errors $\pm\Delta$ for each set of agitators is also presented. For the vessel equipped with single impeller Eq. (5) is applicable within following range of the measurements: $P_{G\text{-}L}/V_L \in$ <160; 5560 W/m^3>; $w_{og} \in$ <1.71×10^{-3}; 8.53×10^{-3} m/s>. The range of an application of this equation for the double impeller systems is as follows: $Re \in$ <7.9×10^4; 16.3×10^4>; $P_{G\text{-}L}/V_L \in$ <540; 4500 W/m^3>; $w_{og} \in$ <1.71×10^{-3}; 6.82×10^{-3} m/s>.

No.	Impeller	A	B	C	±Δ, %
1.	Rushton Turbine	0.029	0.44	0.534	3.7
2.	CD 6	0.028	0.374	0.622	3.3
3.	A 315	0.122	0.432	0.79	5.2

Table 4. The values of the coefficient A and exponents B, C in Eq. (5) for single impeller systems (Kiełbus-Rąpała et al., 2010)

No.	Configuration of impellers		A	B	C	±Δ, %
	lower	upper				
1.	Rushton Turbine	A 315	0.098	0.404	0.689	7
2.	Rushton Turbine	HE 3	0.081	0.374	0.636	6.5
3.	A 315	Rushton Turbine	0.027	0.466	0.575	3
4.	CD 6	Rushton Turbine	0.092	0.334	0.6	2.6

Table 5. The values of the coefficient A and exponents B, C in Eq. (5) for double impeller systems (Kiełbus-Rapała & Karcz, 2010)

3.2 The results for gas-solid-liquid system

On the basis of the experimental studies performed within turbulent regime of the liquid flow the volumetric mass transfer coefficient values for gas-solid-liquid systems agitated by single or double impellers were determined. The results obtained confirm that in the system with solid particles the $k_L a$ coefficient value is strongly affected by operating parameters, and also the presence and concentration of solid phase in the agitated systems.

The results of the measurements of volumetric mass transfer coefficients k_La in two phase gas–liquid and three phase gas–solid–liquid system agitated by single CD 6 impeller are compared in Fig. 8 (Karcz & Kiełbus–Rąpała, 2006). In this Fig. empty points are ascribed to the two-phase system whilst filled points correspond to the three-phase system with the concentration of the solid particles 0.5 (Fig. 8a) and 2.5 mass% (Fig. 8b), respectively.

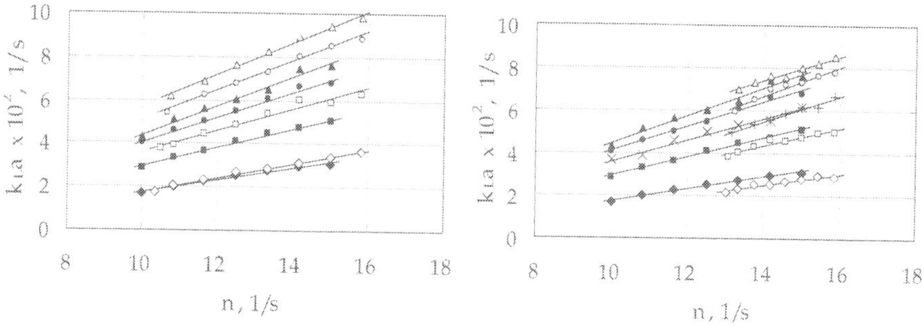

(a) (b)

Fig. 8. Dependence $k_La = f(n)$ for a single impeller CD 6; filled points: $X = 0$ mass %; empty points: a) $X = 0.5$ mass %, b) $X = 2.5$ mass %; various $w_{og} \times 10^3$ m/s: (◊♦) 1.71; (□■) 3.41; (+ ×) 5.12; (○•) 6.82; (Δ▲) 8.53 (Karcz & Kiełbus–Rąpała, 2006)

Analyzing the data in this Fig it could be concluded that in gas-solid-liquid system both superficial gas velocity and impeller speed influenced k_La coefficient value. Similarly to the two-phase system k_La coefficient value depends on the impeller speeds linearly and k_La increases with the increase of n. Also the increase in aeration rate of three-phase system (described by superficial gas velocity w_{og}) causes the increase of the volumetric mass transfer coefficient.

It follows from the results that introducing solids to the system affects volumetric mass transfer coefficient k_La. In the system with 0.5% of solid particles inside, the values of the k_La are significantly higher than in the two-phase system (Fig. 8a). In coalescing system, it can be explained that solids in a small quantity during their movement in the system hit the gas bubbles, which are breaking into the smaller ones and in consequence, the interfacial area increases. However, adding more particles - 2.5 mass %, causes the decrease of k_La values. For lower values of superficial velocity w_{og}, the volumetric mass transfer coefficient values are even lower than those obtained in the gas - liquid system (Fig. 8b).

Results of the k_La coefficient measurements for two double-impeller configurations working in gas-liquid, and gas-solid-liquid systems are compared in next Figs (Kiełbus-Rapała & Karcz, 2009). The data for A 315 (lower)-RT (upper) impeller configuration are presented in Fig. 9, whilst those agitated by CD 6 (lower)-RT (upper) can be found in Fig. 10. In these figures, similarly to the data in Fig. 8 filled points are related to the two-phase system whilst empty points correspond to the three-phase system with the concentration of solid particles 0.5 mass % (Figs 9a, 10a) and 2.5 mass % (Figs 9b, 10b), respectively.

The results for the system agitated by two impellers on a common shaft show the same dependency of the volumetric mass transfer coefficient as was observed in the vessel equipped with single impeller. In the system with 0.5 mass % of solid particles, the values of k_La are significantly higher compared to the data for two-phase system for both configurations of the impellers (Figs. 9a, 10a). It is worth to notice that the differences between the k_La values for the system with or without particles are greater for the A 315–RT configuration (Fig. 9). In the three-phase system stirred by CD 6–RT impellers, volumetric mass transfer coefficients for the lowest values of superficial gas velocity were even lower than those for gas-liquid system. These observations are also confirmed by the literature results concerning single-impeller (Alper et al., 1980; Özbek & Gayik, 2001) or double-impeller (Galaction et al., 2004) systems. For low concentration of solids, the increase of k_La values can be explained by the interaction of particles with gas bubbles. The particles present in the liquid cause an increase in the turbulence on the film surrounding the gas bubbles, promoting thus the bubbles surface renewal and breaking, or leading to a smaller effective diffusion layer (Galaction et al., 2004; Alper et al., 1980; Kluytmans et al., 2003). Solids in low concentration and of small particle size can, during their movement in the system, collide and interact with the gas-liquid interface and, in consequence, the gas bubbles break into smaller ones causing an increase in interfacial area (Kiełbus-Rapała & Karcz, 2009).

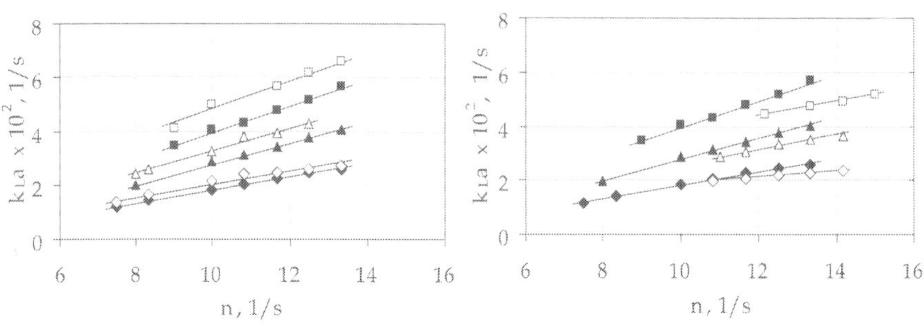

(a) (b)

Fig. 9. Dependence $k_La = f(n)$ for configuration A 315 (lower)–Rushton turbine (upper); filled points: $X = 0$ mass %; empty points: a) $X = 0.5$ mass %, b) $X = 2.5$ mass %; various $w_{og} \times 10^3$ m/s: (◊♦) 1.71; (Δ▲) 3.41; (□■) 6.82 (Kiełbus-Rąpała & Karcz, 2009)

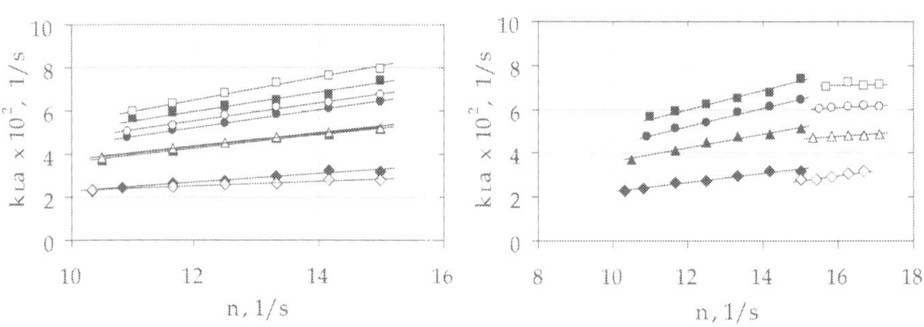

(a) (b)

Fig. 10. Dependence $k_La = f(n)$ for the stirrers design CD 6 (lower)–Rushton turbine (upper); filled points: $X = 0$ mass %; empty points: a) $X = 0.5$ mass %, b) $X = 2.5$ mass %; various $w_{og} \times 10^3$ m/s: (◊♦) 1.71; (Δ▲) 3.41; (○•) 5.12; (□■) 6.82 (Kiełbus-Rąpała & Karcz, 2009)

The investigations of mass transfer process in the system agitated by double-impellers confirm that the volumetric mass transfer coefficient increases, compared to the system without solids, only below a certain level of particles concentration.

Introducing more particles, X = 2.5 mass % into the system causes a decrease of $k_{L}a$ in A 315–RT (Fig. 9b) as well as in CD 6–RT (Fig. 10b) configuration. This tendency (reduction of $k_{L}a$ with increasing solid concentrations) was observed also in literature (Özbek & Gayik, 2001). It was proposed that increasing solid concentration causes an increase of slurry viscosity, resulting in an increase of coalescence processes and, in consequence, decrease of $k_{L}a$ coefficients.

The effect of solids on $k_{L}a$ could be probably associated with the differences in flow patterns generated in the agitated vessel as seen by the significant differences in the $k_{L}a$ values obtained using different impeller designs. Özbek and Gayik (2001) stated that the concentration at which $k_{L}a$ began to decline depended on solids type and size of the particles.

The results obtained for the solid concentration X = 2.5 mass % (Kiełbus-Rapała & Karcz, 2009) differ from those of a single-impeller system (Fig. 8) so the effect of particles with this concentration for the double - impeller system was not the same. Although in both cases, the values of $k_{L}a$ were lower compared to the system with X = 0.5 mass %, $k_{L}a$ coefficients were significantly lower compared to a gas–liquid system at all w_{og} values (Figs 9, 10), whilst in a single-impeller system (Fig. 8b), the volumetric mass transfer coefficient values were lower only for lower values of superficial gas velocity w_{og} (Karcz & Kiełbus-Rapała, 2006).

The effect of the solid phase concentration on the volumetric mass transfer coefficient value for both impeller configurations differ in lower impeller is presented in Fig. 11 (Kiełbus-Rąpała & Karcz, 2009). The values of $k_{L}a$ coefficient, obtained at a constant value of superficial gas velocity for the system with different solid concentration agitated by A 315-RT configuration (Fig. 11a) and CD 6-RT (Fig. 11b) are compared. In this Fig. triangles described data for gas-liquid system, squares - X = 0.5 mass%, circles - X = 2.5 mass%.

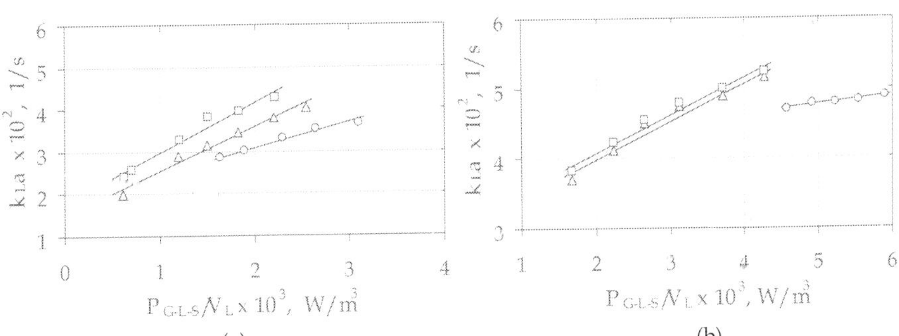

Fig. 11. Effect of the solid concentration on $k_{L}a$ value for double impeller configuration; a) A 315 (lower)–Rushton turbine (upper), b) CD 6 (lower)–Rushton turbine (upper); w_{og} = const = 3.41 × 10^{-3} m/s, X = 0 mass % (triangles); X = 0.5 mass % (squares); X = 2.5 mass % (circles) (Kiełbus–Rąpała & Karcz, 2009)

In the three-phase system agitated by both impeller configurations differ in an upper impeller type significantly lower $k_{L}a$ coefficient value were obtained in the system with solid concentration X = 2.5 mass %. The data for RT-A 315 and RT-HE 3 agitator design are compared in Fig. 12 in which squares concerned system with X = 0.5 mass% whilst triangles : X = 2.5 mass %.

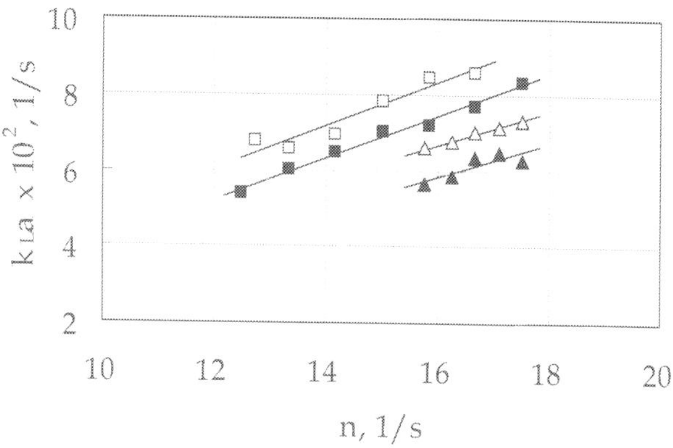

Fig. 12. The dependence of k_La = f(n) for two configurations of the agitator working in gas–solid–liquid system; empty points: RT (lower)–A 315 (upper), filled points: RT (lower)–HE 3 (upper); various values of X: (□■) 0.5 mass %; (Δ▲) 2.5 mass %

The effect of the single agitator type on the volumetric mass transfer coefficient k_La is presented in Fig. 13. Comparing the results obtained for each impeller it can be stated that, in general, the highest values of the k_La coefficient are characterized for Rushton turbine. However, at higher values of the superficial gas velocity w_{og} the better results of the volumetric mass transfer coefficient correspond to CD 6 impeller. The A 315 impeller, which differs in the generated profile of the liquid circulation from the turbines, is less effective for the mass transfer processes in the agitated vessel (Karcz & Kiełbus-Rąpała, 2006).

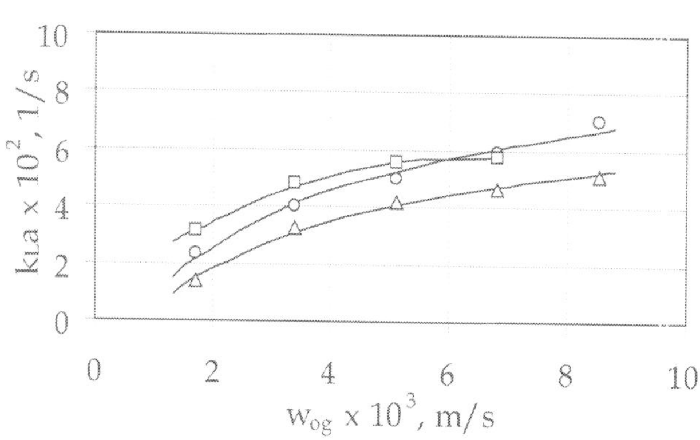

Fig. 13. Comparison of the k_La = f(w_{og}) for different single impellers working in the three-phase system with solid concentration X = 2.5 mass %; (□) RT, n = 13.33 1/s; (○) CD 6, n = 13.33 1/s; (Δ) A 315, n = 12.50 1/s

The effect of upper agitator type on the volumetric mass transfer coefficient value in the gas-solid-liquid system is presented in Fig. 14.

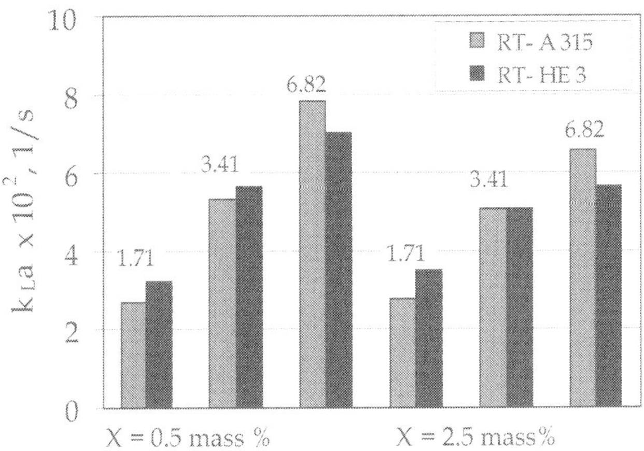

Fig. 14. Comparison of values of k_La coefficient for two impeller configurations, working in gas-solid-liquid systems; $X \neq$ const; n = 15 1/s; various values of superficial gas velocity w_{og} $\times 10^{-3}$ m/s

In the three-phase system that, which agitator ensure better conditions to conduct the process of gas ingredient transfer between gas and liquid phase, depends significantly on the quantity of gas introduced into the vessel. Comparison of values of k_La coefficient for two impeller configurations, working in gas-solid-liquid systems with two different solid concentrations is presented in Fig. 14. At lower value of superficial gas velocity (w_{og} = 1.71 $\times 10^{-3}$ m/s) both in the system with lower solid particles concentration X = 0.5 mass % and with greater one: X = 2.5 mass %, higher of about 20 % values of volumetric mass transfer coefficient were obtained in the system agitated by means of the configuration with HE 3 impeller. With the increase of gas velocity w_{og} the differences in the values of k_La coefficient achieved for two tested impellers configuration significantly decreased. Moreover, in the system with lower solid concentration at the velocity w_{og} = 3.41$\times 10^{-3}$ m/s, in the whole range of impeller speeds, k_La values for both sets of impellers were equal.
Completely different results were obtained when much higher quantity of gas phase were introduced in the vessel. For high value of w_{og} for both system including 0.5 and 2.5 mass % of solid particles, more favourable was configuration Rushton turbine - A 315. Using this set of agitators about 20 % higher values of the volumetric mass transfer coefficient were obtained, comparing with the data characterized the vessel with HE 3 impeller as an upper one (Fig. 12).
The data obtained for three-phase systems were also described mathematically. On the strength of 150 experimental points Equation (2) was formulated:

$$k_La = A\left(\frac{P_{G\text{-}L\text{-}S}}{V_L}\right)^b w_{og}{}^c\left(\frac{1}{1+m_1X^2+m_2X}\right) \tag{6}$$

The values of the coefficient A, m_1, m_2, and exponents B, C in this Eq. are collected in Table 6 for single impeller system and in Table 7 for the configurations of double impeller differ in a lower one. In these tables mean relative error $\pm\Delta$ is also presented. For the vessel equipped with single impeller Eq. (6) is applicable within following range of the measurements: $P_{G\text{-}L\text{-}S}/V_L \in$ <118; 5700 W/m^3>; $w_{og} \in$ <1.71×10^{-3}; 8.53×10^{-3} m/s>; $X \in$ <0.5; 5 mass %>. The range of an application of this equation for the double impeller systems is as follows: $P_{G\text{-}L\text{-}S}/V_L \in$ <540; 5960 W/m^3>; $w_{og} \in$ <1.71×10^{-3}; 6.82×10^{-3} m/s>; $X \in$ <0.5; 5 mass %>.

No.	Impeller	A	b	c	m_1	m_2	$\pm\Delta$, %	Exp. point
1.	Rushton turbine (RT)	0.031	0.43	0.515	-186,67	11.921	6.8	100
2.	Smith turbine (CD 6)	0.038	0.563	0.67	-388.62	23.469	6.6	98
3.	A 315	0.062	0.522	0.774	209.86	-11.038	10.3	108

Table 6. The values of the coefficient A, m_1, m_2 and exponents b, c in Eq. (6) for single impeller systems (Kiełbus-Rąpała et al., 2010)

No.	Configuration of impellers		A	b	c	m_1	m_2	$\pm\Delta$, %
	lower	upper						
1.	CD 6	RT	0.164	0.318	0.665	0.361	8.81	3
2.	A 315	RT	0.031	0.423	0.510	-0.526	-8.31	4

Table 7. The values of the coefficient A, m_1, m_2 and exponents b, c in Eq. (6) for double impeller systems (Kiełbus-Rąpała & Karcz, 2009)

The results of the k_La coefficient measurements for the vessel equipped with double impeller configurations differ in an upper impeller were described by Eq. 7.

$$k_L a = A\left(\frac{P_{G-L-S}}{V_L}\right)^B w_{og}{}^C \frac{1}{1+mX} \tag{7}$$

The values of the coefficient A, m, and exponents B, C in this Eq. for both impeller designs are collected in Table 8. The range of application of Eq. 2 is as follows: $Re \in$ <9.7; 16.8×10^4>; $P_{G\text{-}L\text{-}S}/V_L \in$ <1100; 4950 W/m^3>; $w_{og} \in$ <1.71×10^{-3}; 6.82×10^{-3} m/s>; X ∈ <0; 0.025>.

No.	Configuration of impellers		A	B	C	m	$\pm\Delta$, %
	lower	upper					
1.	Rushton Turbine	A 315	0.103	0.409	0.695	4.10	4.8
2.	Rushton Turbine	HE 3	0.031	0.423	0.510	6.17	6.4

Table 8. The values of the coefficient A and exponents B, C in Eq. (7) (Kiełbus-Rąpała & Karcz 2010)

4. Conclusions

In the multi - phase systems the k_La coefficient value is affected by many factors, such as geometrical parameters of the vessel, type of the impeller, operating parameters in which process is conducted (impeller speed, aeration rate), properties of liquid phase (density, viscosity, surface tension etc.) and additionally by the type, size and loading (%) of the solid particles.
The results of the experimental analyze of the multiphase systems agitated by single impeller and different configuration of two impellers on the common shaft show that within the range of the performed measurements:

1. Single radial flow turbines enable to obtain better results compared to mixed flow A 315 impeller.
2. Geometry of lower as well as upper impeller has strong influence on the volumetric mass transfer coefficient values. From the configurations used in the study for gas-liquid system higher values of k_La characterized Smith turbine (lower)-Rushton turbine and Rushton turbine-A 315 (upper) configurations.
3. In the vessel equipped with both single and double impellers the presence of the solids in the gas-liquid system significantly affects the volumetric mass transfer coefficient k_La. Within the range of the low values of the superficial gas velocity w_{og}, high agitator speeds n and low mean concentration X of the solids in the liquid, the value of the coefficient k_La increases even about 20 % (for single impeller) comparing to the data obtained for gas-liquid system. However, this trend decreases with the increase of both w_{og} and X values. For example, the increase of the k_La coefficient is equal to only 10 % for the superficial gas velocity w_{og} = 5.12 x10^{-3} m/s. Moreover, within the highest range of the agitator speeds n value of the k_La is even lower than that obtained for gas-liquid system agitated by means of a single impeller.
   In the case of using to agitation two impellers on the common shaft k_La coefficient values were lower compared to a gas-liquid system at all superficial gas velocity values.
4. The volumetric mass transfer coefficient increases, compared to the system without solids, only below a certain level of particles concentration. Introducing more particles, X = 2.5 mass % into the system causes a decrease of k_La in the system agitated by both single and double impeller systems.
5. In the gas-solid-liquid system the choice of the configuration (upper impeller) strongly depends on the gas phase participation in the liquid volume:
   -The highest values of volumetric mass transfer coefficient in the system with small value of gas phase init were obtained in the vessel with HE 3 upper stirrer;
   -In the three-phase system, at large values of superficial gas velocity better conditions to mass transfer process performance enable RT-A 315 configuration.

Symbols

a	length of the blade	m
B	width of the baffle	m
b	width of the blade	m
C	concentration	g/dm^3
D	inner diameter of the agitated vessel	m
d	impeller diameter	m

d_d	diameter of the gas sparger	m
d_p	particles diameter	m
e	distance between gas sparger and bottom of the vessel	m
H	liquid height in the agitated vessel	m
h_1	off - bottom clearance of lower agitator	m
h_2	off - bottom clearance of upper agitator	m
i	number of the agitators	
J	number of baffles	
$k_L a$	volumetric mass transfer coefficient	s^{-1}
n	agitator speed	s^{-1}
n_{JSG}	critical impeller speed for gas - solid - liquid system	s^{-1}
P	power consumption	W
R	curvature radius of the blade	m
t	time	s
V_L	liquid volume	m^3
$\dot{V}_g$	gas flow rate	m^3s^{-1}
w_{og}	superficial gas velocity ($= 4\ V_g / \pi D^2)^{\cdot}$	$m\ s^{-1}$
X	mass fraction of the particles	% w/w
Z	number of blades	

Greek Letters

β	pitch of the impeller blade	deg
ρ_p	density of solid particles	kgm^{-3}

Subscripts

G-L	refers to gas - liquid system
G-S-L	refers to gas - solid - liquid system

5. References

Alper, E.; Wichtendahlet, B. & Deckwer, W.D. (1980). Gas absorption mechanism in catalytic slurry reactions. *Chemical Engineering Science,* 35, 217-222

Arjunwadkar, S.J.; Sarvanan, K.; Pandit, A.B. & Kulkarni, P.R. (1998). Gas - liquid mass transfer in dual impeller bio-reactor. *Biochemical Engineering Journal,* 1, 93-96

Bartos, T.M. & Satterfield, C.N. (1986). Effect of finely divided solids on mass transfer between a gas and organic liquid. *AICHE Journal,* 32, 773-780

Brehm, A.; Oguz, H. & Kisakurek, B. (1985). Gas - liquid mass transfer data in an three phase stirred vessel. *Proceedings of the 5th European Conference on Mixing,* pp. 419-425, Wurzburg, Germany, June 10-12, 1985

Brehm, A. & Oguz, H. (1988). Measurement of gas/liquid mass transfer: peculiar effects In aqueous and organic slurries. *Proceedings of the 6th European Conference on Mixing,* pp. 413-420, Pavia, Italy, 24-26.05.1988

Chandrasekaran, K. & Sharma, M.M. (1977). Absorption of oxygen in aqueous solutions of sodium sulfide in the presence of activated carbon as catalyst. Chemical Engineering Science, 32, 669-675

Chapman, C.M.; Nienow, A.W.; Cooke, M. & Middleton, J.C. (1983). Particle - gas - liquid mixing in stirred vessels. Part IV: Mass transfer and final conclusions. *Chemical Engineering Research & Design,* 61, 3, 182-185

Dohi, N.; Matsuda, Y.; Itano, N.; Shimizu, K.; Minekawa, K. & Kawase, Y. (1999). Mixing characteristics in slurry stirred tank reactors with multiple impellers. *Chemical Engineering Communications*, 171, 211-229

Dohi, N.; Takahashi, T.; Minekawa, K. & Kawase, Y. (2004). Power consumption and solid suspension performance of large - scale impellers in gas - liquid - solid three - phase stirred tank reactors. *Chemical Engineering Journal*, 97, 103-114, doi:10.1016/S1385-8947(03)00148-7

Dutta, N.N. & Pangarkar, V.G. (1995). Critical impeller speed for solid suspension in multi - impeller three phase agitated contactors. *Canadian Journal of Chemical Engineering*, 73, 273-283

Fujasova, M.; Linek, V. & Moucha, T. (2007). Mass transfer correlations for multiple - impeller gas - liquid contactors. Analysis of the effect of axial dispersion in gas liquid phases on "local" k_La values measured by the dynamic pressure method in individual stages of the vessel. *Chemical Engineering Science*, 62, 1650-1669, doi:10.1016/j.ces.2006.12.03

Galaction, A.I.; Cascaval, D.; Oniscu, C. & Turnea, M. (2004). Modeling of oxygen transfer in stirred bioreactors for bacteria, yeasts and fungus broths. *Proceedings of the 16th International Congress of Chemical & Processing Engineering, CHISA*, Paper F1.5 Praha, Czech Republic, August 22-26, 2004

Garcia-Ochoa, F. & Gomez, E. (2009). Bioreaktor scale-up oxygen transfer rate in microbial processes. An overview. *Biotechnology Advances*, 27, 153-176.

Gogate, P.R.; Beenackers, A.A.C.M. & Pandit A.B. (2000). Multiple - impeller systems with special emphasis on bioreactors: a critical review. *Biochemical Engineering Journal*, 6, 109-144

Gogate, P.R. & Pandit, A.B. (1999). Survey of measurement for gas-liquid mass transfer coefficient in bioreactors. *Biochemical Engineering Journal*, 4, 7-15

Greaves, M. & Loh, V.Y. (1985). Effect of hight solids concentrations on mass transfer and gas hold-up in three – phase mixing. *Proceedings of 5th European Conference on Mixing*, pp. 451-467, Germany, Wurzburg, 10-12.06.1985

Harnby, N.; Edwards, M.F. & Nienow A.W. (1997). Mixing in the process industries, second edition, Butterworth, Heinemann, ISBN 0 7506 37609, Oxford

Jahoda, M.; Machoň, V.; Veverka, P. & Majiřová, H. (2000). Homogenisation of the liquid in three - phase multiimpeller stirred reactor. *Proceedings of the 14th International Congress of Chemical & Processing Engineering, CHISA*, Paper E4.5, Praha, Czech Republic August 27-31, 2000

Jin, B. & Lant, P. (2004). Flow regime, hydrodynamics, floc size distribution and sludge properties in activated sludge bubble column, air - lift and aerated stirred reactors. *Chemical Engineering Science*, 59, 2379-2388, doi:10.1016/j.ces.2004.01.061

Joosten, G.E.H; Schilderet, J.G.M. & Jansen, J.J. (1977). The influence of suspended solid material on the gas-liquid mass transfer in stirred gas-liquid contactors. Chemical Engineering Science, 32, 563-570

Karcz, J. & Kiełbus—Rąpała, A. (2006). An effect of solids on the volumetric gas - liquid mass - transfer coefficient in mechanically agitated gas - solid - liquid system. *Proceedings of the 17th International Congress of Chemical & Processing Engineering, CHISA*, Praha, Czech Republic, August 27-31, 2006

Kiełbus–Rąpała, A. & Karcz, J. (2009). Influence of suspended solid particles on gas-liquid mass transfer coefficient in a system stirred by double impellers. *Chemical Papers*, 63, 2, 188-196, doi:10.2478/s11696-009-013-y

Kiełbus–Rąpała, A. & Karcz, J. (2010). The influence of stirrers' configuration on the volumetric gas-liquid mass transfer coefficient in multiphase systems. Paper P5.80. *Proceedings of the 19th International Congress of Chemical & Processing Engineering, CHISA*, Praha, Czech Republic, August 28- September 1, 2010

Kiełbus–Rąpała, A., Karcz, J. & Cudak, M. (2010). The effect of physical properties of the liquid phase on the gas-liquid mass transfer coefficient in the two- and three-phase agitated systems. *Proceedings of the 37th International Conference of Slovak Society of Chemical Engineering*, pp. 1310-1318, ISBN 978-80-227-3290-1, Tatranske Matliare, Slovakia, May 24-28, 2010

Kordac, M. & Linek, V. (2010). Efect of solid particle size on mass transfer between gas and liquid. Paper I3.1. *Proceedings of the 19th International Congress of Chemical & Processing Engineering, CHISA*, Praha, Czech Republic, August 28- September1, 2010

Kluytmans, J.H.J.; Wachem, B.G.M;, Kuster, B.F.M. & Schouten, J.C. (2003). Mass transfer in sparged and stirred reactors: influence of carbon particles and electrolyte. *Chemical Engineering Science*, 58, 4719-4728, doi:10.1016/j.ces.2003.05.004

Kralj, F. & Sincic, D. (1984). Hold - up and mass transfer in two- and three - phase stirred tank reactor, *Chemical Engineering Science*, 39, 604-607

Lee, J.C.; Ali, S.S. & Tasakorn, P. (1982). Influence of suspended solids on gas-liquid mass transfer in an agitated tank. *Proceedings of 4th European Conference on Mixing*, pp. 399-408, Noordwijkerhout, The Netherlands, April, 1982

Lemoine, R. & Morsi, B.I. (2005). An algorithm for predicting the hydrodynamic and mass transfer parameters in agitated reactors. *Chemical Engineering Journal*, 114, 9-31

Linek, V.; Moucha, T. & Sinkule, J. (1996). Gas - liquid mass transfer in vessel stirred with multiple impellers - I. Gas - liquid mass transfer characteristics in individual stages. *Chemical Engineering Science*, 51, 3203-3212

Linek, V.; Vacek, V. & Benes, P. (1987). A critical review and experimental verification of correct use of dynamic method for the determination of oxygen transfer in aerated agitated vessels to water, electrolytic solutions and viscous liquids. *Chemical Engineering Journal*, 34, 11-34

Linek, V.; Benes, P. & Vacek, V. (1982). A critical review and experimental verification of correct use of dynamic method for the determination of oxygen transfer in aerated agitated vessels to water, electrolytic solutions and viscous liquids. *Chemical Engineering Journal*, 25, 77-88

Littlejohns, J.V. & Daugulis, A.J. (2007). Oxygen transfer in a gas - liquid system containing solids of varying oxygen affinity. *Chemical Engineering Journal*, 129, 67-74, doi:10.1016/j.cej.2006.11.002

Lu, W.M.; Hsu, R.C. & Chou, H.S. (1993). Effect of solid concentration on gas liquid mass transfer in a machanically agitated three phase. *Journal Chin. I. Chemical Engineering*, 24, 31-39

Lu, W.M.; Wu, H.Z. & Chou, C.Y. (2002). Gas recirculation rate and its influences on mass transfer in multiple impeller systems with various impeller combinations. *The Canadian Journal of Chemical Engineering*, 80, 51-62

Machon, V.; Vlcek, J. & Hudcova, V. (1988). Multi-impeller gas-liquid contactors. *Proceedings of the 6th European Conference on Mixing*, pp. 351-360, Pavia, Italy, 1988, May 24-26

Majiřova, H.; Prokopa, T.; Jagoda, M. & Machoň, V. (2002). Gas hold – up and power input in two - and three - phase dual - impeller stirred reactor. *Proceedings of the 15th International Congress of Chemical & Processing Engineering, CHISA*, Praha, Czech Republic, 2002, August 25-29

Markopoulos, J.; Christofi,C. & Katsinaris, J. (2007). Mass transfer coefficients in mechanically agitated gas-liquid contactors. Chemical Engineering & Technology 30, 7, 829-834, doi: 10.1002/ceat.200600394

Martin, M.; Montes, F.J. & Galan, M.E. (2008). On the contribution of the scales of mixing to the oxygen transfer in stirred tanks. *Chemical Engineering Journal,* 145, 232-241

Mills, D.B.; Bar, R. & Kirwan, D.J. (1987). Effect of solid on oxygen transfer in agitated three phase systems. AICHE Journal, 33, 1542

Moilanen, P.; Laakkonen, M.; Visuri, O.; Alopaeus, V. & Aittamaa, J. (2008). Modelling mass transfer in an aerated 0.2 m^3 vessel agitated by Rushton, Phasejet and Combijet impellers. *Chemical Engineering Journal,* 142, 92-108

Moucha, T.; Linek, V. & Prokopová, E. (2003). Gas hold - up, mixing time and gas - liquid volumetric mass transfer coefficient of various multiple - impeller configurations: Rushton turbine, pitched blade and Techmix impeller and their combinations. *Chemical Engineering Science*, 58, 1839-184, doi:10.1016/S0009-2509(02)00682-6

Nocentini, M.,Fajner, D., Pasquali, G.& Majeli, F. (1993). Gas - liquid mass transfer and hold-up in vessels stirred with multiple Rushton turbines: water and glicerol solution. *Industrial Engineering & Chemical Research*, 32, 19-24

Oguz, H.; Brehm, A. & Deckwer, W.D. (1987). Gas-liquid mass transfer in sparged agitated slurries. *Chemical Engineering Science*, 42, 1815-1822

Özbek, B. & Gayik, S. (2001). The studies on the oxygen mass transfer coefficient in bioreactor. *Proc. Biochem.*, 36, 729-741

Ozkan, O.; Calimli, A.; Berber, R. & Oguz, H. (2000). Effect on inert solid particles at low concentrations on gas - liquid mass transfer in mechanically agitated reactors. *Chemical Engineering Science*, 55, 2737-2740

Paul, E.L; Atiemo-Obeng, V.A & Kresta, S.M. (2004). Handbook of Industrial Mixing, Wiley-Interscience, ISBN 0-471-26919-0, Hoboken, New Yersey

Pinelli, D. (2007). The role of small bubbles in gas-liquid mass transfer in stirred vessels and assessment of a two-fraction model for noncoalescent or moderately viscous liquids. *Chemical Engineering Science*, 62, 3767-3776

Puthli, M.S.; Rathod, V.K. & Pandit, A.B., (2005). Gas - liquid mass transfer studies with triple impeller system on a laboratory scale bioreactor. *Biochemical Engineering Journal*, 23, 25-30, doi:10.1016/j.bej.2004.10.006

Roman, R.V. & Tudose, R.Z., (1997). Studies on transfer processes in mixing vessels: power consumption of the modified Rushton turbine agitators in three - phase systems. *Bioprocess Engineering*, 17, 307-316

Ruthiya, K.C.; Schaaf, J. & Kuster, B.F.M, (2003). Mechanism of physical and reaction enhancement of mass transfer in a gas inducing stirred slurry reactor. *Chemical Engineering Journal,* 96, 55-69, doi:10.1016/j.cej.2003.08.05

Van't Riet, K., (1979). Review of measuring methods and results in nonviscous gas-liquid mass transfer in stirred vessels. *Industrial Engineering Chemical Process Design and Development,* 18, 357-364

Vasconcelos, J.M.T.; Orvalho, S.C.P.; Rodrigues, A.M.A.F. & Alves, S.S. (2000). Effect of Blade shape on the performance of six bladed disc turbine impellers. *Industrial Engineering & Chemical Research*, 39, 203-208

Wu, H. (1995). An issue on applications of a disk turbine for gas-liquid mass transfer. *Chemical Engineering Science*, 50, 17, 2801-2811

Yawalkar, A.A.; Heesink, A.B.M.; Versyeeg, G.F. & Pangarkar, V.G. (2002). Gas-liquid mass transfer coefficient in stirred tank reactors. *Canadian Journal of Chemical Engineering*, 80, 5, 840-848

Yoshida, M.; Kitamura, A.; Yamagiwa, K. & Ohkawa, A. (1996). Gas hold-up and volumetric oxygen transfer coefficient in an aerated agitated vessel without baffles having forward-reverse rotating impellers. *Canadian Journal of Chemical Engineering*, 74, 840-848

Zhang, G.D.; Cai, W.F., Xu, C.J., Zhou & M. (2006). A general enhancement factor model of the physical absorption of gases in multiphase system. *Chemical Engineering Science*, 61, 558-568, doi:10.1016/j.ces.2005.07.035

Zwietering, T.N. (1958). Suspending of solids particles in liquid by agitation. *Chemical Engineering Science*, 8, 244-253.

6

Gas-Liquid Mass Transfer in an Unbaffled Vessel Agitated by Unsteadily Forward-Reverse Rotating Multiple Impellers

Masanori Yoshida[1], Kazuaki Yamagiwa[1],
Akira Ohkawa[1] and Shuichi Tezura[2]
[1]*Department of Chemistry and Chemical Engineering, Niigata University*
[2]*Shimazaki Mixing Equipment Co., Ltd.*
Japan

1. Introduction

Gas-sparged vessels agitated by mechanically rotating impellers are apparatuses widely used mainly to enhance the gas-liquid mass transfer in industrial chemical process productions. For gas-liquid contacting operations handling liquids of low viscosity, baffled vessels with unidirectionally rotating, relatively small sized turbine type impellers are generally adopted and the impeller is rotated at higher rates. In such a conventional agitation vessel, there are problems which must be considered (Bruijn *et al.*, 1974; Tanaka and Ueda, 1975; Warmoeskerken and Smith, 1985; Nienow, 1990; Takahashi, 1994): 1) occurrence of a zone of insufficient mixing behind the baffles and possible adhesion of a scale to the baffles and the need to clean them periodically; 2) formation of large gas-filled cavities behind the impeller blades, producing a considerable decrease of the impeller power consumption closely related to characteristics on gas-liquid contact, i.e., mass transfer; 3) restriction in the range of gassing rate in order to avoid phenomena such as flooding of the impeller by gas bubbles, etc. Neglecting these problems may result in a reduced performance of conventional agitation vessels. Review of the literatures for the conventional agitation vessel reveals that a considerable amount of work was carried out to improve existing type apparatuses. However, a gas-liquid agitation vessel which is almost free of the above-mentioned problems seems not to have hitherto been available. Therefore, there is a need to develop a new type apparatus, namely, an unbaffled vessel which provides better gas-liquid contact and which may be used over a wide range of gassing rates.

As mentioned above, in conventional agitation vessels, baffles are generally attached to the vessel wall to avoid the formation of a purely rotational liquid flow, resulting in an undeveloped vertical liquid flow. In contrast, if a rotation of an impeller and a flow produced by the impeller are allowed to alternate periodically its direction, a sufficient mixing of liquid phase would be expected in an unbaffled vessel without having anxiety about the problems encountered with conventional agitation vessels. We developed an agitator of a forward-reverse rotating shaft whose unsteady rotation proceeds while alternating periodically its direction at a constant angle (Yoshida *et al.*, 1996). Additionally, we designed an impeller with four blades as are longer and narrower and are of triangular sections. The impellers were attached on the agitator shaft to be multiply arranged in an unbaffled vessel with a liquid height-to-diameter ratio of 2:1. This unbaffled vessel agitated by the forward-reverse rotating impellers was applied to an air-water system and then its performance as a gas-liquid contactor was experimentally assessed, with resolutions for the above-mentioned problems being provided (Yoshida *et al.*, 1996; Yoshida *et al.*, 2002; Yoshida *et al.*, 2005).

Liquid phases treated in most chemical processes are mixtures of various substances. Presence of inorganic electrolytes is known to decrease the rate for gas bubbles to coalesce because of the electrical effect at the gas-liquid interface (Marrucci and Nicodemo, 1967; Zieminski and Whittemore, 1971). In many cases, the electrical effect creates different gas-liquid dispersion characteristics, such as decreased size of gas bubbles dispersed in liquid phase without practical changes in their density, viscosity and surface tension (Linek *et al.*, 1970; Robinson and Wilke, 1973; Robinson and Wilke, 1974; Van't Riet, 1979; Hassan and Robinson, 1980; Linek *et al.*, 1987). The present work assesses the mass transfer characteristics in aerated electrolyte solutions, following assessment of those in the air-water system, for the forward-reverse agitation vessel. In conjunction with the volumetric coefficient of mass transfer as viewed from change in power input, which is a typical performance characteristic of gas-liquid contactors, the dependences of mass transfer parameters such as the mean bubble diameter, gas hold-up and liquid-phase mass transfer coefficient were examined. Such investigations including correlation of the mass transfer parameter could quantify enhancement of the gas-liquid mass transfer and predict reasonably the values of volumetric coefficient.

2. Experimental

2.1 Experimental apparatus

A schematic diagram of the experimental set-up is shown in Fig. 1. The vessel was a combination of a cylindrical column (0.25 m inner diameter, D_t, 0.60 m height) made of transparent acrylic resin and a dish-shaped stainless-steel bottom (0.25 m inner diameter and 0.075 m height). The liquid depth, H, was maintained at 0.50 m, which was twice D_t. An impeller is one with four blades whose section is triangular (0.20 m diameter, D_i, Fig. 2), and was used in a multiple manner where the triangle apex of the blade faces downward. Different experiments employed 2-8 impellers; the number is represented as n_i. The impellers were set equidistantly on the shaft in its section between the lower end of the column and the liquid surface. Additionally, the angular difference of position between the blades of one impeller and those of upper and lower adjacent impellers was 45 degree. In the mechanism for transmitting motion used here (Yoshida *et al.*, 2001), when the crank is rotated one revolution, the shaft on which the impellers were attached first rotates up to one-quarter of a revolution in one direction, stops rotating at that position and rotates one-quarter of a revolution in the reverse direction. That is, the angular amplitude of forward-reverse rotation, θ_o, is $\pi/4$. When such a rotation with sinusoidal angular displacement is expressed in the form of a cosine function, the angular velocity of impeller, ω_i, is given by the sine function as

$$\omega_i=2\pi\theta_o N_{fr}\sin(2\pi N_{fr}t) \tag{1}$$

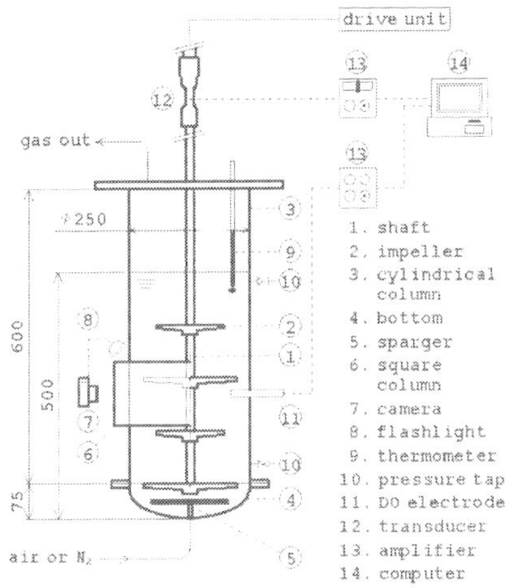

Fig. 1. Schematic flow diagram of experimental apparatus. Dimensions in mm.

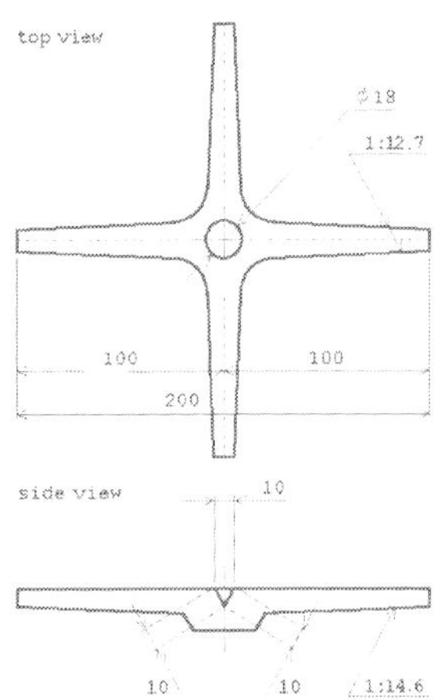

Fig. 2. Structure and dimensions (in mm) of the impeller used.

where N_{fr} is the frequency of forward-reverse rotation and was varied from 1.67 to 6.67 Hz as an agitation rate. A ring sparger with 24 holes of 1.2 mm diameter (the circle passing through the holes' centers is 0.16 m diameter) was used for aeration. The gassing rate ranged from 0.4×10^{-2} to 1.7×10^{-2} m/s in the superficial gas velocity, V_s. Comparative experiments in the unidirectional rotation mode of impeller were undertaken using a conventional impeller, a disk turbine impeller with six flat blades (DT, 0.12 m D_i). DTs were set in a dual configuration on the shaft and a nozzle sparger with a single hole of 7 mm diameter was equipped for the fully baffled vessel. Geometrical conditions such as D_t and H were common to the forward-reverse and unidirectional agitation modes. Sodium chloride solutions of different concentrations (up to 2.0 wt%) were used at 298 K as the liquid phase containing electrolyte. Physical properties of these liquids such as density, ρ, viscosity, μ, and diffusivity, D_L, were approximated by those for water.

2.2 Measuring system for power consumption of impeller

A system measuring unsteady torque of the shaft due to unsteady rotation of the impeller consisted of the fluid force transducing part, impeller displacement transducing part and signal processing part. In the fluid force transducing part, the strain generated during operation in a copper alloy coupling having four strain gauges is recorded continuously. In the impeller displacement transducing part, a switching circuit composed of a light emitting diode and a phototransistor, etc. pulses the rest point in cycles of forward-reverse rotation of impeller, thereby adjusting the frequency of forward-reverse rotation and defining the trigger point of measurements as the rest point. In the signal processing part, the analog signals of voltage from the fluid force and impeller displacement transducers are input into a computer after being digitized to permit calculations of the torque of the forward-reverse rotating shaft. The fluid force transducer detects the strains caused by different forces such as fluid forces acting on the impeller and shaft and inertia forces due to the acceleration of the motions of the impeller and shaft. The fluid force acting on the shaft was found to be negligibly small in analysis, compared with that acting on the impeller. Hence, the moment of the fluid force acting on the impeller, i.e., the agitation torque, can be obtained by subtracting the value measured in air from that in liquid, with the impellers attached.
The time-course curve of instantaneous power consumption, P_m, was obtained by multiplying the instantaneous torque, T_m, measured over one cyclic time of forward-reverse rotation of impeller by the angular velocity of impeller at the corresponding time [Eq. (1)]. The time-averaged power consumption, P_{mav}, that is based on the total energy transmitted in one cycle was graphically determined from the time-course curve of P_m.

$$P_{mav} = \oint P_m \mathrm{d}t/(1/N_{fr}) = \Sigma P_m \Delta t/(1/N_{fr}) \tag{2}$$

The following equation was used to calculate the power consumption for aeration, P_a:

$$P_a = \rho g V_s V_o \tag{3}$$

where g is the acceleration due to gravity and V_o is the liquid volume above the sparger.

2.3 Measuring system for mass transfer parameters

For mass transfer experiments, the physical absorption of oxygen in air by water was used. The volumetric coefficient of oxygen transfer was determined by the gassing-out method with purging nitrogen. The time-dependent dissolved oxygen concentration (DO), C_L, after starting aeration under a given agitation was measured at the midway point of the liquid depth, i.e., the distance 0.25 m above the vessel bottom, using a DO electrode. When there was assumed to be little difference of oxygen concentration between the inlet air and outlet gas, the overall volumetric coefficient based on the liquid volume, K_La_L, was obtained from the following relation:

$$\ln[(C_L^*-C_L)/(C_L^*-C_{Lo})] = -K_La_Lt \tag{4}$$

where C_{Lo} is the initial concentration, C_L^* is the saturated concentration and t is the time. The error in the value of volumetric coefficient due to the response lag of the DO electrode was corrected based on the first-order model using the time constant obtained in response experiments. $K_L a_L$ determined in this way was regarded as the liquid-side volumetric coefficient, $k_L a_L$, because in this system, the resistance to mass transfer on the gas side was negligible compared with that on the liquid side.

In analyzing the time-course of oxygen concentration, a model was used assuming well-mixed liquid phase and gas phase without depletion. Previous researchers including one (Calderbank, 1959) referred for comparison have resolved the difficulty to analyze changes in gas phase by ignoring the depletion of solute, so that gas bubbles are assumed to have the same composition between the inlet and outlet gas streams at all time. It has been demonstrated that the errors inherent in such assumptions are significant and that their effect on evaluation of the volumetric coefficient is considerable (Chapman *et al.*, 1982; Linek *et al.*, 1987), which may underestimate the values of volumetric coefficient. Justification for the assumptions lies in the fact that agreement between the observed values and calculated ones from earlier empirical equations (Van't Riet, 1979; Nocentini *et al.*, 1993) was satisfactory and that the analytical result still preserves the relative order of difference between the agitation modes, making them practical comparison. Therefore, it is to be noted that the values of volumetric coefficient evaluated in this work are confined to the control for comparison and would be required for the reliability to be improved.

Photographs of gas bubbles were taken at the midway point of the liquid depth, i.e., the distance 0.25 m above the vessel bottom. A square column was set around the vessel section where the photographs were taken. The space between the square column and vessel was also filled with water to reduce optical distortion. A point immediately inside the vessel wall was focused on. When a lamp light was collimated through slits to illuminate the vertical plane including that point, bodies within 25 mm inside the vessel wall could be almost in focus. The average value of readings of a scale placed in that space was employed as a measure for comparison. A spheroid could approximate the bubble shape observed on the photographs. By measuring the major and minor axes for at least 100 bubbles photographed, the volume-surface mean diameter, d_{vs}, was calculated. The overall gas hold-up, ϕ_{gD}, based on the gassed liquid volume was determined using the manometric technique (Robinson and Wilke, 1974). The manometer reading was corrected for the difference of dynamic pressure, namely, that of the reading measured in ungassed liquid. When the dispersion is assumed to comprise spherical gas bubbles of size d_{vs}, the gas-liquid interfacial area per unit volume of gassed liquid, a_D, is calculated from the following equation:

$$a_D = 6\phi_{gD}/d_{vs} \tag{5}$$

The liquid-phase mass (oxygen) transfer coefficient, k_L, was separated from the volumetric coefficient based on the liquid volume, $k_L a_L$, using a_D and ϕ_{gD}.

$$k_L = (k_L a_D)/a_D = (k_L a_L)(1-\phi_{gD})/a_D \tag{6}$$

3. Power characteristics of forward-reverse agitation vessel

3.1 Viscous and inertial drag coefficients

The following expression is assumed for the torque of the forward-reverse rotating shaft on which the impellers were attached, i.e., the agitation torque, T_m:

$$T_m = C_d \rho D_i^5 \omega_i |\omega_i| + C_m \rho D_i^5 (\mathrm{d}\omega_i/\mathrm{d}t) \tag{7}$$

where D_i is the diameter of impeller, ω_i is the angular velocity of impeller, ρ is the density of fluid around impeller and t is the time. C_d is the viscous drag coefficient relating to the moment of viscous drag on impeller and C_m is the inertial drag coefficient relating to the moment of inertia force due to the acceleration of fluid motion caused by impeller forward-reverse rotation. These coefficients are expressed in a form of average over one cyclic time of forward-reverse rotation of impeller as follows, respectively, using the coefficients of the fundamental frequency components of sine and cosine obtained by expanding Eq. (7), into which the time-dependence of ω_i [Eq. (1)] was substituted, in Fourier series:

$$C_d=(3\pi/8)[(1/\pi)\oint(T_m/\rho D_i^5\theta_o^2\omega_{fr}^2)\sin(\omega_{fr}t)d(\omega_{fr}t)] \tag{8}$$

$$C_m=\theta_o[(1/\pi)\oint(T_m/\rho D_i^5\theta_o^2\omega_{fr}^2)\cos(\omega_{fr}t)d(\omega_{fr}t)] \tag{9}$$

where ω_{fr} is the angular frequency of the sinusoidal time-course of ω_i and is equal to $2\pi N_{fr}$. Moreover, Eq. (7) for the time-course of T_m is rewritten as follows:

$$T_m=(\rho D_i^5\theta_o^2\omega_{fr}^2)[(8C_d/3\pi)\sin(\omega_{fr}t)+(C_m/\theta_o)\cos(\omega_{fr}t)] \tag{10}$$

The data of agitation torque, T_m, measured in electrolyte solutions of different concentrations when the gassing rate, the agitation rate and the number of impellers were varied were analyzed based on Eq. (10). An example of ungassed and gassed analytical results is shown in Fig. 3. The thin solid line in the figure is for the values calculated from Eq. (10) with the viscous and inertial drag coefficients, C_d and C_m, determined experimentally using Eqs (8) and (9). Good agreements were found between the observed and calculated values, regardless of the conditions with and without aeration. For the difference due to aeration, it was found that the values of gassed T_m were on the whole small compared those of ungassed T_m. Both the resultant C_d and C_m exhibited the low values under the gassed condition in comparison with the ungassed one.

For all systems when the agitation conditions such as the agitation rate, N_{fr}, and the number of impellers, n_i, were varied in electrolyte solutions of different concentrations, C_e, the drag coefficients decreased with increase of the superficial gas velocity, V_s. The dependences of the ratios of gassed coefficients to ungassed ones, C_{dg}/C_{do} and C_{mg}/C_{mo}, characterizing the decrease of the resistance of fluid for the impeller rotation due to aeration, on the agitation conditions were examined. C_{dg}/C_{do} and C_{mg}/C_{mo} decreased with increase of N_{fr}, whereas the coefficient ratios were almost independent of n_i and C_e. The drag coefficients with variation of the aeration and agitation condition in the electrolyte solutions were correlated in the following form:

$$C_d=(0.0024n_i^{0.89})\exp[-(2.3\times0.52)V_s^{0.69}N_{fr}^{0.69}] \tag{11}$$

$$C_m=(0.00032N_{fr}^{-0.06}n_i^{1.00})\exp[-(2.3\times0.31)V_s^{1.07}N_{fr}^{1.07}] \tag{12}$$

The correlation results are shown in Figs 4 and 5 as the relation between C_{dg}/C_{do} and $0.52V_s^{0.69}N_{fr}^{0.69}$ and that between C_{mg}/C_{mo} and $0.31V_s^{1.07}N_{fr}^{1.07}$, respectively. As can be seen from the figures, the observed values of respective drag coefficients were satisfactorily reproduced by Eqs (11) and (12).

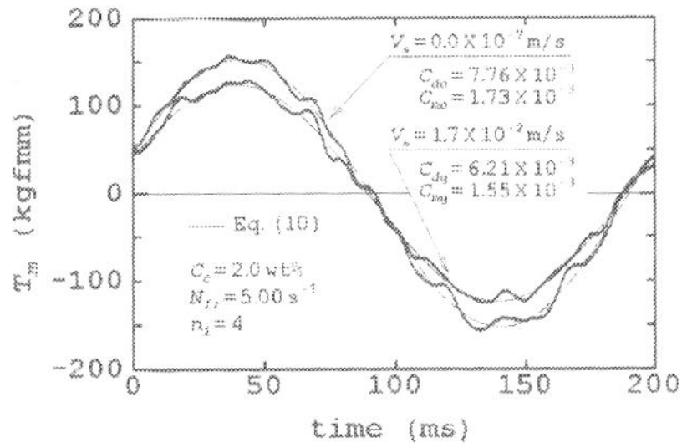

Fig. 3. Time-course of agitation torque, T_m.

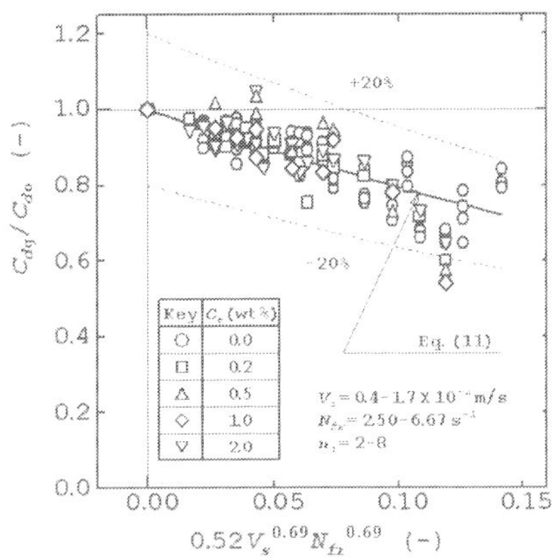

Fig. 4. Relationship between drag coefficient C_{dg}/C_{do} and operating conditions.

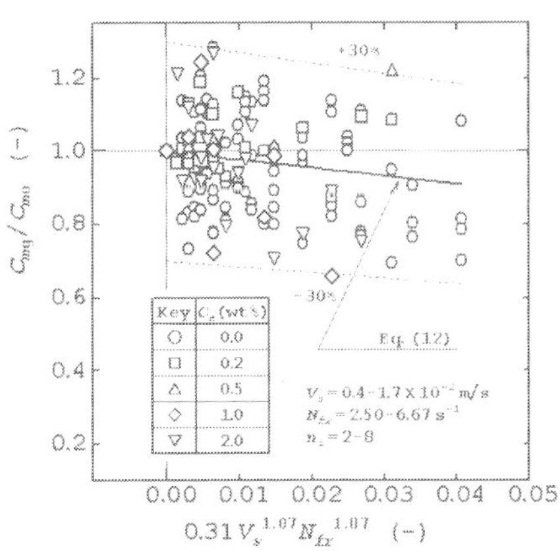

Fig. 5. Relationship between drag coefficient C_{mg}/C_{mo} and operating conditions.

3.2 Power consumption of impeller

The instantaneous power consumption, i.e., the agitation power, P_m, in the cycle of forward-reverse rotation of impeller could be expressed by the following equation as the product of the agitation torque, T_m, [Eq.(10)] and the angular velocity of impeller, ω_i, [Eq.(1)]:

$$P_m=(\rho D_i^5\theta_o^3\omega_{fr}^3)\sin(\omega_{fr}t)[(8C_d/3\pi)\sin(\omega_{fr}t)+(C_m/\theta_o)\cos(\omega_{fr}t)] \quad (13)$$

Using Eq. (13), the time-averaged power consumption, P_{mav}, that is based on the total energy transmitted in one cycle of forward-reverse rotation of impeller is related to the viscous drag coefficient, C_d, as follows:

$$P_{mav}=\oint P_m dt/(2\pi/\omega_{fr})=(4/3\pi)(\rho D_i^5\theta_o^3\omega_{fr}^3)C_d \quad (14)$$

Figure 6 shows an example of the changes in P_m with time. The thin solid line in the figure is for the values calculated from Eq. (13) with the drag coefficients, C_d and C_m, determined experimentally. Agreements between the observed and calculated values were found to be good. According to Eq. (2), the value of P_{mav} was determined by integrating graphically P_m with the time. On the other hand, combined use of Eq. (14) with Eq. (11) enables to calculate P_{mav} as a function of the aeration and agitation conditions such as V_s, N_{fr} and n_i. Figure 7

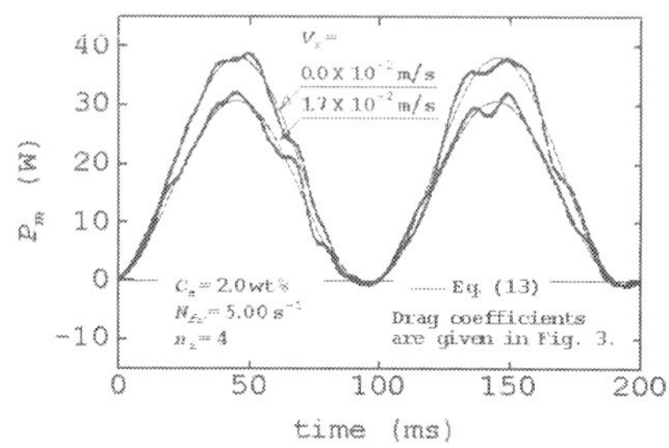

Fig. 6. Time-course of agitation power, P_m.

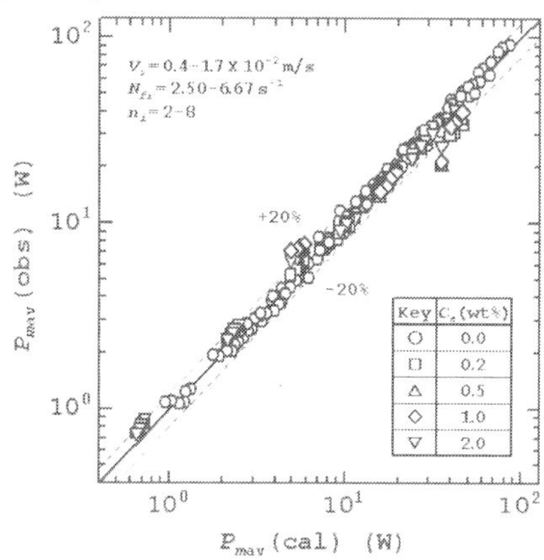

Fig. 7. Comparison of average agitation power, P_{mav}, values observed with those calculated.

compares the P_{mav} values determined experimentally with those calculated from Eq. (14) used with Eq. (11). These equations reproduced the experimental P_{mav} values with an accuracy of ±20 % and was demonstrated to be useful for prediction of the values of the power consumption of the forward-reverse rotating impeller.

Equations (11) and (12) indicate that the power consumption of the forward-reverse rotating impeller in liquid phase where gas bubbles are dispersed is independent of the electrolyte concentration. That is, the power characteristics are perceived to be independent of the dispersing gas bubble size which changes depending on the electrolyte concentration in liquid phase. This result, which is observed also for unidirectionally rotating impellers (Bruijn *et al.*, 1974), would be caused by difficulty for the cavities formed behind the blades of impeller to be affected by small sized gas bubbles dispersed in liquid phase.

4. Mass transfer characteristics of forward-reverse agitation vessel

The differences of the volumetric coefficient of mass transfer when the aeration and agitation conditions were varied were investigated in terms of the power input. A total power input was employed as the sum of the aeration and agitation power inputs. The aeration power input defined as the power of isothermal expansion of gas bubbles to their surrounding liquid was calculated from Eq. (3). For the agitation power input, the average power consumption of impeller calculated from Eqs (14) and (11) was used. Figure 8 shows a typical result of the volumetric coefficient, $k_L a_L$, plotted against the total power input per unit mass of liquid, P_{tw}, with the electrolyte concentration, C_e, and the superficial gas velocity, V_s, as parameters. For any system, $k_L a_L$ tended to increase almost linearly with P_{tw}. The rate of increase in $k_L a_L$ with P_{tw} was practically independent of V_s but differed depending on the conditions with and without electrolyte in liquid phase.

The results for the baffled vessel agitated by the unidirectionally rotating multiple DT impellers examined as a control and those reported by Van't Riet (1979) and Nocentini *et al.* (1993) are also shown in Fig. 8. Although the tendency that the dependence of $k_L a_L$ on P_{tw} becomes larger in existence of electrolyte in liquid phase was common to the forward-reverse and unidirectional agitation modes, the dependence for the former mode was larger than that for the latter one. As a result, favorably comparable $k_L a_L$ values were obtained in the unbaffled vessel agitated by the forward-reverse rotating impellers.

Presence of electrolytes in liquid phase is known to decrease the rate for gas bubbles to coalesce (Marrucci and Nicodemo, 1967; Zieminski and Whittemore, 1971) and to decrease the size of gas bubbles dispersed in liquid phase (Linek *et al.*, 1970; Robinson and Wilke, 1973; Robinson and Wilke, 1974; Van't Riet, 1979; Hassan and Robinson, 1980; Linek *et al.*, 1987). Decreased size of gas bubbles in liquid phase containing electrolyte causes increase of the gas-liquid interfacial area, a_L, which is further enhanced by the tendency for the gas hold-up to increase with decrease of the bubble size. On the other hand, decrease of the bubble size causes decrease of the liquid-phase mass transfer coefficient, k_L, and then k_L is often considered to be a function of the bubble size (Robinson and Wilke, 1974; Hassan and Robinson, 1980). That is, the volumetric coefficient that is the product of a_L and k_L suffers the two counter influences. In the following sections, the mass transfer parameters such as a_L and k_L are addressed for enhancement of the gas-liquid mass transfer in the forward-reverse agitation vessel to be assessed.

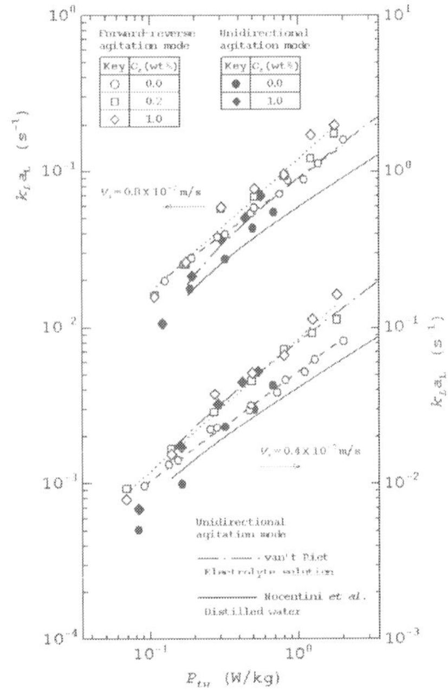

Fig. 8. Comparison of volumetric coefficient, k_La_L, as viewed from change in specific total power input, P_{tw}.

5. Hydrodynamics of forward-reverse agitation vessel

5.1 Mean bubble diameter

The dependence of the size of gas bubbles on aeration and agitation conditions was investigated in terms of the power input. Figure 9 shows a typical relationship between the mean bubble diameter, d_{vs}, and the total power input per unit mass of liquid, P_{tw}, with the electrolyte concentration, C_e, and the superficial gas velocity, V_s, as parameters. For any system, d_{vs} tended to decrease with P_{tw}. The values of d_{vs} at the same level of P_{tw} were almost independent of V_s but differed depending on C_e.

The mean bubble diameter, d_{vs}, was then analyzed with the aeration and agitation conditions. Based on the results shown in Fig. 9, the following functional form was assumed for the empirical equation of d_{vs} with the specific total power input, P_{tw}.

$$d_{vs}=AP_{tw}^{a} \tag{15}$$

The exponent, a, of P_{tw} was obtained from the slope of the straight lines as drawn in Fig. 9. Its value was independent of the electrolyte concentration, C_e. The coefficient, A, changed depending on C_e and its dependence was expressed for the experimental material of this work as follows:

$$A=-1.49C_e^{0.096}+2.95 \tag{16}$$

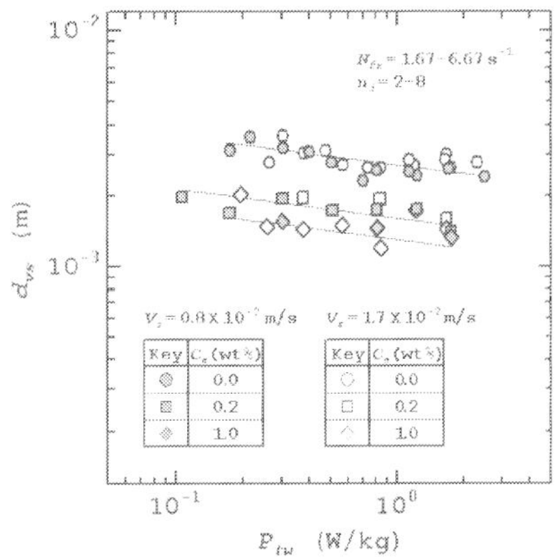

Fig. 9. Relationship between mean bubble diameter, d_{vs}, and specific total power input, P_{tw}.

As a result, the empirical equation of d_{vs} is

$$d_{vs}=(-1.49C_e^{0.096}+2.95)P_{tw}^{-0.12} \tag{17}$$

Figure 10 presents a comparison between d_{vs} values observed and those calculated from Eq. (17). As shown in the figure, d_{vs} could be correlated within approximately 20 %.

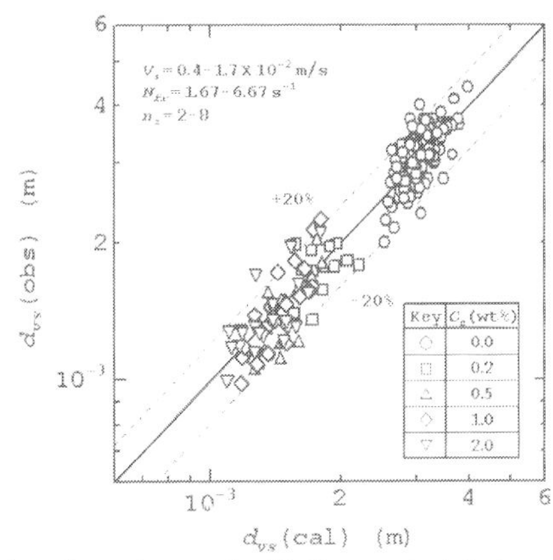

Fig. 10. Comparison of d_{vs} values observed with those calculated.

5.2 Gas hold-up

The dependence of gas hold-up was investigated in relation to the total power input, similarly to that of bubble size. Figure 11 shows a typical relationship between the gas hold-up, ϕ_{gD}, and the total power input per unit mass of liquid, P_{tw}. Although ϕ_{gD} increased with P_{tw}, its values differed depending on the electrolyte concentration, C_e, and the superficial gas velocity, V_s. The gas hold-up, ϕ_{gD}, was then analyzed with the aeration and agitation conditions. Based on the results shown in Fig. 11, the following functional form was inferred for the empirical equation of ϕ_{gD}.

$$\phi_{gD}=BP_{tw}^{b1}V_s^{b2} \tag{18}$$

The exponent, b1, of the specific total power input, P_{tw}, was obtained from the slope of the straight lines as drawn in Fig. 11. The exponent, b2, of the superficial gas velocity, V_s, was determined from the slope of the cross plots. The coefficient, B, was expressed as a function of the electrolyte concentration, C_e, as follows:

$$B=0.629C_e^{0.27}+1.32 \tag{19}$$

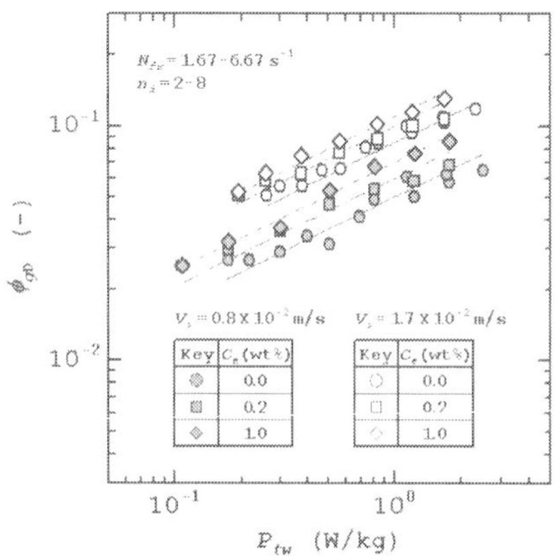

Fig. 11. Relationship between gas hold-up, ϕ_{gD}, and specific total power input, P_{tw}.

Figure 12 shows that ϕ_{gD} could be correlated by the following equation within approximately 30 %.

$$\phi_{gD}=(0.629C_e^{0.27}+1.32)P_{tw}^{0.46}V_s^{0.70} \tag{20}$$

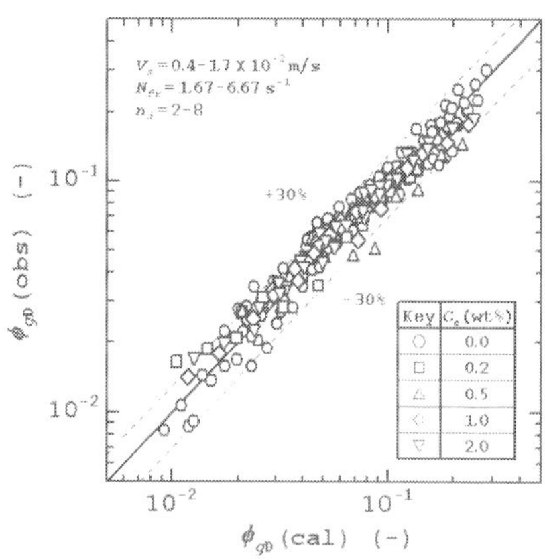

Fig. 12. Comparison of ϕ_{gD} values observed with those calculated.

5.3 Gas-liquid interfacial area

For electrolyte solutions aerated at the velocities ranged in this work, no significant difference in the mean bubble diameter and the gas hold-up determining the magnitude of gas-liquid interfacial area was observed between forward-reverse and conventional agitation vessel. From this, somewhat larger values of volumetric coefficient as illustrated in Fig. 8 in the former vessel may be a reflection of the contribution of forward-reverse rotation of the impeller to increase of the liquid-phase mass transfer coefficient with occurrence of an intensified liquid turbulence in the vicinity of the gas-liquid interface.

6. Mass transfer consideration

6.1 Analysis of mass transfer coefficient

Calderbank and Moo-Young (1961) first examined the liquid-phase mass transfer coefficient, k_L, as a function of the size of particles dispersed in a conventional baffled vessel agitated by a unidirectionally rotating impeller. Subsequently, they elucidated the mechanism of mass transfer by comparing the results with two typical theoretical predictions for single particle. They recommended applications of Froessling's laminar boundary layer theory (Froessling, 1938) and Higbie's penetration theory (Higbie, 1935) to elucidate the respective mechanisms of gas-liquid mass transfer for small (diameter, d_b<0.5 mm) and large (d_b>2.5 mm) gas bubbles. These two theories express the mass transfer characteristics in relation to the rising velocity of gas bubble, V_b, as follows.

Froessling equation:

$$k_L=(2D_L/d_b)+0.55\rho^{1/6}\mu^{-1/6}V_b^{1/2}d_b^{-1/2}D_L^{2/3} \tag{21}$$

Higbie equation:

$$k_L=(2/\pi^{1/2})[D_L/(d_b/V_b)]^{1/2} \tag{22}$$

In these equations, ρ is the liquid density, μ is the liquid viscosity, and D_L is the liquid-phase diffusivity.

The relationship between the liquid-phase mass transfer coefficient, k_L, and the gas bubble size was investigated for the unbaffled vessel agitated by the forward-reverse rotating impellers. Figure 13 shows the plot of k_L against the volume-surface mean bubble diameter, d_{vs}, which is a parameter that characterizes the size of a swarm of gas bubbles. The k_L values for the baffled vessel agitated by the unidirectionally rotating multiple DT impellers (conventional agitation vessel) examined as a control and those reported by Calderbank and Moo-Young (1961) are also shown. The dotted lines for comparison indicate the two theoretical predictions for single bubble with the diameter d_b. Here, the values of V_b that are necessary for calculations from Eqs (21) and (22) were estimated using the equation presented by Tadaki and Maeda (1961). The values of d_{vs} observed in the forward-reverse agitation vessel ranged from 1 to 4 mm. Regarding the difference of k_L caused by the bubble size, the two regions with the boundary of 2.5 mm diameter was common to the forward-reverse and conventional agitation vessel. As can be seen from the figure, k_L values for forward-reverse agitation vessel were noticeably different from those for conventional one. That is, k_L came to exhibit higher values than those by Calderbank and Moo-Young, with the maximum difference being about three times, when the agitation rate, N_{fr}, was increased. For such a peculiar difference of k_L with N_{fr} for bubbles of similar sizes, consideration is necessary in terms of hydrodynamic parameters, taking into account not only the mean bubble diameter but also other important variables such as the rising velocity of a swarm of gas bubbles that changes depending on the gas hold-up. Specifically, examination is required of the mass transfer characteristics as viewed from change in the Reynolds number, which is a parameter characterizing the liquid flow around gas bubble.

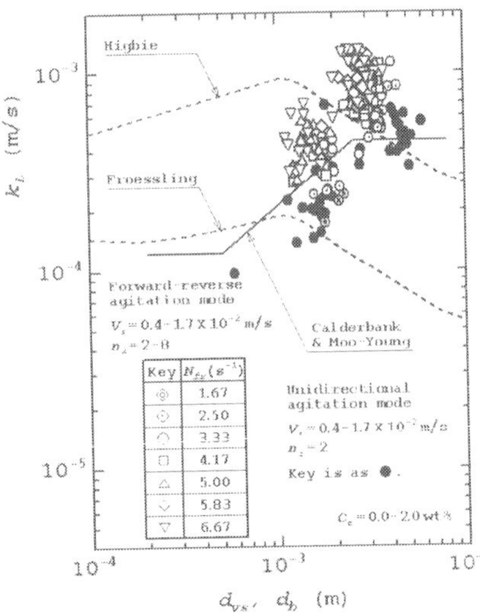

Fig. 13. Relationship between mass transfer coefficient, k_L, and gas bubble size.

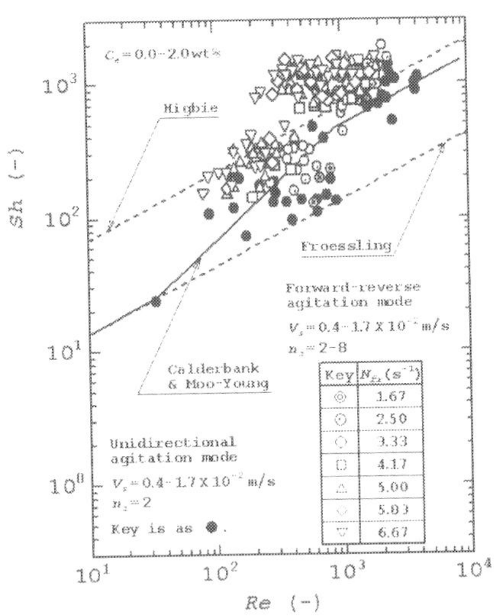

Fig. 14. Relationship between Sh and Re.

6.2 Correlation of Sherwood number

Mass transfer characteristics from bodies in a steadily flowing fluid or from bodies moving steadily in a fluid at rest are expressible using dimensionless terms that characterize transport phenomena involving the size and velocity of the body and the physical properties of the fluid. This has led many researchers to determine the functional form in the following correlation form with dimensionless terms and to evaluate the mass transfer characteristics, i.e., the mechanism of mass transfer, of the intended gas-liquid contactors by comparing results obtained experimentally with the theoretical predictions.

$$Sh = \text{func.}\,(Re, Sc) \tag{23}$$

where Sh is the Sherwood number, Re is the Reynolds number and Sc is the Schmidt number. These dimensionless terms can be given in the following forms, respectively, when the overall mass transfer coefficient for the single bubble with diameter, d_b, and rising velocity, V_b, is approximated by the liquid-phase mass transfer coefficient, k_L.

$$Sh = k_L d_b / D_L \tag{24}$$

$$Re = \rho V_b d_b / \mu \tag{25}$$

$$Sc = \mu / \rho D_L \tag{26}$$

The Froessling and Higbie equations mentioned above are expressed in the forms of Eq. (23), respectively, as
Froessling equation:

$$Sh = 2 + 0.55 Re^{1/2} Sc^{1/3} \tag{27}$$

Higbie equation:

$$Sh = (2/\pi^{1/2}) Re^{1/2} Sc^{1/2} \tag{28}$$

To examine the relationship between the non-dimensionalized k_L, Sh, and Re for the unbaffled vessel agitated by the forward-reverse rotating impellers, d_b in the dimensionless terms was replaced by the mean diameter of a swarm of gas bubbles, d_{vs}, and d_{vs} was regarded as a characteristic length. The mean rising velocity of a swarm of gas bubbles, V_{bs}, as calculated from the following equation when the distribution of gas hold-up was assumed to be uniform on any horizontal section within the vessel, was used for V_b necessary to calculate Re:

$$V_{bs} = V_s / \phi_{gD} \tag{29}$$

where V_s is the superficial gas velocity. ϕ_{gD} is the gas hold-up and the values determined as the average within the vessel were used for calculation from Eq. (29).

Figure 14 shows the relationship between Sh and Re for the forward-reverse agitation vessel. The Sh values for the conventional agitation vessel as a control and those by Calderbank and Moo-Young (1961) are also shown in comparison with the theoretical predictions for single bubble. For the mean bubble diameter, d_{vs}, larger than 2.5 mm, the Sh values for the conventional agitation vessel more closely resembled the values calculated from the Higbie equation than those calculated from the Froessling equation. For d_{vs} smaller than 2.5 mm, decrease of Sh with decrease of Re was remarkable, with Sh exhibiting the values that are intermediary between those from the Higbie and Froessling equations. This fact suggests that the contribution of liquid flow to the mass transfer is reduced considerably if the internal circulation within gas bubble is prevented from becoming fully developed through decreased rising velocity of gas bubble that is attributable to its decreased size, namely, smaller inertial force of rising motion of gas bubble (Tadaki and Maeda, 1963; Sideman *et al.*, 1966). The relationship between *Sh* and *Re* for the forward-reverse agitation vessel under the condition of lower agitation rates exhibited the same tendency as that for the conventional agitation vessel, as shown in the figure. That is, *Sh* under such conditions increased approximately in proportion to $Re^{0.5}$, according to the Higbie equation, for d_{vs} larger than 2.5 mm, and in proportion to *Re* for d_{vs} smaller than 2.5 mm. When the agitation rate, N_{fr}, is increased, an increased gas hold-up and a decreased mean bubble diameter cause *Re*, as defined by Eq. (25), to have lower values. Under the conditions as *Re* decreased or N_{fr} increased, *Sh* for the forward-reverse agitation vessel tended to exhibit higher values than those for the conventional agitation vessel. For the forward-reverse agitation vessel, i.e., unsteady bulk flow type contactor, the dependence of *Sh* on *Re*, characteristically different from that for the conventional agitation vessel, i.e., steady bulk flow type contactor, suggests an effect of unsteady oscillating flow produced by forward-reverse rotation of the impeller on enhancement of the mass transfer. For quantifying *Sh* in this gas-liquid agitation system, further consideration is necessary taking into account a parameter characterizing the unsteady oscillating flow, in addition to *Re* as a measure of the liquid flow around gas bubble.

Many theoretical and experimental studies on the mass transfer between gas bubble and its surrounding liquid in a steady state have given the relation in the form of Eq. (23) as an equation that expresses gas-liquid mass transfer characteristics. The dimensionless terms in Eq. (23) are obtained by non-dimensionalizing the equation of motion determining the liquid flow around gas bubble, the diffusion equation determining the mass transfer between gas bubble and its surrounding liquid and the mass balance equation, respectively, under a condition of steady liquid flow. On the other hand, the time-dependent term in the equation of motion should be considered for analysis of the liquid flow around gas bubble when gas bubble rises in liquid with unsteady oscillating flow produced by forward-reverse rotation of the impeller. The dimensionless terms, the Sherwood number (*Sh*), the Reynolds number (*Re*), the Strouhal number (*St*), and the Schmidt number (*Sc*), which are necessary to express the unsteady mass transfer phenomena, are all derived based on the equation of motion that is non-dimensionalized under a condition of unsteady liquid flow, the dimensionless diffusion equation and the dimensionless mass balance equation. Therefore, *Sh* in this gas-liquid agitation system is given as a function of *Re*, *St* and *Sc*.

$$Sh = \text{func.}\ (Re,\ St,\ Sc) \tag{30}$$

Definitions of *Sh*, *Re* and *Sc* are mentioned above; *St* is defined for single gas bubble as

$$St = f d_b / V_b \tag{31}$$

The diameter, d_b, and the velocity, V_b, for *St* were identical to those for *Re*. The frequency of forward-reverse rotation of impeller, N_{fr}, was taken as a characteristic frequency, *f*.

The experimental results shown in Fig. 14 suggest that the two coexisting liquid flows affect *Sh* in a form of their superposition. Then, the degree of effect would be determined by the relative magnitude of *Re* characterizing the steady slip flow of surrounding liquid with the rising motion of gas bubble and St characterizing the unsteadily oscillating flow of liquid around gas bubble by forward-reverse rotation of the impeller. Furthermore, a value of 1/2, which many researchers (Calderbank, 1959; Yagi and Yoshida, 1975; Nishikawa *et al.*, 1981; Panja and Phaneswara Rao, 1993; Zeybek *et al.*, 1995; Linek *et al.*, 2005) have used for a swarm of gas bubbles in a relatively large size range, is assumed to be adoptable as an exponent indexing the dependence of Sh on Sc. Consequently, Eq. (30) can be concretized in the following functional form:

$$Sh=[\text{func.}\,(Re)+CSt^{c}]Sc^{1/2} \tag{32}$$

Therein, func. (Re) is based on the relation by Calderbank and Moo-Young (1961), which differs depending on the range of mean bubble diameter, as illustrated in Fig. 14, and is given by the following equations.
d_{vs}<2.5 mm:

$$\text{func.}\,(Re)=0.0544Re^{0.90} \tag{33}$$

d_{vs}>2.5 mm:

$$\text{func.}\,(Re)=(2/\pi^{1/2})Re^{1/2} \tag{34}$$

The term in the bracket on the right side of Eq. (32) is expected to express the combined effect of the two liquid flows, namely, the relative influence of the steady slip flow and the unsteady oscillating flow, on the gas-liquid mass transfer.
On the basis of Eq. (32), the relationship between $Sh/Sc^{1/2}$-func. (Re) and St was examined. From the slope of the line and the intercept on the axis in the logarithmic plot, the empirical constants, c and C, were determined for the respective ranges with d_{vs}=2.5 mm as a boundary, and then the following correlation equations were obtained.
d_{vs}<2.5 mm:

$$Sh=[0.0544Re^{0.90}+10.0St^{0.10}]Sc^{1/2} \tag{35}$$

d_{vs}>2.5 mm:

$$Sh=[(2/\pi^{1/2})Re^{1/2}+180St^{0.79}]Sc^{1/2} \tag{36}$$

The result of correlation of Sh is shown in Fig. 15. As shown in the figure, Sh was correlated with Re and St with an accuracy of ±40 %. A positive dependence of Sh on St is considered to indicate that the time-dependence in the flow of liquid around gas bubble, namely, the inertial force caused by the local acceleration produced by forward-reverse rotation of the impeller, contributes to enhancement of the gas-liquid mass transfer. This contribution could be significant when the velocity gradients in the flow of the surrounding liquid, namely, the inertial force caused by the convective acceleration produced by the rising motion of gas bubble, is smaller. Equations (35) and (36) are applicable when Re is 100-2300, that is, the flow of liquid around gas bubble is practically turbulent, for St of up to 0.20.

6.3 Correlation of volumetric coefficient

Correlation of the volumetric coefficient of gas-liquid mass transfer was then made based on the experimental results mentioned above. The volumetric coefficient, k_La_D, based on the gassed liquid volume was intended for correlation. k_La_D is the product of the liquid-phase mass transfer coefficient, k_L, and the gas-liquid interfacial area, a_D.

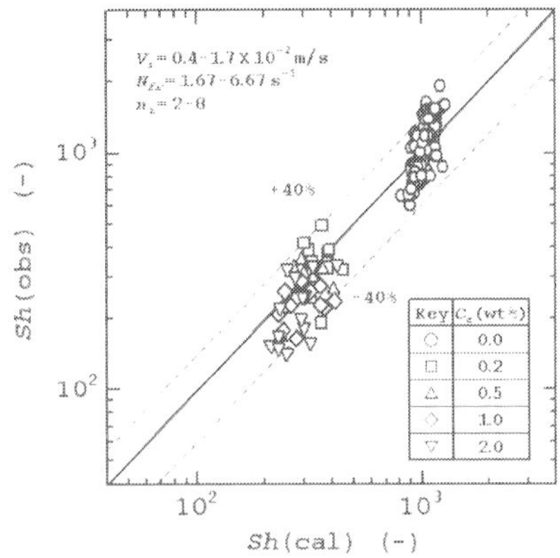

Fig. 15. Comparison of *Sh* values observed with those calculated.

$$k_L a_D = (k_L)(a_D) \tag{37}$$

It is noteworthy that a_D depends on the volume-surface mean bubble diameter, d_{vs}, and the gas hold-up, ϕ_{gD}, according to Eq. (5). Substituting Eq. (5) into Eq. (37) yields the following equation, which gives the relation between $k_L a_D$ and the mass transfer parameters.

$$k_L a_D = 6\phi_{gD} k_L / d_{vs} \tag{38}$$

The empirical equations for d_{vs} and ϕ_{gD} are given respectively as Eqs (17) and (20). In addition, the values of k_L are predicted by combined use of Eq. (35) or (36) and Eqs (17) and (20). Figure 16 shows the observed $k_L a_D$ values along with those obtained by substituting the calculated values of the mass transfer parameters into Eq. (38). As shown in the figure, $k_L a_D$ could be correlated with an accuracy of ±35 %. Moreover, Eqs (17) and (20) are relations in terms of the specific power input and Eqs (35) and (36) are a dimensionless relation. As demonstrated by some researchers (Van't Riet, 1979), the relations based on such correlation forms are expected to be useful for operational design, i.e., scale-up, of the apparatus.

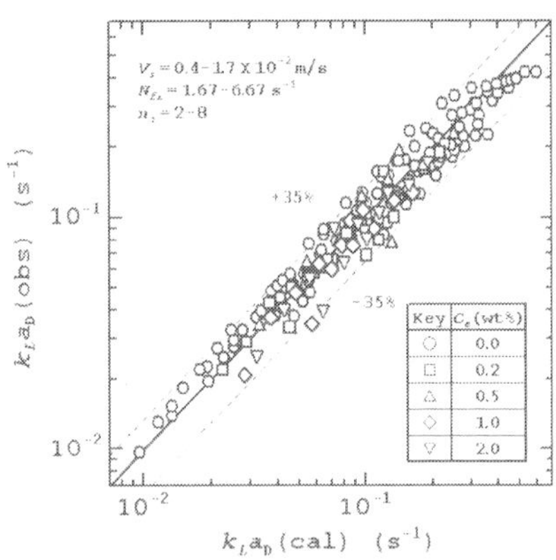

Fig. 16. Comparison of $k_L a_D$ values observed with those calculated.

7. Conclusion

For a gas-sparged, unbaffled vessel agitated by the forward-reverse rotating multiple impellers, the volumetric coefficient of oxygen transfer in air-electrolyte solution systems was correlated in relation to mass transfer parameters such as the mean bubble diameter, gas hold-up and liquid-phase mass transfer coefficient. The effects of gassing rate, agitation rate and the number of impellers on the bubble diameter and hold-up were examined. The empirical equations were obtained to predict their values as a function of the specific total power input, superficial gas velocity and electrolyte concentration in liquid phase. The difference of liquid-phase mass transfer coefficient, non-dimensionalized in a form of the Sherwood number, was related to changes of the Reynolds number and Strouhal number characterizing the mass transfer between gas bubble and its surrounding liquid in the unsteadily oscillating flow.

8. References

Bruijn, W., Van't Riet, K. and Smith, J. M. (1974). Power consumption with aerated Rushton turbines. *Trans. Inst. Chem. Engrs*, Vol. 52, pp. 88-104

Calderbank, P. H. (1959). Physical rate processes in industrial fermentation, Mass transfer coefficient in gas-liquid contacting with mechanical agitation. *Trans. Inst. Chem. Engrs*, Vol. 37, pp. 173-185

Calderbank, P. H. and Moo-Young, M. B. (1961). The continuous phase heat and mass-transfer properties of dispersions. *Chem. Eng. Sci.*, Vol. 16, pp. 39-54

Chapman, C. M., Gibilaro, L. G. and Nienow, A. W. (1982). A dynamic response technique for the estimation of gas-liquid mass transfer coefficients in a stirred vessel. *Chem. Eng. Sci.*, Vol. 37, pp. 891-896

Froessling, N. (1938). On the vaporization of a falling drop. *Gerlands Beitr Geophys*, Vol. 52, pp. 170-216

Hassan, I. T. M. and Robinson, C. W. (1980). Mass transfer coefficient in mechanically agitated gas-aqueous electrolyte dispersions. *Can. J. Chem. Eng.*, Vol. 58, pp. 198-205

Higbie, R. (1935). The rate of absorption of a pure gas into a still liquid during short periods of exposure. *Trans. Am. Inst. Chem. Engrs*, Vol. 31, pp. 365-388

Linek, V., Mayrhoferova, J. and Mosnerova, J. (1970). The influence of diffusivity on liquid phase mass transfer in solutions of electrolytes. *Chem. Eng. Sci.*, Vol. 25, pp. 1033-1045

Linek, V., Vacek, V. and Benes, P. (1987). A critical review and experimental verification of the correct use of the dynamic method for the determination of oxygen transfer in aerated agitated vessels to water, electrolyte solutions and viscous liquids. *Chem. Eng. J.*, Vol. 34, pp. 11-34

Linek, V., Kordac, M. and Moucha, T. (2005). Mechanism of mass transfer from bubbles in dispersions, mass transfer coefficients in stirred gas-liquid reactor and bubble column. *Chem. Eng. Processing*, Vol. 44, pp. 121-130

Marrucci, G. and Nicodemo, L. (1967). Coalescence of gas bubbles in aqueous solutions of inorganic electrolytes. *Chem. Eng. Sci.*, Vol. 22, pp. 1257-1265

Nienow, A. W. (1990). Gas dispersion performance in fermenter operation. *Chem. Eng. Prog.*, No. 2, pp. 61-71

Nishikawa, M., Nakamura, M. and Hashimoto, K. (1981). Gas absorption in aerated mixing vessels with non-Newtonian liquid. *J. Chem. Eng. Japan*, Vol. 14, pp. 227-232

Nocentini, M., Fajner, D., Pasquali, G. and Magelli, F. (1993). Gas-liquid mass transfer and holdup in vessels stirred with multiple Rushton turbines: Water and water-glycerol solutions. *Ind. Eng. Chem. Res.*, Vol. 32, pp. 19-26

Panja, N. C. and Phaneswara Rao, D. (1993). Measurement of gas-liquid parameters in a mechanically agitated contactor. *Chem. Eng. J.*, Vol. 52, pp. 121-129

Robinson, C. W. and Wilke, C. R. (1973). Oxygen absorption in stirred tanks, A correlation for ionic strength effects. *Biotechnol. Bioeng.*, Vol. 15, pp. 755-782

Robinson, C. W. and Wilke, C. R. (1974). Simultaneous measurement of interfacial area and mass transfer coefficients for a well-mixed gas dispersion in aqueous electrolyte solutions. *AIChE J.*, Vol. 20, pp. 285-294

Sideman, S., Hortacsu, O. and Fulton, J. W. (1966). Mass transfer in gas-liquid contacting systems. *Ind. Eng. Chem.*, Vol. 58, pp. 32-47

Tadaki, T. and Maeda, S. (1961). On the shape and velocity of single air bubbles rising in various liquids. *Kagaku Kogaku*, Vol. 25, pp. 254-264

Tadaki, T. and Maeda, S. (1963). On the CO_2 desorption in the gas bubble column. *Kagaku Kogaku*, Vol. 27, pp. 808-814

Takahashi, K. (1994). *Fluid Mixing Techniques for New Materials*, p. 143, IPC Publications Co., Ltd., Tokyo

Tanaka, H. and Ueda, K. (1975). Design of a jar fermentor for filamentous microorganisms. *J. Ferment. Technol.*, Vol. 53, pp. 143-150

Van't Riet, K. (1979). Review of measuring methods and results in nonviscous gas-liquid mass transfer in stirred vessels. *Ind. Eng. Chem. Process Des. Dev.*, Vol. 18, pp. 357-364

Warmoeskerken, M. M. C. G. and Smith, J. M. (1985). Flooding of disc turbines in gas-liquid dispersions, A new description of the phenomenon. *Chem. Eng. Sci.*, Vol. 40, pp. 2063-2071

Yagi, H. and Yoshida, F. (1975). Gas absorption by Newtonian and non-Newtonian fluids in sparged agitated vessels. *Ind. Eng. Chem. Process Des. Dev.*, Vol. 14, pp. 488-493

Yoshida, M., Kitamura, A., Yamagiwa, K. and Ohkawa, A. (1996). Gas hold-up and volumetric oxygen transfer coefficient in an aerated agitated vessel without baffles having forward-reverse rotating impellers. *Can. J. Chem. Eng.*, Vol. 74, pp. 31-39

Yoshida, M., Ito, A., Yamagiwa, K., Ohkawa, A., Abe, M., Tezura, S. and Shimazaki, M. (2001). Power characteristics of unsteadily forward-reverse rotating impellers in an unbaffled aerated agitated vessel. *J. Chem. Technol. Biotechnol.*, Vol. 76, pp. 383-392

Yoshida, M., Watanabe, M., Yamagiwa, K., Ohkawa, A. Abe, M., Tezura, S. and Shimazaki, M. (2002). Behaviour of gas-liquid mixtures in an unbaffled reactor with unsteadily forward-reverse rotating impellers. *J. Chem. Technol. Biotechnol.*, Vol. 77, pp. 678-684

Yoshida, M., Akiho, M., Nonaka, K., Yamagiwa, K., Ohkawa, A. and Tezura, S. (2005). Mixing and mass transfer characteristics of an unbaffled aerated agitation vessel with unsteadily forward-reverse rotating multiple impellers. *Lat. Amer. Appl. Research*, Vol. 35, pp. 37-42

Zeybek, Z., Oguz, H. and Berber, R. (1995). Gas/liquid mass transfer in concentrated polymer solutions. *Chem. Eng. Res. Des.*, Vol. 73, pp. 622-626

Zieminski, S. A. and Whittemore, R. C. (1971). Behavior of gas bubbles in aqueous electrolyte solutions. *Chem. Eng. Sci.*, Vol. 26, pp. 509-520

7

Toward a Multiphase Local Approach in the Modeling of Flotation and Mass Transfer in Gas-Liquid Contacting Systems

Jamel Chahed and Kamel M'Rabet
National Engineering School of Tunis, BP N37, Le Belvédère, 1002, Tunis Tunisia

1. Introduction

This chapter comprises two major parts: The first part is devoted to the development of a kinetic model of flotation based on the theory of mass transfer in gas - liquid bubbly flows. The second part presents some requirements in order to go forward in implementing multiphase local approach in the modeling of flotation and mass transfer in gas-liquid contacting systems.

The common kinetic models applied to the flotation process are first order and they are validated on the basis of experimental analysis. Classical kinetics model of flotation cannot represent the bubble carrying capacity because their solution supposes that for long contact time, the concentration of the slurry have to vanish whatever the rate of the superficial bubbles loading may be.

The bubble carrying capacity, which can be interpreted as a superficial saturation of the bubbles, cannot be represented by simple first order models. However, saturation phenomena have been observed in some flotation devises. For example the experiments of the flotation column of Bensley et al., (1985) show that, for a given flux of bubbles, the recovery of particles in the slurry may still be constant at a relatively low value whatever the height of the flotation column may be. This is similar to what it happens in common gas-liquid mass transfer devices.

Gas-liquid mass transfer theory indicates that the flux of mass through the gas-liquid interface is proportional to the gas-liquid contact area. It indicates also that the resistance to the interfacial transfer of low solubility gases may be described by a first order kinetics law where the main variable is the difference between the local concentration of the gas in the liquid and the concentration at saturation in the same thermodynamic conditions. For long contact time, this difference vanishes: the mass transfer is stopped up when the concentration of saturation in the liquid is reached. A kinetic model inspired from the theory of mass transfer in gas - liquid medium is presented. This model, developed to describe flotation in a bubble column, is interpreted with regard to the effect of the superficial saturation of the bubbles on the kinetics of flotation.

The second part of the chapter is devoted to local analysis and modeling of gas-liquid turbulent flows. The local description of gas-liquid contacting systems (average velocity, turbulence, void fraction etc.) represents an important scientific challenge and creates a major interest in many industrial applications especially when transfer phenomena are in concern. Many experimental results in gas-liquid bubbly flows indicate the important effect of the interfacial interactions on the turbulence of the liquid phase, Lance &Bataille (1991). The eulerian-eulerien model of Chahed et al. (2003) insists on two aspects of the interaction between phases : first the turbulent correlations associated with the added mass force are taken into account in the expression of the force exerted by the liquid on the bubbles and second the Reynolds stress tensor of the continuous phase is separated into two parts: a turbulent part produced by the gradient of mean velocity and a pseudo-turbulent part induced by bubbles displacements, each part is predetermined by a transport equation. This two-fluid model allows the prediction of the turbulence and of the void fraction in basic bubbly flows with moderate void fractions. We present the most prominent results, which are commented in order to specify the basic requirements in the elaboration of CFD codes applied to gas-liquid reactors.

2. Mass transfer approach applied to the modeling of flotation

2.1 First order kinetic model of flotation

In flotation in a bubble column, air bubbles are injected in an aqueous phase (slurry) containing a suspension of non-wettable (hydrophobic) and wettable (hydrophilic) particles. As the bubbles move through the slurry, they collect the hydrophobic particles and carry them to the surface, where they form a stable froth. The froth is removed and the floated particles recovered.

As for the description of the kinetics of chemical transformations, the kinetics of flotation is essentially described by first order models. First order models express that the rate of appearance/disappearance of specie in a process of transformation is proportional to its concentration. Applied to flotation, this model, describes the reduction of the concentration in the liquid during the process of flotation in the form:

$$\frac{dC}{dt} = -kC \tag{1}$$

where C is the liquid concentration (mass of particles per unit volume of the liquid) and k the kinetic coefficient (frequency) assumed to be constant. Broadly speaking, for column flotation where we only consider the collection that occurs in the collection zone, the kinetic models are established in order to represent the number of particles that succeed to have a collision with a bubble and to be effectively attached to the gas liquid interface. The kinetic coefficient of the flotation process is thus formulated as the product of three factors representing the frequencies of collision (P_c), of attachment (P_a) and of non-detachment (P_d):

$$k = P_c\, P_a\, P_d \tag{2}$$

Many models have been developed in order to determine the factors involved in the modeling of flotation kinetics and excellent reviews have been produced establishing the state of the art in the domain, (Tuetja et al., 1994; Zongfu et al., 2000). These studies show that there is a variety of kinetic models used for flotation; but when confronted to the experiments, the disparity of their results may indicate that the mechanisms involved in flotation phenomena are too complex to be described by relatively simple formulations.

Generally speaking, this kind of model may be validated on the basis of experimental analysis provided that the first order behavior is supported by the experimental data, as in the work of Ahmed and Jameson (1985) for example. Nevertheless, it is reported that the phenomenon of flotation may be limited by bubble carrying capacity, (Finch & Dobby, 1990; Yoon et al, 1993), or by the ability of the bubbles to transport large size particles (Nguyen, 2003).

Similar phenomenon of saturation occurs in gas-liquid mass transfer with this difference that the saturation concerns, in this case, the concentration of the liquid where the gas is dissolved. Indeed gas-liquid mass transfer theories indicates that the resistance to the interfacial transfer of low solubility gases may be described by a first order kinetics law where the main variable is the difference between the local concentration of the gas in the liquid and the concentration at saturation in the same thermodynamic conditions. For long contact time, this difference vanishes: the mass transfer is stopped up when the concentration of saturation in the liquid is reached.

Flotation can be seen as interfacial liquid-gas transfer and one may expect that we can draw on mass transfer theory in gas-liquid flows in order to build kinetic models for flotation which would be capable of reproducing the phenomenon of saturation. The application of mass transfer theory to flotation process has been previously suggested by Jameson et al. (1977) and Ityokumbul (1992).The later proposed, for the collection zone of a flotation column, a kinetic model based on a mass balance at the gas-liquid interfaces where the rate of particle attachment to the bubbles is assumed to be proportional to the particle concentration in the liquid and to the presence of non-saturated bubble surface. Chahed & Mrabet (2008) have built on this model and have proposed a more complete formulation. They also tried to comment some reductions of this formulation in comparison with the common first order model of flotation and with the model proposed by Ityokumbul (1992). In the following, we briefly present some aspects of gas-liquid mass transfer theory than we present the results obtained in the formulation of kinetics of flotation in bubble columns based on mass transfer theory.

2.2 Interfacial mass transfer in gas-liquid flow

To focus on the physical significance of the mass transfer phenomena let us analyze the relatively simple two-film model, figure (1). In the two-film model, the rate of mass transfer due to the diffusion of a gas through a gas-liquid interface can be expressed in two ways according as we consider the liquid side film (thickness δ_L) or that on the gas side (thickness δ_G). In the gas side film, the transfer is based on the partial pressure diving force $P_A - P_A^*$ while in the liquid one it is based on the concentration driving force.

$$\frac{dC}{dt} = K_G a(P_A - P_A^*) = K_L a(C_A^* - C_A) \tag{3}$$

where P_A is the partial pressure of the gas and C_A is the concentration of the dissolved gas in the liquid, C_A^* , is the concentration of the gas in the liquid that will be in equilibrium with P_A and P_A^* is the partial pressure of the gas that will be in equilibrium with C_A ; K_G and K_L are respectively the mass transfer coefficients based on the gas and liquid phases ; and a is the interfacial area which represents the interfacial surface per unit volume of the liquid.

The appropriate equilibrium values A_A^* or C_A^* may be given by Henry's law:

$$P_A^* = HC_A \text{ and } C_A^* = \frac{1}{H} P_A \tag{4}$$

Where H is the Henry's constant

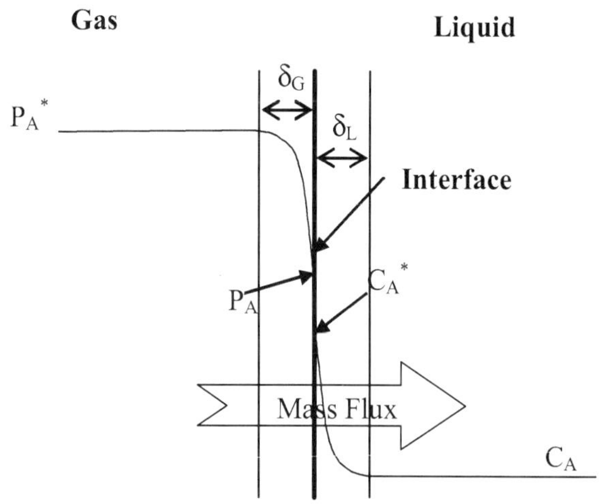

Fig. 1. Schematic representation of double-film mass transfer model

2.3 Mass balance at gas-liquid interfaces in flotation device

The kinetic model of Ityokumbul (1992) is based on a mass balance at the gas-liquid interfaces where the rate of particle attachment to the bubbles is assumed to be proportional to the concentration in the liquid C and to the presence of non-saturated bubble surface. The model considers alos a rate of detachment proportional to the surface load of the bubbles C_b (mass per unit surface). This mass balance is written in the form:

$$\frac{dC_b}{dt} = k_a (C_{mb} - C_b)C - k_d C_b \tag{5}$$

Where C_{mb} represents the maximum surface load of the bubbles that corresponds to the mass of the maximum of particles that bubbles can theoretically carry per surface unit. The model of Ityokumbul introduces two kinetic coefficients: a detachment coefficient k_d which has the dimension of the inverse of time and an attachment coefficient k_a defined so that the product $k_a C$ has the dimension of the inverse of time.

The model has the great advantage of connecting the kinetics of flotation to the presence of interfaces. Indeed, the mass transfer from the liquor is the amount accumulated on the interfaces, so that we can write:

$$\frac{dC}{dt} = -a\frac{dC_b}{dt} = -k_a a(C_{mb} - C_b)C + k_d a C_b \tag{6}$$

Where a is the interfacial area.

In reference to experimental data, Ityokumbul (1992) considers that in ordinary flotation systems, the surface load of the bubbles of the bubbles C_b is still generally small in comparison with the maximum surface load of the bubbles C_{mb}. He deduced that C_b could be neglected and the model (6) degenerates in the form:

$$\frac{dC}{dt} = -k_a a C_{mb} C \tag{7}$$

Which is identical to first order models generally used to describe the kinetics of flotation. Commenting the model of Ityokumbul (1992), Chahed & Mrabet (2008) assume the hypothesis considering that the surface load C_b is generally small as compared to the maximum surface load C_{mb}, allows to neglect C_b in the first term in the rhs of equation (6), but certainly does not allow, in the general case, to neglect the last term. This term represents the rate of detachment of the particles and when neglected, the model degenerates so much so that it becomes impossible to represent the effect of the surface load of bubble on its potential carrying capacity. In order to avoid this loss of generality, Chahed & Mrabet (2008) proposed to keep the term of detachment but they neglected C_b in the first term of the second member of the mass balance (6) against C_{mb}. In these conditions, the model (6) reduces to:

$$\frac{dC_b}{dt} = k_a C_{mb} C - k_d C_b \quad \text{and} \quad \frac{dC}{dt} = k_a a C_{mb}(C_R - C) \tag{8}$$

Where $C_R = \dfrac{k_d C_b}{k_a C_{mb}}$ is a minimum concentration that we can reach under a given operating condition. This model has a similar formulation than the models used in gas-liquid mass transfer. It shows explicitly that the interfacial transfer is proportional to the interfacial area. This outcome is in concordance with the experimental works of Gorain et al. (1997).

The irreducible concentration C_R depends on the superficial loading condition of the bubble interface. By analogy, C_R represents for flotation what the concentration C* (concentration of the gas in the liquid that is in equilibrium with the partial pressure of the gas) represents for gas-liquid mass transfer. Both determine a certain condition of gas-liquid interface equilibrium. So, we have to build a phenomenological model similar to the Henry's law used in gas-liquid mass transfer in order to derive constitutive relations for the irreducible concentration C_R. The expression of C_R suggests that, for a given operative conditions, C_R depends on the loading condition of the bubble surface in comparison to the maximum load that the bubble can potentially carry per unit of surface area. It is also expected that irreducible concentration will depend on the local conditions (slurry concentration, turbulence, superficial tension, contaminants, etc.)

The model proposed by Ityokumbul (1992) corresponds to a reduction of the model (8) in which the detachment is neglected, equation (7). In this case C_R is zero and there is no limit to the minimum slurry concentration. This corresponds to situations where the bubbles are very lightly loaded ($C_b \to 0$). If we consider, in first approximation, that the number of particles accumulated by unit of bubbles surface is proportional to d_P^{-2} (d_P is the diameter of the particle), the maximum load C_{mb} is thus proportional to $\rho_P d_p$ (ρ_P is the density of the particle). This result is concordance with the correlation proposed by Finch and Dobby(1990). On the other hand, for spherical bubbles, the interfacial area is given by:

$$a = \frac{6\alpha}{d_b} \cdot \tag{9}$$

where α is the void fraction that represents the rate of the gas volume and d_b is the bubble diameter. Equation (7) writes:

$$\frac{dC}{dt} = -h\alpha \frac{d_p}{d_b} C \tag{10}$$

Where h is a constant coefficient. For a given void fraction, equation (10) corresponds to the conceptual first collision model published by Sutherland in a famous paper that introduced the notion of flotation kinetics, Sutherland (1948).

2.4 Modeling of flotation in a bubble column

The kinetic model based on gas-liquid mass transfer theory presented above has been applied to two cases of literature experimental data, Chahed & Mrabet (2008). The first set of experiments is due to Bensley et al. (1985) and concerns the flotation of fine coal particles in a bubble column. The experimental data represent the variation of the flotation recovery as a function of the height of the collection zone in the flotation column. These experiments show that, for the reason that coal particles are very hydrophobic, the most important part of the flotation process takes place in the zone where occurs the first contact between the slurry and the fresh bubbles injected at the bottom of the column. It appears that the bubbles are rapidly saturated and the recovery of the bubble column remained less than 80 % whatever the bubble column height may be. Obviously, the classical first order models with a constant coefficient are inadequate for reproducing the flotation kinetics observed in these experiments.

The second experiment concerns the flotation of less hydrophobic particles (flourite), (Yachausti et al., 1988). The authors reported also experimental data representing the variation of the flotation recovery as a function of the height of the collection zone. In this experiment, we remark that the recovery of flourite is improved as the height of the flotation column is increased. It seems that the flotation process is not terminated by a saturation effect, at least in the range of flotation column heights employed in the experiments; correspondingly, the irreducible concentration is not reached in the experiments. Nevertheless, a trend to a limitation of the flotation can be observed for the highest columns. In both experiments, classical first order models are insufficient and the simulations produced by Chahed & Mrabet (2008) showed how it is pertinent to use mass transfer approach to describe the flotation kinetics in bubble columns. The simulation of the first set of experiments shows that, when the irreducible concentration is suitably adjusted from the experiments, the kinetics of flotation can be adjusted and the model reproduces correctly the experimental results, with an irreducible concentration χ_R^* greater than 20% , figure (2). Note that the collection zone has a height L and the liquid moves through the column with an uniform velocity denoted u_p, $\chi_R^* = \frac{C^*}{C_0}$ represents the concentration of the liquid at the outlet of the column normed by the concentration of the slurry at the inlet.

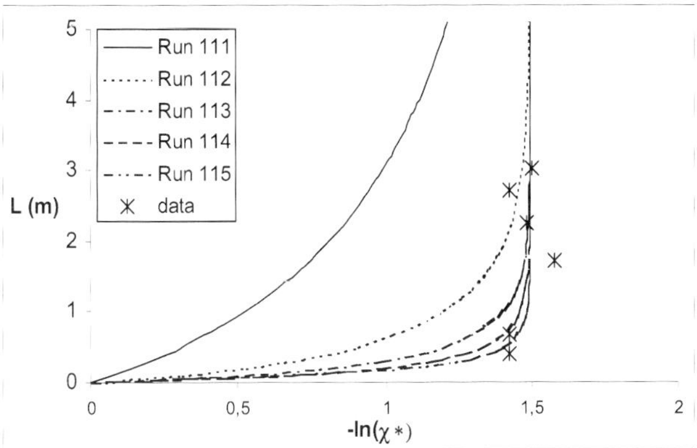

Fig. 2. Effect of the kinetic coefficient on the flotation yield, Simulations with $\chi_R^* = 0.225$, $\frac{k_d}{u_p} = 1$ (Run 111), $\frac{k_d}{u_p} = 5$ (Run 112), $\frac{k_d}{u_p} = 10$ (Run 113), $\frac{k_d}{u_p} = 15$ (Run 114), $\frac{k_d}{u_p} = 20$ (Run 115). Experimental data of coal recovery as function of recovery zone height, (Bensley et al., 1985).

The application of the model to the second set of experiments, with an irreducible concentration of order of $\chi_R^* = 2.5\%$ allows quite good reproduction of the experimental results when the kinetic coefficient is suitably adjusted, figure (3).

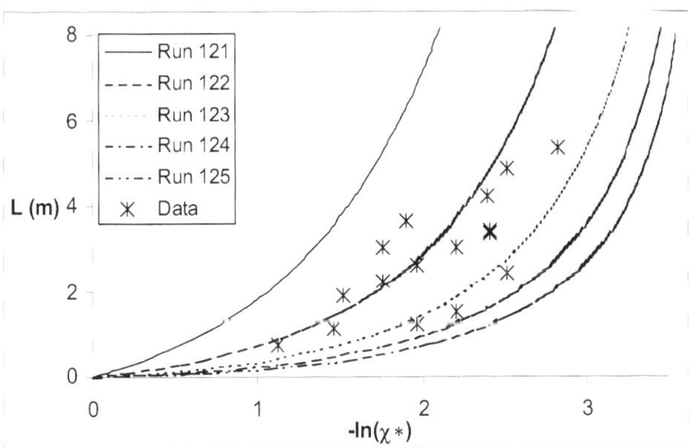

Fig. 3. Effect of the kinetic coefficient on the flotation yield, Simulations with $\chi_R^* = 0.025$, $\frac{k_d}{u_p} = 1$ (Run 121), $\frac{k_d}{u_p} = 2.5$ (Run 122), $\frac{k_d}{u_p} = 5$ (Run 123), $\frac{k_d}{u_p} = 7.5$ (Run 124), $\frac{k_d}{u_p} = 10$ (Run 125). Experimental data of flourite recovery as function of recovery zone height, (Yachausti et al., 1988).

3. Local modeling of transfers in gas-liquid bubbly flows

3.1 Mass transfer in gas-liquid flow

There are number of formulations of mass transfer in gas-liquid flow but all of them come down to a general formulation of the general mass flux, including flotation, in the general form:

$$S_c = k_L a(C^* - C) \quad (11)$$

Where k_L is the transfer coefficient which has dimension of velocity, C^* is the concentration of gas in the liquid at saturation, a is the interfacial area, which is the surface of the interfaces per unit volume, and C_L the concentration far away from the interfaces. In mono-disperse bubbly flows with spherical bubbles, the volumetric interfacial area is proportional to the void fraction and inversely proportional to the bubble diameter ($a = \frac{6\alpha}{d}$) which indicates the important role of the bubble diameter and of its distribution in the computation of the local mass transfer.

The experimental and numerical work carried out in recent decades has made significant advances in local modeling of gas-liquid systems, (Lain et al. 1999; Cockx et al., 2001; Buscaglia et al., 2002; Ayed et al., 2007). These studies show the relevance of the approach and open new perspectives of development. Three important questions are prerequisites for the development of general numerical codes for the modeling of transfers in gas-liquid systems bubbly systems. The first question is related to the modeling of the transfer coefficient, the second issue concerns the turbulence modeling in two-phase gas-liquid flows and the third relates to the prediction of the local distribution of the interfaces, this includes the prediction of the local void fraction and of the bubbles size distributions.

a) Gas-liquid mass transfer coefficients

The formulation of gas-liquid mass transfer coefficients involves local time scales in relation with the mechanisms that control the interfacial transfer. The earliest models still commonly used were proposed by Dankwerts (1951) and Hygbie (1935). For instance the Hygbie model involves time scale related to the bubble displacement. If the formulation of the transfer coefficient is theoretically acceptable for simple two-phase flow situations (spherical bubbles in free ascent in a liquid at rest), the effects of turbulence, the deformation of bubbles and the effects of the contaminants are far to be completely controlled.

A significant progress in the study and modeling of the gas-liquid flows has been achieved and many experimental studies have been carried out during the last decades. The experimental data are used in order to provide suitable correlations of the transfer rate (Aisa et al., 1981; Wanninkhof & Gills, 1999). On the other hand basic experiments had established a local description of the interfacial mass transfer at the gaz-liquid horizontal interface in homogeneous turbulence generated by micro-jets, (George et al., 1994) or by oscillating grid, (Reidel et al., 2004). The characteristic scales of transfers in turbulent gas-liquid systems depend on the scales of turbulence in general, but also depend on the specific scales that characterize the bubble and its movement. The problem is more complex considering that the average and fluctuating fields of gas and of the liquid phases are disturbed by the interfacial exchange.

b) Turbulence in gas-liquid bubbly flows

The available experimental results in gas liquid bubbly flows indicate the important effect of the interfacial interactions between phases on the turbulence structure of the liquid phase. The presence of the dispersed phase in the flow alters considerably the liquid turbulence structure by affecting the whole of the turbulence mechanisms e.g. diffusion, production, dissipation, redistribution (Lance & Bataille, 1991).

In particular, the experiments of homogeneous turbulence with a constant shear (Lance et al., 1991) show that the bubbles, in their random movements, induce a supplementary stretching of the turbulent eddies that leads to a more isotropy of the turbulent fluctuations with a reduction of the turbulent shear. With regard to turbulence modeling, this experimental result is of great consequence: it suggests that we have to go up to turbulence closure level so that we may take into account the effect of the bubbles on the redistribution mechanism. In order to attain this objective, second order turbulence modeling is required and in several recent two-fluid models, the turbulence closures are effectively modeled using second order closure modeling (Lance et al. 1991; Chahed et al., 2003; Zhou, 2001; Lopez de Bertodano et al., 1990).

On the other hand, the turbulence structure of the liquid phase was pointed to have an important role in the phase distribution (Drew & Lahey, 1982); more recently it has been proved that the turbulent contributions of momentum interfacial transfer are important in the phase distribution phenomena, (Chahed et al., 2002). Thus one of the previous questions in the elaboration of efficient prediction tools is the accurate predetermination of the turbulence stress tensors in both liquid and gas phases.

c) Void fraction and bubble size distribution in turbulent bubbly flows

Gas - liquid reactors used in the industrial domain often bring about multiphase dense flows (high void fraction in bubbly flows) where transfer and transformation phenomena occur. In these reactors, often with complex geometries, the distribution of the phases is an important factor: transfer and transformation phenomena are in direct relation with the exchange between phases, therefore with the rate of presence of the interfacial area that materializes the contact between the two phases. Indeed, the interfacial area measures the contacting surface by unit volume and it depends on the distribution of the rate of presence of the phases and on the distribution of the sizes of the bubbles.

3.2 Two-fluid model for turbulent bubbly flows : basic equations and closure issues

All of the considerations presented above led to admit that the development of adequate models of transfers in gas-liquid systems should proceed from a comprehensive approach of two-phase flow modeling. With the current progress in the modeling of turbulent gas-liquid two-phase flows, we may expect that a local phenomenological approach of the problem may be useful for developing more suitable modeling of gas-liquid reactors.

Eulerian two-fluid models are based on a classical averaging of the balance equations that express in each phase the mass and the momentum conservation. We consider turbulent gas-liquid bubbly flow with low solubility of the gas in the liquid so that we neglect the effect of the mass transfer on the dynamic of the gas-liquid flow. We consider that without breakup and coalescence, the bubble diameter is still roughly constant.

We note α the void fraction and $\overline{\overline{\mathbf{u}_L}}$, $\overline{\overline{p_L}}$, ρ_L respectively the average liquid velocity, the pressure and the density of the liquid phase. $\mathbf{g}$ is the gravity acceleration and $\overline{\overline{\mathbf{u}_R}} = \overline{\overline{\mathbf{u}_G}} - \overline{\overline{\mathbf{u}_L}}$

is the relative velocity of the bubbles. With the subscript G the variables are related to the gas phase. We consider stationary incompressible bubbly flows and we neglect the effect of the mass transfer on the dynamic of the gas-liquid flow, the averaged balance equations of mass and momentum in the liquid and in the gas are, (Chahed et al., 2003):

$$\nabla \bullet (1-\alpha)\overline{\overline{\mathbf{u}_L}} = 0 \qquad \nabla \bullet \alpha\overline{\overline{\mathbf{u}_G}} = 0 \tag{12}$$

$$(1-\alpha)\rho_L \frac{D}{Dt}\overline{\overline{\mathbf{u}_L}} = -\nabla \overline{\overline{p_L}} - \nabla \bullet ((1-\alpha)\rho_L \overline{\overline{\mathbf{u}_L^{'}\mathbf{u}_L^{'}}}) + (1-\alpha)\rho_L \mathbf{g} \tag{13}$$

$$0 = -\alpha\nabla \overline{\overline{p_L}} + \mathbf{M}_G \tag{14}$$

$$\begin{aligned}\mathbf{M}_G = &-\frac{3}{4}\rho_L \frac{C_D}{d}\left\|\overline{\overline{\mathbf{u}_R}}\right\|\overline{\overline{\mathbf{u}_R}} - 2\rho_L C_L (1 - f_{LP}(y^{**}))\overline{\overline{\omega_L}} \times \overline{\overline{\mathbf{u}_R}} \\ &- \rho_L C_A \left(\frac{d}{dt}\overline{\overline{\mathbf{u}_G}} - \frac{D}{Dt}\overline{\overline{\mathbf{u}_L}}\right) - \rho_L C_A \frac{1}{\alpha}\nabla.\alpha(\overline{\overline{\mathbf{u}_G^{'}\mathbf{u}_G^{'}}} - \overline{\overline{\mathbf{u}_L^{'}\mathbf{u}_L^{'}}})\end{aligned} \tag{15}$$

Where $\frac{D}{Dt} = \frac{\partial}{\partial t} + (\overline{\overline{\mathbf{u}_L}}.\nabla)$, $\frac{d}{dt} = \frac{\partial}{\partial t} + (\overline{\overline{\mathbf{u}_G}}.\nabla)$

In the equation of the momentum balance, the acceleration and the weight of the gas are neglected in comparison to the force exerted by the liquid on the bubbles $\rho_G << \rho_L$; so the total force exerted on the bubbles is zero as indicated in equation (14). This force contains the non disturbed flow action (pressure term or Tchen force) and the interfacial term $\mathbf{M}_G$.
In the common formulation of the momentum interfacial exchange only the contributions due the average velocities fields of the liquid and the gas phases is considered while the turbulent contributions of the interfacial force are ignored or eventually expressed via a supplementary dispersion term proportional to the void fraction gradient (Lance & Lopez de Bertonado, 1992). In their model Chahed et al. (2002) proposed to take into account beside the average contributions of the interfacial transfer, the turbulent correlations issued from the added mass force and they proved that these turbulent correlations are important in the phase distribution phenomenon, in particular in gas-liquid flow under micro-gravity condition. According to their formulation, the interfacial momentum transfer (4) includes respectively the drag force (drag coefficient C_D), the added mass force (coefficient C_A) that includes the average and turbulent contributions in the liquid and in the gas and the lift force (coefficient C_L) that is expressed with a modified lift coefficient taking into account an eventual wall interaction effect characterised by the function $f_{LP}(y^*)$ where $y^* = \frac{2y_p}{d}$ is the distance y_p from the wall normed by the bubble radius. Considering the expression of the wall force in laminar flow by Antal et al. (1991), a formulation of the function $f_{LP}(y^*)$ is proposed Chahed & Masbernat (2000).
The two-fluid model presented requires closure of the turbulent stress tensors in the liquid and in the gas. The turbulent stress tensor of the gas is related to that of the liquid through a turbulent dispersion model and the turbulent stress tensor of the liquid is computed using a

second order closure of the turbulence developed for bubbly flows, (Chahed & Masbernat, 2003)

3.3 Turbulence modeling

The turbulence modeling is based on the decomposition of the Reynolds stress tensor of the continuous phase into two independent parts: a turbulent part $\overline{\overline{\mathbf{u}'_L\mathbf{u}'_L}}^{(T)}$ produced by the gradient of the mean velocity which also contains the turbulence generated in the bubble's wake (where an equilibrium production-dissipation is assumed) and an irrotational part $\overline{\overline{\mathbf{u}'_L\mathbf{u}'_L}}^{(S)}$ induced by bubbles displacements and controlled by the added mass effects. Each part is computed by a transport equation:

$$\frac{D}{Dt}\overline{\overline{\mathbf{u}'_L\mathbf{u}'_L}}^{(S)} = Diff(\overline{\overline{\mathbf{u}'_L\mathbf{u}'_L}}^{(S)}) + \frac{3}{20}\frac{D}{Dt}\alpha\left\|\overline{\overline{\mathbf{u}_R}}\right\|^2\delta + \frac{1}{20}\frac{D}{Dt}\alpha\overline{\overline{\mathbf{u}_R\mathbf{u}_R}} \tag{14}$$

$$\frac{D}{Dt}\overline{\overline{\mathbf{u}'_L\mathbf{u}'_L}}^{(T)} = Diff(\overline{\overline{\mathbf{u}'_L\mathbf{u}'_L}}^{(T)}) - 2sym\left[\overline{\overline{\mathbf{u}'_L\mathbf{u}'_L}}^{(T)} \bullet \nabla\overline{\overline{\mathbf{u}_L}}\right] + \Phi - \varepsilon\delta \tag{15}$$

In comparison to second order turbulence modeling in single phase flow, the diffusion and redistribution terms in the transport equation of the turbulent part were modified in order to take into account the interfacial effects. These effects are related to a time scale τ_b that characterizes the relative displacement of the bubbles.

$$Diff(\Psi) = \frac{C_{S\Psi}}{1-\alpha}\nabla\left[(1-\alpha)(\tau_t\overline{\overline{\mathbf{u}'_L\mathbf{u}'_L}}^{(T)} + \tau_b\overline{\overline{\mathbf{u}'_L\mathbf{u}'_L}}^{(S)})\nabla\Psi\right] \tag{16}$$

with

$$\tau_t = \frac{trace(\overline{\overline{\mathbf{u}'_L\mathbf{u}'_L}}^{(T)})}{2\varepsilon}, \quad \tau_b = C_R\frac{d}{\left\|\overline{\overline{\mathbf{u}_R}}\right\|}$$

$$\Phi^{(NL)} = -C_1(\tau_t^{-1} + \alpha\tau_b^{-1})\left[\overline{\overline{\mathbf{u}'_L\mathbf{u}'_L}}^{(T)} - \frac{1}{3}trace(\overline{\overline{\mathbf{u}'_L\mathbf{u}'_L}}^{(T)})\right] \tag{17}$$

Where $\Phi^{(NL)}$ is the non-linear part of the redistribution term Φ ($\Phi = \Phi^{(NL)} + \Phi^{(L)}$). The linear part $\Phi^{(L)}$ and the dissipation rate ε are modeled as in single-phase flow. The reduction of the second order closure of the turbulence furnishes the following turbulent viscosity in bubbly flows, (Chahed et al., 2003):

$$\nu_t = \nu_{t0}\frac{(1+\frac{C_{\mu b}}{C_{\mu 0}}\frac{k_S}{k_0})}{(1+\alpha\frac{\tau_t}{\tau_b})} \tag{18}$$

Where $\nu_{t0} = C_{\mu 0}\frac{k_0^2}{\varepsilon}$ is the equivalent single-phase flow turbulent viscosity, $k_0 = \frac{trace(\overline{\overline{\mathbf{u}'_L \mathbf{u}'_L}}^{(T)})}{2}$ and $k_S = \frac{trace(\overline{\overline{\mathbf{u}'_L \mathbf{u}'_L}}^{(S)})}{2}$ are respectively the turbulent and the pseudo-turbulent parts of the kinetic turbulent energy in the liquid phase. The two coefficients $C_{\mu 0}$ and $C_{\mu b}$ depend on the turbulence and the pseudo-turbulence anisotropy. The equation (18) expresses two antagonist interfacial effects on the turbulent viscosity: the bubbles agitation induces in one hand an enhancement of the turbulent viscosity and on the other hand a modification of the characteristic scale of the eddies stretching which can reduces the shear stress.

3.4 Transport equation of concentration

Recall that we consider gas-liquid bubbly flow with low solubility of the gas in the liquid. As a result the effect of the mass transfer on the dynamic of the gas-liquid flow is neglected. Without the source term associated with external inputs or reactive exchanges, we consider that the mass transfer is limited to mass flux through the gas-liquid interface we can write, in these conditions, a transport equation of a concentration of the liquid in general form:

$$\frac{DC}{Dt} = Diff(C) + S_c \tag{19}$$

Where S_c is transfer term through the interface given by equation (11)

3.5 Some results and discussion: Requirements for advanced computation of gas-liquid contacting systems

a) Turbulence modeling

The second order turbulence model presented above predicts correctly the large enhancement of the momentum diffusivity observed in bubbly flow with an important amount of pseudo-turbulence as in mixing layer and wake bubbly flows, (Chahed et al., 2003). On the other hand, the model reproduces the turbulence structure as it is altered by the bubble presence in the boundary layer bubbly flows, figure (4) and pipe bubbly flows figure (5). Figure (4) shows the profiles of the turbulent intensity in the liquid phase produced by the two-fluid model in near wall boundary layer bubbly flow for single phase and two-phase bubblys flows. These profiles are compared to the experimental data of Moursali et al. (1995). Note that y^+ denoted the adimensional distance to the wall $y^+ = \frac{u_* d}{\nu}$ where u_* is the friction velocity, d distance to the wall and ν is the kinematic viscocity of the liquid. Figure (5) shows the profiles of the turbulent shear stress produced by the two fluid model in a pipe single phase and bubbly flows. The numerical results are compared to the experimental data of Serizawa (1992). The results of all of these simulations make clear the mechanisms whereby the attenuation of the turbulence in bubbly flows can occurs: the supplementary stretching induced by the bubbles displacements provokes an attenuation of the shear stress, as a result the turbulence production by the mean velocity gradient is reduced and we can note a diminution of the turbulent intensity as observed in some experiences of wall-bounded bubbly flows.

The decomposition of the Reynolds stress tensor in a turbulent and pseudo-turbulent contributions with specific transport equation for each part makes possible the computation of the specific scales involved in each part. The determination of these scales allows to describe correctly the different effects of the bubbles agitation on the liquid turbulence structure.

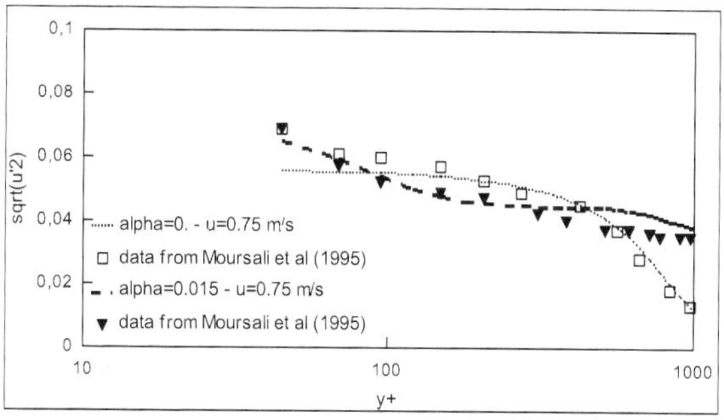

Fig. 4. Turbulent intensity in single-phase and bubbly boundary layer.

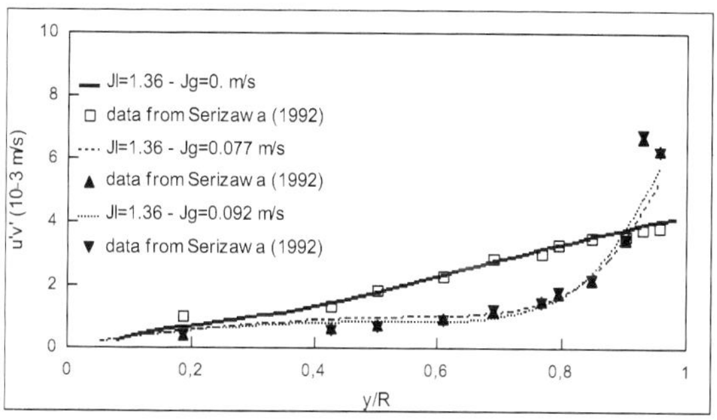

Fig. 5. Turbulent shear stress in single-phase and bubbly pipe flows.

If from a theoretical point of view, second order is an adequate level for turbulence closure in bubbly flows, the implementation of such turbulence models in two-fluid models clearly improves the predetermination of the turbulence structure in different bubbly flow configurations, (Chahed et al., 2002, 2003). Nevertheless, from a practical point of view, second order modeling is still difficult to use and turbulence models based on turbulent viscosity concept, particularly two-equation models, remain widely used in industrial applications. Several two-equation models were developed for turbulent bubbly flows (Lopez de Bertodano et al., 1994; Lee et al., 1989; Morel, 1995; Troshko & Hassan, 2001). All of these models are founded on an extrapolation of single-phase turbulence models by introducing supplementary terms (source terms) in the transport equations of turbulent energy and dissipation rate. In some models, the turbulent viscosity is split into two contributions according to the model of Sato et al. (1981): a "turbulent" contribution induced by shear and a "pseudo-turbulent" one induced by bubbles displacements. To adjust the turbulence models some modifications of the conventional constants are sometimes proposed (Lee et al., 1989; Morel, 1995).

The reduction of second order turbulence modeling developed for two-phase bubbly flows furnish an interpretation of second order turbulence closure in term of turbulent viscosity model. On the basis of this turbulent viscosity model, two-equation turbulence models (k-ε model, (Chahed et al., 1999) and k-ω model (Bellakhel et al., 2004) were developed and applied to homogeneous turbulence in bubbly flows (uniform and with a constant shear). The numerical results clearly show that the model reproduces correctly the effect of the bubbles on the turbulence structure.

The turbulent viscosity formulation (18) keeps the essential of the physical mechanisms involved in second order turbulence modeling. It expresses two antagonist interfacial effects due to the presence of the bubbles on the turbulent shear stress of the liquid phase: the bubbles agitation induces in one hand an enhancement of the turbulent viscosity as compared to and on the other hand a modification of the eddies stretching characteristic scale that causes more isotropy of the turbulence with an attenuation of the shear stress. According as the amount of pseudo-turbulence is important or not, we can expect an increase or a decrease of the turbulent viscosity. As a result, the turbulent shear stress in bubbly flow can be more or less important than the corresponding one in the equivalent single-phase flow. In the case where the turbulent shear stress is reduced, the turbulence production by the mean velocity gradient is lower and we can reproduce, under certain conditions, an attenuation of the turbulence as observed in some wall bounded bubbly flows (Liu and Bankoff, 1990; Serizawa et al., 1992).

Void fraction and bubbles size distributions

The distribution of void fraction is governed by the interfacial forces exerted by the continuous phase on the bubbles as they move throughout the liquid. We have to specify the contributions of the average and fluctuating flow fields to this force. Numerical simulations of upward pipe bubbly flow in micro-gravity and in normal gravity conditions show clearly the role of the turbulence and of the interfacial forces on the void fraction distribution, (Chahed et al., 2002). These numerical simulations are compared to the experimental data of Kamp. et al. (1994). An important result of these experiences is to show that the radial void fraction gradient is inverted according as the gravity is active or not (according as the interfacial momentum transfer associated with the average relative velocity is important or not). Figure (5) shows the profile of void fraction in pipe upward and downward bubbly flows in microgravity and in normal gravity conditions. In micro-gravity condition, the average relative velocity between phases is weak; thus the action of the continuous phase on the bubbles is reduced to the pressure gradient effect (Tchen force) and to the turbulent contributions of the interfacial force. The pressure gradient effect provokes a bubble migration toward the wall and can't explain the experimental void fraction profile. When the turbulent terms issued from the added mass force are introduced, the whole action of turbulence is inverted and the phase distribution prediction is in good agreement with the experimental data.

This result indicates that the effect of the continuous phase turbulence on the phase distribution includes, beside the pressure gradient action (Tchen force), the turbulent contributions of the interfacial forces. Consequently, the accuracy in the predetermination of the turbulence of the dispersed phase is also for importance in the computation of the void fraction distribution. The turbulent stress tensor of the dispersed phase can be related to the liquid one through a turbulent dispersion models, (Hinze, 1975; Csanady, 1963). The recent results issued from numerical simulations can be viewed as a prelude to more progress in this direction.

As compared to the void fraction profile in micro-gravity condition, the prediction of the void fraction distribution in upward and downward bubbly flows in normal gravity conditions clearly shows the effect of the lift force. In upward flow, the lift force is responsible of the near-wall void fraction peaking while in downward flow, the lift force action is inverted and the migration of the bubble toward the centre of the pipe provoked by the global turbulent action is more pronounced than in micro-gravity condition. The adjustment of the coefficients in the expression of the near wall lift force was tested in boundary layer bubbly flow ($\overset{=}{u} = 0.75 m/s$ and $\overset{=}{u} = 1 m/s$) with bubble's diameter between 2.3 and 3.5 mm (the more is the external void fraction the more is the bubble diameter); in these simulations the diameter of the bubbles was adjusted from the experimental data of Moursali et al. (1995). It yields C_L =0.08, y_1^* =1 and y_2^* =1.5. These computations allow us to consider that these coefficients could have a somewhat general character. The value of y_1^* suggests that the position of the void fraction peaking is, for the most part, controlled by lift and wall forces: its value corresponds to the void fraction peaking position observed in the experiences.

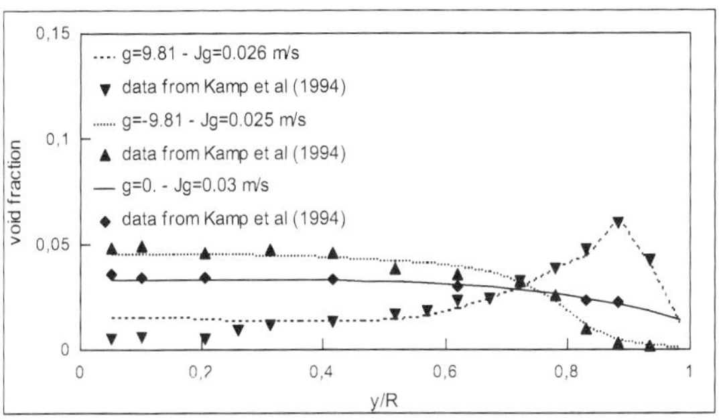

Fig. 5. Void fraction distribution in pipe bubbly flows : upward - downward and in micro-gravity conditions. Data from Kamp et al. (1995)

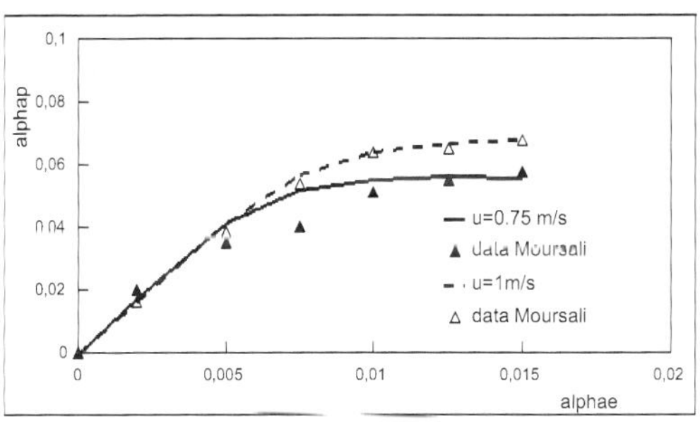

Fig. 6. Amplitude of the near wall void fraction peaking as a function of the external void fraction in boundary layer bubbly flow.

Figure (6) shows that the less is the bubble diameter the more is amplitude of the void fraction peaking near wall. This result is well reproduced by the model for millimetric bubbles: the lift force formulation including the wall effect brings implicitly into account the bubble size. When the bubble's size becomes greater and its shape deviates severely from the sphericity the expression of the force exerted by the liquid should be reviewed. Also on this point, we can expect some progress issued from the numerical simulation. On the other hand.

4. Conclusion

Many industrial processes in chemical, environmental and power engineering employ gas-liquid contacting systems that are often designed to bring about transfer and transformation phenomena in two-phase flows. As for all gas-liquid contacting systems, flotation devices bring into play gas-liquid bubbly flows where the interfacial interactions and exchanges determine not only the dynamics of the system but are, in the same time, the technological reason of the process itself. When applied to flotation, mass transfer approach turns out to be very convenient for representing various behaviors of the flotation kinetics. It allows a more phenomenological approach in the analysis of the interfacial phenomena involved in the flotation process.

From a practical point of view, the development of general models which are able to predict the fields of certain average kinematic properties of both gas and liquid phases and their presence rates in two-phase flows is of great interest for the design, control and improvement of gas-liquid contacting systems. From the scientific point of view, the modeling and simulation of gas-liquid flows set many important questions; in particular the ability to predict the phase distribution in gas-liquid bubbly flows remains limited by the inadequate modeling of the turbulence and of the interfacial forces. Especially in industrial gas-liquid systems characterized by various additional complexities such as : the geometry of the reactor, the hydrodynamic interactions particularly in dense gas-liquid flows (high void fraction), the chemical reactivity, the interfacial area modulation due to the phenomena of rupture and coalescence... All of these issues require new original experiments in order to sustain the modeling effort that aims at developing more general closures for advanced Computational Fluid Dynamics of complex gas-liquid systems.

5. References

Ahmed N. & Jameson G.J. (1985). The effect of bubble size on the rate of flotation of fine particles, *Int. J. of Mineral Processing*, Vol 14, pp. 195-215

Aisa L.; Caussade B.; George J. & Masbernat L. (1981), Echange de gaz dissous en écoulements stratifiés de gaz et de liquide : International Journal of Heat and Mass Transfer vol 24 pp 1005–1018.

Antal S.P; Lahey JR & Flaherty J.E. (1991). Analysis of phase distribution in fully developed laminar bubbly two-phase flow, *Int. J. Multiphase Flow*, 5, pp. 635-652

Ayed H.; Chahed J. & Roig V. (2007). Experimental analysis and numerical simulation of hydrodynamics and mass transfer in a buoyant bubbly shear layer, *AIChE Journal*, 53 (11), pp. 2742-2753

Buscaglia G. C.; Bombardelli F. A. & Garcia M., 2002. Numerical modeling of large scale bubble plumes accounting for mass transfer effects, *Int. J. of Multiphase Flow*, 28, 1763-1785

Bellakhal G.; Chahed J. & Masbernat L. (2004). Analysis of the turbulence structure in homogeneous shear bubbly flow using a turbulent viscosity model, *Journal of Turbulence*, Vol. 5, N°36

Chahed J.; Masbernat L. & Bellakhel G. (1999). k-epsilon turbulence model for bubbly flows, *2nd Int. Symposium On Two-Phase flow Modelling and Experimentation*, Pisa, Italy, May 23-26

Chahed J. & Masbernat L. (2000). Requirements for advanced Computational Fluid Dynamics (CFD) applied to gas-liquid reactors, P*roc. of the Int. Specialized Symp. on*

Fundamentals and Engineering Concepts for Ozone Reactor Design. INSA, Toulouse 1-3 Mars, pp. 307-310

Chahed J.; Colin C. & Masbernat L. (2002) Turbulence and phase distribution in bubbly pipe flow under micro-gravity condition", *Journal of Fluids Engineering*, Vol. 124, pp. 951-956

Chahed J.; Roig V. & Masbernat L. (2003). Eulerian-eulerian two-fluid model for turbulent gas-liquid bubbly flows. *Int. J. of Multiphase flow*. Vol. 29, N°1, pp. 23-49

Csanady G.T. (1963). Turbulent diffusion of heavy particles in the atmosphere" *J. Atm. Sc*, Vol. 20, pp. 201-208

Chahed J. & Mrabet K. (2008). Gas-liquid mass transfer approach applied to the modeling of flotation in a bubble column, *Chem. Eng. Technol*, 31 N°9 pp.1296-1303

Cockx A.; Do-Quang Z., Audic J.M.; Liné A. & Roustan M. (2001). Global and local mass transfer coefficients in waste water treatment process by computational fluid dynamics. *Chemical Engineering and Processing*, Vol. 40, pp. 187-194.

Dankwerts, P.V. (1951). Significance of liquid-film coefficients in gas absorption. *Ind. Eng. Chem.*, Vol. 43, pp. 1460-67

Drew D.A. & Lahey R.T (1982) Phase distribution mechanisms in turbulent low-quality two-phase flow in circular pipe, *J. Fluid Mech.*, Vol. 117, pp. 91-106.

Finch J. A. (1995). Column flotation: a selected review. Part IV: novel flotation devices, *Minerals Engineering*, Vol. 8, N° 6, pp. 587-602

George J.; Minel F. & Grisenti M. (1994). Physical and hydrodynamical parameters controlling gas-liquid mass transfer: *J. Fluid Mechanics*, Vol. 37 pp. 1569-1578.

Gorain B. K.; Franzidis J. P. & Manlapig E. V. (1997). Studies on impeller type, impeller speed and air flow rate in an industrial scale flotation cell. Part 4: Effect of bubble surface area on flotation performance, *Minerals Engineering*, Vol. 10, N° 4, pp. 367-379

Higbie, R. (1935). The rate of absorption of a pure gas into a still liquid during short periods of exposure. *Trans. A.I.Ch.E*, Vol 31, pp. 365-388.

Hinze J. O. (1995). Turbulence, *2nd edition, Mc Graw-Hill*,

Ityokumbul M.T. (1992). A mass transfer approach to flotation column design, *Chemical Engineering Science*, Vol. 13, N° 14, pp. 3605-3612

Jameson G. J.; Nam S. & Moo-Young M. (1977). Physical factors affecting recovery rates in flotation. *Miner. Eng. Sci.* Vol. 9, pp. 103-118

Kamp A.; Colin C. & Fabre J. (1995). The local structure of a turbulent bubbly pipe flow under different gravity conditions, *Proceeding of the Second International Conference on Multiphase Flow*, Kyoto, Japan

Lain S.; Bröder D. & Sommerfeld M. (1999). Experimental and numerical studies of the hydrodynamics in a bubble column, *Chemical Engineering Science*, Vol. 14, pp. 4913-4920

Lance M. & Bataille J. (1991). Turbulence in the liquid phase of a uniform bubbly air water flow, *J. Fluid Mech.*, Vol. 222, pp. 95-118.

Lance M.; Marié J.L. & Bataille J. (1991). Homogeneous turbulence in bubbly flows, *J. Fluids Eng.*, 113, pp. 295-300

Lance M. & Lopez de Bertonado M. (1992). Phase distribution phenomena and wall effects in bubbly two-phase flows, *Third Int. Workshop on Two-Phase Flow Fundamentals*, Imperial College, London, June 15 -19

Liu T.J. & Bankoff S.G. (1990). Structure of air-water bubbly flow in a vertical pipe : I- Liquid mean velocity and turbulence measurements", *Int. J. Heat and Mass Transfer*, vol. 36 (4) pp. 1049-1060

Lopez de Bertodano M.; Lee S. J.; Lahey R. T. & Drew D. A. (1990). The prediction of two-phase turbulence and phase distribution using a Reynolds stress model, *J. of Fluid Eng.*, Vol. 112, pp. 107-113.

Lopez de Bertodano M.; Lee S.J. & Lahey R.T., Jones. O. C. (1994). Development of a k-ε model for bubbly two-phase flow, J. *Fluids Engineering*, Vol. 116, pp. 128-134.

Lee S.J.; Lahey Jr R.T & Jones Jr O.C. (1989). The prediction of two-phase turbulence and phase distribution phenomena using kε model, *Japanese J. of Multiphase Flow*. Vol. 3, pp. 335-368.

Morel C. (1995). An order of magnitude analysis of the two-phase k-ε model, *Int. J. of Fluid Mechanics Research*, Vol. 22 N° 3&4, pp. 21-44.

Moursali E., Marié J.L. & Bataille J. (1995). An upward turbulent bubbly layer along a vertical flat plate, *Int. J. Multiphase Flow*, Vol. 21 N°1, pp. 107-117

Nguyen A. V. (2003). New method and equations determining attachment and particle size limit in flotation, *Int. J. Miner. Process*, Vol. 68, pp. 167-183

Reidel, Boston, McKenna S.P. & Mc Gillis W.R. (2004) : The role of free-surface turbulence and surfactants in air–water gas transfer: *International Journal of Heat and Mass Transfer*, Vol. 47, pp. 539–553.

Rivero M.; Magnaudet J. & Fabre J. (1991). Quelques résultats nouveaux concernant les forces exercées sur une inclusion sphérique par un écoulement accéléré, *C. R. Acad. Sci. Paris*, t.312, serie II, pp. 1499-1506

Serizawa A.; Kataoka I. & Michiyoshi I. (1992). Phase distribution in bubbly flow. *Multiphase Science and Technology*, Vol. 6, Hewitt G. F. Delhaye, J. M., Zuber, N., Eds, *Hemisphere Publ. Corp.*, pp. 257-301.

Sutherland K. L. (1948). Physical chemistry of flotation XI. Kinetics of the flotation process, *J. Phy. Chem.*, Vol. 52, pp. 394-425

Sato Y.; Sadatomi L. & Sekouguchi K. (1981). Momentum and heat transfer in two phase bubbly flow, *Int. J. Multiphase Flow*, Vol. 7, pp. 167-190.

Troshko A. A. & Hassan Y. A. (2001). A two-equation turbulence model of turbulent bubbly flows, *Int. J. Multiphase Flow*, Vol. 27, pp. 1965-2000.

Tuteja R.K.; Spottiswood D.J. & Misra V.N. (1994). Mathematical models of the column flotation process, a review, *Minerals Engineering*, Vol. 7, N°12, pp. 1459-1472

Wanninkhof; R. & McGillis W. R. (1999), A cubic relationship exchange and wind speed. Geophysical Research Letters, Vol. 26, N° 13, pp.1889-1892.

Yachausti R. A.; McKay J. D. & Foot Jr. D. G. (1988). Column flotation parameters - their effects. *Column flotation '88 (K. V. S. Sasty ed.)*, Society of Mining Engineers, Inc. Littleton, CO, pp. 157-172

Yoon R. H.; Mankosa M. J. & Luttrel G. H. (1993). Design and scale-up criteria for column flotation, *XVII International Mineral Processing Congress*, Sydney, Austria. pp. 785-795

Zongfu D.; Fornasiero D. & Ralston J. (2000). Particle bubble collision models - a review, *Advances in Collid and Interface Science*, Vol. 85, pp. 231-256

Zhou, L. X. (2001). Recent advances in the second order momentum two-phase turbulence models for gas-particle and bubble-liquid flows, 4th *International Conference on Multiphase Flow*, paper 602, New Orleans.

8

Mass Transfer in Two-Phase Gas-Liquid Flow in a Tube and in Channels of Complex Configuration

Nikolay Pecherkin and Vladimir Chekhovich
Kutateladze Institute of Thermophysics, SB RAS
Russia

1. Introduction

Successive and versatile investigation of heat and mass transfer in two-phase flows is caused by their wide application in power engineering, cryogenics, chemical engineering, and aerospace industry, etc. Development of new technologies, upgrading of the methods for combined transport of oil and gas, and improvement of operation efficiency and reliability of conventional and new apparatuses for heat and electricity production require new quantitative information about the processes of heat and mass transfer in these systems. At the same time necessity for the theory or universal prediction methods for heat and mass transfer in the two-phase systems is obvious.

In some cases the methods based on analogy between heat and mass transfer and momentum transfer are used to describe the mechanism of heat and mass transfer. These studies were initiated by Kutateladze, Kruzhilin, Labuntsov, Styrikovich, Hewitt, Butterworth, Dukler, et al. However, there are no direct experimental evidences in literature that analogy between heat and mass transfer and momentum transfer in two-phase flows exists. The main problem in the development of this approach is the complexity of direct measurement of the wall shear stress for most flows in two-phase system. The success of the analogy for heat and momentum transfer was achieved in the prediction of heat transfer in annular gas-liquid flow, when the wall shear stress is close to the shear stress at the interface between gas core and liquid film.

Following investigation of possible application of analogy between heat and mass transfer and hydraulic resistance for calculations in two-phase flows is interesting from the points of science and practice.

The current study deals with experimental investigation of mass transfer and wall shear stress, and their interaction at the cocurrent gas-liquid flow in a vertical tube, in channel with flow turn, and in channel with abrupt expansion. Simultaneous measurements of mass transfer and friction factor on a wall of the channels under the same flow conditions allowed us to determine that connection between mass transfer and friction factor on a wall in the two-phase flow is similar to interconnection of these characteristics in a single-phase turbulent flow, and it can be expressed via the same correlations as for the single-phase flow. At that, to predict the mass transfer coefficients in the two-phase flow, it is necessary to know the real value of the wall shear stress.

2. Analogy for mass transfer and wall shear stress in two-phase flow

2.1 Introduction

The combined flow of gas and liquid intensifies significantly the heat and mass transfer processes on the walls of tubes and different channels and increases pressure drop in comparison with the separate flow of liquid and gas phases.

According to data presented in (Kutateladze, 1979; Hewitt & Hall-Taylor, 1970; Collier, 1972; Butterworth & Hewitt, 1977; et al), the methods based on semi-empirical turbulence models and Reynolds analogy are the most suitable for convective heat and mass transfer prediction in two-phase flows. Their application assumes interconnection between heat and mass transfer and hydraulic resistance in the two-phase flow.

Several publications deal with experimental check of analogy between heat and mass transfer and momentum in the two-phase flows. Mass transfer coefficients in the two-phase gas-liquid flow in a horizontal tube are compared in (Krokovny et al., 1973) with mass transfer of a single-phase turbulent flow for the same value of wall shear stress. The mass transfer coefficient in vertical two-component flow was measured by (Surgenour & Banerjee, 1980). Wall shear stress was determined by pressure drop measurements. The experimental study for Reynolds analogy and Karman hypothesis for stratified and annular wave film flows is presented in (Davis et al., 1975). Experimental studies mentioned above prove qualitatively and, sometimes, quantitatively the existence of analogy between heat and mass transfer and wall shear stress.

The main difficulties in investigation of analogy between heat and mass transfer and friction are caused by the measurement of wall shear stress. Determination of friction by measurements of total pressure drop in the two-phase flow can give significant errors at calculation of pressure gradients due to static head and acceleration. Therefore, friction measurements require methods of direct measurement, which allow simultaneous measurement of heat and mass transfer coefficients. Among these methods there is the electrodiffusion method of investigation of the local hydrodynamic characteristics of the single-phase and two-phase flows (Nakoryakov et al., 1973, 1986; Shaw & Hanratty, 1977).

The current study presents the results of simultaneous measurement of mass transfer coefficients and wall shear stress for the cocurrent gas-liquid flow in a vertical tube within a wide alteration range of operation parameters.

2.2 Experimental methods

The experimental setup for investigation of heat and mass transfer and hydrodynamics in the two-phase flows is a closed circulation circuit, Fig. 1. The main working liquid of the electrochemical method for mass transfer measurement is electrolyte solution $K_3Fe(CN)_6 + K_4Fe(CN)_6 + NaOH$; therefore, all setup elements are made of stainless steel and other corrosion-proof materials. Liquid is fed by a circulation pump through a heat exchanger into the mixing chamber, where it is mixed with the air flow. Then, two-phase mixture is fed into the test section. Experiments were carried out with single-phase liquid and with liquid-air mixture in a wide alteration range of liquid and air flow rates and pressure. The test section is a vertical tube with the total length of 1.5 m, inner diameter of 17 mm, and it consists of the stabilization section, the section for visual observation of the flow, and measurement sections. The measurement sections are changeable. They have different design and they are made for investigation of mass transfer and wall shear stress in a straight tube. There is also section for heat transfer study, and the sections for mass transfer measurement in channels of complex configuration.

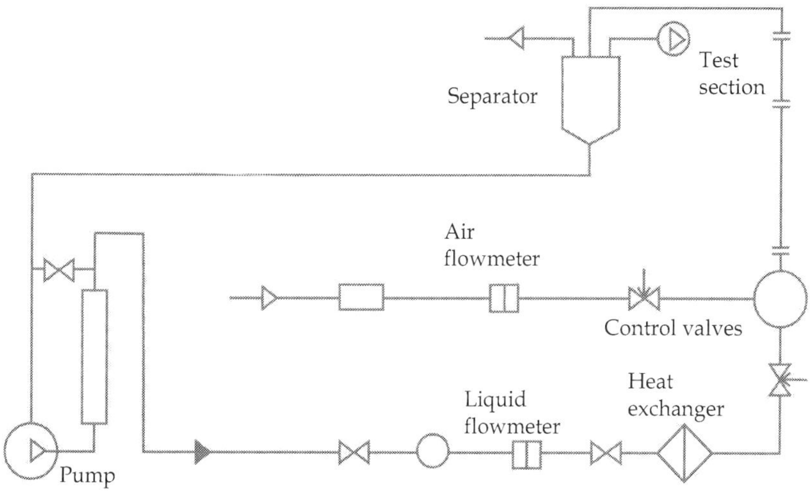

Fig. 1. Experimental setup

The method of electrodiffusion measurement of mass transfer coefficients is described in detail in (Nakoryakov et al., 1973, 1986). The advantage of this method is the fact that it can be used for the measurement of wall shear stress, mass transfer coefficient, and velocity of liquid phase only with the change in probe configuration. When this method is combined with the conduction method local void fraction in two phase flow can be measured. To determine the mass transfer coefficient is necessary to measure current in red-ox reaction $Fe(CN)_6^{3-} + e \Leftrightarrow Fe(CN)_6^{4-}$ on the surface of electrode installed on the wall, Fig. 2-1. The current in a measurement cell (cathode - solution - anode) is proportional to mass transfer coefficient (1)

$$I = kFSC_{\infty} \tag{1}$$

where k is mass transfer coefficient, S is area of probe surface; F is Faraday constant; and C_{∞} is ion concentration of main flow.

Connection between wall shear stress and current is determined by following dependence

$$\tau = A \cdot I^3 \tag{2}$$

where τ is wall shear stress, Pa; I is probe current; A is calibration constant.

Probes for wall shear stress measurements were made of platinum wire with the diameter of 0.3 mm, welded into a glass capillary, Fig. 2-2. The working surface of the probe is the wire end, polished and inserted flash into the inner surface of the channel. The glass capillary is glued into a stainless steel tube, fixed by a spacing washer in the working section. Friction probes were calibrated on the single-phase liquid. The probe for velocity measurements, Fig. 2-3, is made of a platinum wire with the diameter of 0.1 mm, and its size together with glass insulation is 0.15 mm. The incident flow velocity is proportional to the square of probe current $v \sim I^2$.

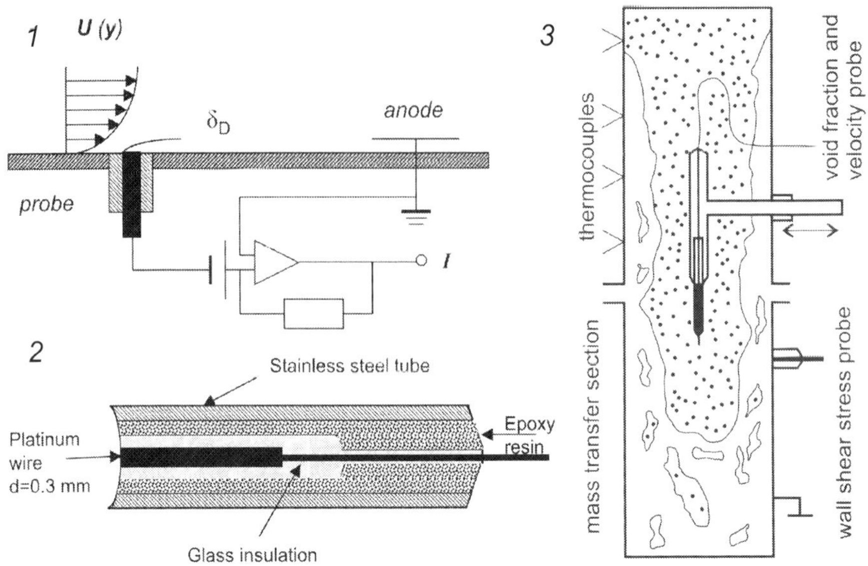

Fig. 2. The electrodiffusion method. *1* - electrochemical cell; *2* - probe for wall shear stress measurement; *3* - scheme of the test section for measurement of the mass transfer coefficient, wall shear stress and liquid velocity

To exclude the effect of entrance region and achieve the fully developed value of mass transfer coefficient, the probe for measurement mass transfer coefficient should be sufficiently long. Theoretical and experimental studies of (Shaw & Hanratty, 1977), carried out by the electrochemical method give the expression for dimensionless length of stabilization

$$L^+ \geq 1.9 \cdot 10^3 Sc^{1/4} \qquad (3)$$

where $L^+ = L\upsilon_*/\nu$, $Sc = \nu/D$ is Schmidt number, and L is probe length. According to (3), the length of mass transfer probe should be not less than 70-100 mm.

2.3 Wall shear stress in two-phase flow in a vertical tube

Experiments on mass transfer and hydrodynamics of the two-phase flow were carried out in the following alteration ranges of operation parameters:

V_{L0}	Superficial liquid velocity	0.5-3 m/s
Re_L	Reynolds number of liquid	8500-54000
G_G	Mass flow rate of air	0.6-35 g/s
Re_G	Reynolds number of air	3000-140000
V_{G0}	Superficial gas velocity at p =0.1 MPa	2-100 m/s
p	Pressure	0.1-1 MPa

Table 1. Experimental conditions

Measurement error for the main parameters: for liquid flow rate it is 2%, for air flow rate it is 4%, for mass transfer coefficient it is 4%, and for wall shear stress with consideration of friction pulsations it is10%.

Experiments were carried out in the slug, annular and dispersed-annular flows. The main purpose of investigations on hydrodynamics of the two-phase flows was measurement of wall shear stress under the same flow conditions as for mass transfer investigations. Moreover, measurement of friction at the flow of gas-liquid mixtures is of a particular interest because there are no direct measurements of local friction in the range of high void fraction for the vertical channels and direct measurement of wall shear stress at high pressures. The friction probe was located at the distance of 60 calibers from the inlet of the test section. There is no effect of stabilization zone length at this distance. The currents of friction and velocity probes were registered simultaneously, Fig. 3. The velocity probe serves simultaneously for void fraction measurement. It is located in the same cross-section of the test section as the friction probe. When this probe is in liquid, its readings correspond to the value of liquid phase velocity. The moments, when the probe current drops to zero, correspond to the gas phase pass.

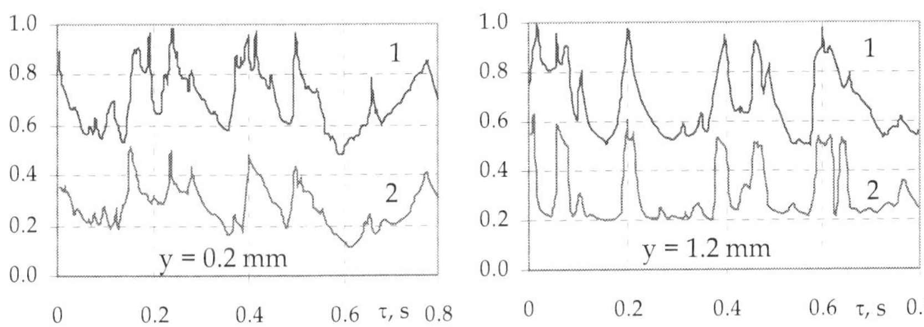

Fig. 3. Oscillograph tracings of wall shear stress (*1*) and liquid velocity in the film (*2*)

Oscillograms in Fig. 3 (left) correspond to distance from the wall y = 0.2 mm. In this position the velocity probe is in liquid during the whole measurement period; void fraction is zero. A synchronous change in the velocity of liquid in the film and wall shear stress is obvious. When the probe moves from the wall, void fraction in the flow core increases, and at the distance of 1–2 mm from the wall it becomes almost equal to one. Fig. 3 (right) corresponds to distance from the wall y = 1.2 mm. Here we can see rare moments, when the velocity probe is in liquid. These moments correspond to wave passing. At these particular moments, wall shear stress increases. Wave passing with simultaneous increase in wall shear stress causes an increase of velocity in a solid layer of the liquid film. Apparently, waves propagate over the film surface under the action of dynamic pressure of gas. The velocity of roll waves on the film surface will depend on wave amplitude and gas velocity. The motion of wave relative to the solid film layer will cause an increase in the velocity gradient in this layer. As a result, an additional shear stress appears on the wall, and it is observed in the form of friction pulsations. In the slug flow friction pulsations are caused by alternation of gas slugs and liquid plugs moving with the velocity of mixture. The level of wall shear stress pulsations depends on the flow conditions and void fraction, and it can reach the value of average friction for low flow rates of liquid. At maximal flow rates of liquid this value approaches the value typical for the single-phase turbulent flow.

Results on wall shear stress measurements under the atmospheric pressure are shown in Fig. 4. The effect of superficial velocities of liquid V_{L0} and gas V_{G0} is shown here. For constant superficial liquid velocities increase in the superficial gas velocity causes a nonlinear increase of wall shear stress, Fig. 4 (a). And for constant superficial gas velocities increase in the superficial liquid velocity results in increase of wall shear stress, Fig. 4 (b).

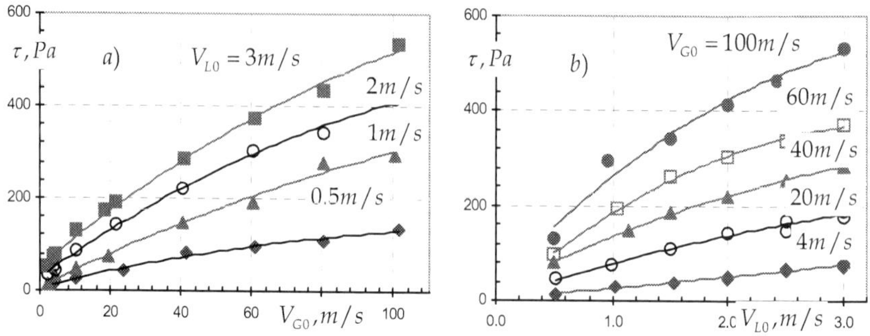

Fig. 4. The dependence of the wall shear stress on the gas superficial velocity (a), and on the liquid superficial velocity (b).

For all studied liquid flow rates at low superficial velocities of gas ($V_{G0} = 2-10$ m/s) wall shear stress depends weakly on pressure. At high velocities of air the effect of pressure on friction becomes significant. A change in pressure causes a change in following values: gas density ρ_G, mass flow rate $\rho_G V_{G0}$, and dynamic pressure $\rho_G V_{G0}^2$.

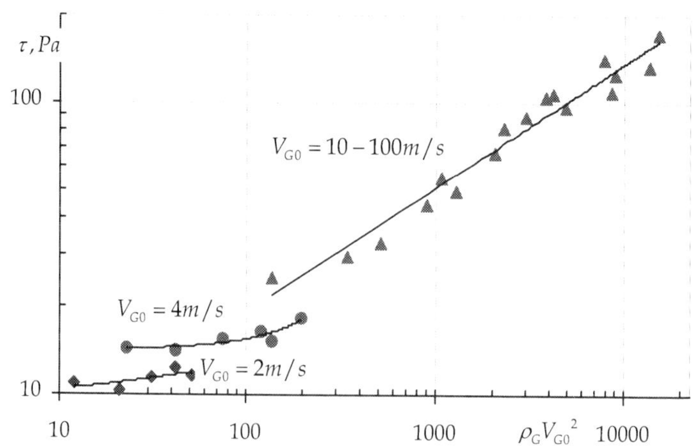

Fig. 5. The influence of the dynamic pressure on wall shear stress

According to analysis of data obtained, the effect of pressure on friction is weak in the bubble and slug flows, when liquid is continuous phase. Pressure effect is significant in the annular and dispersed-annular flows (high air velocities), when gas in flow core is continuous phase. In the last case the liquid film is thin; therefore, wall shear stress is almost equal to friction at the film interface, determined by dynamic pressure of gas, Fig. 5.

The well-known homogeneous model is the simplest model for pressure drop prediction in the two-phase flows. According to this model, the two-phase flow is replaced by the single-phase flow with parameters $\rho_{TP}, \mu_{TP}, V_{TP}$ without slipping between the phases. To determine viscosity of the two-phase mixture there are several relationships; however, since there is some liquid on the tube wall at the two-phase flow without boiling, it is more reasonable to use the liquid phase viscosity instead of μ_{TP}. Experimental data on wall shear stress in the two-phase gas-liquid flow divided by τ_0 - wall shear stress for flow liquid with velocity V_{L0} are shown in Fig. 6 (a) depending on the ratio of superficial velocities of phases. Calculation of relative wall shear stress by the homogeneous model is also shown there. The satisfactory agreement with calculation by the homogeneous model is observed.

Correlations (Lockhart & Martinelli, 1949) are widely used for prediction of pressure drop in two-phase flows. Processing of experimental data in coordinates of Lockhart-Martinelli is shown in Fig. 6 (b) for all studied pressures and liquid and gas flow rates. There is satisfactory agreement of experimental results with Lockhart-Martinelli correlation.

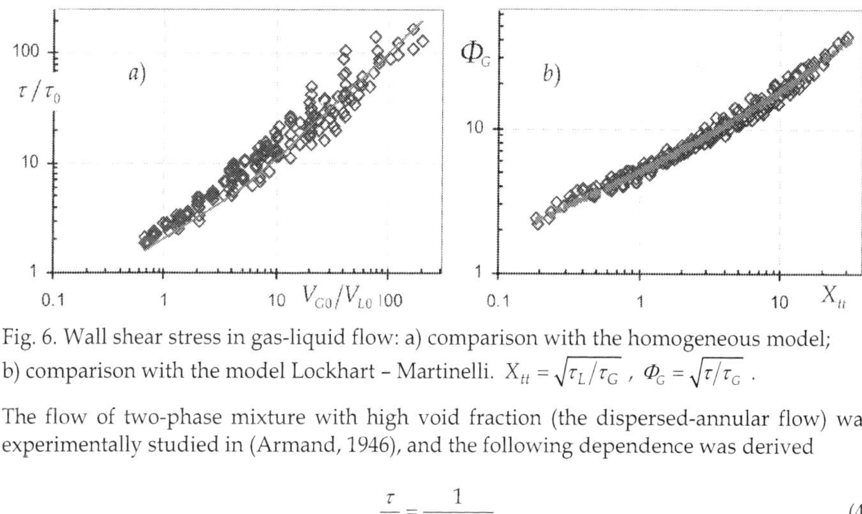

Fig. 6. Wall shear stress in gas-liquid flow: a) comparison with the homogeneous model; b) comparison with the model Lockhart - Martinelli. $X_{tt} = \sqrt{\tau_L/\tau_G}$, $\Phi_G = \sqrt{\tau/\tau_G}$.

The flow of two-phase mixture with high void fraction (the dispersed-annular flow) was experimentally studied in (Armand, 1946), and the following dependence was derived

$$\frac{\tau}{\tau_0} = \frac{1}{(1-\varphi)^n} \tag{4}$$

where τ_0 is friction in the single-phase flow; and φ is void fraction. Equation (4) was obtained with the assumption of the power law for the velocity distribution in the liquid phase. The friction factor in this case is determined by the Blasius equation with actual velocity of liquid phase. Results of our experiments show good agreement with this model.

However, there is a range of operation parameters at low velocities of liquid phase in the bubble flow regime, with an abnormal increase in friction on the tube wall, Fig. 8 (b). Wall shear stress in this area depends not only on the volumetric quality, but also on the distribution of gas bubbles in the cross section of the pipe. Mentioned above models do not predict wall shear stress in such regimes. Therefore, to check the analogy between heat and mass transfer and wall shear stress, it is necessary to measure the coefficients of heat and mass transfer and wall shear stress under the same conditions of the two-phase flow.

2.4 Mass transfer in gas-liquid flow in a vertical tube

Mass transfer on the tube wall at forced two-phase flow was studied by the electrochemical method. In this case mass transfer is identified with ion transfer carried out by the gas-liquid flow between the test electrode (cathode) and referencc electrodes (anode) in the electrochemical cell. In the diffusion limitation regime the diffusion current depends only on the rate of ion supply to the test electrode surface and therefore, it is the quantitative characteristic of mass transfer on a surface, Eq. (1). The diffusion coefficients of reacting ions in the chosen red-ox reaction correspond to Schmidt number $Sc \approx 1500$. Thickness of diffusion boundary layer δ_D, where the main change in concentration of reacting ions occurs, is significantly less than thickness of hydrodynamic boundary layer δ, i.e. $\delta_D/\delta \sim Sc^{-1/3}$. Application of the electrochemical method for mass transfer measurement has an advantage over other known methods (Kottke & Blenke, 1970) - it allows measurement of mass transfer and wall shear stress in one experiment. It is practically important for determination of interconnection between heat and mass transfer and hydrodynamics in the two-phase flows. Moreover, application of the electrochemical method for mass transfer measurement expands significantly the range of physical properties of the studied liquids towards the higher Prandtl numbers. Relatively thin near-wall liquid layer becomes the most important zone of the flow, and this allows us to study the role of the two-phase flow core in the process of heat and mass transfer.

The mass transfer coefficients in the two-phase flow were measured simultaneously with wall shear stress under the conditions shown in Table 1. The plate of the 5-mm width and 100-mm length was used as the probe. The probe length is sufficient for stabilization of the diffusion boundary layer (dimensionless length $L^+ > 4000$).

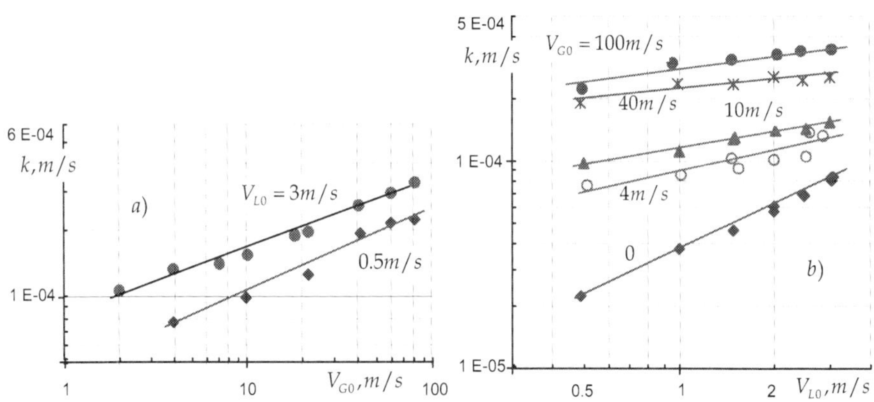

Fig. 7. The dependence of the mass transfer coefficient on the tube wall on superficial gas velocity (a), and superficial liquid velocity (b)

The effect of superficial gas velocity on mass transfer coefficient is shown in Fig. 7 (a). The mass transfer coefficient increases with a rise of superficial velocity of gas. The effect of superficial gas velocity is almost the same for all studied liquid flow rates. The effect of superficial velocity of liquid V_{L0} on mass transfer coefficient k is shown in Fig. 7 (b). The lower line corresponds to the flow of liquid. With an addition of gas into the flow the effect of V_{L0} on k decreases in comparison with the single-phase flow. The effect of volumetric quality $\beta = V_{G0}/(V_{L0}+V_{G0})$ on the relative mass transfer coefficient for the straight tube is shown in Fig. 8 (a). It is obvious that for superficial velocities of liquid phase from 0.5 to 1 m/s the relative mass transfer coefficient depends not only on volumetric quality, but also on liquid flow rate. This ambiguous dependence of mass transfer intensity on the wall is connected with the character of void fraction distribution over the cross-section in the bubble flow. The similar effect of volumetric quality on the relative wall shear stress in the gas-liquid flows in tubes was observed in (Nakoryakov et al., 1973), Fig. 8 (b). It was explained by an increasing in bubble concentration near the wall at low superficial velocities of liquid and additional agitation of near-wall layer. Later it was shown on the basis of simultaneous measurements of wall shear stress and distribution of void fraction and velocity in an inclined flat channel (Kashinsky et al., 2003). At high velocities of liquid the level of these perturbations becomes insignificant on the background of high turbulence of the carrying flow. Under these conditions the relative mass transfer coefficients depend definitely on the value of void fraction and can be calculated by the known models. Figure 8 illustrates that it is impossible to use the known models, for instance, the homogeneous one for calculation of mass transfer coefficients and wall shear stress at low void fraction. Data on heat transfer in the two-phase bubbly flows illustrating an abnormal increase in heat transfer coefficients under similar conditions are also available (Bobkov et al., 1973).

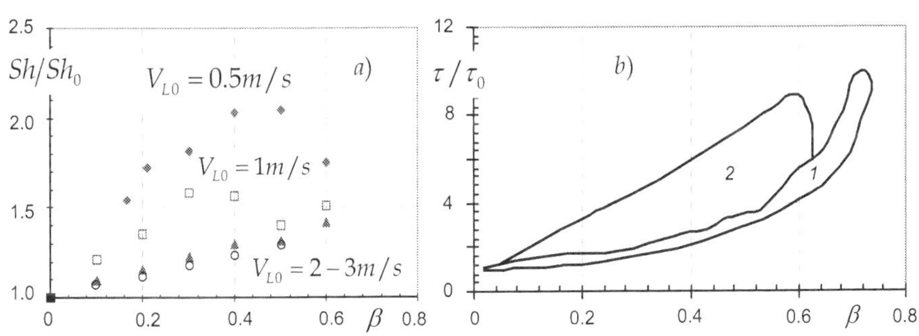

Fig. 8. Effect of volumetric quality on the relative mass transfer coefficient (a) and wall shear stress (b): *1* - homogeneous model; *2* - abnormal increasing of the wall shear stress

The relative mass transfer coefficient is shown in Fig. 9 depending on the ratio of superficial velocities of phases. It is obvious that relative wall shear stress and mass transfer coefficients depend similarly on relative velocity in the whole studied range of operation parameters. In these coordinates there are no deviations observed in the zone of low volumetric quality, Fig. 8. If we compare the relative friction and mass transfer coefficients under the same flow conditions, when inaccuracies of calculation dependences are excluded, we can see their qualitative and quantitative coincidence, Fig. 9.

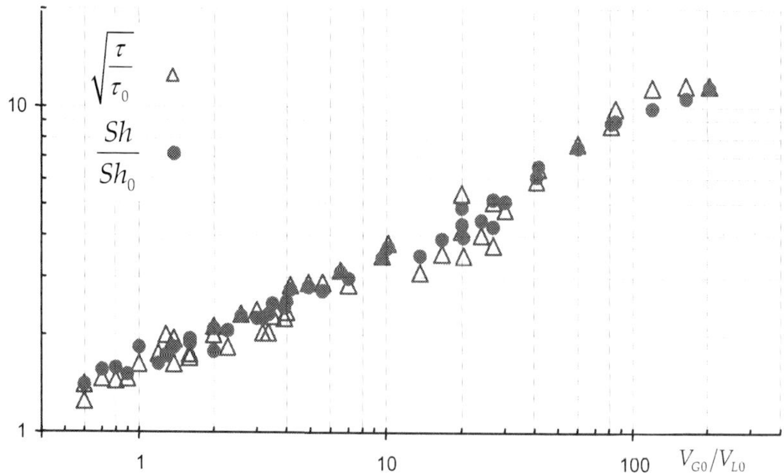

Fig. 9. Comparison of relative mass transfer coefficient and wall shear stress in two-phase flow.

It follows from data in Fig. 9 that

$$\frac{Sh}{Sh_0} = \sqrt{\frac{\tau}{\tau_0}} \tag{5}$$

i.e., connection between wall shear stress and mass transfer in the two-phase flow is the same as in the single-phase flow. Hence, the same dependences as for the single-phase flow can be applied for calculation of mass transfer in the two-phase flow. It is shown in (Chekhovich & Pecherkin, 1987) that relationship (5) is valid also for heat transfer in the two-phase gas-liquid flow.

For convective heat transfer at $\Pr \gg 1$ Kutateladze (1973) has obtained correlation

$$Nu = 0.115\sqrt{\zeta/8}\,\mathrm{Re}\,\mathrm{Pr}^{1/4} \tag{6}$$

Application of (6) for calculations in the two-phase flows is impossible because the specific velocity included into the Reynolds number and friction factor are not determined. However, their product $\sqrt{\zeta/8}\cdot u = v_*$ can be found experimentally from wall shear stress measurements, $v_* = \sqrt{\tau/\rho_L}$. Then $\sqrt{\zeta/8}\cdot \mathrm{Re} = v_* d/\nu' = \mathrm{Re}_*$ and correlation (6) can be applied for the two-phase flow. For mass transfer it can be written as

$$Sh = 0.115\,\mathrm{Re}_*\,Sc^{1/4} \tag{7}$$

where $Sh = \frac{kd}{D}$ is Sherwood number; $Sc = \frac{\nu}{D}$ is Schmidt number, D is diffusion coefficient, ν is kinematic viscosity of liquid phase. Experimental data on mass transfer in the gas-liquid flow at p = 0.1–1 MPa are shown in Fig. 10. The value of friction velocity is determined by measurements of wall shear stress simultaneously with mass transfer coefficients. These data are compared with correlations on convective heat and mass transfer.

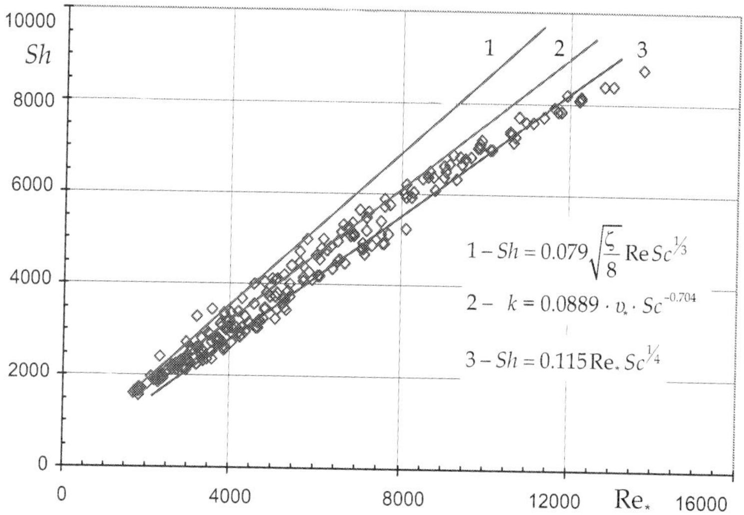

Fig. 10. Comparison of the mass transfer measurements in gas-liquid flow with calculation. 1 - Petukhov, (1967); 2 - Shaw & Hanratty, (1977); 3 -Kutateladze, (1973), Eq. (7).

In the whole range of studied parameters mass transfer coefficients in the two-phase flow coincide with calculation by correlations for the single-phase convective heat and mass transfer at $Pr \gg 1$.

For liquid flows with $Pr \gg 1$ heat and mass transfer occurs via turbulent pulsations penetrating into the viscous sublayer of boundary layer (Levich, 1959; Kutateladze, 1973). Thermal resistance of the turbulent flow core is insignificant. Apparently, the similar mechanism is kept in the two-phase flow. The measure of turbulent pulsations is friction velocity v_*. Since the turbulent core of the boundary layer does not resist to mass transfer, the flow character in the core is not important, either it is the two-phase or the single-phase flow with equivalent value v_*. Apparently, it is only important is that the liquid layer with thickness $\delta^+ > 5$ would be kept on the wall. The above correlations for calculation of mass transfer coefficients differ only by the exponent of Prandtl number, what is caused by the choice of a degree of turbulent pulsation attenuation in the viscous sublayer, (Kutateladze, 1973; Shaw & Hanratty, 1977). Scattering of experimental data on mass transfer in the two-phase flows is considerably higher than difference of calculations by available correlations; thus, we can not give preference to any of these correlations based on these data. It is shown in (Kutateladze, 1979) that the eddy diffusivity at $Pr \gg 1$ changes proportionally to the fourth power of a distance from the wall in the viscous sublayer, therefore, dependence (6) should be considered more grounded.

According to analysis of results shown in Figs. 9–10, mass transfer mechanism in the two-phase flow with a liquid film on the tube wall is similar to mass transfer mechanism in the single-phase flow and can be calculated by correlations for the single-phase convective mass transfer, if the wall shear stress is known.

3. Mass transfer in the channels with complex configuration

3.1 Introduction

Many components of the equipment in nuclear and heat power engineering, chemical industry are subject to erosion and corrosion wear of wetted surfaces. The channels of complex shape such as various junctions, valves, tubes with abrupt expansion or contraction, bends, coils, are affected most. The flow of liquids and gases in these channels is characterized by variations in pressure and velocity fields, by the appearance of zones of separation and attachment, where flow is non-stationary and is accompanied by generation of vortices. Analysis of the conditions in which there are certain items of equipment with two-phase flows, shows that the most typical and dangerous is the impact of drops, cavitation erosion, chemical and electrochemical corrosion (Sanchez-Caldera, 1988).

The process of corrosion wear in general consists of two stages: formation of corrosion products and their entrainment from the surface into the flow. The first stage is determined by the kinetics of the reaction or the degree of mechanical action of the flow on the surface. The supply of corrosion-active impurities to the surface and entrainment of corrosion products into the flow are determined by mass transfer process between the flow and the surface (Sydberger & Lotz, 1982). Due to significant non-uniformity in distribution of the local mass transfer coefficients the areas with increased deterioration appear on internal surfaces. Intensification of mass transfer processes caused by the above reasons can lead to a considerable corrosive wear of equipment parts. Changes in the temperature regimes due to heat transfer intensification result in the appearance of temperature stresses, which affect the reliability of equipment operation and the safety of power units (Poulson, 1991; Baughn et al., 1987). Therefore for safe operation of power plants it is very important to know the location of areas with maximal mass transfer coefficients in the channels with complex configuration and the mass transfer enhancement in comparison with the straight pipelines. The single-phase flow in the bend of various configurations with turn angles 90° and 180° was studied in (Baughn et al., 1987; Sparrow & Chrysler, 1986; Metzger & Larsen, 1986). For this purpose the authors used thin film coating with low melting temperature on internal surface of channels, temperature field measurements, Reynolds analogy for calculations of mass transfer coefficients based on heat transfer measurements, etc. In spite of the fact that two-phase coolants are widely used in cooling systems of various equipment, experimental studies on two-phase flow separation and flow attachment in channels are limited, (Poulson, 1991; Mironov et al., 1988; Lautenschlager & Mayinger, 1989). Intensity of these processes is determined by flow hydrodynamics within thin near-wall layers. Therefore the experimental study of these phenomena should be carried out using the methods which do not distort the flow pattern in the near-wall area in complex channels. The electrochemical method makes it possible to measure local values of wall shear stress and mass transfer rate for single-phase and two-phase flows in the channels with complex configuration.

In this section the results on experimental investigation on distribution of local mass transfer coefficients in single-phase and two-phase cocurrent gas-liquid flow in vertical channels with 90° turn and abrupt expansion are presented. The scheme of the experimental setup is shown on Fig. 1. The scheme of the test sections are presented in Fig. 11.

In a channel with turn flow the liquid or two-phase medium is fed from bottom and changes the flow direction at 90°. To provide fully developed flow straight tube of 20 mm diameter and 2 m long is installed before bend. The channel with the bend is made of two plexiglas sections, sealed with each other by rubber gaskets and pulled together by bolts, Fig. 11. The inner diameter of channel is 20 mm and the relative bending radius is $\tilde{R} = 5$. Fifteen electrochemical probes were installed on the test section: 5 - on inner generatrix, 5 - on middle generatrix, and 5 - on outer generatrix. The probes were installed in the cross-sections with turn angles $\varphi = 10, 28, 45, 63, 80^{\circ}$. One more probe was installed on a straight section of the tube in front of the inlet to the channel. This probe measures the local mass transfer coefficient in a straight tube. The electrochemical probes for measurements of local mass transfer coefficient were made of platinum wire of 0.3 mm in diameter welded into the glass capillary, Fig. 2-2. After probe mounting in test section their working surface was flushed to the internal surface of the channel. The assembled channel was fixed to the flanges of feed and lateral pipelines.

The channel with sudden expansion was made of plexiglass and enabled to visualize the flow, as well as to make photo - and video of the process.

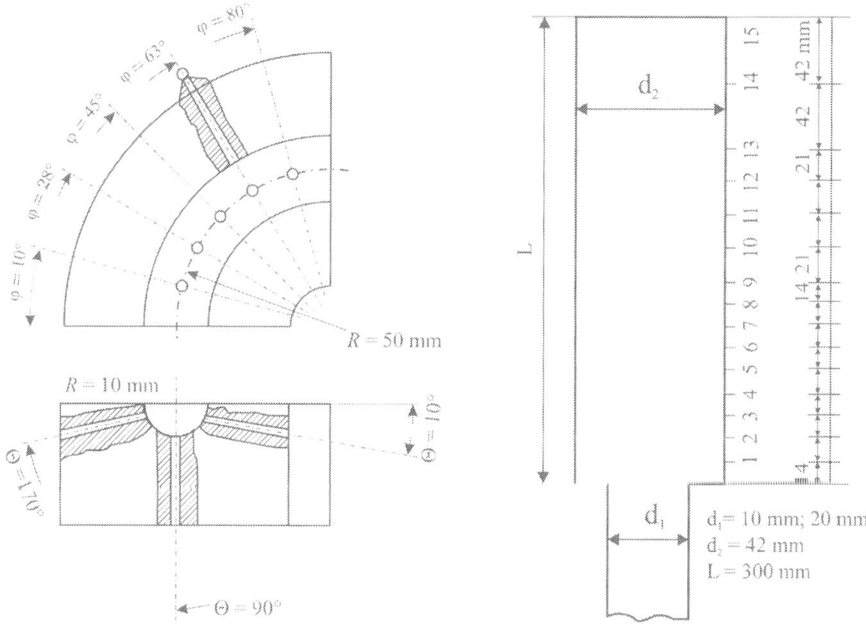

Fig. 11. Scheme of the test sections with turn angle 90° and with abrupt expansion

The inner diameter of the channel was $d_2 = 42$ mm, and the length was $L = 300$ mm. The channel was connected with the stabilization section in such a way that the assembly formed sudden expansion. The stabilization sections were made of two diameters: d_1 = 10 and 20 mm, correspondingly, and ratio $E = d_1/d_2$ was 1:2 and 1:4 (the exact values of E were equal to 0.476 and 0.238), and relative channel length was $L/d_2 = 7.1$.

3.2 Mass transfer in a channel with turn flow

In the experiments on measurements of local mass transfer coefficients on the wall of the channel with the turn flow the volumetric quality β was changed within the range from 0 to 0.6, and liquid superficial velocities from 0.5 to 2.6 m/s. At these parameters the main flow pattern of two-phase mixture is the bubble flow. In certain flow regimes at small liquid flow rates and maximal gas flow rates the slug fluctuating flow was observed. In order to mark out the effect of the flow turn angle the data obtained are presented in the form of ratio of the local mass transfer coefficients in the bend to the local mass transfer coefficient in the straight tube at the same values of the volumetric quality. Figure 12 shows variation of local mass transfer coefficient depending on the turn angle for two values of liquid superficial velocity: 0.5 m/s and 2.6 m/s (Pecherkin & Chekhovich, 2008). Data for the single-phase flow are shown in the same figure. In case of single-phase flow the first probe on inner generatrix ($\varphi = 10°$) indicates approximately the same value as in the straight tube independently of the flow rate. Further, as far as the turn angle increases the mass transfer coefficient diminishes and then slightly increases at the channel outlet. Probably, a significant decrease in mass transfer coefficients is associated with the flow separation in this area. The addition of gas into the liquid flow essentially changes distribution of the local mass transfer coefficient. In the first half of the channel at the turn angles from 10° to 45° the increase in mass transfer coefficients is observed as compared with that in straight tube. The increase in mass transfer coefficients comparing with the straight tube reaches up to 40% at low liquid flow rates, and approximately 20% at high flow rates. At the channel outlet at a horizontal part of the bend the mass transfer coefficients decrease comparing with the straight tube.

On the middle generating line, as a single-phase liquid flows, the intensification reaches 60% at the bend outlet. The mass transfer character in gas-liquid flow is the same as in the single-phase flow. As compared with the straight tube intensification makes up 10-20% at low liquid flow rates and 40-50% at high liquid flow rates depending on volumetric quality.

On the external generating line, for small velocities of single-phase liquid flows at the channel inlet, the mass transfer coefficient remains the same as in a straight tube. At the outlet of the bend mass transfer enhancement reaches 30%. An increase of volumetric quality causes rapid decrease in mass transfer coefficient at the inlet to the channel, and it reaches the minimal value at $\varphi = 10\text{-}30°$, and then smoothly increases downstream to the channel outlet. At high liquid superficial velocities maximal mass transfer coefficients are observed at the turn angles of 50-70° and increase with volumetric quality.

The highest mass transfer enhancement in the single-phase flow is observed at the channel outlet on the middle generatrix. The maximal mass transfer coefficient for these areas can be expressed by the following relation

$$Sh = 0.0287\,\mathrm{Re}^{7/8}\,Sc^{1/4} \qquad (8)$$

Comparison of (8) with correlation for wall mass transfer coefficients in the coil (Abdel-Aziz et al., 2010) shows satisfactory agreement. Clearly expressed local maximum in a two-phase flow is situated on the inner generatrix within the zone of $\varphi = 10\text{–}45°$, and the absolute maximum is observed at the channel outlet on the middle and outer generatrices.

Figure 13 shows the effect of volumetric quality on distribution of local mass transfer coefficients in the bend. The data are presented in the form of ratio of mass transfer coefficients for gas-liquid flow to the mass transfer coefficients for single-phase flow at the same turn angles.

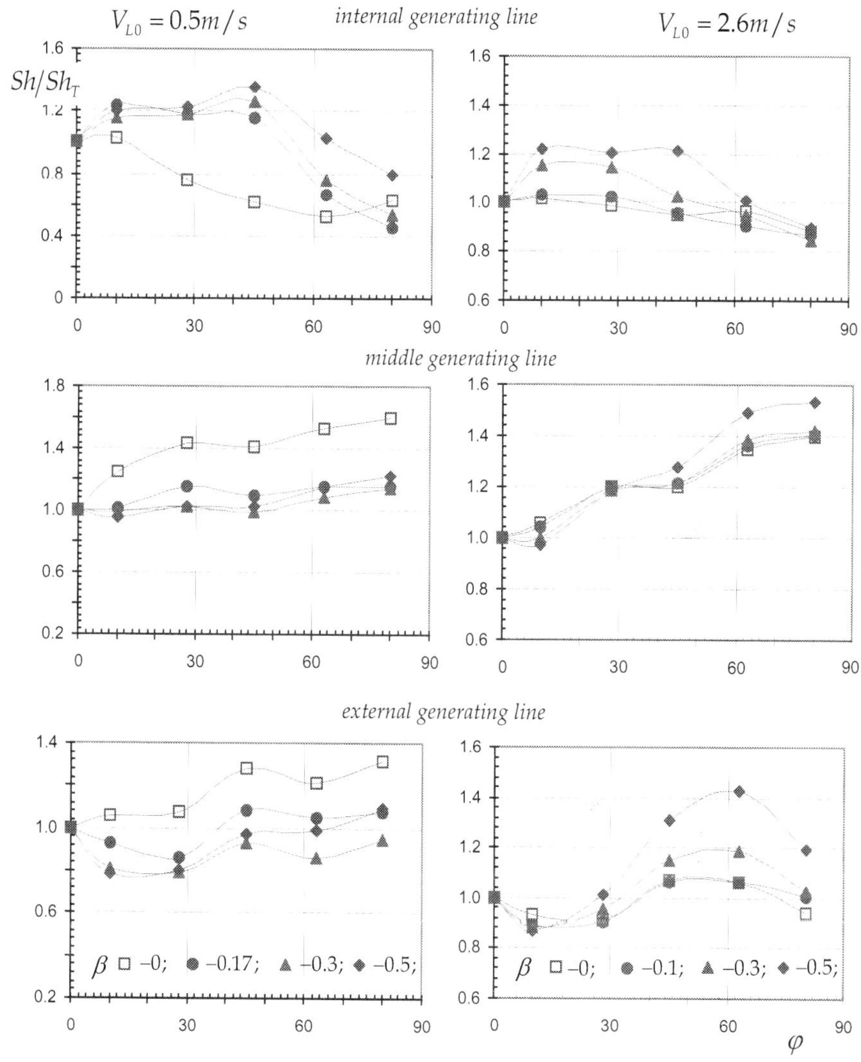

Fig. 12. The influence of the turning angle on the relative mass transfer coefficient in a bend.

At low liquid flow rate, $V_{L0} = 0.5$ m/s, on the inner generatrix at $\varphi = 45°$ mass transfer intensification is 5-fold higher as compared to that for the single-phase flow, Fig. 13 (a).
At higher liquid superficial velocity $V_{L0} = 2.6$ m/s, intensification reaches 60-80% at high volumetric quality, Fig. 13 (b).

void fraction in the channel we observed the foamed flow with large-scale bubbles, while at very high outflow velocities the flow detaches from the channel walls, Fig. 14, right photo. After separation of the flow from the pipe wall the two-phase jet in the center of the channel was observed. Near the outlet from the test section the jet diameter increases, and the certain portion of liquid drops out to the channel walls and flows down as a film or rivulets. The location of flow attachment may move along the channel height depending on the velocity of jet. A decrease in flow rate of one of the components at constant flow rate of another component leads to step-like reverse transition: now the two-phase flow fills up the whole cross-section of the channel along its height.

Figure 15 (a) presents gas flow rates corresponding to transition to the jet flow depending on liquid mass flow rate. The less is liquid flow rate the larger gas flow rate is required to provide the transition to the jet flow. The kind of transition shows the change in the balance of inertial and mass forces in the flow.

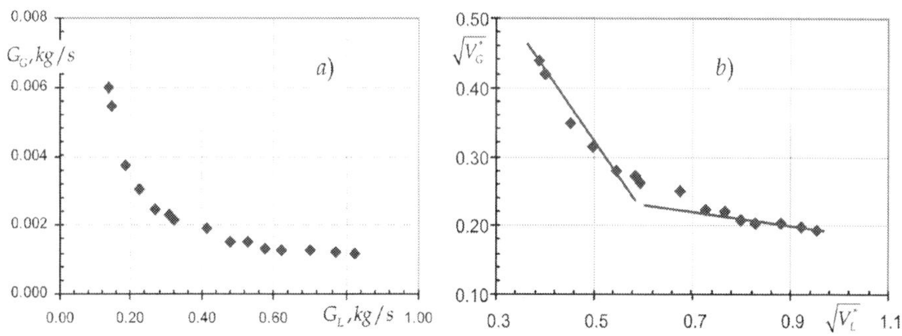

Fig. 15. The correlation between mass flow rate of liquid and gas phases at the boundary of the jet flow transition

The similar phenomenon is observed at counter-current two-phase flow in a vertical tube. Increasing gas flow rate over the critical value causes flooding. Though the flooding mechanisms and mechanisms of transition to jet pattern most likely are different, nevertheless the transition criteria in both cases may be the same. Froude numbers or their combinations may serve as dimensionless criteria to characterize interaction between the gravity forces and inertial forces. Wallis, (1969) proposed the empirical correlation for description of flooding process

$$V_G^{*1/2} + aV_L^{*1/2} = c \tag{9}$$

where $V_G^* = V_G\sqrt{\frac{\rho_G}{gD(\rho_L - \rho_G)}}$; $V_L^* = V_L\sqrt{\frac{\rho_L}{gD(\rho_L - \rho_G)}}$, V_L, V_G are superficial liquid and gas velocities; ρ_L, ρ_G are densities of liquid and gas. We obtained a = 1.02, c = 0.84 for $V_L^* < 0.4$ and a = 0.092, c = 0.29 for $V_L^* > 0.4$, Fig. 15 (b). More detailed investigations are needed to study the regime of two-phase jet flow in a channel with abrupt expansion.

3.3.2 Mass transfer on the wall of a channel with abrupt expansion

The results on measurements of mass transfer coefficients on the wall of channel with abrupt expansion in gas-liquid flow are presented in this section. Fifteen probes were installed to measure the local mass transfer coefficients at the internal surface. Along the initial section of the channel with expansion the probes were installed with the interval of 14 mm, and at the outlet of the channel, where the flow becomes stable, the interval was increased up to 42 mm, Fig. 11. The design and the size of electrochemical probes for measurements of the local mass transfer coefficients were similar to those used for measurements of wall shear stress, Fig. 2 - 2.

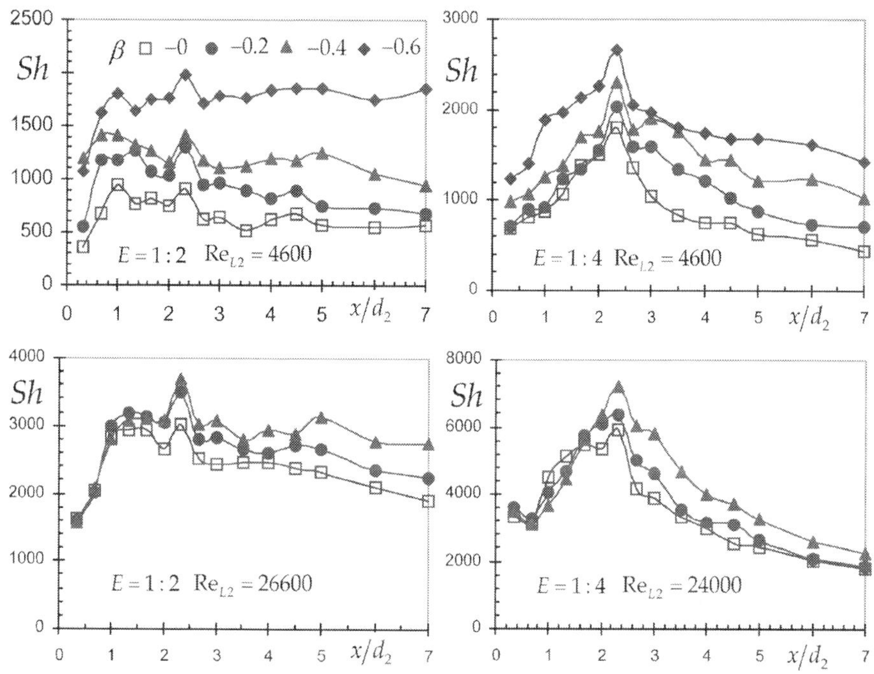

Fig. 16. Distribution of local mass transfer coefficient along the channel with abrupt expansion

Figure 16 represents the dimensionless mass transfer coefficient *Sh* depending on dimensionless length of the channel for various volumetric quality and diameter enlargement (Pecherkin et al., 1998). Distribution of the local mass transfer depends both on liquid velocity and volumetric quality β. Mass transfer coefficient depends on the length in a way similar to that for the single-phase flow, though there may appear local maximums in mass transfer depending on volumetric quality. The effect of volumetric quality becomes apparent in different ways for various liquid flow rates. At low liquid flow rates there are two local maximums at the distances of 1 and 2 channel diameters. In the second half of the channel the mass transfer coefficient practically does not change along the length at $E = 1:2$.

With the increase in liquid flow rate the local maximum is shifted to $x/d_2 = 2.3$. At a distance of up to $x/d_2 = 1$ the effect of volumetric quality diminishes and almost disappears at high liquid flow rates. The mass transfer coefficients at $\beta = 0.1 - 0.4$ practically do not differ from those for single phase flow. The visual observations show that under these conditions in the corners of the channel near the expansion cross-section there are almost no gas bubbles. At high liquid flow rates the local maximum appears at a distance $x/d_2 = 5$, probably due to generation of vortices of another scale. The visual observations show that the flow in this area is of the chaotic nature with no clear stream direction. For the channel with expansion ratio $E = 1:4$ distribution of mass-transfer coefficient has a clearly defined peak. Near the tube inlet the mass transfer coefficient does not differ practically from that for the single-phase flow, while at high liquid flow rates in this zone the local minimum appears as well. The mass transfer coefficient at the maximal point is significantly higher as compared with the channel $E = 1:2$. After reaching the maximum the mass transfer coefficients decrease abruptly, and at the channel outlet they approach the same values as for the channel with expansion ratio $E = 1:2$. Therefore within these distances the flow is almost fully developed and flow regime should depend only on void fraction. The effect of volumetric quality on a relative mass transfer coefficient is shown in Fig. 17.

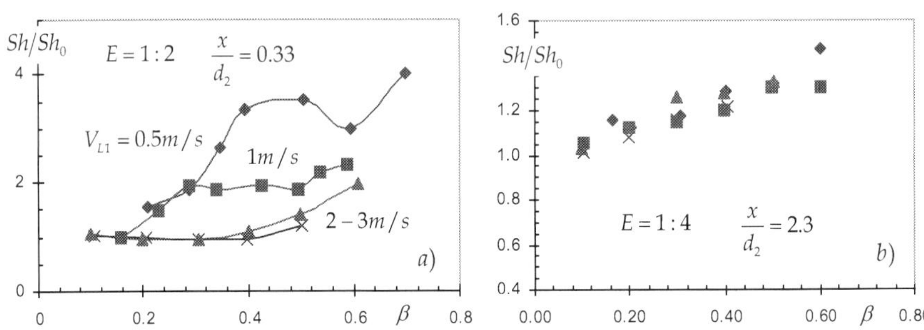

Fig. 17. Effect of volumetric quality on the relative mass transfer coefficient

At a distance of one diameter from the entrance for $E = 1:2$ and high flow rates of fluid there is no influence of volumetric quality. At low liquid flow rates the abnormal increase in mass transfer coefficients was observed. This was concerned with bubbles distribution over the cross-section and vortex flow near the channel inlet. In the vicinity of maximum mass transfer at $x/d_2 = 2.3$ the effect of liquid flow rate on relative mass transfer coefficient is lacking, here mass transfer increases in proportion to the two-phase mixture velocity like in the flow inside the tube.

Experimental results presented in Fig.13 and Fig.17 show, that the relative mass transfer coefficient is effected both by volumetric quality and, to a considerable degree, velocity of the liquid. Similarly to the gas-liquid mixture flow inside the tubes such dependence is explained exceptionally by distribution of void fraction over the tube cross-section. In case of the flow in the channel with abrupt expansion bubble distribution over the cross-section will be significantly effected by vortex zones. The size of these zones and flow intensity inside them depend on liquid velocity.

In the recirculation zone near the expansion cross-section at high liquid flow rates the effect of volumetric quality is not observed at all because this zone is almost free of bubbles. This zone is out of the interest in terms of mass transfer process enhancement. The most important in this respect is the zone between 2-4 calibers. In this area mass transfer maximum is observed within a sufficiently broad range of volumetric quality and flow patterns as well as for various expansion ratios. The maximum in heat and mass transfer in a single-phase flow almost coincides with the flow attachment location. Specific dimension in Nusselt and Reynolds numbers is the diameter of the outflow tube. The data on heat transfer in a single-phase flow in the expansion channels are generalized right in such a way (Zemanic & Dougall, 1970; Krall & Sparrow, 1966), as well as our data on mass transfer in a single-phase flow. In some publications the length of the recirculation zone is used as the reference dimension (Terekhov & Bogatko, 2008).
Processing of the data on maximal mass transfer coefficients in a two-phase gas-liquid flow in the channel with sudden expansion is carried out in (Chouikhi et al., 1987), and correlation has been proposed:

$$Sh_{\max} = A \cdot \mathrm{Re}_{L2}^{\;0.7} \cdot Sc^{1/3} \cdot \left(V_{G2}/V_{L2}\right)^{n} \tag{10}$$

where A, n are functions of the expansion ratio, n changes from 0.05 for $E = 1:2$ to 0.095 for $E = 1:6$, Re_{L2} is Reynolds number calculated by the liquid superficial velocity V_{L2} in a channel of larger diameter, V_{G2} is gas superficial velocity. Substitution of Reynolds number Re_{L2} by the Reynolds number for the tube of a smaller diameter, i.e. Re_{L1} similarly to a single-phase flow allowed us to summarize the data on the maximal mass transfer coefficient (Chouikhi et al., 1987) for all expansion ratios by the following relation within $\pm 15\%$:

$$Sh_{\max} = 0.23 \cdot \left(\mathrm{Re}_{L1}\right)^{0.7} \cdot Sc^{1/3} \cdot \left(V_{G1}/V_{L1}\right)^{0.07} \tag{11}$$

Processing of data on maximal heat and mass transfer coefficients in a single-phase flow has shown that the velocity of outflow from the tube of a smaller diameter identically determines intensity of heat and mass transfer in the area of flow attachment. Hence the attempt was made to process obtained data for the maximal mass transfer coefficient in two-phase flow based on actual velocity of liquid phase V'_{L1} in a channel, from where the two-phase medium outflows. The void fraction was calculated according to Armand (1946) correlation $\varphi = 0.83\beta$. The results obtained are shown in Fig. 18. This graph represents all data obtained for expansion ratios $E = 1:2$ and $E = 1:4$, as well as the data for single-phase flow. The effect of void fraction is identically taken into account by the actual velocity, and in contrast to correlation (10), does not require introduction of additional parameter, which is the ratio of the superficial velocities. The obtained data are satisfactorily generalized by the following correlation

$$Sh_{\max} = 0.252 \cdot \mathrm{Re}'^{\;0.7}_{L1} \cdot Sc^{1/4} \tag{12}$$

The visual observations of the flow and measured data show that flow stabilization after abrupt expansion takes place at the channel outlet. The effect of the expansion ratio on mass transfer coefficients both in the single-phase and two-phase flows disappears and specific

dimension is diameter d_2. Such processing by velocity of mixture $V_{TP2} = V_{L2} + V_{G2}$ in a channel with expansion is presented in Fig. 19.
The data on mass transfer coefficients in the single-phase flow and at high liquid flow rates in two-phase flow are satisfactorily generalized by correlation

$$Sh_2 = 0.135 \cdot \mathrm{Re}_{TP2}{}^{0.75} \cdot Sc^{1/4} \tag{13}$$

Deviations are observed in the area of abnormal effect of volumetric quality on mass transfer at low velocities of liquid phase.

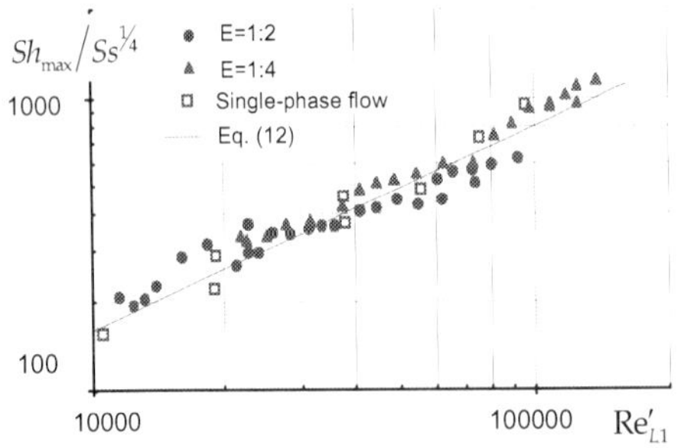

Fig. 18. The maximum coefficient of mass transfer in a channel with abrupt expansion

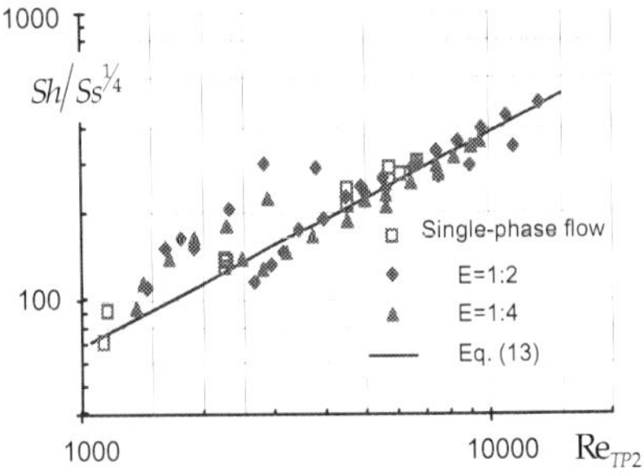

Fig. 19. The mass transfer coefficient in the zone of developed flow

For the studied flow patterns the maximal mass transfer coefficients in the channel with the flow turn of 90° are observed for intermediate and outer generatrices at the channel outlet. And only for lowest liquid flow rates the zone with maximal mass transfer coefficients is situated on the internal surface. The maximal effect of void fraction on the relative mass transfer coefficient is observed at the inner surface and most probably is explained by the higher concentration of gas bubbles. The increase in mass transfer at the outer generatrix of the bend may be related to the shift of velocity maximum towards this area, i.e. with redistribution of phases because of circulation flows.
Distribution of mass transfer coefficients in the channel with sudden expansion in the two-phase flow as a whole is of the same nature as for the single-phase flow. The maximal effect of void fraction is observed from the zone with mass transfer maximum up to the channel outlet. At distance $x/d_2 = 1 \div 1.5$ at high liquid flow rates the effect of void fraction is lacking, because gas practically does not flow into this area due to existence of circulation zones. The absolute values of mass transfer coefficients at the outlet of the channels with various expansion and the same values of volumetric quality are quite close to each other. This confirm the fact that for the given regime parameters the flow becomes almost developed.
The experimental data on the effect of volumetric quality on mass transfer intensity on a wall in the curvilinear channel and in the channel with abrupt expansion, obtained in the present work, may be used at modeling of the erosion- and corrosion wear process in channels of coolant circulation systems of the power and chemical equipment.

4. References

Abdel-Aziz, M.H., Mansour, I.A.S. & Sedahmed, G.H. (2010). Study of the rate of liquid–solid mass transfer controlled processes in helical tubes under turbulent flow conditions, *Chemical Engineering and Processing: Process Intensification*, Vol. 49, Issue 7, pp. 643–648

Armand, A.A. (1946). Resistance to the motion of two-phase system in horizontal pipes, *Izv. VTI*, No. 1, pp. 16-23

Baughn, J. W., Hoffman, M. A., Launder, B. E., Lee, D. & Yap, C. (1989). Heat Transfer, temperature, and velocity measurements downstream of an abrupt expansion in a circular tube at a uniform wall temperture, *J. of Heat Transfer*, Vol. 111, pp. 870–876

Baughn, J., Hoffman, M., Takahashi, R., & Launder, B. (1984). Local heat transfer downstream of an abrupt expansion in circular channel with constant heat flux, *Journal of Heat Transfer*, V. 106, No. 4, pp. 789-796

Baughn, J., Iakovides, H., Jackson, D., & Launder, B. (1987). Local Heat Transfer Measurement in Turbulent Flow around a 180-deg Pipe Bend, *Journal of Heat Transfer*, Vol. 109, No. 1, pp. 43–48

Bel Fdhila R., Suzanne C., & Masbernat L. (1990). Measurement in two-phase bubly flows in complex geometries, *Proc. Int. Sem. Phase-Interface Phenomena in Multiphase Flow*, May 14-18 1990, Dubrovnik, Yugoslavia.

Bobkov, V. P., Ibragimov, M. Kh., Tychinsky, N. A. & Fedotovsky, V. S. (1973). Heat Diffusion at Turbulent Flow of Water with Gas Bubbles, *Inzh. Fiz. Zh.*, Vol. 24, No. 5, pp. 781-789

Chekhovich, V. Yu. & Pecherkin, N. I. (1987-88). Heat and mass transfer and wall shear stress in vertical gas-liquid flow, *Experimental Heat Transfer*, Vol. 1, No. 4, pp. 253–264, ISSN 0891-6152

Chouikhi, S., Patric, M. & Wragg, A. (1987). Mass transfer at pipe walls downstream of abrupt enlargement of diameter in two-phase (gas-liquid) flow, *Proc. 3-d Int. Conf. On Multiphase Flow*, p. 285-294, The Hague Netherlands, May 1987

Collier, J. (1972). *Convective Boiling and Condensation*, McGraw – Hill, London

Davis, E.J., Hung S., & Arciero, S. (1975). An analogy for heat transfer with wavy-stratified gas-liquid flow, *AIChE Journal*, Vol. 21, No. 5, pp. 872–878, Sept, 1975

Heat Transfer in Two-Phase Flow, Butterworth, D. & Hewitt, G. (eds.), Oxford University Press, Oxford, 1977

Hewitt, G.F. & Hall-Taylor N.S. (1970). *Annular Two-Phase Flow*, Pergamon Press, London

Kashinsky O.N., Chinak, A.V., & Kaipova, E.V. (2003). Bubble Gas–Liquid Flow in an Inclined Rectangular Channel, *Thermophysics and Aeromechanics*, Vol. 10, No. 1, pp. 71–78

Kottke, V. & Blenke, H. (1970). Meвmethoden konvektiver Stoffubertragung, *Chem. Ing. Techn.* , Vol. 50, No. 2., pp. 81–90

Krall, K., & Sparrow E. (1966). Turbulent heat transfer in separated, reattached, and redevelopment regions of a circular tube, *Journ. of Heat Transfer*, Vol. 88, No. 1, pp. 131-136

Krokovny, P.M., Kashinskii, O.N., & Burdukov, A.P. (1973). Study of mass transfer in horizontal gas-liquid flow, Inzh. Fiz. Zh., Vol. 38, No. 1, pp. 16-21

Kutateladze, S.S. (1973). *Near-wall turbulence*, Nauka, Novosibirsk, 1973

Kutateladze, S.S. (1979). *Fundamentals of Heat Transfer Theory*, 5th ed., Atomizdat, Moscow

Lautenschlager, G. & Mayinger, F. (1989). Experimental Investigation of Dispersed Flow Heat Transfer in Curved Tubes, *Proc. Int. Symp. Adv. Phase Change Heat Transfer*, pp. 196–202, Chongqinq, China, May 1988, International Academic Publishers

Levich, V.G. (1959). *Physico-chemical hydrodynamics*, Fiz-matgiz, Moscow, 1959

Lockhart, R. & Martinelli, R. (1949). Proposed correlation of data for isothermal two-phase, two – component flow in pipes, *Chem. Eng. Progr.* , Vol. 45, No. 1,pp. 39-48

Metzger, D. & Larsen, D. (1986). Use of Melting Point Surface Coatings for Local Convection Heat Transfer Measurements in Rectangular Channel Flows with 90-deg Turn, *Journal of Heat Transfer,* Vol. 108, No. 1, p. 49

Mironov, Yu.L., Kostomarov, V.M., Vel'kin, V.I., et al. (1988). Experimental Investigation of a Dispersed–Film Flow in a Curvilinear Channel, *Energomashinostroenie,* No. 1, pp. 7–10

Nakoryakov, V.E., Burdukov, A.P., Kashinskii, O.N., & Geshev, P.I., (1986). Elektrodiffuzionnyi metod issledovaniya lokal'noi struktury turbulentnykh techenii (Electrodiffusion Method to Investigate the Local Structure of Turbulent Flows), Novosibirsk: Institute of Thermophysics, Siberian Branch of the USSR Academy of Sciences, 1986

Nakoryakov, V.E., Burdukov, A.P., Pokusaev, B.G., Kuz'min, V.A., et al., (1973). Issledovanie turbulentnykh techenii dvukhfaznykh sred (Investigation of Turbulent Flows of Two-Phase Media), Kutateladze, S.S., Ed., Novosibirsk: Institute of Thermophysics, Siberian Branch of the USSR Academy of Sciences, 1973

Pecherkin, N. I. & Chekhovich, V. Yu. (2008). Mass Transfer in a Two-Phase Flow in a Curvilinear Channel, *Journal of Engineering Thermophysics,* Vol. 17, No. 2, pp. 113–119

Pecherkin N., Chekhovich V., & Nigmatulin B. (1998). Experimental investigation of the distribution of the local mass transfer coefficient on the walls of the channels with a complex geometry. *Proc. Of the Third Intern. Conf. on Multiphase Flow ICMF-98,* Lyon, France, June 8-12, 1998

Petukhov, B.S. (1967). *Heat transfer and resistance in laminar flow of fluid in the pipes,* Energiya, Moscow, 1967

Poulson, B., (1991). Measuring and Modelling Mass Transfer at Bends in Annular Two-Phase Flow, *Chem. Eng. Sci.,* Vol. 46, No. 4, pp. 1069–1082

Sanchez-Caldera, L.E., Griffith, P., & Rabinowicz, E. (1988). The Mechanism of Corrosion-Erosion in Steam Extraction of Power Stations, *J. Eng. Gas Turbines Power,* Vol. 110, No. 2, pp. 180–184

Shaw, D.A. & Hanratty, T.J. (1977). Turbulent mass transfer rates to a wall for large Shmidt numbers, *AIChE Journal,* Vol. 23, No. 1, pp. 28–37

Sparrow, E. & Chrysler, G. (1986). Turbulent Flow and Heat Transfer in Bends of Circular Cross Section: 1-Heat Transfer Experiment, *Journal of Heat Transfer,* Vol. 108, No. 1, pp. 40–47

Surgenour, B. & Banerjee, S. (1980). Wall shear stress and mass transfer in vertical two-component flow, *Proceedings of 30 – th Can. Chem. Eng. Conf.,* pp. 508-523, Edmonton, Canada

Sydberger, T. & Lotz, U. (1982). Relation between Mass Transfer and Corrosion in a Turbulent Pipe Flow, *J. Electrochem. Soc.: Electrochem. Science Technol.,* Vol. 129, No. 2, pp. 276–283

Terekhov, V. I. & Bogatko, T. V. (2008). Effect of boundary layer thickness before the flow separation on aerodynamic characteristics and heat transfer behind an abrupt expansion in a round tube, *Thermophysics and Aeromechanics,* Vol. 15, No. 1, pp. 91-97, March 2008

Wallis, G. B. (1969). *One-Dimensional Two-Phase Flow,* McGraw-Hill, New York

Zemanic, P.P., & Dougall, R.S. (1970). Local heat transfer downstream of abrupt circular channel expansion, *Journal of Heat Transfer,* Vol. 92, No. 1, pp.53–60

Laminar Mixed Convection Heat and Mass Transfer with Phase Change and Flow Reversal in Channels

Brahim Benhamou[1], Othmane Oulaid[1,2],
Mohamed Aboudou Kassim[1] and Nicolas Galanis[2]
[1]*LMFE, CNRST-URAC27, Cadi Ayyad University, Marrakech,*
[2]*Mechanical Eng. Department, Université de Sherbrooke, Sherbrooke,*
[1]*Morocco*
[2]*Quebec, Canada*

1. Introduction

Due to its widespread applications, heat and mass transfer in an air stream with liquid film evaporation or condensation in open channels has received considerable attention in the literature. This kind of flows is present in many natural and engineering processes, such as human transpiration, desalination, film cooling, liquid film evaporator, cooling of microelectronic equipments and air conditioning.

Since the original theory for flow of a mixture of vapour and a non-condensable gas by Nusselt (1916) and its extension by Minkowycz & Sparrow (1966), many theoretical and experimental studies have been published in the literature. These studies deal with different geometric configurations such as a flat plate, parallel-plate channel and rectangular or circular-section ducts.

Heat and mass transfer convection over a flat plate wetted by a liquid film has been investigated by Vachon (1979). He performed an analytical and experimental study of the evaporation of a liquid film streaming along a porous flat plate into a naturally driven airflow. This author established correlations for Nusselt and Sherwood numbers in connection with a combined Grashof number. Ben Nasrallah & Arnaud (1985) investigated theoretically film evaporation in buoyancy driven airflow over a vertical plate heated with a variable heat flux. The solutions of the governing equation have been obtained by means of semi-analytical and finite difference methods. The authors present their results in term of expressions of the wall temperature and mass fraction as well as the local Nusselt and Sherwood numbers. A numerical and experimental analysis has been carried out by Tsay et al. (1990) to explore the detailed heat transfer characteristics for a falling liquid film along a vertical insulated flat plate. Free stream air temperature was set at 30°C and inlet liquid film temperature was taken equal to 30°C or 35°C. The results show that latent heat transfer connected with vaporization is the main cause for cooling of the liquid film. The authors affirm that when the inlet liquid temperature is equal to the ambient one, latent heat transfer due to the film vaporization initiates heat transfer in the film and gas flow. Aguanoun et al (1994; 1998) studied numerically the evaporation of a falling film on an inclined plate heated at a constant temperature in a humid air stream. They considered forced (Agunaoun et al., 1994) or mixed convection (Agunaoun et al., 1998) with several liquid mixtures. A boundary layer type model was adopted. The authors stated that the liquid film-gas interface has approximately the same temperature as the plate in the case of forced convection.

Yan & Soong (1995) considered turbulent heat and mass transfer convection over a wetted inclined plate. The liquid film flow is turbulent and waves less. Their results pointed out that the plate and the gas-liquid interface temperatures are reduced consequently to the increase of the plate's inclination angle, the inlet film thickness or the air velocity. Mezaache & Daguenet (2000) conducted a numerical study of the evaporation of a water film falling on an inclined plate in a forced convection flow of humid air. The plate is insulated or heated by a constant heat flux. The main result of this study was that the enthalpy diffusion term in the energy equation does not influence the film temperature. This term represents the effect of the species diffusion on enthalpy of the humid air mixture. Volchkov et al. (2004) reported a numerical work on both laminar and turbulent forced convection of humid air over an infinite flat plate. The authors aim to establish the validity of the heat-mass transfer analogy. The steam in the humid airflow may condense on the plate whose temperature is lower than that of the airflow. The authors made a major simplification by neglecting the effect of the condensate film. This assumption is justified by the experimental data in the literature which indicate that measurements of the liquid film-air interface temperature in humid-air flows show that it is close to the saturation temperature corresponding to the vapour concentration at the plate. This point is addressed hereafter. Recently, Maurya et al. (2010) developed a numerical analysis of the evaporating flow of a 2-D laminar, developing film falling over an inclined plate, subjected to constant wall heat flux. Their results show that the evaporation process begins only after the growing thermal boundary layer reaches the interface.

Heat and mass transfer in a vertical heated tube with a liquid film falling on its inside walls was treated by Feddaoui et al. (2003). They considered co-current downward flows of liquid water and humid air. The airflow is turbulent while the liquid one is laminar and without surface waves. The authors concluded that better cooling of the liquid film is obtained for higher heat flux or lower inlet liquid flow. Lin et al (1988) analysed numerically combined heat and mass buoyancy effects on laminar forced convection in a vertical tube with liquid falling film. Their results show that large film evaporation rates are obtained for larger tube wall temperature. Convective instability of heat and mass transfer for laminar forced convection in the thermal entrance region of a horizontal rectangular channel has been examined by Lin et al. (1992). The rectangular channel is thermally insulated except for the bottom wall. The latter is maintained at a constant temperature and covered by a thin liquid water film. The thickness of this film is neglected, thus it is treated as a boundary condition for heat and mass transfer. Inlet air temperature was fixed at 20°C. The effects of changes of bottom wall temperature, relative humidity of air at the entrance and the channel aspect ratio are examined. The results show that the convective instability is affected by changes in inlet air relative humidity and temperature and also the channel aspect ratio. Huang et al. (2005) conducted a numerical study on laminar mixed convection heat and mass transfer in a rectangular duct. Two of the duct walls were wetted by a thin liquid water film and maintained at different constant temperatures. The other walls are insulated. Air entered the duct with a constant temperature lower than that of the walls. The authors established that vapour condensation occurred on the wall with lower temperature.

Nelson & Wood (1989) studied the developing laminar natural convection flow in a vertical parallel-plate channel. They used a boundary-layer approximation model to derive a correlation for the heat and mass transfer coefficients in the case of uniform temperature and concentration plates. Yan and co-workers (Yan et al. 1989; Yan & Lin, 1989; Tsay et al. 1990, Yan et al. 1990; Yan 1991; Yan & Lin 1991; Yan et al. 1991; Yan 1993; Yan 1995a; Yan 1995b) investigated the influences of wetted walls on laminar or turbulent mixed convection heat and mass transfer in parallel-plate channels. Constant wall temperature, constant wall heat flux or insulated walls were considered. All of these numerical works were conducted with a boundary-layer type mathematical model. The results of these studies showed that the effects of the water film evaporation on the heat transfer are rather substantial. Two-phase modelling of laminar film condensation from mixtures of a vapor and a non-condensing gas in parallel-plate channels has been studied by Siow et al. (2007). The channel is inclined downward from the horizontal and has an isothermal cooled bottom plate and an insulated upper one. Results for steam-air mixtures are presented and the effects of changes in the angle of inclination, inlet gas mass fraction, airflow Reynolds number and inlet temperature are examined. It was found that an increase in the angle of inclination results in thinner and faster moving liquid films. The authors show that, increasing airflow Reynolds number always produced thinner films and higher Nusselt number.

One of the main matters of the considered problem is the liquid film modelling. Many simplifying assumptions were used in the literature to derive mathematical models for the liquid phase. The boundary layer approximations are often used to derive a simplified model (Chow & Chung 1983; Chang et al. 1986; Lin et al. 1988; Yan & Lin 1989; Yan et al. 1989; He et al. 1998; Mezaache & Daguenet 2000).

Regarding the liquid-gas interface, one of the main hypotheses used in the literature is the so-called Nusselt's approximation (Nusselt, 1916), which neglects the shear stresses along this interface (Suzuki et al. 1983; Shembharkar & Pai 1986; Baumann & Thiele 1990). On the other hand, for accurate estimation of the evaporation process at the liquid-gas interface Maurya et al. (2010) used a numerical model that couples the Volume Of Fluid (VOF) method and the Ghost Fluid Technique (GFT). The VOF conservation method permits the interface tracking, while the GFT enables the authors to implement the interfacial condition (i.e. the prescription of saturation temperature) and the accurate estimation of temperature gradients on either side of the interface. This technique gives a more accurate estimation of the temperature gradients at the wall (and, hence, the heat transfer coefficient) than Nusselt's theory, which ignores the inertial effects.

An interesting way to deal with the liquid film modelling was used by some authors (Lin et al. 1988; Yan 1993; Fedorov et al. 1997; Volchkov et al. 2004; Huang et al. 2005; Azizi et al. 2007; Laaroussi et al. 2009; Oulaid et al. 2010b). These authors assumed an extremely thin liquid film so that it could be treated as a boundary condition. The validity of this assumption has been investigated by Yan (1992; 1993) for both air-water and air-ethanol systems. The author conducted a study of laminar mixed convection with evaporation of a liquid film dripping on the inner walls of a vertical channel. The walls are isothermally heated (Yan, 1993) or heated by a uniform heat flux (Yan, 1992). In each case the author conducted two studies: one with the conservation equations solved both in liquid and gas phases and in the other the film thickness was neglected. By comparing the results of these two studies, the author was able to demonstrate that the assumption of negligible film thickness is valid for low liquid film flow rates.

Many authors assumed constant thermo-physical properties evaluated at a reference temperature T_{ref} and mass fraction ω_{ref} obtained by these expressions: $T_{ref} = (2.T_w + T_{in})/3$ and $\omega_{ref} = (2.\omega_w + \omega_{in})/3$, where T_w , T_{in}, ω_w and ω_{in} are respectively the channel wall and inlet air temperatures and mass fractions. This way of evaluating thermo-physical properties, known as the one-third rule, has been used previously in the literature (Hubbard el al. 1975; Chow & Chung 1983). Chow & Chung (1983) performed a numerical study of evaporation of water in a laminar air stream. The water surface temperature was assumed to be constant and equal to the wet bulb temperature of the free stream. Different airflow temperatures were considered (150-500°C). By comparing the results of their models with variable and constant properties evaluated by the one-third rule, the authors concluded that the latter agree well with the variable-property results, even at high airflow temperatures. Earlier, Hubbard et al. (1975) conducted a numerical study on a single droplet evaporation of octane in stagnant air. The initial temperature of the droplet was 27°C and the air temperature was varied in the range 327-1727°C. By comparing their results using various reference property temperatures and concentrations, they concluded that the one-third rule yields the best agreement with the variable-properties model. The validity of the one-third rule for heat and mass transfer problems has been recently checked by Laaroussi et al. (2009). These authors mentioned that the one-third rule remains valid provided that the vapour mass fraction is small.

Flow reversal was studied analytically for fully developed flow with coupled heat and mass transfer by Salah El-Din (1992) and Boulama-Galanis (2004). These authors presented the criteria of occurrence of this phenomenon. These studies are some of the rare ones concerning flow reversal in combined mixed convection heat and mass transfer. On the other hand, flow reversal was extensively studied in thermal convection problems (Nguyen et al., 2004; Wang et al., 1994; Maré et al., 2008; Nesreddine et al., 1998; Faghri et al., 1980; Salah El-Din, 2001). For instance, Nguyen et al. (2004) studied the flow reversal and the instability of a transient laminar thermal mixed convection in a vertical tube subjected to a uniform time-dependent wall heat flux. The problem was investigated numerically by using a full 3D-transient-model and Boussinesq's assumption. The authors showed that the structures of the flow and thermal field appear to remain stable for Gr_T up to 5.0×10^5 and 10^6, respectively, for opposed and assisted-buoyancy cases. Beyond these critical values, the convergence of the numerical scheme becomes extremely slow and tedious. Thus, the authors were enable to point out any flow transition. Mixed convection with flow reversal in the thermal entrance region of horizontal and vertical pipes was studied numerically by Wang et al. (1994). Their results show that, for ascending flow in a vertical pipe, flow reversal is observed at the pipe centre in the heating case (or near the wall in the cooling case) at relatively high $|Gr/Re|$ with constant Péclet number. The regime of flow reversal has been identified for both heating and cooling cases in the Pe-$|Gr/Re|$ coordinates for a vertical pipe and in the Pe-Ra coordinates for heating in a horizontal pipe. Experimental and numerical studies of thermal mixed convection with flow reversal in coaxial double-duct heat exchangers have been carried out by Maré et al. (2008). Velocity vectors in a vertical heat exchanger for parallel ascending flow of water under conditions of laminar mixed convection have been determined experimentally using the particle image velocimetry technique. Their results show that measured velocity distributions are in very good agreement with corresponding numerical predictions and illustrate the simultaneous existence of flow reversal in the tube and the annulus for both heating and cooling of the fluid in the tube. As far as developing flows with heat and mass transfer are concerned, studies on flow reversal are rare (Laaroussi et al. 2009; Oulaid et al. 2010b). Indeed, the majority of the studies on this problem adopted a parabolic model where axial diffusion of momentum, energy and concentration is not taken into account. This model may not be appropriate for flows with low Péclet number where axial diffusion is not negligible (Nesreddine et al., 1998).

The objective of this paper is to summarize some recent work of the authors on heat and mass transfer in parallel-plate channel with phase change with special emphasis on flow reversal.

2. Problem definition and modelling

The physical problem is a parallel-plate channel (Figure 1). Liquid water film flows on the internal faces of these plates, which are maintained at a uniform temperature T_W or thermally insulated. An upward flow of ambient air enters at constant temperature T_0, relative humidity ϕ_0 and uniform velocity u_0. Steady state conditions are considered and the flow is assumed laminar. Radiation heat transfer, the transfer of energy by inter-diffusion of species, viscous dissipation and the work of the compressive forces are considered negligible. The secondary effects of concentration gradient on the thermal diffusion (Dufour effect) and thermal gradient on the mass diffusion (Soret effect) are neglected (Gebhart & Pera, 1971). Finally, the physical properties are assumed to be constant except for the density in the body forces, which is considered to be a linear function of temperature and mass fraction (Oberbeck-Boussinesq approximation),

$$\rho = \rho_0\left[1-\beta(T-T_0)-\beta^*(\omega-\omega_0)\right] \quad (1)$$

The coefficients of thermal and mass fraction expansion, are defined by,

$$\beta_T = \frac{-1}{\rho_0}\left(\frac{\partial \rho}{\partial T}\right)_{\omega=cste,\ p=cste}$$

$$\beta^* = \frac{-1}{\rho_0}\left(\frac{\partial \rho}{\partial \omega}\right)_{T=cste,\ p=cste} \quad (2)$$

As the gas flow considered here is humid air, which is assumed to be a perfect gas, these coefficients are given by,

$$\beta = 1/T_0$$

$$\beta^* = \frac{M_a - M_v}{(M_a - M_v)\omega + M_v} \quad (3)$$

Mass fraction of water vapour in the humid air ω is low, thus the last expression may be simplified as follows

$$\beta^* = M_a/M_v - 1 \quad (4)$$

The Oberbeck-Boussinesq assumption is considered valid for small temperature and mass fraction differences (Gebhart et al., 1988). The validity of this assumption for simultaneous heat and mass transfer was investigated by Laaroussi et al. (2009). These authors compared the Oberbeck-Boussinesq and variable-density models at relatively high temperatures in a vertical parallel-plate channel laminar mixed convection associated with film evaporation.

Their results showed that the Oberbeck-Boussinesq model works well for temperature and mass fraction differences less than 20K and 0.1kg/kg respectively.
The liquid films are assumed to be extremely thin. This hypothesis, known in the literature as the *zero film thickness model*, allows us to handle only the conservation equations in the gas flow with the appropriate boundary conditions. The liquid film is supposed to be at the imposed wall temperature. As reported in the introduction, the *zero film thickness model* is valid for low liquid flow rates (Yan 1992; Yan 1993).
The following non-dimensional variables are defined

$$X=\frac{x}{D_h},\ Y=\frac{y}{D_h},\ U=\frac{u}{u_0},\ V=\frac{v}{u_0},$$
$$P_m=\frac{p_m}{\rho_0\,u_0^2},\ \Theta=\frac{T-T_0}{T_W-T_0},\ C=\frac{\omega-\omega_0}{\omega_w-\omega_0} \tag{5}$$

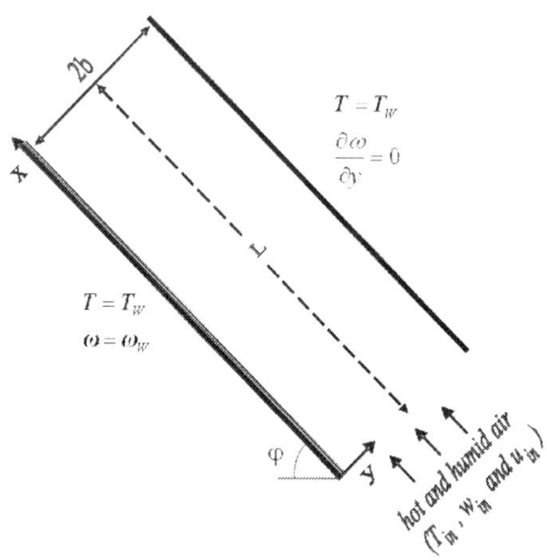

Fig. 1. Sketch of the physical system.

Using these variables and under the above assumptions, the governing equations of the problem are
Continuity equation

$$\frac{\partial U}{\partial X}+\frac{\partial V}{\partial Y}=0 \tag{6}$$

Stream wise momentum equation

$$\left(U\frac{\partial U}{\partial X}+V\frac{\partial U}{\partial Y}\right)=-\frac{\partial P_m}{\partial X}+\frac{1}{Re}\left(\frac{\partial^2 U}{\partial X^2}+\frac{\partial^2 U}{\partial Y^2}\right)+\frac{\sin\varphi}{Re^2}\left(Gr_T\Theta+Gr_M C\right) \tag{7}$$

Span wise momentum equation

$$\left(U\frac{\partial V}{\partial X}+V\frac{\partial V}{\partial Y}\right)=-\frac{\partial P_m}{\partial Y}+\frac{1}{Re}\left(\frac{\partial^2 V}{\partial X^2}+\frac{\partial^2 V}{\partial Y^2}\right)+\frac{\cos\varphi}{Re^2}\left(Gr_T\Theta+Gr_M C\right) \tag{8}$$

Energy equation

$$\left(U\frac{\partial \Theta}{\partial X}+V\frac{\partial \Theta}{\partial Y}\right)=\frac{1}{Re\,Pr}\left(\frac{\partial^2 \Theta}{\partial X^2}+\frac{\partial^2 \Theta}{\partial Y^2}\right) \tag{9}$$

Species conservation equation

$$\left(U\frac{\partial C}{\partial X}+V\frac{\partial C}{\partial Y}\right)=\frac{1}{Re\,Sc}\left(\frac{\partial^2 C}{\partial X^2}+\frac{\partial^2 C}{\partial Y^2}\right) \tag{10}$$

2.1 Boundary conditions

At the channel inlet (X = 0) the airflow velocity is assumed uniform and in the x-direction with constants temperature and mass fraction,

$$U = 1 \text{ and } V = C = \Theta = 0 \tag{11}$$

At the channel exit ($X = 1/2\gamma$), the flow is assumed fully developed. Hence,

$$\frac{\partial U}{\partial X}=\frac{\partial V}{\partial X}=\frac{\partial \Theta}{\partial X}=\frac{\partial C}{\partial X}=0 \tag{12}$$

Four types of conditions on the channel walls are considered here:

- **BC1**: Both of the pates ($Y = 0$ and $Y = 0.5$) are isothermal and wetted by thin liquid water films.
- **BC2**: The lower pate ($Y = 0$) is isothermal and wetted by thin liquid water films, while the upper one ($Y = 0.5$) is isothermal and dry.
- **BC3**: One of the plates ($Y = 0$) is isothermal and wetted by a thin liquid water film while the other ($Y = 0.5$) is thermally insulated and dry.
- **BC4**: both of the plates are thermally insulated and wetted by thin liquid water films.

At the insulated and dry plate, the airflow velocity components as well as the temperature and mass fraction gradients are zero. Thus,

$$U = V = 0, \text{ and } \frac{\partial C}{\partial Y}=\frac{\partial \Theta}{\partial Y}=0 \tag{13a}$$

At the isothermal and dry plate,

$$U = 0,\ V = 0,\ \frac{\partial C}{\partial Y}=\frac{\partial \Theta}{\partial Y}=0 \text{ and } \Theta = 1 \tag{13b}$$

At the isothermal and wet plate, the airflow axial velocity is obviously null (no slip) and its transverse velocity is equal to that of the vapour velocity at the liquid-gas interface. The mass fraction is that of saturation conditions at the plate's temperature. Hence,

$$U = 0,\ V = \pm Ve \text{ and } C = \Theta = 1 \tag{13c}$$

At the insulated and dry plate,

$$U = V = 0, \text{ and } \frac{\partial C}{\partial Y} = \frac{\partial \Theta}{\partial Y} = 0 \tag{13a}$$

At the isothermal and dry plate,

$$U = 0,\ V = 0,\ \frac{\partial C}{\partial Y} = 0 \text{ and } \Theta = 1 \tag{13b}$$

At the insulated and wet plate,

$$U = 0,\ V = \pm Ve,\ \frac{\partial \Theta}{\partial Y} = 0 \text{ and } C = 1 \tag{13d}$$

V_e represents the evaporated liquid film into the air stream or condensed humid air vapour on the wet plate. Its non-dimensional form is (Burmeister, 1993)

$$V_e = \frac{-1}{\text{Re Sc}} \frac{(\omega_w - \omega_0)}{(1 - \omega_w)} \left(\frac{\partial C}{\partial Y} \right)_{Y=0} \tag{14}$$

The mass fraction at the wall ω_w corresponding to the saturation conditions at T_w, is calculated by assuming that air-vapour mixture is an ideal gas mixture, and its expression is given by:

$$\omega_W = \frac{\dfrac{M_v}{M_a}}{\dfrac{M_v}{M_a} + \dfrac{P}{P_{sat}(T_W)} - 1} \tag{15}$$

2.2 Flow, heat and mass transfer parameters

The friction factor at the channel walls is

$$f.\text{Re} = 2 \left(\frac{\partial U}{\partial Y} \right)_{Y=0,Y=0.5} \tag{16}$$

Heat transfer between the wet walls and the humid air is the sum of a sensible and a latent component flux

$$q'' = q''_S + q''_L = -k \left. \frac{\partial T}{\partial y} \right|_{y=0} - \frac{\rho D h_{fg}}{1 - \omega_w} \left. \frac{\partial w}{\partial y} \right|_{y=0} \tag{17}$$

Therefore, the total local Nusselt number is

$$Nu_T = \frac{h D_h}{k} = \frac{q''_T D_h}{k(T_w - T_m)} = Nu_S + Nu_L \tag{18}$$

where

$$Nu_S = -\frac{1}{1-\Theta_m}\left.\frac{\partial \Theta}{\partial Y}\right|_{Y=0} \tag{19}$$

$$\mathrm{Nu_L} = -\frac{\rho\, D\, h_{fg}}{k(1-\Theta_\mathrm{m})}\frac{(\omega_w - \omega_0)}{(1-\omega_w)}\frac{1}{(T_w - T_0)}\left.\frac{\partial \mathrm{C}}{\partial \mathrm{Y}}\right|_{Y=0} \tag{20}$$

The dimensionless bulk temperature is defined as follows

$$\Theta_m = \frac{1}{U_m}\int_0^{0.5} U\ \Theta\ dY \tag{21}$$

The Sherwood number characterizes mass transfer at the air-liquid interface

$$Sh = \frac{h_m D_h}{D} = \frac{\dot{m}}{\rho(\omega_W - \omega_m)}\frac{D_h}{D} \tag{22}$$

At the liquid-gas interface the liquid and vapour mass fluxes are equal. This flux, which is due to convective and diffusion transfer, is given by Burmeister (1993)

$$\dot{m} = \rho_\ell v_\ell = \omega_W \rho\ \ v_e - \rho\, D\left.\frac{\partial \omega}{\partial y}\right|_{y=0} \tag{23}$$

For small mass transfer at the liquid-gas interface, as it is the case here, the following expression for Sherwood number is derived (Oulaid et al. 2010b; Huang et al. 2005)

$$Sh = \frac{-1}{(1-C_m)}\left.\frac{\partial C}{\partial Y}\right|_{Y=0} \tag{24}$$

3. Numerical method

A finite volume method is used for the discretization of the governing equations (Eq. 3-7). The combined convection-diffusion term is calculated with a power-law scheme and an the block-correction method coupled with a line-by-line procedure is used to solve the resulting algebraic equations (Patankar, 1981). The velocity-pressure coupling is treated by the SIMPLER algorithm (Patankar 1980; Patankar 1981). Convergence of this iterative procedure is declared when the relative variations of any dependent variable is less than 10^{-4} and the mass residual falls below 10^{-6} at all the grid points. The grid is non-uniform in both the streamwise and transverse directions with greater node density near the inlet and the walls. Furthermore, different grid sizes were considered to ensure that the solution was grid-independent. Results of the grid sensibility study are given by Ait Hammou et al. (2004) and Kassim et al. (2010). These results show that heat and mass transfer parameters, as well as the local variables, obtained for three grids 100x35, 200x70 and 400x140 vary within less than 5%. Hence, a grid with 100 nodes in the axial direction and 35 nodes in the transverse one is adopted for the results presented here.

Validation of the computer code and the mathematical model has been carried out first for hydrodynamically and thermally developing forced thermal convection (Ait Hammou et al.

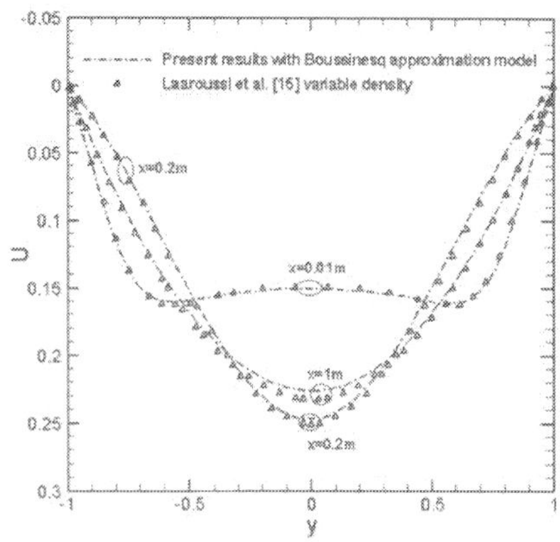

Fig. 2. Comparison between the Oberbeck-Boussinesq model (Oulaid et al. 2010b) and the variable density model (Laaroussi et al., 2009), for $T_W = T_0$=327 K, ω_0=0 and ω_W=0.1.

2004). The results are compared to those obtained by Mercer in the case of a parallel-plate channel with one of the plates isothermal and the other insulated (Kassim et al. 2010)a. Moreover, comparison with the available heat-mass transfer results in the literature for the isothermal wetted parallel-plate channel are satisfactory (Oulaid et al. 2010b).

Figure 2 shows comparison of the velocity profiles obtained by the Oberbeck-Boussinesq model (Oulaid et al. 2010b) and those calculated by the variable-density model (Laaroussi et al., 2009). The discrepancy between the two models is less than 4%.

In view of these successful validations, the computer code, as well as the mathematical model, are considered reliable.

The results presented hereafter are obtained for constant thermo-physical properties evaluated at a reference temperature and mass fraction calculated by the *one-third rule* (Chow and Chung, 1983). As stated in the introduction, this way of evaluating thermo-physical proprieties is believed to be relevant for the present conditions.

4. Uniform wall temperature

In the case of uniform channel walls temperature, both of the plates are subject to the boundary conditions **BC1** (Eq. 13c) for the vertical symmetric channel or **BC2** (Eqs. 13b-c) for the inclined isothermal asymmetrically wetted channel. The plates are maintained at a fixed temperature T_w = 20°C, which is supposed to be the liquid water film temperature. The boundary condition for the film is that of saturated air at T_w, hence ω_w = 14.5 g/kg. The Reynolds number is set at 300 and the channel's aspect ratio is $\gamma = 1/65$ for BC1 and $\gamma = 1/50$ for BC2.

4.1 Flow structure

Figure 3 presents the axial velocity profiles for the vertical symmetric channel. The case of forced convection ($Gr_M = Gr_T = 0$) is also reported to point out the effects of buoyancy forces. Close to the channel entrance (X = 0.15) the velocity profiles for mixed and forced convection are quite close, due to the prevalence of the viscous forces in the developing

boundary layer. As the air moves downstream, these forces become weak and the effect of buoyancy forces becomes clear. As both thermal and solutal Grashof numbers are negative, buoyancy forces act in the opposite direction of the upward flow and decelerate it near the walls. This deceleration produces a flow reversal close to the channel walls at $X = 2.31$. Buoyancy forces introduce a net distortion of the axial velocity profile compared to the case of forced convection. The flow reversal is clear in Figure 4, which show the evolution of the axial velocity, near the plates. Three different temperatures at the channel inlet are represented in this figure: T_0= 30°C (Gr_T= -0.88.10^5 and Gr_M=1.07.10^4), 41°C (Gr_T= -1.71.10^5 and Gr_M= 0) and 50°C (Gr_T= -2.29.10^5 and Gr_M=-1.29.10^4). We notice that the axial velocity takes negative values for the last two cases over large parts of the channel length. Along these intervals, air is flowing in the opposite direction of the entering flow. That change in the flow direction gives rise to a recirculation cell and to the flow reversal phenomenon.
Figure 5 shows the streamlines for the vertical symmetric channel. Two recirculation cells are present close to the channel entrance. Careful inspection of Fig. 5 show that the streamlines contours in the recirculation cells are open near the plates. Indeed, these streamlines are normal to the channel walls. Local velocity is then directed to these walls, as condensation occurs here (Oualid et al., 2010b).

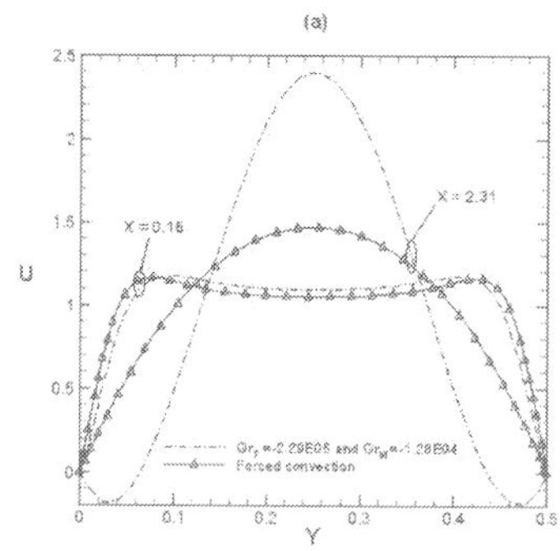

Fig. 3. Axial velocity profiles in the vertical symmetric channel for T_0= 50°C and ϕ_0 = 30% (Oulaid et al., 2010b)

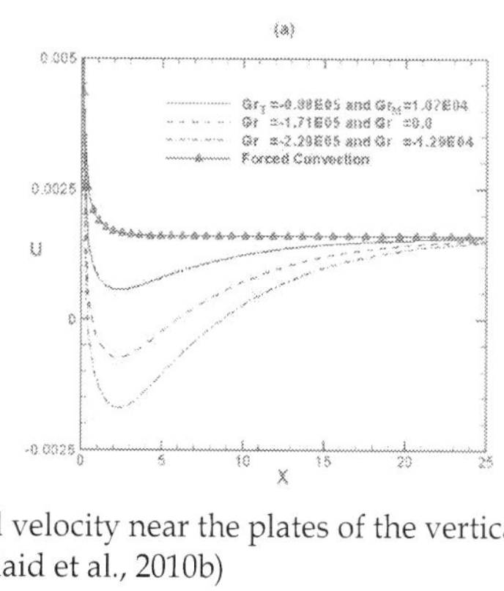

Fig. 4. Evolution of the axial velocity near the plates of the vertical symmetric channel for ϕ_0 = 30% at Y=1.33 10^{-4} (Oulaid et al., 2010b)

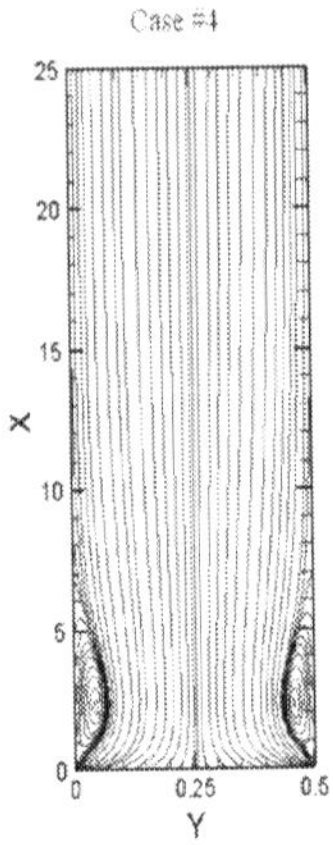

Fig. 5. Streamlines in the vertical symmetric channel for T_0= 41°C and ϕ_0 = 43.25% (Gr_T = -1.71.10^5 and Gr_M= -10^4) (Oulaid et al., 2010b).

For the inclined isothermal asymmetrically wetted channel, the flow structure is represented in Fig. 6 by the axial velocity profiles for different inclination angles. Remember that for this case only the lower plate (Y=0) is wet while the upper one is dry. The maximum of distortion of U is obtained for the vertical channel, for which buoyancy forces takes their maximum value in the axial direction. Fig. 6 show that flow reversal occurs for φ = 60° and

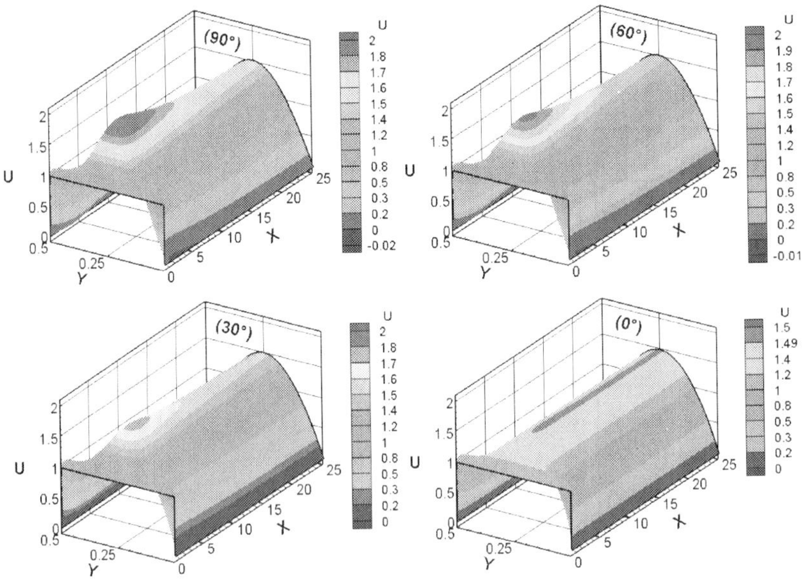

Fig. 6. Axial velocity profiles in the inclined isothermal asymmetrically wetted channel for T_0= 40°C and ϕ_0= 45.5% (Gr_T= -1.64 10^5 and Gr_M= -10^4) (Oulaid et al., 2010d).

90°. This is clearer from Fig. 9, which presents the friction factor f at the lower wet plate in the isothermal asymmetrically wetted channel. Negative values of f occur in the flow reversal region. Streamlines presented in Fig. 8, show the recirculation cells near the lower wet plate, where the airflow is decelerated due to its cooling. It can be seen clearly from Fig. 8 that the streamlines contours in the flow reversal region are not closed. Indeed, close to the lower wet plate, airflow velocity is directed towards the channel wall. This velocity, which is equal to the vapour velocity at the air-liquid interface V_e, is shown in Fig. 9. It is noted that V_e is negative which indicate that water vapour is transferred from airflow towards the wet plate. Thus, this situation corresponds to the condensation of the water vapour on that plate. It is interesting to note that close to the channel entrance, ($X < 4.37$) the magnitude of V_e for forced convection (and the horizontal channel too) is larger than for the inclined channel; while further downstream forced convection results in lower values of V_e magnitude. This inversion in V_e tendency occurs at the end of the flow reversal region ($X = 4.37$). In this region, as the channel approaches its vertical position, buoyancy forces slowdown airflow thus, water vapour condensation diminishes.

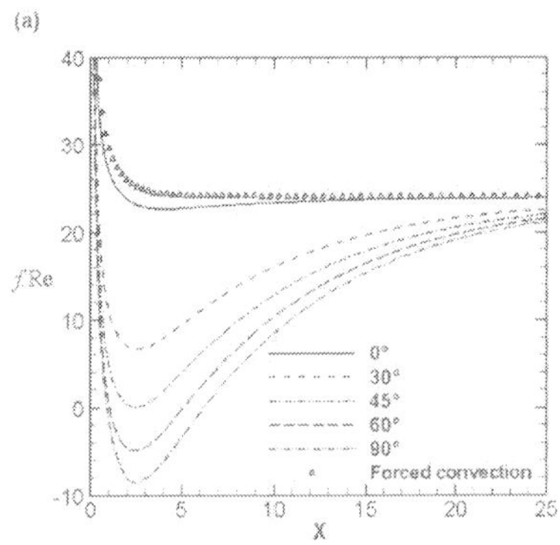

Fig. 7. Axial evolution of the friction factor at the lower wet plate in the isothermal asymmetrically wetted channel for T_0= 40°C and ϕ_0= 45.5% (Gr_T= -1.64 10^5 and Gr_M= -10^4) and different inclination angles (Oulaid et al., 2010d).

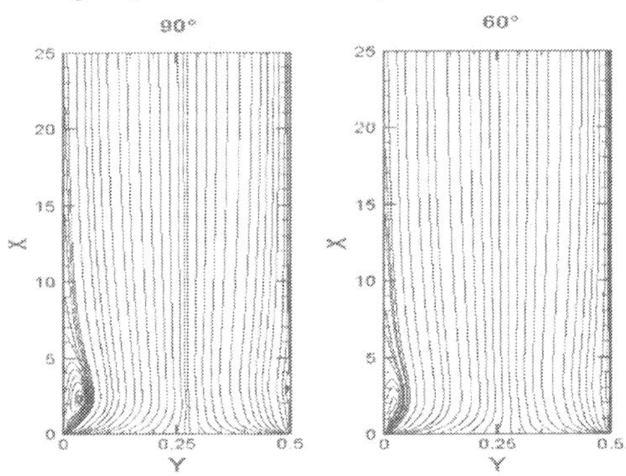

Fig. 8. Streamlines in the isothermal asymmetrically wetted channel for T_0= 40°C and ϕ_0= 45.5% (Gr_T= -1.64 10^5 and Gr_M= -10^4) and different inclination angles (Oulaid et al., 2010d).

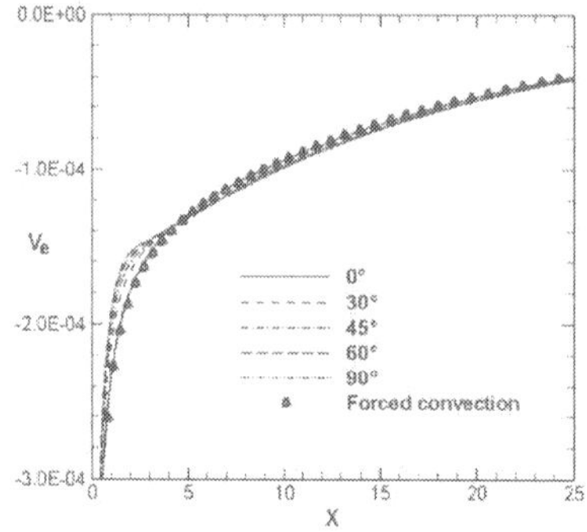

Fig. 9. Vapour velocity at the lower plate of the isothermal asymmetrically wetted channel for T_0= 40°C and ϕ_0= 45.5% (Gr_T= -1.64 10^5 and Gr_M= -10^4) and different inclination angles (Oulaid et al., 2010d).

4.2 Thermal and mass fraction characteristics

Figure 10 presents the evolution of the latent Nusselt number (Nu_L) at the wet plate of the isothermal asymmetrically wetted inclined channel. Nu_L is positive indicating that latent heat flux is directed towards the wet plate. Thus, water vapour contained in the air is condensed on that plate, as shown in Fig. 9. As the air moves downstream, water vapour is removed from the air; thus, the gradient of mass fraction decreases, and that explains the decrease in Nu_L. In the first half of the channel, Nu_L is less significant as the channel approaches its vertical position, due to the deceleration of the flow by the opposing buoyancy forces as depicted above. Close to the channel exit, the buoyancy forces magnitude diminishes; hence, Nu_L takes relatively greater values for the vertical channel (Oulaid et al., 2010a). Figure 11 show the Sensible Nusselt number at the wet plate of the isothermal asymmetrically wetted channel. It is clear that the buoyancy forces diminish heat transfer. This diminution is larger in the recirculation zone. Figure 12 presents Sherwood number at the wet plate of the isothermal asymmetrically wetted channel. The behaviour of Sh resembles to that of Nu_S, as Le ≈ 1 here.

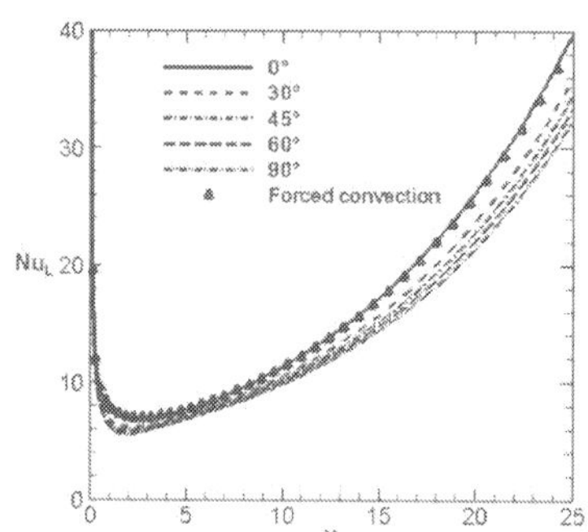

Fig. 10. Latent Nusselt number at the wet plate of the isothermal asymmetrically wetted inclined channel for T_0= 40°C and ϕ_0= 45.5% (Gr_T= -1.64 10^5 and Gr_M= -10^4) and different inclination angles (Oulaid et al., 2010d).

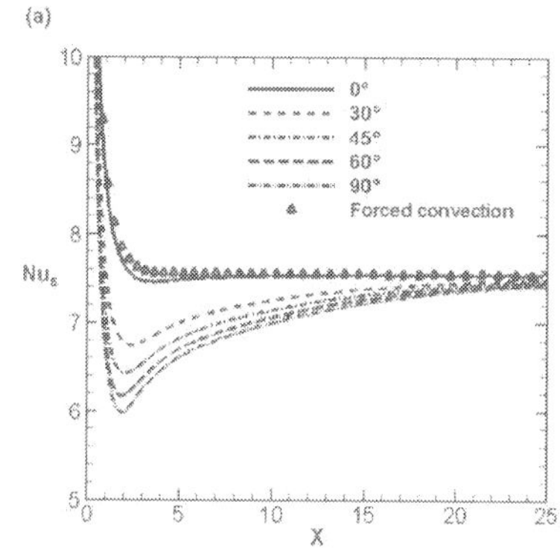

Fig. 11. Sensible Nusselt number Nu_S at the wet plate of the isothermal asymmetrically wetted inclined channel for T_0= 40°C and ϕ_0= 45.5% (Gr_T= -1.64 10^5 and Gr_M= -10^4) and different inclination angles (Oulaid et al., 2010d).

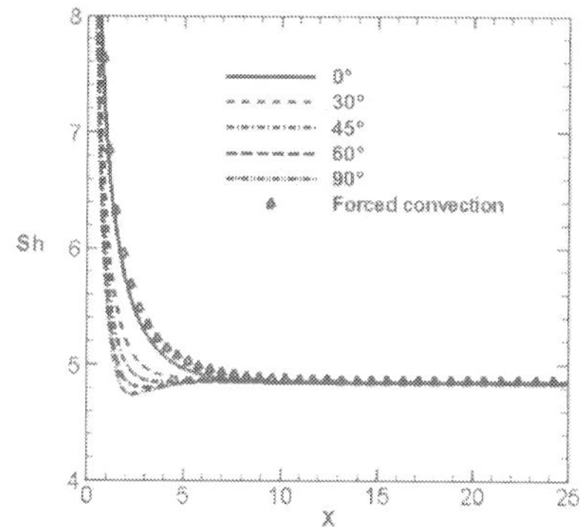

Fig. 12. Sherwood number Sh at the wet plate of the isothermal asymmetrically wetted inclined channel for T_0= 40°C and ϕ_0= 45.5% (Gr_T= -1.64 10^5 and Gr_M= -10^4) and different inclination angles (Oulaid et al., 2010d).

4.3 Flow reversal chart

As stated in the introduction, flow reversal in heat-mass transfer problems was not studied extensively in the literature. This phenomenon is an important facet of the hydrodynamics of a fluid flow and its presence indicates increased flow irreversibility and may lead to the onset of turbulence at low Reynolds number. Hanratty et al. (1958) and Scheele & Hanratty (1962) were pioneers in experimental study of flow reversal in vertical tube mixed convection. These authors have shown that the non-isothermal flow appears to be highly unstable and may undergo its transition from a steady laminar state to an unstable one at rather low Reynolds number. The unstable flow structure has shown, the 'new equilibrium' state that consisted of large scale, regular and periodic fluid motions. The condition of the existence of flow reversal in thermal mixed convection flows were established by many authors for different conditions (Wang et al. 1994; Nesreddine et al. 1998, Zghal et al. 2001;

Behzadmehr et al. 2003). As heat and mass transfer mixed convection is concerned, such studies are rare as depicted in the introduction.

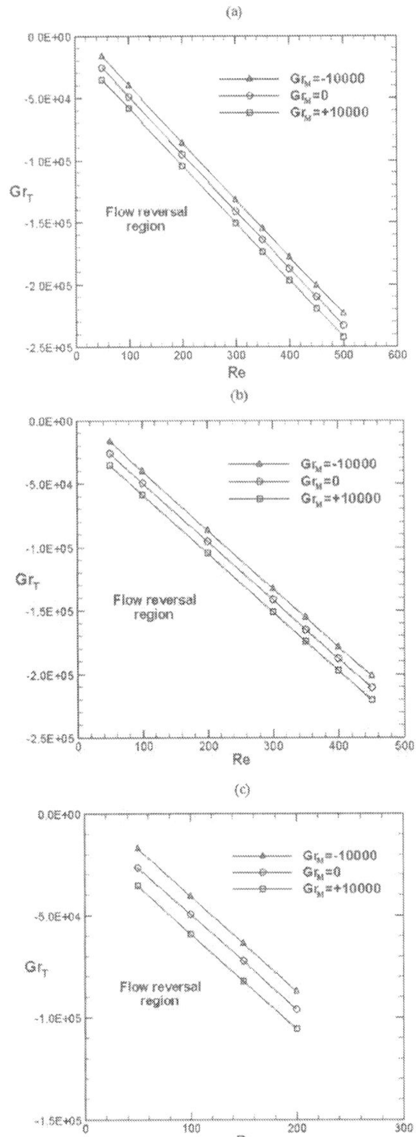

Fig. 13. Flow reversal chart for the vertical symmetric channel (a) $\gamma = 1/35$, (b) $\gamma = 1/50$ and (c) $\gamma = 1/65$. (Oulaid et al. 2010b)

The conditions for the existence of flow reversal was established in the symmetric vertical channel (Oulaid et al., 2010b) and the isothermal asymmetrically wetted inclined channel (Oulaid et al., 2010d). For a given Re we varied T_0 (i.e. Gr_T) at fixed Gr_M (i.e. ϕ_0) in asequence of numerical experiments until detecting a negative axial velocity. All the considered combinations of temperature and mass fraction satisfy the condition for the application of the Oberbeck-Boussinesq approximation, as the density variations do not exceed 10%. These series of numerical experiments enabled us to draw the flow reversal charts for different aspect ratios of the channel (γ = 1/35, 1/50 and 1/65). These flow reversal charts are presented in Figs 13-14. These charts would be helpful to avoid the situation of unstable flow associated with flow reversal. The flow reversal charts are also expected to fix the validity limits of the mathematical parabolic models frequently used in the heat-mass transfer literature (Lin et al., 1988; Yan et al., 1991; Yan and Lin, 1991; Debbissi et al., 2001; Yan, 1993; Yan et al., 1990; Yan and Lin, 1989; Yan, 1995).

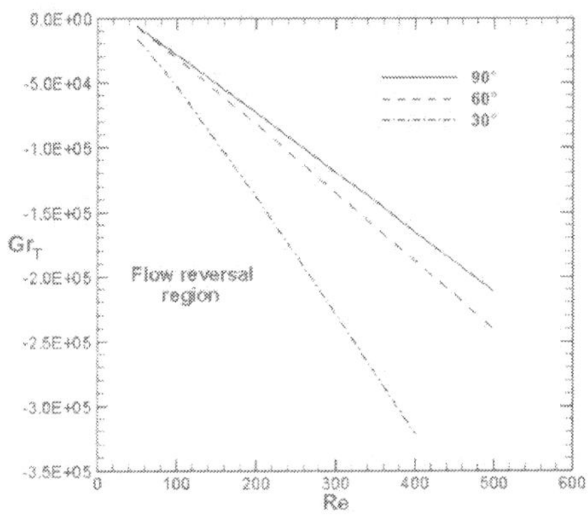

Fig. 14. Flow reversal chart in the isothermal asymmetrically wetted inclined channel for $Gr_M = -10^4$ and $\gamma = 1/65$ (Oulaid et al. 2010d)

5. Asymmetrically cooled channel

For the asymmetrically cooled parallel-plate channel, the plates are subject to the boundary condition **BC3** (i.e. one of the plates is wet and maintained at a fixed temperature T_w = 20°C, while the other is dry and thermally insulated). The Reynolds number is set at 300 and the channel's aspect ratio is γ = 1/130 (L =2m).

5.1 Flow structure

The streamlines for the asymmetrically cooled vertical channel is presented in Fig. 15. This figure shows the recirculation cell, which is induced by buoyancy forces. The dimension of this recirculation cell is more significant than in the case of the isothermal channel (Figs. 5 and 8). The recirculation cell occupies a larger part of the channel and its eye is closer to the channel axis. The flow structure is strongly affected by the buoyancy forces. These forces induce a momentum transfer from the wet plate, where the flow is decelerated, towards the dry plate, where the flow is accelerated (Kassim et al. 2010a).

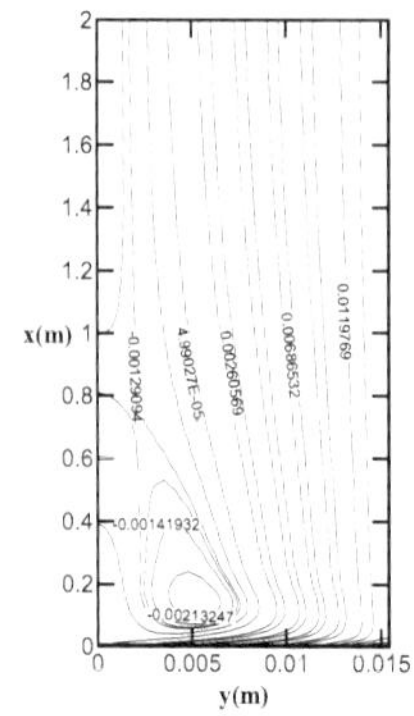

Fig. 15. Streamlines in asymmetrically cooled vertical channel for T_0 =70°C and ϕ_0 = 70% (Gr_T= - 1'208'840 and Gr_M= -670'789) (Kassim et al. 2010a)

5.2 Thermal and mass fraction characteristics

The vapour mass flux at the liquid-air interface is shown in Figure 16. The represented cases correspond to vapour condensation (water vapour contained in airflow is condensed at the isothermal wetted plate in all cases). For ϕ_0 = 10%, phase change and mass transfer at the liquid-air interface is weak, thus condensed mass flux decreases rapidly and stretches to zero. Considering the other cases (ϕ_0 = 30% or 70%) the behaviour of the condensed mass flux is complex. It exhibits local extrema, which are more pronounced as ϕ_0 is increased. Its local minimum occurs at the same axial location of the recirculation cell eye (Fig. 15). Thus, it can be deduced that the increase of the vapour mass flux towards its local maximum is attributed to the recirculation cell. The latter induces a fluid mixing near the isothermal plate and thus increases condensed mass flux. As the recirculation cell switches off, the condensed mass flux decreases due to the boundary layer development.

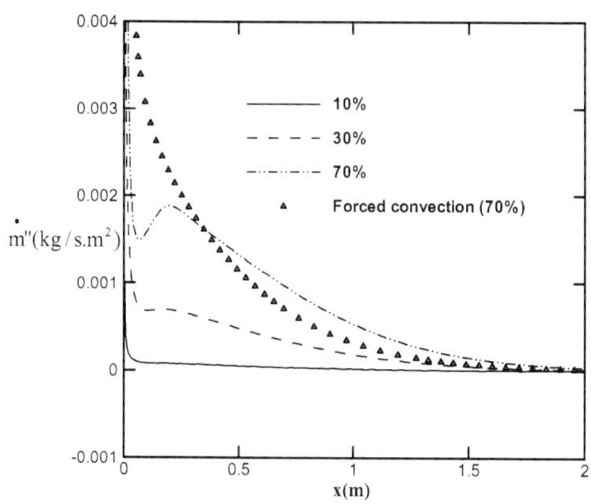

Fig. 16. Vapour mass flux at the wet plate in asymmetrically cooled vertical channel of T_0= 70°C and different inlet humidity ϕ_0= 10% (Gr_T= - 1'180'887 and Gr_M= - 24'359), 30% (Gr_T= - 1'189'782 and Gr_M= - 226'095) and 70% (Gr_T= - 1'208'840 and Gr_M= - 670'789) (Kassim et al. 2010a)

Figure 17 presents axial development of airflow temperature at the channel mid-plane (y= 0.0068m). Airflow is being cooled in all cases as it goes downstream, due to a sensible heat transfer from hot air towards the isothermally cooled plate. The airflow temperature at the channel mid-plane exhibits two local extremums near the channel entrance. These extremums are more pronounced for ϕ_0 = 70%. In this case the local minimum of air temperature is 44.24°C which occurs at x = 0.092m and the local maximum is 46.59°C which occurs at x = 0.208m. These axial locations are closer to that corresponding to local minimum and maximum of the condensed mass flux (Fig. 16). Once again, it is clear that the existenceof local extremums of air temperature at the channel mid-plane is related to the fluid mixing induced by flow reversal near the isothermal wet plate. This fluid mixing increases the condensed mass flux, thus the airflow temperature increases. Indeed, vapour condensation releases latent heat, which is partly absorbed by airflow. Moreover, close to the channel inlet, airflow at the channel mid-plane is cooler as ϕ_0 is increased. In this region the buoyancy forces decelerate the upward airflow and induce flow reversal and thus, increase the air-cooling through sensible heat transfer towards the isothermal plate (Kassim et al. 2010a).

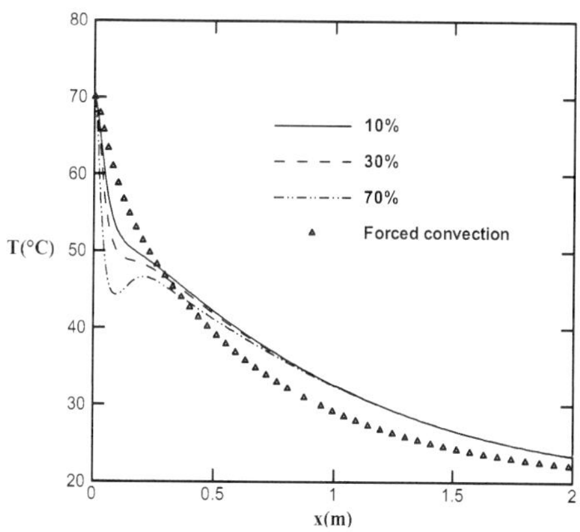

Fig. 17. Airflow temperature at the mid-plane (y = 0.0074m) of the asymmetrically cooled vertical channel for T_0= 70°C and different inlet humidity ϕ_0= 10% (Gr_T= - 1'180'887; Gr_M= - 24'359), 30% (Gr_T= - 1'189'782; Gr_M= - 226'095) and 70% (Gr_T= - 1'208'840; Gr_M= - 670'789) (Kassim et al. 2010a)

Axial evolution of the local latent Nusselt number Nu_L at the isothermal plate is represented in Fig. 18. For ϕ_0 = 10%, Nu_L diminishes and stretches to zero at the channel exit, as phase change and mass transfer at the liquid-air interface is weak (Fig. 16). The axial evolution of Nu_L for ϕ_0 = 30% and 70%, is more complex and exhibits local minimum and maximum. The positions of these extremums, which are the same as for the vapour mass flo rate at the liquid-air interface (Fig. 16), depend on ϕ_0 and are more pronounced for ϕ_0 = 70%. Furthermore, the development of Nu_L and $\dot{m}$ is analogous. Thus, the occurrence of the local extremums of Nu_L is due to the interaction between the vapour condensation and flow reversal as explained above.

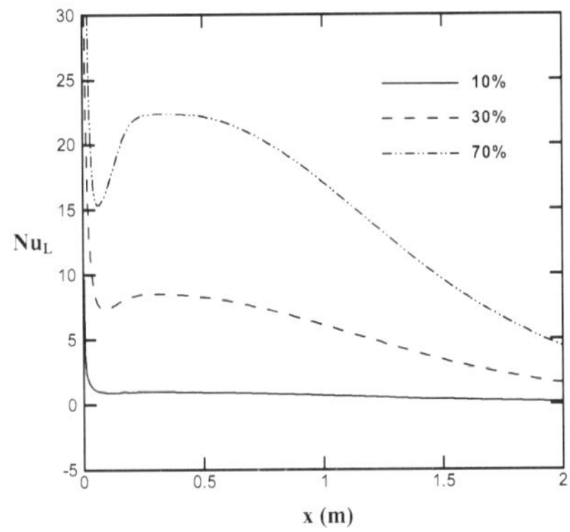

Fig. 18. Latent Nusselt number at the wet plate of the asymmetrically cooled vertical channel for T_0= 70°C and different inlet humidity ϕ_0= 10% (Gr_T= - 1'180'887; Gr_M= - 24'359), 30% (Gr_T= - 1'189'782; Gr_M= - 226'095) and 70% (Gr_T= - 1'208'840; Gr_M= - 670'789) (Kassim et al. 2010a)

6. Insulated walls

The channel walls are subject to the boundary condition **BC4** (i.e., both of the plates are thermally insulated and wet). In this case, an experimental study was conducted and its resuts are compared to the numerical one. Detailed description of the experimental setup is given by Cherif et al. (2010). Only some important aspects of this setup are reported here. The channel is made of two square stainless steel parallel plates (50cm by 50cm) and two Plexiglas rectangular parallel plates (50cm by 5cm). Thus, the channel's aspect ratio is γ = 1/10. The channel is vertical and its steel plates are covered on their internal faces with falling liquid films. In order to avoid dry zones and wet the plates uniformly, very thin tissues support these films. Ambient air is heated through electric resistances and upwards the channel, blown by a centrifugal fan, via an settling box equipped with a honeycomb. Airflow and water film temperature are measured by means of Chromel-Alumel (K-type) thermocouples. For the liquid films, ten thermocouples are welded along each of the wetted plates. For the airflow, six thermocouples are placed on a rod perpendicular to the channel walls. This rod may be moved vertically in order to obtain the temperature at different locations. The liquid flow rate is low and a simple method of weighing is sufficient to measure it. The evaporated mass flux was obtained by the difference between the liquid flow rate with and without evaporation (Cherif et al, 2010; Kassim et al. 2009; Kassim et al. 2010b). The liquid film flow rate was set between 1.55 10^{-4} $kg.s^{-1}.m^{-1}$ and 19.4 10^{-4} $kg.s^{-1}.m^{-1}$. These values are very low compared to the considered mass fluxes in Yan (1992; 1993). Thus, it is expected that the *zero film thickness model* will be valid. The comparison of the numerical and experimental results is conducted for laminar airflow. The Reynolds number is set at 1620 (U_0 = 0.27 m/s).

The airflow temperature is presented in Fig. 19 at three different axial locations. It is clear that the concordance between the experimental measurements and the numerical results is satisfactory. This concordance is excellent at the plates and close to it. Nevertheless, the difference between these results does not exceed 8% elsewhere. It is noted that airflow is cooled as it upwards the channel. This cooling essentially occurs in the vicinity of the wet plates. The wet plates temperature profile is presented in Fig. 20. It should be noted that, in the experimental study, T_w is the water film temperature. The comparison between the measurements and the numerical results is good, as the difference is less than 1.5%. It can be deduced that the assumption of extremely thin liquid film, adopted in the numerical model, is reliable here. On the other hand, it is noted that the liquid film is slightly cooled and then a bit heated in contact with the hot airflow. It is important to remind that air enters the channel at x=0m while the water film enters at x=0.5m. However, the water film temperature remains quasi-constant within 2.5°C. It can be deduced that air is cooled mostly by latent heat transfer associated to water evaporation. The global evaporated mass flux is presented in Fig. 21 with respect to the inlet air temperature T_0. This mass flux is calculated in the numerical study by the following equation,

$$\dot{m}_{ev} = \frac{1}{L}\int_0^L \rho V_e \, dx \qquad (25)$$

Experimentations were performed for three inlet air temperature 30, 35 et 45°C. Fig. 21 shows that the evaporated mass flux increases as the inlet air temperature is increased. This is attributed to the increase of sensible heat transfer from the airflow to the water film, which results in mass transfer from the film to the airflow associated with water evaporation. Meticulous examination of Fig. 21 reveals that numerical calculation predicts well the measured evaporated mass flux for T_0 = 30°C. For larger inlet air temperature, the mathematical model underestimates the evaporated mass flux. Indeed the discrepancy between the calculations and the measurements increases with T_0. It is believed that this is due to the calculation method. Indeed, in Eq 25 the density is considered constant and calculated at the reference temperature (obtained by the *one third rule*). However, global agreement between the calculations and the measurements is found in Fig. 21 as the discrepancy does not exceed 10%.

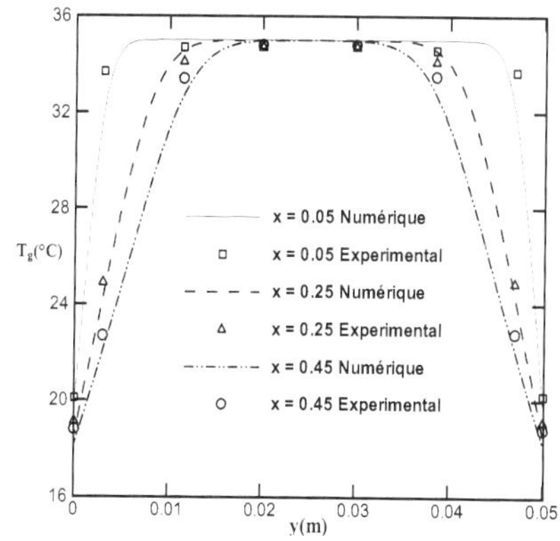

Fig. 19. Airflow temperature profiles at different axial locations for the insulated parallel-plate vertical channel (Kassim et al., 2010b). *Experimental conditions:* u_0 *= 0.27m/s, Re = 1620, water flow rate =1.5 l/h, inlet liquid temperature= 17.7°C, ambient air humidity = 41% and temperature = 18.2 °C, inlet airflow humidity* ϕ_0 *= 16% and temperature* T_0 *= 45°C.*

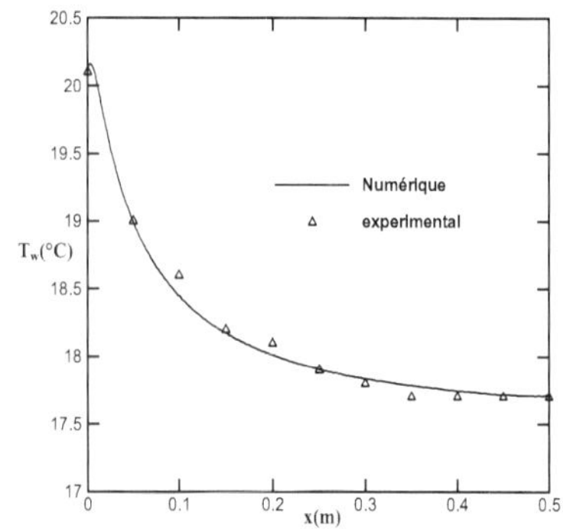

Fig. 20. Wall temperature for the insulated parallel-plate vertical channel (Kassim et al., 2009). Experimental conditions: u_0 = 0.27m/s, Re = 1620, water flow rate = l/h, inlet liquid temperature= 18°C, ambient air humidity = 45% and temperature = 19 °C, inlet airflow humidity ϕ_0 = 16 and temperature T_0 = 45°C.

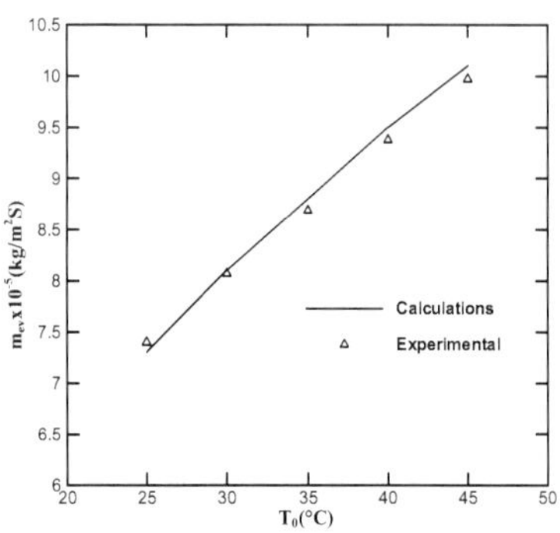

Fig. 21. Evaporated mass flux at the liquid-air interface (Kassim et al., 2009). Experimental conditions: u_0 = 0.27m/s, Re = 1620, water flow rate =1 l/h, inlet liquid temperature= 18°C, ambient air humidity = 45% and temperature = 19 °C.

7. Conclusion

Heat and mass transfer mixed convection in channels, with special emphasis on phase change and flow reversal, is considered. The literature review reveals that the flow reversal phenomenon in such problems was not extensively studied. Some recent numerical and experimental work of the authors are reported. The chanel is a parallel-plate one, which may be isothermally cooled or thermally insulated. Water liquid films stream along one or both of the plates. The effects of buoyancy forces on the flow structure as well as on heat and mass transfer characteristics are analysed. The conditions of the occurrence of flow reversal are particularly addressed. Flow reversal charts, which specify these conditions, are given. The comparison between the numerical and experimental results is satisfactory. However, the numerical study is limited by its hypotheses. Future research may concern more elaborated mathematical models taking into account the variability of the thermo-physical properties and the thickness of the liquid film. On the other hand, as flow reversal may induce flow instability, a transient mathematical model, such as the low Reynolds number turbulence model may be more appropriate. Finally, more experimental investigations of the considered problem is needed.

8. Acknowledgements

The experimental study was conducted at the *Laboratoire d'Energétique et des Transferts Thermiques et Massiques, Faculty of Sciences, Bizerte* (Tunisa) . The financial support of the Morocco-Tunisian Cooperation Program (Grant no MT/06/41) is acknowledged.

9. Nomenclature

b	half width of the channel (m)
C	dimensionless mass fraction, $= (\omega - \omega_0) . (\omega_w - \omega_0)^{-1}$
D	mass diffusion coefficient $[m^2.s^{-1}]$
D_h	hydraulic diameter, = 4b [m]
f	friction factor
g	gravitational acceleration $[m.s^{-2}]$
Gr_M	mass diffusion Grashof number, $= g.\beta^*.D_h^3.(\omega_w - \omega_0).\nu^{-2}$
Gr_T	thermal Grashof number, $= g.\beta.D_h^3.(T_w - T_0).\nu^{-2}$
h	local heat transfer coefficient $[W.m^{-2}.K^{-1}]$
h_m	local mass transfer coefficient $[m.s^{-1}]$
h_{fg}	latent heat of vaporization $[J.kg^{-1}]$
k	thermal conductivity $[W\ m^{-1}\ K^{-1}]$
L	channel height [m]
$\dot{m}$	vapour mass flux at the liquid-gas interface $[kg.s^{-1}.m^{-2}]$
M_a	molecular mass of air $[kg.kmol^{-1}]$
M_v	molecular mass of water vapour $[kg.kmol^{-1}]$
N	buoyancy ratio, $= Gr_M/Gr_T$
Nu_S	local Nusselt number for sensible heat transfer
Nu_L	local Nusselt number for latent heat transfer
p	pressure
P_m	modified dimensionless pressure, $= (p + \rho_0\ g\ x).(\rho_0\ u_0^2)^{-1}$
Pr	Prandtl number, $= \nu/\alpha$
q''	heat flux $[W.m^{-2}]$
Re	Reynolds number, $= u_0.D_h.\nu^{-1}$
R_{Nu}	ratio of latent to sensible Nusselt numbers, $= Nu_L/Nu_S$
Sc	Schmidt number, $= \nu/D$
Sh	Sherwood number
T	temperature [K]
u, v	velocity components $[m.s^{-1}]$
U, V	dimensionless velocity components, $= u/u_0$, v/u_0
V_e	dimensionless transverse vapour velocity at the air-liquid interface.

x, y axial and transverse co-ordinates [m]
X, Y dimensionless axial and transverse co-ordinates, = x/D_h, y/D_h

Greek symbols

α thermal diffusivity [$m^2 s^{-1}$]
β coefficient of thermal expansion,
β^* coefficient of mass fraction expansion,
γ aspect ratio of the channel, = 2b/L
Θ dimensionless temperature, = $(T - T_0)/(T_w - T_0)$
ν kinematic viscosity [$m^2.s^{-1}$]
ρ density [$kg.m^{-3}$]
ϕ relative humidity (%)
φ inclination angle of the channel
ω mass fraction [kg of vapour/ kg of mixture]

Subscripts

a relative to the gas phase (air)
L relative to latent heat transfer
ℓ relative to the liquid phase
m mean value
0 at the inlet
S relative to sensible heat transfer
sat at saturation conditions
v relative to the vapour phase
w at the wall

10. References

Agunaoun A., A. Daif, R. Barriol, & M. Daguenet (1994), Evaporation en convection forcée d'un film mince s´écoulant en régime permanent, laminaire et sans ondes, sur une surface plane inclinée, *Int. J. Heat Mass Transfer*, 37, 2947–2956.

Agunaoun A., A. IL Idrissi, A. Daıf, & R. Barriol (1998) Etude de l'évaporation en convection mixte d'un film liquide d'un mélange binaire s'écoulant sur une plaque inclinée soumise à un flux de chaleur constant, *Int. J. Heat Mass Transfer*, 41, 2197–2210.

Ait Hammou Z., B. Benhamou, N. Galanis & J. Orfi (2004), Laminar mixed convection of humid air in a vertical channel with evaporation or condensation at the wall, *Int. J. Thermal Sciences*, 43, 531-539.

Azizi Y., B. Benhamou, N. Galanis & M. El Ganaoui (2007), Buoyancy effects on upward and downward laminar mixed convection heat and mass transfer in a vertical channel», *Int. J. Num. Meth. Heat Fluid Flow*, 17, 333-353.

Baumann W.W. & Thiele F., (1990) Heat and mass transfer in evaporating two-component liquid film flow, *Int. J. Heat Mass Transfer*, 33, 273–367.

Behzadmehr A., Galanis N., Laneville A., (2003) Low Reynolds number mixed convection in vertical tubes with uniform wall heat flux, Int. J. Heat Mass Transfer 46, 4823–4835.

Ben Nasrallah S. & Arnaud G. (1985) Évaporation en convection naturelle sur une plaque verticale chauffée à flux variable. *Journal of Applied Mathematics and Physics*, 36, 105-119.

Boulama, K., Galanis, N., (2004) Analytical solution for fully developed mixed convection between parallel vertical plates with heat and mass transfer, *J. Heat Transfer*, 126, 381–388.

Burmeister, L.C., (1993) *Convective Heat Transfer*, J. Wiley, New York.

Chang C. J., Lin T. F. & Yan W. M., (1986) Natural convection flows in a vertical open tube resulting from combined buoyancy effects of thermal and mass diffusion, *Int. J. Heat Mass Transfer*, 29, 1453-1552.

Chow, L.C., & Chung, J.N. (1983), Evaporation of water into laminar stream of air and superheated steam, *Int. J. Heat Mass Transfer*, 26, 373-380.

A. S. Cherif, M. A. Kassim, B. Benhamou, S. Ben Jabrallah, S. Harmand, J.P. Courriou "Experimental Study of Mixed Convection Heat and Mass Transfer in a Vertical Channel with Film Evaporation", *Int. Journal Thermal Science*, Submitted June 2010.

Fedorov A. G., R. Viskanta & A. A. Mohamad (1997) Turbulent heat and mass transfer in assymetrically heated, vertical parallel plate channel, *Int. J. Heat and Fluid Flow*, 18, 307–315.

Feddaoui M., Mir A. & Belahmidi E. (2003) Cocurrent turbulent mixed convection heat and mass transfer in falling film of water inside a vertical heated tube, *Int. J. Heat Mass Transfer*, 46, 3497–3509.

Fuji, T., Kato, Y. & Bihara, K. (1977), *Expressions of transport and thermodynamic properties of air, steam and water*, Sei San Ka Gaku Ken Kuu Jo, Report no 66, Kyu Shu University, Kyu Shu, Japan.

Gebhart B. & Pera L. (1971), The nature of vertical natural convection flows resulting from the combined buoyancy effects of thermal and mass diffusion, *Int. J. Heat Mass Transfer*, 14, 2025-2050.

Gebhart B., Jaluria Y., Mahajan R. L. & Sammakia B. (1988), *Buoyancy-induced flows and transport*, Hemisphere Pub. Co., New York.

Hanratty T. J., Rosen E. M., Kabel R. L. (1958), Effect of heat transfer on flow field at low Reynolds numbers in vertical tubes, *Indust. Engrg. Chem.*, 50, 815-820.

He S., An P., Li J. & Jackson J.D. (1998) Combined heat and mass transfer in a uniformly heated vertical tube with water film cooling, *Int. J. Heat Fluid Flow*, 19, 401-417.

Hubbard G.L., Denny, V.E. & Mills, A.F. (1975), Droplet evaporation: effects of transients and variable properties, *Int. J. Heat Mass Transfer*, 18, 1003-1008.

Huang, C.C. Yan, W.M. and Jang, J.H. (2005), Laminar mixed convection heat and mass transfer in a vertical rectangular duct with film evaporation and condensation, *Int. J. Heat Mass Transfer*, 48, 1772-1784.

Kassim M. A., Benhamou B. & Harmand S. (2010a) Combined heat and mass transfer with phase change in a vertical channel, *Computational Thermal Science*, 2, 299-310.

Kassim M. A., Cherif A. S., Benhamou B., Harmand S. et Ben Jabrallah S. (2010b) étude numérique et expérimentale de la convection mixte thermosolutale accompagnant un écoulement d'air laminaire ascendant dans un canal vertical adiabatique. 1er *Colloque International Francophone d'Energétique et Mécanique (CIFEM'2010)*, Saly, 17-19 mai 2010, Senegal.

Kassim M. A., Benhamou B., Harmand S., Cherif A. S., Ben Jabrallah S. (2009) " Etude numérique et expérimentale sur les transferts couplés de chaleur et de masse avec changement de phase dans un canal vertical adiabatique " IXème Colloque Inter-universitaire Franco–Québécois Thermique des systèmes CIFQ2009, 18-20 mai 2009, Lille, France

Laaroussi N., Lauriat G. & Desrayaud G. (2009), Effects of variable density for film evaporation on laminar mixed convection in a vertical channel, *Int. J. Heat Mass Transfer*, 52, 151-164.

Lin T.F., Chang C.J., & Yan W.M. (1988), Analysis of combined buoyancy effects of thermal and mass diffusion on laminar forced convection heat transfer in a vertical tube, *ASME J. Heat Transfer* 110, 337-344.

Lin, J. N., Tzeng, P. Y., Chou, F. C., Yan, W. M. (1992), Convective instability of heat and mass transfer for laminar forced convection in the thermal entrance region of horizontal rectangular channels, *Int. J. Heat Fluid Flow*, 13, 250-258.

Maurya R. S., Diwakar S. V., Sundararajan T., & Das Sarit K. (2010), Numerical investigation of evaporation in the developing region of laminar falling film flow under constant wall heat flux conditions, Numerical Heat Transfer, Part A, 58: 41–64

Maré, T., Galanis, N., Voicu, I., Miriel, J., Sow, O., (2008) Experimental and numerical study of mixed convection with flow reversal in coaxial double-duct heat exchangers, *Exp. Therm. Fluid Sci.*, 32, 1096–1104.

Mezaache H. & Daguenet M. (2000) Etude numérique de l'évaporation dans un courant d'air humide laminaire d'un film d'eau ruisselant sur une plaque inclinée, *Int. J. Thermal Sci.*, 39, 117–129.

Minkowycz, W.J. & Sparrow E.M. (1966), Condensation heat transfer in the presence of non-condensable, interfacial resistance, superheating, variable properties, and diffusion, *Int. J. Heat Mass Transfer*, 9, 1125–1144

Nesreddine, H., Galanis, N. & Nguyen, C.T., (1998) Effects of axial diffusion on laminar heat transfer with low Péclet numbers in the entrance region of thin vertical tubes. Numer. Heat Transfer Part A, 33, 247–266.

Nelson D. J. & Wood B. D. (1989) Combined heat and mass transfer natural convection between vertical parallel plates, Int. J. Heat Mass Transfer, 328, 1779–1789.

Nguyen, C.T., Maiga, S.E., Landry, M., Galanis, N., Roy, G. (2004) Numerical investigation of flow reversal and instability in mixed laminar vertical tube flow, *Int. J. Therm. Sci.*, 43, 797–808.

Nusselt, W. (1916) Die Oberflchenkondensation des Wasserdampfes, *Z. Ver. Deutch. Ing.* 60 541–546. 569–575.

Oulaid O., B. Benhamou, N. Galanis (2010a), Combined buoyancy effects of thermal and mass diffusion on laminar forced convection in a vertical isothermal channel, *Computational Thermal Science*, 2, 125-138.

Oulaid O., Benhamou B., Galanis N. (2010b), Mixed Convection Heat and Mass Transfer with Phase Change and Flow Reversal in Channels, *Int. J. Heat Fluid Flow*, 31, 711-721 (*doi:10.1016/j.ijheatfluidflow.2010.04.007*).

Oulaid O., Benhamou B., Galanis N. (2010c), Simultaneous heat and mass transfer in inclined channel with asymmetric conditions, *14th Int. Heat Transfer Conference IHTC14*, Washigton USA, 8-13 August 2010

Oulaid O., Benhamou B., Galanis N. (2010d), Simultaneous heat and mass transfer in inclined channel with asymmetric conditions, *Journal of Applied Fluid Mechanics, in press.*

Oulaid Othmane (2010), *Transferts couplés de chaleur et de masse par convection mixte avec changement de phase dans un canal*, Ph. D. Thesis, Jointly presented at Cadi Ayyad Univesity Marrakech (Morocco) and University of Sherbrooke (Canada).

Oulaid O., B. Benhamou, N. Galanis (2009), "Effet de l'inclinaison sur les transferts couplés de chaleur et de masse dans un canal" *IXème Colloque Inter-Universitaire Franco-Québécois Thermique des systèmes CIFQ2009*, 18-20 mai 2009, Lille, France

Oulaid O., B. Benhamou (2007), Effets du rapport de forme sur les transferts couples de chaleur et de masse dans un canal vertical, *Actes des 13èmes Journées Internationales de Thermique - JITH07*, pp. 76-80, Albi, France, 28-30 Aout 2007. (http://hal.archives-ouvertes.fr/docs/00/22/66/58/PDF/A20.pdf)

Patankar, S. V. (1980), *Numerical Heat Transfer and Fluid Flow*, Hemisphere/McGraw-Hill.

Patankar, S. V. (1981), A calculation procedure for two-dimensional elliptic situations, *Numerical Heat Transfer*, 4, 409-425.

Salah El-Din, M.M., (1992) Fully developed forced convection in vertical channel with combined buoyancy forces, *Int. Comm. Heat Mass Transfer*, 19, 239–248.

Scheele G. F., Hanratty T. J. (1962), Effect of natural convection on stability of flow in a vertical pipe, J. Fluid Mech., 14, 244-256.

Siow, E. C., Ormiston, S. J., Soliman, H. M. (2007), Two-phase modelling of laminar film condensation from vapour-gas mixtures in declining parallel-plate channels, *Int. J. Thermal Sciences*, 46, 458-466.

Suzuki K., Y. Hagiwara & T. Sato (1983), Heat transfer and flow characteristics of two-component annular flow, *Int. J. Heat Mass Transfer*, 26, 597-605.

Shembharkar T. R. & Pai B. R., (1986) Prediction of film cooling with a liquid coolant, *Int. J. Heat Mass Transfer*, 29, 899-908.

Tsay Y. L., Li T. F. et Yan W. M. (1990) Cooling of falling liquid film through interfacial heat and mass transfer. *Int. J. Multiphase Flow*, 16, 853-865.

Vachon M. (1979) Étude de l'évaporation en convection naturelle. *Thèse de Doctorat*, Université de Poitiers, Poitiers, France.

Volchkov E. P., Terekhov V. V., Terekhov V. I. (2004), A numerical study of boundary layer heat and mass transfer in a forced convection of humid air with surface steam condensation, *Int. J. Heat Mass Transfer*, 47, 1473-1481.

Wang, M., Tsuji, T., Nagano, Y., (1994) Mixed convection with flow reversal in the thermal entrance region of horizontal and vertical pipes, *Int. J. Heat Mass Transfer*, 37, 2305–2319. 656

Yan W.M., Y.L. Tsay & T.F. Lin (1989), Simultaneous heat and mass transfer in laminar mixed convection flows between vertical parallel plate with asymmetric heating, *Int. J. Heat Fluid Flow* 10, 262–269.

Yan W.M. & Lin T.F. (1989), Effects of wetted wall on laminar mixed convection heat transfer in a vertical channel, J. Thermophys. Heat Transfer, 3, 94–96.

Yan W. M., Y. L. Tsay & T. F. Lin (1990), Effects of wetted walls on laminar natural convection between vertical parallel plate with asymmetric heating, *Applied Scientific Research*, 47, 45-64.

Yan W.M. (1991), Mixed convection heat transfer enhancement through latent heat transport in vertical parallel plate channel flows, *Can J. Chem. Eng.*, 69, 1277–1282.

Yan W. M., T. F. Lin & Y. L. Tsay (1991), Evaporative cooling of liquid film through interfacial heat and mass transfer in a vertical channel - I. Experimental study, *Int. J. Heat Mass Transfer*, 34, 1105-1111.

Yan W. M. & Lin T. F. (1991), Evaporative cooling of liquid film through interfacial heat and mass transfer in a vertical channel - II. Numerical study, *Int. J. Heat Mass Transfer* 34, 1113-1124.

Yan W. M. (1992), Effect of film evaporation laminar mixed convection heat and mass transfer in a vertical channel, *Int. J. Heat Mass Transfer*, 35, 3419–3429.

Yan, W.M. (1993), Mixed convection heat transfer in a vertical channel with film evaporation, *Canadian J. Chemical Engineering*, 71, 54-62.

Yan W.M. (1995a), Turbulent mixed convection heat and mass transfer in a wetted channel, *ASME J. Heat Transfer* 117, 229–233.

Yan W. M. (1995b), Effects of film vaporization on turbulent mixed convection heat and mass transfer in a vertical channel, *Int. J. Heat Mass Transfer*, 38, 713-722.

Yan W. M. & Soong C. Y. (1995) Convective heat and mass transfer along an inclined heated plate with film evaporation, Int. J. Heat Mass Transfer, 38, 1261-1269.

10

Liquid-Liquid Extraction With and Without a Chemical Reaction

Claudia Irina Koncsag[1] and Alina Barbulescu[2]
[1]*University of Warwick*
[2]*"Ovidius" University of Constanta*
[1]*United Kingdom*
[2]*Romania*

1. Introduction

The extraction of mercaptans with alkaline solution is accompanied by a second- order instantaneous reaction. As explained in Section 2.2, in this case, the mass transfer coefficients can be calculated as for the physical extraction, since the mass transfer is much slower than the reaction rate.The liquid-liquid extraction is a mass transfer process between two phases. One liquid phase is *the feed* consisting of a *solute* and a *carrier*. The other phase is the *solvent*. The extraction is understood to be a transfer of the solute from the feed to the solvent. During and at the end of the *extraction* process, the feed deprived of solute becomes a *raffinate* and the solvent turns into *extract*. Extraction is a separation process aiming to purify the feed or to recover one or more compounds from it.

The mass transfer mechanism can be described by the well known double film theory, the penetration theory or the surface renewal theory. Especially the stationary double film theory describes most accurately the liquid-liquid extraction. With the means of this theory, the dimensioning of the extraction equipment can be done.

Sometimes, over the physical extraction process, a chemical reaction is superposed. Depending on the reaction rate compared with the mass transfer rate, the process can be considered driven by the mass transfer or by the chemical reaction. Also, in some cases, the chemical reaction has an effect of enhancement for the extraction, contributing to speed up the process. As a consequence, the dimensioning of the equipment is different.

Many studies have been performed in the last decades for the mathematical modelling of the processes. Accurate correlations between physical properties (densities, density difference, interfacial tension), and dimensions involved in the extraction equipment dimensioning: the drop size diameter, the characteristic velocity of the drop and the slip velocity of the phases were worked out. A smaller number of correlations are available for the calculation of the mass transfer coefficients. Some of the elements needed for the dimensioning of the extractors would be determined experimentally, if a certain accuracy is expected. The experiment is compulsory for the mass transfer coefficients when a new type of equipment is used.

The present work exemplifies the theoretical aspects of the liquid-liquid extraction with and without a chemical reaction and the dimensioning of the extractors with original experimental work and interpretations. The experiment involved extraction of acid compounds from sour petroleum fractions with alkaline solutions in structured packing columns. Such an example is useful for understanding the principles of dimensioning the extraction equipment but also offers a set of experimental data for people developing processes in petroleum processing industry. A simple, easy to handle model composed by two equations was developed for the mercaptans (thiols) extraction.

2. Theoretical aspects

The immiscible liquid phases put in contact (the feed and the solvent) form a closed system evolving towards the thermodynamic equilibrium. According to the Gibbs law:

$$l = c + 2 - f = 3 + 2 - 2 = 3 , \qquad (1)$$

the system can be defined by three parameters (l=3), the number of components being c=3 (solvent, solute and carrier), and the phases number f=2. Usually, the parameters taken into account are the temperature (T), the concentration of the solute in the raffinate (x) and the concentration in the extract (y). So, the equilibrium general equation in this case is:

$$y = f(x)_{t=const} \qquad (2)$$

The equilibrium equation can have different forms, but most frequently, if the liquid phases are completely immiscible and the solute concentration is low, the Nernst law describes accurately the thermodynamic equilibrium:

$$y = m \cdot x , \qquad (3)$$

where m is the repartition coefficient of the solute between the two phases. The Nernst law can be applied also at higher concentration of the solute but in a narrow range of concentrations.

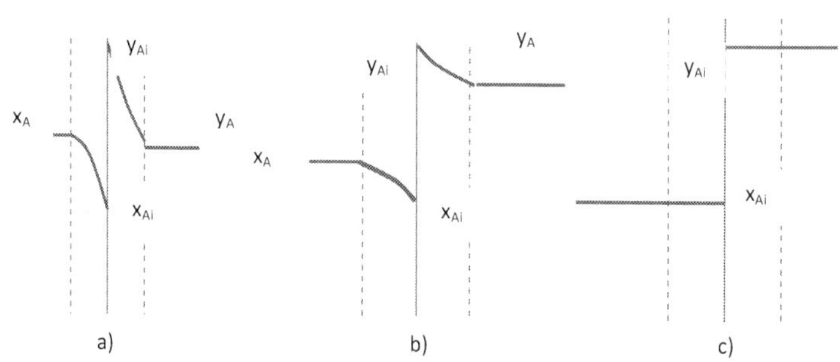

Fig. 1. The evolution of the solute concentration in the vicinity of the interface in a closed system

The double stationary film theory of Whitman leads to very good practical results for the determination of mass transfer coefficients. According to this theory, the phases are separated by an interface and a double film (one of each phase) adheres to this interface. The mass transfer takes place exclusively in this double stationary film by the molecular diffusion mechanism. In the bulk of both phases, the concentration of the solute is considered uniform as a consequence of perfect mixing.

In Fig.1, the evolution in time is presented for a closed system approaching the equilibrium, in the light of double film theory. Notations x_{Ai} and y_{Ai} are for the concentration at the interface in raffinate and extract respectively; x_A and y_A denote the concentration of the solute A in the bulk of the raffinate and of the extract respectively. In Fig.1, the mass transfer is presented in a closed system in evolution from the initial state *a* to the final equilibrium state *c*. The concentrations at the interface are constant and linked by the equilibrium equation since the concentration of the solute in the bulk feed /raffinate decreases and the concentration of the solute in the bulk solvent/ extract increases in time until equalling the equilibrium concentrations. If the system is open, y_A and x_A are constant in time (the regime becomes stationary) and the system is maintained in the state *a*.

2.1 Mass transfer coefficients in physical extraction

In liquid-liquid extraction, the best mechanism describing the mass transfer is the unicomponent diffusion (the solute *A* diffusing in one direction without a counter diffusion). According to Maxwell- Stefan model, the mass transfer rate in the raffinate film is:

$$N_A = \frac{c_R \cdot D_{AR}}{l_R(1 - x_A)_{ml}}(x_A - x_{Ai}) \tag{4}$$

In the Eq.4, N_A is the flow of component *A* transferred from the raffinate through the film to the interface; D_{AR} is the diffusion coefficient of the solute in the raffinate phase; c_R is the total concentration of components in the raffinate, usually expressed as $kmol/m^3$; l_R is the thickness of the raffinate film; $\frac{c_R \cdot D_{AR}}{l_R(1 - x_A)_{ml}}$ denoted with k_R is the partial mass transfer coefficient in the raffinate phase and $1/k_R$ is the resistance to the transfer.

Similarly, Eq.5 describes the mass transfer rate in the extract film, *E* being the notation for "extract":

$$N_A = \frac{c_E \cdot D_{AE}}{l_E(1 \quad y_A)_{ml}}(y_{Ai} - y_A) \tag{5}$$

During a stationary regime, the component A doesn't accumulate in the raffinate film as well as in the extract film; this means that the flux transferred in the raffinate film to the interface equals the flux transferred from the interface into the extract phase:

$$N_A = k_E(y_{Ai} - y_A) = k_R(x_A - x_{Ai}) \tag{6}$$

In Eq.6, $(y_{Ai}$-$y_A)$ and $(x_A$-$x_{Ai})$ are the driving forces of the mass transfer in the extract film and in the raffinate film respectively (related to the partial mass transfer coefficients). These partial driving forces can be read on the axes in the Fig. 2, where the system state is represented by the point A and the equilibrium concentrations at the interface are represented by the point A_i. The arrow AA_i denotes the distance from the actual state of the system to the equilibrium state. But the overall driving force is $(x_A$-$x_{Ae})$, related to the raffinate phase and $(y_{Ae}$- $y_A)$, related to the extract respectively. The overall driving forces refer to the distance from the actual state of the system to an hypothetical state when the actual concentration of the raffinate (x_A) would be in equilibrium with the extract (y_{Ae}), or the actual concentration of the extract (y_A) would be in equilibrium with the raffinate (x_{Ae}).

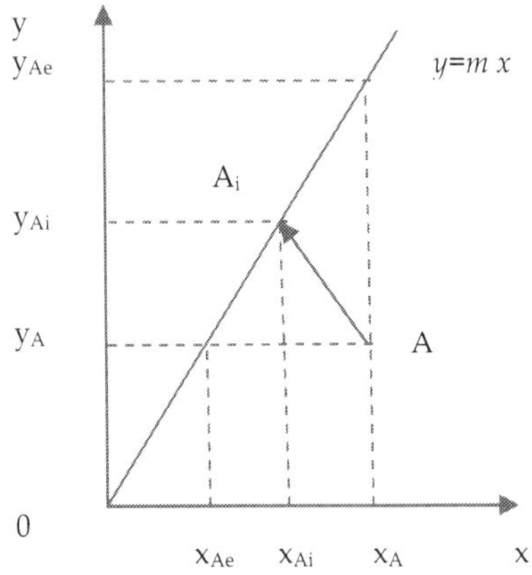

Fig. 2. The representation of the driving forces for the mass transfer (immiscible liquid phases; equilibrium described by Nernst law)

In connection with the overall driving forces, the overall mass transfer coefficients are defined in the equations (7) and (8):

$$N_A = K_R \cdot (x_A - x_{Ae}) \tag{7}$$

$$N_A = K_E \cdot (y_{Ae} - y_A) \tag{8}$$

As seen in Fig.2, the slope of the equilibrium curve (m) can be calculated from geometrical dimensions (Eq.9):

$$m = \frac{y_{Ai} - y_A}{x_{Ai} - x_{Ae}} = \frac{y_{AE} - y_A}{x_A - x_{Ae}} = \frac{y_{AE} - y_{Ai}}{x_A - x_{Ai}} \tag{9}$$

By manipulating the Eq.6-9 and Fig.2, the Eq.(10) and (11) are obtained and would be used for the calculation of the overall mass transfer coefficients K_R and K_E [m·s^{-1}] when the partial coefficients k_R and k_E are known:

$$\frac{1}{K_R} = \frac{1}{k_R} + \frac{1}{m \cdot k_E} \tag{10}$$

$$\frac{1}{K_E} = \frac{1}{k_E} + \frac{m}{k_R} \tag{11}$$

More often, the mass transfer coefficients are not related to the raffinate/ extract phases but more important, to the continuous and the dispersed phase. The extraction system is in fact an emulsion: one of the phases is in form of droplets and the other one is continuous. Which one is discontinuous, depends on the volume ratio of the phases and on the interfacial phenomena. Sometimes, the raffinate is the dispersed phase, at other times it is the continuous phase. This is why, the equations (10) and (11) are re-written in terms of *overall volumetric mass transfer coefficients* for the dispersed phase (*d*) and for the continuous phase (*c*), $K_d a$ and $K_c a$ [s^{-1}], as the interfacial area a [m^2/m^3] is included in their value:

$$\frac{1}{K_d \cdot a} = \frac{1}{k_d \cdot a} + \frac{\rho_d}{\rho_c \cdot m \cdot k_c \cdot a} \tag{12}$$

$$\frac{1}{K_c \cdot a} = \frac{1}{k_c \cdot a} + \frac{\rho_c \cdot m}{\rho_d \cdot k_c \cdot a} \tag{13}$$

Eq. (12) and (13) are written for the raffinate as dispersed phase. The partial coefficients in Eq. (12) and (13) can be calculated from the diffusivity coefficients D_{AR}, D_{AE} and the thickness of the double film: l_R and l_E (Eq.4 and 5). Since for the diffusivity, there are a few accurate correlations (the most used is the correlation Wilke & Chang, 1955) or simple experiments to perform, measuring the double film thickness is more complicated.
Alternatively, the partial coefficients can be calculated with criterial equations; for example, the most used correlation for the calculation of the partial coefficient for the continuous phase is Eq.14 (Treybal, 2007):

$$Sh_c = 0.725 \cdot Sc_c^{0.42} \cdot Re_c^{0.57} \cdot (1-\varphi) \ , \tag{14}$$

where:

- the partial mass transfer coefficient in the continuous phase film is included in the Sherwood criterion ($Sh_c = k_d d_{32} / D_c$, d_{32} being the medium Sauter diameter of the drops and D_c - diffusivity of the reactant A in the continuous phase);
- Sc_c is the Schmidt criterion for the continuous phase, $Sc_c = \mu_c / D_c$;
- Re_c is Reynolds number for continuous phase, $Re_c = \rho_c d_{32} V_{slip} / \mu_c$ (V_s- slip velocity of phases);
- φ is the dispersed phase hold-up.
- The following notions: Sauter mean diameter (d_{32}), slip velocity(V_{slip}) and dispersed phase hold up (φ), will be explained in section 2.3.

The correlation recommended by Laddha and Degaleesan (1974) for the partial coefficient for the discontinuous phase is Eq.15:

$$k_d = 0.023 \cdot V_{slip} \cdot Sc_c^{-0.5} \tag{15}$$

In practice, the calculations are done in reverse order: the overall coefficients are determined in experimental studies, as explained in Section 2.3, then the partial coefficients are calculated from Eq.(10) and (11). From these partial coefficients one can calculate the thickness of the double film. In the extreme case when the solvent has a high affinity for the solute A, much higher than the raffinate, it is accepted that $K_E \approx k_E$.
Knowing the overall global coefficients for a certain system is crucial, because they can't be avoided at the equipment dimensioning.

2.2 Mass transfer coefficients in chemical extraction

Let's consider a reaction in the liquid-liquid system:

$$A+q\cdot B\rightarrow \text{Products}$$

The first phase contains the component A which diffuses from the first phase into the second one containing B, reacting with B in that phase. Then, products diffuse in the same phase 2. Reactions in liquid-liquid systems can be classified from kinetically point of view as slow, fast and instant (Sarkar *et al,* 1980).

The equation describing the diffusion of the reactant A simultaneously with the chemical reaction is (Astarita, 1967):

$$D_A\cdot\nabla^2 c_A = \mathrm{u}\cdot\nabla \mathrm{c_A} + \frac{\partial \mathrm{c_A}}{\partial t} + \mathrm{v_{RA}} \qquad (16)$$

The term on the left hand side of the Eq.16 represents the molecular diffusion of the component A through the film of phase 1, The terms on the right hand sides have the following meaning: the first one describes the transport by convection through the same film, the second one is the accumulation of A in the film and the third represents the reaction rate. The Eq. 16 can be simplified in the conditions of the double film theory, where the diffusion direction of A is perpendicular to the interface (direction x), eddies are inexistent in the film and component A doesn't accumulate in the film:

$$D_A\cdot\frac{d^2 c_A}{dx^2} = v_{RA} \qquad (17)$$

The Eq.17 can be detailed for both reactants:

$$D_A\cdot\frac{d^2 c_A}{dx^2} = \frac{dc_A}{dt} \qquad (18)$$

$$D_B\cdot\frac{d^2 c_B}{dx^2} = \frac{dc_B}{dt} \qquad (19)$$

Fast and instant reactions

In case of fast and instant reactions, the reaction takes place in the plane located in the film of phase 2 (phase containing the component B). The component A diffuses through the film 1 to the interface then from interface to the reaction plane (see Fig. 3 a). In Fig.3 a, a particular case of fast reaction: the irreversible instantaneous reaction is illustrated; in this case, both reactants diffuse to the reaction plane, where their concentrations equals to zero. The term "instantaneous" is idealised since the reaction rate is always finite, but in this case, the mass transfer rate is much lower than the reaction rate, so the process is entirely controlled by the diffusion mechanism.

Taking into account the position of the reaction plane (at the distance λ from the interface) and the stoechiometric coefficient of the reaction q, the Eq.18 and 19 considering their equality, and integrating, the Eq. 20 is obtained:

$$D_A\cdot\left(\frac{dc_A}{dx}\right)_{x=\lambda} = \frac{1}{q}\cdot D_B\cdot\left(\frac{dc_B}{dx}\right)_{x=\lambda}. \qquad (20)$$

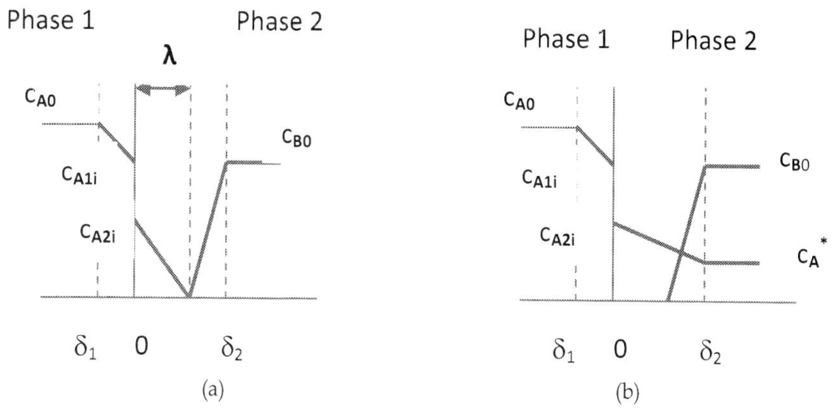

Fig. 3. Profiles of reactants concentration at the extraction with a chemical reaction: a - instantaneous irreversible reaction taking place in phase film 2; b - slow reaction taking place in the film phase 2

By integrating Eq.20 between the limits $x=\lambda$ and $x=l$, (l- the film thickness), it results:

$$q \cdot D_A \cdot \frac{c_{A2i}}{\lambda} = D_B \cdot \frac{c_{B0}}{l-\lambda} \tag{21}$$

In Eq.21, c_{A2i} is the concentration of A at the interface on the film's 2 side and c_{B0} is the concentration of B in bulk of the phase 2. The Eq. 21 can be re- written in another form:

$$\lambda = l \cdot \frac{q \cdot D_A \cdot c_{A2i}}{q \cdot D_A \cdot c_{A2i} + D_B \cdot c_{B0}} \tag{22}$$

$\frac{l}{\lambda}$ is in fact the ratio between the the overall mass transfer coefficient with a chemical reaction $K{\cdot}a$, and the overall mass transfer coefficient at the physical extraction (without a chemical reaction), $K^0{\cdot}a$:

$$\frac{K \cdot a}{K^0 \cdot a} = \frac{l}{\lambda} = 1 + \frac{D_B}{D_A} \cdot \frac{c_{B0}}{q \cdot c_{A2i}}, \tag{23}$$

So, the overall mass transfer coefficient in the case of instant reaction is proportional to the coefficient for the physical extraction. It means that the coefficient at the extraction with instant chemical reaction depends on hydrodynamics in the same extent as that for physical extraction.

For instantaneous irreversible reactions, the enhancement factor E_i is defined (Pohorecki, 2007) by Eq.24:

$$E_i = \frac{Q(instantaneous\,reaction)}{Q(physical)} \tag{24}$$

where: Q_c is the continuous phase volumetric flow, [$m^3 \cdot s^{-1}$]
Q_d- the dispersed phase volumetric flow, [$m^3 \cdot s^{-1}$]
B_{max} is the flooding capacity, [$m^3 \cdot m^{-2} \cdot s^{-1}$]; considering the flow in the free cross-sectional area of the column. The flooding capacity B_{max} is in fact the sum of the flooding velocities of phases; it depends on the physical properties of the system: the density (ρ_c and ρ_d), the viscosity (μ_c and μ_d) and the interfacial tension σ. The flooding capacity can be predicted following extensive studies, as exemplified in Section 3.
k- the flooding coefficient, with values from 0.4 (dispersion column) to 0.8 (column equipped with structured packing); this coefficient would be kept as high as possible, in order to increase the mass transfer rate and the processing capacity of the column.

The flooding capacity is experimentally determined for each type of column. It consists in derangements of the countercurrent flow bringing about entrainment of one phase in the other one, or the impossibility for one flow to enter the column. There are three main mechanisms of flooding:

- The phase inversion provoked by the excessive increasing of dispersed phase flow;
- The entrainment of the drops in the continuous phase when the flow of continuous phase increases too much;
- The flooding due to the contaminants at the interface creating instability of the interface of even inversion of the phases.

According to authors after Hanson (1971), the diameter calculation is made using the concept of *slip velocity*, V_{slip}, which is the velocity of the dispersed phase related to the continuous phase; the *slip velocity* is in fact, the sum of linear velocities of the phases, not in the free cross-sectional area but in the actual cross-sectional area, taking into account the internal parts of the column and the dispersed phase holdup, ϕ. According to this definition, in the dispersion column (without any internal parts, such as trays or packing), the slip velocity is:

$$V_{slip} = \frac{V_d}{\phi} + \frac{V_c}{1-\phi} \tag{31}$$

The dispersed phase holdup is the fraction occupied by the drops in the free cross sectional area of the column. In fact, the holdup is not uniformly distributed in the column because the drops are of various dimensions, with an irregular shape, oscillating. For approximate calculations, a mean diameter would be taken into consideration. The most usual expression of this is the Sauter mean diameter, d_{32}:

$$d_{32} = \frac{\sum n_i \cdot d_i^3}{\sum n_i \cdot d_i^2} \tag{32}$$

The mean diameter d_{32}, is correlated with the holdup ϕ and the interfacial area a:

$$a = \frac{6 \cdot \phi}{d_{32}} \tag{33}$$

Empirical correlations can be used to correlate Sauter mean diameter d_{32} with physical properties of the system. It can be calculated with the formula recommended by Seibert & Fair (1988) and verified by Iacob & Koncsag (1999) on systems with high interfacial tension:

$$d_{32} = 1.15\left(\frac{\sigma}{\Delta\rho \cdot g}\right)^{0.5} \tag{34}$$

where $\Delta\rho$- density difference of phases, g -gravitational constant.

It is difficult to determine the slip velocity, but it can be easier done but an easier by correlating it with the singular drop's characteristic velocity, V_K, which can be determined experimentally. V_K is defined as the ratio between the distance travelled by the singular drop in the column and the time of this trip. The singular drop's characteristic velocity is uninfluenced by the presence of other drops but is influenced by the presence of the internal parts of the equipment. For example, in the case of packed column, V_K can be correlated with the physical properties of the liquid-liquid system and the geometrical characteristics of the packing:

$$V_K = 0.637\left(\frac{a_p \cdot \rho_c}{\varepsilon^3 \cdot g \cdot \Delta\rho}\right)^{-0.5} \tag{35}$$

There are correlations between the slip velocity and the singular drop's characteristic velocity. A simple correlation is the Pratt- Thornton equation which is valid for rigid (non-oscillating) drops and low values of the holdup (Thornton, 1956):

$$V_{slip} = V_K \cdot (1-\phi) \tag{36}$$

For higher values of holdup, a more accurate equation would be used (Misek, 1994):

$$V_{slip} = V_K \cdot (1-\phi)\exp(a\phi) \tag{37}$$

2.3.2 The column height

For extractors with countercurrent flow of the phases, the active height is calculated on the basis of the *mass transfer unit* notion. In the hypothesis of the plug flow, the height of the column is:

$$H = [NTU]_{od} \cdot [HTU]_{od} = [NTU]_{oc} \cdot [HTU]_{oc} \tag{38}$$

where $[NTU]_{od}$ and $[NTU]_{oc}$ are the number of transfer units relative to the dispersed and to the continuous phase respectively, when expressing the mass transfer rate as the overall mass transfer coefficients. $[HTU]_{od}$ and $[HTU]_{oc}$ are the height of the transfer unit relative to the overall mass transfer coefficient in the same phases.

The height of the transfer unit is the height of the column which ensures the decreasing by *e* (=2.71...) of the driving force defined as in the Eq.7 and 8, taking into account which is the dispersed phase and the continuous one.

$[NTU]_{od}$ and $[NTU]_{oc}$ are calculated taking into account the equilibrium data. The relationships given in Eq.39 and 40 are related to the extract (*E*) and the raffinate (*R*) and it is to see which equation applies to the continuous phase or the dispersed phase (e.g. the raffinate can be dispersed phase in one application and continuous phase in another one):

$$[NTU]_{oR} = \int_{x_1}^{x_2} \frac{dx}{x - x_e} \tag{39}$$

$$[NTU]_{oE} = \int_{y_1}^{y_2} \frac{dy}{y - y_e} \tag{40}$$

where:

x_1, x_2, y_1, y_2 are the solute concentration in the raffinate (x) and in the extract respectively (y) in the flow entering (1) or exiting (2) the column.

x_e , y_e are the solute concentrations in raffinate and in the extract respectivelly, in equilibrium conditions in every point along the column.

$[HTU]_{od}$ and $[HTU]_{oc}$ are experimentally found by dividing the active height of the column (the height of the column where the dispersed phase and the continuous phase co-exists) by the $[NTU]_{od}$ or $[NTU]_{oc}$.

Let's express the mass transfer for the dispersed phase. The volumetric overall mass transfer coefficients related to the dispersed phase $K_{od}a$ are correlated with the height of the mass transfer unit $[HTU]_{od}$ and the superficial velocity of the dispersed phase v_d (which is defined as the volumetric flow divided by the cross- sectional area of the column):

$$K_{od} \cdot a = \frac{v_d}{HTU_{od}} \tag{41}$$

In practice, the height of the column is calculated starting with the experimental determination of mass transfer coefficients, continuing by the calculation of HTU with Eq.41 and finally, applying the Eq. 38.

3. Experimental data

The theoretical aspects presented here are very general. In fact, an engineer needs mathematical models for the dimensioning of the industrial equipment, specific for a given type of extractor. Here we present the process of the model's development, using a large database, partially relying on our original experiment, partially on data from literature, for the dimensioning of a packed column.

The original experimental data were obtained in a 76 mm diameter column with structured packing type Sulzer SMV 350 Y. The specific area of the packing was: a_p= 340 m^2/m^3 and the void fraction ε = 0.96. The packing bed was made of 4 structured packing elements with a total height of 840 mm. The detailed description of the pilot plant was presented in a previous work (Koncsag & Barbulescu, 2008). Another type of ordered packing -corrugated metal gauze- was used in a mass transfer study at laboratory scale, also described in (Koncsag & Barbulescu, 2008). The experiment at laboratory scale was performed in another installation including a glass column with the internal diameter of 3 cm and an active height of 70 cm. The handicraft packing was made of corrugated metal gauze and had the following geometric characteristics: ε = 0.98 and a_p = 60 m^2/ m^3. Taking into account the small opening of the spiral, the drops are forced to detour and the tortuosity of their motion increases; as a consequence, the residence time of the drops in the column increases and the mass transfer improves comparing with the simple dispersion column. Also, the experiment was performed for the dispersion column (the column unpacked).

So, three sets of data were obtained: for the Sulzer SMV350 packing, for the corrugated metal gauze packing and without packing (Table1).

The studied systems were: water- gasoline, NaOH solution 20%wt - gasoline and carbon tetrachloride- water. These systems were very different from the viewpoint of density, interfacial tension and viscosity. The first and the third systems are usually taken into account in the hydrodynamic studies and the second one is common in the purifying of hydrocarbon streams. The results of the flooding tests are expressed as pairs of limiting superficial velocities of the phases (continuous and dispersed), in flooding conditions: V_{cf}, V_{df}.

Other experimental data from the literature were connected to the results of the original experiment (Table 1), in order to have a larger database for the mathematical model (Table 2). The data were chosen for hydrocarbon - water systems (gasoline- water or toluene- water) and for very different types of packing: Raschig rings of different size (Crawford & Wilke, 1951); Norton ordered packing (Seibert & Fair, 1988) and Intalox saddles (Seibert *et al*, 1990).

Flooding superficial velocities for the system gasoline- water, d_{32} = 0.0052 m, ρ_c=996 kg/m^3, ρ_d=740 kg/m^3, μ_c= 0.000993 kg/m·s, σ= 52.0·10^{-3}N/m, v_K= 0.055 m/s				
1	2	3	4	5
0.88	0.84	0.69	0.32	0.21
0.17	0.27	0.43	0.84	1.00
Flooding superficial velocities for the system gasoline- 20%NaOH solution, d_{32} = 0.0047 m, ρ_c=1220 kg/m^3, ρ_d=740 kg/m^3, μ_c= 0.00366 kg/m·s, σ= 78.6 10^{-3}N/m, v_K= 0.067 m/s				
1	2	3	4	5
0.80	0.68	0.43	0.32	0.27
0.11	0.17	0.84	1.17	1.31
Flooding superficial velocities for the system water - CCl_4 at 5°C, d_{32} = 0.0035 m, ρ_c=1610 kg/m^3, ρ_d=996kg/m^3, μ_c= 0.00123 kg/m·s, σ= 47.3·10^{-3}N/m, v_K= 0.0604 m/s				
1	2	3	4	5
0.64	0.62	0.58	0.35	0.24
0.17	0.27	0.37	1.17	1.43

Table 1. Experimental data at the hydrodynamic test in the column equipped with Sulzer packing (a_p= 340 m^2/m^3, ε=0.96, D_{col} = 0.076 m)

The second part of the experiment consisted of a mass transfer study concerning the extraction of mercaptans from petroleum fractions with alkaline solutions, a process encountered in the oil processing industry.

The raw material was the gasoline enriched in mercaptans (ethanethiol or 1-propanethiol or 1-buthanethiol) and pumped into the column where it forms the dispersed phase; after the extraction and the coalescence, the gasoline exits the column at the top, as a refined phase. The solvent - the continuous phase- is in fact a caustic solution (NaOH) with concentration in range of 5-15% wt. The continuous phase enters the column free of mercaptans and exits as an extract enriched in the said mercaptans. Samples of feed and refined phase are collected and the mercaptans concentration is analyzed by a volumetric method using AgNO3. The concentration of the mercaptans in the extract is calculated by material balance. The volumetric overall mass transfer coefficients were calculated byEq.37, 38 and 40 and are presented in Table 3.

Flooding superficial velocities for the system gasoline- water in case of the contactor equipped with Raschig rings 1/ 2" (a_p= 310 m^2/m^3, ε=0.71, D_{col} = 0.305 m) ,d_{32} = 0.0052 m, ρ_c=996 kg/m^3, ρ_d=740 kg/m^3 , μ_c= 0.000993 kg/m· s, σ= 42.4 ·10^{-3}N/m, v_K= 0.037 m/s					
Superficial velocities of phases, m/s	1	2	3	4	
$10^2 \cdot V_{df}$	0.88	0.58	0.38	0.13	
$10^2 \cdot V_{cf}$	0.25	0.44	0.60	0.97	
Flooding superficial velocities for the system gasoline- water in case of the contactor equipped with Raschig rings 1" (a_p= 195 cm^2/cm^3, ε=0.74, D_{col} = 0.305 cm); d_{32} = 0.0052 m, ρ_c=996 kg/m^3, ρ_d=740 kg/m^3 , μ_c= 0.000993 kg/m· s, σ= 42.4 ·10^{-3}N/m, v_K= 0.05 m/s					
Superficial velocities of phases, m/s	1	2	3	4	5
$10^2 \cdot V_{df}$	1.64	1.10	0.88	0.67	0.36
$10^2 \cdot V_{cf}$	0.54	0.83	1.00	1.23	1.69
Flooding superficial velocities for the system toluene- water in case of the contactor equipped with Norton packing (a_p= 213 m^2/m^3, ε=0.97, D_{col} = 0.425 m);d_{32} = 0.0055 m, ρ_c=996 kg/m^3, ρ_d=864 kg/m^3, μ_c= 0.00089 kg/m· s, σ= 30.0 ·10^{-3}N/m, v_K= 0.0497 m/s					
Superficial velocities of phases, m/s	1	2	3	4	5
$10^2 \cdot V_{df}$	1.13	0.93	0.72	0.60	0.30
$10^2 \cdot V_{cf}$	0.9	1.01	1.14	1.30	1.60
Flooding superficial velocities for the system toluene- water in case of the contactor equipped with Intalox saddles No25 IMTP (a_p= 226 m^2/m^3, ε=0.95, D_{col} = 0.102 m); d_{32} = 0.0055 m, ρ_c=996 kg/m^3, ρ_d=864 kg/m^3, μ_c= 0.00089 kg/m· s, σ= 30.0 ·10^{-3}N/m, v_K= 0.0477 m/s					
Superficial velocities of phases, m/s	1	2	3	4	5
$10^2 \cdot V_{df}$	1.44	1.32	0.93	0.72	0.51
$10^2 \cdot V_{cf}$	0.79	1.05	1.40	1.58	1.75

Table 2. Experimental data from literature concerning the hydrodynamic tests in the columns equipped with other type of packing

4. Discussion

The original experimental data about the flooding- linked to the column capacity- (Table 1) were compared with the predicted data from older models in the literature for packed columns, in order to see if they are satisfactory models or should be improved. If not, a new model would be developed.

4.1 The Crawford-Wilke model

A good old model is the Crawford- Wilke correlation curve (Crawford & Wilke, 1951). At the beginning, this model correlated a total of 160 experimental points for a large range of random packing but relatively few liquid systems.

	NaOH conc.	Dispersion column: Dispesed phase velocity, V_d ($x10^2$,m/s)	Dispersion column: $K_{od}a$ ($x10^3$,s^{-1})	Laboratory packed column: Dispesed phase velocity, V_d ($x10^2$,m/s)	Laboratory packed column: $K_{od}a$ ($x10^3$,s^{-1})	Pilot packed column: Dispesed phase velocity, V_d ($x10^2$, m/s)	Pilot packed column: $K_{od}a$ ($x10^3$,s^{-1})
Buthanethiol	5%	0.17	0.91	0.17	1.25	0.21	3.0
		0.23	1.06	0.25	1.43	0.43	4.5
		0.33	1.16	0.33	2.05	0.68	6.5
	10%	0.17	0.89	0.17	1.56	0.21	3.1
		0.23	1.20	0.25	1.93	0.43	4.8
		0.33	1.38	0.33	2.87	0.68	7.1
	15%	0.17	1.49	0.17	1.93	0.21	3.1
		0.23	1.74	0.25	2.47	0.43	3.7
		0.33	2.19	0.33	2.54	0.68	7.0
Propanethiol	5%	0.13	1.64	0.13	1.69	0.21	5.4
		0.20	1.66	0.20	2.17	0.32	7.8
		0.26	1.77	0.26	2.62	0.68	13.1
	10%	0.13	1.53	0.13	1.94	0.21	6.3
		0.20	1.98	0.20	2.23	0.32	10.2
		0.26	2.08	0.26	2.32	0.68	13.8
	15%	0.13	2.34	0.13	2.85	0.21	7.1
		0.20	2.65	0.20	3.08	0.32	11.2
		0.26	2.73	0.26	3.94	0.68	16.1
Ethanethiol	5%	0.17	2.78	0.17	6.02	0.21	8.3
		0.23	3.66	0.23	6.27	0.32	11.4
		0.33	4.31	0.33	8.34	0.58	18.2
	10%	0.17	5.06	0.17	6.45	0.21	9.3
		0.23	5.36	0.23	8.26	0.32	13.3
		0.33	8.21	0.33	13.87	0.58	22.2
	15%	0.17	5.77	0.17	7.84	0.21	10.2
		0.23	6.43	0.23	10.30	0.43	19.2
		0.33	11.67	0.33	14.46	0.68	28.6

Table 3. Experimental data at the mercaptans extraction

This model started from the premise that the sum of the square roots of the flooding velocity of both phases is a constant. This seemed reasonable for a large range of packing and systems, even that it has no a theoretical basis. The sum: $(V_{cf}^{1/2}+V_{df}^{1/2})^2$ = constant, is correlated with the physical properties of the liquids and the characteristics of the packing (a_p, ε):

$$\frac{(V_{cf}^{\frac{1}{2}}+V_{df}^{\frac{1}{2}})\cdot\rho}{a_p\cdot\mu_c}=f\left[\left(\frac{\sigma}{\rho_c}\right)^{0.2}\left(\frac{\mu_c}{\Delta\rho}\right)\left(\frac{a_p}{\varepsilon}\right)\right] \qquad (42)$$

The variables of Eq.42 are plotted in Figure 4.

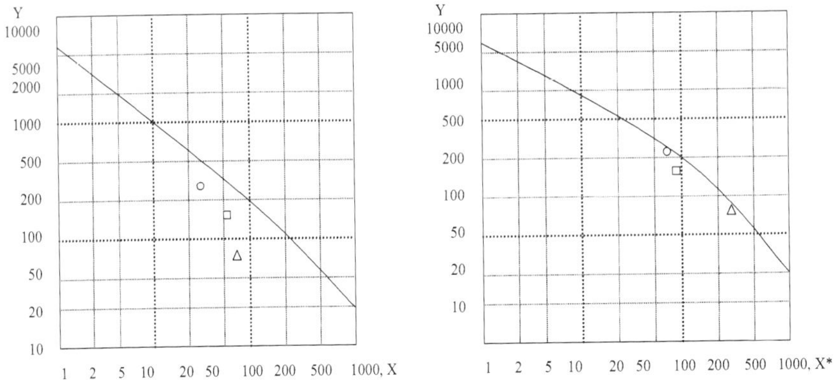

(a) (b)

Fig. 4. The correlation between the flooding velocity of the phases - the Crawford-Wilke model. Legend: a- the original correlation ($X=\frac{(V_{cf}^{1/2}+V_{dj}^{1/2})\cdot\rho}{a_p\cdot\mu_c}$, $Y=\left(\frac{\sigma}{\rho_c}\right)^{0.2}\left(\frac{\mu_c}{\Delta\rho}\right)^{1.0}\left(\frac{a_p}{\varepsilon}\right)^{1.5}$); b- the modified correlation ($X^*=([V]_{cf}^{1/2}+k_1[V]_{df}^{1/2})\cdot\rho)/(a_p\cdot\mu_c)$; square for the system water- gasoline, circle for the system NaOH solution- gasoline and triangle for the system CCl_4- water

As one can see from Figure 4.a, our experimental points are pretty far from the original curve. Other authors (Nemunaitis *et al* 1971) reported the same. They found that flooding occurred at loading only 20% of those predicted by the flooding correlation of Crawford and Wilke.
An explanation could be that the assumption of the sum $(V_{cf}^{1/2}+V_{df}^{1/2})$ constancy is not true. Even Crawford and Wilke (1951) expressed their doubts, however they considered this hypothesis as reasonable.
The sum $(V_{cf}^{1/2}+V_{df}^{1/2})$ *= constant indicates that the curves* $V_{df}^{1/2}$ *vs* $V_{cf}^{1/2}$ *are lines with a slope of (–1):*

$$V_{cf}^{1/2}+V_{df}^{1/2}=k_2 \qquad (43)$$

Eq. 43 does not describe correctly the flooding line. The sum $(V_{cf}^{1/2}+V_{df}^{1/2})$ should be modified as follows: $(V_{cf}^{1/2}+k_1V_{df}^{1/2})$ with respect to the real slope of the flooding line. In this case, $k_1 = -1/m'$ where (m') is the real slope of the flooding line. For the systems studied in my experiment, it was found:

- k_1= 1.17 - for Water-Gasoline; *(m' = -0.85)*
- k_1 = 2.44- for NaOH sol.20%wt- Gasoline ; *(m'=-0.41)*
- k_1 = 2.63- for CCl_4 - Water; *(m'= -0.38)*

As one can see, slope m' is in fact very different from the original (-1). This wasn't obvious when Crawford and Wilke established their model, because the liquid- liquid systems taken into account at that time were (all) low interfacial tension systems. In the present work, very different systems have been considered, proceeding from present authors' original experimental data as well as from other authors' data, e.g. (Nemunaitis *et al* 1971,), (Watson *et al,* 1971), (Seibert & Fair, 1998), (Seibert *et al.*1990). It is to say that the scientific literature is very poor in flooding data in case of liquid-liquid countercurrent contactors.

In order to fit all these data and to have a model responding to systems with very different physical properties, the Crawford- Wilke correlation should be modified as follows: on the X- axis would appear the expression $X^* = \dfrac{(V_{cf}^{1/2} + k_1 \cdot V_{df}^{1/2}) \cdot \rho}{a_p \cdot \mu_c}$ instead of the expression $X = \dfrac{(V_{cf}^{1/2} + V_{df}^{1/2}) \cdot \rho}{a_p \cdot \mu_c}$.

The k_1 constant should be calculated with Eq.43 as recommended by Watson *et al* (1975):

$$k_1 = 0.466 \cdot \rho_c^{+0.5} \cdot \sigma^{+0.25} \cdot \mu_c^{+0.25} \cdot \varepsilon^{+0.5} \cdot d_{32}^{-1} \tag{44}$$

From Fig. 4b, one can see that it is a better accordance of experimental data with the Crawford- Wilke model when the real slope of the flooding line is considered. The maximum error of the modified model for our experimental points was 24%, lower than that reported by the authors of the original correlation: 35% for their own data.

4.2 The Seibert- Fair model

This model was developed after Eighties, when structure packing type was introduced, but it can be also applied to the random type packing as well. The models presented previously were empirical but this one is analytical, starting from the following assumption: the drops are rigid and spherical, the drop size can be represented by a Sauter mean diameter d_{32}, the axial mixing of the continuous and of the dispersed phase can be neglected. The authors consider a drop traveling at an *angle of ascent* θ, in order to avoid the packing surface in its path. In a spray column, the mean angle of ascent is 90° but in a packed column the angle is smaller and depends on the drop size and the packing specific area. The packing increases the droplet velocity and the path length. This can be expressed as a *tortuosity factor* ξ:

$$\xi = a_p \cdot d_{32} / 2 \tag{45}$$

Manipulation of the classic equations of hydrodynamics and the use of the maximum theoretical holdup value of 0.52 lead to the final expression of the model:

$$1.08 \cdot V_{cf} + [\cos\left(\frac{\pi\xi}{4}\right)]^{-2} \cdot V_{df} = 0.192 \cdot \varepsilon \cdot V_{s0} \tag{46}$$

where V_{s0} is the slip velocity of the singular drop in the dispersion column defined by hydrodynamic Eq. 47:

$$V_{s0} = [(\Delta\rho g d_{32}) / (3\rho_c C_D)]^{0.5} \tag{47}$$

The dispersed phase holdup at flooding point ϕ_f depends on the the phases flow ratio and by consequence, on their superficial velocities ratio (V_d and V_c). Seibert & Fair (1988) proposed the empirical Eq.48:

$$\phi_f = \frac{V_d \cdot \left[\cos\left(\frac{\pi \cdot \xi}{4}\right)\right]^{-2}}{\varepsilon \cdot V_{s0} \exp\left(-1.92 \cdot \phi_f\right) - \frac{V_c}{1-\phi_f}} \tag{48}$$

This correlation is used especially for the prediction of the holdup at flooding, by trial-error method.

Legend: square for the system water- gasoline, circle for the system NaOH solution-gasoline and triangle for the system CCl_4- water.

The Fair- Sibert model was verified by its authors and gave good results for their own data obtained on a small diameter (100 mm) column (Seibert & Fair 1988). At larger scale: 400 mm diameter, the model gave bigger errors (Seibert *et al.* 1990). Also, the model was verified with data obtained in the original experiment (Koncsag & Stratula 2002) giving a maximum error of 26.4%. The parity plot for the original data is shown in Figure 5.

It seems to be a good accordance of the original data with the model but in fact the errors are systematic; for example, in the case of CCl_4- water system, the errors go continuously from negative to positive values. This could be explained either by non- reliable data or by a non-reliable model. The authors of the present work tend to consider a non- reliable model as long as the authors of the model themselves had a parity plot which indicated exclusively negative values for the standard errors (Seibert *et al.*1990).

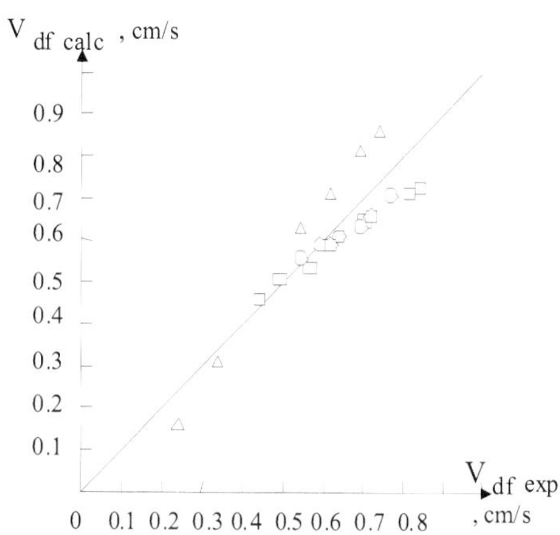

Fig. 5. The parity plot: $V_{df\ calc}$ (Seibert -Fair model) vs. original experimental values

5. Modelling the removal of mercaptans from liquid hydrocarbon streams in structured packing columns

5.1 Calculating the diameter

The data from Tables 1 and 2 were processed by dimensional analysis, with the goal of finding a mathematical model, which gives the limiting capacity of the contactor, expressed as *slip velocity of the phases* in flooding conditions.

The relevant list of dimensional parameters and dimensional constant describing this process is:

$$d_{32}\,,\ \sigma\,,\ \rho_c\,,\ \mu_c\,,\ V_{slip}\,,\ a_p\,,\ g \tag{49}$$

According to the dimensional analysis theory, this set of 7 dimensional parameters and constants, reduces to a set of only 4 dimensionless numbers. These numbers are chosen by us according to their physical significance for the extraction process:

$$Re,\ C_D\,,\ Ar\,,\ \Pi_4 \tag{50}$$

where Re is the Reynolds number, proportional to the ratio between the inertial force and the viscous force, $(\rho_c \cdot d_{32} \cdot V_{slip}) / \mu_c$;

- C_D - the drag coefficient, proportional to the ratio between the gravitational force and the inertial force, $(g \cdot \Delta\rho \cdot d_{32}) / (\rho_c \cdot V_{slip}^2)$;
- *Ar*- the Archimedes number, proportional to the ratio between the gravitational force and the viscous force, $(g \cdot \rho_d \cdot d_{32}{}^3 \cdot \Delta\rho / \mu_c^2)$;
- Π_4 - a dimensionless number, which combines the geometrical characteristics of the packing and of the drops, $s \cdot (1-\phi) \cdot \dfrac{\varepsilon}{a_p}$, s being the characteristic surface of the mean drop ($= 6 / d_{32}$) and ϕ being the hold-up of the dispersed phase in the column.

So, the criterial equation for the process would be:

$$Re^a \cdot C_D^b \cdot Ar^c \cdot \Pi_4{}^d = \text{const.} \tag{51}$$

The linear velocity which intervenes in the Re and C_D numbers is *the slip velocity*.

By processing the data from Tables 1 and 2, the exponents *a, b, c, d* and the constant in the Eq.51 were found and the general model is the following:

$$Re \cdot C_D \cdot Ar^{-0.6} \cdot \Pi_4 = 2.97 \tag{52}$$

The average deviation of this model is 7.7% and the absolute maximum error is 11.8%.

Replacing the dimensionless numbers with their definition correlations, the Eq.52 turns into Eq.53 for the calculation of the slip velocity in flooding conditions:

$$V_{slip} = 0.33 \cdot \Delta\rho^{0.4} \cdot \rho^{-0.6} \cdot d_{32}^{-0.8} \cdot \mu_c^{0.2} \cdot \left(\frac{\varepsilon}{a_p} \right) \cdot (1-\phi) \tag{53}$$

Or, taking into consideration the Eq. 36, the characteristic velocity of the singular drop is:

$$V_K = 0.33 \cdot \Delta\rho^{0.4} \cdot \rho^{-0.6} \cdot d_{32}^{-0.8} \cdot \mu_c^{0.2} \cdot \left(\frac{\varepsilon}{a_p} \right) \tag{54}$$

The new model can serve for dimensioning the liquid- liquid countercurrent contactor equipped with packing: random or ordered. The steps of the dimensioning are:

- the calculation of the characteristic velocity (Eq.54);
- the choice of the desired ratio of the phases (V_d/V_c);
- the calculation of the dispersed phase holdup by trial- error method (Eq.48);
- the calculation of slip velocity (Eq.53 applied to the flooding conditions, ϕ_f);
- the prediction of the throughputs limit (V_{df} and V_{cf}), knowing their ratio ($V_d / V_c = V_{df} / V_{cf}$;
- the calculation of the diameter with Eq. 30, where $B_{max} = V_{df} + V_{cf}$.

5.2 Calculating the active height of the column

The extraction of mercaptans with alkaline solution is accompanied by a second- order instantaneous reaction. As explained in Section 2.2, in this case, the mass transfer coefficients can be calculated as for the physical extraction, since the mass transfer is much slower than the reaction rate.

The calculation of the active height of the column is performed with Eq.38-41. The main difficulty consists of calculating the integrals (Eq.39 and 40) because one should know the concentrations profiles along the column. For systems following the Nernst law (Eq.3) and for very high values of the extraction factor (defined by Eq.56), the number of the transfer units NTU_{od} can be calculated from the number of theoretical stages *NTT*, with the Eq.55:

$$\frac{NTT}{NTU_{od}} = \frac{1 - \frac{1}{E}}{lnE} \tag{55}$$

NTT can be found graphically in a McCabe-Thiele- type construction.
The extraction factor *E* is defined by the Eq.56:

$$E = m \cdot \frac{S}{A} \tag{56}$$

where S/A is the solvent- to- feed gravimetric ratio.

As seen in Eq.56, the phase ratio refers to the solvent and feed and not to dispersed and continuous, as in diameter calculation. In mass transfer, the direction of the transfer is very important. Always, the direction is from the feed to the solvent, whatever the dispersed phase is.

The data used in modelling the extraction of the buthanethiol, propanethiol and ethanethiol are given in Table 3.

A model of the same type for all the data in Table 3 was developed, taking into account the factors determining the mass transfer rate: c- the concentration of NaOH solution (%wt), the geometrical characteristics of the packing (ε, a_p), the dispersed phase (the feed) superficial velocity and the acidity of the mercaptans:

$$K_{od} \cdot a \cdot 10^3 = \alpha_i \left(0.95 + \frac{c}{100}\right)^{A_1} \left(\frac{a_p}{\varepsilon}\right)^{A_2} \left(10^2 \cdot V_d\right)^{A_3}, i = 1,2,3 \quad (57)$$

In Eq.57, the coefficients α_1, α_2, α_3 correspond to different mercaptans, respectively buthanethiol, propanethiol and ethanethiol.

The inferior limit value for a_p in Eq.57 is 0.01 m^2/m^3, being assigned to the unpacked column.

Taking the logarithms in the previous formula and denoting by $Y = \ln(K_{od} \cdot a \cdot 10^3)$, $A_0 = ln\,\alpha$, $X_1 = ln\left(0.95 + \frac{c}{100}\right)$, $X_2 = \ln(\frac{a_p}{\varepsilon})$, $X_3 = \ln(10^2 \cdot V_d)$, $C_i = ln\alpha_i, i = 1,2,3,$

the model becomes:

$$Y = A_0 + A_1X_1 + A_2X_2 + A_3X_3 + \xi, \quad (58)$$

where Y is the dependent variable, X_1, X_2, X_3 are independent (explicative) variables and ξ is the specification error of the model. In a equivalent form, the system can be written:

$$y = xA + \xi\ . \quad (59)$$

Using the least squares method, the solution of Eq.58 is:

$$\begin{pmatrix} \hat{C}_1 \\ \hat{C}_2 \\ \hat{C}_3 \\ \hat{A}_1 \\ \hat{A}_2 \\ \hat{A}_3 \end{pmatrix} = \begin{pmatrix} 1.442 \\ 2.048 \\ 2.867 \\ 4.119 \\ 0.091 \\ 0.837 \end{pmatrix} \quad (60)$$

In what follows, $\hat{y}_t$ is the estimated (computed) value of y_t and e_t- the residual, i.e $e_t = y_t - \hat{y}_t$, t = 1...81.

Remark. A distinction has to be made between the specification error of the model, ξ_t which is and remains unknown and the residual, which is known.

The variance of the error $\sigma^2(\varepsilon)$ can be estimated by:

$$\hat{\sigma}^2(\varepsilon) = \frac{ee'}{n-k-1} = 0.059\ , \quad (61)$$

where: n = 81 is the observations number, k = 3 is the number of explicative variables and e is the vector containing the residuals, et , t = 1...81.

Making the calculus, it can be seen that the residuals sum is zero. Since the residual variance is closed to zero, the residuals in the model are very small.

From 60, the fitting quality is measured using the determination coefficient, R^2:

$$R^2 = 1 - \frac{\sum_{t=1}^{27} e_t^2}{\sum_{t=1}^{81} (y_t - \overline{y})^2} = 0.925 \quad (62)$$

and the modified determination coefficient, $\overline{R}^2$:

$$\overline{R}^2 = 1 - \frac{n-1}{n-k-1}\left(1 - R^2\right) = 0.922 \quad (63)$$

Since these values are closed to 1, the fitting quality is very good.

I. Tests on the model coefficients

i. It was verified if the explicative variables have significant contributions to the explanation of the dependent variable, by testing the hypothesis:

$$H_0 : A_i = 0, i = 0,1,2,3, \quad (64)$$

at the significance level $\alpha = 5\%$.
It has been done by a *t* test, that rejected the null hypothesis H_0, so the model coefficients are significant.

ii. The global significance of the model has been tested by a *F* test , for which the null hypothesis is:

$$H_0 : A_0 = A_1 = A_2 = A_3 = C_1 = C_2 = C_3 = 0$$

The hypothesys H_0 was rejected, at significance level $\alpha = 5\%$
From i. and ii. it results that the model coefficients are well chosen.

II. Tests on the errors

We saw that the residual sum is zero and the residual variance is 0.059.
We complete the information on the errors distribution providing the results concerning their normality, homoscedasticity and correlation.

i. Normality test
In order to verify the normality of the errors, the well-known Kolmogotrov-Smirnov test has been used, as well as the Jarque Bera test (Barbulescu & Koncsag, 2007).
Both tests lead us to accept the normality hypothesis.

ii. Homoscedasticity test
The test Bartlett is used to verify the errors homoscedasticity.
The hypothesis which must be tested is:

H_0: the errors have the same variance.

First, the selection values are divided in i = 3 groups, each of them containing n_i =27 data, and the test statistic is calculated by the formula:

$$X^2 = \frac{-\sum_{j=1}^{i} n_j \ln \frac{s_j^2}{s^2}}{1 + \frac{1}{i-1}\sum_{j=1}^{i}\left(\frac{1}{n_j} - \frac{1}{n}\right)} \quad (65)$$

where s_1^2, s_2^2, s_3^2, s are respectivelly the selection variance of the groups and of the sample.
The hypothesis H_0 is accepted at the significance level $\alpha = 5\%$ since $X^2 = 0.1138 < 5.991 = \chi^2(2)$, where $\chi^2(2)$ is the value given in the tables of the repartition χ^2 with two degrees of freedom.

iii. Correlation test

In order to determine if there exists a correlation of first order between the errors, the test Durbin Watson is used, for which the statistics test is defined by Barbulescu & Koncsag (2007):

$$DW = \frac{\sum_{t=2}^{81}(e_t - e_{t-1})^2}{\sum_{t=1}^{81} e_t^2} = 1.268$$

Since DW=1.268<d_1 (the critical value in the Durbin- Watson tables) , it results that the errors are correlated at the first order.

6. Conclusions

The result of this work consists on a model for the calculation of the industrial scale column serving to the extraction of mercaptans from hydrocarbon fractions with alkaline solutions. The work is based on original experiment at laboratory and pilot scale. It is a simple, easy to handle model composed by two equations.
The equation for the slip velocity, linked to the throughputs limit of the phases and finally linked to the column diameter, shows the dependency of the column capacity on the physical properties of the liquid- liquid system and the geometrical characteristics of the packing:

$$V_{slip} = 0.33 \cdot \Delta\rho^{0.4} \cdot \rho^{-0.6} \cdot d_{32}^{-0.8} \cdot \mu_c^{0.2} \cdot \left(\frac{\varepsilon}{a_p}\right) \cdot (1-\phi)$$

It is recommended for the usual commercial packing having a_p in range of 195-340 m^2/m^3 and ε in range of 0.74-0.96 and for liquid-liquid systems with interfacial tension in the range of 30-80 $\cdot$ 10^{-3} N/m. The average deviation of the model is 7.7% and the error's maximum maximorum is 11.8%.
The equation for the mass transfer coefficients at the extraction of different mercaptans is linked to the calculation of the active height of the column:

$$K_{od} \cdot a \cdot 10^3 = \alpha_i \left(0.95 + \frac{c}{100}\right)^{A_1} \left(\frac{a_p}{\varepsilon}\right)^{A_2} \left(10^2 \cdot V_d\right)^{A_3}, i = 1,2,3$$

where A_1=4.119; A_2= 0.091; A_3=0.835. α has different values for buthanethiol, propanethiol and ethanethiol respectively:1.442; 2.0867; 2.867.
The residual sum is zero and the residual variance is 0.059, so the accuracy of the model is very good. The fitting quality is confirmed by the high values of the determination coefficients. The model is satisfactory also points of view of statistics, since its coefficients are significant and the errors have a normal repartition and the same dispersion.
The model works for all type of packing, structured or bulk.

7. Nomenclature

a - the interfacial area, m^2/m^3
a_p – packing specific area, m^2/m^3
Ar- Archimedes number, $g \cdot \rho_d \cdot d_{32}{}^3 \cdot \Delta\rho / \mu_c{}^2$, dimensionless
A_0, A_1, A_2, A_3- constants, dimensionless
c - the concentration of NaOH solution, % wt
C_D– drag coefficient, $(g \cdot \Delta\rho \cdot d_{32}) / (\rho_c \cdot V_{slip}{}^2)$, dimensionless
d_{32}– Sauter mean diameter of drops, $(\Sigma n_i d_i{}^3)/(\Sigma n_i d_i{}^2)$, m
D- diffusivity, m^2/s
D_c-column diameter, m
E- extraction factor, dimensionless
g- gravitational constant, m/s^2
H- active height of the column, m
HTU- height of mass transfer unit, m
k- partial mass transfer coefficient
k- reaction rate constant
K- overall mass transfer coefficient
$K_{od} \cdot a$- overall volumetric mass transfer coefficient related to the dispersed phase, s^{-1}
m- repartition coefficient, Nernst law
NTU- number of mass transfer units, dimensionless
Re-Reynolds number, $(\rho \cdot d_{32} \cdot v_{slip}) / \mu$, dimensionless
s- characteristic surface of the mean drop, $6 / d_{32}$
Sc-Schmidt criterion, $Sc=\mu/D$, dimensionless
Sh-Sherwood criterion, $Sh= k_d \cdot d_{32} /D$, dimensionless
V_{cf} ,V_{df}– superficial velocities of the continuous phase and the dispersed phase respectively, m/s
V_K- characteristic velocity of drops, m/s
V_{slip} – slip velocity of phases, m/s
α- coefficient, dimensionless
ε – void fraction of the packing, m^3/m^3
μ – viscosity, kg/m.s
ρ – density, kg/m^3
$\Delta\rho$– density difference of the phases, kg/m^3
σ – interfacial tension, N/m
ϕ - holdup of the dispersed phase, m^3/m^3
Π_4 - dimensionless number, $s \cdot (1 - \phi) \cdot \varepsilon / a_p$

Subscripts:
c- continuous phase
d-dispersed phase
D- drag (coefficient)
E-extract
f- in flooding conditions

i- at interface
o- overall
p- packing
R-raffinate
0-single drop
Superscripts:
0-in absence of chemical reaction

8. Acknowledgement

The research was supported for the second author by the national authority CNCSIS-UEFISCSU under research grant PNII IDEI 262/2007.

9. References

Astarita, G. (1967). *Mass Transfer With Chemical Reaction*, Elsevier Publishing Co, Amsterdam

Barbulescu A. & Koncsag C.(2007). A new model for estimating mass transfer coefficients for the extraction of ethanethiol with alkaline solutions in packed columns, *Appl. Math. Modell*, Elsevier, 31(11), 2515-2523, ISSN 0307-904X

Crawford, J.W. & Wilke, C.R. (1951). Limiting flows in packed extraction columns, *Chem. Eng. Prog*, 47, 423-431, ISSN 0360-7275

Ermakov, S.A.; Ermakov, A.A.; Chupakhin, O.N.&Vaissov D.V.(2001). Mass transfer with chemical reaction in conditions of spontaneous interfacial convection in processes of liquid extraction, *Chemical Engineering Journal*, 84(3), 321-324, ISSN 1385-8947

Godfrey, J.C. & Slater M.J (1994). *Liquid- Liquid Extraction Equipment*, John Wiley &Son, ISBN 0471941565, Chichester

Hanson, C.(1971). *Recent Advances in Liquid-Liquid Extraction*, Pergamon-Elsevier, ISBN 9780080156828, Oxford, New York

Iacob,L. & Koncsag, C.I. (1999). *Hydrodynamic Study of the Drops Formation and Motion in a Spray Tower for the Systems Gasoline -NaOH Solutions*, XIth Romanian International Conference On Chemistry and Chemical Engineering (RICCCE XI), Section: Chemical Engineering, Bucureşti, p.131 (on CD)

Koncsag, C.I.& Stratula, C. (2002). Extractia lichid- lichid în coloane cu umplutură structurată.Partea I: Studiul hidrodinamic, *Revista de Chimie*, 53 (12), 819-823, ISSN0034-7752

Koncsag, C.I. (2005). Models predicting the flooding capacity of the liquid- liquid extraction columns equipped with structured packing, Proceedings of the *7th World Congress of Chemical Engineering*, Glasgow,UK, ISBN 0 85295 494 8 (CD ROM)

Koncsag, C.I. & Barbulescu,A.(2008). Modelling the removal of mercaptans from liquid hydrocarbon streams in structured packing columns, *Chemical Engineering and Processing*, 47, 1717-1725, ISSN0255-2701

Laddha, G.S.& Dagaleesan, T.E.(1976). *Transport Phenomena in Liquid-Liquid Extraction*, Tata-McGraw-Hill, ISBN 0070966885, New Delhi

Misek, T. (1994). *Chapter 5. General Hydrodynamic Design Basis for Columns*, în Godfrey, J.C. & Slater M.J.*Liquid- Liquid Extraction Equipment*, John Wiley &Son, ISBN 0471941565 Chichester

Nemunaitis, S.R., Eckert, J.S., Foot E.H. & Rollison, L.H. (1971). Packed liquid-liquid extractors, *Chem. Eng. Prog.*, 67(11), 60-64, ISSN 0360-7275

Pohorecki, R.(2007). Effectiveness of interfacial area for mass transfer in two-phase flow in microreactors, *Chem. Eng. Sci.*, 62, 6495-6498, ISSN 0009-2509

Pratt, H.R.C. (1983). Interphase mass transfer, *Handbook of Solvent Extraction*, Wiley/Interscience, ISBN 0471041645, New York

Thornton, J.D.(1956). Spray Liquid-Liquid Extraction Column:Prediction of Limiting Holdip and Flooding Rates, *Chem. Eng. Sci.*, 5, 201-208, ISSN 0009-2509

Treybal, R.E. (2007). *Liquid Extraction*, Pierce Press, ISBN 978-1406731262, Oakland, CA

Sarkar, S., Mumford, C.J. & Phillips C.R. (1980). Liquid-liquid extraction with interpase chemical reaction in agitated columns.1.Mathematical models., *Ind. Eng. Chem. Process Des. Dev.*, 19, 665-671, ISSN 0196-4305

Seibert, A.F. & Fair, J.R. (1988) Hydrodynamics and mass transfer in spray and packed liquid-liquid extraction columns, *Ind. Eng. Chem. Res.*, 27, 470-481, ISSN 0888-5885

Seibert, A.F. Reeves, B.E. and Fair, J.R.(1990), Performance of a Large-Scale Packed Liquid-Liquid Extractor, *Ind. Eng. Chem. Res.*, 29 (9), 1907-1914, ISSN 0888-5885

Watson, J.S , McNeese, L.E. ,Day, J& Corroad, P.A.(1975). Flooding rates and holdup in packed liquid-liquid extraction columns, *AIChEJ*, 21(6), 1980-1986, ISSN 0001-1541

Wilke, C.R & Chang, P.(1955). Correlation of diffusion coefficients in dillute solutions, *AIChEJ*,1 (2), 264-270, ISSN 0001-1541

Zhu S.L.& Luo G.S. in Proceedings of the International Solvent Extraction Conference ISEC'96, *Value Adding Through Solvent Extraction,* ISBN 073251250, University of Melbourne, p.1251-1255

11

Modeling Enhanced Diffusion Mass Transfer in Metals during Mechanical Alloying

Boris B. Khina[1] and Grigoriy F. Lovshenko[2]
[1]*Physico-Technical Institute, National Academy of Sciences of Belarus*
[2]*Belorussian National Technical University*
Minsk, Belarus

1. Introduction

The subject of this Chapter is an urgent cross-disciplinary problem relating to both Mass Transfer and Materials Science, namely enhanced, or abnormal diffusion mass transfer in solid metals and alloys under the action of periodic plastic deformation at near-room temperatures. This phenomenon takes place during the synthesis of advanced powder materials by mechanical alloying (MA) in binary and multi-component systems, which is known as a versatile means for producing far-from equilibrium phases/structures possessing unique physical and chemical properties such as supersaturated solid solutions and amorphous phases (Benjamin, 1992; Koch, 1992; Koch, 1998; Ma & Atzmon, 1995; Bakker et al., 1995; El-Eskandarany, 2001; Suryanarayana, 2001; Suryanarayana, 2004; Zhang, 2004; Koch et al., 2010).

Along with MA, this phenomenon is relevant to other modern processes used for producing bulk nanocrystalline materials by intensive plastic deformation (IPD) such as multi-pass equal-channel angular pressing/extrusion (ECAP/ECAE) (Segal et al., 2010; Fukuda et al., 2002), repetitive cold rolling (often termed as accumulative roll bonding-ARB, or folding and rolling-F&R) (Perepezko et al., 1998; Sauvage et al., 2007; Yang et al., 2009), twist extrusion (Beygelzimer et al., 2006) and high-pressure torsion (HPT) using the Bridgman anvils (X.Quelennec et al., 2010). It is responsible for the formation of metastable phases such as solid solutions with extended solubility limits during IPD, demixing of initial solid solutions or those forming in the course of processing, and is considered as an important stage leading to solid-state amorphization in the course of MA.

In these and similar situations, the apparent diffusion coefficients at a room temperature, which are estimated from experimental concentration profiles, can reach a value typical of a solid metal near the melting point, $D \sim 10^{-8}$-10^{-7} cm^2/s, and even higher.

Mechanical alloying (MA) was discovered by J.S.Benjamin in early 1970es as a means for producing nickel-base superalloys with dispersed fine oxide particles (Benjamin & Volin, 1974). Later it was found that MA brings about the formation of non-equilibrium structures in many metal-base systems, such as supersaturated solid solutions, amorphous and quasicrystalline phases, nanograins etc., and it had acquired a wide use for the synthesis of novel metallic and ceramic materials. MA of powder mixtures is performed in attritors, vibratory and planetary mills and other comminuting devices where particle deformation occurs during incidental ball-powder-ball and ball-powder-wall collisions. An important advantage of MA is cost efficiency since the alloys are produced without furnaces and other high-temperature equipment while its main drawback is contamination of the final product because of wear of the balls and inner surface of a milling device.

On the first stage of MA, particle fracturing and cold welding over juvenile surfaces bring about the formation of composite particles, which contain interweaved lamellas of dissimilar metals (if both of the initial components are ductile) or inclusions of a brittle component in a matrix of a ductile metal. This substantially increases the contact area of the starting reactants. After that, within a certain milling time depending on the energy input to the comminuting device, which is characterized by the ball acceleration reaching 60-80g in modern industrial-scale planetary mills (Boldyrev, 2006)), non-equilibrium phases are formed. These transformations occur due to plastic deformation of composite particles, which brings about generation of non-equilibrium defects in the metals and enhanced solid-state diffusion mass transfer. Hence, the latter is virtually the most important phenomenon responsible for metastable phase transformations during MA.

Despite vast experimental data accumulated in the area of MA, a deep understanding of the complex underlying physicochemical phenomena and, in particular, deformation-enhanced solid-state diffusion mass transfer, in still lacking. As outlined in (Boldyrev, 2006), this situation hinders a wider use of cost and energy efficient MA processes and the development of novel advanced materials and MA-based technologies for their production. Further development in this promising and fascinating area necessitates a new insight into the mechanisms of deformation-enhanced diffusion, which is impossible without elaboration of new physically grounded models and computer simulation. As a first step, it seems necessary to review the known viewpoints on this intricate phenomenon.

In this Chapter, analysis of the existing theories/concepts of solid-state diffusion mass transfer in metals during MA is performed and a new, self-consistent model is presented, which is based on the concept of generation of non-equilibrium point defects in metals during intensive periodic plastic deformation. Numerical calculations within the frame of the developed model are performed using real or independently estimated parameter values (Khina et al., 2004; Khina et al., 2005; Khina & Formanek, 2006).

2. Brief analysis of existing concepts

Different models that are used in the area of MA can be divided into three large groups: mechanistic, atomistic and macrokinetic ones. The mechanistic models (Maurice & Courtney, 1990; Magini & Iasona, 1995; Urakaev & Boldyrev, 2000a; Urakaev & Boldyrev, 2000b; Chattopadhyay et al., 2001; Lovshenko & Khina, 2005) consider the mechanics of ball motion and incidental ball-powder-ball and ball-powder-wall collisions in a milling device. The concept of elastic (Hertzian) collision is employed. This approach permits estimating the maximal pressure during collision, energy transferred to the powder, the collision time, strain and strain rate of the powder particles, local adiabatic heating and some other parameters, which can be used for assessing the physical conditions under which deformation-enhanced diffusion and metastable phase formation occur in the particles in the course of MA. This approach was used by the authors for evaluating the MA parameters for a vibratory mill of the in-house design (Lovshenko & Khina, 2005). However, these and similar models all by itself cannot produce any information about the physics of defect formation, enhanced diffusion mass transfer and non-equilibrium structural and phase transformations in metals and alloys under mechanical deformation.

Atomistic models employ molecular statics and molecular dynamics simulations (MDS) (Lund & Schuh, 2004a; Lund & Schuh, 2004b; Odunuga et al., 2005; Delogu & Cocco, 2005). They permit studying ordering and disordering processes, atomic intermixing, i.e. diffusion over a small (nanometric) scale and non-equilibrium phase transitions, e.g., amorphization, in crystalline solids under the influence of an external mechanical force (the so-called "mechanically driven alloys"). In MDS, considered are individual nanosized particles or thin films. However, these models are poorly linked both to external conditions, i.e. the processing regimes in a milling device, and to macroscopic physicochemical parameters that are measured, directly or indirectly, basing on the experimental results or are known in literature, such as diffusion coefficients. Besides, it should be born in mind that since MDS is typically performed over a relatively small-size matrix using periodic boundary conditions, generalization of the obtained results to a macroscopic scale is not always well justified and hence one should use them with caution when interpreting the experimental data on MA.

Macrokinetic models of MA occupy a position in between the mechanistic and atomistic approaches and are based on the results of the latter. They can give important information on the physicochemical mechanisms of non-equilibrium phase and structure formation and deformation-enhanced diffusion during MA, which is necessary for optimization of existing and development of novel MA-based technologies and MA-produced materials, link the mechanical parameters of MA to the transformation kinetics in an individual particle, and bridge the existing gap between the two aforesaid approaches. However, such models are least developed as compared with the mechanistic and atomistic ones, which is connected with the problem complexity. Up to now, two basic concepts are know.

Most elaborated is a semi-quantitative concept according to which the dominated role in the mechanochemical synthesis belongs to fracturing of initial reactant particles with the formation of juvenile surfaces during collisions in a comminuting device (Butyagin, 2000; Delogu & Cocco, 2000; Delogu et al., 2003; Butyagin & Streletskii, 2005). In this case, crystal disordering occurs in surface layers whose thickness is several lattice periods. In the contact of juvenile surfaces of dissimilar particles during collision, co-shear under pressure brings about the so-called "reactive intermixing" on the atomic level, which leads to the formation of a product (i.e. chemical compound) interlayer. Here, the most important factor is a portion of the collision energy transferred to the reactant particles per unit contact surface area, which was estimated in the above cited works. In our view, such a mechanism of interaction is typical of mechanical activation and mechanochemical synthesis in inorganic systems where the reactant particles (salts, oxides, carbonates etc.) are hard and brittle, and the dominating process during collisions is brittle fracture over cleavage planes.

In binary and multicomponent metal base systems, unlike brittle inorganic substances, the main process during mechanical alloying is plastic deformation of composite (lamellar) particles formed on earlier stages due to fracturing and cold welding of initial pure metal particles. The formation of solid solutions, metastable (e.g., amorphous) and stable (e.g., intermetallic) phases in the course of MA is impossible without intermixing on the atomic level in the vicinity of interfaces in composite particles (boundaries of lamellas of pure metals), i.e. without diffusion. Thus, the second macrokinetic concept of MA outlines the role of deformation-induced solid-state diffusion mass transfer (Schultz et al., 1989; Lu & Zhang, 1999; Zhang & Ying, 2001; Ma, 2003), which is less developed in comparison with the "reactive intermixing" model referring to the area of inorganic mechanochemistry.

It should be noted that the phenomenon of abnormal (enhanced) non-equilibrium diffusion mass transfer under intensive plastic deformation (IPD) was experimentally observed in bulk metals at different regimes of loading, from ordinary mechanical impact to shock-wave (explosion) processing in a wide rage of temperature, strain ε and strain rate $\dot{\varepsilon}$ (Larikov et al., 1975; Gertsriken et al., 1983; Arsenyuk et al., 2001a; Arsenyuk et al., 2001b; Gertsriken et al., 1994; Gertsriken et al., 2001), and at ultrasound processing (Kulemin, 1978). The apparent diffusion coefficient, which is calculated from the time dependence of the diffusion zone width, was found to increase by many orders of magnitude and approach a value typical of a metal in the pre-melting state. Extensive experimental investigation performed in the above cited works using a wide range of techniques such are autoradiography, X-ray analysis, Mossbauer spectroscopy and other methods have demonstrated that IPD of bimetallic samples (a metal specimen clad with another metal) in binary substitutional systems brings about the formation of supersaturated solid solutions. The penetration depth of atoms from a surface layer into the bulk material reaches several hundred microns and the concentration of alloying element can be large: 18% Al in copper and up to 10% Cu in aluminum in the Cu-Al bimetallic couple at 300 K and $\dot{\varepsilon}$ =120 s^{-1} (Gertsriken et al., 1994).

However, an adequate explanation of the enhanced non-equilibrium diffusion mass transfer phenomenon in crystalline solids under IPD has not been developed so far. Moreover, the very role of diffusion in MA is a subject of keen debates in literature: in particular, a series of mutually contradicting papers was published in journal "Metal Science and Heat Treatment" (Farber, 2002; Skakov, 2004; Gapontsev & Koloskov, 2007; Skakov, 2007; Shtremel', 2002; Shtremel', 2004; Shtremel', 2007). In (Farber, 2002), the physical factors that could be responsible for the acceleration of solid-state diffusion, e.g., generation of non-equilibrium point defects during deformation, were described in detail on a qualitative level but no calculations nor even simple numerical estimates were given. Experimental data on the formation of supersaturated solid solutions at MA were reviewed in (Skakov, 2004; Skakov, 2007) and a qualitative hypothesis was presented. In (Gapontsev & Koloskov, 2007), a model for enhanced diffusion is presented wherein the disclinations (i.e. triple grain junctions) act as sources and sinks of non-equilibrium vacancies during IPD thus giving rise to intensive diffusion fluxes of vacancies across grains, which, it turn, promote the diffusion of alloying atoms. On the other hand, in (M.A.Shtremel', 2002; M.A.Shtremel', 2004) simple numerical estimates based on the classical theories of diffusion and plastic deformation, which can not account for the process-specific factors acting in the conditions of MA, were used to support an opposite viewpoint that atomic diffusion plays an insignificant and even negative role in the formation of solid solutions and intermetallics during MA. It is speculated that the basic reason of alloying during IPD is not diffusion mass transfer but "mechanical intermixing of atoms" at shear deformation (Shtremel', 2004; Shtremel', 2007) but the physical meaning of this term is not explained; the author of the cited papers did not present any theories nor numerical estimates to support this concept. Different viewpoints on the role of atomic diffusion and deformation-generated point defects in the structure formation in alloys under IPD have been recently reviewed in (Lotkov et al., 2007).

As was noted earlier (Khina & Froes, 1996), this situation is determined by insufficient theoretical knowledge of the physical mechanisms underlying the deformation-enhanced diffusion mass transfer during MA on the background of extensive experimental data accumulated in this area. Unfortunately, this statement is still valid now to a large extent. The absence of a comprehensive macrokinetic model is a constraint on the way of a further development of novel materials and technologies based on MA and other IPD techniques.

In several theoretical works employing the macrokinetic approach, mathematical models of deformation-induced diffusion mass transfer during MA considered only diffusion along curved dislocation lines (the so-called dislocation-pipe diffusion) (Rabkin & Estrin, 1998) or a change of geometry of an elementary diffusion couple in a composite (lamellar) particle because of deformation (Mahapatra et al., 1998); in the latter case, traditional diffusion equation (the Fick's law) was used. In these attempts, the role of deformation-generated point defects was not included. Besides, the whole processing time of powders in a milling device was considered as the time of diffusion (from 1 h in (Rabkin & Estrin, 1998) to 50 h in (Mahapatra et al., 1998)) although it is known from mechanistic models that at MA the collision time, during which deformation-induced diffusion occurs, is substantially (by several orders of magnitude) shorter than intervals between collisions (Benjamin, 1992; Suryanarayana, 2001; Suryanarayana, 2004; Maurice & Courtney, 1990; Chattopadhyay et al., 2001; Lovshenko & Khina, 2005). In (Mahapatra et al., 1998), the numerical value of the main parameter, viz. volume diffusion coefficient, was taken at an elevated temperature, which was varied arbitrary (in the range 505-560 K for binary system Cu-Zn and up to 825 K for system Cu-Ni) to attain agreement with experimental data. This is motivated by particle heating during ball-powder-ball collisions, although it is known that during a head-on collision, which provides maximal pressure, strain and strain rate of particles, a local temperature rise is small (~10 K) for most of the milling devices, and the temperature quickly decreases to the background level due to high thermal conductivity of metals (Maurice & Courtney, 1990; Lovshenko & Khina, 2005). The main conclusion from modeling performed in (Mahapatra et al., 1998) is trivial: to achieve agreement between the calculations obtained using the Fickian equation and the experimental data on the alloying degree reached at a long processing time, the diffusion coefficient must have a value typical of that at a high temperature. This fact is known for many years: the effective (i.e. apparent) diffusion coefficient at IPD exceeds the equilibrium value at the processing temperature by several orders of magnitude.

There are several models of abnormal solid-state diffusion at shock loading of a bimetallic specimen, which are based on extended non-equilibrium thermodynamics (Sobolev, 1997; Buchbinder, 2003). Fast diffusion in metals caused by a propagating shock wave is described using the hyperbolic telegrapher equation, i.e. the equation of a decaying elastic wave. Within this concept, in the left-hand side of the Fickian diffusion equation, the second time-derivative, $\partial^2C/\partial t^2$, is included along with term $\partial C/\partial t$, where C is concentration. This brings about a final propagation velocity of the concentration disturbance (Buchbinder, 2003). But in this approach, mass transfer is considered as occurring in a structureless continuum (a fluid), and a physical mechanism responsible for fast diffusion in a crystalline solid is not revealed. In (Bekrenev, 2002), a similar situation is analyzed by introducing a drift term into the right-hand side of the diffusion equation to describe the motion of solute atoms in the field of an external force. However, this term was not analyzed in detail.

Models for diffusion demixing of a solid solution or intermetallic compound in the course of MA have been developed (Gapontsev & Koloskov, 2007; Gapontsev et al., 2000; Gapontsev et al., 2002; Gapontsev et al., 2003) which consider the formation of non-equilibrium vacancies in grain boundaries and their diffusion into grains. The vacancy flux directed into grains brings about an oppositely directed diffusion flux of solute atoms, which ultimately results in demixing of this stable or metastable phase. In its physical meaning, this model refers to a case when diffusion processes in a lamellar particle has already completed and a uniform product phase (metastable or equilibrium) has formed, and further milling brings about decomposition of the MA product. It should be noted that cyclic process of formation and decomposition of an amorphous or intermetallic phase was observed during prolonged

ball milling in certain systems (El-Eskandarany et al., 1997; Courtney & Lee, 2005). In these models, grain boundaries (Gapontsev et al., 2000; Gapontsev et al., 2002; Gapontsev et al., 2003) or disclinations (triple grain junctions) (Gapontsev & Koloskov, 2007) can act as vacancy sources when the deformation proceeds via grain boundary sliding and rotational modes. This corresponds to a situation when the size of grains in the particle has reduced to nanometric. Similar deformation mechanisms operate at superplastic deformation of micron and submicron grained alloys at elevated temperatures where accommodation of grains takes place via grain boundary diffusion (Kaibyshev, 2002) and vacancies arising in the boundary may penetrate into grains. However, as noted in (Shtremel', 2007), a mechanism via which disclinations can generate vacancies is not described in (Gapontsev & Koloskov, 2007), and estimates for the vacancy generation rate are not presented in (Gapontsev et al., 2000; Gapontsev et al., 2002; Gapontsev et al., 2003). Besides, the interaction of vacancy flux in a grain with edge dislocations, which can substantially reduce the vacancy concentration, is not considered, i.e. it is implied that nonograins, whose typical size in the powders processed by MA lies within 20-100 nm, do not contain dislocations. But experimental observations using high-resolution transmission electron microscopy have revealed that dislocation density in nanograined Ni (20-30 nm) obtained by IPD (particularly, accumulative roll bonding) is very high, $\sim 10^{12}$ cm^{-2} (Wu & Ma, 2006).
Thus, models (Gapontsev & Koloskov, 2007; Gapontsev et al., 2000; Gapontsev et al., 2002; Gapontsev et al., 2003) can be considered as incomplete and relating to a distant stage of MA where nanorgains of a solid solution or intermetallic compound have already been formed via a certain physical mechanism which was not considered in these works.
An idea of solid solution formation during MA by the "shear-drift diffusion" (Foct, 2004) or "trans-phase dislocation shuffling" (Raabe et al., 2009; Quelennec et al., 2010) has been proposed, which the authors of these works base upon the certain outcomes of atomistic simulations (Bellon & Averback, 1995). Since this viewpoint has acquired a certain use in literature, it is necessary to analyze it in detail. It is implied that during plastic deformation, which in crystalline solids is produced by dislocations gliding over glide planes, dislocations can cross the phase boundary. At large strains or strain rates the dislocation glide may occur over intersecting glide planes. According to the above concept, this results in "trans-phase dislocation shuffling" of groups of atoms at the boundary. In other words, after several dislocations gliding over different planes have crossed the phase boundary between dissimilar metals, say the A/B boundary, a group of atoms A originally located in phase A near the interface appears inside phase B (see Fig.9 in (Raabe et al., 2009)).
Although this qualitative concept seems clear and simple from the viewpoint of the classical dislocation theory and continuum mechanics, it contradicts the existing theories of plastic deformation of bulk polycrystalline materials, both coarse-grained (with micron-sized grains) and nanograined. During plastic deformation of polycrystals with grain size of the order of 1-100 μm, the dislocations that glide from an intragrain source (a Frank-Read source) towards a grain boundary under the action of shear stress cannot "burst" through the boundary (Meyers & Chawla, 2009): they accumulate near the latter forming the so-called pile-ups where the number of piled dislocations is $\sim 10^2$-10^3. The arising elastic stress activates a Frank-Read source in the adjacent grain, which results in macroscopic deformation revealing itself in a step-like displacement of the grain boundary. This theory results in the known Hall-Petch equation which shows a good agreement with numerous experimental data. Gliding dislocations can really cross a phase boundary, but only in the case of a coherent (or at least semi-coherent) interface between a matrix and a small-sized

inclusion of a strengthening phase, which has formed during ageing of precipitation-hardening alloys. This brings about matched co-deformation of the matrix and precipitate, or shearing of a particle, which is accompanied with interface steps formation (Argon, 2008). In (Raabe et al., 2009), it is speculated that dislocations can cross a phase boundary in a nanograined material under a high shear stress. It is argued that a <111> texture formed at co-deformation of two fcc phases (Cu and Ag) during drawing of Cu-5 at.% Ag-3 at.% Nb wires with a maximal true strain of 10.5 brings about the matching of highly stressed slip systems in both phases, and this consideration is used as an argument for the dislocation shuffling concept. However, the existence of same orientation of grains is a necessary but insufficient condition. A boundary between dissimilar metals formed by cold welding during MA is typically a non-coherent high-angle one and, in a special case, it may convert into a low-angle boundary due to grain-boundary sliding and grain rotation during IPD. But the authors of the cited work did not propose a physical mechanism via which a non-coherent boundary could convert into a coherent or at least a semi-coherent one.

In alloys with submicrocrystalline/nanograined structure, where the Hall-Petch law is not valid, deformation proceeds via different mechanisms. In nanosized grains the traditional Frank-Read sources of dislocations cannot operate because of lack of space. Then the leading dislocation in a pile-up formed near a grain boundary will interact with the interface, which results in absorption of dislocation by the latter, i.e. conversion into a grain-boundary dislocation accompanied with its core spreading (delocalization) (Segal et al., 2010). This results in grain-boundary sliding and may bring about rotational modes of deformation. In this case, triple grain junctions (disclinations) and non-equilibrium grain boundaries act as sources of dislocations for an adjacent grain (Segal et al., 2010; Gutkin et al., 2001; Bobylev et al., 2009; Gutkin et al., 2005), which results in deformation of the material as a whole. The dislocations (actually, dislocations semiloops) emitted by the grain boundary are blown like bubbles into the next grain by the shear strain; a similar phenomenon was observed at low-angle grain/subgrain boundaries during hot deformation of traditional (with micron-sized grains) metals and alloys (see Figs.3.12 and 3.13 in (Levitin, 2006)).

This concept is supported by atomistic modeling (Wang et al., 2008) for a Cu-Nb system: "a single mixed dislocation, from either Cu or Nb, cannot cross the interface even at resolved shear stresses in excess of 1.0 GPa". Thus, a non-coherent grain/phase boundary is typically an impermeable obstacle for dislocation glide in both micro and nanograined crystalline materials. Hence, in (Raabe et al., 2009), where work (Wang et al., 2008) was cited, the results of molecular-dynamics simulations performed in the latter were misinterpreted.

So, the qualitative concept of "shear-drift diffusion" or "dislocation shuffling" does not correspond to a real situation in MA and thus appears to be physically meaningless.

From the above brief analysis it is seen that, despite a number of attempts, a physically grounded model for the formation of (supersaturated) solid solutions in metallic systems during IPD and, in particular, MA has not been developed so far. This stage of MA is of primary scientific and technological importance since the goal of MA is the production of a far-from-equilibrium product possessing advanced properties.

3. Formulation of the model

3.1 Physical background and basic assumptions

As a physical situation, we consider an individual lamellar particle formed at an initial stage of MA due to fracturing of cold welding of initial metal particles, and separate a unit structural element, viz. diffusion couple "metal B (phase 2)-metal A (phase 1)" where diffusion mass transfer occurs during MA. In binary metal systems, diffusion in normal conditions proceeds mainly via a substitutional (vacancy) mechanism. Since the directions of incidental ball-powder-ball and ball-powder-wall collisions in a milling device are chaotic, we neglect a change of the diffusion-couple geometry, and reduce the role of plastic deformation only to the formation of defects in the crystal lattice of both phases.

It has been demonstrated experimentally that in many substitutional systems under IPD the enhanced solid-state diffusion mass transfer in a wide temperature range is dominated by volume rather than grain boundary diffusion (Larikov et al., 1975; Gertsriken et al., 1983; Arsenyuk et al., 2001a; Arsenyuk et al., 2001b; Gertsriken et al., 1994; Gertsriken et al., 2001). Since we consider early stages of MA, when the initial concentration gradient at the phase boundary 2/1 is very high, it seems reasonable to assume that phase transformation (e.g., formation of an intermetallic compound) at this interface does not occur, i.e. only diffusion of atoms A and B across the initial boundary can take place. This physical assumption is based on the results of works (Khusid & Khina, 1991; Desre & Yavari, 1990; Desre, 1991; Gusak et al., 2001) where it has been demonstrated using both kinetic (Khusid & Khina, 1991; Gusak et al., 2001) and thermodynamic (Desre & Yavari, 1990; Desre, 1991) considerations that in the field of a sharp concentration gradient in a binary metallic system nucleation of an equilibrium phase, e.g., intermetallic compound, is suppressed and can occur only after the gradient decreases in the course of diffusion to a certain critical level.

Volume diffusion in substitutional alloys (at close diameters of A and B atoms) can be strongly influenced by the formation of non-equilibrium vacancies and also by generation of interstitial atoms whose diffusion rate is high. In a number of works on IPD, the phenomenon of enhanced diffusion is attributed to the interstitial mechanism (Larikov et al., 1975; Skakov, 2004): highly mobile interstitial atoms (say, of sort A), which in binary substitutional solid solutions normally diffuse via a vacancy mechanism, are formed at the A/B interface due to "pushing" of lattice atoms A into interstices of crystal lattice B by the shear stress. After that, they rapidly diffuse in metal B in the field of an external force, e.g., a pressure jump generated by an incidental collision during MA or by explosion at shock-wave processing of metals. This mechanism is similar to the ballistic effects observed in reactor materials under radiation (Bullough et al., 1975; Murphy, 1987; Mansur, 1979) or at ion-beam processing of metals. However, molecular-dynamics simulation has revealed that for this mechanism to operate, local pressure at the phase boundary must be ~100 GPa (Gusak & Bushin, 1996), which is typical of explosion processing, whereas the mechanistic modeling of MA in ball mills has demonstrated that the maximal pressure during collisions is ~1 GPa (Maurice & Courtney, 1990; Lovshenko & Khina, 2005), i.e. substantially lower. That is why this mechanism is not relevant to "traditional" MA/IPD processes and hence is not considered here. Besides, within the above concept it is not specified what happens with the crystal lattice of metal B near the interface when foreign atoms A are "pushed" into interstitial positions of the former in a mass quantity.

In traditional polycrystalline materials with micron-size grains, dislocation glide is the dominating mechanism of plastic deformation (Nabarro et al., 1964; Novikov, 1983; Hirth & Lothe, 1982). As noted earlier, nanocrystalline materials with a grain size of 20-30 nm produced by IPD feature a high density of lattice dislocations (Wu & Ma, 2006). Molecular-dynamics simulations performed for copper (Vo et al., 2008) have unambiguously shown that even at very high strain rates (up to 10^{10} s^{-1}), which correspond to shock-wave processing of metals, dislocations account for 90% of the strain at the grain size of 20 nm, and the role of grain boundary sliding, which corresponds to the remaining 10%, decreases with strain. This outcome of modeling is supported by experimental observations (Gurao & Suwas, 2009): the major contribution to plastic strain at IPD (rolling to 90% reduction in thickness) of nanocrystalline Ni with a grain size of about 20 nm occurs through normal dislocation slip. It is known that during deformation of metals via the dislocation mechanism, excess point defects (vacancies and self-interstitials) are produced due to interaction of gliding dislocations (Nabarro et al., 1964; Novikov, 1983). Hence, it this model we consider namely this route of point defect generation.

3.2 Derivation of the model equations

Let us define the concentration of species in ratio to the density of lattice sites N_0 (Voroshnin & Khusid, 1979): $c_k = N_k/N_0$, where N_k is a number of k-th species per unit volume. Here the species in phases 1 and 2 are lattice atoms A and B, vacancies v and interstitial atoms A and B, which are denoted as A_i and B_i. Since $N_0 = N_B + N_A + N_v$, then $c_B + c_A + c_v = 1$, and thus the overall sum including the concentration of interstitials will exceed unity.
The diffusion mass transfer equation looks as (Adda & Philibert, 1966; Gurov et al., 1981):

$$\partial c_k / \partial t = -\,div\, J_k + F_k\,,\quad k \equiv A,B,v,A_i,B_i \tag{1}$$

where F_k is the sink/source term for k-the species and J_k is the diffusion flux. Neglecting the gradient of the lattice site density $\partial N_0/\partial x$, let us write the expression for diffusion fluxes of lattice atoms A and B and vacancies v taking into account the interconnection of fluxes via the cross-term, or off-diagonal interdiffusion coefficients following the classical theory of diffusion in solids (Voroshnin & Khusid, 1979; Adda & Philibert, 1966; Gurov et al., 1981):

$$J_k = -\sum_n D_{kn}\, grad\, c_n\,,\quad \sum_k J_k = 0\,,\ k,n \equiv A,B,v, \tag{2}$$

where D_{kn} are elements of the matrix of interdiffusion coefficients.
For diffusion fluxes of atoms and vacancies v, interdiffusion coefficients D_{kn} that appear in Eq. (2) can be determined using the theory of diffusion in solid solutions with non-uniform vacancy distribution (Gurov et al., 1981). We take into account that self-diffusion coefficient D^* in quasi-equilibrium conditions, i.e. at an anneal without external influences, is estimated as $D^* = f_c D_v c_v^0$, where c_v^0 is the equilibrium vacancy concentration at a given temperature, D_v is the vacancy diffusion coefficient, $D_v >> D^*$, and $f_c \approx 1$ is the correlation factor. Then in the case when non-equilibrium vacancies are present in a solid solution, i.e. $c_v >> c_v^0$, the diffusion coefficient of atoms will increase proportionally to term c_v/c_v^0. Hence we have

$$\begin{aligned}
&D_{BB} = D_B{}^*(g_{BB} - g_{AB})c_v / c_v^0, D_{AA} = D_A{}^*(g_{AA} - g_{BA})c_v / c_v^0,\\
&D_{Bv} = -D_B{}^* c_B / c_v^0, D_{vB} = D_{AA} - D_{BB},\\
&D_{vv} = D_v + D_{AA}, D_v = \left[c_B D_B{}^* + c_A D_A{}^*\right] / c_v^0\\
&g_{kk} = 1 + \partial(\ln\gamma_k)/\partial(\ln c_k), g_{kj} = \left(c_k / c_j\right)\partial(\ln\gamma_k)/\partial\left(\ln c_j\right), k,j \equiv A,B,\ k \neq j,
\end{aligned} \tag{3}$$

where D_B^* и D_A^* are the self-diffusion coefficients of atoms A and B in the given phase (1 or 2), g is the thermodynamic factor and γ is the activity coefficient. As a common first approximation, the solid solutions are considered to be ideal and then $g_{kk}=1$, $g_{kj}=0$.

As seen from Eqs. (3), increasing the vacancy concentration above c_v^0 substantially raises the diffusion coefficient. This corresponds to an increase in the pre-exponential factor D_0 in the Arrhenius formula $D^* = D_0\exp[-E/(RT)]$ at an almost unchanged activation energy E. This is connected with the vacancy diffusion mechanism in substitutional solid solutions where E is the energy barrier height for an atomic jump on the adjacent vacant position while D_0 is related to the number of available vacant sites, the latter being proportional to c_v.

Excess point defects are generated during deformation, i.e. at short-time collisions during MA in a ball mill. Relaxation of point defects towards equilibrium concentration during intervals between collisions occurs by two basic mechanisms: (i) interaction with edge components of dislocation loops, which act as volume-distributed sinks; in this case intersitials A_i and B_i become lattice atoms of the corresponding sort, and (ii) vacancy-interstitial annihilation, as a result of which an interstitial atom becomes a lattice one. To describe the rates of these processes we use certain results of the theory of defects in irradiated alloys (Bullough et al., 1975; Murphy, 1987; Mansur, 1979). The equilibrium concentration of self-interstitials in crystalline metals is typically assumed to be negligibly small. Then the rate of vacancy-interstitial annihilation is proportional to the deviation of the vacancy concentration from the equilibrium value times the concentration of self-interstitials. The rate of point defect adsorption by edge dislocations is proportional to the density of the latter, the diffusion coefficient of point defects and their concentration.

Then the equations for concentrations of diffusing species, viz. atoms B, vacancies and interstitials B_i and A_i in each phase are formulated as following:

$$\frac{\partial c_B}{\partial t} = \frac{\partial}{\partial x}\left(D_{BB}\frac{\partial c_B}{\partial x} + D_{Bv}\frac{\partial c_v}{\partial x}\right) - P_i c_{Bi} + \zeta_{Bi} D_{Bi} c_{Bi}\rho_e + K_{iv}(c_v - c_v^0)c_{Bi}\,, \tag{4}$$

$$\frac{\partial c_v}{\partial t} = \frac{\partial}{\partial x}\left(D_{vB}\frac{\partial c_B}{\partial x} + D_{vv}\frac{\partial c_v}{\partial x}\right) + P_v - \zeta_v D_v (c_v - c_v^0)\rho_e - K_{iv}(c_v - c_v^0)(c_{Bi} + c_{Ai})\,, \tag{5}$$

$$\frac{\partial c_{Bi}}{\partial t} = \frac{\partial}{\partial x}\left(D_{Bi}\frac{\partial c_{Bi}}{\partial x}\right) + P_i c_B - \zeta_{Bi} D_{Bi} c_{Bi}\rho_e - K_{iv}(c_v - c_v^0)c_{Bi}\,, \tag{6}$$

$$\frac{\partial c_{Ai}}{\partial t} = \frac{\partial}{\partial x}\left(D_{Ai}\frac{\partial c_{Ai}}{\partial x}\right) + P_i c_A - \zeta_{Ai} D_{Ai} c_{Ai}\rho_e - K_{iv}(c_v - c_v^0)c_{Ai}\,, \tag{7}$$

$$c_B + c_A + c_v = 1\,. \tag{8}$$

Here D_{Bi} and D_{Ai} are the diffusion coefficients of interstitial atoms, P_i and P_v are the generation rates of interstitials and vacancies per unit volume of an alloy, ρ_e is the density of edge dislocations, ζ_{Ai}, ζ_{Bi} and ζ_v are dimensionless coefficients describing the efficiency of edge dislocations as sinks point defects, $\zeta_k \approx 1$, $k \equiv A_i, B_i, v$ (Bullough et al., 1975; Murphy, 1987; Mansur, 1979), and K_{iv} is the vacancy-interstitial recombination rate per unit volume.

Unlike radiation damage of reactor materials, where Frenkel pairs and collision cascades are formed due to knocking lattice atoms out of their regular positions by high-energy particles, the basic mechanism of non-equilibrium point-defect generation in metals under plastic deformation is jog dragging by gliding screw dislocations. The jogs are formed on gliding edge and screw dislocations during intersection with the forest dislocations, i.e. those not involved in the active glide system. Jogs on edge dislocations move together with the dislocation while those on screws act as obstacles for the dislocation glide. Under shear stress, a screw dislocation bends and drags the jogs, which produce excess vacancies or interstitial atoms depending on the jog sign (Nabarro et al., 1964; Novikov, 1983). The defect production rates P_v and P_i are described using the Hirsch-Mott theory (Nabarro et al., 1964):

$$P_v = (\dot{\varepsilon} b / 2)(f_v \xi \rho / 2)^{1/2}, \quad P_i = (\dot{\varepsilon} b / 2)(f_i \xi \rho / 2)^{1/2}, \tag{9}$$

Here ρ is the total dislocation density, $\rho_s \approx \rho_e = \rho/2$ where ρ_s is the density of screws (Novikov, 1983), ξ is the fraction of forest dislocations (usually $\xi \approx 0.5$), b is the Burger vector length, $\dot{\varepsilon}$ is the strain rate, f_v and f_i are the fractions of vacancy and interstitial producing jogs, $f_v + f_i = 1$, and typically $f_v > f_i$ (Nabarro et al., 1964; Novikov, 1983).
The recombination-rate coefficient K_{iv} in a binary solid solution is determined as

$$K_{iv} = (4\pi r_0 / \omega)(D_i + D_v), \tag{10}$$

where r_0 is the capture radius, $\omega = a_0^3$ is the average volume of a crystal cell, a_0 is the lattice period. The capture radius is usually determined as $r_0 = b/2$ (Bullough et al., 1975; Murphy, 1987; Mansur, 1979), where the b is the Burgers vector. In the theory of radiation-induced diffusion it is considered that the diffusion coefficient of self-interstitials is by several orders of magnitude higher than that of vacancies (Murphy, 1987; Mansur, 1979). For modeling we assume that D_{Ai}, $D_{Bi} \sim 10^3 D_v$, where D_v is defined by formulas (3).

3.3 Initial and boundary conditions

To complete the problem, diffusion mass transfer equations (4)-(7) should be supplemented with relevant initial and boundary conditions. For definiteness, let phase 2 (initially pure metal B) be on the left (with respect to the direction of the coordinate axis x) and phase 1 (initially pure metal A) on the right. The thickness of the diffusion couple L consists of two half-thicknesses of the corresponding metal layers. Then the initial conditions look as

$$c_{Bi}(x) = c_{Ai}(x) = 0, \; c_v(x) = c_v^0 \text{ at } 0 < x < L, \; c_B(x) = 1 \text{ at } 0 < x < g, \; c_B(x) = 0 \text{ at } g < x < L \tag{11}$$

where g=L/2 is the coordinate of the 2/1 interface.
The Neumann-type boundary conditions to Eqs. (4)-(7) at x=0 and x=L are

$$J_k(x = 0, L) = 0, \quad k \equiv B, v, B_i, A_i. \tag{12}$$

For vacancy diffusion, an additional boundary condition is posed at the 2/1 interface:

$$c_v(x = g) = c_v^0. \tag{13}$$

Condition (13) is determined by the fact that the interface between starting pure metals A and B is incoherent and consists of grain-boundary dislocations (Kosevich et al., 1980) and hence can act as a localized sink for non-equilibrium point defects.
Since problem (3)-(13) describing diffusion in substitutional alloys in the conditions of IPD is substantially nonlinear, it can be solved only numerically. For this purpose, a computer procedure is developed employing a fully implicit finite-difference scheme, which is derived using the versatile integration-interpolation method (Kalitkin, 1978).

4. Parameter values for modeling

We consider a deformation-relaxation cycle with parameters typical of MA in vibratory mill "SPEX 8000" with oscillation frequency ω=20 Hz, then the cycle duration is $t_c = (2\omega)^{-1}$ = 0.025 s. Deformation, when point defect are generated, occurs during collisions whose duration is t_d~10^{-4} s, and a lower-level estimate for strain rate is $\dot{\varepsilon}$ ~10 s^{-1} (Maurice & Courtney, 1990; Lovshenko & Khina, 2005). Then the interval between collisions, when relaxation occurs (P_i=P_v=0), is $t_r = t_c - t_d$ = 0.0249 s. The thickness of metal layers in lamellar particles is 0.5-0.05 μm; we assume L=0.1 μm. The dislocation density is ρ~10^{11}-10^{12} cm^{-2}; we take $\rho=10^{11}$ cm^{-2}. Since $f_v > f_i$ (Nabarro et al., 1964; Novikov, 1983), we suppose f_v/f_i=2.

The following model binary system is considered: Al (metal B, phase 2)-Cu (metal A, phase 1). At collisions during MA in a vibratory mill, a local temperature rise in particles is small: 10 K for Al and about 20 K for Cu (Maurice & Courtney, 1990), and thus its influence on diffusion coefficients D_B^*, D_A^* is negligible. We take a constant process temperature, T_{MA} = 100° C = 373 K, which corresponds to an industrial milling device with a water-cooled shell. The equilibrium vacancy concentration in both phases is determined as

$$c_v^0 = \exp[-\Delta H_v^f / (k_B T)], \tag{14}$$

where ΔH_v^f is the vacancy formation enthalpy and k_B is the Boltzmann constant.

The values of E and D_0 for calculating self-diffusion coefficients D_B^*, D_A^*, the Burgers vector, vacancy formation enthalpy and the estimated parameter values at T_{MA} are listed in Table 1. It is seen that in equilibrium conditions parameters D* and c_v^0 in both phases are very small.

species/ phases	phase 2 (Al)			phase 1 (Cu)			ΔH_v^f, eV	$c_v^0(T_{MA})$	b, cm
	D_0, cm^2/s	E, kJ/mol	$D^*(T_{MA})$, cm^2/s	D_0, cm^2/s	E, kJ/mol	$D^*(T_{MA})$, cm^2/s			
Al	1.71	142.3	$2.0\cdot10^{-20}$	0.3	196.8	$8.6\cdot10^{-29}$	0.76	$5.4\cdot10^{-11}$	$2.86\cdot10^{-8}$
Cu	0.647	135	$7.8\cdot10^{-20}$	0.2	196.4	$6.6\cdot10^{-29}$	1.17	$1.6\cdot10^{-16}$	$2.55\cdot10^{-8}$

Table 1. Parameters D_0 and E for self and impurity diffusion in Al and Cu (Brandes & Brook, 1992), the vacancy formation enthalpy (Bokshtein, 1978), Burgers vector (Brandes & Brook, 1992) and the parameters values calculated for T_{MA}=373 K

5. Numerical results and discussion

The results of computer simulations with the above described parameters are presented in Figs. 1-3 for different situations. In the first case, boundary condition (13) was not considered, i.e. phase boundary 2/1 was assumed to be permeable for vacancy diffusion and did not work as a sink. This refers to a hypothetical situation of a coherent phase boundary, which for some reason retains during deformation. Here, relaxation of non-equilibrium point defects occurs via vacancy-interstitial annihilation and interaction with volume-distributed sinks (edge dislocations). In this case, atoms B (Al) diffuse into phase 1 (Cu-base solid solution), i.e. the concentration profile is asymmetrical (see Fig. 1 (a)).

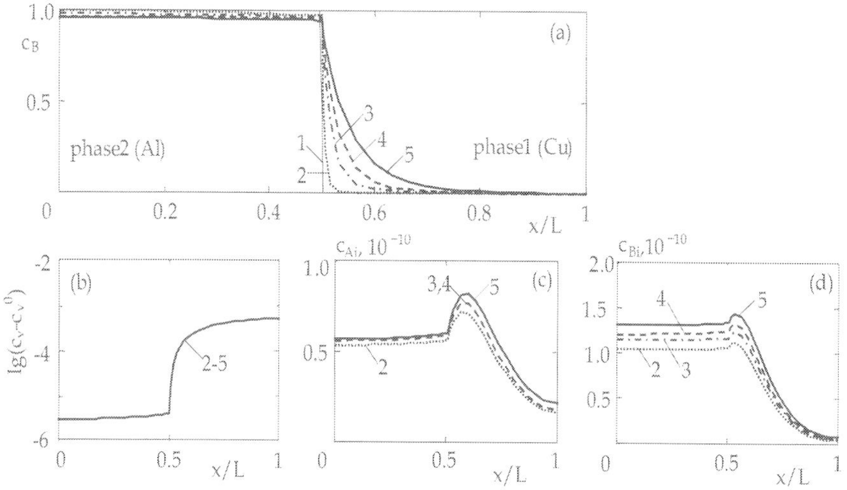

Fig. 1. Calculated concentration profiles of atoms B (a), vacancies (b) and interstitial atoms A_i (c) and B_i (d) when the 2/1 interface is not a sink for vacancies: 1, t=0; 2, t=250 s (10000 cycles); 3, t=1250 s (50000 cycles); 4, t=2000 s (80000 cycles); 5, t=4000 s (160000 cycles)

As a result of interplay between vacancy generation, diffusion and relaxation, which occur at different rates in phases 1 and 2, the concentration of non-equilibrium vacancies appears to be very high (Fig. 1 (b)), and in phase 1 it is substantially larger than in phase 2, $c_v^{(1)} > c_v^{(2)}$, where superscripts denote the phase numbers. The steady-state vacancy profile with a steep gradient near the interface x=g (i.e. x/L=0.5) is established after a short time, about 200 s. As seen from Eq. (4), at a large value of term $|D_{Bv}\partial c_v/\partial x|$ atoms B will diffuse in the direction opposed to that of the diffusion flux of vacancies, which brings about noticeable diffusion alloying of copper with aluminum (metal B) near the interface at a relatively short time, 4000 s. The concentration of interstitials A_i and B_i is several orders of magnitude lower than that of vacancies (Fig 1, (c and d)), which is due to high annihilation rates in phases 1 and 2 and a large vacancy concentration. Since away from the 2/1 interface $c_v^{(1)} > c_v^{(2)}$, the concentration of interstitials A_i and B_i in phase 1 is lower than in phase 2. The peaks in the concentration profiles of interstitials in phase 1 near the 2/1 interface (see Fig 1, (c and d)) are attributed to the existence of a steady-state profile of vacancies. Thus, acceleration of diffusion in the conditions of MA is connected not only with a substantial increase of the partial diffusion coefficients, D_{BB} and D_{AA}, due to a high vacancy concentration (see Eqs. (3)) but also with interaction of diffusion fluxes of atoms and vacancies via off-diagonal terms D_{Bv}, D_{vB}. Such a cross-term effect is sometimes referred to as the "inverse Kirkendall effect".

Results of simulations for a more realistic case, when the 2/1 interface acts as localized sink for excess vacancies, i.e. boundary condition (13) is accounted for, are shown in Fig. 2.

Unlike the previous case, diffusion of atoms A (Cu) from phase 1 into phase 2 (Al-base solid solution) occurs (see Fig. 2 (a)). As in the previous situation, very high concentration of vacancies is observed inside both of the phases, and $c_v^{(1)} > c_v^{(2)}$. Since the equilibrium vacancy concentration is sustained at the interface, steep vacancy concentration gradients arise in phases 1 and 2 near the boundary, so the vacancy flux in each phase is directed to the interface and $|\partial c_v^{(1)}/\partial x|_{x=g+0} >> |\partial c_v^{(2)}/\partial x|_{x=g-0}$ (Fig. 2 (b)). The steady-state profile of vacancies is established after a longer time, about 1000 s. As seen from Eqs. (4),(5), a counter-current flux of vacancies accelerates the diffusion of lattice atoms while a co-current flux retards it. In the given situation, all of the above brings about alloying of phase 2 with atoms A (Cu) within a relatively short time, t=4000 s. A small peak of atoms A is seen at the interface at a large time (curve 5 in Fig. 2 (a)). Thus, interaction of vacancy fluxes, which arise due to vacancy generation under periodic IPD, with the phase boundary can have a selective influence on diffusion of different atoms (A and B) during MA. The concentration of interstitials A_i and B_i inside both of the phases are small as compared with vacancies (see Fig. 2 (c and d)). This is due to fast recombination of the former with excess vacancies whose concentration in phase 1 away from the interface is much higher than in phase 2.

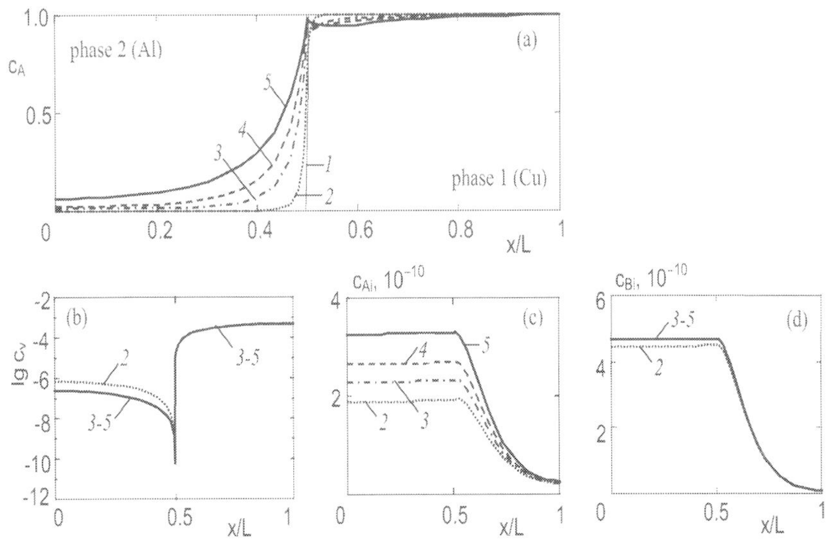

Fig. 2. Calculated concentration profiles of atoms A (a), vacancies (b) and interstitial atoms A_i (c) and B_i (d) when the 2/1 interface acts as a localized sink for vacancies: 1, t=0; 2, t=250 s (10000 cycles); 3, t=1250 s (50000 cycles); 4, t=2000 s (80000 cycles); 5, t=4000 s (160000 cycles)

Therefore, along with the accumulation of excess vacancies, increase in the diffusion coefficients and cross-term linking between the vacancy and atomic fluxes, important is the interaction of fluxes of non-equilibrium vacancies with the phase boundary. This factor has not previously received due attention in literature.

It should be noted that in the above situation the diffusion coefficients of atoms in phases 1 and 2 attain high values due to a high vacancy concentration while at the interface, where $c_v=c_v^0$, D_{AA} and D_{BB} fall down to equilibrium, i.e. very low values. Thus, the interface acts as a diffusion barrier. But diffusion over grain boundaries proceeds substantially faster than in the volume at the same temperature. So, in the third case we consider that at x=g the values of D_{BB} and D_{AA} correspond to the grain-boundary diffusion coefficients for corresponding pure metals. Using the data presented in (Mishin, 2001) for metals with fcc lattice (both Cu

and Al), we obtain the following estimates at T_{MA}=373 K: for Cu (metal A) $D_{AA}(x=g) \sim 10^{-16}$-$10^{-17}$ cm²/s, for Al (metal B) $D_{BB}(x=g) \sim 10^{-12}$ cm²/s.
The results of simulation for this situation are presented in Fig. 3.

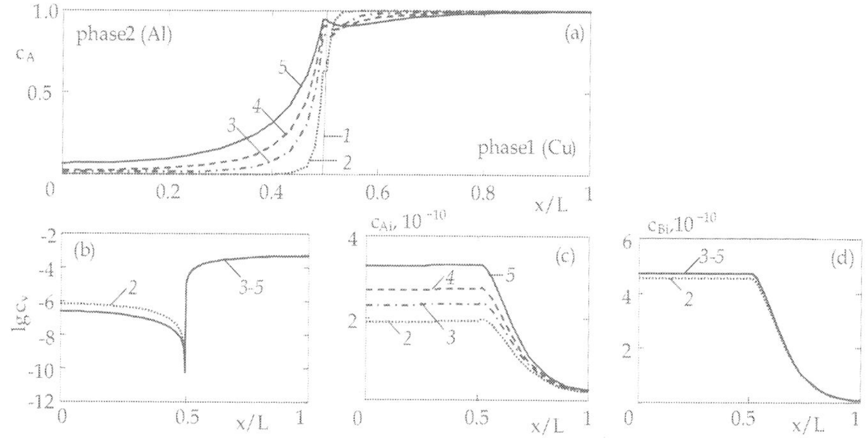

Fig. 3. Same is Fig. 2 but the values of diffusion coefficients of atoms A and B at the 2/1 interface correspond to the grain boundary diffusion

Since the barrier for atomic diffusion at the phase boundary is lower, the degree of alloying of phase 2 (Al) with atoms B (Cu), i.e. total mass of the latter in the solid solution, has increased in comparison with the previous case. The peak of atoms A at the interface becomes more pronounced (Fig. 3 (a)); the concentration profiles of point defects (Fig. 3 (b-d)) change insignificantly. Thus, interaction of diffusion fluxes of both vacancies and atoms with the interface is important in the enhancement of diffusion during MA.
The observed formation of solid solution within a short time of periodic IPD due to enhanced diffusion, and the revealed features of the process qualitatively agree with numerous experimental data on MA. In particular, asymmetric concentration profiles were observed in miscible (Fe-Ni, Fe-Co, Fe-Mn) and immiscible (Fe-Cu) systems as results of MA (Cherdyntsev & Kaloshkin, 2010). The calculated vacancy concentration semi-quantitatively agrees with experimental data for copper after ECAP and ARB (Ungar et al., 2007): at the dislocation density of $4\cdot10^{11}$ cm^{-2} the value of c_v reached $5\cdot10^{-6}$ inside grains and up to 10^{-3} in grain boundaries.
It should be noted that in this work we have used lower-level estimates for strain rate, which, according to the outcomes of mechanistic models, in the conditions of MA in a high-speed vibratory mill can reach 10^4 s^{-1} (Maurice & Courtney, 1990; Lovshenko & Khina, 2005). It should be noted that at dynamic deformation regimes, i.e. high strain rates ($\dot{\varepsilon}$ >10 s^{-1}), the point defect production rates may be substantially higher (Popov et al., 1990) than predicted by the Hirsch-Mott theory (Eq. (9)). Besides, according to the theory of non-equilibrium grain boundaries, i.e. those in metals under the conditions of IPD (Segal et al, 2010), the grain-boundary diffusion coefficients will be orders of magnitude higher that the values used in this work. With all of these factors taken into account, diffusion mass transfer under the action of periodic IPD will proceed still faster.

6. Conclusion

Thus, a consistent model of non-equilibrium enhanced solid-state diffusion in binary metallic systems during mechanical alloying has been developed, which accounts for the generation of non-equilibrium point defects due to jog dragging by gliding screw dislocations during plastic deformation at incidental collisions in a milling device and defect relaxation during time intervals between collisions. Numerical simulations using realistic values of the diffusion parameters for a model system (Cu-Al) at 100 °C have demonstrated a possibility of considerable alloying within a short time of MA (Khina et al., 2005; Khina & Formanek, 2006). The acceleration of diffusion mass transfer is connected with the interplay and competition of several factors: (i) accumulation of non-equilibrium vacancies due to their fast generation during plastic deformation and slow relaxation in the intervals between collisions, (ii) increase of partial diffusion coefficients due to excess vacancies, (iii) cross-link interaction of diffusion fluxes via off-diagonal diffusion coefficients, and (iii) interaction of the diffusion fluxes of both vacancies and atoms with the phase boundary. The role of interstitial atoms is minor in comparison with non-equilibrium vacancies.

Fast diffusion that results in the formation of a zone of supersaturated solid solution around the interface, along with accumulation of excess vacancies inevitably increases the free energy of the alloy and causes a distortion of the crystal lattice. This may ultimately bring about a non-equilibrium phase transition such as solid-state amorphization: for example, in pure copper the latter can occur at c_V=0.077 (Fecht, 1992).

One of the prospective directions of further research is combining the developed model with a theory for dislocation evolution during IPD to obtain a comprehensive picture of structure formation. Also, it seems interesting to unite this approach with the theory of solid-state amorphization (Gusak at al., 2001) to get an opportunity to predict metastable phase transitions during MA.

7. References

Benjamin, J.S. (1992). Fundamentals of mechanical alloying. *Materials Science Forum*, Vol. 88-90, pp. 1-18, ISSN 1662-9752

Koch, C.C. (1992). The synthesis of non-equilibrium structures by ball-milling. *Materials Science Forum*, Vol. 88-90, pp. 243-262, ISSN 1662-9752

Koch, C.C. (1998). Intermetallic matrix composites prepared by mechanical alloying-a review. *Materials Science and Engineering A*, Vol. 244, No. 1, pp. 39-48, ISSN 0921-5093

Ma, E. & Atzmon, M. (1995). Phase transformations induced by mechanical alloying in binary systems. *Materials Chemistry and Physics*, Vol. 39, pp. 249-267, ISSN 0254-0584

Bakker, H.; Zhou, G.F. & Yang, H. (1995). Mechanically driven disorder and phase transformations in alloys. *Progress in Materials Science*, Vol. 39, No. 3, pp. 159-241, ISSN 0079-6425

El-Eskandarany, M.S. (2001). *Mechanical Alloying for Fabrication of Advanced Engineering Materials*, Noyes Publications, ISBN 0-8155-1462-X, Norwich, NY

Suryanarayana, C. (2001). Mechanical alloying and milling. *Progress in Materials Science*, Vol. 46, No. 1-2, pp. 1-184, ISSN 0079-6425

Suryanarayana, C. (2004). *Mechanical Alloying and Milling*, Marcel Dekker, Inc., ISBN 0-8247-4103-X, New York, NY

Zhang, D.L. (2004). Processing of advanced materials using high-energy mechanical milling. *Progress in Materials Science*, Vol. 49, No. 3-4, pp. 537-560, ISSN 0079-6425

Koch, C.C.; Scattergood, R.O.; Youssef, K.M.; Chan, E. & Zhu, Y.T. (2010). Nanostructured materials by mechanical alloying: new results on property enhancement. *Journal of Materials Science*, Vol. 45, No. 17, pp. 4725-4732, ISSN 0022-2461

Segal, V.M.; Beyerlein, I.J.; Tome, C.N.; Chuvildeev, V.N. & Kopylov, V.I. (2010). *Fundamentals and Engineering of Severe Plastic Deformation*, Nova Science Publishers, Inc., ISBN 978-1-61668-190-6, Hauppauge, NY

Fukuda, Y.; Oh-ishi, K.; Horita, Z. & Langdon, T.G. (2002). Processing of a low-carbon steel by equal-channel angular pressing. *Acta Materialia*, Vol. 50, No. 6, pp. 1359-1368, ISSN 1359-6454

Perepezko, J.H.; Sieber, H.; Sagel, A. & Fecht, H.-J. (1998). Synthesis of an amorphous Zr-Al-Ni-Cu alloy with large supercooled liquid region by cold-rolling of elemental foils. *Acta Materialia*, Vol. 46, No. 12, pp. 4233-4241, ISSN 1359-6454

Sauvage, X.; Dinda, G.P. & Wilde, G. (2007). Non-equilibrium intermixing and phase transformation in severely deformed Al/Ni multilayers. *Scripta Materialia*, Vol. 56, No. 3, pp. 181-184, ISSN 1359-6462

Yang, D.; Hodgson, P. & Wen, C. (2009). The kinetics of two-stage formation of $TiAl_3$ in multilayered Ti/Al foils prepared by accumulative roll bonding. *Intermetallics*, Vol. 17, No. 9, pp. 727-732, ISSN 0966-9795

Beygelzimer, Y.; Orlov, D.; Korshunov, A.; Synkov, S.; Varyukhin, V.; Vedernikova, I.; Reshetov, A.; Synkov, A.; Polyakov, L. & Korotchenkova, I. (2006). Features of twist extrusion: Method, structures & material properties. *Solid State Phenomena*, v.114, pp. 69-78, ISSN 1662-9779

Quelennec, X.; Menand, A.; Le Breton, J.M.; Pippan, R. & Sauvage, X. (2010). Homogeneous Cu-Fe supersaturated solid solutions prepared by severe plastic deformation. *Philosophical Magazine*, Vol. 90, No. 9, pp. 1179-1195, ISSN 1478-6435.

Benjamin, J.S. & Volin, T.E. (1974). The mechanism of mechanical alloying. *Metallurgical Transactions*, Vol. 5, No. 8, pp. 1924-1934

Boldyrev, V.V. (2006). Mechanochemistry and mechanical activation of solids. *Russian Chemical Reviews*, Vol. 75, No. 3, pp. 177-189

Khina, B.B.; Solpan, I. & Lovshenko, G.F. (2004). Modelling accelerated solid-state diffusion under the action of intensive plastic deformation. *Journal of Materials Science*, Vol. 39, No. 16-17, pp. 5135-5138, ISSN 0022-2461

Khina, B.B.; Lovshenko, G.F.; Konstantinov, V.M. & Formanek, B. (2005). Mathematical model of solid-state diffusion at periodic plastic deformation. *Metal Physics and Advanced Technologies*, Vol. 27, No. 5, pp. 609-623, ISSN 1024-1809 (in Russian)

Khina, B.B. & Formanek, B. (2006). Mathematical modeling of solid-state diffusion during mechanical alloying. *Defect and Diffusion Forum*, Vol. 249, pp. 105-110, ISSN 1662-9507

Maurice, D.R. & Courtney, T.H. (1990). The physics of mechanical alloying: a first report. *Metallurgical Transactions A*, Vol. 21, No. 2, pp. 289-303

Magini, A. & Iasona, A. (1995). Energy transfer in mechanical alloying (overview). *Materials Transactions (JIM)*, Vol. 36, No. 3, pp. 123-133

Urakaev, F.Kh. & Boldyrev, V.V. (2000a). Mechanism and kinetics of mechanochemical processes in comminuting devices: I. Theory. *Powder Technology*, Vol. 107, No. 1-2, pp. 93-107, ISSN 0032-5910

Urakaev, F.Kh. & Boldyrev, V.V. (2000b). Mechanism and kinetics of mechanochemical processes in comminuting devices: II. Applications of the theory and experiment. *Powder Technology*, Vol. 107, No. 3, pp. 197-206, ISSN 0032-5910

Chattopadhyay, P.P.; Manna, I.; Talapatra, S. & Pabi, S.K. (2001). A mathematical analysis of milling mechanics in a planetary ball mill. *Materials Chemistry and Physics*, Vol. 68, No. 1-3, pp. 85-94, ISSN 0254-0584

Lovshenko, G.F. & Khina, B.B. (2005). Estimation of factors influencing mechanochamical transformations during mechanical alloying in a vibratory mill. *Journal of Friction and Wear*, Vol. 26, No. 4, pp. 434-445 ISSN 1068-3666 (in Russian)

Lund, A.C. & Schuh, C.A. (2004a). Molecular simulation of amorphization by mechanical alloying. *Acta Materialia*, Vol. 52, No. 8, pp. 2123-2132, ISSN 1359-6454

Lund, A.C. & Schuh, C.A. (2004b). Topological and chemical arrangement of binary alloys during severe deformation. *Journal of Applied Physics*, Vol. 95, No. 9, pp. 4815-4822, ISSN 0021-8979

Odunuga, S.; Li, Y.; Krasnochtchekov, P.; Bellon, P. & Averback, R.S. (2005). Forced chemical mixing in alloys driven by plastic deformation. *Physical Review Letters*, Vol. 95, No. 4, 045901 (4 pp.), ISSN 0031-9007

Delogu, F. & Cocco, G. (2005). Molecular dynamics investigation on the role of sliding interfaces and friction in the formation of amorphous phases. *Physical Review B*, Vol. 71, No. 14, 144108 (11 pp.), ISSN 1098-0121

Butyagin, P. (2000). Mechanochemical synthesis: mechanical and chemical factors. *Journal of Materials Synthesis and Processing*, Vol. 8, No. 3/4, pp. 205-211, ISSN 1064-7562

Delogu, F. & Cocco, G. (2000). Relating single-impact events to macrokinetic features in mechanical alloying processes. *Journal of Materials Synthesis and Processing*, Vol. 8, No. 5/6, pp. 271-277, ISSN 1064-7562

Delogu, F.; Orru, R. & Cao. G. (2003). A novel macrokinetic approach for mechanochemical reactions. *Chemical Engineering Science*, Vol. 58, No. 3-6, pp. 815-821, ISSN 0009-2509

Butyagin, P.Yu. & Streletskii, A.N. (2005). The kinetics and energy balance of mechanochemical transformations. *Physics of the Solid State*, Vol. 47, No. 5, pp. 856-862, ISSN 1063-7834

Schultz, R.; Trudeau, M.; Hout, J.Y. & Van Neste, A. (1989). Interdiffusion during the formation of amorphous layers by mechanical alloying. *Physical Review Letters*, Vol. 62, No. 24, pp. 2849-2852, ISSN 0031-9007

Lu, L. & Zhang, Y.F. (1999). Influence of process control agent on interdiffusion between Al and Mg during mechanical alloying. *Journal of Alloys and Compounds*, Vol. 290, No. 1-2, pp. 279-283, ISSN 0925-8388

Zhang, D.L. & Ying, D.Y. (2001). Solid state reactions in nanometer scaled diffusion couples prepared using high energy ball milling. *Materials Science and Engineering A*, Vol. 301, No. 1, pp. 90-96, ISSN 0921-5093

Ma, E. (2003). Amorphization in mechanically driven material systems. *Scripta Materialia*, Vol. 49, No. 10, pp. 941-946, ISSN 1359-6462

Larikov, L.N.; Fal'chenko, V.M.; Mazanko, V.F.; Gurevich, S.M.; Harchenko, G.K. & Ignatenko, A.I. (1975). Abnormal acceleration of diffusion at pulse loading of metals. *Soviet Physics Doklady*, Vol. 221, No. 5, pp. 1073-1075 (in Russian)

Gertsriken, D.S.; Mazanko, V.F. & Fal'chenko, V.M. (1983). Certain regularities of atom migration in the conditions of high-strain rate plastic deformation. *Metal Physics*, Vol. 5, No. 4, pp. 74-90, ISSN 1024-1809 (in Russian)

Arsenyuk, V.V.; Gertsriken, D.S.; Mazanko, V.F.; Tyshkevich, V.M. & Fal'chenko, V.M. (2001a). Effect of defects of crystal structure on the mobility of atoms in metals at impact compression. *Metal Physics and Advanced Technologies*, Vol. 23, No. 7, pp. 927-937, ISSN 1024-1809(in Russian)

Arsenyuk, V.V.; Gertsriken, D.S.; Mazanko, V.F.; Tyshkevich, V.M. & Fal'chenko, V.M. (2001b). Specific features of atom distribution in metals under pulse loading. *Metal Physics and Advanced Technologies*, Vol. 23, No. 9, pp. 1203-1212, ISSN 1024-1809 (in Russian)

Gertsriken, D.S.; Ignatenko, A.I.; Mazanko, V.F.; Tyshkevich, V.M. & Fal'chenko, V.M. (1994). Specific features of interation of iron and copper with different elements at pulse deformation of metals. *Metal Physics and Advanced Technologies*, Vol. 16, No. 12, pp. 44-52, ISSN 1024-1809 (in Russian)

Gertsriken, D.S.; Mazanko, V.F.; Tyshkevich, V.M. & Fal'chenko, V.M. (2001). *Mass Transfer in Metals at Low Temperatures in the Conditions of External Actions*, Institute of Metal Physics of the Ukrainian Academy of Sciences Publ., Kiev, Ukraine (in Russian)

Kulemin, A.V. (1978). *Ultrasound and Diffusion in Metals*, Metallurgiya, Moscow (in Russian)

Farber. V.M. (2002). Contribution of diffusion processes to structure formation in intense cold plastic deformation of metals. *Metal Science and Heat Treatment*, Vol. 44, No. 7-8, pp. 317-323, ISSN 0026-0673

Skakov, Yu.A. (2004). High-energy cold plastic deformation, diffusion, and mechanochemical synthesis. *Metal Science and Heat Treatment*, Vol. 46, No. 3-4, pp. 137-145, ISSN 0026-0673

Gapontsev, V.L. & Koloskov, V.M. (2007). Nonequilibrium vacancy-stimulated diffusion (induced diffusion) as the main mechanism of activated alloy formation. *Metal Science and Heat Treatment*, Vol. 49, No. 11-12, pp. 503-513, ISSN 0026-0673

Skakov, Yu.A. (2007). Comments on the paper of V.L.Gapontsev and V.M.Koloskov "Nonequilibrium vacancy-stimulated diffusion (induced diffusion) as the main mechanism of activated alloy formation". *Metal Science and Heat Treatment*, Vol. 49, No. 11-12, pp. 514-516, ISSN 0026-0673

Shtremel', M.A. (2002). Participation of diffusion in the processes of mechanical alloying. *Metal Science and Heat Treatment*, Vol. 44, No. 7-8, pp. 324-327, ISSN 0026-0673

Shtremel', M.A. (2004). In what direction does diffusion go? (Letter to the Editor). *Metal Science and Heat Treatment*, Vol. 46, No. 3-4, pp. 146-147, ISSN 0026-0673

Shtremel', M.A. (2007). Review of the paper of V. L. Gapontsev and V. M. Koloskov "Nonequilibrium vacancy-stimulated diffusion (induced diffusion) as the main mechanism of activated alloy formation". *Metal Science and Heat Treatment*, Vol. 49, No. 11-12, pp. 517-518, ISSN 0026-0673

Lotkov, A.I.; Baturin, A.A.; Grishkov, V.N. & Kopylov, V.I. (2007). Possible role of crystal structure defects in grain structure nanofragmentation under severe cold plastic deformation of metals and alloys. *Physical Mesomechanics*, Vol. 10, No. 3-4, pp. 179-189

Khina, B.B. & Froes, F.H. (1996). Modeling mechanical alloying: advances and challenges. *Journal of Metals (JOM)*, Vol. 48, No. 7, pp. 36-38, ISSN 1047-4838

Rabkin, E. & Estrin, Y. (1998). Pipe diffusion along curved dislocations: an application to mechanical alloying. *Scripta Materialia*, Vol. 39, No. 12, pp. 1731-1736, ISSN 1359-6462

Mahapatra, T.K.; Das, D.; Manna, I. & Pabi, S.K. (1998). Mathematical modelling of the mechanical alloying kinetics. *Acta Materialia*, Vol. 46, No. 10, pp. 3501-3510, ISSN 1359-6454

Sobolev, S.L. (1997). Local non-equilibrium transport models. *Physics Uspekhi*, Vol. 40, No. 10, pp. 1043-1054

Buchbinder, G.L. (2003). Mass transfer in field of fast-moving deformation disturbance. *International Journal of Heat and Mass Transfer*, Vol. 46, No. 19, pp. 1665-1673, ISSN 0017-9310

Bekrenev, A.N. (2002). Mass transport in metals under intensive impulse reactions. *Journal of Physics and Chemistry of Solids*, Vol. 63, No. 9, pp. 1627-1631, ISSN 0022-3697

Gapontsev, V.L.; Kesarev, A.G.; Kondrat'ev, V.V. & Ermakov, A.E. (2000). Demixing of nanocrystalline alloys at generation of non-equilibrium vacancies at grain boundaries. *The Physics of Metals and Metallography*, Vol. 89, No. 5, pp. 8-13, ISSN 0031-918X

Gapontsev, V.L.; Kesarev, A.G. & Kondrat'ev, V.V. (2002). Theory of diffusion phase transformation in nanocrystalline alloys at intensive plastic deformation. I. Stage of formation of concentration inhomogeneity near grain boundaries. *The Physics of Metals and Metallography*, Vol. 94, No. 3, pp. 5-10, ISSN 0031-918X

Gapontsev, V.L.; Razumov, I.K. & Gornostyrev, Yu.N. (2003). Theory of diffusion phase transformation in nanocrystalline alloys at intensive plastic deformation. II. Demixing of ideal solid solutions. *The Physics of Metals and Metallography*, Vol. 96, No. 4, pp. 5-15, ISSN 0031-918X

El-Eskandarany, M.S.; Akoi, K.; Sumiyama, K. & Suzuki, K. (1997). Cyclic crystalline-amorphous transformations of mechanically alloyed $Co_{75}Ti_{25}$. *Applied Physics Letters*, Vol. 70, No. 13, pp. 1679-1681, ISSN 0003-6951

Courtney, T.H. & Lee, J.K. (2005). Cyclical phase transformations and dynamic equilibrium in mechanical alloying. *Philosophical Magazine*, Vol. 85, No. 2-3, pp. 153-170, ISSN 1478-6435

Kaibyshev, O.A. (2002). Fundamental aspects of superplastic deformation. *Materials Science and Engineering A*, Vol. 324, No. 1-2, pp. 96-102, ISSN 0921-5093

Wu, X.-L. & Ma, E. (2006). Dislocations in nanocrystalline grains. *Applied Physics Letters*, Vol. 88, No. 23, 231911 (3 pp.), ISSN 0003-6951

Foct, J. (2004). An attempt to extract common behaviour from heterogeneous results about mechanical alloying. *Journal of Materials Science*, Vol. 39, No. 16-17, pp. 5011-5017, ISSN 0022-2461

Raabe, D.; Ohsaki, S. & Hono, K. (2009). Mechanical alloying and amorphization in Cu-Nb-Ag in situ composite wires studied by TEM and atom probe tomography. *Acta Materialia*, Vol. 57, No. 17, 5254-5263, ISSN 1359-6454

Quelennec, X.; Menand, A.; Le Breton, J.M.; Pippan, R. & Sauvage, X. (2010). Homogeneous Cu-Fe supersaturated solid solutions prepared by severe plastic deformation. *Philosophical Magazine*, Vol. 90, No. 9, pp. 1179-1195, ISSN 1478-6435

Bellon, P. & Averback, R.S. (1995). Nonequilibrium roughening of interfaces in crystals under shear: application to ball milling. *Physical Review Letters*, Vol. 74, No. 10, pp. 1819-1822, ISSN 0031-9007

Meyers, M.A. & Chawla, K.K. (2009). *Mechanical Behavior of Materials*, Cambridge University Press, ISBN 978-0-521-86675-0, Cambridge, UK

Argon, A. (2008). *Strengthening Mechanisms in Crystal Plasticity*, Oxford University Press, ISBN 978-0-19-851600-2, Oxford, UK

Gutkin, M.Yu.; Ovid'ko, I.A. & Pande, C.S. (2001). Theoretical models of plastic deformation processes in nanocrystalline materials. *Reviews on Advanced Materials Science*, Vol. 2, pp. 80

Bobylev, S.V.; Mukherjee, A.K. & Ovid'ko, I.A. (2009). Transition from plastic shear into rotation deformation mode in nanocrystalline metals and ceramics. *Reviews on Advanced Materials Science*, Vol. 19, No1-2, pp. 103-113

Gutkin, M.Yu.; Ovid'ko, I.A. & Skiba, N.V. (2005). Emission of partial dislocations from triple junctions of grain boundaries in nanocrystalline materials. *Journal of Physics D: Applied Physics*, Vol. 38, No. 21, pp. 3921-3925, ISSN 0022-3727

Levitin, V. (2006). *High Temperature Strain of Metals and Alloys: Physical Fundamentals*, Wiley-VCH, ISBN: 3-527-31338-9, Weinheim

Wang, J.; Hoagland, R.G.; Hirth, J.P. & Misra, A. (2008). Atomistic modeling of the interaction of glide dislocations with "weak" interfaces. *Acta Materialia*, Vol. 56, No. 19, pp. 5685-5693, ISSN 1359-6454

Khusid, B.M. & Khina, B.B. (1991). Kinetic model for intermetallic compound formation during interdiffusion in a binary system. *Physical Review B*, Vol. 44, No. 19, pp. 10778-10793, ISSN 1098-0121

Desre, P.J. & Yavari, A.R. (1990). Suppression of crystal nucleation in amorphous layers with sharp concentration gradients. *Physical Review Letters*, Vol. 64, No. 13, pp. 1533-1536, ISSN 0031-9007

Desre, P.J. (1991). Effect of sharp concentration gradients on the stability of a two-component amorphous layer obtained by solid state reaction. *Acta Metallurgica et Materialia*, Vol. 39, No. 10, pp. 2309-2315, ISSN 1359-6454

Gusak, A.M.; Hodaj, F. & Bogatyrev, A.O. (2001). Kinetics of nucleation in the concentration gradient. *Journal of Physics: Condensed Matter*, Vol. 13, No. 12, pp. 2767-2787, ISSN 0953-8984

Bullough, R.; Eyre, B.L. & Krishan, K. (1975). Cascade damage effects on the swelling of irradiated materials. *Proceedings of the Royal Society of London*, Vol. A346, No. 1644, pp. 81-102

Murphy, S.M. (1987). Spatial instability in dislocation structure under irradiation. *Europhysics Letters*, Vol. 3, No. 12, pp. 1267-1272

Mansur, L.K. (1979). Effect of point defect trapping and solute segregation on radiation-induced sweeling and creep. *Journal of Nuclear Materials*, Vol. 83, pp. 109-127, ISSN 0022-3115

Gusak, A.M. & Bushin, I.N. (1996). Modeling the motion of point defects in the shock wave front. *Metal Physics*, Vol. 18, No. 9, pp. 68-72, ISSN 1024-1809 (in Russian)

Nabarro, F.R.N.; Basinski, Z.S. & Holt, D.B. (1964). The plasticity of pure single crystals. *Advances in Physics*, Vol. 50, pp. 193-323

Novikov, I.I. (1983). *Defects of Crystal Structure in Metals*, Metallurgiya, Moscow (in Russian)

Hirth, J.P. & Lothe, J. (1982). *Theory of Dislocations*, second Edition, John Wiley & Sons, Inc., New York, NY

Vo, N.Q.; Averback, R.S.; Bellon, P.; Odunuga, S. & Caro, A. (2008). Quantitative description of plastic deformation in nanocrystalline Cu: Dislocation glide versus grain boundary sliding. *Physical Review B*, Vol. 77, No. 13, 134108 (9 pp.), ISSN 1098-0121

Gurao, N.P. & Suwas, S. (2009). Deformation mechanisms during large strain deformation of nanocrystalline nickel. *Applied Physics Letters*, Vol. 94, No. 19, 191902 (3 pp.), ISSN 0003-6951

Voroshnin, L.G. & Khusid, B.M. (1979). *Diffusion Mass Transfer in Multicomponent Systems*, Nauka i Tehnika Publ., Minsk, Belarus (in Russian)

Adda, Y. & Philibert, J. (1966). *Diffusion in Solids*, Presses Universitaires de France, Paris, Vol. 1 (in French)

Gurov, K.P.; Kartashkin, B.A. & Ugaste, U.E. (1981). *Mutual Diffusion in Multiphase Metallic Systems*. Nauka, Moscow (in Russian)

Kosevich, V.M.; Ievlev, V.M.; Palatnik, L.S. & Fedorenko, A.I. (1980). *Structure of Intercrystalline and Interphase Boundaries*, Metallurgiya, Moscow (in Russian)

Kalitkin, N.N. (1978). *Numerical Methods*, Nauka, Moscow (in Russian)

Brandes, E.A. & Brook, G.B. (Ed.) (1992). *Smithells Metals Reference Book*, seventh edition, Butterworth-Heinemann, ISBN 0 7506 3624 6, Oxford, UK

Bokshtein, B.S. (1978). *Diffusion in Metals*, Metallurgiya, Moscow (in Russian)

Mishin, Y. (2001). 50 years of grain boundary diffusion: what do we know about it today? *Defect and Diffusion Forum*, Vol. 194-199, pp. 1113-1126, ISSN 1662-9507

Cherdyntsev, V.V. & Kaloshkin, S.D. (2010). On the kinetics of phase and structural transformations upon mechanical alloying. *The Physics of Metals and Metallography*, Vol. 109, No. 5, pp. 492-504, ISSN 0031-918X

Ungar, T.; Schafler, E.; Hanak, P.; Bernstorff, S. & Zehetbauer, M. (2007). Vacancy production during plastic deformation in copper determined by in situ X-ray diffraction. *Materials Science and Engineering A*, 2007, Vol. 462, No. 1-2, pp. 398-401, ISSN 0921-5093

Popov, L.E.; Starenchenko, V.A. & Shalygin, I.I. (1990). Intensity of point defect generation at plastic deformation. *The Physics of Metals and Metallography*, No. 6, pp. 31-36, ISSN 0031-918X

Fecht, H.J. (1992). Defect-induced melting and solid-state amorphization. *Nature*, Vol. 356, No. 6365, pp. 133-135, ISSN 0028-0836.

12

Mass Transfer in Steelmaking Operations

Roberto Parreiras Tavares
Federal University of Minas Gerais
Brazil

1. Introduction

The productivity of steelmaking processes, including production and refining of liquid steel, depends on the mass transfer rates. Due to the high temperatures involved in the processing of liquid steel, the rate controlling step of the processes is usually a mass transfer step.

In steelmaking operations, different situations of mass transfer can occur, depending on the phases involved,

- liquid-liquid mass transfer, in the case of reactions involving liquid steel and slag;
- liquid-gas mass transfer, when a gas is injected into or onto liquid steel;
- liquid-solid mass transfer, when solid particles are injected into liquid steel to promote refining reactions.

In all these situations, the evaluation of the mass transfer coefficient and the identification of the factors that affect its value are very important tasks. Different techniques have been developed to evaluate the mass transfer coefficient. These techniques vary according to the phases involved in the system being analysed.

In the present chapter, the mass transfer coefficients will be defined. Then, the different techniques that have been adopted to evaluate the mass transfer coefficient will be presented and briefly discussed. Finally, a case study, analysing the mass transfer rate during decarburization in the RH degasser, will be described and discussed.

2. Mass transfer coefficient

2.1 Definition

The mass transfer rate of a certain component A between two phases can be expressed by the following relationship

$$j_A = -D_A \cdot \left.\frac{\partial C_A}{\partial x}\right|_{x=0} .S \qquad (1)$$

where

j_A = mass transfer rate of component A (kg/s);

S = interface area (m^2);

D_A = diffusion coefficient of component A in a certain phase (m^2/s);

$\left.\frac{\partial C_A}{\partial x}\right|_{x=0}$ = concentration gradient of component A at the interface between the two phases (kg/m^4).

In equation (1), the mass transfer rate is defined only in terms of diffusion. At the interface between two fluid phases, there is also an additional contribution to the mass transfer rate due to bulk flow.

The application of equation (1) in the evaluation of mass transfer rates presents some difficulties. It is then usually necessary to resort to experimental data to determine the mass transfer rate. To obtain experimental data, a mass transfer coefficient, defined by equation (2), is normally used.

$$j_A = k.\ S.\ (C_A^o - C_A^s) \tag{2}$$

$$k = \frac{-D_A \cdot \left.\frac{\partial C_A}{\partial x}\right|_{x=0}}{(C_A^o - C_A^s)} \tag{3}$$

Combining equations (1) and (2), an expression for the mass transfer coefficient can be obtained.

It is also useful to define a volumetric mass transfer coefficient, which is the product of the mass transfer coefficient and the interface area.

$$k_V = k.S \tag{4}$$

where k_V is the volumetric mass transfer coefficient (m^3/s).

In steelmaking systems, when there is chemical reaction and component A is transferred from one phase to the other, the concentration of A at the interface is close to the equilibrium concentration, since the mass transfer is usually the rate controlling step in these situations.

2.2 Evaluation

The value of the mass transfer coefficient can be deduced based on first principles. In this case, it is necessary to model the flow of the phases and to determine the concentration gradient at the interface. Then, equation (3) is applied and the mass transfer coefficient is determined. The values of the mass transfer coefficient are then expressed in terms of correlations involving dimensionless numbers (Sherwood, Reynolds, Grashof and Schmidt). This method is normally applied to simple situations, usually in one dimensional flow. In more complex situations, this method can also be adopted but it will be necessary to solve the Navier-Stokes equations (laminar or turbulent form, according to the kind of flow) and the mass conservation equations for the species involved.

In steelmaking processes, the mass transfer coefficient is usually estimated based on experimental results. Due to the costs and difficulties associated to experiments with liquid steel, physical modelling of the industrial processes, using water to simulate liquid steel, is usually adopted to investigate mass transfer rates.

Recently, Singh et al (2009) investigated mass transfer between slag and metal in a physical model of a BOF converter. Different oils were used to simulate liquid slag. To evaluate the mass transfer rate, the distribution coefficient of benzoic acid between water and oil was determined. Assuming that the mass transfer rate between the two fluids owing to mixing of the phases was of first order, the authors estimated the mass transfer rate constant, which is related to the mass transfer coefficient.

Wei et al (2007) used the dissolution of sodium chloride powder to determine the mass transfer coefficient in a physical model of a RH degasser.
A different approach was adopted by Sakaguchi & Ito (1995) to analyse the mass transfer process in gas stirred vessels under reduced pressure. In this investigation, the desorption of CO_2 from aqueous solutions was used to deduce the volumetric mass transfer coefficient, assuming that the process is controlled by mass transfer. A similar technique was used by Maruoka et al (2010) to investigate the effect of the bottom bubbling condition on surface reaction on an oxygen-water system, simulating refining of steel in a ladle. In this case, the desorption of oxygen was monitored during the experiments.
In industrial systems, the mass transfer coefficient is usually obtained by curve fitting to experimental data. A mathematical model, based on mass transfer control, is proposed. Variations of concentration of a certain component as a function of time are predicted and compared to experimental data. A mass transfer coefficient that leads to the best agreement between experimental data and theoretical predictions is then determined (Kitamura et al, 2009; Hamano et al, 2004).
In the next section, a case study, analysing mass transfer rate during decarburization in the RH degasser will be presented and discussed.

3. Case study: decarburization in the RH degasser

3.1 Introduction

The RH process is a secondary refining process that can simultaneously attain significant levels of removal of interstitial elements, such as nitrogen, hydrogen and carbon elements, from liquid steel.
In this process, the melt circulation and the decarburization rates play a very significant role in determining the productivity of the equipment. In previous investigations, the melt circulation rate in the RH process has been evaluated by means of mathematical modeling (Park et al, 2001; Park et al, 2000), physical modeling (Kamata et al, 1998; Seshadri & Costa, 1986; Nakanishi et al, 1975) and also using data obtained in industrial plants (Kondo et al, 1989; Kuwabara et al, 1988). The effects of different parameters, such as argon flow rate, snorkels diameters, pressure in the vacuum chamber and position of the argon injection points, on the melt circulation rate have been analyzed in these investigations.
The decarburization rate in the RH process has also been studied using data from plant trials (Inoue et al, 1992; Yamaguchi et al, 1992; Kishimoto et al, 1993; Takahashi et al, 1995), lab scale experiments (Inoue et al, 1992; Kishimoto et al, 1993) and water models (Inoue et al, 1992; Seshadri et al, 2006). Mathematical models have been proposed to evaluate the decarburization kinetics. The roles of the different reaction sites, interfaces between molten steel and argon bubbles injected into the snorkel, free surface of the molten steel in the vacuum chamber and bubbles of carbon monoxide, have been analyzed (Takahashi et al, 1995). It is general agreement that, at low carbon concentrations, the decarburization rate is controlled by the carbon mass transfer, and that a first-order equation can describe the decarburization kinetics (Inoue et al, 1992; Yamaguchi et al, 1992; Takahashi et al, 1995).
In this case study, the melt circulation rate and the decarburization reaction in the RH process have been studied by means of physical modeling. The effects of the gas flow rate and of the diameter of the nozzles used in the gas injection have been analyzed. The decarburization reaction of liquid steel was simulated using a reaction of CO_2 desorption from caustic solutions (Guo & Irons, 1998).

3.2 Methodology

The experiments with the physical model were divided in two parts, one to evaluate the melt circulation rate and the other to determine the decarburization rate. These experiments are described below, after a description of the physical model and of the experimental set-up.

3.2.1 Experimental set-up

A schematic view of the physical model of the RH degasser is depicted in Fig. 1.

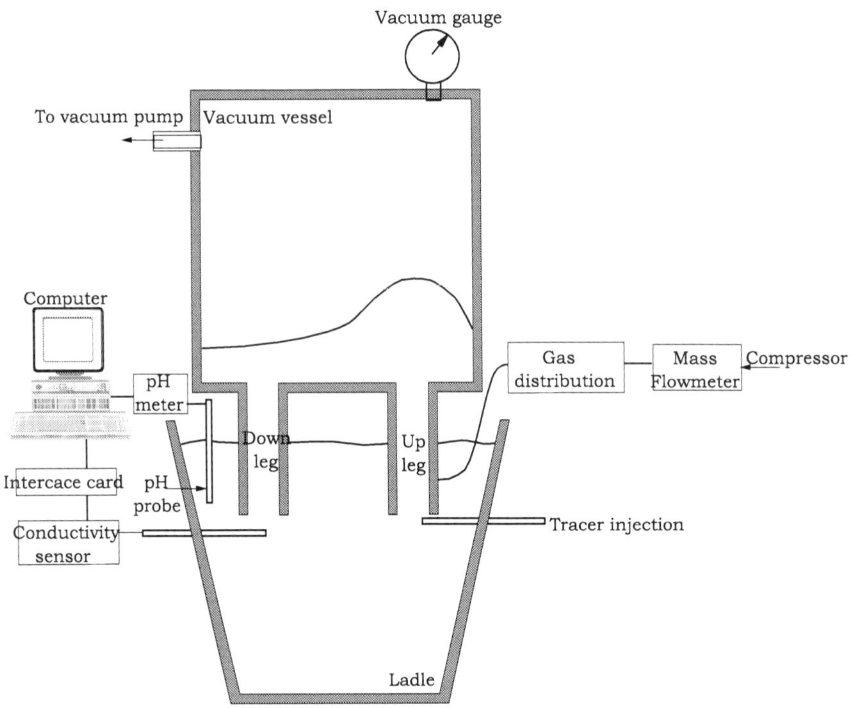

Fig. 1. Schematic view of the experimental set-up

Table 1 presents the dimensions of the physical model and the main characteristics of the gas injection system. These dimensions correspond to a 1:5 scale of an industrial RH degasser.

During the experiments, air supplied by a compressor was injected in the upleg snorkel. The flowrate was measured by a mass flowmeter and controlled manually. There were no individual measurements of flowrate for each injection nozzle. To equally distribute the gas flow rate among the nozzles, the air was first injected in the central region of a small chamber. This chamber was connected to each nozzle using pipes with the same length and diameter.

The pressure in the vacuum chamber was controlled by a system of valves and monitored by a pressure gauge.

The levels of water in the ladle and in the vacuum chamber were controlled and kept constant in all the experiments.

Parameters	Value
Ladle:	
- upper diameter (m)	0.720
- lower diameter (m)	0.648
- height (m)	0.750
- liquid level (m)	0.655
Vacuum chamber:	
- diameter (m)	0.415
- height (m)	0.700
- liquid level (m)	0.090
- pressure (Pa - gauge)	- 2700
Snorkels:	
- length (m):	0.312
- diameter (m)	0.120
- depth of immersion (m)	0.120
- distance between centers (m)	0.300
Gas injection:	
- flowrate (STP l/min)	50 - 500
- nozzles:	
- number	10
- diameters (mm)	1.0, 1.5, 2.8
- position (below liquid level) (m)	0.04

Table 1. Characteristics of the RH physical model

3.2.2 Melt circulation rate

To determine the melt circulation rate, the method used by Seshadri et al (2006) was adopted. In this procedure, a solution of potassium chloride is injected in the upleg snorkel in the form of a pulse. The concentration of KCl in the downleg snorkel is then continuously monitored. Since the electrical conductivity of water is affected by the presence of KCl, its concentration was calculated based on the signal generated by the electrical conductivity sensor, using a calibration curve previously determined.

Fig. 2 illustrates typical curves representing the variation of concentration of KCl in the downleg snorkel as a function of time. The circulation rate was estimated using the following equation

$$\Gamma = \frac{V}{A/\Delta C}\,\rho \qquad (5)$$

where

Γ is the circulation rate (kg/s);

V is the volume of water in ladle (m^3),

A is the area under the curve during the first peak (kg.s/m^3) (see Fig. 2),

ΔC is the variation of concentration of KCl at the end of the experiment (kg/m^3),

ρ is the density of the circulating fluid (kg/m^3).

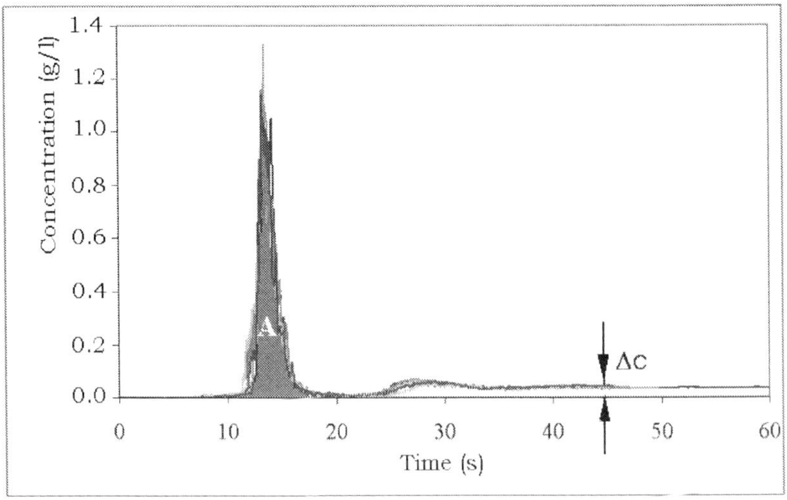

Fig. 2. Typical curves obtained in the experiments to evaluate the melt circulation rate

To guarantee the reliability of the experiments, the mass balance of KCl was verified in all the tests. Ten experiments were carried for each experimental condition.

3.2.3 Decarburization rate

The desorption of CO_2 from caustic solutions was used to simulate the decarburization reaction of liquid steel (Guo & Irons, 1998). In these experiments, CO_2 was injected into an aqueous solution of sodium hydroxide (NaOH) with a concentration of 0.01 g.mol/l, until the pH reached a preset value (approximately 6.5). At this point, the operation of the RH physical model was started, with the injection of air at the upleg snorkel. This injection leads to CO_2 desorption and to an increase in the pH of the circulating fluid. The variation of the pH was monitored during the entire test at a frequency of 0.33 Hz. A calibration curve was used to convert pH values into aqueous CO_2 concentration. Different locations for the pH sensor were adopted, without significant differences in the results. Air flowrates, ranging from 50 to 500 STP l/min, were used in the tests. Nozzles with diameters of 1.0, 1.5, 2.2 and 2.8 mm were also tested in the simulations. The remaining conditions were kept constant in all the experiments. Three experiments were run for each experimental condition. Fig. 3 shows a typical variation of aqueous CO_2 concentration during a test to evaluate the decarburization rate.

3.3 Results and discussion

3.3.1 Melt circulation rate

The variation of the melt circulation rate as a function of the gas flow rate for the different diameters of the injection nozzles is presented in Fig. 4. For a given diameter of the nozzle, the circulation rate tends to increase when the gas flow rate increases. Above a certain flow rate, the circulation rate remains approximately constant or shows a slight tendency of reduction. This kind of behavior has also been determined in other works (Park et al, 2001; Seshadri & Costa, 1986; Seshadri et al, 2006).

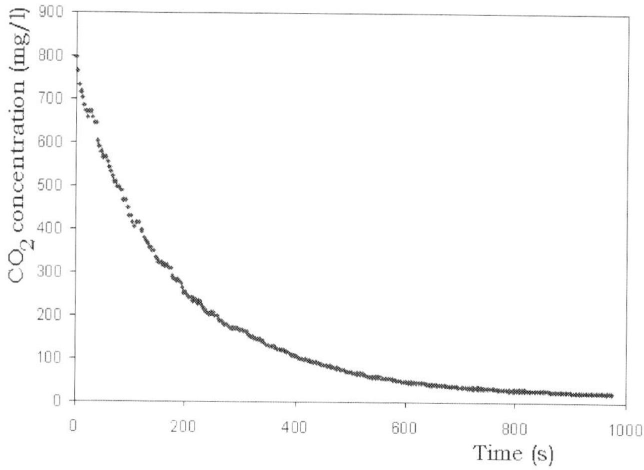

Fig. 3. Typical variation of aqueous CO_2 concentration as a function of time during a test

An increase in the diameter of the injection nozzles leads to higher circulation rates. According to Park et al (2001), this effect can be analysed considering the variation of the plume area in the upleg snorkel. To confirm that, during the experiments, images of the cross section of the upleg snorkel were obtained using a video camera. These images were post processed to identify the area occupied by the plume. Fig. 5 illustrates the original and the post processed images taken from the upleg snorkel. In Fig. 5, the black area represents the plume.

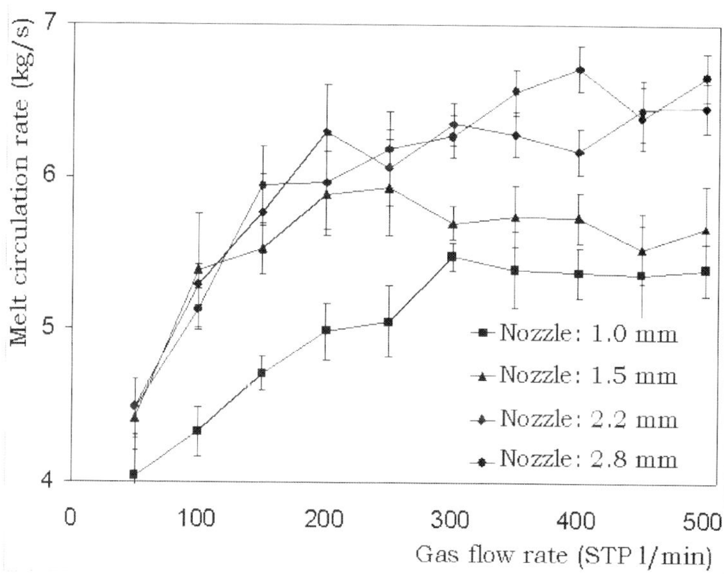

Fig. 4. Variation of the melt circulation rate with gas flow rate for different nozzles diameters

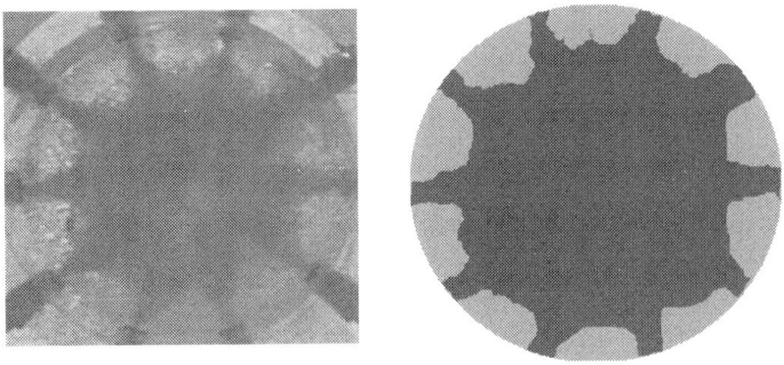

Nozzle: 1.0 mm

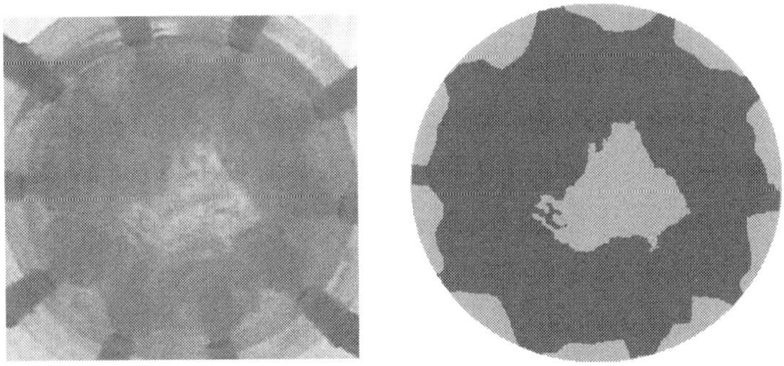

Nozzle: 2.8 mm

a) b)

Fig. 5. a) Image taken from the cross section of the upleg snorkel. b) Post processed image identifying the plume

Fig. 6 presents the results of circulation rate and the percentage of the area of the upleg snorkel that is occupied by the plume, for the experiments with nozzles of 1.5 and 2.8 mm in diameter. Similar trend was observed for the other nozzles. Based on these results, it is possible to analyze the variation of the circulation rate in terms of the variation of the plume area in the upleg snorkel. This supports the theoretical analysis carried out by Park et al (2001). Fig. 7 presents a graph of the circulation rate as a function of the plume area, showing a linear relationship between these two variables, within the conditions tested in the present work.

In an attempt to generalize the results for the circulation rate, a dimensionless circulation rate was defined as:

$$\text{Dimensionless Circulation Rate} = \frac{V_{ds}^2}{g.d_o} \tag{6}$$

where

V_{ds} is the velocity of the fluid in the downleg snorkel (m/s),

g is the acceleration due to gravity (m/s^2),

d_o is the diameter of the nozzle (m).

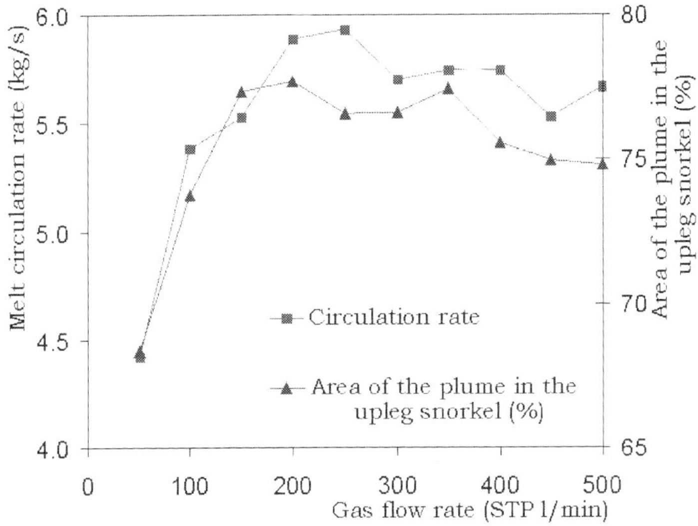

Nozzle: 1.5 mm

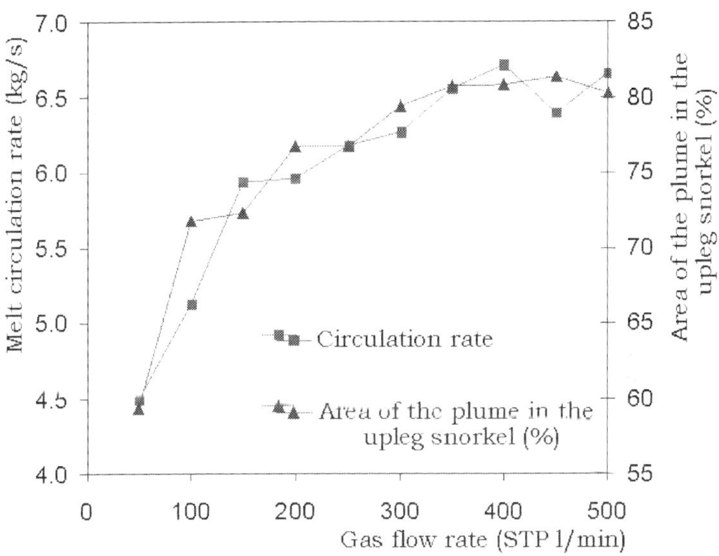

Nozzle: 2.8 mm

Fig. 6. Circulation rate and area of the plume in the upleg snorkel as a function of gas flow rate for two different nozzle diameters

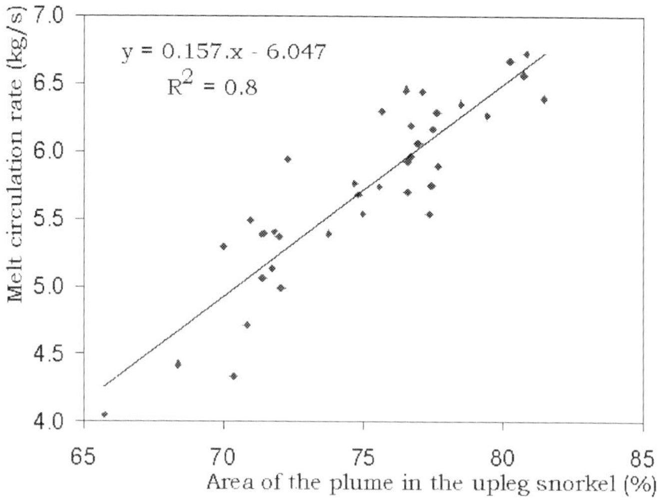

Fig. 7. Variation of the melt circulation rate with the plume area in the upleg snorkel

This parameter was correlated to the modified Froude number defined as

$$Fr_m = \frac{\rho_g\, Q^2}{d_o^5\, g\, (\rho_l - \rho_g)} \tag{7}$$

where

ρ_g is the density of the gas injected (air) (kg/m^3),
Q is volumetric flow rate of the gas injected in the upleg snorkel (m^3/s),
ρ_l is the density of the liquid (water) (kg/m^3).

The results for different nozzles are presented in Fig. 8. It can be seen that there is a logarithmic relationship between the two parameters and that one single equation fit the results for the different nozzles. The relation between these two dimensionless numbers is reasonable, since the modified Froude number is related to the geometry of the plume formed in the upleg snorkel (Sato et al, 2004).

3.3.2 Decarburization rate

In the model proposed by Yamaguchi et al (1992), the variation of the concentration of aqueous CO_2 (shown in Fig. 3) can be expressed by a first order reaction rate given by

$$-\frac{dCO_{2_{aq}}}{dt} = K(C_{CO_{2aq}} - C^{eq}_{CO_{2aq}}) \tag{8}$$

where

C_{CO2aq} is the concentration of aqueous CO_2 in the ladle (mg/l),
t is the time (s),
K is the reaction constant (s^{-1}),
C^{eq}_{CO2aq} is the concentration of aqueous CO_2 in equilibrium with the gas phase.

In this model, the rate constant is given by

$$K = \frac{1}{W\left(\frac{1}{\Gamma} + \frac{1}{ak_c\rho}\right)} \tag{9}$$

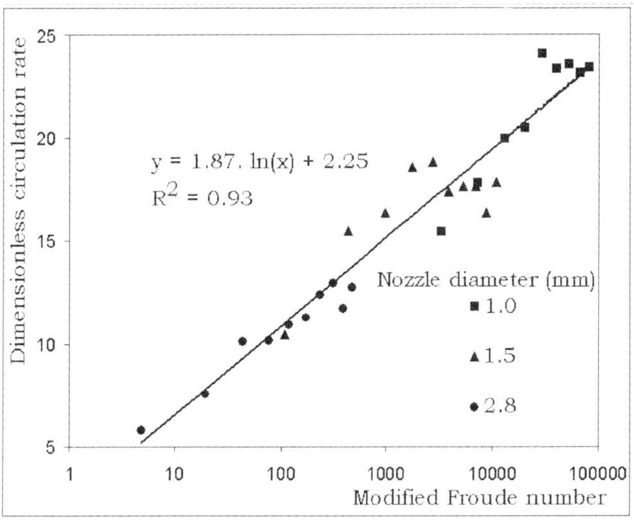

Fig. 8. Relation between the dimensionless circulation rate and the modified Froude number

where
W is the mass of water in the ladle (kg),
Γ is the circulation rate (kg/s),
a is the reaction area (m^2),
k_c is the mass transfer coefficient (m/s),
ρ is the density of water (kg/m^3).

The following assumptions were made in the development of this model:
- the reaction rate is controlled by mass transfer in the liquid,
- the liquid is perfectly mixed in the ladle and in the vacuum chamber and
- the amount of liquid in the ladle is much larger than that in the vacuum chamber.

All these assumptions seem very reasonable in the present investigation.

Equation (8) can be integrated to give

$$-\ln\frac{(C_{CO_{2aq}} - C^{eq}_{CO_{2aq}})}{(C^{i}_{CO_{2aq}} - C^{eq}_{CO_{2aq}})} = kt \tag{10}$$

where $C^{i}_{CO_{2aq}}$ is the initial concentration of aqueous CO_2.

A plot of the experimental data is presented in Fig. 9. This graph indicates that the results are very well reproduced by equation (10). The value of K is given by the slope of the straight line.
The product $a.k_c$ can be estimated using equation (9). This product represents the volumetric mass transfer coefficient.
Fig. 10 presents the variation of the logarithm of the volumetric mass the transfer coefficient as a function of the logarithm of the gas flow rate (Q). Sakaguchi and Ito(15) also calculated the volumetric mass transfer coefficient for gas stirred vessels under reduced pressure. They expressed the mass transfer coefficient as a function of the gas flow rate. According to their analysis, the mass transfer coefficient was expected to be proportional to the 0.744 power of the gas flow rate (although their experimental results indicated a power of 0.717), when the vacuum chamber pressure and the temperature are constant. The straight line with a slope of 0.744 is also indicated in figure 10.

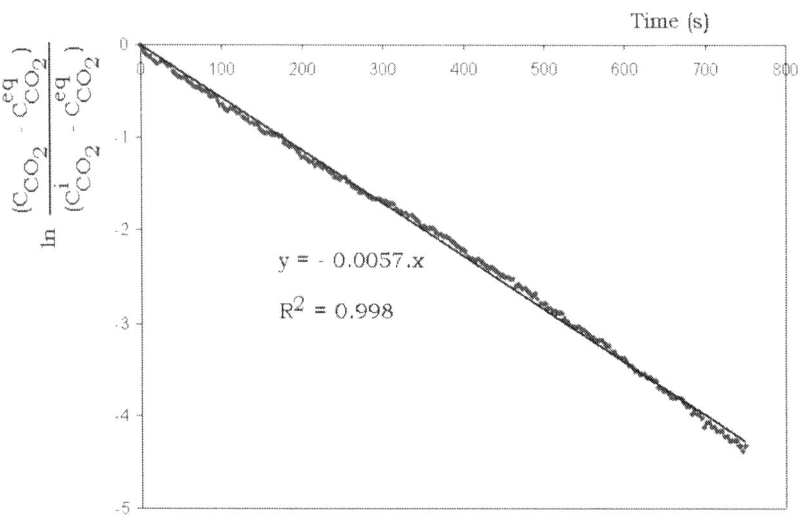

Fig. 9. Experimental data plot according to equation (10)

In the region of low flow rates, the slopes of the curves are similar to that indicated by Sakaguchi & Ito (1995). At higher flow rates, particularly for the nozzles with diameters of 1.5 and 2.8 mm, there is an increase in the slope above a certain flow rate. The behavior observed in Fig. 10 can be analyzed considering that there are two sites for the reaction to occur. One is in the interface between the gas and the liquid in the upleg snorkel. The other site is in the vacuum chamber and is usually associated to the splash that occurs there. In their study, Sakaguchi & Ito (1995) admitted that the contribution of the mass transfer at the free surface could be neglected. Apparently, the slope of 0.744 is valid for situations where the contribution of the mass transfer at the free surface is small. When this contribution is more significant, i.e. when the splash in the vacuum chamber is more pronounced, there is a tendency of the slope to increase. This is probably the cause of the increase in the slope observed in Fig. 10, especially for nozzle diameters of 1.5 and 2.8 mm.

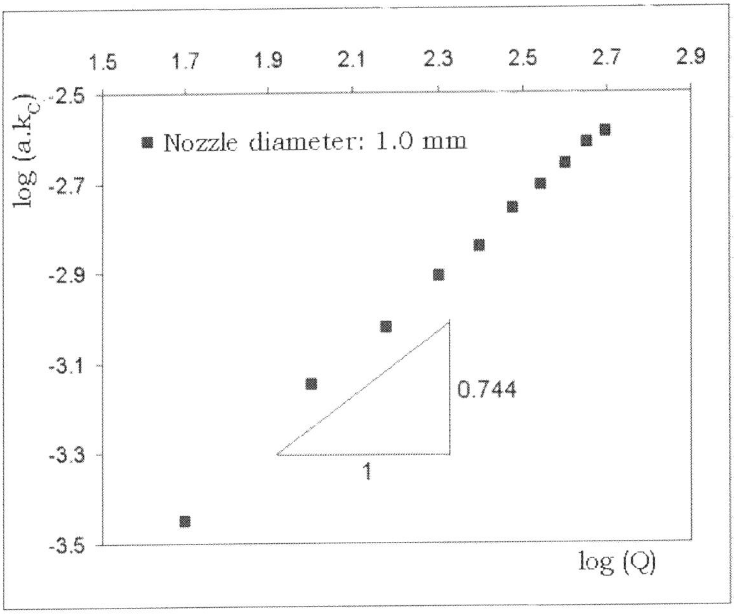

a) Nozzle diameter: 1.0 mm

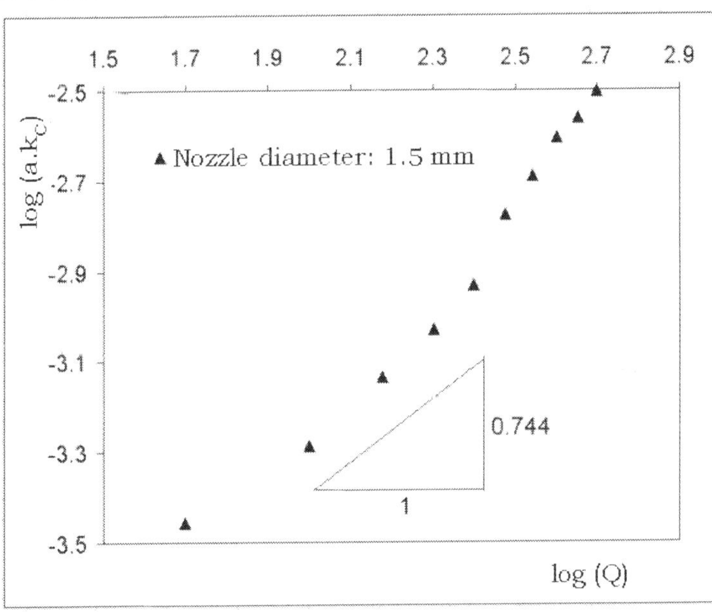

b) Nozzle diameter: 1.5 mm

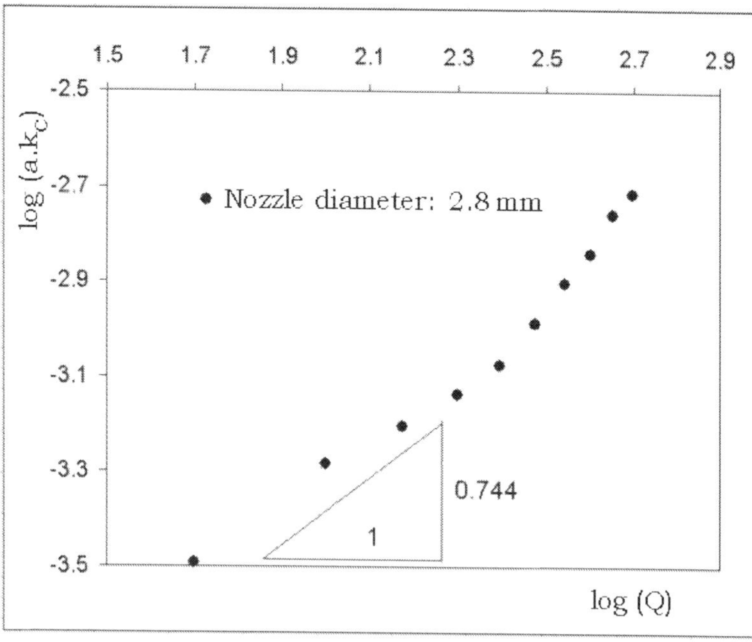

c) Nozzle diameter: 2.8 mm

Fig. 10. Relation between the gas flow rate and the volumetric mass transfer coefficient for the three nozzle diameters

In a previous work developed with a copper converter, Themelis & Schmidt (1967) related the volumetric mass transfer coefficient to the nozzle Reynolds number. They determined a linear relationship between the mass transfer coefficient and the Reynolds number in the range of 2,000 to 9,000. The same straight line fit the results for different nozzle diameters.

Fig. 11 shows the values of the mass transfer coefficient as a function of the nozzle Reynolds number obtained in the present work. In the low Reynolds region (Re < 15,000), the effect of the nozzle diameter is relatively small and the results indicate a linear relationship between the mass transfer coefficient and the Reynolds number. For Reynolds number above 15,000, the linear trend remains, but the slopes of the lines for the different nozzles increase. The magnitude of this increase varies according to the nozzle diameter. For the 1.0 mm nozzle, the variation is very small. For the other two nozzles, the variation is more noticeable.

The behavior observed in Fig. 11 can also be analyzed considering the two reaction sites. The relevance of each site changes according to the Reynolds number and to the nozzle diameter. For the 1.0 mm nozzle, the jet coming from the nozzle disrupts in very small bubbles, with large surface area. In this case, the site in the upleg snorkel has a significant role even when the splash in the vacuum chamber is pronounced. For the nozzles with bigger diameters, particularly for those with 2.8 mm diameter, the effect of the splash becomes more relevant at higher Reynolds number, especially when a swinging motion of the bath surface is observed (Sato et al, 2004). This effect is probably responsible for the increase in the slopes of the lines observed in Fig. 11.

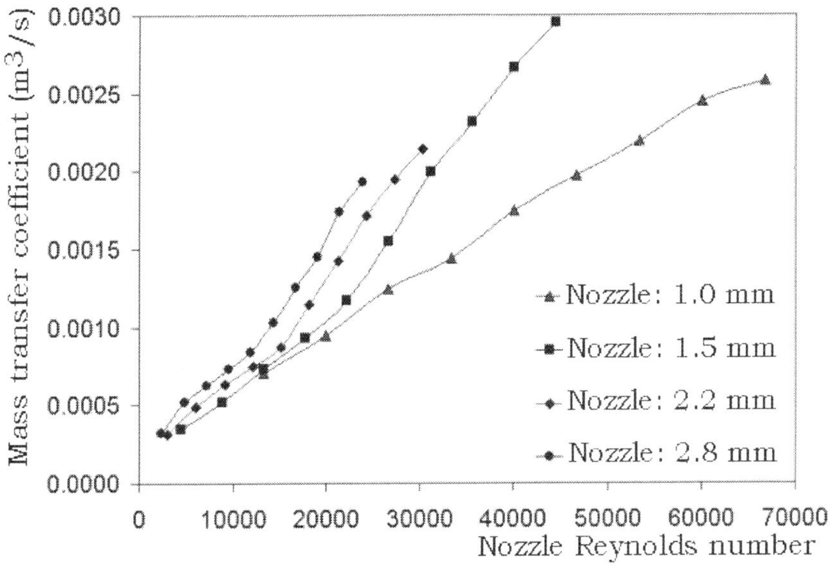

Fig. 11. Volumetric mass transfer coefficient as a function of the nozzle Reynolds number

Fig. 12 depicts images of the vacuum chamber, when different nozzles are used. The nozzle Reynolds number is approximately the same in three pictures. The splash is more pronounced for the 2.8 mm nozzle diameter and is certainly leading to the higher volumetric mass transfer coefficients observed in Fig. 11.

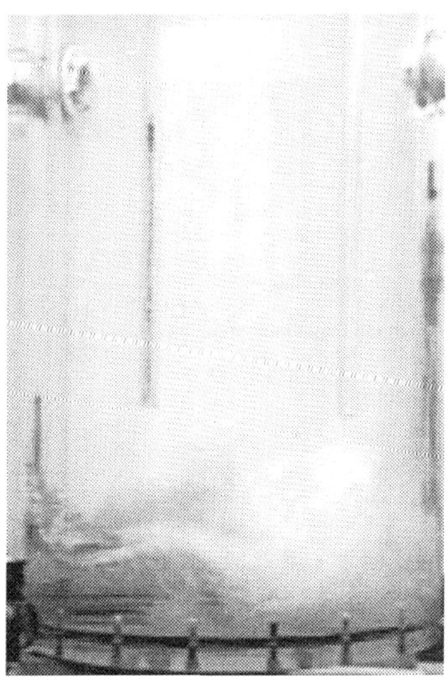

a) Nozzle: 1.0 mm

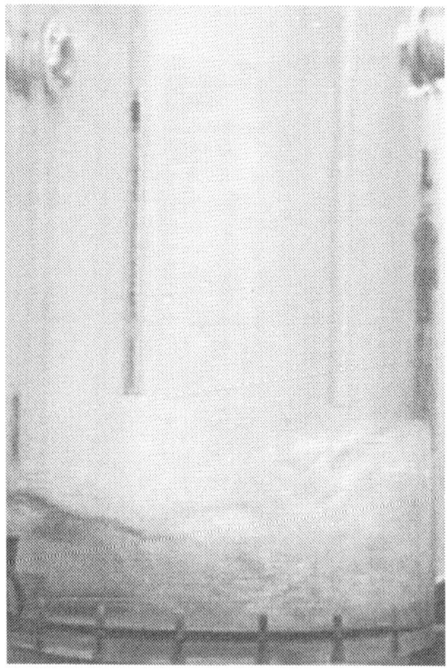

b) Nozzle: 1.5 mm

c) Nozzle: 2.8 mm.

Fig. 12. Images of the vacuum chamber when different nozzles are used. Nozzle Reynolds number ≅ 20,000

4. Conclusions

Mass transfer plays a significant role in determining the rate of steelmaking operations. Therefore, the evaluation of the mass transfer coefficient and the identification of the factors that affect the mass transfer rate are very important tasks. After defining the mass transfer coefficients and briefly discussing the techniques applied in their evaluation, a case study, analysing decarburization in the RH degasser was presented.

In this case study, a physical model was used to study the circulation rate and the kinetics of decarburization in a RH degasser. The effects of the gas flow rate and of the diameters of the nozzles used in the gas injection were investigated. The decarburization of liquid steel was simulated using a reaction of desorption of CO_2 from caustic solutions.

The results showed that the circulation rate increases with an increase in the diameter of the nozzles and in the gas flow rate. The effect of the gas flow rate becomes less significant at higher flow rates. A relationship between a dimensionless circulation rate and the modified Froude number was determined. This relationship fit the results for all nozzle diameters tested.

The kinetics of the reaction follows a first order equation and is controlled by mass transfer in the liquid phase. The reaction rate constant was affected by the gas flow rate and nozzle diameter. An increase in the gas flow rate lead to an acceleration of the reaction. For a given flow rate, the smaller nozzle tend to give higher reaction rates.

A volumetric mass transfer coefficient was calculated based on the rate constants and on the circulation rate. The logarithm of the mass transfer coefficient showed a linear relationship with the logarithm of the gas flow rate. The slope of the line was found to vary according to the relevance of the reaction at the free surface in the vacuum chamber.

A linear relationship between the volumetric mass transfer coefficient and the nozzle Reynolds number was also observed. Again, the slopes of the lines changed according to the relative importance of the two reaction sites, gas-liquid interface in the upleg snorkel and in the vacuum chamber (mainly due to the splash). At higher Reynolds number, the reaction in the vacuum chamber tends to be more significant.

5. Acknowledgments

The financial support of FAPEMIG in the form of a research grant to R. P. Tavares (Process No. TEC - PPM-00197-09) is gratefully acknowledged.

6. References

Guo, D. & Irons, G.A. (1998). Water Modeling of Vacuum Decarburization in a Ladle, Proceedings of the 1998 Steelmaking Conference Proceedings, pp. 601-607, 1-886362-26-2.

Hamano, T.; Horibe, M. & Ito, K. (2004). The Dissolution Rate of Solid Lime into Molten Slag Used for Hot-metal Dephosphorization. ISIJ International, 44, 2, 263–267, 0915-1559.

Inoue, S.; Furuno, Y.; Usui, T. & Miyahara, S. (1992). Acceleration of Decarburization in RH Vacuum Degassing Process, ISIJ International, 32, 1, 120-125, 0915-1559.

Kamata, C.; Matsumura, H; Miyasaka, H.; Hayashi, S.; Ito, K. (1998). Cold Model Experiments on the Circulation Flow in RH Reactor Using a Laser Doppler Velocimeter, Proceedings of the 1998 Steelmaking Conference, (1998), pp. 609-616, 1-886362-26-2.

Kishimoto, Y.; Yamaguchi, K.; Sakuraya, T. & Fujii, T. (1993). Decarburization Reaction in Ultra-Low Carbon Iron Melt Under Reduced Pressure, ISIJ International, 33, 3, 391-399, 0915-1559.

Kitamura, S-Y; Miyamoto, K.I.; Shibata, H.; Maruoka, N. & Matsuo, M. (2009). Analysis of Dephosphorization Reaction Using a Simulation Model of Hot Metal Dephosphorization by Multiphase Slag. ISIJ International, 49, 9, 1333–1339, 0915-1559.

Kondo, H.; Kameyama, K; Nishikawa, H.; Hamagami, K. & Fujji, T. (1989). Comprehensive refining process by the Q-BOP-RH Route for Production of Ultra-Low Carbon Steel, Iron & Steelmaker, 16, 10, 34-38.

Kuwabara, T.; Umezawa, K; Mori, K & Watanabe, H. (1988). Investigation of Decarburization Behavior in RH-Reactor and its Operation Improvement, Transactions of ISIJ, 28, 4, 305-314, 0021-1583.

Maruoka, N.; Lazuardi, F.; Nogami, H.; Gupta, G.S. & Kitamura, S-Y. (2010) Effect of Bottom Bubbling Conditions on Surface Reaction Rate in Oxygen–Water System. ISIJ International, 50, 1, 89 -94, 0915-1559

Nakanishi, K.; Szekely, J. & Chang, C.W. (1975). Experimental and Theoretical Investigation of Mixing Phenomena in the RH-Vacuum Process, Ironmaking & Steelmaking, 2, 2, 115-124, 0301-9233.

Park, Y-G.; Yi, K-W & Ahn, S-B. (2001). The Effect of Operating Parameters and Dimensions of the RH System on Melt Circulation Using Numerical Calculations, ISIJ International, 41, 5, 403-409, 0915-1559.

Park, Y-G.; Doo, W-C; Yi, K-W & Ahn, S-B. (2000). Numerical Calculation of Circulation Flow Rate in the Degassing Rheinstahl-Heraeus Process, ISIJ International, 40, 8, 749-755, 0915-1559.

Sakaguchi, K. & Ito, K. (1995). Measurement of the Volumetric Mass Transfer Coefficient of Gas-Stirred Vessel under Reduced Pressure, ISIJ International, 35, 11, 1348-1353, 0915-1559.

Sato, T.; Bjurström, M.; Jönsson, P. & Iguchi, M. (2004). Swinging Motion of Bath Surface Induced by Side Gas Injection, ISIJ International, 44, 11, 1787-1792, 0915-1559.

Seshadri, V. Costa, S.L.S. (1986). Cold Model Studies of RH Degassing Process, Transactions of ISIJ, 26, 2, 133-138, 0021-1583.

Seshadri, V.; Silva, C.A.; Silva, I.A.; Vargas, G.A. & Lascosqui, P.S.B. (2006). Decarburization Rates in RH-KTB Degasser of the CST Steel Plant (Companhia Siderúrgica de Tubarão, Vitória, Brazil) Through a Physical Modeling Study, Ironmaking & Steelmaking, 33, 1, 34-38, 0301-9233.

Singh, V.; Lenka, S.N.; Ajmani, S.K.; Bhanu, C. & Pathak, S. (2009). A Novel Bottom Stirring Scheme to Improve BOF Performance through Mixing and Mass Transfer Modelling. ISIJ International, 49, 12, 1889-1894, 0915-1559

Takahashi, M.; Matsumoto, H. & Saito, T. (1995). Mechanism of Decarburization in RH Degasser, ISIJ International, 35, 12, 1452-1458, 0915-1559.

Themelis, N.J. & Schmidt, P.R. (1967). Transactions of AIME, 239 , 1313, ISSN.

Wei, J-H; Jiang, X-Y.; Wen, L-J. & Li, B. (2007). Mass Transfer Characteristics between Molten Steel and Particles under Conditions of RH-PB(IJ) Refining Process. ISIJ International, 47, 3, 408–417, 0915-1559.

Yamaguchi, K.; Kishimoto, Y.; Sakuraya, T.; Fujii, T.; Aratani, M. & Nishikawa, H. (1992). Effect of Refining Conditions for Ultra Low Carbon Steel on Decarburization Reaction in RH Degasser, ISIJ International, 32, 1, 126-135, 0915-1559.

13

Effects of Surface Tension on Mass Transfer Devices

Honda (Hung-Ta) Wu[1] and Tsair-Wang Chung[2]
[1]*Center of General Education, Chungyu Institute of Technology*
[2]*Department of Chemical Engineering/R&D Center for Membrane Technology, Chung-Yuan Christian University*
Taiwan, ROC

1. Introduction

Fluid flow resulted from the gradient of surface tension usually called as Marangoni effect or surface tension effect, and the induced convection was called as Marangoni convection. Earlier studies about Marangoni effect were to discuss and analyze the disturbed phenomena in the gas-liquid interface. The phenomenon of the so called "tears and wine" was first studied by Carlo Marangoni in 1865. The Benard cells resulted from the gradient of temperature were another instance of Marangoni convections. Nowadays, the surface tension effect was extensively applied in many fields. For example, the nanostructure changed as a result of Marangoni effect in enhanced laser nanopatterning of silicon. Besides, to avoid spotting in silicon wafers, the matter of low surface tension was blown over the wet wafer to lead the gradient of surface tension and to dry wafer surface by the induced Marangoni effect. Marangoni effect was also utilized in dyeing works. The dyes or pigments were floated on the surface of the basic medium, and then they moved toward the diffusion direction by Marangoni effect. Finally, the surface was covered by paper or cloth to take a print.

On the basis of small disturbance analysis, the interfacial disturbances can be divided into stable, stability and instability state. The stable state means that the fluid flowed phenomenon is not affected by Marangoni effect. The studies about stability state were always focused on critical Marangoni number or neutral stability curve. The instability state could be subdivided into stationary and oscillatory instabilities, and they were known as Marangoni instability. The regular hexagonal pattern of convective cells, such as Benard cells, was formed by heating from below or cooling from above, and which was the typical stationary instability, that is, the Marangoni convections with regular convection were called as stationary instability; however, the Marangoni convection with irregular convection was called as oscillatory instability. In general, the mass transfer performance can be enhanced by the Marangoni instability or so called interfacial disturbance. Therefore, studies about mass transfer affected by interfacial disturbance were focused on performance enhancement. Both of stationary instability or oscillatory instability can be called as interfacial disturbance in these studies.

Mentioned above, Marangoni instability or interfacial disturbance can be resulted from the gradient of surface tension. Since fluids are the indispensable element for mass transfer devices, fluid flow affected by surface tension and effect of Marangoni instability on mass transfer were discussed in recent years. Generally speaking, the reason for the induced Marangoni convection could be divided into artificial and spontaneous Marangoni convection. For example, the disturbance induced by surface additive injected into absorption system could be called as artificial Marangoni instability; however the spontaneous Marangoni instability could be produced by some composed components in the distillation, extraction, bubble columns and so on. The Marangoni effect could be occurred in the gas-liquid and liquid-liquid contacting systems or mass transfer devices, such as packed distillation column, falling film absorber, absorption process with chemical reaction, two-phase flow system, liquid jets system and so on.

In addition to the gradient of surface tension, the liquid fluid with continuous phase is an important reason to trigger the Marangoni effect so much that the liquid fluid with continuous phase can be observed in the mass transfer devices mentioned above. Therefore, the purpose of this chapter is to discuss effects of Marangoni instability on mass transfer devices. Besides, some experimental results are present to describe effects of Maranfoni effect on absorption performance. The interfacial disturbance and surface stress were also observed and calculated to analyze mass transfer performance for water vapor absorbed by triethylene glycol (TEG) solution in packed bed absorber. Described above, the phenomena of Marangoni effect in the thin liquid film, thinker liquid layer, and mass transfer devices were elucidated in the first. Secondly, the definitions related to artificial and spontaneous Marangoni convections were described. And then effects of interfacial disturbance resulted from the gradient of surface tension on the performance of mass transfer devices were discussed. Finally, the summary of this chapter was described in the conclusion.

2. Marangoni effect in thin liquid film, thinker liquid layer, and mass transfer devices

2.1 Thin liquid film

Fluid flow driven by the gradient of surface tension had been called as Marangoni effect, and the surface of liquid thin film was always inhomogeneous or wavy in the microview. As shown in Fig. 1, the horizontal coordinate toward the thinner region is assumed to be positive x, that is the direction of $+x$, and the section of between real line and dotted line can be regarded as a cellular convection in the interface. Since the concentration in the thinner region is higher than that in the thicker region, the concentration gradient, eq. 2, is greater than zero for the gradient of surface tension, eq. 1.

$$\frac{d\gamma}{dX} = \frac{\partial \gamma}{\partial C_{AL}} \frac{\partial C_{AL}}{\partial X} \tag{1}$$

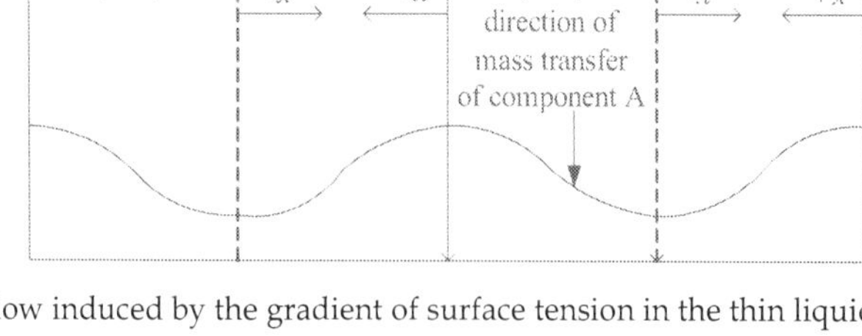

Fig. 1. Fluid flow induced by the gradient of surface tension in the thin liquid film

$$\frac{\partial C_{AL}}{\partial X} > 0 \tag{2}$$

where the symbol γ is surface tension, and C_{AL} is the concentration of solute in liquid phase. Mentioned above, the direction of fluid flow is dominated by the gradient of surface tension with respect to the concentration of liquid solution, that is $\partial\gamma / \partial C_{AL}$.

a. $\frac{\partial \gamma}{\partial C_{AL}} < 0$

If the gradient of surface tension with respect to concentration is less than zero (negative), the gradient of surface tension (eq. 1) will be negative. The liquid will flow from thinner region to thicker region. Compared with liquid flowing on the supported surface, such as packing surface, the gas-liquid contacting area is reduced by the contraction of liquid film on packing surface, which leads to the less mass transport. Therefore, the phenomenon was called as "Marangoni negative system".

b. $\frac{\partial \gamma}{\partial C_{AL}} > 0$

If the gradient of surface tension with respect to concentration is greater than zero (positive), the gradient of surface tension will be positive. The liquid will flow from thicker region to thinner region. Since the fluid flow under this condition makes liquid film flowing homogeneously on the supported surface, the gas-liquid contacting area is larger than the"Marangoni negative system". The mass transfer performance is always better for this system, and the phenomenon is called as "Marangoni positive system".

Extended from the concept of Marangoni effect acting on thin liquid film, effect of surface tension on mass transfer performance of packed distillation column was investigated by Patberg et al., 1983. Since the surface tension of feeding solution was almost not changed while contacting with the reflux, Fig. 2 (a) showed the liquid was subject to the path of the shortest distance and the lowest resistance. Flow phonmenon in Fig. 2 (a) was resulted from Marangoni negative or neutral system in packed distillation column. Therefore, the poor distilling performance was due to the bad efficiency of packing wetted. On the opposite, the solution on the button of packing could be drawn by the feeding solution on the top of packing due to the surface tension of feeding solution increased by the reflux. Therefore, Fig. 2 (b) showed the solution flowing more homogeneously over the packing material. Since the wetting efficiency of packing material is good for mass transfer under the condition of Fig. 2 (b), the mass transfer performance of packed distillation column is better than Fig. 2 (a). This can be called as Marangoni positive system in the packed distillation column. In addition, Patberg et al., 1983 also found that the interface refreshment was affected by the smaller packing and the lower liquid flow rates more significantly. Patberg et al., 1983 assumed that the shear stress was equal to the largest possible surface tension difference divided by an assumed creeping height, which resulted in the constant shear stress and constant thickness of creeping film. To achieve a more detailed approximation, the creeping film phenomenon (Fig. 3) for packed distillation column was proposed by Dijkstra & Drinkenburg, 1990 to discuss effects of surface tension on wetted area and mass transfer. The numerical results showed that Marangoni effect was more significant in lower Biot number (Buoyancy effect), and the creeping height was increased with the increased Marangoni number. Finally, the Marangoni effect resulted from evaporation of acetone affected mass transfer flux for the acetone-water system was also demonstrated by Dijkstra & Drinkenburg, 1990.

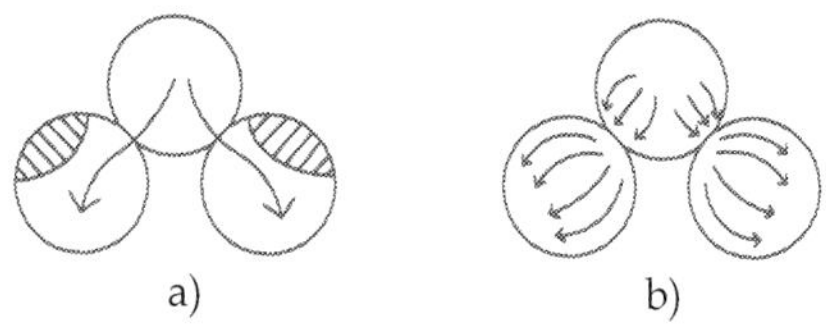

Fig. 2. Schematic diagram of liquid flow over packing under the conditions of (a) negative or neutral system (b) positive system. (referred from Patberg et al., 1983)

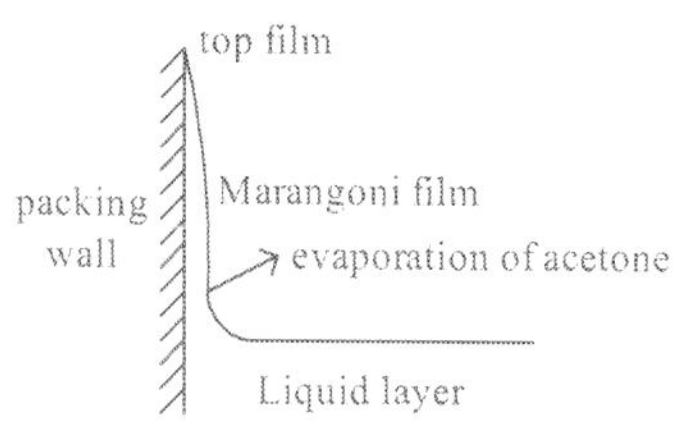

Fig. 3. Schematic diagram of the phenomenon of creeping film. (referred from Dijkstra & Drinkenburg, 1990)

2.2 Liquid layer

Marangoni convection or Marangoni instability was usually resulted from the gradient of surface tension in the thinker liquid layer. In addition to the interfacial disturbance resulted from heating the bottom of liquid layer, the interfacial disturbance also can be induced by the gradient of concentration, such as chemisorptions of carbon dioxide by monoethanolamine (MEA) solution. Brian et al., 1967 proposed the chemisorptions mechanism for carbon dioxide absorbed by MEA solution as follows:

$$NH_3 + CO_2 \rightarrow NH_3COOH \tag{3}$$

$$NH_3COO^- + H^+ + NH_3 \rightarrow NH_4^+ \tag{4}$$

The absorption efficiency of carbon dioxide could be enhanced by the induced interfacial disturbance in the system. In order to analyze effects of surface tension on cellular convection, the chemisorptions for the components of H_2S-MEA-H_2O and CO_2-MEA-H_2O were investigated by Buzek, 1983. Absorption of H_2S by MEA solution was an instantaneous and irreversible reaction, and the mass transfer resistance in the gas phase was negligible. Since the liquid surface and its vicinity were occupied by the only ionized products, there was no concentration gradient responsible for cellular convection. Although the mass transfer resistance in the gas phase was still negligible for absorption of CO_2 by MEA solution, the rate of chemical reaction between MEA solution and CO_2 was finite. The gradient of interfacial tension could be resulted from nonuniform interfacial distribution of reactant and product. Therefore, the cellular convection could be resulted from absorption of CO_2 by MEA solution due to the gradient of interfacial tension. For the chemisorptions, Kaminsky et al., 1998 proposed the model of energy-balance equation, and the results showed that the mass transfer rate between phases was increased by the induced interfacial disturbance. Besides, to discuss the influences of surfactant solutions spreading on hydrophilic surfaces affected by Marangoni effect, Cachile et al., 1999 used nonionic surfactants, such as $C_{12}E_4$ and $C_{12}E_{10}$, in elthylene glycol (EG) and diethylene glycol (DEG) to deposit on the surface of oxidized silicon wafer. Cachile et al., 1999 found that the spreading of surfactant solutions on hydrophilic surfaces and the structure of the instability pattern were dominated by the mobility of pure surfactant and the relative humidity, especially for that higher than 80%. In recent years, Marangoni convections were also discussed in the systems of solute evaporating from a liquid phase to an inert phase, surfactant transport from an aqueous to an organic phase, and absorption and desorption of carbon dioxide into and from organic solvents by Colinet et al., 2003, Lavabre et al., 2005, and Sun, 2006 respectively.

In general, the interfacial disturbance resulted from spontaneous mass transfer is insignificant, and it is difficult to observe by naked eyes. Therefore, some studies compared mass transfer data with and without Marangoni effect to show influence of surface tension on mass transfer performance. On the other hand, some studies used the disturbed phenomena in the macro view or established the disturbed model to deduce interfacial disturbance resulted from the gradient of surface tension. Mentioned above, scaling up the interfacial phenomena from micro view and proving by experimental data under the conditions without violating scientific theory is one way to realize interfacial phenomena affected by the Marangoni effect.

In order to observe and realize the interfacial phenomena resulted from the gradient of surface tension for the absorption system, the water drop was instilled on the surface of TEG solution to observe the interfacial disturbance and calculate the surface stress. The schematic diagram for observing water drop instilled on the surface of TEG solution is shown in Fig. 4. Since the disturbed phenomena for water drop instilled on different concentrations of TEG solutions are similar, only water drop instilled on 95 wt. %. TEG solution is shown to describe the interfacial disturbance, such as Fig. 5 (a), (b) and (c). As shown in Fig. 5, the microscope with the software of image processing was used to observe the interfacial phenomena. The water drop can be called as the spreading liquid and the TEG solution can be called as the supporting liquid during the process of instilling water drop on the surface of the TEG solution. Since the surface tension of water drop was greater than that of TEG solution, the contraction of water drop inward was occurred by the induced interfacial stress, as shown in Fig. 5 (a) and (b). The results showed that the rate of instantaneous contraction for the interfacial contour was faster than dissolution of water drop into TEG solution. And then the drop diverged gradually due to mutual dissolution between water and TEG, as shown in Fig. 5 (c). In addition, the longitudinal gradient of surface tension made the disturbed behavior around the peripheral region of water drop, which could be called as the interfacial instability and the instability lasted from 30s to 40s.

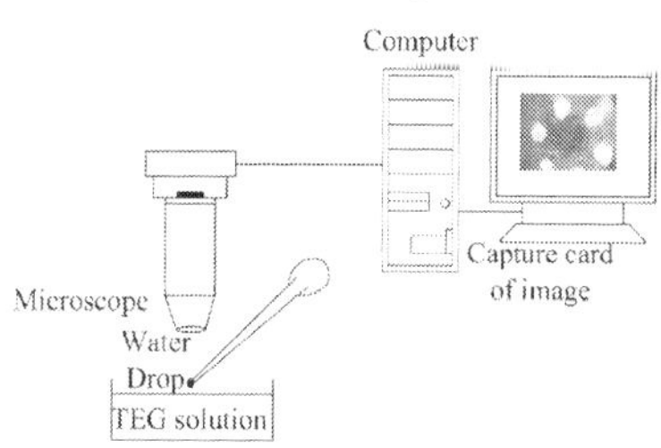

Fig. 4. The observed system of water drop instilled on the surface of TEG solutions

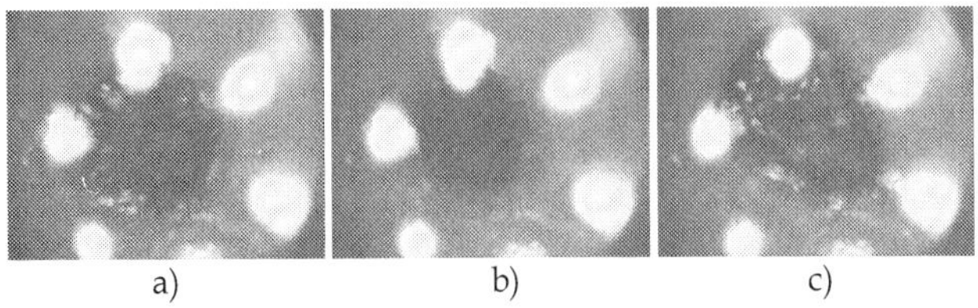

Fig. 5. Images of water drop instilled on surface of 95 wt. %. TEG solution (a) the start of water drop on the TEG solution, (b) the contraction of water drop, (c) divergence of water drop on TEG surface

The interfacial stress was calculated and the relationship between interfacial stress and concentration of TEG solution was drawn after the images of water drop instilled on the surface of TEG solutions were captured. The schematic diagram of water drop on the TEG surface is shown in Fig. 6, and the assumptions of homogeneous water film and plug flow is made for the contraction of water drop in this system. Mentioned above, the interfacial stress can be deduced as follows:

$$dF = dma \tag{5}$$

$$dF = d\rho V_{drop} a \tag{6}$$

where the symbol F is the interfacial stress, m is the mass of liquid drop, V is the volume of liquid drop, ρ is the density of liquid drop, and a is the acceleration of leading edge of liquid drop. Assuming the acceleration maintained a constant at that instant.

$$dF = a \cdot d(\rho V_{drop}) \tag{7}$$

$$\because V_{drop} = 2\pi r \times \omega \tag{8}$$

ω =the thickness of liquid film

Eq. 7 is replaced by eq. 8, and the interfacial stress can be obtained from eq. 9.

$$F = a\rho\omega 2\pi \times \int_{r1}^{r2} r dr \tag{9}$$

On the basis of eq. 9, the interfacial stress resulted from the gradient of surface tension can be calculated, and the relationship between interfacial stress and concentration of TEG solution is shown in Fig. 7. As known, the surface tension of TEG solution is decreased with the increased concentration of TEG solution. The surface tension difference between water and TEG solution should be greater for the higher TEG concentration, which leads to the stronger interfacial stress. Fig. 7 also shows that the interfacial stress increases dramatically for the concentration higher than 93 wt. %. TEG solution. Therefore, the absorption performance of water vapor absorbed by TEG solution could be increased more significant as TEG concentration greater than 93 wt. %, and the deduction is consistent with experimental results by Wu and Chung, 2006. Although the interfacial stress is insignificant for lower concentration, the interfacial instability resulted from longitudinal gradient of surface tension around the peripheral region of water drop is still being. The interfacial stress and Marangoni instability resulted from the enough difference of surface tension between spreading and supporting liquids was demonstrated, and the disturbed phenomena described above could also be helpful for explaining why the performance of mass transfer devices affected by the Marangoni effect.

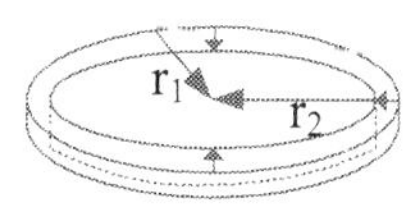

Fig. 6. The schematic diagram of water drop contracted inward on the surface of TEG solution

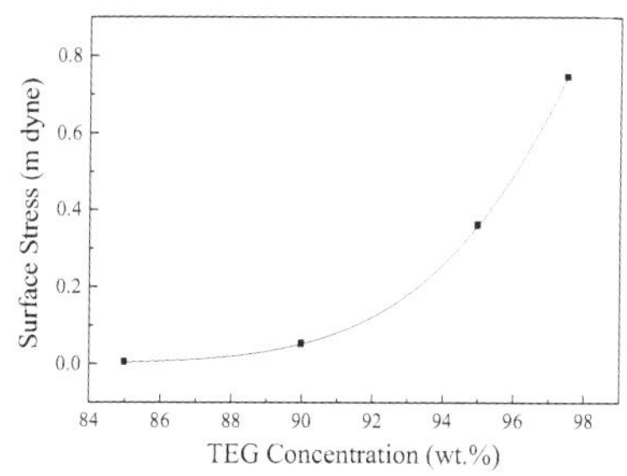

Fig. 7. Effects on TEG concentration on interfacial stress

2.3 Mass transfer devices

In addition to Benard cell resulted from the gradient of surface tension in the liquid layer, studies related to Marangoni effect were almost devoted to packed distillation column and liquid-liquid contacting system before 1990. For example, Bakker et al., 1967 defined the ratio *F* of the measured concentration to the calculated concentration to analyze effect of driving force on the ratio *F* for the liquid-liquid extraction system. The ratio *F* was increased with the increased driving force. Bakker et al., 1967 deduced that the discrepancy between measured and calculated concentration could be attributed to the interfacial movement. The components and the changed range of ratio *F* for the system are shown in Table 1. Besides, Moens & Bos, 1972 used pool column to investigate effect of surface tension on surface renewal. The relationship between stabilizing index, $M = -(d\gamma/dx)(x-x^*)$, and number of transfer units N_{og} was used to analyze mass transfer performance affected by the gradient of surface tension. Roll cells were observed only for stabilizing index *M* greater than 5 dn/cm, and N_{og} was not decreased beyond 0.15 dn/cm. Moens & Bos, 1972 concluded that the surface was renewed by the longitudinal gradient of surface tension. The entering liquid spread over the interface and moved towards outlet rapidly under the condition of positive *M*, which led interfacial velocity and mass transfer coefficient to be increased. However, the entering liquid did not spread over the interface under the condition of negative *M*. As a result of limiting spread of liquid, the insignificant surface renewal and the limited Marangoni effect could be derived for this condition. For absorption system, the surface additive could be added to absorption system to induce interfacial disturbance. For example, n-octanol was added to the aqueous solution of lithium bromide to induce interfacial disturbance by Kashiwagi et al., 1993 in the falling-film system. Both of adding n-octanol vapor and adding saturated n-octanol to the aqueous solution of lithium bromide were performed by Kashiwagi et al., 1993. The results showed that absorption of steam was enhanced by the induced Marangoni effect. On the other hand, sodium lauryl sulfate (SLS) and cetyltrimethyl ammonium bromide (CTMAB) were used as surfactant respectively by Vazquez et al., 1996 to test the performance of carbon dioxide absorbed by water. Experimental results showed that the performance of carbon dioxide absorbed by water could be enhanced by the convection-inducing liquid, 20-100 wt. % aqueous solution of methanol, ethanol, 2-propanol, and the mass transfer coefficient would be reduced with the increased surfactant concentration.

Similar to Patberg et al., 1983, Proctor et al., 1998 also discussed effects of surface tension on packed distillation column. The difference between them is that the experimental parameters, include different scale of packed distillation column and liquid flow rate were performed by Proctor et al., 1998. Effects of surface tension on mass transfer performance for the small-scale packed distillation column were consistent with previous studies. However, the extra surface was produced by spray and small drops for the larger scale column in the negative system, and then the mass transfer performance was better for the negative system at heavier loading. For absorption of carbon dioxide, liquid water, monoethanolamine (MEA), and metheldiethanolamine (MDEA) aqueous solution were often used as absorbent solutions to absorb carbon dioxide in the open studies. For example, aqueous solution of MDEA was used to absorb carbon dioxide by Zhang et al., 2003 to discuss the discrepancy of absorption rate between experimental data and kinetics model, and hence they thought that the enhanced absorption rate could be attributed to Marangoni effect resulted from the elevated partial pressure of carbon dioxide. In addition, some studies related to Marangoni effect in the recent years can also be found from absorption of CO_2 and NH_3 absorbed by NaOH and water in the falling film and bubble absorption systems, as shown in Table 1.

Mentioned above, the gradient of surface tension could be formed by mass transfer in the interface, and then the Marangoni instability could be induced by the gradient in the mass transfer device with continuous liquid phase. Therefore, the packed-bed absorber with continuous liquid phase was tested by Wu et al., 2001 to discuss effects of Marangoni convection on mass transfer performance of water vapor absorbed by TEG solution. Since the surface tension of absorbent solution was depend on concentration and temperature, the stabilizing index (*M*-index) was established with respect to the differentiation of concentration. On the basis of dimensional analysis and *M*-index, the empirical mass transfer correlation with M-index was established in eq. 10.

$$\frac{k_{x}a \cdot M \cdot d_p^{\;2}}{\mu_L} = 1\times10^{-7}\cdot\left(\frac{\rho_L \cdot L \cdot d_p}{\mu_L}\right)^{0.5}\left(\frac{\mu_L}{D_L \cdot \rho_L}\right)^{1/3}\left(\frac{L}{G}\right)^{1.55}(M-index)^{0.25}. \tag{10}$$

where k_{xa} is the mass transfer coefficient in the liquid phase, M is the molecular weight of the transferred matter, d_p is the diameter of the transferred matter, μ_L is the viscosity of liquid fluid, the term in the first parentheses is the Reynolds number, the term in the second parentheses is the Schmidt number, L is the liquid flow rate, G is the gas flow rate, and *M-index* is the Marangoni-index. The difference between experimental mass transfer coefficients and predicted by eq. 10 is about 7%, which is better than that predicted by the empirical mass transfer correlation without *M*-index. The results mean that mass transfer phenomena and performance should be affected by Marangoni effect under the process of water vapor absorbed by TEG solution.

Method of Mass Transfer	Authors	Response Value	Changed Range	Components
Extraction	Bakker et al., 1967	F	1~3	acetic acid from water to isobutyl alcohol
Distillation (Packed Column)	Proctor et al., 1998	H_{og}(mm) (height of transfer units)	20-100	n-propanol/water
Distillation (Pool Column)	Moens & Bos, 1972	N_{og} (number of transfer units)	2.55-3.65	n-heptane/methyl-cyclohexane overall gas
Absorption (Falling Film)	Kashiwagi et al., 1993	Na (kg/m²s)	0.9-1.9	steam absorption by 58.3 %wt. LiBr solution
Absorption (Falling Film)	Zanfir et al., 2005	conversion, %	40-100	CO_2 absorbed by NaOH
Absorption (Pool absorber)	Vazquez et al., 1996	k_l (m/s)	6.6-7.8×10^{-5}	CO_2 absorbed by water
Absorption (Packed Absorber)	Zhang et al., 2003	N (kmol/m^2s)	1.229-26.699×10^{-6}	CO_2 absorbed by MDEA
Absorption (Bubble)	Kim et al., 2006	m (g/s) absorption rate	0.3-2.8	NH_3 absorbed by water

Table 1. Response value and the changed range for different mass transfer devices

3. Artificial and spontaneous marangoni convections

Researches about Marangoni effect can be categorized into experimental operation and numerical simulation. For the experimental operation, some studies compared experimental data to demonstrate that mass transfer performance affected by Marangoni effect, and the others observed or analyzed surface velocity and interfacial properties resulted from the gradient of surface tension to show effect of interfacial disturbance on mass transfer. Researches about Marangoni effect discussed by numerical simulation can also be categorized as follows. One is to simulate Marangoni effect resulted from the gradient of surface tension in the mass transfer system, and show that the performance is affected by Marangoni effect; the other is discuss the roll cells resulted from Marangoni instability and to analyze the induced interfacial disturbance by dimensionless numbers based on mass transfer principle and linear stability analysis. According to the collected references, the studies about interfacial disturbance discussed by numerical simulation are beyond 70 percent. Half of the other studies are to investigate effect of Marangoni effect on mass and heat transfer performance by practical experimental data; and the rest is to analyze and discuss Marangoni convection by the observed technology. The difference of study number shows that it is not easy to design a pilot engineering device accompanied with surface tension effect. The designer not only need to have the ability to design mass or heat transport device, but also need to have the ability to make the Marangoni effect occurring in the mass transfer device. Furthermore, studies about transfer performance affected by Marangoni effect in mass transfer devices and image observation during the process of mass transfer were not increased in recent years, which causes it is difficult to find the relevant paper for Marangoni effect occurring in the mass transfer devices. However, heat and mass transport engineering and drying of chip and semiconductor affected by Marangoni effect have been demonstrated in the open literatures. This is why the subject of effects of surface tension on mass transfer devices was selected to discuss in this chapter; however, it still need more hands to fill the gap in the literature. The purpose of this chapter is to discuss effect of Marangoni effect on mass transfer devices, and hence most of the descriptions are focused on the mass transfer enhancement affected by Marangoni effect. Some results obtained from numerical simulation are used to assist the descriptions about interfacial behaviors.

Mass Transfer Device	Method	Times of Mass Transfer Enhancement	Authors(year)
liquid-liquid system	100 ml water + 0.002-0.05 g ionic and non-ionic surfactants	1-7 times (compared with the absence of surfactant)	Agble & Mendes, 2000
falling film absorber	saturated n-octanol vapor was supplied to the absorber	increase 20% (mass flux, kg/m^{2}s)	Kashiwagi et al., 1993
plane absorption system (two concentric absorption cell)	methanol, ethanol, 1-propanol, 2-propanol or acetone (20-100% wt aqueous solution) was deposited at the surface of water liquid	3-4 times (compared with the absence of surfactant)	Vazquez et al., 1996
plane absorption system (two concentric absorption cell)	methanol, ethanol or n-propanol, (0-100 % wt.) was deposited at the surface of water liquid	3-4 times (compared with the absence of surfactant)	Lu et al., 1996
plane absorption system (two concentric absorption cell)	2-ethyl-1-hexanol was deposited at the surface of water liquid	1-4 times (compared with the absence of surfactant)	Kim et al., 1996
plane absorption system (two concentric absorption cell)	ethanol was added to the surface of liquid water	2-5 times (compared with the absence of surfactant)	Lu et al., 1997
plane absorption system (two concentric absorption cell)	ethanol was added to the surface of TEG solution and ethanol vapor was added to absorption system	increase 15-60% (removal efficiency, %)	Yang et al., 2008
plane absorption system	ethanol was added to the surface of absorbent solution (triethylene glycol)	increase 5-17% (mass transfer coefficient, mol/m^{2}min)	Wu et al., 2008

Table 2. Mass transfer devices and the method to result in interfacial disturbance for the artificial Marangoni convection

Mass Transfer Device	Properties	Purpose	Authors
Wetted wall column	solutal Marangoni effect experimental data	to discuss the intensity of interfacial disturbance for solutes transferring	Maroudas & Awistowski, 1964
Wetted wall column	solutal Marangoni effect experimental data	to show that absorption of carbon dioxide into monoethanolamine affected by interfacial turbulence	Brian et al., 1967
Liquid-liquid extraction	solutal Marangoni effect experimental data	to analyze the relationship between mass transfer data and driving force across liquid-liquid interfaces	Bakker et al., 1967
horizontal liquid layer	solutal Marangoni effect numerical simulation	to develop the transient models of transfer processes based on the transient age distributions	Chung et al., 1971
Packed distillation column	solutal Marangoni effect experimental data	to estimate influence of driving force on the efficiency of distillation column	Moens, 1972
Liquid-jet and wetted wall column	solutal Marangoni effect experimental data	to discuss mass transfer enhancement affected by interfacial disturbance for desorbing surface-active solute	Imaishi et al., 1982
Packed distillation column	solutal Marangoni effect experimental data	to discuss effect of positive and negative driving force on different packings	Patberg et al. (1983)
Pilot wetted wall	solutal Marangoni effect numerical simulation	to discuss mass transfer enhancement by the model of creeping film	Dijkstra et al., 1990
Liquid layer with finite deep	solutal Marangoni effect numerical simulation	to study Marangoni instability for chemisorptions	Warmuzinski & Tanczyk, 1991
Packed rectification column	solutal Marangoni effect experimental data	to discuss effect of positive and negative systems on rectification efficiency	Martin & Perez, 1994

Horizontal liquid layer	thermal Marangoni effect numerical simulation	to analyze effect of viscosity and deformable free surface on stationary thermocapillary convection	Kalitzova et al., 1996
Horizontal liquid layer	thermal Marangoni effect experimental data numerical simulation	to study effect of Marangoni number on steady and oscillatory thermocapillary flow	Kamotani et al., 1996
Packed distillation column	solutal Marangoni effect experimental data	to discuss effect positive and negative driving force on mass transfer performance	Proctor et al., 1998
Quiescent gas-Liquid contactor and gas-liquid channel	solutal Marangoni effect experimental data	to show the mass-transfer performance enhanced by interfacial turbulence and to observe interfacial convection by schlieren photography	Sun et al., 2002

Table 3. Some studies related to spontaneous Marangoni convections

Generally speaking, the interfacial disturbance can be divided into artificial and spontaneous Marangoni convection. In order to enhance mass transport, the interfacial disturbance resulted from the added surfactants is called as artificial Marangoni convection. In contrast with artificial Marangoni convection, the gradient of interfacial tension resulted from the process of mass transfer is called as spontaneous Marangoni convection. Some studies related to artificial and spontaneous Marangoni convection are listed in Table 2 and 3.

3.1 Artificial Marangoni convection

By means of the difference of surface tension between spreading and supporting liquids, the artificial Marangoni convection can be induced by the added surfactant or solution on the surface of supporting liquid. In addition, the Marangoni convection could be produced by injecting a few of volatile solute into solvent or adding surfactant vapor to mass transfer system in the process of gas-liquid contacting, and then the mass transfer performance could be enhanced. Except for numerical simulation, the searched papers discussed about mass transfer enhancement by artificial Marangoni convection are shown in Table 2. The artificial Marangoni convections could be occurred in the device with continuous liquid phase, such as falling film absorber, plate absorption system, and liquid-liquid contacting system. For example, the concept of larger difference of surface tension between vapor and absorbent solution can be utilized to produce imbalanced surface tension on liquid surface of falling film system. Once the vapor or the droplet is condensed on liquid surface, the Marangoni convection or wavy surface can be resulted from the imbalanced surface tension. The absorption performance could be enhanced by the artificial Marangoni convection, such as the saturated n-octanol vapor was added to the falling film absorber by Kashiwagi et al., 1993, and the ethanol vapor was added to the absorption system by Yang et al., 2008. Vazquez et al., 1996, Lu et al., 1997, and Kim et al., 1996 used capillary tube to deposit liquid drops of methanol, ethanol, and n-propanol respectively on the surface of liquid water to enhance mass transfer performance for two concentric absorption system, and the mass transfer enhancement was also shown in Table 2. In addition, aqueous solutions of ionic and non-ionic surfactants were added to the liquid-liquid system respectively to discuss mass transfer enhancement by Agble & Mendes, 2000.

In addition to the interfacial disturbance induced by vapor condensation and liquid drop, the liquid ethanol was used to produce interfacial disturbance in the plate absorption system by authors of this chapter based on the higher volatility and the lower surface tension for liquid ethanol with the properties of high volatility and low surface tension was used to produce interfacial disturbance in the plate absorption system by authors of this chapter. As shown in Fig. 8, the working solutions used to absorb water vapor in the absorption system included triethylene glycol (TEG) and diethylene glycol (DEG) solutions. Pure ethanol was added to the absorbent solution up to 5 wt. % for each experimental run. In order to make humid to be carried by air, pure water was poured into the flask A. Air humidity can be controlled by air flow rate and numbers of flask. After the humidity attained equilibrium in the system, TEG solution with the added ethanol was injected into the absorption cell by liquid valve. Humidity and temperature were measured in the entrance and exit of the absorption cell, and then the mass transfer coefficient were calculated to discuss mass transfer coefficient changed with time and mass transfer performance affected by artificial Marangoni effect. The solution was regenerated at 80°C after experimental operation. Fig. 9 and 10 shows the scheme of mass transfer coefficient changed with time for water vapor absorbed by TEG and DEG solutions respectively. Compared Fig. 9 with Fig. 10, the mass transfer coefficient of water vapor absorbed by DEG solution is slightly greater than that by TEG solution. The mass transfer coefficient for addition of ethanol is greater than that without addition of ethanol, and the mass transfer coefficient is leveled off after 240s. Since the more ethanol evaporates from glycol solution to air phase at the beginning of absorption process, the induced interfacial disturbance should be stronger for the beginning. As also shown in Fig. 9 and 10, the mass transfer enhancement is significant before 150 sec, and then the mass transfer coefficients with and without addition of ethanol are closer. Therefore, the mass transfer performance enhanced by the induced interfacial disturbance can be demonstrated by comparing mass transfer coefficient with and without addition of ethanol in this study.

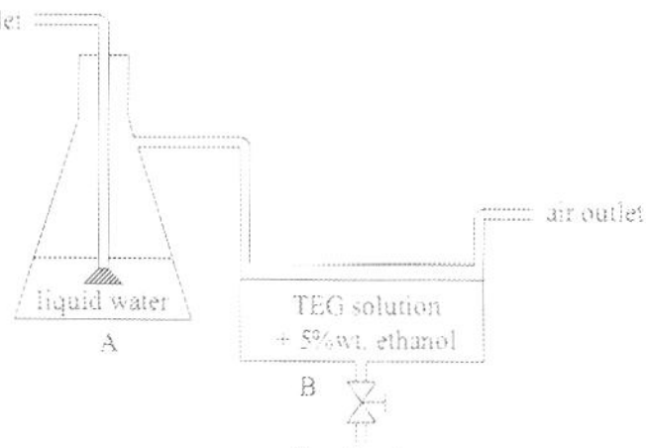

Fig. 8. Plane absorption system with addition of ethanol

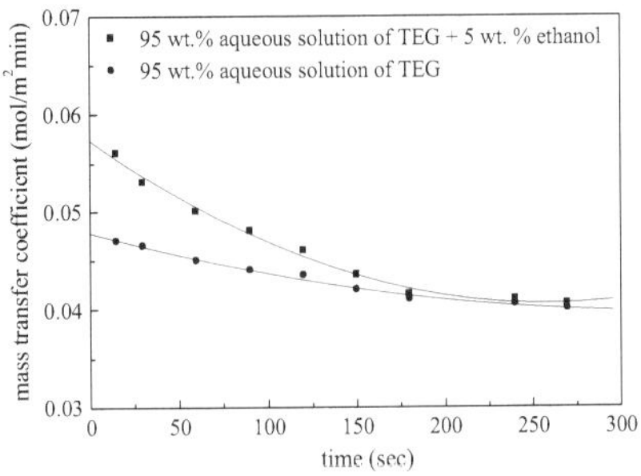

Fig. 9. Mass transfer coefficient for water vapor absorbed by TEG solution in plane absorption system

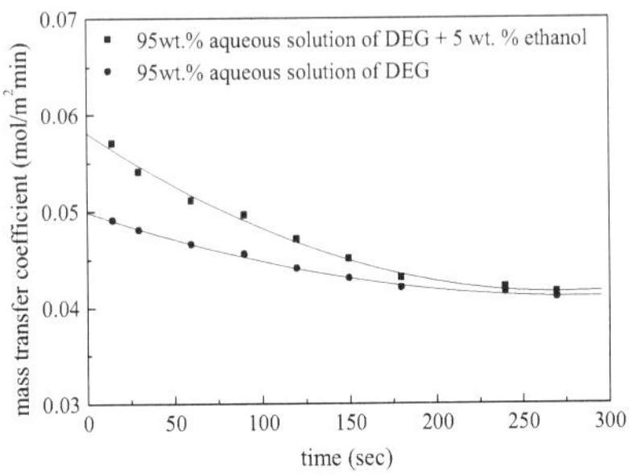

Fig. 10. Mass transfer coefficient for water vapor absorbed by DEG solution in plane absorption system

3.2 Spontaneous Marangoni convection

Table 2 and Table 3 show that the artificial Marangoni convection can be applied into falling-film absorption system, plane-absorption system, and liquid-liquid contacting system; however, the spontaneous Marangoni convection was occurred in the system with fluid circulation or chemical reaction, such as packed distillation column or chemisorptions. Since the difference of surface tension between feed liquid and reflux is larger enough to result in the gradient of surface tension in distillation column, the interface would be disturbed, renewed or accelerated by the gradient. For the spontaneous Marangoni convection, the interfacial instability for falling-film absorption system, packed distillation column, and the system of horizontal liquid layer heated from bottom were often analyzed by mass transfer fundamental and linear stability analysis. Based on the collected references, just some studies discussed effects of spontaneous Marangoni convection on mass transfer performance by practical experimental data; the most studies analyzed and discussed interfacial instability by numerical simulation. Table 3 lists some studies to elucidate effect of Marangoni effect on mass transfer devices. For the studies of spontaneous Marangoni convection performed by experimental operation, the mass transfer data affected by spontaneous Marangoni convection could be compared with that without spontaneous Marangoni convection or the theoretical data, and the results showed that the mass transfer data affected by spontaneous Marangoni convection were greater than that without spontaneous Marangoni convection or the theoretical data, such as Bakker et al., 1967, Moens, 1972, Patberg et al., 1983, Martin & Perez, 1994, Proctor et al., 1998, and Sun et al., 2002 in Table 3. In addition, most studies attributed the discrepancy between experimental data and predicted results to that the Marangoni effect was not considered into traditional mass transfer theory. For the studies with numerical simulation, some studies discussed effects of Marangoni number and other dimensionless number on interfacial instability for the gradient of surface tension resulted from temperature, such as Kalitova et al., 1996 and Kamotani et al., 1996. Some studies devoted to analyze solutal Marangoni instability resulted from chemisorptions, such as absorption of carbon dioxide by MEA solution. The relevant models were set and solved by numerical method to analyze effects of surface tension on mass transfer, such as Dijkstra et al., 1990 and Warmuzinski & Tanczyk, 1991. The amount of studies related to Marangoni effect is much greater for discussing by numerical simulation; however, establishment of experimental system and confirmation of experimental data are the way to promote engineering and science technology. Therefore, such field still needs more scholars to make effort in future.

4. Marangoni effect in the mass transfer devices and mass transfer performance affected by Marangoni effect

Table 1 shows mass transfer devices and their performance affected by Marangoni effect. As shown in Table 1 and Table 2, Marangoni effect was often discussed for the devices of packed-distillation column, falling-film absorber, two-concentric absorption system, and liquid-liquid contacting system. The dependent variables H_{og} and N_{og} were usually used to discuss mass transfer performance for packed-distillation column, the dependent variables mass transfer coefficient (k_l or k_g) and mass transfer flux (N) were usually used to discuss mass transfer enhancement for absorption system, and the factor F was usually used to discuss the difference of transfer performances with and without Marangoni effect. Since effects of surface tension on performances of mass transfer devices were emphasized in this chapter, introduction of mass transfer devices and effects of surface tension on mass transfer performance are elucidated for packed-distillation column, two-concentric absorption cell, falling film absorber, and liquid-liquid contact system respectively.

4.1 Packed-distillation column

A typical packed distillation column is shown in Fig. 11. The purpose of distillation column is to separate miscible liquids by boiling points of mixture components. In general, a distillation device consists of a distillation column, a condenser, a reboiler, reflux tube, and a heat source. In order to provide contacting area between liquid and vapor phases, the packed-bed or the tray column can be selected. The difference between the packed-bed column and the tray column is that the surface area for packed-bed column is continuous and the surface area for the tray column is discrete. Since the Marangoni effect could be induced from the continuous liquid phase, the packed-bed column was discussed in this chapter. Liquid flows down the packed bed, and vapor upflows to contact with liquid phase in the countercurrent. The vapor was cooled and condensed in the condenser, and the liquid was reboiled in the reboiler. Once the contacting time is provided enough for gas and liquid pgases, the matter with the property of volatile or low boiling point can be obtained in the top of condenser, and the heavier matter can be obtained in the bottom of condenser.

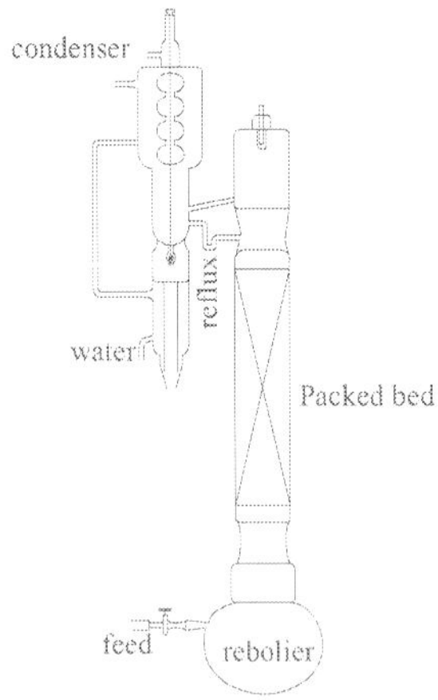

Fig. 11. Schematic diagram of packed-bed distillation column

For the gradient of surface tension, Marangoni effect in the packed-bed distillation column can be divided into positive and negative system. For example, a component of low surface tension transferred from a liquid phase to a gas phase may increase surface tension of the transferred spot on the surface of liquid layer, and then the liquid surrounding the spot is drawn to the spot. The flow phenomenon driven by this kind of surface tension gradient may spread over the packing well in packed-bed column and increase mass transfer performance. Therefore, the system making more packing surface wetted by liquid is called as positive system for the packed-bed distillation column. In the opposite case, if a component of high surface tension transfers from a liquid phase to a gas phase, surface tension of the transferred spot will be decreased. The induced stress is directed from the spot to the surrounding liquid, which leads the wetted surface to be contrasted. Since the mass transfer performance would be decreased with the decreased contact area between gas and liquid phases, such system is called as negative system. In addition, Moens & Bos, 1972 pointed out that the surface renewal effects could be caused by the longitudinal gradient of surface tension for the pool distillation column, that is, evaporation of the component of low surface tension would accompany with the increased surface tension in the direction of liquid flow. Since the liquid flow would be accelerated along the interface and the mass transfer performance would be enhanced by the surface renewal, such a system for promoting surface renewal could be called as a positive system for the pool distillation column. In contrast with the positive system, the surface tension would be decreased in the direction of liquid flow by transferring the component of high surface tension from a liquid phase to a gas phase. The flow velocity would be retarded, and the surface renewal of pool column would be decreased under this condition. Since the mass transfer performance was increased insignificantly with the increased driving force, the system with bad surface renewal was called as negative system for the pool distillation column. As shown in Fig. 12, Moens & Bos, 1972 and Patberg et al., 1983 demonstrated that the mass transfer

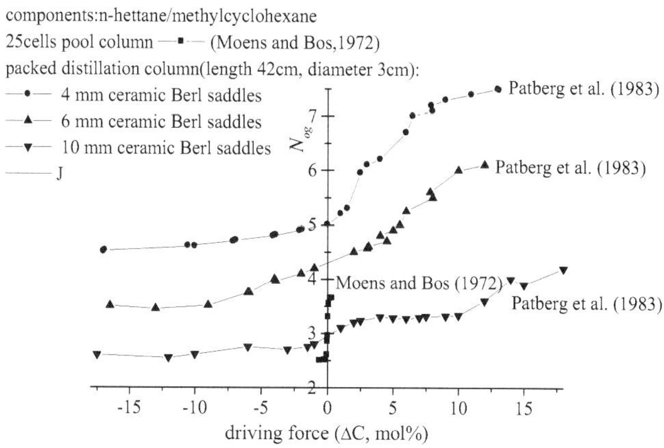

Fig. 12. Effect of driving force on N_{og} for n-heptane/methylcyclohexane

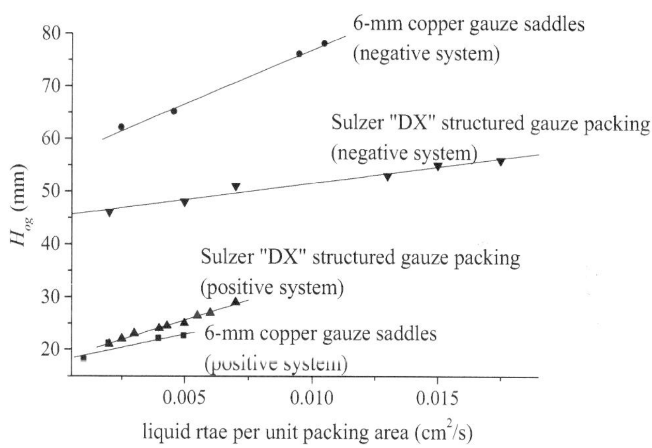

Fig. 13. Effect of surface tension gradient on HTU Values for different packings. (data source: Proctor et al., 1998)

performances were increased significantly for the positive system, especially for the smaller packing. Since the size of pool distillation column established by Moens & Bos, 1972 is larger than that of Patberg et al., 1983, the mass transfer performance seems to be increased more significantly across zero driving force. Besides, the relationship between height of transfer unit and liquid rate was established by Proctor et al., 1998 for packed distillation column, as shown in Fig. 13. The results also demonstrated that the mass transfer performance of positive system was better than that of negative system. Since the specific surface area of 6-mm copper gauze saddles is larger than that of Sulzer "DX" structured gauze packing, the mass transfer performance for that packed with 6-mm copper gauze saddles is slightly higher than that packed with Sulzer "DX" structured gauze packing under the positive condition. However, the mobility of n-propanol on the surface of Sulzer "DX" structured gauze packing is better than that of 6-mm copper gauze saddles so much that the detrimental effects under the condition of negative system may be overcome partly to lead the better mass transfer performance for Sulzer "DX" structured gauze packing.

4.2 Two-concentric absorption cell

The schematic diagram of two-concentric absorption cell is shown in Fig. 14. The absorbent liquid is injected in the bottom of this system. The liquid flows upwardly along the center of the inner cylinder, and then flows on the plane surface of the inner cylinder. In order to induce interfacial disturbance, the liquid of low surface tension could be fed on the surface of absorbent liquid by the capillary tube. If the interfacial disturbance was considered to be induced by surfactant vapor, the saturated vapor is the better choice to inject from inlet of surfactant vapor. The distance between the fed liquid and the surface of absorbent liquid is as close as possible to avoid Marangoni effect interfered by gravity. Inlet and outlet of the treated air are usually mounted in the opposite sides of capillary tube. The gas can be single component or mutil-components. In order to ensure effective contact for gas and liquid phases, some fixed blades are suggested to mount in the absorption system.

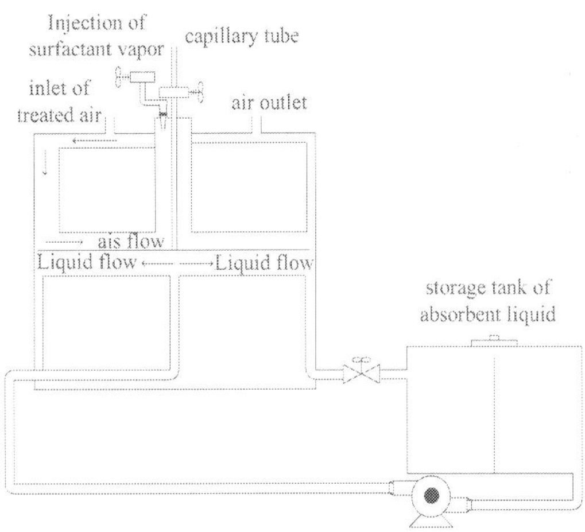

Fig. 14. Schematic diagram of two-concentric absorption system

As known, the gradient of surface tension could be arisen from transferring a component across interface. Thus the interfacial disturbance resulted from the gradient could be occurred in some separation processes, such as distillation, absorption, and extraction. However, it is not easy to discuss effects of interfacial behaviors on mass transfer process because of other interferences, such as buoyancy, gravity, viscosity, and etc. In order to control the intensity of the induced Marangoni convection, the interfacial convection was induced by adding surfactant liquid of low surface tension at the interface. As shown in Fig. 14, if the difference of surface tension between the absorbent solution (supporting liquid) and the fed liquid (spreading liquid) is large enough, the tangential stress at the interface will be resulted in. Therefore, the Marangoni convection will be produced by the stress in the region between liquid surface and underlying liquid. In general, the spreading liquid with low surface tension was usually added on the surface of supporting liquid to make Marangoni convection artificially, and hence the absorption performance was enhanced with the increased effective area between gas and liquid phases or with the promoted surface renewal. For example, the mass transfer performance of water vapor absorbed by solution of lithium bromide is increased with the increased concentrations of 2-ethyl-1-hexanol in the range from 10 to 100 ppm, as shown in Fig. 15. Fig. 15 also shows that the efficiency of water vapor removed by the solution of lithium chloride is better for the addition of ethanol into gas stream than into working solution. In addition, to discuss effect of surface additives on interfacial disturbance quantitatively, some surface additives were added to absorbent solution, such as aqueous solutions of anionic sodium lauryl sulfate (SLS), anionic sodium dodecyl sulfate (SDS), aqueous solution of cationic cetyltrimethylammonium bromide (CTMAB), and aqueous solution of dodecyl trimethyl ammonium chloride (DTMAC). Lu et al., 1997 showed that mass transfer performance of carbon dioxide absorbed by water could be enhanced by interfacial disturbance resulted from addition of ethanol; however, the mass transfer coefficient was decreased with the increased surfactant concentration, as shown in Fig. 16. Therefore, effects of anionic and cationic surfactants on mass transfer performance of absorption system were demonstrated by Lu et al., 1997 and Vazquez et al., 2000.

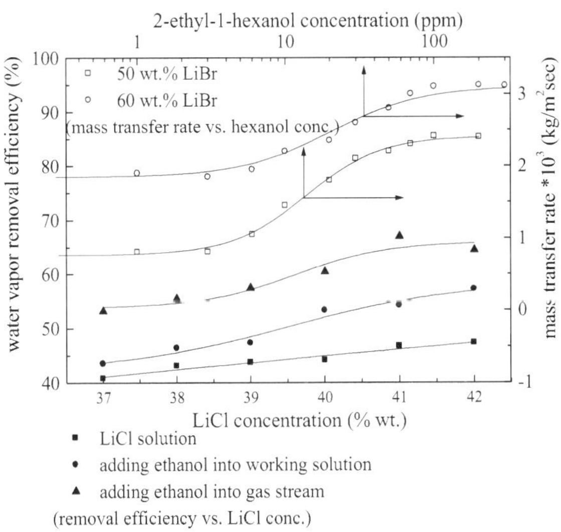

Fig. 15. Effects of the induced interfacial disturbance on mass transfer performance. (data source: Kim et al., 1996 and Yang et al., 2008)

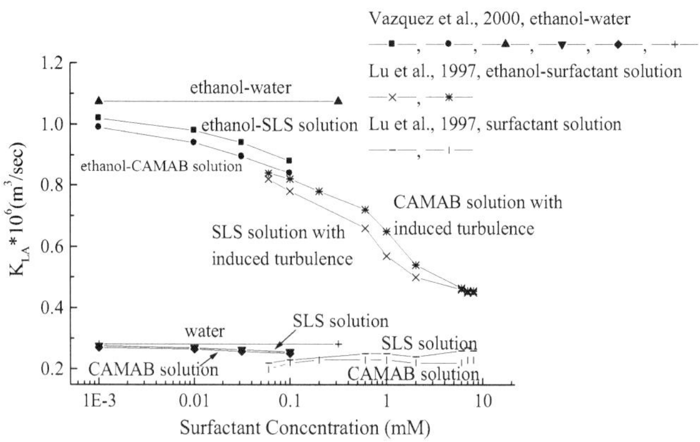

Fig. 16. Effects of the induced Marangoni convection and surfactant concentration on mass transfer performance. (data source: Lu et al., 1997 and Vazquez et al., 2000)

4.3 Falling-film absorber

The common falling-film absorption systems are shown in Fig. 17 and Fig. 18. The advantage of the device in Fig. 17 is that the contacting time or distance between liquid and gas can be adjusted easily; however, the large-scale contacting area between gas and liquid phases, suh as the device in Fig. 18, is suitable for observing interfacial behaviors during absorption process. The falling-film absorption system shown in Fig. 17 is mainly consisted of two-concentric annulus pipes, cap, and some flow controller. The absorbent liquid is introduced into the bottom of the falling-film system. The absorbent liquid flows up the inside of the inner annulus pipe, and then is distributed by a cap to form liquid film. Gas inlet and outlet can be designed in the top or bottom of the system. If the gas inlet is in the top of the system, the gas and liquid will flow in the cocurrent. Oppositely, if the gas inlet is in the bottom of the system, the gas and liquid will flow in the countercurrent. While the liquid film flows down the outside of the inner annulus pipe, pollutant in the gas phase is absorbed by the liquid film between cap and gas outlet. The thickness of liquid film can be determined by the width between cap and inner annulus pipe and the liquid flow rate. For the falling-film absorption system shown in Fig. 18, the absorbent liquid may be introduced into the system by the slit-shape distributor or liquid nozzle so much that the thickness of liquid film can be determined by the liquid distributor and liquid flow rate. Similar to the device in Fig. 17, the depositions of gas inlet and outlet determine gas and liquid flowing in the cocurrent or countercurrent.

Similar to the packed-bed distillation column, the gradient of surface tension can be formed by transferring a solute from a liquid phase to a gas phase, and promoting surface renewal by Marangoni convection. If the difference of surface tension between solute and solvent is large enough, the gradient of surface tension will be formed around the spot where the solute evaporates or desorbs. The interfacial disturbance is induced by the gradient of surface tension at the interface so much that the desorbed solute is called as surface-active solute. For example, the liquid-phase mass transfer coefficients with and without interfacial disturbance in the falling film absorption system were compared by Imaishi et al., 1982, and the results showed that desorption performance would be enhanced with the increased concentration of solute, as shown in Fig. 19. Kashiwagi et al., 1993 also demonstrated that the mass transfer performance would be enhanced by addition of vapor of low surface tension in the falling film system, and the mass transfer enhanced by the surfactant concentration was shown in Fig. 19. The difference of activation of Marangoni convection between Kashiwagi et al., 1993 and Imaishi et al., 1982 is that Kashiwagi et al., 1993 added vapor of low surface tension to induce Marangoni convection artificially, and Imaishi et al., 1982 used the surface-active solute desorbed from absorbent liquid to result in Marangoni convection spontaneously. Whatever the activated method of Marangoni convection was used, the mass transfer enhancements were demonstrated by experimental results how to activate the Marangoni convection. Besides, the solutal Marangoni effect can also be resulted from chemisorptions, such as carbon dioxide absorbed by aqueous solution of monoethanolamine Therefore, mass transfer enhancement for carbon dioxide absorbed by MEA solution was demonstrated by Brian et al., 1967, and the mass transfer enhancement affected by MEA concentration was shown in Fig. 19. Since the operating conditions were different for these three data point of Brian et al., 1967, the trend differed with other studies was not focused here.

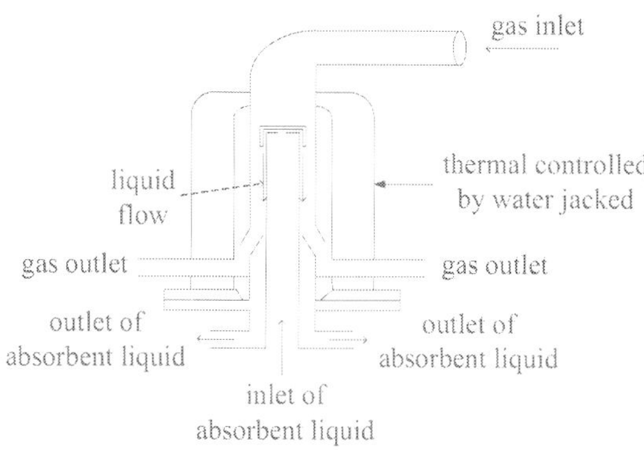

Fig. 17. Schematic diagram of falling-film absorption system consisted of two-concentric pipes. (referred from Imaishi et al., 1982)

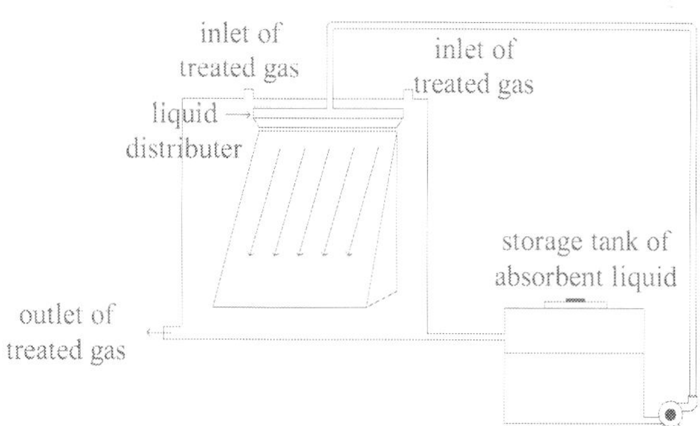

Fig. 18. Schematic diagram of falling-film absorption system consisted of the inclined plane

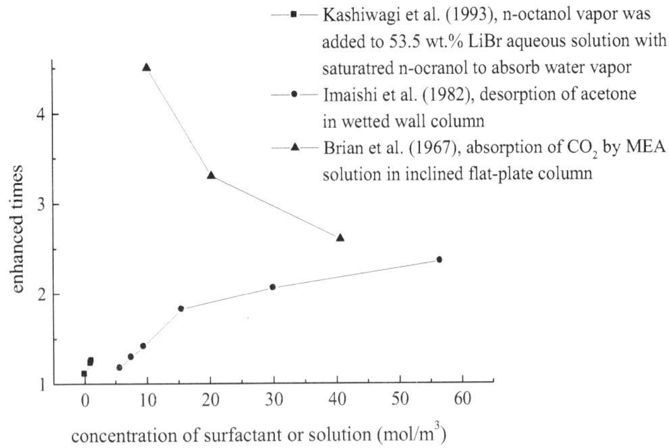

Fig. 19. Effect of concentration of surfactant or solution on mass transfer enhancement

4.4 Liquid-liquid contact system

In order to investigate effect of Marangoni instability on mass transfer performance or interfacial phenomena induced by the gradient of surface tension for liquid-liquid contact system, some kinds of liquid-liquid contact devices were designed. Based on the direction of fluid flow, the liquid-liquid contact system can be generally categorized into vertical and horizontal liquid-liquid mass transfer devices. The schematic diagram of vertical liquid-liquid contact device is shown in Fig. 20. The heavier liquid is introduced at A, and comes into contact with the lighter liquid in the cocurrent. Then the heavier liquid flows over the rod C, and leaves at the valve D. The lighter liquid is introduced at B and leaves the device at valve E. Described above, transport of solute between phases is occurred in the section between F and G. The horizontal liquid-liquid contact devices can also be divided into dynamic and static liquid-liquid contact devices, as shown in Fig. 21 and 22 respectively. As shown in Fig. 21, the heavier and the lighter liquids are introduced at the inlet A and B, and the divider C is used to smooth out the flow. The length between D and E decides the exposure time for transferring solute between phases. The heavier liquid leaves at valve F while the lighter liquid leaves at valve G after the exposure time. In order to observe periodic Marangoni instability in liquid-liquid mass transfer device, transport of a surfactant, such as cetyltrimethylammonium bromide, from an aqueous to an organic phase in the static liquid-liquid contact system, as shown in Fig. 22, was performed by Lavabre et al., 2005. In general, the organic solution was placed in the bottom of the beaker, and then the aqueous solution was introduced on the top of the organic solution. The advantage of this system is to make the visualization of interfacial disturbance easier.

For the system of transferring a solute from one liquid phase to another liquid phase, development of interfacial instability depends on the differences of surface tension between liquid A and surfactant and that between surfactant and liquid B. Fig. 23 shows that a solute diffuses from liquid A to liquid B. Assuming that the surface tension of solute is less than those of liquid A and liquid B, and then the interfacial tension will be decreased with the increased concentration of solute. If a solute is transferred from bulk liquid A to the point S, a small disturbance near point S will be resulted from the gradient of interfacial tension. The underlying liquid is brought with the small disturbance to the interface, and the interfacial instability is promoted. Furthermore, the solute diffuses from interface to bulk liquid B, and then the interfacial tension at point S increases with the decreased concentration of solute. Since the gradient of interfacial tension resulted from solute transferring from interface to liquid B is opposite to that transferring from liquid A to interface, the disturbance induced by transferring solute from interface to liquid B will be opposite to that transferring from liquid A to interface. Therefore, the interfacial instability will depend on the differences of surface tension between surfactant and liquid A and that between surfactant and liquid B. Mentioned above, transferring a solute from one liquid phase to another liquid phase might cause the gradient of concentration in the interface, and then the gradient of surface tension may be formed by the gradient of concentration. Once the difference of surface tension between surfactant and liquid A is differ from that between surfactant and liquid B significantly, the interfacial disturbance will be occurred by the solutal Marangoni effect. For example, Maroudas & Sawistowskis, 1964 used the F-factor to discuss transfer of solute across liquid/liquid system in the horizontal, and the F-factor was used to show the intensity of interfacial disturbance. Since the F-factor is greater than unity in the dynamic liquid/liquid system especially for the initial 500s, the spontaneous interfacial disturbance was verified for transferring phenol between carbon tetrachloride and water, as shown in Fig. 24. Besides, the mass transfer flux was used by Agble & Mendes, 2000 to discuss effect of surfactants on interfacial mass transfer for liquid-liquid contact system. As a result of the presence of ionic surfactant, such as dodecyl trimethyl ammonium bromide, the molar flux was enhanced by the induced Marangoni convection especially for the initial 2000s, as shown in Fig. 24. In contrast with nonionic surfactant, such as ATLAS G 1300, and pure water, the molar fluxes were lower than that addition of an ionic surfactant, and the molar fluxes seemed to be equal for the nonionic surfactant and pure water. The result means that Marangoni convection could be dampened by the addition of a nonionic surfactant.

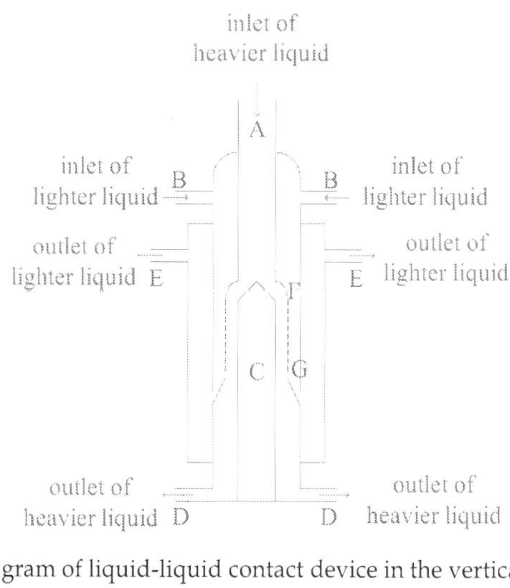

Fig. 20. Schematic diagram of liquid-liquid contact device in the vertical. (referred from Bakker et al., 1967)

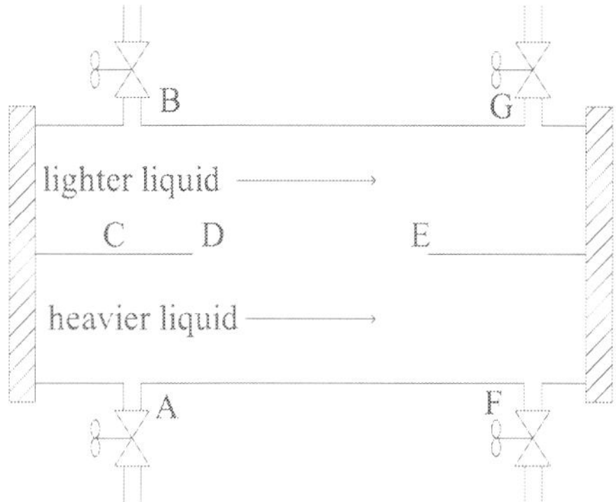

Fig. 21. Schematic diagram of dynamic liquid-liquid contact device in the horizontal. (referred from Maroudas & Sawistowski, 1964)

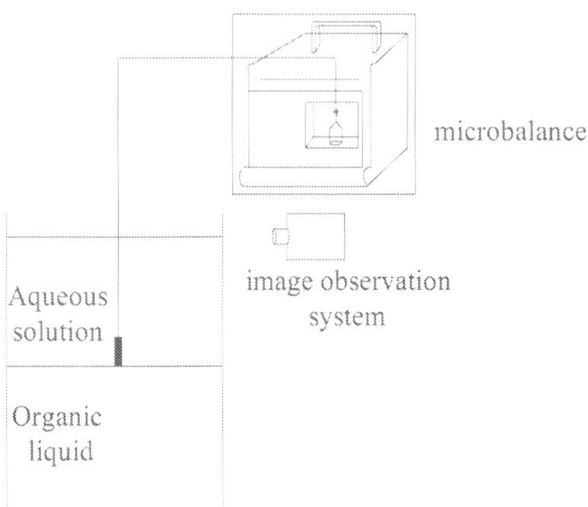

Fig. 22. Schematic diagram of static liquid-liquid contact device in the horizontal. (referred from Lavabre et al. 2005)

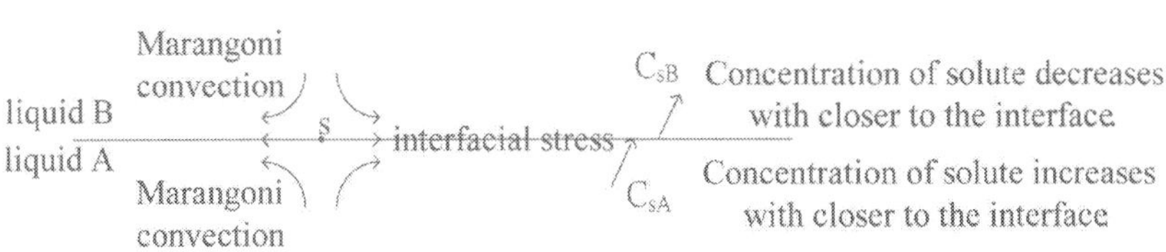

Fig. 23. Schematic diagram of Marangoni convection resulted from liquid-liquid contact system. (Referred from Miller & Neogi, 1985)

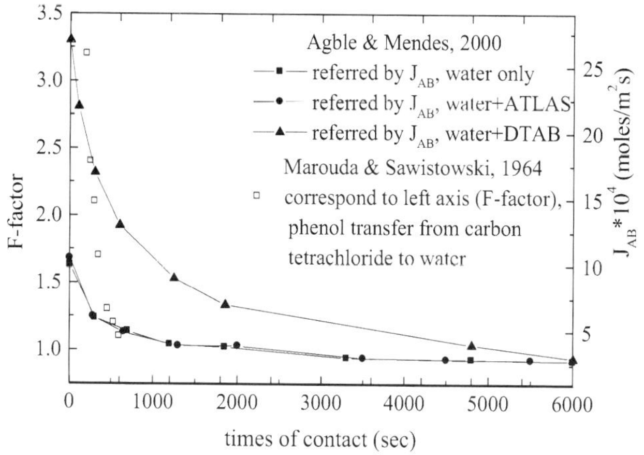

Fig. 24. Effects of contact time of mass transfer enhancement

5. Conclusions

The surface tension not only can apply for daily life but also can apply for engineering and scientific technology. For example, the surface tension can be applied into eye drops, ball pen, condensed liquid drop, capillary phenomenon, liquid lens, and etc. for the daily life. Besides, the surface tension can also be applied into semiconductor drying, artificial rainfall, and mass transfer enhancement for engineering and scientific technology. Since the effective area can be increased by interfacial disturbance, the surface tension plays an important role between phases. For example, the gradient of surface tension was used to induce interfacial disturbance to improve mass transfer efficiency by some studies. The purpose of this chapter is also to discuss the performance of mass transfer devices affected by the Marangoni effect, which is resulted from the gradient of surface tension.

First, the fluid flow in thin liquid film, thinker liquid layer, and mass transfer devices was introduced, and then the observation of interfacial disturbance resulted from liquid drop instilled on the surface of TEG solution was presented by the author of this chapter. The disturbed phenomena of liquid drop on the surface of TEG solution was used to explain successfully why the Marangoni effect occurred in the absorber with continuous liquid phase. In addition, artificial and spontaneous Marangoni convection were described in this chapter. Table 2 and Table 3 show some studies related to artificial and spontaneous Marangoni convection in the mass transfer devices. Evaporation of ethanol was used by author to form the gradient of surface tension, and then the mass transfer performance increased with the induced interfacial disturbance was demonstrated by the experimental results. Finally, the interfacial disturbance resulted from the gradient of surface tension and performance affected by the interfacial disturbance for these devices, such as packed-bed distillation column, two-concentric absorption cell, falling-film absorber, and liquid-liquid contact system, were described and analyzed.

Mentioned above, the surface tension is an important physics at the interface. The formed gradient of surface tension not only changes effective area between gas and liquid phases, but also affects the mass transfer mechanism. The results always make the traditional mass transfer fundamental underestimate the practical mass transfer data. In order to describe the interfacial disturbance resulted from the gradient of surface tension more precisely; some models and relevant governing equations had been established and solved to explain mass transfer affected by Marangoni effect. However, the phenomena of fluid flow dominated by Marnngoni effect seem to be still in the stage of speculation. Since the interfacial convection induced by the gradient of surface tension is difficult to observe by naked eyes, development of computational fluid dynamic can be suggested to describe effects of surface tension on fluid flow of the mass transfer devices in future. In order to promote the application of the Marangoni effect in the engineering and scientific technology, more researches discussed about Marangoni effect and more data confirmed the interfacial disturbance will be expected to fill the gap of this filed by more excellent scholars.

6. References

Agble, D. & Mendes-Tatsis, M. A. (2000). The Effect of Surfactants on Interfacial Mass Transfer in Binary Liquid-Liquid System. *International Journal of Heat and Mass Transfer*, 43, 1025-1034, ISSN: 0017-9310.

Bakker, C. A. P.; Vlissingen, Fentener, F. H. & Beek, W. J. (1967). The Influence of the Driving Force in Liquid-Liquid Extraction—a Study of Mass Transfer with and without Interfacial Turbulence under Well-Defined Conditions. *Chemical Engineering Science*, 22, 1349-1355, ISSN: 0009-2509.

Brian, P. L. T.; Vivian, J. E. & Matiatos, D. C. (1967). Interfacial Turbulence during the Absorption of Carbon Dioxide into Monoethanolamine. *AIChE J.*, 13(1), 28-36, ISSN: 0001-1541.

Buzek, J, (1983). Some Aspects of the Mechanism of Cellular Convection. *Chemical Engineering Science*, 38(1), 155-160, ISSN: 0009-2509.

Cachile, M.; Cazabat, A. M.; Bardon, S.; Valignat, M. P. & Vandenbrouck, F. (1999). Spontaneous Spreading of Surfactant Solutions on Hydrophilic Surfaces. *Colloids and Surfaces A: Physicochemical and Engineering Aspects*, 159, 47-56, ISSN: 0927-7757.

Chung, B. T. F.; Fan, L. T. & Hwang, C. L. (1971). Surface Renewal and Penetration Models in the Transient State. *AIChE J.*, 17(1), 154-160, ISSN: 0001-1541.

Colinet, P.; Joannes L.; Iorio C. S.; Haut, B.; Bestehorn, M.; Lebon, G. & Legros, J. C. (2003). Interfacial Turbulence in Evaporating Liquids: Theory and Preliminary Results of the ITEL-MASTER 9 Sounding Rocket Experiment. *Adv. Space Res.*, 32(2), 119-127, ISSN: 0273-1177.

Dijkstra, H. A. & Drinkengburg, A. A. H. (1990). Enlargement of Wetted Area and Mass Transfer due to Surface Tension Gradients: the Creeping Film Phenomenon. *Chemical Engineering Science*, 45(4), 1079-1088, ISSN: 0009-2509.

Imaishi, N.; Suzuki, Y.; Hozawa, M. & Fujinawa, K. (1982). Interfacial Turbulence in Gas-Liquid Mass Transfer. *International Chemical Engineering*, 22(4), 659-665, ISSN: 0020-6318.

Kalizova-Kurteva, P. G.; Slavtchev, S. G. & Kurtev, I. A. (1996). Stationary Marangoni Instability in a Liquid Layer with Temperature-Dependent Viscosity and Deformable Free Surface. *Microgravity sci. technol.*, IX/4, 257-263, ISSN: 0938-0108.

Kaminsky V. A.; Vyaz'min A. V.; Kulov N. N. & Dil'man V. V. (1998). Marangoni effect in the Presence of Bulk Turbulence. *Chemical Engineering Science*, 53(19), 3347-3353, ISSN: 0009-2509.

Kamotani, Y. & Masud, J. (1996). Oscillatory Convection Due to Combined Buoyancy and Thermocapillarity. *Journal of Thermophysics and Heat Transfer*, 10(1), 102-108, ISSN: 0887-8722.

Kashiwagi, T.; Rie, Dong-ho; Kurosawa, S.; Nomura, T. & Omata, K. (1993). Marangoni Effect on the Process of Steam Absorption into a Falling Film of an Aqueous Solution of LiBr. *Heat Transfer-Japanese Research*, 22(4), 355-371, ISSN: 0096-0802.

Kim, J.-K.; Jung, J. Y.; Kim, J. H.; Kim, M.-G.; Kashiwagi, T & Kang, Y. T. (2006). The Effect of Chemical Surfactants on the Absorption Performance during NH_3/H_2O Bubble Absorption Process. *International Journal of Refrigeration*, 29, 170-177, ISSN: 0140-7007.

Kim, Kwang J; Berman, Neil S. & Wood, Byard D. (1996). Absorption of Water Vapor into LiBr Solutions with 2-Ethyl-1-Hexanol. *AIChE J.*, 42(3), 884-888, ISSN: 0001-1541.

Lavabre, D.; Pradines, V; Micheu, Jean-Claude & Pimienta, V. (2005) Periodic Marangoni Instability in Surfactant (CATB) Liquid/Liquid Mass Transfer. *J. Phys. Chem. B*, 109, 7582-7586, ISSN: 1089-5647.

Lu, Hsin-Hsen; Yang, Yu-Min; Maa & Jer-Ru (1996). Effect of Artificially Provoked Marangoni Convection at a Gas/Liquid Interface on Absorption. *Ind. Eng. Chem. Res.*, 35, 1921-1928, ISSN: 0888-5885.

Lu, Hsin-Hsen; Yang, Yu-Min & Maa, Jer-Ru (1997). On the Induction Criterion of the Marangoni Convection at the Gas/Liquid Interface. *Ind. Eng. Chem. Res.*, 36, 474-482, 0888-5885

Maroudas, N. G. & Sawistowski, H (1964). Simultaneous Transfer of Two-Solutes across Liquid-Liquid Interfaces. *Chemical Engineering Science*, 19, 919-931, ISSN: 0009-2509.

Martin, Martin & Perez, Chica (1994). The Influence of Surface Tension on the Volumetric Mass Transfer Coefficient in Rectification. *International Chemical Engineering*, 34(1), 76-81, ISSN: 0020-6318.

Miller, C. A. & Neogi, P. (1985). *Interfacial Phenomena Equilibrium and Dynamic Effects*, Marcel Dekker, ISBN: 0-8247-7490-6, New Work.

Moens, F. P. & Bos, R. G. (1972). Surface Renewal Effects in Distillation. *Chemical Engineering Science*, 27, 403-408, ISSN: 0009-2509.

Patberg, W. B.; Koers, A.; Steenge, W. D. E. & Drinkenburg A. A. H. (1983). Effectiveness of Mass Transfer in a Packed Distillation Column in Relation to Surface Tension Gradients. *Chemical Engineering Science*, 38(6), 917-923, ISSN: 0009-2509.

Proctor, Sebastian J.; Biddulph, Michael W. & Krishnamurthy, Krish R. (1998). Effects of Marangoni Surface Tension Forces on Modern Distillation Packings. *AIChE J.*, 44(4), 831-835, ISSN: 0001-1541.

Sun, Z. F. (2006). Onset of Rayleigh-Benard-Marangoni Convection in Gas-Liquid Mass Transfer with Two-Phase Flow: Comparison of Measured Results with Theoretical Results. *Ind. Eng. Chem. Res.*, 45, 6325-6329, ISSN: 0888-5885.

Sun, Z. F.; Wang, S. Y. & Miao, Y. Z. (2002). Absorption and Desorption of Carbon Dioxide into and from Organic Solvents: Effects of Rayleigh and Marangoni Instability. *Ind. Eng. Chem. Res.*, 41, 1905-1913, ISSN: 0888-5885.

Vazquez, G.; Antorrena, G.; Navaza, J. M. & Santos, V. (1996). Absorption of CO_2 by Water and Surfactant Solutions in the Presence of Induced Marangoni Effect. *Chemical Engineering Science*, 51(12), 3317-3324, ISSN: 0009-2509.

Vazquez, G; Antorrena, G. & Navaza, J. M. (2000). Influence of Surfactant Concentration and Chain Length on the Absorption of CO_2 by Aqueous Surfactant Solutions in the Presence and Absence of Induced Marangoni Effect. *Ind. Eng. Chem. Res.*, 39, 1088-1094, ISSN: 0888-5885.

Warmuzinski, Krzysztof & Tanczyk, Marek (1991). Marangoni Instability during the Absorption of Carbon Dioxide into Aqueous Solutions of Monoethanolamine. *Chem. Eng. Process.*, 30, 113-121, ISSN: 0255-2701.

Wu, Honda & Chung Tsair-Wang (2006). Analysis of Effeft of Gas-Liquid Interfacial Disturbance on Mass Transfer Performance. *Surface Review and Letters*, 13 (2 & 3), 279–286, ISSN: 0218-625X.

Wu, Honda & Chung, Tsair-Wang (2008). Influences for the Addition of Ethanol to the Absorption System on the Interfacial Disturbances and Mass Transfer Performance. *Ind. Eng. Chem. Res.*, 47, 7397-7404, ISSN: 0888-5885.

Wu, Honda; Chung, Tsair-Wang & Lai, Ming-Hong (2001). Effects of Marangoni Convection on the Mass Transfer Performance in a Packed-Bed Absorber. *Ind. Eng. Chem. Res., 40*, 885-891, ISSN: 0888-5885.

Yang, Nai-Hsuan; Chen, Yi-Jen; Liao, Chien-Chin & Chung, Tsair-Wang (2008). Improved Absorption in Gas-Liquid Systems by the Addition of a Low Surface Tension Component in the Gas and/or Liquid Phase. *Ind. Eng. Chem. Res.*, 47, 8823-8827, ISSN: 0888-5885.

Zanfir, M.; Gavriilidid, A; Wille, Ch. & Hessel, V. (2005). Carbon Dioxide Absorption in a Falling Film Microstructured Reactor: Experiments and Modeling. *Ind. Eng. Chem. Res.*, 44, 1742-1751, ISSN: 0888-5885.

Zhang, X.; Wang, J.; Zhang, C.-F.; Yang, Y.-H. & Xu, J.-J. (2003). Absorption Rate into a MDEA Aqueous Solution Blended with Piperazine under a High CO_2 Partial Pressure. *Ind. Eng. Chem. Res.*, 42, 118-122, ISSN: 0888-5885.

14

Overall Mass-Transfer Coefficient for Wood Drying Curves Predictions

Rubén A. ANANIAS[1], Laurent CHRUSCIEL[2], André ZOULALIAN[3], Carlos SALINAS-LIRA[4] and Eric MOUGEL[5]

[1]University of Bio-Bio, Faculty of Engineering, Department of Wood Engineering 1202 Av. Collao, 4081112, Concepción

[2,3]University Henri Poincaré Nancy I, Faculty of Sciences and Technology, Lermab Bld. des Aiguillettes, 54506, Vandoeuvre, Nancy

[4]University of Bio-Bio, Faculty of Engineering, Department of Mechanical Engineering 1202 Av. Collao, 4081112, Concepción-

[5]University Henri Poincaré Nancy I, ENSTIB-Lermab 27 rue du Merle blanc, 88051, Epinal

[1,4]Chile

[2,3,5]France

1. Introduction

Wood is a biological material, hygroscopic, anisotropic and highly heterogeneous. To predict the wood drying rate, the drying kinetics need to be described. The wood drying kinetics can be described by three periods. Except in the first drying period, where the internal resistance of the water transfer can be neglected, it is necessary to take into account the internal transfer resistance during the second and third drying periods. It is sufficient to combine both the internal and external resistances, expressing the drying rate according to an overall mass transfer coefficient K. The driving force can be a physical characteristic of water in either the gas or solid phase. Karabagli et al. (1997) and Chrusciel et al. (1999), then Ananías et al. (2009a, 2009b) opted for an absolute average wood moisture content difference. The overall mass transfer coefficient is characterized by four operating parameters: wood thickness, air velocity, air temperature and air relative humidity. In general, K includes the internal resistance of the moisture movement through wood and the external resistance of the transfer of moisture from wood surface to the air. This chapter is a review of model permitting the determination of wood drying rate represented by an overall mass transfer coefficient, noted K, and a driving force expressed as the difference between the average wood moisture content and the equilibrium wood moisture content.

2. Wood drying kinetics

The drying rate Φ of wood can be represented by the derivative of the absolute wood moisture content ($-d\bar{x}/dt$), then

$$\Phi = \frac{-M_0}{A}\left(\frac{d\bar{x}}{dt}\right) \tag{1}$$

Also, the drying rate can be characterized by the transfer resistance of the gas phase (Bramhall 1979a, 1979b). It is:

$$\Phi = k_G \cdot (P_i - P_v) \tag{2}$$

The transfer coefficient k_G can be estimated by correlating with values reported in the literature, but the evaluation of Φ requires the knowledge of the partial pressure P_i. The relationship between this pressure and the average wood moisture content is not evident, and this relationship is applicable only during the first drying period, because P_i, the vapor pressure, is only equal to the saturated vapor pressure of the liquid water at the interface temperature.

Van Meel (1958) proposed a characteristic drying curve to represent the three drying phases with a relationship between the reduced drying rate Φ^+ and the non-dimensional parameter φ

$$\varphi = \frac{\bar{x} - x^*}{\bar{x}_C - x^*} \tag{3}$$

Where $\bar{x}$, $\bar{x}_C$, x^* indicate the average wood moisture content, the wood moisture content at the end of the first drying period, and the wood moisture content at the end of drying (equilibrium moisture content), respectively.

The equilibrium moisture content x^* can be determined by Simpson correlations (Siau 1984):

$$x^* = \left[\frac{w_1 \cdot w_2 \cdot RH}{1 + w_1 \cdot w_2 \cdot RH} + \frac{w_2 \cdot RH}{1 - w_2 \cdot RH}\right] \cdot \frac{18}{w_3} \tag{4}$$

Where RH is the relative humidity

$$RH = \frac{P_v}{P_s} \tag{5}$$

In wich Pv and Ps are defined as:

$$P_v = \left\{P_{sw} - \frac{\left(1.3332 \cdot 10^5 - P_{sw}\right) \cdot \left(T - T_w\right)}{1546 - 1.44 \cdot T_w}\right\} \tag{6}$$

$$P_s = 133.32 \cdot e^{\left(51.29 - \frac{6651}{T_K} - 4.53 \cdot \ln(T_K)\right)} \tag{7}$$

$$P_{Sw} = 133.32 \cdot e^{\left(51.29 - \frac{6651}{T_{Kw}} - 4.53 \cdot \ln(T_{Kw})\right)} \tag{8}$$

And w_1, w_2 and w_3 are coefficients as a function of temperatures T (°C)

$$w_1 = 4.737 + 0.0477 \cdot T - 0.0005 \cdot T^2 \tag{9}$$

$$w_2 = 0.7095 + 0.0017 \cdot T - 5.5534 \cdot 10^{-6} \cdot T^2 \tag{10}$$

$$w_3 = 223.385 + 0.6942 \cdot T + 0.0185 \cdot T^2 \tag{11}$$

Φ^+ represents the relation between the drying rate and the maximum drying rate obtained during the first drying period. The maximum drying rate can be estimated by equation 2, where the partial vapor pressure Pi is replaced by the saturated pressure of the liquid water at the interface temperature. However, Ananias et al. (2009a) used the heat and mass transfer analogy to estimate the maximal drying rate using the equation:

$$\Phi_{MAX} = \frac{h \cdot (T - T_W)}{\Delta h_v} \tag{12}$$

Where h is the external heat transfer coefficient, which can be estimated by various correlations according to the wood drier geometry:

$$h = 0.023 \cdot \left(\frac{v \cdot d_H}{\upsilon}\right)^{\frac{4}{5}} \cdot \left(\frac{v \cdot \rho \cdot C_{Pa}}{\lambda}\right)^{\frac{1}{3}} \cdot \frac{\lambda}{d_H} \tag{13}$$

In wich υ, ρ, λ and C_{Pa} are the air properties defined as following (Jumah et al. 1997):

$$\upsilon = \frac{\mu}{\rho} \tag{14}$$

$$\mu = 1.691 \cdot 10^{-5} + 4.984 \cdot 10^{-8} \cdot T_K - 3.187 \cdot 10^{-11} \cdot T_K^2 + 1.319 \cdot 10^{-14} \cdot T_K^3 \tag{15}$$

$$\rho = \frac{353}{T_K} \tag{16}$$

$$\lambda = 0.02425 + 7.889 \cdot 10^{-5} \cdot T_K - 1.79 \cdot 10^{-8} \cdot T_K^2 - 8.57 \cdot 10^{-12} \cdot T_K^3 \tag{17}$$

$$C_{Pa} = 1.0092 - 4.0403 \cdot 10^{-2} \cdot T_K + 6.1759 \cdot 10^{-4} \cdot T_K^2 - 4.097 \cdot 10^{-17} \cdot T_K^3 \tag{18}$$

With d_H and Δh_v calculated as:

$$d_H = \frac{4 \cdot S}{W_P} \tag{19}$$

$$\Delta h_V = 2503 - 2.43 \cdot T_K \tag{20}$$

Van Meel's characteristic drying curve requires the knowledge of the maximum drying rate and the average wood moisture content $\overline{x}_C$, corresponding to the transition between the first and second drying periods.

The concept of a characteristic drying curve is variously validated by the works of Moyne (1984), Basilico (1985), Moser (1992), Keey (1994), Martin et al. (1995), Pang (1996a) in convective wood drying at both high and very high temperatures, but the first drying period is rarely observed. For parameter φ values below 1, these authors have demonstrated that a functional relationship for Φ might be developed in terms of φ.

Except during the first drying period, where the internal resistance of the water transfer can be neglected, it is necessary to take into account the internal transfer resistance during the second and third drying periods. It is sufficient to combine both the internal and external resistances, expressing the drying rate according to an overall mass transfer coefficient K. In this stage, the driving force can be a physical characteristic of water in either the gas or solid phase. Karabagli et al. (1997) and then Chrusciel et al. (1999) opted for an absolute average wood moisture content difference. Consequently, the drying rate is expressed as:

$$\Phi = K \cdot (\bar{x} - x^*) \tag{21}$$

In general, K includes the internal resistance of the moisture movement through the wood and the external resistance of the transfer of moisture from the wood surface to the air.
Ananias et al. (2009a) applied the drying model proposed by Van Meel and justified the phenomenological drying model recommended by Karabagli et al. (1997) and Chrusciel et al. (1998). According to Ananias et al. (2009a), these curves do not show the period where the drying rate is constant nor the case when the initial wood moisture content is of the order of 100 %. The first drying period is not observed. This observation is also reported by other authors (Basilico 1985, Moser 1992), who suggest that the absence of the first drying period is due to the low initial wood moisture content values. However Keey et al. (2000) propose that with hardwoods and the heartwood of sapwoods, the critical moisture content is probably the initial value.
Ananias et al (2009a) affirm that the drying rate decreases during the second and third drying period, and estimate that the relationship between the drying rate and the driving force ($\bar{x}$ -x*) is substantially linear. They believe that under constant operating conditions (temperature, humidity and air velocity), the overall mass- transfer coefficient K (equation 21) is substantially constant, justifying the phenomenological law proposed by Karabagli et al. (1997) to express the drying rate. They have not observed a period of constant drying rate even when the initial wood moisture constant was high. Ananias et al (2009a) show that the highest point of the drying curves depends remarkably on the wood thickness, air velocity and air temperature (Fig. 1)
By extrapolation of these linear relationships in the limit value of Φ^+ equal to 1, Ananias et al. (2009a) obtain the critical value of the average wood moisture content, $\bar{x}_C$. They show that the critical wood moisture content depends on all of the studied operating parameters and that they are in a range of absolute wood moisture content between 56.3 and 138.3 %.
According to Ananias et al. (2009a), the $\bar{x}_C$ values cannot be known a priori. Van Meel's representation does not lead to a unique characteristic drying curve. From these representations, they only obtained the existence of a considerably linear relationship between the drying rate and the driving force ($\bar{x}$ -x*), which confirms the hypothesis of Karabagli et al. (1997) for the overall transfer coefficient K: that an overall mass transfer coefficient remains constant during the kiln drying cycle when the operating conditions temperature, humidity relative and air velocity are also maintained constant.
Ananias et al (2009a) show that for all the kiln-drying runs and for the analyzed operating conditions, the average relative error is below 7 %, justifying the selection of the phenomenological drying rate modeling and the hypothesis of a constant overall mass-transfer coefficient. However, since the critical moisture content of the Van Meel's characteristic drying curve and the overall mass-transfer coefficient K depend on the operating conditions, the operating conditions will have a varying influence on the internal and external resistance.
The influence of the operating parameters is demonstrated by Chrusciel et al. (1999). They show that the global mass transfer coefficient K could be represented by two partial mass transfer coefficients, it is, K_g in the gas phase (air) and K_s in the solid phase (wood). Then the water vapor flow rate exchanged between wood and the drying air is estimated by:

$$\phi = K_g \cdot A \cdot (H_S - H) = K_S \cdot A \cdot (x_C - x_S) \quad (22)$$

Where (H_S-H) is the specific humidity slope between the air interface that is in equilibrium with the wood surface and the drying air flow. And (x_C-x_S) is the wood moisture content slope between the wood-core and the wood-surface.

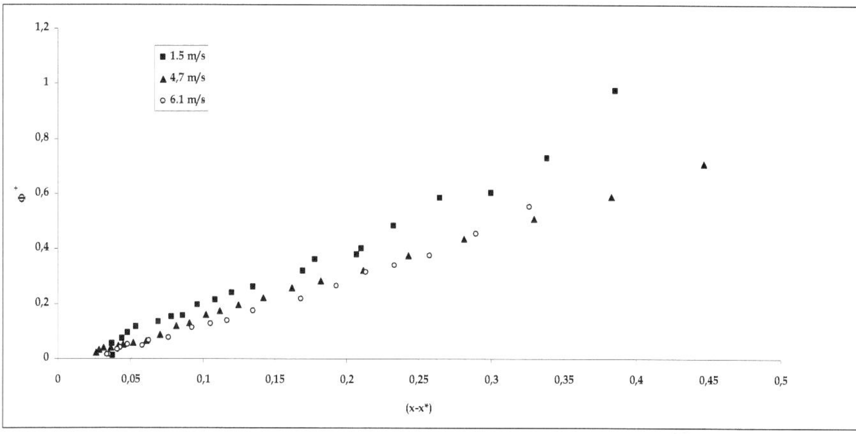

Fig. 1. Effect of air velocity on wood drying rate (After Ananias et al. 2009a)

Suppossing that x_C is not too different from the average x-value and linking the specific humidity (H_S) to the wood moisture content at the interface (x_S) by a proportional coefficient (*m*), the following equation is obtained

$$H_S = m \cdot x_S \quad (23)$$

Assuming that at the end of the drying cycle, H_S=H and x_S=x*, then the relation between air humidity and wood moisture content is

$$H = m \cdot x^* \quad (24)$$

The above equations (22)-(24) then becomes

$$\frac{1}{K} = \frac{1}{K_S} + \frac{1}{m \cdot K_G} \quad (25)$$

Or

$$\frac{1}{K} = R_I + R_E \quad (26)$$

According to Ananias et al (2009a), the internal resistance R_I is proportional to wood thickness and the external resistance R_E depends on air velocity according to the law of type v^{-n}, where the exponent n varies between 0.5 and 1.

Chrusciel et al. (1999) have shown that the inverse of K, which can be assimilated to a global mass transfer resistance, is a decreasing exponential function of the residual air desiccation ratio z (z = 1 - RH/100).

It is assumed that the global mass transfer resistance 1/K is the sum of two mass transfer resistances: one characterizing the solid phase (wood) and the second one the gas phase (wet air). It is supposed that only the resistance in the gas phase depends on the relative humidity RH and the air velocity v of the air while only the resistance in the solid phase depends on the wood thickness.
It is hard to determine separately the influence of the air temperature on the gas resistance from the one on the solid phase. It is also difficult to distinguish the influence of the air relative humidity from the one of the air temperature on the global mass transfer coefficient. However, it has been demonstrated that the variations of K as a function of the air temperature can be represented by an Arrhenius law. It is then supposed that the influence of T on the two local mass transfer resistances is quite similar. All these considerations permit to propose a general correlation for the global mass transfer resistance:

$$\frac{1}{K} = \alpha + \beta \cdot \exp\left(\frac{-z}{x_{PSF} - x^*}\right) \tag{27}$$

and

$$\alpha = a_0 \cdot \exp\left(\frac{c_0}{T_K}\right) \cdot e \tag{28}$$

$$\beta = b_0 \cdot \exp\left(\frac{c_0}{T_K}\right) \cdot v^{-p} \tag{29}$$

In relation (27), $x_{PSF} - x^*$ is the gap between the fiber saturation point and the equilibrium moisture content at the temperature T. According to the many authors such as Lartigue and Puiggali (1987) or Nadler et al. (1985), when z tends to 1 (so RH tends to 0 %), x* tends to a low but positive value (about 1 %) so that the ratio $z / (x_{PSF} - x^*)$ is defined when T is lower than the water desorption temperature (about 103 °C at 1 atm). The limit of validity of equation (27) is reached when z is very closed from 0 (so when RH tends to 100 %) even if some authors such as Babiak and Kudela (1995) underline that the fiber saturation point can hardly be defined as the value of x* at RH = 100 %.
Consequently, the overall mass-transfer coefficient K can be expressed in general, according to the operating parameters, as:

$$\frac{1}{K} = a_0 \cdot \exp\left(\frac{c_0}{T_K}\right) \cdot e + b_0 \cdot \exp\left(\frac{c_0}{T_K}\right) \cdot v^{-n} \cdot \exp\left[\frac{(RH - 1)}{(\overline{x}_{FSP} - x^*)}\right] \tag{30}$$

In the equation 30, RH is the relative humidity of kiln drying and $\overline{x}_{FSP}$ is the wood moisture content at the fiber saturation point, taken equal to 0.3.
If the exponent n is maintained at 0.8, then the three coefficients in equation 30, a_0, b_0 and c_0 are determined by optimization from the calculated values of the overall mass-transfer coefficients by minimizing the relative error function between the optimized values and the calculated values obtained with equation 30. The values of these three coefficients obtained for optimization are: a_0= 0.12 (m.s.kg^{-1}), b_0= 23.9 (m^2.kg^{-1}) and c_0= 2683 (K).
Fig 2 compares the calculated values and the optimized values for the overall mass-transfer coefficient. Equation 30 represents all the experimental determinations with an average relative error equal at 15.3 % (and a standard deviation equals at 11 %). This correlation is

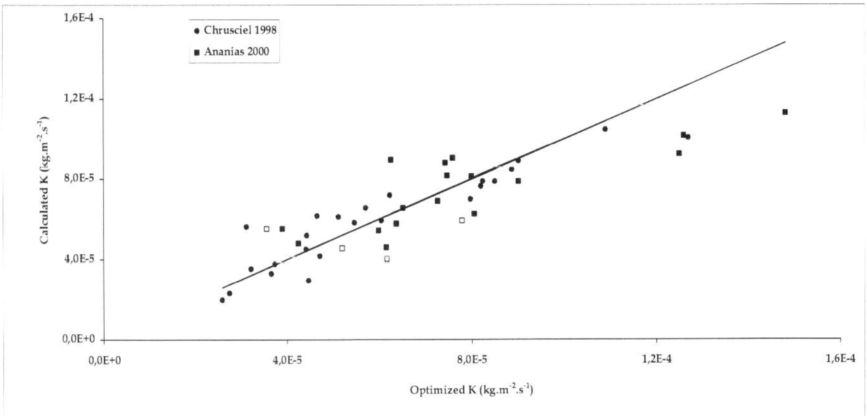

Fig. 2. Comparison of the optimized and calculated values of overall mass transfer coefficient (After Ananias et al. 2009a)

valid for two kinds of lumber studied (spruce and beech), and is satisfactory for a whole kiln-drying schedule.

3. Mathematical model

The model states that the drying rate is a linear function of the drying potential, the moisture-content difference ($\bar{x}$ -x*), and a constant coefficient of proportionality, which is the overall mass-transfer coefficient. This hypothesis has been verified in a previous study (Ananías et al. 2009a).

The model further assumes that the mass and enthalpy transfer takes place unidirectionally, the initial moisture content is homogenous, and the air distribution through the stack is uniform. Heat losses and temperature changes throughout the stack are considered negligible. The model requires the initial values of moisture content and temperature for each sub-system (wood and air) to be known, as well as the overall mass (*K*) and heat-transfer (*h*) coefficients.

The model equations (Karabagli et al. 1997) are from the mass balance of water in drying air,

$$G.(W_{out} - W_{in}) = K.A.(\bar{x} - x^*) \tag{31}$$

the mass balance of water in the wood,

$$-M_0.\frac{d\bar{x}}{dt} = K.A.(\bar{x} - x^*) \tag{32}$$

the enthalpy balance over the drying air,

$$G.\begin{bmatrix} \{C_{pa}.T_{out} + W_{out}.(\Delta h_o + C_{pV}.T_{out})\} - \\ \{C_{pa}.T_{in} + W_{in}.(\Delta h_o + C_{pV}.T_{in})\} \end{bmatrix} = K.A.\Delta h_V.(\bar{x} - x^*) - h.A.(T - T_w) \tag{33}$$

and the enthalpy balance for the wood,

$$M_0 \cdot \left(C_{pS} + C_{pL} \cdot \bar{x}\right) \cdot \frac{dT_W}{dt} = -G \cdot \begin{bmatrix} C_{pa} \cdot (T_{out} - T_{in}) - (C_{pL} \cdot T_W) \cdot (W_{out} - W_{in}) + \\ C_{pv} \cdot (W_{out} \cdot T_{out} - W_{in} \cdot T_{in}) \\ -\Delta h_0 \cdot (W_{out} - W_{in}) \end{bmatrix} \quad (34)$$

The four equations have been solved as an initial-value problem in a previous study (Ananias et al. 2001, Broche et al. 2002). If equation 32 is solved by means of a finite-difference method, then we can calculate the theoretical wood moisture content at any time ($\bar{x}^{j+1}$). In rearranging this equation, the following relation is obtained:

$$k = -\frac{d\bar{x}}{(\bar{x} - x^*)} \quad (35)$$

On making the transient terms discrete, the following equation is found:

$$k = -\frac{\bar{x}^{j+1} - \bar{x}^{j}}{\dfrac{\bar{x}^{j+1} - x^{*j+1} + \bar{x}^{j} - x^{*j}}{2}} \quad (36)$$

Rearranging this equation to find moisture content at any time ($\bar{x}^{j+1}$), we get:

$$\bar{x}^{j+1} = \frac{(2-k) \cdot \bar{x}^{j}}{2+k} + \frac{k \cdot (x^{*j} + x^{*j+1})}{2+k} \quad (37)$$

Note that $\bar{x}^{j}$ and x^{*j} are experimental values and are related to the following error function:

$$E = 100 \cdot \left| \frac{\left(\bar{x}_{exp} - \bar{x}_{cal}\right)}{\bar{x}_{exp}} \right| \quad (38)$$

Finally,

$$K = -\frac{k \cdot M_0}{A \cdot \Delta t} \quad (39)$$

Since the model assumes that coefficient *K* remains constant during drying, it is necessary to dry under constant conditions.

4. Drying curves predictions

We predicted kiln-drying curves, using Spruce (*Picea abies*), Beech (*Fagus sylvatica*) and chilean coigüe (*Nothofagus dombeyi*). The corresponding drying curves are presented in Figures 3-5.

There were minor differences between the experimental and calculated moisture content of the wood. The magnitude of the overall mass-transfer coefficients K is in the range of 0.43×10^{-5} kg/m^{2}s in Chilean coigüe of 38-mm thickness to 12.5×10^{-5} kg/m^{2}s in Spruce of 18-mm thickness (Table 1).

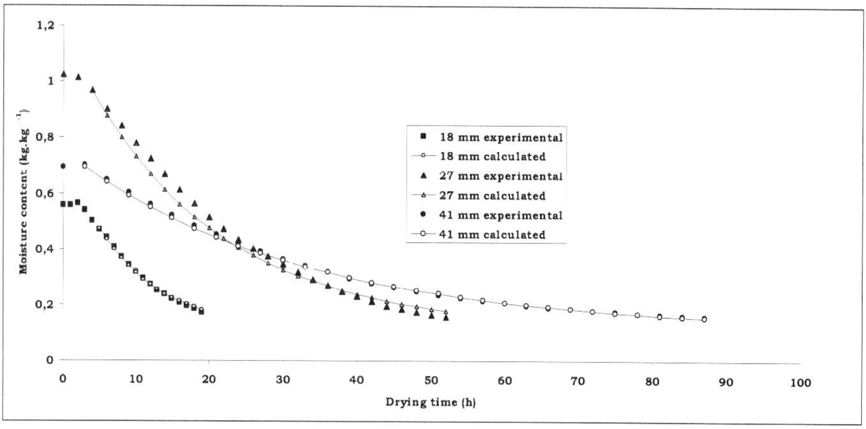

Fig. 3. Kiln drying curves of Spruce wood (After Ananias et al. 2009a)

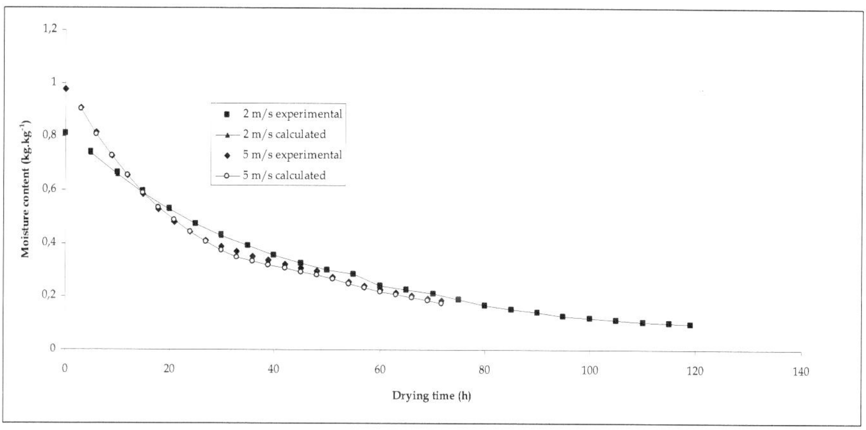

Fig. 4. Kiln drying curves of Beech wood (After Ananias et al 2009a)

Wood Species	e (mm)	T (°C)	Tw (°C)	v (m/s)	K.10^5 (kg/m^2.s)	Reference
Spruce	18	70	50	3	12.5	Ananias et al. 2009a
Spruce	27	70	50	3	7.48	
Spruce	41	70	50	3	6.39	
Beech	30	70	50	2	5.21	
Beech	30	70	50	5	7.81	
Coigüe	38	60	44	2.5	0.43	Ananias et al. 2009b

Table 1. Overall-mass transfer coefficient

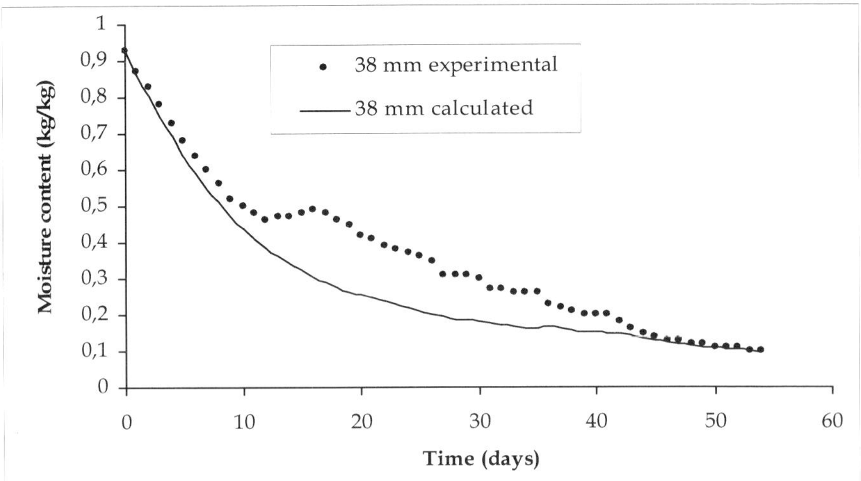

Fig. 5. Kiln drying curve of Chilean coigüe

5. Conclusion

Many low-temperature conventional wood drying curves can be predicted by a constant overall mass-transfer coefficient. The model presented has been tested on different wood drying schedules and the results obtained are very satisfactory. For these reasons, it is suggested that it can be successfully used for drying schedule optimization at industrial scale.

6. List of symbols

A	Transfer surface (m^2)
a	Wood width (m)
a_b	Stickers width (m)
a_o, b_o, c_o, n	Model constants
C_{pa}	Air specific heat (J/kg.°C)
C_{pL}	Water-Liquid specific heat (J/kg.°C)
C_{pv}	Water-Vapor specific heat (J/kg.°C)
d_H	Hydraulic diameter (m)
e	Wood thickness (mm)
G	Air flow rate (kg/s)
H	Specific humidity (kg/kg)
h	Overall heat-transfer coefficient ($W/m^2.K$)
K	Overall mass-transfer coefficient ($kg/m^2.s$)
K_S	Partial mass-transfer coefficient in air-phase ($kg/m^2.s$)
K_g	Partial mass-transfer coefficient in solid-phase ($kg/m^2.s$)
k	Mass-transfer coefficient (non-dimensional)
k_G	Mass-transfer coefficient ($kg/m^2.s.Pa$)
l	Wood length (m)
Mo	Wood dry mass (kg)

Pi	Partial pressure at the interface (Pa)
P_v	Partial pressure (Pa)
P_S	Saturation pressure (Pa)
RH	Relative humidity (kg/kg)
S	Wood surface (m^2)
t	Drying time (h)
T	Air temperature (°C)
T_K	Air temperature (K)
T_w	Wet-bulb temperature (°C)
T_{Kw}	Wet-bulb temperature (K)
v	Air velocity (m/s)
W_P	Wetted perimeter (m)
$\bar{x}$	Moisture content (kg/kg)
$\bar{x}_C$	Critical moisture content (kg/kg)
x_i	Initial moisture content (kg/kg)
$\bar{x}_{PSF}$	Fiber saturation point (kg/kg)
x^*	Equilibrium moisture content (kg/kg)
z	Residual air dessication ratio [/]
Δh_0	Heat of vaporization at T= 0 °C *(J/kg)*
Δh_V	Heat of vaporization *(J/kg)*
Φ	Drying rate [kg/m^2s]
Φ_{MAX}	Maximum drying rate ($kg/m^2.s$)
Φ^+	Reduced drying rate
φ	Non dimensional parameters
u	Air dynamic viscosity (kg/m.s)
υ	Air cinematic viscosity (m^2/s)
ρ	Air density (kg/m^3)
λ	Air thermal conductivity (W/m.K)

7. References

Ananías, R. A.; Mougel, E. & Zoulalian, A. (2009a) Introducing an overall mass-transfer coefficient for prediction of drying curves at low temperature drying rates. *Wood Science and Technology* 43(1): 43-56.

Ananías, R.A.; Broche, W.; Alvear, M.; Salinas, C. & Keey, R.B. (2009b) Using an overall mass-transfer coefficient for prediction of drying of Chilean coigüe. *Wood and fiber Science* 41(4):426-432.

Ananías, R. A.; Broche, W.; Salinas, C. & Ruiz, P. (2001) Drying modeling of Chilean coigüe. Part 1. Theoretical aspects. (In Spanish, abstract in English). *Maderas. Ciencia y tecnología* 3 (1-2):27-34.

Ananías, R. A. (2000) *Modelisation du séchage convectif basse température et optimisation du séchage du hêtre vis a vis des problèmes de discoloration.* (In French, Abstract in English). Thèse de doctorat, Université Henri Poincaré, Nancy 1, France.

Babiak, M. & Kudela, J. (1995) A contribution to the definition of the fibre saturation point. *Wood Science and Technology* 29(3): 217-226.

Basilico, C. (1985) *Le séchage convectif à haute température du bois massif. Etude des mécanismes de transfert de chaleur et de masse.* (In French, Abstract in English). Thèse de doctorat, Institut National Polytechnique de Lorraine, Nancy, France

Bramhall, G. (1979a) Mathematical model for lumber drying. I - Principles involved. *Wood Science* 12 (1):14-21.

Bramhall, G. (1979b) Mathematical model for lumber drying. II. The model. *Wood Science* 12 (1): 22-31.

Broche, W.; Ananías, R.A.; Salinas, C. & Ruiz, P. (2002) Drying modeling of Chilean coigüe. Part 2. Experimental results. (In Spanish, abstract in English). *Maderas. Ciencia y tecnología* 4(2):69-76.

Chrusciel, L.; Mougel, E.; Zoulalian, A. & Meunier, T. (1999) Characterisation of water transfer in a low temperature convective wood drier: influence of the operating parameters on the mass transfer coefficients. *Holz als Roh- und Werkstof* 57: 439-445.

Chrusciel, L. (1998). *Etude de l'association d'une colonne d'absorption à un séchoir convectif à bois basse température. Influence de l'absorbeur sur la cinétique et la qualité du séchage..* (In French, Abstract in English). Thèse de doctorat, Université Henri Poincaré, Nancy 1, France.

Jumah, R.Y.; Mujumdar, A.S.; Raghavan, G.S.V. 1997. A mathematical model for constant and intermittent batch drying of grains in a novel rotating jet spouted bed. In mathematical modeling and numerical technique in drying technology. Ed. By I. Turner & A.S. Mujumdar. Dekker, Inc. N. York, pp. 339-380.

Karabagli, A.; Mougel, E.; Chrusciel, L. & Zoulalian, A. (1997) Study of a low temperature convective wood drier. Influence of some operating parameters on drier modelling and on the quality of dried wood. *Holz als Roh- und Werkstof* 55:221-226.

Keey, R. B.; Langrish, T. A. G. & Walker, J. C. F. (2000) *Kiln-Drying of Lumber.* Springer Science. N. York

Keey, R. B. (1994) Heat and mass transfer in kiln drying. Proceeding of the 4th IWDC, Rotorua, New Zealand, pp.22-44.

Lartigue, C. & Puiggali, J.R. (1995). Caractéristiques des pins des Landes à la compréhension des phénomènes de séchage. . (In French, Abstract in English). Actes du 2ème colloque sciences et industries du bois, A.R.Bo.Lor, Nancy, France, pp.57-64.

Martin, M.; Perré, P. & Moser, M. (1995) La perte de température à travers la charge: Intérêt pour le pilotage d'un séchoir à bois à haute température. (In French, Abstract in English). *International Journal of Heat and Mass Transfer* 38 (6): 1075-1088.

Moser, M. (1992) Le séchage convectif à haute température. Observation des mécanismes à deux échelles: La planche et la pile. (In French, Abstract in English). Thèse de doctorat, Institut National Polytechnique de Lorraine, Nancy, France.

Moyne, C. (1984) *Contribution à l'étude du transfert simultané de chaleur et de masse au cours du séchage sous vide d'un bois résineux.* (In French, Abstract in English). Thèse de doctorat, Institut National Polytechnique de Lorraine, Nancy. France.

Nadeau, J. P. & Puigalli, J. R. (1995) *Séchage: des processus physiques aux procédés industriels.* . (In French). Lavoisier, Paris, France.

Nadler, K.C.; Choong, E.T. & Wetzel D.M. (1985). Mathematical modelling of the diffusion of water in wood during drying. *Wood and Fiber Science* 17 (3): 404-423.

Pang, S. (1996a) Development and validation of a kiln-wide model for drying of softwood lumber. Proceeding of the 5th IWDC, Quebec, Canada, pp. 103-110.

Pang, S. (1996b) External heat and mass transfer coefficients for kiln drying timber. *Drying Technology* 14(3/4):859-871.

Salin, J. G. (1996) Prediction of heat and mass trasfer coefficient for individual boards and board surfaces. A review. Proceeding of the 5th IWDC, Quebec, Canada, pp. 49-58.

Siau, J.F. 1984. *Transport processes in wood.* Springer-verlag. Berlín.

Van Meel, D. A. (1958) Adiabatic convection batch drying with recirculation of air. *Chemical Engineering Science* 9(1958):36-44.

15

Transport Phenomena in Paper and Wood-based Panels Production

Helena Aguilar Ribeiro[1], Luisa Carvalho[1,2], Jorge Martins[1,2] and Carlos Costa[1]
[1]Universidade do Porto – Faculdade de Engenharia, Laboratory for Process, Environmental and Energy Engineering,
[2]Instituto Politécnico de Viseu, Wood Engineering Department, Portugal

1. Introduction

1.1 A brief historical perspective of paper and wood-based materials

The pulp and paper industry is a vital manufacturing sector that meets the demands of individuals and society. Paper is an essential part of our culture and daily lives, as it is used to store and share information, for packaging goods, personal identification, among other end uses. In an age of computers and electronic communication, paper is still envisaged as one of the most convenient and durable option of data storage, and a material of excellence for artists and writers. It is not surprising that the birth of modern paper and printing industry is commonly marked from the increasing demand for books and important documents in the 15th century. In 2008 the Confederation of European Paper Industries (CEPI) reported a global world paper production of 390.9 million tonnes covering a wide range of graphic paper grades, household and sanitary, packaging and other carton board grades (CEPI, 2010). The CEPI member countries account for 25.3% of the world paper and board production, slightly above North America (24.5%) but far behind Asia (40.2%). In volume terms, graphic paper grades account for 48% of the Western European paper production, packaging paper grades for some 41%, and hygiene and utility papers for 11% (CEPI, 2010). Additionally, forecasts indicate that from 1998 to 2015 there will be an increase of 2.8% in the consumption of paper and board globally. It is clear, therefore, that despite the growth of alternatives to paper like electronic media, several paper grades will still play an important role in our lives. Moreover, other materials used in a day-to-day basis derive from wood fibres extracted from a diversity of arboraceous species. As an example, "wood-based panels" (WBP) - a general term for a variety of different board products which have an impressive range of engineering properties (Thoemen, 2010) - are used in a wide range of applications, from non-structural to structural applications, outdoor and indoor, mostly in construction and furniture, but also in decoration and packaging. The large-scale industrial production of wood composites started with the plywood industry in the late 19th century. A number of new types of wood based panels have been introduced since that time as hardboard, particleboard, Medium Density Fibreboard (MDF), Oriented Strand Board (OSB), LVL-Laminated Veneer Lumber and more recently LDF (Light MDF) and HDF (High Density Fibreboard). The production of wood-based panels is still an important part of the world's total volume of wood production. In 2009, FAO (Food and Agriculture Organization of the United Nations) reported that a total of 255 million m^3 was produced in the world (Europe 29.7%, Asia 43.9%, North America 18.3% and others 2.5%). In case of MDF the production in Europe was 19.1 million m^3 (Wood Based Panels International, 2010).

1.2 Research and development in a high-tech industry: major advances and concerns

Research, development and innovation are the key to many of the challenges paper and wood-based materials industry are facing today. In the last decades, substantial development work has been undertaken to improve the pulp and paper qualities of today, taking into account features such as printability, press runnability, sheet opacity/low grammage and barrier properties. Modern paper machines are giant tailor-made units that carry out the two major steps of papermaking: dewatering and consolidation of a wet paper web made of cellulose fibres, chemical additives and water. In fact, the production of paper is mainly a question of removing as much water as possible from the pulp at the lowest possible cost. During papermaking, water removal takes place in three stages, namely in the wire, press and drying sections of the paper machine. In the first stage, water content is reduced from 99% down to about 80% using gravitational force or with the aid of suction boxes. In the press section, the dewatering process continues by mechanical pressure, increasing the paper web dryness to about 35-50%. The paper then enters the drying section, which is comprised of several rotating heated cylinders, and most of the remaining water is evaporated from the paper. At this stage, the dryness of the web has increased up to about 90-95%. Even though the water removal in the drying section is relatively modest, this is by far the most energy demanding stage of the web consolidation process, making mechanical dewatering a much more cost-effective process than evaporation. Also, the demand for higher productivity led to a significant increase in the speed of the paper machine, which in its turn results in higher water content after the press section, thus increasing the effort put in the dryer section. As a result, a considerable emphasis has been given over the last thirty years, by researchers and paper makers, to the development of more efficient press sections. In the 80's, a new concept arised with the development of the so-called *extended nip presses*, which includes the terms *high impulse presses*, *long-nip presses*, *wide-nip presses* and *shoe presses*, the common feature to all being the increased contact time between the paper web and the pressing element, thus leading to a significant higher dryness (Pikulik, 1999). In some emerging techniques such as *press drying*, the *Condebelt process* and more recently *impulse drying*, higher levels of dryness are possible. Moreover, the implementation of these methods showed to significantly reduce the dimensions of the paper machine dryer section and the use of steam while allowing to obtain a drier and stronger sheet at the end of the press section. In summary, the overall-aim of developments in the press section has been to improve the energy efficiency of web consolidation and paper properties.

Similar technological advances have been undertaken in the field of wood-based panels, which are produced from particles (as particleboard or OSB), fibres (as MDF, softboard or hardboard) or veneers (as plywood or LVL), using a thermosetting resin, through a hot pressing process. The hot-pressing operation is the final stage of its manufacturing process, where fibres/particles are compressed and heated to promote the cure of the resin. This operation is the most important and costly in the manufacture of wood-based panels. In the last decade, the technology for the production of wood-based panels had an important change in response to ever changing markets. The international research in this field is driven by improvements in quality (better resistance against moisture and better mechanical resistance) and cost reduction by energy savings (shorter pressing times) as well as the use of more cost effective raw materials (cheaper and alternative raw materials, reuse and recycling) (Carvalho, 2008). Environmental regulations and legislation regarding VOCs (volatile organic compounds) emissions, in particular formaldehyde, are important driving forces for technological progresses. Although panel product emissions have been dramatically reduced over the last decades, the recent reclassification of formaldehyde by IARC (International Agency for Research and Cancer) as "carcinogenic to humans", is forcing panels manufacturers, adhesive suppliers and researchers to develop systems that lead to a decrease in its emissions to levels as low as those present in natural wood (Athanassiadou et al., 2007).

2. Heat and mass transfer phenomena in porous media

2.1 Introduction

Many problems in scientific and industrial fields as diverse as petroleum engineering, agricultural, chemical, textiles, biomedical and soil mechanics, involve multiphase flow and displacement processes in a heterogeneous porous medium. These processes are mainly controlled by the pore space morphology, the interplay between the viscous and capillary forces, and the contact angles of the fluids with the surface of the pores. Estimating the capillary pressure and relative fluid permeabilities across the porous media can therefore be very complex, especially if the medium is deformable as is the case of paper and wood-based panels. In fact, the most important process in paper production is dewatering of the cellulose fibre suspension, which has a concentration less than 1% entering the forming section of the paper machine. In particular, the wet pressing of paper - or other wood based materials - may be envisaged as the simultaneous flow of two fluids, water and a mixture of air and water vapour, in a deformable porous medium. The following sections address the drying processes of paper and MDF, with special emphasis in the dewatering and consolidation mechanisms involved in the press section. Here, a deep knowledge of the interactions between heat and water is of utmost importance to control and optimize this operation in order to improve paper/MDF quality and to reduce the operational costs. The development of theoretical models based on the many physical, chemical and mechanical phenomena that are involved in this operation, constitutes an attempt to understand and quantify the most diverse interacting transfer mechanisms (simultaneous heat and mass transfer with phase change, and the rheological behaviour of the fibrous material).

2.2 Foundations of flow analysis in compressible porous media

2.2.1 Consolidation mechanisms involved

As previously mentioned, the production of paper and wood-based materials, such as MDF, is mainly a question of consolidation of the fibrous network by removing as much water or gas (air + water vapour) as possible from the interstitial void space. For instance, in the pressing process in a roll press, the paper web is squeezed together with one or more press felts between two rolls exerting a mechanical pressure on both materials (Fig. 1). During the compression phase water will flow from the paper web into the felt forced by a positive hydraulic pressure gradient. At the end of the press nip, when load is being released, the hydraulic pressure gradient will become negative, which may result in some rewetting caused by the back-flow of water and air from the felt to the paper web. Furthermore, if applying a heated press roll an energy flow from the roll to the paper web will be established at the moment the web makes contact with the press roll. Depending on the temperature and pressure conditions imposed to the paper web/felt sandwich steam may be generated inside the paper web and ultimately induce web delamination, which occurs when the force dissipated by the flow of steam generated inside the paper web is larger than its z-directional strength (Larsson et al., 1998; Orloff et al., 1998). It has been shown, however, that proper temperature/pressure control in the press nip may prevent steam generation inside the paper web. Moreover, the ability of pulp fibres to form fibre-to-fibre bonds during the consolidation process is an important characteristic, which strongly influences the structural and mechanical properties of paper and wood-based materials in general. It depends mainly on wood species, and/or pulping method, fines content, amount of bonding agents (additives, resins), chemical modification of fibres, refining and ultimately on the pressing conditions (Skowronski, 1987). In fact, when high temperature pressing conditions are employed, fibre flexibility and conformability are improved, which may explain the higher sheet densification levels observed under such intense operating conditions.

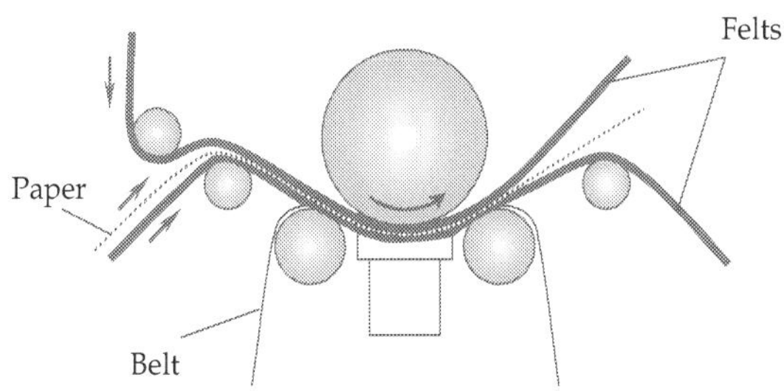

Fig. 1. Press nip of a shoe pressing machine (Aguilar Ribeiro, 2006).

The thermal softening of the fibre's cell wall material is thus partially responsible for the increased mat consolidation and sheet density, but it also induces a significant drop in air and water permeability as the fibrous material dries and consolidates. Since the flow of water and air encounters different cumulative flow resistances across the thickness of the web, the final density profiles may show some signs of stratification, e.g. nonuniform z-direction density profiles. This is influenced by several factors such as the permeability of the pressing head contacting the fibrous material, the temperature/pressure conditions of the pressing event, the web moisture content and fibre's properties, and the uniformity of pressure application.

2.2.2 Hydraulic and structural pressures generated during compression of a wet web: factors affecting the governing mechanisms of water removal

According to Szikla, the role of various factors in dynamic compression of paper is greatly influenced by the moisture ratio of the web, suggesting different governing mechanisms over different ranges of moisture ratio and/or density (Szikla, 1992). In order to remove water by compaction from a web, the mechanical stiffness of the structure must be overcome and water must be transported. The mechanical stiffness of a fibrous mat is influenced by its moisture content, reaches its maximum when all the water has been removed from the web, and decreases continuously as the moisture content increases. Therefore, the pressure carried by the mechanical stiffness of a saturated web during the compression phase of a pressing event cannot be higher than the pressure measured at the same density when an unsaturated web is pressed. The two values may be close to each other as long as significant water transport does not take place in the unsaturated web. The experimental results obtained by Szikla (1992) for 50 g.m^{-2} paper sheets of mechanical or chemical pulps under dynamic load and ingoing moisture ratios in the range 2.0-4.0 kg H_20/kg dry fibres, showed that an increase in chemical pulp beating resulted in higher contribution from hydraulic pressure; an increase in fibre's stiffness, the removal of fines and a decrease in compression rate all lowered the hydraulic pressure. His results also showed that flow in the inter-fibre voids plays an important role in the dynamic compression behaviour of wet fibre mats. When the moisture ratio of the web is high and the compression is fast, as in paper machines, most of the compression force is balanced by the hydraulic pressure that builds up in the layers of the web close to the impermeable pressing surface. This is the case for low grammage paper (e.g. 40-50 g.m^{-2}). The role of hydraulic pressure in balancing the compression force decreases as the compaction of the web increases.

Regarding the mechanisms of dynamic compression of wet fibre mats, the following conclusions can be drawn from the work of Szikla (1992):

- The mechanical stiffness of the structure must be overcome and water must be transported in order to bring about compression of a wet fibre mat. According to this, the force balance prevailing in pressing can be written in the following form:

$$P_t = P_{mec} + P_{flow} \tag{1}$$

where P_t is the total compressing pressure, P_{mec} the pressure carried by the mechanical stiffness of the mat, and P_{flow} the pressure required to transport water;

- The load applied to a wet fibre mat is carried partly by the structure and partly by the water in the interstices of the structure. The structure is formed by fibre material and water. Water located in the lumen of the fibre wall and bound to external surfaces is an integral part of the structure. The pressure carried by the structure is often called structural pressure, P_{st}, and the load carried by the water hydraulic pressure, P_h. The pressure carried by the mechanical stiffness of a fibre mat constitutes only a part of the structural pressure. Another part of the structural pressure is a result of water transport within the fibre material. According to this classification, the force balance can be written in the following form:

$$P_t = P_{st} + P_h = (P_{mec} + P_{fh}) + P_h \tag{2}$$

where P_{fh} is the structural pressure due to water transport within the fibre material. The structural pressure is equal to the pressure carried by the mechanical stiffness of the fibre material only when water transport within the fibre material is negligible. On the other hand, in most paper sheets there are large density ranges over which the pressure generated by the water transport within the fibre material plays a dominant role in forming the structural pressure.

Quantitatively, Terzaghi's principle has to be used carefully in the case of highly deformable pulp fibre networks, as it applies rigorously only to solid undeformable particles with point-like contact points. In a deformable porous material the hydraulic pressure is only effective on a share (1-α) of the area A (Fig. 2). So being, the stress balance may be written as:

$$P_t A = P_{st} A + P_h A(1-\alpha) \Leftrightarrow P_t = P_{st} + P_h(1-\alpha) \tag{3}$$

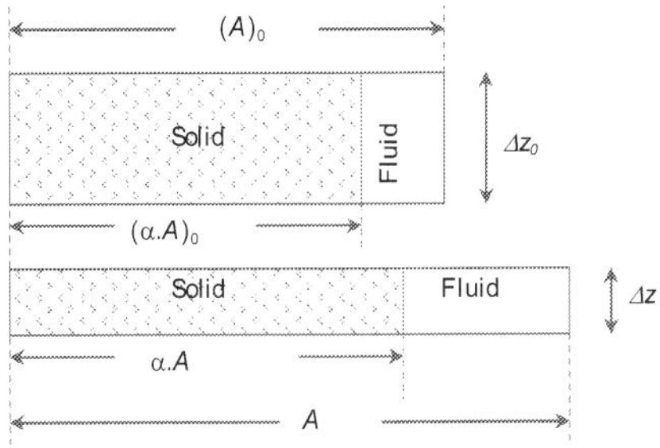

Fig. 2. Schematic diagram of the compression of a deformable porous medium (Δz_0 and Δz are the initial and final thickness of the fibrous material, respectively).

In conclusion, the dynamic compressing force is balanced in the paper web by the following factors: (i) the flow resistance in the inter-fibre channels; (ii) the flow resistance within the fibres (intra-fibre water); (iii) and the mechanical stiffness of the fibre material.

2.3 Fundamentals of wet pressing and high-intensity drying processes: simultaneous heat and mass transfer

2.3.1 Wet pressing

It is convenient to think of wet pressing as a one-dimensional volume reduction process, with the fibrous matrix and water assumed to be a more or less homogeneous continuum. However, when visualized in the microscope (Fig. 3), wet pressing is a far more complex process which combines important mechanical changes in the fibre network with three-dimensional, highly unsteady, two-phase flow through a rapidly collapsing interconnected porous network.

In wet pressing, volume reduction, fluid flow, and static water pressure gradients are intimately interrelated. Classical Fluid Mechanics states that the static water pressure is reduced in the direction of flow by conversion into kinetic energy (water velocity). Some of the total energy available at each layer is lost to friction with the surrounding fibre and by microturbulence in the narrowing flow paths. This loss is associated with fluid shear stresses. However, the water-filled fibre network should not really be considered a continuous confined system (e.g. water flowing in a pipe).

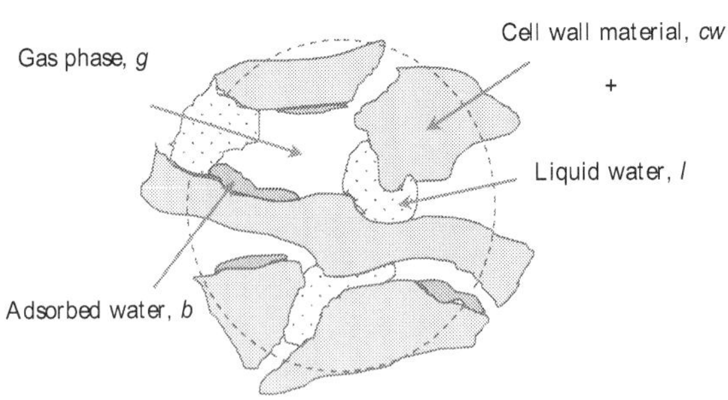

Fig. 3. Micro-scale constituents of paper and MDF (in this case free liquid water should not be considered) (Aguilar Ribeiro, 2006).

The local velocity vector changes direction frequently as the water is forced to take a tortuous path across the collapsing fibre network. Despite the simplifications offered by classical fluid mechanics, it seems safe to say that the static water pressure is highest at the smooth roll surface (if referred to a roll press of a paper machine - or in a lab-scale platen press, as shown in Fig. 4), where water is not in motion relative to the fibres, and lowest at the felted side of the paper, where water velocity is highest. In a roll press the largest static water pressure gradient is not directly downward - it is oriented slightly upstream and, coupled with a significantly higher in-plane sheet permeability, must create some longitudinal water flow component towards the nip entrance.

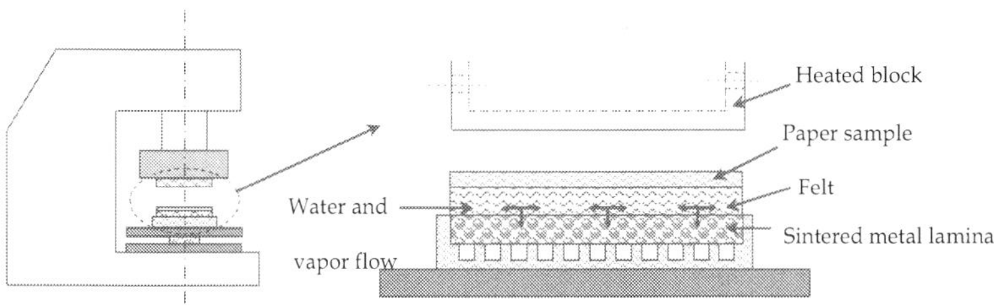

Fig. 4. Schematic drawing of a lab-scale platen press for paper consolidation experiments. The inset shows the water flow pattern and the paper/felt sample arrangement within the press nip (Aguilar Ribeiro, 2006).

Wahlström showed that water is removed from the paper web in the converging part of the press nip due to web compression, but that part of the expressed water returns into the web on the outgoing side of the nip due to capillary forces (Wahlström, 1960). Another important pillar of wet pressing theories has been the division of the total applied load into hydraulic and structural components. The sum of the two pressure components has been considered to be constant in the z-direction and equal to the total applied pressure in the pressing machine but the contribution of these components is considered to change in that, progressing in the direction of water flow hydraulic pressure decreases and structural pressure increases. This idea of separating the two components of pressure, originating from Terzaghi (1943), was applied to the compression behaviour of paper webs by Campbell (1947). Later on, Carlsson's and his co-workers' studies (Carlsson et al., 1977) revealed the important role of water held within fibres in wet pressing, showing that water present in the intra-fibre voids must make a significant contribution to the structural pressure. Only the water in the inter-fibre voids is responsible for the hydraulic pressure. The rest of the structural pressure is the result of mechanical stiffness.
However, it was gradually realized that the original definition of hydraulic and structural pressures was oversimplified. Classical wet pressing theory separates the total applied pressure into only two components - static water pressure and the network compressive stress (usually called mechanical pressure). The stress associated with the fluid drag force - here called fluid shear stress - and the static water pressure drop always appear together; one cannot exist without the other. The vertical component of fluid shear stress should be added to the fibre network stress to obtain the total compressive stress acting at each layer of the fibrous web. Fluid stress is maximum at the outflow side of the paper and nonexistent at the smooth press roll side. It is also nonexistent after the point of zero hydrodynamic pressure since water flow has ceased. Fluid shear stress also has an in-plane component

which must be taken into account when considering paper properties. Although the value of Terzaghi's principle as a tool for quantitative predictions has been questioned by Kataja et al. (1995), it still constitutes the basis of our understanding of wet pressing. Consequently, the operations of press nips are traditionally divided into two categories. In the first case, the press nip is considered to be compression-controlled. Here, the mechanical stress in the fibre network is the dominating factor, and the maximum web dryness is determined by the applied pressure, and is independent of the pressing time. On the other hand, the nip is considered to be flow-controlled when the viscous resistance between water and fibres controls the amount of dewatering. Here, web dryness increases with the residence time at the nip, and the fluid flow is proportional to the pressure impulse which is the product of pressure and time. Schiel's work (1969) led to the conclusion that for many cases the problem was not in applying enough press load (this wouldn't bring much dryness improvement), but in applying enough pressing time. Wahlström also coined the well known terms "pressure-controlled" and "flow-controlled" pressing as a way to denote whether the water removal was restricted by fibre compression response or by fluid flow resistance inside the paper sheet (Fig. 5). It was then concluded that the moisture content of a wet sheet leaving a press nip depends both on the compressibility of the solid fibrous skeleton and on the resistance to flow in the porous space (Wahlström, 1960).

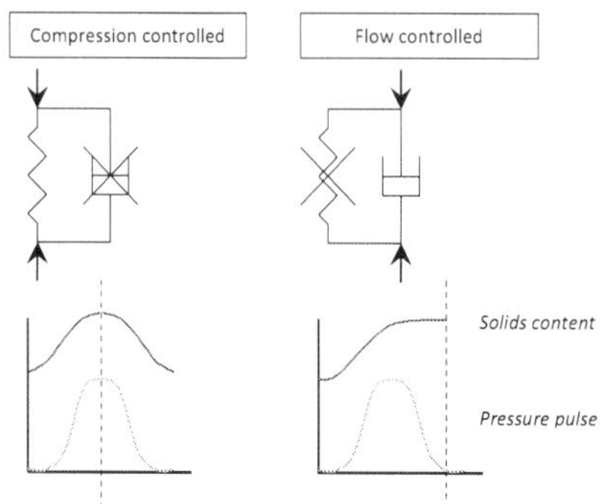

Fig. 5. Schematic drawing of compression-controlled and flow-controlled press nips for an applied roll-like pressure profile on a paper machine (adapted from Carlsson et al., 1982; Aguilar Ribeiro, 2006).

As a consequence of the applicability of Terzaghi's principle to flow-controlled press nips, the web layers closer to the felt in a paper machine are compacted first, with the higher density at the sheet-felt interface. MacGregor (1983) described this phenomenon as stratification and its existence has been observed in laboratory experiments (Burns et al., 1990; Szikla and Paulapuro, 1989a, 1989b; Szikla, 1992). Yet, a recent study performed by Lucisano shows opposing evidence to the existence of a density profile as it was previously reported by several authors. When trying to characterize the delamination process by the changes in transverse permeability and solidity profiles he found no evidence that wet pressing, and even impulse pressing (see section 2.3.3), induced stratification in non-delaminated sheets and concluded that the parabolic solidity profiles observed were due to capillary forces present during oven drying and not a result of hydrodynamic forces induced onto the fibres during the pressing event (Lucisano, 2002).

2.3.2 Batch and continuous hot pressing of medium-density fiberboard (MDF)

MDF, as other wood-based panels, can be manufactured using batch (single or multidaylight) or continuous presses. Steam injection, platen and/or radio-frequency or micro-waves can be used as heating systems. The most common type is the batch press with heated plates (multidaylight), but in the last decade batch presses are being substituted by continuous presses with moving belts. Continuous presses have heating zones along their length and are more efficient than batch presses for thin MDF, allowing to attain line speeds of 120 m/min (Irle & Barbu, 2010).

The consolidation of MDF panels is therefore achieved through hot-pressing. The thermal energy is used to promote the cure of the thermosetting adhesive and soften the wood elements, and the mechanical compression is needed to increase the area of contact between the wood elements to allow the possibility of adhesive bond formation. The hot-pressing process should be regarded as a process of simultaneous mass and heat transfer. However, other mechanisms are also important as they are tightly coupled with heat and mass transfer: the rheological behaviour and the adhesive polymerisation reaction. The material's rheological behaviour, affected by the development of adhesive bonds among fibres, as resin cures, will determine the formation of a density profile in the thickness direction of the MDF panel. These mechanisms are also dependent on temperature and moisture distributions and have direct influence on heat and mass transfer across the mattress porous structure.

Fig. 6. Continuous press in a particleboard plant (courtesy from Sonae Indústria, Portugal).

In MDF, the mat of fibres forms a capillary porous material in which voids between fibres contain a mixture of air and steam. In addition, liquid water may be adsorbed onto the fibres surface. During the hot-pressing process, heat is transported by conduction from the hot platen to the surface. This leads to a rapid rise in temperature, vaporising the adsorbed water in the surface and thus increasing the total gas pressure. The gradient between the surface and the core results in the flow of heat and vapour towards the core of the mattress, therefore increasing its pressure. As a consequence, a positive pressure differential is established from the interior towards the lateral edges, and then a mixture of steam and air will flow through the edges. So, the most important mechanisms of heat and mass transfer involved are (Pereira et al., 2006):

i. Heat transfer by conduction due to temperature gradients and by convection due to the bulk flow of gas: conduction follows Fourier's law;
ii. The gaseous phase (air + water vapour) is transferred by convection; each component is transferred by diffusion and convection in the gas phase. Diffusion follows Fick's law and the gas convective flow obeys Darcy's law: the driving force for gas flow is the total pressure gradient, and diffuse flow is driven by the partial pressure gradient of each component;
iii. The migration of water in the adsorbed phase occurs by molecular diffusion due to the chemical potential gradient of water molecules within the adsorbed phase;
iv. phase change of water from the adsorbed to the vapour state and vice-versa.

Heat transfer by conduction: Heat is transferred through the interface plate/mat to the interior by conduction and will be used to resin polymerisation and to remove water present in the mat as bound water. To evaporate this water it is necessary to supply energy equal to the sum of the water latent heat of vaporisation and the heat of wetting (or sorption) sufficient to break hydrogen bonds between water and wood constituents.

Heat transfer by convection: Convection occurs because the heat transferred from the hot platens causes the vaporisation of moisture, increasing the water vapour pressure. A gradient of vapour partial pressure is formed across the board thickness, causing a convective flow of vapour towards the mat centre. On the other hand, the increase of gas pressure will cause a horizontal pressure gradient that will create a flow of heat by convection to the edges. When the temperature of the medium exceeds water ebullition point, imposed by the external pressure, the horizontal pressure gradient becomes the more important driving force (Constant et al., 1996). However, it is not necessary to attain the ebullition point of free water to have a vapour flow. Any change in temperature will affect the EMC (equilibrium moisture content) of wood and so the vapour partial pressure in the voids (Humphrey & Bolton, 1989). Also, if the vapour is cooled, it will condense, liberating the latent heat and a rapid rise of temperature will occur. So, there is also a phase change associated with the bulk flow, which imparts the temperature change (Kamke, 2004). This condensation will happen continuously from the surface to the core and not as a discrete event, which complicates the modelling of this system.

Heat transfer by radiation: Heat transfer by radiation is usually neglected, since for the relatively lower range of temperatures (< 200 °C), it would be insignificant compared with conduction and convection. However, during press closing and before the platen makes contact with the mat, as well as during the first instants of pressing while mat density is relatively low, heat transfer by radiation can be a significant part of the total heat transferred (Humphrey & Bolton, 1989). On the other hand, on the exposed edges the heat is continuously transported to the surroundings by radiation (Zombori, 2001).

Other heat sources: The other possible sources are the exothermal reaction of the resin cure and the heat of compression. The contribution of the heat of compression is generally neglected. Bowen (1970) estimated that its contribution for heat transfer was around 2%. The contribution of the exothermic polymerisation of the resin depends on the reaction rate and condensation enthalpy.

Mass transfer by convection: In WBP hot-pressing, it is generally assumed that moisture content is below the FSP (fibre saturation point) and so water is present as vapour in cell lumens and voids between particles/fibres, and bound water in cell walls (Kavvouras, 1977; Humphrey, 1982; Carvalho et al., 1998; Carvalho et al., 2003; Zombori, 2001; Thoemen & Humprey, 2006; Pereira et al., 2006). Two main phases are then considered, the gaseous phase (air + water vapour) and the bound water; local thermodynamic equilibrium is also assumed. The gaseous phase is transferred by convection due to a gas pressure gradient (bulk flow) and the water vapour is transferred by diffusion. The bulk flow occurs in response to a gas pressure gradient caused by the vaporisation of moisture present in the mat. Diffusion inside the mat during hot-pressing includes vapour diffusion and bound water diffusion. The driving force for the diffusive flow of vapour is the partial pressure gradient. The convective and diffusive fluxes occur simultaneously, but it is widely accepted that convective gas flow is the predominant mass transfer mechanisms during hot-pressing (Denisov et al., 1975; Thoemen & Humphrey, 2006).

Mass transfer by diffusion: The migration of water in the adsorbed phase occurs by molecular diffusion and follows Fick's first law with the chemical potential gradient of water molecules within the adsorbed phase as the driving force to diffusive flux. This is a slow process and thus it is often considered negligible by some authors (Carvalho et al., 2003) in comparison with steam diffusion. Zombori and others (2002) studied the relative significance of these mechanisms and they found that the diffusion is negligible during the short time associated to the hot-pressing process. The adsorbed water and steam are then related by a sorption equilibrium isotherm.

Capillary transport: At press entry the moisture content of the furnish is relatively low (generally below 14%) and although a possible presence of liquid water brought by the adhesive (water content around 50%) and capillary condensation in some tiny pores, it is generally assumed that the whole mat is below the FSP (Kavvouras, 1977; Humphrey, 1982; Zombori, 2001; Thoemen & Humprey; 2006). In case of particleboard, the moisture content at the press entry might be 11%, while the particle moisture content before resin blending could be around 2-4%. During blending, considerable quantities of water are added with the resin (water content around 50%), and so unless the equilibrium is achieved by the furnish before entering the press (in that case, the water will be adsorbed in the cell walls of wood) some capillary translation might occur (Humphrey & Bolton, 1989). In case of MDF, the fibre drying after the resin spraying in the blow-line results in the decrease of moisture and it is reasonable to consider that the equilibrium will be attained before the hot-pressing, and thus the water will be adsorbed in the fibres (Carvalho, 1999). There is also a possibility of capillary condensation in tiny pores. In case of WBPs, the relative humidity does not exceed 90% (Humphrey, 1984, Kamke and Casey, 1988) and considering a temperature of 115 °C, inside the mat, the maximum pore diameter filled with water will be 0.007 μm. This will correspond to capillary pressures of 14.6 to 20 kPa, which are an order of magnitude less than the predicted maximum vapour pressure differential between the centre and the edges of board (at atmospheric pressure). So, even if some fine capillaries do fill by capillary condensation, it is unlikely that capillary translation of liquid will occur (Carvalho, 1999).

2.3.3 The concept of impulse drying: application to paper production

The most obvious goal driving the development of high-intensity pressing and drying techniques is the quest for higher drying rates and more efficient mechanisms of water removal. One such process seems to be impulse drying which combines wet pressing and drying into a single operation. Impulse drying has been postulated to be economically advantageous since it uses less energy than conventional drying because the increased amount of water removed in the improved press section may not need to be evaporated in the dryer section, which now may use less heated cylinders. Designing more compact and shorter paper machines would mean substantial savings in investments. The concept of impulse drying was first suggested in a Swedish patent application by Wahren (1978). Instead of conducting heat through thick steel dryer cylinders, heat was transferred rapidly from a hot surface to the paper web using a high pressure pulse. The high heat flow to the paper web generates steam in the vicinity of the paper web surface and the idea was that the formed steam would pass right through the paper web and drag the remaining free liquid water towards a "permeable surface" (the felt) on the other side of the paper web, which would result in extremely high water removal rates and energy efficiencies. According to Arenander and Wahren (1983), this could be explained if the following mechanisms would take place during the pressing/drying event:

i. In the first part of the nip, the wet web is subjected to a compressive load and heat is transferred from the heated surface into the proximate layers of the web. The initial part of the drying event may be considered as a consolidation strategy which enhances dewatering by volume reduction and temperature effects on fibres compressibility and water viscosity;
ii. If the boiling point of water at the actual hydraulic pressure is reached, some steam is generated near the hot surface; steam could only expand towards the felt due to the steam pressure gradient established between the upper and lower surfaces of the paper web; at this moment, the voids in the web are completely or partially filled with water, except for the steam pressurized layers close to the hot surface;
iii. If the steam actually flows through the sheet, it may drag some interstitial water out of the web and into the felt (Fig. 7); moreover, water in the fibres walls and lumens is transferred into the inter-fibre space, becoming accessible to removal either by steam rushing through the web or evaporation.

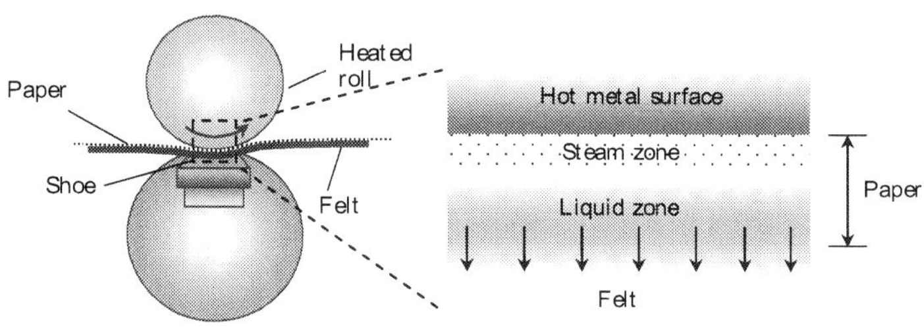

Fig. 7. Design of a shoe press nip. The inset shows a vapour front displacing liquid water in an impulse drying event, as suggested by Arenander and Wahren (1983).

The concept of impulse drying today is somewhat different to Wahren's idea, which consisted in pressing the paper web at a high pressure and high temperature over a short dwell time. Typical operating parameters would be a peak of 2-8 MPa, a temperature of 150-480 °C and a dwell time of 5-15 ms. Temperatures of 200-350 °C and lower average pressures are now being used (Metso, 2010). The contact time in the press nip is 15-50 ms depending on the machine speed and the press nip length. A development of impulse drying is to increase the dwell time even further, to super-elongated press nips, to take full advantage of the effects of high pressing temperatures. For effective dewatering and densification of the paper web, it was therefore proposed that impulse drying should be used in the form of a longer nip dwell time or a so-called shoe press (Metso, 2010). In light of this, the heat and mass transfer mechanisms operating in such a complex event will be further addressed throughout the present manuscript.

2.3.3.1 Hot and superhot pressing, evaporative dewatering, steam assisted displacement dewatering: experimental highlights

The high heat fluxes and water removal rates experienced during the impulse drying event suggest that the mechanisms that control dewatering differ substantially from those of conventional pressing and drying operations. Explanations for such high water removal rates are manifold, but an analysis of the literature published in the field suggests that three modes of water removal can take part in the impulse event:

Hot and superhot pressing, i.e., dewatering by volume reduction, enhanced by temperature effects on network compressibility and water viscosity. The water inside the paper web is considered to be in the liquid state, even if temperature exceeds 100 °C.

Evaporative dewatering, in which thermal energy is used to evaporate water. Here, two modes of liquid-vapour phase change are considered: traditional evaporation or drying, and flashing. Drying refers to the water removal process in which thermal energy is used to overcome the latent heat of evaporation of the liquid phase. Flashing or flash evaporation is another mode of removing liquid water from a solid matrix in which water is exposed to a pressure lower than the saturation pressure at its temperature. In the press nip, water is kept in the liquid state and sensible heat is stored as superheat, which is then converted into latent heat of vaporisation upon nip opening - liquid water is flashed to vapour. The theory of a flash evaporation at the final stage of the impulse drying event was suggested by several authors to explain the dewatering process in impulse drying (Macklem & Pulkowski, 1988; Larsson et al., 2001).

Steam-assisted displacement dewatering, in which liquid water is displaced by the action of a vapour phase. According to some authors (Arenander & Wahren, 1983; Devlin, 1986) the resulting steam pressure hypothetically developed in the initial stage of the pressing event is expected to act as the driving force for water removal, displacing the free liquid water from the wet web to the felt (Fig. 7).

The two main opposing theories to explain the high heat fluxes observed in impulse drying - flashing evaporation and steam-assisted displacement dewatering - found experimental evidence in the works developed by Devlin (1986), Lavery (1987), Lindsay and Sprague (1989), and more recently Lucisano (2002) and Aguilar Ribeiro (2006). Lucisano et al. (2001) performed an investigation of steam forming during an impulse drying event by measuring the transient temperature profiles of wet paper webs subjected to a compressive load in a heated platen press. The initial temperature of the platen press was set from 150 to 300 °C and the length of the applied pressure pulse varied from 100 ms to 5 s. In light of their findings, they advanced that for faster compression rates - as those used in impulse drying - the web stratification induced an increase in the hydraulic pressure which, in its turn, would tend to shift the boiling point of water and prevent steam generation. In summary, the authors believe that for short pulses the hydraulic pressure in most of the sheet is high enough to prevent steam generation and water is present in the liquid phase until the pressure is released. Also, with platen temperatures greater than 200 °C and nip dwell times shorter than 500 ms, they observed a sudden increase in temperature when pressure was released from the paper samples. The same qualitative trends were observed by Aguilar Ribeiro (2006) when experiments were conducted with more realistic pressing conditions (pressing dwell times reaching down to 75 ms and pressure profiles resembling more those used in real press machines) - Fig. 8. The results show that platen temperatures below 150°C did not induce steam generation as the temperature inside the web remained under 100°C.

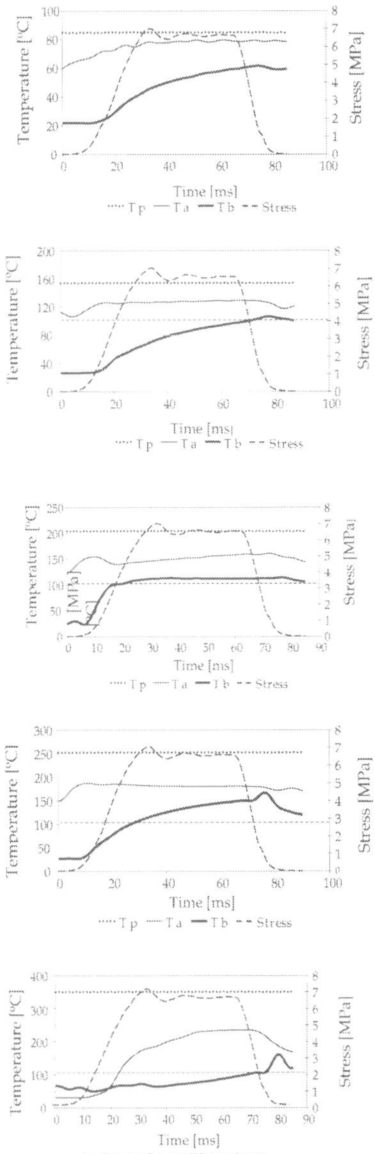

Fig. 8. Internal web temperatures during press drying of 60 g.m^{-2} hardwood unsaturated paper samples. The hot plate temperature was set to 80, 150, 200, 250 and 350 °C and the nip dwell time was 75 ms. T_p is the platen temperature and, T_a and T_b refer to the temperature at the platen/paper and paper/felt interfaces. (Aguilar Ribeiro, 2006).

However, at 200°C and higher temperatures, a sudden increase of web temperature was recorded when the mechanical load was released. This suggests that thin paper sheets tend to exhibit phase change at the end of the press pulse. The flashing of water as pressure is relieved at the end of the nip is seen as a rapid temperature decrease down to 100 °C. When even shorter pulses (30 ms) were applied to thin sheets no clear evidence of such temperature increase was found, except for temperatures of 250°C or above. As suggested before, this may occur as the increasing compression rate causes an increase in the hydraulic pressure, which may imply almost no in-nip steam generation. According to Lucisano's experimental results, this type of flashing phenomena might only be seen for pulse lengths well beyond those encountered in industrial pressing conditions. Lucisano et al. proposed that this mechanism should be termed "flashing-assisted displacement dewatering" since it differs from the steam-assisted displacement of liquid water originally proposed by Wahren (1982) because of the different driving force (Lucisano & Martin, 2006).

Despite some similarities in the heat and mass transfer mechanisms involved in the consolidation process of paper and MDF, there are in fact significant technological differences in what concerns the operating conditions of the corresponding industrial pressing/drying units. Table 1 gives an overview of the typical operating conditions for high-intensity pressing and drying of paper and MDF.

		Paper			MDF
		Press drying	Condebelt drying	Impulse drying	Hot pressing
Mechanical pressure	MPa	0.1 - 0.4	0.02 - 0.5	1 - 5	3 - 4
Temperature of pressing	°C	100 - 250	120 - 180	150 - 500	190 - 220
Dwell time	ms	200 - 300	250 - 10 000	5 - 50; 15 - 100	-
Maximal outgoing dryness	%	45	-	> 50	-
Ingoing moisture content	%	-	-	-	11
Machine speed	m/min	-	100	> 800	7 - 8
Web initial thickness	mm	0.7 - 0.8			40 - 50
Web final thickness	mm	< 0.1			15 - 20

Table 1. Typical operating conditions for continuous high-intensity pressing and drying of paper (adapted from Aguilar Ribeiro, 2006) and MDF (Pereira et al., 2006; Carvalho, 1999; Irle, M. & Barbu M., 2010).

3. Modelling of the high-intensity drying processes

3.1 Introduction

The transport mechanisms in high-intensity drying processes are by nature very complex: modelling and simulation of transport mechanisms in a rigid porous medium pose many problems and the situation is even more complicated when the medium is compressible, such as paper and wood-based materials like, for instance, MDF. Moreover, the coupling between heat and mass transfer is strong, making the material description complicated. The following sections present a brief description of the main heat and mass transfer models that constitute the basis of the development of more complex models used to explain what happens at high-intensity pressing conditions of highly deformable porous materials, such as paper and MDF. Although special emphasis is given to the main driving mechanisms of water removal (temperature and pressure) it is also worth mention the fundamental role of the fibre network consolidation process, which is here addressed in terms of a structural analysis similar to that used for composite materials.

3.2 Mechanical models applied to dewatering processes of compressible fibrous networks

3.2.1 Elasticity, viscoelasticity and plasticity of fibrous composites: paper and MDF

A paper sheet is basically a multiphase material composed of moisture, fibres, voids, and chemical additives, bonded together in a complex network. Thus, it may be considered a composite material, with fibres tending to lie predominantly in the plane of the sheet. Wood itself may be thought of as a natural composite consisting of cellulose fibres interconnected by a primarily lignin binder.

Low and medium-density solids, such as paper and MDF, can therefore be assembled as random networks of fibres, the contact points of which may be bonded together and, according to some authors, mechanical and thermal properties of these materials have much in common with those of cellular materials - honeycombs and foams (Gibson & Ashby, 1988). The question now is to know the preferred mode of deformation experienced by paper and MDF during drying/pressing operations, and how it can be modelled.

The rheological behaviour of paper or MDF in the course of a pressing event is quite complex: the stresses developed due to densification can be relaxed, blocked in the solid structure, released or originate elastic/plastic deformations. These physical processes are tightly coupled with temperature and humidity distributions; the density profile affects the heat and steam/liquid fluxes across the mattress porous structure. During MDF hot pressing, as the resin cures, it is expected that an increase of stress relaxation take place, because of the formation of a network structure that promotes the development of a uniform distribution of stresses (Carvalho et al., 2003). At the beginning of press closure, the compression of the mat is linear. A yield point is reached, when fibre to fibre contact is made from bottom to top of the mat and wide spread fibre bending occurs (Kamke, 2004). From this point on the compression is nonlinear due to the collapse of cell wall. The fibres begin to compress and lumen starts to diminish. The fibre mat behaves as a viscoelastic material and this behaviour is influenced by temperature, moisture content and time. During the hot-pressing event, it can be considered that the MDF mat responds with elastic strain, delayed elastic strain and viscous strain. The elastic stress is immediately recovered after the removal of stress. The delayed elastic strain is also recoverable but not immediately; in addition, the viscous strain is not recoverable upon removal of the stress (Kamke, 2004). So, the four-element Burger model is frequently used to model this behaviour (Fig. 9a). Pereira et al. (2006) used the burger model for modelling the continuous pressing of MDF. However, irreversible changes of the cell wall and mat structure that happen instantaneously upon loading are not represented by the Burger model. So, to account for both viscoelastic behaviour and the instantaneous but irreversible deformation, Thoemen and Humphrey (2003) considered a modified Burger model with a plastic and micro fracture element in series (represented by a spring that operates only in one direction) - Fig. 9b. Carvalho et al. (2006) and Zombori (2001) considered the Maxwell body (Fig. 9c) as an alternative simplified model, due to excessive solution time of their global models for the hot-pressing of MDF and OSB, respectively.

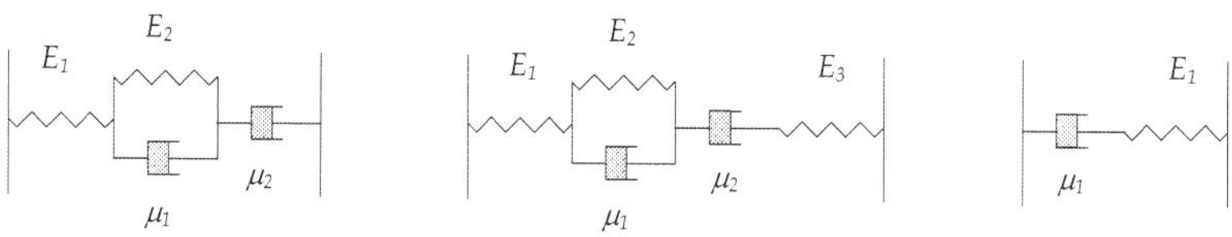

(a) Burger model (b) Modified Burger model (c) Maxwell model

Fig. 9. Mechanical analogues used to describe the rheological behaviour of MDF throughout a pressing/drying event (Carvalho, 1999).

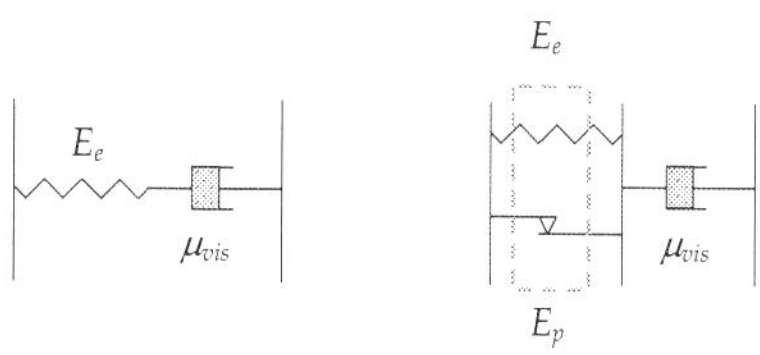

$$\varepsilon = f(\sigma, t) \tag{4}$$

$$\frac{d\varepsilon}{dt} = \underbrace{\left(\frac{\partial \varepsilon}{\partial \sigma}\right)\frac{d\sigma}{dt}}_{\text{elastoplasticity}} + \underbrace{\left(\frac{\partial \varepsilon}{\partial t}\right)_{\sigma}}_{\text{viscosity}} \tag{5}$$

(a) Maxwell model (b) Modified Maxwell model (visco/elastoplastic model)

Fig. 10. Mechanical analogue of the so-called modified Maxwell unit representing the "visco/elastoplastic" model proposed by Aguilar Ribeiro (2006) for the transverse compression of paper in a press nip. E_e and E_p represent the elastic and plastic moduli of the composite material, respectively.

The rheological behaviour of wet paper samples is somehow similar to MDF mats. When subjected to low loads, paper behaves as a linear elastic material, but under very high loads its stress-strain curve shows a hysteresis. This phenomenon is due to a plastic strain of the web during compression, giving rise to a nonlinear stress-strain curve. In order to model the deformation of paper and MDF during compression, an approach to cellular materials theories is presented in Section 3.2.2. Meanwhile, a brief description of the rheological model for paper is presented herein, taking into account the flow resistance of intra- and extra-fibre water, the contribution of water vapour and air to the elastic modulus of the network, and the fibre rheology itself. At the risk of some simplification, the proposed model may be represented by a mechanical analogue consisting of three elements coupled in series and in parallel (Fig. 10). One element is purely elastic (recoverable), the other is plastic (the unrecoverable strain is associated with the structure of the material), and the third element is viscous (time-dependent). In summary, the elastic and plastic elements characterize the structural integrity of the medium, and the viscous elements the degree to which the medium components move or change position with time when subjected to a stress. In a press nip, paper deforms in three stages. At first, the volume is reduced; as higher pressure is imposed, the fibre network starts to take up load and deformation is achieved through elastic-plastic buckling; at some critical deformation of the fibre network, further deformation can only be achieved by "crushing" of the fibres (Rodal, 1989; Gibson & Ashby, 1988). Bearing this in mind, Aguilar Ribeiro (2006) proposed a modified Maxwell

model to describe the nonlinear densification of paper in the pressing section of a paper machine, using simple arrangements of springs, dampers and "dry-friction" elements; the mechanical model has been referred to as "visco/elastoplastic" (Fig. 10).
Considering that strain is only a function of two variables, i.e. time (t) and stress (σ), the governing equation for the modified Maxwell model, which defines paper's behaviour when subjected to a dynamic stress, is given by Eq. (4). It clearly states that the total deformation (ε) of the material may be separated into elastoplastic and viscous deformation. Starting from Hooke's law and taking the derivative form of the equation, it follows that

$$\sigma = E \cdot \varepsilon \Rightarrow \frac{\partial \sigma}{\partial \varepsilon} = E + \varepsilon \frac{\partial E}{\partial \varepsilon} \tag{6}$$

As for the viscous constituent of the modified Maxwell unit, it can be represented by Newton's law of viscosity:

$$\left(\frac{\partial \varepsilon}{\partial t}\right)_{\sigma} = \frac{\sigma}{\mu_{vis}} \tag{7}$$

where μ_{vis} is the viscous parameter of the network, including the viscosity of its constituents (Aguilar Ribeiro, 2006).
Finally, the governing differential equation for a modified Maxwell element, describing the compression behaviour of paper, may be expressed as

$$\frac{d\varepsilon}{dt} = \frac{1}{E + \varepsilon \frac{\partial E}{\partial \varepsilon}} \frac{d\sigma}{dt} + \frac{\sigma}{\mu_{vis}} \tag{8}$$

A similar approach may be defined for MDF. In this case, a linear viscoelastic behaviour may be assumed, as suggested by Carvalho (1999) – the first term on the right-hand side of Eq. (8) is simplified to $\frac{1}{E}\frac{d\sigma}{dt}$. The following section presents a brief description of the cellular solids theory to estimate the elasticity modulus of the composite materials (E), paper and MDF.

3.2.2 Application of cellular solids theory to paper and wood-based materials

The applicability of the cellular material compression theories to describe the nonlinearity of the transverse compression of solid wood has been demonstrated by several authors (Wolcott et al., 1989; Lenth & Kamke, 1996a; Lenth & Kamke, 1996b; Easterling et. al, 1982). This same approach has been used to describe the consolidation of MDF and paper by Carvalho (1999) and Aguilar Ribeiro (2006), respectively.
In a press nip of a paper machine, as the fibre network becomes compacted, a hydraulic pressure builds up in the water held within the fibre walls (intra-fibre water) and a part of it is driven out (Carlsson, 1983; Vomhoff, 1998). Thus, only a part of the structural stress is due to the mechanical stiffness of the fibre network (Szikla & Paulapuro, 1989a, 1989b). The main resistance is due to the flow of the intra-fibre water within and out of the fibre walls, which is inherently a viscous phenomenon. The rheology of the fibre network can therefore be expected to be rate-dependent, and a model of the fibre network rheology should capture the mechanical stiffness and the stress due to the flow resistance inside the fibres. From a

composite point of view, the "elastoplastic" modulus of a paper sheet (E) is therefore seen as the sum of the contributions of both solid and fluid phases. Consequently, the following relation based on the longitudinal rule of mixtures can be written,

$$E = E_s\phi_s + E_g\phi_g + E_l\phi_l \tag{9}$$

where ϕ_s, ϕ_g and ϕ_l represent the volume fractions of the solid, gas and liquid phases in the composite material.

In addition, it is known that for cellular materials, the elasticity modulus (E) depends both on strain and relative density of the solid (ρ_{rel} - defined as the ratio of the apparent density of the porous material (cell wall material + additives + adsorbed water), ρ, to the real density of the solid of which it is made, ρ_s). This relation has been described by Gibson and Ashby (1988) for closed cellular solids as

$$E = C_2 E_{cw} \rho_{rel}^{\alpha} \tag{10}$$

where C_2 is a constant, E_{cw} the elasticity modulus of the fibre cell wall material, and α a parameter which is a function of the material structure ($1.5 < \alpha < 3.0$). Bearing in mind the water removal mechanisms occurring in a paper pressing event, Aguilar Ribeiro (2006) considered the "open-celled foam" version of Eq. (10) to estimate the elasticity modulus of the fibrous solid structure (Maiti et al., 1984). The solid matrix is therefore assumed to consist of open interconnected cells through which fluids (liquid water, air and water vapour) flow as a consequence of material deformation:

$$E = C_2 E_{cw} \rho_{rel}^{2} \tag{11}$$

Finally, it is worth mention that Eq. (9) is derived from Hooke's law for linear elasticity, and this is not valid for the paper and MDF consolidation process since these materials exhibit nonlinear behaviour. As such, the corresponding compressive stress-strain curves are represented by the modified Hooke's law, which takes into account the linear as well as the nonlinear mechanical response of the material by introducing an additional nonlinearising term, $\Phi(\varepsilon, \rho_{rel})$ (Gibson & Ashby, 1988; Maiti et. al, 1984). Eq. (9) may now be written in its final form as

$$E = C_2 E_s \rho_{rel}^{2} \Phi(\varepsilon, \rho_{rel}) + E_g\phi_g + E_l\phi_l \tag{12}$$

where E_s accounts for the fibre cell wall material, adsorbed water on the surface of the cellulose fibres, and any possible additives or resins used in the fibre stock preparation.

3.3 Heat and mass transfer models

3.3.1 Simultaneous heat and mass transfer models for paper and MDF

A rigorous model for high-temperature pressing of paper and other wood-based materials will involve the simultaneous solution of a wet pressing model and a heat transfer model. Such a model will be extremely complex and highly non-linear, especially if phase-change phenomena are to be included.

Heat and mass transfer models for MDF: batch, continuous and HF pressing

Since the eighties, several models have been published in the literature for the batch process, mostly for particleboard. However, these models have inherent limitations, either because they are one-dimensional or do not couple all the phenomena involved in the process. The first models that were developed for the hot-pressing of particleboard attempted to describe only simultaneous heat and mass transfer (Kamke & Wolcott, 1991). The improvement of computer performance induced the development of two or three-dimensional models, although with some limiting simplifications, namely treating the problem as pseudo-steady-state (Humphrey & Bolton, 1989) or simply predicting the behaviour of a single variable (Hata et al., 1990). For MDF, a three-dimensional unsteady-state model was presented (Carvalho & Costa, 1998), describing the heat and mass transfer. A global model, integrating all the mechanisms involved (rheological behaviour and resin polymerisation reaction) was also presented later (Carvalho et al., 2003). Almost at the same time, a combined stochastic deterministic model was developed by Zombori (2001, 2002) to characterise the random mat formation and the physical mechanisms during the hot-pressing of OSB. A two-dimensional model of heat and mass transport within an oriented strand board (OSB) mat, during the hot pressing process, was presented also by Fenton et al. (2003). This model was later combined with a model to predict mat formation and compression (Painter el al., 2006a) and another to predict the mechanical properties (Painter et al., 2006b). The global model was also used in a genetic algorithm to carry out an optimisation study of batch OSB manufacturing (Painter et al., 2006b). Dai and Yu (2004, Dai et al., 2007) presented a model that provides a mathematical description of the coupled physical phenomenon in hot-pressing of OSB. This model was then validated with experimental data (temperature and gas pressure inside the mat) (Dai et al., 2007).

As for the continuous pressing, the description of the phenomena involved corresponds to the modelling of a porous and heterogeneous media in movement. The main difficulty associated to this type of problems is the choice of the reference system to make easy the numerical solution of the equations of conservation of mass, energy and momentum. While the batch process represents an unsteady state problem, the continuous process can be described as a steady-state process using the press as the reference system. Thoemen and Humphrey (1999, 2001, 2003) presented an analytical model that is based on thermodynamics, rheological concepts and numerical solution used by Humphrey. This model accounts for combined heat and mass transfer, adhesive cure, mat densification and stress relaxation. Lee (2006) presented an optimisation of OSB manufacturing that focused on the continuous pressing process, but did not consider mechanical strength of the panel. Pereira et al. (2006) presented a three-dimensional model for the continuous pressing of MDF. A comprehensive description of the mechanisms involved, as well as the equations and the numerical method used for solving this problem were described. The set of equations for the conservation of energy, mass and momentum was deduced in an Eulerian reference system (taking the press as reference), then changed to a moving reference system (taking the mat as reference) and finally solved as an unsteady-state problem.

An alternative to the hot-platen heating is the use of high frequency heating, which has the advantage of reducing the duration of the pressing cycle, the platen temperature and the post-curing time, for the same resin formulation. The reported work on modelling of HF heating in the production of wood-based panels is scarce. Pereira et al. (2004) presented an electromagnetic heating model that was coupled with a three-dimensional model for heat and mass transfer and resin polymerisation previously presented. This dynamic model was used to predict the evolution of the local variables related to heat and mass transfer (temperature and moisture content), as well as the variable connected to the electromagnetic behaviour (dielectric properties of the mat).

Heat and mass transfer models for paper: high-intensity processes and impulse drying

Phase-change problems in which a phase boundary moves have received much attention in recent years, and impulse drying of paper is just an example. Here, and according to some authors (Ahrens, 1984; Pounder, 1986), the vapour-liquid boundary moves not only because of phase-change but also the liquid is driven out by the generated vapour pressure. Impulse drying is also related to another set of moving boundary problems involving phase displacement in porous media. Pounders (1986) and Ahrens (1984) presented a model for high-intensity drying of paper, which could be applied to impulse drying. The drying process is idealized in the sense that paper is divided in different zones comprising different amounts of fibre, liquid water and water vapour. The model is based on solving the conservation equations of heat and mass in the different zones, combining equations which describe the applied pressure and the physical properties of liquid and vapour, as well as equations describing the thermal properties, compressibility and permeability of paper. From their study it was found that in many cases the model predicted a higher degree of water removal than that observed in press drying experiments. Later, Lindsay (1991) proposed a model in which vapour and liquid are assumed to be in equilibrium in a two-phase zone between the dry zone (close to the hot surface of the press machine) and the wet zone (near the felt). Heat transfer is then governed by evaporation occurring at the dry interface and condensation at the wet interface, thus predicting an almost constant temperature profile within the two-phase zone. Experimental temperature profiles used for comparison in his study showed similar behaviour, a plateau of almost constant temperature. In addition, Lindsay found that the model could predict the heat fluxes during impulse drying, showing the basic features found in the experimental investigations, but they were somehow overestimated.

Unlike the earlier authors, Riepen (2000) proposed a model in which conduction and convective heat transfer was considered. His model includes the coupling of a wet pressing model and a heat transfer model, and it describes the transfer of mass and energy through the paper, providing the possibility of studying flash expansion during nip opening. According to Riepen's model, the steam formed could increase the hydraulic pressure and thus the hydraulic pressure gradient across the paper thickness, which is the driving force in the impulse drying dewatering process (Riepen, 2000).

In an effort to use a simple model describing the dewatering process in impulse drying, Nilsson and Stenström (2001) treated paper as a two-phase medium of cellulose fibres and water where the two components constitute a homogeneous matrix. The model is based on solving the energy equations for a compressible medium and is limited to the compression phase of the impulse drying event, without phase change and not considering the structural aspects of the fibrous network. According to the authors, the good agreement between the predicted temperature profiles and the experiments performed at elevated temperatures reinforces the assumption that impulse drying is a process with enhanced wet pressing due to the increased web temperature followed by flashing of superheated water. In addition, they advanced that heat conduction and convection are always present regardless of the temperature and pressure (and the transport mechanisms involved in impulse drying can occur simultaneously in different parts of the web), although the convective heat transfer is low and of minor importance (Nilsson & Stenström, 2001).

Gustafsson and Kaul (2001) presented a general model of wet pressing at high temperature, in which the fibrous network rheology was described by the model suggested by Lobosco and Kaul (2001). The model includes the rate dependency of the fibre network stress and thus takes into account the flow resistance of the intra-fibre water, something not done in previous works. The capability of the model to predict changes in the solids content and structural properties of the fibre web as it passes through the press nip was good within the range of temperatures and linear loads studied by the authors.

More recently, a similar approach was used to describe the consolidation of paper in an impulse drying event taking into account its nonlinear behaviour to compression resembling that of cellular structures (Aguilar Ribeiro, 2006; Aguilar Ribeiro & Costa, 2007a). Paper is seen as a medium composed of three phases, solid, liquid and gas (air and water vapour) in thermodynamic equilibrium; and three main heat and mass transfer mechanisms are assumed: heat conduction, convective heat and mass transfer, and phase change of water either from the adsorbed or liquid state into vapour. A brief description of the model is presented hereafter, and a more detailed explanation can be found elsewhere (Aguilar Ribeiro, 2006; Aguilar Ribeiro & Costa, 2007a; Aguilar Ribeiro & Costa, 2007b).

The development of a mathematical model for any system involves the formulation of equations describing the physics of the process based on the fundamental laws of conservation of matter and energy. For high temperature pressing, five coupled equations have to be solved to calculate the heat and mass transfer in the wet fibre web: the continuity equations for free liquid water (if it exists), water vapour, gas phase and energy; and one equilibrium equation (moisture sorption isotherm or one describing the liquid/vapour saturation). Finally, the mechanical behaviour of the mat to compression is described by one equation that relates the vertical position in the web thickness direction and local deformation, one to determine the stress development inside the mat, and one equation for the total thickness of the web. The description of the model presented hereafter does not intend to be exhaustive, but rather give some insight into the main constitutive equations that govern the heat and mass transfer phenomena during a press/drying event of a wet fibre mat, such as paper or MDF. In particular, this work focus the compression of a paper web as it passes through the nip of an impulse drying unit of a paper machine. A more complete and detailed description of the models can be found elsewhere (Aguilar Ribeiro, 2006; Aguilar Ribeiro & Costa, 2007a; Aguilar Ribeiro & Costa, 2007b; Carvalho, 1999).

3.3.1.1 The energy conservation equation

The first term in Eq. (13) refers to the accumulation of heat in the wet web followed by the terms concerning the liquid/vapour phase change, which equal the conductive and convective heat fluxes in the three main directions, x, y and z (machine, cross-machine and thickness direction, respectively):

$$\left(\rho C_p \frac{\partial T}{\partial t} + C_p T \frac{\partial \rho}{\partial t}\right) + \dot{M}\left(\Delta H_{vap} + Q_l\right) = -\left(\frac{\partial \varphi_{t_x}}{\partial x} + \frac{\partial \varphi_{t_y}}{\partial y} + \frac{\partial \varphi_{t_z}}{\partial z}\right) \tag{13}$$

where T is the temperature inside the mat, ρ and C_p are the density and the specific heat capacity of the mat, $\dot{M}$ the amount of vaporised water per unit time and volume, ΔH_{vap} the latent heat of vaporisation of water, Q_l the differential heat of sorption for the wood-water system, and φ_{t_i} ($i = x, y, z$) is the total heat flux in the three main directions of the mat defined as:

$$\varphi_{t_i} = -\lambda_i \frac{\partial T}{\partial i} + \left(\rho_g v_{g_i} C_{p_g} + \rho_l v_{l_i} C_{p_l}\right) T \tag{14}$$

where λ_i is the thermal conductivity of the mat in a given i direction, ρ_g , ρ_l , C_{p_g} and C_{p_l} the specific densities and heat capacities of both gas and liquid phases, and finally v_{g_i} and v_{l_i} represent the superficial velocity of gas and liquid in the three main directions defined by Darcy's law.

3.3.1.2 The mass conservation equations: liquid and gas phases

Water vapour phase

To establish the mass balance for water vapour, two possible scenarios may be drawn: (i) *below the fibre saturation point:* at this stage there is no free liquid water in the web, which means that the vapour formed comes entirely from the vaporisation of the adsorbed water on the cellulose fibres (this is the case of MDF, in which only adsorbed water is present); (ii) *above the fibre saturation point*: as adsorbed water is assumed to be retained on the surface of the cellulose fibres, the liquid water is considered to be in equilibrium with its saturated vapour. The water vapour mass balance may then be expressed as:

$$\left(\dot{M}_b+\dot{M}_l\right)=\frac{\partial\left(\phi_g\rho_v\right)}{\partial t}+\left(\frac{\partial\varphi_{v_x}}{\partial x}+\frac{\partial\varphi_{v_y}}{\partial y}+\frac{\partial\varphi_{v_z}}{\partial z}\right) \tag{15}$$

where $\dot{M}_b$ and $\dot{M}_l$ are the amount of adsorbed or liquid water vaporised per unit time and volume, ϕ_g is the volume fraction of gas (air and water vapour) in the entire mat, ρ_v the specific density of vapour and φ_{v_i} is the total water vapour flux in the i direction defined as:

$$\varphi_{v_i}=\rho_v v_{g_i}-\phi_g D_i^{eff}\frac{\partial\rho_v}{\partial i}=\frac{MM_v}{R}\left[\frac{P_v}{T}v_{g_i}-\phi_g D_i^{eff}\frac{\partial}{\partial i}\left(\frac{P_v}{T}\right)\right] \tag{16}$$

D_i^{eff} is the effective diffusivity of water vapour in the pores of the mat for a given i direction, MM_v the molar mass of water, R the universal gas constant and P_v the vapour partial pressure.

Gas phase (air + water vapour)

Following the same approach used for the vapour mass balance, the conservation equation for the gas phase may be written as:

$$\left(\dot{M}_b+\dot{M}_l\right)=\frac{\partial\left(\phi_g\rho_g\right)}{\partial t}+\left(\frac{\partial\varphi_{g_x}}{\partial x}+\frac{\partial\varphi_{g_y}}{\partial y}+\frac{\partial\varphi_{g_z}}{\partial z}\right) \tag{17}$$

where ρ_g is the specific density of the gas phase, and the gas flux in the i direction (φ_{g_i}) is given by:

$$\varphi_{g_i}=\rho_g v_{g_i}=\rho_g\left(-\frac{K_{g_i}}{\mu_g}\frac{\partial P_g}{\partial i}\right) \tag{18}$$

and K_{g_i} and μ_g are the permeabilities and the viscosity of the gas phase in the porous medium.

Liquid phase

When the material is above the fibre saturation point, the amount of vapour generated comes from the liquid phase in equilibrium with its saturated vapour (in this case, $\dot{M} = \dot{M}_l$ and $\dot{M}_b = 0$). Therefore, the amount of vaporised liquid is defined by the liquid/vapour equilibrium relation, given by the water vapour saturation pressure ($P^{saturation}$) at a temperature T:

$$P_v = P^{saturation} = f(T) \;\Rightarrow \frac{\partial P_v}{\partial t} = \frac{\partial P^{saturation}}{\partial t} = \left(\frac{\partial P^{saturation}}{\partial T}\right)\left(\frac{\partial T}{\partial t}\right) \tag{19}$$

The mass balance for the liquid phase is then expressed as:

$$\frac{\partial(\phi_l \rho_l)}{\partial t} = -\left(\frac{\partial \varphi_{l_x}}{\partial x} + \frac{\partial \varphi_{l_y}}{\partial y} + \frac{\partial \varphi_{l_z}}{\partial z}\right) - \dot{M}_l \tag{20}$$

where ρ_l is the specific density of the liquid phase, and the water flux in the i direction (ϕ_{l_i}) is given by:

$$\varphi_{l_i} = \rho_l v_{l_i} = -\frac{\rho_l K_{l_i}}{\mu_l}\left(\frac{\partial P_g}{\partial i} - \frac{\partial P_c}{\partial i}\right) \tag{21}$$

and K_{l_i}, μ_l and P_c are the permeabilities, the viscosity and the capillary pressure of the liquid water in the porous medium.

On the other hand, if no liquid water is present in the mat (this is the case for MDF), the vapour results exclusively from the vaporisation of adsorbed water which is now in equilibrium with the vapour in the pores (in this case, $\dot{M} = \dot{M}_b$ and $\dot{M}_l = 0$). Therefore, the water/vapour equilibrium may be described by a sorption isotherm as that derived by Nadler et al. (1985). The relative humidity, *RH*, and the moisture ratio at the equilibrium (here defined as MR_{eq}) are related by the following equation:

$$MR_{eq} = f(RH) = f\left(\frac{P_v}{P^{saturation}}\right) \tag{22}$$

Assuming that the amount of water being vaporised per unit time and volume is given by the variation of the moisture ratio of the web,

$$\dot{M}_b = -\frac{\partial m_b}{\partial t} = -\frac{\partial}{\partial t}\left(\frac{m_{dry} \cdot MR_{eq}}{V}\right) = -\frac{\partial}{\partial t}\left(\rho_{dry} \cdot MR_{eq}\right) \tag{23}$$

where ρ_{dry} is the apparent density of the dry solid defined as the mass of the fibre cell wall material and additives, m_{dry}, per unit volume V.

Finally, if no liquid water is present in the web, Eq. (19) may be rewritten as follows:

$$\frac{\partial P_v}{\partial t} = \frac{P^{saturation}\left(\dfrac{\partial MR_{eq}}{\partial t}\right)}{\left(\dfrac{\partial MR_{eq}}{\partial RH}\right)_T} = -\frac{\dfrac{P^{saturation}}{\rho_{dry}}\left(\dot{M}_b + MR_{eq}\dfrac{\partial \rho_{dry}}{\partial t}\right)}{\left(\dfrac{\partial MR_{eq}}{\partial RH}\right)_T} \tag{24}$$

3.3.1.3 The momentum conservation equation

The governing differential equations describing the consolidation of the web take into account the viscoelastic behaviour of the mat under compression, as explained in section 3.2. According to the mat discretisation grid, used to model the entire mat (Fig.11), the vertical position for each *ijk* representative volume element is given by Eq. (25a).

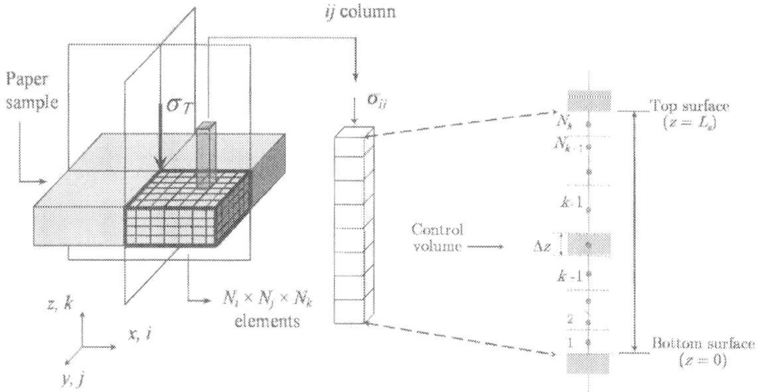

Fig. 11. Schematic representation of the mechanical model in compression, with discretisation in the thickness direction of the web (here divided in $N_i \times N_j \times N_k$ control volumes) (Aguilar Ribeiro, 2006).

$$\frac{\partial z_{ijk}}{\partial t} = \frac{\Delta z_0}{2}\left(\frac{\partial \varepsilon_{ijk}}{\partial t} + 2\sum_{l=1}^{k-1}\frac{\partial \varepsilon_{ijl}}{\partial t}\right) \tag{25a}$$

with

$$\frac{\partial \varepsilon_{ijk}}{\partial t} = \frac{1}{E_{ijk} + \varepsilon_{ijk}\dfrac{\partial E_{ijk}}{\partial \varepsilon_{ijk}}}\frac{\partial \sigma_{ijk}}{\partial t} + \frac{\sigma_{ijk}}{\mu_{ijk}} \tag{25b}$$

The applied stress in every *ij* column of the web (σ_{ij}) may be expressed in terms of the local deformation (ε_{ij}):

$$\frac{\partial \sigma_{ij}}{\partial t} = E_{ij}\frac{\partial \varepsilon_{ij}}{\partial t} - \frac{E_{ij}}{\mu_{ij}}\sigma_{ij} \tag{26a}$$

with

$$E_{ij} = \frac{N_k}{\displaystyle\sum_{k=1}^{N_k}\frac{1}{E_{ijk} + \varepsilon_{ijk}\dfrac{\partial E_{ijk}}{\partial \varepsilon_{ijk}}}} \tag{26b}$$

$$\mu_{ij} = \frac{N_k}{\sum_{k=1}^{N_k} \frac{1}{\mu_{ijk}}} \tag{26c}$$

where E_{ij} and μ_{ij} are the elastic and viscous parameters defined for each ij column. Again, for the entire mat it will be assumed the same rheological behaviour, *i.e.* the total deformation of the fibrous network (ε_T) is described by the viscoelastic constitutive relation:

$$\frac{\partial \varepsilon_T}{\partial t} = \frac{1}{E_T}\frac{\partial \sigma_T}{\partial t} + \frac{\sigma_T}{\mu_T} \tag{27a}$$

$$E_T = \frac{1}{N_i N_j}\sum_{ij} E_{ij} \tag{27b}$$

$$\mu_T = \frac{1}{N_i N_j}\sum_{ij} \mu_{ij} \tag{27c}$$

where the subscript "T" refers to the entire structure. The initial and boundary conditions, as well as the physical, mechanical and thermodynamic properties of the web, required to solve this set of partial differential equations are described elsewhere (Aguilar Ribeiro, 2006; Aguilar Ribeiro & Costa, 2007a; Aguilar Ribeiro & Costa, 2007b; Carvalho, 1999).

4. Technological barriers and future perspectives

Although the knowledge concerning the phenomena involved in high temperature pressing of paper, including impulse drying, has increased tremendously over the years, modelling and simulation of the process must me continued. Improvements and extensions are required to describe the post-nip stage of the process and to clarify the amount and severity of the flash expansion. As the highest temperature occurs at the paper side in contact with the heated cylinder of the press machine, the application of coatings with special thermal properties (high thermal mass) at this interface would influence the interface temperature as well as the distribution inside the paper sheet and most probably reduce the effects of flash expansion.

Concerning wood-based panels, advances particularly in the fields of adhesive formulation, production technology, as well as online measuring and control techniques, have triggered a technology push in the manufacture of these materials (Thoemen, 2010). However, in the majority of production lines, the scheduling of the press cycle is still done empirically, based on a trial and error methodology in the production line. Models are important tools for the scheduling of the press cycle as well as for the prediction of the final product properties, but technological barriers still exist for the direct application in industry. Despite the advances on computer modelling in other areas during the last two decades, the full potential of today's models have not been exploited yet, and not all the models meet the needs of the industry (COST Action E49, 2004). The increase in complexity (real world conditions, raw material variability due to the use of wood residues and recycled wood) and the need for faster numerical solutions (real time simulator) might well be the driving forces for the use of other numerical strategies, including the use of artificial intelligence methods (genetic algorithms, neural networks, fractals) and distributed agents methods.

5. References

Aguilar Ribeiro, H. (2006). Study of the rheology of deformable porous media: application to paper production. PhD thesis, University of Porto, Portugal.

Aguilar Ribeiro, H. & Costa, C.A.V. (2007a). Modeling and Simulation of the Hot-Pressing Process in Paper Production: A Heat- and Mass-Transfer Analysis. *Industrial & Engineering Chemistry Research*, 46, 24, 8205-8219.

Aguilar Ribeiro, H. & Costa, C.A.V. (2007b). Modelling and simulation of the Nonlinear Behavior of Paper: A Cellular Materials Approach. *Chemical Engineering Science*, 62, 6696-6708.

Ahrens, F. (1984). Fundamentals of drying. Technical Report Project 3470, The Institute of Paper Chemistry, USA.

Arenander, S. & Wahren, D. (1983). Impulse drying adds new dimension to water removal. *Tappi Journal*, 66, 9, 123-126.

Athanassiadou, E.; Tsiantzi, S. & C. Markessini (2007). Towards composites with formaldehyde emission at natural wood levels, In: *COST Action E49 'Measurement and Control of VOC Emissions from Wood-Based Panels'*, Braunschweig, Germany.

Bowen, M.E. (1970). Heat transfer in particleboard during hot-pressing. PhD dissertation, Colorado State University, Corvallis, USA.

Burns, J.R.; Connors, T.E. & Lindsay, J.D. (1990). Dynamic measurement of density gradient development during wet pressing. *Tappi Journal*, 73, 4, 107- 113.

Campbell, W.B. (1947). The physics of water removal. *Pulp and Paper Magazine of Canada*, 48, 3, 103-109.

Carlsson, G. (1983). Some Fundamental Aspects of the Wet Pressing of Paper. PhD thesis, Royal Institute of Technology, Sweden.

Carlsson, G.; Lindström, T. & B. Norman (1982). Some basic aspects on wet pressing of paper, In: *IPC International Water Removal Symposium*, 5-12, CPPA, Vancouver, Canada.

Carlsson, G.; Lindström, T. & Söremark, C. (1977). Expression of Water from Cellulosic Fibers under Compressive Loading, *Transactions of the British Paper and Board Industry Federation Symposium on Fiber – Water Interactions in Papermaking*, 389 – 409, British Paper and Board Industry Federation, Oxford, UK.

Carvalho L. (1999). Estudo da Operação de prensagem do Aglomerado de Fibras de Média Densidade: Prensa descontínua de pratos quentes (Study of the hot-pressing of Medium Density Fiberboard (MDF): batch platen press). PhD Thesis in Chemical Engineering, Faculty of Engineering of University of Porto, Portugal.

Carvalho, L.H.; Costa, M.R.N. & Costa, C.A.V. (2003). A Global Model for the Hot-Pressing of Medium Density Fiberboard (MDF). *Wood Sci. and Technol.*, 37, 241-258.

Carvalho, L.M. & Costa, C.A.V. (1998). Modeling and simulation of the hot-pressing process in the production of medium density fiberboard (MDF). *Chem. Eng. Comm.*, 170, 1-21.

Carvalho, L.M. (2008). Modelling of the Hot-Pressing Process, In: *Core Document of COST E34 "Bonding of Timber"*, Lignovisionen issue 18, 197-201, Universität für Bodenkultur, ISSN 1681-2808, Wien, Austria.

Confederation of European Paper Industries (2010). Key Statistics 2009 - European Pulp and Paper Industry. CEPI, Brussels.

Constant, T.; Moyne, C. & Perré, P. (1996). Drying with internal heat generation: theoretical aspects and application to microwave heating. *AIChE J.*, 42, 2, 359-368.

COST Action E49 (Processes and Performance of Wood-based Panels) "Memorandum of Understanding", 2004.

Dai, C. & Yu, C. (2004). Heat and mass transfer in wood composite panels during hot-pressing: Part I. A physical-mathematical model. *Wood and Fibre Sci.*, 36, 4, 585-597.

Dai, C.; Yu, C.; Xu C. & He, G. (2007). Heat and mass transfer in wood composite panels during hot pressing: Part 4. Experimental investigation and model validation. *Holzforshung*, 61, 83–88.
Denisov, O.B.; Anisov, P. & Zuban P.E. (1975). Untersuchung der Permeabilität von Spanvliesen. *Holztechnologie*, 16, 1, 10-14.
Devlin, C.P. (1986). An Investigation of the Mechanism of High-Intensity Paper Drying. PhD thesis, Institute of Paper Chemistry, USA.
Easterling, K.E.; Harrysson, R.; Gibson, L.J. & Ashby, M.F. (1982). On the mechanics of balsa and other woods. *Proceedings of the Royal Society of London*, Vol. A383, No. 1784 (September, 1982), 31-41.
Fenton, T.; Budman, H.; Pritzker, M.; Bernard, E. & Broderick, G. (2003). Modeling of Oriented Strandboard Pressing. *Ind. Eng. Chem. Res.*, 42, 5229-5238.
Gibson, L. & Ashby, M. (1988). *Cellular Solids: Structure and Properties*. Pergamon Press, ISBN 0521495601, Oxford.
Gustafsson, J.-E. & Kaul, V. (2001). A general model of deformation and flow in wet fibre webs under compression. *Nordic Pulp and Paper Research Journal*, 16, 2, 149-155.
Hata, T.; Kawai, S. & Sasaki, H. (1990). Computer simulation of temperature behavior in particle mat during hotpressing and steam injection pressing. *Wood Sci. and Technol.*, 24, 65.
Humphrey, P.E. & Bolton, A.J. (1989). The Hot Pressing of Dry-formed Wood-based Composites. Part II: A Simulation Model for Heat and Moisture Transfer, and Typical Results. *Holzforschung*, 43, 3, 199-206.
Humphrey, P.E. (1982). Physical aspects of wood particleboard manufacture. Ph.D. Thesis, University of Wales, UK.
Irle, M. & Barbu M. (2010). Wood-Based Panel Technology, In: *Wood-based Panels: An Introduction to Specialists*, H. Thoemen, M. Irle & M. Sernek (Ed.), 1-94, Brunel University Press, ISBN 978-1-902316-82-6, London, England.
Kamke, F.A. (2004). Physics of Hot-pressing, In: *Fundamentals of Composite Processing: Proceedings of a Workshop*, 3-18, USDA Forest Service, General Technical Report FPL-GRT-149, Madison WI, U.S.A.
Kamke, F.A. & Casey, L.J. (1988). Fundamentals of flakeboard manufacture: internal-mat conditions. *For. Prod. J.*, 38, 6, 38- 44.
Kamke F.A. & Wolcott, M.P. (1991). Fundamentals of flakeboard manufacture: wood-moisture relationships. *Wood Sci. and Technol.*, 25, 57.
Kataja, M.; Kirmanen, J. & Timonen, J. (1995). Hydrostatic and structural pressure in compressed paper webs and press felts. *Nordic Pulp and Paper Research Journal*, 10, 3, 162-166.
Kavvouras, P.K. (1977). Fundamental process variable in particleboard manufacture, Ph.D. Thesis, University of Wales, UK.
Larsson, H. & Stenström, S. (1998). Critical pressure control of delamination in impulse drying. *Tappi Journal*, 81, 7, 117-122.
Larsson, H.; Nilsson, J. & Stenström, S. (2001). Experiments in a pilot-scale impulse unit - Part I: Heat transfer and water removal rates. *Nordic Pulp and Paper Research Journal*, 16, 156-161.
Lavery, H.P. (1987). Impulse drying of linerboard, *Proceedings of the TAPPI Papermakers Conference*, pp. 43-49, Atlanta, GA, April 1987, TAPPI Press, Norcross, GA.
Lee, C.; Budman, H. & Pritzker, M. (2006). Simulation and Optimisation of the Continuous Oriented Strand Board Pressing Process. *Ind. Eng. Chem. Res.*, 45, 6, 1974-1988.
Lenth, C.A. & Kamke, F.A. (1996a). Investigations of flakeboard mat consolidation. Part I. Characterizing the cellular structure. *Wood and Fiber Sci.*, 28, 2, 153-167.

Lenth, C.A. & Kamke, F.A. (1996b). Investigations of flakeboard mat consolidation. Part II. Modeling mat consolidation using theories of cellular materials. *Wood and Fiber Sci.*, 28, 3, 309-319.
Lindsay, J.D. & Sprague, C.H. (1989). MIPPS: A numerical moving-boundary model for impulse drying. *Journal of Pulp and Paper Science*, 15, 4, J135-141.
Lindsay, J.D. (1991). Advances in the numerical modeling of impulse drying, In: *Tappi Engineering and Management in the Pulp and Paper Industry*, M. J. Coleman (Ed.), 665-677, Tappi Press, Atlanta, GA, USA.
Lobosco, V. & Kaul, V. (2001). An Elastic/Viscoplastic model of the fibre network stress in wet pressing: Part I. *Nordic Pulp and Paper Research Journal*, 16, 1, 1217.
Lucisano, M.F.C. (2002). On Heat and Paper: From Hot Pressing to Impulse Technology. PhD thesis, Royal Institute of Technology, Sweden.
Lucisano, M.F.C. & Martin, A.R. (2006). Visualization of liquid-vapor phase change phenomena in impulse technology. *Tappi Journal*, 5, 6, 15-21.
Lucisano, M.F.C.; Mazzatorta, P. & Martinez, D.M. (2001). On the mechanism of steam forming during impulse pressing of wet paper webs. *Nordic Pulp and Paper Research Journal*, 16, 4, 355-361.
MacGregor, M.A. (1983). A description of sheet stratification caused by wet pressing. *Tappi Journal*, 66, 6, 53-57.
Macklem, E.A. & Pulkowski, J.H. (1988). Impulse drying - a pressing/flashing drying phenomenon, *Proceedings of the TAPPI Engineering Conference*, pp. 447-453, Chicago, IL, September 1988, TAPPI Press, Norcross, GA.
Maiti, S.K.; Gibson, L.J. & M.F. Ashby (1984). Deformation and energy absorption diagrams for cellular solids. *Act. Metal*, 32, 11, 1963-1975.
Metso Paper (2010). http://www.metso.com/pulpandpaper/MPwUpRunning.nsf/WebWID/WTB-090603-2256F-67CFF?OpenDocument (December 13, 2010).
Nilsson, J. & Stenström, S. (2001). Modelling of heat transfer in hot pressing and impulse drying of paper. *Drying Technology: An International Journal*, 19, 10, 2469-2485.
Orloff, D.I.; Patterson, T.F. & Parviainen, P.M. (1998). Opening the operating window of impulse drying - Part II: Pressure differential as a source of delamination. *Tappi Journal*, 81, 8, 195-203.
Painter, G.; Budman, H. & Pritzker, M. (2006a). Prediction of oriented strand board properties from mat formation and compression operating conditions. Part 1. Horizontal density distribution and vertical density profile. *Wood Sci. and Technol.*, 40, 139–158.
Painter G.; Budman, H. & Pritzker, M. (2006b). Prediction of oriented strand board properties from mat formation and compression operating conditions. Part 2: MOE prediction and process optimization. *Wood Sci. and Technol.*, 40, 291-307.
Pereira, C.; Carvalho, L.H. & Costa, C.A.V. (2004). High Frequency Heating of Medium Density Fiberboard (MDF): Theory and Experiment. *Chem. Eng. Sci.*, 59, 4, 735-745.
Pereira, C.; Carvalho, L.H. & Costa, C.A.V. (2006). Modeling the continuous hot-pressing of MDF. *Wood Sci. and Technol.*, 40, 308-326.
Pikulik, I. (1999). The Effects of Shoe Pressing on Fine Paper Properties. *Tappi Journal*, 82, 11 88-92.
Pounder, J. (1986). A Mathematical Model of High Intensity Paper Drying. PhD thesis, The Institute of Paper Chemistry, USA.
Riepen, M. (2000). An inside view on impulse drying phenomena by modelling, *Proceedings of the 12th International Drying Symposium IDS2000*, ISBN 9780444504227, August 2000, Elsevier Science Ltd., Noordwijkerhout, The Netherlands.
Rodal, J.J.A. (1989). Soft-nip calendering of paper and paperboard. *Tappi Journal*, 72, 5, 177-186.

Schiel, C. (1969). Optimizing the nip geometry of transversal-flow presses. *Pulp and Paper Magazine of Canada*, 70, 3, T71-T76.

Skowronski, J. & Bichard, W. (1987). Fibre-to-fibre bonds in paper. Part I. Measurement of bond strength and specific bond strength. *Journal of Pulp and Paper Science*, 13, 5, 165-169.

Szikla, Z. (1992). On the Basic Mechanisms of Wet Pressing. PhD thesis, Helsinki University of Technology, Finland.

Szikla, Z. & Paulapuro, H. (1989a). Changes in z-direction density distributions of paper in wet pressing. *Journal of Pulp and Paper Science*, 15, 1, J11-J17.

Szikla, Z. & Paulapuro., H. (1989b). Compression behaviour of fibre mats in wet pressing.In: Fundamentals of Papermaking. In: *Transactions of the Ninth Fund. Res. Symp.*, C. F. Baker and V. W. Punton (Ed.), 587-624, Mechanical Engineering Publications, London, UK.

Terzaghi, K. (1943). *Theoretical Soil Mechanics*. John Wiley Sons, ISBN 0471853054, New York.

Thoemen, H., (2010). Preface In: *Wood-based Panels: An Introduction to Specialists*, H. Thoemen, M. Irle & M. Sernek (Ed.), 123-176, Brunel University Press, ISBN 978-1-902316-82-6, London, England.

Thoemen, H. & Humprey, P.E. (1999). The Continuous pressing process for WBPs: an analytical model. *Proceedings of The Third European Panel Products Symposium*, pp. 18-30, Llandudno, Wales, UK.

Thoemen, H. & Humprey, P. (2001). Hot pressing of wood-based composites: selected aspects of the physics investigated by means of a simulation, *Proceedings of The Fifth European Panel Products Symposium*, pp. 18-30, Llandudno, Wales, UK.

Thoemen, H. & Humprey, P.E. (2003). Modeling the continuous pressing process for wood-based composites. *Wood and Fibre Sci.*, 35, 456-468.

Thoemen, H. & Humprey, P.E. (2006). Modeling the physical processes relevant during hot pressing of wood-based composites. Part I. Heat and Mass transfer. *Holz Roh-Werkst.*, 35, 456-468.

Vomhoff, H. (1998). Dynamic Compressibility of Water-Saturated Fibre Networks and Influence of Local Stress Variations in Wet Pressing. PhD thesis, Royal Institute of Technology, Sweden.

Wahlström, B. (1960). A long term study of water removal and moisture distribution on a newsprint machine press section - Part I. *Pulp and Paper Magazine of Canada*, 61, 8, T379-T401.

Wahren, D. (1978). Förfarande och anordning för konsolidering och torkning av en fuktig porös bana, Swedish Patent no. 7803672-0.

Wahren, D. (1982). Method and apparatus for the rapid consolidation of moist porous web, USA Patent no. 4,324,613.

Wolcott, M.P.; Kasal, B.; Kamke, F.A. & Dillard, D.A. (1989). Modeling wood as a polymeric foam: An application to wood-based composite manufacture, *Proceedings of the Third Joint ASCE-ASME Mechanics Conference*, pp. 53—60, San Diego, LaJolla, CA, July 1989. University of California.

Wood-based Panels International (2010). Particleboard survey part 1, In: *Wood-based Panels International*, issue 5.

Zombori, B. (2001). Modeling the transient effects during the Hot-Pressing of Wood-based Composites. PhD Thesis, Faculty of the Virginia Polytechnic Institute and State University, Blacksburg, Virginia, USA.

Zombori, B.; Kamke, F.A. & Watson, L.T. (2002). Simulation of the internal conditions during the hot-pressing process. *Wood and Fibre Sci.*, 35,1, 2-23.

16

Control of Polymorphism and Mass-transfer in Al_2O_3 Scale Formed by Oxidation of Alumina-Forming Alloys

Satoshi Kitaoka, Tsuneaki Matsudaira and Masashi Wada
Japan Fine Ceramics Center
Japan

1. Introduction

Thermal barrier coatings (TBCs) are widely used for hot section components of gas turbine engines to protect the underlying metals from the high operating temperatures; they serve to both increase the engine efficiency and improve the durability of components. TBC systems typically consist of a Ni-based superalloy substrate, an alumina forming-alloy bond coat and an yttria-stabilized zirconia topcoat. When TBCs are exposed to high temperatures in oxidizing environments, a thermally grown oxide (TGO) develops on the bond coat surface underneath the top coat. Fracture of TBCs progresses in the vicinity of the TGO, which attains a critical thickness during thermal cycling operations (Evans et al., 2001). Thus, suppressing oxidation of the bond coat is anticipated to enhance the durability of TBCs.

Alumina-forming alloys tend to form metastable Al_2O_3 polymorphs such as gamma- and theta-phases in oxidizing environments at 1100-1450 K (Brumm et al., 1992, Tolpygo et al., 2000). The thermodynamically stable alpha-Al_2O_3 typically forms at longer oxidation times and/or higher temperatures. Metastable oxide scales consisting of the gamma- and theta-phases contain lattice defects so that they hardly act as a protective layer to further oxidation of the alloys compared with a alpha-Al_2O_3 scale (Brumm et al., 1992). It is also well known that the transformation of the alpha-phase from metastable polymorphs involves large changes in volume (~13%) and morphology.

Metastable oxides tend to form on the bond coat during coating with the topcoat in an oxygen-containing atmosphere at high temperatures, so that they are present at the interface between the bond coat and topcoat. The transformation of these oxides to the stable alpha-phase at higher operating temperatures promotes topcoat spalling. Thus, in order to enhance the durability of TBCs, the as-processed TBC microstructures should contain a thin layer of alpha-Al_2O_3 at the interface with no the metastable oxides. The standard grit-blasting procedure that is used in practical applications to prepare the bond coat for topcoat deposition promotes the formation of alpha-Al_2O_3. However, it also results in severe contamination of the coating surface and accelerates oxidation of the bond coat (Tolpygo et al., 2001). Some studies have revealed that a pre-oxidation step that forms a alpha-Al_2O_3 TGO on the bond coat without grit blasting reduces further oxidation of the bond coat and improves the durability of the TBC (Tolpygo et al., 2005, Nijdam et al., 2006, Matsumoto et al., 2006, 2008). Although it is generally preferable to perform pre-oxidation above 1450 K to produce a thin alpha-Al_2O_3 oxide and avoid producing metastable oxides, high temperature treatment degrades the bond coat or causes excessive oxidation. Thus, pre-oxidation should be carried out at lower temperatures and with shorter oxidizing times to form a thin alpha-Al_2O_3 layer. Matsumoto et al. (2006, 2008) reported that pre-oxidation of CoNiCrAlY under a low oxygen partial pressure (P_{O2}) and at lower temperature of 1323 K for 4 h, for which thermodynamic calculations predict that only alpha-$(Al,Cr)_2O_3$ will be produced, resulted in the formation of a dense alpha-Al_2O_3 scale with large grains, which functioned as an excellent protective layer against further oxidation of the alloy. They also reported that the theta-phase remained when $(Co,Ni)(Al,Cr)_2O_4$ was produced by oxidation under a higher P_{O2}, despite using the same oxidizing time as that under the lower P_{O2}. These results suggest that reducing the P_{O2} in pre-oxidized environments accelerates the polymorph transformation to the alpha-phase, but the extent of this acceleration and its mechanism remain unclear.

It is well known that adding chromium or iron to alumina-forming alloys accelerates the transformation from metastable polymorphs to alpha-Al_2O_3 (Brumm et al., 1992, Pint et al., 1997, Peng et al., 2003). Because these isovalent ions, which are larger than Al^{3+}, can dissolve into both the metastable phases and the alpha-phase, they partially compensate for the dilatational stress field of the dislocation and transformation (Clarke, 1998). This may be why the transformation is accelerated by adding chromium or iron. By contrast, the transformation is retarded when gamma-Al_2O_3 powders are substituted with 0.1-0.5 mol% of divalent cations such as Co^{2+} or Ni^{2+} (Odaka et al., 2008). Doping gamma-Al_2O_3 with 5 mol% Ni^{2+} also increases the hydrothermal stability relative to non-doped gamma-Al_2O_3 (Nagano et al, 2008). They may be due to their poor solvencies in the alpha-phase, although the divalent cations are larger than Al^{3+} ions. Thus, it is anticipated that the polymorphism of the Al_2O_3 scale can be controlled by controlling the P_{O2} during the pre-oxidation step. In other words, the transformation is accelerated under lower P_{O2}, when both aluminum and chromium in the alloy are simultaneously oxidized to form alpha-$(Al,Cr)_2O_3$, but it may be retarded under higher P_{O2}, when cobalt and nickel in the alloy are oxidized in addition to aluminum and chromium, resulting in the formation of $(Co,Ni)(Al,Cr)_2O_4$. In this chapter, control of Al_2O_3 polymorph formation was verified by the oxidation of specific components in CoNiCrAlY under a thermodynamically determined P_{O2}, resulting in the rapid formation of a dense, smooth alpha-Al_2O_3 scale (Kitaoka et al., 2010).

Suppressing mass-transfer through the alpha-Al_2O_3 scale is also anticipated to further improve the durability of the TBCs. When oxidation of the alumina forming alloys takes place through alumina scale under higher oxygen partial pressures (P_{O2}), such as in air, an oxygen potential gradient is generally induced in the direction opposite to the aluminum potential gradient in accordance with the Gibbs-Duhem equation. Thus, the alumina scale is grown on the alloys by the inward grain boundary diffusion of oxygen and the outward grain boundary diffusion of aluminum, resulting in the development of the grain boundary ridges on the scale surfaces. By contrast, ridges do not form when oxidation is carried out through alumina scale in a low P_{O2} environment, such as in a purified argon flow (Nychka et al., 2005). The ridge formation mechanism (i.e., mass-transfer through the alumina scale) is thus thought to depend on the P_{O2}.

The inherent effectiveness of monolithic alumina as a barrier to oxygen permeation has been estimated directly by measuring oxygen permeation through a polycrystalline alumina wafer exposed to oxygen potential gradients at high temperatures, where the each surface of the wafer is deliberately subjected to different P_{O2} values (Matsudaira et al., 2008, Wada et al., 2008, Kitaoka et al., 2009, Volk et al., 1968, Courtright et al., 1992, Ogura et al., 2001). When the potential gradients are produced by a combination of high P_{O2} values, oxygen permeates mainly via grain boundary diffusion of aluminum through aluminum vacancies from the lower to the higher P_{O2} surface, resulting in the formation of grain boundary ridges on the latter surface (Matsudaira et al., 2008, Wada et al., 2008, Kitaoka et al., 2009). In other words, O_2 molecules are adsorbed onto the surface at higher P_{O2} and subsequently dissociate into oxygen ions (forming alumina), whereas oxygen ions on the opposite surface at a lower P_{O2} are desorbed by association into O_2 molecules (decomposition of alumina). Conversely, under an oxygen potential gradient generated by a combination of low P_{O2} values, oxygen permeation occurs by grain boundary diffusion of oxygen through oxygen vacancies from the higher to the lower P_{O2} surface. In this case, very little ridge formation occurs at grain boundaries because of the very low aluminum flux (Kitaoka et al., 2009). Thus, the main diffusing species leading to oxygen permeation through the grain boundaries depends on the P_{O2} values producing the oxygen potential gradients. This readily explains the absence of grain boundary ridges on alumina scale formed under low P_{O2} conditions, such as in a purified argon flow (Nychka et al., 2005).

Many studies have focused on oxygen grain boundary diffusion in polycrystalline alumina using either secondary ion mass spectroscopy (SIMS) (Plot et al., 1996, Nakagawa et al., 2007, Messaoudi et al., 1998) or nuclear reaction analysis (NRA) (Heuer, 2008) to determine ^{18}O depth profiles after high temperature exchange with ^{18}O-enriched oxygen. Messaoudi et al. (1998) determined oxygen grain boundary diffusion coefficients during transport through a growing alumina scale. This was done by oxidizing alumina-forming alloys in a $^{16}O_2$ atmosphere, further oxidizing them in $^{18}O_2$ and then determining the ^{18}O distribution in the alumina scale using SIMS. The oxygen grain boundary diffusion coefficients measured by this procedure were larger than those determined from extrapolated diffusion data for polycrystalline alumina annealed in a homogeneous environment without any oxygen potential gradients (Plot et al., 1996, Nakagawa et al., 2007, Heuer, 2008). The corresponding activation energies (Messaoudi et al., 1998) derived using the double oxidation technique were smaller than those obtained by annealing (Plot et al., 1996, Nakagawa et al., 2007, Heuer, 2008). Kitaoka et al. (2009) have reported that the oxygen grain boundary diffusion coefficients, which were determined from the oxygen permeation coefficients of undoped polycrystalline alumina wafers exposed to oxygen potential gradients at high temperatures, decreased with increasing P_{O2}, and the corresponding activation energies were similar to those for the actual alumina scales formed on the alloys (Messaoudi et al., 1998). Thus, oxygen grain boundary diffusivity in polycrystalline alumina under oxygen potential gradients is apparently different from that in a homogeneous environment.

On the other hand, there has been only two reports of aluminum lattice diffusion coefficients determined using SIMS (Paladino et al., 1962, Le Gall et al., 1994). In this case, the appropriate tracer, ^{26}Al, has a very low specific activity and an extremely long half-life of 7.2×10^5 years, making it very difficult to perform radiotracer diffusion experiments. Recent studies have used Cr^{3+} instead of Al^{3+} as the diffusion species because of the extremely long half-life of the ^{26}Al isotope. Cr^{3+} is a suitable substitute for Al^{3+} because it has a similar ionic radius and is isovalent. For example, Bedu-Amissah et al. used a diffusion couple of alumina and chromia to measure grain boundary diffusion coefficients from Cr^{3+} concentration profiles obtained using an X-ray microprobe (Amissah et al., 2007). Kitaoka et al. (2009) have recently estimated aluminum grain boundary diffusion coefficients from the oxygen permeation coefficients of undoped polycrystalline alumina wafers exposed to oxygen potential gradients at high temperatures (Kitaoka et al., 2009). The aluminum grain boundary diffusion coefficients were found to increase with increasing P_{O2}, in an inverse relationship to the P_{O2} dependence of the oxygen grain boundary diffusion coefficients.

Alumina-forming alloys contain small quantities of oxygen-reactive elements (REs) (e.g., Y, La, Ti, Zr, and Hf) to improve the oxidation-resistance of the alloys. These REs segregate at grain boundaries in growing alumina scales during oxidation of the alloys. The obvious difficulty is that it is not known what the diffusing species along the grain boundaries in the scales might be. Nevertheless, the REs are speculated to inhibit the scale growth by effectively blocking the outward grain-boundary diffusion of aluminum due to an ionic size misfit, since the ionic sizes of the REs are larger than Al^{3+} (Nychka et al., 2005, Pint et al., 1998). However, during long-duration, high-temperature oxidation, REs that segregated at grain boundaries were found to diffuse toward the scale surfaces together with aluminum, resulting in the precipitation of RE-rich particles on the surfaces (Pint et al., 1998). This casts a doubt on the conjecture that the REs can effectively control the movement of aluminum. The coexistence of various kinds of REs further complicates the interpretation of the experimental results. On the other hand, rare earth doping can significantly increase the high-temperature creep resistance of polycrystalline alumina (Matsunaga et al., 2003, Ikuhara et al., 2001, Yoshida et al., 2002). Several studies have suggested that segregation of large dopant elements changes the grain boundary environment by 'site blocking' critical oxygen diffusion pathways (Amissah et al., 2007, Wang et al., 1999, Cho et al., 1999, Cheng et al., 2008, Priester, 1989, Korinek et al., 1994) and/or by strengthening grain boundaries by enhancing bond strengths in the vicinity of dopant ions (Yoshida et al., 2002, Buban et al., 2006).

As mentioned above, when undoped polycrystalline alumina wafers were exposed to oxygen potential gradients at high temperatures, the main diffusing species depended on P_{O2}. This phenomenon will be useful for elucidating how much the migration of oxygen and aluminum is affected by REs segregated at the grain boundaries.

In this chapter, the effect of lutetium doping on oxygen permeability in polycrystalline alumina wafers exposed to steep oxygen potential gradients was evaluated at high temperatures to investigate the mass-transfer phenomena through the scale (Matsudaira et al., 2010). It is well known that lutetium doping can significantly improve high-temperature creep resistance in polycrystalline alumina (Matsunaga et al., 2003, Ikuhara et al., 2001, Yoshida et al., 2002); therefore, it is also expected to retard mass-transfer in alumina under oxygen potential gradients.

2. Control of polymorphism in Al_2O_3 scale

2.1 Thermodynamic prediction of stable oxides

The P_{O2} dependence of equilibrium amounts and components of stable condensed oxide species are first estimated thermodynamically when CoNiCrAlY (Co-28%Ni-21%Cr-16%A-0.3%Y (in atomic %)) is exposed to an oxidizing environment at a high temperature. The equilibrium calculation was conducted using FactSage_free-energy minimization computer code together with a database for the Co-Ni-Cr-Al-Y-O system that mainly consists of the SGTE database and some solution models of condensed phases. The Redlich-Kister-Muggianu polynomial model was applied to solid solution phases such as the gamma-phase (alloy) and (Co,Ni)O, and multi-sublattice formalism to the beta-phase (alloy) and $(Co,Ni)(Al,Cr)_2O_4$. Subregular (Degterov et al., 1996) and sublattice (Ansara et al., 1997) models were used for $(Al,Cr)_2O_3$ and the gamma'-phase, respectively. Initial conditions for the calculation were 1 mol of CoNiCrAlY, a total pressure of 10^5 Pa (equal to the sum of partial pressures of O_2 and Ar), and a temperature of 1323 K.

Figure 1 shows the equilibrium amounts of the condensed species as a function of the equilibrium P_{O2} at 1323K. alpha-$(Al,Cr)_2O_3$ is thermodynamically stable under the equilibrium P_{O2} below 10^{-9} Pa. On the contrary, $(Co,Ni)(Al,Cr)_2O_4$ and (Co,Ni)O are stable under the P_{O2} above 10^{-9} Pa and 10^{-5} Pa, respectively. All the yttrium in the alloy is also oxidized to produce about 10^{-3} mol of $Al_5Y_3O_{12}$ in the P_{O2} range shown in Fig. 1.

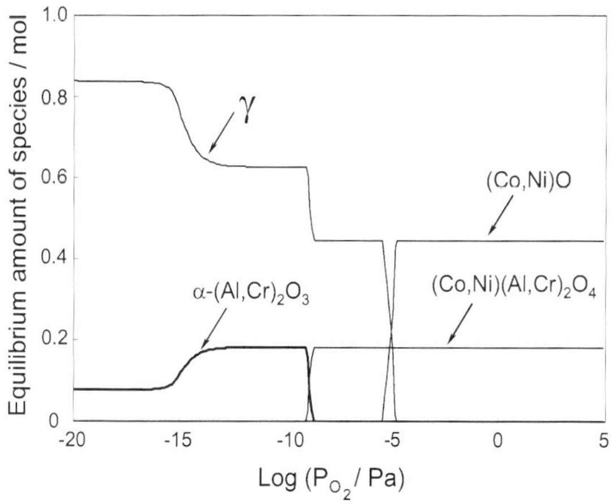

Fig. 1. Equilibrium amount of species as a function of partial pressure of O_2 (P_{O2}) at 1323 K for Co-0.28Ni-0.21Cr-0.16Al-0.003Y (at %). γ-phase represents a metallic phase.

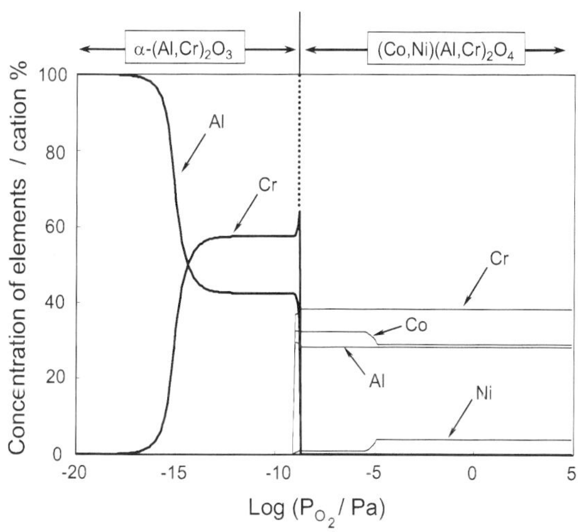

Fig. 2. Equilibrium concentration of each element in the solid solution phases of alpha-$(Al,Cr)_2O_3$ and $(Cr,Ni)(Al,Cr)_2O_4$ as a function of P_{O2} at 1323 K for Co-0.28Ni-0.21Cr-0.16Al-0.003Y (at %).

The equilibrium concentration of each element in the solid solution phases of alpha-$(Al,Cr)_2O_3$ and $(Cr,Ni)(Al,Cr)_2O_4$ as a function of P_{O2} at 1323 K for the CoNiCrAlY alloy is

shown in Fig. 2. For a P_{O2} below 10^{-9} Pa, where alpha-$(Al,Cr)_2O_3$ is thermodynamically stable, the concentration of chromium dissolved in the oxide increases sharply with an increase in P_{O2} above 10^{-17} Pa, up to about 60 cation % for a P_{O2} above 10^{-13} Pa. By contrast, for a P_{O2} above 10^{-9} Pa, at which $(Co,Ni)(Al,Cr)_2O_4$ is stable, both cobalt and nickel dissolve in the complex oxide in addition to aluminum and chromium.
In the actual system, it is speculated that initially amorphous and/or gamma-alumina, which can dissolve in oxides consisting of other elements such as chromium, and/or cobalt and nickel in the alloy, are produced at lower temperatures during heating in the oxidation treatment. This is followed by transformation into the theta-phase, and finally conversion into thermodynamically stable phases such as alpha-alumina and/or $(Co,Ni)(Al,Cr)_2O_4$ while holding at a higher temperature. Thus, the oxidation of the alloy under a P_{O2} at which both aluminum and chromium are oxidized and at which neither cobalt nor nickel is oxidized will accelerate transformation into the alpha-alumina phase. According to Figs. 1 and 2, a suitable P_{O2} for promoting the polymorph transformation lies below 10^{-9} Pa at 1323 K. By contrast, the transformation will be retarded in a P_{O2} above 10^{-9} Pa. In this chapter, control of Al_2O_3 polymorph formation was verified by the oxidation of CoNiCrAlY under a thermodynamically determined P_{O2} of 10^{-14} Pa at 1323 K, compared with oxidation under P_{O2} of 1 and 10^5 Pa (Kitaoka et al., 2010).

2.2 Experimental procedures

CoNiCrAlY alloy (Co-28%Ni-21%Cr-16%Al-0.3%Y (in atomic %)) was coated by vacuum plasma spraying to a thickness of 150 micrometer on a 20 × 40 × 2.5-mm grit-blasted substrate of Inconel 738LC and the alloy surface was then polished to Ra = 0.25 micrometer. Each sample was introduced into a V-notch with a radius of curvature below 20 micrometer on the substrate side to prepare the cross-section of the sample after the subsequent oxidation treatment by brittle fracture in liquid N_2. The V-notched samples were ultrasonically cleaned in acetone and dried at 393 K for 1 h. Each sample was set in a furnace, the internal atmosphere of which was replaced with either O_2, commercial Ar, or P_{O2} controlled Ar using a solid-state electrochemical oxygen pump system (Matsumoto et al., 2006, 2008). The gas flow rate was 3×10^{-4} m^3/min. The sample was heated to 1323 K at a rate of 5 K/min, and then held at the temperature for 10, 200, and 600 min, where the P_{O2} values were 10^5 Pa for O_2, 10 Pa for the commercial Ar, and 10^{-14} Pa for the controlled Ar. After that, the samples were cooled to room temperature at a rate of 5 K/min.
Crystalline phases in the scale formed by oxidation were identified by X-ray diffraction (XRD). Identification of Al_2O_3 polymorphs in the scale such as the theta- and alpha-phases was also performed by photostimulated Cr^{3+} luminescence spectroscopy (PSLS). The surface morphology and cross-sections of the scales were measured by scanning electron microscopy (SEM). Elemental depth profiles of the scale were measured by secondary ion mass spectrometry (SIMS).

2.3 Characteristics of oxide scales

The crystalline phases in the oxide scales, which were identified by XRD and PSLS, are summarized in Table 1. The higher the P_{O2} used for oxidation was, the longer the metastable theta-phase survived. The P_{O2} dependence for the formation of the oxide phases in the scale on oxidation time is in good agreement with the thermodynamic predictions shown in Fig. 1. This demonstrates that the oxidation of the alloy under a precisely regulated P_{O2} can be used to control the polymorphism of the Al_2O_3 scale.

Time/ min	Po_2 / Pa		
	10^{-14}	10	10^5
10	α-Al_2O_3	θ + α-Al_2O_3 (Co,Ni)(Al,Cr)$_2$O$_4$	θ + α-Al_2O_3 (Co,Ni)(Al,Cr)$_2$O$_4$
200	↓	α-Al_2O_3 (Co,Ni)(Al,Cr)$_2$O$_4$	↓
600	↓	↓	α-Al_2O_3 (Co,Ni)(Al,Cr)$_2$O$_4$

Table 1. Crystalline phases in the oxide scales.

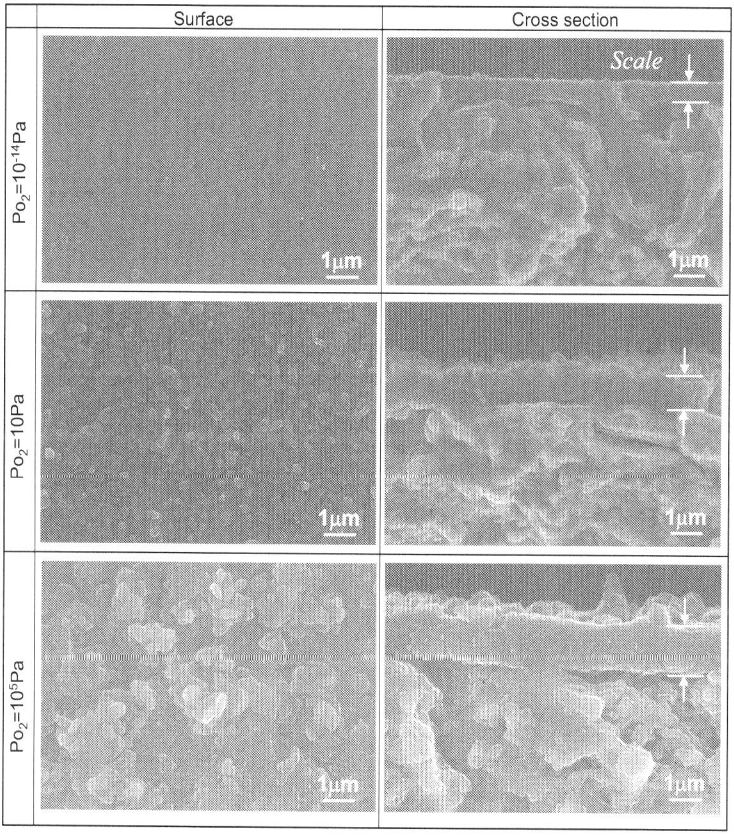

Fig. 3. SEM micrographs of the surfaces and cross-sections of the samples oxidized at 1323 K for 600 min under P_{O2} of 10^{-14}, 10, and 10^5 Pa.

The SEM micrographs of the surfaces and cross-sections of the samples oxidized at 1323 K for 600 min under P_{O2} of 10^{-14}, 10, and 10^5 Pa, respectively, are shown in Fig. 3. The surface

of the oxide scale formed under a P_{O2} of 10^{-14} Pa is relatively smooth and its thickness is about 1 micrometer. On the other hand, the higher the P_{O2} for the oxidation is, the larger the oxide crystals are which are exposed on the scales, increasing the density of surface irregularities. The scale thickness increases with an increase in P_{O2} for the oxidation: the scale thickness for oxidation under a P_{O2} of 10^5 Pa is at least twice that under a P_{O2} of 10^{-14} Pa. Some of the crystals grown on the oxide scales under the higher P_{O2} are considered to be $(Co,Ni)(Al,Cr)_2O_4$, as shown in Fig. 1 and Table 1. It is well known that the morphology of theta-Al_2O_3 consists of blade-like crystals (known as whiskers). In addition, when theta-Al_2O_3 survives for a long time at high temperatures, this oxide crystal grows outward about an order of magnitude faster than alpha-Al_2O_3 (Tolpygo et al., 2000). Therefore, since the theta-phase exists longer under a higher P_{O2}, the oxide has longer whiskers than those transformed earlier, resulting in the formation of an oxide scale with a rougher surface.
Figure 4 shows the SIMS depth profiles of selected elements through the CoNiCrAlY coats of the samples oxidized at 1323 K for 600 min under P_{O2} of 10^{-14} and 10^5 Pa, respectively. For the oxidation under a P_{O2} of 10^{-14} Pa (Fig. 4(a)), chromium, cobalt, and nickel are concentrated near the surface of the scale, which consists of only the crystalline alpha-Al_2O_3 phase, and high-purity alpha-Al_2O_3 is formed near the scale side of the interface between the scale and alloy. Chromium in the scale formed under a lower P_{O2} should be oxidized to form a solid solution of alpha-$(Al,Cr)_2O_3$, whereas both cobalt and nickel detected in the subsurface should segregate as metals, as shown in Figs. 1 and 2. For oxidation under a P_{O2} of 10^5 Pa (Fig. 4(b)), the concentrations of chromium, cobalt, and nickel in the scale are considerably higher than those under a P_{O2} of 10^{-14} Pa, and such a high-purity alpha-Al_2O_3 layer evidently does not exist at the interface between the scale and alloy.

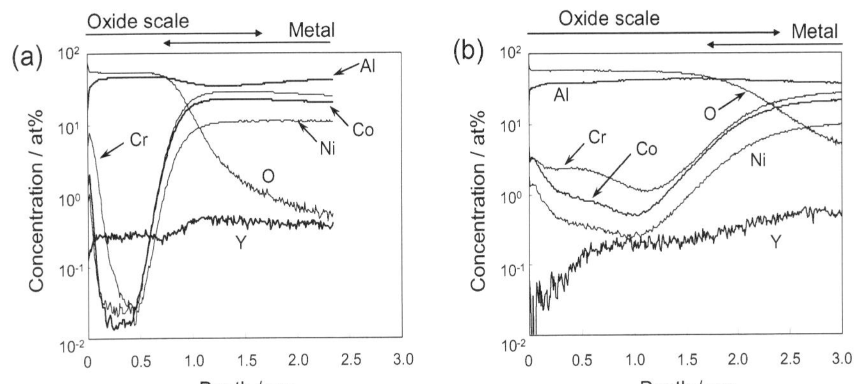

Fig. 4. SIMS depth profiles of selected elements through the CoNiCrAlY coats of the samples oxidized at 1323 K for 600 min under a P_{O2} of (a) 10^{-14} and (b) 10^5 Pa.

We have evaluated the oxygen permeability of polycrystalline alpha-Al_2O_3 wafers exposed to steep oxygen potential gradients at high temperatures to investigate complicated mass-transfer phenomena through the alpha-Al_2O_3 scale formed on the alloy, as discussed later (Matsudaira et al., 2008, 2010, Wada et al., 2008, Kitaoka et al., 2009). Diffusion of aluminum and oxygen species, which were responsible for the oxygen permeation along the grain boundaries of alpha-Al_2O_3, was found to be strongly dependent on P_{O2}, forming oxygen potential gradients.

When the wafer was subjected to potential gradients caused by a combination of low P_{O2} values, oxygen permeation primarily occurred by grain boundary diffusion of oxygen through oxygen vacancies from the higher P_{O2} surface to the lower P_{O2} surface. Grain boundary ridges were hardly formed on the surfaces under higher P_{O2} because of the very low aluminum flux. Thus, oxidation of CoNiCrAlY through the alpha-Al_2O_3 scale under a P_{O2} of below 10^{-14} Pa is thought to be mainly controlled by inward grain boundary diffusion of oxygen, because oxidation progressed without grain boundary ridges in similar to oxidation under purified argon (Nychka et al., 2005). Nevertheless, chromium, cobalt, and nickel are concentrated near the scale surface formed by oxidation under a P_{O2} of 10^{-14} Pa, as shown in Fig. 4(a). The reason for the segregation of these elements near the scale surface is discussed below.

Figure 5 shows the thermodynamic equilibrium phase boundary (solid line) between alpha-$(Al,Cr)_2O_3$ and $(Cr,Ni)(Al,Cr)_2O_4$ as a function of T^{-1}. Lower oxidation temperature results in a larger stability region for $(Co,Ni)(Al,Cr)_2O_4$. Broken line (A) in Fig. 5 indicates the transition of P_{O2} in the furnace as the temperature increased during oxidation treatment under a P_{O2} of 10^{-14} Pa at 1323 K, corresponding to the testing conditions of Fig. 4(a). The segregation of both cobalt and nickel near the scale surface shown in Fig. 4(a) seems to be caused by initial oxidation during temperature increase to produce $(Co,Ni)(Al,Cr)_2O_4$, followed by reduction and decomposition to cobalt, nickel, and alpha-$(Al,Cr)_2O_3$. According to Fig. 2, the surface segregation of chromium may be thermodynamically promoted by reducing the solubility of chromium ions in the alpha-phase with decreasing oxygen chemical potential in the scale from the scale surface to the interface between the scale and the alloy.

In TBC systems, if a topcoat such as yttria-stabilized zirconia is coated on the pre-oxidized bond coat of CoNiCrAlY, where metallic cobalt and nickel are segregated near the surface of the alpha-$(Al,Cr)_2O_3$ scale on the alloy (Fig. 4(a)), these segregated metals will react with alpha-$(Al,Cr)_2O_3$ in the scale to produce $(Co,Ni)(Al,Cr)_2O_4$ in oxidizing environments at high temperatures, promoting the spalling of TBCs. If the oxidation of the alloy is carried out under a P_{O2} exactly controlled according to broken line (B) in Fig. 5, which indicates the transition of P_{O2} in the furnace when the temperature is increasing, production of $(Co,Ni)(Al,Cr)_2O_4$ at low temperatures will be inhibited. In other words, although the thickness of the scale formed along line (B) in Fig. 5 will be similar to that formed along line (A) in Fig. 5, the surface segregation of cobalt and nickel in the alpha-$(Al,Cr)_2O_3$ scale will be suppressed.

The SIMS depth profiles of cobalt and nickel through the CoNiCrAlY coats of the samples oxidized at a holding temperature of 1323 K under a P_{O2} of 10^{-14} Pa are shown in Fig. 6. Lines (a) and (b) in Fig. 6 are when the temperature was increased to 1323 K according to the P_{O2} along line (A) in Fig. 5 and then held at 1323 K for 10 and 600 min, respectively. Line (c) in Fig. 6 is when the temperature was increased up to 1323 K according to the P_{O2} along line (B) in Fig. 5 and then held for 600 min. When the samples were treated during oxidation under the P_{O2} along line (A) in Fig. 5, only varying the holding time at 1323 K, the concentration depths of both cobalt and nickel near the scale surface are constant and independent of the holding time, as shown by lines (a) and (b) of Fig. 6. Because the oxidation treatments use the same P_{O2} transition and heating rate when the temperature was increased, the amount of $(Co,Ni)(Al,Cr)_2O_4$ produced at lower temperature was thought to be constant and did not depend on the holding time at 1323K. As shown in Fig. 6(c), when P_{O2} during the temperature increase in the oxidation treatment is reduced in the manner indicated by line (B) in Fig. 5, concentrations of cobalt and nickel at the top surface of the scale are decrease to about 1/10 those under the P_{O2} indicated by line (A) in Fig. 5. The lower P_{O2} during the temperature increase in the oxidation treatment is, the lower surface concentrations of these elements are, and monolithic alpha-$(Al,Cr)_2O_3$ scale will certainly form. It is expected that the adherence between the topcoat and bond coat will be considerably improved by controlling the P_{O2} transition during the temperature increase, resulting in further improvement in the durability of TBC systems.

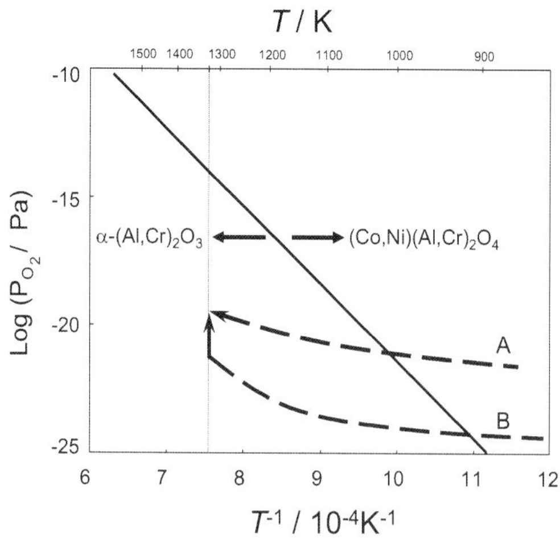

Fig. 5. Thermodynamic equilibrium phase boundary line (solid line) between alpha-$(Al,Cr)_2O_3$ and $(Cr,Ni)(Al,Cr)_2O_4$ as a function of T^{-1}. The broken lines A and B in Fig. 5 indicate the transition of P_{O2} in the furnace during the temperature increase in the oxidation treatment under a P_{O2} of 10^{-14} Pa at 1323 K.

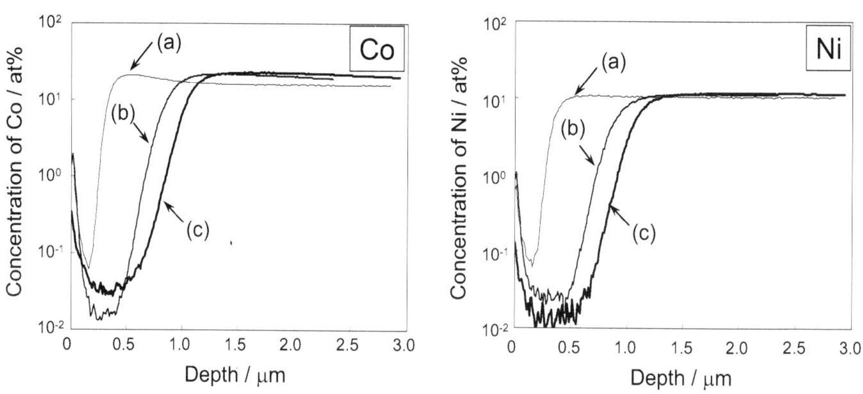

Fig. 6. SIMS depth profiles of Co and Ni through the CoNiCrAlY coats of the samples oxidized at a holding temperature of 1323 K under a P_{O2} of 10^{-14} Pa. Lines (a) and (b) in Fig. 8 are when the temperature was increased to 1323 K according to P_{O2} along line A in Fig.5 and then held for 10 and 600 min, respectively. Line (c) in Fig. 6 is when the temperature was increased to 1323 K according to P_{O2} along line B in Fig. 5 and then held for 600 min.

3. Mass-transfer of Al_2O_3 polycrystals under oxygen potential gradients

3.1 Experimental procedures

3.1.1 Materials

Commercial, high-purity alumina powder (TM-DAR, Taimei Chemicals Co., Ltd., Nagano, Japan, purity > 99.99 wt%) was used for the undoped alumina. Lutetia-doped powders (0.2 mol% of Lu_2O_3) were also prepared by mixing the alumina powder and an aqueous solution of lutetium nitrate hydrate ($Lu(NO)_3 \cdot xH_2O$ (>99.999%), Sigma-Aldrich Co., MO, USA) and subsequent drying to remove the water solvent. Each powder was molded by a uniaxial press at 20 MPa and then subjected to cold isostatic pressing at 600 MPa. The green compacts were pressureless sintered in air at 1773 K for 5 h. Wafers with dimensions of diameter 23.5×0.25 mm were cut from the sintered bodies and then polished so that their surfaces had a mirror-like finish. The relative density of the wafers was 99.5% of the theoretical density. All the wafers had similar microstructures with an average grain size of about 10 micrometer.

3.1.2 Oxygen permeability constants

Figure 7 shows a schematic diagram of the oxygen permeability apparatus. A polycrystalline alpha-Al2O3 wafer was set between two alumina tubes in a furnace. Platinum gaskets were used to create a seal between the wafer and the Al_2O_3 tubes by loading a dead weight from the top of the upper tube. A gas-tight seal was achieved by heating at 1893-1973 K under an Ar gas flow for 3 hrs or more. After that, a P_{O2} of oxygen included as an impurity in the Ar gas was monitored at the outlets of the upper and lower chambers that enclosed the wafer and the Al_2O_3 tubes using a zirconia oxygen sensor at 973K. The partial pressure of water vapor (P_{H2O}) was measured at room temperature using an optical dew point sensor. These measured P_{O2} and P_{H2O} were regarded as backgrounds. Then, pure O_2 gas or Ar gas containing either 1-10 vol% O_2 or 0.01-1 vol% H_2 was introduced into the upper chamber at a flow rate of 1.67×10^{-6} m^3/s. A constant flux for oxygen permeation was judged to be achieved when the values of the P_{O2} and P_{H2O} monitored in the outlets became constant.

When either O_2 gas or the Ar/O_2 mixture was introduced into the upper chamber and Ar was introduced into the lower chamber to create an oxygen gradient across the wafer, oxygen permeated from the upper chamber to the lower chamber. The P_{O2} values in the lower chamber at the experimental temperatures were calculated thermodynamically from the values measured at 973 K. The calculated values were almost the same as those at 973 K. On the other hand, when the Ar/H_2 mixture was introduced into the upper chamber and Ar was introduced into the lower chamber, a tiny amount of oxygen in the Ar permeated from the lower chamber to the upper chamber and reacted with H_2 to produce water vapor. As a result, the P_{H2O} in the upper chamber increased while the H_2 partial pressure (P_{H2}), which was measured at room temperature by gas chromatography, in the upper chamber decreased. The increase of P_{H2O} in the upper chamber was comparable to the reduction of P_{O2} in the lower chamber in terms of oxygen, and the P_{H2O} in the lower chamber remained constant during the permeation tests; thus, hydrogen permeation from the upper chamber to the lower chamber was negligibly small in comparison with the oxygen permeation in the opposite direction. The P_{O2} values in the upper chamber at the experimental temperatures were estimated thermodynamically from the P_{H2O} and P_{H2} measured at room temperature.

The oxygen permeability constant, PL, was calculated from the difference between the P_{O2} estimated thermodynamically in one chamber (which had a lower P_{O2} than that in another chamber) and the background in the lower P_{O2} chamber using [20), 22), 23)]

$$PL = \frac{C_p \cdot Q \cdot L}{V_{st} \cdot S}, \quad (1)$$

where C_p is the concentration of permeated oxygen (P_{O2}/P_T, where P_T = total pressure), Q is the flow rate of the test gases, V_{st} is the standard molar volume of an ideal gas, S is the permeation area of the wafer, and L is the wafer thickness.

The wafer surfaces exposed to oxygen potential gradients at 1923 K for 10 hrs were observed by scanning electron microscopy (SEM) combined with energy dispersive spectroscopy (EDS), and X-ray diffraction (XRD). The volume of the grain boundary ridges formed on the surfaces by the oxygen potential gradients was measured by 3D laser scanning microscopy, and was compared with the total amount of the oxygen permeated in the wafer.

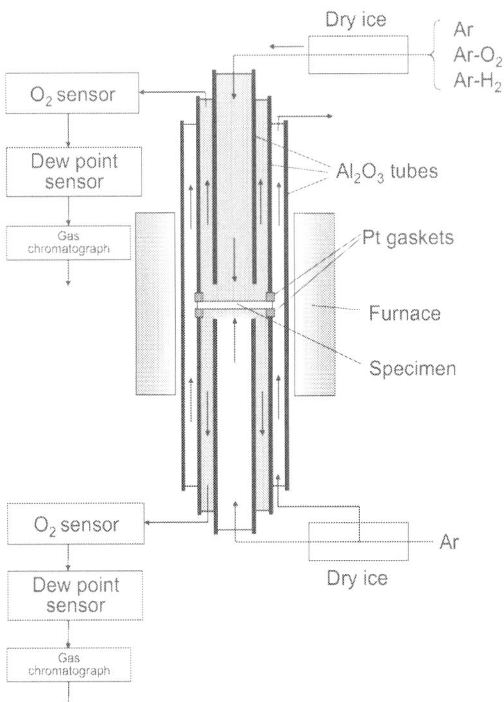

Fig. 7. Schematic diagram of the gas permeability apparatus.

3.1.3 Determination of grain boundary diffusion coefficients

(a) Fluxes of charged particles

The charged particle flux is described as

$$J_i = -Z_i \left(\frac{C_i D_i}{RT} \right) \frac{\partial \eta_i}{\partial x}, \quad (2)$$

where Z_i is the charge of the diffusing particle, C_i is the molar concentration per unit volume, D_i is the diffusion coefficient, R is the gas constant, T is the absolute temperature, x is a space coordinate, and η_i is the electrochemical potential.
The flux of oxygen that permeates through the wafer is equal to the sum of J_{Al} and J_O,

$$J_{TO} = J_{Al} + J_O = -\left(\frac{{Z_{Al}}^2}{Z_O} C_{Al} D_{Al} + Z_O C_O D_O\right)\frac{(t_{h^\cdot} + t_{e'})}{RT}\cdot\frac{\partial \mu_O}{\partial x} \quad (3)$$

where t_i is the transport number and μ_O is the oxygen chemical potential.
Integrating Eq. (3) from x = 0 to x = L gives

$$\int_0^L J_{TO} dx = -\frac{(t_{h^\cdot} + t_{e'})}{2}\left(\frac{{Z_{Al}}^2 C_{Al}}{Z_O}\int_{P_{O_2}(I)}^{P_{O_2}(II)} D_{Al} d\ln P_{O_2} + Z_O C_O \int_{P_{O_2}(I)}^{P_{O_2}(II)} D_O d\ln P_{O_2}\right) \quad (4)$$

Equation (4) is applicable to the case of ideal oxygen permeation when there is no interaction between electrons and holes, or when either electrons or holes exclusively participate (Kitaoka et al., 2009, Matsudaira et al., 2010).

(b) Oxygen grain boundary diffusion

The flux of oxygen that permeates through the wafer is postulated to be equal only to J_O. It is also assumed that oxygen permeates only through reactions between defects, in which both oxygen vacancies and electrons participate. In these reactions, dissociative adsorption of O_2 molecules is assumed to progress on the surface exposed to the higher P_{O2} (i.e., P_{O2}(II)) as follows.

$$1/2O_2 + V_O^{\cdot\cdot} + 2e' \rightarrow O_O^X \quad (5)$$

Oxygen ions migrate through oxygen vacancies from the P_{O2}(II) side to the lower P_{O2} side (i.e., P_{O2}(I)), and oxygen vacancies and electrons diffuse in the opposite direction to the oxygen flux. The inverse reaction to Eq. (5) proceeds on the P_{O2}(I) surface, and oxygen ions recombine to produce O_2 molecules.
If the diffusing species migrate mainly along the grain boundaries of polycrystalline Al_2O_3, the grain boundary diffusion coefficient of oxygen related to Eq. (5), is written as

$$D_{Ogb}\delta = \frac{D_{V_{Ogb}^{\cdot\cdot}}}{C_{Ob}S_{gb}}\left(\frac{1}{4K_{V_{Ogb}^{\cdot\cdot}}}\right)^{1/3} {P_{O_2}}^{-1/6} = -\frac{A_O}{6C_{Ob}S_{gb}} {P_{O_2}}^{-1/6} \quad (6)$$

where D_{Ogb} is the grain boundary diffusion coefficient of oxygen, δ is the grain boundary width, C_{Ob} is the molar concentration of oxygen per unit volume, S_{gb} is the grain boundary density, which is determined from the average grain size in the Al_2O_3. $D_{V_{Ogb}^{\cdot\cdot}}$ is the grain boundary diffusion coefficient of an oxygen vacancy and $K_{V_{Ogb}^{\cdot\cdot}}$ is the equilibrium constant of reaction (5) that occurs at grain boundaries. Assuming that $t_{e'}$ = 1 and $D_{Ogb} >> D_{Algb}$, and inserting Z_O = -2 and Eq. (6) into Eq. (4) gives

$$\int_0^L J_{TO}dx = A_O(P_{O_2}(II)^{-1/6} - P_{O_2}(I)^{-1/6}) = 4PL \quad (7)$$

If the constant A_O is determined experimentally using Eq. (7), $D_{Ogb}\delta$ for a certain P_{O2} can be estimated from Eq. (6).

(c) Aluminum grain boundary diffusion

The flux of oxygen that permeates through the wafer is premised to be equal only to J_{Al}. Oxygen permeation is also assumed to occur by reactions in which both aluminum vacancies and holes participate. O_2 molecules are absorbed on the surface exposed to $P_{O2}(II)$ as follows.

$$1/2O_2 \rightarrow O_O^X + 2/3V_{Al}''' + 2h^{\bullet} \quad (8)$$

Aluminum vacancies move from the $P_{O2}(II)$ side to the $P_{O2}(I)$ side, and aluminum ions and holes migrate in the opposite direction. Finally, the inverse reaction of (8) occurs on the $P_{O2}(I)$ surface, and oxygen ions recombine to produce an O_2 molecule.
In a similar way to Section 3.1.3(b), the grain boundary diffusion coefficient of aluminum, D_{Algb}, is obtained as follows.

$$D_{Algb}\delta = \frac{D_{V_{Algb}'''}}{C_{Alb}S_{gb}}\left(\frac{K_{V_{Algb}'''}}{9}\right)^{3/8} P_{O_2}{}^{3/16} = \frac{A_{Al}}{12C_{Alb}S_{gb}} P_{O_2}{}^{3/16} \quad (9)$$

C_{Alb} denotes the molar concentration of aluminum per unit volume, $D_{V_{Algb}'''}$ is the grain boundary diffusion coefficient of aluminum vacancies, $K_{V_{Algb}'''}$ is the equilibrium constant of reaction (8) that occurs at the grain boundaries. If it is assumed that $t_{h^{\bullet}}=1$ and $D_{Algb} >> D_{Ogb}$, then substituting $Z_{Al} = +3$ and $Z_O = -2$ into Eq. (4) gives

$$\int_0^L J_{TO}dx = A_{Al}(P_{O_2}(II)^{3/16} - P_{O_2}(I)^{3/16}) = 4PL \quad (10)$$

If the experimental value of A_{Al} is obtained using Eq. (10), $D_{Algb}\delta$ for a certain P_{O2} can be calculated from Eq. (9).

3.2 Oxygen permeation

Figure 8 shows the temperature dependence of oxygen permeability constant of polycrystalline Al_2O_3 (non-doped and doped with 0.2 mol% Lu_2O_3) exposed to oxygen potential gradients (ΔP_{O2}). The solid and open symbols indicate data for specimens exposed under $P_{O2}(II)/P_{O2}(I)$= 1 Pa/10^{-8} Pa and 10^5 Pa/1 Pa, respectively. The other lines are data from the literature under a similar ΔP_{O2} as that for the open symbols. The oxygen permeability constants are found to increase with increasing temperature, such that they are proportional to T^{-1}, in a similar manner as the data from the literature. The oxygen permeability constants tend to decrease with increasing purity of Al_2O_3. For $P_{O2}(II)/P_{O2}(I)$ = 10^5 Pa/1 Pa, the oxygen permeability constants of the lutetia-doped wafer are similar to those of the undoped wafer. Although the slopes of the curves for $P_{O2}(II)/P_{O2}(I)$ = 1 Pa/10^{-8} Pa are the same for both samples, they are markedly different from those for $P_{O2}(II)/P_{O2}(I)$= 10^5 Pa/1 Pa. Furthermore, the permeability constants obtained for $P_{O2}(II)/P_{O2}(I)$ = 1 Pa/10^{-8} Pa are clearly reduced by lutetia doping. These results suggest that the effect of lutetia doping on the oxygen permeation and the corresponding permeation mechanism vary depending on the oxygen potential gradients.

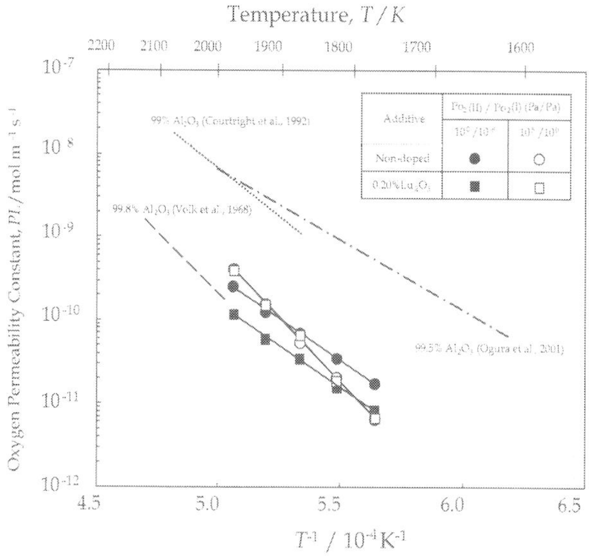

Fig. 8. Temperature dependence of oxygen permeability constant of polycrystalline Al_2O_3 (non-doped and doped with 0.2 mol% Lu_2O_3) exposed to oxygen potential gradients (ΔP_{O2}). The solid and open symbols indicate data for specimens exposed under $P_{O2}(II)/\ P_{O2}(I) = 1$ Pa/10^{-8} Pa and 10^5 Pa/1 Pa, respectively. The other lines are data from the literature under a similar ΔP_{O2} as that for the open symbols.

Because the oxygen permeability constants of a single-crystal Al_2O_3 wafer were lower than the measurable limit of this system (below 1×10^{-12} mol·m^{-1}s^{-1} at 1773 K), the oxygen permeation is thought to occur preferentially through the grain boundaries for the polycrystalline Al_2O_3 (Matsudaira et al., 2008). Furthermore, the oxygen permeability constants of the polycrystalline wafers were inversely proportional to the wafer thickness. According to Eq.(2), therefore, the oxygen permeation is considered to be controlled by diffusion in the wafer, not by interfacial reaction between the wafer surfaces and ambient gases.

Figure 9 shows the effect of P_{O2} under a steady state in the upper chamber on the oxygen permeability constants of polycrystalline alumina (undoped and doped with 0.20 mol% Lu_2O_3) at 1923 K, where the P_{O2} in the lower chamber is constant at about 1 Pa. For P_{O2} values of less than 10^{-3} Pa, the oxygen permeability constants decrease with increasing P_{O2} for both the undoped and lutetia-doped wafers. The slopes of the curves correspond to a power constant of n = -1/6, which is applicable to the defect reaction given in Eq. (5) and is related to $P_{O2}(I)$ in accordance with Eq. (7), since $P_{O2}(II) >> P_{O2}(I)$. O_2 molecules are assumed to permeate mainly by grain boundary diffusion of oxygen through the oxygen vacancies from the higher to the lower P_{O2} surface. When the doping level is 0.2 mol%, the oxygen permeability constants are about three times smaller than for undoped alumina, although the slopes of the curves are similar. Thus, lutetium doping seems to suppress the mobility of oxygen without changing the oxygen diffusion mechanism. On the other hand, the oxygen permeability constants for all the polycrystals for P_{O2} values above 10^3 Pa in the upper chamber are similar to each other and increase with increasing P_{O2}, as shown in Fig. 9. Their slopes correspond to a power constant of n = 3/16 that suggests participation in the defect reaction given in Eq. (8) and $P_{O2}(II)$ in accordance with Eq. (10), since $P_{O2}(II) >> P_{O2}(I)$. Under potential gradients generated by P_{O2} values above approximately 10^3 Pa, O_2 molecules seem to permeate mainly by grain boundary diffusion of aluminum through aluminum vacancies from the lower to the higher P_{O2} surface. In this case, the lutetium segregated at grain boundaries would be expected to have little effect on the diffusivity of aluminum.

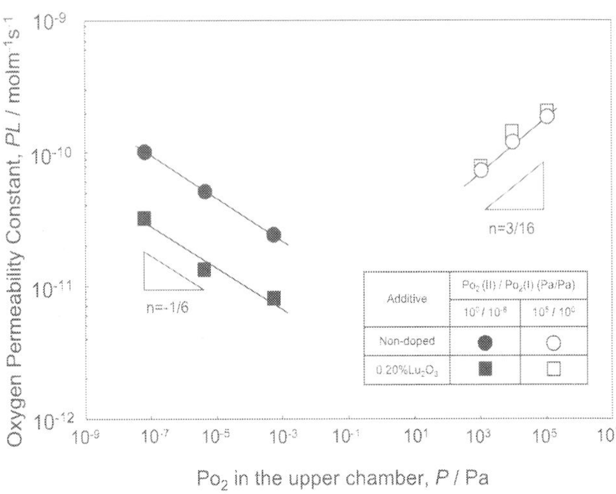

Fig. 9. Effect of P_{O2} in the upper chamber on the oxygen permeability constants of polycrystalline alumina (non-doped and doped with 0.2 mol% Lu_2O_3) at 1923 K. The solid symbols indicate data for specimens exposed to a ΔP_{O2} between about $P_{O2}(II)$ = 1 Pa in the lower chamber and a much lower P_{O2} ($P_{O2}(I)$) in the upper chamber. The open symbols indicate data for specimens exposed to a ΔP_{O2} between $P_{O2}(I)$ = 1 Pa in the lower chamber and a much higher P_{O2} ($P_{O2}(II)$) in the upper chamber.

Figure 10 shows SEM micrographs of the surfaces and cross-sections of non-doped polycrystalline alumina exposed at 1923 K for 10 h under ΔP_{O2} with $P_{O2}(II)/ P_{O2}(I)$= 1 Pa/10^{-8} Pa and 10^5 Pa/1 Pa. For $P_{O2}(II)/ P_{O2}(I)$= 1 Pa/10^{-8} Pa, grain boundary grooves are observed on both the surfaces, of which morphology is similar to that formed by ordinary thermal etching. The oxygen potential gradients with combination of the lower P_{O2} values hardly affect the surface morphological change. The absence of the grain boundary ridges suggests that the migration of aluminum was scarcely related to the oxygen permeation. This surface morphology supports the oxygen permeation mechanism with n = -1/6 as shown in Fig. 9. For $P_{O2}(II)/ P_{O2}(I)$= 10^5 Pa/1 Pa, grain boundary ridges with heights of a few micrometers can be seen on the P_{O2}(II) surface, while deep crevices are formed at the grain boundaries on the P_{O2}(I) surface. The total volume of the grain boundary ridges, measured by 3D laser scanning microscopy, was consistent with the volume of alumina that should be produced given the observed amount of oxygen permeation (Kitaoka et al., 2009). This result provides adequate support for an oxygen permeation mechanism with n = 3/16, as shown in Fig. 9.

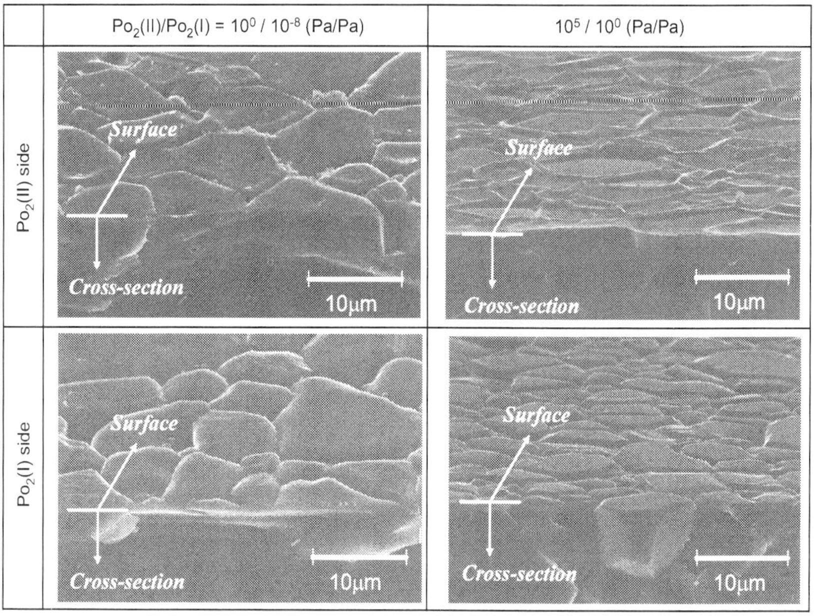

Fig. 10. shows SEM micrographs of the surfaces and cross-sections of non-doped polycrystalline alumina exposed at 1923K for 10h under ΔP_{O2} with $P_{O2}(II)/P_{O2}(I)$=1 Pa/10^{-8} Pa and 10^5 Pa/1 Pa.

Figure 11 shows SEM micrographs of the surfaces and cross-sections of polycrystalline alumina doped with 0.2 mol% Lu_2O_3 exposed at 1923 K for 10 h under ΔP_{O2} with $P_{O2}(II)/P_{O2}(I)$= 1 Pa/10^{-8} Pa and 10^5 Pa/1 Pa. Figure 12 shows top-view SEM images of the surfaces corresponding to Fig. 11. In the case of $P_{O2}(II)/ P_{O2}(I)$ = 1 Pa/10^{-8} Pa, as shown in Fig. 11, shallow grain boundary grooves, similar to those produced by conventional thermal etching, are observed on both surfaces, as in the case of undoped alumina. In addition, as seen in Fig. 12, a large number of particles with diameters of about 1 micrometer are uniformly distributed at the grain boundaries on both surfaces. The distribution of the particles was maintained during exposure under the oxygen potential gradient at 1923 K. These particles were identified as $Al_5Lu_3O_{12}$ by XRD and EDS and had already precipitated at the grain boundaries by reaction of alumina grains with excess lutetium during sintering the sample. The remainder of the added lutetium should then become segregated at the grain boundaries. This implies that the lutetium species scarcely migrates, remaining in the wafer during oxygen permeation, and inhibiting the mobility of oxygen from the region of higher P_{O2} to the region of lower P_{O2} (Fig. 9).

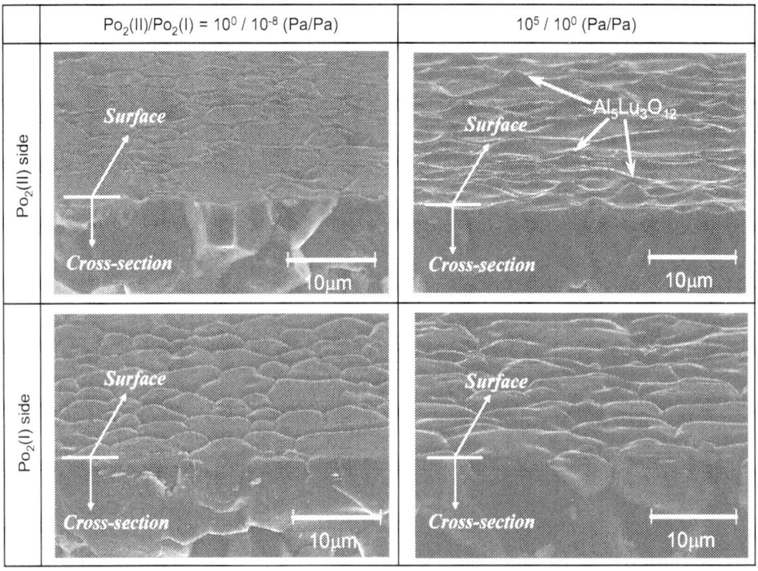

Fig. 11. SEM micrographs of the surfaces and cross-sections of polycrystalline alumina doped with 0.2 mol% Lu_2O_3 exposed at 1923 K for 10 h under ΔP_{O2} with $P_{O2}(II)/ P_{O2}(I)= 1$ Pa/10^{-8} Pa and 10^5 Pa/1 Pa.

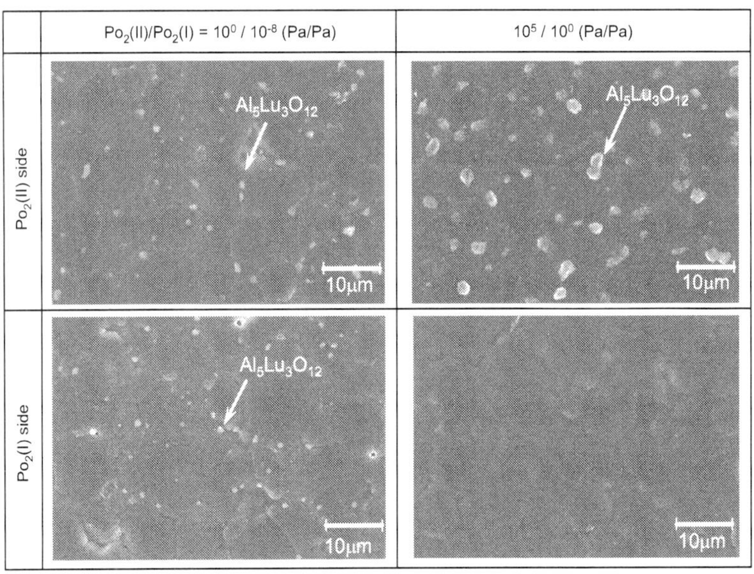

Fig. 12. SEM micrographs of the surfaces of polycrystalline alumina doped with 0.2 mol% Lu_2O_3 exposed at 1923K for 10h under ΔP_{O2} with $P_{O2}(II)/P_{O2}(I)$=1 Pa/10^{-8}Pa and 10^5Pa/1Pa.

For $P_{O2}(II)/ P_{O2}(I) = 10^5$ Pa/1 Pa, Fig. 11 reveals that the grain boundaries on the higher P_{O2} surface are raised to a height of a few micrometer, while deep trenches are formed at the grain boundaries on the lower P_{O2} surface, similar to the case for undoped alumina. Furthermore, as seen in Fig. 12, the higher P_{O2} surface exhibits $Al_5Lu_3O_{12}$ particles with diameters of several micrometer, but such particles are not found on the opposite surface. This can be explained by a migration of both lutetium and aluminum from the lower to the higher P_{O2} region.

3.3 Grain boundary diffusion coefficients

The grain boundary diffusion coefficients of oxygen and aluminum ($D_{gb}\delta$) are estimated from the oxygen permeability constants shown in Fig.9 by the procedure described in Section 3.1.3. Figure 13 shows $D_{gb}\delta$ for oxygen and aluminum in polycrystalline alumina (undoped and doped with 0.20 mol% Lu_2O_3) as a function of the equilibrium partial pressure of oxygen in the upper chamber at 1923 K. Values of oxygen diffusion coefficients

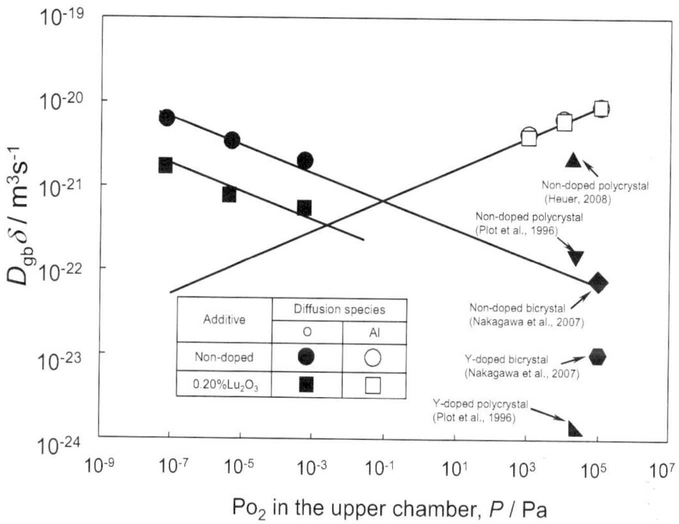

Fig. 13. $D_{gb}\delta$ of oxygen and aluminum in polycrystalline alumina (non-doped and doped with 0.2 mol% Lu_2O_3) as a function of the equilibrium partial pressures of oxygen in the upper chamber at 1923 K. The solid and open symbols indicate the $D_{gb}\delta$ of oxygen and aluminum, respectively.

taken from the literature (Plot et al., 1996, Nakagawa et al., 2007, Heuer, 2008) are also shown in Fig. 13. They were determined using an ^{18}O isotopic tracer profiling technique for bicrystalline or polycrystalline alumina annealed in a homogeneous environment in the absence of an oxygen potential gradient, and their P_{O2} values on the abscissa correspond to those in the annealing environments. The data for refs. (Nakagawa et al., 2007, Heuer, 2008) are estimated by extrapolating to 1923 K. For lutetia-doped polycrystalline alumina, there are unfortunately no data for oxygen grain boundary diffusion coefficients determined by the tracer profiling technique, but some measurements have been carried out on yttria-doped alumina. On the other hand, it has been reported that creep resistance in polycrystalline alumina was improved remarkably by doping to only 0.05-0.1 mol% with oxides such as Lu_2O_3 and Y_2O_3 in a similar effect on the creep resistance to each other (Ikuhara et al., 2001). Thus, the grain boundary coefficients for oxygen in yttria-doped alumina (polycrystal and bicrystal) are shown in Fig. 13 for reference.

The $D_{gb}\delta$ value for oxygen is seen to decrease with increasing P_{O2}, whereas the value for aluminum increases for both undoped and lutetia-doped alumina. Increasing the doping level to 0.2 mol% lutetia causes an approximately three times reduction in $D_{gb}\delta$, while maintaining the slope of the curve. In contrast, the $D_{gb}\delta$ value of aluminum is unaffected by lutetia doping. Thus, lutetia doping has the effect of reducing the mobility of oxygen along the grain boundaries, but has little influence on the diffusion of aluminum.

For the undoped alumina, the line extrapolated to higher P_{O2} for $D_{gb}\delta$ of oxygen is consistent with previous reported data obtained using SIMS (Plot et al., 1996, Nakagawa et al., 2007), but deviates widely from data using NRA (Heuer, 2008). There is a thermal equilibrium level of defects such as Schottky pairs (Buban et al., 2006) or Frenkel pairs (Heuer, 2008) in alumina held in uniform environments at high temperatures. As shown in Figs. 9-13, the oxygen potential gradients through the wafer seem to result in the formation of new defects such as oxygen vacancies for lower P_{O2} ranges and aluminum vacancies for higher P_{O2} ranges, in addition to the thermally induced defects. Because $D_{gb}\delta$ for oxygen and aluminum are proportional to the concentration of their respective vacancies, the dominant defects in the wafer are probably oxygen vacancies for lower P_{O2} values and aluminum vacancies for higher P_{O2} values. Therefore, the extrapolated line in Fig. 8 may correspond to the SIMS data (Plot et al., 1996, Nakagawa et al., 2007), where the concentration of oxygen vacancies induced by the oxygen potential gradient for the higher P_{O2} ranges is asymptotic to that under thermal equilibrium. Nevertheless, the reason why the NRA result deviates so much cannot be ascertained based on the descriptions given in the paper (Heuer, 2008).

As mentioned above, elements such as yttrium and lutetium that were segregated at the grain boundaries of alumina by addition of only 0.05-0.1 mol% Ln_2O_3 effectively retarded oxygen grain boundary diffusivity, creep deformation and final-stage sintering under uniform environments (Nakagawa et al., 2007, Ikuhara et al., 2002, Yoshida et al., 2002, 2007, Watanabe et al., 2003). Retardation of such mass transfer can be explained by a 'site-blocking' mechanism (Amissah et al., 2007, Wang et al., 1999, Cho et al., 1999, Cheng et al., 2008, Priester, 1989, Korinek et al., 1994) and/or grain boundary strengthening (Yoshida et al., 2002, Buban et al., 2006). Under the oxygen potential gradients used in this study, it was found that oxygen diffusivity was unaffected by 0.05 mol% lutetia-doping (Matsudaira et al., 2010), and even for 0.2 mol% doping, the retardation was small compared to the effect in uniform environments. This may be related to the generation of a large number of oxygen vacancies in the vicinity of the grain boundaries under an oxygen potential gradient, despite the fact that Lu^{3+} is isovalent with Al^{3+}.

As mentioned in the Introduction, Bedu-Amissah et al. measured Cr^{3+} diffusion in alumina under a Cr^{3+} concentration gradient (Amissah et al., 2007). From the chromium diffusion profile, they found that yttrium doping retards cation diffusion in the vicinity of the grain boundary, reducing $D_{gb}\delta$ by at least one order of magnitude (Amissah et al., 2007). In contrast, the $D_{gb}\delta$ value of aluminum under oxygen potential gradients is unaffected by lutetia doping. Thus, lutetia doping has little influence on the diffusion of aluminum along grain boundaries. This may be related to the generation of a large number of aluminum vacancies around grain boundaries under an oxygen potential gradient, which reduces the effect of 'site-blocking' and/or grain boundary strengthening, resulting in outward diffusion of both lutetium and aluminum, as shown in Figs. 11 and 12.

4. Conclusions

The oxidation of the CoNiCrAlY alloy under a P_{O2} of 10^{-14} Pa at 1323 K, during which both aluminum and chromium in the alloy were oxidized and elements such as cobalt and nickel were not oxidized, accelerated the transformation from metastable theta-Al_2O_3 to stable alpha-Al_2O_3, resulting in the formation of a dense, smooth alpha-$(Al,Cr)_2O_3$ scale. The surface concentrations of cobalt and nickel in the scale, which was evolved by formation of $(Co,Ni)(Al,Cr)_2O_4$ during the temperature increase and subsequent reduction and decomposition of the oxide at a higher temperature, could be effectively reduced by decreasing the P_{O2} during the temperature rise in the oxidation treatment. By contrast, oxidation at a higher P_{O2} required a longer time for the transformation and $(Co,Ni)(Al,Cr)_2O_4$ was also produced in the scale with a rougher surface.

The oxygen permeability of undoped and lutetia-doped polycrystalline alpha-alumina wafers that were exposed to oxygen potential gradients (ΔP_{O2}) was evaluated at high temperatures to investigate the mass-transfer phenomena through the alumina scale. The main diffusion species during oxygen permeation through the alumina grain boundaries was found to depend on P_{O2} values, which created ΔP_{O2}. Under ΔP_{O2} generated by low P_{O2} values, where oxygen permeation occurred by oxygen diffusion from regions of higher to low P_{O2}, segregated lutetium at the grain boundaries suppressed only the mobility of oxygen in the wafers, without affecting the oxygen permeation mechanism. By contrast, under ΔP_{O2} generated by high P_{O2} values, where oxygen permeation proceeded by aluminum diffusion from regions of lower to higher P_{O2}, lutetium had little effect on aluminum diffusion and migrated together with aluminum, resulting in precipitation and growth of $Al_5Lu_3O_{12}$ particles on the higher P_{O2} surface.

5. References

Evans A. G., Mumm D. R., Hutchinson J. W., Meier G. H. & Pettit F. S. (2001). Mechanisms Controlling the Durability of Thermal Barrier Coatings. *Prog. Mater. Sci.*, 46, 505-553

Brumm M. W. & Grabke H.J. (1992). The Oxidation Behavior of NiAl - I. Phase Transformations in the Alumina Scale During Oxidation of NiAl and NiAl-Cr Alloys. *Corrosion Science*, 33, 1677-1690

Tolpygo V.K. & Clarke D.R. (2000). Microstructural Study of the Theta-Alpha Transformation in Alumina Scales Formed on Nickel-Aluminides. *Microscopy of Oxidation*, 17, 59-70

Tolpygo V.K., Clarke D.R. & Murphy K. S. (2001). The Effect of Grit Blasting on the Oxidation Behavior of Platinum-Modified Nickel-Aluminide Coating. *Metall. Mater. Trans.*, 32A, 1467-1478

Tolpygo V.K. & Clarke D.R. (2005). The Effect of Oxidation Pre-Treatment on the Cyclic Life of EB-PVD Thermal Barrier Coatings with Platinum-Aluminide Bond Coats. *Surf. Coat. Technol.* 200, 1276-1281

Nijdam T.J. & Sloof W.G. (2006). Combined Pre-Annealing and Pre-Oxidation Treatment for the Processing of Thermal Barrier Coatings on NiCoCrAlY Bond Coatings. *Surf. Coat. Technol.*, 201, 3894-3900

Matsumoto M., Hayakawa K., Kitaoka S., Matsubara H., Takayama H., Kagiya Y. & Sugita Y. (2006). The Effect of Preoxidation Atmosphere on Oxidation Behavior and Thermal Cyclic life of Thermal barrier Coatings. *Mater. Sci. Eng.* A 411, 119-125

Matsumoto M., Kato T., Hayakawa K., Yamaguchi N., Kitaoka S. & Matsubara H. (2008). The Effect of Pre-Oxidation Atmosphere on the Durability of EB-PVD Thermal Barrier Coatings with CoNiCrAlY Bond Coats. *Surf. Coat. Technol.*, 202, 2742-2748

Pint A., Treska M. & Hobbs L.W. (1997). The Effect of Various Oxide Dispersions on the Phase Compositui n and Morphology of Al_2O_3 Scales Grown on beta-NiAl. *Oxid. Met.*, 47, 1-20

Peng X., Clarke D.R. & Wang F. (2003). Transient-Alumina Transformations during the Oxidation of Magnetron-Sputtered CoCrAl Nanocrystalline. *Oxid. Met.*, 60, 225-240

Clarke D.R. (1998). Epitaxial Phase Transformations in Aluminum Oxide. *Phys. Stat. Sol.*, 166, 183-196

Odaka A., Yamaguchi T., Fujita T., Taruta S. & Kitajima K. (2008). Cation Dopant Effect on Phase Transformation and Microstructural Evolution in M^{2+}-Substituted gamma-Alumina Powders. *J. Mater. Sci.*, 43, 2713-2720

Nagano T., Sato K., Saitoh T. & Takahashi S. (2009). Hydrothermal Stability of Mesoporous Ni-Doped gamma-Al_2O_3. *J. Ceram. Soc. Jpn.*, 117, 832-835

Kitaoka S., Kuroyama T., Matsumoto M., Kitazawa R. & Kagawa Y. (2010). Control of Polymorphism in Al_2O_3 Scale Formed by Oxidation of Alumina-forming Allys. *Corrosion Science*, 52, 429-434

Nychka J.A. & Clarke D.R. (2005). Quantification of Alumminum Outward Diffusion During Oxidation of FeCrAl Alloys. *Oxid. Metals*, 63, 325-352

Matsudaira T., Wada M., Kitaoka S., Asai T., Miyachi Y. & Kagiya Y. (2008). Gas Permeability of Oxide Ceramics at Ultra-high Temperatures. *J. Soc. Mater. Sci. Jpn.*, 57, 532-538

Wada M., Matsudaira T. & Kitaoka S. (2008). Oxygen Permeability and Morphological Stability of Alumina Ceramics under O_2 Potential Gradients at Ultra-High Temperatures. *AMTC Letters*, 1, 34-35

Kitaoka S., Matsudaira T. & Wada M. (2009). Mass-Transfer Mechanism of Alumina Ceramics under Oxygen Potential Graddients at High Temperatures. *Mater. Trans.*, 50, 1023-1031

Volk H.F. & Meszaros F.W. (1968). *Ceramic Micro Structures their Analysis Significance & Production*, ed. by R. M. Fullhrath and J. A. Pask, John Wiley and Sons, New York

Courtright E.L. & Prater J.T. (1992). *Oxygen Permeability of Several Oxides above 1200 deg.C*, US DOE Rep. PNL-SA-20302

Ogura Y., Kondo M., Morimoto T., Notomi A. & Sekigawa T. (2001). Oxygen Permeability of Y_2SiO_5. *Mater. Trans.*, 42, 1124-1130.

Plot D., Gall M.L., Lesage B., Huntz A.M. & Monty C. (1996). Self-diffusion in Alpha-Al_2O_3 IV. Oxygen Grain-boundary Self-diffusion in Undoped and Yttria-doped Alumina Polycrystals.*Philos. Mag. A*, 73, 935-949

Nakagawa T., Sakaguchi I., Shibata N., Matsunaga K., Mizoguchi T., Yamamoto T., Haneda H. & Ikuhara Y. (2007). Yttrium Doping Effect on Oxygen Grain Boundary Diffusion in Alpha-Al_2O_3. *Acta Mater.*, 55, 6627-6633

Messaoudi K., Huntz A.M. & Lesage B. (1998). Diffusion and Growth Mechanism of Al_2O_3 Scales on Ferritic Fe-Cr-Al Alloys. *Mater. Sci. Eng. A*, 247, 248-262

Heuer A.H. (2008). Oxygen and Aluminum Diffusion in Alpha-Al_2O_3 : How Much Do We Really Understand ?. *J. Eur. Ceram. Soc.*, 28, 1495-1507

Paladino A.E. & Kingery W.D. (1962). Aluminum Ion Diffusion in Aluminum Oxide. *J. Chem. Phys.*, 37, 957-962

Le Gall M., Lesage B. & Bernardini J. (1994). Self-diffusion in Alpha-Al_2O_3 I. Aluminum Diffusion in Single Crystals. *Philos. Mag. A*, 70, 761-773

Amissah K.B., Rickman J.M., Chan H.M. & Harmer M.P. (2007). Grain-boundary Diffusion of Cr in Pure and Y-Doped Alumina. *J. Am. Ceram. Soc.*, 90, 1551-1555

Pint B.A., Garratt-Reed A.J. & Hobbs L.W. (1998). Possible Role of the Oxygen Potential Gradient in Enhancing Diffusion of Foreign Ions on Alpha-Al_2O_3 Grain Boundaries. *J. Am. Ceram. Soc.* 81, 305-314

Matsunaga K., Tanaka T., Yamamoto T. & Ikuhara Y. (2003). First-principles Calculations of Intrinsic Defects in Al_2O_3. *Phys. Rev. B*, 68, 085110

Ikuhara Y., Yoshida H. & Sakuma T. (2001). Impurity Effects on Grain Boundary Strength in Structural Ceramics. *Mater. Sci. Eng.*, A319-321, 24-30

Yoshida H., Ikuhara Y. & Sakuma T. (2002). Grain Boundary Electronic Structure Related to the High-temperature Creep Resistance in Polycrystalline Al_2O_3. Acta Mater., 50, 2955-2966

Wang C.M., Cargill III G.L., Harmer M.P., Chan H.M. & Cho J. (1999). Atomic Structural Environment of Grain Boundary Segregated Y and Zr in Creep Resistant Alumina from EXAFS. *Acta Mater.*, 47, 3411-3422

Cho J., Wang C.M., Chan H.M., Rickman J.M. & Harmer M.P. (1999). Role of Segregating Dopants on the Improved Creep Resistance of Aluminum Oxide. *Acta Mater.*, 47, 4197-4207

Cheng H., Dillon S.J., Caram H.S., Rickman J.M., Chan H.M. & Harmer M.P. (2008). The Effect of Yttrium on Oxygen Grain-boundary Transport in Polycrystalline Alumina Measured Using Ni Maker Particles. *J. Am. Ceram. Soc.*, 91, 2002-2008

Priester L. (1989). Geometrical Speciality and Special Properties of Grain Boundaries. *Rev. Phys.*, 24, 419-438

Korinek S.L. & Dupau F. (1994). Grain Boundary Behavior in Superplastic Mg-doped Alumina with Yttria Codoping. *Acta. Metall. Mater.*, 42, 293-302

Buban J.P., Matsunaga K., Chen J., Shibata N., Ching W.Y., Yamamoto T. & Ikuhara Y. (2006). Grain Boundary Strengthening in Alumina by Rare Earth Impurities. *Science*, 311, 212-215

Matsudaira T., Wada M., Saitoh T. & Kitaoka S. (2010). The Effect of Lutetium Dopant on Oxygen Permeability of Alumina Polycristals under Oxygen Potential Gradients at Ultra-high Temperatures. *Acta Mater.*, 58, 1544-1553

Degterov S. & Pelton A.D. (1996). Critical Evaluation and Optimization of the Thermodynamic Properties and Phase Diagrams of the $CrO-Cr_2O_3$, $CrO-Cr_2O_3-Al_2O_3$, and $CrO-Cr_2O_3-CaO$ Systems. *J. Phase Equilibria*, 17, 476-487

Ansara I., Dupin N., Lukas H.L. & Sundman B.J. (1997). Thermodynamic Assessment of the Al-Ni System. *Alloys and Compounds*, 247, 20-30

Watanabe S., Yoshida H. & Sakuma T. (2003). Densification Behavior and Grain Boundary Diffusivity in Cation-doped Alumina. *Key Engineering Materials*, 247, 67-70

Yoshida H., Morita K., Kim B.N., Hiraga K., Yamamoto T. & Sakuma T. (2007). estimation of Grain Boundary Diffusivity in Cation-doped Polycrystalline Alumina. *Mater Sci Forum*, 558-559, 997-1002

17

Mass Transfer Investigation of Organic Acid Extraction with Trioctylamine and Aliquat 336 Dissolved in Various Solvents

Md Monwar Hossain
United Arab Emirates University (Company)
United Arab Emirates

1. Introduction

Organic acids have been used in producing biodegradable polymeric materials (polylactate) and they are also being considered for manufacture of drugs, perfumes and flavours as raw materials. Therefore the production of high purity organic acids is very important. They can be produced by chemical methods. However, fermentation technology has proven to be the best alternative being more energy efficient and having potential. To allow production and separation simultaneously. The major part of the production cost accounts for the cost of separation from very dilute reaction media where productivity is low due to the inhibitory nature of many organic acids. The current method of extraction/separation is both expensive and environmentally unfriendly. Therefore, there is great scope for development of an alternative technology that will offer increased productivity, efficiency, economic and environmental benefits. One of the promising technologies for recovery of organic acids from fermentation broth is reactive liquid-liquid extraction (Tamada and King, 2001, Dutta et al., 2006). However, common organic solvents when used alone show low distribution coefficients and do not give efficient separation. Reactive liquid-liquid extraction (RLLE) utilizes a combination of an extractant (also known as carrier) and diluents to intensify the separation through simultaneous reaction and extraction. Thus this method provides high selectivity and enhances the recovery. RLLE has been applied in many analytical, industrial, environmental and metallurgical processes (Parthasarathy et al., 1997; Klassen, et al., 2005; Kumar et al., 2001; Urtiaga et al., 2005; Carrera et al., 2009). In most of these applications one of these following solvents: kerosene, toluene/mixtures of kerosene and methyl isobutyl ketone (MIBK), hexane/decanol/octanol or any solvent system with similar toxic characteristics have been examined. These solvents have been proven to separate the "target" component from the aqueous solutions containing it. However, they have the issues of sustainability, health and safety, operator-friendliness and environmental impact. Therefore, efforts are devoted to determine a solvent that will partially or fully address these issues. In this chapter, a new, non-traditional solvent is examined for its ability to separate a specific component by applying the reactive extraction. Lactic acid (an organic acid) is chosen as the specific component (as a model for all other organic acids), experiments are presented to show its capacity and finally the analysis is extended to include the mass transfer processes in microporous hollow-fiber membrane module (HFMM). In the next few paragraphs lactic acid is described with the processes of production and ongoing research in the development of techniques to separate it from the production media. From the methods one is selected (i.e. RLLE) and the new solvent system that has the potential to overcome the disadvantages of the currently practiced solvent, is examined.

Lactic acid (2-hydroxypropanoic acid, $CH_3CHOHCOOH$) is a colorless, organic liquid. It has a variety of applications in the food, chemical, pharmaceutical and cosmetic industries [Hong, et al., 2002]. The Food and Drug Administration (FDA) have approved lactic acid and its salts to be GRAS (Generally Recognized as Safe) [Lee, et al., 2004]. It can be converted to a polylactic acid used for the synthesis of biodegradable materials [Coca, et al., 1992]. As well as being environmentally friendly, there is a growing demand; due to strict environmental laws being legislated for biodegradable polymers as a substitute for conventional plastic materials. Biodegradable copolymers are also used for the production of new materials with biomedical applications such as drug delivery systems [Choi, and Hong, 1999].

Lactic acid is typically produced via either chemical synthesis or the fermentation of whey or another in-expensive carbon source [Lee, et al., 2004]. Due to the increasing cost of the common raw material for the chemical synthesis, the efficient production of lactic acid through fermentation has become increasingly important [Han, et al., 2000; Heewsink, et al., 2002; Drioli, et al., 1996; Hano, et al., 1993; Siebold, et al., 1995]. As mentioned earlier, an economical and efficient method for the recovery from fermentation broth is vital as the overall cost of production is dominated by the cost of recovery [Han, et al., 2000; Drioli, et al., 1996].

The production of most organic acids from fermentation media are subject to product inhibition as the reaction proceeds [Hano, et al., 1993; Hong and Hong, 1999; Yuchoukov, et al., 2005]. Hence, the separation of the organic acid as it is being produced is highly desirable. The extractive fermentation, *in situ* application of the solvent extraction technique, keeps the product concentration in the broth at a low level and suppresses the product inhibition by continuously removing them from a fermentation broth [Siebold, et al., 1995; Yankov et al., 2005; Frieling and Schugerl, 1999].

Various methods for the extraction of lactic acid have been reported such as precipitation, ion exchange process, adsorption, diffusion dialysis, microcapsules, esterification and hydrolysis, reactive extraction as well as a simulated moving bed process (Hong, et al., 2002; Tik, et al., 2001; Tong, et al., 1999; Ju, and Verma, 1994; Gong, et al., 2006; Sun et al., 2006). These methods have several disadvantages including high cost, and they produce large volumes of waste, require multiple steps, and operate with low efficiency under practical conditions. As mentioned earlier, the RLLE method using microporous Hollow Fibre Membrane Contactor (HFMC) may potentially overcome many of the disadvantages and provide a better alternative for the recovery of lactic acid (Wasewar, et al., 2002; Datta and Henry, 2006; Schlosser, 2001; Lin, and Chen, 2006). In a recent review, a process based on RLLE in HFMM has been found to be competitive from the process, economic and environmental points of view (Sun, et al., 2006; Joglekar, et al., 2006; Datta, et al., 2006). The advantages of the membrane mass transfer process over the conventional systems are (Lin, and Chen, 2006; Sun, et al., 2006; Joglekar, et al., 2006; Datta, et al., 2006):

- Selectivity and flexibility of extraction
- *in situ* application to reduce any inhibitory effect
- Reduction of number of steps (improved productivity)
- Use of operator and environmentally-friendly organic system
- Minimal dispersion of phases (less contamination)
- Recycle of extracting media and generation of smaller wastes

- Lower temperature operation requiring less energy
- Ability to operate on identical density systems
- Availability of large-scale module (i.e. easy scale up methods).

Amine compounds have been found useful as extractants for the separation of organic acids (Tamada and King, 2000; Kertesz, and Schlosser, 2005). They provide high efficiency and selectivity. Secondary, tertiary and quaternary amines and their mixtures have been employed for this purpose. An organic solvent is required to dissolve the reaction product, and a diluent is required to control the viscosity and to stop formation of any third phase. Active polar and proton-donating diluents as alcohols have been shown to be the most suitable diluents for amines as they show high distribution coefficient. The reaction mechanism changes with the combination of the extractant and solvent type. But the mass transfer equations and analysis of the processes involved are similar to those developed in the following section.

To understand and explore more of this process the main aims were set

- to determine a less toxic, environmentally-friendly solvent and a carrier or a mixture of carriers for extraction of lactic acid and the effect of conditions (temperature and pH) similar to the fermentation,
- to discuss the results of the experiments on liquid-liquid extraction
- to develop a mathematical model for the mass transfer processes in a small pilot-scale contactor
- to evaluate the performance of the less toxic solvent for extraction under fermentation conditions (i.e. in presence of salts and lactose).
- to compare the results of the hollow-fibre experiments.

The results show that the new system has the potential to overcome some of the disadvantages mentioned above. More research is required to optimise the experimental conditions, to develop a more comprehensive mathematical model including extraction and re-extraction and obtain performance data with "real" (rather than synthetic system) system.

In the next section, mathematical modelling is presented for liquid-liquid extraction and mass transfer processes in a commercially available membrane module (i.e. HFMM). Rather than a comprehensive approach a simple analysis is proposed to provide an understanding of the mass transfer phenomena in a small pilot-scale module.

2. Modeling of equilibrium and mass transfer

2.1 Liquid-liquid extraction

The reaction of undissociated lactic acid (HLA) with a carrier (B) dissolved in the solvent gives a reaction complex (BHLA) which remains largely in the organic phase and may be represented by (Juang and Huang, 1997; Datta, et al., 2006; Yuchoukov, et al., 2005):

$$(\mathrm{HLA})_{aq} + (\mathrm{B})_{org} \Leftrightarrow \left(\mathrm{BH^{+}LA^{-}}\right)_{org} \quad (1)$$

A simple 1:1 stoichiometry (the molar ratio of organic acid to that of extractant) has been proposed. However, this depends on the type of the organic acid and its ionic state, the type of the extractant and the type of the solvent (Uslu et al., 2009). The reaction mechanism does not change the mass transfer processes.

For reactive extractions microporous hollow fiber membrane contactors, in various configurations have been evaluated (Klassen et al., 2005; Yang, and Cussler, 2000; Ren et al.,

2005; Tong et al., 1998; Juang, et al., 2000; Prasad and Sirkar, 1988). Typically two modules are used, one for the extraction and the other for the re-extraction or stripping process.

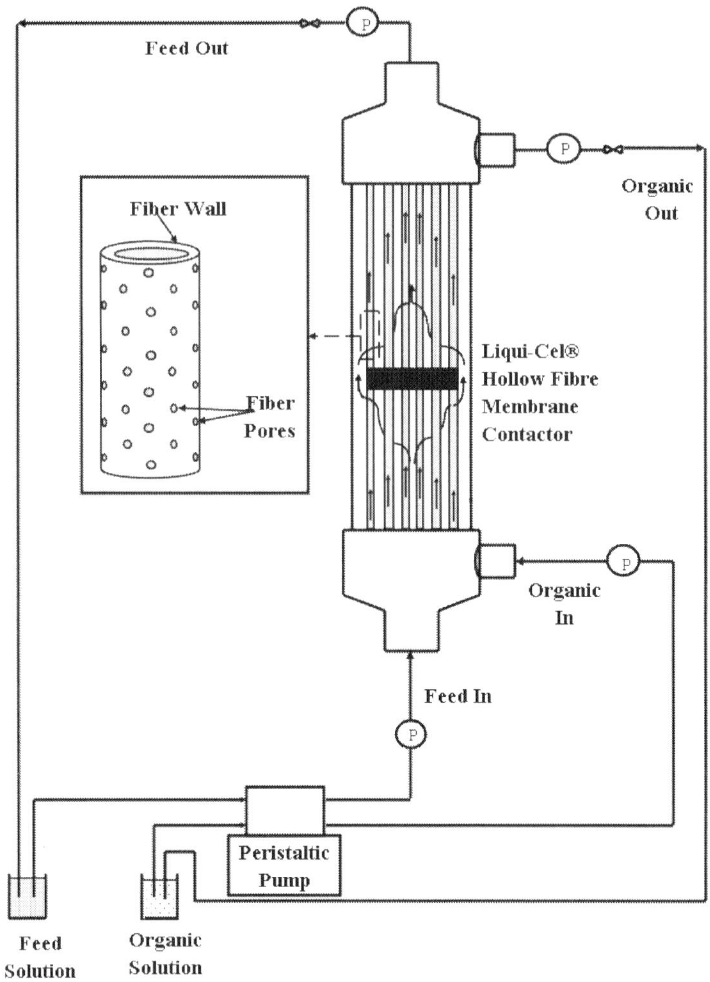

Fig. 1. A schematic diagram of the mass transfer operation of the hollow-fibre membrane contactor.

These are commercially available modules, e.g. a module with catalogue No. 5PCM-218, obtained from Separation Products Division, Hoechst Celanese Corporation, Charlotte, NC, USA, have been extensively used for mass transfer operation. The contactor has a shell-and-tube configuration with a total of 10,000 polypropylene hollow fibers (Celgard X-30, 240 μm ID, 300 μm, OD, length 15 cm) potted in polyethylene in a polypropylene case of 6 cm ID. The surface area of the contactor is 1.4 m^2. The hollow fiber module is usually set up as shown in Figure 1. When these modules are used, aqueous and organic solutions flow continuously, one through the lumen side of the fibre and the other through the shell side. The both the solutions get into contact through the pores of the wall. Phase entrainment is avoided by applying a little higher pressure on the aqueous side. The pressure difference between the phases is between 0.2 - 0.3 bar and it has been reported that it has no influence on the mass transfer processes.

These contactors have been used in various process configurations such as hollow fibre contained liquid membranes, HFCLM (Yang et al., 2003; Dai et al., 2000), hollow fibre supported liquid membrane, HFSLM, (Yang and Kocherginsky, 2006; Rathore, et al., 2001), non-dispersive solvent extraction, NDSX, Ortiz, et al., 2004) and hollow fibre renewal liquid membranes, HFRLM, (Ren et al., 2008). The main difference between these configurations is that the contacting pattern of the liquid phases are different. In a newly developed mass transfer operation known as emulsion pertraction, PERT, a single module is used for extraction and stripping simultaneously (Klaassen and Jansen, 2001; Klassen et al., 2005). The emulsion consists of an organic solvent with a dissolved extractant phase with droplets of strip liquid dispersed in it. The phases are separated by the hydrophobic membrane surface. The contact between the water phase and emulsion phase occurs at the pore mouth. The organic acid-extractant complex diffuses through the pores and on the other side of the membrane the extractant is regenerated by strip liquid. The analysis below is not applicable to the PERT process, it is devoted to extraction in a single module.

2.2 Mass transfer in a hollow fibre membrane contactor

A schematic of the transport mechanism of solute molecules from an aqueous feed side to the organic side through hollow-fibre wall is shown in Figure 2 (Hossain and Mysuria, 2008). The mass transfer processes can be described by the solute transport through the resistances from the aqueous feed (inside the fibre) to the organic phase (shell side). The steps considered for the mass transport and reactive extraction, the solute (lactic acid) molecules:

- are transported from the feed solution to the feed-pore interface and can be expressed by Eq. (2).
- at the interface the reaction between the solute and the carrier takes place (Eq.1) to form a solute-carrier complex. The equilibrium concentrations at the interface in the aqueous and organic phases can be related by an apparent distribution coefficient (DE), given by Eq. (3).
- The diffusion of the solute-carrier complex through the pores of the hollow-fibers filled with the organic phase and this can be expressed by Eq. (4).
- The final step is the transport through the solvent boundary layer at the outer end of the pore mouth and this step can be expressed as in Eq. (5).

The following assumptions have been considered for writing the model equations:

- The system works at isothermal conditions.
- Equilibrium is reached at the interfaces of the aqueous and organic phases.
- The curvature of the interfaces does not affect significantly the processes.
- The distribution coefficient of the solute is considered to be constant with the conditions used
- Uniform pore size along the entire length of the contactor.
- The mass transfer processes in the boundary layer is described by the diffusion model.
- The phases are immiscible and the pores are wetted by the organic phase only.

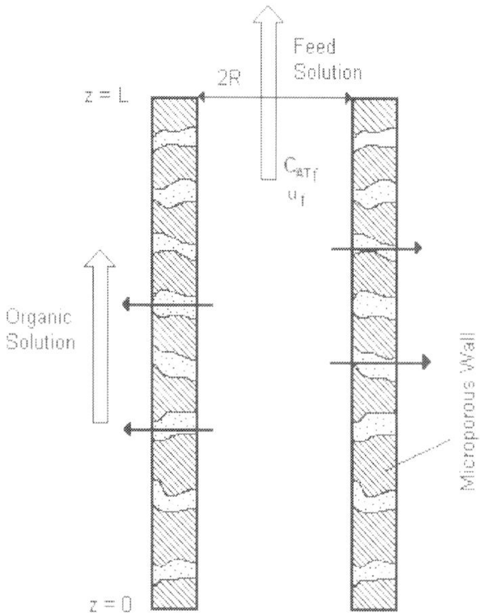

Convective flux = $-u_f\left(\frac{V}{A}\right)_{in}\frac{\partial C^m_{LAf}}{\partial z}$

The interfacial reaction:

$(HLA)_{aq} + (B)_{org}^{-} \Leftrightarrow (BH^{+}LA^{-})_{org}$

$\left(BH^{+}LA^{-}\right)$ diffuses through the wall to the shell side (the diffusive flux)

Flux = $-K_{of}\left(C^m_{LAf} - \frac{C^m_{LAo}}{D_E}\right)$

Fig. 2.(a) A schematic of the mass transfer processes in the membrane module.

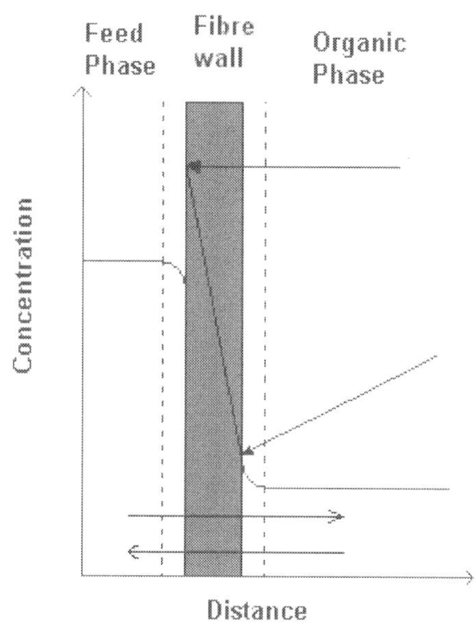

Fig. 2.(b) Concentration profiles of lactic acid in the feed phase (fibre side), in membrane wall and in the organic phase (shell side).

The transport equations are

$$N_{LAf} = k_f\left(C_{LAf} - C_{LAfi}\right) \tag{2}$$

$$D_E = C_{LAOi} / C_{LAfi} \tag{3}$$

$$N_{LAm} = k_{mf}\left(C_{LAfi} - C_{LAOi}\right) \tag{4}$$

$$N_{LAo} = k_o\left(C_{LAOi} - C_{LAO}\right) \tag{5}$$

where k_f , k_{mf} and k_o are the mass transfer coefficient in the feed side, on the membrane and in the organic side, respectively. The concentrations C_{LAf} and C_{LAfi} are the total lactic acid concentrations in the bulk and at the interface, respectively. The concentrations C_{LAOi} and C_{LAO} are the concentration of lactic acid at the membrane-organic interface and in the bulk organic phase, respectively.
Combining the above equations the flux in the system at the steady state is obtained as:

$$N_{LAf} = K_{of}\left(C_{LAf} - C_{LAO}/D_E\right) \tag{6}$$

where K_{of} is the overall mass transfer coefficient of the process. K_{of} is related to the individual transfer coefficients by the following equation:

$$\frac{1}{K_{of}} = \frac{1}{k_f} + \frac{1}{D_E k_{mf}} + \frac{1}{k_O D_E} \tag{7}$$

In order to calculate the overall mass transfer coefficient from the above equation the individual mass transfer coefficients have to be known in addition to the distribution ratio of lactic acid between the aqueous-organic solutions. There are many correlations available in the literature for calculating the individual mass transfer coefficients (Lin, and Chen, 2006; Bringas, et al., 2009; Coelhoso et al. 2000). Each of the correlation is based on the specific experimental conditions and equipment set-up used to develop the correlation. So the assumption of the correlation needs to be matched for its appropriateness before applying to any other system. The calculations from the correlations will be discussed later. In the section below an approximate solution is presented for hollow-fibre membrane modules to evaluate the overall mass transfer coefficient from an analysis of concentration versus time data.

2.3 Approximate solution for the mass transfer model

The membrane modules are operated in recycling mode as the percentage extraction in once-through operation is small. In the recycling mode, it is considered that the feed and the organic solutions are circulated through the fiber side and shell side of the module, respectively.
The mathematical model consists of two mass balance equations, Eq. (8) and (9) that defines the change in solute concentration (i) in the module and (ii) in the feed tank, where aqueous solution is continuously circulated.

$$\left(\frac{V}{A}\right)_{in} \frac{\partial C^{m}_{LAf}}{\partial t} = -u_f\left(\frac{V}{A}\right)_{in} \frac{\partial C^{m}_{LAf}}{\partial Z} - K_{of}\left(C^{m}_{LAf} - \frac{C^{m}_{LAO}}{D_E}\right) \tag{8}$$

$$\frac{dc_{LAf}^{T}}{dt}=\frac{q_f}{v^T}\left(C_{LAf}^{m}\Big|_{z=L}-C_{LAf}^{m}\Big|_{z=0}\right) \tag{9}$$

where $\left(\frac{V}{A}\right)_{in}$ is the ratio of the volume to inner area of mass transfer in the fibers, L is the length of the fiber, u_f is the linear velocity, q_f is the feed flow rate and v^T is the tank volume. The superscripts m and T refer to the membrane module and tank, respectively.

The ratio, $\left(\frac{V}{A}\right)_{in}$, for the feed solution circulating along the inside of the fiber is equal to di/4, where di is the inner radius of each fiber. The factor, $\left(\frac{V}{A}\right)_{in}\frac{1}{L}$, is a small number for this type of contactor, i.e. approx. 4 x 10^{-4}. Using this and assuming a slow rate of change of solute concentration in the module, Eqn (8) can be simplified to

$$\frac{dc_{LAf}^{m}}{dZ}=-\frac{1}{u_f\left(\frac{V}{A}\right)_{in}}K_{of}\left(C_{LAf}^{m}/L-C_{LAO}^{m}/D_E\right) \tag{10}$$

An approximate solution of the model equations form is given by

$$Ln\left\{\frac{C_{LAi}}{(1+1/D_E)C_{LAf}-\frac{C_{LAi}}{D_E}}\right\}=\frac{q_f}{v^T}\left\{1-\exp\left(-BK_{of}\right)\right\}\ t \tag{11}$$

where B has been defined by the following equation:

$$B=\frac{\left(1+\frac{1}{D_E}\right)L}{\left(\frac{V}{A}\right)_{in}u_f} \tag{12}$$

The overall mass transfer coefficient (K_{of}) can be determined from the value of the slope of the linear plots of the left hand side (LHS) of equation (11) versus t (time). The LHS of equation (11) requires (i) the experimental values of the concentrations and (ii) the partition coefficient of the solute. We also need to calculate B from equation (12) which requires the velocity inside the fibre, module characteristics (volume, mass transfer area) and the partition coefficient.

As mentioned earlier in Equation (7), the overall mass transfer coefficient can be calculated using the individual mass transfer coefficients: mass transfer coefficient in the aqueous side (k_f), membrane mass transfer coefficient (k_{mf}) and the mass transfer coefficient in the organic side (k_o). In most of the hollow fibre systems, the aqueous phase containing the "target" component is allowed to flow (under laminar conditions) through the inside of the hollow fibres. The mass transfer coefficient in the aqueous side can be calculated using an equation of the following form (Skelland, 1974):

$$Sh = (1.5 \text{ or } 1.62) \times (Gz)^{0.33} \tag{13}$$

where Sh and Gz are, respectively the Sherwood and Graetz numbers. The values of k_f reported (Bringas et al., 2009) within the Renolds number range, Re (0.08-200) are 1.8 x 10 $^{-6}$ m/s to 9.5 x 10 $^{-4}$ m/s. It is noted that the equation will be different (if the organic flow is allowed through the fibres) as used in Lin and Chen (2006).

For the membrane mass transfer coefficient (k_{mf}), the following equations have been suggested (Schlosser, 2001; Prasad and Sirkar, 1988):

$$k_{mf} = D_m.\varepsilon/\sigma.\tau \tag{14}$$

where σ is the thickness of the hollow fibre, τ is the membrane totuosity, ε is the membrane porosity and D_m is the molecular diffusivity of the carrier-solute complex. The values estimated (Bringas et al.; 2009) using the reported values of D_m are: 1.1 x 10 $^{-8}$ m/s to 1.5 x 10 $^{-5}$ m/s. Other equations have also been used by researchers in this field (de Haan, et al., 1989; Hu and Wiencek, 1998).

For the organic phase flowing through the shell side of the contactor, the value of (k_o) can be estimated using the following equation (Yang and Cussler, 1986):

$$Sh = 1.25\ (\mathrm{Re}.d_h/L)^{0.93}.(Sc)^{0.33} \tag{15}$$

Where d_h is the hydraulic diameter, L is the effective length of the fibres. The dimensionless numbers: Re and Sc represent the Reynolds and Schmidt numbers, respectively. The values of k_o reported (Bringas et al., 2009) are in the range 7.4 x 10 $^{-7}$ m/s to 6.5 x 10 $^{-5}$ m/s.

A detailed discussion of various systems and the estimated values of the mass transfer coefficients are available in Bringas et al. (2009). It is concluded that the estimation by any empirical correlation mentioned in the literature require a description of the specific system, the characteristics of the fibres and contactor, the physical and transport properties of the system under study, the operating conditions and the assumption that the specific system is at least similar to the one where the correlation was developed.

The degree of extraction, E(%), for the HFM experiments is measured by the extraction efficiency which is defined by the following equations:

$$E(\%) = \frac{C_{LAi} - C_{LAf}}{C_{LAi}} \times 100 \tag{16}$$

where C_{LAi} and C_{LAf} are the initial and final concentration values of the feed solution in the recirculation system, respectively.

3. Equilibrium experiments

3.1 Effect of organic solvent

In the mass transfer analysis of the process the value of the distribution coefficient (DE) is required. It is worthwhile to know the range of values obtained for the extraction of organic acids. The physical extraction of lactic acid with pure solvent (no carrier added) is less and these values for various organic phases are listed in Table 1.

Solvent	Distribution Coefficient, DE(-)
Tri-butyl phosphate (TBP)	0.82
Oleyl Alcohol (OlA)	0.25
Decanol (D)	0.43
Hexane (H)	0.25
Dodecane (DD)	0.25
Shellsol TK (STK)	0.11
Oleic Acid (OA)	0.67
Sunflower oil	0.11

Table 1. The values of the distribution coefficient of lactic acid at pH 2.4.

The data in Table 1 suggest that TBP is the optimum solvent for extraction of lactic acid. TBP is followed by oleic acid, decanol, oleyl alcohol, hexane, dodecane. Shellsol TK and sunflower oil gave the lowest distribution ratio of lactic acid. It is evident that active solvents (with functional groups), e.g. TBP, oleic acid have greater extraction power than the inactive solvents, e.g. shellsol TK, dodecane. The active solvent can assist the solvation of the lactic acid molecules and enhance the solubility of the lactic acid complex (Wasewar, et al., 2002). The complex formed with the functional group of the active solvent may be more stable and soluble in the organic phase and thus allow greater extraction compared to those solvents without any functional groups (Tamada, and King, 2000). Although this result with active solvent is good, other points like solvent loss (due to solubility in the aqueous phase) and environmental effects should also be considered before the final selection of the solvent. It has been reported that the solvents containing phosphorus-bonded oxygen atoms (like tributyl phosphate) are not favourable from the points of water solubility and environmental considerations (Matsumoto, et al., 2003).
As mentioned earlier that the reaction with the carrier molecules can enhance the values of DE. The values of DE increased when extraction is performed with adding carrier in the solvent phase. These values at natural pH (2.4) of lactic acid and at room temperature are presented in Table 2 for the extraction using tributyl phosphate as solvent. At this low pH the values of the distribution coefficient were moderate to high for all the carriers. The organic system of 10 wt% TOA - 90 wt% TBP gave a distribution coefficient as 4.88. This value increased with the increase in the concentration of TOA for a fixed feed concentration.

10 wt% Carrier in TBP	Distribution Coefficient, DE (-)
2-octyl amine	4.26
Trioctyl amine	4.88
Aliquat 336	1.22

Table 2. Effects of carrier in tri-butyl phosphate on lactic acid extraction at pH 2.4.

Oleic acid, a component of sunflower oil, did not perform as well when carrier was added to it. However, the best organic solvent, TBP, is expensive and is not recognized as a GRAS (generally regarded as safe) solvent. Sunflower oil is a non-toxic, cheap and can be regarded as an environmentally friendly solvent. It is worth mentioning that the solvents used in this study are less toxic to the fermentation media compared to those reported in the study of organic acid extraction (Wasewar, et al., 2002).

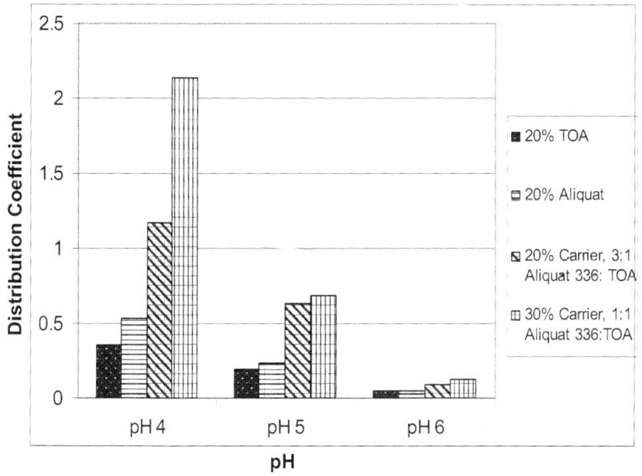

Fig. 3. Effect of feed solution pH on reactive extraction of lactic acid.

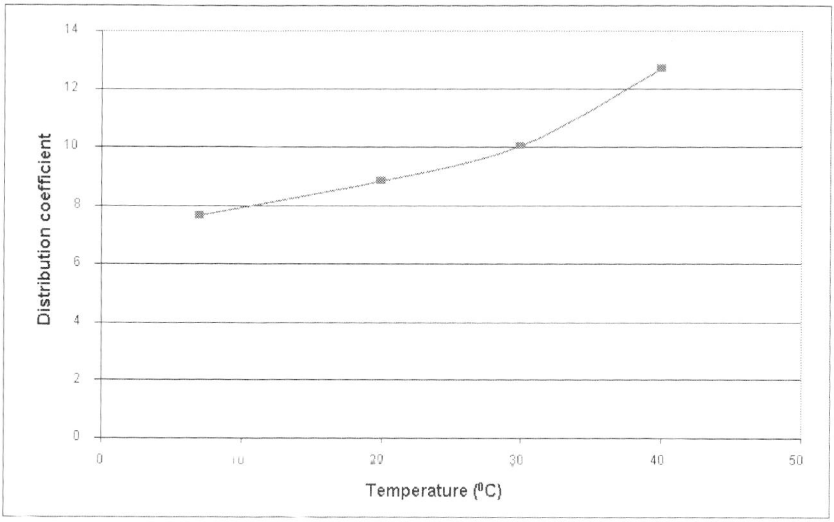

Fig. 4. Effect of temperature on the extraction of lactic acid with sunflower oil.

The values of the distribution coefficient also depend on the extraction condition such as feed solution pH (Canari and Eyal, 2003) and temperature. The values for extraction of 0.2M lactic acid with sunflower oil (as diluent) and Aliquat 336/TOA (as carriers) is shown in Figure 3. The distribution ratio decreased considerably with the increase in aqueous solution pH range 4-6 (Kyuchoukov, et al., 2006). The magnitude of the distribution ratio was greater with Aliquat 336 than for TOA at higher pH. The two extractants when combined also gave considerably higher distribution ratios than the individual extractant. The synergistic effect is apparent in these results, as reported in the literature (Hong, and Hong, 1999; Kyuchoukov, et al., 2001; Yankov, et al., 2005). However, this synergistic effect is much more apparent at lower pH (pH 4) than at higher pH (pH 6). The organic phase which comprised of 15% Aliquat 336 and 15% TOA give the highest distribution ratios at all pH values. Since pH used for fermentative production of lactic acid is approx. 6, the values at high pH are useful in the analysis of mass transfer processes. This will determine their applicability in the *in situ* recovery of lactic acid and thus reduce its the inhibitory effect to enhance its productivity.

The values of DE also depend on the operating temperature. As shown in Figure 4 the temperature (both phases at the same temperature) increased the value of distribution coefficient. Hence, better extraction could be expected at temperatures higher than the room temperature (Martak, J. and Schlosser, 2004). This trend is a positive result, as the optimal temperature for lactic acid fermentation could be in the range 30-38°C.

3.2 Hollow fibre experiments

For the analysis of the mass transfer processes the experiments in hollow-fibre were chosen and a typical base case of the operating conditions are presented in Table 3. The effects of the flow rate, the fermentation media, the carrier, the operating temperature with the solvents: TBP and sunflower oil are discussed below.

Feed Solution	
Concentration of lactic acid (M):	0.2
Volume (L)	0.5
pH (-):	2.4-4.5
Flow rate (L/h):	12-13
Organic solution	
Concentration of carrier in the solvent (%w/w):	5,10
Volume (L):	0.5
Flow rate (L/h):	9-10

Table 3. The operating conditions for mass transfer experiments in the HFMC.

3.2.1 Effect of flow rate

The maximum extraction percentage of approx. 93% was obtained with 10% TOA-TBP organic phase within a processing time of 2 - 3 h (Figure 5). The extraction profiles for three different flow rates were identical except for the profile of 8.33 mL/s that attained lower than the other two organic systems. This could be due to the experimental error (as is observed in Figure 5) that occurred during the initial period of monitoring. The effect of flow rate in the tested range is small and suggests that the overall extraction is not significantly affected by the external mass transfer of lactic acid (i.e. the aqueous side resistance).

3.2.2 Effect of solvent type

Oleyl alcohol (OA) is non- toxic to acid producing bacteria and the results for extraction of lactic acid with this solvent are presented. The percentage extraction increased with time and then attained a plateau at approx. 93% within 2-3 h (Figure 6). With OA the extraction rate was initially slower, increased with time and finally attained approx. 84% (which is within 10% of that for TBP). The lower efficiency and slower rate could be attributed to the high viscosity of oleyl alcohol resulting in a decreased mass transfer rate and requiring longer time to complete the process compared to the less viscous solvent systems viscous.

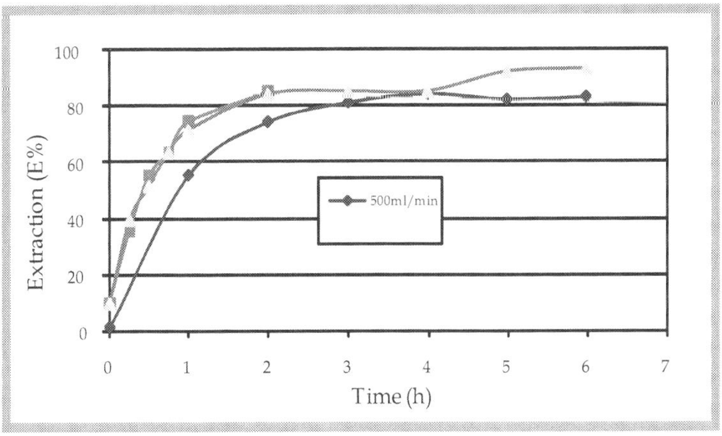

Fig. 5. Effect of flow rate on the extraction of lactic acid with TOA-TBP system.

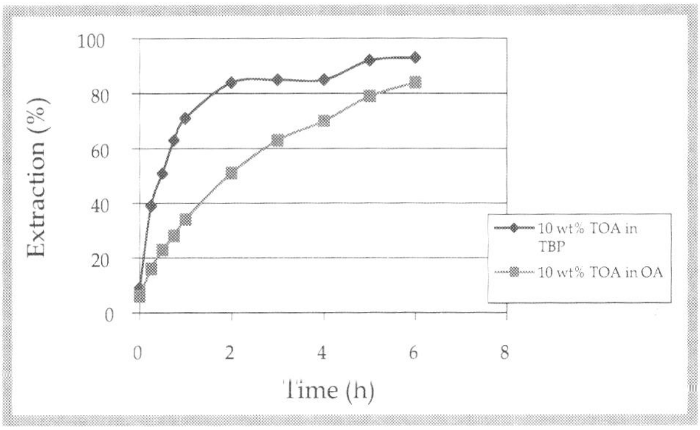

Fig. 6. Effect of the solvent type with TOA as a carrier in the extraction of lactic acid.

3.2.3 Effect of fermentation media

The synthetic fermentation broths (containing lactose and salts) and the pure lactic acid solution gave nearly the same extraction efficiency of about 90% as shown in Figure 7. The variation of the solution pH, i.e. a pH of 4.5 for the synthetic fermentation broth (compared to pH 2.4 for pure lactic acid) and presence of salts did not significantly affect the extraction efficiency. These results compare well with those in the literature (Tik, et al., 2001; Yankov, et al., 2005).

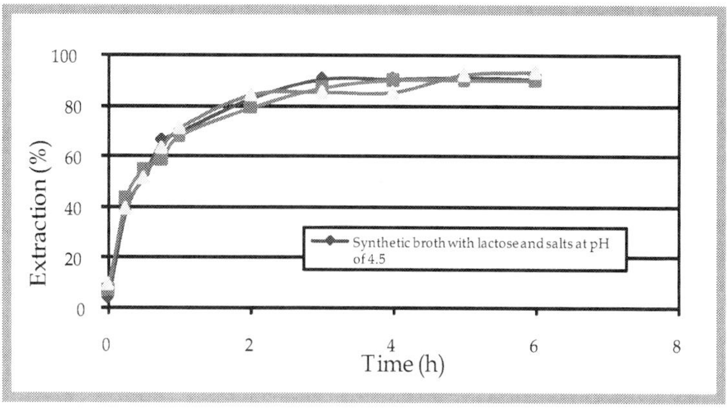

Fig. 7. Effect of fermentation broth on the extraction of lactic acid at two pH.

Because the above-mentioned solvents are not considered suitable (although many studies in the literature have used them) for large scale applications due to cost, health and safety and environmental reasons, further hollow-fibre experiments were conducted with sunflower oil.

3.3 Hollow-fibre extraction with sunflower oil

3.3.1 Effect of carrier

The presence of carrier (either as a single solvent or combination with others) in the solvent had a large effect on the efficiency of the extraction. All of the trials reached steady state after a period of 45-60 mins. The run with no carrier (only sunflower oil) performed poorly with only 9% extraction efficiency (Figure 8). The experimental run with 30% carrier, which consisted of

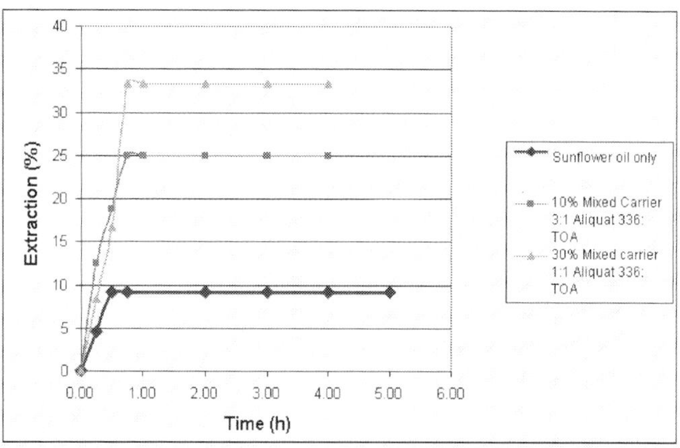

Fig. 8. Extraction with TOA-sunflower oil at feed pH 4.5.

half Aliquat 336 and half TOA, performed the best, and reached an extraction efficiency of 33%. This percentage extraction is low but was obtained at conditions similar to the fermentation conditions; so selective separation of lactic acid from the fermentation can be achieved without addition of any chemicals to adjust the pH of the broth. Thus *in situ* application would be possible without interruption of the fermentation. This continuous removal of lactic acid is expected to reduce its inhibitory effect and eventually the productivity of lactic acid can be increased.

3.3.2 Comparison with other solvent (TBP)

The extraction results of lactic acid at pH 6 with sunflower oil and TBP are compared in Figure 9. The carrier used was a mixture of 15% Aliquat 336 and 15% TOA. As expected TBP performed better; TBP achieved 50% extraction efficiency compared to 33% with sunflower oil. It is surprising to see that sunflower oil, having a much lower distribution coefficient than TBP, achieved a moderately good performance. Comparing the cost, industrial acceptability and environmental benefits sunflower oil could be a very good candidate as a single solvent or in combination with other high performing solvents.

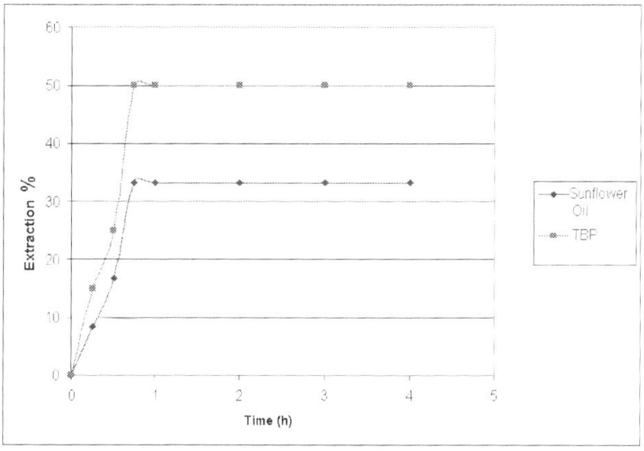

Fig. 9. Comparison of Extractions: sunflower oil vs. TBP.

3.3.3 Effect of temperature on extraction with sunflower oil

The temperature has an effect on the extraction. This is observed in Figure 10, the experiment performed at room temperature reached 21%, whereas the run at 30.5 °C reached 28% extraction using a 10% TOA in sunflower oil (lactic acid feed pH 6). The increase in temperature increased the distribution of the organic phase and the diffusivity of the carrier-acid complex (as the viscosity of the two phases are decreased at higher temperature). These effects allowed faster mass transport through the pores (Frieling & Schugerl, 1999) and achieved greater percentage extraction. Therefore, the amount of extracted lactic acid can be further increased, by applying (i) higher temperature, (ii) lower aqueous solution pH, (iii) increased TOA concentration, and (iv) a mixed solvent system of sunflower oil and TBP. It is to be noted that temperatures more than 45 °C is not allowable in the hollow-fibre contactors used in these experiments. Also, fermentation of lactic acid is usually carried out below 40 °C as higher temperatures may make bacteria ineffective.

3.3.4 Optimal operating conditions

The results of extraction at the optimal operating parameters are presented in Figure 11. The conditions are: pH 5.0, at 35 °C, with an organic phase: a mixture of 15% Aliquat 336 and 15% TOA and a binary solvent system of 35% TBP and 35% sunflower oil. It is seen in Figure 11 that approx. 70% of lactic acid was extracted within 4h of contact time. The extraction rate was still increasing gradually and would have reached higher value at a longer operating time. It is also observed (from the comparison of the profiles in Figure 10 and 11) that the extraction rate decreases at these conditions (higher pH and larger composition of carrier and solvents). This could be because the distribution of lactic acid decreases at higher pH and at with larger composition of other solvents the diffusivity decreases (because of higher viscosity). But the extraction process goes on continuously. This is the effect of the mixed carrier being used in this experiment.

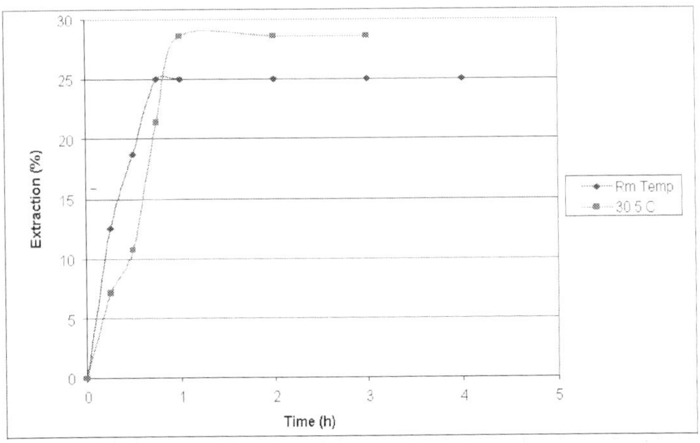

Fig. 10. Effect of temperature on extraction with 10% TOA in sunflower oil.

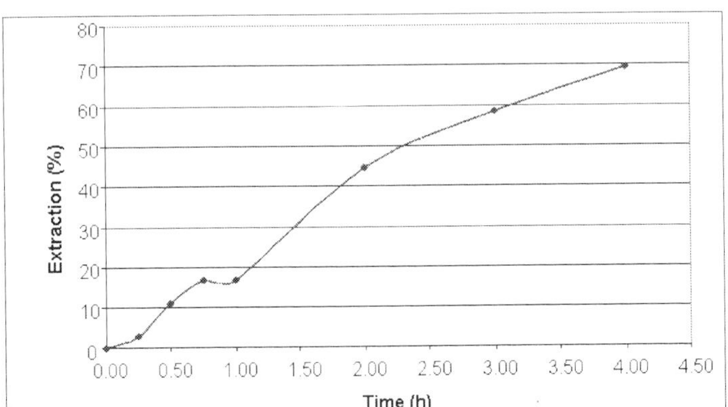

Fig. 11. Extraction (%) with a mixed carrier and mixed solvent system.

3.4 Overall mass transfer coefficient

The overall transfer coefficient (K_{of}) was calculated from the slope of the plot of the left-hand side of Eq. (11) versus time. The value of the distribution coefficient obtained experimentally for each organic phase was used in the calculation. The experimental data during the initial period (approx. 60-70 mins) showed good correlation ($R^2 = 0.9$ at least) and were considered for the calculation. A plot of the LHS of Eq. (11) versus the extraction time with various organic phases is shown in Figure 12. From the best fit line of the plot of LHS versus time, the slope was calculated and with the value of B from Eq. (12), the overall mass transfer coefficient has been calculated.

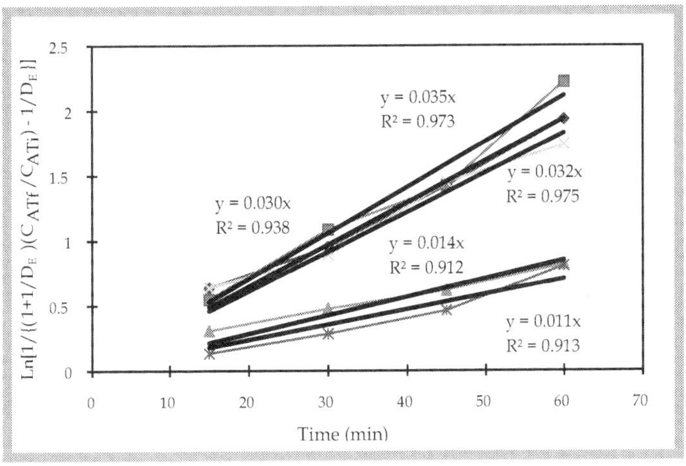

Fig. 12. A plot of LHS of Eqn. (11) versus time (min) for the overall mass transfer coefficient.

The values of the mass transfer coefficients are in the range (0.4 -2.3) x 10^{-5} cm/s (shown in Table 3). This is in consistent with the values of the distribution coefficient (Figure 9), i.e. the mass transfer rate is much faster in tributyl phosphate than in the other two organic phases. The values of overall mass transfer coefficient reported in the literature are range: (0.2 - 5) x 10^{-5} cm/s (Juang and Huang, 1997; Coelhoso et al., 2000; Frieling, and Schugerl, 1999). Considering the fact that this analysis is based on the extractions only (simultaneous re-extraction was not considered in this analysis), the values obtained in this study compare very well. The other advantage of the approximate method used in the present analysis is that the estimation of the mass transfer coefficient is based on semi-analytical equation and requires no correlation and additional data from any other experiments or literature. The model is based on many assumptions such as plug flow in both the aqueous feed and the organic phase, this has been reported in the literature in many extraction processes.

The knowledge of the analysis of results in terms of mass transfer coefficient could be useful in estimating the membrane area for a large-scale fermentative production process (as a preliminary excercise) where lactic acid can be removed as it is produced and thus enhancing the productivity.

In order to test the prediction capability of the model, more experimental results are required in a wide range of conditions with a variety of process solutions (especially with real process solutions obtained under industrial conditions). This is left for future work.

Organic phase	Feed solution	Overall mass transfer coefficient (K_{of}) cm/s x 10^5
10% TOA in TBP (250 ml/min)	Aqueous lactic acid (0.2M)	1.82
10% TOA in TBP (750 ml/min)	Aqueous lactic acid (0.2M)	2.30
10% TOA in OA (250 ml/min)	Aqueous lactic acid (0.2M)	0.52
10% TOA in TBP (250 ml/min)	Synthetic fermentation broth	1.57
15% TOA + 15% Aliquat 336 in sunflower oil	Aqueous lactic acid (0.2M)	0.4

Table 4. The values of the overall mass transfer coefficient (K_{of}) for lactic acid extraction.

4. Conclusions

- The values of the distribution coefficient of lactic acid with solvents such as tri-butyl phosphate (TBP, the best), oleic acid, oleyl alcohol, shellsol TK and sunflower oil are small for physical extraction (solvent only).
- These values can be increased by reactive extraction with tri-octylamine (TOA/Aliquat 336 as the carrier) dissolved in any of the above solvents. The best organic system TOA-TBP gave a high distribution ratio at the natural pH 2.4. An extraction of approx. 93% can be achieved in the hollow fibre module within 2-3 h using 10% (w/w) TOA-TBP.
- The use of less toxic solvents such as oleyl alcohol (instead of TBP or its mixture with TBP) would give a lower percentage extraction with the final value around 83%.
- The presence of the components of the fermentation media, such as, lactose and salts do not significantly affect the percentage extraction in the membrane module.
- Extractions with these two organic systems are considered less suitable for large scale production because of cost, toxicity and potentially harmful environmental effects.
- For industrial application an organic phase with sunflower oil, is more attractive because of its cost (cheaper by an order of magnitude), non-toxicity and environmental benefits. It has proven to be effective at operating conditions (pH and temperature) similar to the fermentative production of lactic acid and shows stability with the commercially available hollow fibre modules.
- With sunflower oil-TOA/Aliquat 336 as the organic phase the percentage extraction is lower (30-35%) at room temperature. The extraction can be enhanced by operating at higher temperature.
- The percentage removal at this rate (at feed flow rate of approx. 12-13 L/h) from the fermentation media should reduce the product inhibition (lactic acid) effects and would improve the productivity by allowing the process over a longer period of time.

Notation

A	surface area (m^2)
B	defined in Equation (12)
C	concentration of lactic acid (mmol/L)
di, do	inner diameter, outer diameter of a hollow fiber (cm)
D_E	distribution coefficient of lactic acid defined in Eq. (3)
D_m	diffusion coefficient of lactic acid defined in Eq. (14)
d_h	hydraulic radius of lactic acid defined in Eq. (15)

E (%)	percent of extraction of lactic acid, defined in Eq. (17)
H+	hydrogen ion
HLA	undissociated lactic acid
Gz	Graetz number defined in Eq. (13)
K_{of}	mass transfer coefficient in Equation (6), (cm/s)
L	length of the fiber (cm)
n_f	number of fibers
N	steady state flux, defined in Equation (6)
Re	Reynolds number defined in Eq. (15)
Sc	Schmidt number defined in Eq. (15)
Sh	Sherwood number defined in Eq. (13)
t	time, s
v	flow rate, L/s
u	linear velocity in the hollow fibers (cm/s)
V	volume (L)
Z	axial distance in the module (cm)

Subscripts

aq	aqueous phase
f	feed solution
fi	feed side at aqueous-organic interface
fo	organic side at aqueous-organic interface
in	based on inside fibre diameter
LAf	final lactic acid concentration
LAi	initial lactic acid concentration
oi	organic side interface
o, org	organic phase

Superscripts

m	module
T	tank

Greek symbols

e	membrane porosity
σ	thickness of the hollow fibres
τ	tortuosity of the pores

5. Acknowledgement

The financial assistance of the University of Auckland Staff Research Grant, Auckland, New Zealand, is gratefully acknowledged. The author is also thankful to the United Arab Emirates University, Al Ain, UAE, for their overall support.

6. References

Bringas, E.; San Roman, MF. (2009). An overview of the mathematical modelling of liquid membrane separation processes in hollow fibre contactors. *J. Chem Tech Biotechnol.*, 84, 1583-1614.

Canari R.; Eyal. A. M. (2003). Selectivity in monocarboxylic acids extraction from their mixture solutions using an amine-based extractant: effect of pH. *Ind. Eng. Chem. Res.*, 42, 1307- 1314.

Carrera, J. A.; Bringas, E.; San Roman, M.F.; Ortiz, I. (2009). Selective membrane alternative to the recovery of zinc from hot-dip galvanizing effluents. *J. Membr. Sci.*, 326, 672-680.

Choi, J.; Hong, W. H. (1999). Recovery of lactic acid by batch distillation with chemical reactions using ion exchange resin, *J. Chem. Eng. Japan*. 32, 184-189.

Coca, J.; Pazos, C; San-Martin, M. 1992. Reactive extraction of lactic Acid with Alamine336 in the presence of salts and lactose, *J. Chem. Tech. Biotech.* 54, 1-6.

Coelhoso, I.M.; Cardoso, M.M.; Viegas, R.M.C.; Crespo, J.P.S.G. (2000). Transport mechanisms and modelling in liquid membrane contactors. *Sep. Purfc Technol.*, 19, 183-197.

Datta, R.; Henry, M. (2006). Lactic acid: Recent advances in products, processes and technologies – a review. *J. Chem. Tech. Biotech*, 81, 1119-1129.

Dai, X. P.; Yang, Z. F.; Luo, R. G.; Sirkar, K. K. (2001).Lipase-facilitated separation of organic acids in a hollow-fibre contained liquid membrane module. *J. Membr. Sci.*, 171, 183-196.

de Haan, A.B.; Bartels, P.V.; de Graauw, J. (1989). Extraction of metals from waste water: Modeling the mass transfer in a supported liquid membrane process. *J. Membr. Sci.*, 281-297.

Drioli, E.; Giorno, L.; Spicka, P. (1996). Downstream processing of lactic acid by membrane based solvent extraction. *Sep. Sci. Technol.*, 31, 2159-2169.

Frieling, P. von; Schugerl, K. (1999). Recovery of lactic acid from aqueous model solutions and fermentation broths. Proc.Biochem., 34, 685-696.

Gong, C.; Luo, G. S.; Yang, W. W.; Wu, F. Y. (2006). Separation of organic acids by newly developed polysulfone microcapsules containing Trioctylamine. *Sep. Purif. Technol.*, 48, 235-243.

Han, D. H.; Hong, Y. K.; Hong, W. H. (2000). Separation characteristics of lactic acid in reactive extraction and stripping, *Korean J. Chem. Eng.* 17, 528-533.

Hano, T.; Matsumoto, M.; Ohtake, T. (1993). Recovery of lactic acid by extraction from fermentation broth. *Solvent Extraction in the Process Industries*, 2, 1025-1031.

Heewsink, A. B. M.; Pangarkar, V. G.; Versteeg, G. F.; Wasewar, K. L. (2002). Equilibria and kinetics for reactive extraction of lactic acid using Alamine 336 in decanol, *J. Chem. Tech. Biotech.*, 77, 1068-1075.

Hong, Y. K.; Hong, W. H. (1999). Reactive extraction of lactic acid with mixed tertiary amine extractants. *Biotechnol. Techniques*, 13, 915-918.

Hong, W. H.; Kim, Y. J.; Wonzy, G. (2002). Effect of recycle and feeding method on batch reactive recovery system of lactic Acid. *Korean J. Chem. Eng.* 19, 808-814.

Hossain, Md. M.; Dean, J; (2008). Modelling and analysis of reactive extraction of penicillin G in a hollow-fibre membrane module. *Sep. Purific. Technol.*, 62(2), 437-443.

Hossain, Md M.; Maisuria, J. L. (2008). Effects of the Organic phase, Fermentation media and Operating conditions on Lactic acid Extraction. *Biotechnol. Prog.*, 24(3), 757-765.

Hu, S.B.; Wiencek. (1998). Emulsion-liquid-membrane extraction of copper using a hollow-fiber contactor. *AIChE J.*, 44, 570-581.

Joglekar, H. G.; Rahman, I.; Babu, S.; Kulkarni, B. D.; Joshi, A. (2006). Comparative assessment of downstream processing options for lactic acid. *Sep. Purif. Technol.*, 52, 1-17.

Ju, L.K.; Verma, A. (1994). Characteristics of lactic acid transport in supported liquid membranes. *Sep. Sci. Technol.*, 29, 2299-2313.

Juang, R.-S.; Huang, R.-H. (1997). Kinetic studies on lactic acid extraction with amine using a microporous membrane-based stirred cell. *J. Membr. Sci.*, 129, 185-196.

Juang, R. S.; Chen, J. D.; Huan, H. C. (2000). Dispersion-free membrane extraction: case studies of metal ion and organic acid extraction. *J. Membr. Sci.*, 165, 59-73.

Klaassen, R.; Jansen, A. E. (2001). The membrane contactor: environmental applications and possibilities. *Environ. Prog.*, 20, 37-43.

Klaassen, R.; Feron, P. H. M.; Jansen, A. E. (2005). Membrane contactors in industrial applications. *Chem. Eng. Res. Des.*, 83, 234-246.

Kertesz, R.; Schlosser, S. (2005). Design and simulation of two phase hollow fibre c.ontactors for simultaneous membrane based solvent extraction and stripping of organic acids and bases. *Sep. Purif. Technol.*, 41, 275-287.

Kumar, A.; Haddad, R.; Sastre A. M. (2001). Integrated membrane process for gold recovery from hydrometallurgical solutions. *AIChE J.*, 47, 328 - 340.

Kyuchoukov, G., Marinova, M., Molinier, J., Albet, J.; Malmary, G. (2001). Extraction of lactic acid by means of a mixed extractant. *Ind. Eng. Chem. Res.*, 40, 5623-5639.

Kyuchoukov, G.; Labbaci., A.; Albet, J.; Molinier, J. (2006). Simultaneous influence of active and "Inert" diluents on the extraction of lactic acid by means of Tri-*n*-octylamine (TOA) and Tri-*iso*-octylamine (TIOA). *Ind. Eng. Chem. Res.*, 45 503-510

Lee, H. J.; Koo, Y. M.; Wang, N. H. L. (2004). Separation of lactic acid from acetic acid using a four-zone SMB. *Biotechnol. Prog.* 20, 179-192.

Lin, S.-H.; Chen, C.-N. (2006). Simultaneous reactive extraction separation of amino acids from water with D2EHPA in hollow fiber contactors. *J. Membr. Sci.*, 280, 771-780.

Martak, J. and Schlosser, S. Ionic liquids in pertraction and extraction of organic acids, in: B. Lenarcik and R. Wodzki (Eds.), Proc. 19th symp. Ars Separatoria (http://www.ars_separatoria.chem.uni.torun.pl/MATERIAL Y4.htm), Zloty Potok (PL), 2004, p106.

Matsumoto, M.; Takahashi, T.; Fukushima, K. (2003). Synergistic extraction of lactic acid with alkylamine and tri-n-butyl phosphate: effects of amines, diluents and temperature. *Sep. Purif. Technol.*, 33, 89-93.

Ortiz, I.; Bringas, E.; San Roman, M. F.; Urtiaga, A. M. (2004). Selective separation of zinc and iron from spent pickling solutions by membrane-based solvent extraction: Process viability. *Sep. Sci. Technol.*, 39, 2441-2455.

Pathasarathy, N.; Pelletier, M.; Buffle, J. (1997). Hollow fibre based supported liquid membrane: a novel analytical system for trace metal analysis. *Anal. Chim. Acta*, 350, 183-195.

Prasad, R., Sirkar, K.K. (1988). Dispersion-free solvent extraction with microporous hollow-fiber modules. *AIChE J.*, 34, 177-188.

Rathore, N. S.; Sonawane, J. V.; Kumar, A.; Venugopalan, A. K.; Singh, R. K.; Bajpal, D. D. (2001). Hollow fibre supported liquid membrane: a novel technique for separation and recovery of plutonium from aqueous acidic wastes. *J. Membr. Sci.*, 189, 119-128.

Ren, Z.; Zhanh, W.; Li, H.; Lin, W. (2009). Mass transfer characteristics of citric acid extraction by hollow fibre renewal liquid membrane. *Chem. Eng. J.*, 146, 220-226.

Schlosser, S; Sabolova, E. Mass-transfer characteristics of hollow-fibre contactors for pertraction and membrane based extraction of organic acids. Proceedings of

Engineering with Membranes, Vol I, paper No. PA.4-17, p313, June 3-6, 2001, Granada, Spain.

Siebold, M.; Frieling, P. v.; Joppien, R.; Rindfleisch, D.; Schugerl, K.; Roper, H. (1995). Comparison of the production of lactic acid by three different lactobacilli and its recovery by extraction and electrodialysis. *Proc. Biochem.*, 30, 81-95.

Skelland, A.H.P., Diffusional Mass Transfer, Wiley, New York (1974).

Sun, Q.; Wang, W.; Zhao, Ma, H.; Sakata, S. (2006). Extraction and purification of lactic acid from fermentation broth by esterification and hydrolysis method. *Sep. Purif. Technol.*, 49, 43-48.

Tamada, J. A.; King, C. J. (2000). Extraction of carboxylic acids with amine extractants. 2. Chemical interactions and interpretation of data. *Ind. Eng. Chem. Res.*, 29, 1327-1333.

Tik, N.; Bayraktar, E.; Mehmetoglu, U. (2001). In situ reactive extraction of lactic acid from fermentation media. *J. Chem. Technol. Biotechnol.* 76, 764-768.

Tong, Y.; Hirata, M.; Takanashi, H.; Hano, T. (1999). Back extraction of lactic acid with microporous hollow fiber membrane. *J. Membr. Sci.*, 157, 189-198.

Tong, Y.; Hirata, M. Takanashi, H.; Hano, T.; Kuboto, F.; Goto, M. (1998). Extraction of lactic acid from fermented broth with microporous hollow fibre membranes. *J. Membr. Sci.*, 143, 81-91.

Urtiaga, A.M.; Abellan, M. J., Irabien, J. A.; Ortiz, I. (2005). Membrane contactors for the recovery of metallic compounds. Modelling of copper recovery from WPO processes. *J. Membr. Sci.*, 257, 161-170.

Uslu, H.; Kirbaslar, I.; Wasewar, K. L. (2009). Reactive extraction of levulinic acid by Amberlite LA-2 extractant. J. Chem Eng. Data. 54, 712-718.

Wasewar, K.; Leesink, L.; Veersteeg, A. B. M.; Pangarkar, V. G. (2002). Reactive extraction of lactic acid using Alamine 336 in MIBK: equilibria and kinetics. *J. Biotechnol.*, 97, 59-68.

Yang, M.C.; Cussler, E.L. (1986). Designing hollow fiber contactors. *AIChE J.*, 32, 1910-1916.

Yang, Q.; Kochergnisky, N. M. (2006). Copper recovery and spent ammoniacal etchant regeneration based hollow fibre supported liquid membrane technology: from bench-scale to pilot-scale tests. *J. Membr. Sci.*, 286, 301-309.

Yang, X. J.; Fane, A. G.; Soldenhoff, K. (2003). Comparison of liquid membrane processes for metal separations: permeability, stability, and selectivity. *Ind. Eng. Chem. Res.*, 42, 392-403.

Yankov, D., Molinier, J., Kyuchoukov, G., Albet, J. and Malmary, G. (2005). Improvement of the lactic acid extraction. Extraction from aqueous solutions and simulated fermentation broth by means of mixed extractant and TOA, partially loaded with HCl. *Chem. Biochem. Eng. Quarterly*, 19, 17-24.

Yuchoukov, G.; Yankov, G.; Albet, J.; Molinier, J. (2005). Mechanism of lactic acid extraction with quaternary ammonium chloride (Aliquat 336). *Ind. Eng.Chem. Res*,. 44, 5739-5746.

18

New Approaches for Theoretical Estimation of Mass Transfer Parameters in Both Gas-Liquid and Slurry Bubble Columns

Stoyan NEDELTCHEV and Adrian SCHUMPE
Institute of Technical Chemistry, TU Braunschweig
Germany

1. Introduction

Bubble columns (with and without suspended solids) have been used widely as chemical reactors, bioreactors and equipment for waste water treatment. The key design parameters in bubble columns are:

- gas holdup;
- gas-liquid interfacial area;
- volumetric liquid-phase mass transfer coefficient;
- gas and liquid axial dispersion coefficients;

Despite the large amount of studies devoted to hydrodynamics and mass transfer in bubble columns, these topics are still far from being exhausted. One of the essential reasons for hitherto unsuccessful modeling of hydrodynamics and mass transfer in bubble columns is the unfeasibility of a unified approach to different types of liquids. A diverse approach is thus advisable to different groups of gas-liquid systems according to the nature of liquid phase used (pure liquids, aqueous or non-aqueous solutions of organic or inorganic substances, non-Newtonian fluids and their solutions) and according to the extent of bubble coalescence in the respective classes of liquids. It is also necessary to distinguish consistently between the individual regimes of bubbling pertinent to a given gas-liquid system and to conditions of the reactor performance.

The mechanism of mass transfer is quite complicated. Except for the standard air-water system, no hydrodynamic or mass transfer characteristics of bubble beds can be reliably predicted or correlated at the present time. Both the interfacial area a and the volumetric liquid-phase mass transfer coefficient k_La are considered the most important design parameters and bubble columns exhibit improved values of these parameters (Wilkinson et al., 1992). For the design of a bubble column as a reactor, accurate data about bubble size distribution and hydrodynamics in bubble columns, mechanism of bubble coalescence and breakup as well as mass transfer from individual bubbles are necessary. Due to the complex nature of gas-liquid dispersion systems, the relations between the phenomena of bubble coalescence and breakup in bubble swarms and pertinent fundamental hydrodynamic parameters of bubble beds are still not thoroughly understood.

The amount of gas transferred from bubbles into the liquid phase is determined by the magnitude of k_La. This coefficient is an important parameter and its knowledge is essential for the determination of the overall rate of chemical reaction in heterogeneous systems, i.e. for the evaluation of the effect of mass transport on the overall reaction rate. The rate of interfacial mass transfer depends primarily on the size of bubbles in the systems. The bubble size influences significantly the value of the mass transfer coefficient k_L. It is worth noting that the effects of so-called tiny bubbles (d_s<0.002 m) and large bubbles (d_s≥0.002 m) are opposite. In the case of tiny bubbles, values of mass transfer coefficient increase rapidly as the bubble size increases. In the region of large bubbles, values of mass transfer coefficient decrease slightly as the bubble diameter increases. However, such conclusions have to be employed with caution. For the sake of correctness, it would therefore be necessary to distinguish strictly between categories of tiny and large bubbles with respect to the type of liquid phase used (e.g. pure liquids or solutions) and then to consider separately the values of liquid-phase mass transfer coefficient k_L for tiny bubbles (with immobile interface), for large bubbles in pure liquids (mobile interface) and for large bubbles in solutions (limited interface mobility).

The axial dispersion model has been extensively used for estimation of axial dispersion coefficients and for bubble column design. Some reliable correlations for the prediction of these parameters have been established in the case of pure liquids at atmospheric pressure. Yet, the estimations of the design parameters are rather difficult for bubble columns with liquid mixtures and aqueous solutions of surface active substances.

Few sentences about the effect of high pressure should be mentioned. Hikita et al. (1980), Öztürk et al. (1987) and Idogawa et al. (1985a, b) in their gas holdup experiments at high pressure observed that gas holdup increases as the gas density increases. Wilkinson et al. (1994) have shown that gas holdup, k_La and a increase with pressure. For design purposes, they have developed their own correlation which relates well k_La and gas holdup. As the pressure increases, the gas holdup increases and the bubble size decreases which leads to higher interfacial area. Due to this reason, Wilkinson et al. (1992) argue that both a and k_La will be underestimated by the published empirical equations. The authors suggest that the accurate estimation of both parameters requires experiments at high pressure. They proposed a procedure for estimation of these parameters on the basis of atmospheric results. It shows that the volumetric liquid-phase mass transfer coefficient increases with pressure regardless of the fact that a small decrease of the liquid-side mass transfer coefficient is expected. Calderbank and Moo-Young (1961) have shown that the liquid-side mass transfer coefficient decreases for smaller bubble size. The increase in interfacial area with increasing pressure depends partly on the relative extent to which the gas holdup increases with increasing pressure and partly on the decrease in bubble size with increasing pressure.

The above-mentioned key parameters are affected pretty much by the bubble size distribution. In turn, it is controlled by both bubble coalescence and breakup which are affected by the physico-chemical properties of the solutions used. On the basis of dynamic gas disengagement experiments, Krishna et al. (1991) have confirmed that in the heterogeneous (churn-turbulent) flow regime a bimodal bubble size distribution exists: small bubbles of average size 5×10^{-3} m and fast rising large bubbles of size 5×10^{-2} m. Wilkinson et al. (1992) have proposed another set of correlations by using gas holdup data obtained at pressures between 0.1 and 2 MPa and extensive literature data.

The flow patterns affect also the values of the above-mentioned parameters. Three different flow regimes are observed:

- homogeneous (bubbly flow) regime;
- transition regime;

- heterogeneous (churn-turbulent) regime.

Under common working conditions of bubble bed reactors, bubbles pass through the bed in swarms. Kastanek et al. (1993) argue that the character of two-phase flow is strongly influenced by local values of the relative velocity between the dispersed and the continuous phase. On the basis of particle image velocimetry (PIV), Chen et al. (1994) observed three flow regimes: a dispersed bubble regime (homogeneous flow regime), vortical-spiral flow regime and turbulent (heterogeneous) flow regime. In the latter increased bubble-wake interactions are observed which cause increased bubble velocity. The vortical-spiral flow regime is observed at superficial gas velocity u_G=0.021-0.049 m/s and is composed of four flow regions (the central plume region, the fast bubble flow region, the vortical-spiral flow region and the descending flow region) from the column axis to the column wall. According to Koide (1996) the vortical-spiral flow region might occur in the transition regime provided that the hole diameter of the gas distributor is small. Chen et al. (1994) have observed that in the fast bubble flow regime, clusters of bubbles or coalesced bubbles move upwards in a spiral manner with high velocity. The authors found that these bubble streams isolate the central plume region from direct mass exchange with the vortical-spiral flow region. In the heterogeneous flow regime, the liquid circulating flow is induced by uneven distribution of gas holdup. At low pressure in the churn-turbulent regime a much wider range of bubble sizes occurs as compared to high pressure. At low pressure there are large differences in rise velocity which lead to a large residence time distribution of these bubbles. In the churn-turbulent regime, frequent bubble collisions occur.

Deckwer (1992) has proposed a graphical correlation of flow regimes with column diameter and u_G. Another attempt has been made to determine the flow regime boundaries in bubble columns by using u_G vs. gas holdup curve (Koide et al., 1984). The authors recommended that if the product of column diameter and hole size of the distributor is higher than 2×10^{-4} m^2, the flow regime is assumed to be a heterogeneous flow regime. In the bubble column with solid suspensions, solid particles tend to induce bubble coalescence, so the homogeneous regime is rarely observed. The transition regime or the heterogeneous regime is usually observed.

In some works on mass transfer, the effects of turbulence induced by bubbles are considered. The flow patterns of liquid and bubbles are dynamic in nature. The time-averaged values of liquid velocity and gas holdup reveal that the liquid rises upwards and the gas holdup becomes larger in the center of the column.

Wilkinson et al. (1992) concluded also that the flow regime transition is a function of gas density. The formation of large bubbles can be delayed to a higher value of superficial gas velocity (and gas holdup) when the coalescence rate is reduced by the addition of an electrolyte. Wilkinson and Van Dierendonck (1990) have demonstrated that a higher gas density increases the rate of bubble breakup especially for large bubbles. As a result, at high pressure mainly small bubbles occur in the homogeneous regime, until for very high gas holdup the transition to the churn-turbulent regime occurs because coalescence then becomes so important that larger bubbles are formed. The dependence of both gas holdup and the transition velocity in a bubble column on pressure can be attributed to the influence of gas density on bubble breakup. Wilkinson et al. (1992) argue that many (very) large bubbles occur especially in bubble columns with high-viscosity liquids. Due to the high rise velocity of the large bubbles, the gas holdup in viscous liquids is expected to be low, whereas the transition to the churn-turbulent regime (due to the formation of large bubbles) occurs at very low gas velocity. The value of surface tension also has a pronounced influence on bubble breakup and thus gas holdup. When the surface tension is lower, fewer large bubbles occur because the surface tension forces oppose deformation and bubble breakup (Otake et al., 1977). Consequently, the occurrence of large bubbles is minimal due to bubble breakup especially in those liquids that are characterized with a low surface tension and a low liquid viscosity. As a result, relatively high gas holdup values are to be expected for such liquids, whereas the transition to the churn-turbulent regime due to the formation of large bubbles is delayed to relatively high gas holdup values.

1.1 Estimation of bubble size

The determination of the Sauter-mean bubble diameter d_s is of primary importance as its value directly determines the magnitude of the specific interfacial area related to unit volume of the bed. All commonly recommended methods for bubble size measurement yield reliable results only in bubble beds with small porosity (gas holdup≤0.06). The formation of small bubbles can be expected in units with porous plate or ejector type gas distributors. At these conditions, no bubble interference occurs. The distributions of bubble sizes yielded by different methods differ appreciably due to the different weight given to the occurrence of tiny bubbles. It is worth noting that the bubble formation at the orifice is governed only by the inertial forces. Under homogeneous bubbling conditions the bubble population in pure liquids is formed by isolated mutually non-interfering bubbles.

The size of the bubbles leaving the gas distributor is not generally equal to the size of the bubbles in the bed. The difference depends on the extent of bubbles coalescence and break-up in the region above the gas distributor, on the distributor type and geometry, on the distance of the measuring point from the distributor and last but not least on the regime of bubbling. In coalescence promoting systems, the distribution of bubble sizes in the bed is influenced particularly by the large fraction of so-called equilibrium bubbles. The latter are formed in high porosity beds as a result of mutual interference of dynamic forces in the turbulent medium and surface tension forces, which can be characterized by the Weber number We. Above a certain critical value of We, the bubble becomes unstable and splits to bubbles of equilibrium size. On the other hand, if the primary bubbles formed by the distributor are smaller than the equilibrium size, they can reach in turbulent bubble beds the equilibrium size due to mutual collisions and subsequent coalescence. As a result, the mean diameter of bubbles in the bed again approaches the equilibrium value. In systems with suppressed coalescence, if the primary bubble has larger diameter than the equilibrium size, it can reach the equilibrium size due to the break-up process. If however the bubbles formed by the distributor are smaller than the equilibrium ones the average bubble size will remain smaller than the hypothetical equilibrium size as no coalescence occurs. Kastanek et al. (1993) argue that in the case of homogeneous regime the Sauter-mean bubble diameter d_s increases with superficial gas velocity u_G.

The correct estimation of bubble size is a key step for predicting successfully the mass transfer coefficients. Bubble diameters have been measured by photographic method, electroresistivity method, optical-fiber method and the chemical-absorption method. Recently, Jiang et al. (1995) applied the PIV technique to obtain bubble properties such as size and shape in a bubble column operated at high pressures.

In the homogeneous flow regime (where no bubble coalescence and breakup occur), bubble diameters can be estimated by the existing correlations for bubble diameters generated from perforated plates (Tadaki and Maeda, 1963; Koide et al., 1966; Miyahara and Hayashi, 1995) or porous plates (Hayashi et al., 1975). Additional correlations for bubble size were developed by Hughmark (1967), Akita and Yoshida (1974) and Wilkinson et al. (1994). The latter developed their correlation based on data obtained by the photographic method in a bubble column operated between 0.1-1.5 MPa and with water and organic liquids. In electrolyte solutions, the bubble size is generally much smaller than in pure liquids (Wilkinson et al., 1992).

In the transition regime and the heterogeneous flow regime (where bubble coalescence and breakup occur) the observed bubble diameters exhibit different values depending on the measuring methods. It is worth noting that the volume-surface mean diameter of bubbles measured near the column wall by the photographic method (Ueyama et al., 1980) agrees well with the predicted values from the correlation of Akita and Yoshida (1974). However, they are much smaller than those measured with the electroresistivity method and averaged over the cross-section by Ueyama et al. (1980).

When a bubble column is operated at high pressures, the bubble breakup is accelerated due to increasing gas density (Wilkinson et al., 1990), and so bubble sizes decrease (Idogawa et al., 1985a, b; Wilkinson et al., 1994). Jiang et al. (1995) measured bubble sizes by the PIV technique in a bubble column operated at pressures up to 21 MPa and have shown that the bubble size decreases and the bubble size distribution narrows with increasing pressure. However, the pressure effect on the bubble size is not significant when the pressure is higher than 1.5 MPa.

The addition of solid particles to liquid increases bubble coalescence and so bubble size. Fukuma et al. (1987a) measured bubble sizes and rising velocities using an electro-resistivity probe and showed that the mean bubble size becomes largest at a particle diameter of about 0.2×10^{-3} m for an air-water system. The authors derived also a correlation.

For pure, coalescence promoting liquids, Akita and Yoshida (1974) proposed an empirical relation for bubble size estimation based on experimental data from a bubble column equipped with perforated distributing plates. The authors argue that their equation is valid up to superficial gas velocities of 0.07 m/s. It is worth noting that Akita and Yoshida (1974) used a photographic method which is not very reliable at high gas velocities. The equation does not include the orifice diameter as an independent variable, albeit even in the homogeneous bubbling region this parameter cannot be neglected.

For porous plates and coalescence suppressing media Koide and co-workers (1968) derived their own correlation. However, the application of this correlation requires exact knowledge of the distributor porosity. Such information can be obtained only for porous plates produced by special methods (e.g. electro-erosion), which are of little practical use, while they are not available for commonly used sintered glass or metal plates. Kastanek et al. (1993) reported a significant effect of electrolyte addition on the decrease of bubble size. According to these authors, for the inviscid, coalescence-supporting liquids the ratio of Sauter-mean bubble diameter to the arithmetic mean bubble diameter is approximately constant (and equal to 1.07) within orifice Reynolds numbers in the range of 200-600. It is worth noting that above a certain viscosity value (higher than 2 mPa s) its further increase results in the simultaneous presence of both large and extremely small bubbles in the bed. Under such conditions the character of bubble bed corresponds to that observed for inviscid liquids under turbulent bubbling conditions. In such cases, only the Sauter-mean bubble diameter should be used for accurate bubble size characteristics. Kastanek et al. (1993) developed their own correlation (valid for orifice Reynolds numbers in between 200 and 1000) for the prediction of Sauter-mean bubble diameter in coalescence-supporting systems.

According to it, the bubble size depends on the volumetric gas flow rate related to a single orifice, the surface tension and liquid viscosity.
The addition of a surface active substance causes the decrease of Sauter-mean bubble diameter to a certain limiting value which then remains unchanged with further increase of the concentration of the surface active agent. It is frequently assumed that the addition of surface active agents causes damping of turbulence in the vicinity of the interface and suppression of the coalescence of mutually contacting bubbles. It is well-known fact that the Sauter-mean bubble diameters corresponding to individual coalescent systems differ only slightly under turbulent bubbling conditions and can be approximated by the interval $6\text{-}7\times10^{-3}$ m.

1.2 Estimation of gas holdup

Gas holdup is usually expressed as a ratio of gas volume V_G to the overall volume (V_G+V_L). It is one of the most important parameters characterizing bubble bed hydrodynamics. The value of gas holdup determines the fraction of gas in the bubble bed and thus the residence time of phases in the bed. In combination with the bubble size distribution, the gas holdup values determine the extent of interfacial area and thus the rate of interfacial mass transfer. Under high gas flow rate, gas holdup is strongly inhomogeneous near the gas distributor (Kiambi et al., 2001).
Gas holdup correlations in the homogeneous flow regime have been proposed by Marrucci (1965) and Koide et al. (1966). The latter is applicable to both homogeneous and transition regimes. It is worth noting that the predictions of both equations agree with each other very well. Correlations for gas holdup in the transition regime are proposed by Koide et al. (1984) and Tsuchiya and Nakanishi (1992). Hughmark (1967), Akita and Yoshida (1973) and Hikita et al. (1980) derived gas holdup correlations for the heterogeneous flow regime. The effects of alcohols on gas holdup were discussed and the correlations for gas holdups were obtained by Akita (1987a) and Salvacion et al. (1995). Koide et al. (1984) argues that the addition of inorganic electrolyte to water increases the gas holdup by 20-30 % in a bubble column with a perforated plate as a gas distributor. Akita (1987a) has reported that no increase in gas holdup is recognized when a perforated plate of similar performance to that of a single nozzle is used. Öztürk et al. (1987) measured gas holdups in various organic liquids in a bubble column, and have reported that gas holdup data except those for mixed liquids with frothing ability are described well by the correlations of Akita and Yoshida (1973) and Hikita et al. (1980). Schumpe and Deckwer (1987) proposed correlations for both heterogeneous flow regime and slug flow regime in viscous media including non-Newtonian liquids. Addition of a surface active substance (such as alcohol) to water inhibits bubble coalescence and results in an increase of gas holdup. Grund et al. (1992) applied the gas disengagement technique for measuring the gas holdup of both small and large bubble classes. Tap water and organic liquids were used. The authors have shown that the contribution of small class bubbles to k_La is very large, e.g. about 68 % at u_G=0.15 m/s in an air-water system. Grund et al. (1992) suggested that a rigorous reactor model should consider two bubble classes with different degrees of depletion of transport component in the gas phase. Muller and Davidson (1992) have shown that small-class bubbles contribute 20-50 % of the gas-liquid mass transfer in a column with highly viscous liquid. Addition of solid particles to liquid in a bubble column reduces the gas holdup and correlations of gas holdup valid for transition and heterogeneous flow regimes were proposed by Koide et al. (1984), Sauer and Hempel (1987) and Salvacion et al. (1995).

Wilkinson et al. (1992) have summarized some of the most important gas holdup correlations and have discussed the role of gas density. The authors reported also that at high pressure gas holdup is higher (especially for liquids of low viscosity) while the average bubble size is smaller. Wilkinson et al. (1992) determined the influence of column dimensions on gas holdup. Kastanek et al. (1993) reported that at atmospheric pressure the gas holdup is virtually independent of the column diameter provided that its value is larger than 0.15 m. This information is critical to scale-up because it determines the minimum scale at which pilot-plant experiments can be implemented to estimate the gas holdup (and mass transfer) in a large industrial bubble column. Wilkinson et al. (1992) reached this conclusion for both low and high pressures and in different liquids.

Wilkinson et al. (1992) argues that the gas holdup in a bubble column is usually not uniform. In general, three regions of different gas holdup are recognized. At the top of the column, there is often foam structure with a relatively high gas holdup, while the gas holdup near the sparger is sometimes measured to be higher (for porous plate spargers) and sometimes lower (for single-nozzle spargers) than in the main central part of the column. The authors argue that if the bubble column is very high, then the gas holdup near the sparger and in the foam region at the top of the column has little influence on the overall gas holdup, while the influence can be significant for low bubble columns. The column height can influence the value of the gas holdup due to the fact that liquid circulation patterns (that tend to decrease the gas holdup) are not fully developed in short bubble columns (bed aspect ratio<3). All mentioned factors tend to cause a decrease in gas holdup with increasing column height. Kastanek et al. (1993) argues that this influence is negligible for column heights greater than 1-3 m and with height to diameter ratios above 5.

Wilkinson et al. (1992) have shown that the influence of the sparger design on gas holdup is negligible (at various pressures) provided the sparger hole diameters are larger than approximately $1\text{-}2\times10^{-3}$ m (and there is no maldistribution at the sparger). In high bubble columns, the influence of sparger usually diminishes due to the ongoing process of bubble coalescence. Wilkinson et al. (1992) argue that the relatively high gas holdup and mass transfer rate that can occur in small bubble columns as a result of the use of small sparger holes will not occur as noticeably in a high bubble column. In other words, a scale-up procedure, in which the gas holdup, the volumetric mass transfer coefficient and the interfacial area are estimated on the basis of experimental data obtained in a pilot-plant bubble column with small dimensions (bed aspect ratio<5, D_c<0.15 m) or with porous plate spargers, will in general lead to a considerable overestimation of these parameters. Shah et al. (1982) reported many gas holdup correlations developed on the basis of atmospheric data and they do not incorporate any influence of gas density.

In the case of liquid mixtures, Bach and Pilhofer (1978), Godbole et al. (1982) and Khare and Joshi (1990) determined that gas holdup does not decrease if the viscosity of water is increased by adding glycerol, carboxymethyl cellulose (CMC) or glucose but passes through a maximum. Wilkinson et al. (1992) assumes that this initial increase in gas holdup is due to the fact that the coalescence rate in mixtures is lower than in pure liquids. The addition of an electrolyte to water is known to hinder coalescence with the result that smaller bubbles occur and a higher gas holdup than pure water.

The addition of solids to a bubble column will in general lead to a small decrease in gas holdup (Reilly et al., 1986) and the formation of larger bubbles. The significant increase in gas holdup that occurs in two-phase bubble columns (due to the higher gas density) will also occur in three-phase bubble columns. A temperature increase leads to a higher gas

holdup (Bach and Pilhofer, 1978). A change in temperature can have an influence on gas holdup for a number of reasons: due to the influence of temperature on the physical properties of the liquid, as well as the influence of temperature on the vapor pressure.
Akita and Yoshida (1973) proposed their own correlation for gas holdup estimation. The correlation can be safely employed only within the set of systems used in the author's experiments, i.e. for systems air (O_2, He, CO_2)-water, air-methanol and air-aqueous solutions of glycerol. The experiments were carried out in a bubble column 0.6 m in diameter. The clear liquid height ranged between 1.26 and 3.5 m. It is worth noting that the effect of column diameter was not verified. Hikita and co-workers (1981) proposed another complex empirical relation for gas holdup estimation based on experimental data obtained in a small laboratory column (column diameter=0.1 m, clear liquid height=0.65 m). Large set of gas-liquid systems including air-(H_2, CO_2, CH_4, C_3H_8, N_2)-water, as well as air-aqueous solutions of organic liquids and electrolytes were used. For systems containing pure organic liquids the empirical equation of Bach and Pilhofer (1978) is recommended. The authors performed measurements in the systems air-alcohols and air-halogenated hydrocarbons carried out in laboratory units 0.1-0.15 m in diameter, at clear liquid height > 1.2 m. Hammer and co-workers (1984) proposed an empirical correlation valid for pure organic liquids at low superficial gas velocities. The authors pointed out that there is no any relation in the literature that can express the dependence of gas holdup on the concentration in binary mixtures of organic liquids. The effect of the gas distributor on gas holdup can be important particularly in systems with suppressed bubble coalescence. The majority of relations can be employed only for perforated plate distributors, while considerable increase of gas holdup in coalescence suppressing systems is observed in units with porous distributors. Kastanek et al. (1993) argue that the distributor geometry can influence gas holdup in turbulent bubble beds even in coalescence promoting systems at low values of bed aspect ratio and plate holes diameter.
The gas holdup increases with decreasing surface tension due to the lower rise velocity of bubbles. The effect of surface tension in systems containing pure liquids is however only slight. Gas holdup is strongly influenced by the liquid phase viscosity. However, the effect of this property is rather controversial. The effect of gas phase properties on gas holdup is generally of minor importance and only gas viscosity is usually considered as an important parameter. Large bubble formation leads to a decrease in the gas holdup. Kawase et al. (1987) developed a theoretical correlation for gas holdup estimation. Godbole et al. (1984) proposed a correlation for gas holdup prediction in CMC solutions.
Most of the works in bubble columns dealing with gas holdup measurement and prediction are based on deep bubble beds (Hughmark, 1967; Akita and Yoshida, 1973; Kumar et al., 1976; Hikita et al., 1980; Kelkar et al., 1983; Behkish et al., 2007). A unique work concerned with gas holdup ε_G under homogeneous bubbling conditions was published by Hammer et al. (1984). The authors presented an empirical relation valid for pure organic liquids at $u_G \leq 0.02\ m \cdot s^{-1}$. Idogawa et al. (1987) proposed an empirical correlation for gas densities up to $121\ kg \cdot m^{-3}$ and u_G values up to $0.05\ m \cdot s^{-1}$. Kulkarni et al. (1987) derived a relation to compute ε_G in the homogeneous flow regime in the presence of surface–active agents. By using a large experimental data set, Syeda et al. (2002) have developed a semi–empirical correlation for ε_G prediction in both pure liquids and binary mixtures. Pošarac and Tekić (1987) proposed a reliable empirical correlation which enables the estimation of gas holdup in bubble columns operated with dilute alcohol solutions. A number of gas holdup correlations were summarized by Hikita et al. (1980). Recently, Gandhi et al. (2007) have

proposed a support vector regression-based correlation for prediction of overall gas holdup in bubble columns. As many as 1810 experimental gas holdups measured in various gas–liquid systems were satisfactorily predicted (average absolute relative error: 12.1%). The method is entirely empirical.

In the empirical correlations, different dependencies on the physicochemical properties and operating conditions are implicit. This is primarily because of the limited number of liquids studied and different combinations of dimensionless groups used. For example, the gas holdup correlation proposed by Akita and Yoshida (1973) can be safely employed only within the set of systems used in the authors' experiments (water, methanol and glycerol solutions). The effect of column diameter D_c was not verified and the presence of this parameter in the dimensionless groups is thus only formal. In general, empirical correlations can describe ε_G data only within limited ranges of system properties and working conditions. In this work a new semi–theoretical approach for ε_G prediction is suggested which is expected to be more generally valid.

1.3 Estimation of volumetric liquid-phase mass transfer coefficient

The volumetric liquid-phase mass transfer coefficient is dependent on a number of variables including the superficial gas velocity, the liquid phase properties and the bubble size distribution. The relation for estimation of k_La proposed by Akita and Yoshida (1974) has been usually recommended for a conservative estimate of k_La data in units with perforated-plate distributors. The equation of Hikita and co-workers (1981) can be alternatively employed for both electrolytes and non-electrolytes. However, the reactor diameter was not considered in their relation. Hikita et al. (1981), Hammer et al. (1984) and Merchuk and Ben-Zvi (1992) developed also a correlation for prediction of the volumetric liquid-phase mass transfer coefficient k_La.

Calderbank (1967) reported that values of k_L decrease with increasing apparent viscosity corresponding to the decrease in the bubble rise velocity which prolongs the exposure time of liquid elements at the bubble surface. The k_L value for the frontal area of the bubble is higher than the one predicted by the penetration theory and valid for rigid spherical bubbles in potential flow. The rate of mass transfer per unit area at the rear surface of spherical-cap bubbles in water is of the same order as over their frontal areas. For more viscous liquids, the equation from the penetration theory gives higher values of k_L than the average values observed over the whole bubble surface which suggests that the transfer rate per unit area at the rear of the bubble is less than at its front.

Calderbank (1967) reported that the increase of the pseudoplastic viscosity reduces the rate of mass transfer generally, this effect being most substantial for small bubbles and the rear surfaces of large bubbles. The shape of the rear surface of bubbles is also profoundly affected. Evidently these phenomena are associated with the structure of the bubble wake.

In the case of coalescence promoting liquids, almost no differences have been reported between k_La values determined in systems with large- or small-size bubble population. For coalescence suppressing systems, it is necessary to distinguish between aqueous solutions of inorganic salts and aqueous solutions of surface active substances in which substantial decrease of surface tension occurs. Values of k_La reported in the literature for solutions of inorganic salts under conditions of suppressed bubble coalescence are in general several times higher than those for coalescent systems. On the other hand, k_La values observed in the presence of surface active agents can be higher or lower than those corresponding to pure water. No quantitative relations are at present available for prediction of k_La in solutions containing small bubbles. The relation of Calderbank and Moo-Young (1961) is considered the best available for the prediction of k_L values. It is valid for bubble sizes greater than 2.5×10^{-3} m and systems water-oxygen, water-CO_2 and aqueous solutions of glycol or polyacrylamide-CO_2. For small bubbles of size less than 2.5×10^{-3} m in systems of aqueous solutions of glycol-CO_2, aqueous solutions of electrolytes-air, waxes-H_2 these authors proposed another correlation. An exhaustive survey of published correlations for k_La and k_L was presented by Shah and coworkers (1982). The authors stressed the important effect of both liquid viscosity and surface tension. Kawase and Moo-Young (1986) proposed also an empirical correlation for k_La prediction. The correlation developed by Nakanoh and Yoshida (1980) is valid for shear-thinning fluids.

In many cases of gas-liquid mass transfer in bubble columns, the liquid-phase resistance to the mass transfer is larger than the gas-phase one. Both the gas holdup and the volumetric liquid-phase mass transfer coefficient k_La increase with gas velocity. The correlations of Hughmark (1967), Akita and Yoshida (1973) and Hikita et al. (1981) predict well k_La values in bubble columns of diameter up to 5.5 m. Öztürk et al. (1987) also proposed correlation for k_La prediction in various organic liquids. Suh et al. (1991) investigated the effects of liquid viscosity, pseudoplasticity and viscoelasticity on k_La in a bubble column and they developed their own correlation. In highly viscous liquids, the rate of bubble coalescence is accelerated and so the values of k_La decrease. Akita (1987a) measured the k_La values in inorganic aqueous solutions and derived their own correlation. Addition of surface-active substances such as alcohols to water increases the gas holdup, however, values of k_La in aqueous solutions of alcohols become larger or smaller than those in water according to the kind and concentration of the alcohol (Salvacion et al., 1995). Akita (1987b) and Salvacion et al. (1995) proposed correlations for k_La prediction in alcohol solutions.

The addition of solid particles (with particle size larger than 10 μm) increases bubble coalescence and bubble size and hence decreases both gas holdup and k_La. For these cases, Koide et al. (1984) and Yasunishi et al. (1986) proposed correlations for k_La prediction. Sauer and Hempel (1987) proposed k_La correlations for bubble columns with suspended particles. Sada et al. (1986) and Schumpe et al. (1987) proposed correlations for k_La prediction in bubble columns with solid particles of diameter less than 10 μm. Sun and Furusaki (1989) proposed a method to estimate k_La when gel particles are used. Sun and Furusaki (1989) and Salvacion et al. (1995) showed that k_La decreases with increasing solid concentration in gel-particle suspended bubble columns. Salvacion et al. (1995) showed that the addition of alcohol to water increases or decreases k_La depending on the kind and concentration of the alcohol added to the water and proposed a correlation for k_La including a parameter of retardation of surface flow on bubbles by the alcohol.

1.4 Estimation of liquid-phase mass transfer coefficient

The liquid-phase mass transfer coefficients k_L are obtained either by measuring k_La, gas holdup and bubble size or by measuring k_La and a with the chemical absorption method. Due to the difficulty in measuring distribution and the averaged value of bubble diameters in a bubble column, predicted values of k_L by existing correlations differ. Hughmark (1967), Akita and Yoshida (1974) and Fukuma et al. (1987b) developed correlations for k_L prediction. In the case of slurry bubble columns, Fukuma et al. (1987b) have shown that the degrees of dependence of k_L on both bubble size and the liquid viscosity are larger than those in a bubble column. Schumpe et al. (1987) have shown that low concentrations of high density solids of size less than 10 μm increase k_L by a hydrodynamic effect on the liquid film around the bubbles.

For pure liquids and large bubbles ($d_s \geq 0.002$ m), Higbie's (1935) relation based on the penetration theory of mass transfer can be used as the first approximation yielding qualitative information on the effect of fundamental physico-chemical parameters (viscosity, density, surface tension) on k_L values. All these parameters influence both the size of the bubbles (and consequently also their ascending velocity) and the hydrodynamic situation at the interface (represented by an appropriate value of liquid molecular diffusivity). Kastanek et al. (1993) proposed their own correlation for calculation of k_L.

Values of k_L decrease with increasing apparent viscosity corresponding to the decrease in bubble rise velocity which prolongs the exposure time of liquid elements at the bubble surface. According to Calderbank (1967), k_L for the frontal area is 1.13 times higher than the one predicted by the penetration theory and valid for spherical bubbles in potential flow. In the case of water, the rate of mass transfer per unit area at the rear surface of spherical-cap bubbles is of the same order as over their frontal areas. For more viscous liquids, the transfer rate per unit area at the rear of the bubble is less than at its front.

Calderbank (1967) reported that in general the increase of pseudoplastic viscosity reduces the rate of mass transfer, this effect being most substantial for small bubbles and the rear surfaces of large bubbles. The shape of the rear surface of bubbles is also profoundly affected. According to Calderbank (1967), these phenomena are associated with the structure of the bubble wake. Calderbank and Patra (1966) have shown experimentally that the average k_L obtained during the rapid formation of a bubble at a submerged orifice is less than the value observed during its subsequent ascent. According to the authors, this is a consequence of the fact that if the rising bubbles are not in contact with each other the mean exposure time of liquid elements moving round the surface of a rising bubble must be less than the corresponding exposure time during its formation.

Large bubbles ($d_s > 2.5\times10^{-3}$ m) have greater mass transfer coefficients than small bubbles ($d_s < 2.5\times10^{-3}$ m). Small "rigid sphere" bubbles experience friction drag, causing hindered flow in the boundary layer sense. Under these circumstances the mass transfer coefficient is proportional to the two-thirds power of the diffusion coefficient (Calderbank, 1967). For large bubbles ($>2.5\times10^{-3}$ m) form drag predominates and the conditions of unhindered flow envisaged by Higbie (1935) are realized. The author assumed unhindered flow of liquid round the bubble and destruction of concentration gradients in the wake of the bubble.

Griffith (1960) suggested that the mass transfer coefficient for the region outside a bubble may be computed if one knows the average concentration of solute in the liquid outside the bubble, the solute concentration at the interface and the rate of solute transfer. Leonard and Houghton (1963) reported that the k_L values for pure carbon dioxide bubbles dissolving in water is proportional to the square of the instantaneous bubble radius for diameters in the range $6\text{-}11\times10^{-3}$ m where the rise velocity appeared to be independent of size. Leonard and Houghton (1961) found that for bubbles with diameters below 6×10^{-3} m mass transfer seems to have an appreciable effect upon the velocity of rise, indicating that surface effects predominate in this range of sizes. Hammerton and Garner (1954) argue that there is a simple hydrodynamic correspondence between bubble velocity and mass transfer rate. According to Leonard and Houghton (1963) k_L is not only a function of bubble diameter but is also a function of the distance from the point of release. The variation of k_L with distance from the release point indicates that the rate is a function of time after release or some other related variable such as bubble size or hydrostatic pressure. Baird and Davidson (1962) observed a time dependence for carbon dioxide bubbles in water, but only for bubbles larger than 25×10^{-3} m in diameter, the explanation being that the time dependence was due to the unsteady state eddy diffusion into the turbulent wake at the rear of the bubble. Davies and Taylor (1950) developed a relation for k_L prediction in potential flow around a spherical-cap bubble. The authors argue that the bubble shape becomes oblate spheroidal for bubble sizes below 15×10^{-3} m.

Leonard and Houghton (1963) reported that the effect of inert gas is to reduce somewhat the mass transfer rate by about 20-40 % and to introduce more scatter in the calculated values of k_L, presumably because of the smaller volume changes. Gas circulation is also involved. The effect of an inert gas is to reduce the specific absorption rate, presumably by providing a gas-film resistance that may be affected by internal circulation.

Leonard and Houghton (1963) argue that there is a detectable decrease of k_L with increasing distance from the release point during absorption, the reverse appearing to be true for desorption. The addition of surfactant can reduce mass transfer without affecting the rise velocity. Mass transfer from single rising bubbles is governed to a large extent by surface effects, particularly at the smaller sizes.

The theory of isotropic turbulence can be used also for k_L prediction (Deckwer, 1980). The condition of local isotropy is frequently encountered. The theory of local isotropy gives information on the turbulent intensity in the small volume around the bubble. Turbulent flow produces primary eddies which have a wavelength or scale of similar magnitude to the dimensions of the main flow stream. These large primary eddies are unstable and disintegrate into smaller bubbles until all their energy is dissipated by viscous flow. When the Reynolds number of the main flow is high most of the kinetic energy is contained in the large eddies but nearly all of the dissipation occurs in the smallest eddies. If the scale of the main flow is large in comparison with that of the energy-dissipating eddies a wide spectrum of intermediate eddies exist which contain and dissipate little of the total energy. The large eddies transfer energy to smaller eddies in all directions and the directional nature of the primary eddies is gradually lost. Kolmogoroff (1941) concludes that all eddies which are much smaller than the primary eddies are statistically independent of them and the properties of these small eddies are determined by the local energy dissipation rate per unit mass of fluid. For local isotropic turbulence the smallest eddies are responsible for most of the energy dissipation and their time scale is given by Kolmogoroff (1941). Turbulence in the immediate vicinity of a bubble affects heat and mass transfer rates between the bubble and the liquid and may lead to its breakup.

Kastanek et al. (1993) suggested that the mass transfer in the turbulent bulk-liquid region is accomplished by elementary transfer eddies while in the surface layer adjacent to the interface turbulence is damped and mass transfer occurs due to molecular diffusion. In agreement with the theory of isotropic turbulence, the authors represented the contact time as the ratio of the length of elementary transport eddy to its velocity at the boundary between the bulk liquid and diffusion layer. Kastanek et al. (1993) argue that the rate of mass transfer between the gaseous and the liquid phase is decisively determined by the rate of energy dissipation in the liquid phase.

Kastanek (1977) and Kawase et al. (1987) developed a theoretical model for prediction of volumetric mass transfer coefficient in bubble columns. It is based on Higbie's (1935) penetration theory and Kolmogoroff's theory of isotropic turbulence. It is believed that turbulence brings up elements of bulk fluid to the free surface where unsteady mass transfer occurs for a short time (called exposure or contact time) after which the element returns to the bulk and is replaced by another one. The exposure time must either be determined experimentally or deduced from physical arguments. Calderbank and Moo-Young (1961) and Kawase et al. (1987) developed a correlation relating the rate of energy dissipation to turbulent mass transfer coefficient at fixed surfaces.

1.5 Estimation of gas-phase mass transfer coefficient

There is a lack of research in the literature on the estimation of the volumetric gas-phase mass transfer coefficients k_Ga. On the basis of chemical absorption and vaporization experiments Metha and Sharma (1966) correlated the k_Ga values to the molecular diffusivity in gas, the superficial gas velocity and static liquid height. Botton et al. (1980) measured k_Ga by the chemical absorption method in a SO_2 (in air)-Na_2CO_3 aqueous solution system in a wide range of u_G values. Cho and Wakao (1988) carried out experiments on stripping of five organic solutes with different Henry's law constants in a batch bubble column with water and they proposed two correlations (for single nozzle and for porous tube spargers) for k_Ga prediction. Sada et al. (1985) developed also correlation for k_Ga prediction. In the case of slurry bubble columns, the authors measured k_Ga by using chemical adsorption of lean CO_2 into NaOH aqueous solutions with suspended $Ca(OH)_2$ particles and they developed a correlation. Its predictions agree well with those observed by Metha and Sharma (1966) and Botton et al. (1980) in a bubble column.
The gas-phase mass transfer coefficient k_G decreases with increasing pressure due to the fact that the gas diffusion coefficient is inversely proportional to pressure (Wilkinson et al., 1992). In the case of bubble columns equipped with single nozzle and porous tube spargers the k_Ga value can be calculated by the correlation of Cho and Wakao (1988).

1.6 Estimation of interfacial area

The specific gas-liquid interfacial area varies significantly when hydrodynamic conditions change. Several methods exist for interfacial area measurements in gas-liquid dispersions. These are photographic, light attenuation, ultrasonic attenuation, double-optical probes and chemical absorption methods. These methods are effective under certain conditions only. For measuring local interfacial areas at high void fractions (more than 20 %) intrusive probes (for instance, double optical probe) are indispensable (Kiambi et al., 2001).
Calderbank (1958) developed a correlation for the specific interfacial area in the case of non-spherical bubbles. Akita and Yoshida (1974) derived also their own empirical equation for the estimation of the interfacial area. Leonard and Houghton (1963) related the interfacial area to the bubble volume using a constant shape factor, which would be 4.84 for spherical bubbles. In the case of perforated plates, Kastanek et al. (1993) reported a correlation for the estimation of the interfacial area. Very frequently the reliability of estimation of the specific interfacial area depends on the accuracy of gas holdup determination.

2. Effect of bubble shape on mass transfer coefficient

Deformed bubbles are generally classified as ellipsoids or spherical caps (Griffith, 1960; Tadaki and Maeda, 1961). The shapes and paths of larger non-spherical bubbles are generally irregular and vary rapidly with time, making the exact theoretical treatment impossible. Bubbles greater than 1.8×10^{-2} m in diameter assume mushroom-like or spherical-cap shapes and undergo potential flow. Calderbank (1967) argue that the eccentricity decreases with increasing viscosity accompanied by the appearance of "tails" behind small bubbles and of spherical indentations in the rear surfaces.

Calderbank et al. (1970) developed a new theory for mass transfer in the bubble wake. Their work with aqueous solutions of glycerol covers the bubble size range $0.2\text{-}6.0\times10^{-2}$ m and includes the various bubble shapes as determined by the bubble size and the viscosity of the Newtonian liquid. Calderbank and Lochiel (1964) measured the instantaneous mass transfer coefficients in the liquid phase for carbon dioxide bubbles rising through a deep pool of distilled water. Redfield and Houghton (1965) determined mass transfer coefficients for single carbon dioxide bubbles averaged over the whole column using aqueous Newtonian solutions of dextrose. Davenport et al. (1967) measured mass transfer coefficients averaged over column lengths of up to 3 m for single carbon dioxide bubbles in water aqueous solutions of polyvinyl alcohol and ethyl alcohol, respectively. Angelino (1966) has reported some shapes and terminal rise velocities for air bubbles in various Newtonian liquids. Liquid-phase mass transfer coefficients for small bubbles rising in glycerol have been determined by Hammerton and Garner (1954) over bubble diameters ranging from 0.2×10^{-2} m to 0.6×10^{-2} m. Barnett et al. (1966) reported the liquid-phase mass transfer coefficients for small CO_2 bubbles ($0.5\text{-}4.5\times10^{-3}$ m) rising through pseudoplastic Newtonian liquids. This bubble size range was extended to $3\text{-}50\times10^{-3}$ m in the data reported by Calderbank (1967). Astarita and Apuzzo (1965) presented experimental results on the rising velocity and shapes of bubbles in both purely viscous and viscoeleastic non-Newtonian pseudoplastic liquids. According to Calderbank et al. (1970) bubble shapes observed in distilled water vary from spherical to oblate spheroidal ($0.42\text{-}1.81\times10^{-2}$ m) to spherical cap ($1.81\text{-}3.79\times10^{-2}$ m) with increasing bubble size. Over the size range ($4.2\text{-}70\times10^{-3}$ m) the bubbles rise with a zigzag or spiral motion and between bubble diameters of 7×10^{-3} m (Re=1800) and 18×10^{-3} m (Re=5900) an irregular ellipsoid shape is adopted and the bubble pulsates about its mean shape. Over the bubble size range $1.8\text{-}3.0\times10^{-2}$ m a transition from irregular ellipsoid to spherical cap shape occurs and surface rippling is much more evident. For bubble sizes greater than 3×10^{-2} m the bubbles adopt fully developed spherical cap shapes and exhibit little surface rippling. These spherical cap bubbles rise rectilinearly.

Calderbank et al. (1970) developed theory of mass transfer from the rear of spherical-cap bubbles. The authors argue that the overall mass transfer coefficients enhance by hydrodynamic instabilities in the liquid flow round bubbles near the bubble shape transition from spherical cap to oblate spheroid. Calderbank et al. (1970) reported that for bubble sizes $1\text{-}1.8\times10^{-2}$ m a shape transition occurs, the bubble rear surface is gradually flattening and becoming slightly concave as the bubble size is increased. The onset of skirting is accompanied by a flattening of the bubble rear surface. The authors argue that the bubble eccentricity decreases with increasing Newtonian liquid viscosity, though there is a tendency towards convergence at large bubble sizes.

Davidson and Harrison (1963) indicated that the onset of slug flow occurs approximately at bubble size/column diameter>0.33. In the case of spherical-cap bubbles it is expected that there will be appreciable variations between front and rear surfaces. Behind the spherical-cap bubble is formed a torroidal vortex.

Calderbank et al. (1970) reported that a maximum value of k_L occurs shortly before the onset of creeping flow conditions and corresponds to a bubble shape transition from spherical cap to oblate spheroid. This shape transition and the impending flow regime transition results in instabilities in the liquid flow around the bubble resulting in k_L enhancement. The results of Zieminski and Raymond (1968) indicate that for CO_2 bubbles a maximum of k_L occurs at bubble size of 3×10^{-3} m which they attribute to a progressive transition between circulating and rigid bubble behavior.

Calderbank (1967) stated that the theory of mass transfer has to be modified empirically for dispersion in a non-isotropic turbulent field where dispersion and coalescence take place in different regions. Coalescence is greatly influenced by surfactants, the amount of dispersed phase present, the liquid viscosity and the residence time of bubbles. The existing theories throw little light on problems of mass transfer in bubble wakes and are only helpful in understanding internal circulation within the bubble. The mass-transfer properties of bubble swarms in liquids determine the efficiency and dimensions of the bubble column.

If the viscous or inertial forces do not act equally over the surface of a bubble they may cause it to deform and eventually break. A consequence of these dynamic forces acting unequally over the surface of the bubble is internal circulation of the fluid within the bubble which induces viscous stresses therein. These internal stresses also oppose distortion and breakage.

3. New approach for prediction of gas holdup (Nedeltchev and Schumpe, 2008)

Semi-theoretical approaches to quantitatively predict the gas holdup are much more reliable and accurate than the approaches based on empirical correlations. In order to estimate the mass transfer from bubbles to the surrounding liquid, knowledge of the gas-liquid interfacial area is essential. The specific gas–liquid interfacial area, defined as the surface area available per unit volume of the dispersion, is related to gas holdup ε_G and the Sauter–mean bubble diameter d_s by the following simple relation:

$$a = \frac{6\varepsilon_G}{d_s} \tag{1}$$

Strictly speaking, Eq. (1) (especially the numerical coefficient 6) is valid only for spherical bubbles (Schügerl et al., 1977).

The formula for calculation of the interfacial area depends on the bubble shape. Excellent diagrams for bubble shape determination are available in the books of Clift et al. (1978) and Fan and Tsuchiya (1990) in the form of log–log plots of the bubble Reynolds number Re_B vs. the Eötvös number Eo with due consideration of the Morton number Mo. A comparison among the experimental conditions used in our work and the above–mentioned standard plots reveals that the formed bubbles are no longer spherical but oblate ellipsoidal and follow a zigzag upward path as they rise. Vortex formation in the wake of the bubbles is also observed. The specific interfacial area a of such ellipsoidal bubbles is a function of the number of bubbles N_B, the bubble surface S_B and the total dispersion volume V_{total} (Painmanakul et al., 2005; Nedeltchev et al., 2006a, b, 2007a):

$$a = \frac{N_B S_B}{V_{total}} = \frac{N_B S_B}{AH} \tag{2}$$

where A denotes the column cross–sectional area. The number of bubbles N_B can be deduced from the bubble formation frequency f_B and bubble residence time (Painmanakul et al., 2005):

$$N_B = f_B \frac{H}{u_B} = \frac{Q_G}{V_B} \frac{H}{u_B} \tag{3}$$

where Q_G is the volumetric gas flow rate, u_B is the bubble rise velocity and V_B is the bubble volume. The substitution of Eq. (3) into Eq. (2) yields:

$$a = \frac{f_B S_B}{A u_B} \tag{4}$$

The bubble rise velocity u_B can be estimated from Mendelson's (1967) correlation:

$$u_B = \sqrt{\frac{2\sigma_L}{\rho_L d_e} + \frac{g d_e}{2}} \tag{5}$$

This equation is particularly suitable for the case of ellipsoidal bubbles.
The volume of spherical or ellipsoidal bubbles can be estimated as follows:

$$V_B = \frac{\pi d_e^3}{6} = \frac{4}{3}\pi \left(\frac{l}{2}\right)^2 \frac{h}{2} \tag{6}$$

If some dimensionless correction factor f_c due to the bubble shape differences is introduced, then Eqs. (1) and (4) might be considered equivalent:

$$\frac{6\varepsilon_G}{d_s} = f_c \frac{f_B S_B}{A u_B} \tag{7}$$

Rearrangement of Eq. (7) yields:

$$\varepsilon_G = f_c \frac{d_s f_B S_B}{6 A u_B} \tag{8}$$

The surface S_B of an ellipsoidal bubble can be calculated as follows (Nedeltchev et al., 2006a, b, 2007a):

$$S_B = \pi \frac{l^2}{2} \left[1 + \left(\frac{h}{l}\right)^2 \frac{1}{2e} \ln \frac{(1+e)}{(1-e)} \right] \tag{9}$$

where e is the bubble eccentricity. It can be calculated as follows:

$$e = \sqrt{1 - \left(\frac{h}{l}\right)^2} \tag{9a}$$

An oblate ellipsoidal bubble is characterized by its length l (major axis of the ellipsoid) and its height h (minor axis of the ellipsoid). The ellipsoidal bubble length l and height h can be estimated by the formulas derived by Tadaki and Maeda (1961) and Terasaka et al. (2004):
for $2<Ta<6$:

$$l = \frac{d_e}{1.14 Ta^{-0.176}} \tag{10a}$$

$$h = 1.3 d_e Ta^{-0.352} \tag{10b}$$

for 6<Ta<16.5:

$$l = \frac{d_e}{1.36 Ta^{-0.28}} \tag{11a}$$

$$h = 1.85 d_e Ta^{-0.56} \tag{11b}$$

where

$$Ta = Re_B Mo^{0.23} \tag{12}$$

$$Re_B = \frac{d_e u_B \rho_L}{\mu_L} \tag{13}$$

$$Mo = \frac{g \mu_L^4}{\rho_L \sigma_L^3} \tag{14}$$

It is worth noting that the major axis of a rising oblate ellipsoidal bubble is not always horizontally oriented (Yamashita et al., 1979). The same holds for the minor axis of a rising oblate ellipsoidal bubble, i.e. it is not necessarily vertically oriented (Akita and Yoshida, 1974). Equations (10a)-(14) were used to calculate both l and h values under the operating conditions examined. The Morton number Mo is the ratio of viscosity force to the surface tension force. The Tadaki number Ta characterizes the extent of bubble deformation; the Ta values fell always in one of the ranges specified above. This fact can be regarded as an additional evidence that the bubbles formed under the operating conditions examined are really ellipsoidal.

The above correlations (Equations (10a)–(14)) imply that one needs to know a priori the bubble equivalent diameter d_e. Very often in the literature is assumed that d_e can be approximated by the Sauter–mean bubble diameter d_s. The latter was estimated by means of the correlation of Wilkinson et al. (1994):

$$\left(\frac{g \rho_L d_s^2}{\sigma_L}\right) = 8.8 \left(\frac{u_G \mu_L}{\sigma_L}\right)^{-0.04} \left(\frac{\sigma_L^3 \rho_L}{g \mu_L^4}\right)^{-0.12} \left(\frac{\rho_L}{\rho_G}\right)^{0.22} \tag{15}$$

Equation (15) implies that the bubble size decreases as the superficial gas velocity u_G or the gas density ρ_G (operating pressure P) increase. The calculated d_s values for all liquids examined imply an ellipsoidal shape. Equation (5) along with Eq. (15) (for d_s estimation) was used also to calculate the bubble Reynolds number Re_B (Eq. (13)) needed for the estimation of both l and h values.

The bubble equivalent diameter d_e of an ellipsoidal bubble can be also calculated from Eq. (6) by assuming a sphere of equal volume to the volume of the ellipsoidal bubble:

$$d_e = \left(l^2 h\right)^{1/3} \tag{16}$$

Estimating the characteristic length of ellipsoidal bubbles with the same surface–to–volume ratio (the same d_s value as calculated from Eq. (15)) required an iterative procedure but led to only insignificantly different values than simply identifying the equivalent diameter d_e with d_s when applying Eqs. (10a–b) or (11a–b). In other words, the differences between bubble diameters estimated by Eq. (15) and Eq. (16) are negligibly small.

Liquid	D_c [m]	Gas Sparger	Gases Used	P [MPa]	ρ_L [$kg \cdot m^{-3}$]	μ_L [10^{-3} Pa·s]	σ_L [10^{-3} $N \cdot m^{-1}$]
Acetone	0.095	D4	air	0.1	790	0.327	23.1
Anilin	0.095	D4	air	0.1	1022	4.4	43.5
Benzene	0.095	D4	air	0.1	879	0.653	28.7
1–Butanol	0.095 0.102	D1, D2 D4	N_2, air, He	0.1–4.0	809	2.94	24.6
Carbon tetrachloride	0.095	D4	air, He, H_2, CO_2	0.1	1593	0.984	26.1
Cyclohexane	0.095	D4	air	0.1	778	0.977	24.8
Decalin	0.102	D1, D2	N_2, He	0.1–4.0	884	2.66	32.5
1,2-Dichloroethane	0.095	D4	air	0.1	1234	0.82	29.7
1,4-Dioxane	0.095	D4	air	0.1	1033	1.303	32.2
Ethanol (96 %)	0.102	D1, D2, D3	N_2, He	0.1–4.0	793	1.24	22.1
Ethanol (99 %)	0.095	D4	air	0.1	791	1.19	22.1
Ethyl acetate	0.095	D4	air	0.1	900	0.461	23.5
Ethyl benzene	0.095	D4	air	0.1	867	0.669	28.6
Ethylene glycol	0.095 0.102	D1, D2, D4	N_2, air, He	0.1–4.0	1112	19.9	47.7
Gasoline	0.102	D1	N_2	0.1–4.0	692	0.464	21.6
LigroinA (b. p. 90–110 °C)	0.095	D4	air	0.1	714	0.470	20.4
Ligroin B(b. p. 100–140°C)	0.095	D4	air	0.1	729	0.538	21.4
Methanol	0.095	D4	air	0.1	790	0.586	22.2
Nitrobenzene	0.095	D4	air	0.1	1203	2.02	38.1
2-Propanol	0.095	D4	air	0.1	785	2.42	21.1
Tap water	0.095 0.102	D1, D2 D3, D4	N_2, air, He	0.1–4.0	1000	1.01	72.74
Tetralin	0.095	D4	N_2, air, CO_2	0.1	968	2.18	34.9
Toluene	0.095 0.102	D1, D2, D4	N_2, air, He	0.1–4.0	866	0.58	28.5
Xylene	0.095	D4	N_2, air, He, H_2, CO_2	0.1	863	0.63	28.4

Table 1. Properties of the organic liquids and tap water (293.2 K)

Our semi-theoretical approach is focused on the derivation of a correlation for the correction term f_c introduced in Eq. (7). Many liquids covering a large spectrum of physicochemical properties, different gas distributor layouts and different gases at operating pressures up to 4 MPa are considered (Nedeltchev et al., 2007a). As many as 386 experimental gas holdups were obtained in two bubble columns. The first stainless steel column (D_c=0.102 m, H_0=1.3 m) was equipped with three different gas distributors: perforated plate, 19 × Ø 1×10^{-3} m (D1), single hole, 1 × Ø 4.3×10^{-3} m (D2) and single hole, 1 × Ø 1×10^{-3} m (D3) (Jordan and Schumpe, 2001). In the second plexiglass column (D_c=0.095 m, H_0=0.85 m) the gas was always introduced through a single tube of 3×10^{-3} m in ID (D4) (Öztürk et al., 1987). The ε_G values were measured in 21 organic liquids, 17 liquid mixtures and tap water (see Tables 1 and 2).

Liquid Mixture	Key Fig. 9	D_c [m]	Gas Sparger	Gas Used	P [MPa]	ρ_L [kg·m^{-3}]	μ_L [10^{-3} Pa·s]	σ_L [10^{-3} N·m^{-1}]
Benzene/Cyclohexane 6.7 %	A	0.095	D4	air	0.1	865	0.634	27.6
Benzene/Cyclohexane 13.4 %	B	0.095	D4	air	0.1	854	0.628	26.9
Benzene/Cyclohexane 31.5 %	C	0.095	D4	air	0.1	834	0.631	26.2
Benzene/Cyclohexane 54 %	D	0.095	D4	air	0.1	814	0.672	25.4
Benzene/Cyclohexane 78.5 %	E	0.095	D4	air	0.1	797	0.772	24.9
Benzene/Cyclohexane 90 %	F	0.095	D4	air	0.1	787	0.858	24.9
Glycol 22.4 %/Water	G	0.095	D4	air	0.1	1043	2.32	53.8
Glycol 60 %/Water	H	0.095	D4	air	0.1	1072	5.6	52.0
Glycol 80 %/Water	I	0.095	D4	air	0.1	1091	9.65	51.0
Toluene/Ethanol 5 %	J	0.095	D4	air	0.1	863	0.578	27.6
Toluene/Ethanol 13.6 %	K	0.095	D4	air	0.1	859	0.587	27.3
Toluene/Ethanol 28 %	L	0.095	D4	air	0.1	852	0.616	25.5
Toluene/Ethanol 55 %	M	0.095	D4	air	0.1	838	0.731	25.0
Toluene/Ethanol 73.5 %	N	0.095	D4	air	0.1	823	0.961	24.2
Toluene/Ethanol 88.5 %	O	0.095	D4	air	0.1	807	1.013	23.3
Toluene/Ethanol 94.3 %	P	0.095	D4	air	0.1	800	1.103	22.7
Toluene/Ethanol 97.2 %	Q	0.095	D4	air	0.1	796	1.135	22.2

Table 2. Properties of the liquid mixtures (293.2 K)

In both tables are listed the different combinations of liquids, gases, gas distributors and operating pressures that have been used. It is worth noting that in a 0.095 m in ID bubble column equipped with a sparger D4 every liquid or liquid mixture was aerated with air. Table 1 shows that in the case of few liquids (carbon tetrachloride, tetralin, toluene and xylene) some other gases have been used. It should be mentioned that in the case of 0.102 m in ID bubble column no air was used (only nitrogen and helium).

The gas holdups ε_G in 1–butanol, ethanol (96 %), decalin, toluene, gasoline, ethylene glycol and tap water were recorded by means of differential pressure transducers in the 0.102 m stainless steel bubble column operated at pressures up to 4 MPa. The following relationship was used:

$$\varepsilon_G = \frac{\Delta P_{\text{no gas}} - \Delta P_{\text{gas}}}{\Delta P_{\text{no gas}}} \tag{17}$$

where ΔP is the pressure difference between the readings of both lower (at 0 m) and upper (at 1.2 m) pressure transducers. The subscript "no gas" denotes the pressure difference at the clear liquid height H_0, whereas the subscript "gas" denotes the pressure difference at the aerated liquid height H. The gas holdups ε_G in all other liquids and liquid mixtures were estimated by visually observing the dispersion height under ambient pressure in the 0.095 m in ID bubble column. The upper limit (transitional gas velocity) of the homogeneous regime (transition gas velocity u_{trans}) was estimated by the formulas of Reilly et al. (1994).

Most of the gas–liquid systems given in Tables 1 and 2 were characterized with Tadaki numbers Ta lower than 6 and thus Eqs. (10a–b) for the estimation of both bubble length l and bubble height h were applied. Only in the case of ethylene glycol–(helium, air and nitrogen), 1–butanol–(helium and air) and decalin–helium the Ta values exceeded 6 and then Eqs. (11a–b) were used.

It was found that the dimensionless correction factor f_c can be correlated successfully to both the Eötvös number Eo and a dimensionless gas density ratio:

$$f_c = 0.78 Eo^{-0.22}\left(\frac{\rho_G}{\rho_G^{ref}}\right)^{0.07} = 0.78\left(\frac{g(\rho_L - \rho_G)d_e^2}{\sigma_L}\right)^{-0.22}\left(\frac{\rho_G}{1.2}\right)^{0.07} \tag{18}$$

where ρ_G^{ref} is the reference gas density (1.2 kg·m^{-3} for air at ambient conditions: 293.2 K and 0.1 MPa). All experimental gas holdup data (386 points) were fitted with an average error of 9.6%. The dimensionless gas density ratio is probably needed because the correlation of Wilkinson et al. (1994) was derived for pressures up to 1.5 MPa only, whereas the present data extend up to the pressure of 4 MPa. It is worth noting that Krishna (2000) also used such a dimensionless gas density ratio for correcting his correlations for large bubble rise velocity and dense–phase gas holdup.

Figure 1 illustrates the decrease of the product $f_c(\rho_G/1.2)^{-0.07}$ with increasing Eo. At smaller bubble sizes (with shapes approaching spheres), Eo will be lower and thus f_c higher (gradually approaching unity). It is worth noting that most of the liquids are characterized with Eo values in a narrow range between 2 and 8.

Figure 2 illustrates that the correction factor f_c increases with gas density ρ_G (operating pressure) leading to bubble shrinkage. For example, the correction term f_c decreases in the following sequence: toluene > ethanol > decalin > 1–butanol > ethylene glycol. The smallest bubble size is formed in toluene, whereas the largest bubble size is formed in the case of ethylene glycol. When very small (spherical) bubbles are formed, the correction factor f_c should be equal to unity and both expressions for the interfacial area should become identical (see Eq. (7)).

Figures 3 and 4 exhibit that the experimental gas holdups ε_G measured in 1–butanol, decalin and toluene at pressures up to 4 MPa can be predicted reasonably well irrespective of the gas distributor type.

The same result holds for ethylene glycol and tap water (see Fig. 5). The successful prediction of gas holdups in ethylene glycol should be regarded as one of the most important merits of the presented method since the viscosity is much higher than that of the other liquids.

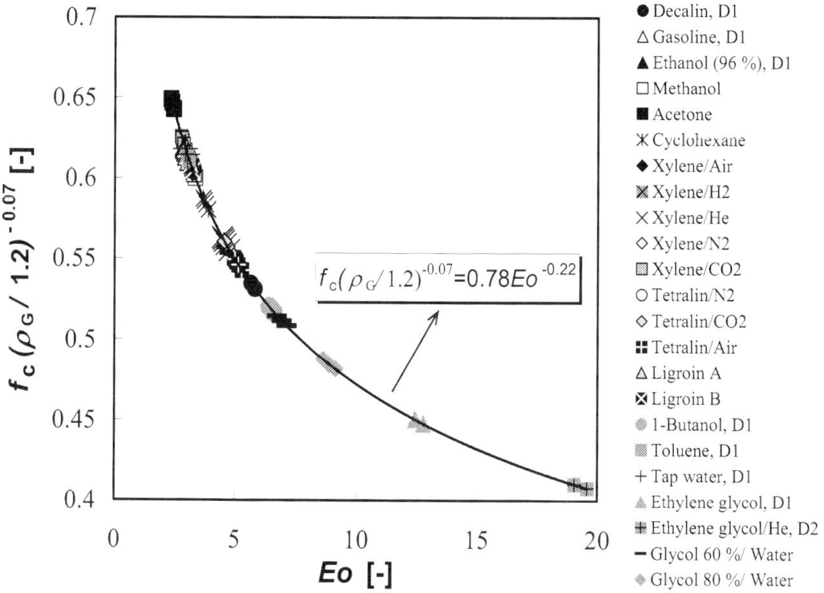

Fig. 1. Product $f_c(\rho_G/1.2)^{-0.07}$ as a function of Eo for 12 organic liquids, 2 liquid mixtures and tap water at ambient pressure (Gas distributor: D4 unless specified in the legend)

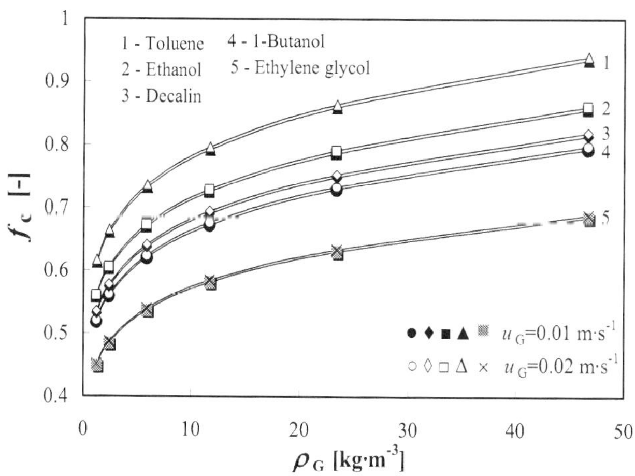

Fig. 2. Correction factor f_c as a function of gas density ρ_G in five organic liquids (Gas distributor: D1; D_c: 0.102 m)

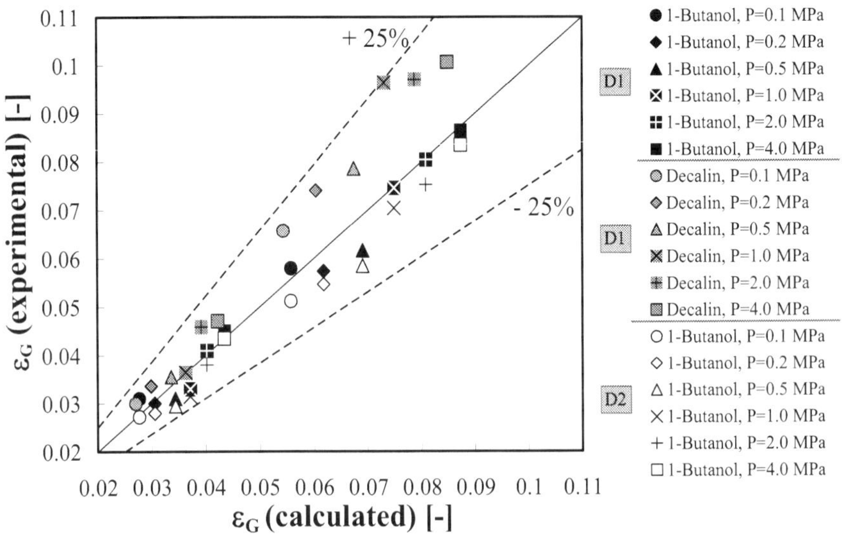

Fig. 3. Parity plot for gas holdups in 1–butanol and decalin sparged with nitrogen through gas distributors D1 and D2 at various pressures

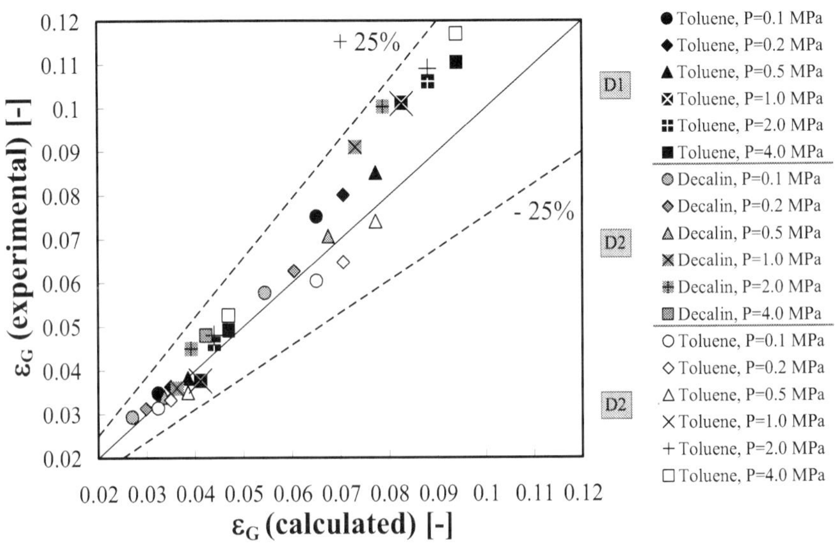

Fig. 4. Parity plot for gas holdups in toluene and decalin sparged with nitrogen through gas distributors D1 and D2 at various pressures

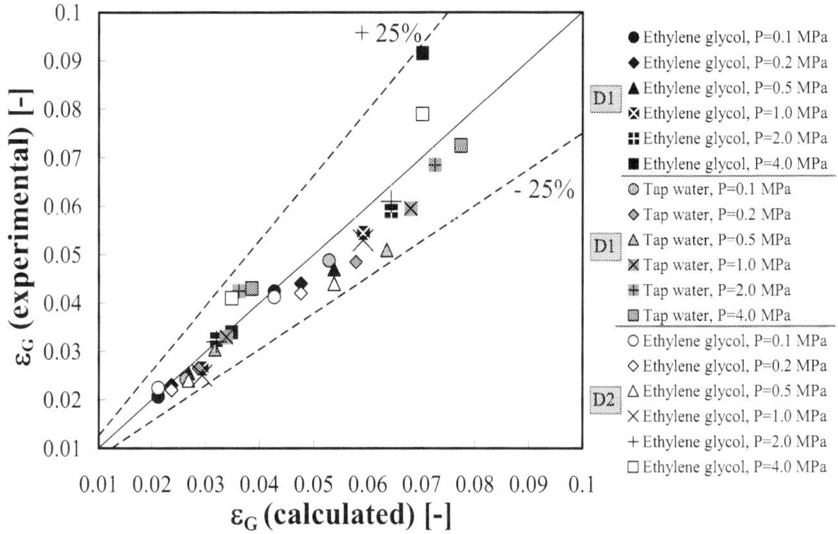

Fig. 5. Parity plot for gas holdups in ethylene glycol and tap water aerated with nitrogen through gas distributors D1 and D2 at various pressures

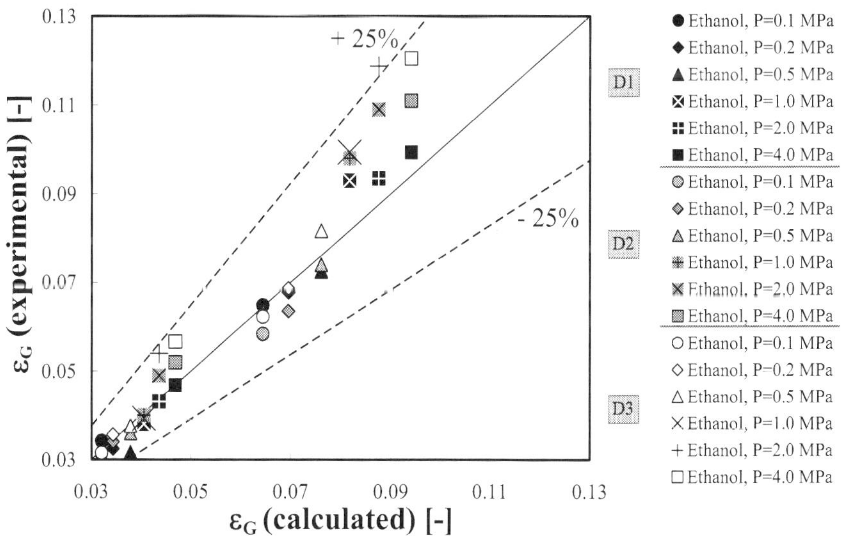

Fig. 6. Parity plot for gas holdups in ethanol (96 %) sparged with nitrogen through gas distributors D1–D3 at various pressures

Figure 6 shows, for ethanol (96 %) as an example, that the gas distributor type is not so important. The same holds for 1–butanol, decalin and toluene (Figs. 3–5) and tap water. This fact is in agreement with the work of Wilkinson et al. (1992) who stated that once the hole size of the gas distributor is greater than $1–2\times10^{-3}$ m, then it has no significant effect on the gas holdup.

Eight organic liquids and tap water have been aerated not only with air or nitrogen but also with other gases (helium, hydrogen and carbon dioxide). Figure 7 shows that the developed model is capable of predicting satisfactorily the experimental gas holdups at these operating conditions.

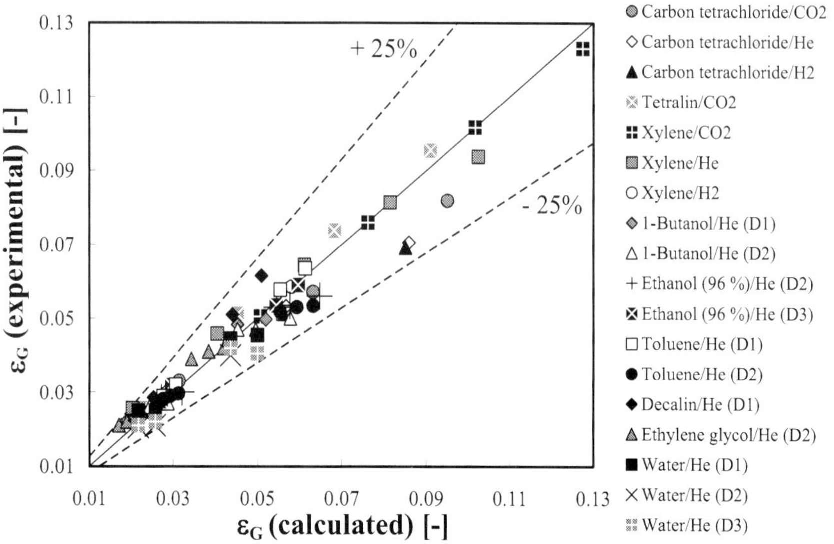

Fig. 7. Parity plot for gas holdups ε_G in 8 organic liquids and tap water aerated with other gases (helium, hydrogen and/or carbon dioxide) at ambient pressure. Gas distributor: D4 unless specified in the legend

Figure 8 shows that the model predicts reasonably well the experimental gas holdups measured in 15 organic liquids at ambient pressure. This fact should be regarded as further evidence that by the introduction of a correction term the presented method becomes generally applicable.

Table 2 and Figs. 3–8 reveal that our approach is applicable not only to tap water but also to organic liquids covering the following ranges of the main physicochemical properties: $692\leq(\rho_L/\text{kg m}^{-3})\leq1593$, $0.327\times10^{-3}\leq(\mu_L/\text{Pa s})\leq 19.9\times10^{-3}$, $20.4\times10^{-3}\leq(\sigma_L/\text{N m}^{-1})\leq47.7\times10^{-3}$.

Figure 9 exhibits that the proposed method for gas holdup prediction along with the new correction factor (Eq. (18)) is also applicable to various liquid mixtures. Table 2 shows that the examined liquid mixtures cover the following ranges of the main physicochemical properties: $787\leq(\rho_L/\text{kg m}^{-3})\leq1091$, $0.578\times10^{-3}\leq(\mu_L/\text{Pa s})\leq9.65\times10^{-3}$ and $22.2\times10^{-3}\leq(\sigma_L/\text{N}\cdot\text{m}^{-1})$ $\leq53.8\times10^{-3}$.

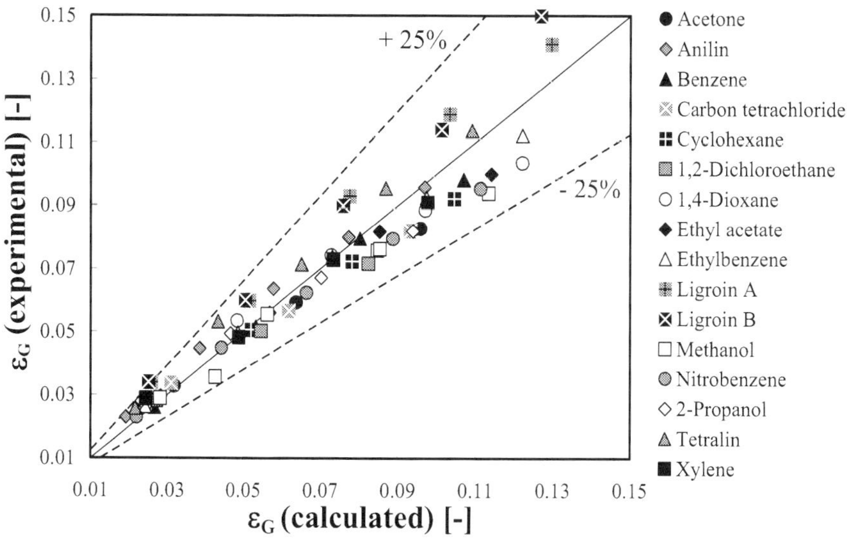

Fig. 8. Parity plot for gas holdups in 15 organic liquids aerated with air by means of gas distributor D4 (D_c=0.095 m) at ambient pressure

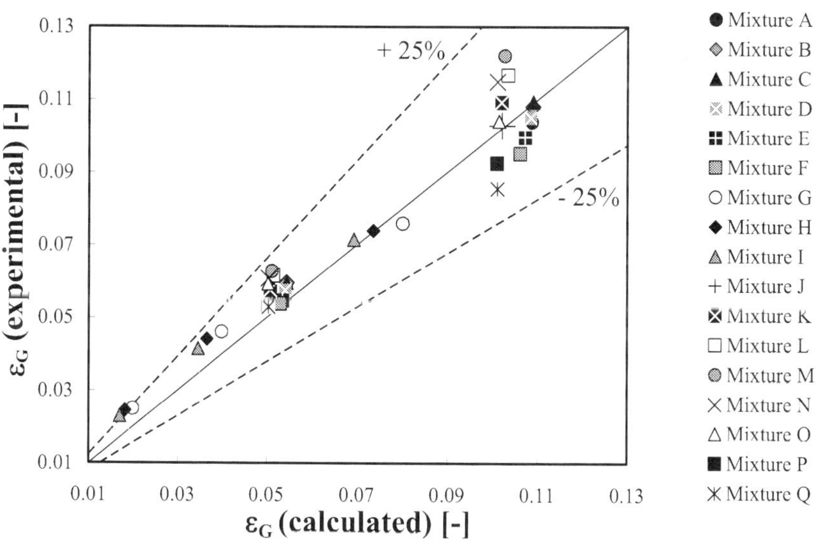

Fig. 9. Parity plot for gas holdups in 17 liquid mixtures sparged with air (gas distributor D4, D_c=0.095 m) at ambient pressure. The legend keys are given in Table 2

Nedeltchev and Schumpe (2008) have shown that in the homogeneous flow regime the semi-theoretical approach improves the gas holdup predictions and turns out to be the most reliable one (since it produces better gas holdup predictions than the empirical correlations of both Hammer et al. (1984) and Akita and Yoshida (1974)). It is worth underlying that our approach is applicable at $u_G \leq u_{trans}$, where the transition gas velocity u_{trans} can be predicted by the correlations of Reilly et al. (1994). In this work, the u_{trans} values were always less than 0.04 $m \cdot s^{-1}$.

4. New approach for prediction of volumetric liquid-phase mass transfer coefficient (Nedeltchev et al., 2006a, b, 2007a)

The k_La value can be predicted if one knows how to estimate both the liquid–side mass transfer coefficient k_L and the interfacial area a. These parameters depend on the bubble diameter. Both k_L and a are closely related to the bubble bed hydrodynamics. k_L incorporates the effects of the complex liquid flow field surrounding the rising gas bubbles. The interfacial area a inherently reflects the system bubble behavior.

Gas bubbles are nonspherical, except when interfacial tension and viscous forces are much more important than inertial forces (Clift et al., 1978). Most of the bubble sizes in this work fall in the range of $1.4–6 \times 10^{-3}$ m. Such bubbles are no longer spherical and follow a zigzag or helical upward path. Viscous drag is augmented by vortex formation in the wake, and velocity of rise remains fairly constant over the bubble size range (Miller, 1974).

As a rough approximation, Higbie (1935) assumed that the average time of exposure for mass transfer (called contact time or exposure time for mass transfer) can be estimated as follows:

$$t_c = \frac{\text{Bubble surface}}{\text{Rate of surface formation}} \tag{19}$$

In our approach all theoretical k_La calculations were based on this definition of the gas–liquid contact time t_c. The latter characterizes the residence time of the micro eddies (responsible for mass transfer in the liquid film) at the interface, i.e. at the bubble surface. It is practically impossible to measure directly the contact time t_c, so usually it is estimated from correlations for the bubble surface and the rate of surface formation. The contact time is the time required for the bubble to rise through its equivalent spherical diameter. Higbie's (1935) model predicts a decrease in k_L with increase in diameter when in fact it increases quite rapidly. Higbie's theory does not predict an effect of distance from the point of release. The simple model of Higbie (1935) in which the contact time is approximated by macro-scale parameters fails for turbulent bubbling conditions where bubble swarms exist. Higbie's (1935) model assumes direct contact between the gas phase and the bulk liquid.

The contact time takes part in the evaluation of the liquid-phase mass transfer coefficient:

$$k_L = \sqrt{\frac{4D_L}{\pi t_c}} \tag{20}$$

The equation is valid explicitly for rigid spherical bubbles. Higbie (1935) postulated that a gas bubble moving through a liquid splits the liquid at its advancing tip. The penetration theory assumes unsteady–state absorption of a gas by a fluid element adjacent to the surface. The element moves at a uniform velocity from the front of the bubble to the rear as penetration into it occurs. Timson and Dunn (1960) showed that the bubble surface increases when the spherical bubble is deformed into an ellipsoid. This leads to higher contact time and thus lower k_L value.

In order to calculate the volumetric liquid–side mass transfer coefficient k_La one also needs to know how to calculate the interfacial area a. The formula for its calculation depends on the bubble shape. The specific interfacial area a is a function of the number of bubbles N_B, the bubble surface area S_B and the total dispersion volume V_{total} (see Eq. (2)).

The theoretically calculated k_La values were obtained as a product of both Eq. (20) (using Eq. (19) for the contact time) and Eq. (2). Experimentally, on the other hand, it is much easier to measure the product of k_L and a than the individual values. In the homogeneous flow regime there is a narrow bubble size distribution and the researchers use frequently the Sauter–mean bubble diameter d_s for their mass transfer calculations. The bubbles rise uniformly in nearly straight lines and have roughly uniform bubble size (Krishna, 2000; Lucas et al., 2005). Therefore, it is a reasonable simplification in the homogeneous flow regime to use the mean bubble diameter and disregard the bubble size distribution.

For the sake of theoretical prediction of the k_La values by means of Eqs. (2), (19) and (20) one needs to calculate the bubble surface S_B, the rate of surface formation R_{sf} and the number of bubbles N_B. The calculation of the bubble surface S_B depends on the bubble shape (Painmanakul et al., 2005). Two excellent diagrams for bubble shape determination are available in the books of Clift et al. (1978) and Fan and Tsuchiya (1990) in the form of log–log plots of the bubble Reynolds number Re_B vs. the Eötvös number Eo with due consideration of the Morton number Mo. A comparison among the experimental conditions used in this work and the above–mentioned standard plots reveals that the formed bubbles are no longer spherical but oblate ellipsoidals that follow a zigzag upward path as they rise. Vortex formation in the wake of the bubbles is also observed.

An oblate ellipsoidal bubble is characterized by its length l (major axis of the ellipsoid) and its height h (minor axis of the ellipsoid). The surface area S_B of an ellipsoidal bubble is calculated on the basis of Eqs. (9) and (9a) (Fan and Tsuchiya 1990; Nedeltchev et al., 2006a, b). In order to calculate both ellipsoidal bubble length l and height h, the correlations of Terasaka et al. (2004) (for $2<Ta<6$) were used (see Eqs. (10a-b)). The Tadaki numbers Ta fell always in the range specified above. However, the correlations of Terasaka et al. (2004) imply that one needs to know a priori the bubble equivalent diameter d_e.

Very often in the literature is assumed that d_e can be approximated by the Sauter–mean bubble diameter d_s. Bubble shape, motion and any tendency for the interface to ripple, fluctuate or otherwise deform are all related to bubble size. In turn, bubble size is determined by the physical characteristics of the system and the operating conditions. The d_s value was estimated by means of the correlation of Wilkinson et al. (1994) which is one of the most frequently cited in the literature. This equation implies that the bubble size decreases as superficial gas velocity u_G or gas density ρ_G (operating pressure P) increase. The calculated d_s values for all liquids examined correspond to an ellipsoidal shape.

The bubble equivalent diameter d_e of an ellipsoidal bubble was calculated by assuming a sphere of volume equal to the volume of the ellipsoidal bubble (see Eqs. (6) and (16)). Estimating the characteristic length of ellipsoidal bubbles with the same surface–to–volume ratio (the same d_s value as calculated from Wilkinson et al.'s (1994) correlation) required an iterative procedure but led to only insignificantly different values than simply identifying the equivalent diameter d_e with d_s when applying the correlations for both l and h (see Terasaka et al., 2004).

The equivalent bubble diameter d_e is needed also for the calculation of the bubble rise velocity u_B. The latter was estimated by means of the correlation of Mendelson (1967). This equation is particularly suitable for the case of ellipsoidal bubbles. Mendelson's (1967) correlation for u_B along with Wilkinson et al.'s (1994) relationship for d_s estimation were also used to calculate the bubble Reynolds number Re_B (Eq. (13)) needed for the estimation of both l and h values.

The bubble rise velocity u_B and both the bubble length l and height h of an ellipsoidal bubble take part in the calculation of the rate of surface formation R_{sf} (Higbie, 1935). For oblate ellipsoidal bubbles (see Nedeltchev et al., 2006a, b):

$$R_{sf} = \pi\sqrt{\frac{l^2+h^2}{2} - \frac{(l-h)^2}{8}}u_B \tag{21}$$

Equation (21) is needed for the calculation of the contact time t_c (Eq. 19) and thus the liquid–side mass transfer coefficient k_L (Eq. 20).

The number of bubbles N_B was deduced from the bubble residence time (aerated liquid height H divided by the bubble rise velocity u_B) and the bubble formation frequency f_B (Painmanakul et al., 2005; Nedeltchev et al., 2006a, b). The latter was expressed as volumetric flow rate Q_G divided by the bubble volume V_B (see Eqs. (3) and (6)). The interfacial area can be calculated as a function of bubble frequency f_B, bubble surface S_B, bubble rise velocity u_B and cross section (see Eq. (4)). Therefore, the theoretical k_La values for ellipsoidal bubbles can be calculated as a product of k_L value (estimated by means of Eqs. (5), (9), (10a-b), (20) and (21)) and a value (estimated by means of Eq. (4)).

The theoretical k_La values have to be multiplied with some correction factor f_c in order to match satisfactorily the experimental k_La values. Specifically, the accurate k_La values for ellipsoidal bubbles should be calculated by means of the following formula:

$$k_La = f_c\sqrt{\frac{4D_LR_{sf}}{\pi S_B}}\frac{f_BS_B}{Au_B} \tag{22}$$

The introduction of the correction factor f_c could be attributed to the fact that even when Eq. (19) is modified for oblate ellipsoidal bubbles it does not take into account the effect of the bubble wake and surface disturbances. Due to these supplementary effects on the mass transfer rate some additional term should be introduced.

Nedeltchev et al. (2006a) have illustrated that at pressures up to 1 MPa a good prediction of the experimental k_La values is achieved when the correction factor is expressed as a function only of the Eötvös number Eo:

$$f_c = 0.185Eo^{0.737} \tag{23}$$

where

$$Eo = \frac{g(\rho_L - \rho_G)d_e^2}{\sigma_L} \tag{24}$$

The Eötvös number Eo represents the gravitational force–to–surface tension force ratio. As mentioned above, the bubble shape depends on the Eo value. Eqs. (23) and (24) show that as ρ_G increases and especially as bubble size reduces, Eo and thus f_c become lower.

The above–described theoretical method was applied to conduct a systematic comparison between predicted and experimental k_La results over a wide range of physicochemical properties, which are characteristic for gas–liquid systems. Both pure organic liquids and liquid mixtures as well as tap water were considered. In such a way, the capability of the correction method to predict the numerous experimental k_La data available in the literature was assessed. Following the introduction of an additional correction term accounting for the gas density effect, the predicted k_La values at high pressures were improved by as many as 18 % (in terms of the maximum relative error).

Volumetric liquid–phase mass transfer coefficients k_La measured in 18 organic liquids, 14 liquid mixtures and tap water at 293. 2 K (Tables 1 and 2) have been considered. Results for 1–butanol, ethanol (96 %), toluene, decalin and tap water were reported by Jordan and Schumpe (2001) for pressures P of 0.1, 0.2, 0.5, 1, 2 and 4 MPa, respectively. The k_La data in the other pure and organic liquids were measured at ambient pressure by Öztürk et al. (1987). These authors presented also some k_La data in tap water. All k_La data in the present approach refer to the dispersion volume.

The bubble column used by Jordan and Schumpe (2001) had an inner diameter of 0.1 m and a height of 2.4 m. Three different gas distributors were used: perforated plate, 19 × Ø 1×10^{-3} m (D1), single hole, 1 × Ø 4.3×10^{-3} m (D2) and single hole, 1 × Ø 1×10^{-3} m (D3). The clear liquid height was set at 1.3 m. Jordan and Schumpe (2001) measured their k_La values with a dynamic oxygen desorption technique. The oxygen desorption from the liquid was traced with an optical probe (MOPS, Comte, GmbH, Hannover, Germany) based on fluorescence extinction by oxygen. By dissolving the fluorophore in the liquid, rather than fixing it to the tip of the glass fiber, an instantaneous sensor response was achieved. It is worth noting that the D_L values for oxygen were either measured (Öztürk et al., 1987) or calculated by the correlation of Schumpe and Lühring (1990).

Öztürk et al. (1987) carried out their k_La experiments in a jacketed glass bubble column of 0.095 m in ID. The clear liquid height was set at 0.85 m. A single tube of Ø 3×10^{-3} m ID was used as the gas distributor (D4). Air, nitrogen, hydrogen or helium were employed as the gas phase.

The k_La values were measured by dynamic oxygen absorption or desorption methods. The oxygen fugacity in the liquids was measured with a polarographic oxygen electrode (WTW–EO 90) inserted horizontally at half the dispersion height. The electrode response time was 3 s in water and less in most organic liquids. For absorption runs, oxygen was desorbed by sparging nitrogen. After disengagement of all nitrogen bubbles, a preadjusted air flow was fed by switching two magnet valves, and the increase in oxygen fugacity was recorded. For desorption runs, oxygen–free inert gas (nitrogen, hydrogen or helium) was sparged into air–saturated liquid.

In the present work only k_La data obtained in the homogeneous flow regime of bubble column operation were analyzed. The upper boundary (so–called transitional gas velocity) of this flow regime was determined by the formulas of Reilly et al. (1994). According to the bubble shape diagrams presented by Clift et al. (1978) and Fan and Tsuchiya (1990) the bubbles formed under all operating conditions were oblate ellipsoidals.

By means of a nonlinear regressional routine applied to the experimental k_La data the following new correction factor was derived:

$$f_c = 0.124 Eo^{0.94}\left(\frac{\rho_G}{\rho_G^{ref}}\right)^{0.15} = 0.124\left(\frac{g(\rho_L - \rho_G)d_e^2}{\sigma_L}\right)^{0.94}\left(\frac{\rho_G}{1.2}\right)^{0.15} \quad (25)$$

where ρ_G^{ref} is the reference gas density (1.2 kg m^{-3} for air at ambient conditions: 293.2 K and 0.1 MPa). 263 experimental k_La values are fitted with an average relative error of 10.4 %. It is worth noting that this correction factor f_c combines the individual corrections of both k_L and a due to the ellipsoidal shape of bubbles. The average relative error without the gas density correction term is 14.9 %.

The dimensionless gas density ratio is the additional correction term that has not been reported hitherto. Its introduction is needed because the correlation of Wilkinson et al. (1994) was tested only up to 1.5 MPa. In our work we correlate k_La data at pressures of 2 and 4 MPa that could not be fitted without this term. Moreover, the correlations of Terasaka et al. (2004) have been derived under ambient pressure. To the best of our knowledge, these equations have not been validated under high pressure. It is worth noting that such a dimensionless gas density correction factor has been also used by Krishna (2000) for correcting his correlations for large bubble rise velocity and dense–phase gas holdup. The different correction factors reported in our previous papers (Nedeltchev et al., 2006a, b) are due to both lower number of liquids considered and lower operating pressures (up to 1 MPa).

Figs. 10a–10c show the parity plots of k_La values measured in four organic liquids aerated with three different distributors (D1, D2 and D3) at pressures up to 4 MPa.

It is clear that by means of the new correction factor (Eq. (25)) the k_La values can be predicted reasonably well (within ± 20 %). It is worth noting that the k_La values obtained in

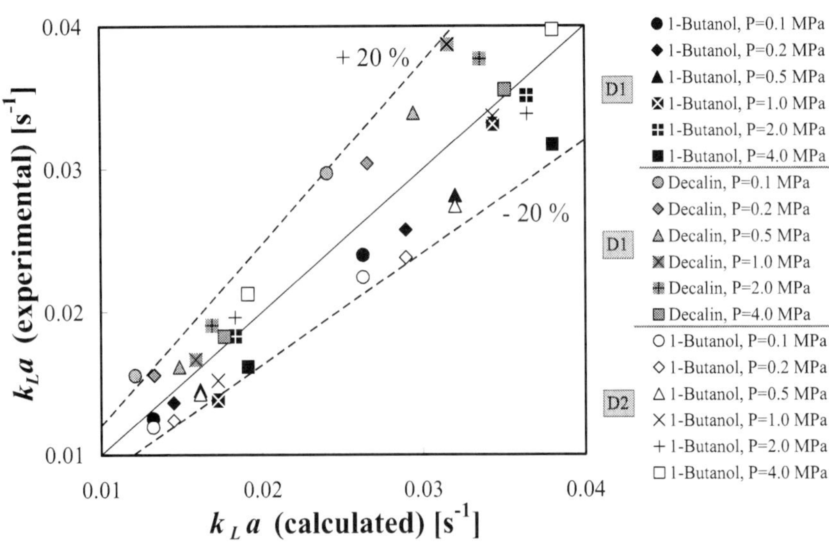

Fig. 10a. Parity plot of k_La values measured in 1–butanol and decalin at pressures up to 4 MPa (Gas distributors: D1 and D2)

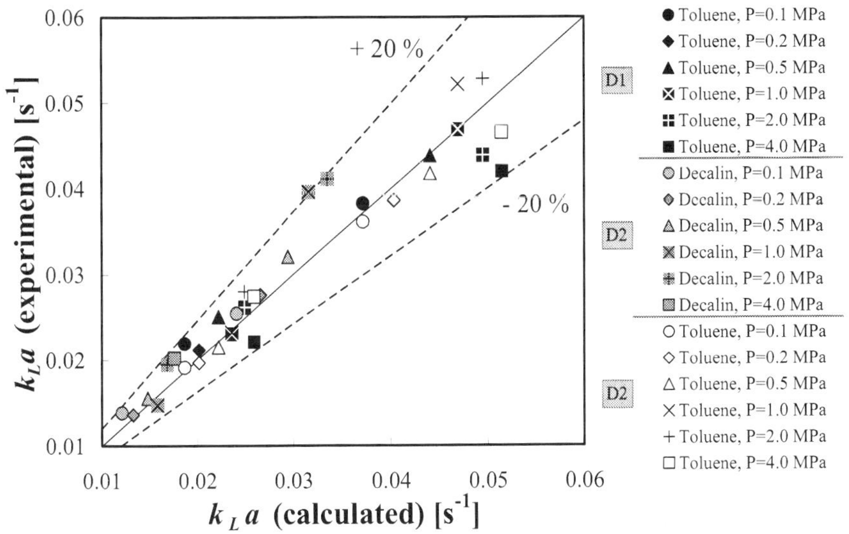

Fig. 10b. Parity plot of k_La values measured in toluene and decalin at pressures up to 4 MPa (Gas distributors: D1 and D2)

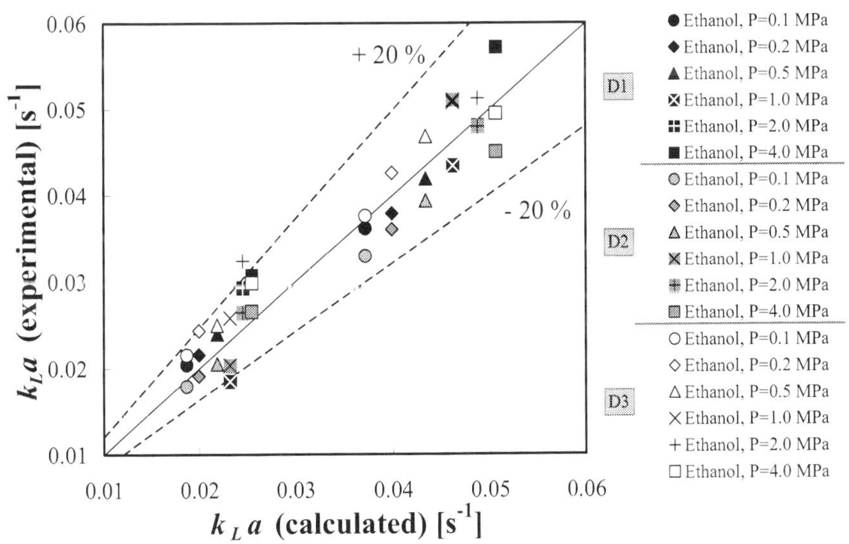

Fig. 10c. Parity plot of k_La values measured in ethanol (96 %) at pressures up to 4 MPa (Gas distributors: D1, D2 and D3)

tap water aerated with gas spargers D1 and D4 can also be predicted satisfactorily by means of Eq. (25). One of the merits of our work is the successful prediction of k_La data at very high pressure (up to 4 MPa). The presented theoretical approach predicts satisfactorily the k_La values measured by making use of four different gas distributors (D1–D4). It is known that the sparger has no substantial influence on both the gas holdup and mass transfer, provided that the diameter of the sparger holes is not too small (Wilkinson et al., 1994).

Fig. 11a exhibits that the correction method predicts satisfactorily the k_La values measured in 15 different organic liquids aerated with gas sparger D4 at ambient pressure. In the case of xylene and tetralin not only air (nitrogen) but also helium (He) and hydrogen were used. Ethanol (96 %) was also sparged with helium.

The new correction method applies not only to pure organic liquids but also to liquid mixtures. Fig. 11b shows that the predicted k_La values are in reasonable agreement with the experimental results. The keys for the different organic mixtures used are given in Table 2. All 14 organic mixtures were aerated with a gas sparger D4. As in some other parity plots, high k_La values tend to be overpredicted. This indicates the onset of coalescence at high gas velocity (flow regime transition). In Fig. 2b are included also predicted and experimental k_La values obtained in ethanol (96 %), toluene and 1–butanol aerated with helium (He) at different pressures by using different gas spargers (D2 and D3). These data demonstrate again that the developed theoretical approach along with the new correction factor (Eq. (25)) is valuable and capable of predicting numerous experimental k_La data.

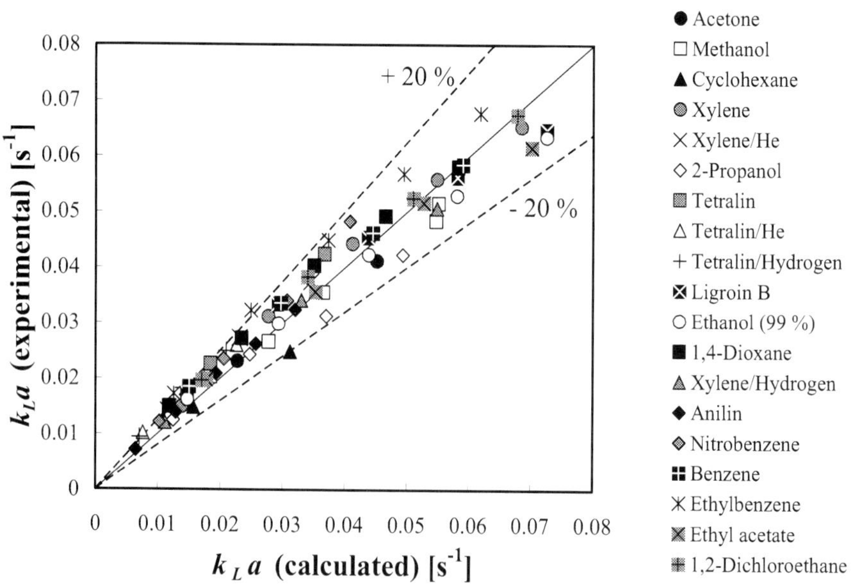

Fig. 11a. Parity plot of k_La values measured in 15 different organic liquids at ambient pressure (Gas distributor: D4)

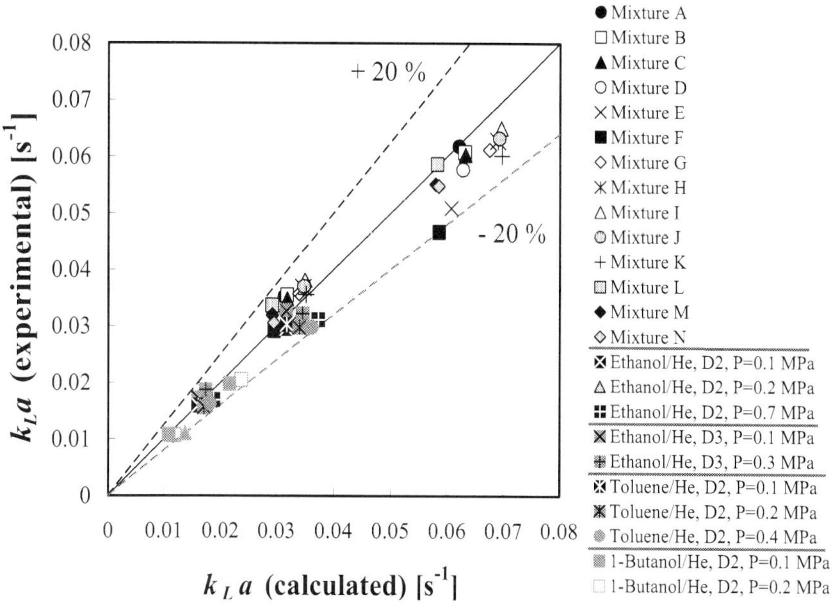

Fig. 11b. Parity plot of k_La values measured in 14 different liquid mixtures (P=0.1 MPa, sparger D4) as well as in ethanol, toluene and 1–butanol aerated with helium (He) at different pressures and gas spargers (D2, D3). The legend keys for the liquid mixtures are given in Table 2.

Fig. 12 shows the product $f_c(\rho_G/1.2)^{-0.15}$ (see Eq. (25)) as a function of Eo. This product increases with Eo for 13 organic liquids at ambient pressure. The trend holds also for different gases (air, nitrogen, helium and hydrogen). The $f_c(\rho_G/1.2)^{-0.15}$ values fall in the range of 0.3–1.1 which corresponds to Eo values from 2 to 10. It is worth noting that the f_c values for the liquids and operating conditions examined in this work are always lower than unity (0.3–0.8). In the case of the liquid mixtures specified in Table 2 the $f_c(\rho_G/1.2)^{-0.15}$ vs. Eo relationship follows the same trendline as the one shown in Fig. 12. For the sake of brevity this figure is omitted.

Nedeltchev et al. (2006a) have shown that the values of the correction factor (Eq. (23)) fall in line with the predictions of the correlation of Wellek et al. (1966) for the bubble aspect ratio E:

$$E - 1 = 0.163Eo^{0.757} \tag{26}$$

where

$$E = \frac{l}{h} \tag{27}$$

This similarity holds also for Eq. (25). Figure 13 exhibits that the $f_c(\rho_G/1.2)^{-0.15}$ trendline at Eo≤5 is very close to the predictions of Eq. (26). Beyond this Eo value the values of the product $f_c(\rho_G/1.2)^{-0.15}$ become systematically higher than (E-1) values.

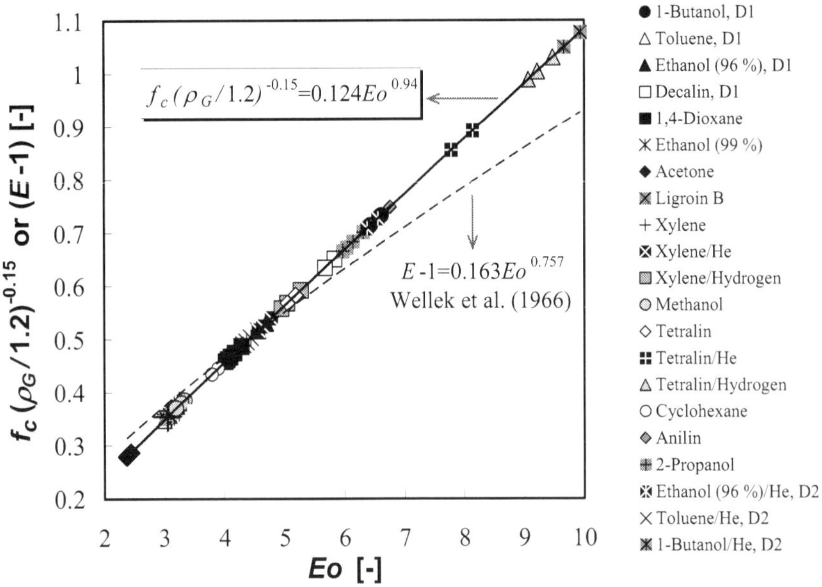

Fig. 12. Product $f_c(\rho_G/1.2)^{-0.15}$ as a function of the Eötvös number Eo for 13 different organic liquids at ambient pressure (Gas distributor: D4 unless otherwise mentioned)

It was found that the correction factors f_c gradually decrease with the increase of gas density ρ_G (operating pressure). As the superficial gas velocity u_G increases, a gradual f_c reduction is also observed. Both trends are illustrated for four organic liquids in Fig. 13. The decreasing f_c trends are explainable in terms of Eq. (25). The new correction factor f_c is primarily dependent on the Eötvös number Eo which involves the bubble equivalent diameter d_e. As u_G or ρ_G increases, the bubble size decreases (according to the correlation of Wilkinson et al. (1994)) leading to lower Eo and thus lower f_c (this trend is stronger than the effect of the gas density ratio).

Larger ellipsoidal bubbles are characterized with higher f_c values since larger wakes or vortices are formed behind them which enhance the mass transfer coefficient (see Lochiel and Calderbank, 1964; Miller, 1974) and thus the correction factor f_c becomes closer to unity. In addition, with larger ellipsoidal bubbles more surface disturbances occur which also enhance the mass transfer characteristics. Fan and Tsuchiya (1990) argue that larger ellipsoidal bubbles (larger Re_B) will have larger shedding frequency of vortex pairs from a bubble which will increase the mass transfer coefficient across the base of an ellipsoidal bubble and thus the overall liquid–phase mass transfer coefficient. It is worth noting that Eq. (20) is valid at very high bubble Reynolds numbers Re_B (Calderbank, 1967). It means that in the same liquid larger ellipsoidal bubbles will have higher Re_B and thus their correction factors f_c will be higher (closer to unity). The correction factor given in Eq. (25) supports this explanation. Lochiel and Calderbank (1964) have demonstrated that liquid drops (ethyl acetate drops in water and water drops in isobutanol) with larger diameter (higher Re_B) are characterized with higher correction factors.

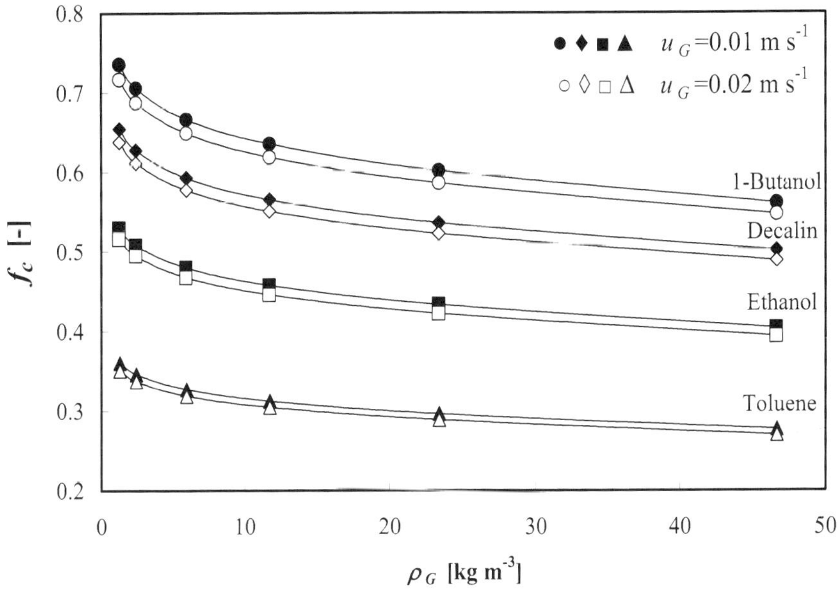

Fig. 13. Effects of both superficial gas velocity u_G and gas density ρ_G (operating pressure P) on the correction factor f_c. Gas distributor: D1–perforated plate, 19 × Ø 1×10^{-3} m, triangular pitch

The presented correction of the penetration theory is applicable in the homogeneous flow regime of bubble column operation. The upper boundary (so–called transitional gas velocity) of this flow regime was determined by the formulas of Reilly et al. (1994).

Only in the case of two other organic liquids, carbon tetrachloride and ethylene glycol (reported by Öztürk et al., 1987), the presented correction method was not capable of fitting the experimental k_La data satisfactorily. Table 1 shows that the correction method is applicable to organic liquids with densities ρ_L in the range of 714–1234 kg m^{-3} and viscosities μ_L in the range of 0.327–2.94×10^{-3} Pa s. Carbon tetrachloride has a higher liquid density (ρ_L=1593 kg m^{-3}), whereas ethylene glycol has a higher liquid viscosity (μ_L=19.94×10^{-3} Pa s).

In order to predict the k_La values (referred to liquid volume) at high temperature conditions, Nedeltchev et al. (2010) used another definition of the specific gas-liquid interfacial area a which is referred to unit liquid volume (see Lemoine et al., 2008):

$$a = \frac{6\varepsilon_G}{(1-\varepsilon_G)d_s} \qquad (28)$$

It is worth noting that experimental gas holdups were used for calculating those interfacial areas. Two different elevated temperatures T (323 and 343 K) and three different operating pressures P (0.1, 0.2 and 0.5) were considered, respectively.

In the case of ellipsoidal bubbles, it is reasonable to assume that the correction factor f_c is proportional to the interphase drag coefficient C_D and the bubble eccentricity term (E−1) (taking into account the distortion from the perfect spherical shape). According to Olmos et al. (2003) and Krishna and van Baten (2003) the drag coefficient C_D for ellipsoidal bubbles is a function of Eo:

$$C_D = \frac{2}{3}\sqrt{Eo} \tag{29}$$

The drag coefficient C_D accounts for the bubble-bubble interactions and bubble-induced turbulence. Raymond and Zieminski (1971) reported that there is a similarity in the behavior of the mass transfer coefficient and drag coefficient. The decrease of the mass transfer coefficient corresponds to an increase of the drag coefficient (due to friction drag for small bubbles or form drag for large bubbles) in the case of straight chain alcohols. According to Raymond and Zieminski (1971) the drag coefficient is proportional to the minor semi-axis of the ellipsoidal bubble. The authors reported also some dependence of the mass transfer coefficient on the bubble shape. Wellek et al. (1966) argue that the term (E−1) can be also correlated to the Eötvös number (see Eq. (26)).

Nedeltchev et al. (2010) expressed the correction factor f_c for prediction of k_La values (referred to liquid volume) as a product of both C_D and (E−1) as well as dimensionless temperature ratio:

$$f_c = B\, C_D (E-1)\left(\frac{T}{273.15}\right) \tag{30}$$

where B is proportionality constant.

The use of dimensionless temperature ratio is needed since the correlations for both C_D and (E−1) as well as Sauter-mean bubble size d_s (Eq. (15)), geometrical characteristics of the ellipsoidal bubbles (Eqs. (10a) and (10b)) and bubble rise velocity (Eq. (5)) were derived at ambient temperature and thus it is not certain whether they remain valid at elevated temperatures. It is worth noting that Cockx et al. (1995) introduced a correction term for the volumetric interfacial area which is also a function of the bubble eccentricity. Talvy et al. (2007a) have proven that the local bubble interfacial area is sensitive to the bubble shape. For bubble sizes above 3×10^{-3} m they have reported correction factors above 1.1. Talvy et al. (2007b) have shown that the ellipsoidal shape of the bubbles seems to be significant in estimation of the drag coefficient. When these correlations are improved and become applicable at various temperatures, then the dimensionless temperature ratio in Eq. (30) might become useless. The substitution of Eqs. (26) and (29) into Eq. (30) yields:

$$f_c = 0.109B\, Eo^{1.257}\left(\frac{T}{273.15}\right) \tag{31}$$

By means of a non-linear regressional analysis applied to 167 k_La values (referred to liquid volume), the constant of proportionality B was estimated as 1.183. All k_La values were successfully fitted with an average relative error of 9.3 %.

The above-mentioned approach was also applied to slurry bubble columns operated in the homogeneous regime (Nedeltchev et al., 2007b). Six different liquid-solid systems (water/activated carbon, water/aluminium oxide, tetralin/aluminium oxide, 0.8 M sodium sulfate solution/kieselguhr (diatomaceous earth), ligroin (petroleum ether)/polyethylene and ligroin/polyvinylchloride) were considered. 85 experimental k_La values were fitted with a mean error of 19 %.

5. Conclusions

New approaches for the prediction of both gas holdups and volumetric liquid-phase mass transfer coefficients in gas-liquid and slurry bubble columns operated in the homogeneous regime have been developed. In these models the shape of the bubbles plays a major role. In the model for gas holdup prediction, the bubble surface and geometrical characteristics of the bubble (its length and height) are considered. In the model for k_La prediction the same parameters along with the circumference of the ellipsoidal bubbles (taking part in the rate of surface formation) are involved. In the process of evaluation of the mass transfer coefficient, it is made clear that Higbie's (1935) equation should be corrected due to the non-spherical shape of the bubbles. New correction factors were derived. The Eötvös number Eo plays a major role in them. Mass transfer coefficients at elevated temperatures can also be predicted by this approach.
As many as 23 organic liquids, 17 liquid mixtures and 6 liquid/solid systems were tested and the two major mass transfer parameters were predicted successfully.

6. Acknowledgment

Dr. Stoyan Nedeltchev expresses his gratitude to both DAAD and the Alexander von Humboldt Foundation (Germany) for the postdoctoral fellowships that made this research work possible. The financial support of the European Commission (7th Framework Programme, Grant Agreement No. 221832) for the continuation of this work is also gratefully acknowledged.

7. Nomenclature

A	= cross–sectional area of the reactor	[m^2]
a	= specific interfacial area (referred to dispersion volume)	[m^{-1}]
B	= constant in Eqs. (30) and (31)	[–]
C_D	= drag coefficient	[–]
D_c	= column diameter	[m]
d_e	= bubble equivalent diameter	[m]
D_L	= molecular diffusivity	[$m^2 \cdot s^1$]
d_s	= Sauter–mean bubble diameter	[m]
e, E	= bubble eccentricity (9a) or (27)	[–]
Eo	= Eötvös number, Eq. (24)	[–]
f_B	= bubble formation frequency	[s^{-1}]
f_c	= correction factor	[–]
g	= gravitational acceleration	[$m \cdot s^{-2}$]
h	= height of an ellipsoidal bubble	[m]
H	= aerated liquid height	[m]
H_0	= clear liquid height	[m]
k_G	= gas-phase mass transfer coefficient	[$m \cdot s^{-1}$]
k_Ga	= volumetric gas-phase mass transfer coefficient	[s^{-1}]
k_L	= liquid-phase mass transfer coefficient	[$m \cdot s^{-1}$]
k_La	= volumetric liquid-phase mass transfer coefficient	[s^{-1}]
l	= length (width) of an ellipsoidal bubble	[m]
Mo	= Morton number, Eq. (14)	[–]

N_B = number of bubbles in the dispersion [–]
P = operating pressure [Pa]
ΔP = pressure difference between the readings of both lower (at 0 m) and upper (at 1.2 m) pressure transducers [Pa]
Q_G = gas flow rate [$m^3 \cdot s^{-1}$]
R_{sf} = rate of surface formation [$m^2 \cdot s^{-1}$]
Re = Reynolds number [–]
Re_B = bubble Reynolds number, Eq. (13) [–]
S_B = bubble surface [m^2]
Ta = Tadaki number, Eq. (12) [–]
T = operating temperature [K]
t_c = contact time [s]
u_B = bubble rise velocity [$m \cdot s^{-1}$]
u_G = superficial gas velocity [$m \cdot s^{-1}$]
u_{trans} = transition gas velocity [$m \cdot s^{-1}$]
V_B = bubble volume [m^3]
V_G = gas volume [m^3]
V_L = liquid volume [m^3]
V_{total} = dispersion volume [m^3]
We = Weber number [–]

8. Greek letters

ε_G = gas holdup [–]
μ_L = liquid viscosity [$Pa \cdot s$]
ρ_G= gas density [$kg \cdot m^{-3}$]
ρ_G^{ref} = reference gas density (air at ambient conditions) [$kg \cdot m^{-3}$]
ρ_L = liquid density [$kg \cdot m^{-3}$]
σ_L = liquid surface tension [$N \cdot m^{-1}$]

9. References

Akita, K. & Yoshida, F. (1973). Gas Holdup and Volumetric Mass Transfer Coefficient in Bubble Columns, *Ind. Eng. Chem. Process Des. Dev.*, Vol. 12, 76-80

Akita, K. & Yoshida, F. (1974). Bubble Size, Interfacial Area and Liquid-Phase Mass Transfer Coefficient in Bubble Column, *Ind. Eng. Chem. Process Des. Dev.*, Vol. 13, 84-91

Akita, K. (1987a). Effect of Electrolyte on Mass Transfer Characteristics in Gas Bubble Column, *Kagaku Kogaku Ronbunshu,* Vol. 13, 166-172

Akita, K. (1987b). Effect of Trace Alcohol on Mass Transfer in Gas Bubble Column, *Kagaku Kogaku Ronbunshu,* Vol. 13, 181-187

Angelino, H. (1966). Hydrodynamique des grosses bulles dans les liquides visqueux, *Chem. Eng. Sci.*, Vol. 21, 541-550

Astarita, G. & Apuzzo, G. (1965). Motion of Gas Bubbles in Non-Newtonian Liquids, *AIChE J.*, Vol. 11, 815-820

Bach, H. F. & Pilhofer, T. (1978). Variation of Gas Holdup in Bubble Columns with Physical Properties of Liquids and Operating Parameters of Columns, *Ger. Chem. Eng.*, Vol. 1, 270

Baird, M. H. I. & Davidson J. F. (1962). Gas Absorption by Large Rising Bubbles, *Chem. Eng. Sci.*, Vol. 17, 87-93

Barnett, S. M.; Humphrey A. E. & Litt, M. (1966). Bubble Motion and Mass Transfer in Non-Newtonian Fluids, *AIChE J.*, Vol. 12, 253-259

Behkish, A.; Lemoine, R., Sehabiague, L., Oukaci, R. & Morsi, B. I. (2007). Gas Holdup and Bubble Size Behavior in a Large-Scale Slurry Bubble Column Reactor Operating with an Organic Liquid Under Elevated Pressures and Temperatures, *Chem. Eng. J.*, Vol. 128, 69-84

Botton, R.; Cosserat, D. & Charpentier, J. C. (1980). Mass Transfer in Bubble Columns Operating at High Gas Throughputs, *Chem. Eng. J.*, Vol. 20, 87-94

Calderbank, P. H. (1958). Physical Rate Process in Industrial Fermentation. Part I: The Interfacial Area in Gas-Liquid Contacting with Mechanical Agitation, *Trans. Inst. Chem. Engrs.*, Vol. 36, 443-463

Calderbank, P. H. & Moo-Young, M. B. (1961). The Continuous-Phase Heat and Mass-Transfer Properties of Dispersions, *Chem. Eng. Sci.*, Vol. 16, 39-54

Calderbank, P. H. & Lochiel A. C. (1964). Mass Transfer Coefficients, Velocities and Shapes of Carbon Dioxide Bubbles in Free Rise Through Distilled Water, *Chem. Eng. Sci.*, Vol. 19, 485-503

Calderbank, P. H. & Patra R. P. (1966). Mass Transfer in the Liquid Phase During the Formation of Bubbles, *Chem. Eng. Sci., Vol.* 21, 719-721

Calderbank, P. H. (1967). Gas Absorption from Bubbles. *Trans. Instn. Chem. Engrs.*, Vol. 45, CE209-CE233.

Calderbank, P. H.; Johnson, D. S. L. & Loudon, J. (1970). Mechanics and Mass Transfer of Single Bubbles in Free Rise Through Some Newtonian and Non-Newtonian Liquids, *Chem. Eng. Sci.*, Vol. 25, 235-256

Chen, R. C.; Reese, J. & Fan, L.-S. (1994). Flow Structure in a Three-Dimensional Bubble Column and Three-Phase Fluidized Bed, *AIChE J.*, Vol. 40, 1093-1104

Cho, J. S. & Wakao, N. (1988). Determination of Liquid-Side and Gas-Side Volumetric Mass Transfer Coefficients in a Bubble Column, *J. Chem. Eng. Japan*, Vol. 21, 576-581

Clift, R.; Grace, J. R. & Weber, M. (1978). *Bubbles, Drops and Particles*, Academic Press, New York, U. S. A.

Cockx, A.; Roustan, M., Line, A. & Hebrard, G. (1995). Modelling of Mass Transfer Coefficient k_L in Bubble Columns, *Trans. Instn. Chem. Engrs.*, Vol. 73(A), 617-631

Davenport, W. G.; Richardson, F. D. & Bradshaw, A. V. (1967). Spherical Cap Bubbles in Low Density Liquids, *Chem. Eng. Sci.*, Vol. 22, 1221-1235

Davidson, J. F. & Harrison, D. (1963). *Fluidised Particles*, Cambridge University Press, UK.

Davies, R. M. & Taylor, G. I. (1950). *Proc. Roy. Soc.*, Vol. A200, 375

Deckwer, W.-D. (1980). On the Mechanism of Heat Transfer in Bubble Column Reactors. *Chem. Eng. Sci.*, Vol. 35, 1341-1346.

Deckwer, W.-D. (1992). *Bubble Column Reactors*, John Wiley and Sons, Chichester, UK.

Fan, L.-S. & Tsuchiya, K. (1990). *Bubble Wake Dynamics in Liquids and Liquid–Solid Suspensions*, Butterworth–Heinemann Series in Chemical Engineering, Stoneham, U.S.A.

Fukuma, M.; Muroyama, K. & Yasunishi, A. (1987a). Properties of Bubble Swarm in a Slurry Bubble Column, *J. Chem. Eng. Japan,* Vol. 20, 28-33

Fukuma, M.; Muroyama, K. & Yasunishi, A. (1987b). Specific Gas-Liquid Interfacial Area and Liquid-Phase Mass Transfer Coefficient in a Slurry Bubble Column, *J. Chem. Eng. Japan,* Vol. 20, 321-324

Gandhi, A. B.; Joshi, J. B., Jayaraman, V. K. & Kulkarni, B. D. (2007). Development of Support Vector Regression (SVR)-Based Correlation for Prediction of Overall Gas Hold-Up in Bubble Column Reactors for Various Gas–Liquid Systems, *Chem. Eng. Sci.* Vol. 62, 7078-7089

Godbole, S. P.; Honath M. F. & Shah, Y. T. (1982). Holdup Structure in Highly Viscous Newtonian and Non-Newtonian Liquids in Bubble Columns, *Chem. Eng. Commun.,* Vol. 16, 119-134

Godbole, S. P.; Schumpe, A., Shah, Y. T. & Carr, N. L. (1984). Hydrodynamics and Mass Transfer in Non-Newtonian Solutions in a Bubble Column, *AIChE J.,* Vol. 30, 213-220

Griffith, R. M. (1960). Mass Transfer from Drops and Bubbles, *Chem. Eng. Sci.,* Vol. 12, 198-213

Grund, G.; Schumpe, A. & Deckwer, W.-D. (1992). Gas-Liquid Mass Transfer in a Bubble Column with Organic Liquids, *Chem. Eng. Sci.,* Vol. 47, 3509-3516

Hammer, H.; Schrag, H., Hektor, K., Schönau, K., Küsters, W., Soemarno, A., Sahabi U. & Napp, W. (1984). New Subfunctions on Hydrodynamics, Heat and Mass Transfer for Gas/Liquid and Gas/Liquid/Solid Chemical and Biochemical Reactors, *Front. Chem. Reac. Eng.,* 464-474

Hammerton, D. & Garner, F. H. (1954). *Trans. Instn. Chem. Engrs.,* Vol. 32, S18

Hayashi, T.; Koide K. & Sato, T. (1975). Bubbles Generated from Porous Plate in Organic Solutions and Aqueous Solutions of Organic Substances, *J. Chem. Eng. Japan,* Vol. 8, 16-20

Higbie, R. (1935). The Rate of Absorption of Pure Gas into a Still Liquid During Short Periods of Exposure, *Trans. AIChE,* Vol. 31, 365-389

Hikita, H.; Asai, S., Tanigawa, K., Segawa, K. & Kitao, M. (1980). Gas Holdup in Bubble Columns, *Chem. Eng. J.,* Vol. 20, 59-67

Hikita, H.; Asai, S., Tanigawa, K., Segawa, K. & Kitao, M. (1981). The Volumetric Liquid-Phase Mass Transfer Coefficient in Bubble Columns, *Chem. Eng. J.,* Vol. 22, 61

Hughmark, G. A. (1967). Holdup and Mass Transfer in Bubble Columns, *Ind. Eng. Chem. Process Des. Dev.,* Vol. 6, 218-220

Idogawa, K.; Ikeda, K., Fukuda, T. & Morooka, S. (1985a). Behavior of Bubbles in a Bubble Column under High Pressure for Air-Water System, *Kagaku Kogaku Ronbunshu,* Vol. 11, 253-258

Idogawa, K.; Ikeda K., Fukuda, T. & Morooka, S. (1985b). Effects of Gas and Liquid Properties on the Behavior of Bubbles in a Bubble Column under High Pressure, *Kagaku Kogaku Ronbunshu,* Vol. 11, 432-437

Idogawa, K.; Ikeda, K., Fukuda, T. & Morooka, S. (1987). Effect of Gas and Liquid Properties on the Behaviour of Bubbles in a Column Under High Pressure, *Intern. Chem. Eng.,* Vol. 27, 93-99

Jiang, P.; Lin, T.-J., Luo, X. & Fan, L.-S. (1995). Flow Visualization of High Pressure (21 MPa) Bubble Columns: Bubble Characteristics, *Trans. Inst. Chem. Eng.,* Vol. 73, 269-274

Jordan, U. & Schumpe, A. (2001). The Gas Density Effect on Mass Transfer in Bubble Columns with Organic Liquids, *Chem. Eng. Sci.*, Vol. 56, 6267-6272

Kastanek, F. (1977). The Volume Mass Transfer Coefficient in a Bubble Bed Column, *Collect. Czech. Chem. Commun.*, Vol. 42, 2491-2497

Kastanek, F.; J. Zahradnik, Kratochvil J. & Cermak, J. (1993). *Chemical Reactors for Gas-Liquid Systems*, Ellis Horwood, Chichester, West Sussex, UK

Kawase, Y. & Moo-Young, M. (1986). Influence of Non-Newtonian Flow Behaviour on Mass Transfer in Bubble Columns with and without Draft Tubes, *Chem. Eng. Commun.*, Vol. 40, 67-83

Kawase, Y.; Halard, B. & Moo-Young, M. (1987). Theoretical Prediction of Volumetric Mass Transfer Coefficients in Bubble Columns for Newtonian and Non-Newtonian Fluids, *Chem. Eng. Sci.*, Vol. 42, 1609-1617

Kelkar, B. G.; Godbole, S. P., Honath, M. F., Shah, Y. T., Carr, N. L. & Deckwer, W.-D. (1983). Effect of Addition of Alcohols on Gas Holdup and Backmixing in Bubble Columns, *AIChE J.*, Vol. 29, 361-369

Khare, A. S. & Joshi, J. B. (1990). Effect of Fine Particles on Gas Holdup in Three-Phase Sparged Reactors, *Chem. Eng. J.*, Vol. 44, 11

Kiambi, S. L.; Duquenne, A. M., Bascoul, A. & Delmas, H. (2001). Measurements of Local Interfacial Area: Application of Bi-Optical Fibre Technique, *Chem. Eng. Sci.*, Vol. 56, 6447-6453

Koide, K.; Hirahara, T. & Kubota, H. (1966). Average Bubble Diameter, Slip Velocity and Gas Holdup of Bubble Swarms, *Kagaku Kogaku*, Vol. 30, 712-718

Koide, K.; Kato, S., Tanaka, Y. & Kubota, H. (1968). Bubbles Generated from Porous Plate, *J. Chem. Eng. Japan*, Vol. 1, 51-56

Koide, K.; Takazawa, A., Komura M. & Matsunaga, H. (1984). Gas Holdup and Volumetric Liquid-Phase Mass Transfer Coefficient in Solid-Suspended Bubble Columns, *J. Chem. Eng. Japan*, Vol. 17, 459-466

Koide, K. (1996). Design Parameters of Bubble Column Reactors With and Without Solid Suspensions, *J. Chem. Eng. Japan*, Vol. 29, 745-759

Kolmogoroff, A. N. (1941). *Dokl. Akad. Nauk SSSR*, Vol. 30, 301

Krishna, R.; Wilkinson, P. M. & Van Dierendonck, L. L. (1991). A Model for Gas Holdup in Bubble Columns Incorporating the Influence of Gas Density on Flow Regime Transitions, *Chem. Eng. Sci.*, Vol. 46, 2491-2496

Krishna, R. (2000). A Scale-Up Strategy for a Commercial Scale Bubble Column Slurry Reactor for Fischer-Tropsch Synthesis, Oil and Gas Science and Techn.-Rev. IFP, Vol. 55, 359-393

Krishna, R. & van Baten, J. M. (2003). Mass Transfer in Bubble Columns, Catalysis Today 79-80, 67-75

Kulkarni, A.; Shah, Y. T. & Kelkar, B. G. (1987). Gas Holdup in Bubble Column with Surface-Active Agents: a Theoretical Model, *AIChE J.*, Vol. 33, 690-693

Kumar, A.; Degaleesan, T. T., Laddha, G. S. & Hoelscher, H. E. (1976). Bubble Swarm Characteristics in Bubble Columns, *Can. J. Chem. Eng.*, Vol. 54, 503-508

Lemoine, R.; Behkish, A., Sehabiague, L., Heintz, Y. J., Oukaci, R. & Morsi, B. I. (2008). An Algorithm for Predicting the Hydrodynamic and Mass Transfer Parameters in Bubble Column and Slurry Bubble Column Reactors, Fuel Proc. Technol., Vol. 89, 322-343

Leonard, J. H. & Houghton, G. (1961). *Nature* (London), Vol. 190, 687

Leonard, J. H. & Houghton, G. (1963). Mass Transfer and Velocity of Rise Phenomena for Single Bubbles, *Chem. Eng. Sci.*, Vol. 18, 133-142

Lochiel, C. & Calderbank, P. H. (1964). Mass Transfer in the Continuous Phase Around Axisymmetric Bodies of Revolution, *Chem. Eng. Sci.*, Vol. 19, 471-484

Lucas, D.; Prasser, H.-M. & Manera, A. (2005). Influence of the Lift Force on the Stability of a Bubble Column, Chem. Eng. Sci., Vol. 60, 3609-3619

Marrucci, G. (1965). Rising Velocity of a Swarm of Spherical Bubbles, *Ind. Eng. Chem. Fund.*, Vol. 4, 224-225

Mendelson, H. D. (1967). The Prediction of Bubble Terminal Velocities from Wave Theory, *AIChE J.*, Vol. 13, 250-253

Merchuk, J. C. & Ben-Zvi, S. (1992). A Novel Approach to the Correlation of Mass Transfer Rates in Bubble Columns with Non-Newtonian Liquids, *Chem. Eng. Sci.*, Vol. 47, 3517-3523

Metha, V. D. & Sharma, M. M. (1966). Effect of Diffusivity on Gas-Side Mass Transfer Coefficient, *Chem. Eng. Sci.*, Vol. 21, 361-365

Miller, D. N. (1974). Scale-Up of Agitated Vessels Gas-Liquid Mass Transfer, *AIChE J.*, Vol. 20, 445-453

Miyahara, T. & Hayashi, T. (1995). Size of Bubbles Generated from Perforated Plates in Non-Newtonian Liquids, *J. Chem. Eng. Japan*, Vol. 28, 596-600

Muller, F. L. & Davidson J. F. (1992). On the Contribution of Small Bubbles to Mass Transfer in Bubble Columns Containing Highly Viscous Liquids, *Chem. Eng. Sci.*, Vol. 47, 3525-3532

Nakanoh, M. & Yoshida, F. (1980). Gas Absorption by Newtonian and Non-Newtonian Liquids in a Bubble Column, *Ind. Eng. Chem. Process Des. Dev.*, Vol. 19, 190-195

Nedeltchev, S.; Jordan, U. & Schumpe, A. (2006a). Correction of the Penetration Theory Applied to the Prediction of k_La in a Bubble Column with Organic Liquids, *Chem. Eng. Tech.*, Vol. 29, 1113-1117

Nedeltchev, S.; Jordan, U. & Schumpe, A. (2006b). A New Correction Factor for Theoretical Prediction of Mass Transfer Coefficients in Bubble Columns, *J. Chem. Eng. Japan*, Vol. 39, 1237-1242

Nedeltchev, S.; Jordan, U. & Schumpe, A. (2007a). Correction of the Penetration Theory Based On Mass-Transfer Data from Bubble Columns Operated in the Homogeneous Regime Under High Pressure, *Chem. Eng. Sci.*, Vol. 62, 6263-6273

Nedeltchev, S. & Schumpe, A. (2007b). Theoretical Prediction of Mass Transfer Coefficients in a Slurry Bubble Column Operated in the Homogeneous Regime, *Chem. & Biochem. Eng. Quarterly*, Vol. 21, 327-334

Nedeltchev, S. & Schumpe, A. (2008). A New Approach for the Prediction of Gas Holdup in Bubble Columns Operated Under Various Pressures in the Homogeneous Regime, *J. Chem. Eng. Japan*, Vol. 41, 744-755

Nedeltchev, S.; Jordan U. & Schumpe, A. (2010). Semi-Theoretical Prediction of Volumetric Mass Transfer Coefficients in Bubble Columns with Organic Liquids at Ambient and Elevated Temperatures, *Can. J. Chem. Eng.*, Vol. 88, 523-532

Olmos, E., Gentric, C. & Midoux, N. (2003). Numerical Description of Flow Regime Transitions in Bubble Column Reactors by a Multiple Gas Phase Model, *Chem. Eng. Sci.*, Vol. 58, 2113-2121

Otake, T.; Tone, S., Nakao, K. & Mitsuhashi, Y. (1977). Coalescence and Breakup of Bubbles in Liquids, *Chem. Eng. Sci.*, Vol. 32, 377-383

Öztürk, S; Schumpe, A. & Deckwer, W.-D. (1987). Organic Liquids in a Bubble Column: Holdups and Mass Transfer Coefficients, *AIChE J.*, Vol. 33, 1473-1480

Painmanakul, P.; Loubière, K., Hébrard, G., Mietton-Peuchot, M. & Roustan, M. (2005). Effect of Surfactants on Liquid–Side Mass Transfer Coefficients, *Chem. Eng. Sci.*, Vol. 60, 6480-6491

Pošarac, D. & Tekić, M. N. (1987). Gas Holdup and Volumetric Mass Transfer Coefficient in Bubble Columns with Dilute Alcohol Solutions, *AIChE J.*, Vol. 33, 497–499

Raymond, D. R. & Zieminski, S. A. (1971). Mass Transfer and Drag Coefficients of Bubbles Rising in Dilute Aqueous Solutions. *AIChE J.*, Vol. 17, 57-65

Redfield, J. A. & Houghton, G. (1965). Mass Transfer and Drag Coefficients for Single Bubbles at Reynolds Numbers of 0.02-5000, *Chem. Eng. Sci., Vol.* 20, 131-139

Reilly, I. G.; Scott, D. S., de Bruijn, T. J. W., Jain, A. K. & Piskorz, J. (1986). Correlation for Gas Holdup in Turbulent Coalescing Bubble Columns, *Can. J. Chem. Eng.* 64, 705-717

Reilly, I. G.; Scott, D. S., De Bruijn, T. J. W. & MacIntyre, D. (1994). The Role of Gas Phase Momentum in Determining Gas Holdup and Hydrodynamic Flow Regimes in Bubble Column Operations, *Can. J. Chem. Eng.*, Vol. 72, 3-12

Sada, E.; Kumazawa, H., Lee, E. & Fujiwara, N. (1985). Gas-Liquid Mass Transfer Characteristics in Bubble Columns with Suspended Sparingly Soluble Fine Particles, *Ind. Eng. Chem. Process Des. Dev.*, Vol. 24, 255-261

Sada, E.; Kumazawa, H., Lee, E. & Iguchi, T. (1986). Gas Holdup and Mass Transfer Characteristics in a Three-Phase Bubble Column, *Ind. Eng. Chem. Process Des. Dev.*, Vol. 25, 472-476

Salvacion, J. L.; Murayama, M., Ohtaguchi, K. & Koide, K. (1995). Effects of Alcohols on Gas Holdup and Volumetric Liquid-Phase Mass Transfer Coefficient in Gel-Particle Suspended Bubble Column," *J. Chem. Eng. Japan,* Vol. 28, 434-442

Sauer, T. & Hempel, D.-C. (1987). Fluid Dynamics and Mass Transfer in a Bubble Column with Suspended Particles, *Chem. Eng. Technol., Vol.* 10, 180-189

Schumpe, A. & Deckwer, W.-D. (1987). Viscous Media in Tower Bioreactors: Hydrodynamic Characteristics and Mass Transfer Properties, *Bioprocess Eng.*, Vol. 2, 79-94

Schumpe, A.; Saxena, A. K. & Fang, L. K. (1987). Gas/Liquid Mass Transfer in a Slurry Bubble Column, *Chem. Eng. Sci.*, Vol. 42, 1787-1796

Schumpe, A. & Lühring, P. (1990). Oxygen Diffusivities in Organic Liquids at 293.2 K, J. Chem. and Eng. Data, Vol. 35, 24-25

Schügerl, K.; Lucke, J. & Oels, U. (1977). Bubble Column Bioreactors, *Adv. Biochem. Eng.*, Vol. 7, 1-84

Shah, Y. T.; Kelkar, B. G. & Deckwer, W.-D. (1982). Design Parameters Estimation for Bubble Column Reactors, *AIChE J.*, Vol. 28, 353-379

Suh, I.-S.; Schumpe, A., Deckwer, W.-D. & Kulicke, W.-M. (1991). Gas-Liquid Mass Transfer in the Bubble Column with Viscoelastic Liquid, *Can. J. Chem. Eng.*, Vol. 69, 506-512

Sun, Y. & Furusaki, S. (1989). Effect of Intraparticle Diffusion on the Determination of the Gas-Liquid Volumetric Oxygen Transfer Coefficient in a Three-Phase Fluidized Bed Containing Porous Particles, *J. Chem. Eng. Japan*, Vol. 22, 556-559

Syeda, S. R.; Afacan, A. & Chuang, K. T. (2002). Prediction of Gas Hold–Up in a Bubble Column Filled with Pure and Binary Liquids, *Can. J. Chem. Eng.*, Vol. 80, 44-50

Tadaki, T. & Maeda, S. (1961). On Shape and Velocity of Single Air Bubble Rising in Various Liquids, *Kagaku Kogaku*, Vol. 25, 254-264

Tadaki, T. & Maeda, S. (1963). The Size of Bubbles from Single Orifice, *Kagaku Kogaku*, Vol. 27, 147-155

Talvy, S.; Cockx, A. & Line, A. (2007a). Modeling of Oxygen Mass Transfer in a Gas–Liquid Airlift Reactor, AIChE J., Vol. 53, 316-326

Talvy, S.; Cockx, A. & Line, A. (2007b). Modeling Hydrodynamics of Gas–Liquid Airlift Reactor," AIChE J., Vol. 53, 335-353

Terasaka, K.; Inoue, Y., Kakizaki, M. & Niwa, M. (2004). Simultaneous Measurement of 3–Dimensional Shape and Behavior of Single Bubble in Liquid Using Laser Sensors, *J. Chem. Eng. Japan*, Vol. 37, 921-926

Timson, W. J. & Dunn, C. J. (1960). Mechanism of Gas Absorption from Bubbles Under Shear, *Ind. & Eng. Chem.*, Vol. 52, 799-802

Tsuchiya, K. & Nakanishi, O. (1992). Gas Holdup Behavior in a Tall Bubble Column with Perforated Plate Distributors, *Chem. Eng. Sci.*, Vol. 47, 3347-3354

Ueyama, K.; Morooka, S., Koide, K., Kaji, H. & Miyauchi, T. (1980). Behavior of Gas Bubbles in Bubble Columns, *Ind. Eng. Chem. Process Des. Dev.*, Vol. 19, 592-599

Wellek, R. M., Agrawal, A. K. & Skelland, A. H. P. (1966). Shape of Liquid Drops Moving in Liquid Media, AIChE J., Vol. 12, 854-862

Wilkinson, P. M. & van Dierendonck, L. L. (1990). Pressure and Gas Density Effects on Bubble Breakup and Gas Holdup in Bubble Columns, *Chem. Eng. Sci.*, Vol. 45, 2309-2315

Wilkinson, P. M.; Spek A. P. & Van Dierendonck, L. L. (1992). Design Parameters Estimation for Scale-Up of High-Pressure Bubble Columns, *AIChE J.*, Vol. 38, 544-554

Wilkinson, P. M.; Haringa, H. & Van Dierendonck, L. L. (1994). Mass Transfer and Bubble Size in a Bubble Column under Pressure, *Chem. Eng. Sci.*, Vol. 49, 1417-1427

Yamashita, F.; Mori Y. & Fujita, S. (1979). Sizes and Size Distributions of Bubbles in a Bubble Column, *J. Chem. Eng. Japan*, Vol. 12, 5-9

Yasunishi, A.; Fukuma, M. & Muroyama, K. (1986). Hydrodynamics and Gas-Liquid Mass Transfer Coefficient in a Slurry Bubble Column with High Solid Content, *Kagaku Kogaku Ronbunshu*, Vol. 12, 420-426

Zieminski, S. A. & Raymond, D. R. (1968). Experimental Study of the Behaviour of Single Bubbles, *Chem. Eng. Sci.*, Vol. 23, 17-28

19

Influence of Mass Transfer and Kinetics on Biodiesel Production Process

Ida Poljanšek and Blaž Likozar
University of Ljubljana
Slovenia

1. Introduction

Biodiesel is produced by the transesterification of large branched triglycerides (TG) (usually vegetable oils) into smaller, generally straight-chain molecules of alkyl (most often methyl) esters in the presence of a catalyst. Di- and monoglycerides (DG and MG) are intermediates and glycerol (G) is the side product. The three reactions are consecutive and reversible.

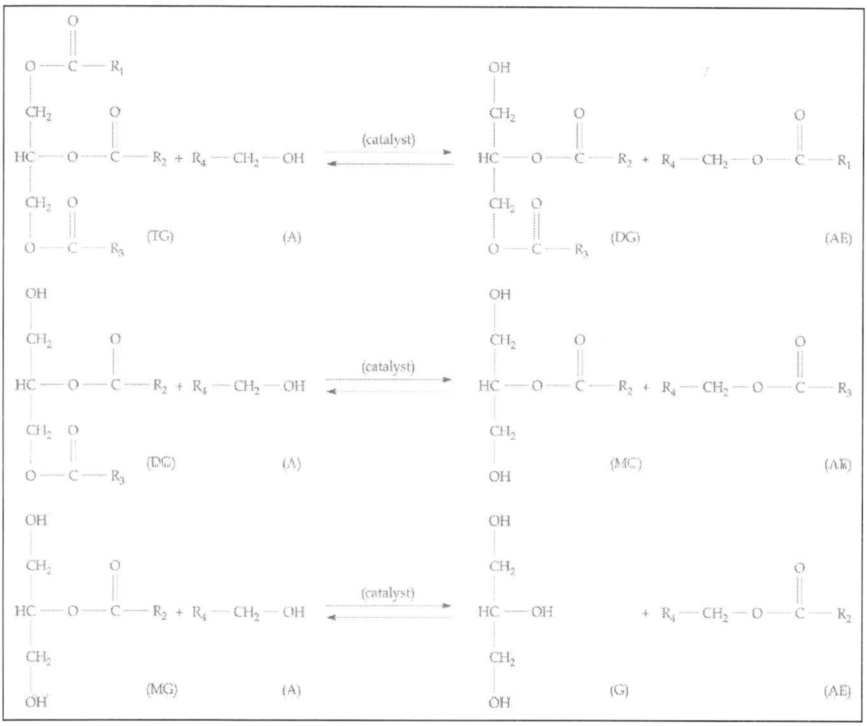

Fig. 1. Reaction scheme of triglyceride transesterification to glycerol and alkyl ester

2. Mass transfer-determined rate of biodiesel production process

2.1 Batch reactors

The reaction system in a batch reactor may be considered as a pseudo-homogeneous one with no mass transfer limitations (Marjanovic et al., 2010). Nonetheless, a reaction mechanism consisting of an initial mass transfer-controlled region followed by a kinetically controlled region is generally proposed (Noureddini & Zhu, 1997). Recently, there is an increased interest in new technologies related to mass transfer enhancement (Leung et al., 2010). Biodiesel production process may be catalyzed by acids and bases, and these influence mass transfer in a batch reactor. Lewis acid catalysts are active for both esterification and transesterification, but the reaction is very slow due to mass transfer limitations between methanol and oil phase (Hou et al., 2007). Experiments may be conducted at ambient temperature to study mass transfer limitations, indicated by the presence of a triglyceride induction period, during the acid-catalyzed transesterification reaction (Ataya et al., 2008b). The immiscibility of methanol and vegetable oil leads to a mass transfer resistance in the transesterification of vegetable oil (Guan et al., 2009). Likewise, the use of meso-structured supports is shown as a factor improving the catalytic performance as compared with macro-porous sulfonic acid-based resins, likely due to an enhancement of the mass transfer rates of large molecules, such as triglycerides, within the catalyst structure (Melero et al., 2010). However, acid exchange resins deactivation in the esterification of free fatty acids is always present in the system (Tesser et al., 2010).

The conventional base-catalyzed transesterification is characterized by slow reaction rates at both initial and final reaction stages limited by mass transfer between polar methanol/glycerol phase and non-polar oil phase (Zhang et al., 2009). If using specific catalysts, the homogeneous single phase is formed at 3:1 methanol to oil molar ratio and the mass transfer resistance between the methanol/triglyceride phases disappears (Tsuji et al., 2009). However, the methanol in the system is not effectively used for the reaction due to interface mass transfer resistance (Kai et al., 2010). Meanwhile, the process model indicates that the transesterification reaction is controlled by both mass transfer and reaction (Liu et al., 2010). Transesterification was performed in a 30 L reactor by Sengo et al. (2010), under previously optimized conditions and a yield of 88% fatty acid methyl esters was obtained after 90 min of reaction time, due to mass transfer limitations. Thus, the transesterification reaction is initially mass transfer-limited because the two reactants are immiscible with each other, and later because the glycerol phase separates together with most of the catalyst (sodium or potassium methoxide) (Cintas et al., 2010).

The sigmoid kinetics of the process is explained by the mass transfer controlled region in the initial heterogeneous regime, followed by the chemical reaction-controlled region in the pseudo-homogenous regime. The mass transfer is related to the drop size of the dispersed (methanol) phase, which reduces rapidly with the progress of the methanolysis reaction (Stamenkovic et al., 2008). It is observed that droplet size has a major influence on reaction end point and that the reaction is mass transfer-limited. This observation is confirmed by developing a mass transfer-based reaction model using the data from the batch reactor which agrees with results from other researchers (Slinn & Kendall, 2009). Biodiesel fuel yields increase with the addition of sodium dodecyl sulfonate as surface active agent because the mass transfer rates of protons and methanol to the oil phase through the oil−methanol interface are increased with increasing interfacial area (Furukawa et al., 2010).

Analogously may be reasoned when a solid catalyst is present in the process. The sigmoid process kinetics is explained by the initial triglyceride mass transfer controlled region, followed by the chemical reaction controlled region in the latter reaction period. The triglyceride mass transfer limitation is due to the small available active specific catalyst surface, which is mainly covered by adsorbed molecules of methanol. In the later phase, the adsorbed methanol concentration decreases, causing the increase of both the available active specific catalyst surface and the triglyceride mass transfer rate, and the chemical reaction rate becomes smaller than the triglyceride mass transfer rate (Veljkovic et al., 2009). A kinetic model can also be expressed as three significant controlled regions, i.e., a mass transfer-controlled region in the internal surface of a heterogeneous catalyst, an irreversible chemical reaction-controlled region in the pseudo-homogenous fluid body and a reversible equilibrium chemical reaction-controlled region near to the transesterification equilibrium stage (Huang et al., 2009). The methanolysis process using calcium hydroxide catalyst is also shown to involve the initial triglyceride mass transfer-controlled region, followed by the chemical reaction controlled region in the later period. The triglyceride mass transfer limitation is caused by the low available active specific catalyst surface due to the high adsorbed methanol concentration. Both the triglyceride mass transfer and chemical reaction rates increase with increasing the catalyst amount (Stamenkovic et al., 2010).

Influence of mass transfer on the production of biodiesel may be observed through mixing variation as the use of different mixing methods (magnetic stirrer, ultrasound and ultra-turrax) results in different conversions for the transesterification of rape oil with methanol in both acidic and basic systems (Lifka & Ondruschka, 2004a; Lifka & Ondruschka, 2004b). A reaction mechanism for sunflower oil is proposed involving an initial region of mass transfer control followed by a second region of kinetic control. The initial mass transfer-controlled region is not significant using 600 rpm (Vicente et al., 2005). The mechanism of *Brassica carinata* oil methanolysis also involves an initial stage of mass transfer control, followed by a second region of kinetic control. However, the initial mass transfer-controlled step is negligible using an impeller speed of at least 600 rpm (Vicente et al., 2006). In the case of crude sunflower oil, mass transfer limitation is effectively minimized at agitation speeds of 400–600 rpm with no apparent lag period (Bambase et al., 2007).

Optimization of mechanical agitation and evaluation of the mass transfer resistance is essential in the oil transesterification reaction for biodiesel production. The KOH-catalyzed transesterification of sunflower oil with methanol was studied by Frascari et al. (2009) in batch conditions in a 22 L stirred reactor in order to develop criteria for the energetic optimization of mechanical agitation in the biodiesel synthesis reaction, obtain preliminary information on the decantation of the reaction products, and evaluate the influence of the mass transfer resistance under different mixing conditions. An evaluation of the reaction and mass transfer characteristic times shows that the optimized tests are characterized by a not negligible mass transfer resistance (Frascari et al., 2009).

The tests conducted with one single static mixer at a 1.3 m/s superficial velocity (Reynolds number, Re = 1490) result in a profile of sunflower oil conversion versus time equivalent to that obtained in the best-performing test with mechanical agitation, indicating the attainment of a reaction run not affected by mass transfer limitations. In an evaluation of the energy requirement for the attainment of the alcohol/oil dispersion, the static mixer tests perform better than those with mechanical agitation (17 vs. 35 J/kg of biodiesel, in the reaction conditions without mass transfer constraints) (Frascari et al., 2008).

In the case of increasing ultrasound intensity, the observed mass transfer and kinetic rate enhancements are due to the increase in interfacial area and activity of the microscopic and macroscopic bubbles formed when ultrasonic waves of 20 kHz are applied to a two-phase reaction system (Colucci et al., 2005). The high yield under the ultrasonic irradiation condition is due to high speed mixing and mass transfer between the methanol and triolein as well as the formation of a micro-emulsion resulting from the ultrasonic cavitation phenomenon (Hanh et al., 2008). Cavitation results in conditions of turbulence and liquid circulation in the reactor which can aid in eliminating mass transfer resistances. The cavitation may be used for intensification of biodiesel synthesis (esterification) reaction, which is mass transfer-limited reaction considering the immiscible nature of the reactants, i.e., fatty acids and alcohol (Kelkar et al., 2008). A certain degree of conversion attributed to heterogeneity of the system, which adds to mass transfer resistances under conventional approach, appears to get eliminated due to ultrasound (Deshmane et al., 2009). The high yield for the crude cottonseed oil biodiesel under the ultrasonic irradiation condition is also attributed to the efficacy of cavitation, which can enhance the mass transfer between the methanol and crude cottonseed oil (Fan et al., 2010).

At three temperatures studied by Stamenkovic et al. (2008), the mass transfer coefficients of triglycerides into alcohol phase (at good confidence interval values) ranged from 1.40 (± 0.01) $\times 10^{-7}$ to 1.45 (± 0.01) $\times 10^{-6}$ m s^{-1}, consistent with the reported literature values of approximately 10^{-7}–10^{-3} m s^{-1} (Frascari et al., 2009; Klofutar et al., 2010). From these values, the specific activation energies of first-order triglyceride mass transfer (E_a) were estimated, and again a good fit was obtained. Table 1 shows the (average) values of the mass transfer coefficients at different reaction temperatures. Also, it has to be noted that in Table 1 the mass transfer coefficient values using the mixing rate of 700 min^{-1} are 10.3-times higher than those reported in the literature using mechanical agitation of 100 min^{-1}. This data was used to determine the activation energy of the mass transfer coefficients. The estimated mass transfer coefficients for sunflower oil and KOH catalyst reported in the literature (Frascari et al., 2009; Klofutar et al., 2010) were compared at similar mixing rates. Lower mass transfer coefficients were reported for lower temperatures, since the mass transfer coefficients obtained by Frascari et al. (2009) at 60 °C were greater than that obtained by Klofutar et al. (2010) at 40 °C and 50 °C. The mass transfer coefficient ($k_c a$) determined by Ataya et al. (2007) was much lower than that reported in the other literature. In the case of Liu et al. (2010), the apparent mass transfer coefficient value is not representative of the maximal coefficient that can be reached by reacting molecules before the reaction will occur, since the determination of mass transfer parameters was performed without acknowledging the experimental regime and the coefficient was consequentially unusually high.

Table 2 shows the calculated effective activation energies and pre-exponential factors of mass transfer coefficients (E_a and k_{c0}) for different reaction conditions and the corresponding literature sources. Similar activation energy values were obtained for the methanolysis of sunflower oil regardless of different impeller speeds (Stamenkovic et al., 2008; Klofutar et al., 2010). The disagreement was quite small in both cases, and as a result, the proposed mass transfer model adequately described the results from the experiments. The value of E_a, corresponding to the mass transfer in the case of the reactions of canola oil, was much lower and was considered to indicate less temperature-dependent behaviour of the coefficient. In this sense, the mass transfer of triglycerides to alcohol phase to give diglycerides, monoglycerides and glycerol is not much more favourable upon temperature increase, because of the poorer miscibility of canola oil-originating triglycerides and alcohol, which consequently involves comparably greater mass transfer resistance in the direction of alcohol phase at high temperatures in the case of canola oil than in the case of sunflower oil. Consequently, the mass transfer step (the mass transfer of triglyceride into alcohol) may be considered rate-determining for higher temperatures in the case of canola oil in comparison to sunflower oil. According to the k_c values at higher temperatures (50 °C), the mass transfer from triglyceride phase to alcohol phase was slower for canola oil than for sunflower oil.

Reaction temperature /°C	N /min^{-1}	k_c or k_ca	Oil	Catalyst	Literature
20	500	7.28×10^{-7} s^{-1}	Canola	H_2SO_4	(Ataya et al., 2007)
10	200	1.45×10^{-7} m s^{-1}	Sunflower	KOH	(Stamenkovic et al., 2008)
20	200	3.02×10^{-7} m s^{-1}	Sunflower	KOH	(Stamenkovic et al., 2008)
30	200	1.45×10^{-6} m s^{-1}	Sunflower	KOH	(Stamenkovic et al., 2008)
10	200	1.40×10^{-7} m s^{-1}	Sunflower	KOH	(Stamenkovic et al., 2008)
20	200	3.02×10^{-7} m s^{-1}	Sunflower	KOH	(Stamenkovic et al., 2008)
30	200	1.30×10^{-6} m s^{-1}	Sunflower	KOH	(Stamenkovic et al., 2008)
45	Variable	1.67×10^{-7} s^{-1}	Palm	Lipase	(Al-zuhair et al., 2009)
60	100	5.30×10^{-5} m s^{-1}	Sunflower	KOH	(Frascari et al., 2009)
60	200	1.20×10^{-4} m s^{-1}	Sunflower	KOH	(Frascari et al., 2009)
60	250	1.60×10^{-4} m s^{-1}	Sunflower	KOH	(Frascari et al., 2009)
60	300	2.00×10^{-4} m s^{-1}	Sunflower	KOH	(Frascari et al., 2009)
60	400	2.80×10^{-4} m s^{-1}	Sunflower	KOH	(Frascari et al., 2009)
60	700	5.50×10^{-4} m s^{-1}	Sunflower	KOH	(Frascari et al., 2009)
40	500	4.00×10^{-6} m s^{-1}	Sunflower	KOH	(Klofutar et al., 2010)
50	500	1.70×10^{-5} m s^{-1}	Sunflower	KOH	(Klofutar et al., 2010)
40	500	6.92×10^{-6} m s^{-1}	Canola	KOH	(Klofutar et al., 2010)
50	500	1.18×10^{-5} m s^{-1}	Canola	KOH	(Klofutar et al., 2010)
40	500	7.83×10^{-5} m s^{-1}	Sunflower Canola	KOH	(Klofutar et al., 2010)
50	500	2.04×10^{-4} m s^{-1}	Sunflower Canola	KOH	(Klofutar et al., 2010)
65	900	1.15×10^{-1} m s^{-1}	Soybean	$Ca(OCH_3)_2$	(Liu et al., 2010)

Table 1. Mass transfer parameters for the triglyceride transesterification reaction

N /min^{-1}	C_c /wt. %	E_a /kJ mol^{-1}	k_{c0} /m s^{-1}	Oil	Catalyst	Literature
200	1	81.8	1.54×10^{8} m s^{-1}	Sunflower	KOH	(Stamenkovic et al., 2008)
200	1	79.2	5.06×10^{7} m s^{-1}	Sunflower	KOH	(Stamenkovic et al., 2008)
500	1	121.7	8.10×10^{12} m s^{-1}	Sunflower	KOH	(Klofutar et al., 2010)
500	1	45.2	2.38 m s^{-1}	Canola	KOH	(Klofutar et al., 2010)
500	1	80.4	2.04×10^{7} m s^{-1}	Sunflower Canola	KOH	(Klofutar et al., 2010)

Table 2. Activation energies and pre-exponential factors of mass transfer coefficients

To study the effect of a solvent on mass transfer, experiments were performed by Ataya et al. (2006) at ambient temperature to investigate mass transfer during the transesterification reaction of canola oil with methanol (CH_3OH) to form fatty acid methyl esters by use of a sodium hydroxide (NaOH) base catalyst. Small conversions, at ambient conditions, accentuate the effects of mass transfer on the transesterification reaction. The influence of mass transfer is indicated by the increased reaction rate resulting from stirring a two-phase reaction mixture and changing a two-phase reaction to a single-phase reaction through the addition of a solvent (Ataya et al., 2006).

Symbol	Value	Description and units	Literature
$[TG]$	Variable	Triglyceride concentration in dispersed phase /kmol m^{-3}	/
$[TG]_i$	$[TG]_d/D$ or $[TG]$	Interface triglyceride concentration /kmol m^{-3}	/
$[TG]_d$	ρ_{TG}/M_{TG}	Dispersed phase triglyceride concentration /kmol m^{-3}	/
$[DG]$	Variable	Diglyceride concentration in dispersed phase /kmol m^{-3}	/
$[MG]$	Variable	Monoglyceride concentration in dispersed phase /kmol m^{-3}	/
$[G]$	Variable	Glycerol concentration in dispersed phase /kmol m^{-3}	/
$[A]$	Variable	Alcohol concentration in dispersed phase /kmol m^{-3}	/
$[AE]$	Variable	Alkyl ester concentration in dispersed phase /kmol m^{-3}	/
$[TG]_0$	0	Initial triglyceride concentration in dispersed phase /kmol m^{-3}	/
$[DG]_0$	0	Initial diglyceride concentration in dispersed phase /kmol m^{-3}	/
$[MG]_0$	0	Initial monoglyceride concentration in dispersed phase /kmol m^{-3}	/
$[G]_0$	0	Initial glycerol concentration in dispersed phase /kmol m^{-3}	/
$[A]_0$	ρ_A/M_A	Initial alcohol concentration in dispersed phase /kmol m^{-3}	/
$[AE]_0$	0	Initial alkyl ester concentration in dispersed phase /kmol m^{-3}	/
M_{TG}	871.55	Triglyceride molecular mass /kg $kmol^{-1}$	(Klofutar et al., 2010)
M_{DG}	611.73	Diglyceride molecular mass /kg $kmol^{-1}$	(Klofutar et al., 2010)
M_{MG}	351.91	Monoglyceride molecular mass /kg $kmol^{-1}$	(Klofutar et al., 2010)
M_G	92.09	Glycerol molecular mass /kg $kmol^{-1}$	/
M_A	32.05	Alcohol molecular mass /kg $kmol^{-1}$	/
M_{AE}	291.87	Alkyl ester molecular mass /kg $kmol^{-1}$	(Klofutar et al., 2010)
ρ_{TG}	Variable	Triglyceride density /kg m^{-3}	(Hilal et al., 2004)
ρ_{DG}	Variable	Diglyceride density /kg m^{-3}	(Hilal et al., 2004)
ρ_{MG}	Variable	Monoglyceride density /kg m^{-3}	(Hilal et al., 2004)
ρ_G	Variable	Glycerol density /kg m^{-3}	(Hilal et al., 2004)
ρ_A	Variable	Alcohol density /kg m^{-3}	(Hilal et al., 2004)
ρ_{AE}	Variable	Alkyl ester density /kg m^{-3}	(Hilal et al., 2004)
k_c	$k_{c0}\exp(-E_a/(RT))\ d_{ref}/d$	Mass transfer coefficient /m s^{-1}	(Klofutar et al., 2010)
a	$6\phi/d$	Specific surface area /m^{-1}	/
k_1	$A_1\exp(-E_{a1}/(RT))$	Triglyceride transesterification forward reaction rate constant /m^3 $kmol^{-1}$ s^{-1}	(Klofutar et al., 2010)
k_2	$A_2\exp(-E_{a2}/(RT))$	Diglyceride transesterification forward reaction rate constant /m^3 $kmol^{-1}$ s^{-1}	(Klofutar et al., 2010)
k_3	$A_3\exp(-E_{a3}/(RT))$	Monoglyceride transesterification forward reaction rate constant /m^3 $kmol^{-1}$ s^{-1}	(Klofutar et al., 2010)
k_4	$A_4\exp(-E_{a4}/(RT))$	Triglyceride transesterification backward reaction rate constant /m^3 $kmol^{-1}$ s^{-1}	(Klofutar et al., 2010)
k_5	$A_5\exp(-E_{a5}/(RT))$	Diglyceride transesterification backward reaction rate constant /m^3 $kmol^{-1}$ s^{-1}	(Klofutar et al., 2010)
k_6	$A_6\exp(-E_{a6}/(RT))$	Monoglyceride transesterification backward reaction rate constant /m^3 $kmol^{-1}$ s^{-1}	(Klofutar et al., 2010)
T	Variable	Time /s	/

Table 3. Descriptions and numerical values of the symbols in Equations (1)–(7)

Experiments were also performed by Ataya et al. (2007) at ambient temperature to investigate the effects of mass transfer during the transesterification reaction of canola oil with methanol (CH_3OH) to form fatty acid methyl esters using a sulfuric acid (H_2SO_4) catalyst at a CH_3OH/oil molar ratio of 6:1. Experiments at ambient conditions result in reaction rates that are slow enough to permit the effects of mass transfer on the transesterification reaction to become more evident than at higher temperatures. The influence of mass transfer was investigated by comparing a mixed versus quiescent two-phase reaction and changing a two-phase reaction to a single-phase reaction through the addition of a solvent, tetrahydrofuran (Ataya et al., 2007). The presence of tetrahydrofuran minimizes the mass transfer problem normally encountered in heterogeneous systems (Soriano et al., 2009). Vegetable oil such as corn, sunflower, rapeseed, soybean, and palm oil may also be completely transesterified into biodiesel fuel in short time because of high mass transfer rate in the homogeneous solution formed by adding environment-friendly solvent of dimethyl ether (Guan et al., 2007). The feasibility of fatty acid methyl ester as a co-solvent used to increase the mass transfer between oil and methanol was investigated by Park et al. (2009).

The solution to the new system model acknowledging mass transfer and kinetics is presented in Fig. 2–6. The whole [TG], [DG], [MG], [G], [A] and [AE] versus t set ofsolutions for arbitrary conditions and initial $[TG]_0$, $[DG]_0$, $[MG]_0$, $[G]_0$, $[A]_0$ and $[AE]_0$ were obtained using the fourth-order Runge-Kutta method in the form described by Klofutar et al. (2010). The value of each parameter of Equations (2)–(7) is given in Table 3. The symbols which are not explained in Table 3 are gas constant (R), temperature (T), distribution coefficient of triglyceride in continuous and dispersed phase (D) (Hilal et al., 2004), pre-exponential factors (A_i) (Klofutar et al., 2010), activation energies (E_{ai}) (Klofutar et al., 2010), dispersed phase drop size (d) (Klofutar et al., 2010), reference dispersed phase drop size (d_{ref}) (Klofutar et al., 2010), and dispersed phase volume fraction (ϕ) (Klofutar et al., 2010).

$$\begin{aligned} TG + A &\underset{k_4}{\overset{k_1}{\rightleftarrows}} DG + AE \\ DG + A &\underset{k_5}{\overset{k_2}{\rightleftarrows}} MG + AE \\ MG + A &\underset{k_6}{\overset{k_3}{\rightleftarrows}} G + AE \end{aligned} \tag{1}$$

$$\frac{d[TG]}{dt} = k_c a([TG]_i - [TG]) - k_1[TG][A] + k_4[DG][AE] \tag{2}$$

$$\frac{d[DG]}{dt} = k_1[TG][A] - k_2[DG][A] - k_4[DG][AE] + k_5[MG][AE] \tag{3}$$

$$\frac{d[MG]}{dt} = k_2[DG][A] - k_3[MG][A] - k_5[MG][AE] + k_6[G][AE] \tag{4}$$

$$\frac{d[G]}{dt} = k_3[MG][A] - k_6[G][AE] \quad (5)$$

$$\frac{d[A]}{dt} = -k_1[TG][A] - k_2[DG][A] - k_3[MG][A] + k_4[DG][AE] + k_5[MG][AE] + k_6[G][AE] \quad (6)$$

$$\frac{d[AE]}{dt} = k_1[TG][A] + k_2[DG][A] + k_3[MG][A] - k_4[DG][AE] - k_5[MG][AE] - k_6[G][AE] \quad (7)$$

The examination of Fig. 2 reveals the sigmoid curve for the production of methyl esters and glycerol. This indicates the low reaction rate or the delay at the beginning which is followed by a sudden surge and finally a lower rate as the reactions approach equilibrium. This is the typical behaviour for autocatalytic reactions or reactions with changing mechanisms. Since the transesterification reaction of triglycerides is not known to be an autocatalytic reaction, a second possibility is hypothesized as a mass transfer-controlled region (low rate) followed by a kinetics-controlled region (high rate) and a final low-rate region as the equilibrium is approached. This hypothesis was in more detail discussed in the previous paragraphs and will be supported with the simulated data in the following figures.

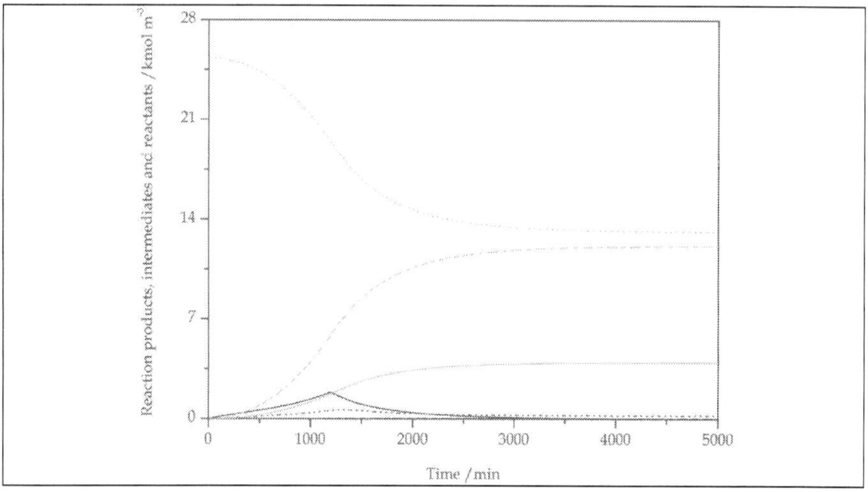

Fig. 2. The composition of the reaction mixture in dispersed phase during the transesterification of sunflower oil with catalyst concentration of 1 wt. % at 0 °C and Reynolds number (Re) of 49.2 (impeller speed of 100 rpm); (——) triglycerides; (·········) diglycerides; (— — —) monoglycerides; (········) glycerol; (·········) alcohol; (— — —) alkyl esters; the parameters for simulation were obtained from the literature (Klofutar et al., 2010)

Consequently, the effect of mixing was studied. In this transesterification reaction, the reactants initially form a two-phase liquid system. The rate is diffusion-controlled and the poor diffusion between the phases results in a low rate. As methyl esters are formed, theyact as the mutual solvent for reactants, intermediates and products and a single-phase system is formed. This was substantiated with simulations by differing impeller speeds showing that the low-rate region is practically absent when using 200 rpm (Re = 98.4) or more (Fig. 3).

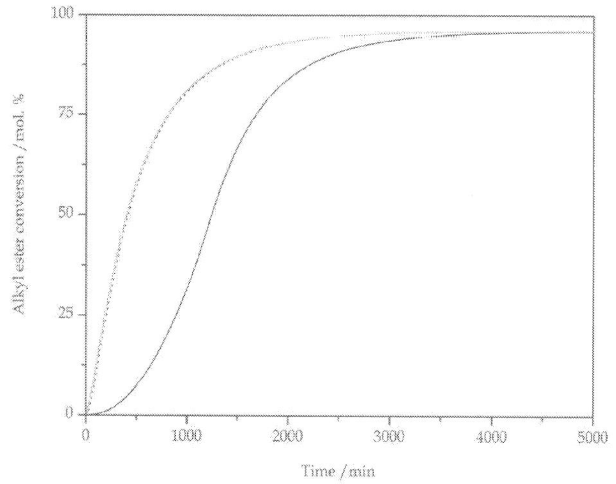

Fig. 3. The effect of mixing intensity and time on the overall conversion of sunflower oil with catalyst concentration of 1 wt. % to alkyl esters at 0 °C; (——) Re = 49.2 (N = 100 rpm); (·········) Re = 98.4 (N = 200 rpm); (– – –) Re = 147.6 (N = 300 rpm); (——) Re = 196.8 (N = 400 rpm); (·········) Re = 246.0 (N = 500 rpm); (– – –) Re = 295.2 (N = 600 rpm); (——) Re = 344.4 (N = 700 rpm); (·········) Re = 393.6 (N = 800 rpm); (– – –) Re = 442.8 (N = 900 rpm); the parameters for simulation were obtained from the literature (Klofutar et al., 2010)

Results revealed that during the very early stages of the reaction, the mixture is separated into two phases. However, as the process was continued, a single phase was observed at the time corresponding either to inflection point or maximum in the concentration−time diagram. Fig. 3 summarizes this delay or the low-rate region as a function of the mixing intensity at a constant reaction temperature of 0 °C. As expected, this time lag decreased as the mixing intensity was increased and reached a practically constant value of less than 1 min for Re greater than 98.4.

Subsequently, the effect of temperature was studied. The temperature dependency of the overall alkyl ester formation reaction rate is presented in Fig. 4 and 5 at two different mixing intensities. This dependency is similar to the effect of the mixing intensity and has to be analyzed separately for the previously hypothesized mass transfer- and kinetics-controlled regions. The time of the mass transfer-controlled region is shortened as temperature is increased (Fig. 4 and 5) which is due to the higher energy level of molecules resulting in more fruitful diffusion into continuous phase. The improved solubility of triglycerides in alcohol at elevated temperatures is also partially responsible for this behaviour. Mass transfer-controlled region is reduced from 1200 min to about 180 min as temperature is increased from 0 °C to 10 °C at Re = 98.4 (Fig. 4). At higher mixing intensities, mass transfer-controlled region is short and this effect is not significant (Fig. 5).

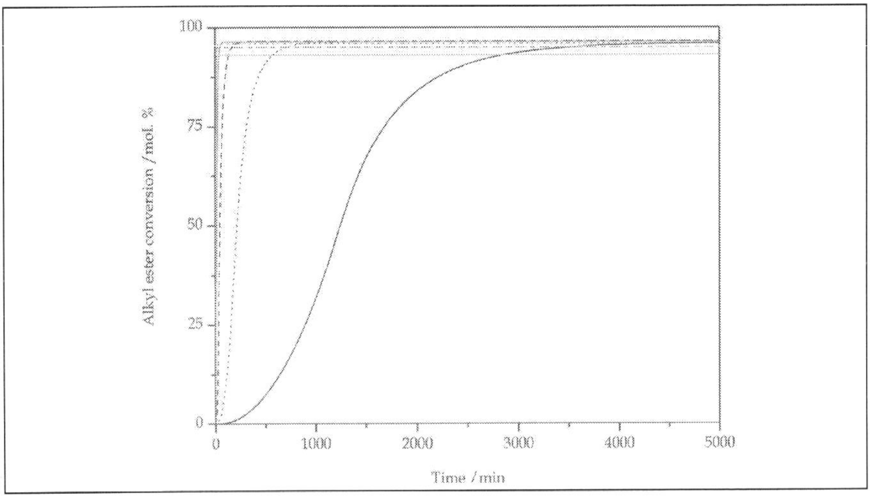

Fig. 4. The effect of temperature and time on the overall conversion of sunflower oil with catalyst concentration of 1 wt. % to alkyl esters at Re = 49.2 (N = 100 rpm); (——) 0 °C; (·········) 10 °C; (– – –) 20 °C; (———) 30 °C; (·········) 40 °C; (— — —) 50 °C; (·········) 60 °C; the parameters for simulation were obtained from the literature (Klofutar et al., 2010)

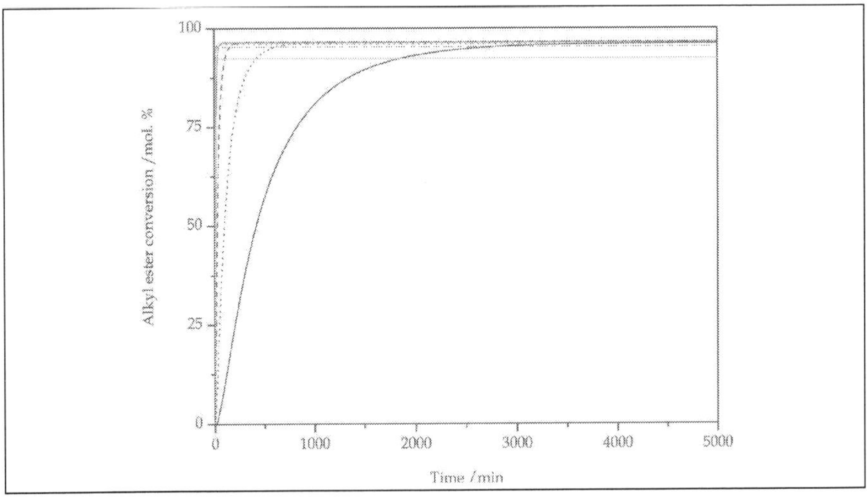

Fig. 5. The effect of temperature and time on the overall conversion of sunflower oil with catalyst concentration of 1 wt. % to alkyl esters at Re = 98.4 (N = 200 rpm); (——) 0 °C; (·········) 10 °C; (– – –) 20 °C; (———) 30 °C; (·········) 40 °C; (— — —) 50 °C; (·········) 60 °C; the parameters for simulation were obtained from the literature (Klofutar et al., 2010)

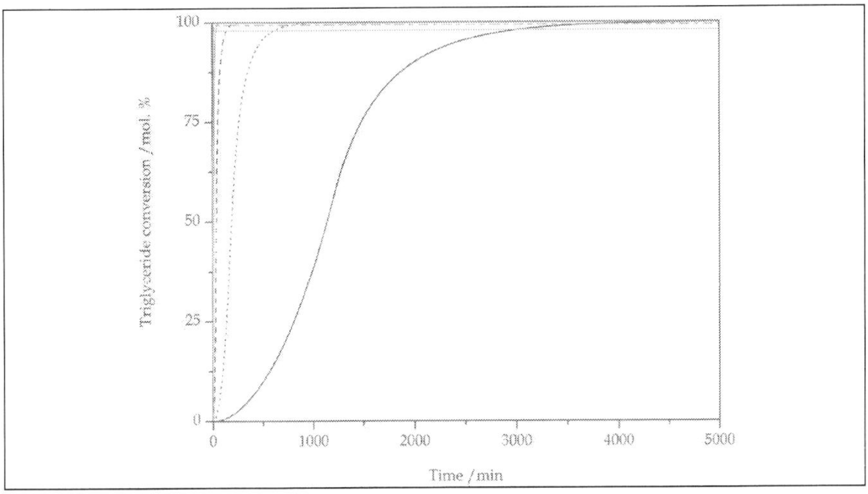

Fig. 6. The effect of temperature on the triglycerides conversion in sunflower oil with catalyst concentration of 1 wt. % at Re = 49.2 (N = 100 rpm); (——) 0 °C; (·········) 10 °C; (– – –) 20 °C; (———) 30 °C; (·········) 40 °C; (— — —) 50 °C; (·········) 60 °C; the parameters for simulation were obtained from the literature (Klofutar et al., 2010)

The effects of impeller speed and temperature are thus predominant in influencing the mass-transfer-determined rate of the biodiesel production process in batch reactors. Twenty-one simulations of reactions were carried out all in all varying temperature and impeller speed using 1 wt. % of potassium hydroxide concentration in sunflower oil. Fig. 6 represents the triglyceride conversion change with time for the reactions using 100 rpm at 0 °C, 10 °C, 20 °C, 30 °C, 40 °C, 50 °C and 60 °C. In all cases, triglyceride conversions were very low during the beginning of reactions, which involved a low methyl ester production rate at this stage. The conversions then increased and at the end remained approximately constant as equilibrium was approached. Noureddini & Zhu (1997) also observed these three regions of different rate behaviour. As mentioned, this behaviour is typical for the processes with changing mechanisms. The observed low-rate region was again due to the immiscibility of sunflower oil and methanol during the first stages of the reactions. As mentioned, in this region, the rate is controlled by mass transfer.

2.2 Semi-batch reactors

In a semi-batch reactor, the immiscibility of oil in alcohol (canola oil and methanol were studied) provides a mass transfer challenge in the early stages of the transesterification of oil (specifically, canola oil) in the production of fatty acid methyl esters (biodiesel) (Dube et al., 2007).

2.3 Continuous reactors

Continuous reactor technologies enhance reaction rate, reduce molar ratio of alcohol to oil and energy input by intensification of mass transfer and heat transfer and in situ product separation, thus achieving continuous product in a scalable unit (Qui et al., 2010).

One of the possible continuous reactors for biodiesel production process is an oscillatory flow reactor. An oscillatory motion is superimposed upon the net flow of the process fluid, creating flow patterns conducive to efficient heat and mass transfer, whilst maintaining plug flow (Harvey et al., 2003).

Another possibility is a continuous tubular reactor, in which at low conversion values (< 25−35%), the system has mass transfer limitations due to the immiscibility of the oil−methanol system (Busto et al., 2006).

Yet another possibility is a packed bed reactor, in which kinetics and mass transfer of free fatty acids esterification with methanol are a key pre-treatment in biodiesel production. The collected experimental data may be interpreted by means of a mono-dimensional packed bed reactor model in which the external mass transfer limitation (fluid-to-particle) is accounted for (Santacesaria et al., 2007a). Experiments were performed by Ataya et al. (2008a) to study the mass-transfer limitations during the acid-catalyzed transesterification reaction of triglyceride with methanol (CH_3OH) to fatty acid methyl ester (biodiesel). The rate constant at two-phase conditions (largest velocity, smallest packing particle size, and maximum pressure gradient) are comparable to that obtained at single-phase conditions, indicating that the mass-transfer limitations for two-phase experiments can be effectively overcome using a liquid-liquid packed bed reactor. The diminished mass transfer is explained by the formation of a new interfacial area between the two liquid phases, caused by the droplets being momentarily deformed into an elongated non-spherical shape as they pass through the openings between the solid particles of the packed bed (Ataya et al., 2008a).

The fourth possibilities are a slurry reactor and a loop reactor. Both the well-stirred slurry reactor and spray tower loop reactor show liquid−solid phase mass transfer limitations (Santacesaria et al., 2007b). Different configurations and dynamics of single air/alcohol gas-liquid compound drops in vegetable oil may largely improve mass transfer in a slurry reactor (Duangsuwan et al., 2009). The kinetic model that is developed on the basis of several batch runs is also able to simulate the behaviour of dynamic tubular loop reactor, providing that the external mass transfer resistance is properly accounted for. The mass transfer coefficient is satisfactorily modelled using correlations available in literature (Tesser et al., 2009).

The fifth possibility is a continuous reactor with static mixing. A novel continuous static reactor concept improves mass transfer in two-phase chemical reactions between one and two orders of magnitude (Massingill et al., 2008).

The sixth possibility is a film reactor. This reactor is a co-current, constant diameter (0.01 m), custom-made packed column where the mass transfer area between the partially miscible methanol-rich and vegetable oil-rich phases is created in a non-dispersive way, without the intervention of mechanical stirrers or ultrasound devices (Narvaez et al., 2009).

The seventh possibility is a continuous reactor with counter flow. In the latter, the excess methanol is subjected to a mass transfer from the liquid phase into the gas phase, which is withdrawn through the head of the reactor and condensed in an external condenser unit (Iglauer & Warnecke, 2009).

The eighth possibility is a micro-structured reactor. Micro-structured reactors have an equivalent hydraulic diameter up to a few hundreds of micrometers and, therefore, provide high mass and heat transfer efficiency increasing the reactor performance drastically, compared to the conventional one. Particular attention is given to the identification of the parameters that control the flow pattern formed in micro-capillaries regarding the mass transfer efficiency (Kashid & Kiwi-Minsker, 2009). The homogeneous state in the micro-tube should be a benefit for the transesterification of waste cooking oil due to the enhancement of the mass transfer between oil and methanol (Guan et al., 2010).

The ninth possibility is a membrane reactor. A new alternative technology, using hydrophobic porous membranes, can be used to prevent bulk mixing of the two phases and facilitate contact and mass transfer of species between the two phases (Sdrula, 2010).

2.4 Bioreactors

One of the possible organisms used for biodiesel production process are fungi. The initial reaction rate is increased notably (204%) with oil pre-treatment on the cells before catalyzing the reaction, which is possibly due to the improved mass transferring of substrates (Zeng et al., 2006). The accumulated glycerol influences whole cell stability through mass transfer limitation only, while the accumulated methyl ester influences whole cell stability through both mass transfer limitation and product inhibition (Li et al., 2008). Halim et al. (2009) studied continuous biosynthesis of biodiesel from waste cooking palm oil in a packed bed reactor; investigating optimization using response surface methodology and mass transfer, specifically the effect of mass transfer in the packed bed bioreactor has been studied extensively. Models for fatty acid methyl ester yield are developed for cases of reaction control and mass transfer control. The results show very good agreement compatibility between mass transfer model and the experimental results obtained from immobilized lipase packed bed reactor operation, showing that in this case the fatty acid methyl ester yield is mass transfer-controlled (Halim et al., 2009). A kinetic model was developed by Al-Zuhair et al. (2009) to describe a bioreactor system, taking into consideration the mass transfer resistances of the reactants. The experimental results were used to determine the kinetic parameters of the proposed model and to determine the effect of mass transfer (Al-Zuhair et al., 2009).

Sim et al. (2009) studied the effect of mass transfer and enzyme loading on the biodiesel yield and reaction rate in the enzymatic transesterification of crude palm oil. Efforts in minimizing mass transfer effects in enzymatic transesterification of crude palm oil in a biphasic system have always been the compromise between enzyme loading and agitation speed, therefore, effect of enzyme loading and agitation speed on fatty acid methyl ester productivity in terms of intrinsic and external mass transfer limitations and the effective reaction time may be determined using factorial design. Graphical plots of experimental results reveal that the mass transfer effect for the transport of reactant from bulk liquid to immobilized lipase and within the intra-particle of immobilized lipase are absent at 150 rpm and 6.65% enzyme loading (Sim et al., 2009). In the case of continuous process, circulation and long-term continuous system are investigated for development of efficient mass transfer system (Lee et al., 2010). The aim of the study by Sotoft et al. (2010) was to determine reaction enthalpy for the enzymatic transesterification and to elucidate the mass transfer and energetic processes taking place. Although it is possible to determine thermodynamic properties such as reaction enthalpy and reaction rate, the difficulty in actually measuring the true non-mass transfer-limited reaction kinetics is exposed by the high time resolution of isothermal calorimetry (Sotoft et al., 2010).

Another possibility are bacteria, for which employment of immobilization seems to result in a decrease in the maximum rate (v_{max}) and an increase in the Michaelis constant (K_M), most likely due to the mass transfer resistance arising from formation of micelles during the lipase immobilization process (Liu & Chang, 2008).

Yet another possibility are algae, for which maximizing cost efficient mass transfer of CO_2 to cells in an aqueous environment is not a trivial task for large-scale liquid culture systems as is anticipated for outdoor algal cultures (Xu et al., 2009).
However, in all these organisms the reaction catalysts are the enzymes themselves, critical aspects of these being mass transfer limitations, use of solvents and water activity together with process considerations and evaluation of possible reactor configurations, if industrial production with enzymes is to be carried out (Fjerbaek et al., 2009). However, these heterogeneous acid- and enzyme-catalyzed systems still suffer from serious mass transfer limitation problems and therefore are not favourable for industrial application. Nevertheless, a few latest technological developments that have the potential to overcome the mass transfer limitation problem such as oscillatory flow reactor, ultrasonication, microwave reactor and co-solvent use are being studied (Lam et al., 2010).

2.5 Downstream processing

To remove the unwanted side products (e.g. glycerol) from the wanted main product (biodiesel), adsorption may be used. Industrial adsorption units with beads (3.18 mm silica beads were studied) suffer from mass transfer limitations inside the pellet pores, and for the particle size investigated by Yori et al. (2007), the breakthrough point (output concentration per input concentration of 1%, $C/C_0 = 0.01$) was located at about one-half of the time of full saturation. Assuming a linear isotherm gives erroneous results; fitting the experimental breakthrough curves produces underestimated values of the Henry's adsorption constant and of the mass transfer resistances. Accordingly, breakthrough curves are fairly well predicted using an irreversible isotherm, a shrinking-core adsorption model, and common correlations for the mass transfer coefficients (Yori et al., 2007). To remove the unwanted glycerol from the wanted biodiesel, glycerol may also be used as a substrate in an enzyme reaction. Volpato et al. (2009) studied the effects of oxygen volumetric mass transfer coefficient and pH on lipase production by *Staphylococcus warneri* EX17. The principal objectives of this study were to evaluate the kinetics of lipase production by *Staphylococcus warneri* EX17 under different oxygen volumetric mass transfer coefficients (k_La) and pH conditions in submerged bioreactors, using glycerol (a biodiesel by-product) as a carbon source (Volpato et al., 2009). To remove the unwanted glycerol from the wanted biodiesel, glycerol may also be used as a reactant in a chemical reaction for the production of dichloropropanol. The reaction is conducted at high agitation speed in order to avoid mass transfer limitation between glycerol and hydrochloric acid gas (Song et al., 2009).
To remove the unwanted side products (e.g. salts) from the dispersed phase after removing glycerol (mostly water), desalination may be used. A competent grasp of thermodynamics and heat and mass transfer theory, as well as a proper understanding of current desalination processes, is essential for ensuring beneficial improvements in desalination processes (Semiat, 2008). To remove the unwanted salts from the dispersed phase, adsorption may also be used. Kinetic experiments show that the rate of mass transfer in the adsorbent/liquid binary system is high (Carmona et al., 2009).
Mass transfer also plays an important role during biodiesel's final application as a fuel, usually for an engine. For example, engine internal processes are usually studied by means of exergy balances based on engine indicating data, which provides information about the impact of biodiesel blending on the amount of fuel exergy exchanged through heat, work and mass transfer (Bueno et al., 2009). Also, the improved physical (mass transfer, filtering of C-containing species) and chemical (reaction kinetics) processes during hydrocarbon-selective catalytic reduction over powders compared to monoliths leads to better initial

catalyst activity, but it also accelerates catalyst deactivation which lead to increased diffusion limitations (Sitshebo et al., 2009).

3. Kinetics-determined rate of biodiesel production process

3.1 Influence of lipid and alcohol on reaction rate

The transesterification of oils (triglycerides) with simple alcohols in the presence of a catalyst has long been the preferred method for preparing biodiesel. As already mentioned the initial mass transfer-controlled region is not significant using the impeller speed of 600 rpm, high temperature, or the molar ratio of alcohol to oil higher than 6:1.

The general, overall apparent reaction of the transesterification of triglycerides with alcohol is reversible and every triglyceride molecule can react with three molar equivalents of alcohol (usually methanol) to produce glycerol and three fatty acid alkyl ester molecules. Fatty acid originating chains vary in composition depending on the source of triglycerides. The reaction of transesterification is highly dependent on oil quality, especially in terms of free fatty acid and water content. Today, most of biodiesel comes from the transesterification of edible resources, such as animal fats, vegetable oils, and even waste cooking oils, in the conditions of alkaline and acid catalysis. Conventional oils, such as canola oil, sunflower oil, and soybean oil, are usually employed for the biodiesel synthesis under acidic conditions (Santacesaria et al., 2007b). In novel process studies, the use of tin chloride, lipases, and supercritical alcohol without catalyst is being investigated, using conventional oils, such as canola oil, sunflower oil, soybean oil, and palm oil. In most industrial biodiesel processes, in which biodiesel is synthesized under alkaline conditions, oils such as canola oil, sunflower oil, soybean oil, palm oil, olive oil, *Brassica carinata* oil, and *Pongamia pinnata* oil were used (Noureddini & Zhu, 1997; Vicente et al., 2006; Bambase et al., 2007; Stamenkovic et al., 2008) Canola oil is the preferred oil feedstock for the biodiesel production in most of Europe,

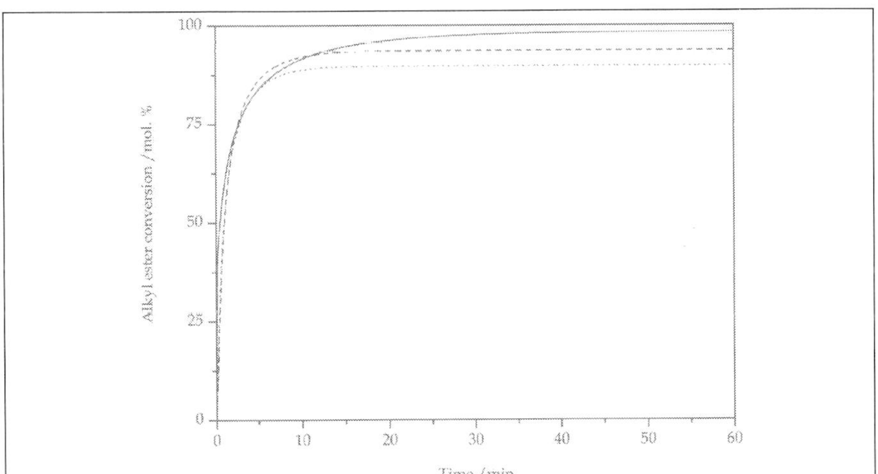

Fig. 7. The effect of lipid on the alkyl ester conversion in oil with catalyst concentration of 1 wt. % at Re → ∞ and 65 °C; (——) *Brassica carinata* oil; (·········) sunflower oil; (— — —) canola oil; the parameters for simulation were obtained from the literature (Vicente et al., 2006, Klofutar et al., 2010)

Reaction temperature /°C	N / min^{-1}	k_1 /m^3 $kmol^{-1}$ min^{-1}	k_2 /10^1 m^3 $kmol^{-1}$ min^{-1}	k_3 /10^{-1} m^3 $kmol^{-1}$ min^{-1}	k_4 /10^1 m^3 $kmol^{-1}$ min^{-1}	k_5/10^1 m^3 $kmol^{-1}$ min^{-1}	k_6 /10^{-2} m^3 $kmol^{-1}$ min^{-1}	Oil	C_c /wt. %	Literature
25	600	0.07	0.02	2.20	0.03	0.01	1.60	Sunflower	0.50	(Vicente et al., 2005)
35	600	0.20	0.17	2.70	0.10	0.22	1.10	Sunflower	0.50	
45	600	0.37	0.38	3.20	0.19	0.44	0.77	Sunflower	0.50	
55	600	0.80	1.05	3.40	0.60	1.59	0.35	Sunflower	0.50	
65	600	1.50	2.30	4.00	1.37	4.14	0.26	Sunflower	0.50	
25	600	0.11	0.03	2.60	0.04	0.02	1.30	Sunflower	1.00	
35	600	0.30	0.25	3.70	0.15	0.31	0.77	Sunflower	1.00	
45	600	0.50	0.70	4.50	0.31	1.00	0.30	Sunflower	1.00	
55	600	1.55	2.05	6.10	0.85	2.25	0.12	Sunflower	1.00	
65	600	3.06	3.25	5.40	2.39	5.75	0.09	Sunflower	1.00	
25	600	0.15	0.04	4.90	0.07	0.04	0.40	Sunflower	1.50	
35	600	0.40	0.43	6.00	0.20	0.47	0.05	Sunflower	1.50	
45	600	0.74	1.19	7.40	0.50	1.26	0.06	Sunflower	1.50	
55	600	2.05	3.01	8.30	1.09	2.95	0.01	Sunflower	1.50	
65	600	4.00	5.50	9.10	2.70	6.55	0.01	Sunflower	1.50	
25	600	0.05	0.02	1.98	0.03	0.01	1.10	*B. carinata*	0.50	(Vicente et al., 2006)
35	600	0.22	0.16	2.52	0.11	0.23	1.00	*B. carinata*	0.50	
45	600	0.40	0.40	2.70	0.20	0.44	0.67	*B. carinata*	0.50	
55	600	0.75	1.01	3.90	0.68	1.70	0.05	*B. carinata*	0.50	
65	600	1.30	2.20	4.08	1.45	4.24	0.26	*B. carinata*	0.50	
25	600	0.11	0.03	3.30	0.05	0.03	1.00	*B. carinata*	1.00	
35	600	0.28	0.24	3.72	0.14	0.32	0.44	*B. carinata*	1.00	
45	600	0.55	0.69	4.20	0.30	1.01	0.30	*B. carinata*	1.00	
55	600	1.59	2.15	6.00	0.97	2.39	0.11	*B. carinata*	1.00	
65	600	3.10	3.27	4.98	2.45	5.95	0.06	*B. carinata*	1.00	
25	600	0.20	0.06	4.68	0.06	0.08	0.50	*B. carinata*	1.50	
35	600	0.40	0.47	6.00	0.26	0.49	0.04	*B. carinata*	1.50	
45	600	0.80	1.01	6.72	0.50	1.30	0.03	*B. carinata*	1.50	
55	600	2.90	2.95	7.98	1.12	2.89	0.02	*B. carinata*	1.50	
65	600	4.13	5.55	9.18	2.82	6.39	0.01	*B. carinata*	1.50	
40	500	0.02	0.01	0.70	0.00	0.00	0.42	Sunflower	1.00	(Klofutar et al., 2010)
50	500	0.08	0.02	0.97	0.00	0.01	0.88	Sunflower	1.00	
40	500	0.04	0.02	0.65	0.01	0.03	0.47	Canola	1.00	
50	500	0.09	0.05	1.56	0.01	0.07	0.61	Canola	1.00	

Table 4. Reaction rate constants for the triglyceride transesterification reaction with KOH catalyst

partly because canola produces more oil per unit of land area compared to other oil sources. *Brassica carinata*, a native plant of Ethiopia, is a promising alternative oilseed crop for the biodiesel production in the Mediterranean area. It generally has high levels of undesirable glucosinolates and erucic acid which, however; do not affect biodiesel purity, maximum conversion and purification step.

The effect of canola, sunflower, and *Brassica carinata* oil on alkyl (methyl) ester conversion with the catalyst concentration of 1 wt. % at 65 °C is presented in Fig. 7. Under the mentioned conditions, the maximal conversion was achieved with the biodiesel production process using *Brassica carinata* oil.

As already mentioned, the maximum conversion in transesterification equilibrium stage depends on the oil itself, especially in terms of the composition of fatty acids bonded to glycerol. Generally, a vegetable oil with lower iodine value is more suitable for biodiesel production. From this point of view, steric interactions must be considered as an important factor in attempts to achieve the highest equilibrium conversion. In Table 4 and 5 the influence of lipid on the calculated reaction rate constants for the forward and backward reactions is presented.

Reaction temperature /°C	N /min^{-1}	k_1 /10^{-1} m^3 kmol^{-1} min^{-1}	k_2 /10^{-1} m^3 kmol^{-1} min^{-1}	k_3 /10^{-1} m^3 kmol^{-1} min^{-1}	k_4 /10^{-1} m^3 kmol^{-1} min^{-1}	k_5 /m^3 kmol^{-1} min^{-1}	k_6 /10^{-2} m^3 kmol^{-1} min^{-1}	Oil	C_c /wt. %	Literature
50	300	0.50	2.15	2.42	1.10	1.23	0.70	Soybean	0.20	(Noureddini & Zhu, 1997)
25	400	0.07	0.20	1.37	0.01	0.01	0.54	Sunflower	0.50	(Bambase et al., 2007)
40	400	0.22	0.65	2.28	0.03	0.03	1.28	Sunflower	0.50	
60	400	0.50	1.00	1.03	1.35	0.41	5.99	Sunflower	0.25	
60	400	0.90	3.48	4.88	0.09	0.13	3.80	Sunflower	0.50	
60	400	2.31	4.49	8.77	0.17	0.11	6.32	Sunflower	1.00	
60	600	1.32	3.23	8.61	0.20	0.25	6.06	Sunflower	0.50	

Table 5. Reaction rate constants for the triglyceride transesterification reaction with NaOH catalyst

3.2 Influence of temperature on reaction rate

When agitation speeds are in the range of 400−600 rpm and mass transfer limitations are practically eliminated, temperature becomes the most influential factor affecting the apparent rate of transesterification as system becomes kinetically controlled (Vicente et al., 2005; Vicente et al., 2006; Bambase et al., 2007; Klofutar et al., 2010). The effect of temperature on the transesterification of canola oil (Klofutar et al., 2010), sunflower oil (Vicente et al., 2005; Bambase et al., 2007; Klofutar et al., 2010), and *Brassica carinata* oil

(Vicente et al., 2006) was also studied and the influence of temperature on the calculated reaction rate constants for forward and backward reactions is presented in Table 4 and 5. It was observed that temperature had a noticeable effect on transesterification reactions.

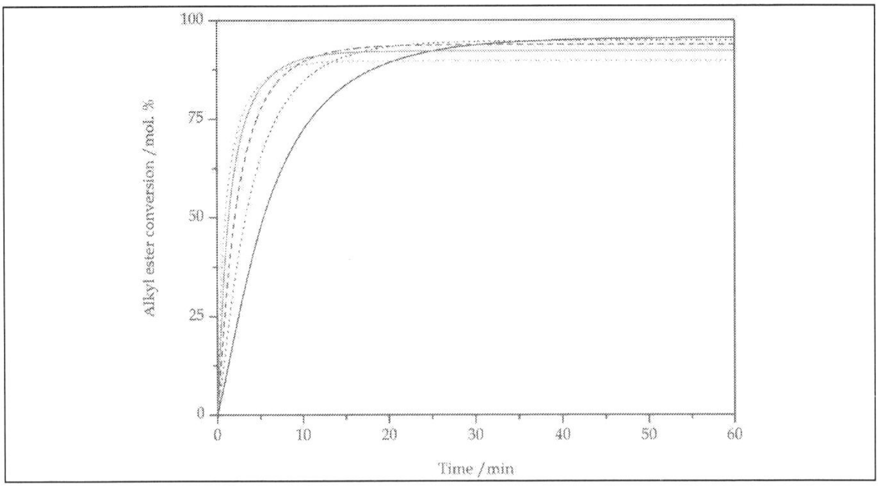

Fig. 8. The effect of temperature on the alkyl ester conversion in sunflower oil with catalyst concentration of 1 wt. % at Re → ∞; (——) 45 °C; (·········) 50 °C; (– – –) 55 °C; (——) 60 °C; (·········) 65 °C; the parameters for simulation were obtained from the literature (Klofutar et al., 2010)

Lower temperature slows down the rate of reactions prolonging the reaction time required to achieve the maximal alkyl ester production. On the other hand, higher reaction temperature results in lower maximal alkyl ester production due to reversible reaction steps (Fig. 8). All reaction rate constants increase with increased temperature, but the backward reaction rate constants increase more intensively in comparison to the forward reaction rate constants. Fig. 8 shows the effect of temperature on the alkyl (methyl) ester conversion in sunflower oil with the catalyst concentration of 1 wt. % (per weight of oil), where KOH was used as the catalyst and the methanol to oil molar ratio of 6:1 was employed at 45 °C, 50 °C55 °C, 60 °C, and 65 °C. The apparent rate of esters production increased with the increase in reaction temperature. After 1 h of reaction time or even sooner, the maximal extent of reaction is approached. The temperature of reaction mixture dictates the maximum conversion itself and the rate of transesterification which affects the reaction time required to achieve maximum conversion. From Fig. 8 it follows that after 5 min the conversion of sunflower oil to methyl ester was about 47 mol. %, 65 mol. %, 77 mol. %, 82 mol. %, and 84 mol. % at 45 °C, 50 °C, 55 °C, 60 °C, and 65 °C, respectively. After 60 min the conversion was above 96 mol. %, 95 mol. %, 94 mol. %, 92 mol. %, and 90 mol. % at 45 °C, 50 °C, 55 °C, 60 °C, and 65 °C, respectively.

The obtained reaction rate constants show that the transesterification of sunflower oil is quite sensitive to temperature. The results obtained from examining kinetics suggest that in the temperature range studied, the rates of the forward reactions of transesterification dominate. For the global reaction, TG + 3 A → G + 3 AE, with 1 wt. % of KOH as the catalyst (per weight of oil) and the methanol to oil molar ratio of 6:1, the activation energy for the forward reaction is lower than the activation energy of the backward reaction and therefore AE formation is favoured at moderately high temperatures. Overall, the results for sunflower oil suggest that the rates of the forward reactions dominate over the backward ones. A similar trend regarding the activation energies as for the case of the sunflower oil (Klofutar et al., 2010) was obtained for the activation energies of the transesterification of soybean oil by Noureddini & Zhu (1997), who obtained the maximal value of E_a for the reaction of the transesterification of diglycerol.

3.3 Influence of catalyst on reaction rate

The base-catalyzed transesterification of vegetable oils proceeds faster than the acid-catalyzed reactions.

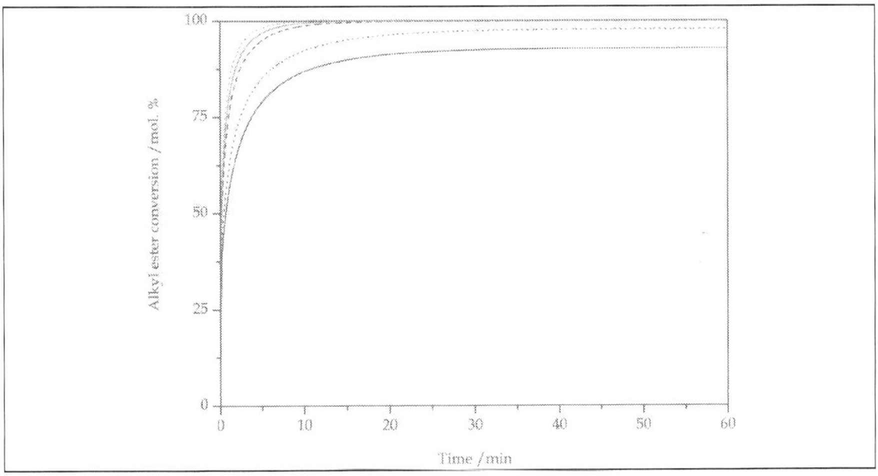

Fig. 9. The effect of catalyst (content) on the alkyl ester conversion in sunflower oil at Re → ∞ and 65 °C; (——) 0.5 wt. %; (·········) 1.0 wt. %; (– – –) 1.5 wt. %; (———) 2.0 wt. %; (·········) 2.5 wt. %; the parameters for simulation were obtained from the literature (Vicente et al., 2005)

A common catalyst for transesterification reactions is potassium hydroxide. An advantage of this catalyst is that at the end of reactions the mixture can be neutralized with phosphoric acid, whereas the products of neutralization reaction can be used as an artificial fertilizer.

The common acid-catalyzed processes of the biodiesel production from low-cost raw materials involve the use of mineral acids; this is due to the large amounts of the edible resources with free fatty acids, which are not compatible with the use of base-catalyzed methods. The acid-catalyzed processes and the corresponding kinetic studies included the use of sulphuric acid as the catalyst (Santacesaria et al., 2007b), whereas in another study of Santacesaria et al. (2007a), a sulphonic active group-functionalized copolymer matrix gel was applied as a catalyst as well.

It must be pointed out, however; that major drawbacks - such as reactor corrosion and the substantial generation of by-products and waste materials, including the salts formed as a result of mineral acid neutralization, which must be disposed of in the environment - represent insurmountable problems for the mineral-acid-catalyzed process.

The kinetics of biodiesel production can usually be differentiated by the mechanism of transesterification process. As mentioned before, in the novel kinetic studies of biodiesel production process, the use of tin chloride as a catalyst, lipases as biocatalysts, and supercritical alcohol without catalyst is being investigated.
In most industrial biodiesel processes, however; the oil as a reactant is added to methoxide (methanol and an alkali metal base) and mixed for some time at a temperature just below the boiling point of methanol, followed by settling and washing. This type of process represents the most investigated area of research in the field of base-catalyzed biodiesel production kinetics (Noureddini & Zhu, 1997; Vicente et al., 2006; Bambase et al., 2007; Stamenkovic et al., 2008; Slinn & Kendall, 2009). These base-catalyzed processes and the corresponding kinetic studies include the use of either sodium (Noureddini & Zhu, 1997; Bambase et al., 2007; Slinn & Kendall, 2009) or potassium (Vicente et al., 2006; Stamenkovic et al., 2008, Klofutar et al., 2010) hydroxide as the catalyst.
Fig. 9 shows the effect of catalyst loading on the process of the methanolysis reactions in the catalyst concentration range 0.5–2.5 wt. % at 65 °C. As expected, the alkyl ester production rate increases with catalyst concentration. A lower rate of transesterification is observed at a lower concentration of catalyst (0.5 wt. %) with the swift reaction termination at 89% equilibrium conversion. It is believed that, at this point, all of the available hydroxide catalyst had been saponified. Bambase et al. (2007) observed the same effect of catalyst concentration on reaction system. Increasing the catalyst concentration to 1.5 wt. %, 2.0 wt. % or even 2.5 wt. % speeds up the reactions, giving a more rapid apparent rate of alkyl ester production during the initial period of process although the conversion maximum does not differ much once a critical catalyst content is reached. It does not seem necessary to use the catalyst concentration higher than 1.5 wt. %, since adding surplus hydroxide catalyst leads to the formation of excess soaps resulting in the very low recovery of alkyl (usually methyl) esters during phase separation. At the end, the optimal catalyst loading is highly dependent on the quality of oil, especially in terms of free fatty acid and water content.

4. Conclusion

Fatty acid alkyl (usually methyl) esters, known as biodiesel, have a great potential as the substitutes for conventional diesel fuel. Rudolph Diesel is said to have already conducted experiments with vegetable oils as fuel for his engine in the late 1890s. In recent years, the biodiesel production from vegetable oils and animal fats has gained much attention, because of its eco-friendly nature, liquid state, and easy portability. However, this green fuel production is problematic because it comes into direct competition with the production of food, consequently causing the prices of the latter to increase. We should thus be careful not to extend the production of biofuel so much that it would significantly contribute to the loss of cultivated land. Biodiesel is also more expensive than fossil fuels, which limits its applications. Biofuels appear to be the only realistic short term source of renewable energy as an alternative to the fossil fuels in transportation sector. Their liquid nature and compatibility with traditional fuels are their main strengths. The Renewable Energy Directive (RED) of the European Union requires biofuels (biodiesel and bioethanol) to reach 10% of total automotive fuel consumption by the year 2020. Biofuels should contribute to a 35% CO_2 reduction at the introduction of the new directive, aiming to reach 50% by 2017 (60% for new production plants). On the other hand, diesel continues its strong growth,

while gasoline is in decline. At the same time, the production of biodiesel is growing and it is expected to reach around 18 million tons of biodiesel per year in EU-27 by 2020. This forecast represents around 50% compliance with the EU directive target. The recent suggestion for an EU target of 10% by 2020 is considered ambitious by most analysts.
The wide-spread popularity of biodiesel in recent years has also led to the founding of biodiesel-powered electrical plant in Slovenia in August 2010.

5. References

Al-Zuhair, S.; Dowaidar, A. & Kamal, H. (2009). Dynamic modeling of biodiesel production from simulated waste cooking oil using immobilized lipase. *Biochemical Engineering Journal*, 44, 2-3, (May 2009) (256-262), 1369-703X

Ataya, F.; Dube, M.A. & Ternan, M. (2006). Single-phase and two-phase base-catalyzed transesterification of canola oil to fatty acid methyl esters at ambient conditions. *Industrial & Engineering Chemistry Research*, 45, 15, (July 2006) (5411-5417), 0888-5885

Ataya, F.; Dube, M.A. & Ternan, M. (2007). Acid-catalyzed transesterification of canola oil to biodiesel under single- and two-phase reaction conditions. Energy & Fuels, 21, 4, (July-August 2007) (2450-2459), 0887-0624

Ataya, F.; Dube, M.A. & Ternan, M. (2008a). Transesterification of canola oil to fatty acid methyl ester (FAME) in a continuous flow liquid-liquid packed bed reactor. *Energy & Fuels*, 22, 5, (September-October 2008) (3551-3556), 0887-0624

Ataya, F.; Dube, M.A. & Ternan, M. (2008b). Variables affecting the induction period during acid-catalyzed transesterification of Canola oil to FAME. *Energy & Fuels*, 22, 1, (January-February 2008) (679-685), 0887-0624

Bambase, M.E.; Nakamura, N.; Tanaka, J. & Matsumura M. (2007). Kinetics of hydroxide-catalyzed methanolysis of crude sunflower oil for the production of fuel-grade methyl esters. *Journal of Chemical Technology and Biotechnology*, 82, 3, (March 2007) (273-280), 0268-2575

Bueno, A.V.; Velásquez, J.A. & Milanez, L.F. (2009). Effect of soybean oil ethyl ester/diesel fuel blends on engine efficiency. *International Journal of Vehicle Design*, 50, 1-4, (May 2009) (229-247), 0143-3369

Busto, M.; D'Ippolito, S.A.; Yori, J.C.; Iturria, M.E.; Pieck, C.L.; Grau, J.M. & Vera C.R. (2006). Influence of the axial dispersion on the performance of tubular reactors during the noncatalytic supercritical transesterification of triglycerides. *Energy & Fuels*, 20, 6, (November 2006) (2642-2647), 0887-0624

Carmona, M.; Valverde, J.L.; Perez, A.; Warchol, J. & Rodriguez, J.F. (2009). Purification of glycerol/water solutions from biodiesel synthesis by ion exchange: sodium removal Part I. *Journal of Chemical Technology and Biotechnology*, 84, 5, (May 2009) (738-744), 0268-2575

Cintas, P.; Mantegna, S.; Gaudino, E.C. & Cravotto, G. (2010). A new pilot flow reactor for high-intensity ultrasound irradiation. Application to the synthesis of biodiesel. *Ultrasonics Sonochemistry*, 17, 6, (August 2010) (985-989), 1350-4177

Colucci, J.A.; Borrero, E.E. & Alape, F. (2005). Biodiesel from an alkaline transesterification reaction of soybean oil using ultrasonic mixing. *Journal of the American Oil Chemists Society*, 82, 7, (July 2005) (525-530), 0003-021X

Deshmane, V.G.; Gogate, P.R. & Pandit, A.B. (2009). Ultrasound-assisted synthesis of biodiesel from palm fatty acid distillate. *Industrial & Engineering Chemistry Research*, 48, 17, (September 2009) (7923-7927), 0888-5885

Duangsuwan, W.; Tuzun, U. & Sermon, P.A. (2009). Configurations and dynamics of single air/alcohol gas-liquid compound drops in vegetable oil. *Chemical Engineering Science*, 64, 13, (July 2009) (3147-3158), 0009-2509

Dube, M.A.; Tremblay, A.Y. & Liu, J. (2007). Biodiesel production using a membrane reactor. *Bioresource Technology*, 98, 3, (February 2007) (639-647), 0960-8524

Fan, X.H.; Wang, X. & Chen, F. (2010). Ultrasonically assisted production of biodiesel from crude cottonseed oil. *International Journal of Green Energy*, 7, 2, (March 2010) (117-127), 1543-5075

Fjerbaek, L.; Christensen, K.V. & Norddahl, B. (2009). A review of the current state of biodiesel production using enzymatic transesterification. *Biotechnology and Bioengineering*, 102, 5, (April 2009) (1298-1315), 0006-3592

Frascari, D.; Zuccaro, M.; Pinelli, D. & Paglianti, A. (2008). A pilot-scale study of alkali-catalyzed sunflower oil transesterification with static mixing and with mechanical agitation. *Energy & Fuels*, 22, 3, (May-June 2008) (1493-1501), 0887-0624

Frascari, D.; Zuccaro, M.; Paglianti, A. & Pinelli, D. (2009). Optimization of mechanical agitation and evaluation of the mass-transfer resistance in the oil transesterification reaction for biodiesel production. *Industrial & Engineering Chemistry Research*, 48, 16, (August 2009) (7540-7549), 0888-5885

Furukawa, S.; Uehara, Y. & Yamasaki, H. (2010). Variables affecting the reactivity of acid-catalyzed transesterification of vegetable oil with methanol. *Bioresource Technology*, 101, 10, (May 2010) (3325-3332), 0960-8524

Guan, G.Q.; Kusakabe, K.; Sakurai, N. & Moriyama, K. (2007). Synthesis of biodiesel fuels at room temperature in the presence of dimethyl ether. *Chemistry Letters*, 36, 12, (December 2007) (1408-1409), 0366-7022

Guan, G.; Kusakabe, K.; Sakurai, N. & Moriyama, K. (2009). Transesterification of vegetable oil to biodiesel fuel using acid catalysts in the presence of dimethyl ether. *Fuel*, 88, 1, (January 2009) (81-86), 0016-2361

Guan, G.Q.; Teshima, M.; Sato, C.; Son, S.M.; Irfan, M.F.; Kusakabe, K.; Ikeda, N. & Lin, T.J. (2010). Two-phase flow behavior in microtube reactors during biodiesel production from waste cooking oil. *AIChE Journal*, 56, 5, (May 2010) (1383-1390), 0001-1541

Halim, S.F.A.; Kamaruddin, A.H. & Fernando, W.J.N. (2009). Continuous biosynthesis of biodiesel from waste cooking palm oil in a packed bed reactor: Optimization using response surface methodology (RSM) and mass transfer studies. *Bioresource Technology*, 100, 2, (January 2009) (710-716), 0960-8524

Hanh, H.D.; Dong, N.T.; Starvarache, C.; Okitsu, K.; Maeda, Y. & Nishimura, R. (2008). Methanolysis of triolein by low frequency ultrasonic irradiation. *Energy Conversion and Management*, 49, 2, (February 2008) (276-280), 0196-8904

Harvey, A.P.; Mackley, M.R & Seliger, T. (2003). Process intensification of biodiesel production using a continuous oscillatory flow reactor. *Journal of Chemical Technology and Biotechnology*, 78, 2-3, (February-March 2003) (338-341), 0268-2575

Hilal, S.H.; Karickhoff, S.W. & Carreira, L.A. (2004). Prediction of the solubility, activity coefficient and liquid/liquid partition coefficient of organic compounds. *QSAR & Combinatorial Science*, 23, 9, (November 2004) (709-720), 1611-020X

Hou, X.L.; Qi, Y.Q.; Qiao, X.G.; Wang, G.F.; Qin, Z.F. & Wang, J.G. (2007). Lewis acid-catalyzed transesterification and esterification of high free fatty acid oil in subcritical methanol. *Korean Journal of Chemical Engineering*, 24, 2, (March 2007) (311-313), 0256-1115

Huang, K.; Xu, Q.L.; Zhang, S.P.; Ren, Z.W. & Yan, Y.J. (2009). Multi-step controlled kinetics of the transesterification of crude soybean oil with methanol by Mg(OCH3)(2). *Chemical Engineering & Technology*, 32, 10, (October 2009) (1595 1604), 0930-7516

Iglauer, S. & Warnecke, H.J. (2009). Simulation and experimental validation of two-phase flow in an aerosol-counter flow reactor using computational fluid dynamics. *Chemical Engineering & Technology*, 32, 6, (June 2009) (939-947), 0930-7516

Kai, T. ; Mataki, Y. ; Nakazato, T. & Takanashi, H. (2010). Optimization of reaction conditions of two-step batch operation for biodiesel fuel production using KOH catalyst. *Journal of Chemical Engineering of Japan*, 43, 1, (January 2010) (90-94), 0021-9592

Kashid, M.N. & Kiwi-Minsker, L. (2009). Microstructured reactors for multiphase reactions: State of the art. *Industrial & Engineering Chemistry Research*, 48, 14, (July 2009) (6465-6485), 0888-5885

Kelkar, M.A.; Gogate, P.R. & Pandit, A.B. (2008). Intensification of esterification of acids for synthesis of biodiesel using acoustic and hydrodynamic cavitation. *Ultrasonics Sonochemistry*, 15, 3, (March 2008) (188-194), 1350-4177

Klofutar, B.; Golob, J.; Likozar, B.; Klofutar, C.; Zagar, E. & Poljansek, I. (2010). The transesterification of rapeseed and waste sunflower oils: Mass-transfer and kinetics in a laboratory batch reactor and in an industrial-scale reactor/separator setup. *Bioresource Technology*, 101, 10, (May 2010) (3333-3344), 0960-8524

Lam, M.K.; Lee, M.T. & Mohamed, A.R. (2010). Homogeneous, heterogeneous and enzymatic catalysis for transesterification of high free fatty acid oil (waste cooking oil) to biodiesel: A review. *Biotechnology Advances*, 28, 4, (July-August 2010) (500-518), 0734-9750

Lee, J.H.; Kim, S.B.; Park, C.; Tae, B.; Han, S.O. & Kim, S.W. (2010). Development of batch and continuous processes on biodiesel production in a packed-bed reactor by a mixture of immobilized Candida rugosa and Rhizopus oryzae lipases. *Applied Biochemistry and Biotechnology*, 161, 1-8, (May 2010) (365-371), 0273-2289

Leung, D.Y.C.; Wu, X. & Leung, M.K.H. (2010). A review on biodiesel production using catalyzed transesterification. *Applied Energy*, 87, 4, (April 2010) (1083-1095), 0306-2619

Li, W.; Du, W.; Liu, D.H. & Yao, Y. (2008). Study on factors influencing stability of whole cell during biodiesel production in solvent-free and tert-butanol system. *Biochemical Engineering Journal*, 41, 2, (September 2008) (111-115), 1369-703X

Lifka, J. & Ondruschka, B. (2004a). Influence of mass transfer on the production of biodiesel. *Chemie Ingenieur Technik*, 76, 1-2, (January 2004) (168-171), 0009-286X

Lifka, J. & Ondruschka, B. (2004b). Influence of mass transfer on the production of biodiesel. *Chemical Engineering & Technology*, 27, 11, (November 2004) (1156-1159), 0930-7516

Liu, C.H. & Chang, J.S. (2008). Lipolytic activity of suspended and membrane immobilized lipase originating from indigenous Burkholderia sp C20. *Bioresource Technology*, 99, 6, (April 2008) (1616-1622), 0960-8524

Liu, X.J.; Piao, X.L.; Wang, Y.J. & Zhu, S. L. (2010). Model study on transesterification of soybean oil to biodiesel with methanol using solid base catalyst. *Journal of Physical Chemistry A*, 114, 11, (March 2010) (3750-3755), 1089-5639

Marjanovic, A.V.; Stamenkovic, O.S.; Todorovic, Z.B.; Lazic, M.L. & Veljkovic, V.B. (2010). Kinetics of the base-catalyzed sunflower oil ethanolysis. *Fuel*, 89, 3, (March 2010) (665-671), 0016-2361

Massingill, J.L.; Patel, P.N.; Guntupalli, M.; Garret, C. & Ji, C. (2008). High efficiency nondispersive reactor for two-phase reactions. *Organic Process Research & Development*, 12, 4, (July-August 2008) (771-777), 1083-6160

Melero, J.A.; Bautista, L.F.; Iglesias, J.; Morales, G.; Sanchez-Vazquez, R. & Suarez-Marcos, I. (2010). Biodiesel production over arenesulfonic acid-modified mesostructured catalysts: Optimization of reaction parameters using response surface methodology. *Topics in Catalysis*, 53, 11-12, (July 2010) (795-804), 1022-5528

Narvaez, P.C.; Sanchez, F.J. & Godoy-Silva, R.D. (2009). Continuous methanolysis of palm oil using a liquid-liquid film reactor. *Journal of the American Oil Chemists Society*, 86, 4, (April 2009) (343-352), 0003-021X

Noureddini, H. & Zhu, D. (1997). Kinetics of transesterification of soybean oil. *Journal of the American Oil Chemists Society*, 74, 11, (November 1997) (1457-1463), 0003-021X

Park, J.Y.; Kim, D.K.; Wang, Z.M. & Lee, J.S. (2009). Fast biodiesel production with one-phase reaction. *Applied Biochemistry and Biotechnology*, 154, 1-3, (May 2009) (246-252), 0273-2289

Qiu, Z.Y.; Zhao, L.N. & Weather, L. (2010). Process intensification technologies in continuous biodiesel production. *Chemical Engineering and Processing*, 49, 4, (April 2010) (323-330), 0255-2701

Santacesaria, E.; Tesser, R.; Di Serio, M.; Guida, M.; Gaetano, D. & Agreda, A.G. (2007a). Kinetics and mass transfer of free fatty acids esterification with methanol in a tubular packed bed reactor: A key pretreatment in biodiesel production. *Industrial & Engineering Chemistry Research*, 46, 15, (July 2007) (5113-5121), 0888-5885

Santacesaria, E.; Tesser, R.; Di Serio, M.; Guida, M.; Gaetano, D.; Agreda, A.G. & CamMarchota, F. (2007b). Comparison of different reactor configurations for the reduction of free acidity in raw materials for biodiesel production. *Industrial & Engineering Chemistry Research*, 46, 25, (December 2007) (8355-8362), 0888-5885

Sdrula N. (2010). A study using classical or membrane separation in the biodiesel process. *Desalination*, 250, 3, (January 2010) (1070-1072), 0011-9164

Semiat R. (2008). Energy issues in desalination processes. *Environmental Science & Technology*, 42, 22, (November 2008) (8193-8201), 0013-936X

Sengo, I.; Gominho, J.; d'Orey, L.; Martins, M.; d'Almedia-Duarte, E.; Pereira, H. & Ferreira-Dias, S. (2010). Response surface modeling and optimization of biodiesel production from Cynara cardunculus oil. *European Journal of Lipid Science and Technology*, 112, 3, (March 2010) (310-320), 1438-7697

Sim, J.H.; Kamaruddin, A.H. & Bhatia, S. (2009). Effect of mass transfer and enzyme loading on the biodiesel yield and reaction rate in the enzymatic transesterification of crude palm oil. *Energy & Fuels*, 23, 9, (September 2009) (4651-4658), 0887-0624

Sitshebo, S.; Tsolakis, A. & Theinnoi, K. (2009). Promoting hydrocarbon-SCR of NOx in diesel engine exhaust by hydrogen and fuel reforming. *International Journal of Hydrogen Energy*, 34, 18, (September 2009) (7842-7850), 0360-3199

Slinn, M. & Kendall, K. (2009). Developing the reaction kinetics for a biodiesel reactor. *Bioresource Technology*, 100, 7, (April 2009) (2324-2327), 0960-8524

Song, S.H.; Lee, S.H.; Park, D.R.; Kim, H.; Woo, S.Y.; Song, W.S.; Kwon, M.S. & Song, I.K. (2009). Direct preparation of dichloropropanol from glycerol and hydrochloric acid gas in a solvent-free batch reactor: Effect of experimental conditions. *Korean Journal of Chemical Engineering*, 26, 2, (March 2009) (382-386), 0256-1115

Soriano, N.U.; Venditti, R. & Argyropoulos, D.S. (2009). Biodiesel synthesis via homogeneous Lewis acid-catalyzed transesterification. *Fuel*, 88, 3, (March 2009) (560-565), 0016-2361

Sotoft, L.F.; Westh, P.; Christensen, K.V. & Norddahl, B. (2010). Novel investigation of enzymatic biodiesel reaction by isothermal calorimetry. *Thermochimica Acta*, 501, 1-2, (March 2010) (84-90), 0040-6031

Stamenkovic, O.S.; Todorovic, Z.B.; Lazic, M.L.; Veljkovic, V.B. & Skala D.U. (2008). Kinetics of sunflower oil methanolysis at low temperatures. *Bioresource Technology*, 99, 5, (March 2008) (1131-1140), 0960-8524

Stamenkovic, O.S.; Veljkovic, V.B.; Todorovic, Z.B.; Lazic, M.L.; Bankovic-Ilic, I.B. & Skala D.U. (2010). Modeling the kinetics of calcium hydroxide catalyzed methanolysis of sunflower oil. *Bioresource Technology*, 101, 12, (June 2010) (4423-4430), 0960-8524

Tesser, R.; Casale, L.; Verde, D.; Di Serio, M. & Santacesaria, E. (2009). Kinetics of free fatty acids esterification: Batch and loop reactor modeling. *Chemical Engineering Journal*, 154, 1-3, (November 2009) (25-33), 1385-8947

Tesser, R.; Di Serio, M.; Casale, L.; Sannino, L.; Ledda, M. & Santacesaria, E. (2010). Acid exchange resins deactivation in the esterification of free fatty acids. *Chemical Engineering Journal*, 161, 1-2, (July 2010) (212-222), 1385-8947

Tsuji, T.; Kubo, M.; Shibasaki-Kitakawa, N. & Yonemoto, T. (2009). Is excess methanol addition required to drive transesterification of triglyceride toward complete conversion?. *Energy & Fuels*, 23, 12, (December 2009) (6163-6167), 0887-0624

Vicente, G.; Marchtinez, M.; Aracil, J. & Esteban, A. (2005). Kinetics of sunflower oil methanolysis. *Industrial & Engineering Chemistry Research*, 44, 15, (July 2005) (5447-5454), 0888-5885

Vicente, G.; Marchtinez, M. & Aracil, J. (2006). Kinetics of Brassica carinata oil methanolysis. *Energy & Fuels*, 20, 4, (July 2006) (1722-1726), 0887-0624

Veljkovic, V.B.; Stamenkovic, O.S.; Todorovic, Z.B.; Lazic, M.L.; & Skala D.U. (2009). Kinetics of sunflower oil methanolysis catalyzed by calcium oxide. *Fuel*, 88, 9, (September 2009) (1554-1562), 0016-2361

Volpato, G.; Rodrigues, R.; Heck, J.X. & Ayub, M.A.Z. (2009). et al.Title: Effects of oxygen volumetric mass transfer coefficient and pH on lipase production by Staphylococcus warneri EX17. *Biotechnology and Bioprocess Engineering*, 14, 1, (February 2009) (105-111), 1226-8372

Xu, L.; Weathers, P.J.; Xiong, X.R. & Liu, C.Z. (2009). Microalgal bioreactors: Challenges and opportunities. *Engineering in Life Sciences*, 9, 3, (June 2009) (178-189), 1618-0240

Yori, J.C.; D'Ippolito, S.A.; Pieck, C.L. & Vera, C.R. (2007). Deglycerolization of biodiesel streams by adsorption over silica beds. *Energy & Fuels*, 21, 1, (January-February 2007) (347-353), 0887-0624

Zhang, Y.; Stanciulescu, M. & Ikura, M. (2009). Rapid transesterification of soybean oil with phase transfer catalysts. *Applied Catalysis A-General*, 366, 1, (September 2009) (176-183), 0926-860X

Zeng, J.; Du, W.; Liu, XY.; Liu, D.H. & Dai, L.M. (2006). Study on the effect of cultivation parameters and pretreatment on Rhizopus oryzae cell-catalyzed transesterification of vegetable oils for biodiesel production. *Journal of Molecular Catalysis B-Enzymatic*, 43, 1-4, (December 2006) (15-18), 1381-1177

20

Condensation Capture of Fine Dust in Jet Scrubbers

M.I. Shilyaev and E.M. Khromova
Department of heating and ventilation,
Tomsk State University of Architecture and Building
Russia

1. Introduction

Dust particles with a size of hundredths and tenths of a micron can be hardly separated from the cleaned gas flow in the inertial dust catchers without their preliminary agglomeration. This processing can be carried out in the sound field; as a result of this processing the small particles become significantly larger because of acoustic coagulation (according to (Uzhov & Valdberg, 1972), soot particles with a size of 0.03 μm enlarge to 5 μm) or because of condensation of water vapors on these particles. In the last case efficient dust capture is provided by a rise of inertia of particles, where water vapors condense during particle motion in the working section of apparatus, and, hence, by a rise of probability of their collision with droplets of irrigating fluid (Fig. 1), for instance, in jet scrubbers. Thus, at soot capture from cracked gases in the counter-flow jet scrubber (Fig. 2) the efficiency of about 90% was achieved at high inlet moisture content in gases (about 1 kg/kg of dry gases) (Uzhov & Valdberg, 1972, Shilyaev et al., 2006). The conditions of intensive vapor condensation on particles should be achieved already at the inlet; and reliable prognosis and optimization of dust capture require the adequate mathematical models. It can be scarcely determined in experiment. Some generalizing experimental data on condensation dust capture in different devices of the wet type are shown in (Uzhov & Valdberg, 1972, Valdberg & Savitskaya, 1993). However, it is shown in (Shilyaev et al., 2008) that in these devices it is wrong to generalize dust capture efficiency vs. moisture content by a single dependence (as it was done in the mentioned papers) because the processes of dust capture there are determined by different mechanisms (Shilyaev et al., 2006).

The authors failed to find well-grounded theoretical publications on combined heat and mass transfer and condensation dust capture on droplets in jet scrubbers. The attempt of such modeling is shown in (Pazhi & Galustov, 1984), one of the chapters of this book presents one-dimensional model, but unfortunately, it is shown with numerous misprints and it is difficult to understand. According to equations, the moisture content and thermodynamics of the vapor-gas flow were not considered in general in connection with heat and mass transfer with particles and droplets. Therefore, well-illustrated correlation between calculations and experimental data on dust capture causes some doubts. At that, there is no the description of experimental conditions and required physical, operation, and geometrical parameters.

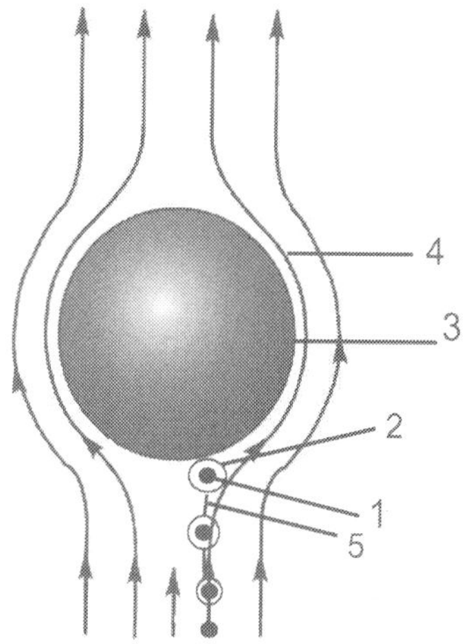

Fig. 1. The scheme of interaction of a particle with condensate on the surface with a droplet of irrigating fluid: 1 - particle; 2 - condensate on particle surface; 3 - droplet; 4 - stream line of the vapor-gas flow; 5 - formation trajectory of a particle with condensate on the surface

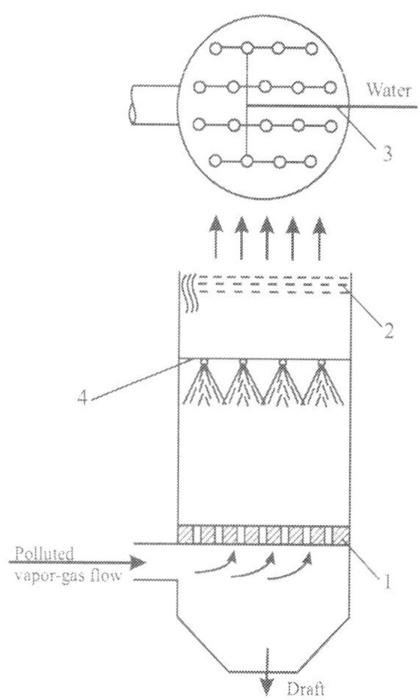

Fig. 2. The scheme of the counter-current hollow jet scrubber: 1 - gas-distributing grid; 2 - droplet-catcher; 3 - water collector; 4 - jets

The above-mentioned data prove the necessity of more complete and accurate statement of the discussed problem.

The current paper represents the main points and results of implementation of the physical-mathematical model on condensation capture of submicron particles in jet scrubbers, occurring by the inertial mechanism, proved by the experimental data of (Uzhov & Valdberg, 1972).

2. Main assumptions and equations of the model

The current model was based on the model of heat and mass transfer of fluid droplets with the vapor-gas flow in irrigation chamber (Shilyaev & Khromova, 2008, Shilyaev et al., 2008). It was modified by addition of the equation of particle motion, continuity equation for their "spread" density (mass concentration), which considers the effect of collision of particles with condensate on the surface, and equations of heat and mass transfer of new formations with the vapor-gas flow. At that, we have considered a change in the mass of irrigating fluid droplets both via mass transfer with the vapor-gas flow and precipitation of formations on their surface. The equation of droplet heat transfer considers both the latent heat of phase transition on droplet surface and heat transfer of new formations due to their precipitation on droplets. The equation of moisture content considers the mass fluxes from new formations together with the mass fluxes caused by fluid evaporation (vapor condensation) from (on) droplets. The continuity equation for the "spread" density of irrigating fluid takes into account a change in its mass both due to total phase transition on droplets and formations and due to absorption of formations by droplets at their collision. The scheme of collision of droplet and particle with condensate on the surface is shown in Fig. 1.

In general we consider the motion and heat and mass transfer of the three-phase four-component heterogeneous medium of fluid droplets–monodispersed submicron solid particles–vapor-gas mixture in the hollow scrubbers with jet irrigation in the framework of the model of various-temperature and various-velocity continua. At that, considering the small size of particles with condensate on the surface, their motion with the vapor-gas mixture is assumed dynamically equilibrium, but temperature-unbalanced. This model takes into account condensation of vapors on droplets and particles or evaporation depending on local thermodynamics parameters in the vapor-gas flow. The effect of precipitation of particles with condensate on the surface on droplets is taken into account with the help of capture coefficient, determined by the empirical formula of Langmuir-Blodgett. It is known that the higher the content of vapor in gas, the more efficient the process of fine particle capture in the scrubber. Therefore, the current model takes into account the effect of increased moisture content on mass transfer of droplets and particles with condensate on their surface by the correction for the Stefan flow. Volumetric concentrations of droplets and particles are assumed low, what is typical for operation of scrubbers and irrigation chambers of air-conditioning units; therefore, the equations of liquid and solid phase motion in the framework of continuum model are turned to the motion equations for separate droplets and particles. At that, we consider that the droplets and submicron particles are monodispersed, with some, for instance, mean-mass diameter.

The model presentation of the problem is shown in Fig. 3. This schematic view of irrigation chamber was chosen to formulate the model equations in more general form, considering that the chambers can be horizontal (β=0) or vertical ($\beta=\pi/2$) or even inclined at unconditioned angle. The counter-flow jet scrubbers are vertical, irrigation chambers of air-conditioning units are horizontal (predominantly) and vertical (sometimes), Venturi tubes are horizontal, vertical and inclined at unconditioned angle (in battery arrangement). In Venturi scrubber dust is captured in the diffuser (diverging) part of Venturi tube, which should be taken into account in the equation of the vapor-gas flow velocity. At that, the scheme can be counter-flow a) and direct-flow b), what is determined by supply of fluid and vapor-gas flow. In the counter-flow scheme a) fluid is fed from one side with the following initial parameters: Θ_0 is droplet temperature, $q = Q_f/Q$ is irrigation coefficient (Q_f and Q are volumetric flow rates of fluid and vapor-gas flow), δ_{d0} is mean-mass size of droplets generated by jets and calculated via the jet parameters (Uzhov & Valdberg, 1972, Pazhi & Galustov, 1984, Vitman et al., 1962), V_{d0} is the initial velocity of droplets also calculated via jet parameters. The vapor-gas mixture with small particles is fed from the opposite side with the following initial parameters: T_{00}, d_0, U_0 are temperature, moisture content and velocity of the vapor-gas flow. In direct-flow scheme b) fluid and vapor-gas mixture and particles are fed from one side. At interaction the particles with condensate on the surface collide with droplets, absorbed by them and entrained into sludge.

Let's write down model equations in the general vector form.

The equation of droplet motion with consideration of their mass variability because of condensation or evaporation and absorption of formations by droplet is

$$\frac{d\vec{V}_d}{d\tau} = \vec{g} + \vec{R}_d - \frac{\vec{V}_d}{m_d}\frac{dm_d}{d\tau}, \tag{1}$$

where $\vec{V}_d, m_d, \vec{R}_d$ is velocity vector, droplet mass and vector of aerodynamic resistance acting on the droplet per a unit of its mass, and $\vec{g}$ is vector of gravity acceleration, τ is time.

The motion equation for particles with condensate on the surface is (equilibrium is assumed because of the small size of formations)

$$\vec{V}_{pf} = \vec{U}, \tag{2}$$

where $\vec{V}_{pf}$ and $\vec{U}$ are velocity vectors of formations and vapor-gas flow.

The continuity equation for irrigating fluid is

$$\frac{d\rho_d}{d\tau} = \frac{\rho_d}{m_d}\frac{dm_d}{d\tau} - \rho_d \nabla \vec{V}_d, \tag{3}$$

where ρ_d is "spread" density of droplets.

The equation of mass transfer of fluid droplet with the flow of vapor-gas mixture due to evaporation-condensation and with particles due to their precipitation on droplets is

$$\frac{dm_d}{d\tau} = \frac{2\pi M_1 K_s D\delta_d}{RT}(P_1 - P_{1d})\Phi + \rho_{pf} V_r \frac{\pi\delta_d^2}{4}\eta_{Stk}, \tag{4}$$

or

$$\frac{dm_d}{d\tau} = \beta_d \pi\delta_d^2(\rho_1 - \rho_{1d}) + \rho_{pf} V_r \frac{\pi\delta_d^2}{4}\eta_{Stk}, \tag{4'}$$

where M_1 is molecular mass of vapor (for steam M_1=18 kg/kmole), D is coefficient of vapor diffusion in mixture, δ_d is current size of the droplet, R=8.314·10³ kJ/kmole·K is universal gas constant, T is absolute temperature of the vapor-gas flow, P_1 and P_{1d} are partial pressures of vapors far from the droplet and on its surface, corresponding to saturation temperature, which equals temperature of droplet surface Θ_s, K_s is correction for the Stefan flow at high moisture contents (Uzhov & Valdberg, 1972, Amelin, 1966):

$$K_s = 1 + \frac{P_1 + P_{1d}}{2B},$$

$B = P_1 + P_2$ is barometric pressure, P_2 is partial pressure of dry gas in the flow, and Φ is Frossling correction for the inertia of the flow past the droplet (Shilyaev et al., 2006):

$$\Phi = 1 + 0,276\mathrm{Re}_d^{0,5} \cdot \mathrm{Sc}^{0,33},$$

Re_d is Reynolds number of the flow past the droplet:

$$\mathrm{Re}_d = \frac{V_r \delta_d \rho}{\mu},$$

Sc is Schmidt number:

$$\mathrm{Sc} = \frac{\mu}{\rho D},$$

$V_r = \left|\vec{V}_d - \vec{U}\right|$ is absolute value of relative droplet velocity, ρ and μ are density and dynamic viscosity of the vapor-gas mixture, ρ_{pf} is "spread" density of particle formations with condensate on the surface, η_{Stk} is coefficient of formation capture by fluid droplets, β_d is mass transfer of a droplet by concentration difference of vapors, ρ_1 and ρ_{1d} are partial densities (mass concentrations of vapor) far from the droplet and near its surface, calculated from state equations

$$\rho_1 = \frac{P_1 M_1}{RT}, \quad \rho_{1d} = \frac{P_{1d} M_1}{R\Theta_s}.$$

The equation of mass transfer of particle formation with condensate on the surface with the flow of vapor gas mixture (Maxwell equation) is

$$\frac{dm_{pf}}{d\tau} = \frac{2\pi M_1 K_s D \delta_{pf}}{RT}\left(P_1 - P_{1p}\right), \tag{5}$$

or

$$\frac{dm_{pf}}{d\tau} = \beta_{pf} \pi \delta_{pf}^2 \left(\rho_1 - \rho_{1p}\right), \tag{5'}$$

where m_{pf} is mass of formation, δ_{pf} is diameter of formation, P_{1p} is partial pressure of vapors near the formation surface, calculated by its temperature T_{pf}, corresponding to saturation temperature, β_{pf} is coefficient of formation mass transfer by concentration difference, and ρ_{1p} is partial density of vapors near the formation surface, calculated from the equation of state

$$\rho_{1p} = \frac{P_{1p} M_1}{R T_{pf}} .$$

The continuity equation for particles with condensate on the surface with consideration of their precipitation on droplets of irrigating fluid is

$$\frac{d\rho_{pf}}{d\tau} = \frac{\rho_{pf}}{m_{pf}} \frac{dm_{pf}}{d\tau} - \rho_{pf} \nabla \vec{U} - \rho_{pf} V_r \frac{\pi \delta_d^2}{4} \eta_{Stk} \frac{\rho_d}{m_d} , \tag{6}$$

where $\rho_d / m_d = n_d$ is calculating droplet concentration in the vapor-gas flow.

The continuity equation for dry particles with consideration of precipitation of particle formation with condensate on the surface on droplets of irrigating fluid is

$$\frac{d\rho_p}{d\tau} = -\rho_p \nabla \vec{U} - \rho_p V_r \frac{\pi \delta_d^2}{4} \eta_{Stk} \frac{\rho_d}{m_d} , \tag{7}$$

where ρ_p is "spread" density of dry particles in the flow.

The equation for moisture content of the vapor-gas flow is

$$\frac{dd}{d\tau} = \frac{W}{(1 - \varepsilon_{dp}) \rho_2} , \tag{8}$$

where d is moisture content made up of the mass of fluid vapors per one kilogram of dry part of gas in the mixture,

$$\rho_2 = \frac{P_2 M_2}{RT} ,$$

ρ_2 is partial density of dry gases in the flow, M_2 is molecular mass of dry gases (for air M_2=29 kg/kmole), ε_{dp} is total volumetric concentration of droplets and particles in the flow, as usual ε_{dp}<<1, W is the total mass withdrawal (condensation), income (evaporation of droplets and formations) of fluid vapors in the flow per a time unit in a volume unit of the vapor-gas mixture:

$$W = -\left[\frac{dm'_d}{d\tau} \frac{\rho_d}{m_d} + \frac{dm_{pf}}{d\tau} \frac{\rho_{pf}}{m_{pf}} \right] ,$$

where

$$\frac{dm'_d}{d\tau} = \frac{2\pi M_1 K_s D \delta_d}{RT} (P_1 - P_{1d}) \Phi ,$$

or

$$\frac{dm'_d}{d\tau} = \beta_d \pi \delta_d^2 (\rho_1 - \rho_{1d}) .$$

The equation for velocity of the vapor-gas mixture along chamber axis x (Shilyaev & Khromova, 2008) is

$$U = U_0 \frac{T}{T_{00}} \frac{K+d}{K+d_0}, \tag{9}$$

where $K = M_1/M_2$.

The equation of heat transfer of a droplet with the vapor-gas mixture flow is

$$c_d m_d \frac{\mathrm{d}\Theta}{\mathrm{d}\tau} = -\alpha_d \pi\delta_d^2(\Theta_s - T) + r_f \frac{\mathrm{d}m_d'}{\mathrm{d}\tau} + c_{pf}\rho_{pf}V_r \frac{\pi\delta_d^2}{4}\eta_{Stk}T_{pf}, \tag{10}$$

where Θ is mean-mass temperature of droplet, Θ_s is temperature of droplet surface, c_d is heat capacity of droplet together with absorbed particles, then we will assume it equal to fluid heat capacity c_f because of low concentration of particles, r_f is specific heat of the phase transition evaporation-condensation, then we will assumed $c_{pf}= c_f$.

The equation of heat transfer between particle formations with condensate on the surface and the flow of vapor-gas mixture is

$$c_{pf} m_{pf} \frac{\mathrm{d}T_{pf}}{\mathrm{d}\tau} = -\alpha_{pf}\pi\delta_{pf}^2(T_{pf} - T) + r_f \frac{\mathrm{d}m_{pf}}{\mathrm{d}\tau}, \tag{11}$$

where a_{pf} is heat transfer of formation.

The equation for the temperature of vapor-gas mixture is

$$\rho \frac{\mathrm{d}cT}{\mathrm{d}\tau} = \alpha_d \pi\delta_d^2(\Theta_s - T)\frac{\rho_d}{m_d} + \alpha_{pf}\pi\delta_{pf}^2(T_{pf} - T)\frac{\rho_{pf}}{m_{pf}}. \tag{12}$$

where c is heat capacity of the vapor-gas mixture.

Equations (1)-(12) should be solved under the following initial conditions:

$$\text{at } \tau = 0 \quad \vec{V}_d = \vec{V}_{d0}, \vec{U} = \vec{U}_0, \rho_d = \rho_{d0}, m_d = m_{d0}, m_{pf} = m_{p0}, \rho_{pf} = \rho_{p0}, d = d_0, \Theta = \Theta_0, T = T_{00} \tag{13}$$

The model equations can be solved in the stationary statement in the Euler coordinate system for direct flow (assignment of initial conditions for the droplets of irrigating fluid, vapor-gas mixture and particles on one side of the apparatus at x=0) and for counter flow (assignment of initial conditions for the droplets of irrigating fluid on one side of the apparatus at x=0, and for the vapor-gas mixture and particles on another side at x=l). At that $\frac{\partial}{\partial t} = 0$ and

$$\frac{\mathrm{d}_i(\ldots)}{\mathrm{d}\tau} = \vec{V}_i \nabla(\ldots), \tag{14}$$

where index i determines substantial derivative. If axis x coincides with apparatus axis and direction of the vapor-gas flow, relationship (14) is

$$\frac{d\vec{V}_d}{d\tau} = \vec{V}_d \nabla \vec{V}_d\,,\ \frac{d\rho_d}{d\tau} = \vec{V}_d \nabla \rho_d\,,\ \frac{dm_d}{d\tau} = \vec{V}_d \nabla m_d\,,\ \frac{dm_{pf}}{d\tau} = \vec{U} \nabla m_{pf} = U\frac{\partial m_{pf}}{\partial x}\,,$$

$$\frac{d\rho_{pf}}{d\tau} = \vec{U}\nabla\rho_{pf} = U\frac{\partial \rho_{pf}}{\partial x}\,,\ \frac{d\rho_2}{d\tau} = \vec{U}\nabla\rho_2 = U\frac{\partial\rho_2}{\partial x}\,,\ \frac{dd}{d\tau} = \vec{U}\nabla d = U\frac{\partial d}{\partial x}\,,\ \frac{d\Theta}{d\tau} = \vec{V}_d\nabla\Theta\,,$$

$$\frac{dT_{pf}}{d\tau} = \vec{U}\nabla T_{pf} = U\frac{\partial T_{pf}}{\partial x}\,,\ \frac{d(cT)}{d\tau} = \vec{U}\nabla(cT) = U\frac{\partial (cT)}{\partial x}\,.$$

The efficiency of dust particle capture is determined by expression:

$$\eta = 1 - \frac{1}{h\rho_{p0}U_0}\int_0^h (\rho_p U)_{x=l}\,dy\,, \tag{15}$$

where h is chamber height (width).
In equation (1) and then, if we assume droplet spherical,

$$m_d = \frac{\pi\delta_d^3}{6}\rho_f\,, \tag{16}$$

where ρ_f is density of irrigating fluid, and δ_d is droplet diameter. With consideration of gravity forces in the coordinate system of Fig. 3 the velocity vector of irrigating fluid droplets has two components V_{dx} and V_{dy}. The strength of aerodynamic drag $\vec{R}_d$, effecting the droplet, per a unit of its mass is:

$$\vec{R}_d = \tilde{\xi}_d \frac{\left(\vec{V}_d - \vec{U}\right)}{\tau_d}\,, \tag{17}$$

where

$$\tau_d = \frac{\rho_f \delta_d^2}{18\mu}\,. \tag{18}$$

In (17) $\tilde{\xi}_d = \xi_d/\xi_s$ is relative coefficient of droplet resistance, ξ_d is actual coefficient of droplet resistance, and $\xi_s = 24/\mathrm{Re}_d$ is Stokes coefficient of droplet resistance.
The density of vapor-gas flow is determined as the sum of partial densities of mixture components:

$$\rho = \rho_1 + \rho_2\,. \tag{19}$$

The dynamic viscosity of vapor-gas flow is calculated through the partial pressure of binary mixture components (Shilyaev & Khromova, 2008), or on the basis of molecular-kinetic theory of gases (Shilyaev et al., 2008).
Let's take relative resistance coefficient $\tilde{\xi}_d$ by the known approximation dependence, valid for a wide range of Reynolds numbers $\mathrm{Re}_d = 0{,}1 - 3\cdot 10^5$ (Shilyaev et al., 2006):

$$\tilde{\xi}_d = 1 + 0{,}197\mathrm{Re}_d^{0,63} + 2{,}6\cdot 10^{-4}\mathrm{Re}_d^{1,38}\,. \tag{20}$$

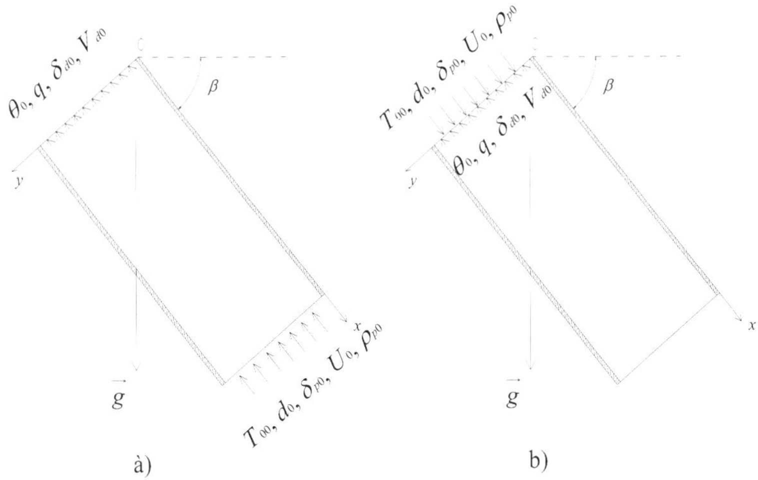

Fig. 3. The schemes of irrigation chambers: a) - counter flow; b) - direct flow

Let's take capture coefficient η_{Stk}, determining efficiency of precipitation of particle formations with condensate on their surface on droplets of irrigating fluid by empirical dependence of Langmuir-Blodgett with the correction of N.A. Fuks for the capture effect (Shvydkiy & Ladygichev, 2002):

$$\eta_{\mathrm{Stk}} = \left(\frac{\mathrm{Stk}}{\mathrm{Stk}+0,5}\right)^2 + 2,5\frac{\delta_{pf}}{\delta_d}, \tag{21}$$

where Stk is Stokes number:

$$\mathrm{Stk} = \tau_\delta \frac{V_r}{\delta_d}, \quad \tau_\delta = \frac{\rho_f \delta_{pf}^2}{18\mu}. \tag{22}$$

The second summand in equation (21) determines the capture effect.

In equation (4') β_d is coefficient of droplet mass transfer with the vapor-gas flow by concentration difference of vapors $\Delta\rho = \rho_1 - \rho_{1d}$, $\rho_{1d} = P_{1d}(\Theta_s)M_1/R\Theta_s$ (in calculations we will assume $\Theta_s \approx \Theta$, what is reasonable, as it is shown in (Shilyaev & Khromova, 2008), for droplets with the size of up to 600-800 μm):

$$\beta_d = 2\frac{D}{\delta_d}K_s\Phi. \tag{23}$$

In equation (5) m_{pf} is mass of formation:

$$m_{pf} = \frac{\pi\delta_{pf}^3}{6}\rho_{pf}^0 \approx \frac{\pi\delta_{pf}^3}{6}\rho_f. \tag{24}$$

where ρ_{pf}^{0} is averaged density of formation, then, we will assume $\rho_{pf}^{0} \approx \rho_{f}$.
In equation (5') β_{pf} is coefficient of formation mass transfer with the vapor-gas flow:

$$\beta_{pf} = 2\frac{D}{\delta_{pf}}K_{s}. \tag{25}$$

In equation (10) we determine the coefficient of droplet heat transfer a_d by Drake formula:

$$\mathrm{Nu}_{d} = \frac{\alpha_{d}\delta_{d}}{\lambda} = 2 + 0,495\mathrm{Re}_{d}^{0,55}\mathrm{Pr}^{0,33} \quad , \tag{26}$$

Nu_d is Nusselt number of heat transfer, λ is coefficient of heat conductivity for the vapor-gas mixture calculated by partial pressures (Shilyaev & Khromova, 2008) or on the basis of molecular-kinetic theory of gases for the binary mixture (Shilyaev et al., 2008), $\mathrm{Pr} = \frac{\mu c}{\lambda}$, where c is heat capacity of the vapor-gas flow:

$$c\rho = c_{1}\rho_{1} + c_{2}\rho_{2}, \tag{27}$$

then

$$c = \frac{c_{1}d + c_{2}}{1 + d}, \tag{28}$$

since by equation

$$d = \frac{\rho_{1}}{\rho_{2}}, \tag{29}$$

c_1 and c_2 are heat capacities of vapor and dry gas (for steam c_1=1.8 kJ/kg·K, for air c_2=1.005 kJ/kg·K).
The second summand in the right part of (10) is latent heat of phase transition of droplet evaporation or condensation of vapors on the droplet, the third summand in the right part is heat transferred by formations on the droplet. Let's determine heat transfer coefficient of formation a_{pf} by equation

$$\alpha_{pf} = 2\frac{\lambda}{\delta_{pf}} \tag{30}$$

because of its small size.
Equation (12) describes a change in enthalpy per a time unit in a volume unite of the vapor-gas flow because of its convective heat transfer with droplets of irrigating fluid and formations.
The initial condition for mass concentration of fluid is set as

$$\rho_{d0} = q\rho_{\mathrm{f}}\frac{U_{0}}{V_{d0}} \quad . \tag{31}$$

The condition of vapor condensation on particles follows from the equation of mass transfer, for instance, (5'): $\rho_1 - \rho_{1p} > 0$, then $dm_{pf}/d\tau > 0$. This condition can be written at τ=0 in the form $\left(\frac{P_1 M_1}{RT_{00}} - \frac{P_{1p}(T_{00}) M_1}{RT_{00}}\right)_{\tau=0} > 0$, or $P_1 - P_{1p} > 0$.

Since from state equation $P_1 = B\frac{d_0}{K+d_0}$, $P_{1p} = B\frac{d_{1p}}{K+d_{1p}}$, d_{1p} is moisture content determined by temperature at chamber inlet T_{00}, the condition for the beginning of liquid vapor condensation on particles takes form

$$d_0 > K\frac{a}{1-a}, \tag{32}$$

where

$$a = \frac{P_{1p}(T_{00})}{B}. \tag{33}$$

For instance, for steam and air at B=101325 Pa, T_{00}=333 K (60 ^{0}C), P_{1p}=0.199·10^5 Pa, K=18/29=0,621. Then, a=0.1964 and d_0 > 0.152 kg/kg of dry air. Therefore, condition (32) should be taken into account at realization of the above problem.

3. Comparison of calculation results with experimental data

Results of model implementation for the experimental data on soot capture in the jet scrubber by the method of methane electric cracking from cracked gases are shown in Fig. 4. On the basis of data from (Uzhov & Valdberg, 1972), we managed to determine approximately the physical parameters of cracked gases through the comparison with the molecular weights of the known gases: $M_g = 11.24\, kg/kmole$ is molecular mass; $c_g = 2.4\, kJ/kg \cdot K$ is specific heat capacity at constant pressure; coefficients of dynamic viscosity are

$$\mu_g = 6{,}47 \cdot 10^{-6}\left(\frac{T}{T_0}\right)^{1,7}, \text{ Pa·s}, \; T_0 = 273\,\text{K}, \tag{34}$$

and coefficients of heat conductivity are

$$\lambda_g = 1{,}34 \cdot 10^{-2}\left(\frac{T}{T_0}\right)^{1,7}, \; W/m \cdot K, \tag{35}$$

coefficient of steam diffusion in cracked gases is:

$$D_0 = 13{,}1 \cdot 10^{-6}\left(\frac{T}{T_0}\right)^{3/2}, \text{ m}^2/\text{s}. \tag{36}$$

Calculations have been carried out for the following conditions (Uzhov & Valdberg, 1972, Table XIII.1, p.221):

- inlet gas temperature - $t_0 = 160 - 180^0\,$C;

- outlet gas temperature – $t_{out} = t(H) = 50-55^0$ C;
- inlet velocity of vapor-gas flow – $U_0 = 0,25$ m/s;
- irrigation coefficient – $q = 7,1\cdot 10^{-3}$ m³/m³;
- inlet soot concentration – $\rho_{p0} = 1.72\,g/m^3\left(2.8\,g/m^3\left(normalconditions\right)\right)$;
- outlet soot concentration – $\rho_{pout} = \rho_p(H) = 0.356\,g/m^3\left(0.425g/m^3\left(normalconditions\right)\right)$;
- inlet water temperature – $\theta_0 = 20^0$ C;
- scrubber diameter – $D = 3$ m;
- scrubber height – $H = 12,75$ m;
- water pressure on jets (evolvent) – $P_f = 300$ kPa;
- jet nozzle diameter – $d_n = 12$ mm;
- density of cracked gases under normal conditions – $\rho_g = 0,51 kg/m^3\left(normalconditions\right)$.

Approximated calculation of the size of irrigating fluid droplets by (Uzhov & Valdberg, 1972)gives the values of mean-mass diameter δ_{d0}=700 µm and initial velocity of droplets $V_{d0} = 24,5$ m/s. The estimate of moisture content difference by empirical data of (Uzhov & Valdberg, 1972) allowed determination of $\Delta d = 0,849$ kg/kg of dry air for the given experiment. Exhaustive search of inlet moisture contents (there is no d_0 in experimental data) for determination of experimental value $\Delta d \approx 0.85$ kg/kg of dry gas in calculations allows us to take $d_0 = 0.93$ kg/kg of dry gas. In calculations we have also taken the initial size of soot particles $\delta_{p0} = 0.1$ µm.
Let's determine the efficiency values by the ratio of mass flow rates of particles at the scrubber outlet and inlet by formula (15) with consideration of dependence (9):

$$\eta = 1 - \frac{\rho_p(H)\cdot U(H)}{\rho_{p0}\cdot U_0} = 1 - \frac{0,356}{1,72}\frac{T(H)}{T_{00}}\frac{K + d(H)}{K + d_0} =$$

$$= 1 - \frac{0,356}{1,72}\frac{325,2}{443}\frac{\left(\frac{18}{11,24} + 0,081\right)}{\left(\frac{18}{11,24} + 0,93\right)} = 0,899 \ \mathbf{(89,9\%)},$$

where $d(H)$ and d_0 are taken by calculation because there is no experimental value of velocity $U(H)$ in [1], $\rho_p(H)$ and ρ_{p0} are real concentrations of particles at scrubber outlet and inlet, and temperatures $T(H)$ and T_{00} are assumed average from data presented.
The theoretical value of efficiency for the given version of calculation is η=89.3 %, and the diagrams in Fig. 4 prove that.
Calculation results on parameters described by the suggested model are shown in Fig. 4. According to the diagrams, the "spread" density (mass concentration) of dry particles increases drastically at first, then it starts decreasing slowly. An increase is caused by a fast reduction in velocity of the vapor-gas flow because of a significant withdrawal of vapors via their condensation of droplets and particles; then particles with condensation on the surface are entrapped by droplets and dust concentration in the flow decreases. In this case the size of particles increases by the factor of 3.5; i.e., their mass increases by the factor of 43.

Calculated outlet gas temperatures differ significantly from the experimental ones, and we suppose that this is connected with uncertainty of assignment of initial moisture content and averaging of temperature within 20^0 C from the measured values.

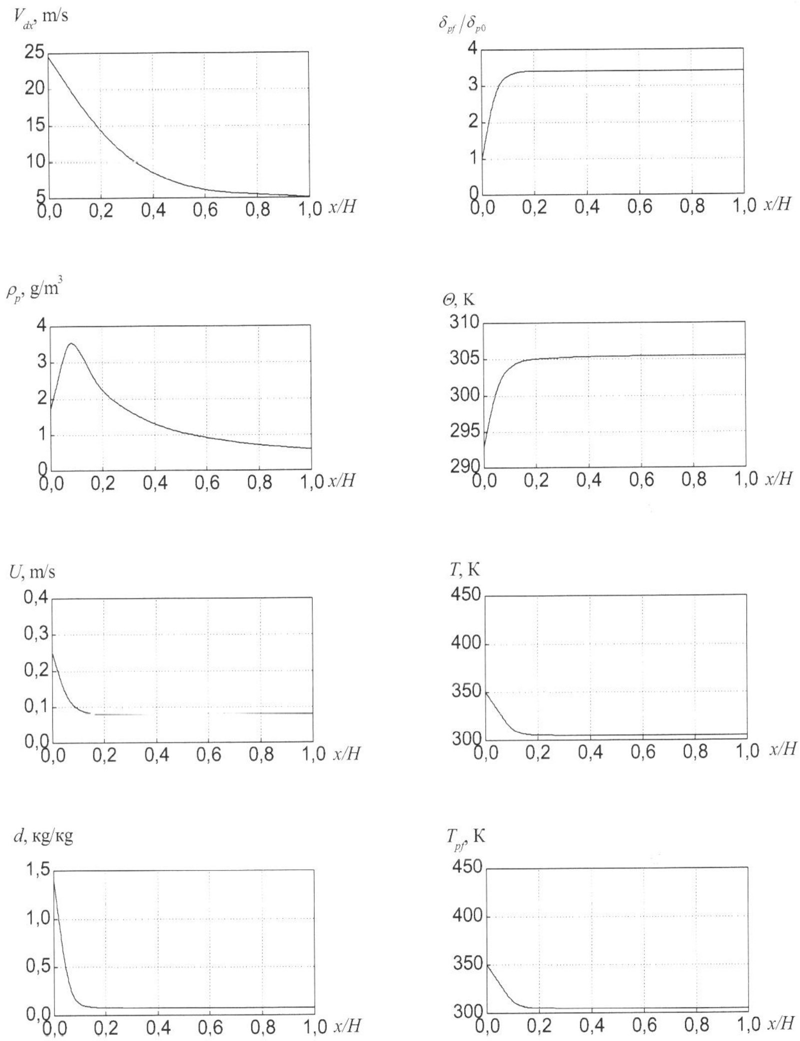

Fig. 4. Results of model calculations: H=12.75 m, q=7.1 $\cdot 10^{-3}$ m^3/m^3, δ_{d0}=7$\cdot 10^{-4}$ m, V_{d0}=24.5 m/s, Θ_0=293 K, T_{00}=443 K, d_0=0.93 kg/kg, U_0=0.25 m/s, δ_{d0}=10^{-7} m, ρ_{p0}=1.72 g/m^3

As we can see, the theoretical results correlate well with the experimental values, what proves model efficiency.

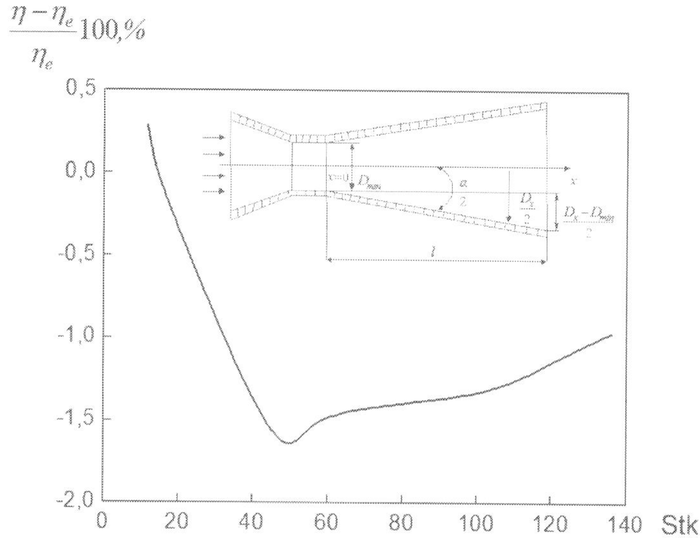

Fig. 5. Comparison of model with experimental data under isothermal conditions for Venturi scrubber: η was determined by formula (15) and η_e - by formula (37)

To assure additionally efficiency of the model, the process of dust capture on water droplets from air was calculated with the use of this model under the isothermal conditions (at t=20 ^{0}C) without mass transfer in a standard Venturi tube (Uzhov & Valdberg, 1972). Calculation results are compared in Fig. 5 with the experimental data described by the known dependence of fractional efficiency on Stokes number (Uzhov & Valdberg, 1972)

$$\eta_e = 1 - \mathrm{e}^{-q \cdot 10^{-3} b \sqrt{\mathrm{Stk}}}, \quad \mathrm{Stk} = \frac{\rho_p^0 \delta_{p0}^2}{18\mu} \frac{V_{r0}}{\delta_{d0}}, \tag{37}$$

at q=0.5·10^{-3} m^3/m^3, b=1.5 (b is constructive parameter: b=1.25-1.56 (Uzhov & Valdberg, 1972, Shilyaev et al., 2006)). Difference in a wide range of Stokes numbers Stk does not exceed 2 %. In calculations velocity of the vapor-gas flow was determined by formula

$$U = U_0 \frac{T}{T_{00}} \frac{K + d}{K + d_0} \left(\frac{D_{min}}{D_x} \right)^2, \tag{38}$$

where U_0 is velocity of the vapor-gas flow in the throat of tube with diameter D_{min}, D_x is the current diameter of diffuser:

$$D_x = D_{min} + 2x\mathrm{tg}\frac{\alpha}{2},$$

α is diffuser angle, and x is coordinate along the tube axis. The size of fluid droplets, fed into the tube throat, was calculated by formula of Nukiyama-Tanasava (Shilyaev et al., 2006, Shvydkiy & Ladygichev, 2002):

$$\delta_{d0} = \frac{0,585}{V_{r0}}\sqrt{\frac{\sigma_f}{\rho_f}} + 53,4\left(\frac{\mu_f}{\sqrt{\rho_f \sigma_f}}\right)^{0,45} q^{1,5}\text{, м,} \quad (39)$$

where σ_f is coefficient of surface tension of fluid (for water σ_f=0.072 N/m), ρ_f is fluid density (for water ρ_f=10^3 kg/m^3), μ_f is coefficient of dynamic viscosity of fluid (for water μ_f=10^{-3} Pa·s at t=20 ^{0}C), $V_{r0} = \left|V_g - V_{d0}\right|$, V_g is gas velocity in the throat of Venturi tube, and V_{d0} is velocity of droplets in the throat of Venturi tube, assumed equal to 4-5 m/s. The density of particles was taken conditionally $\rho_p^0 = 10^3$ kg/m^3. The diameter of tube throat was taken D_{min}=0.1 m, length of diffuser part was l=1 m, and angle was α=6^0.

4. Condensation effect of single particle enlargement in irrigation chamber

Results of model (Shilyaev et al., 2008) implementation together with mass transfer equation for a single submicron droplet (5') under the condition of fluid vapor condensation on it (32) are shown in Figs. 6-9 for the air-water system (calculations have been carried out at T_{00}=333 K, δ_{d0}=500 μm, q=0.001 m^3/m^3, Θ_0=293 K, U_0=3 m/s, V_{d0}=12 m/s; q is coefficient of irrigation; $V_{d0}, U_0, \Theta_0, T_{00}$ are inlet velocities and temperatures of irrigating fluid droplets and vapor-gas flow; δ_{d0} is initial size of irrigating fluid droplets; δ_0 is initial size of submicron droplet; d_0 is moisture content at the inlet to the chamber; and l is chamber length). The effect of collision between submicron droplet and irrigating fluid droplet was not taken into account.

According to Figs. 6 and 7, at high moisture contents the condensation effect is very strong and inverse to initial size δ_0. The droplet size for δ_0=0.1 μm increases by the factor of 450 up to 45 μm, for δ_0=0.01 μm it increases by 4500 times up to the same size. These formations can be efficiently captured even independently in vortex drop catchers.

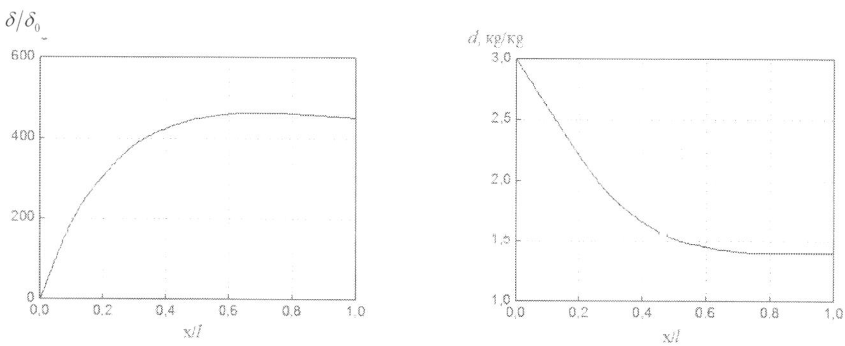

Fig. 6. Condensation of fluid vapors in a vertical chamber in direct flow on droplet with size δ_0=10^{-7} m: l=2 m, d_0=3 kg/kg of dry air

Results of calculations under outstanding conditions of condensation (32) are shown in Fig. 8. In this case critical value is d_0=0.15 kg/kg of dry air. According to the figure, the droplet with initial size δ_0=0.1 μm evaporates along the whole chamber and disappears almost at the chamber inlet.

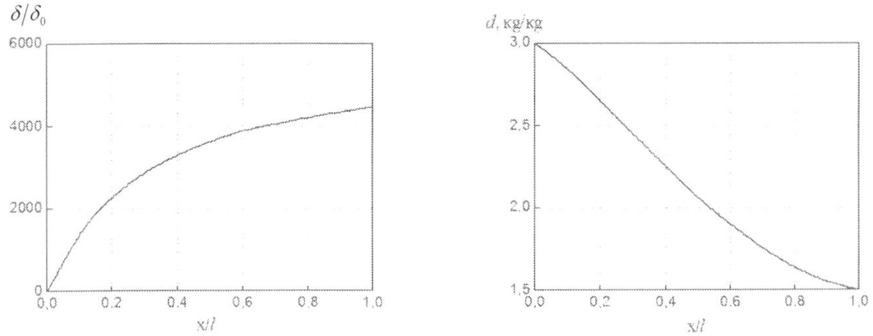

Fig. 7. Condensation of fluid vapors in a vertical chamber in direct flow on droplet with size $\delta_0=10^{-7}$ m: l=1m, d_0=3 kg/kg of dry air

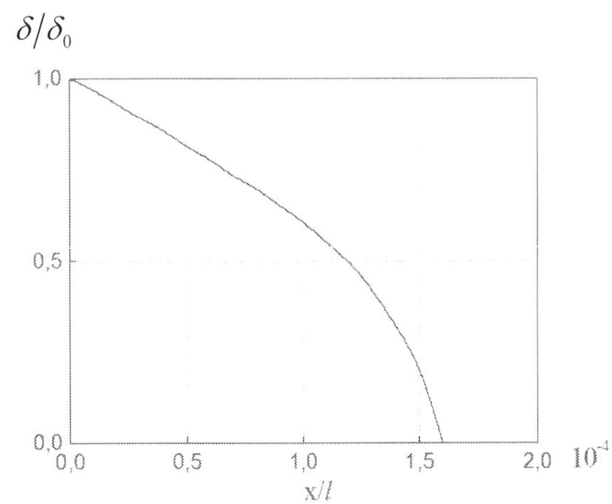

Fig. 8. Droplet evaporation in the vertical chamber at the direct flow: $\delta_0=10^{-7}$ m, l=1m, d_0=0.15 kg/kg of dry air

Calculation results for condition (32) satisfied at the inlet are shown in Fig. 9. The process of mass transfer between particle and flow starts from condensation. The droplet size at initial moisture content d_0=0.17 kg/kg of dry air increases until the middle of chamber length and becomes equal to 4 μm (by the factor of 40), then it starts evaporating and at the distance of 0.7 l it disappears turning to vapor.

Therefore, condensation processes in irrigation chambers under some certain conditions can effect positively the efficiency of submicron particle capture, but these conditions can be achieved only on the basis of adequate mathematical models similar to the suggested one including model equations (Shilyaev et al., 2008) combined by mass and heat balance, heat and mass transfer equations of particles under the conditions of their absorption by fluid droplets at the motion along the chamber.

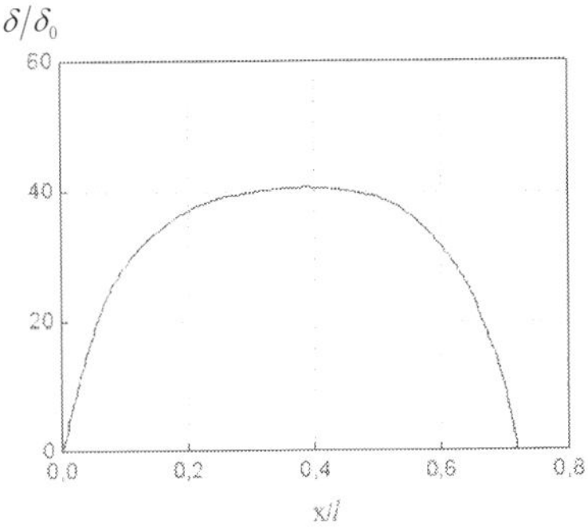

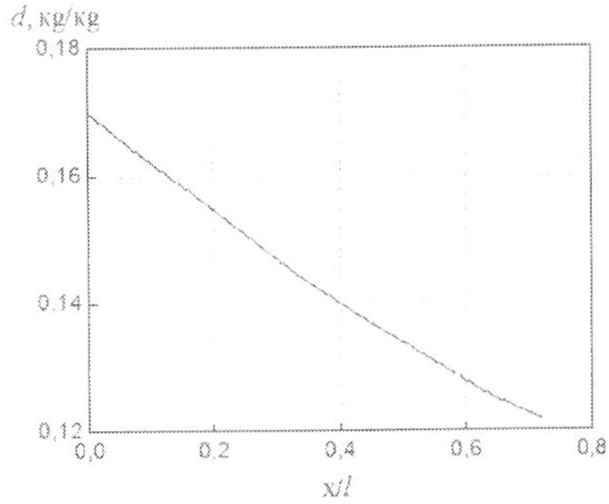

Fig. 9. Condensation is evaporation of a droplet in the vertical chamber at direct flow: $\delta_0=10^{-7}$ m, l=1m, d_0=0.17 kg/kg of dry air

Let's determine the average velocity of vapor at its condensation on a droplet from balance relationship

$$\frac{\Delta m}{\Delta \tau} = \pi \bar{\delta}^2 \bar{w}_v \bar{\rho}_v = \frac{\pi}{6}\left(\delta_d^3 - \delta_0^3\right)\frac{\rho_f}{\Delta \tau}, \tag{40}$$

where Δm is droplet mass increase during time $\Delta\tau$ of its passing along distance l, for instance, the chamber length; $\bar{w}_v$ is vapor velocity to the surface of condensation (droplet); $\bar{\rho}_v$ is the average vapor density on distance l near the droplet surface calculated by its temperature equal to the temperature of saturation; δ_0 and δ_d are initial and final diameters of droplet; ρ_f is droplet density; $\bar{\delta}$ is average size of a droplet on distance l. Time is $\Delta\tau = l/U$, where U is velocity of the vapor-gas flow along the chamber axis.

If we assume $\bar{\delta} = \sqrt{\delta_d \delta_0}$, from (40) we can obtain

$$\bar{w}_v \approx \frac{1}{6}\frac{\delta_d^{\,2}}{\delta_0}\frac{U}{l}\frac{\rho_f}{\bar{\rho}_v}. \tag{41}$$

In equation (41) we neglect summand δ_d^2/δ_0, since $\delta_0 << \delta_d$. It can be seen from (41) that velocity $\bar{w}_v$ is reverse to the initial size of droplets. This regularity can be also obtained directly from the equation of droplet mass transfer:

$$\frac{dm}{d\tau} = \pi\delta^2 w_v \rho_v = \beta\pi\delta^2\left(\rho_{1d} - \rho_1\right). \tag{42}$$

It follows from (42) that

$$w_v = \beta\frac{\rho_{1d} - \rho_1}{\rho_{1d}}, \tag{43}$$

but since we can assume for small droplets, as it was already mentioned, $\beta = 2\frac{D}{\delta}$, it follows from equation (43) that $w_v \sim \frac{1}{\delta}$.

Let's estimate with the help of formula (41) condensation rates for Figs. 6 and 7:
at $l = 0.1\text{m}$

$$\delta_{d1}/\delta_{01} = 450\,,\ \delta_{01} = 10^{-1}\mu\text{m}\,,\ \delta_{d1} = 45\,\mu\text{m}\,;$$

at $l = 1\text{m}$

$$\delta_{d2}/\delta_{02} = 4500\,,\ \delta_{02} = 10^{-2}\mu\text{m}\,,\ \delta_{d2} = 45\,\mu\text{m}\,.$$

Assuming in (41) that values of $U/\overline{\rho}_v$ differ slightly in these two considered cases, we obtain

$$\frac{\overline{w}_{v1}}{\overline{w}_{v2}} = \left(\frac{\delta_{d1}}{\delta_{d2}}\right)^2 \frac{\delta_{02}}{\delta_{01}}\,. \tag{44}$$

Along distance $l = 0{,}1\text{м}$ $\frac{\delta_{d1}}{\delta_{01}} = 1700\,,\ \delta_{02} = 10^{-2}\mu\text{m}\,,\ \delta_{d2} = 17\,\mu\text{m}\,;\ \frac{\delta_{d2}}{\delta_{02}} = 170\,,\ \delta_{01} = 10^{-1}\mu\text{m}\,,$ $\delta_{d1} = 17\,\mu\text{m}$.

As we can see, in two considered cross-sections of chamber two calculation versions give the same final size of droplets: in the first case it is 45 μm and in the second it is 17 μm.
Thus, it follows from (44) in connection with $\delta_{d1} \approx \delta_{d2}$ that

$$\frac{\overline{w}_{v1}}{\overline{w}_{v2}} = \frac{\delta_{02}}{\delta_{01}}\,. \tag{45}$$

Relationship (45) proves the fact that the diffusion mechanism of small particle deposition on large droplets is insignificant because of small diffusion velocities of vapors at condensation on their surfaces, and it can be neglected; simultaneously it is very important for small droplets. This conclusion correlates with formula of B.V. Deryagin and S.S. Dukhin (Uzhov & Valdberg, 1972)

$$\eta_d = \frac{144\pi\mu D\left(\rho - \rho_v\right)}{g\rho_p^0\rho_{da}\delta_d\left(\delta_{d0}^2 - \delta_0^2\right)}\,. \tag{46}$$

Here μ is dynamic viscosity of vapor-gas flow, ρ_p^0 is density of particles, ρ_{da} is density of dry air in the vapor-gas mixture, and η_d is capture efficiency of particles with size δ_0 due to the diffusion effect.
According to calculations by formula (46), at $\delta_0 << \delta_d$ the efficiency of submicron dust deposition is low (Shilyaev et al., 2006).

5. Parametrical analysis of condensation capture of fine dust in Venturi scrubber

The Venturi scrubber (VS) is the most common type of wet dust collector for efficient gas cleaning from dust particles even of a micron size. Together with dust capture the absorption and thermal processes can occur in VS. The VS is used in various industries:

ferrous and non-ferrous metallurgy, chemistry and oil industry, production of building materials, power engineering, etc. The construction of VS includes combination of irrigated Venturi tube and separator (drop catcher). The Venturi tube has gradual inlet narrowing (converging cone) and gradual outlet extension (diffuser). A pinch in cross-section of Venturi tube is called a "throat". The operation principle of VS is based on catching of dust particles, absorption or cooling of gases by droplets of irrigating fluid dispersed by the gas flow in Venturi tube. Usually the gas velocity in the throat of scrubber tube is 30-200 m/s, and specific irrigation is 0.1-6.0 l/s^3. In the current section we are considering optimization of possible application of Venturi scrubber for fine dust capture under condensation conditions on the basis of the suggested physical-mathematical model.

Results of calculation on the basis of suggested model for VS are shown in Figs. 10 and 11. According to Fig. 10, at low moisture contents (almost dry air) with a rise of initial particle concentration the efficiency of their capture increases slightly and with an increase in moisture content it decreases (Fig. 10a). At that high efficiency of dust capture can be achieved al low particle concentrations and high moisture contents at the VS inlet (Fig. 10b). Dependence of dust capture efficiency on diffuser angle of Venturi tube α is shown in Fig. 11a, and it is obvious that for the given case the optimal is $\alpha \approx 7.7^0$. For any other case this optimal angle can be calculated by the model.

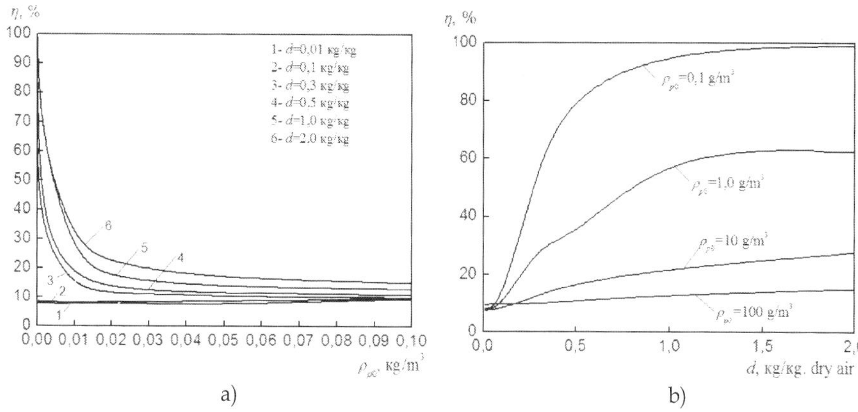

Fig. 10. Effect of initial particle concentration and moisture content on dust capture efficiency: V_0=5 m/s, Θ_0=293 K, ρ_p^0=10^3 kg/m^3, q=0.5·10^{-3} m^3/m^3, U_0= 160 m/s, T_{00}=333 K, a=6^0, l=1m, δ_0=10^{-7} m

Dust capture efficiency vs. relative diffuser length is shown in Fig. 11b, and it can be seen that the optimal length of diffuser tube, which provides the required dust capture efficiency, can be determined with the help of the model. Thus, for this case at required efficiency η=99 % the length of diffuser should be l=1 m.

According to Fig. 12, efficiency depends significantly on the flow velocity in the tube throat and irrigation coefficient. Calculations were carried out for diagram a) at following parameters: l=1 m, V_{d0}=5 m/s, δ=0.1 μm, Θ_0=293 K, a=6^0, ρ_{p0}=1 g/m^3, ρ_p^0=10^3 kg/m^3, q=2 l/m^3, U_0= 80 m/s, T_{00}=303 K, d_0=0.01193 kg/kg of dry air.

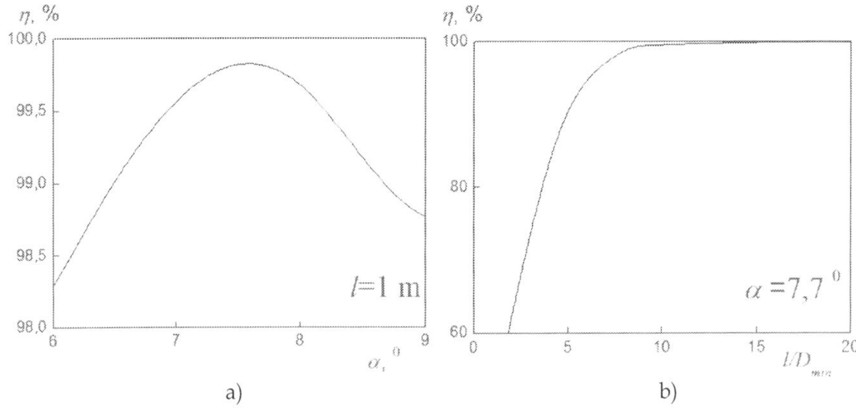

Fig. 11. The effect of diffuser angle (a) and diffuser length (b) on dust capture efficiency: V_0=5 m/s, Θ_0=293 K, $\rho_p^0=10^3$ kg/m^3, $q=10^{-3}$ m^3/m^3, U_0= 80 m/s, T_{00}=333 K, ρ_{p0}=1 g/m^3, d_0=0.5 kg/kg of dry air, $\delta_0=10^{-6}$ m

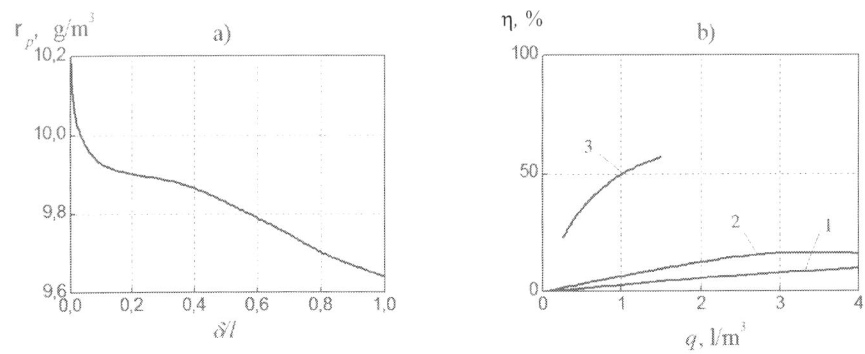

Fig. 12. Calculation results: a) distribution of particle concentration along the diffuser; b) efficiency of particle capture depending on irrigation coefficient: 1 - U_0= 80 m/s, d_0=0.01193 kg/kg of dry air; 2 - U_0= 100 m/s, d_0=0.01193 kg/kg of dry air; 3 - U_0= 100 m/s, d_0=0.5 kg/kg of dry air, other parameters are the same as for Fig. a)

6. Comparison of direct-flow and counter-flow apparatuses of condensation capture of fine dust

It is interesting to compare specific power inputs for gas cleaning from fine dust under the conditions of condensation of particle capture on fluid droplets in the direct-flow and counter-flow apparatuses as well as their sizes under the same conditions. For this purpose let's compare the counter-flow jet scrubber (CJS) and Venturi scrubber (VS) under the same flow rates of cracked gases cleaned from soot particles, corresponding to experimental data of (Uzhov & Valdberg, 1972) for CJS.

Comparative calculations were carried out for the following data. For CJS: q=7.1 $\cdot 10^{-3}$ m^3/m^3, δ_{d0}=700 μm, V_{d0}=24.5 m/s, Θ_0=293 K, T_{00}=443 K, d_0=0.93 kg/kg of dry air, U_0=0.25 m/s, δ_0=0.12 μm, ρ_{p0}=1.72 g/m^3, D=3.0 m, H=2; 3; 6; 9; 11; and 12.5 m. For VS the diameter of tube throat D_{min} was determined from the equilibrium equation for volumetric flow rates of the vapor-gas flow at the inlets of the compared apparatuses. Thus, at U_0=80 m/s D_{min}=0.17 m, at U_0=160 m/s D_{min}=0.12 m. Angle α was varied as well as scrubber length l. Initial size of droplets δ_{d0} for VS was calculated by Nukiyama-Tanasava formula (39) depending on (U_0 – V_{d0}), fluid density ρ_f, q, coefficient of fluid surface tension σ_f (for water σ_f=0.072 N/m); the value of initial velocity of droplets in tube throat V_{d0} was set 4.0 m/s.

Calculation results are generalized in Figs. 13 and 14 for optimal angle α=7.7^0, corresponding to maximal efficiency of dust capture. According to the figures, with an increase in the relative length of Venturi tube and relative height of CJS, the efficiency increases significantly, but the higher l/D_{min} and H/D, the less expressive is this growth. According to Fig. 13b, the efficiency growth is caused firstly by enlargement of "formations" (particles with condensate on their surface). Deceleration of efficiency growth depending on converging cone length and scrubber height is caused by a decrease in particle concentration in the flow and reduction in probability of collisions between "formations" and fluid droplets. The suggested model provides a possibility to determine the optimal length of Venturi tube l or scrubber height H for the required efficiency of dust capture.

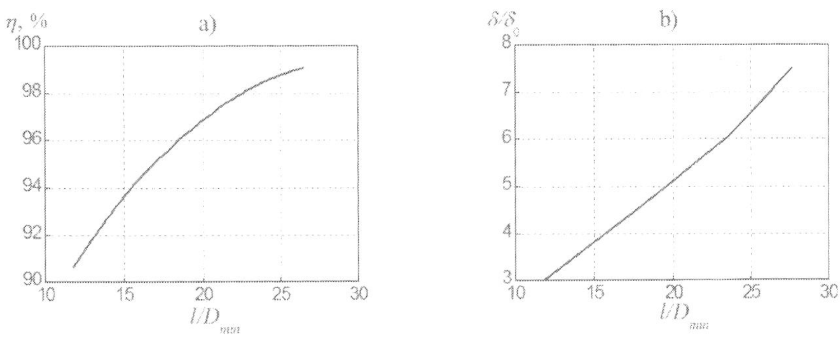

Fig. 13. Results of calculations by the model for Venturi scrubber: q=7.1 $\cdot 10^{-3}$ m^3/m^3, V_{d0}=4.0 m/s, Θ_0=293 K, T_{00}=443 K, d_0=0.93 kg/kg of dry air, U_0=80 m/s, δ_0=1.2$\cdot 10^{-7}$ m, ρ_{p0}=1.72 g/m^3, D_{min}=0.17 m, a=7.7^0

Calculation results on the relative size of "formations" and CJS efficiency under the same conditions as for VS, corresponding to experimental data for CJS on soot capture from cracked gases (Uzhov & Valdberg, 1972), are shown in Fig. 15 depending on the initial temperature of droplets. The height of experimental CJS was H=12.7 m, and diameter was D=3 m. It can be seen from the figure that with a decrease in droplet temperature at the inlet Θ_0 efficiency increases significantly. Thus, an increase in Θ_0 from 293 K (20 ^{0}C) to 278 K (5 ^{0}C) increases efficiency by 8 %. This important result proves the fact that the same experimental efficiency $\eta \approx 90$ % can be obtained at significantly less height of the scrubber. Thus, according to calculations, at Θ_0=278 K (5 ^{0}C) this value of efficiency can be achieved at height $H \approx 4$-5 m instead of 12.7 m, what reduces the dimensions and specific quantity of metal of the whole construction. The point in Fig. 15b indicates the experimental value of efficiency, and this means that model operability is proved well by the experiment.

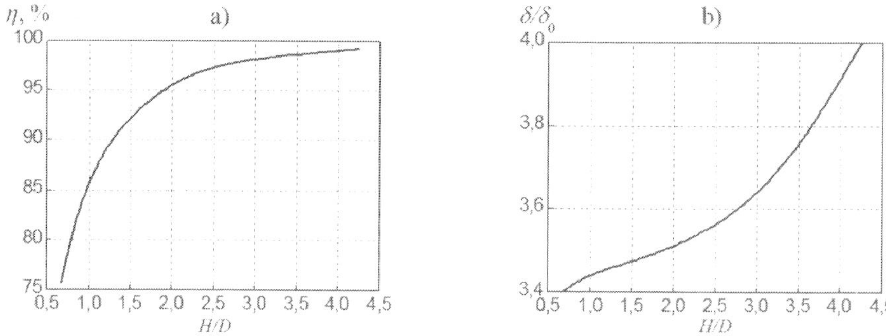

Fig. 14. Results of calculations for CJS: q=7.1·10^{-3} m^3/m^3, δ_{d0}=7·10^{-4} m, V_{d0}=24.5 m/s, Θ_0=293 K, T_{00}=443 K, d_0=0.93 kg/kg of dry air, U_0=0.25 m/s, δ_0=1.2·10^{-7} m, ρ_{p0}=1.72 g/m^3, D=3.0 m

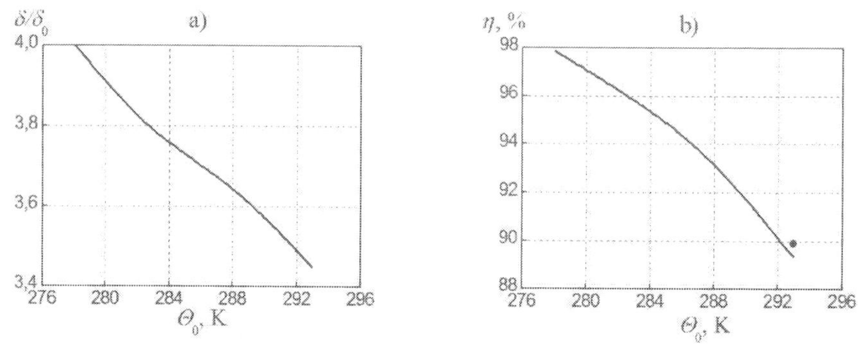

Fig. 15. Results of calculations by the model for CJS: H=12.75 m, q=7.1·10^{-3} m^3/m^3, δ_{d0}=7·10^{-4} m, V_{d0}=24.5 m/s, T_{00}=443 K, d_0=0.93 kg/kg of dry air, U_0=0.25 m/s, δ_0=10^{-7} m, ρ_{p0}=1.72 g/m^3.

According to analysis, for similar required efficiency, the direct-flow dust catchers (in this case they are VS), despite their advantage by dimensions over the counter-flow apparatuses, require higher power inputs for gas cleaning, determined by pressure drops in apparatuses. Actually, for VS the coefficient of hydraulic resistance can be estimated by formula (Shilyaev et al., 2006)

$$\xi_{t.v} = \xi_{d.t}\left(1 + 0{,}63 q^{0,7}\frac{\rho_f}{\rho}\right), \qquad (47)$$

where $\xi_{d.t}$ is resistance coefficient of the dry Venturi tube, it is assumed to be 0.12–0.15, ρ is gas density. Under our conditions for estimate calculation we assume q=7.1·10^{-3} m^3/m^3, ρ_f=10^3 kg/m^3, $\xi_{d.t}$=0.12, $\rho \approx \rho_0 \dfrac{273}{T_m}$, T_m=397.6 K, where T_m=0.5(T_{00}+T_{out}), T_{00}=443 K, T_{out}=352.2 K, and ρ_0 (273 K)=0.51 kg/m^3 (Uzhov & Valdberg, 1972). Then, ρ=0.35 kg/m^3.

Substituting these data into (47), we obtain $\xi_{t.v} = 6{,}9$. Hence, at U_0=80 m/s the pressure drop on the Venturi tube is $\Delta P_{t.v} = \xi_{t.v}\rho\frac{U_0^2}{2} = 7728$ Pa, and at U_0=160 m/s $\Delta P_{t.v} = 30912$ Pa.

This resistance exceeds the resistance of CJS, where the main part of energy is spent for fluid spraying, and hydraulic resistance is low (as usual, the velocity of cleaned gas does not exceed 1 m/s). At that, the same energy is spent for fluid spraying in Venturi tube. Thus, specific energy spent for fluid spraying is

$$\Delta P_f = qP_f,$$

where P_f is pressure of fluid fed to the spraying jets, equal to 300-400 kPa. Thus, in our case we obtain

$$\Delta P_f = 7{,}1\cdot10^{-3}\left(300-400\right)\,10^3 = \left(2130-2840\right)\ \text{J/m}^3.$$

Moreover, to estimate the total dimensions of VS, the sizes of droplet catcher should be added to the sizes of Venturi tube and power inputs for overcoming of hydraulic resistance should be taken into account.

All the above mentioned proves the fact that the counter-flow schemes of condensation dust capture are in preference to the direct-flow ones.

7. Conclusion

Therefore, the physical-mathematical model of heat and mass transfer and condensation capture of fine dust in scrubbers was formulated, and its efficiency was determined. The suggested model can be used for preliminary calculations and estimation of the most rational determining parameters of apparatuses, which provide efficient gas cleaning.

8. References

Amelin, A.G. (1966). *Theoretical Foundations of Fog Formation at Vapor Condensation.* - Moscow: Khimiya.

Pazhi, D.G., Galustov, V.S. (1984). *Foundations of Liquids Spraying Technology.* - Moscow: Khimiya.

Shilyaev, M.I., Khromova, E.M. (2008). Simulation of heat and mass transfer in spray chambers. *Theoretical Foundations of Chemical Engineering*, Vol. 42, No. 4, P. 404-414.

Shilyaev, M.I., Khromova, E.M., Tumashova, A.V. (2008). Physical-mathmatical model of heat and mass transfer process in jet irrigation chambers at high moisture contents. *Izv. Vuziv. Stroitelstvo*, No. 6, P. 75-81.

Shilyaev, M.I., Shilyaev, A.M., Grischenko, E.P. (2006). *Calculation Methods for Dust Catchers.* - Tomsk: Tomsk State University of Architecture and Building.

Shilyaev, M.I., Shilyaev, A.M., Khromova, E.M., Doroshenko, Yu.N. (2008). About condensation mechanisms of dust capture intensification in CJS and DC. *Izv. Vuzov. Stroitelstvo*, No.4, P. 61-67.

Shvydkiy, V.S., Ladygichev, M.G. (2002). *Gas Cleaning: Reference Book.* - Moscow: Teploenergetik.

Uzhov, V.N., Valdberg, A.Yu. (1972). *Gas Cleaning by Wet Filters.* - Moscow: Khimiya.

Valdberg, A.Yu., Savitskaya, N.M. (1993). Calculation of dust capture at condensation operation of scrubbers. *Theoretical Foundations of Chemical Engineering*, Vol. 27, No. 5, P. 526-530.

Vitman, L.A., Katsnelson, B.D., Paleev, I.I. (1962). *Liquid Spraying by Jets*. - Moscow-Leningrad: Gosenergoizdat.

21

Mass Transfer in Filtration Combustion Processes

David Lempert, Sergei Glazov and Georgy Manelis
Institute of Problems of Chemical Physics of Russian Academy of Sciences
Russian Federation

1. Introduction

Wave combustion is one of wide-spread regimes of chemical reactions progress in the systems with the enthalpy excess. Combustion waves in porous medium have some special features, that let consider them as especial kind of combustion processes. Usually one denominates the filtration combustion (FC) as the oxidation of any solid combustible at gaseous oxidizer filtration. The presence of two phase states, intensive heat- and mass exchange between these two phase states, a constant countercurrent flow of solid and gas phases complicate considerably theoretical description of FC wave, as well as experimental results explication. In such systems one has to consider not only heat and concentration fields, but also the gas flow dynamics and heterogeneous reactions peculiarities. Besides it a huge difference between densities of components provides the necessity of common consideration of processes with appreciably different characteristic rates. Anyway due to some peculiarities filtration combustion waves remain very attractive objects for industrial application.

Combustion regimes with heat accumulation occupy an especial place in wave combustion processes. A typical example it is the combustion of a solid fuel at gas oxidizer filtration, when the combustion front direction coincides with the gas flow one (Aldushin et al., 1999; Hanamura et al., 1993; Salganskii et al., 2008). In coordinates, cohered with the combustion front (zone of the exothermic transformation) this process may be considered as the interaction of gas and condensed flows, coming from the opposite direction, passing through the chemical reactions zone, and being transformed in this zone with the change of both chemical content and physical-chemical properties (Fig.1).

The presence of high temperature area with an intensive interphase mass-transfer processes between counter-current phases flows forms a zone structure. In each zone there are physical and chemical processes depending on corresponding conditions (temperature, medium properties, reagents concentration etc.). Space separation of zones supplies an accumulation of either, one or another substance in the definite zone accordingly to his physical-chemical properties, and provides possibility of some useful components extraction. These peculiarities allow to realize some industrial processes in extremely effective and a low-price regime, basing on heat-effectiveness of combustion wave. Examples of FC processes industrial application are known. It is waste extermination using superadiabatic combustion (Manelis et al., 2000; Brooty & Matcowsky, 1991) underground oil recovery (Chu, 1965; Prato, 1969), metallurgical burden agglomeration (Voice & Wild,

1957; Zhu-lin, 2006) oxidative catalyst regeneration (Kiselev, 1988), self-propagating high temperature synthesis (Merzhanov & Borovinskaya, 1975; Novozhilov, 1992) etc. These processes are typical examples of FC with counter-current flow and superadiabatic overheat.

We have to notice that in this paper we consider heterogeneous combustion only. We do not consider the FC of gases where preliminary mixed gaseous fuel and oxidizer burns in porous heated medium (Babkin, 1993), because in these systems heterogeneous processes are not determinative.

Due to the wave structure the heat, released in chemical reactions, transfers intensively to source materials with no use of outside heat-exchange devices, only because of extremally intensive interphase heat-exchange while gas filtration. The heat accumulation may be so considerable that combustion temperature can exceed by several times the adiabatic temperature, when it calculated assuming that the initial temperature of any portion of reacting compounds is equal to the ambient temperature. That is why sometimes one uses the terms «superadiabatic heating», «superadiabatic regime of filtration combustion», or simply «superadiabatic combustion».

The term «superadiabatic» seems disputable at first glance, however any heat recuperation from combustion products to initial substances can increase the adiabatic temperature (Wainberg, 1971) of the mixture. Really due to an intensive interphase heat exchange in such system the temperature of initial interacting compounds is far higher than the ambient temperature and may approach the combustion front temperature. Anyway the term «superadiabatic» has been used during many years and we guess one should not replace it.

Just in superadiabatic regime the effectiveness of the heat recuperation may be maximally high, whereby namely when the solid combustible contains enough high amount of an inert material, and when the gaseous oxidizer contains enough high fraction of inert gas component (Salganskii et al., 2008). It is due to the FC process organization - inert components are very effective heat carriers, thus both combustible and oxidizer can be overheated maximally before they enter into the zone of chemical reactions. Solid combustible is heated due to gaseous combustion products, while gaseous oxidizer - due to ash residue and solid inert material.

The most interesting peculiarity of combustion waves in such systems is the independence of the stationary combustion wave temperature on the value of the reaction heat release (if it is a positive value). After ignition the temperature in the combustion front increases until the heat input (due to exothermic reactions) is equal to the side heat losses. Minimizing side heat losses the thermal equilibrium is reached at very high temperature, enough for considerable increase of chemical reaction rates. So, heat losses in FC processes play more important role than in case of classic combustion waves, because in the case of FC the heat losses determine to more considerable degree the temperature in the reaction zone Temperature profile of such combustion wave is shown schematically in Fig.2. Due to an intensive heat exchange between source reagents and combustion products the released energy is accumulated mainly close to the combustion zone. If the mixture has a small heat release value (e.g. a mixture of carbon with a high amount of an inert material) the FC process will accumulate the heat energy with a lower rate and therefore it will reach the stationary regime longer.

At conditions of counter-current flows of combustible and oxidizer the combustion rate (that is very important characteristic) is determined mostly not with the heat transfer rate, but with the rate of reagents supply into the combustion zone (that is with the filtration rate).

Besides, before the combustion zone (in Fig. 1 and 2 - to the right of the high temperature area where the main chemical exothermic reactions run with highest effectiveness) the reducing zone exists with high amount of combustible and rather high temperature, that results in complete gaseous oxidizer consumption. Behind the combustion zone (in Fig. 1 and 2 - to the left of the high temperature area), contrariwise, there is a hot zone with high content of oxidizer, that provides the completeness of the material burning.

In view of the aforesaid, it is obvious that the FC process is very attractive for industry, particularly when it is needed

- To burn cheaply a material containing small amount of combustible
- To obtain high combustion temperatures,
- To provide maximal fullness of solid fuel burning,
- To get space separation of zones (heating, pyrolysis, evaporation, oxidation, condensation, cooling etc.) in solid porous fuel.

Hereby the energy outlay may be minimal due to effective heat recuperation in FC waves.

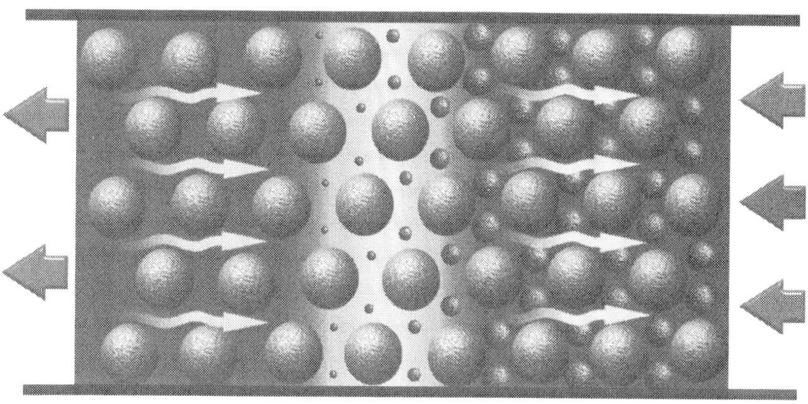

Fig. 1. Schema of combustion wave with superadiabatic heating. The solid combustible material - small balls, while the inert material - big balls. The solid material flow - right to left, the gas flow - left to right. High-temperature zone - the area with more light background

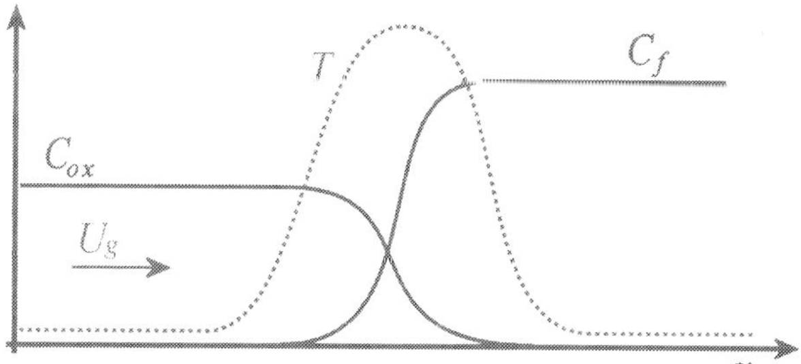

Fig. 2. Temperature and concentration profiles of the combustion wave in case of equal heat capacities of the flows of condensed and gaseous phases

2. Peculiarity of the physical and chemical structure of FC

2.1 The simplest case of FC process

There are many possibilities to realize mass transfer in FC processes. The simplest case in one-dimensional approximation is the chemical interaction of counter-current of solid fuel flow with gaseous oxidizer (being filtrating through the solid material) flow when a single combustion product forms. We are expecting the presence of both inert material in the solid fuel and other gaseous components (that do not participate in chemical reactions, e.g. nitrogen) in gaseous oxidizer. Hereby, depending on the phase state of the combustion product, this product is added to the respective flow through the reaction zone. For example, at carbon oxidation the combustion product is gaseous carbon dioxide, while at aluminum oxidation it is solid aluminum oxide. So, we have an interphase mass transfer of either solid fuel to gaseous product (Fig. 3b) or gaseous oxidizer to solid product (Fig. 3b). In both cases the whole redox process and the summary heat release are concentrated in the single reaction zone.

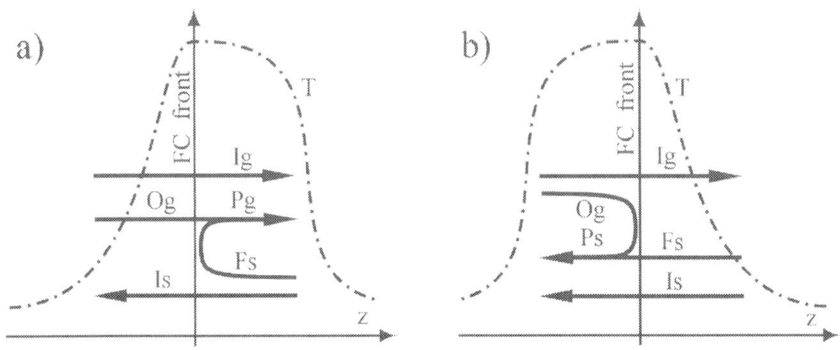

Fig. 3. Mass flows through the reaction front in cases of: a) gaseous products (*Pg*) and b) solid products (*Ps*). *Og* and *Ig* - gaseous oxidizer and inert; *Fs* and *Is* - solid fuel and inert substances, correspondingly

Let's presume that the temperature level in the reaction zone is enough high, it allows to consider this zone width being negligible small in comparison with the warming-up zone of the combustion wave. Besides we presume that the interphase heat-transfer at the filtration process is so effective that the difference between temperatures of solid and gaseous phase is negligible. Then depending on real conditions (combustible concentration in the solid mixture and oxygen concentration in gaseous oxidizer) the heat structure of the FC wave may be either like the curve in Fig.4a ("reaction trailing" structure), or like the curve in Fig.4b ("reaction leading" structure). The type of the heat structure is determined with the ratio of heat capacities of counter-current solid and gas flows through the reaction front (Aldushin et al.,1999; Salganskii et al., 2008). The heat, released in combustion, is removed with the gas flow in the case of the reaction trailing structure, while in the case of the reaction leading structure it is removed with the solid material flow. These two heat flows determine the type of the profile of the FC wave. It is possible that two these heat flows are equal, it provides a symmetric profile of the combustion wave and maximal heat accumulation in the combustion wave [Aldushin et al.,1999]. In this case the heat of chemical reactions is removed with both solid material and gas. In all considered cases an intensive interphase heat-transfer results in the accumulation of all released heat near the combustion front. If the reactor is long enough, all products leave it at the initial temperature. Continuous heat energy accumulation results in the expansion of the warming-up zone in the direction either of the solid material or gas flow depending on the type of the heat structure of the FC wave. When side heat losses exist, a stationary profile of combustion wave can form. When side heat losses are negligible, a stationary process is possible at uncompleted heat-transfer only, in this case either gas or solid material leaves at hot temperature.

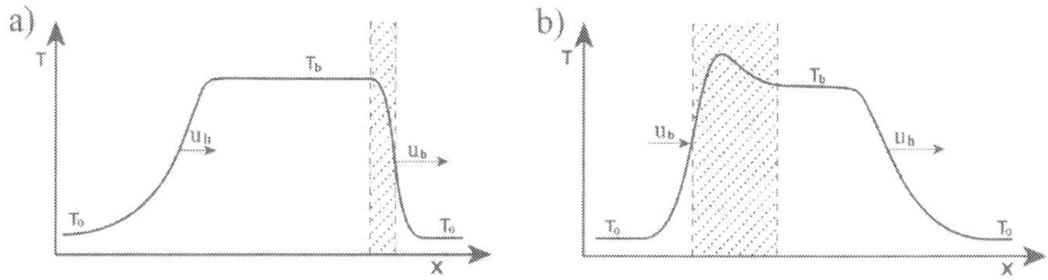

Fig. 4. Temperature profiles of combustion wave in case of there is no heat losses: a) - reaction leading heat structure, b) - reaction trailing structure. Hatchs indicate zones of chemical reactions

2.2 Attended processes of evaporation and condensation

The heat structure of the FC wave determines conditions of compounds heating at combustion wave propogation, and all accompanying physical and chemical processes. For example, the presence of an additional volatile component in the solid fuel (besides the combustible itself and an inert material) results in the localization of the zone of this component concentration (evaporation - condensation) in the region of the fuel warming-up (Fig. 5a). The main heat release, providing the existence of whole FC wave structure, takes place in the combustion front. Evaporation process occurs due to convective heat flow from the combustion front. Mass transfer of the vaporized component with the gas flow takes place before the area of condensation. If the convective heat flow from the combustion front is higher than heat losses for the evaporation, the zone of the accumulation of the vaporized component expands. If there are side heat losses the expansion of this zone ends sooner or later, and further all processes set moves stationary as a batch.

In the case of the reaction leading structure, the evaporation zone is situated near the combustion front, which determine and provide the FC wave structure. Therefore considerable heat expenses for the evaporation may decrease the combustion front temperature, and surely it has an influence on all characteristics of FC waves. In the case of reaction trailing structure, the heat expenses for the component evaporation decrease the temperature in the region of warming-up, not in the combustion front, therefore these heat expenses do not influence the value of heat release in the combustion front. It is an extraordinary peculiarity of these regimes of the FC. The zone of condensation of vaporized component is situated a bit farther along the gas flow. The condensation process is accompanied with some heat release, therefore in this case there is not mass transfer only, but heat transfer from one zone to another one too.

Typical example of vaporized component presence is the fuel moisture. Due to superadiabatic heating it is possible to organize the FC regimes where high content of moisture does not prevent propagation of stable combustion wave (Salganskaya, 2008).
It is not necessary that the condensation of the vaporized component occurs always to its accumulation in the determined reactor zone. For example, the water condensation occurs to an aerosol forming. The higher size of drops of the liquid, the easier they sediment on the initial solid material during the filtration process. Temperature gradients in the FC wave may be very high. In this case a high rate of the gas cooling occurs to forming very small drops (less than 10^{-6} m), which sediment badly under filtration and may be removed (as a fog) from the reactor with the gas flow. Thus, it is rather simple to organize the extraction of a volatile component from the source solid material.

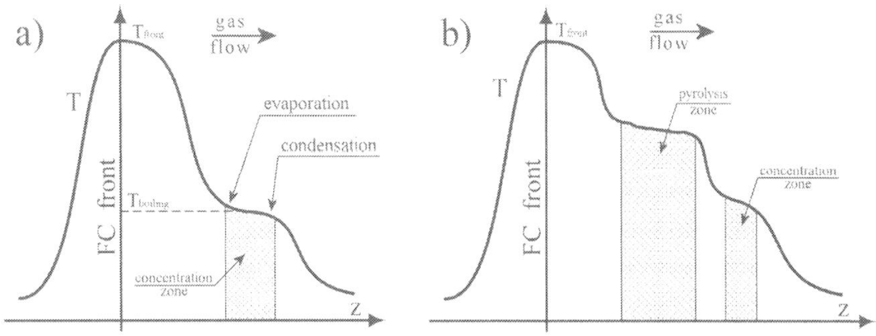

Fig. 5. Heat structure of the FC wave, propagating through a porous solid fuel: (a) – in case of an evaporating component, and (b) – in case of pyrolytic decomposition of the fuel

2.3 Peculiarities of filtration combustion of carbonic systems

Layer burning of carbonic fuel has been used long since, and many systems of gas generators, industrial furnaces work still using this process. The combustion of porous burden containing solid carbonic fuel and incombustible material at air or another oxygen-containing gaseous oxidizer filtration is of great interest for industrial application in processes of solid fuel burning optimization, as well as for developing environmentally friendly methods for different combustible wastes recycling.
Heterogeneous carbon oxidation is a complicated and multistage process. The final product are carbon dioxide and monoxide. There is no sure answer which one of these two oxides is the primary product of the carbon particles oxidation, and which one forms already in the gas phase. It is so difficult to find out it because as soon monoxide forms it may be oxidized immediately to dioxide, while dioxide at rather high temperature may be reduced to monoxide above carbon surface. Currently most part of researchers guess that in result of heterogeneous processes two oxides form together (Lizzio et al., 1990; Bews et al., 2001; Chao'en & Brown, 2001). Oxidation mechanism and the quantitative ratio of formed oxides depend on conditions (temperature, pressure etc.) as well as on properties of carbon particles surface.
At the interaction of the main components of FC in counter-current flows of solid fuel and gaseous oxidizer, a zone structure forms, each zone differs from another one in temperature and reagents concentrations. In the main zone of heat release (combustion front) carbon is

oxidized to **CO** and **CO_2**. In case of "reaction leading" wave structure solid combustion products near combustion front stay in oxygen medium at high temperature, that's why here carbon burns completely. However it is possible that oxygen is not expended completely because a quickly gas flow cooling behind the combustion front may occur to oxidation reactions deceleration.

In case of "reaction trailing" wave structure the appearance of mass transfer is entirely different. Solid products, leaving the combustion front, cool abruptly. Hereby regimes with incomplete carbon combustion are possible. Contrariwise, gaseous combustion products get through high-temperature area with big amount of hot carbon. It leads to complete oxygen exhaust, as well as to forming the zone of endothermic reactions, where carbon dioxide may be reduced to monoxide:

$$CO_2(g) + C(s) = 2\,CO(g) \quad -172\ kJ$$

Besides if water steam there is in gaseous oxidizer (steam-air gasification), other very important reaction proceeds in the same zone on the carbon surface:

$$H_2O(g) + C(s) = CO(g) + H_2(g) \quad -131.2\ kJ.$$

These reactions proceed with considerable rate only at enough high temperature, therefore they decrease local temperature in the hottest places. Hereby two combustible gases appear in gaseous combustion products: an additional carbon monoxide, and considerable amount of hydrogen (at steam-air gasification up to 30 vol.%). So, depending on conditions the FC of carbonic systems can proceed by considerably different ways, and with different results.

These peculiarities of the heat structure of the FC waves at carbonic systems combustion have to be considered at industrial realization of technologies based on superadiabatic condition regimes.

2.4 Attended processes of thermal decomposition at filtration combustion wave

The structure of the FC waves may be rather complicated. The main heat release in the combustion front determines the common temperature level. When components predisposed to thermal decomposition there are in the solid fuel, a new zone forms in the combustion wave structure: zone of corresponding chemical processes. For example, if there is calcium carbonate (chalk, buhr) in carbonic fuel, during the heating it will decompose in a varying degree, dependently on temperature. Hereby solid combustion product (quicklime) remains in the burden, while gaseous carbon dioxide removes together with other gaseous combustion products. Fig.6a shows the results of the thermodynamic calculations of the equilibrium **$CaCO_3 \leftrightarrow CaO + CO_2$** at the pressure 1 atm in air medium at temperatures since 800 up to 1200 K. On the other hand if for example there is copper oxide **CuO** in the burden, **CuO** begins to decompose (Fig.6b) at high temperature (higher than 1400 K) and an additional oxygen appears in gaseous combustion products, then this oxygen reacts immediately with the fuel. In this case the combustion wave structure is complicated because of two new zones (the zone of **CuO** decomposition and an additional zone of the fuel oxidation) appearance. Hereby in each zone individual physical and chemical processes proceed accordingly the temperature level and reagents concentration.

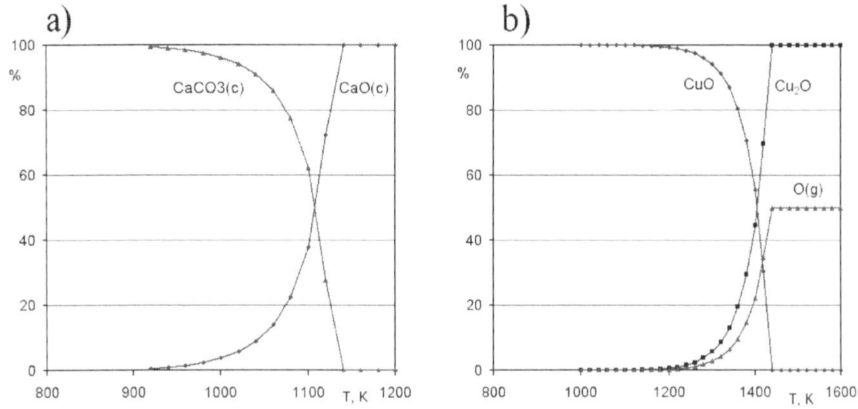

Fig. 6. Thermodynamic equilibrium in systems containing **$CaCO_3$** (a) and **CuO** (b)

In this system mass transfer may be too complicated. Details of the temperature profile of the complex combustion wave reflect all processes with heat release and heat absorption.

2.5 Filtration combustion of fuel able to pyrolytic decomposition

Filtration combustion of organic fuel is a particular case of combustion wave with thermal decomposition processes. Being heated theses fuels usually pyrolyze forming liquid and gaseous products, as well as coke residue. Typical examples are organic fuels: wood, peat, natural coals etc.

In this case the heat wave structure is complicated - a new zone of thermal decomposition appears before the combustion front (Fig. 5.b). Pyrolysis proceeds in the zone of solid fuel warming-up where no oxygen presents. Usually thermal effect of pyrolysis is rather small in comparison with heat release in the combustion front.

Usually solid coke residue, pyrolysis tars, and gaseous destruction products form during the pyrolysis. Then the coke falls into the combustion zone and burns there. At relatively high temperature pyrolysis tars stay in gas state and move with the gas flow and gaseous pyrolysis products from pyrolysis zone into the region with lower temperature. There the pyrolysis tars, which is a mixture of different hydrocarbons, condense. It provides appearance of zone of liquid products accumulation, like the zone of the volatilile components accumulation, but with the only difference - the origin of the products accumulated in these zones is different.

The content of pyrolysis tars is rather complicated and it may be different depending on the nature and properties of the material under pyrolysis, as well as on the rate and intensiveness of the heating. There are thousands of organic substances in pyrolysis tars, among them many toxic substances. The worth of these tars is not considerable because in order to obtain any goods (e.g. motor-fuel) it is necessary to organize rather complex chemical processes. So, at this stage it is appropriate to burn pyrolysis tars and to obtain heat or electric energy. However we have to consider the possibility to develop technology of liquid fuel producing from non-petrolic source, moreover this source may free, even have a negative price (if one utilizes some kinds of organic waste).

Pyrolysis tars, which condense in gas flow at its cooling, form aerosol by the same way as volatile components do. And by the same way pyrolysis tars may be removed (as small fog drops) together with the gas flow from the reactor (Salganskii et al., 2010). Unlike moisture and other incombustible components, pyrolysis tars are combustible and may be burnt in presence of gaseous oxidizer.

3. Characteristics of filtration combustion of some metal-containing systems

Investigations on FC processes showed (Manelis et al., 2006) that this process may be successfully used for some metal extraction, namely metals, which can form relatively volatile products (products of oxidation as well as of reduction), because even at their low concentration in the gas phase the may be removed together with gas flow, shifting the thermodynamic equilibrium to the needed direction. The most interesting is the realization of FC in superadiabatic regime for extraction less-common metals from unconventional sources - poor ores, burrows etc.

Mass transfer of different metal derivatives in the FC waves may be successfully realized because the pressure of saturated steams of some metals themselves and some of their derivatives at temperatures from 800 to 1200°C (typical temperature for FC processes) is enough for their extraction. As objects of this kind of mass transfer may be considered some free metals (**Zn, Cd, Hg, As, Se, Tl, Ta**) as well as some oxidized forms (trioxides of molybdenum and rhenium, oxides of selenium, tellur, tantalum, tungsten hydroxides). New possibily appears to develop effective technologies for extraction valuable metals from unconventional sources.

All physico-chemical processes said above, which can realize mass transfer and extraction of valuable metals, may be realized without using filtration combustion, that is by known methods, but only in superadiabatic regime of FC due to maximal level of heat recuperation, and therefore due to maximal heat efficiency, it is possible to realize the same processes with minimal energetic expenses, that is maximally effectively from an economic point of view.

Naturally, mass transfer of relatively volatile substances from the reaction zone is accompanied with incessant processes of evaporation (as the zone of this substance staying is heated) and condensation (as steams of this substance falls into the zone with lower temperature). So, when a few products move from the initial mixture they may be separated spatially depending on their volatility, adsorption coefficients etc. Fig.7 demonstrates that in FC of mixture where, besides fuel and inert material, additionally iron, zinc, and cadmium (iron is not volatile, cadmium volatility is far higher than zinc volatility) present , the iron concentration does not change, while concentration of zinc and cadmium change so manner, that there is an incessant accumulation of these metals in determined places. The zone of cadmium maximal accumulation is farther from the combustion front than the zone of zinc maximal accumulation.

The fact that in filtration combustion process the whole reaction zone anytime is separated on two parts - oxidation zone and reducing one, is very useful if one considers filtration combustion regime as a way for metals mass transfer. All reactor volume is not uniform, there are zones with different temperatures and different redox nature of gaseous phase there. The zone left to combustion front (Fig. 3a) is the oxidizing zone, right to combustion front (Fig. 3b) - reducing zone.

This peculiarity should be used for the optimization of processes of different metals extraction. For example, when we extract molybdenum (**MoO_3** is far more volatile is individual metal) we have to organize a combustion process in "reaction leading" mode (Fig. 4a). In this case Mo-containing products form in the oxidizer zone (naturally at rather high temperature though lower than combustion front temperature) relatively volatile **MoO_3** which moves together with the gas flow behind combustion front.

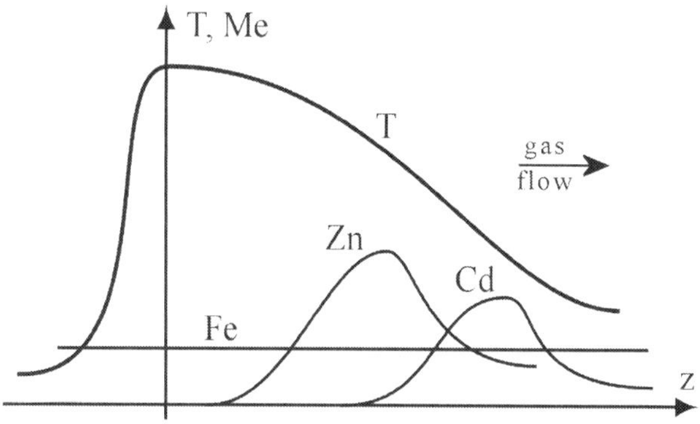

Fig. 7. Zones of metals accumulation separation in the wave of filtration combustion

When we want to realize mass transfer of metals having rather volatile reducing forms (e.g. free metals such as **Zn, Cd**) we have to organize FC in regime with "reaction trailing" structure. Then this compounds reduce with carbon monoxide before the combustion front in hot reducing zone (Fig.4b) and metal vapour moves together with gas flow and may be extracted or at least accumulated in burden portions left of combustion front.
A correct choice of combustion regime for realization of mass transfer of the giving metal may be obtained preliminary from results of thermodynamic calculations of equilibrium concentrations (e.g. using the code TERRA (Trusov, 2002). For example, we are representing results of thermodynamic estimation of the system containing metallurgy tailing containing high amount of iron and zinc. We looked for the possibility to extract useful metals from secondary heavytonnage source (there are million tons of this kind of tailing in Russia only), which can not be recycled with economic effect using traditional technologies. One of real samples has been investigated, it contains (mass.%%): **Fe**-28.4; **Zn**-12.05; **Ca**-5; **Si**-2.65; **Mn** -1.26; **Pb**-1.07; **Mg**-0.86; **Al**-0.2; **Cr**-0.16; **Cu**-0.11 and **P**-0.037. Thermodynamic analysis considered atmosphere pressure and temperature from about 500 till 1300°C with different oxygen concentration. As result we got the listing of possible reaction products and their equilibrium concentration in the given conditions. It gave the first resumes and ideas. It was shown that zinc and lead are the most interesting for their extraction using FC processes. **Zn** and **Pb** forms the most volatile substances. In oxidizing zone (Fig.8) practically all zinc stays in condensed phase (as **ZnO(c)**), so it is too hard to extract zinc using the regime "reaction trailing" structure of combustion wave. Changing the gas content in direction to **CO** excess, **Zn**-containing substances begin to be reduced starting from determined temperature and form free metal that moves to the gas phase (vapor pressure of **Zn** is 0.00002 MPa at 1200 K and 0.0001 MPa at 1300 K). If initial coal portion in the mixture increases (that is the ratio **O/Zn** decreases) **Zn** vaporizes at lower temperature (compare Fig 8b and 8c), and therefore it makes process of **Zn** extraction easier.

As for lead, unlike **Zn** even in oxidation zone there are enough **Pb**-containing substances (different oxides) in the gas phase, at temperature higher than ~1000°C practically a half of **Pb** is already in the gas phase, up to~1400°C mainly in the forms of **Pb2O(g)** and **PbO(g)**. Change of gas medium properties (in reducing medium) gaseous **Pb** appears beginning from ~800°C, and by ~1200°C it remains practically the only Pb-containing gaseous product (vapor pressure of **Pb** is 0.00016 MPa at 1200 K, and 0.0028 MPa at 1300 K). Unlike the case with **Zn**, systems, containing **Pb**, do not change with the change of reducing potential (compare curves on Fig. 8b and 8c, they are practically the same.

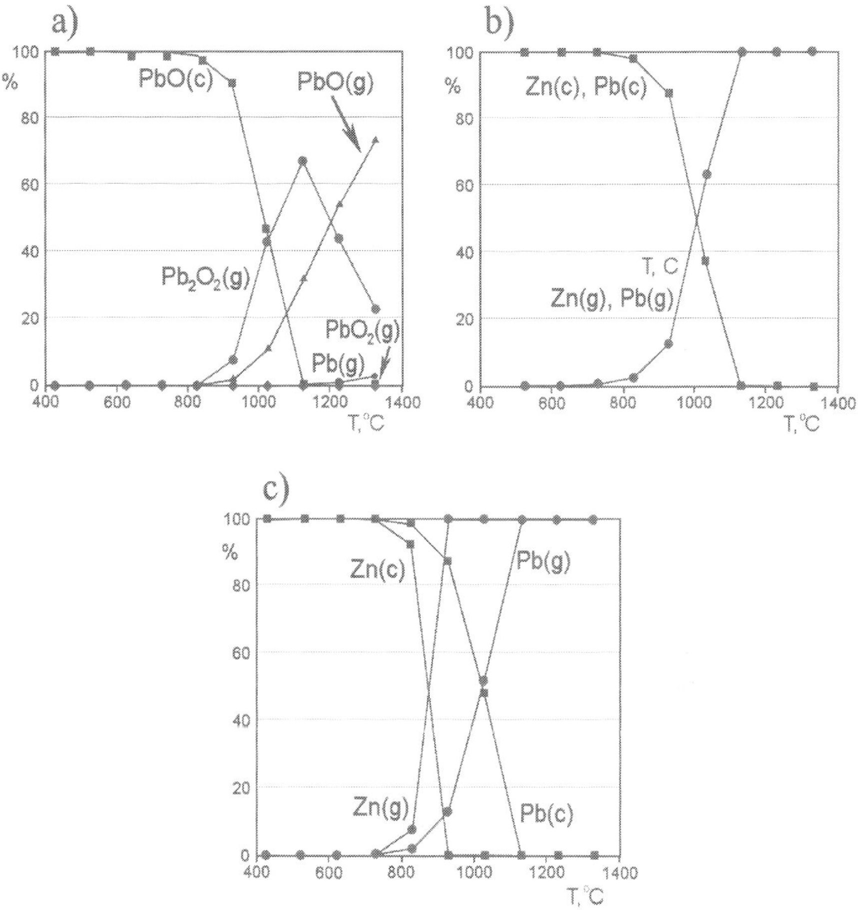

Fig. 8. Main substances, containing **Zn** and **Pb**
a) in oxidizing gas medium with oxygen excess,
b) in reducing gas medium, where the most part of carbon is in **CO**,
c) in reducing gas medium, where the most part of carbon is in carbon itself

Thermodynamic analysis shows that other metals, represented in the tailing under consideration, practically do not form gaseous products neither in oxidizer zone nor in reducing one. So, it was theoretically shown that using FC it is possible to extract **Zn** and **Pb** from that tailing. **Zn** may be extracted when relatively high temperature there is in the reducing zone (that is at "reaction trailing" wave structure) while **Pb** may be extracted using both regimes - "reaction leading" or "reaction trailing" wave structures. In case of "reaction leading" wave structure lead oxides will leave, while in case of "reaction trailing"- gaseous lead.

Above in section 3 we discussed only systems containing a few metal in mixture with organic fuel, inert material and air, that is systems **C-N-H-O-Me**. But there are few metals and metal oxides, having rather high vapor pressure at temperatures lower 1200-1300° C (at higher temperatures it is hard to organize industrial processes that would be economically acceptable). The listing of such metals includes mainly **Zn, Cd, Hg, Te, Tl, Se, As, W, Re, Pb.** However this listing may be considerably expanded if we introduce chlorine or flourine into the system, that is if we consider the possibility of mass transfer of gaseous chlorides and fluorides (chlorides and fluorides of few tens metals are rather volatile at temperature lower than 1200°C). So, potential abilities of FC processes would be strongly enlarged, but we will meet another serious drawback - presence of chlorine, moreover fluorine, creates problems in environmental protection. We only recently began to investigate FC processes of systems with chlorine (Balabaeva, 2009) and we do not consider these systems in this paper. We have to notice a special case of fuel, metal sulphides, e.g. molybdenum sulphide (**MoS_2**), arsenic sulphide (**As_2S_3**). It was shown (see infra section 3.1) that at FC of mineral molybdenite concentrate (**MoS_2**) it is possible to organize an effective FC process with no additional fuel, hereby relatively high volatile molybdenum trioxide removes from the initial burden. Some examples of the effectiveness of the FC process are described below (sections 3.1-3.3).

3.1 Trioxide molybdenum extraction from mineral molybdenite

Experimental testing of **MoO_3** mass transfer possibility at FC of molybdenite has been realized using flotation concentrate of mineral molybdenite (by ~50% **Mo**). FC process needs a good gas penetrability of the solid reaction mixture, therefore the initial material has to be granulated in order to provide a rather quickly and stable air flow through the burden. Flotation concentrate of mineral molybdenite has been granulated with bentonite clay, obtained granules (2-4 mm diameter) contained by 40-50% **Mo**. FC process was realized in a model vertical quartz reactor length ~500 mm, diameter ~20 mm. The reactor had a system of preliminary ignition from an outside thermal source, which was installed before (lower) the main burden mass. Reaction temperature was measured using a set of thermocouples, installed in a few places in the burden in different distances from the combustion start. After the mixture is burnt and the residue is cooled a few samples from different places in different distance from the combustion start have been analyzed. Sublimated products were analyzed too.

Depending on combustion conditions (air flow rate, burden content etc.) the combustion front velocity was between 3 and 7 mm/min. For burden with no additional coal (if molybdenite content in the mixure is rather high, the heat of its combustion is enough to support a stable filtration combustion front without additional fuel) and no inert component the increase of air flow rate from 0.6 to 1.1 m/s increases combustion front velocity from 3.8 to 6.6 mm/min, hereby the maximal temperature in the combustion front increase too (from

900 to 1300°C). Addition of coal (up to 20%) into the initial mixture changes the character of the temperature rise - after its strong increase there is a period of slow rise up to the maximal value.
Addition of the inert material changes the combustion temperature at the same gas flow rate. In this case temperature profile has a practically symmetric form, which characterize the optimal superadiabatic regime, maximal temperature was about 1100°C.
Chemical analysis showed that by burning molybdenite in FC regime it is possible to obtain molybdenum trioxide of high purity. Hereby we did not notice undesirable caking of the burden. Almost half (56-57%) of obtained molybdenum trioxide did not leave the reactor and remained in the calcine. After processing all this molybdenum trioxide can be moved in a solution form by the water ammonia, and simply extracted.

3.2 Filtration combustion of mixtures containing used catalysts with molybdenum

Aluminosilicate catalysts doped with nickel, and containing 11-12% $\mathbf{MoO_3}$ (spent catalysts are tonnage wastes after industrial petrochemical synthesis processes) have been investigated as a secondary source for molybdenum recycling. In order to burn this material it is already necessary to add a fuel into the burden because the catalyst itself does not contain any combustible compounds. In case of this material burning in the FC regime, one does not need to add another solid inert compound, because the catalyst itself serves as inert material (it contains up to 80% inert aluminosilicate). Granules with catalyst and bentonite clay (similar those describing in section 3.1), mixted with coal (at different ratio) were burnt in the same maner as granules with molybdenite were executed. It was found that at coal content 3-10% in the solid mixture the filtration combustion process goes in stable regime [3]. At constant air flow (about 0.7 m/s) the increase of coal content in the initial mixture from 3 to 10% results in practically linear increase of maximal temperature in combustion front (from 600 up to 1340°C). Gas flow rate increase do not influence practically the maximal temperature. Increase of the coal content as well as the air flow rate rises the combustion front velocity practically linear.
During combustion process a white deposit forms at «cool» reactor walls, far from hot zone this deposit becomes gray. Chemical analysis showed that it is mainly molybdenum trioxide. Totaly 78% of molybdenum went to sublimate. Optical microscope tests in transmitted and polarized light showed that products of the combustion of **Mo**-containing systems, taken from reactor walls, are transparent crystal powders containing needle-like particles of lenth 50 μm to 1-2 mm and width 10-20 μm. Analysis shows that samples of the sublimate crystals taken out of burden zone contains higher than 95% $\mathbf{MoO_3}$.
It was shown that at the end of combustion process the content of $\mathbf{MoO_3}$ increases on approaching to the final end of the burden, but it is always lower than molybdenum concentration in the initial mixture, that proves once more that mass transfer of **Mo**-containing derivatives left the initial reaction mixture. An interesting regularity was observed: maximal yield of **Mo** into sublimate occurs at coal content 5-6% (combustion front temperature is 850-950 °C), the increase of coal fraction up to 8-10%, and correspondingly, combustion front temperature up to 1000-1340°C makes the yield considerably lower. It seems that at coal content increase the heat structure of the combustion wave changes from "reaction leading" (at low coal content) to "reaction trailing" (at coal content increase). In the last case the reducing zone becomes more heated and molybdenum oxides may be reduced there to metal molybdenum which is not volatile in these temperatures. Another possible reason of $\mathbf{MoO_3}$ yield fall at T = 1000-1340°C may be any change of crystal structure

of aluminosilicate cage of catalyst structure at intensive heating and as result an aggravation of conditions for $\mathbf{MoO_3}$ mass transfer.

3.3 Filtration combustion of mixtures containing zinc

Experimantal investigations of FC and mass transfer of **Zn**-containing products have been carried out with a wide set of model mixtures, including zinc as a free metal as well as zinc oxide (**ZnO**). It was shown that by the control of the filtration combustion process organization (varying control parameters in wide interval, that is changing the coal content from 10 to 40%, and air flow rate from 0.01 to 1.8 m/s) it is possible to optmize FC regime with reaction leading structure (**ZnO** extraction) as well as regime with reaction trailing structure, where the reducing of ZnO to free Zn is possible due to endothermic reactions

$$\mathbf{ZnO(s) + CO(g) = Zn(s) + CO_2(g)} \quad - 66 \text{ kJ},$$

with further zinc sublimation:

$$\mathbf{Zn(s) = Zn(g)} \quad - 130 \text{ kJ}.$$

Stable combustion regimes for the systems under investigation were found at air flow rate higher than 0.06 - 0.11 m/s. Hereby temperatures of combustion front at the tested interval of controlling parameters reached 690 - 1300 C and increased with the fuel content rise as well as with air flow rate increase. For all samples under consideration combustion front velocities were between 0.7 and 17.5 mm/min and they increased with the air flow rate rise at the constant fuel content.

It was shown that the depth of **ZnO** extraction to the sublimate depends on initial zinc content in granules of Zn-containing burden component. During the burning of systems, containing many **ZnO** (10-30%), we found that the concentration of **ZnO** in the burnt mass decreased from the start point of combustion to the end point if the air flow rate was rather small (lower than 0.1 m/s); at higher air flow rates (0.4 - 1.7 m/s), conversely, **ZnO** concentration increased from the start point of combustion to the end point. It proves that **Zn**-containing combustion products move through the burden. Increase of combustion front temperature rises the level of **Zn** extraction from the system and this yield was between 3 and 34% depending on many factors.

For systems with small zinc content in the burden (2-4%) the level of zinc extraction depends on air flow rate and on fuel content, at small flow rates the effectiveness of **Zn** extraction reached sometimes 100% or so.

4. Principal advantage of filtration combustion over traditional methods of substances extraction

We have to stress once more: all processes of the mass transfer described above may be executed in simple stove regime, that is heating all reaction mass in the media of necessary gas (depending on the type of the substance to be extracted, this gas may be air, carbon monoxide, hydrogen etc), but in this case we have to waste energy in several times higher than it is needed to proceed the same chemical reaction in superadiabatic regime using the method of filtration combustion. Incessant heat recuperation in filtration combustion wave and the fact that only a small part of all reaction mass is heated at the moment and the most part of the thermic energy moves along the reaction mass and do not disperse outside provide high energetic efficiency of the process. By this way it is possible to realize mass transfer of different kinds of compounds able to form volatile products. Spending of energy for the mass transfer of compound to be extracted by the FC method are considerably lower than necessary spending for heating all reactive mass to obtain the same extraction effect. Moreover, filtration combustion process may become not energy-consuming, but even energy-generating if further one uses combustible gaseous products (**CO**, **H_2**) as fuel for other industrial goals (heating, eclectic energy obtaining etc.).

5. References

Aldushin, A.P.; Rumanov, I. E. & Matkowsky B. J. (1999). Maximal energy accumulation in a superadiabatic filtration combustion wave, *Combustion and Flame*, Vol.118, pp.76-90,

Babkin V.S. Filtration combustion of gases. (1993). Present state of affairs and prospects, *Pure Appl. Chem*, Vol.65, 335-344,

Balabaeva, E.M.; Korshunova, L.A.; Manelis, G.B.; Polianchik, E.V. & Tsvetkov, M.V. (2009).Chlorine neutralisation with Ca-containing sorbents at solid fuel gasification. *Alternative energetics and ecology*, Vol. 76, N 8, pp. 190-194 (Rus)

Bews, I.M.; Hayhurst, A.N.; Richardson S.M. & Taylor, S.G. (2001). The Order, Arrhenius Parameters, and Mechanism of the Reaction Between Gaseous Oxygen and Solid Carbon. *Combustion and Flame*. Vol. 124, pp. 231-245,

Brooty, M.R. & Matcowsky, B.J. (1991). *Combust. Sci. Technol.* Vol. 80. pp. 231- 264,

Chao'en, Li; Brown, T.C. (2001). Carbon oxidation kinetics from evolved carbon oxide analysis during temperature-programmed oxidation. *Carbon*, Vol. 39, pp. 725-732,

Chu, C. (1965). The Vaporization-condensation Phenomenon in a Linear Heat Wave, *Society of Petroleum Engeneers Journal*, Vol.5, N.3, pp. 196-210,

Hanamura, K.; Echigo, R. & Zhdanok, S. (1993). Superadiabatic Combustion in Porous Media, Int.J.*Heat and Mass transfer*, Vol.36, pp. 3201-3209,

Kiselev, O.V.; Matros, Yu.Sh. & Chumakova, N.A. (1988). Phenomenon of thermal front propagation in catalyst layer. *Proceedings of «Propagation of Thermal Waves in Heterogeneous Media»*, pp. 145-203, Novosibirsk, 1988, Nauka, (in Russian),

Lizzio, A.A.; Jiang, H. & Radovic, L.R. (1990). On the kinetics of carbon (char) gasification: reconciling models with experiments, *Carbon*, Vol. 28, N1. pp. 7-9,

Manelis, G.B.; Polianchik, E.V. & Fursov V.P. (2000). Energetic technology of burning basing on the phenomenon of superadiabatic heating, *Chemistry for Sustainable Development.*, Vol. 8., N 4, pp. 537-545 (Rus),

Manelis G. et al. (2006). Rus.Patent, N 2278175, June 20.

Merzhanov, A.G. & Borovinskaya, I.P. (1992).*Combust. Sci. Technol.*, lo, 195-201 (1975) ;

Novozhilov, B.V. (1992). Non-linear SHS phenomena: Experiment, theory, numerical modeling. *Pure & Appl. Chem.*, Vol. 64, N. 7, pp. 955-964.

Prato, M. (1969). The heat efficiency of thermal recovery process. *Journal Petroleum Technology*, Vol.21,. No. 3, pp.323-330,

Salganskaya, M.V.; Glazov, S.V.; Salganskii, E.A.; Kislov, V.M. , Zholudev A.F. & Manelis, G.B. (2008). Filtration Combustion of Humid Fuels. *Russian Journal of Physical Chemistry* B, Vol. 2, No. 1, pp. 71-76,

Salganskii, E.A.; Kislov, V.M.; Glazov, S.V.; Zheludev, A. F. & Manelis, G. B. (2008). Filtration Combustion of a Carbon-Inert Material System in the Regime with Superadiabatic Heating, *Combustion, Explosion, and Shock Waves*, Vol. 44, N. 3, pp. 273-280,

Salganskii, E.A.; Kislov, V.M.; Glazov, S.V.; Zholudev, A.F. & Manelis, G.B. (2010). Specific Features of Filtration Combustion of a Pyrolized Solid Fuel. *Combustion, Explosion, and Shock Waves*, Vol. 46, 2010, N. 5. (in press)

Trusov, B.G. (2002). Program System TERRA for Simulation Phase and Thermal Chemical Equilibrium. *Proceedings of «XIV Intern. Symp. on Chemical Thermodynamics»*, pp. 483-484, St-Petersburg, 2002, July, Russia,

Voice, E. W. & Wild, R. (1957). *Iron Coal Trade Review*, Vol. 175, pp.841-850,

Wainberg F.J. (1971). Combustion temperatures: The future? *Nature*, Vol.233, September, 24,

Zhu-lin, Liu; Chen, Zi-lin & Tang Le-yun. (2006). Experimental Study on Optimization of Sintering Technology, *J. Iron & Steel*, Vol. 41(5), pp.15-19.

22

Mass Transfer in Hollow Fiber Supported Liquid Membrane for As and Hg Removal from Produced Water in Upstream Petroleum Operation in the Gulf of Thailand

U. PANCHAROEN[1], A.W. LOTHONGKUM[2] and S. CHATURABUL[3]
[1]*Department of Chemical Engineering, Faculty of Engineering Chulalongkorn University, Bangkok 10330*
[2]*Department of Chemical Engineering, Faculty of Engineering King Mongkut's Institute of Technology Ladkrabang, Bangkok 10520*
[3]*Salamander Energy (Bualuang) Limited, Bangkok 10120 Thailand*

1. Introduction

This chapter describes the merits of using hollow fiber supported liquid membrane (HFSLM), one of liquid membranes in supported structures, and the step-by-step mass transfer in removing arsenic (As) and mercury (Hg) ions from produced water in upstream petroleum operation from the Gulf of Thailand. Apart from the necessary fundamentals, the contents comply existing information and data based on our up-to-date publications in journals.

Arsenic (As) and mercury (Hg) are naturally trace components in petroleum reservoir. In certain Gulf of Thailand fields, the concentration levels vary widely and Hg, in particular, is drawing local statutory attention since its concentration is found typically higher than those found in the rest of the global petroleum production areas. A number of hypotheses have suggested the origin of As and Hg in the petroleum reservoirs laid underneath the Gulf. But one common fact agrees that mercury predominantly presents in an elemental form Hg(0) with the rest in an inorganic form ($HgCl_2$), organic forms (CH_3HgCH_3 and $C_2H_5HgC_2H_5$) and an organo-ionic form ($ClHgCH_3$). For arsenic, it presents as As(III) and As(V) (Korte & Fernando, 1991). The As(III) is normally in the form of H_3AsO_3 whereas As(V) is in either $H_2AsO_4^-$ or $HAsO_4^{2-}$. Arsenic and mercury are grouped in the most hazardous metals since arsenic compounds are carcinogens, both severe acute and chronic toxicity, while mercury can evaporate in soil or water; short-term exposure results in kidney damage; and a lifetime of exposure can lead to impairments in neurological functioning (U.S. EPA, 1984).

In the Gulf of Thailand, petroleum development and upstream production have been very active recently following an increasing domestic energy demand and a soaring global oil price. Accordingly, numbers of leading and national oil companies have been gathering in the Gulf for an interest of petroleum exploration and production. Figure (1) presents 2009 updated petroleum concessionaire map in the Gulf of Thailand where total gross production per day has been reported around 680,000 barrels of oil equivalent [DMF, 2010] and hundreds of thousands of barrels daily for the produced water are generated. Trend of produced water production is increasing dramatically according to an exponential expansion of petroleum development to meet high domestic demand and maturity of production reservoir itself.

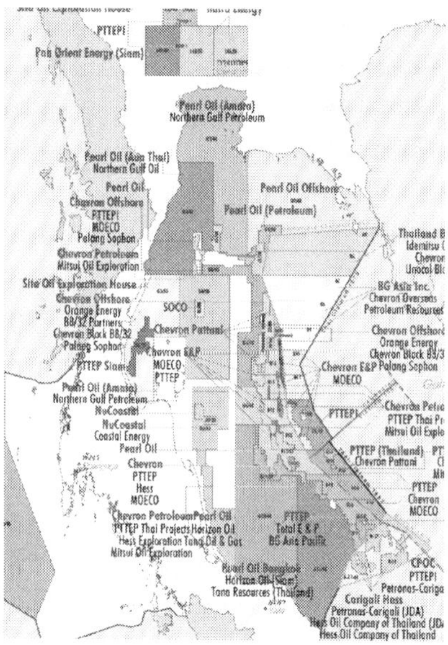

Fig. 1. Concessionaire map in the Gulf of Thailand (DMF, 2009)

To handle this massive daily generated amount of produced water, the oil company operators in the Gulf undertake an enduring path to pursue a "zero discharge" program enforced by Thai local authorities by injection of the produced water back into the formations. Nevertheless, overboard discharge of the produced water into the ocean is still necessitated when the re-injection facilities are in breakdown state or for some production fields where uncertainties in geological formations to receive such great volume of produced water hinder the feasibility. Pollution and biological impact to local marine environment is of concern when overboard discharge of arsenic and mercury contaminating produced water into the ocean. Various aspects of the potential environmental effects of arsenic and mercury have been assessed by local environmental control authorities. According to the overboard discharge limits made to reference with the Ministry of Industry and the Ministry of Natural Resources and Environment, Thailand, it is a must to remove arsenic and mercury from offshore waste discharges to no greater than 250 ppb and 5 ppb, respectively.

The conventional methods to treat heavy-metal-contaminated water are precipitation/co-precipitation, coagulation, electrolysis, reverse osmosis, carbon adsorption, membrane filtration, solvent extraction, ion exchange, etc (Arpa et al., 2000; Dabrowski et al., 2004; Van der Vaart et al., 2001). Nevertheless, those are reported to be ineffective at a very low concentration separation. Precipitation/co-precipitation is the most commonly used process.

It uses chemical additives to transform dissolved contaminants into insoluble solid which is then removed from the water phase by clarification or filtration. Adsorption is an alternative to precipitation/co-precipitation. However, it poses some constraints as it is effective for treating the water which has only mercury as a contaminant. The adsorption is likely to be affected by characteristics of the media and non-mercury contaminants (U.S. EPA, 2007).

Chevron Thailand Exploration and Production Ltd., one of the major oil company operators in the Gulf of Thailand, has applied a continuous chemical treatment process to comply with these regulations. However, the on-line As and Hg monitors have been problematic to achieve the desired discharge concentrations (Gallup & Strong, 2007). Apart from that, the use of chemical treatment has raised concerns from operation point of view. For instance, sodium hydroxide and ferric chloride in the process cause severe corrosion with the associated metal made pipeworks. Costly super alloy material is the only viable option for construction material of high pressure rating process but high capital cost. High toxicity from the chemicals deteriorates the circumstance by exposing personnel operators with health hazards when loading and unloading of the chemicals to the process.

To cope with the problems met by conventional methods and chemical treatment process, the removal by hollow fiber supported liquid membrane (HFSLM) has been applied. The modules of hollow fiber can be put in series after the conventional operation to scavenge the remaining arsenic/mercury down to the environmentally acceptable permits. The HFSLM technique is relatively recent. It can supplement or can replace the conventional separation techniques if the concentrations of As and Hg ions in produced water are very low and hardly handled by those conventional methods. To date based on the concept of mass transfer by using HFSLM with either a single extractant or a synergistic one, we published our remarkable works on effective removal of dilute arsenic/mercury concentration from the produced water in accordance with the discharge legislation.

2. Liquid membrane

The membrane is typically fabricated with hydrophobic microporous hollow fibers. With an organic extractant embedded in the micro pores of hydrophobic membrane, the membrane, therefore, acts as a semi barrier between two aqueous phases from contact but still allows the target component to diffuse across the membrane freely while simultaneously impedes the diffusion of other components. Thus, the advantage over solvent extraction method by the use of liquid membrane is non dispersive mass transfer. The flow rates of the phases can be varied independently over a wide range without any problem of flooding, loading, channeling or entrainment.

The system, shown in Figure (2), usually consists of feed phase, liquid membrane phase and stripping phase. The feed is a phase which consists of a mixture of components including the target component. The stripping phase is a phase which preferentially receives the target component from the feed through liquid membrane diffusion. The mechanism involves solvent extraction and a membrane-based mass transfer for the removal of the specific component from feed phase. It is renowned for its combined ability of extraction, diffusion, and stripping of the target component and can be treated as a simultaneous multistage extraction and stripping process (Mohapatra & Manchanda, 2008).

Common forms of liquid membrane are bulk, emulsion and supported liquid membranes. Each has different pros and cons, for example, bulk liquid membrane is easier to prepare and operate but offers less surface to volume ratio than others. Emulsion liquid membrane

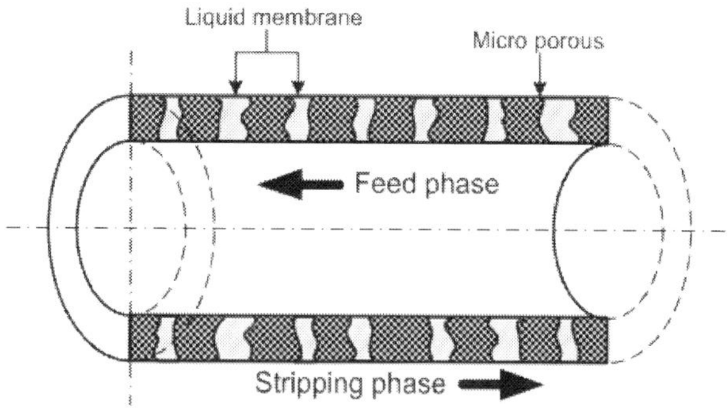

Fig. 2. The microporous hollow fiber liquid membrane

offers large surface to volume ratio but poor stability. Supported liquid membrane draws more industrial interests as it provides large surface to volume ratio to allow fast mass transfer while stability can be very well maintained. Moreover, supported liquid membrane is sustainable for a continuous flow operation. It has gained a broad range of separation applications and also demonstrated a promise in the removal of metal ions, such as nickel, uranium, chromium, copper, etc from the effluent streams (Lothongkum et al., 2009; 2009; Usapein et al., 2009; Pancharoen et al., 2005; Ramakul et al., 2009; 2007; 2006; 2006; 2004; 2003). The hollow fiber supported liquid membrane is, therefore, deployed in many of our study researches.

2.1 Diffusion transport in hollow fiber supported liquid membrane

HFSLM, or membrane contactor in many literatures, creates an immiscible layer between feed and stripping phases from an organic extractant in microporous hollow fibers. The target component reacts with the extractant at the feed-membrane interface to form complex species. Subsequently, the complex species diffuse across the liquid membrane (organic phase) to react with the stripping solution at the opposite interface of the membrane then are stripped into the stripping phase. Thus, the target component can be extracted and stripped simultaneously in a single step. The transportation rate of the target component is driven by the concentration gradient between feed and stripping phases. In this phenomenon of diffusion transport, it can be either a simple facilitated transport or a coupled facilitated transport. The simple facilitated transport occurs when the transport is independent of any other ions. It normally takes place in an application of neutral species extraction. As for ionic species extraction, the coupled transport occurs to maintain the solution electroneutrality (Bringas et al., 2009). The driving force to determine the transport rate is dependent on types and concentrations of co-ions in the feed.

Figure (3) schematically explains the transport of each case. The scheme depicts (A) as the target component, (B) co-ion, (C) the organic extractant, and $(\overline{AC}$ or $\overline{BC}$ or $\overline{ABC})$ the organic complex. In our study, the target component (A) can be arsenic and/or mercury ions. The straightforward mechanism is observed with the simple facilitated transport (Figure (3-a)) since the organic complex $(\overline{AC})$ is produced from the reaction between (A) and (C). Then, $(\overline{AC})$ is decomposed at the interface between the liquid membrane and the stripping phase, and (A) is recovered. The coupled facilitated transport can be classified into co-transport (Figure (3-b)) and counter-transport (Figure (3- c)). For the coupled facilitated co-transport, the extractant reacts with the target component (A) and co-ion (*B*) to form the organic complex ($\overline{ABC}$). The ($\overline{ABC}$) diffuses across the liquid membrane to the stripping interface where both the target component and co-ion are simultaneously recovered. This mechanism has co-ion transporting along with the target component from feed phase to the stripping phase.

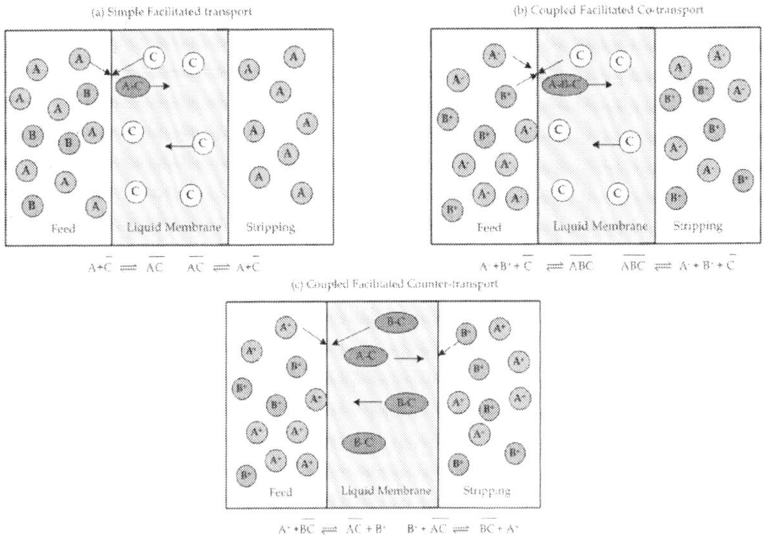

Fig. 3. Schematic pictures of liquid membrane systems (adopted from Bringas et al., 2009)

This coupled facilitated co-transport is common for neutral and basic extraction as schematized in Equation (1) for the reaction with neutral organic extractant, and Equation (2) for the reaction with basic organic extractant. The aqueous pH phase or hydrogen ion (H^+) in the system depicts as (H).
Coupled Co-transport, **Neutral** organic extractant

$$A^{+}_{(aq)} + B^{-}_{(aq)} + \overline{C}_{(org)} \Leftrightarrow \overline{ABC}_{(org)} \tag{1}$$

Coupled Co-transport, **Basic** organic extractant

$$A^{n+}_{(aq)} + mB^{-}_{(aq)} + (m-n)H^{+}_{aq} + \overline{(m-n)C_{(org)}} \Leftrightarrow \overline{(A_n B_m{}^{(m-n)-})((m-n)CH^{+})_{(org)}} \tag{2}$$

The coupled facilitated counter-transport has reverse mechanism from the co-transport. The co-ion (B) transports from the stripping phase to feed phase, against the transport direction of the target component (A). The mechanism starts with the reaction between (A) and the organic extractant in terms of the organic complex ($\overline{BC}$) to form organic complex ($\overline{AC}$) and release co-ion (B) to feed phase. The organic complex ($\overline{AC}$), subsequently, diffuses across the liquid membrane to the stripping interface where the target component is released to

the stripping phase and co-ion is recovered to the organic extractant. The common case of the counter-transport is the reaction using acidic extractant as schematized by Equation (3).
Coupled Counter-transport, **Acidic** organic extractant

$$A^{+}_{(aq)} + \overline{HC_{(org)}} \Leftrightarrow \overline{AC_{(org)}} + H^{+}_{(aq)} \tag{3}$$

Uphill transport where the target component can be transported across the membrane against the concentration gradient of the target component is usually observed from the coupled transport. The uphill effects can continue until the target component diffuses across the liquid membrane to the stripping solution as long as the driving force in the coupled transport system is maintained. The driving force is often acquired from aqueous pH or H^{+} and/or co-ion (B) concentration gradient. Generally, research works in our laboratory shows that the removal of very dilute concentration of arsenic or mercury ions from produced water follows the mechanism of the coupled facilitated transport, and the uphill effects against the target component concentration is usually observed.
Key parameters to represent the performance of liquid membrane system are indicated in terms of the percentage of extraction (%), the percentage of stripping or recovery (%) and the distribution ratio (D). The percentage of extraction is the performance indicator which quantitatively identifies how much the target component, i.e., arsenic or mercury ions in our case is removed from produced water by the organic extractant. The percentage of recovery identifies how much the target component is recovered to the stripping solution. For the distribution ratio, it is a measurement of how much the target component is transferred to the next phase. It equals the concentration of the target component in the membrane phase divided by its concentration in the aqueous phase. Equations (4), (5) and (6) are provided for a better understating of these key parameters.

$$Extraction\,(\%) = \frac{(C_{f\mathbf{0}} - C_f)}{C_f} \tag{4}$$

$$Stripping\,(\%) = \frac{C_s}{C_{f\mathbf{0}}} \tag{5}$$

$$D = \frac{C_m}{C_{aq}} \tag{6}$$

2.2 HFSLM operation set up

The hollow fiber supported liquid membrane system is composed of two gear pumps, two variable flow rate controllers, two rotameters and four pressure gauges. The hollow fiber module is Celgard® microporous polypropylene fibers woven into fabric and wrapped around a central-tube feeder to supply the shell side fluid. The properties of the hollow fiber module are shown in Table (1).
The one-through mode of a single-module HFSLM operation is shown in Figure (4). The selected organic extractant is filled in and subsequently circulated in shell and tube sides of the module for at least 20 minutes to ensure that the extractant is embedded in micro pores

Properties	Descriptions
Material	Polypropylene
Number of fibers	10,000
Module length	20.3 cm
Module diameter	6.3 cm
Porosity	30%
Pore size	0.05 mm
Contact area	1.4 m^2
Area per unit volume	29.3 cm^2/cm^3
Fiber ID	240 μm
Fiber OD	300 μm

Table 1. Properties of the hollow fiber module

of the fibers. The feed produced water and the stripping solution are then pumped counter-currently into the tube side and shell side of the HFSLM, respectively. A 10-mL sample is taken at the end of each separation cycle (about 40 min) from the feed and the stripping solutions to analyze the concentrations of arsenic and/or mercury ions by the inductively coupled plasma spectroscopy (ICP).

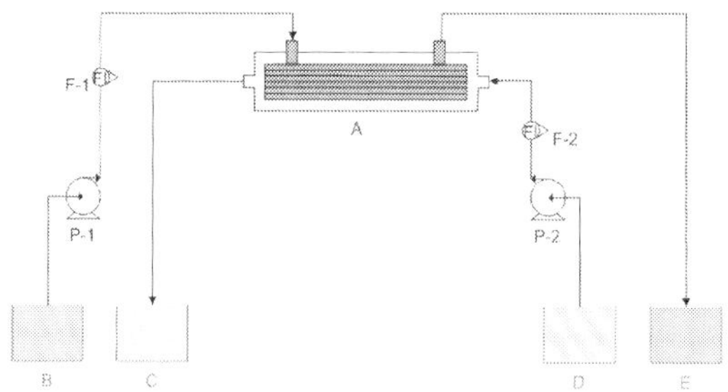

Fig. 4. Schematic counter-current flow diagram of one-through-mode separation by hollow fiber supported liquid membrane: A-HFSLM; B-inlet feed reservoir; C-outlet stripping solution reservoir; D-inlet stripping solution reservoir; E-outlet feed reservoir

3. Mass transfer mechanism in HFSLM

The mass transfer in HFSLM is the net diffusion of mass of arsenic/mercury ions across immiscible liquid membrane phase from feed phase to the stripping phase. The direction of mass transfer depends on the driving force across the liquid membrane. The driving force can be the concentration gradient of arsenic/mercury concentration itself or co-ion (B)/aqueous pH or H^+ concentration for the uphill transport in the case of the coupled facilitated transport as discussed earlier.

In engineering, the mass transfer coefficient is defined to quantify the mass transfer between the phases. It is a diffusion rate constant that relates among the mass transfer rate, mass transfer area and the concentration gradient. To determine the mass transfer coefficient for arsenic/mercury ions diffusion through HFSLM, the relationship between the mass transfer coefficient and permeability coefficient (P) is deployed. The permeability coefficient depends on mass transfer resistances and is reciprocal to the mass transfer coefficients as follows (Rathore et al., 2001).

$$\frac{1}{P}=\frac{1}{k_f}+\frac{r_i}{r_{lm}}\cdot\frac{1}{Dk_m}+\frac{r_i}{r_o}\frac{1}{k_s} \tag{7}$$

where

$$r_{lm}=\frac{r_o-r_i}{\ln(\frac{r_o}{r_i})} \tag{8}$$

Three mass transfer resistances in Equation (7) are in accordance with three steps of the transport mechanisms: first the resistance when arsenic/mercury ions diffuse across the feed interface; second the resistance when complex ions of arsenic/mercury diffuse across liquid membrane; and third the diffusion resistance across the stripping interface. The mass transfer resistance in the stripping interface can be disregarded since the mass transfer coefficient of the stripping solution (k_s) is higher than that of the feed phase (k_f) (Uedee et al., 2008) based on the following assumptions:

1. The film at feed interface is much thicker than that at the stripping interface. This is because of a combination of dense arsenic/mercury ions in the feed and co-ions from buffer solution at the feed interface, while there are only stripping ions and very lean arsenic/mercury ions at the stripping interface. From Equations (9) and (10), thick feed interfacial film (l_{if}) makes the mass transfer coefficient of the feed phase (k_f) much lower than that of the stripping solution (k_s).

Feed mass transfer coefficient

$$k_f=\frac{D}{l_{if}} \tag{9}$$

Stripping mass transfer coefficient

$$k_s=\frac{D}{l_{is}} \tag{10}$$

2. The arsenic/mercury concentration difference between the feed solution (C_f) and the feed-membrane interface (C_f*) is higher than the difference between the membrane-stripping interface (C_s*) and the membrane solution (C_s) (Porter, 1990). At equal flux by Equation (11), k_f is, therefore, much lower than k_s.

$$J=k_f(C_f-C_f^*)=k_s(C_s^*-C_s) \tag{11}$$

3. Only a stripping ion is used in the stripping side and it is directly in contact with the liquid membrane providing rapid dissolution. Thus, we can discard the third resistance in Equation (7) and obtain Equation (12).

$$\frac{1}{P}=\frac{1}{k_f}+\frac{r_i}{r_{lm}}\cdot\frac{1}{Dk_m} \tag{12}$$

The permeability coefficient (P) is also proposed by Danesi (Danesi, 1984) with the following Equations:

$$-V_f\ln\left(\frac{C_f}{C_{f,0}}\right)=AP\frac{\beta}{\beta+1}t\,. \tag{13}$$

where $$\beta = \frac{Q_f}{PL\varepsilon\pi Nr_i} \tag{14}$$

The permeability coefficient is calculated from the slope of the plot between $-V_f ln\left(\frac{C_f}{C_{f,0}}\right)$ against t.

Theoretical model equations for mass transfer mechanism through hollow fiber supported liquid membrane system are defined for better understanding the system and to predict the behavior of the system under different operating condition applications. The accurate model is very useful to scale up the system for industrial applications. Developments of the models, for example, diffusion model, plug flow model, graphical model have been proposed by many researchers. One of the diffusion models which attracts interests of our research group is the model proposed by Huang (Huang et al., 2008) on Fick's law of mass transfer.

The following assumptions are made to simplify the model equations:

1. Pseudo steady state system with a constant operating temperature and constant physical properties.
2. The membrane phase is completely immiscible with the feed and stripping phases.
3. The volume of the liquid membrane phase can be neglected with reference to the bulk volume of feed phase or the stripping phase.
4. The reaction rates at both feed and stripping interfaces are rapid and can ignore the associated reaction resistances.

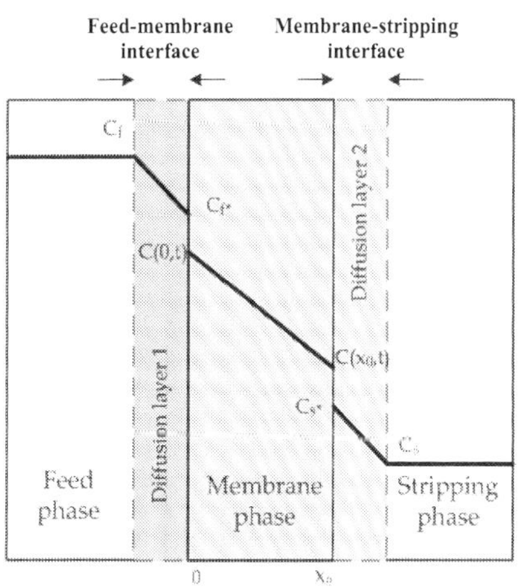

Fig. 5. The proposed steps in Huang's diffusion model

The diffusion transport at each step is described by Fick's law of diffusion as follows:

1. The target component diffuses across the interfacial layer between the feed and membrane.

$$j_f = k_f(C_f - C_f^*) \tag{15}$$

2. The target component reacts with the organic extractant to form the organic complex.

$$j_{r1} = k_1 C_f^* - k_{-1} C(0,t) \tag{16}$$

3. The organic complex diffuses across the membrane to the stripping interface.

$$j_{m1} = -D_e \frac{\partial C(0,t)}{\partial x} \tag{17}$$

$$j_{m2} = -D_e \frac{\partial C(x_0,t)}{\partial x} \tag{18}$$

4. The organic complex is decomposed at the stripping interface and the target component is released. Simultaneously, the organic extractant is regenerated and returns to the membrane phase.

$$j_{r2} = k_{-2} C(x_0,t) - k_2 C_s^* \tag{19}$$

5. The released target component diffuses across the interfacial layer between the membrane and is recovered to the stripping phase.

$$j_s = k_s(C_s^* - C_s) \tag{20}$$

where, k_f and k_s are obtained from the solution of Equation (7).
D_e is the effective diffusivity from Wilke-Chang's correlation, defined as follows

$$D_e = \frac{D\varepsilon'}{\tau'} \tag{21}$$

$$D = 7.4x10^{-8} \frac{(\varphi M_w)^{0.5} T}{\mu V_A^{0.6}} \tag{22}$$

By pseudo steady state assumption, the concentrations of the target component in the feed phase, membrane phase and stripping phase are achieved from the material balance equation.

$$V_f C_{f0} + V_s C_{so} = V_f C_f + V_s C_s + V_m C_m \tag{23}$$

$$C_m = C_f K_{ex} = C_s K_s \tag{24}$$

V_m can be neglected by the assumption that the membrane volume is very small relative to the bulk volume of feed and the stripping volume. The calculated results using the aforementioned proposed model equations show good agreement with the experimental results from various extraction applications via HFSLM by our group.

4. HFLSM applications for arsenic and mercury removal

Types of the extractants and stripping solutions are pivotal to the success of As and/or Hg separation. Iberhan and Wisniewski (Iberhan and Wisniewski, 2002) extracted As(III) and As(V) using the organic extractants of Cyanex 925, Cyanex 301 and a mixture between Cyanex 925 and Cyanex 301 at different volumetric ratios. The result showed that Cyanex 301 provided higher extraction of As(III) than As(V). The mixture of Cyanex 925 and Cyanex 301 helped remove As(V) significantly, while pure Cyanex 925 could extract As(V) a little better than As(III). Fabrega and Mansur (Fabrega & Mansur, 2007) extracted Hg(II) from HCl solution by Aliquat 336 dissolved in commercial Kerosene Exxol D-80. Mercury was almost extracted within 5 min at pH ≥ 1 and was satisfactorily stripped out about 99% by using thiourea as the stripping solution. Chakrabarty (Chakrabarty et al., 2010) found that by using triocthyamine (TOA) as the extractant in liquid membrane, Hg(II) could be extracted from pure solution more than lignosulfonate-mixed solution.

Knowledge from the past researches elicited our research group to progress further for the effective removal of arsenic/mercury ions from the produced water via HFSLM. Sangtumrong (Sangtumrong et al., 2007) simultaneously separated Hg(II) and As(III) ions from chloride media via HFSLM by TOA dissolved in toluene as the extractant and NaOH as the striping solution. Prapasawat (Prapasawat et al., 2008) used Cyanex 923 dissolved in toluene to separate As(III) and As(V) ions from sulphate media with water as the stripping solution via HFSLM. It was found that more As(V) could be extracted than As(III). Uedee (Uedee et al., 2008) obtained 100% extraction and 97% recovery of Hg(II) ions from chloride media via HFSLM using TOA dissolved in kerosene as the extractant and NaOH as the stripping solution. Recently, Pancharoen (Pancharoen et al., 2009; 2010) separated arsenic and mercury ions from the produced water from different gas fields in the Gulf of Thailand by HFSLM. Aliquat 336 dissolved in kerosene was a proper extractant with 91% arsenic extraction for the produced water without mercury contamination.

The discussions of the results from our research group are provided in the following sub topics (1) selective arsenic removal, (2) selective mercury removal and (3) simultaneous arsenic and mercury removal.

4.1 Selective arsenic removal

A number of researchers sought the organic extractants to remove arsenic ions, mostly they worked in a lab scale using synthetic feeds. Cyanex 301, Cyanex 923, Cyanex 925, a mixture of Cyanex 301 and Cyanex 925, tri-n-butylphosphate (TBP), hydrophobic glycol and hydroxamic acids were used to remove As(III) and As(V) from sulfuric acid solution by liquid-liquid extraction. The acidic reagent, Cyanex 301, could extract As(III) better than As(V). Cyanex 925 and Cyanex 923 were found more suitable for As(V) extraction than As(III) (Wisniewski, 1997; Meera et al., 2001).

Arsenic in the produced water is predominantly in the species of arsenite As(III) and arsenate As(V). The As(III) normally occurs as H_3AsO_3 and $H_2AsO_3^-$complexes. While As(V) occurs as H_3AsO_4, $H_2AsO_4^-$, $HAsO_4^{2-}$ and AsO_4^{3-}. But for the produced water from the Gulf of Thailand, As(III) usually presents in un-dissociated neutral H_3AsO_3 and As(V) presents in dissociated anion complexes of $H_2AsO_4^-$ and $HAsO_4^{2-}$ (Wilson et al., 2007). The As(V) takes majority part of total arsenic in the produced water where the pH is found in the range of 6-6.5. It appears that the removal of arsenic from produced water has to deal with spectroscopic range of arsenic in the water. Therefore, three different types of potential organic extractants for arsenic removal have been investigated in which the summary resultsare provided herewith.

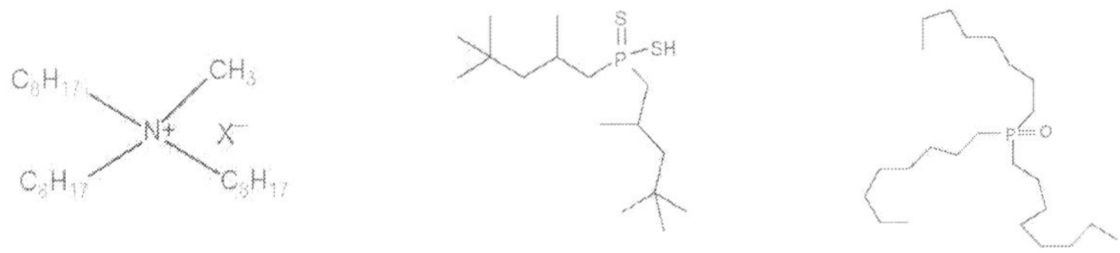

(a) (b) (c)

Fig. 6. Chemical structures of the organic extractants: (a) Aliquat 336, (b) Cyanex 301, and (c) Cyanex 923 (Mohapatra and Manchanda, 2008)

4.1.1 Arsenic extraction using aliquat 336 (basic extractant)

Aliquat 336 (CH_3R_3NCl) is the basic organic extractant which has an ability to react with both dissociated forms ($H_2AsO_4^-$ and $HAsO_4^{2-}$) and un-dissociated form (H_3AsO_3) of the arsenic ions. The transport mechanism for arsenic removal is driven by the concentration gradient of hydroxide ion (OH-) in counter with arsenic ion transport direction. For clear illustration, the schematic transport mechanism is provided in Figure (7) for the extraction of $H_2AsO_4^-$ using Aliquat 336. The extraction reaction and recovery reaction are demonstrated in Equations (25) - (30). Term X^- in the equations is denoted as Cl^- in feed and membrane phases for the first cycle, and OH^- (a counter ion in the stripping phase) will take over the place of Cl^- in the next cycles (Porter, 1990).

Extraction reactions of dissociated arsenic forms by Aliquat 336

$$H_2AsO_4^{\,-} + \overline{CH_3R_3N^+ \cdot X^-} \rightleftharpoons \overline{(CH_3R_3N^+)(H_2AsO_4^{\,-})} + X^- \quad (25)$$

$$HAsO_4^{\,2-} + \overline{2(CH_3R_3N^+ \cdot X^-)} \rightleftharpoons \overline{(CH_3R_3N^+)_2(HAsO_4^{\,2-})} + 2X^- \quad (26)$$

Extraction reactions of un-dissociated arsenic forms by Aliquat 336

$$H_3AsO_3 + \overline{CH_3R_3NCl} \rightleftharpoons \overline{(CH_3R_3NCl)(H_3AsO_3)} \quad (27)$$

Recovery reactions of arsenic-Aliquat 336 complex

$$\overline{(CH_3R_3N)(H_2AsO_4)} + NaOH \rightleftharpoons \overline{CH_3R_3N \cdot OH} + NaH_2AsO_4 \quad (28)$$

$$\overline{(CH_3R_3N)_2(HAsO_4)} + 2NaOH \rightleftharpoons \overline{2CH_3R_3N \cdot OH} + Na_2HAsO_4 \quad (29)$$

$$\overline{(CH_3R_3NCl)(H_3AsO_3)} + 3NaOH \rightleftharpoons \overline{CH_3R_3NCl} + Na_3AsO_3 + 3H_2O \quad (30)$$

Using 0.75 M (35% v/v) Aliquat 336 as the organic extractant and 0.5 M NaOH as the stripping solution, successful reduced arsenic to meet the permissible limit (< 250 ppb). The percentage of the recovery of arsenic ions increased with the concentration of sodium hydroxide up to 0.5 M. After 3-cycle operation, 91% of arsenic extraction from the produced water and 72% of arsenic recovery were achieved (Pancharoen et al., 2009).

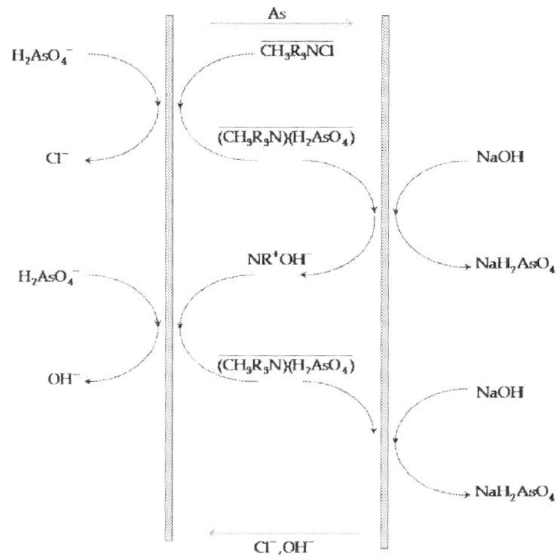

Fig. 7. Schematic representation of $H_2AsO_4^-$ coupled transport with Aliquat 336

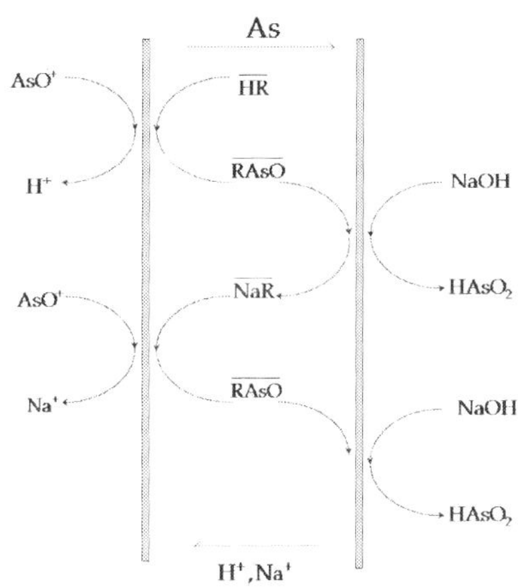

Fig. 8. Schematic representation of AsO^+ coupled transport with Cyanex 301

4.1.2 Arsenic extraction using cyanex 301 (acidic extractant)

Cyanex 301 is an acidic organic extractant which is recommended as an effective extractant for cation arsenic such as AsO^+ from the dissolution of H_3AsO_3 (Iberhan & Wisniewski, 2002). The transport mechanism is schematically illustrated in Figure (8). The process is driven by the hydrogen ion in counter with arsenic ion transport. The extraction and recovery reactions are described in the following equations:

$$AsO^+ + \overline{HR} \rightleftharpoons \overline{RAsO} + H^+ \tag{31}$$

$$\overline{RAsO} + NaOH \rightleftharpoons \overline{NaR} + HAsO_2 \tag{32}$$

For arsenic extraction using Cyanex 301 via HFSLM, the results were reported with relatively high percentage of extraction but with very low percentage of recovery (Pancharoen et al., 2009). The poor performance of recovery is explained by the strong bond that arsenic ions (AsO^+) make with Cyanex 301 and it is difficult for the stripping solution to break the bond (Iberhan & Wisniewski, 2002). This finding does not promote the favor of using Cyanex 301 as it offers an effective extraction while obtaining unacceptable recovery.

4.1.3 Arsenic extraction using cyanex 923 (neutral extractant)

Cyanex 923 or trialkylphosphine oxides (TRPO) is a neutral extractant which is recommended to use for the removal of un-dissociated neutral arsenic ions (Wisniewski, 1997; Meera et al., 2001). In the produced water, $HAsO_2$ is the neutral species of As(III) and H_3AsO_4 is the neutral species of As(V). The extraction of As(III) and As(V) by Cyanex 923 is proposed in Equations (33) and (34), respectively.

$$HAsO_2 + H^+ + \overline{TRPO} \rightleftharpoons \overline{HAsO_2 \cdot H^+ \cdot TRPO} \tag{33}$$

$$H_3AsO_4 + H^+ + \overline{TRPO} \rightleftharpoons \overline{H_3AsO_4 \cdot H^+ \cdot TRPO} \tag{34}$$

Using water as the stripping solution in the study, the recovery reactions are described as follows:

$$\overline{HAsO_2 \cdot H^+ \cdot TRPO} + H_2O \rightleftharpoons HAsO_2 + H^+ + \overline{TRPO} \tag{35}$$

$$\overline{H_3AsO_4 \cdot H^+ \cdot TRPO} + H_2O \rightleftharpoons H_3AsO_4 + H^+ + \overline{TRPO} \tag{36}$$

All equations above can be presented in the schematic diagram of arsenic transport as shown in Figure (9). Our work (Prapasawat et al., 2008) reported the study of using Cyanex 923 (30% v/v) diluted in toluene as the organic extractant and water as the stripping solution, the maximum arsenic extraction was 38% for As(III) species and 45% for As (V) species. Poor arsenic extraction performance was observed from Cyanex 923. This should be attributed to the low contribution of neutral arsenic in the feed and Cyanex 923 can work with neutral species only.

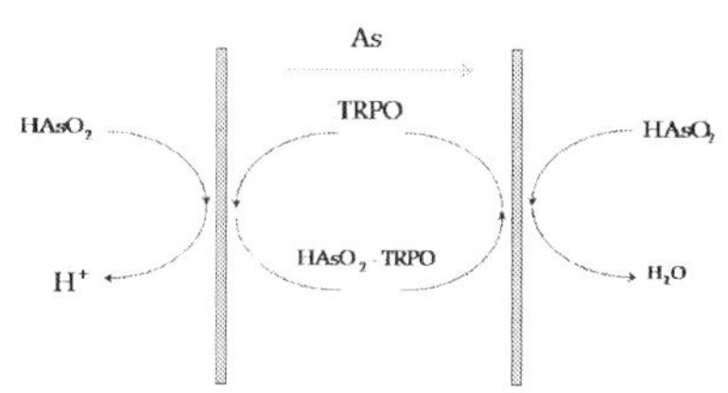

Fig. 9. Schematic representation of neutral As(III) with Cyanex 923

Of all three investigated extractants, Aliquat 336 attains high percentages of extraction and recovery of arsenic ions. This is due to its ability to react with both dissociated forms

($H_2AsO_4^-$ and $HAsO_4^{2-}$) and un-dissociated form (H_3AsO_3) of the arsenic which takes the majority part of total arsenic contribution in the produced water.

4.2 Selective mercury removal

Most of mercury in the produced water is an elemental form Hg(0) with the rest of inorganic form such as $HgCl_2$ or Hg(II). To remove mercury, the elementary mercury is normally taken by a chemical treatment process using an oxidant, ferric ions and flocculent to form a removable sludge containing mercury, which is known as an effective solution for the removal of mercury of high concentration. The residual mercury after the chemical treatment will be diluted and extracted via HFSLM subsequently. Uedee (Uedee et al., 2008) revealed that high extraction and recovery performances of mercury using HFSLM could be constantly maintained under the dilute mercury concentration system.

Normally, the mercury species after chemical treatment are in the form of Hg(II). The existence of Hg(II) comes from inorganic mercury $HgCl_2$ originally in the produced water and the undesired conversion of elementary mercury in the chemical treatment process. For the latter, the oxidation reduction potential is the contributory factor for the conversion. If the oxidation reduction potential exceeds the controlled limit, the elementary mercury is often converted to ionic mercury form of Hg(II) resulting in seriously degradation in the overall treatment process (Frankiewicz & Gerlach, 2000).

Sangtumrong (Sangtumrong et al., 2007) and Uedee (Uedee et al., 2008) removed Hg(II) from Hg(II) contaminated synthetic chloride water using tri-n-octylamine (TOA) by HFSLM. Pancharoen (Pancharoen et al., 2010) succeeded a similar work but used the produced water as the feed. The results corresponded closely; implying that the predominant Hg(II) species in the produced water is valid.

TOA is a basic organic extractant and its chemical structure can be referred to Figure (6). TOA in toluene is found to be the most selective mercury extractant (Sangtumrong et al., 2007; Uedee et al., 2008). However, feed pretreatment is necessary in order to deprotonate the Hg(II) of neutral $HgCl_2$ to anion form which is suitable for the function of the basic extractant (Ramakul & Panchareon, 2003). Equation (37) shows the Hg(II) deprotonation by HCl.

$$HgCl_2 + 2HCl \rightarrow HgCl_4^{2-} + 2H^+ \tag{37}$$

Subsequent to the feed pretreatment, mercury ions in the form of $HgCl_4^{2-}$ will react with the organic extractant (TOA, shown as R_3N) to form the complex species as seen in Equation (38):

$$HgCl_4^{2-} + 2H^+ + 2\overline{R_3N} \rightleftharpoons \overline{(R_3NH)_2 \cdot HgCl_4} \tag{38}$$

The mercury complex species diffuse to the opposite side of the liquid membrane by the concentration gradient and react with NaOH, a stripping solution. The $HgCl_4^{2-}$-ions are recovered to the stripping phase, shown in Equation (39):

$$\overline{(R_3NH)_2 \cdot HgCl_4} + 2OH^- \rightleftharpoons 2\overline{R_3N} + HgCl_4^{2-} + 2H_2O \tag{39}$$

After the stripping reaction, TOA is diffused back to the feed-membrane interface according to its concentration gradient, and again TOA is reacted with $HgCl_4^{2-}$ ions from the feed.

Thus, the transport mechanism of Hg(II) ions in the produced water through liquid membrane can be illustrated in Figure (10).

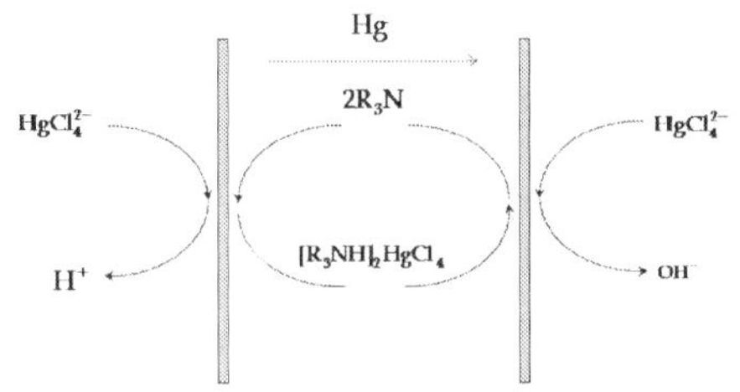

Fig. 10. Co-transport scheme of $HgCl_4^{2-}$ by TOA extractant

Pancharoen (Pancharoen et al., 2010) found that the highest percentages of extraction and recovery of 99.8% and 62%, respectively were achieved in 300 min by a 6th-cycle operation, pH of the feed solution of 2.5, 2% (v/v) TOA, 0.5 M NaOH using 50 mL/min of feed and stripping solutions.

4.3 Simultaneous arsenic and mercury removal

A successive attempt on simultaneous removal of arsenic and mercury from the produced water was investigated. The focused species to be extracted were dissociated As(V) as $H_2AsO_4^-$ and Hg(II) as $HgCl_2$ since they were key contaminated arsenic/mercury in the produced water. To enhance the separation of arsenic and mercury, the synergistic extraction by using the mixture of the organic extractant was examined. Equation (40) defines the synergistic extent in terms of synergistic coefficient (R) relating to the distribution coefficients (Luo et al., 2004).

$$R = \frac{D_{max}}{(D_1 + D_2)} \tag{40}$$

D_{max} is the maximum distribution coefficient or the distribution ratio of the synergistic system to extract the specifed ions, and (D_1 + D_2) is the summation of the distribution coefficient from each single extraction system. The greater synergistic coefficient means that the mixture of the extractant has synergistic effect on arsenic/mercury extraction. Figure (11) shows a comparative plot of the maximum percentages of the extraction of arsenic and mercury ions from the produced water against the different extractants. The sequences of the percentages of extraction are as follows.

As: Aliquat 336+Cyanex 471 > Aliquat 336 > Bromo-PADAP > Cyanex 471 > Cyanex 923,

Hg: Aliquat 336+Cyanex 471 > Aliquat 336 > Cyanex 923 > Bromo-PADAP ≈ Cyanex 471

It was reported that the mixture of 0.22 M Aliquat 336 and 0.06 M Cyanex 471 provided the highest extraction of both arsenic and mercury. The calculated synergistic coefficient (R) to arsenic ions of Cyanex 471 was 2.8; the value greater than 1 indicated that the mixture of Aliquat 336 and Cyanex 471 had the synergistic effect on arsenic extraction.

Among the stripping solutions used in this work, i.e., NaOH, DI water, HNO_3 and H_2SO_4, thiourea (NH_2CSNH_2) was found to be the best stripping solution for arsenic and mercury. Thiourea with large anion in the structure was strong enough to strip mercury complex ion from Aliquat 336, which was composed of a large organic cation associated with a chloride

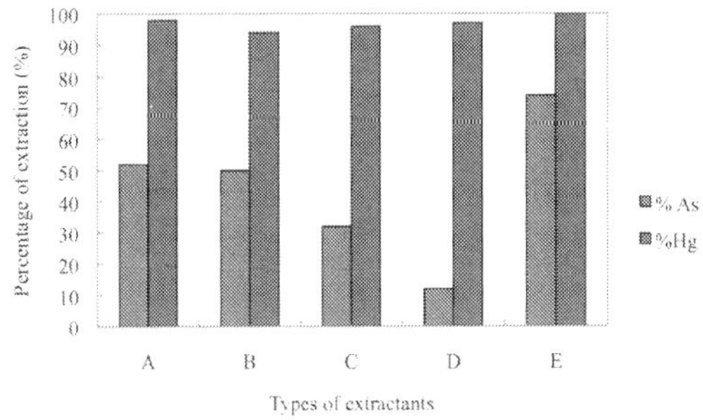

Fig. 11. The maximum percentages of arsenic and mercury ions extraction from produced water against types of the extractants: (A) 0.22 M Aliquat 336, (B) 0.002 M Bromo-PADAP, (C) 0.06 M Cyanex 471, (D) 0.51 M Cyanex 923, (E) 0.22 M Aliquat 336 + 0.06 M Cyanex 471

ion. Moreover, water in aqueous thiourea also contributed to the recovery performance according to the report of using water as the stripping solution for As(III) and As(V) separation (Prapasawat et al., 2008). Using thiourea, no trace of the precipitates was observed unlike NaOH which produced the precipitates with Hg resulting in membrane fouling and poor transport performance in the membrane phase. The discharge concentrations of mercury and arsenic in the produced water to the environment complied with the legislation limits determined by the Ministry of Industry within 1-cycle separation and 3-cycle separation, respectively.

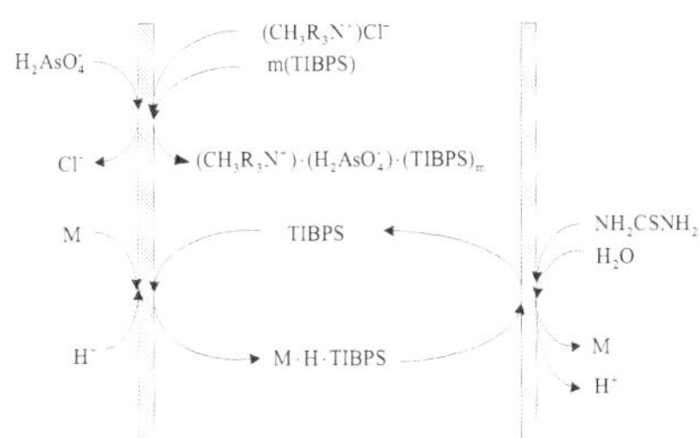

Fig. 12. Schematic extraction and stripping mechanisms of $H_2AsO_4^-$ by the synergistic extractant of Aliquat 336 and Cyanex 471 (TIBPS) with thiourea as the stripping solution

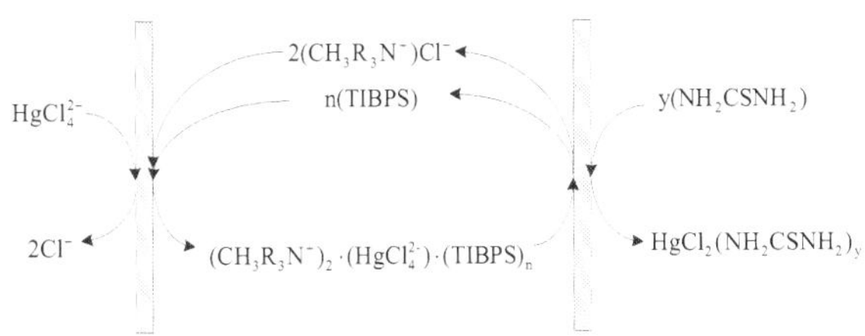

Fig. 13. Schematic extraction and stripping mechanisms of $HgCl_4^{2-}$ by the synergistic extractant of Aliquat 336 and Cyanex 471 (TIBPS) with thiourea as the stripping solution

The descriptive illustrations of the extraction and stripping mechanisms of arsenic/mercury ions by the mixture of Aliquat 336 (($CH_3R_3N^+$)Cl^-) and Cyanex 471 (TIBPS) as the synergistic extractant, and thiourea (NH_2CSNH_2) as the stripping solution, are shown in Figures (12) and (13).
The synergistic extraction reactions are shown in Equations (41) - (44).
Extraction by Aliquat 336

As(V) $$H_2AsO_4^- + \overline{CH_3R_3N^+Cl^-} \rightleftharpoons \overline{(CH_3R_3N^+)\cdot(H_2AsO_4^-)} + Cl^- \quad (41)$$

Hg(II) $$HgCl_4^{2-} + \overline{2CH_3R_3N^+Cl^-} \rightleftharpoons \overline{(CH_3R_3N^+)_2\cdot(HgCl_4^{2-})} + 2Cl^- \quad (42)$$

Extraction by the mixture of Aliquat 336 and Cyanex 471 (TIBPS)

As(V) $$H_2AsO_4^- + \overline{CH_3R_3N^+} + m\overline{(TIBPS)} \rightleftharpoons \overline{(CH_3R_3N^+)\cdot(H_2AsO_4^-)\cdot(TIBPS)_m} \quad (43)$$

Hg(II) $$HgCl_4^{2-} + 2\overline{(CH_3R_3N^+)} + n\overline{(TIBPS)} \rightleftharpoons \overline{(CH_3R_3N^+)_2\cdot(HgCl_4^{2-})\cdot(TIBPS)_n} \quad (44)$$

where, the stoichiometric coefficients of m and n were calculated from the distribution coefficients of the relevant components at various concentrations of the extractant used.
Extraction by Cyanex 471 (TIBPS) for un-dissociated arsenics

As (III &V) $$M + H^+ + \overline{TIBPS} \rightleftharpoons \overline{M\cdot H\cdot TIBPS} \quad (45)$$

where, M stands for H_3AsO_3 (As(III)) or H_3AsO_4 (As(V)). Cyanex 471 (TIBPS) is the neutral organic extractant and effective for un-dissociated ions (Wisniewski, 1997) including un-dissociated arsenic such as H_3AsO_3 (As(III)) and H_3AsO_4 (As(V)) in the produced water. Accordingly, it is regarded as an enhancement to arsenic extraction, on top of primarily focused species of dissociated $H_2AsO_4^-$.

4.3.1 Distribution coefficients and extraction equilibrium constants

Subject to the mass transport analysis, the following terms of the extraction equilibrium constant (K_{ex}) and the distribution ratio are expressed by Equations (46) – (49).
The extraction equilibrium constants (K_{ex}) of arsenic and mercury ions:

$$K_{ex,As} = \frac{[\overline{(CH_3R_3N^+)\cdot(H_2AsO_4^-)\cdot(TIBPS)_m}]}{[H_2AsO_4^-][\overline{CH_3R_3N^+}][\overline{TIBPS}]^m} \quad (46)$$

$$K_{ex,Hg} = \frac{[\overline{(CH_3R_3N^+)_2\cdot(HgCl_4^{2-})\cdot(TIBPS)_n}]}{[HgCl_4^{2-}][\overline{CH_3R_3N^+}]^2[\overline{TIBPS}]^n} \quad (47)$$

The distribution coefficients (D) for arsenic and mercury extractions by the mixture of Aliquat 336 and Cyanex 471 (TIBPS):

$$D_{As} = \frac{\overline{[(CH_3R_3N^+)\cdot(H_2AsO_4^-)\cdot(TIBPS)_m]}}{[H_2AsO_4^-]} = K_{ex,As}\overline{[CH_3R_3N^+]}\overline{[TIBPS]}^{\,m} \tag{48}$$

$$D_{Hg} = \frac{\overline{[(CH_3R_3N^+)_2\cdot(HgCl_4^{2-})\cdot(TIBPS)_n]}}{[HgCl_4^{2-}]} = K_{ex,Hg}\overline{[CH_3R_3N^+]}^{\,2}\overline{[TIBPS]}^{\,n} \tag{49}$$

The distribution coefficients (D) of arsenic and mercury from the extraction by HFSLM, shown in Table (2), are estimated from Equations (48) and (49). The increase of the distribution coefficient indicates the enhancement of the extractability. From Table (2), the distribution coefficients increased with the concentration of Cyanex 471. The maximum distribution coefficients of arsenic and mercury were attained at 0.06 M Cyanex 471 and 0.07 M Cyanex 471, respectively.

Cyanex 471 (M)	Distribution coefficients	
	Arsenic	Mercury
0.02	0.63	-
0.04	1.13	4.52
0.05	1.32	5.57
0.06	1.47	6.59
0.07	-	8.72

Table 2. The distribution coefficients at Cyanex 471 concentration of 0.02 - 0.07 M mixed with 0.22 M Aliquat 336 (0.5 M NaOH as the stripping solution)

The distribution coefficients in Equations (48) and (49) were rewritten as follows:

$$\log D_{As} = \log(K_{ex,As}\cdot\overline{[CH_3R_3N^+]}) + m\log\overline{[TIBPS]} \tag{50}$$

$$\log D_{Hg} = \log(K_{ex,Hg}\cdot\overline{[CH_3R_3N^+]}^{\,2}) + n\log\overline{[TIBPS]} \tag{51}$$

The stoichiometric coefficients (m and n) were calculated from the plots of log D_{As} and log D_{Hg} against log [TIBPS]. The linear relationships with slopes m = 0.7917 or 4/5 for arsenic extraction and n = 1 for mercury extraction were observed. The slopes, m and n, were substituted in the synergistic extraction Equations (43) and (44). The extraction equilibrium constants of arsenic ions ($K_{ex,As}$) and mercury ($K_{ex,Hg}$) were determined by Equations (46) and (47). The equilibrium constant of mercury (1,622 (L/mol^3)) was much higher than that of arsenic (62.7 (L/mol$^{9/5}$) suggesting that the extraction of mercury was higher than arsenic, which was in accordance with the results obtained from the study.

4.3.2 Permeability coefficients

The permeability coefficients of arsenic and mercury, which related to the concentration of Cyanex 471 from 0.02 - 0.07 M, were obtained from Equations (13) and (14) and the slopes (AP $\beta/(\beta+1)$) of the plot between $-V_f \ln(C_f / C_{f,o})$ versus t in Figure (14). From Table (3), it could be observed that the permeability coefficients increased when the concentration of Cyanex 471 increased. The permeability coefficients of mercury were higher than those of arsenic, implying higher mass transfer or higher extraction of mercury ions.

Cyanex 471 (M)	P x 10^3 (cm/s) Arsenic	P x 10^3 (cm/s) Mercury
0.02	5.47	-
0.04	8.90	33.98
0.05	9.81	40.94
0.06	11.54	48.37
0.07	-	53.14

Table 3. The permeability coefficients at Cyanex 471 concentration of 0.02 - 0.07 M mixed with 0.22 M Aliquat 336

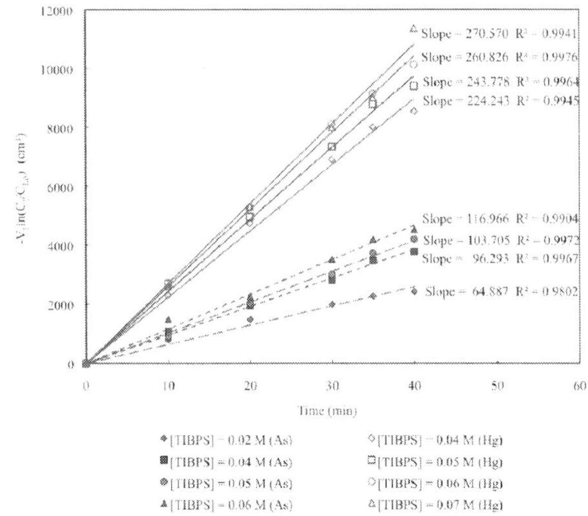

Fig. 14. Plot of $-V_f \ln (C_f / C_{f,o})$ of arsenic and mercury ions in feed solution against time with different concentrations of Cyanex 471 mixed with 0.22 M Aliquat 336 for synergistic organic extractant

4.3.3 Mass transfer coefficients

Equations (52) and (53) were defined assuming the stripping reactions of arsenic and mercury were instantaneous and no contribution of resistance in the stripping phases.

$$\frac{1}{P_{As}} = \frac{1}{k_i} + \frac{r_i}{r_{lm}} \cdot \frac{1}{K_{ex,As} k_m \overline{[CH_3R_3N^+]} \overline{[TIBPS]}^{4/5}} \tag{52}$$

$$\frac{1}{P_{Hg}} = \frac{1}{k_i} + \frac{r_i}{r_{lm}} \cdot \frac{1}{K_{ex,Hg} k_m \overline{[CH_3R_3N^+]}^{2} \overline{[TIBPS]}} \tag{53}$$

where, P_{As} and P_{Hg} were the permeability coefficients of arsenic and mercury, respectively. The organic-phase and aqueous-phase mass transfer coefficients (k_m and k_i) of arsenic were determined from the plot of 1/P against $1/([CH_3R_3N^+][TIBPS]^{4/5})$. The slope and the ordinate were $(r_i/r_{lm})(1/K_{ex}k_m)$ and $1/k_i$. Accordingly, k_m and k_i of arsenic were obtained at 1.02×10^{-4} and 0.0392 cm/s. Similarly, the values of mercury were obtained at 1.27×10^{-5} and 2.210 cm/s from the plot of 1/P against $1/([CH_3R_3N^+]]^2\ [TIBPS])$. According to the calculated k_m and k_i where the organic-phase mass transfer coefficient was lower than the aqueous-phase mass transfer coefficient, thus, the mass transfer within membrane was the rate controlling step of this extraction system.

5. Conclusions

The outstanding feature of HFSLM technique is its ability to treat metal ions of a very low concentration that are hardly treated or impractical by the conventional techniques. Therefore, based on the concept of mass transfer we, for example, deployed the HFSLM to remove low concentrations of arsenic and mercury in the produced water. The result was successfully achieved by using the HFSLM in conjunction with the selection of the suitable single or synergistic organic extractant, and the stripping solution. The removal of mercury contamination was very effective. In many cases, the mercury content could be reduced to meet the permissible discharge limit to the environment from only single extraction cycle. The removal of arsenic was found inferior to the mercury. More than one extraction cycles were required to treat arsenic in the produced water. Of all the potential organic extractants in the study for arsenic treatment, Aliquat 336 was the best extraction since it could extract both dissociated and un-dissociated arsenic under a basic or weak acidic condition.

Synergism extraction was much of interest aiming for the simultaneous removal of arsenic and mercury. By using the mixture of 0.06 M Cyanex 471 and 0.22 M Aliquat 336, it promoted the synergistic effect on arsenic extraction. The calculated synergistic coefficient to arsenic ions was 2.8, implying 2.8 times of arsenic being extracted by the synergistic extractant more than by the single organic extractant. The 0.1 M thiourea was found to be the best stripping solution. The arsenic/mercury contaminated in the effluent produced water complied with the legislation limit after 1-cycle extraction for mercury and 3-cycle extraction for arsenic. Mass transfer during the whole extraction process starting from feed phase through the stripping phase was followed the Fick's law diffusion mass transport. The parameters describing mass transport were determined, i.e., (1) equilibrium constant (K_{eq}), (2) distribution ratio (D), (3) permeability coefficient (P), (4) mass transfer coefficient in the feed phase stagnant layer (k_f) and (5) mass transfer coefficient in the immobilized liquid membrane (k_m). In this extraction via HFSLM system for the simultaneous removal of arsenic and mercury, the rate controlling step was found at the membrane phase.

6. Acknowledgments

Sincere thanks go to the Thailand Research Fund (TRF), the Office of Small and Medium Enterprises Promotion (OSMEP) and Chulalongkorn University for granting research funds;

Ratchadaphiseksomphot Endownment Fund (FW012A). Contributions and efforts by the students to run the experiments are very much appreciated.

7. Notation

A	Membrane area (cm^2)
C	Concentration within the membrane (mol / cm^3)
C_{aq}	Concentration in the aqueous phase (mol / cm^3)
C_f	Concentration in the feed phase (mol / cm^3)
C_{f0}	Initial concentration in the feed phase (mol / cm^3)
C^*_f	Concentration at the feed interface with liquid membrane (mol / cm^3)
C_m	Concentration in the liquid membrane phase (mol / cm^3)
C_s	Concentration in the stripping phase (mol / cm^3)
C_{s0}	Initial concentration in the stripping phase (mol / cm^3)
C^*_s	Concentration at the stripping interface with the liquid membrane (mol / cm^3)
D	Distribution ratio
D_e	Effective diffusivity (cm^2/s)
D_0	Free bulk diffusion coefficient (cm^2/s)
HFSLM	Hollow Fiber Supported Liquid Membrane
J	Flux (mol/cm^2.s)
j_f	Flux across the feed interface film (mol/cm^2.s)
j_{m1}	Flux at feed-liquid membrane interface (mol/cm^2.s)
j_{m2}	Flux at liquid membrane-stripping interface (mol/cm^2.s)
j_{r1}	Flux following the extraction reaction (mol/cm^2.s)
j_{r2}	Flux following the recovery or stripping reaction (mol/cm^2.s)
j_s	Flux across the stripping interface film (mol/cm^2.s)
K_e	Equilibrium constant of the interfacial reaction in the feed phase
K_s	Equilibrium constant of the interfacial chemical reaction in stripping phase
K_f	Dimensionless form of mass transfer coefficient in the feed phase
K_s	Dimensionless form of mass transfer coefficient in the stripping phase
k_1	Forward interfacial reaction rate constant in the feed phase (m/s M)
k_{-1}	Backward interfacial reaction rate constant in the feed phase (m/s M)
k_2	Forward interfacial reaction rate constant in the stripping phase (m/s M)
k_{-2}	Backward interfacial reaction rate constant in the stripping phase (m/s M)
k_f	Mass transfer coefficient in the extraction boundary layer (cm/s)
k_s	Mass transfer coefficient in the stripping boundary layer (cm/s)
L	Length of the hollow fiber (cm)
l_{if} and l_{is}	Thickness of feed interface film and stripping interface film (cm)
M_w	Molecular weight (g/mol)
N	Numbers of hollow fibers in the module
m, n	Stoichiometric coefficient
P	Permeability coefficient (cm/s)
Q_f	Volumetric flow rate of feed solution (cm^3/s)

R	Synergy coefficient
r_i	Internal radius of the hollow fiber (cm)
r_o	External radius of the hollow fiber (cm)
r_{lm}	Log-mean radius of the hollow fiber
T	Absolute temperature (K)
V_A	Molar volume of the target component at its normal boiling temperature (cm^3/mol)
V_f	Volume of the feed phase (cm^3)
V_{f0}	Initial volume of the feed phase (cm^3)
V_m	Volume of the membrane phase (cm^3)
V_{s0}	Initial volume of the stripping phase (cm^3)
V_s	Volume of the stripping phase (cm^3)
x_0	Membrane thickness (cm)
ε	Porosity of the hollow fiber (%)
φ	Association factor of solvent (dimensionless)
μ	Viscosity (cP)
Subscripts	
aq	Aqueous
f	Feed phase
m	Membrane phase
org	Organic
s	Stripping phase

8. References

Arpa, C.; Basyilmaz, E.; Bektas, S.; Genç, O. & Yurum, Y. (2000). Removal of Hg, Cd and Pb from waste water, In: *J. Fuel Process. Technol.*, Vol. 68., pp. 111-120

Bringas, E.; San Rom´an, MF.; Irabien, JA. & Ortiz, I. (2009). An overview of the mathematical modelling of liquid membrane separation processes in hollow fiber contactors, In: *J. Chem Technol Biotechnol*, Vol. 84., pp. 1583–1614

Chakrabarty, K.; Saha, P. & Ghoshal, A.K. (2010). Simultaneous separation of mercury and lignosulfonate from aqueous solution using supported liquid membrane, In: *J. Membr. Sci.*, Vol. 346 ., pp. 37-44

Dabrowski, A.; Hubicki, Z.; Podkoscielny, P. & Robens, E. (2004). Selective removal of the heavy metal ions from waters and industrial wastewaters by ion-exchange method, In: *J. Chemosphere*, Vol. 56 ., No. 2., pp. 91-106

Danesi, P.R. (1984). A simplified model for the coupled transport of metal ions through hollow-fiber supported liquid membranes, In: *J. Membrane Sci.*, Vol. 20., pp. 231-248

Department of Mineral Fuels (DMF), Ministry of Energy, Thailand. (2010). http://www.dmf.go.th / index.php?act=service&sec=year Production, dated 20-July-10

Department of Mineral Fuels (DMF), Ministry of Energy, Thailand. (2009). http://www.dmf.go.th/ file/Concess270309.png

Fábrega, F.d.M. & Mansur, M.B. (2007). Liquid-liquid extraction of mercury(II) from hydrochloric acid solutions by Aliquat 336, In: *J. Hydrometallurgy*, Vol. 87., pp. 83-90

Frankiewicz, T.C. & Gerlach, J. (2000). Removal of hydrocarbons, mercury and arsenic from oil-field produced water, In: *U.S. patent number 6117333*

Gallup, D.L. & Strong, J. B. (2007). Removal of Mercury and Arsenic from Produced Water, Chevron Corporation, pp. 1-9

Huang, D.; Huang, K.; Chen, S. & Liu, S. (2008). Rapid Reaction-Diffusion Model for the Enantioseparation of Phenylalanine across Hollow Fiber Supported Liquid Membrane, In: *J. Sep. Sci. and Tech.*, Vol. 43., pp. 259–272

Iberhan, L. & Wisniewski, M. (2002). Extraction of arsenic(III) and arsenic(V) with Cyanex 925, Cyanex 301 and their mixtures, In: *J. Hydrometallurgy*, Vol. 63., pp. 23-30

Korte, N.E. & Feernando, Q. (1991). A review of arsenic(III) in groundwater, In: *Criti. Rev. Environ. Sci. Technol.*, Vol. 21., pp. 1-39

Lothongkum, A.W.; Khemglad, Y.; Usomboon, N. & Pancharoen, U. (2009). Selective recovery of nickel ions from wastewater of stainless steel industry via HFSLM, In: *J. Alloy Compd.*, Vol. 476., pp. 940-949

Lothongkum, A.W.; Ramakul, P.; Sasomsub, W.; Laoharochanapan, S. & Pancharoen, U. (2009). Enhancement of uranium ion flux by consecutive extraction via hollow fiber supported liquid membrane, In: *J. Taiwan Inst. Chem. Eng.*, Vol. 40., pp. 518-523

Luo F., Li D. & Wei P. (2004). Synergistic extraction of zinc(II) and cadmium(II) with mixtures of primary amine N1923 and neutral organophosphorous derivatives, In: *J. Hydrometallurgy*, Vol. 73., pp. 31-40

Meera, R., Francis, T. & Reddy, M.L.P. (2001). Studies on the liquid-liquid extraction of mercury(II) from acidic chloride solutions using Cyanex 923, In: *J. Hydrometallurgy*, Vol. 61., pp. 97-103

Mohapatra, P.K. & Manchanda, V.K. (2008). Liquid Membrane-Based Separations of Actinides, In: *Handbook of Membrane Separations*, ISBN-13: 978-0-8493-9549-9, CRC Press., Taylor & Francis Group

Pancharoen, U.; Somboonpanya, S.; Chaturabul, S. & Lothongkum, A.W. (2010). Selective removal of mercury as $HgCl_4^{2-}$ from natural gas well produced water by TOA via HFSLM, In: *J. Alloys and Compounds*, Vol. 489., pp. 72-79

Pancharoen, U.; Poonkum, W. & Lothongkum, A.W. (2009). Treatment of arsenic ions from produced water through hollow fiber supported liquid membrane, In: *J. Alloys and Compounds*, Vol. 482., pp. 328-334

Pancharoen, U.; Ramakul, P. & Patthaveekongka, W. (2005). Purely Extraction and Separation of Mixture of Cerium(IV) and Lanthanum(III) Via Hollow Fiber Supported Liquid Membrane, In: *J. Ind. Eng. Chem.*, Vol. 11., No. 6., pp. 926-931

Porter, M.C., (1990). *Handbook of Industrial Technology Membrane*, Noyes Publications, ISBN: 0815512058, New Jersey, USA

Prapasawat, T.; Ramakul, P.; Satayaprasert, C.; Pancharoen, U. & Lothongkum, A.W. (2008). Separation of As(III) and As(V) by hollow fiber supported liquid membrane based on the mass transfer theory, In: *Korean J. Chemical Engineering*, Vol. 25., No. 1., pp. 158-163

Ramakul, P.; Supajaroon, T.; Prapasawat, T.; Pancharoen, U. & Lothongkum, A.W. (2009). Synergistic separation of yttrium ions in lanthanide series from rare earths mixture via hollow fiber supported liquid membrane, In: *J. Chem. Eng.*, Vol. 15., pp.224-228

Ramakul, P.; Prapasawad, T.; Pancharoen, U. & Pattaveekongka, W. (2007). Separation of radioactive metal ions by hollow fiber supported liquid membrane and permeability analysis, In: *J. China. Inst. Chem. Engrs.*, Vol. 38., pp. 489-494

Ramakul, P.; Pattaweekongka, W. & Pancharoen, U. (2006). Mass transfer modeling of membrane carrier system for extraction of Ce(IV) from sulfate media using hollow fiber supported liquid membrane, In: *Korean J. Chem. Eng.*, Vol. 23., No. 1., pp. 85-92

Ramakul, P.; Songkun, E.; Pattaweekongka, W.; Hronec, M. & Pancharoen, U. (2006). Permeation study on the hollow fiber supported liquid membrane for the extraction of Cobalt(II), In: *Korean J. Chem. Eng.*, Vol. 23., No. 1., pp. 117-123

Ramakul, P.; Nakararueng, K. & Panchareon, U. (2004). One-through Selective Separation of Copper, Chromium and Zinc Ions by Hollow Fiber Support Liquid Membrane, In: *Korean J. Chem. Eng.*, Vol. 21., No. 6., pp. 1212-1217

Ramakul, P. & Panchareon, U. (2003). Synergistic extraction of mixture of lanthanum and neodymium by hollow fiber supported liquid membrane, In: *Korean J. Chem. Eng.*, Vol. 20., No. 4., pp. 724-730

Rathore, N. S.; Rathore, N.S.; Sonawane, J.V.; Venugopalan, A. K. & Shukla, J.P. (2001). Hollow fiber supported liquid membrane: a novel technique for separation and recovery of plutonium from aqueous acidic wastes, In: *J. Membr. Sci.*, Vol. 189., pp. 119-128

Sangtumrong, S.; Ramakul, P.; Satayaprasert, C.; Pancharoen, U. & Lothongkum, A.W. (2007). Purely separation of mixture of mercury and arsenic via hollow fiber supported liquid membrane, In: *J. Ind. and Eng. Chem.*, Vol. 13., No. 5., pp. 751-756

Uedee, E.; Ramakul, P.; Pancharoen, U. & Lothongkum, A.W. (2008). Performance of hollow fiber supported liquid membrane on the extraction of mercury (II) ions, In: *Korean J. Chem. Eng.*, Vol. 25., No. 6., pp. 1486-1494

Usapein, P.; Lothongkum, A.W.; Ramakul, P. & Pancharoen, U. (2009). Efficient transport and selective extraction of Cr(VI) from waste pickling solution of the stainless steel-cold rolled plate process using Aliquat 336 via HFSLM, In: *Korean J. Chem. Eng.*, Vol. 26 ., No. 3., pp. 791-798

U.S. Environmental Protection Agency (EPA). (2007). Treatment Technologies for Mercury in Soil, Waste, and Water, pp. ES-4

U.S. Environmental Protection Agency (EPA). (1984). EPA-600/8-83-021F: Health Assessment Document for Inorganic Arsenic, pp. 10-28

Van der Vaart, R.; Akkerhuis, J.; Feron, P. & Jansen, B. (2001), Removal of mercury from gas streams by oxidative membrane gas absorption, In: *J. Membr. Sci.*, Vol. 187., pp. 151-157

Wilson, S.D.; Kelly, W.R.; Holm, T.R. & Talbott, J.L. (2007). Arsenic removal in water treatment facilities: survey of geochemical factors and pilot plant experiments, Midwest Technology Assistance Center, pp. 10–11

Wisniewski, M. (1997) Extraction of arsenic from sulphuric acid solutions by Cyanex 923, In: *J. Hydrometallurgy.*, Vol. 46., pp. 235-241.

23

Mass Transfer in Fluidized Bed Drying of Moist Particulate

Yassir T. Makkawi[1] and Raffaella Ocone[2]
[1]*Chemical Engineering & Applied Chemistry, Aston University, Birmingham B4 7ET,*
[2]*Chemical Engineering, Heriot-Watt University, Edinburgh EH14 4AS,*
UK

1. Introduction

Bubbling fluidized bed technology is one of the most effective means for the interaction between solid and gas flow, mainly due to its good mixing and high heat and mass transfer rate. It has been widely used at a commercial scale for drying of grains such as in pharmaceutical, fertilizers and food industries. When applied to drying of non-porurs moist solid particles, the water is drawn-off driven by the difference in water concentration between the solid phase and the fluidizing gas. In most cases, the fluidizing gas or drying agent is air. Despite of the simplicity of its operation, the design of a bubbling fluidized bed dryer requires an understanding of the combined complexity in hydrodynamics and the mass transfer mechanism. On the other hand, reliable mass transfer coefficient equations are also required to satisfy the growing interest in mathematical modelling and simulation, for accurate prediction of the process kinetics.
This chapter presents an overview of the various mechanisms contributing to particulate drying in a bubbling fluidized bed and the mass transfer coefficient corresponding to each mechanism. In addition, a case study on measuring the overall mass transfer coefficient is discussed. These measurements are then used for the validation of mass transfer coefficient correlations and for assessing the various assumptions used in developing these correlations.

2. Two phase model of fluidization

The first model to describe the essential hydrodynamic features in a bubbling fluidized bed, usually referred to as the simple two phase model, was proposed in the early fifties of the last century by Toomey and Johnstone (1952). The model assumes that all the gas in excess of the minimum fluidization velocity, U_{mf}, passes through the core of the bed in the form of bubbles. The rest of the gas, usually referred to as emulsion gas, was described to pass through a dense solid phase surrounding the bubbles, at a low velocity close or equal to U_{mf}. Later experimental investigations on bubbles formation and rise in two and three dimensional fluidized beds, utilizing conventional photographing and x-ray imaging techniques, have shown a rather more complicated flow pattern of gas around bubbles. A more accurate model, describing the movement of gas/solid and pressure distribution around a rising bubble was then proposed by Davidson and Harrison (1963). This model describes the gas flow through a three dimensional fluidized bed mainly in a spherical or semi-spherical shape bubbles through the core, however, depending on the emulsion gas velocity; the region around the bubble may be surrounded by a cloud as a result of emulsion gas circulation between the dense solid phase and the core of the bubble. This can be schematically described as shown in Fig. 1. The existence of a cloud around fast rising bubbles has been later verified experimentally by a number of researchers. Most recently, Makkawi and Ocone (2009), utilizing Electrical Capacitance Tomography (ECT) imaging have further confirmed the existence of cloud around a single isolated bubble rising through a fluidized bed as shown in Fig. 2. In terms of mass transfer, the existence of cloud and gas circulation between the bubble and its surrounding have a significant contribution to the overall mass transfer mechanism in a bubbling fluidized bed dryer as will be discussed later.

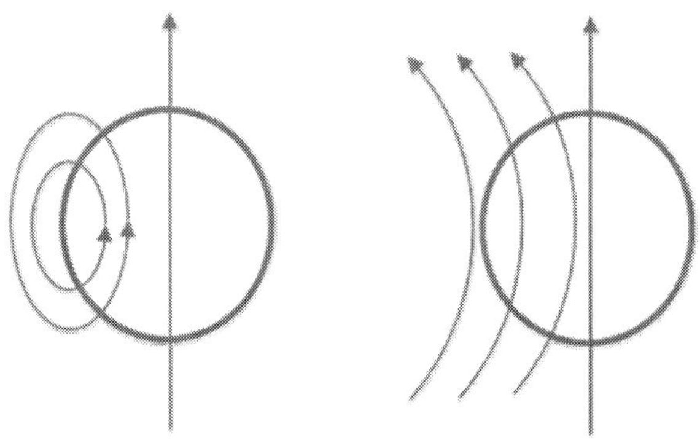

(a) fast rising bubble (b) slow rising bubble

Fig. 1. Proposed gas streamlines in and out of a single rising bubble as described in

3. Mass transfer mechanisms

With the confirmed existence of different phases in a bubbling fluidized bed, it is postulated that in a bubbling fluidized bed dryers different mechanisms can regulate the mass transfer process, depending on the bubbles characteristics and the degree of water content in the bed. The different phases, which all contribute to the removal of moisture from the wet particles, are the bubble phase, its surrounding cloud and the dense annular solid phase.

The most widely used mass transfer model of Kunii and Levenspiel (1991) expresses the overall mass transfer in a bubbling bed in terms of the cloud-bubble interchange and dense-cloud interchange. The cloud-bubble interchange is assumed to arise from the contribution of circulating gas from the cloud phase and in and out of the bubble, usually referred to as throughflow, in addition to the diffusion from a thin cloud layer into the bubble. The dense-cloud interchange is assumed to arise only from diffusion between the dense phase and the cloud boundary. Kunii and Levenspiel (1991) also suggested additional mass transfer resulting from particles dispersed in the bubbles, however, recent advanced imaging technique, have shown bubble free particles in most cases as will be demonstrated later.

For particles of about 500 μm, some researchers assume that the transfer is of a purely diffusional nature, and thus neglect the contribution of bubble throughflow. However, Walker [1975] and Sit and Grace [1978] pointed out that, pure diffusional model may significantly underestimate, the overall mass transfer coefficient. Kunii and Levenspiel (1991) reported that the true overall mass transfer coefficient may fall closer to either of the acting mechanisms depending on the operating conditions (particle size, gas velocity, etc.). They suggested accounting for the first mechanism by summing the diffusional and throughflow, and adding those to the second mechanism in a similar fashion as for additive resistances.

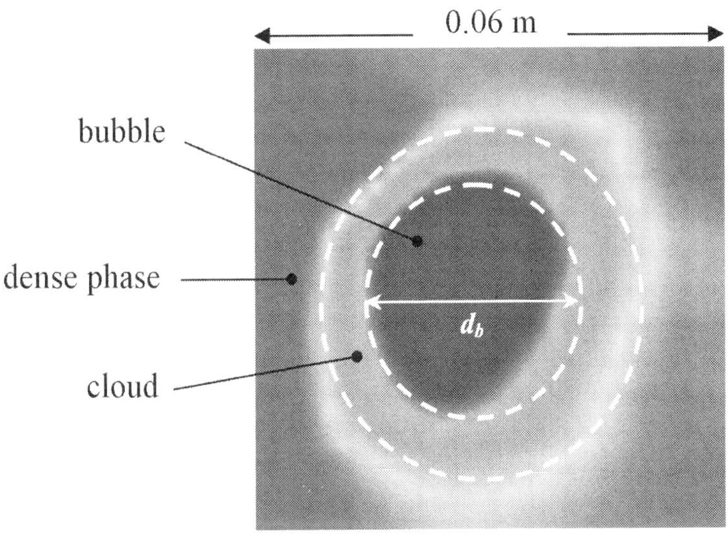

Fig. 2. Dense-cloud and cloud-bubble phases demonstrated in a typical ECT image of an isolated rising clouded bubble in a fluidized bed

4. Mass transfer coefficient from literature

Because of the growing interest on modelling as a tool for effective research and design, researchers on bubbling fluidized bed drying or mass transfer in general are nowadays seeking to validate or develop new mass transfer coefficient equations required for accurate prediction of the process kinetics. Recently, different mass transfer coefficients for drying in fluidized beds have been reported in the literature, most of them are based on the two phase model of fluidization.

Ciesielczyk and Iwanowski (2006) presented a semi-empirical fluidized bed drying model based on cloud-bubble interphase mass transfer coefficient. To predict the generalized drying curve for the solid particles, the interchange coefficient across the cloud-bubble boundary was given by:

$$K_{cb} = 4.5\left(\frac{u_{mf}}{d_b}\right) + 5.85\left(\frac{\mathfrak{D}^{0.5} g^{0.25}}{d_b^{1.25}}\right) \tag{1}$$

where the bubble diameter, d_b, is given by modified Mori and Wen (1975) model for the bubble diameter as follows:

$$\frac{d_{b,m}-d_b}{d_{b,m}-d_{b,0}}=\exp\left(-\frac{0.12H_{mf}}{D_c}\right) \tag{2}$$

where $d_{b,o}$ and $d_{b,m}$ are the initial bubble diameter at the distributor level and at its maximum size respectively, and given by:

$$d_{b,o}=0.376\left(U-U_{mf}\right)^2 \tag{3}$$

$$d_{b,m}=1.636\left[D_c\left(U-U_{mf}\right)\right]^{0.4} \tag{4}$$

According to Davidson and Harrison (1963), the first term in the right side of Eq. 1 is assumed to represent the convection contribution as a result of bubble throughflow. The second term arises from the diffusion across a limited thin layer where the mass transfer takes place. Using area based analysis, Murray (1965) suggested that the first term on the right side of Eq. 1 to be reduced by a factor of 3, which then gives:

$$K_{cb}=1.5\left(\frac{u_{mf}}{d_b}\right)+5.85\left(\frac{\mathfrak{D}^{0.5}g^{0.25}}{d_b^{1.25}}\right) \tag{5}$$

Ciesielczyk and Iwanowski (2006) have shown satisfactory agreement between the above outlined correlation and experimentally determined drying rate and mass transfer coefficient for group B particles of Geldart classification.

Kerkhof (2000) discussed some modeling aspect of batch fluidized bed drying during thermal degradation of life-science products. In this model, it is assumed that the contribution from particle raining or circulating in and out from a bubble is important, therefore the cloud-bubble interphase exchange, given by Eq. 1 above, was combined with the dense-cloud exchange in addition to contribution from the particle internal diffusion to give an overall bed mass transfer coefficients,

$$Sh_{bed}=Sh_{pb}+Sh_{db} \tag{6}$$

where the first sherwood number, Sh_{pb}, represents the mass transfer added from the particles dispersed in the bubble and expressed in terms of the mass transfer coefficient for a single particle, given by,

$$k_{pb}=\frac{\mathfrak{D}}{d_p}\left(2+0.664\,\mathrm{Re}_p^{0.5}\,Sc^{0.33}\right) \tag{7}$$

The second Sherwood number, Sh_{pb}, represents the combined cloud-bubble and dense-cloud exchanges and given in terms of a single mass transfer coefficient, k_{db}, as follows,

$$\frac{1}{k_{db}}=\frac{1}{k_{dc}}+\frac{1}{k_{cb}} \tag{8}$$

where the cloud-bubble mass transfer coefficient, k_{cb}, was given earlier in Eq. 1 in terms of interchange coefficient. Note that the interchange coefficient is expressed as a rate constant $(1/s)$, which can then be multiplied by the bubble volume per unit area to give the mass transfer coefficient in (m/s) as follows:

$$k = \frac{d_b}{6} K \tag{9}$$

The dense-cloud mass transfer coefficient, k_{dc}, which appear in Eq. 8 was adopted from Higbie penetration model, which is expressed in terms of the bubble-cloud exposure time and the effective diffusivity as follows:

$$k_{dc} = 2\left(\frac{\mathfrak{D}_e \varepsilon_{mf}}{\pi t}\right)^{0.5} \tag{10}$$

where $t = \frac{d_c}{u_b}$ is the exposure time between the bubble and the cloud. Kerkhof (2000) made two simplifications to Eq. 10; first, it is assumed that the cloud thickness is negligible, therefore the bubble diameter can be replacement for the cloud diameter (i.e. $d_c \approx d_b$), second, it is assumed that the effective diffusivity is better approximated by the gas molecular diffusivity (i.e. $\mathfrak{D}_e \approx \mathfrak{D}$). Accordingly, Eq. 10 reduces to

$$k_{dc} = 6.77\left(\frac{\mathfrak{D} \varepsilon_{mf} u_b}{d_b^3}\right)^{0.5} \tag{11}$$

Recently, Scala (2007) experimentally studied the mass transfer around a freely active particle in a dense fluidized bed of inert particles. The results suggested that the mass transfer coefficient for a single particle is best correlated by a modified Foressling (1938) equation for Sherwood number,

$$Sh = 2\varepsilon_{mf} + 0.7\,\mathrm{Re}_{p,mf}^{0.5}\, Sc^{0.33} \tag{12}$$

where $\mathrm{Re}_{p,mf}$ is the Reynolds number expressed in terms of the voidage ε_{mf} (i.e. $= \rho u_{mf} d_p / \mu \varepsilon_{mf}$). The above correlation was found to be independent of the fluidization velocity or regime change from bubbling to slugging. Accordingly, Scala (2007) concluded that in a dense bubbling bed the active particle only reside in the dense phase and never enters the bubble phase, hence it has no direct contribution to the bubble-dense phase interchanges. This contradicts the observation noted by Kunii and Levenspiel (1991) and others (e.g. Kerkhof, 2000; Agarwal, 1978), were it is assumed that the contribution of particles dispersed in the bubble should not be neglected. Agarwal (1978) claimed that the particles do circulate in and out of the bubble with 20% of the time residing within the bubble phase.

Clearly, despite of the considerable effort on developing fluidized bed mass transfer coefficients, there still remain uncertainties with respect to the assumptions used in developing these coefficients.

5. Characteristic drying rate profiles

Early experimental observations on fluidized beds suggest that the mass transfer at the single particle level generally occurs at two different drying regimes; one at which the free moisture, either at the particle surface or within large pours, is rapidly withdrawn at a constant rate, followed by a slower rate regime at which the process is controlled by slow diffusion from the fine pores to the particle surface. These are usually referred to as "constant rate" and "falling rate" respectively. The moisture content at the transition between these two regimes is called the critical moisture content. Fig. 3 illustrates the characteristic drying rate curve as function of time and moisture content. There is an argument that these drying curves are in fact oversimplification of the process, and such profiles may change considerably with respect to particle size and material type. Keey (1978) pointed that the drying rate at the beginning of the process may not be constant at all, or at least changes to a small degree, therefore, he recommended calling this as "initial drying period" instead of the commonly used term "constant drying". The same applies to the "falling rate" regime, where it is preferred to call it "second drying period". The existence of the critical moisture content point, on the other hand, is true in most cases.

For non-pours particles, regardless of the material type, the drying process occurs at a single regime, where the moisture residing at the particle surface is rapidly withdrawn, driven by difference in moisture concentration. To a great extent, this resembles free water diffusion into a moving air stream. Fig. 4 shows an example of this behaviour during drying of wet glass beads in a bubbling fluidized bed using air at ambient conditions.

Here it is clear that the drying rate falls exponentially within the first 15 minutes, after which the drying process ends. This confirms a single drying regime rapidly driven by the difference in moisture content between the fluidizing air and particle. Using the data in Fig. 4, one can obtain the water concentration in the fluidized bed as a function of the drying time by the integration of the drying curve function, $F(t)$, such that

$$w_t = w_o - \int_0^t F(t)dt \tag{13}$$

where w_o is the initial water content.

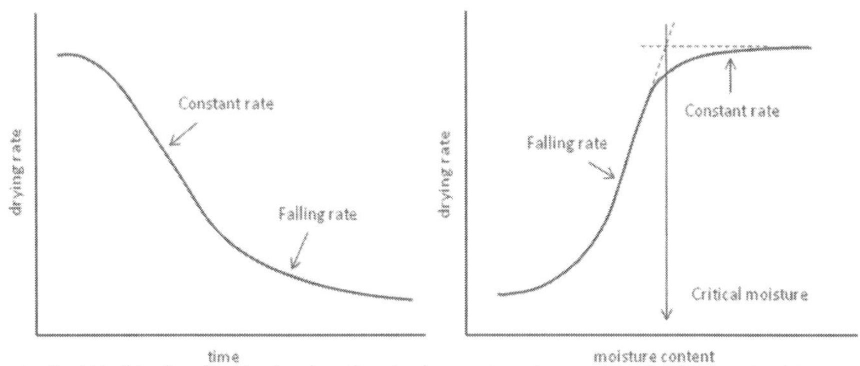

Fig. 3. Characteristic drying curves for moist particles

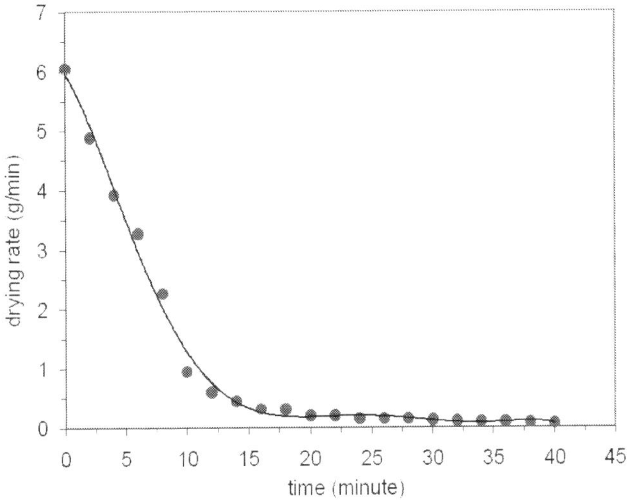

Fig. 4. Drying rate profile for moist glass beads in a bubbling fluidized bed using ambient air (Makkawi and Ocone, 2009).

6. Case study

Experiments have been carried out with the primary objective to measure the mass transfer coefficient for a drying process in a conventional bubbling fluidized bed. This required detailed knowledge of the fluidized bed hydrodynamics and drying rate. For this purpose, non-porous wet solid particles of glass beads were contained in a vertical column and fluidized using air at ambient temperature. The fluidising air was virtually dry and obtained from a high-pressure compressor. An advanced imaging ECT sensor was used to provide dynamic information on the fluidized bed material distribution. The sensor was connected to a data acquisition unit and a computer. The air outlet temperature and its relative humidity were recorded using a temperature/humidity probe. Since the air condition at the inlet of the fluidization column was constant and completely independent of the bubbling bed operating conditions, only one probe was installed at the freeboard (air exit). The detailed experimental set-up is shown in Fig. 5.

6.1 Experimental procedure and materials

The fluidization was carried out in a cast acrylic column, 13.8 cm diameter and 150 cm high. The column was transparent, thus allowing for direct visual observation. A PVC perforated gas distributor with a total of 150 holes (~1.8% free area), was placed 24 cm above the column base. The upstream piping was fitted with pressure regulator, moisture trap, valve and three parallel rotamaters. A one-step valve was connected before the moisture trap and was used as the upstream main flow controller. The particles used were ballotini (non-porous glass beads) with a mean diameter of 125 μm and a density of 2500 kg/m^3 (Geldart A/B mixture). The detailed physical properties of the particles are given in Table 1. Distilled water at ambient condition was used to wet the particles. A variable speed granule shaker was utilised to produce the final wetted mixture.

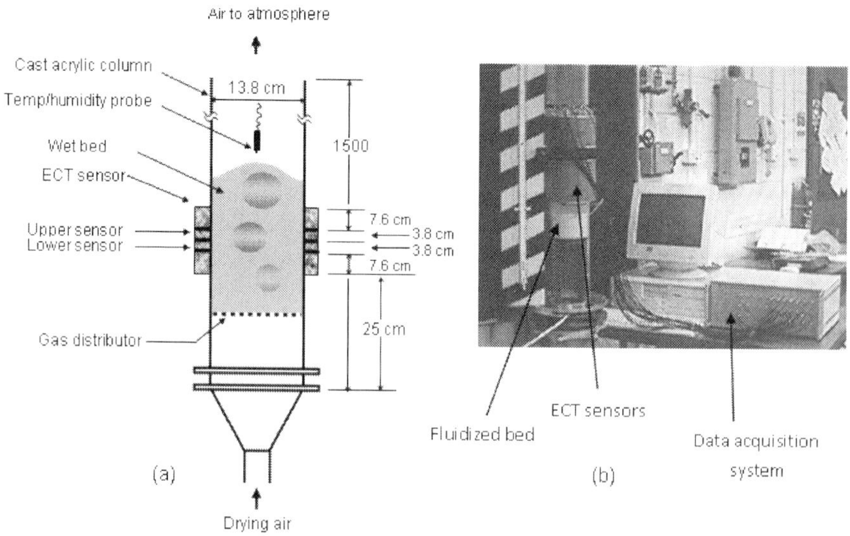

Fig. 5. Experimental set up (a) Schematic of the fluidized bed (b) A photograph of the installation

The Electrical Capacitance Tomography imaging system used (ECT from Process Tomography Limited, Manchester, UK), consisted of two adjacent sensor rings each containing 8 electrode of 3.8 cm length. All electrodes were connected to the computer through a data acquisition system. The PC was equipped with custom communication hardware and software that allow for online and off-line dynamic image display. The system is capable of taking cross-sectional images of the bed at two adjacent levels simultaneously at 100 frames per second. Further details about the ECT system used in this study and its application to fluidization analysis can be found in Makkawi et al. (2006) and Makkawi and Ocone (2007).

Geldart Group	A/B
Particle size range (μm)	50 - 180
Mean particle diameter (μm)	125
Particle density (kg/m^3)	2500
Sphericity	≥ 80%
Pores	< 0.02 nm
Material	Pure soda lime glass ballotini.
Chemical composition	SiO_2=72%, Na_2O=13%, CaO=9%, MgO=4%, Al_2O_3=1%, K_2O & Fe_2O_3=1%
Commercial name	Glass beads - type S, Art. 4500
Electric permittivity	~3.1

Table 1. Physical and chemical properties of the dry particles

The exit air quality was measured using a temperature and humidity probe (Type: Vaisala HMI 31, Vantaa, Finland, measuring range: 0-100% RH, -40-115 C°). The probe was hung by a connecting wire inside the fluidized bed freeboard approximately 10 cm above the maximum expanded bed height.

The experimental procedure employed was completely non-intrusive. The experiment commenced by weighting a total of 4.5 kg of a dry ballotini mixture and placing it in a granule shaker after being wetted by distilled water. The shaker was firmly clamped and operated continuously for at least 25 minutes to ensure an even distribution of water content. Distilled water was used to eliminate any possible interference with the ECT signal (ECT works for non-conducting materials only). The wetted particles were then loaded into the fluidization column. Prior to commencement of drying, the ECT sensor was calibrated for two extreme cases. This was carried out by sliding the ECT sensor up to the freeboard to calibrate for the empty bed case, and down to the static bed area to calibrate for the packed bed case. It should be mentioned that, because the water content was limited to a maximum of 45 ml (1% moisture on dry solid weight basis), the possible changes in the particle/air permittivity during the drying process would be negligible. Further details on the sensitivity of the ECT system to moisture content can be found in Chaplin and Pugsley (2005) and Chaplin et al., (2006). The wet bed material was fluidized at the required air flow rate. This was carefully adjusted to ensure the bed operation at the single bubble regime. The temperature and relative humidity were recorded at 2 minutes intervals. Simultaneously, and at the 5 minutes intervals, a segment of 60 seconds ECT data were recorded. At the same time, the expanded bed height during fluidization was obtained from visual observations. Finally, the drying rate was obtained from the measured air flow rate and temperature/humidity data at inlet and outlet using psychometric charts and mass balance calculations. The recorded ECT data were further processed off-line and loaded into in-house developed MATLAB algorithm to estimate the bubble characteristics.

The above described procedure was repeated for the three different operating conditions summarized in Table 2. To ensure data reproducibility, each operating condition was repeated three times, making a total of nine experiment tests.

Experimental unit	Operating conditions		
Fluidization column	Diameter = 13.8 cm, height = 150 cm, material: cast acrylic, equipped with a Perforated PVC plate of 150 holes, each of 2 mm dia.		
Dry particles	d_p = 125 μm, ρ_p = 2500 kg/m^3, Material: glass		
Fluidization fluid	Air at ambient condition (~20° C)		
Static bed height	20 cm		
	Exp. 1	Exp. 2	Exp. 3
Fluidization velocity (m/s)	0.35	0.47	0.47
Initial water content (wt%)	1.0	1.0	0.5

Table 2. Summary of experimental operating conditions

6.2 Measurement of mass transfer coefficient

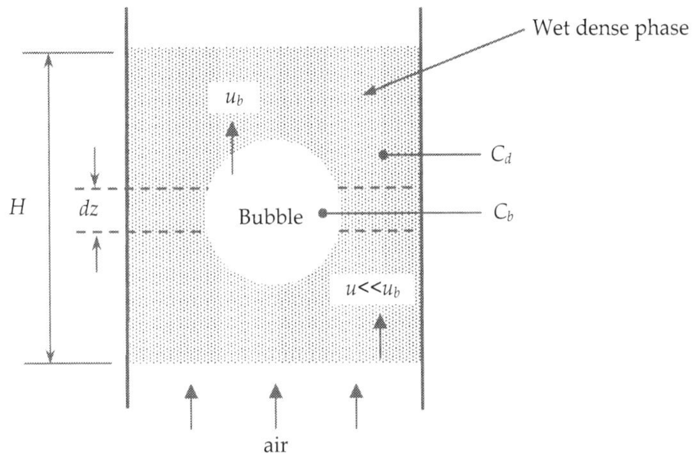

Fig. 6. Schematic representation of the method used in experimental calculation of the overall mass transfer coefficient

Considering a section of the bed as shown in Fig. 6, the overall mass transfer coefficient between the bubble phase and the surrounding dense phase, k_{db}, can be defined by the following rate equation:

$$-u_b \frac{dC_b}{dz} = \left(\frac{S_b}{V_b}\right) k_{db}(C_d - C_b) \tag{14}$$

where C_b is the water concentration in the bubble phase, C_d is the concentration in the surrounding dense phase, u_b, S_b and V_b are characteristic features of the bubble representing the rising velocity, the interphase area and the volume, respectively. For moisture-free inlet air, Eq. 14 is subject to the following boundary conditions:

$$C_b = (C_{in})_{air} = 0 \text{ at } z = 0 \text{ and } C_b = (C_{out})_b \text{ at } z = H \tag{15}$$

where H is the expanded bed height. The bubble moisture content at the outlet $(C_{out})_b$ can be given by:

$$(C_{out})_b = \frac{\text{(drying rate)}}{\text{(bubble mass flow rate)}} = \frac{m_{air}(C_{out} - C_{in})_{air}}{m_b} \tag{16}$$

where m_{air} and m_b are the mass flow rate of the fluidising air and bubbles respectively. Because of the assumption that the bubbles rise much faster than the gas through the dense phase and the inlet air was virtually dry, Eq. 16 reduces to:

$$(C_{out})_b = (C_{out})_{air} \tag{17}$$

where $(C_{out})_{air}$ is obtained from the measured temperature and humidity at the bubbling bed surface.

For a spherical bubble, S_b/V_b ratio appearing in Eq. 14 reduces to $6/d_b$, where d_b is the bubble diameter. It should be mentioned that for a perforated distributor (such as the one used in this experiment), coalescence of bubbles mainly takes place at a few centimetres above the distributor, therefore, the entrance effects are neglected and the bubble characteristics are assumed independent of height (this was confirmed from the ECT images).

Finally, assuming that the water concentration in the dense phase is uniform and remains unchanged during the bubble rise ($C_d = w_{water}/w_{bed}$) and integrating Eq. 14 from $z=0$ to $z=H$, the mass transfer coefficient is obtained as follows:

$$k_{db} = -\left(\frac{d_b u_b}{6H}\right)\ln\left[\frac{C_d-(C_{out})_b}{C_d}\right] \tag{18}$$

where d_b and u_b and are the bubble diameter and velocity respectively.

The experimental measurement of the overall mass transfer coefficient, k_{db}, as a function of the water concentration in the bed is shown in Fig. 7. The values of k_{db} are found to fall within the range of 0.0145-0.021 m/s. It is interesting to note that this range is close to the value one can obtain from the literature for the mass transfer coefficient from a free water surface to an adjacent slow moving ambient air stream (~0.015 m/s) (Saravacos and Z. Maroulis, 2001).

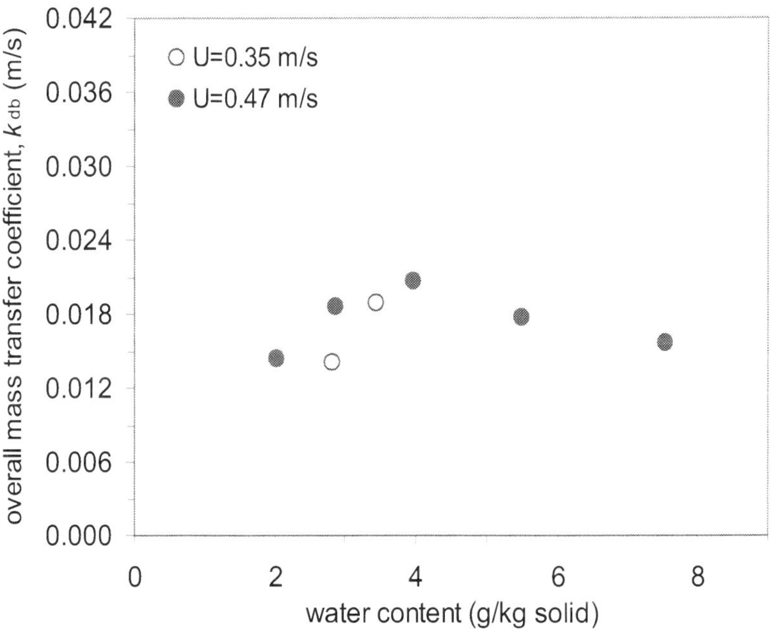

Fig. 7. Experimentally measured overall mass transfer coefficient

6.3 Measurement of bubble characteristics

Experimental determination of the overall mass transfer requires knowledge of the bubble diameter and velocity (see Eq. 18). Using the ECT, the diameter and velocity of the bubbles in a gas-solid fluidized bed can be obtained. The distinct lowering of the solid fraction when the bubble passes across the sensor area, as shown in Figs 11 and 12, allows for identification of the bubble events in a given time and space. The bubble velocity was then calculated from the delay time determined from a detailed analysis of the signal produced by the two adjacent sensors, such that:

$$u_b = \frac{\delta}{\Delta t_b} \tag{19}$$

where $\Delta t_b = t_{b2} - t_{b1}$, t_{b1} and t_{b2} represent the time when the bubble peak passes through the lower and upper level sensors respectively, and δ represents the distance between the centre of the two sensors, which is 3.8 cm. The method is demonstrated for a typical ECT data in Fig. 8.

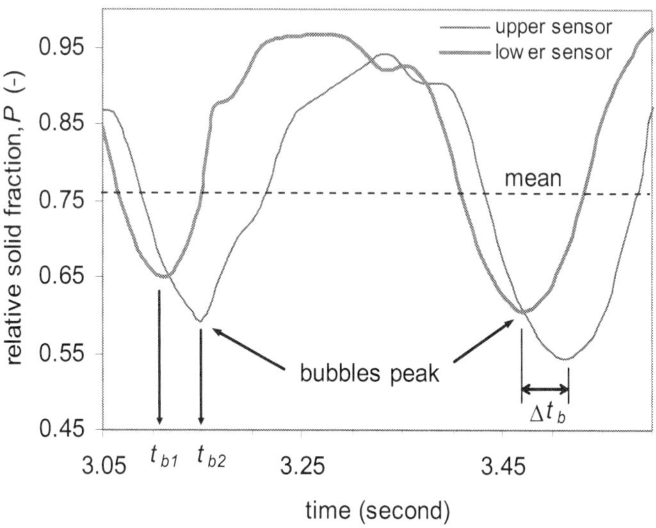

Fig. 8. Estimation of bubble velocity from ECT data

The bubble diameter was obtained from the ECT data of relative solid fraction at the moment of bubble peak across the sensor cross-section. From this, the bed voidage fraction (the fraction occupied by bubbles) was calculated as follows:

$$d_b = D\gamma \tag{20}$$

where γ=(1 - P) is the bubble fraction, P is the relative solid fraction (i.e. packed bed: P = 1; empty bed:: P = 1; empty bed: P = 0) and D is the bed/column diameter. This procedure is demonstrated for a typical ECT data in Fig.9. Further details on the application of twin-plane ECT for the measurements of bubble characteristics in a fluidized bed can be found in Makkawi and Wright (2004).

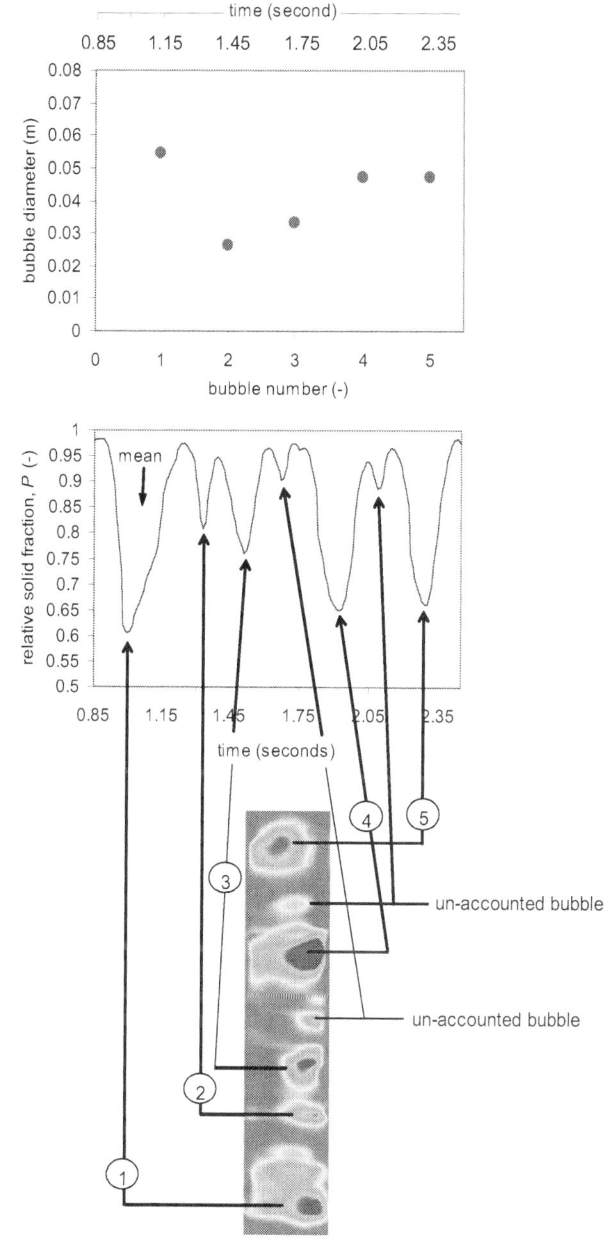

Fig. 9. Estimation of bubble diameter from ECT measurement (a) bubble diameter (b) ECT solid fraction (b) ECT slice images

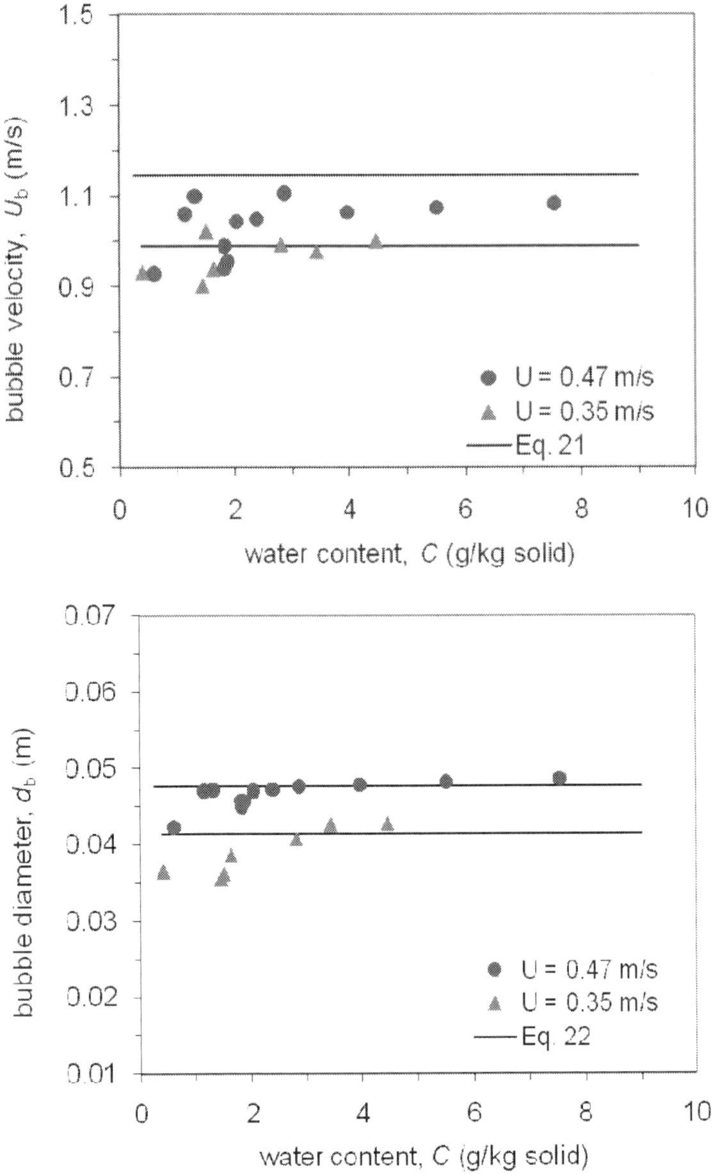

Fig. 10. Variation of the bubble velocity and bubble diameter during the drying process

Fig. 10 shows the measured bubble velocity and bubble diameter as a function of the water content in the bed. These measurements were taken at different time intervals during the drying process. Each data point represents the average over 60 seconds. Both parameters vary slightly within a limited range. These hydrodynamic observations suggest that the bubble characteristics almost remain independent of the water content, at least within the range of operating conditions considered here. This is due to the fact that the initial water content in the bed was not significant enough to cause considerable hydrodynamic changes. Among the available correlations from the literature, the following equations have been found to provide good matches with the experimental measurements:

Bubble velocity:

$$u_b = \psi\left(U - U_{mf}\right) + \alpha\left[0.711\left(gd_b\right)^{0.5}\right] \tag{21}$$

This a modified form of Davidson and Harrison [13] equation, where $\psi = 0.75$ and $\alpha = 3.2D_c^{1/3}$ are correction factors suggested by Werther(1991) [17] and Hilligardt and Werther(1986).

Bubble diameter:

$$d_b = 0.652\left[D_o - \exp\left(-0.3z/D_c\right)\right] + \frac{0.347D_o}{n_o^{0.4}}\exp\left(-0.3z/D_c\right) \tag{22}$$

where

$$D_o = \left[\frac{\pi}{4D_c^2}\left(U - U_{mf}\right)\right]^{0.4} \tag{23}$$

U_{mf} used in Eqs 21 and 23 was given by:

$$U_{mf} = \frac{d_p^2\left(\rho_p - \rho_g\right)g}{150\mu}\frac{\varepsilon_{mf}^3\varphi_p}{\left(1 - \varepsilon_{mf}\right)} \tag{24}$$

For the operating condition given in Table 2, Eq. 24 gives U_{mf} =0.065 m/s, which closely matches the measured value of 0.062 m/s. Despite the fact that Eqs. 21-24 were all originally developed for dry bed operations; they seem to provide a reasonable match with the experimental measurements made here under wet bed condition. This is not surprising, since the water content in the bed was relatively low as discussed above. The expanded bed height, used in the experimental estimation of the overall mass transfer coefficient (Eq. 18), is shown in Fig. 11. Limited increase in the bed expansion as the water is removed from the bed can be noticed

6.4 Combined hydrodynamics and mass transfer coefficient model

In this analysis we assume that mass transfer occurs in two distinct regions: at the dense-cloud interface and at the cloud-bubble interface. The overall mass transfer may be dense-cloud controlled; cloud-bubble controlled or equally controlled by the two mechanisms depending on the operating conditions. The following theoretical formulations of these acting mechanisms are mainly based on the following assumptions:

i. The fluidized bed operates at a single bubble regime.
ii. The bubbles are spherically shaped.
iii. The bubble rise velocity is fast ($u_b > 5U_{mf}/\varepsilon_{mf}$).

iv. The bubbles size and velocity are independent of height above the distributor
v. The contribution of particles presence within the bubble is negligible.
vi. The contribution of the gas flow through the dense phase (emulsion gas) is assumed to be negligible.

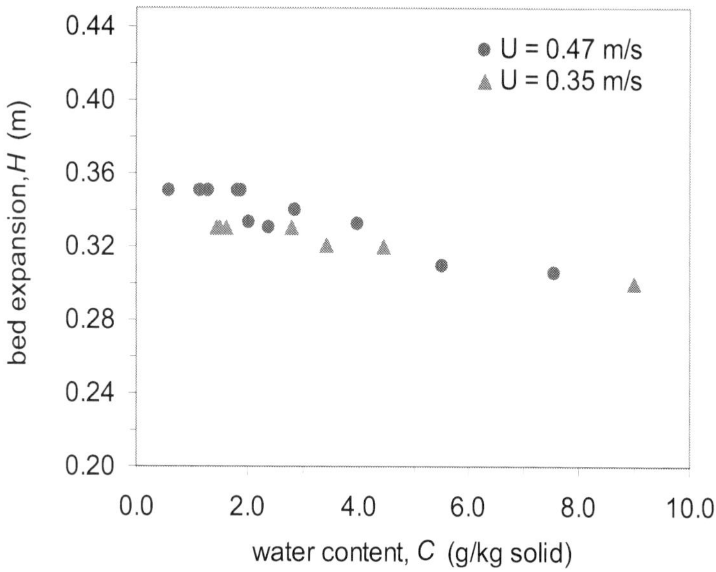

Fig. 11. Variation of expanded fluidized bed height during the drying process

As evident from the tomographic analysis of the bubble characteristics shown in Figs 2, 9 and 12, assumptions (i)-(v) are to a great extent a good representation of the actual bubbling behaviour considered here.

In this study, we assume the mass transfer at the bubble-cloud interface arises from two different contributions:

1. Convection contribution as a result of bubble throughflow, which consists of circulating gas between the bubble and the cloud, given by

$$k_q = 0.25U_{mf} \tag{25}$$

2. Diffusion across a thin solid layer (cloud), given by

$$k'_{cb} = 0.975\mathfrak{D}^{0.5}\left(\frac{g}{d_b}\right)^{0.25} \tag{26}$$

The addition of both acting mechanisms gives the total cloud-bubble mass transfer coefficient,

$$k_{cb} = 0.25U_{mf} + 0.975\mathfrak{D}^{0.5}\left(\frac{g}{d_b}\right)^{0.25} \tag{27}$$

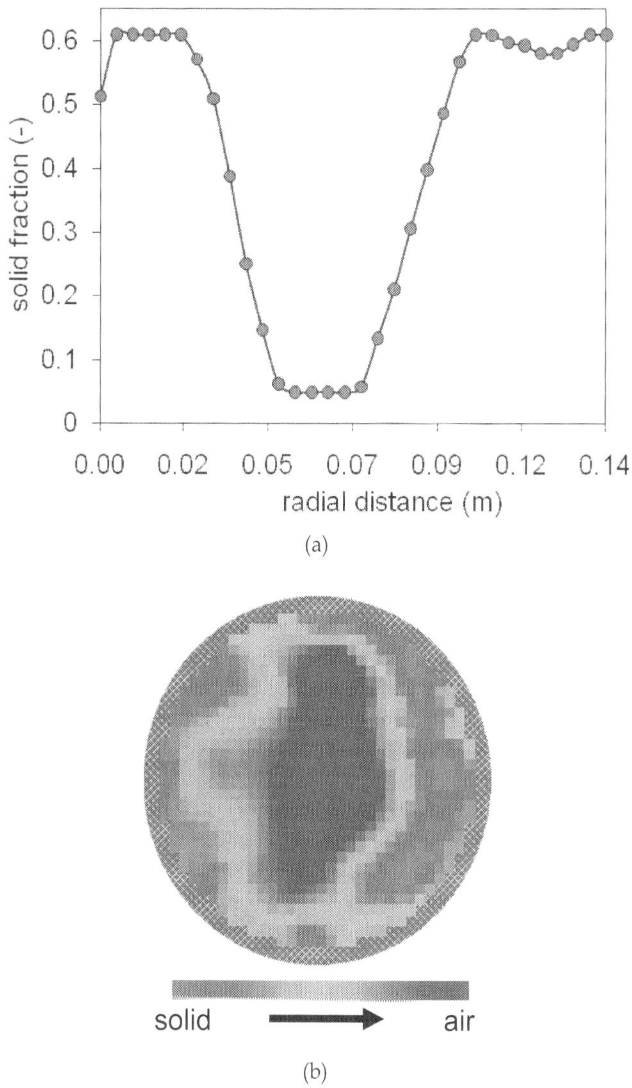

Fig. 12. Cross-sectional tomographic imaging during bubble passage across the sensor demonstrating negligible solid within the bubble core (a) radial solid concentration (b) contour of solid distribution

This suggest that the cloud-bubble interchange is indirectly proportional to the particle size (U_{mf} increases with increasing d_p) and inversely proportional to d_b. Note that the above equation reduces to the same formulation given earlier for the exchange coefficient (Eq. 5) after dividing by the bubble volume per unit area.

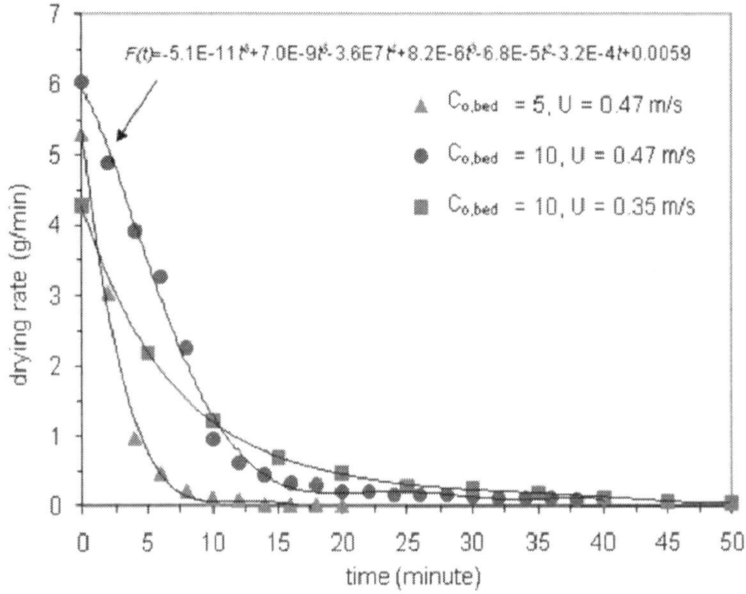

Fig. 13. Drying rate curves for the three conducted experiments

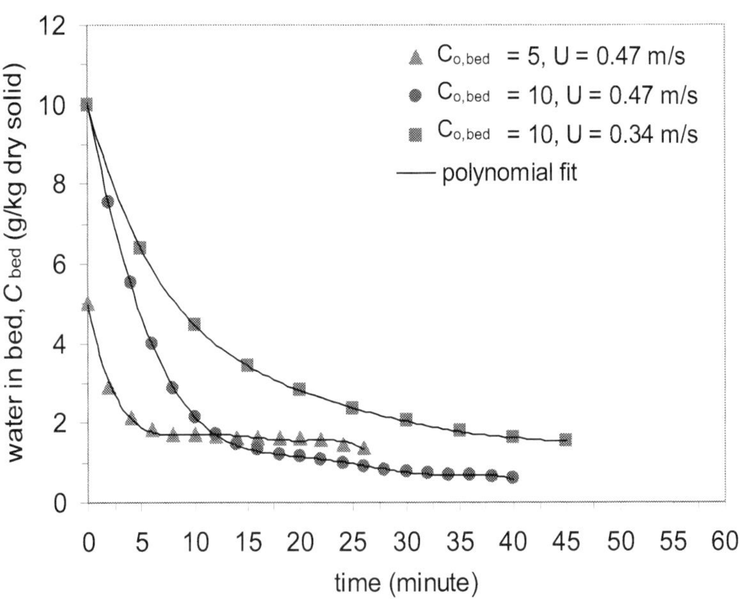

Fig. 14. Variation of water content during drying

The mass interchange coefficient across the dense-cloud boundary can be given by Higbie penetration mode (Eq. 10). For the special case discussed here, the tomographic images of the bubbles and its boundaries suggest that the cloud diameter (the outer ring in Fig. 2) is always within the range of 1.2-1.8 bubble diameters. Therefore, assuming that $d_c \sim 1.5 d_b$, and after multiplying by bubble volume per unit area, the dense-cloud mass transfer coefficient can be given by:

$$k_{dc} = 0.92\left(\frac{\mathfrak{D}\varepsilon_{mf}u_b}{d_b}\right)^{0.5} \tag{28}$$

For an equally significant contribution from cloud-bubble and dense-cloud interchanges, Kunii and Levenspiel (1991) suggested adding both contributions in analogy to parallel resistances, such that the overall bed mass transfer coefficient (k_{db}) is given by:

$$\frac{1}{k_{db}} = \frac{1}{k_{dc}} + \frac{1}{k_{cb}} \tag{29}$$

Substituting Eq. 25 and 26 into Eq. 27 yields the overall mass transfer coefficient as follows:

$$k_{db} = \frac{A_1 B_1}{d_b^{0.5}\left(A_1 + B_1\right)} \tag{30}$$

where

$$A_1 = 0.975\mathfrak{D}^{0.5}\left(d_b g\right)^{0.25} + 0.25U_{mf}d_b^{0.5} \tag{31}$$

$$B_1 = 0.92\left(\mathfrak{D}\varepsilon_{mf}u_b\right)^{0.5} \tag{32}$$

In this model, the bubble diameter and velocity are obtained from the correlations given in Eqs 21 and 22 respectively, and U_{mf} is given by Eq. 24.

6.5 Drying rate

The drying rate curves for the three experiments conducted are shown in Figure 13. The curve fitting is used to obtain the water content in the bed at various times. From this figure, it may be concluded that the drying time is directly proportional to the initial water content, and inversely proportional to the drying air flow rate. For instance, at an air velocity of 0.47 m/s, this time was reduced by half when reducing the initial water content from $C_{o,bed}$ =10% to $C_{o,bed}$ =5%, while at the initial water content of $C_{o,bed}$ =10, this time was ~35% longer when reducing the air velocity from 0.47 m/s to 0.33 m/s. The water concentration in the bed as a function of the drying time is shown in Fig. 14. This was obtained from the integration of the drying curve function as given earlier in Eq. 13.

6.6 Comparison with literature data

Walker 1975) Sit and Grace (1978) measured the mass transfer coefficient in a two-dimensional fluidized bed. The technique employed involves the injection of ozone ozone-rich bubble into an air-solid fluidized bed. Patel et al. (2003) reported numerical prediction of mass transfer coefficient in a single bubbling fluidized bed using a two fluid model based on kinetic theory of granular flow. Comparison between the above mentioned literature and the experimental data obtained in this study is shown in Fig. 15. Taking into consideration the differences in the experimental set-up and operating conditions, the agreement with our measurement appears satisfactory for the particle size considered in this study.

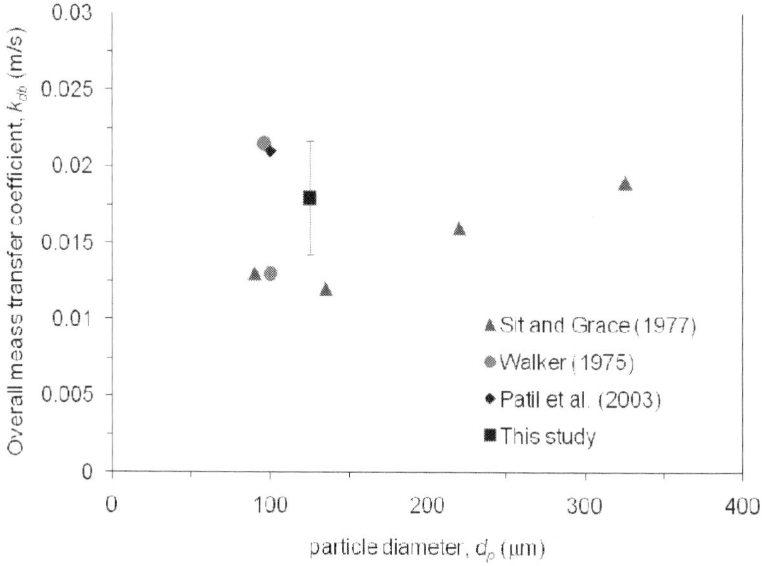

Fig. 15. Experimental overall mass transfer coefficient in comparison with other previously reported results

6.7 Comparison of experimental and theoretical prediction

Fig. 16 compares the measured mass transfer coefficient with the theoretical predictions obtained from the formulations given in section 6.4. The boundaries for the overall mass transfer coefficient are given by: (i) a model accounting for cloud-bubble and dense-bubble diffusion contribution as well as the bubble through flow convective contribution, giving the lower limit (Eq. 30) and (ii) a model accounting for the cloud-bubble contribution, giving the upper limit (Eq. 27). The results also suggest that, within the operating conditions considered here, the drying may well be represented by a purely diffusional model, controlled by either the resistance residing at the dens-cloud interface, or the cloud-bubble interface.

Finally, Table 3 shows the numerical values of the various mass transfer contributions obtained from Eqs 25-32. It is shown that the estimated diffusional resistances, as well as the contribution from the bubble throughflow, are all of the same order of magnitude. Previously, Geldart (1968) argued that the bubble throughflow is not important for small particles and may be neglected. According to our analysis, this may well be the case here. However, generalization of this conclusion should be treated with caution especially when dealing with larger particles.

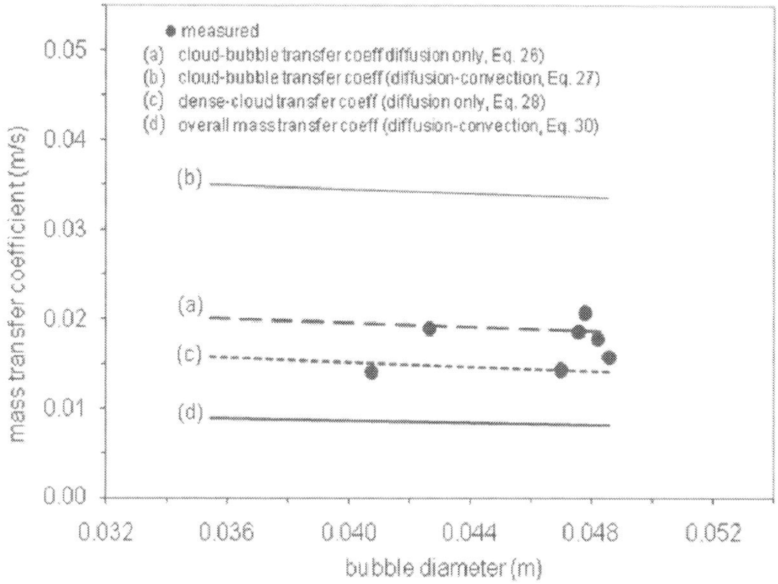

Fig. 16. Comparison between experimental measurement and various theoretical models for mass transfer coefficients

Experimental				Theoretical		
Gas velocity	bubble characteristics		Overall mass transfer coeff.	Dense-cloud interchange (diffusion only)	Cloud-bubble interchange (diffusion only)	Bubble throughflow
U (m/s)	d_b (m)	u_b (m/s)	k_{db} (m/s)	k_{dc} (m/s)- Eq. 28	k_{cb} (m/s)- Eq. 27	k_q (m/s)- Eq. 25
0.35	0.04	0.99	0.0145	0.0178	0.0194	0.015

Table 3. The measured overall mass transfer coefficient for one selected operating condition in comparison to the theoretical predictions of various contributions.

6.8 Conclusion

Mass transfer coefficient in a bubbling fluidized bed dryer has been experimentally determined. This work is the first to utilise an ECT system for this purpose. The ECT allowed for quantification of the bubble diameter and velocity, as well as providing new insight into the bubble-cloud-dense boundaries.

The measured overall mass transfer coefficient was found to be in the range of 0.045-0.021 m^2/s. A simple hydrodynamic and mass transfer model, based on the available correlations

was used to predict the mass transfer coefficient in a bubbling fluidized bed. Despite the complexity of the process, and the number of assumption employed in this analysis, the model based on pure diffusional mass transfer seems to provide satisfactory agreement with the experimental measurements.
This work set the scene for future experimental investigations to obtain a generalised correlation for the mass transfer coefficient in fluidized bed dryer, particularly that utilizes the ECT or other similar imaging techniques. Such a correlation is of vital importance for improved fluidized bed dryer design and operation in its widest application. A comprehensive experimental program, covering a wider range of operating conditions (particle size, gas velocity, water content, porous/non-porous particles) is recommended.

7. Nomenclature

A	[m^2]	column/bed cross-sectional area
A_1, B_1	[m$^{1.5}$s^{-1}]	parameters defined in Eqs. 15, 16 respectively
C_d, C_b	[-]	water concentration in the dense and bubble phases respectively, kg/kg
d, D	[m]	diameter
$\mathfrak{D}, \mathfrak{D}_e$	[m^{2}s^{-1}]	molecular and effective diffusivity respectively
g	[ms^{-2}]	gravity acceleration constant
H	[m]	expanded bed height
k_{db}	[ms^{-1}]	overall mass transfer coefficient (between dense and bubble phases)
k_{cb}	[ms^{-1}]	mass transfer coefficient between cloud and bubble phases
k_{dc}	[ms^{-1}]	mass transfer coefficient between dense and cloud phases
m	[kgs^{-1}]	mass flow rate
P	[-]	relative solid fraction
P	[-]	relative solid fraction
U	[ms^{-1}]	superficial gas velocity
u	[ms^{-1}]	velocity
V	[(m^3)	volume
w	[g]	bed water content
z	[m]	axial coordinate

Greek symbols

ε	[-]	bed voidage
γ	[-]	bubble fraction
ρ	[kg.m^{-3}]	density
δ	[m]	distance between the centre of the two ECT sensors
φ_p	[-]	particle sphericity

Subscripts

b	bubble
c	cloud
d	dense
mf	minimum fluidization
p	particle

8. References

Agarwal, P.K. (1987). The residence phase of active particles in fluidized beds of smaller inert particles. Chemical Engineering Science 42, 2481–2483

Chaplin G. and Pugsley, T. (2005). Application of electrical capacitance tomography to the fluidized bed drying of pharmaceutical granule. *Chemical Engineering Science 60,* 7022.

Chaplin, G., pugsley, T., van der, L., Kantzas, A., Winter's, C. (2005). The dynamic calibration of an electrical capacitance tomography sensor applied to the fluidised bed drying of pharmaceutical granule, *Measurement Science and Technology 16,* 281.

Chiba, T. and Kobayashi, H. (1970). *Chemical Engineering Science 25,*1375.

Ciesielczyk, W. and Iwanowski, J. (2006). Analysis of Fluidized Bed Drying Kinetics on the Basis of Interphase Mass Transfer Coefficient, Drying *Technology 24,* 1153-1157.

Davidson, J. and Harrison, D. (1963). *Fluidized Particles, Cambridge University Press,* Cambridge, England.

Frossling, N. (1938). The evaporation of falling drops. Gerlands Beitrage zur Geophysik 52, 170–216 (in German).

Geldart, D. (1986). *Gas Fluidization Technology,* John Wiley and Sons, Chichester, UK.

Higbie, R., in *Fluidization Engineering (1991),* Second edition (Eds: D. Kunii and O. Levenspiel), Butterworth-Heinemann, Boston, USA.

Hilligardt, K. and Werther, J. (1986). Local bubble gas-holdup and expansion of gas/solid fluidized beds. *Ger. Chem. Eng.*, 9, 215-221. *German Chemical Engineering* 9, 215.

Keey, R. B., Introduction to industrial drying operations (1978), Pergamon Press Ltd., Headington Hall, Oxford, England.

Kerkhof, P. J. A. M. (2000). Some modeling aspects of (batch) fluid-bed drying of life-science products. *Chemical Engineering and Processing, 39,* 69.

Kunii, D. and Levenspiel, O. (1991). *Fluidization Engineering.* Second edition, Butterworth-Heinemann, Boston.

Makkawi, Y. and Ocone, R. (2007). Integration of ECT measurements with hydrodynamic modelling of conventional gas-solid bubbling bed. *Chemical Engineering Science.* 4304-4315.

Makkawi, Y., Wright, P. C., Ocone, R. (2006). The effect of friction and inter-particle cohesive forces on the hydrodynamics of gas-solid flow: a comparative analysis of theoretical predictions and experiments. *Powder Technology* 163. 69-79.

Makkawi, Y. and Wright, P. C. (2004). Electrical Capacitance Tomography for conventional fluidized bed measurements- remarks on the measuring technique. *Powder Technology 148.* 142-157.

Mori, S. and Wen, C. Y. (1975). Estimation of bubble diameter in gaseous fluidized beds. AIChE Journal, 21: 109–115.

Murray, J. D. (1965). On the mathematics of fluidization Part 2. Steady motion of fully developed bubbles. Journal of Fluid Mechanics, 22, 57-80.

Patil, D. J., van Sint Annaland, M., Kuipers, J. A. M. (2003). Gas Dispersion and Bubble-to-Emulsion Phase Mass Exchange in a Gas-Solid Bubbling Fluidized Bed: A Computational and Experimental Study, International Journal of Chemical Engineering, 1-20.

Saravacos G. and Maroulis, Z. *(2001). Transport properties of foods,* CRC Press.

Scala, F. (2007). Mass transfer around freely moving active particles in the dense phase of a gas fluidized bed of inert particles, Chemical Engineering Science, 62(16), 4159-4176 *Chemical Engineering Science 62*, 4159.

Sit, S. P. and Grace, J. R. (1978). Interphase mass transfer in aggregative fluidized beds. *Chemical Engineering Science* , *33*, 1115.

Toomey, R.D. and Johnstone, H.F., Gaseous Fluidization of Solid Particles. Chem. Eng Prog. 48, 220-226, 1952

Walker, B. V. (1975). *Transaction Institute of Chemical Engineers*, *53*, 225.

Werther, J., in *Fluidization Engineering (1991)*, Second edition (Eds: D. Kunii and O. Levenspiel), Butterworth-Heinemann, Boston, USA

24

Simulation Studies on the Coupling Process of Heat/Mass Transfer in a Metal Hydride Reactor

Fusheng Yang and Zaoxiao Zhang
Xi'an Jiaotong University
P.R.China

1. Introduction

Many substances can react with hydrogen under certain conditions, and the products are generally called hydrides. The binary hydrides can be classified into ionic hydrides, covalent hydrides and stable complex hydrides (Berube et al., 2007). The metal hydrides (MH), which feature metallic bonding between hydrogen and host material, belong to the second category and are investigated here. Since the successful development of $LaNi_5$ and TiFe as hydrogen storage materials in 1970s, studies on the metal hydrides have attracted many attentions.

According to (Sandrock & Bowman, 2003), two properties among all have been found crucial in MH applications:

1. The easily reversed gas-solid chemical reaction, which is expressed as follows,

$$M + \frac{x}{2} H_2 \leftrightarrow MH_x + \Delta H \tag{1}$$

Here M denotes a certain kind of metal or alloy, while MH_x is the metal hydride as product. ΔH, the enthalpy change during hydriding/ dehydriding reaction, is generally 30~40kJ/ mol H_2.

2. The well known Van't Hoff equation relating plateau pressure to temperature.

$$\ln P_e = \frac{\Delta H}{R_g T} - \frac{\Delta S}{R_g} \tag{2}$$

For the hydriding/dehydriding reaction, there exists a phase during which the stored amount of H_2 varies a lot while the equilibrium pressure almost keeps constant, the phase is generally called "plateau". The equilibrium pressure in this phase, as shown in Equation (2), depends on the temperature. ΔH and ΔS are the enthalpy and entropy changes during hydriding/dehydriding reaction, respectively. Both parameters take quite different values for various MH materials, thus a wide range of operation temperature and pressure can be covered.

Because of the unique properties mentioned above, MHs can be applied for a number of uses, e.g. hydrogen storage (Kaplan et al., 2006), heat storage (Felderhoff & Bogdanovic, 2009), thermal compression (Murthukumar et al., 2005; Kim et al., 2008; Wang et al., 2010), heat pump (Qin et al., 2008; Paya et al., 2009; Meng et al., 2010), gas separation and purification (Charton et al, 1999). Generally such systems share the common advantages of being environmentally benign, compact and flexible for various operating conditions. It is noteworthy that for any practical application, the reactor where MHs are packed plays an important role in the whole system. Besides the basic function of holding MH materials, the reactor should also facilitate good heat and mass transfer. Therefore the analysis and optimization of MH reactor are very important, and numerical simulation has become a powerful tool for that purpose as the development of computers.

The modeling and simulation of hydriding/dehydriding process in the MH reactor started early. In 1980s, a simple 1-D mathematical model considering heat conduction and reaction kinetics was popular in use, see the pioneer work of (El Osery, 1983; Sun & Deng, 1988). Later (Choi & Mills, 1990) incorporated the classical Darcy's law into the 1-D model for the calculation of hydrogen flow, which added to the completeness of the model concerning the description of multiple physics. Moreover, the treatment also introduced the notion of dealing with the hydriding/dehydriding process in a MH reactor as reactive flow in porous media, which marked a great progress. Then the notion was further developed by (Kuznetsov and Vafai, 1995; Jemni & Ben Nasrallah, 1995a; Jemni and Ben Nasrallah, 1995b). These authors formulated 2-D mathematical models based on the volume averaging method (VAM), which is classical in the study of porous media. The coupling process of porous flow, heat conduction and convection, reaction kinetics were described in the models. In a different way, (Lloyd et al., 1998) derived the model equations for a representative element volume from the basic conservation law, which are similar in form to those obtained by the VAM. Since then the theoretical frame to model the dynamic process in a MH reactor has been established, although still more details were taken into account in recent studies, such as the temperature slip between gas and solid phases (Nakagawa et al., 2000), 3-D description (Mat et al., 2002), the effect of radiative heat transfer (Askri et al., 2003).

Unfortunately, till now most studies are concentrated on the domain of reaction bed while the effect of vessel wall is ignored, and isotropic physical properties are generally assumed in the bed, which is not necessarily the case. In this investigation, a general mathematical model for the MH reactor was formulated and numerically solved by the finite volume method. The effects of vessel wall as well as the anisotropic physical properties were discussed thereafter for a tubular reactor.

2. Mathematical model

2.1 A general picture

Like most gas-solid reactions, the actual hydriding/dehydriding process could be very complicated. According to (Schweppe et al., 1997), the hydriding reaction proceeds in several steps on the scale of MH particles:

1. Transport of H_2 molecules in the inter-particle gas phase;
2. Physisorption of H_2 molecules on the particle surface;
3. Dissociation of physisorbed H_2 into H atoms;
4. Interface penetration of H atoms to the subsurface;
5. Diffusion of H atoms in the hydride (also termed β phase) layer;
6. Formation of the hydride at the α/β interface, α is the solid solution with relatively small amount of hydrogen;
7. Diffusion of H atoms in the α phase.

For dehydriding reaction, the steps are largely similar yet proceed in reversed order.

Fortunately, we don't have to deal with all the above details in modeling the MH reactor. In the general frame for the model, step 1 is described by the porous flow, and the rest steps including the microscopic mass transport are simply incorporated in a lumped kinetic expression where only two superficial parameters matter, namely the activation energy E and pre exponential factor k. Obviously, most microscopic details are dropped in such a treatment, yet the simplification proves acceptable in the macroscopic description of coupling process in the MH reactor.

2.2 Reactor geometry and operation

The tubular type reactor is developed early and widely used, especially for the heat pump and thermal compression applications, as reviewed by (Yang et al.,2010 a). Therefore our investigation was focused on such a reactor, and the schematic was shown in Fig.1.

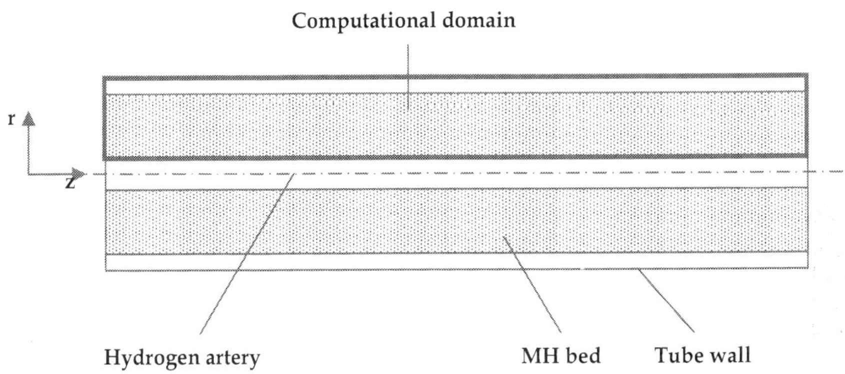

Fig. 1. The schematic of the tubular type reactor for investigation

As can be seen, a central artery is used in this type of reactor for radial hydrogen flow. MH particles ($LaNi_5$ for this investigation, which is commonly used) are packed in the annular space between the artery and the tube wall, while the heat exchange of the reaction bed with heat source/sink can be conducted through the external surface of tube wall. Aluminum foam was supposed to be inserted in the bed for heat transfer enhancement. The upper part of the reactor including both MH bed and tube wall was taken as the computational domain for symmetry. The length of the reactor is 0.5m, the radius of the artery and the bed thickness are 0.003 and 0.0105m, respectively. The thickness of wall is 0.0015m.

2.3 The set of model equations

Before formulating the model equations, the assumptions were made as follows:

1. The physical properties of the reaction bed, including the thermal conductivity, the permeability, the heat capacity etc., are constant during the reaction.
2. The gas phase is ideal from the thermodynamic view.
3. There is no temperature slip between the solid phase and the gas phase, which is also termed "local thermal equilibrium" (Kuznetzov and Vafai, 1995). The common temperature is defined as T_b here.

4. The radiative heat transfer can be neglected due to the moderate temperature range in discussion.

The model equations include,

The mass equation for gas phase (the continuity equation):

$$\frac{\partial \varepsilon_v \rho_g}{\partial t} + \nabla(\rho_g \vec{U}) = -\dot{m} \cdot M_g \tag{3}$$

The mass equation for solid phase:

$$\frac{\partial \varepsilon_{MH} \rho_{MH}}{\partial t} = \dot{m} \cdot M_g \tag{4}$$

Where the mass source term resulting from the hydriding/dehydriding reactions was expressed as,

$$\dot{m} = \frac{\varepsilon_{MH} \cdot \rho_{MH}}{M_{MH}} \cdot \left[\frac{H}{M}\right]_{sat} \cdot \frac{dX}{dt} \tag{5}$$

The momentum equation for the gas phase takes the form of Darcy's law,

$$\vec{U} = \frac{K}{\mu} \nabla P_g \tag{6}$$

For a porous system composed of spherical particles, the permeability K could be calculated according to the Carman-Kozeny correlation,

$$K = \frac{d_p^{\,2} \cdot \varepsilon_v^{\,3}}{180 \cdot (1 - \varepsilon_v)^2} \tag{7}$$

The energy equation for the bulk bed including both gas and solid phases is written as,

$$\frac{\partial \rho_b c_{pb} T_b}{\partial t} + \nabla \cdot \left(\rho_g c_{pg} \vec{U} T_b \right) = \nabla(\lambda_{eff} \nabla T_b) + \dot{m} \cdot \Delta \mathrm{H} \tag{8}$$

The heat capacity of the bulk bed is calculated as follows,

$$\rho_b c_{pb} = \sum_i \varepsilon_i \rho_i c_{pi} \tag{9}$$

Where ε_i, ρ_i and c_{pi} denote the corresponding properties for an individual phase, e.g. MH, hydrogen gas or the materials added (Aluminum foam here). Because multiple mechanisms and complex geometry are involved in the particle-scale heat transfer process (Sun & Deng, 1990), the correlation of effective thermal conductivity to relevant explicit properties is not quite accurate, not to say general. Therefore, the citation of measured value of λ_{eff} seems more practical in the modeling studies, although the nonlinearity of the actual system may not be well reflected after such simplification.

Besides the basic conservation equations given above, still some other equations are needed to close the model, i.e. the P-c-T equations and reaction kinetic equations.

The P-c-T equations are used to describe the relationship of pressure P, hydrogen concentration c and temperature T under the equilibrium state. Van't Hoff equation, namely the equation (2) is the most commonly used one. It features simple expression and few parameters involved (just ΔH and ΔS). The values of ΔH and ΔS for many MH materials can be found in the literature. However, the equation merely covers the plateau phase, in addition the plateau slope and hysteresis are not well reflected. Therefore other P-c-T expressions were presented to cover the full range of hydriding/dehydriding reaction with higher accuracy, e.g. the polynomial equations (Dahou et al., 2007) and the modified Van't Hoff equations (Nishizaki et al., 1983; Lloyd et al., 1998). In this investigation, a modified Van't Hoff equation was adopted(Nishizaki et al., 1983),

$$P_{e,a} = \exp(A - \frac{B}{T_b} + (\theta + \theta_0) \cdot \tan(\pi * (X - 1/2)) + \frac{\beta}{2}) \tag{10a}$$

$$P_{e,d} = \exp(A - \frac{B}{T_b} + (\theta - \theta_0) \cdot \tan(\pi * (X - 1/2)) - \frac{\beta}{2}) \tag{10b}$$

As mentioned in section 2.1, most microscopic details are dropped in a realistic reaction kinetic equation, whose integral form can be generally written as follows,

$$f_1(X) = k \cdot f_2(T) \cdot f_3(P) \cdot t \tag{11a}$$

The equivalent differential form of the kinetic equation can be obtained by simple manipulation of equation (11a),

$$\frac{dX}{dt} = k \cdot f_2(T) \cdot f_3(P) \cdot g_1(X) \tag{11b}$$

Equation (11a) is more used in the experimental determination of the kinetic parameters and reaction mechanism, while equation (11b) is preferred in the modeling of hydriding/dehydriding process in a MH reactor. The specific expression of f_1 or g_1 depends on the reaction mechanism, see Table 1(Li et al., 2004). Among all the expressions, the ones suggesting shrinking core, diffusion control or nucleation & growth mechanisms are widely applied in the kinetic study of MHs. The Arrhenius expression is often adopted as f_2. A few expressions are available for f_3 according to (Ron, 1999), and a so-called normalized pressure dependence expression was recommended. However, some authors argued that f_3 should be related to the reaction mechanism(Forde et al., 2007).

In this investigation, the kinetic equations for $LaNi_5$ are those recommended by (Jemni and Ben Nasrallah, 1995a; Jemni and Ben Nasrallah, 1995b),

$$\frac{dX}{dt} = k_a \cdot \exp(-\frac{E_a}{R_g T_b}) \cdot \ln(\frac{P_g}{P_{e,a}}) \cdot (1 - X) \tag{12a}$$

$$\frac{dX}{dt} = k_d \cdot \exp(-\frac{E_d}{R_g T_b}) \cdot (\frac{P_g - P_{e,d}}{P_{e,d}}) \cdot X \tag{12b}$$

The detailed information about the parameters in equation (10) and (12) are referred to the original papers.

Mechanism	$g_1(X)$	$f_1(X)$
Nucleation& growth	$(1/n)\cdot(1-X)\cdot[-\ln(1-X)]^{1-n}$	$[-\ln(1-X)]^{n}$
Branching nucleation	$X\cdot(1-X)$	$\ln[X/(1-X)]$
Chemical reaction	$(1/n)\cdot(1-X)^{n}$ $(1/2)\cdot(1-X)^{3}$ $(1-X)^{2}$	$1-(1-X)^{n}$ $(1-X)^{-2}-1$ $(1-X)^{-1}-1$
1-D diffusion	$(1/2)\cdot X^{-1}$	X^2
2-D diffusion	$(1-X)^{1/2}\left[1-(1-X)^{1/2}\right]^{-1}$	$\left[1-(1-X)^{1/2}\right]^{2}$
3-D diffusion	$(3/2)(1-X)^{2/3}\left[1-(1-X)^{1/3}\right]^{-1}$ $(3/2)\left[(1-X)^{-1/3}-1\right]^{-1}$	$\left[1-(1-X)^{1/3}\right]^{2}$ $1-2X/3-(1-X)^{2/3}$

Table 1. Part of f_1/g_1 expressions for MH reaction kinetics(Li et al., 2004)

2.4 Initial and boundary conditions

The initial reacted fraction for hydriding and dehydriding were uniform throughout the reactor. The temperature of the reactor was equal to that of inlet fluid, and the system was assumed under the P-c-T equilibrium.

The boundary conditions of MH reactors can be classified into three types (Yang et al, 2008; Yang et al., 2009): adiabatic wall (or symmetry boundary), heat transfer wall and mass transfer boundary.

For the adiabatic wall (or symmetry boundary):

$$\frac{\partial T_b}{\partial z}\Big|_{z=0}=0\ ,\ \frac{\partial P_g}{\partial z}\Big|_{z=0}=0 \tag{13a}$$

$$\frac{\partial T_b}{\partial z}\Big|_{z=L}=0\ ,\ \frac{\partial P_g}{\partial z}\Big|_{z=L}=0 \tag{13b}$$

For the heat transfer wall:

$$\lambda_{eff}\frac{\partial T_b}{\partial r}\Big|_{r=r_o}=h(T_b-T_f)\ ,\ \frac{\partial P_g}{\partial r}\Big|_{r=r_o}=0 \tag{14}$$

where T_f is varied along the axial direction of tubular reactor and can be calculated as follows,

$$h(T_b-T_f)=q_f c_{pf}\frac{\partial T_f}{\partial z} \tag{15}$$

For the mass transfer boundary through which hydrogen enters or leaves the reactor, a Danckwerts' boundary condition (Yang et al., 2010b) is applied to make sure that the flow rate is continuous across the boundary,

$$\begin{cases} \lambda_{eff} \dfrac{\partial T_b}{\partial r}|_{r=r_i} = \rho_{g,in} \vec{U} c_{pg} (T_{in} - T_b) & \text{(for hydriding)} \\ \dfrac{\partial T_b}{\partial r}|_{r=r_i} = 0 & \text{(for dehydriding)} \end{cases} \tag{16}$$

The pressure is simply set as follows,

$$P_g |_{r=r_i} = P_{ex} \tag{17}$$

3. Numerical solutions and validation

3.1 The solution based on FVM

In the field of computational fluid dynamics (CFD), finite volume method (FVM) is widely applied and many commercial packages are based on this method, such as FLUENT, CFD-ACE, CFX, etc. The method uses the integral form of the conservation equation as the starting point. The generic conservation equation is (Tao, 2001):

$$\frac{\partial(\rho\phi)}{\partial t} + div\left(\rho\phi\vec{U}\right) = div(\Gamma grad\phi) + S \tag{18}$$

From left to right, the terms were called non-steady term, convection term, diffusion term and source term, respectively. These terms could be integrated using different "schemes". In this investigation, implicit Euler scheme, 1-order upwind differencing scheme (UDS) and central difference scheme (CDS) were applied for the first 3 terms. The source term S resulting from the hydriding/dehydriding reaction was obtained explicitly in a time step.

The integration is implemented for a number of small control volumes (CVs) in the computational domain. A type-B grid was adopted for the discretization of the domain (Tao, 2001), which defines the boundaries first and then the nodal locations.

It is noteworthy that the distributed and anisotropic physical properties can be easily incorporated into the FVM based solution procedure. Firstly, we could discretize the computational domain so that a certain boundary of grids and the true boundary (e.g. the interface separating the reaction bed and the vessel wall) overlap, see Fig.2.

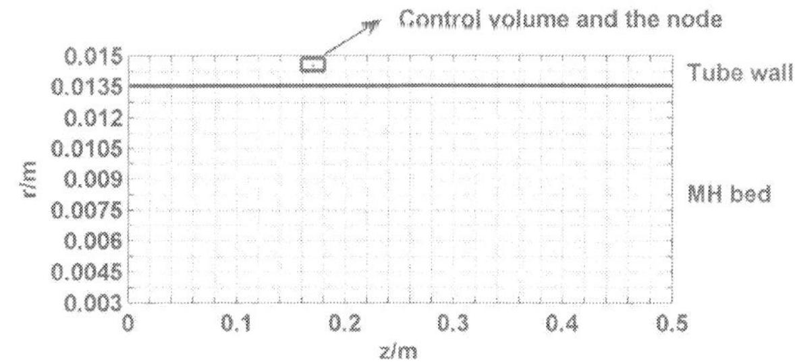

Fig. 2. The discretization of the computational domain

Next, the physical properties should be set according to the positions of the grids. Porosity and heat capacity are scalar quantities defined at the nodes and could be specified easily, see Table 2. Obviously, in the region of tube wall, the governing equations including flow, heat transfer and reaction kinetics degenerate into a simple heat conduction equation. On the contrary, thermal conductivity and permeability, which are tensors defined on the boundaries of CV, should be dealt with carefully. For the boundaries in the domain of reaction bed or vessel wall, the settings are similar to those for scalar properties, yet should be implemented in both axial and radial directions. For the true boundary, the physical properties are obtained by a certain averaging of those properties on both sides. A harmonic averaging is found appropriate to keep a constant flow over the boundary(Tao, 2001) and was applied.

	Reaction bed	Tube wall
Volume fraction of gas(namely porosity) ε_g	0.438	10^{-5}
Volume fraction of MH ε_{MH}	0.462	0
Volume fraction of Al foam ε_{Al}	0.1	0
Volume fraction of wall material ε_w	0	0.99999
Thermal conductivity λ/W/(m·K)	λ_{eff} =7.5	λ_w
Permeability K/m^2	5.8×10^{-13}	10^{-35}(basically impermeable)

Table 2. The physical properties for the wall materials

The governing equations were solved in a segregated manner, and the algorithm is similar to the SIMPLE type method widely applied in the computation of incompressible flow. However, for the compressible flow in the investigation, pressure exerts influences on both velocity and the density of fluid, which should be considered in the solution procedure. The main steps are listed as follows,

1. The increment of reacted fraction in this time step is calculated explicitly by the kinetic equations from the initial pressure P, temperature T and reacted fraction X, thus the source terms in the mass equation and energy equation are obtained accordingly.
2. The gas density ρ^* and velocity U^* are calculated respectively by the state equation for ideal gas and Darcy's law from the initial P and T, while they do not solve the simultaneous continuity equation for the gas.
3. To fulfill the solution of the continuity equation, some correction, namely ρ' and U' should be conducted based on ρ^* and U^*. The expressions of ρ' and U' with regard to P', which denotes the correction of present pressure P^*, could be found according to state equation for ideal gas and Darcy's law.
4. The expressions of ρ' and U' obtained in step 3 are substituted into the continuity equation, and the pressure correction P' is solved as the primitive variable.
5. Use P' to correct the pressure P^*, the density ρ^* and the velocity U^*.
6. Repeat steps 2-5 until the computation of flow field converges, which is assumed after a certain tolerance achieved.
7. The velocity U obtained above is substituted into the energy equation for the solution of temperature T.
8. Repeat steps 2-7 until the computations of both flow and heat transfer converge.
9. Enter the next time step, repeat steps 1-8 till the required time elapses.

After discretization of the domain and integration of the differential equations, a few sets of algebraic equations were formulated. An alternative direction implicit (ADI) method was used to solve them. The time step was 0.01s, while convergence was assumed when the error of P and T were respectively lower than 10^{-3}Pa and 10^{-3}K.

3.2 Validation by the literature data

(Laurencelle and Goyette, 2007) have carried out extensive experimental studies on a MH reactor packed with $LaNi_5$, and the data they reported were used to validate our model. The so-called "small" reactor has an internal diameter of 6.35mm and a length of 25.4mm. 1g of $LaNi_5$ powder was stored and hydriding followed by dehydriding experiment was conducted. The initial pressures for the two sequential processes were 0.6 and 0.0069MPa, respectively. The reacted fraction and system pressure predicted by the model were compared with the experimental data in Figs. 3. As can be seen, the simulation results show satisfactory agreement with the experimental data, thus the model can be used for the study of hydriding/dehydriding processes in a MH reactor.

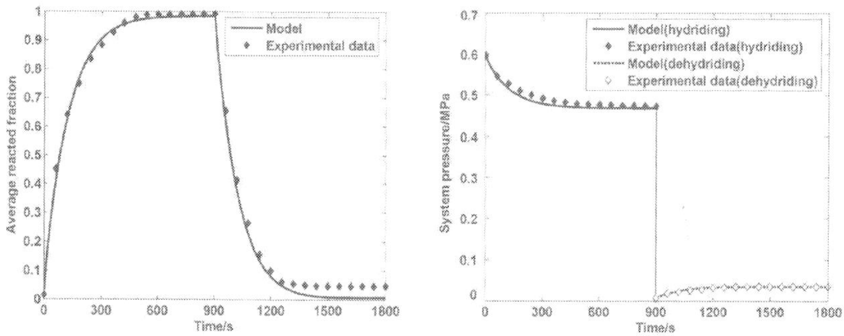

Fig. 3. The comparison of simulated results and the experimental data from the literature

4. Discussions based on the model

In the engineering practice, stainless steel, brass and aluminum are the materials used most frequently for the construction of MH reactors (Murthukumar et al., 2005; Kim et al., 2008; Qin et al., 2008; Paya et al., 2009). Therefore they were considered here, and the reactors using them as wall material are referred to as reactors 1, 2 and 3 respectively. Many factors should be taken into account for the use of a certain wall material, i.e. strength, corrosion and heat transfer. The former two are irrelevant concerning the hydriding/dehydriding processes, thus would not be elaborated here. To conduct the numerical simulation, some physical properties of the wall material should be known beforehand and are listed in Table 3.

	Density ρ_w/kg/m3	Thermal conductivity λ_w/W/(m·K)	Specific heat capacity c_{pw}/J/(kg·K)
Stainless steel(316L)	7959	13.3	488
Brass	8530	121.6	390
Aluminum	2699	237	897

Table 3. Some relevant physical properties of wall materials

The reference operation conditions for the hydriding/dehydriding processes in the MH reactors were specified in Table 4. Water was assumed to be the heat transfer fluid and the number of heat transfer unit (NTU) was set to be 1.

	Hydriding	Dehydriding
Exerted pressure P_{ex}/MPa	0.4	0.8
Initial reacted fraction X	0.2	0.8
Initial bed temperature T_b/K	293	353
Convection heat transfer coefficient h/W/(m^2·K)	1500	1500
Fluid inlet temperature T_f /K	293	353
Fluid mass flow rate q_f/kg/s	0.0168	0.0168
Fluid specific heat capacity c_{pf} /J/(kg·K)	4200	4200

Table 4. The operation conditions for hydriding/dehydriding reaction in the MH reactor

The grid independence test was carried out before further work conducted. 3 sets of grids (10×8, 25×16, 50×24) were applied to simulate the hydriding process of reactor 1 respectively, and Fig.4 shows the comparison of the results. An asymptotic tendency was found when using denser grid, and the 25×16 grid proved to be adequate in obtaining enough accuracy.

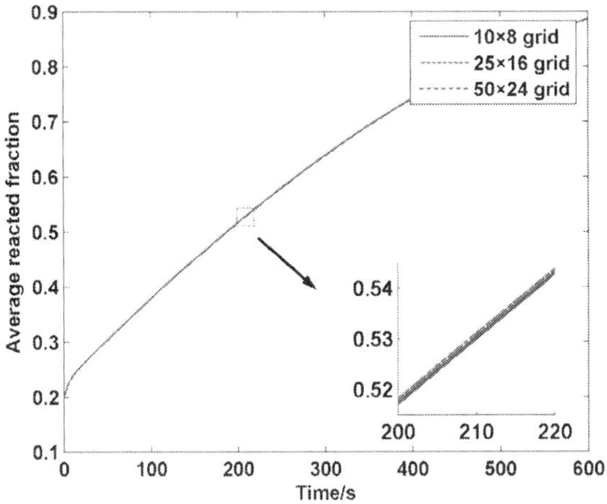

Fig. 4. The simulation results for grid independence test

4.1 General characteristics in a tubular reactor

Firstly the reaction and transport characteristics were analyzed for a tubular reactor (reactor 1) under the reference conditions. The temperature contours during hydriding process were shown in Fig.5. As can be seen, the reactor temperature rises from the initial value of 293K due to the exothermic reaction. The top left corner region($z \to 0$, $r \to r_o$) close to the fluid inlet is better cooled and the corresponding temperature is low, while the peak temperature of around 311K appears in the bottom right region ($z \rightarrow L$, $r \rightarrow r_i$), which is far from both the heat transfer wall and the fluid inlet.

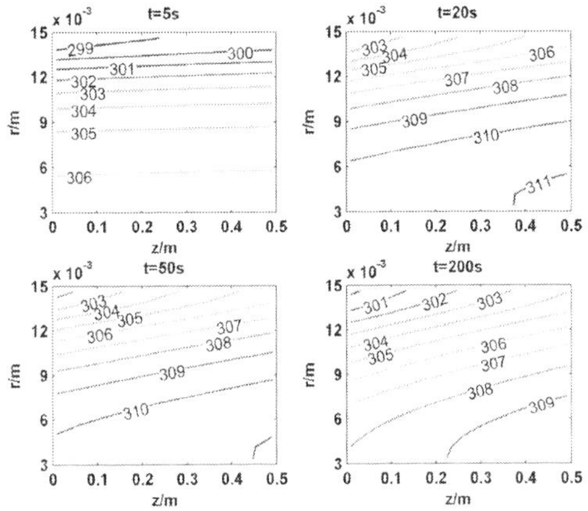

Fig. 5. The temperature contours in reactor 1 at the moment of 5, 20, 50, 200s during hydriding process (K)

The pressure contours in the reactor were shown in Fig.6. From the mass transfer boundary on the bottom side($r \rightarrow r_i$), the pressure decreases slightly across the reactor by a few hundred Pa, thus the resistance against mass transfer is small. Through the pressure gradient we can also determine the direction of hydrogen flow. It is almost radial initially, while as time elapses an axial flow from right to left can be recognized in the region close to the heat transfer wall($r \rightarrow r_o$).

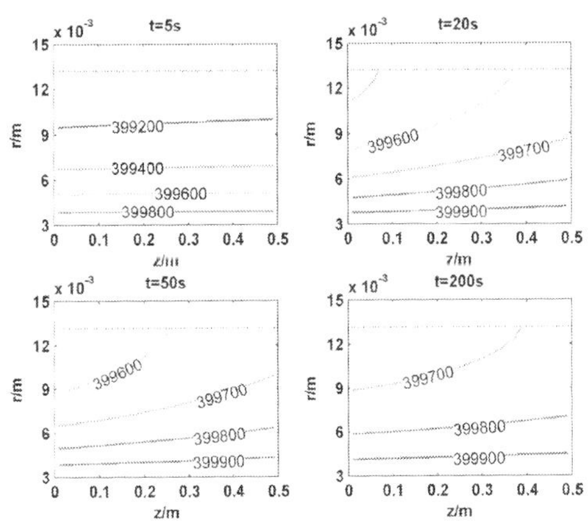

Fig. 6. The pressure contours in reactor 1 at the moment of 5, 20, 50, 200s during hydriding process (Pa)

The contours for reacted fraction were shown in Fig.7. Obviously, the hydriding reaction proceeds most rapidly in the top left corner, while in the bottom right corner region the process is very sluggish. The variation in hydriding rate also explains the appearance of axial flow: the hydrogen tends to move towards the region absorbing it more quickly. Consider the distribution of temperature, pressure and reacted fraction, the effect of heat and mass transfer on the reaction can be assessed. As is well known, low temperature and high pressure are favorable for the hydriding process. Therefore, the distribution of reacted fraction is basically consistent with the temperature distribution, suggesting that heat transfer controls the actual reaction rate under the given conditions.

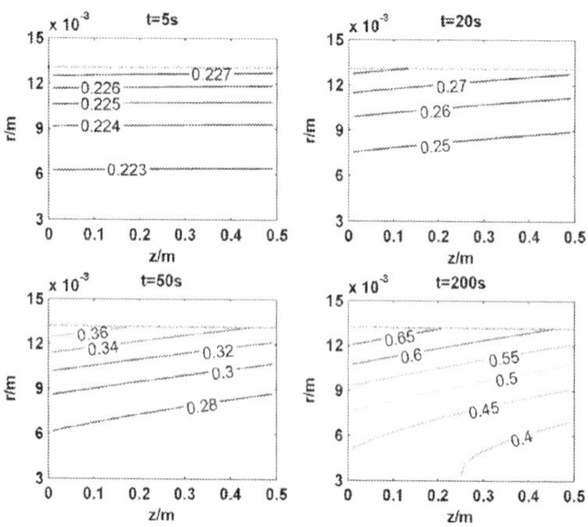

Fig. 7. The contours of reacted fraction in reactor 1 at the moment of 5, 20, 50, 200s during hydriding process

For dehydriding process, the phenomena is basically similar to that in hydriding process, thus the corresponding statements are skipped.

4.2 The effect of wall materials

The reactor depicted in section 2.2 is called a thin-wall reactor, which is better used in heat pump applications. Because in that case the operation pressure is moderate and sufficient heat transfer is more important. The hydriding rates for the reactors 1, 2 and 3 are shown in Fig.8.

As can be seen, more rapid reaction is achieved for the reactors with brass and aluminum as wall material, while the difference between the two is hardly discernable. The phenomena is attributed to the varied thermal conductivity of the materials, see Table 2. Since higher thermal conductivity implies smaller resistance for heat transfer through the wall, the reaction heat from hydriding process could be more effectively removed from the reaction bed of reactors 2 and 3. Thereby, large temperature rise in the bed, which reduces the driving force of reaction and hinders the process, is less likely to occur. The explanation is also supported by the comparison of fluid outlet temperature of the 3 reactors, see Fig.9. With the same mass flow rate and inlet temperature for the fluid, the fluid outlet temperature for reactors 2 and 3 is higher than that for reactor 1 during most of the reaction time, suggesting a larger power of heat release.

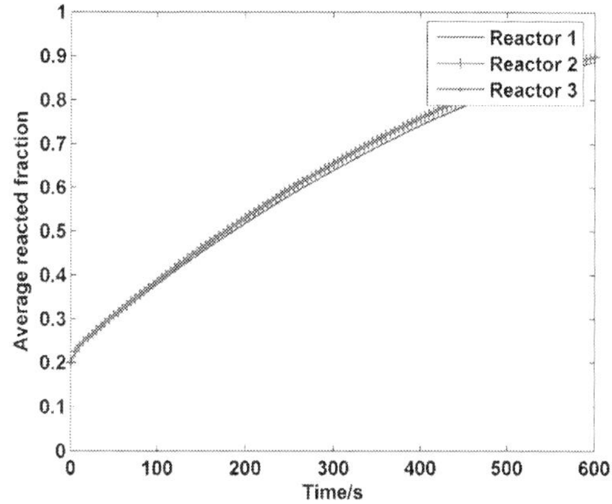

Fig. 8. The simulated hydriding rates for reactors 1, 2 and 3 under reference conditions

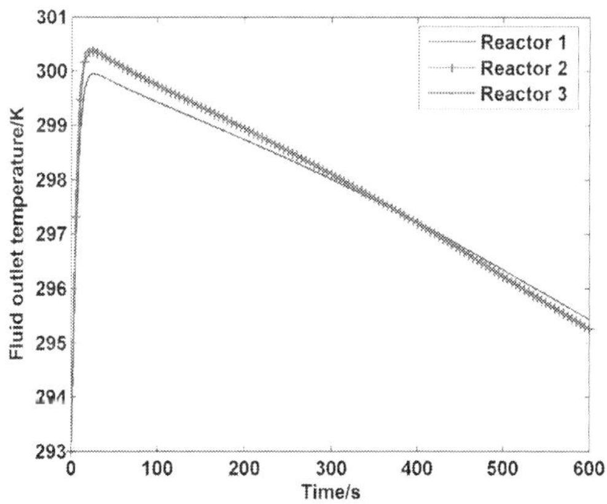

Fig. 9. The simulated fluid outlet temperatures for reactors 1, 2 and 3 under reference conditions

Then a thick-wall reactor preferable in the compression applications was investigated. The wall thickness is increased from 1.5 to 3mm. As shown in Fig.10, due to the large thickness, the effect of wall gains significance. The comparison result is qualitatively similar to that for thin-wall reactor, yet greater improvement in hydriding rate was observed for the reactors 2 and 3.

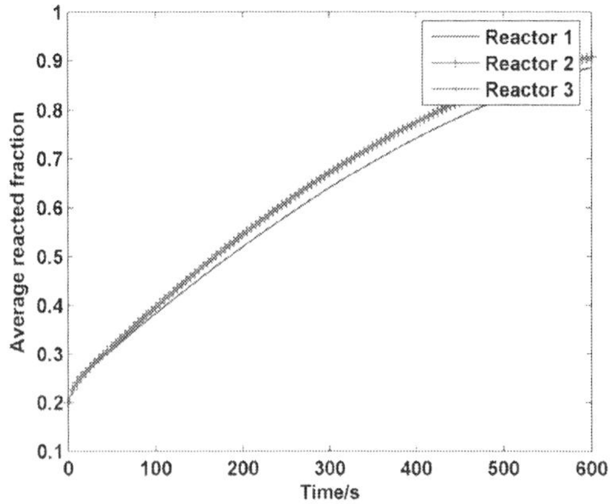

Fig. 10. The simulated hydriding rates for reactors 1, 2 and 3 with larger wall thickness

The effect of wall material on the reaction rate depends not only on the thickness, but also on the operation conditions, especially those related to the transport and accumulation of the reaction heat. The simulation results for the case that air is used as the heat transfer fluid were shown in Fig.11.

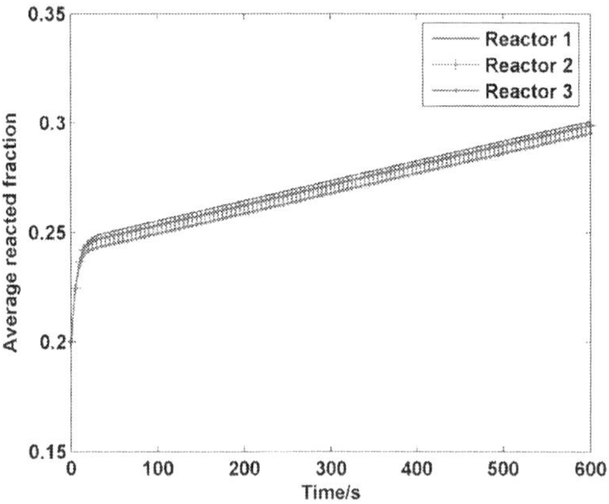

Fig. 11. The simulated hydriding rates for reactors 1, 2 and 3 while using air as heat transfer fluid

Under this condition, a typical convection heat transfer coefficient of 50W/(m^2·K) was set. As can be seen, the hydriding rate gets much slower for all the 3 reactors, because the heat exchange between the bed and fluid is not as sufficient as that for a reactor using water. Moreover, it was found that the highest reaction rate is obtained by reactor 1 rather than reactors 2 and 3, which seems unreasonable considering only the heat transfer. The comparison of the heat capacity of wall for the 3 reactors in Fig.12 explained the simulation results.

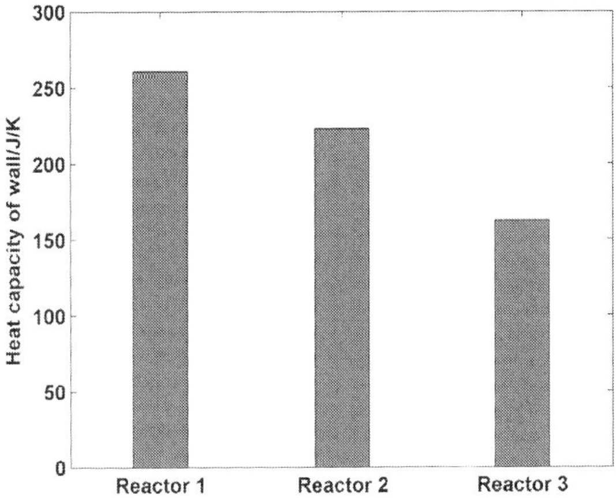

Fig. 12. The heat capacities for the tube wall of reactors 1, 2 and 3

Due to poor external heat transfer when using the air, the reaction heat can no longer be released rapidly and will be accumulated in the bed, even for the reactors 2 and 3. The notion was proved by the temperature contours in reactor 3 at certain moments, see Fig.13.

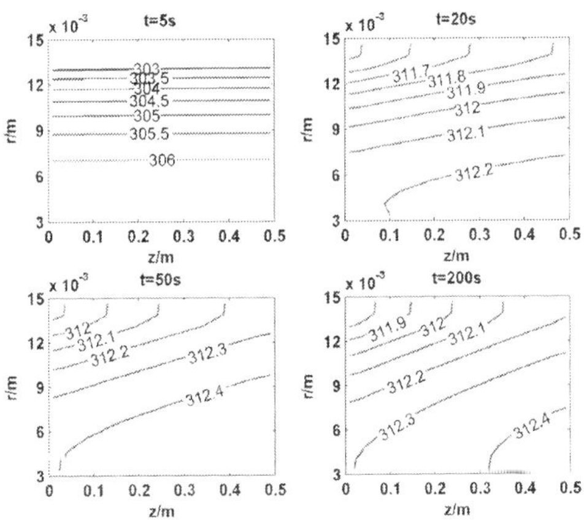

Fig. 13. The temperature contours of reactor 3 at the moment of 5, 20, 50 and 200s while using air as heat transfer fluid

Because of the heat accumulation, the reactor temperature is kept on a high level soon after the hydriding reaction starts. In this case, the temperature rise of the reactor depends more on the total heat capacity. Larger heat capacity limits the temperature fluctuation in the bed during the initial stage, and thus favors the proceeding of reaction. After a certain time elapses and the steady temperature distribution formed, such effect disappears and the same reaction rates were observed for all the 3 reactors, as shown in Fig.11.

For the dehydriding process, the conclusions are largely the same as those drawn above. Thus only the simulation results for the thin wall reactor was shown for the sake of conciseness, see Fig.14.

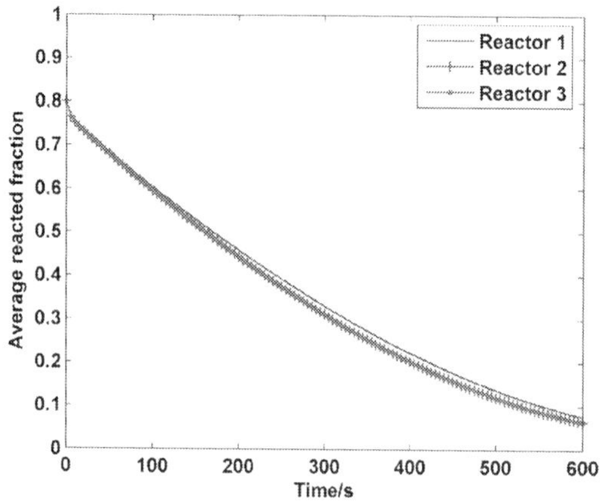

Fig. 14. The simulated dehydriding rates for reactors 1, 2 and 3 under reference conditions

4.3 The effect of anisotropic physical properties

In the randomly packed MH bed, the physical properties are largely isotropic on condition that the effect of gravity and the pulverization phenomena are ignored. However, if the compact with both MH and some solid matrix (e.g. expanded natural graphite, ENG) is used in the reactor bed for heat transfer enhancement, anisotropic properties could be observed, as reported by (Sanchez et al., 2003; Klein & Groll, 2004). The radial thermal conductivity of the compact is significantly higher than the axial one, and the effect of such variation in a realistic tubular reactor has not yet been analyzed.
Here the sample M24 (Klein & Groll, 2004), which has proper porosity and mass fraction of MH was considered. The porosity of the bed is 0.427, the volume fraction for ENG and MH are 0.124 and 0.449, respectively. The radial thermal conductivity of the compact is 17.4W/(m*K), while the axial one was not reported in the original paper. However, the property of a similar compact called IMPEX (Sanchez et al., 2003) could be used as approximation(~1W/(m*K)).
Under the reference conditions in Table 4, the hydriding processes were simulated for reactor 1 using M24 or another sample, which was supposed to have an isotropic thermal conductivity of 17.4W/(m*K). The simulated hydriding rates in both cases were compared and hardly any difference was recognized. It can be easily understood that the axial thermal conductivity works on the hydriding rate through its effect on the temperature distribution in the MH bed. However, even for the maximum temperature difference along axial direction, which lies in the neighborhood of tube wall, such effect is rather insignificant, see Fig.15. Simulations were also carried out for the dehydriding processes in reactors 1, 2 and 3, and the above phenomena repeated. Therefore, it can be concluded that the axial thermal conductivity is much less important than the radial one in a tubular reactor.

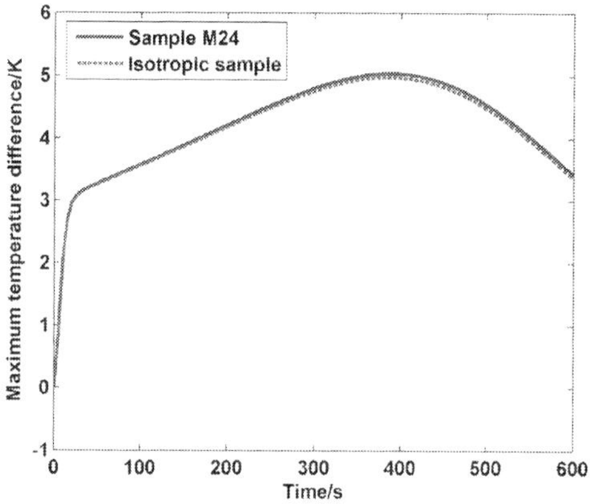

Fig. 15. The maximum temperature difference along axial direction for the reactors using sample M24 and isotropic sample

It is convenient to compare the reaction and transport characteristics of the reactors where different measures are taken for heat transfer enhancement, i.e. Aluminum foam and ENG compact. As shown in Fig.16, faster reaction rate was achieved in the latter reactor under reference conditions, although its axial thermal conductivity is even lower than that of reactor using Aluminum foam.

The temperature in the reactor using ENG compact was indicted in Fig.17. More uniform distribution of temperature was found when compared with Fig.5. The reaction driving force is determined by bed pressure and equilibrium pressure(largely determined by temperature), as can be seen in equation (12). Considering the small pressure drop in the reactor, we can conclude that uniform temperature results in uniform reaction, which is favorable for the operation of MH reactor. In a word, MH reactor using ENG compact shows superior performance to the one using Aluminum foam.

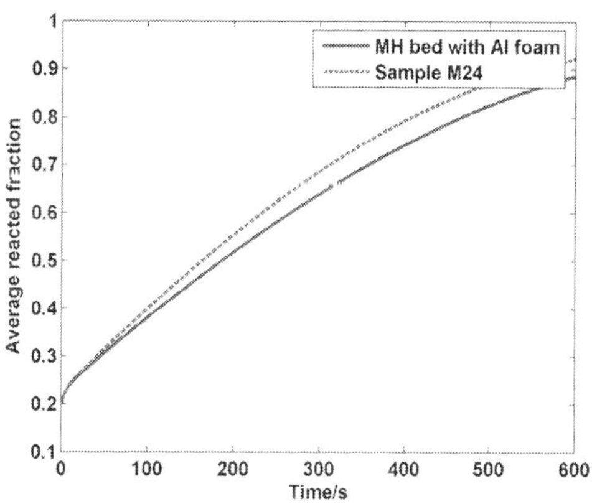

Fig. 16. The simulated hydriding rates for the reactor 1 using Aluminum foam or ENG compact

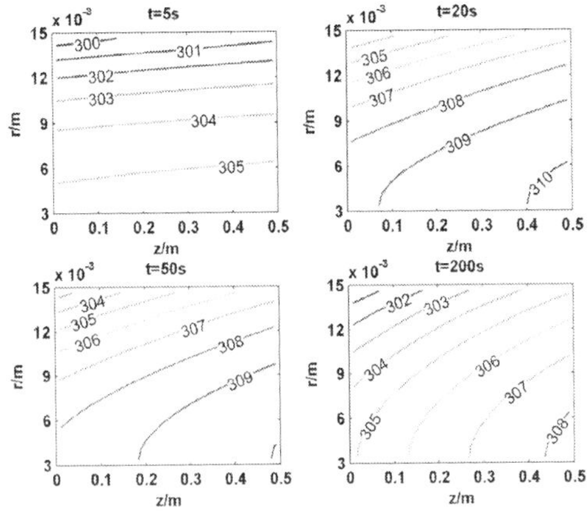

Fig. 17. The temperature contours in reactor 1 using ENG compact at the moment of 5, 20, 50, 200s during hydriding process (K)

5. Conclusion

A mathematical model for the hydriding/dehydriding process in a realistic MH reactor was established, in which the distributed and anisotropic physical properties could be easily handled. Numerical solution of the model was implemented by the FVM procedure, and the model was validated using the literature data. The model was applied for the simulation study of a tubular type MH reactor packed with $LaNi_5$, and the following conclusions were drawn:

1. Heat transfer rather than mass transfer controls the actual reaction rate for the MH reactor under discussion.
2. The effect of wall material on the hydriding/dehydriding rate of a MH reactor depends on two factors, i.e. the thermal conductivity and heat capacity.
3. When the external convection heat transfer is sufficient, the effect of wall depends more on its thermal conductivity, larger λ suggests better heat transfer and more rapid reaction.
4. When air is used as the heat transfer fluid, heat capacity of the wall becomes important as the external convection heat transfer gets poor. The reactor with larger heat capacity shows smaller temperature fluctuation and slightly higher reaction rate in the initial stage.
5. In comparison to the radial thermal conductivity, the axial one of the MH bed is basically unimportant in a tubular reactor. Therefore, the measures merely achieving significant heat transfer enhancement in radial direction, e.g. ENG compact, is recommended for use.

6. Nomenclature

A	parameter in P-c-T equation	
B	parameter in P-c-T equation	$[K]$

c_p	specific heat capacity	$[J/(kg \cdot K)]$
d_p	diameter of particles	[m]
E	activation energy	$[J/mol]$
f	function	
g	function	
h	convective heat transfer coefficient	$[W/(m^2 \cdot K)]$
ΔH	reaction heat	$[J/molH_2]$
$\left[\frac{H}{M}\right]$	mole ratio of stored hydrogen to host metal	
k	reaction rate constant	$[s^{-1}]$
K	permeability	$[m^2]$
$\dot{m}$	source term from reaction	$[\text{mol}/(m^3 \cdot \text{s})]$
L	length of the tubular reactor	$[m]$
M	molecular weight	$[kg/mol]$
P	pressure	$[Pa]$
q	mass flow rate	$[kg/\text{s}]$
r	r-coordinate in the radial direction	$[m]$
R_g	general gas constant	$[J/(mol \cdot K)]$
S	source term	
t	time	$[s]$
T	temperature	$[K]$
$\vec{U}$	gas velocity	$[m/s]$
X	reacted fraction	
z	z-coordinate in the axial direction	$[m]$

Greek symbols

β	hysteresis factor in P-c-T equation	
Γ	coefficient for diffusion-like process	
ε	volume fraction	
θ	plateau flatness factor in P-c-T equation	
θ_0	plateau flatness factor in P-c-T equation	
λ	thermal conductivity	$[W/(m \cdot K)]$
μ	dynamic viscosity	$[Pa \cdot s]$
ρ	density	$[kg/m^3]$
φ	general scalar quantity	

Subscripts

a	absorption, namely hydriding

Al	Aluminum foam
b	bulk
d	desorption, namely dehydriding
e	equilibrium
eff	effective
ex	exerted
f	heat transfer fluid
g	hydrogen gas
i	inner
in	inlet
MH	metal hydride
o	outer
sat	saturated
w	wall

7. References

Askri, F.; Jemni, A. & Ben Nasrallah, S. (2003). Study of two-dimensional and dynamic heat and mass transfer in a metal-hydrogen reactor. *Int. J Hydrogen Energy*, Vol.28, No.5 (537-557), ISSN: 0360-3199.

Berube, V.; Radtke, G.; Dresselhaus, M. & Chen, G. (2007). Size effects on the hydrogen storage properties of nanostructured metal hydrides: A review. *Int. J Energy Research*, Vol.31, No.6-7 (637-663), ISSN: 1099-114X.

Charton, S.; Corriou, J.P. & Schweich, D. (1999). Modeling of hydrogen isotopes separation in a metal hydride bed. *Chemical Engineering Science*, Vol.54, No.1 (103-113), ISSN: 0009-2509.

Choi,H. & Mills, A.F. (1990). Heat and mass transfer in metal hydride beds for heat pump applications. *Int. J Heat Mass Transfer*, Vol.33, No.6 (1281-1288), ISSN: 0017-9310.

Dhaou, H.; Askri, F.; Ben Salah, M.; Jemni, A.; Ben Nasrallah, S. & Lamloumi, J. (2007). Measurement and modeling of kinetics of hydrogen sorption by $LaNi_5$ and two related pseudo-binary compounds. *Int. J Hydrogen Energy*, Vol.32, No.5 (576-587), ISSN: 0360-3199.

El Osery, I.A. (1983). Theory of the computer code ret1 for the calculation of space-time dependent temperature and composition properties of metal hydride hydrogen storage beds. *Int. J Hydrogen Energy*, Vol.8, No.3 (191-198), ISSN: 0360-3199.

Felderhoff, M. & Bogdanovic, B. (2009). High temperature metal hydrides as heat storage materials for solar and related applications. *Int. J Molecular Sciences*, Vol.10, No.1 (325-344), ISSN: 1422-0067.

Forde, T.; Maehlen, J.P.; Yartys, V.A.; Lototsky, M.V. & Uchida, H. (2007). Influence of intrinsic hydrogenation/dehydrogenation kinetics on the dynamic behavior of metal hydrides: A semi-empirical model and its verification. *Int. J Hydrogen Energy*, Vol.32, No.8 (1041-1049), ISSN: 0360-3199.

Jemni,A. & Ben Nasrallah, S. (1995). Study of two-dimensional heat and mass transfer during absorption in a metal-hydrogen reactor. *Int. J Hydrogen Energy*, Vol.20, No.1 (43-52), ISSN: 0360-3199.

Jemni, A. & Ben Nasrallah, S. (1995). Study of two-dimensional heat and mass transfer during desorption in a metal-hydrogen reactor. *Int. J Hydrogen Energy*, Vol.20, No.11 (881-891), ISSN: 0360-3199.

Kaplan, Y.; Ilbas, M.; Mat, M.D.; Demiralp, M. & Veziroglu, T.N. (2006). Investigations of thermal aspects of hydrogen storage in a $LaNi_5$-H_2 reactor. *Int. J Energy Research*, Vol.30, No.6 (447-458), ISSN: 1099-114X.

Kim, J.K.; Park, I.S.; Kim, K.J. & Gawlik, K. (2008). A hydrogen-compression system using porous metal hydride pellets of $LaNi_{5-x}Al_x$. *Int. J Hydrogen Energy*, Vol.33, No.2 (870-877), ISSN: 0360-3199.

Klein, H.P. & Groll, M. (2004). Heat transfer characteristics of expanded graphite matrices in metal hydride beds. *Int. J Hydrogen Energy*, Vol.29, No.14 (1503-1511), ISSN: 0360-3199.

Kuznetsov, A.V.; Vafai, K. (1995). Analytical comparison and criteria for heat and mass transfer models in metal hydride packed beds. *Int. J Heat Mass Transfer*, Vol.38, No.15 (2873-2884), ISSN: 0017-9310.

Laurencelle, F & Goyette, J. (2007). Simulation of heat transfer in a metal hydride reactor with aluminium foam. *Int. J Hydrogen Energy*, Vol.32, No.14 (2957-2964), ISSN: 0360-3199.

Li, Q.; Lin, Q.; Chou, K.C. & Jiang, L.J. (2004). A study on the hydriding-dyhydriding kinetics of $Mg_{1.9}Al_{0.1}Ni$. *J Materials Science*, Vol.39, No.1 (61-65), ISSN: 1573-4803.

Lloyd, G.; Razani, A. & Feldman, K.T. Transitional reactor dynamics affecting optimization of a heat-driven metal hydride refrigerator. *Int. J Heat Mass Transfer*, Vol.41, No.3 (513-527), ISSN: 0017-9310.

Mat, M.D.; Kaplan, Y. & Aldas, K. (2002). Investigation of three-dimensional heat and mass transfer in a metal hydride reactor. *Int. J Energy Research*, Vol.26, No.11 (973-986), ISSN: 1099-114X.

Meng, X.Y.; Bai, F.F.; Yang, F.S.; Bao, Z.W. & Zhang, Z.X. (2010). Study of integrated metal hydrides heat pump and cascade utilization of liquefied natural gas cold energy recovery system. *Int. J Hydrogen Energy*, Vol.35, No.13 (7236-7245), ISSN: 0360-3199.

Murthukumar, P.; Prakash Maiya, M. & Srinivasa Murthy, S. (2005). Experiments on a metal hydride based hydrogen compressor. *Int. J Hydrogen Energy*, Vol.30, No.8 (879-892), ISSN: 0360-3199.

Nakagawa, T.; Inomata, A.; Aoki, H. & Miura, T. (2000). Numerical analysis of heat and mass transfer characteristics in the metal hydride bed. *Int. J Hydrogen Energy*, Vol.25, No.4 (339-350), ISSN: 0360-3199.

Nishizaki, T.; Miyamoto, K. & Yoshida, K. (1983). Coefficients of performance of hydride heat pumps. *J Less-Common Metals*, Vol.89, No.2 (559-566), ISSN: 0022-5088.

Paya, J.; Linder, M.; Laurien, E. & Corberan, J.M. (2009). Dynamic model and experimental results of a thermally driven metal hydride cooling system. *Int. J Hydrogen Energy*, Vol.34, No.7 (3173-3184), ISSN: 0360-3199.

Qin, F.; Chen, J.P.; Lu, M.Q.; Chen, Z.J.; Zhou, Y.M. & Yang, K. (2007). Development of a metal hydride refrigeration system as an exhaust gas-driven automobile air conditioner. *Renewable Energy*, Vol. 32, No.12 (2034-2052), ISSN: 0960-1481.

Ron, M. (1999). The normalized pressure dependence method for the evaluation of kinetic rates of metal hydride formation/decomposition. *J Alloys and Compounds*, Vol.283, No.1-2 (178-191), ISSN: 0925-8388.

Sanchez, A.R.; Klein, H.P.; Groll, M. (2003). Expanded graphite as heat transfer matrix in metal hydride beds. *Int. J Hydrogen Energy* , Vol.28, No.5 (515-527), ISSN: 0360-3199.

Sandrock, G. & Bowman, R.C. (2003). Gas-based hydride applications: recent progress and future needs. *J Alloys and Compounds*, Vol.356-357, No.1-2 (794-799), ISSN: 0925-8388.

Schweppe, F.; Martin, M.; Fromm, E. (1997). Model on hydride formation describing surface control, diffusion control and transition regions. *Journal of Alloys and Compounds*, Vol.261, No.1-2 (254-258), ISSN: 0925-8388.

Sun, D.W. & Deng, S.J. (1988). Study of the heat and mass transfer characteristics of metal hydride beds. *J. Less- Common Metals*, Vol.141, No.1 (37-43), ISSN: 0022-5088.

Sun, D.W. & Deng, S.J. (1990). Theoretical descriptions and experimental measurements on the effective thermal conductivity in metal hydride powder beds. *J Less-Common Metals*, Vol.160, No.2 (387-395), ISSN: 0022-5088.

Tao, W.Q. (2001). *Numerical heat transfer*(in Chinese, 2nd edition). Xi'an Jiaotong University Press, ISBN: 978-7-5605-1436-9, Xi'an, P.R.China.

Wang, Y.Q.; Yang, F.S.; Meng, X.Y.; Guo, Q.F.; Zhang, Z.X.; Park, I.S.; Kim, S.W. & Kim, K.J. (2010). Simulation study on the reaction process based single stage metal hydride thermal compressor. Int. J Hydrogen Energy, Vol.35, No.1 (321-328), ISSN: 0360-3199.

Yang, F.S.; Meng, X.Y.; Deng, J.Q.; Wang, Y.Q. & Zhang, Z.X. (2008). Identifying heat and mass transfer characteristics of metal- hydrogen reactor during adsorption - Parameter analysis and numerical study. Int. J Hydrogen Energy, Vol.33, No.3 (1014-1022), ISSN: 0360-3199.

Yang, F.S.; Meng, X.Y.; Deng, J.Q; Wang, Y.Q. & Zhang, Z.X. (2009). Identifying heat and mass transfer characteristics of metal hydride reactor during adsorption - Improved formulation about parameter analysis. Int. J Hydrogen Energy, Vol.34, No.4 (1852-1861), ISSN: 0360-3199.

Yang, F.S.; Wang, G.X.; Zhang, Z.X.; Meng, X.Y. & Rudolph, V. (2010). Design of the metal hydride reactors- A review on the key technical issues. Int. J Hydrogen Energy, Vol.35, No.8 (3832-3840), ISSN: 0360-3199.

Yang, F.S.; Wang, G.X.; Zhang, Z.X. & Rudolph, V. (2010). Investigation on the influences of heat transfer enhancement measures in a thermally driven metal hydride heat pump. Int. J Hydrogen Energy, Vol.35, No.18 (9725-9735), ISSN: 0360-3199.

25

Mass Transfer around Active Particles in Fluidized Beds

Fabrizio Scala
Istituto di Ricerche sulla Combustione - CNR
Italy

1. Introduction

Fluidized beds are extensively used in a number of gas-solid applications where significant heat and/or mass transfer rates are needed. The design and modelling of such processes requires the precise knowledge of the heat and mass transfer coefficients around immersed objects in the fluidized bed. Thus, it is not surprising that since the early spreading of the fluidized bed technology a considerable experimental and theoretical activity on this topic has been reported in the literature, mostly focused on the heat transfer coefficient. A more limited effort was dedicated to the estimation of the mass transfer coefficient, because of the inherent difficulty of measuring this quantity in the dense environment of a fluidized bed. Unfortunately, a fluidized bed is one of those cases where the analogy between heat and mass transfer does not hold, so that measured heat transfer coefficients cannot be used to estimate mass transfer rates under similar operating conditions. In fact, the bed particles represent an additional path to heat transfer around an immersed object, while they only result in a decrease of the available volume for gas mass transfer (except for the very particular case when the bed particles can adsorb one of the transferred components). Strictly speaking, an analogy exists between mass transfer and the gas-convective contribution to heat transfer in a fluidized bed. On the other hand, the particle-convective contribution to heat transfer (and also the radiative one, if relevant) has not an analogous mechanism in a mass transfer process.

In this review paper we will focus our attention to the mass transfer coefficient around freely moving active particles in the dense phase of a fluidized bed. This case represents most situations of practical interest, whereas the case of a fixed object (with respect to a reference system bound to the reactor walls) is less frequently encountered in mass transfer problems, contrary to the heat transfer case. With active particle we mean a particle that is exchanging mass with the gas phase, because either a chemical reaction or a physical process (phase change) is taking place in or at the surface of the particle. Finally, we will mainly consider the case of mass transfer between the gas and one or few active particles dispersed in a fluidized bed of inert particles, as opposed to the case where the entire bed is made of active particles. This configuration is important for a number of processes like combustion and gasification of carbon particles, and most typically the inert particle size is smaller than the active particle size.

In the next sections we will thoroughly review the experimental and theoretical work available in the literature on mass transfer in the dense phase of fluidized beds, showing the main achievements and the limitations for the estimation of the mass transfer coefficient. On the other hand, only few review papers addressing (partially) this topic have appeared in the literature (La Nauze, 1985; Agarwal & La Nauze, 1989; Ho, 2003; Yusuf et al., 2005), so that a more complete review of the previous literature available on mass transfer in fluidized beds is considered to be useful.

A convenient way to analyze and compare mass transfer data is to use the particle Sherwood number defined as $Sh = k_g \cdot d_a / D$. This quantity represents the average dimensionless gas concentration gradient of the transferring species at the active particle surface.

2. Mass transfer around isolated spheres in a gas flow

Before focusing on the dense phase of a fluidized bed we will briefly describe the mass transfer problem around an isolated sphere in a gas flow, as this is the starting point for further discussion on mass transfer in fluidized beds. This problem is relevant for particles or drops flowing in a diluted gas stream, like in spray-dry or entrained flow applications. It is important to note that in this case each particle moves isolated from the other particles and the analogy between heat and mass transfer processes around the particle is valid. An exact solution to the set of equations describing the boundary layer problem with mass and/or heat transfer around a sphere in a gas flow is not available, so that empirical or semi-empirical correlations are required to describe the experimental results.

Experimental data of mass and heat transfer coefficient for this system are mostly derived from evaporation of single liquid drops in a gas flow, due to the simplicity and accuracy in performing the measurements. In its pioneering experimental and theoretical work Frössling (1938) first proposed to correlate the mass transfer data (in the Reynolds number range of 2 to 1300) with the following expression derived by dimensional analysis:

$$Sh = 2.0 + K \cdot Re^{1/2} \cdot Sc^{1/3} \tag{1}$$

where $Re = \rho_g \cdot U \cdot d_a / \mu_g$, $Sc = \mu_g / D \cdot \rho_g$, and K is a constant, whose value was estimated to be 0.552. The first term on the right hand side represents mass transfer in stagnant conditions (diffusive term), while the second one accounts for the enhancement of mass transfer caused by the gas flowing around the particle (convective term). This expression is consistent with the theoretical requirement that Sh = 2 at Re = 0. It must be highlighted that the use of Eq. 1 (or similar ones) is based on the assumption that a steady boundary layer develops around the particle enabling the use of a steady-state mass transfer approach.

Ranz & Marshall (1952) used Eq. 1 to correlate both their own and previous mass and heat transfer data, and suggested a value K = 0.60 (for $0 < Re < 200$). Successively, Rowe et al. (1965) also correlated with Eq. 1 their own and others' data available to that date and obtained K = 0.69 (for $20 < Re < 2000$). This value of K is probably the most reliable one and with this value Eq. 1 is able to predict the heat and mass transfer data around an isolated sphere in a gas flow with a remarkable accuracy. Recently, Paterson & Hayhurst (2000) gave further theoretical background to this expression.

3. Mass transfer around active spheres in a fluidized bed: experimental data and correlations

In a fluidized bed the active particles are surrounded by a dense bed of inert particles and two different effects occur that influence the mass transfer process. First, the inert particles decrease the gas volume available for mass transfer around the active particle. Second, the presence of the fluidized particles alters the gas fluid-dynamics and the formation of the boundary layer around the active particle. These two effects must be taken into account when interpreting the experimental data.

It is obvious that the experimental technique based on the evaporation of liquid drops for the measurement of the mass transfer coefficient is not feasible in the dense phase of a fluidized bed. Different techniques have been actually used in fluidized beds and reported in the literature, and they mostly belong to three categories: sublimation of solid particles, liquid evaporation from porous particles and combustion of carbon particles. A fourth technique has been recently reported based on chemical reaction on the surface of catalyst spheres. In the following we will examine these four groups separately, indicating advantages and drawbacks of each technique. The only other works found in the literature using a technique not belonging to these four groups to estimate the mass transfer coefficient are those reported by Hsu & Molstad (1955) and by Richardson & Szekely (1961) who studied the adsorption of carbon tetrachloride by a fluidized bed entirely made of activated carbon granules. These early studies, however, showed limited success and will not be examined further.

3.1 Sublimation of solid particles

This technique is based on the determination of the sublimation rate of one or more solid particles in the fluidized bed by the measurement either of their weight change or of the concentration of the sublimating component in the gas phase. Calculation of the mass transfer coefficient requires the knowledge of the vapour pressure and of the diffusion coefficient of the sublimating component at the operating temperature. If appreciable heat effects are associated to the sublimation process, the active particle temperature must be either independently measured during the tests or estimated with a heat balance coupled to the mass balance around the particle.

Most of the experimental data obtained with this technique have been collected using naphthalene as the sublimating component. This substance is conveniently available, non-toxic, easily mouldable, and sublimates at low but detectable rates at temperatures close to the ambient one. Further advantages are the possibility to measure the naphthalene vapour concentration by means of a flame ionization or infrared analyzer, and the small heat effect, so that the active particle temperature can be safely assumed to be close to the bed temperature. This technique was first applied to fluidized beds by Resnick & White (1949) and Chu et al. (1953). These authors used shallow beds composed of all active particles. To extend the range of the studies to smaller particles and deeper beds without approaching saturation in the gas phase, van Heerden (1952) diluted few naphthalene spheres in a bed of carborundum, coke or fly ash particles. In examining the experimental results, this author noted that Sherwood numbers below the theoretical minimum of 2 were obtained at low Reynolds numbers. This result was explained by the reduced volume available for diffusion because of the presence of the inert particles, and the use of an effective diffusion coefficient through the bed interstices was suggested. Hsiung & Thodos (1977) diluted few naphthalene spheres in a bed of inert particles of the same size and density. The inert particles were beads of styrene divinylbenzene copolymer, which were claimed not to adsorb appreciably naphthalene vapour after an initial exposure. The experimental results were correlated by the following expression (rearranged here in terms of the Sherwood number):

$$Sh = \left(0.040 \cdot Re_{mf} + 2.12 \cdot Re_{mf}^{0.41} + 0.62/Re_{mf}^{0.51}\right) \cdot Sc^{1/3} \tag{2}$$

where $Re_{mf} = \rho_g \cdot U_{mf} \cdot d_a / \mu_g$ is the Reynolds number at the minimum fluidization condition. An interesting outcome of this investigation was that Sh appeared to be independent of the total fluidization velocity, but only depended on the minimum fluidization velocity and on the particle size (and also on the gas properties). Prins et al. (1985) and de Kok et al. (1986) extended the investigation to the case where the inert particle size was different from that of the active particle. In particular, the size ratio between the active and the inert particles was varied in the range 3-200. These authors pointed out at two possible problems arising with the use of this technique: the risk of mechanical attrition of the active particles and of adsorption of naphthalene vapour by the bed material. Both these two problems would lead to an enhancement of the apparent mass transfer rate and, in turn, to an overestimation of the mass transfer coefficient. Careful selection of the operating conditions ($T \approx 65°C$) and of the inert material (glass beads and dense alumina) could minimize these problems. The experimental results were in line with results of Hsiung & Thodos (1977). In particular, no effect of the fluidizing velocity on the mass transfer coefficient was observed over the complete range of bed materials. The experimental data of this work and those of Hsiung & Thodos (1977) were correlated by the following empirical expression (rearranged here in terms of the Sherwood number):

$$Sh = \left(0.105 + 1.505 \cdot (d_a/d_i)^{-1.05}\right) \cdot (d_a/d_i) \cdot \left[(1-\varepsilon_{mf})/\varepsilon_{mf}\right] \cdot Re_{mfi}^{1-m} \cdot Sc^{1/3} \tag{3}$$

where $m = 0.35 + 0.29 \cdot (d_a/d_i)^{-0.50}$, and the Reynolds number at minimum fluidization was referred to the inert particle diameter $Re_{mfi} = \rho_g \cdot U_{mf} \cdot d_i / (1-\varepsilon_{mf}) \cdot \mu_g$. Coelho & Guedes de Carvalho (1988) made similar experiments with somewhat heavier naphthalene particles, and substantially confirmed results by Prins et al. (1985).

Pal'chenok & Tamarin (1985) and Tamarin et al. (1985) measured the mass transfer coefficients of single particles composed of a naphthalene spherical shell placed over a core made of different materials in order to vary the total density of the active particles. Bed materials of different size and density were used (glass spheres, corundum and fire clay). Experimental results indicated that for beds of large particles ($d_i > 0.5$ mm) the mass transfer coefficient was independent of the fluidization velocity. However, in the beds of fine particles the mass transfer coefficient slightly increased with the fluidization velocity and then levelled off at a maximum value. These authors proposed the following correlation for the maximum mass transfer coefficient:

$$Sh = 0.117 \cdot (d_i/d_a)^{0.13} \cdot (\rho_i/\rho_a)^{-0.15} \cdot Ar_i^{0.39} \cdot Sc^{1/3} \tag{4}$$

where the Archimedes number of the inert particles is $Ar_i = (\rho_i - \rho_g) \cdot \rho_g \cdot g \cdot d_i^3 / \mu_g^2$. Successively, Palchonok et al. (1992) suggested a new correlation valid over a wider range of particle sizes, for the special case $d_i = d_a$:

$$Sh = 2.0 \cdot \varepsilon_{mf} + 0.117 \cdot Ar_i^{0.39} \cdot Sc^{1/3} \tag{5}$$

A more general correlation was also presented by Palchonok (1998):

$$Sh = \left[\left(0.009 \cdot Ar_i^{0.5}\right) + \left(d_i/d_a\right)^{2/3} \cdot \left(2.0 \cdot \varepsilon_{mf} + 0.117 \cdot Ar_i^{0.39} - 0.009 \cdot Ar_i^{0.5}\right)\right] \cdot Sc^{1/3} \quad (6)$$

Joulié et al. (1986), Joulié & Rios (1993) and Joulié et al. (1997) carried out experiments of sublimation of naphthalene spheres with inert particles of two different densities (sand and zeolite). In their calculations they took into account the heat of sublimation and the difference between the active particle surface temperature and that of the bed. The experimental data were correlated using the following empirical expression:

$$Sh = 0.165 \cdot Ar_i^{0.55} \cdot \left(d_a/d_i\right)^{0.81} \cdot \left(\rho_a/\rho_i\right)^{2.70} \cdot \left[\left(U_{opt} - U_{mf}\right)/U_{mf}\right]^{1.91} \quad (7)$$

where U_{opt} represents the optimal gas velocity at which the maximum sublimation rate is found (U_{opt} was approximately 2 ~ 3 times U_{mf}). All the experiments correlated by Eq. 7 were carried out at $U = U_{opt}$. These authors, however, did not check the absence of adsorption effects by the bed materials used in their experiments. Oka et al. (1995) also carried out experiments of sublimation of naphthalene spheres in a fluidized bed of inert particles (corundum), and proposed a Frössling-type correlation:

$$Sh = 2.0 \cdot \varepsilon_{mf} + 0.97 \cdot Re^{0.60} \cdot Sc^{1/3} \quad (8)$$

The range of variation of the fluidization velocity and of the inert particle size in their experiments was rather small. This was true also for data reported by Donsì et al. (1998 and 2000), who found that the mass transfer coefficient was unaffected by the fluidization velocity for a particular bed particle size. They, however, correlated their data with a Frössling-type expression:

$$Sh = 2.0 + a \cdot \left(Re/\varepsilon\right)^b \cdot Sc^{1/3} \quad (9)$$

which predicts a significant influence of the fluidization velocity. In Eq. 9, a = 0.0086 and b = 1.34 according to Donsì et al. (1998), while a = 0.707 and b = 0.5 according to Donsì et al. (2000).

On the whole, the naphthalene sublimation technique appears to be fairly accurate in measuring the mass transfer coefficient of active particles in a fluidized bed, provided experiments are carefully performed. The main sources of inaccuracy are: mechanical attrition of the naphthalene particles; adsorption of naphthalene vapour by the inert solids; errors in the measurement of the mass loss of naphthalene particles (due to extraction, weighting and re-injection of the particles in the bed); errors in the measurements of the naphthalene vapour concentration in the gas (discontinuous or inaccurate analyzers); errors in the estimation of the vapour pressure and diffusion coefficient of naphthalene; incorrectness of the assumptions for the bulk gas naphthalene vapour concentration and for the naphthalene particle surface temperature; change of particle size during the sublimation process.

Other solid substances have also been proposed as an alternative sublimating component, like p-dichlorobenzene (Riccetti & Thodos, 1961), ice (Joulié et al., 1986) and dry ice (Schlichthaerle & Werther, 2000). The use of these substances, however, appears to be more complicated than naphthalene. Besides the previously reported inaccuracies, in fact, these substances are more difficult to mould, and sublimation is often accompanied by appreciable heat effects.

3.2 Liquid evaporation from porous particles

This technique is based on the measurement of the evaporation rate of a liquid from the surface of one or more porous particles in the fluidized bed. As for the previous technique, calculation of the mass transfer coefficient requires the knowledge of the vapour pressure and of the diffusion coefficient of the evaporating component at the operating temperature. In addition, if appreciable heat effects are associated to the evaporation process, a heat balance must be coupled to the mass balance (or particle temperature measured independently), complicating the calculation of the mass transfer coefficient.

Kettenring et al. (1950), Riccetti & Thodos (1961), Bradshaw & Myers (1963), Petrovic & Thodos (1966, 1967), Wilkins & Thodos (1969), Yoon & Thodos (1972), Schlünder (1977) and Ciesielczyk (1996) studied the mass transfer coefficient in fluidized bed composed of all active particles, by means of porous particles filled either with water, nitrobenzene, n-decane, n-dodecane or n-tetradecane. Ziegler & Brazelton (1964) and Ziegler & Holmes (1966) studied the evaporation of water from a single fixed sphere immersed in a fluidized bed of seven different materials. Recently Tsotsas (1994a and 1994b) studied the evaporation of water from fluidized beds of all aluminium silicate spheres. This author pointed out that because of the porous nature of the particles particular attention should be paid for the additional contribution of particle-side kinetics, especially for the second drying period (falling-rate), as opposed to the first drying period (constant-rate) when the particle surface is still wet. Kozanoglu et al. (2001) analyzed the evaporation of water at reduced pressure in fluidized beds of silica gel or millet particles.

Up to now the only works available in the literature where this technique was applied to few freely moving active particles within a fluidized bed of inert particles are those reported by Vanderschuren & Delvosalle (1980) and Delvosalle & Vanderschuren (1985), and by Cobbinah et al. (1987). The first authors estimated with the aid of a simple model the mass transfer coefficient in beds of refractory silica and alumina particles, where wet particles were dried in beds of dry particles of the same material and size. Interestingly, the mass transfer coefficient was found not to vary with the fluidization velocity, but only with the particle nature and size. Cobbinah et al. (1987), instead, dried wet coarse alumina spheres within fluidized beds of fine sand. These authors also found that the mass transfer coefficient was independent of the fluidization velocity. However, this happened only for fluidization velocities large enough so that no segregation of the coarse and fine particles occurred.

The advantages of this technique are the availability of cheap particles and liquids, the easy measurement of the gas concentration of the evaporating component and the constancy of the active particle diameter with time. On the other hand, apart from possible inaccuracies in measurements and in parameters evaluation, several severe drawbacks are present for a practical application. First, the heat balance must be typically solved together with the mass balance or the particle temperature must be measured as a consequence of the non-negligible heat of evaporation. Second, internal resistances to mass transfer in a porous particle may add serious errors in the estimation of the external mass transfer coefficient, if they are not properly taken into account. Third, as it was the case for the sublimation technique, the evaporating component can be adsorbed over the inert bed particles. Finally, the presence of a liquid layer on the surface of the active particles may disturb smooth fluidization by the action of capillary forces.

3.3 Combustion of carbon particles

This technique is based on the determination of the combustion rate of one or more carbon particles burning in a fluidized bed. This can be accomplished by the measurement of the weight change of the carbon particles and/or of the concentration of CO_2 and CO in the gas phase during combustion. The assumption is made that the carbon particle temperature and size are large enough so that the combustion rate is controlled by external mass transfer of O_2 towards the carbon surface. Alternatively, the intrinsic carbon reactivity and intraparticle mass transfer resistance must be properly considered in the calculations and separately quantified. Carbon particles which leave after combustion a coherent ash layer should be avoided for the experiments, as this would add a further resistance to mass transfer of O_2. Calculation of the mass transfer coefficient requires the knowledge of the diffusion coefficient of O_2 at the operating temperature. In addition, the heat balance must be solved together with the mass balance (or the particle temperature must be measured during the experiment) as a consequence of the appreciable heat effects of combustion. A significant additional complication is associated with the two following still unsolved questions: what are the primary combustion products at the carbon surface, CO_2, CO or both of them and in which relative proportion? Where is CO oxidized to CO_2, nearby or far from the carbon particle? These two issues affect both the mass transfer coefficient and the particle temperature (a large fraction of the total heat release may be associated with CO combustion). These issues have been reviewed in detail by La Nauze (1985), Prins (1987), and Agarwal & La Nauze (1989).

The first attempts to estimate the mass transfer coefficient during the fluidized bed combustion of carbon particles were reported by Avedesian & Davidson (1973), Chakraborty & Howard (1981), Pillai (1981), Tamarin et al. (1982) and Ross & Davidson (1982). These authors used the carbon particle burn-out time data to determine the particle Sherwood number in their experiments. As noted by La Nauze (1985), this method provides only the average Sherwood number during the burning time and does not allow any variation of Sh with the particle diameter. Further, the data analysis was based on a number of questionable assumptions regarding the carbon chemical kinetics, the CO oxidation reaction and the particle temperature. Avedesian & Davidson (1973), on the basis of their experimental results and of the estimation of the active particle Reynolds number (with the gas characteristic velocity in the dense phase assumed to be equal to U_{mf}), suggested that convective effects can be assumed to be negligible and that Sh for the burning particle should be close to its lower asymptotic value, i.e. at stagnant conditions. The authors proposed that in these conditions the limiting value of 2 should be multiplied by the bed voidage in the dense phase to account for the presence of the inert particles. If the two-phase fluidization theory (Davidson & Harrison, 1963) is assumed to hold, then $Sh = 2 \bullet \varepsilon_{mf}$. This result would be valid only in the case that the inert particle size is smaller than the active particle one. Basu et al. (1975) and Basu (1985) suggested that this value should be refined on account of the change of the local voidage with the radial distance from the active particle surface. This would result in a slightly larger value of Sh in stagnant conditions, because the voidage increases from the average bed value ε_{mf} far from the carbon particle to unity at the particle surface. Agarwal et al. (1988a), however, noted that given the poor accuracy of existing experimental data and the uncertainty in the estimation of physical properties, this refinement would be difficult to be verified. Chakraborty & Howard (1981) and Pillai (1981) proposed respectively the two Frössling-type correlations:

$$Sh = 2.0 \cdot \varepsilon + 0.69 \cdot Re^{1/2} \cdot Sc^{1/3} \tag{10}$$

$$Sh = \varepsilon \cdot \left(2.0 + 0.69 \cdot Re^{1/2} \cdot Sc^{1/3}\right) \tag{11}$$

They basically modified Eq. 1 (using the value suggested by Rowe et al. (1965) for the constant K) by introducing the mean bed voidage to account for the shielding effect of the particulate phase surrounding the burning particle. The arbitrary choice of the average bed quantities ε and Re, instead of the dense phase ones ε_{mf} and Re_{mf}, was not justified by the authors nor validated by their experimental data.

The first work in which the mass transfer coefficient was determined from combustion rate data was that reported by La Nauze & Jung (1982, 1983a) and Jung & La Nauze (1983). These authors performed experiments where single petroleum coke particles were burned in a fluidized bed and the mass loss of carbon was measured as a function of time. An independent measure of the chemical rate coefficient and of the particle temperature was available. Experimental results showed that particles with a size larger than 3 mm burned under external diffusion control and closely followed the shrinking sphere model. The authors correlated their results (assuming that no CO escaped the particle boundary layer) with a modification of Eq. 10:

$$Sh = 2.0 \cdot \varepsilon + 0.69 \cdot \left(Re/\varepsilon\right)^{1/2} \cdot Sc^{1/3} \tag{12}$$

The authors considered U/ε rather than U as the characteristic gas velocity relevant for convection around the burning particle. This correlation still predicts a large influence of the fluidization velocity, which however was not demonstrated in their experiments. In addition, the effect of the bed inert particle size was not taken into account. In a later paper La Nauze & Jung (1983b) suggested to change the first term from $2 \bullet \varepsilon$ to $2 \bullet \varepsilon_{mf}$, on the basis of theoretical considerations. Additional experiments were also presented by La Nauze & Jung (1985, 1986), in which the oxygen-nitrogen fluidizing gas was substituted with an oxygen-helium mixture. In this condition the oxygen diffusivity was increased by a factor of about 2.5 times and, correspondingly, the mass transfer coefficient was found to be around two times larger.

Prins (1987) reported a comprehensive work where the burning rate of single graphite spheres in a fluidized bed was measured by following the CO_2 and CO concentrations in the flue gas. The author carefully considered and measured the effect of the burning particle temperature, CO/CO_2 primary ratio and CO oxidation, intraparticle oxygen penetration, and attrition, on the particle combustion rate. On the basis of the experimental results, it was concluded that a previously proposed correlation (Eq. 3) was able to correctly predict the mass transfer coefficient for the burning particle.

Guedes de Carvalho et al. (1991) measured with the same technique the burning rate of batches of coke or char particles in fluidized beds of sand or Pt doped catalyst beads at high temperature and at different pressures. In these conditions the combustion rate of carbon was controlled by external diffusion, and the intrinsic kinetics and intraparticle diffusion resistances were neglected. The particle temperature was not measured, but assumed to be 100 K above that of the bed. Experimental results indicated that the mass transfer coefficient was independent of the fluidization velocity. Comparing the combustion rates measured alternatively with beds of sand and of catalyst beads, the authors interpreted the experimental data as an evidence of CO being the only primary product of carbon oxidation, followed by CO oxidation outside the particle boundary layer. On this basis, previously published mass transfer coefficient data were questioned and re-interpreted under the above assumption.

Salatino et al. (1998) measured the burning rate of single coal and biomass char particles in a fluidized bed of sand. The mass transfer coefficient was estimated on the basis of a simplified particle combustion model. The experimental results showed that Sh was not influenced by the fluidization velocity and oxygen inlet concentration, but only by the active and the inert particle size. The authors stressed the possible influence on the apparent burning rate of attrition and fragmentation of the char particle combined with fines post-combustion within the bed. Sh can be overestimated up to a factor of two (for porous and fragile chars, like those resulting from biomass devolatilization) if these phenomena are not taken into account. In a later paper, Scala et al. (2006) highlighted the significance of the shape of non-spherical particles on the correct evaluation of the exposed particle surface and of the mass transfer coefficient.

Paterson (2000) and Hayhurst (2000) noted that the implicit assumption of equimolar counter-diffusion of gaseous reactants and products around an active particle (which is typically made for the calculation of the mass transfer coefficient from experimental data) might not always be valid. For example, if CO is the only primary product of carbon combustion and it is further oxidized away from the carbon particle, assuming equimolar counter-diffusion (i.e. neglecting the Stefan flow) would lead to a 10 - 20% underestimation of the real mass transfer coefficient. The only case when the equimolar counter-diffusion approach can be safely used independently of the reaction stoichiometry is the case of very dilute gaseous reactant concentration. However, in this analysis it was assumed that the problem could be schematized as a pseudo-binary system and the presence of the other relevant gaseous species was not taken into account. Recently, Scala (2010a) considered the mass transfer coefficient for oxygen reacting with a spherical carbon particle in an atmosphere of O_2, N_2, CO_2, CO, and H_2O and analytically solved the complete set of Stefan-Maxwell equations under the assumption of negligible homogeneous reaction in the boundary layer. Results showed that under typical combustion conditions the use of the equimolar counter-diffusion mass transfer coefficient can lead to errors up to 10%.

Hayhurst & Parmar (2002) measured the burning rate of single freely moving graphite spheres and coal char particles in a fluidized bed of sand by following the CO_2 and CO concentrations in the flue gas. At the same time, the particle temperature was measured by a very thin and flexible thermocouple. The CO/CO_2 ratio as primary combustion products at the carbon surface was also estimated. The experimental results were quite scattered, and showed a slight decrease of Sh with the bed temperature (explained by the change of several physical properties of the system) and no clear trend with the fluidization velocity. The increase of Sh with the sand particle size was mainly attributed to the change of the minimum fluidization velocity. The experimental Sh data were correlated by the Frössling-type expression:

$$Sh = 2.0 \cdot \varepsilon_{mf} + 0.61 \cdot Re_p^{0.48} \cdot Sc^{1/3} \tag{13}$$

where $Re_p = \rho_g \cdot U_p \cdot d_a / \mu_g$, and U_p is the gas velocity in the particulate phase that was estimated from expressions available in the literature. Alternatively, the authors suggested as a good approximation a modification of Eq. 12, where the bed values ε and Re are substituted by the dense phase ones ε_{mf} and Re_{mf}:

$$Sh = 2.0 \cdot \varepsilon_{mf} + 0.69 \cdot \left(Re_{mf}/\varepsilon_{mf}\right)^{1/2} \cdot Sc^{1/3} \tag{14}$$

More recently Dennis et al. (2006) showed that the interpretation of their own and of previous combustion rate measurements can suggest both a linear and a square root dependence of Sh from the active particle diameter, because of the scatter of data. The authors questioned the latter dependence (which is consistent with a Frössling-type expression) and preferred a linear one. This choice was justified by invoking the presence of a gas cushion underneath the active particle and a heap of defluidized sand resting on its upper surface.

Scala (2009) recently proposed a new indirect experimental technique to measure the product CO/CO_2 ratio at the surface of spherical coal char particles during fluidized bed combustion, based on the measurement of the burning rate of a single particle under low oxygen concentration conditions. In these conditions two advantages are obtained: the boundary layer mass transfer coefficient can be calculated without the need to account for high mass transfer rate and/or possible non-equimolar counter-diffusion corrections; heat effects are very limited and the char particle temperature can be assumed to be approximately equal to that of the bed. Scala (2010b) extended this technique and measured the primary CO/CO_2 ratio at the surface of coal char particles at different bed temperatures, oxygen concentrations, fluidization velocities, and inert bed particle sizes. In addition, the actual sphericity and temperature of the particle, as independently measured during the experiments, were taken into account in analyzing the data. Results showed that under all the experimental conditions investigated carbon was completely oxidized to CO_2 within the particle boundary layer. The experiments confirmed that the char particles burned under boundary layer diffusion control in the temperature range 800-900°C. The author concluded that single particle burning rate experiments can be used to estimate the particle Sherwood number in fluidized beds only if a high-reactivity fuel is used and if attrition can be assumed to be negligible. In fact, significant carbon attrition during the experiments would lead to a fictitious increase of the measured apparent particle Sherwood number.

In summary, despite the large amount of experimental data gained with the use of this technique, still some uncertainties exist on the accuracy of the estimated mass transfer coefficient values. Apart from possible errors in the measurements and in the parameters evaluation, the major limitations in the use of this technique are: the possible influence of intrinsic kinetics and intraparticle diffusion in the porous carbon on the overall combustion rate and on the particle conversion pattern; the correct evaluation of the carbon particle temperature; the assumption of the CO/CO_2 primary ratio and of the CO oxidation location; the influence of attrition and fragmentation of the carbon particles on the apparent combustion rate and on the particle number and size evolution; the change of carbon particle size with time by combustion; the use of non-spherical carbon particles.

3.4 Chemical reaction on the surface of catalyst particles

Recently, Venderbosch et al. (1998) proposed the oxidation reaction of CO with oxygen over a Pt catalyst as a model reaction to be used for the determination of mass transfer coefficients in gas-solid systems. It was shown with experiments in a fixed bed and in a riser that this reaction is completely mass transfer controlled in the temperature range 450-500°C, even for very small active particles (< 100 μm) at low CO bulk concentrations. The authors highlighted a number of advantages of this technique like the easy preparation and stability of the catalyst particles and the absence of side reactions. They, however, did not apply the technique to fluidized beds.

Basically, this technique is based on the determination of the CO combustion rate on the surface of one or more catalyst particles dispersed in a fluidized bed. This can be accomplished by measuring the concentration of CO_2 and/or CO in the gas phase. Calculation of the mass transfer coefficient requires the knowledge of the diffusion coefficient of CO at the operating temperature. In addition, if high CO concentrations are used the heat balance must be solved together with the mass balance (or the particle temperature must be measured during the experiments) as a consequence of the appreciable heat effects of combustion.

Scala (2007) applied this technique to measure the mass transfer coefficient around freely moving active particles under bubbling/slugging fluidized bed conditions in a lab-scale reactor. In this work the mass transfer coefficient around one or few attrition-resistant Pt catalyst spheres immersed in an inert bed of sand was measured by following the CO oxidation reaction at 450°C at different fluidization velocities, catalyst sphere sizes and inert bed particle sizes. The experiments were performed using very low CO inlet concentrations (< 1000 ppm). In these conditions heat effects were negligible and the catalyst particle temperature could be assumed to be equal to that of the bed. Experimental results showed that Sh is not influenced by the fluidization velocity and by a change of regime from bubbling to slugging, whereas it increases with a square root dependence with the minimum fluidization velocity and with the active particle size. These results strongly suggest that the active particles only reside in the dense phase and never enter the bubble/slug phase. In addition, the reported 1/2 dependence of Sh on both U_{mf} and d_a is a double (independent) evidence of the applicability of the Frössling-type analysis to active particles dispersed in a fluidized bed. In particular, experimental data were excellently fitted by the following correlation (similar to Eq. 14, but with a slightly different coefficient):

$$\mathrm{Sh} = 2.0 \cdot \varepsilon_{mf} + 0.70 \cdot \left(\mathrm{Re}_{mf}/\varepsilon_{mf}\right)^{1/2} \cdot \mathrm{Sc}^{1/3} \tag{15}$$

A further interesting outcome of this work was that mass transfer around active particles in a fluidized bed of inert particles can be safely studied in a small lab-scale apparatus, since it appears not to be influenced by the bubble/slug fluid-dynamics, provided the local dense phase conditions are the same.

On the whole, this technique appears to be suitable for the accurate determination of mass transfer coefficients in fluidized beds in the temperature range 450-500°C. In fact, it allows to overcome most of the difficulties and uncertainties associated with other available techniques. The following advantages over the other techniques can be identified: a) only the heterogeneous oxidation reaction $CO + 1/2 O_2 = CO_2$ at the catalyst particle surface is active and no parallel homogeneous or side reactions are present; b) the reaction is completely controlled by external mass transfer and no influence of intrinsic kinetics or intraparticle diffusion is present; c) at low CO concentrations heat effects are negligible and the catalyst particle temperature can be assumed to be equal to that of the bed; d) the CO conversion degree can be easily and accurately calculated by measuring CO and/or CO_2

concentrations at the reactor outlet; e) both CO and CO_2 are not adsorbed significantly by typical inert bed materials; f) at low CO concentrations, the Sherwood number can be calculated without the need to account for high mass transfer rate and/or non-equimolar counter-diffusion corrections; g) the Pt catalyst can be easily prepared and is very stable with respect to the oxidation reaction; h) the catalyst particles can be made with a spherical shape and keep a constant size during the experiments; i) by careful selection of the catalyst support material the particle attrition rate in the fluidized bed can be made negligible; j) properties of the different gaseous species and solid particles can be easily and accurately estimated. On the contrary, apart from possible inaccuracies in measurements and in parameters evaluation, the most critical issue is the selection of an attrition-resistant catalyst support. The occurrence of significant catalyst attrition would lead to an overestimation of the Sherwood number because of the continuous generation of new catalytic surface within the bed. Finally, if high CO concentration are used in the experiments, the catalyst particle temperature should be measured or estimated, and the high mass transfer rate and/or non-equimolar counter-diffusion corrections should be taken into account.

4. Mass transfer around active spheres in a fluidized bed: modelling studies

Contrary to the large amount of experimental work, only few theoretical investigations on mass transfer of active particles in fluidized beds of inert particles have appeared in the literature.

Tamarin (1982) applied the steady-state boundary layer theory for flow past a sphere to describe mass transfer to an active particle in a fluidized bed of inert particles. The average velocity gradient and the average tangential stress at the particle surface were determined with suitable simplifying assumptions. Correction to the gas velocity near the active particle was introduced to account for fluctuations of the impingement velocity and possible stagnant zones near the points of contact of adjacent particles. By relating the average stress to the particle weight, the following expression was obtained:

$$Sh = 0.248 \cdot \left(d_a / d_i\right)^{1/2} \cdot Ar_i^{1/3} \cdot Sc^{1/3} \qquad (16)$$

La Nauze & Jung (1983b) and La Nauze et al. (1984) questioned the suitability of a steady-state approach to describe mass transfer around an active particle in a fluidized bed. They proposed an unsteady-state model where mass transfer in the dense phase was assumed to occur because of a gas convective component (due to gas percolating through the bed at minimum fluidization condition) and a particle convective component (due to packets of bed particles coming into contact with the active surface, and whose motion is induced by the bubbles). Solution of the unsteady equation with a suitable estimation of the gas renewal frequency at the surface of the particle gave:

$$Sh = 2 \cdot \varepsilon_{mf} + \left[\left(4 \cdot \varepsilon_{mf} \cdot d_a / \pi \cdot D\right) \cdot \left(U_{mf} / \varepsilon_{mf} + U_b\right)\right]^{1/2} \qquad (17)$$

The authors also suggested that when the active particle size becomes comparable to the inert particle size the particle convective component can be neglected and the term U_b can be dropped from Eq. 17. Guedes de Carvalho & Coelho (1986) noted that in the derivation of Eq. 17 La Nauze and coworkers should have used the molecular diffusivity instead of the effective diffusivity, so that the correct equation should read:

$$Sh = 2 \cdot \varepsilon_{mf} + \varepsilon_{mf} \cdot \left[\left(4 \cdot d_a / \pi \cdot D \right) \cdot \left(U_{mf} / \varepsilon_{mf} + U_b \right) \right]^{1/2} \tag{18}$$

They also noted that in this formulation the gas renewal frequency by packets was likely to be overestimated, because a continuous stream of bubbles near the active particle was assumed. This last point was partly addressed in an alternative formulation by La Nauze & Jung (1985, 1986), where the particle convective gas renewal frequency was directly related to the bubble frequency. By matching the model with experimental data a proportionality parameter was estimated giving:

$$Sh = 2 \cdot \varepsilon_{mf} + \left[\left(4 \cdot \varepsilon_{mf} \cdot d_a / \pi \cdot D \right) \cdot \left(U_{mf} / \varepsilon_{mf} + 6.93 \cdot U_b \cdot \varepsilon_b \cdot d_a / d_b \right) \right]^{1/2} \tag{19}$$

It must be noted, however, that the points raised by Guedes de Carvalho & Coelho (1986) still apply to this later model formulation.

Coelho & Guedes de Carvalho (1988) presented a steady-state model of mass transfer around a large active particle in a packed or fluidized bed of smaller particles. A continuum approach was adopted to describe the flowfield and a potential flow solution was obtained. The assumption was made that no hydrodynamic boundary layer establishes around the particle. Transverse dispersion was considered to be more suitable to describe mass transfer rather than molecular diffusion. An analytical solution was obtained for the particular case of a thin concentration boundary layer around the sphere, with suitable simplifying assumptions. A numerical solution was necessary for the more general equation, and an empirical approximation was derived by matching the numerical model results and bridging the two exact solutions for the stagnant and thin boundary layer cases. For a fluidized bed, assuming that the active particle resides only in the dense phase, the expression becomes (Guedes de Carvalho et al., 1991):

$$Sh = \varepsilon_{mf} / \tau \cdot \left[4 + 0.576 \cdot Pe^{0.78} + 1.28 \cdot Pe + 0.141 \cdot \left(d_i / d_a \right) \cdot Pe^2 \right]^{1/2} \tag{20}$$

where the Peclet number was defined as $Pe = U_{mf} \cdot d_a \cdot \tau / \varepsilon_{mf} \cdot D$, and τ represents the bed tortuosity. In a later paper, Guedes de Carvalho & Alves (1999) examined the more general case where longitudinal dispersion is non-negligible, under the same assumptions of the previous model. The numerical solution was again approximated by an empirical expression representing the product of the solution for advection plus molecular diffusion and the enhancement brought about by convective dispersion. For a fluidized bed the expression becomes:

$$Sh = \varepsilon_{mf} / \tau \cdot \left(4 + 4/5 \cdot Pe^{2/3} + 4/\pi \cdot Pe \right)^{1/2} \cdot \left[1 + 1/9 \cdot \left(d_i / d_a \right) \cdot Pe \right]^{1/2} \tag{21}$$

All the models presented so far assume that the active particle resides only in the dense phase of the fluidized bed. Agarwal (1987) questioned this assumption and, basing on the particle movement patterns under the action of the raising bubbles, developed a model to estimate the probability (p) of the particles being in the dense phase. The author claimed that up to 20% of the life-time of the active particle may be associated to the bubble phase, depending on the density of the particle. On the basis of this particle circulation model Agarwal et al. (1988a, 1988b) developed a mass transfer model for a large active particle in a bed of smaller inert particles. They assumed that the active particle resides alternately in the bubble and emulsion phases, so that:

$$Sh = p \cdot Sh_e + (1-p) \cdot Sh_b \quad (22)$$

where the relevant Sherwood numbers in the emulsion and bubble phases were estimated from a boundary layer model as:

$$Sh_e = K_p \cdot (C_D/8)^{1/3} \cdot (\tau/\varepsilon_{mf} \cdot Re_{mf})^{2/3} \cdot Sc^{1/3} \quad (23)$$

$$Sh_b = 2 + K_p \cdot (C_D/8)^{1/3} \cdot Re_b^{2/3} \cdot Sc^{1/3} \quad (24)$$

where $Re_b = \rho_g \cdot 3 \cdot U_{mf} \cdot d_a / \mu_g$ and $K_p = 0.69 \cdot \left[(1+Re \cdot Sc)^{1/3} - 1\right] / (Re \cdot Sc)^{1/3}$. The drag coefficient C_D and the parameter K_p were evaluated using Re_{mf} and Re_b respectively in the emulsion and bubble phase. Though the assumption underlying this model is somewhat fascinating, it must be noted that no direct evidence exists that (non-gas-emitting) active particles enter the bubble phase during their motion in the fluidized bed.

5. Comparison of available correlations with experimental data

The available empirical/theoretical correlations reported in the previous sections were checked for their accuracy in predicting the experimental mass transfer data of Scala (2007), which we consider to be the most complete and accurate reported in the literature to date. As a preliminary screening, all the correlations which predict a (direct or indirect) influence of the total fluidization velocity on Sh were discarded (Eqs 8-13, 17-19), as they are incompatible with the experimental data. Moreover, Eq. 5 predicts no influence of the active particle size on Sh and Eqs 4 and 6 predict a decrease of Sh with the active particle size, contrary to the experimental data. These correlations were also discarded. Finally, eq. 22 is based on the assumption that the active particle resides partly in the bubble phase, contrary to the experimental evidence, and therefore was not considered further.

Figures 1 and 2 report the comparison of the remaining correlations with the experimental data. As regards the empirical and semi-empirical correlations (Figs 1A and 2A), Eq. 15 excellently fits the experimental data. It is noted that Eq. 14 gives practically the same results as Eq. 15, and fits equally well the experimental Sh data. The purely empirical correlation by Prins et al. (1985), Eq. 3, fits very well the experimental Sh data at varying active particle size (Fig. 2A), but fits worse the Sh data at varying inert particle size (Fig. 1A). It must be noted, however, that for $d_i > 700$ μm the experiments are outside the range of validity of Eq. 3, as reported by the authors in their paper. Eqs 2 (Hsiung & Thodos, 1977) and 7 (Joulié et al., 1997) both significantly overpredict the Sherwood number. In Eq. 7 we have conservatively considered $U_{opt} = 2 \bullet U_{mf}$. Should the ratio U_{opt} / U_{mf} be > 2, the discrepancy of the predicted Sh with the experimental data would have been even larger. It is also worth to note that in Eq. 7 (but also in Eq. 4) the active particle density is explicitly present. In the experiments performed by Scala (2007) this variable was varied in a relatively limited range ($1050 < \rho_a < 1950$ kg/m^3), and within this range it appeared not to influence significantly Sh. It is our opinion that if particle segregation is avoided, the active particle density has no importance on mass transfer, but this speculation needs further experimental confirmation.

In Figs 1B and 2B the available theoretical correlations are compared with the experimental data. Equation 16 (Tamarin, 1982) significantly overpredicts the data. The two models by Guedes de Carvalho and coworkers (Eqs 20 and 21) give practically the same results and underpredict the data, except for the largest inert particle sizes. In these equations a value of $\tau = 1.414$ was used as suggested by the authors. A change of the bed tortuosity value in the range $1.0 < \tau < 1.414$, however, gave only a very limited benefit to the comparison with the experimental data. On the whole, it appears that no available fluidized bed mass transfer theoretical model is able to correctly predict the Sherwood number.

6. Conclusions

Four experimental techniques have been used so far to measure the mass transfer coefficient around active particles dispersed in a fluidized bed. The technique based on liquid evaporation from porous particles appears to be affected by severe limitations and its use is not recommended. The other three techniques, namely sublimation of solid particles, combustion of carbon particles, and chemical reaction on the surface of catalyst particles, all appear to be suitable for measuring Sh, provided experiments are carefully performed.

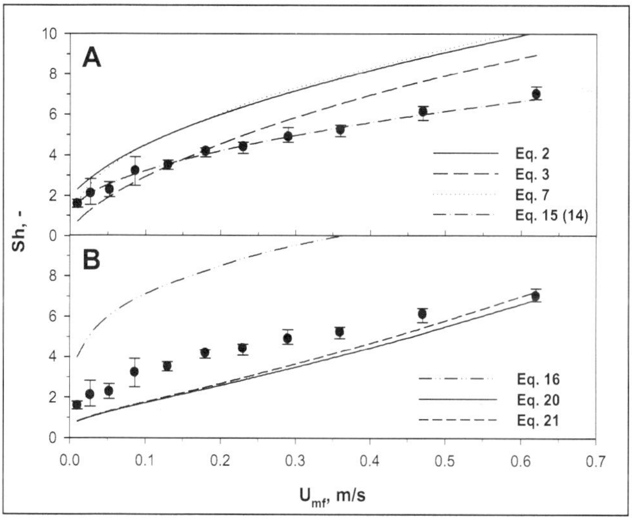

Fig. 1. Comparison of Sherwood number correlations with experimental data taken from Scala (2007). A) Empirical and semi-empirical correlations. B) Theoretical correlations. Average Sherwood number as a function of minimum fluidization velocity ($d_a = 4.6$ mm)

In particular, the most critical issues are connected to the need of minimizing particle attrition during the experiments, and of properly accounting for temperature differences between the active particle and the bed due to heat effects. For the technique based on the combustion of carbon particles a further critical aspect is the assumption of the CO/CO_2 primary ratio and of the CO oxidation location. Among these three techniques, we consider that based on chemical reaction on the surface of catalyst particles to be particularly simple and accurate, and for this reason we recommend its use whenever possible.

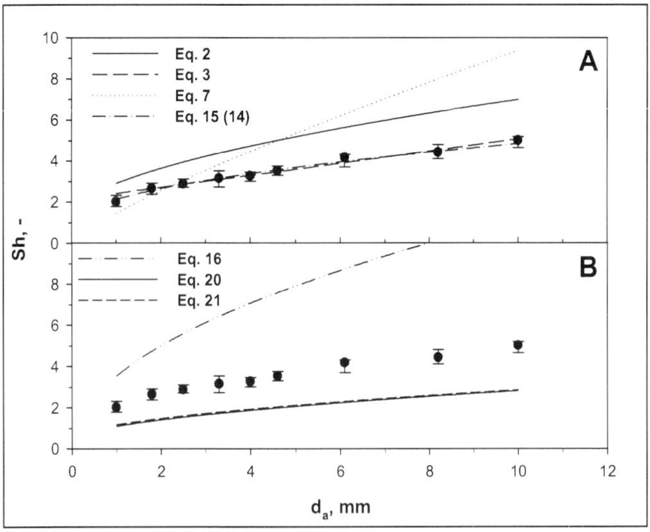

Fig. 2. Comparison of Sherwood number correlations with experimental data taken from Scala (2007). A) Empirical and semi-empirical correlations. B) Theoretical correlations. Average Sherwood number as a function of active particle size (500 µm < d_i < 600 µm)

Comparison of experimental data reported by Scala (2007) with empirical and semi-empirical correlations available in the literature showed that a Frössling-type correlation (Eq. 15) was able to fit excellently the data. The empirical correlation (Eq. 3) proposed by Prins et al. (1985) also fitted satisfactorily the data, but only when d_i < 700 µm. On the contrary, no available theoretical model could correctly predict the Sherwood number.

These results strongly suggest that in order to analyze the dependencies of Sh on the operating variables a Frössling-type analysis is the only one based on sound physical grounds. The first diffusive term accounts for mass transfer around the active particle in stagnant conditions. Contrary to the fixed bed case, however, this term has no strict physical meaning here, since a fluidized bed cannot exist in stagnant conditions (U = 0). So this term must be considered only as an asymptotic behaviour. For a free sphere in stagnant conditions Sh = 2. Two effects must be taken into account when we consider a fluidized bed. First, the inert bed particles decrease the volume available for mass transfer, which depends on the average void fraction around the active particle. This value can be assumed as a first approximation equal to ε_{mf}. A second effect, discussed by Coelho & Guedes de Carvalho (1988), implies that an effective diffusion coefficient D/τ should be used, where τ is the bed tortuosity accounting for the hindering effect of the granular bed on gas diffusion. So in principle the diffusive term should written as: $Sh = 2 \bullet \varepsilon_{mf} / \tau$. It must be noted, however, that to our knowledge no estimation of tortuosity exists for the dense phase of a fluidized beds. Given the theoretical rather than practical significance of this term, it is suggested here that $Sh = 2 \bullet \varepsilon_{mf}$ is a reasonable approximation for the diffusive term.

As regards the convective term, theoretical considerations suggest a 1/2 dependence on the Reynolds number and a 1/3 dependence on the Schmidt number. The first point here is the discrimination of the relevant particle Reynolds number to be used. On the basis of the experimental findings, the relevant gas velocity around the active particle is the dense phase gas interstitial velocity. In most operating conditions this velocity can be safely approximated with U_{mf}/ε_{mf}. So the correct Reynolds number to be used in a Frössling-type expression should be Re_{mf}/ε_{mf}. Analysis of experimental data reported by Scala (2007) strongly support this point. In addition, the reported 1/2 dependence of Sh on both U_{mf} and d_a is an evidence of the soundness of this approach and also of the applicability of the Frössling-type analysis to a fluidized bed.

In spite of the relatively well understood behaviour of Sh with the main operating variables, several issues still remain open to debate, namely the effect of the active particle shape and density on mass transfer coefficient, and the value of Sh for the case $d_a/d_i < 1$.

7. Notation

a	constant in Eq. 9, -
Ar	Archimedes number, -
b	constant in Eq. 9, -
C_D	drag coefficient in Eqs 23 and 24, -
d	particle diameter, m
D	diffusion coefficient, m^2/s
g	gravity acceleration, m/s^2
k_g	mass transfer coefficient, m/s
K	constant in Eq. 1, -
K_p	parameter in Eqs 23 and 24, -
m	parameter in Eq. 3, -
p	probability of active particles being in the dense phase (Eq. 22), -
Pe	Peclet number, -
Re	Reynolds number, -
Sc	Schmidt number, -
Sh	Sherwood number, -
T	temperature, K
U	fluidization velocity, m/s

Greek letters

ε	bed voidage, -
ρ	density, kg/m^3
τ	bed tortuosity, -
μ	viscosity, $kg/m{\cdot}s$

Subscripts

a	active particles
b	bubble or bubble phase
e	emulsion phase (Eqs 22 and 23)
g	gas
i	inert bed particles
mf	at minimum fluidization conditions
mfi	at minimum fluidization conditions referred to inert particles (Eq. 3)
opt	at optimal conditions (Eq. 7)

p particulate phase (Eq. 13)

8. References

Agarwal, P.K. (1987). The residence phase of active particles in fluidized beds of smaller inert particles. *Chemical Engineering Science*, 42, 2481-2483.

Agarwal, P.K.; Mitchell, W.J. & La Nauze, R.D. (1988a). Transport phenomena in multi-particle systems-III. Active particle mass transfer in fluidized beds of inert particles. *Chemical Engineering Science*, 43, 2511-2521.

Agarwal, P.K.; Mitchell, W.J. & La Nauze, R.D. (1988b). Mass transfer processes around burning char particles in fluidized beds. *Proceedings of the Combustion Institute*, 22, 279-286.

Agarwal, P.K. & La Nauze, R.D. (1989). Transfer processes local to the coal particle: a review of drying, devolatilization and mass transfer in fluidized bed combustion. *Chemical Engineering Research and Design*, 67, 457-480.

Avedesian, M.M. & Davidson, J.F. (1973). Combustion of carbon particles in a fluidised bed. *Transactions of the Institution of Chemical Engineers*, 51, 121-131.

Basu, P.; Broughton, J. & Elliott, D.E. (1975). Combustion of single coal particles in fluidized beds, In: *Institute of Fuel Symposium Series no. 1: Fluidised Combustion*, pp. A3-1-A3-10, Institute of Fuel, London.

Basu, P. (1985). The effect of radial distribution of voidage on the burning rate of a carbon sphere in a fluidized bed. *Chemical Engineering Communications*, 39, 297-308.

Bradshaw, R.D. & Myers, J.E. (1963). Heat and mass transfer in fixed and fluidized beds of large particles. *AIChE Journal*, 9, 590-595.

Chakraborty, R.K. & Howard, J.R. (1981). Combustion of char in shallow fluidized bed combustors: influence of some design and operating parameters. *Journal of the Institute of Energy*, 54, 48-54.

Chu, J.C.; Kalil, J. & Wetteroth, W.A. (1953). Mass transfer in a fluidized bed. *Chemical Engineering Progress*, 49, 141-149.

Ciesielczyk, W. (1996). Analogy of heat and mass transfer during constant rate period in fluidized bed drying. *Drying Technology*, 14, 217-230.

Cobbinah, S.; Laguerie, C. & Gibert, H. (1987). Simultaneous heat and mass transfer between a fluidized bed of fine particles and immersed coarse porous particles. *International Journal of Heat and Mass Transfer*, 30, 395-400.

Coelho, M.A.N. & Guedes de Carvalho, J.R.F. (1988). Transverse dispersion in granular beds part II - mass transfer from large spheres immersed in fixed or fluidised beds of small inert particles. *Chemical Engineering Research and Design*, 66, 178-189.

Davidson, J.F. & Harrison, D. (1963). *Fluidised particles*. Cambridge University Press, Cambridge, UK.

de Kok, J.J.; Stark, N.L. & van Swaaij, W.P.M. (1986). The influence of solids specific interfacial area on gas-solid mass transfer in gas fluidised beds, In: *Fluidization V*, Østergaard, K & Sorensen, A. (Eds.), pp. 433-440, Engineering Foundation, New York,.

Delvosalle, C. & Vanderschuren, J. (1985). Gas-to-particle and particle-to-particle heat transfer in fluidized beds of large particles. *Chemical Engineering Science*, 40, 769-779.

Dennis, J.S.; Hayhurst, A.N. & Scott, S.A. (2006). The combustion of large particles of char in bubbling fluidized beds: the dependence of Sherwood number and the rate of burning on particle diameter. *Combustion & Flame*, 147, 185-194.

Donsì, G.; Ferrari, G. & De Vita, A. (1998). Analysis of transport phenomena in two component fluidized beds, In: *Fluidization IX*, Fan, L.-S. & Knowlton, T.M. (Eds.), pp. 421-428, Engineering Foundation, New York.

Donsì, G.; Ferrari, G. & De Vita, A. (2000). Heat and mass transport phenomena between fluidized beds and immersed spherical objects. *Recents Progress en Genie des Procedes*, 14, 213-220.

Frössling, N. (1938). The evaporation of falling drops (in German). *Gerlands Beiträge zur Geophysik*, 52, 170-216.

Guedes de Carvalho, J.R.F. & Coelho, M.A.N. (1986). Comments on mass transfer to large particles in fluidized beds of smaller particles. *Chemical Engineering Science*, 41, 209-210.

Guedes de Carvalho, J.R.F.; Pinto, M.F.R. & Pinho, M.C.T. (1991). Mass transfer around carbon particles burning in fluidised beds. *Transactions of the Institution of Chemical Engineers*, 69, 63-70.

Guedes de Carvalho, J.R.F. & Alves, M.A.M. (1999). Mass transfer and dispersion around active sphere buried in a packed bed. *AIChE Journal*, 45, 2495-2502.

Hayhurst, A.N. (2000). The mass transfer coefficient for oxygen reacting with a carbon particle in a fluidized or packed bed. *Combustion & Flame*, 121, 679-688.

Hayhurst, A.N. & Parmar, M.S. (2002). Measurement of the mass transfer coefficient and Sherwood number for carbon spheres burning in a bubbling fluidized bed. *Combustion & Flame*, 130, 361-375.

Ho, T.C. (2003). Mass Transfer (Chapter 11), In: *Handbook of Fluidization and Fluid-Particle Systems*, Yang, W.C. (Ed.), pp. 287-307, Dekker, New York.

Hsiung, T.H. & Thodos, G. (1977). Mass transfer in gas-fluidized beds: measurement of actual driving forces. *Chemical Engineering Science*, 32, 581-592.

Hsu, C.T. & Molstad, M.C. (1955). Rate of mass transfer from gas stream to porous solid in fluidized beds. *Industrial and Engineering Chemistry*, 47, 1550-1559.

Joulié, R.; Rios, G.M. & Gibert, H. (1986). Sublimation of pure substances in gas fluidized beds at atmospheric pressure. *Drying Technology*, 4, 111-135.

Joulié, R. & Rios, G.M. (1993). Theoretical analysis of heat and mass transfer phenomena during fluidized bed sublimation. *Drying Technology*, 11, 157-182.

Joulié, R.; Barkat, M. & Rios, G.M. (1997). Effect of particle density on heat and mass transfer during fluidized bed sublimation. *Powder Technology*, 90, 79-88.

Jung, K. & La Nauze, R.D. (1983). Sherwood numbers for burning particles in fluidized beds, In: *Fluidization IV*, Kunii, D. & Cole, S.S. (Eds.), pp. 427-434, Engineering Foundation, New York.

Kettenring, K.N.; Manderfield, E.L. & Smith, J.M. (1950). Heat and mass transfer in fluidized systems. *Chemical Engineering Progress*, 46, 139-145.

Kozanoglu, B.U.; Vilchez, J.A.; Casal, J. & Arnaldos, J. (2001). Mass transfer coefficient in vacuum fluidized bed drying. *Chemical Engineering Science*, 56, 3899-3901.

La Nauze, R.D. & Jung, K. (1982). The kinetics of combustion of petroleum coke particles in a fluidized-bed combustor. *Proceedings of the Combustion Institute*, 19, 1087-1092.

La Nauze, R.D. & Jung, K. (1983a). Combustion kinetics in fluidized beds, In: *Proceedings of the 7th International Conference on Fluidized Bed Combustion*, pp. 1040-1053, ASME, New York.

La Nauze, R.D. & Jung, K. (1983b). Mass transfer of oxygen to a burning particle in a fluidized bed, In: *Proceedings of the 8th Australasian Fluid Mechanics Conference*, pp. 5C.1-5C.4, Newcastle, NSW.

La Nauze, R.D.; Jung, K. & Kastl, J. (1984). Mass transfer to large particles in fluidized beds of smaller particles. *Chemical Engineering Science*, 39, 1623-1633.

La Nauze, R.D. & Jung, K. (1985). Further studies of combustion kinetics in fluidized beds, In: *Proceedings of the 8th International Conference on Fluidized Bed Combustion*, pp. 65-73, ASME, New York.

La Nauze, R.D. (1985). Fundamentals of coal combustion in fluidised beds. *Chemical Engineering Research and Design*, 63, 3-33.

La Nauze, R.D. & Jung, K. (1986). Mass transfer relationships in fluidized-bed combustors. *Chemical Engineering Communications*, 43, 275-286.

Oka, S.N.; Ilić, M.S.; Vukašinović, B.N. & Komatina, M.S. (1995). Experimental investigations of mass transfer between single active particle and bubbling fluidized bed, In: *Proceedings of the 13th International Conference on Fluidized Bed Combustion*, pp. 1419-1425. ASME, New York.

Pal'chenok, G.I. & Tamarin, A.I. (1985). Mass transfer at a moving particle in a fluidized bed of coarse material. *Journal of Engineering Physics*, 47, 916-922.

Palchonok, G.I.; Dolidovich, A.F.; Andersson, S. & Leckner, B. (1992). Calculation of true heat and mass transfer coefficients between particles and a fluidized bed, In: *Fluidization VII*, Porter, O.E. & Nicklin, D.J. (Eds.), pp. 913-920, Engineering Foundation, New York.

Palchonok, G.I. (1998). *Heat and mass transfer to a single particle in fluidized bed.* Ph.D. Thesis, Chalmers University of Technology, Sweden.

Paterson, W.R. & Hayhurst, A.N. (2000). Mass or heat transfer from a sphere to a flowing fluid. *Chemical Engineering Science*, 55, 1925-1927.

Paterson, W.R. (2000). Mass transfer to, and reaction on, a sphere immersed in a stationary or flowing gas. *Chemical Engineering Science*, 55, 3567-3570.

Petrovic, L.J. & Thodos, G. (1966). Evaporation from alumina in fixed and fluid beds. *British Chemical Engineering*, 11, 1039-1042.

Petrovic, L.J. & Thodos, G. (1967). Effectiveness factors for mass transfer in fluidized systems, In: *Proceedings of the International Symposium on Fluidization*, Drinkenburg, A.A.H. (Ed.), pp. 586-598, Netherlands University Press, Amsterdam.

Pillai, K.K. (1981). The influence of coal type on devolatilization and combustion in fluidized beds. *Journal of the Institute of Energy*, 54, 142-150.

Prins, W.; Casteleijn, T.P.; Draijer, W. & van Swaaij, W.P.M. (1985). Mass transfer from a freely moving single sphere to the dense phase of a gas fluidized bed of inert particles. *Chemical Engineering Science*, 40, 481-497.

Prins, W. (1987). *Fluidized bed combustion of a single carbon particle.* Ph.D. Thesis, Twente University, The Netherlands.

Ranz, W.E. & Marshall Jr., W.R. (1952). Evaporation from drops. *Chemical Engineering Progress*, 48, (part I) 141-146 & (part II) 173-180.

Resnick, W. & White, R.R. (1949). Mass transfer in systems of gas and fluidized solids. *Chemical Engineering Progress*, 45, 377-390.

Richardson, J.F. & Szekely, J. (1961). Mass transfer in a fluidised bed. *Transactions of the Institution of Chemical Engineers*, 39, 212-222.

Riccetti, R.E. & Thodos, G. (1961). Mass transfer in the flow of gases through fluidized beds. *AIChE Journal*, 7, 442-444.

Ross, I.B. & Davidson, J.F. (1982). The combustion of carbon particles in a fluidised bed. *Transactions of the Institution of Chemical Engineers*, 60, 108-114.

Rowe, P.N.; Claxton, K.T. & Lewis, J.B. (1965). Heat and mass transfer from a single sphere in an extensive flowing fluid. *Transactions of the Institution of Chemical Engineers*, 43, T14-T31.

Salatino, P.; Scala, F. & Chirone, R. (1998). Fluidized-bed combustion of a biomass char: the influence of carbon attrition and fines postcombustion on fixed carbon conversion. *Proceedings of the Combustion Institute*, 27, 3103-3110.

Scala, F.; Chirone, R. & Salatino, P. (2006). Combustion and attrition of biomass chars in a fluidized bed. *Energy & Fuels*, 20, 91-102.

Scala, F. (2007). Mass transfer around freely moving active particles in the dense phase of a gas fluidized bed of inert particles. *Chemical Engineering Science*, 62, 4159-4176.

Scala, F. (2009). A new technique for the measurement of the product CO/CO_2 ratio at the surface of char particles burning in a fluidized bed. *Proceedings of the Combustion Institute*, 32, 2021-2027.

Scala, F. (2010a). Calculation of the mass transfer coefficient for the combustion of a carbon particle. *Combustion & Flame*, 157, 137-142.

Scala, F. (2010b). Fluidized bed combustion of single coal char particles: an analysis of the burning rate and of the primary CO/CO_2 ratio. Submitted for publication.

Schlichthaerle, P. & Werther J. (2000). Influence of the particle size and superficial gas velocity on the sublimation of pure substances in fluidized beds of different sizes. *Drying Technology*, 18, 2217-2237.

Schlünder, E.U. (1977). On the mechanism of mass transfer in heterogeneous systems - in particular in fixed beds, fluidized beds and on bubble trays. *Chemical Engineering Science*, 32, 845-851.

Tamarin, A.I. (1982). Mass transfer between the gas and solid particles in a fluidized bed. *Journal of Engineering Physics*, 41, 1346-1350.

Tamarin, A.I.; Galershteyn, D.M.; Shuklina, V.M. & Zabrodskiy, S.S. (1982). Convective transport between a burning coal particle and an air-fluidized bed of inert particles. *Heat Transfer - Soviet Research*, 14, 88-93.

Tamarin, A.I.; Palchyonok, G.I. & Goryunov, K.E. (1985). Heat and mass transfer of model particles in a fluidized bed of inert material. *Heat Transfer - Soviet Research*, 17, 136-141.

Tsotsas, E. (1994a). Discrimination of fluid bed models and investigation of particle-to-gas mass transfer by means of drying experiments. *Chemical Engineering and Processing*, 33, 237-245.

Tsotsas, E. (1994b). From single particle to fluid bed drying kinetics. *Drying Technology*, 12, 1401-1426.

Vanderschuren, J. & Delvosalle, C. (1980). Particle-to-particle heat transfer in fluidized bed drying. *Chemical Engineering Science*, 35, 1741-1748.

Van Heerden, C. (1952). Some fundamental characteristics of the fluidized state. *Journal of Applied Chemistry*, 2, S7-S17.

Venderbosch, R.H.; Prins, W. & van Swaaij, W.P.M. (1998). Platinum catalyzed oxidation of carbon monoxide as a model reaction in mass transfer measurements. *Chemical Engineering Science*, 53, 3355-3366.

Wilkins, G.S. & Thodos, G. (1969). Mass transfer driving forces in packed and fluidized beds. *AIChE Journal*, 15, 47-50.

Yoon, P. & Thodos, G. (1972). Mass transfer in the flow of gases through shallow fluidized beds. *Chemical Engineering Science*, 27, 1549-1554.

Yusuf, R.; Melaaen, M.C. & Mathiesen, V. (2005). Convective heat and mass transfer modeling in gas-fluidized beds. *Chemical Engineering and Technology*, 28, 13-24.

Ziegler, E.N. & Brazelton, W.T. (1964). Mechanism of heat transfer to a fixed surface in a fluidized bed. *Industrial and Engineering Chemistry Fundamentals*, 3, 94-98.

Ziegler, E.N. & Holmes, J.T. (1966). Mass transfer from fixed surfaces to gas fluidized beds. *Chemical Engineering Science*, 21, 117-122.

26

Mass Transfer Phenomena and Biological Membranes

Parvin Zakeri-Milani and Hadi Valizadeh
Faculty of Pharmacy, Drug Applied Research Center,
Research Center for Pharmaceutical Nanotechnology,
Tabriz University of Medical Sciences, Tabriz
Iran

1. Introduction

Mass transfer is the net movement of mass from one location to another in response to applied driving forces. Mass transfer is used by different scientific disciplines for different processes and mechanisms. It is an important phenomena in the pharmaceutical sciences; drug synthesis, preformulation investigations, dosage form design and manufacture and finally ADME (absorption, distribution, metabolism and excretion) studies. In nature, transport occurs in fluids through the combination of advection and diffusion. Diffusion occurs as a result of random thermal motion and is mass transfer due to a spatial gradient in chemical potential or simply, concentration. However the driving force in convective mass transport is the spatial gradient in pressure (Fleisher, 2000). On the other hand, there are other variables influencing mass transfer like electrical potential and temperature which are important in pharmaceutical sciences. In a complex system mass transfer may be driven by multiple driving forces. Mass transfer exists everywhere in nature and also in human body. In fact in the body, mass transport occurs across different types of cell membranes under different physiological conditions. This chapter is aimed at reviewing transport across biological membranes, with an emphasis on intestinal absorption, its model analysis and permeability prediction.

2. Transport across membranes

Biomembrane or biological membrane is a separating amphipathic layer that acts as a barrier within or around a cell. The membrane that retains the cell contents and separates the cell from surrounding medium is called plasma membrane. This membrane acts as a lipid bilayer permeability barrier in which the hydrocarbon tails are in the centre of the bilayer and the electrically charged or polar headgroups are in contact with watery or aqueous solutions. There are also protein molecules that are attached to or associated with the membrane of a cell. Generally cell membrane proteins are divided into integral (intrinsic) and peripheral (extrinsic) classes. Integral membrane proteins containing a sequence of hydrophobic group are permanently attached to the membrane while peripheral proteins are temporarily attached to the surface of the cell, either to the lipid bilayer or to integral proteins. Integral proteins are responsible for identification of the cell for recognition by other cells and immunological behaviour, the initiation of intracellular responses to external molecules (like pituitary hormones, prostaglandins, gastric peptides,...), moving substances into and out of the cell (like ATPase,...). Concerning mass transport across a cell, there are a number of different mechanisms, a molecule may simply diffuses across, or be transported by a range of membrane proteins (Washington et al., 2000, Lee and Yang, 2001).

2.1 Passive transport

Lipophilic drug molecules with low molecular weight are usually passively diffuses across the epithelial cells. Diffusion process is driven by random molecular motion and continues until a dynamic equilibrium is reached. Passive mass transport is described by Fick,s law which states that the rate of diffusion across a membrane (R) in moles s-1 is proportional to the concentration difference on each side of the membrane:

$$R=(Dk/h).A.\Delta C \quad (1)$$

Where D is the diffusion coefficient of the drug in the membrane, k is the partition coefficient of the drug into the membrane, h is the membrane thickness, A is the area of membrane over which diffusion is occurring, and ΔC is the difference between concentrations on the outside and the inside of the membrane. However it should be noted that the concentration of drug in systemic blood circulation is negligible in comparison to the drug concentration at the absorption surface and the drug is swept away by the circulation. Therefore the driving force for absorption is enhanced by maintaining the large concentration gradient throughout the absorption process. The diffusion coefficient of a drug is mainly influenced by two important factors, solubility of the drug and its molecular weight. For a molecule to diffuse freely in a hydrophobic cell membrane it must be small in size, soluble in membrane and also in the aqueous extracellular systems. That means an intermediate value of partition coefficient is needed. On the other hand, it is necessary for a number of hydrophilic materials, to pass through the cell membranes by membrane proteins. These proteins allow their substrates to pass into the cell down a concentration gradient, and act like passive but selective pores. For example for glucose diffusion into the cell by hexose transporter system, no energy is expended and it occurs down a concentration gradient. This process is called non-active facilitated mass transport (Sinko, 2006, Washington et al., 2000).

2.2 Active transport

In the cell membrane there are a group of proteins that actively compile materials in cells against a concentration gradient. This process is driven by energy derived from cellular metabolism and is defined as primary active trasport. The best-studied systems of this type are the ATPase proteins that are particularly important in maintaining concentration gradients of small ions in cells. However this process is saturable and in the presence of extremely high substrate concentration, the carrier is fully applied and mass transport rate is limited. On the other hand cells often have to accumulate other substances like amino acids and carbohydrates at high concentrations for which conversion of chemical energy into electrostatic potential energy is needed. In this kind of active process, the transport of an ion is coupled to that of another molecule, so that moving an ion out of the membrane down the concentration gradient, a different molecule moves from lower to higher concentration.

Depending on the transport direction this secondary active process is called symport (same directions) or antiport (opposite directions). Important examples of this process are absoption of glucose and amino acids which are coupled to transporter conformational changes driven by transmucosal sodium gradients (Lee and Yang, 2001).

2.3 Endocytic processes

All the above-mentioned mass transport mechanisms are only feasible for small molecules, less than almost 500 Dalton. Larger objects such as particles and macromolecules are absorbed with low efficiency by a completely different mechanism. The process which is called cytosis or endocytosis is defied as extending the membrane and enveloping the object and can be divided into two types, pinocytosis and phagocytosis. Pinocytosis (cell drinking) occurs when dissolved solutes are internalized through binding to non-specific membrane receptors (adsorptive pinocytosis) or binding to specific membrane receptors (receptor-mediated pinocytosis). In some cases, following receptor-mediated pinocytosis the release of undegraded uptaken drug into the extracellular space bounded by the basolateral membrane is happened. This phenomenon called transytosis, represents an important pathway for absorption of proteins and peptides. On the other hand phagocytosis (cell eating) occurs when a particulate matter is taken inside a cell. Although phagocytic processes are finding applications in oral drug delivery and targeting, it is mainly carried out by the specialized cells of the mononuclear phagocyte systems or reticuloendothelial system and is not generally relevant to the transport of drugs across absorption barriers (Lee and Yang, 2001, Fleisher, 2000, Washington et al., 2000).

2.4 Pore transport

The aqueous channels which exist in cell membranes allow very small hydrophilic molecules such as urea, water and low molecular weight sugars to be transported into the cells. However because of the limited pore size (0.4 nm), this transcellular pathway is of minor importance for drug absorption (Fleisher, 2000, Lee and Yang, 2001).

2.5 Persorption

As epithelial cells are sloughed off at the tip of the villus, a gap in the membrane is temporarily created, allowing entry of materials that are not membrane permeable. This process has been termed persorption which is considered as a main way of entering starch grains, metallic ion particles and some of polymer particles into the blood.

3. Intestinal drug absorption

Interest has grown in using in vitro and in situ methods to predict in vivo absorption potential of a drug as early as possible, to determine the mechanism and rate of transport across the intestinal mucosa and to alert the formulator about the possible windows of absorption and other potential restrictions to the formulation approach. Single-pass intestinal perfusion (SPIP) model is one of the mostly used techniques employed in the study of intestinal absorption of compounds which provides a prediction of absorbed oral dose and intestinal permeability in human. In determination of the permeability of the intestinal wall by external perfusion techniques, several models have been proposed (Ho and Higuchi, 1974, Winne, 1978, Winne, 1979, Amidon et al., 1980). In each model, assumptions must be made regarding the convection and diffusion conditions in the experimental system which affects the interpretation of the resulting permeabilities. In addition, the appropriateness of the assumptions in the models to the actual experimental situation must be determined. Mixing tank (MT) model or well mixed model has been previously used to describe the hydrodynamics within the human perfused jejunal segment based on a residence time distribution (Lennernas, 1997). This model has also been used in vitro to simulate gastrointestinal absorption to assess the effects of drug and system parameters on drug absorption (Dressman et al., 1984). However complete radial mixing (CRM) model was used to calculate the fraction dose absorbed and intestinal permeability of gabapentine in rats (Madan et al., 2005). Moreover these two models (MT and CRM) were utilized to develop a theoretical approach for estimation of fraction dose absorbed in human based on a macroscopic mass balance approach (MMBA) (Sinko et al., 1991). Although these models have been theoretically explained, their comparative suitability to be used for experimental data had not been reported. The comparison of proposed models will help to select the best model to establish a strong correlation between rat and human intestinal drug absorption potential. In this section three common models for mass transfer in single pass perfusion experiments (SPIP) will be compared using the rat data, we obtained in our lab. The resulting permeability values differ in each model, and their interpretation rests on the validity of the assumptions (valizadeh et al., 2008).

4. Mass transfer models

Three models are described that differ in their convection and diffusion assumptions (Fig 1).

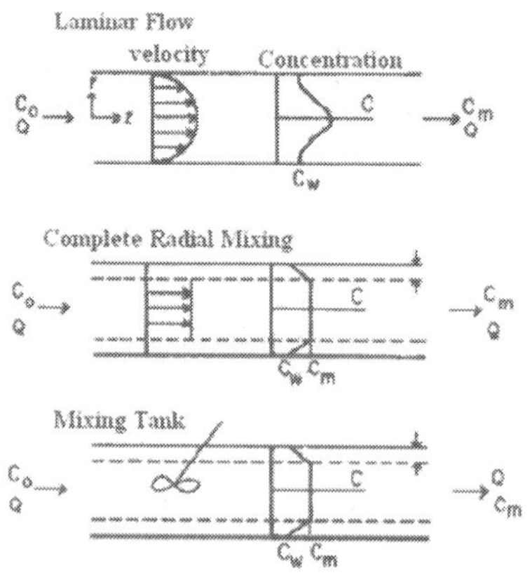

Fig. 1. Velocity and concentration profiles for the models. The concentration profiles are also a function of z except for mixing tank model (Amidon et al., 1980)

These models are the laminar flow, complete radial mixing (diffusion layer) for convective mass transport in a tube and the perfect mixing tank model. It is convenient to begin with the solute transport equation in cylindrical coordinates (Sinko et al., 1991, Elliott et al., 1980, Bird et al., 1960):

$$v^*_z \frac{\partial C}{\partial z^*} = Gz(\frac{1}{r^*}\frac{\partial}{\partial r^*} r^* \frac{\partial C}{\partial r^*}) \quad (2)$$

Where, Z* = Z / L, r* = r / R, υ^*_z = v_z / Vm, Gz = πDL/2Q , R = radius of the tube, L = length of the tube, Vm = maximum velocity, Q = perfusion flow rate

This relationship is subject to the first-order boundary condition at the wall:

$$\left.\frac{\partial C}{\partial r^*}\right|_{r^*=1} = -P^*_w C_w \quad (3)$$

where P^*_w = Pw R/D = the dimensionless wall permeability.

The main assumptions achieving Eq. 1 are: (a) the diffusivity and density are constant; (b) the solution is dilute so that the solvent convection is unperturbed by the solute; (c) the system is at steady state (∂C/∂t = 0); (d) the solvent flows only in the axial (z) direction; (e) the tube radius, R, is independent of Gz; and (f) axial diffusion is small compared to axial convection (Bird et al., 1960). The boundary condition (Eq. 2) is true for many models having a tube wall but does not describe a carrier transport of Michaelis-Menten process at the wall, except at low solute concentrations.

4.1 Complete radial mixing model

For this model the velocity profile as with the plug flow model is assumed to be constant. In addition, the concentration is assumed to be constant radially but not axially. That is, there is complete radial but not axial, mixing to give, uniform radial velocity and concentration profiles. With these assumptions, the solution is written as:

$$C_m/C_0 = \exp(-4 P^*_{eff} Gz) \quad (4)$$

where P^*_{eff} replaces P^*_w (Ho and Higuchi, 1974, Winne, 1978, Winne, 1979). Since no aqueous resistance is included in the model directly, the wall resistance is usually augmented with a film or diffusion layer resistance. That is, complete radial mixing occurs up to a thin region or film adjacent to the membrane. In this model the aqueous (luminal) resistance is confined to this region. Hence, the wall permeability includes an aqueous or luminal resistance term and can be written as:

$$P^*_{eff} = \frac{P^*_w P^*_a}{P^*_w + P^*_a} \quad (5)$$

where P^*_w is the true wall permeability and P^*_a, is the effective aqueous permeability. The aqueous permeability often is written as:

$$P_a = D/\delta \quad (6)$$

Or

$$P^*_a = R/\delta \quad (7)$$

where δ is the film thickness and represents an additional parameter that needs to be determined from the data to obtain P_w^*. For typical experiments, P_a^* or R/δ is an empirical parameter, since the assumed hydrodynamic conditions may not be realistic at the low Reynolds numbers. The complete radial mixing model also can be derived from a differential mass balance approach (Ho and Higuchi, 1974) and often is referred to as the diffusion layer model. The Calculated P_{eff}^* values for tested drugs and the corresponding plot are shown in Table 2 and Fig. 2 respectively.

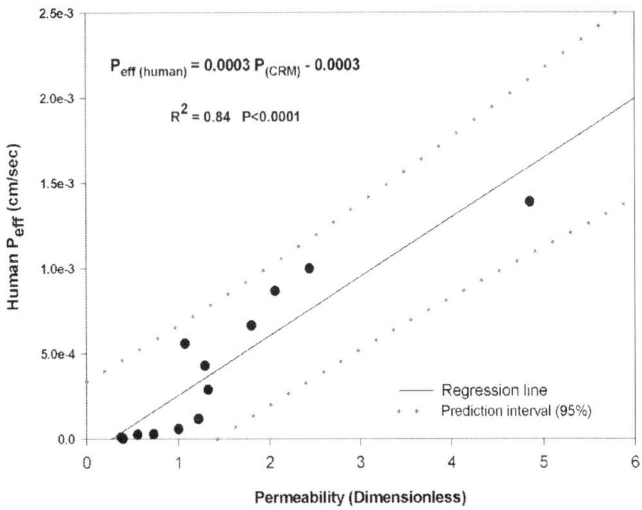

Fig. 2. Plot of dimensionless permeability values vs human P_{eff} values in complete radial mixing model

4.2 Laminar flow model

For flow of a newtonian fluid in a cylindrical tube, the exit concentration of a solute with a wall permeability Pw is given by (Amidon et al., 1980):

$$C_m/C_o = \sum_{n=1}^{\infty} M_n \exp(-\beta_n^2 G_z) \tag{8}$$

Where, Cm = "cup-mixing" outlet solute concentration from the perfused length of intestine,

$$\text{Co = inlet solute concentration; Gz} = \pi DL/2Q; \tag{9}$$

Gz is Graetz number, the ratio of the mean tube residence time to the time required for radial diffusional equilibration.

D = solute diffusivity in the perfusing fluid
L = length of the perfused section of intestine
Q = volumetric flow rate of perfusate = πR2(υ)
R = radius of perfused intestine
(υ) = mean flow velocity

Both the Mn and βn in Eq. 7 are functions of P_w^*, the dimensionless wall permeability,

$$P_w^* = \frac{P_w R}{D} \tag{10}$$

From the form of the solution it appears that Gz is the only independent variable and that the solution is an implicit function of P_w^*. Since P_w^* (or Pw) is the parameter of interest, Eq. 4 is not in a convenient form for its determination.
We now define:

$$\frac{1}{P_{eff}^*} = \frac{1}{P_w^*} + \frac{1}{P_{aq}^*} = \frac{1}{{}^{\circ}P_w^*} + \frac{1}{{}^{\circ}P_{aq}^*} \tag{11}$$

$$P_{eff}^* = \frac{\ln[(C_m/C_0)]_{exp}}{-4Gz} \tag{12}$$

$${}^{\circ}P_{aq}^* = \frac{\ln[(C_m/C_0)]_0}{-4Gz} \tag{13}$$

$$[(C_m/C_0)]_0 = \sum_{n=1}^{5} {}^{o}M_n \exp(-{}^{\circ}\beta_n^2\, Gz) \tag{14}$$

where the superscript o denotes the sink condition (Graetz solution), the superscript * denotes dimensionless quantities [Eq. 8] and subscripts exp stands for experimental condition. The wall permeability is determined in the following manner: First the ${}^{\circ}P_{aq}^*$ is calculated using Eqs. 9 , 11, 14 and Table 1.

(n)	${}^{\circ}\beta_n$	${}^{\circ}M_n$
1	2.7043	0.81905
2	6.6790	0.09752
3	10.6734	0.03250
4	14.6711	0.01544
5	18.6699	0.00878

Table 1. Coefficients, ${}^{\circ}M_n$ and exponents, ${}^{\circ}\beta_n$ for the Graetz solution, equation (12), (sink conditions) (Elliott et al., 1980)

Then the P_{eff}^* is calculated from the experimental results using Eq. 8 and 11 at the third step the value of ${}^{\circ}P_w^*$ is found out from Eq. 10 and finally the value of ${}^{\circ}P_w^*$ is multiplied by the correction factor in Fig 3 to obtain P_w^*.

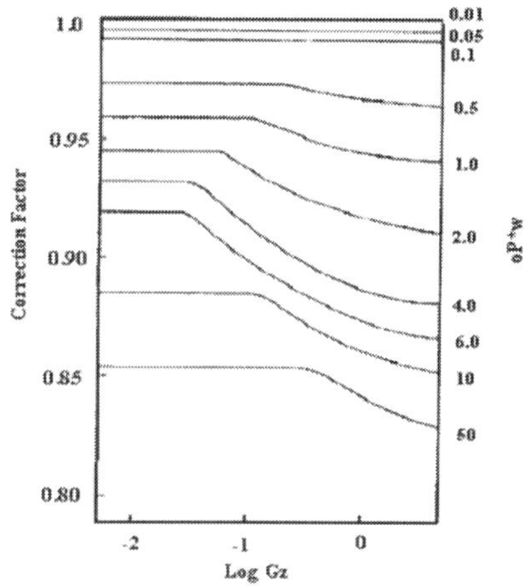

Fig. 3. Correction factor to obtain exact wall permeability (P_w^*) given the estimated wall Permeability ($^{\circ}P_w^*$) and value of Gz. (Elliott et al., 1980)

All calculations were performed for our data in SPIP model. The Gz values were calculated based on equation 8, using the compound diffusivity, length of intestine and flow rate of perfusion which are shown in Table 2. The average value of Gz was found to be 3.34×10^{-2} ($\pm$ 8.6×10^{-3}). It seems that there are limitations for the use of laminar flow model in determination of the dimensionless wall permeability of highly permeable drugs. For instance a negative value of ibuprofen dimensionless wall permeability was obtained based on laminar flow model because of the high P^*_{eff} value of ibuprofen in comparison with its calculated P^*_{aq} sink value and as a result the drug was excluded from correlation plot. Table 2 also represents the obtained dimensionless rat gut wall permeabilities (P_w^*) for tested compounds. The plot of P_w^* versus the observed human intestinal permeability values is shown in Fig. 4.

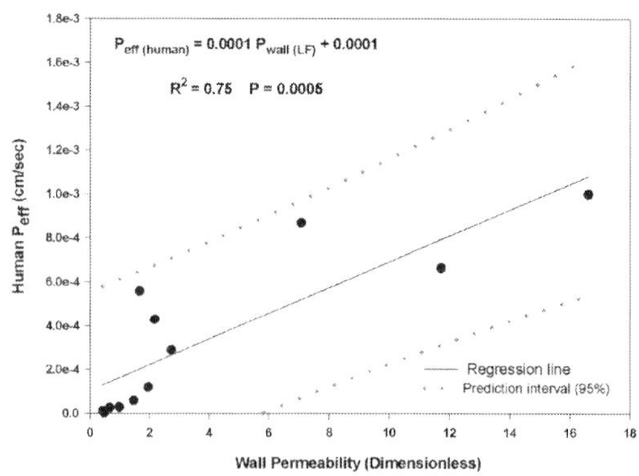

Fig. 4. Plot of dimensionless rat gut wall permeability values vs human Peff values in laminar flow model

4.3 Mixing tank model

This model takes the next step and assumes that both radial and axial mixing are complete. The aqueous resistance again is believed to be confined to a region (film) next to the membrane where only molecular diffusion occurs, and the rest of the contents are well mixed (perfect mixer). This model is described most easily by a mass balance on the system: (mass/time)inlet - (mass/time)outlet = (mass/time)absorbed or:

$$QC_0 - QC_m = (2\pi RL)(P'_{eff})C_m \tag{15}$$

where 2πRL is the area of the mass transfer surface (cylinder) of length L and radius R, P'_{eff} is the permeabilily or mass transfer coefficient of the surface, and C_m is the concentration in the tube (which is constant and equal to the outlet concentration by the perfect mixing assumption). From Eq. 15 it is obtained:

$$\frac{C_0 - C_m}{C_m} = P'_{eff}\frac{2\pi RL}{Q} \tag{16}$$

$$C_0/C_m = 1 + 4P'^{*}_{eff}Gz \tag{17}$$

As with the complete radial mixing model, P^*_{eff} contains additional parameter $P'^{*}_a = R/\delta'$ that must be estimated from the data, The P'^{*}_a and P'^{*}_{eff} values for the mixing tank model differ from those for the complete radial mixing model by nature of the different hydrodynamic assumptions (Amidon et al., 1980). While this model is not appropriate to most perfusion experiments, it is useful to compare its ability for correlation of mass transfer data with other models. As a matter of fact the P^*_{eff} for our data was calculated on the basis of assumptions of mixing tank model. The data and representative plot for this model are shown in Table 2 and Fig. 5 respectively (Valizadeh et al. 2008).

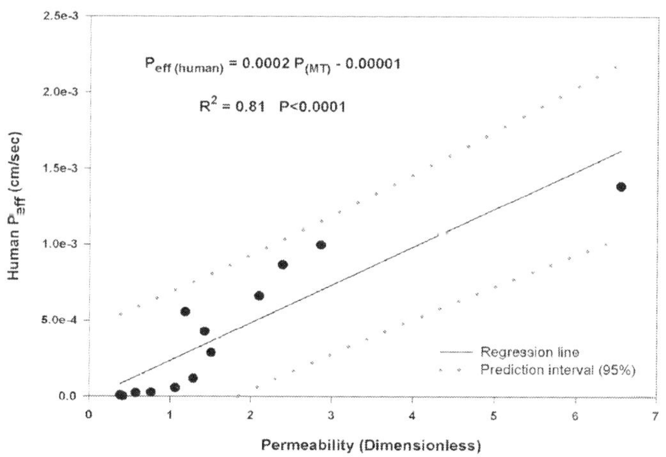

Fig. 5. Plot of dimensionless permeability values vs human P_{eff} values in mixing tank model

Compound	P^*_{wall} (±SD) (LF)	P^*_{eff} (±SD) (MT)	P^*_{eff} (±SD) (CRM)	Graetz no.	Rat no.	Diffusivity[a] (×10^{-6} m^2/sec)
Atenolol	0.41± 0.00	0.38±0.00	0.37±0.00	3.53E-02	1	7.70
				3.46E-02	2	
				2.59E-02	3	
Cimetidine	1.46± 0.07	1.06±0.03	0.99±0.02	3.32E-02	1	8.70
				4.68E-02	2	
				3.98E-02	3	
Ranitidine	0.67± 0.32	0.57±0.25	0.55±0.02	2.99E-02	1	7.40
				3.16E-02	2	
				2.16E-02	3	
Antipyrine	1.65± 0.13	1.18±0.06	1.07±0.04	5.34E-02	1	9.92
				3.56E-02	2	
				4.45E-02	3	
Metoprolol	1.94± 1.35	1.28±0.62	1.21± 0.56	2.01E-02	1	4.98
				1.39E-02	2	
				1.68E-02	3	
Piroxicam	11.70 ± 14.4	2.09±1.18	1.80±0.92	2.84E-02	1	7.92
				3.56E-02	2	
				2.74E-02	3	
Propranolol	2.72± 1.8	1.50±0.61	1.32±0.48	3.46E-02	1	7.70
				3.98E-02	2	
				5.19E-02	3	
Carbamazepine	2.17± 0.35	1.42±0.14	1.29±0.12	3.71E-02	1	8.70
				3.94E-02	2	
				3.47E-02	3	
Furosemide	0.98± 0.69	0.76±0.47	0.72±0.44	2.92E-02	1	8.22
				2.36E-02	2	
				2.58E-02	3	
Hydrochlorothiazide	0.46± 0.26	0.41±0.22	0.39±0.21	4.07E-02	1	9.26
				4.24E-02	2	
				3.82E-02	3	
				4.15E-02	4	
Ibuprofen	-------	6.54±0.53	4.85±0.54	3.82E-02	1	7.40
				2.49E-02	2	
				2.76E-02	3	
Ketoprofen	7.07±3.97	2.38±0.52	2.06±0.40	3.40E-02	1	8.42
				3.02E-02	2	
				4.53E-02	3	
				2.72E-02	4	
Naproxen	16.59± 15.8	2.85±0.55	2.43±0.41	3.26E-02	1	8.55
				3.26E-02	2	
				2.92E-02	3	
				2.96E-02	4	

[a] Diffusivities were calculated using 2D structure of compounds applying he method proposed by Heyduk et al (Hayduk and Laudie, 1974)

Table 2. Dimensionless permeabilities determined based on three mass transfer models

The calculated dimensionless wall permeability values were in the range of 0.37 - 4.85, 0.38-6.54 and 0.41-16.59 for complete radial mixing, mixing tank and laminar flow models respectively. It is clear that drugs with different physicochemical properties belonging to all four biopharmaceutical classes were enrolled in the study. Atenolol a class III drug (high soluble-low permeable) showed lowest effective permeability value in all three investigated models. It is also shown that there is only a small difference in the calculated atenolol permeability coefficients in three models. However this variation becomes more salient for high permeable drugs; i.e. class I (high soluble-high permeable) and class II (low soluble-high permeable) drugs especially in term of permeability in laminar flow model. For instance the observed mean permeability values for naproxen, a class II drug, are 2.43, 2.85 and 16.59 in CRM, MT and LF models respectively. Therefore it seems that in comparison to other model laminar flow model provides larger values for highly permeable drugs in comparison to the other models. However the ranking order for intestinal absorption of tested drugs is almost the same in other evaluated models. In addition it seems that it would be possible to classify drugs correctly by the resulting values. Fig. 2, 4 and 5 demonstrate the obtained correlations for investigated models. It is seen that the plots of rat permeability versus human P_{eff} values, present rather high linear correlations with intercepts not markedly different from zero (R^2= 0.81, P <0.0001 for MT, R^2= 0.75, P =0.0005 for LF, R^2= 0.84, P <0.0001 for CRM). The permeabilities differ for the various models. The permeabilities resulting from application of the other models can be interpreted if it is assumed that the laminar flow permeability measures the wall permeability. The permeability values for the complete radial mixing model are lower than the laminar flow model since this model assumes radial mixing, which leads to lower estimated luminal (aqueous) resistance values and a higher estimated membrane resistance (lower permeability value). However, the usual interpretation of the complete radial mixing model recognizes that the permeability value includes an aqueous resistance. While the permeabilities in mixing tank model, which takes the final step in assuming both radial and axial mixing, were expected to be the lowest among all models, they were in the range between permeabilities in complete radial mixing and dimensionless wall permeabilities. Although theoretically laminar flow model has been established to a reasonable approximation in external perfusion studies, based on the results of correlations of this study, it seems the hydrodynamics in normal physiological situation clearly are more complex and need more investigation to choose from proposed models. Therefore it is concluded that all investigated models work relatively well for our data despite fundamentally different assumptions. The wall permeabilities fall in the order laminar flow > mixing tank > complete radial mixing. Based on obtained correlations it is also concluded that although laminar flow model provides the most direct measure of the intrinsic wall permeability, it has limitations for highly permeable drugs such as ibuprofen and the normal physiological hydrodynamics is more complex and finding real hydrodynamics require further investigations.

5. Prediction of human intestinal permeability using SPIP technique

Previous studies have shown that the extent of absorption in humans can be predicted from single-pass intestinal perfusion technique in rat (Salphati et al., 2001, Fagerholm et al., 1996), however, in this section (Zakeri-Milani et al., 2007) we compare the quantitative differences between permeabilities in human and rat models directly using a larger number of model drugs with a broad range of physicochemical properties for both high and low permeability classes of drugs. In fact more poorly absorbed drugs (cimetidine and ranitidine) have been included in the present work and therefore it is likely that the obtained equations will give a more reliable prediction of the human intestinal permeability and fraction of dose absorbed than previously reported equations. Single-pass intestinal perfusion studies in rats were performed using established methods adapted from the literature. Briefly, rats were anaesthetized using an intra peritoneal injection of pentobarbital (60 mg/kg) and placed on a heated pad to keep normal body temperature. The small intestine was surgically exposed and 10 cm of jejunum was ligated for perfusion and cannulated with plastic tubing. The cannulated segment rinsed with saline (37°C) and attached to the perfusion assembly which consisted of a syringe pump and a 60 ml syringe was connected to it. Care was taken to handle the small intestine gently and to minimize the surgery in order to maintain an intact blood supply. Blank perfusion buffer was infused for 10 min by a syringe pump followed by perfusion of compounds at a flow rate of 0.2 ml/min for 90 min. The perfusate was collected every 10 min in microtubes. The length of segment was measured following the last collection and finally the animal was euthanitized with a cardiac injection of saturated solution of KCl. Samples were frozen immediately and stored at -20°C until analysis. Effective permeability (P_{eff}) (or better named practical permeability, since the effective area of segment is not considered in the calculation) was calculated using following equation (Eq.18) according to the parallel tube mode:

$$P_{eff} = -Q \ln(C_{out}/C_{in})/2\pi rl \quad (18)$$

In which C_{in} is the inlet concentration and C_{out} is the outlet concentration of compound which is corrected for volume change in segment using phenol red concentration in inlet and outlet tubing. Q is the flow rate (0.2 ml/min), r is the rat intestinal radius (0.18 cm) and l is the length of the segment. It has been demonstrated that in humans at a Qin of 2-3 ml/min, P_{eff} is membrane-controlled. In the rat model the Q_{in} is scaled to 0.2 ml/min, since the radius of the rat intestine is about 10 times less than that of human. In 1998 Chiou and Barve (Chiou and Barve, 1998) reported a great similarity in oral absorption (F_a) between rat and human; however they have used an in vivo method, quite different from in situ techniques, that can give an idea of the absorption from the entire GI tract, therefore the significance of rat jejunal permeability values for predicting the human F_a has not been tested in that report. In the present study the obtained P_{eff} values ranged between 2×10^{-4} cm/sec to 1.6×10^{-5} cm/sec and showed a high correlation (R^2=0.93, P<0.0001) with human P_{eff} data for passively absorbed compounds (Fig 6) confirming the validity of our procedure. This correlation was weakened when the actively transported compounds (cephalexin and α methyl dopa) were added to the regression (R^2=0.87, P<0.0001).

The plot of predicted vs observed human P_{eff} values presents a high linear correlation with intercept not markedly different from zero (R^2= 0.93, P <0.0001) (Zakeri-Milani et al., 2007). According to previously reported equations by Salphati et al (Salphati et al., 2001) in the ileum and Fagerholm et al (Fagerholm et al., 1996) in the jejunal segment, the slopes for the same correlation between two models were 6.2 and 3.6 respectively. However based on our results for larger set of compounds including more low-permeable drugs the rat P_{eff} values were on average 11 times lower than those in human. The species differences and the differences in effective absorptive area might be the reasons for the lower permeability values in the rat model. In addition, any changes in the intestinal barrier function during the surgery might be a main reason for obtaining different results in literature concerning intestinal permeability of drugs. A strong correlation was observed between rat permeability data and fraction of oral dose absorbed in human fitting to chapman type equation; $F_{a\,(human)} = 1 - e^{-38450 P_{eff\,(rat)}}$ ($R^2 = 0.91$, $P<0.0001$) (Fig. 7).

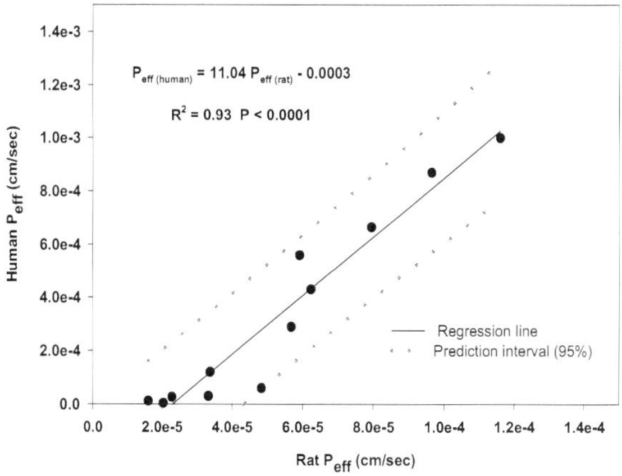

Fig. 6. Plot of P_{eff} rat vs P_{eff} human

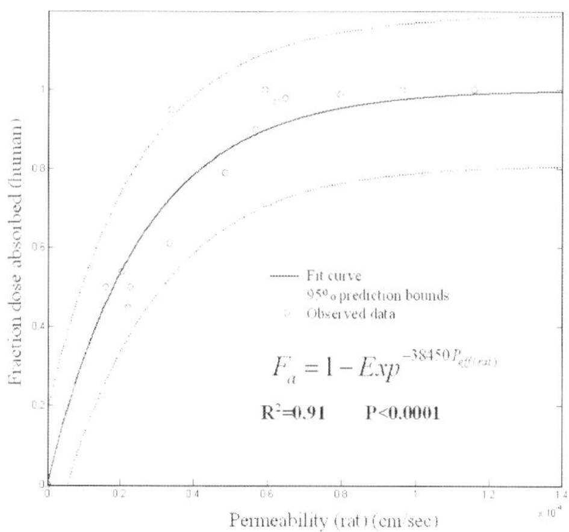

Fig. 7. Plot of rat P_{eff} vs human F_a

The same fitting using human intestinal permeability gives a lower correlation coefficient. The comparison of rat Peff and intestinal absorption in man (F_a) showed that rat P_{eff} values greater than 5.9×10^{-5} cm/sec corresponds to $F_a \approx 1$ while rat P_{eff} values smaller than 3.32×10^{-5} cm/sec corresponds to F_a values lower than 0.6. Corresponding estimates in human are >

0.2×10^{-4} cm/sec and $<0.03\times10^{-4}$ cm/sec, respectively. Moreover the predicted and observed human F_a (%) are linearly correlated ($R^2 = 0.92$, $P < 0.0001$). The rank order for P_{eff} values in rat was compared with those of human P_{eff} and F_a (Zakeri-Milani et al., 2007). The spearman rank correlation coefficients (rs) were found to be 0.96 and 0.91 respectively. Based on the obtained results, it is concluded that in situ perfusion technique in rat could be used as a reliable technique to predict human gastrointestinal absorption extent following oral administration of a drug. However, to render our observation more reliable, it seems that using larger number of compounds belonging to all four biopharmaceutical classes, i.e., different solubility and permeability properties (Lobenberg and Amidon, 2000) especially drugs with low permeability must be tested.

6. Biopharmaceutics classification system using rat P_{eff} as a surrogate for human P_{eff}

In 1995 Amidon et al. devised a biopharmacetics classification system (BCS) to classify drugs based on their aqueous solubility and intestinal permeability, two fundamental properties governing drug absorption (Amidon et al., 1995). This system divides active moieties into four classes: class I (high permeability, high solubility), class II (high permeability, low solubility), class III (low permeability, high solubility) and class IV (low permeability, low solubility). For highly permeable drugs the extent of fraction dose absorbed in human is considered to be more than 90% as defined by US Food and Drug Administration (FDA) (Lennernas and Abrahamsson, 2005, Zakeri-Milani et al., 2009a). The classification of drug solubility is based on the dimensionless dose number (D0) which is the ratio of drug concentration in the administered volume (250 ml) to the saturation solubility of the drug in water. If a drug has dose/solubility ratio less than 250 ml over the pH range from 1 to 7.5 it is classified as highly soluble drug compound (Kasim et al., 2004). BCS classification can help pharmaceutical companies to save a significant amount in development time and reduce costs. This classification provides a regulatory tool to substitute in vivo bioequivalence (BE) studies by in vitro dissolution tests. In fact for immediate-release (IR) solid oral dosage forms containing rapidly dissolving and easily permeating active ingredients bioequivalence studies may not be required because they act like a solution after oral administration. Therefore dissolution rate has a negligible impact on bioavailability of highly soluble and highly permeable (BCS Class I) drugs. As a result, various regulatory agencies including the United States Food and Drug Administration (FDA) now allow bioequivalence of formulations of BCS Class I drugs to be demonstrated by in vitro dissolution (often called a biowaiver) (Takagi et al., 2006) . Waivers for class III drugs have also been recommended (Blume and Schug, 1999, Yu et al., 2002) . Moreover BCS provides distinct rules for determining the rate-limiting factor in the gastrointestinal drug absorption process. As a result it could be helpful in the selection of candidate drugs for full development, prediction and clarification of food interactions, choice of formulation principle and the possibility of in vitro-in vivo correlation in the dissolution testing of solid formulations (Lennernas and Abrahamsson, 2005, Fleisher et al., 1999). Although permeability classification of drugs would be ideally based on human jejunal permeability data, such information is available for only a small number of drugs. Therefore in this section a new classification is presented which is based on a correlation between rat and human intestinal permeability values. However first the calculation of used parameters is explained.

7. Dose number calculation

Dose number is a criterion for solubility (Do) which is defined as the ratio of dose concentration to drug solubility. It is calculated as follows:

$$D_o = \frac{M / V_o}{C_s} \tag{19}$$

Where (C_s) is the solubility, (M) is the maximum dose strength, and (V_o) is the volume of water taken with the dose (generally set to be 250 mL). The values of solubility and maximum dose strength of tested compounds are listed in table 3 . Dose number would be as unity (Do = 1), when the maximum dose strength is soluble in 250 ml of water and the drug is in solution form throughout the GI tract. This criterion is extended to 0.5 for borderline classification, considering the average volume of fluid (500 ml) under fed conditions (Zakeri-Milani et al., 2009b).

8. Dissolution number calculation

Dissolution number refers to the time required for drug dissolution which is the ratio of the intestinal residence time to the dissolution time, which includes solubility (C_s), diffusivity (D), density (ρ), initial particle radius (r_0) of a compound and the intestinal transit time (T_{si}) (Zakeri-Milani et al., 2009b, Varma et al., 2004).

$$Dn = \left(\frac{3D}{r_0^2}\right)\left(\frac{C_s}{\rho}\right)\langle T_{si}\rangle = \frac{\langle T_{si}\rangle}{\langle T_{diss}\rangle} \tag{20}$$

where ρ and T_{si} are generally considered to be 1200 mg/cm^3 and 199 min respectively.

$$T_{diss} = \frac{\rho h r_0}{3DC_s} \tag{21}$$

9. Absorption number calculation

This is the ratio of permeability (P_{eff}) and the gut radius (R) times the residence time in the small intestine which can be written as ratio of residence time and absorption time (Zakeri-Milani et al., 2009b, Varma et al., 2004).

$$An = \frac{P_{eff}}{R} * \langle T_{si}\rangle = \frac{\langle T_{si}\rangle}{\langle T_{abs}\rangle} \tag{22}$$

For calculation the R value of 1.7 cm and the predicted human P_{eff} (based on rat P_{eff}) were used.

10. Absorption time calculation

This parameter is proportional to P_{eff} through the following equation (Zakeri-Milani et al., 2009b, Varma et al., 2004).

$$T_{abs} = \frac{R}{P_{eff}} \tag{23}$$

11. Absorbable dose calculation

Absorbable dose is the amount of drug that can be absorbed during the period of transit time, when the solution contacting the effective intestinal surface area for absorption is saturated with the drug (Zakeri-Milani et al., 2009b, Varma et al., 2004).

$$D_{abs} = P_{eff} C_s A < T_{si} > \tag{24}$$

In this equation A is the effective intestinal surface area for absorption. If the small intestine is assumed to be a cylindrical tube with a radius of about 1.5 cm and length of 350 cm, the available surface area and volume are 3297 cm^2 and 2473 ml, respectively. In reality, the actual volume is around 600 ml and the effective intestinal surface area is then estimated to be about 800 cm^2 assuming the same ratio. Drugs were classified to the BCS on the basis of dose number (Do) and rat jejunal permeability values, which are taken as indicative of fundamental properties of drug absorption, solubility and permeability. On the basis of the relationship between human and rat intestinal permeability (Zakeri-Milani et al., 2009a, Zakeri-Milani et al., 2007) , rat P_{eff} values greater than 5.09×10^{-5} cm/sec corresponds to F_a > 85 % while P^{eff} values smaller than 4.2×10^{-5} cm/sec corresponds to Fa values lower than 80 %. Therefore, as it can be seen in Fig 8 a cutoff for highly permeable drugs, $P_{eff\ rat} = 5.09\times10^{-5}$ cm/sec with a border line cutoff of 4.2×10^{-5} cm/sec can be set. Drugs with permeability in the range of 4.2-5.09e^{-5} cm/sec were considered as borderline drugs. The intersections of dashed lines drawn at the cutoff points for permeability and dose/solubility ratio divide the plane in Fig. 8 into four explicitly defined drug categories (I – IV) and a region of borderline.

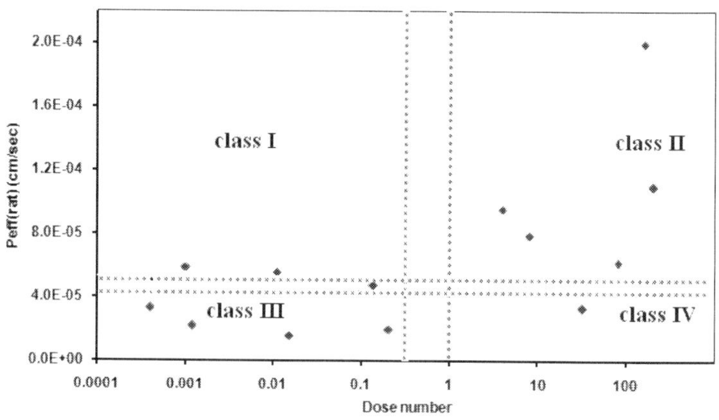

Fig. 8. Plot of Dose number vs rat P_{eff} values representing the four classes of tested compounds

The biopharmaceutical properties of a drug determine the pharmacokinetic characteristics as below:

Class I, Do <0.5, $P_{eff\ (rat)} > 5.09\times10^{-5}$ cm/sec

The drugs in this category are highly soluble and highly permeable and are ideal candidates for oral delivery. These drugs are characterized by the high An, high Dn and low Do, showing that they are in solution form throughout the intestine and is available for permeation. Therefore the rate of absorption of drugs in this class is controlled only by gastric emptying. Examples of this category include antipyrine and propranolol.

Class II, Do > 1, $P_{eff\ (rat)} > 5.09\times10^{-5}$ cm/sec

Class II drugs have high lipophilicity and therefore are highly permeable across the GI membrane, primarily by passive transport. These drugs are characterized by mean absorption time less than mean dissolution time, and thus gastric emptying and GI transit are important determinants of drug absorption (Varma et al., 2004). These drugs are expected to have a dissolution-limited absorption and an IVIVC is expected (Lennernas and Abrahamsson, 2005). Low dissolution rate of these molecules limit the concentration at the site of absorption thereby leading to less passive diffusion. Therefore formulation plays an important role in the rate and extent of intestinal absorption of such drugs. Although there are methods to enhance the solubility of class II drugs (Valizadeh et al., 2004, Valizadeh et al., 2007), incorporation of polar groups into the chemical backbone, salt generation and prodrug approaches are the primary methods for improving deliverability during lead optimization.. This class includes drugs such as ketoprofen, naproxen, piroxicam and carbamazepine.

Class III, Do <0.5, $P_{eff\ (rat)} < 4.2\times10^{-5}$ cm/sec

The absorption of class III drugs is limited by their intestinal permeability and no IVIVC should be expected. These drugs are either having unfavorable physicochemical properties leading to less intrinsic permeability and/or are strong substrates to efflux transporters and/or gut wall metabolic enzymes (Varma et al., 2004). Therefore the rate and extent of intestinal absorption may be controlled by drug molecule properties and physiological factors rather than pharmaceutical formulation properties (Yu et al., 2002). They must possess optimum lipophilicity in order to permeate the lipophilic epithelial cell membranes lining the gastrointestinal tract. Thus for highly polar compounds, administration of less polar, more lipophilic prodrugs may improve absorption. Balance between the hydrophilicity and lipophilicity should be maintained during incorporation of lipophilic groups into the structure. Atenolol, hydrochlorothiazide and ranitidine are examples of drugs in this group.

Class IV, Do > 1, $P_{eff\ (rat)} < 4.2\times10^{-5}$ cm/sec

Low and variable absorption for these drugs is anticipated because of the combined limitation of solubility and permeability. Formulation may improve the bioavailability of these drugs. However they are compromised by their poor intestinal membrane permeability. These drugs are more likely susceptible to P-gp efflux and gut metabolism, as the concentration of the drug in the enterocytes at any given time will be less to saturate the transporter (Varma et al., 2004). Strategies to improve both solubility and permeability should be worked out for these molecules, which may not be an easy task. However, obtaining this type of quality information will certainly improve drug design and help in optimizing candidates with "brick-like" properties.

Borderline Class, 0.5 <Do <1 or $4.2\times10^{-5} < P_{eff\ (rat)} < 5.09\times10^{-5}$ cm/sec

In this region, bordered by the dashed lines of the four cutoff points, the predictions become more uncertain for drugs lying. Cimetidine which is supposed to be in class III, has been classified in this region. All in all, 13 of 15 test drugs (87%) are correctly classified with respect to their rat Peff values, however, metoprolol, a drug with high permeability, was classified as a low permeability drug in the presented plot (False negative). Furthermore there are some more fundamental parameters describing oral drug absorption. These parameters include absorption number, dissolution number, absorption time and dissolution time (Varma et al., 2004). There is also an extra parameter named absorbable dose which was calculated to propose the absorption limiting steps in oral absorption of tested drugs. Three dimensionless parameters (Do, An and Dn) which were shown in Table 3 can be used to qualitative classification of drugs. The four BCS classes of drugs were defined as below on the basis of these three parameters. For easy comparison Table 3 was set in which the dimensionless parameters for each class of drugs were compared.

Compound	D_{abs} Calculated (mg)	T_{diss} calculated (min)	T_{abs} Calculated (min)	D_n Calculated	A_n calculated	D_o calculated	C_s (mg/ml)	Dose (mg)	Mean P_{eff} (*10^5 cm/s)
Antipyrine	3519359	0.01	76.5	11784.9	2.58	0.001	1000 [a]	250[a]	5.9 ± 0.2
Propranolol	109621	0.65	81.1	302.1	2.44	0.011	33 [a]	90 [a]	5.6 ± 2.0
Carbamazepine	38	1915.2	71.2	0.10	2.78	80.0	0.01 [a]	200 [a]	6.2 ± 0.6
Ibuprofen	150	2249.7	18.0	0.08	10.99	160.0	0.01 [a]	400 [a]	20 ± 2.2
Ketoprofen	328	395.8	41.1	0.50	4.81	4.0	0.05 [b]	50 [a]	9.6 ± 1.8
Naproxen	81	1947.3	33.1	0.10	5.98	200.0	0.01 [b]	500 [a]	11 ± 0.2
Piroxicam	26	4204.7	52.1	0.04	3.79	8.0	0.005 [a]	10 [a]	7.9 ± 4.0
Metoprolol	1448424	0.03	185.9	5917.4	1.06	0.0004	1000 [a]	100 [a]	3.3 ± 1.5
Furosemide	14	2025.8	190.6	0.09	1.03	32.0	0.01 [a]	80 [a]	3.3 ± 2.0
Cimetidine	15841	3.1	102.0	62.0	1.94	0.133	6 [c]	200 [a]	4.8 ± 0.1
Atenolol	196	0.81	36326.8	242.6	0.005	0.015	26.5 [a]	100 [a]	1.6 ± 0.02
Ranitidine	560303	0.02	480.6	8800.8	0.41	0.001	1000 [a]	300 [a]	2.2 ± 1.0
Hydrochlorothiazide	360	17.9	747.0	11.0	0.26	0.200	1 [a]	50 [a]	2.0 ± 1.0

Table 3. Dose, solubility and calculated oral drug absorption parameters for tested compounds (Zakeri-Milani et al., 2009b)

Class	Dimensionless parameters	Permeability	Solubility
I	A_n↑* D_n↑ D_o↓	High	High
II	A_n↑ D_n↓ D_o↑	High	Low
III	A_n↓ D_n↑ D_o↓	Low	High
IV	A_n↓ D_n↓ D_o↑	Low	Low

*symbols↓ and↑ represent low and high quantity for parameters

Table 4. Qualitative classification of drugs based on dimensionless parameters

Absorption limiting step	Examples	Comments	Condition
No limited	Antipyrine, Propranolol, Cimetidine	There is no limitation in drug absorption since all three parameters are in acceptable range.	T_{diss}< 50 min $P_{eff\ rat}$ > 4.2×10^{-5} D_{abs} >> Dose
Dissolution limited	Ketoprofen, Piroxicam	Although solubility itself imparts to poor dissolution, the dissolution here mainly refers to particle size. The absolute bioavailability increases with increasing dose.	T_{diss} > 199 min $P_{eff\ rat}$ > 4.2×10^{-5} D_{abs} >> Dose
Solubility limited	Ibuprofen, Carbamazepine, Naproxen	Solubility-limited absorption occurs mainly when a high dose saturates part of the gut. The absolute bioavailability does not increase with increasing dose.	T_{diss} > 199 min $P_{eff\ rat}$ > 4.2×10^{-5} D_{abs} < Dose
Permeability limited	Ranitidine, Atenolol, Metoprolol, Hydrochlorothiazide	This limiting step is considered for highly soluble drugs dosed in solutions: assume no precipitation occurs. The absolute bioavailability increases with increasing dose.	T_{diss}< 50 min $P_{eff\ rat}$ < 4.2×10^{-5} D_{abs} >> Dose
Dissolution-permeability-solubility-limited	Furosemide	Drug absorption is limited by all steps including solubility, permeability and dissolution	T_{diss} > 199 min $P_{eff\ rat}$ < 4.2×10^{-5} D_{abs} < Dose

Table 5. Absorption limiting steps and their corresponding conditions

This classification is in accordance with quantitative classification model which was given in the first part of current section, i.e. all compounds lie in the same class as did in quantitative classification. For example atenolol with a Do = 0.015 (low), An = 0.005 (low) and Dn = 242 (high) is classified in class III which is in agreement with above-mentioned QBCS. Again metoprolol with An of 1.06 lies in class III as it did before in quantitative model. However this is a false negative result, since it was known to have a high permeability belonging to class I. Another interesting aspect of using these dimensionless parameters is to determine the absorption limiting steps which was summarized as a framework in Table 5. As it was mentioned before, the mean small intestinal transit time was found to be 199 minutes with a standard deviation of 78 minutes (Yu, 1999, Zakeri-Milani et al., 2009b). This means that as a worst case, the small intestinal transit in some individuals may be only 43 minutes (mean small intestinal transit time - 2 × standard deviation). The time of 50 minutes was used as a reference time of dissolution to determine if the dissolution is fast enough to permit complete dissolution in the small intestine (Yu, 1999). The $P_{eff\ (rat)}$ was set at 4.2×10^{-5} cm/sec which based on our correlations, corresponds to over 80% of dose absorbed. Table 3 provides distinguishing conditions under which each limiting case occurs. Considering these conditions, antipyrine and propranolol meet the criteria for no-limited absorption. All of these three drugs belong to class I. However cimetidine a drug which was false positive in our previous quantitative and qualitative classification lies in no-limited class again. On the other hand based on dissolution time, permeability and absorbable dose for furosemide, a drug of class IV, its absorption would be limited by all three parameters. Therefore it takes place in the last class of Table 5. Furthermore drugs with low permeability which have a high absorbable dose and low dissolution time such as ranitidine and hydrochlorothiazide (class III), are classified in permeability-limited category. Finally the drugs of remaining class of BCS (class II) are divided in two groups based on their relative values of dimensionless parameters. All of these drugs have high dissolution time (Table 3), but regarding the absorbable dose, their absorption could be dissolution or solubility-limited. For instance, piroxicam and ketoprofen lie in dissolution-limited class, while naproxen is placed in solubility-limited category. According to obtained results and proposed classification for drugs, it is concluded that drugs could be categorized correctly based on dose number and their P_{eff} values in rat model using SPIP technique. This classification enables us to remark defined characteristics for intestinal absorption of all four classes using suitable cutoff points for both dose number and rat effective intestinal permeability values. Therefore the classification of drugs using their intestinal permeability values in rats can help pharmaceutical companies to save a significant amount in development time and reduce costs. Moreover it could be as a regulatory tool to substitute in vivo bioequivalence (BE) studies by in vitro dissolution tests. However this work relies on only 13 compounds which their P_{eff} values in rat were measured and to confirm the proposed classification the larger data set is needed.

12. Biopharmaceutical classification of drugs using intrinsic dissolution rate (IDR) and rat intestinal permeability

The solubility and dissolution rate of active ingredients are of major importance in preformulation studies of pharmaceutical dosage forms (Valizadeh et al., 2007, Valizadeh et al., 2004, Barzegar-Jalali et al., 2006, Zakeri-Milani et al., 2009a). The formulation characteristics including shelf life, process behavior, and even the bioavailability are affected by physicochemical properties of drug molecules (Haleblian and McCrone, 1969). The intrinsic dissolution rate (IDR) has been used to characterize solid drugs for many years. For example it could be used to understand the relationship between the dissolution rate and crystalline form and also to study the effects of surfactants and pH on the solubilization of poorly soluble drugs (Amidon et al., 1982, Yu et al., 2004, Zakeri-Milani et al., 2009a). IDR is generally defined as the dissolution rate of a pure drug substance under the condition of constant surface area, agitation or stirring speed, pH and ionic strength of the dissolution medium. The true intrinsic dissolution rate may be better described as the rate of mass transfer from the solid surface to the liquid phase. The apparatus for intrinsic dissolution testing was originally developed by John Wood which enables the calculation of the dissolution rate per centimeter squared of the intrinsic ingredients of pharmaceutical products (Levy and Gumtow, 1963, Nelson, 1958). It has been suggested that it might be feasible to use IDR to classify drugs instead of solubility (Yu et al., 2004). The reason is that, just like permeability, IDR is a rate phenomenon instead of an equilibrium phenomenon. Therefore it might correlate better with in vivo drug dissolution rate than solubility, although for drugs having either extremely high or low dose, discrepancies may exist between the solubility and IDR methods since dose is considered in the classification of solubility while intrinsic dissolution does not consider the effect of dose. In the present study the intrinsic dissolution rate and rat intestinal permeability (using SPIP technique) were measured for drugs with different physicochemical properties. The suitability of IDR-permeability for biopharmaceutical classification of drugs was evaluated.

13. Procedure of IDR measurement

A quantity of 100 mg of each drug was compressed at an average compression force of 7.84 MPa for 1 minute to make non-disintegrating compacts using die and punch with diameter of 6 mm. The surface area of the compacts was 0.2826 cm^2. The improved method of wood et al was used for disk dissolution studies (Wood et al., 1965). Compacts were placed in a molten beeswax-mold in such a way that only one face could be in contact with dissolution medium. Dissolution study was conducted using USP II dissolution apparatus using 900mL of phosphate buffer (pH=6.8) at temperature of 37°C ± 1°C as the dissolution media with paddle rotating at 100 rpm. Samples were collected through 0.45-μm syringe filters over a period of 8 hours for low-soluble and 20 minutes for highly soluble drugs. Sampling time intervals were 30 min and 2 min respectively. All studies were carried out in triplicate. Absorbances were determined in triplicate using a UV-Vis spectrophotometer at the maximum absorbance wavelength for each active tested. The cumulative amount dissolved per surface unit of the compact was plotted against time for each vessel. The slope of the linear region ($R^2 \geq 0.95$) was taken as intrinsic dissolution rate. IDR is easily calculated by

$$G = (dw/dt)(1/S) = DCs/h \tag{25}$$

where G is intrinsic dissolution rate ($mg/min/cm^2$); dw is the change in drug dissolved (mg); dt is the change in time (minutes); S is the surface area of the compact (cm^2); D is diffusion coefficient (cm^2/sec); Cs is solubility (mg/cm^3) and h is stagnant layer thickness (cm) (Zakeri-Milani et al., 2009a).

14. Solubility studies

Solubilities were determined in at least triplicates by equilibrating excess amount of drugs in phosphate buffer solutions (pH=6.8). The samples were kept in thermostated water bath at 37°C and shaked at a rate of 150 rpm for 24 hours. The absorbances of filtered and suitably diluted samples were measured with an UV-VIS spectrophotometer at the maximum absorbance wavelength for each active tested. The solubilities were calculated using calibration curves determined for each drug (Zakeri-Milani et al., 2009a). Current BCS guidance defines an API as "highly soluble" when the highest dose recommended is soluble in 250 mL or less of aqueous media over the pH range of 1.2 to 7.5 (Gupta et al., 2006). However the pH 6.8 is scientifically justified over pH 7.4 (Gupta et al., 2006). In order to set a condition for BCS classification of compounds and since small intestine is the major site for drug absorption, where the pH is about 6.8, IDR measurements were conducted in pH 6.8.

The presence of sink condition in dissolution medium during the experiment is upholded by comparison of the final concentration of drugs and their solubility in dissolution medium. Classification of tested drugs based on their intestinal permeability and IDR for human and rat is shown in Fig. 9 and Fig. 10 respectively. Drugs are scientifically identified based on their solubility and human intestinal permeability. Since human intestinal permeability could be predicted with precise using the rat effective permeability values, the same classification can be constructed utilizing the solubility and rat intestinal permeability values. IDR is a parameter which could be used easily to characterize the pure drug substance. The determination of this parameter allows labs to screen experimental drug formulations and to understand their behavior under different bio-physical conditions. Table 6 shows the obtained solubility and IDR values in the present work for tested drugs.

Compound	Drug class					IDR (mg cm^{-2} min^{-1})	Solubility (mg/l)	Wavelength (nm)
	Dissolution based	BDDCS	BCS	This work**	This work*			
Antipyrin	I	I	I	I	I	56.79	683271.6	243
Metoprolol	I	I	I	III	I	34.64	779580.8	274
Propranolol	II	I	I	I	I	16.596	71797.17	288
Verapamil	I	I	I	I	I	16.192	71602.64	274
Ketoprofen	II	I	I	II	II	0.6348	2121.80	261
Naproxen	II	II	II	II	II	0.388	1604.45	262
Carbamazepine	IV	II	II	II	II	0.0355	164.59	285
Ibuprofen	-	II	II	II	II	0.2844	1315.41	222
Piroxicam	-	II	II	II	II	0.0739	157.64	353
Atenolol	III	III	III	III	III	3.449	16868.14	224
Cimetidine	-	III	III	III	III	7.2	46276.68	219
Ranitidine	III	III	III	III	III	42.18	>1000000	228
Furosemide	IV	IV	IV	IV	IV	0.58	1464.42	277

* proposed class based on IDR and human intestinal permeability
**proposed class based on IDR and rat intestinal permeability

Table 6. Experimental wavelength, Solubility, intrinsic dissolution rate (IDR), and respective class of tested compounds using different approaches

The IDR results on tested drugs are in agreement with previously reported values (Yu et al., 2004). In the present study the obtained rat Peff values showed a high correlation (R^2=0.93, P<0.0001) with human Peff data for passively absorbed compounds confirming the validity of our procedure (Zakeri-Milani et al., 2007). It was found that a strong correlation was observed between rat permeability data and fraction of oral dose absorbed in human (R^2=

0.91, P<0.0001). The same correlation for human intestinal permeability data and fraction of oral dose absorbed gives a lower correlation coefficient (R^2= 0.81, P<0.0001). However according to obtained equations, the permeabilities of 0.0000509 and 0.000047 cm/sec in rat and human respectively corresponds to Fa=85% which are set as cut-off points for highly permeable drugs. On the other hand, IDR correlates with the BCS solubility classification with 1-2 mg/min/cm² as a class boundary. It is seen that antipyrin, ranitidine and metoprolol with IDRs of 56.79, 42.18 and 34.64 mg/cm²/min respectively have the higher values in comparison to others whereas carbamazepine and piroxicam have the lowest intrinsic dissolution rate in the series (IDR=0.035 and 0.07 mg/cm²/min respectively). This order is almost the same for solubility of mentioned drugs. However in the case of permeability this arrangement is not expected. The reason is that the investigated drugs belong to all four biopharmaceutical classes. That means a drug with high IDR value may belong to high or low permeability classes. In the present study passively absorbed drugs are classified based on their intrinsic dissolution rates and human intestinal permeability values (Zakeri-Milani et al., 2009a).. IDR was expected to correlate more closely with in vivo dissolution dynamics of drug than solubility. Therefore it could be used to correct assignment of a drug to a specific BCS class. This classification is presented in Fig 9 and Fig 10 (Zakeri-Milani et al., 2009a).

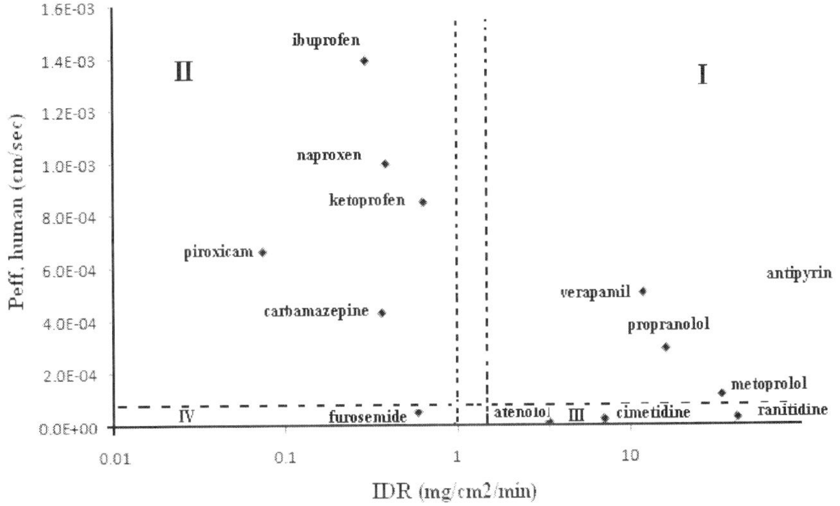

Fig. 9. Classification of tested drugs based on their human intestinal permeability and IDR

Based on this classification, drugs are placed in four explicitly defined categories (I-IV) which are made by intersections of dashed lines drawn at the cutoff points for permeability and IDR. These classes are characterized as below:

Class I: $_{Peff,rat}$ > 5× 10^{-5} (cm/sec) or $P_{eff,human}$ > 4.7× 10^{-5} (cm/sec) , IDR > 2 (mg/min/cm²)

Examples of the compounds of this category include propranolol, metoprolol, verapamil and antipyrin which exhibit a high dissolution and absorption. However according to intestinal permeability estimates in rat, metoprolol is assigned in class III.

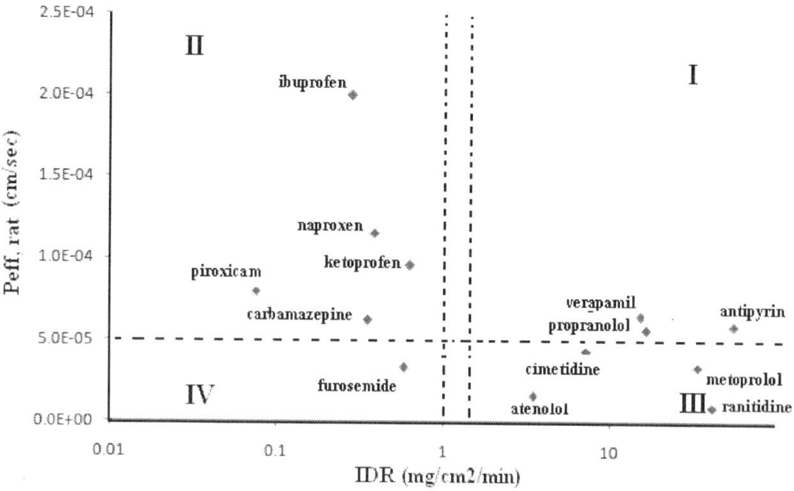

Fig. 10. Classification of tested drugs based on their rat intestinal permeability and IDR

Class II: $P_{eff,rat} > 5\times 10^{-5}$ (cm/sec) or $P_{eff,human} > 4.7\times 10^{-5}$ (cm/sec), IDR < 1 (mg/min/cm^2)
Drugs like ketoprofen, naproxen, piroxicam, ibuprofen and carbamazepine are included in this category. Class II drugs have a high absorption but a low dissolution therefore absorption is limited primarily by drug dissolution in the gastrointestinal tract (Amidon et al., 1995).

Class III: $P_{eff,rat} < 5\times 10^{-5}$ (cm/sec) or $P_{eff,human} < 4.7\times 10^{-5}$ (cm/sec) , IDR > 2 (mg/min/cm^2)
Class III drugs, have high dissolution and low absorption. In vivo permeability is rate limiting step for drug absorption . Examples are atenolol, ranitidine and cimetidine.

Class IV: $P_{eff,rat} < 5\times$ 10-5 (cm/sec) or $P_{eff,human} < 4.7\times 10^{-5}$ (cm/sec) , IDR < 1 (mg/min/cm^2)

Furosemide is an example of drugs of this category which exhibit a lot of problems for effective oral administration. From the obtained results it is provided that the presented classification based on IDR and human intestinal permeability of drugs is in high agreement with previously introduced classification and most of the compounds are placed in correct categories they belong to (Amidon et al., 1995). Although using the rat intestinal permeability values instead of human intestinal permeability, metoprolol was almost misclassified, considering non-feasibility of using human in intestinal perfusion studies, which is the major difficulty in assigning drugs to BCS classes, it may be suggested that determined intestinal permeability of drugs in rats could be used as a criterion for biopharmaceutical classification of compounds. On the other hand, it was proposed that a biopharmaceutics drug disposition classification system (BDDCS) based on extent of drug metabolism could provide an alternative simple method to assign drugs in class I for a waiver of in vivo bioequivalence studies (Takagi et al., 2006, Benet et al., 2008, Wu and Benet, 2005). According to this classification highly metabolized drugs exhibit high permeability. Therefore a drug is considered to be class I if it is highly soluble and highly metabolized. However this definition excludes drugs that have high absorption but are excreted unchanged in to bile and urine (Takagi et al., 2006). Comparison of our results with

BDDCS classification (≥50% being defined as extensive metabolism) of drugs (Wu and Benet, 2005) shows high agreement (92% and 85% using human and rat intestinal permeability respectively) in classification of tested compounds (Table 6). Another classification system namely dissolution-based classification was developed by Papadopoulou et al (Papadopoulou et al., 2008) using mean intestinal transit time (MITT), mean dissolution time (MDT) and mean absorption time (MAT). The comparison of this classification with our results is also shown in Table 6. However in dissolution-based classification propranolol and carbamazepine are classifies as class II and class IV drugs respectively which are expected to be assigned in class I and II respectively as was shown in other classifications in Table 6. It seems that the presented classification could be used to waive in vivo bioavailability and bioequivalence studies for immediate release solid oral dosage forms which allows pharmaceutical companies to forego clinical bioequivalence studies, if their drug product meets the required specification. However at the time being, our attempt is to introduce some thermodynamic parameters as a surrogate for permeability measurements.

15. References

Amidon, G. E., Higuchi, W. I. & Ho, N. F. (1982). Theoretical and experimental studies of transport of micelle-solubilized solutes. *J Pharm Sci,* 71, 77-84.

Amidon, G. L., Kou, J., Elliott, R. L. & Lightfoot, E. N. (1980). Analysis of models for determining intestinal wall permeabilities. *J Pharm Sci,* 69, 1369-73.

Amidon, G. L., Lennernas, H., Shah, V. P. & Crison, J. R. (1995). A theoretical basis for a biopharmaceutic drug classification: the correlation of in vitro drug product dissolution and in vivo bioavailability. *Pharm Res,* 12, 413-20.

Barzegar-Jalali, M., Nayebi, A. M., Valizadeh, H., Hanaee, J., Barzegar-Jalali, A., Adibkia, K., Anoush, M. & Sistanizad, M. (2006). Evaluation of in vitro-in vivo correlation and anticonvulsive effect of carbamazepine after cogrinding with microcrystalline cellulose. *J Pharm Pharm Sci,* 9, 307-16.

Benet, L. Z., Amidon, G. L., Barends, D. M., Lennernas, H., Polli, J. E., Shah, V. P., Stavchansky, S. A. & Yu, L. X. (2008). The use of BDDCS in classifying the permeability of marketed drugs. *Pharm Res,* 25, 483-8.

Bird, R. B., Stewart, W. E. & Lightfoot, E. N. (1960). *Transport phenomena,* New York, Wiley.

Blume, H. H. & Schug, B. S. (1999). The biopharmaceutics classification system (BCS): class III drugs - better candidates for BA/BE waiver? *Eur J Pharm Sci,* 9, 117-21.

Chiou, W. L. & Barve, A. (1998). Linear correlation of the fraction of oral dose absorbed of 64 drugs between humans and rats. *Pharm Res,* 15, 1792-5.

Dressman, J. B., Fleisher, D. & Amidon, G. L. (1984). Physicochemical model for dose-dependent drug absorption. *J Pharm Sci,* 73, 1274-9.

Elliott, R. L., Amidon, G. L. & Lightfoot, E. N. (1980). A convective mass transfer model for determining intestinal wall permeabilities: laminar flow in a circular tube. *J Theor Biol,* 87, 757-71.

Fagerholm, U., Johansson, M. & Lennernas, H. (1996). Comparison between permeability coefficients in rat and human jejunum. *Pharm Res,* 13, 1336-42.

Fleisher, D. 2000. Biological transport phenomena in the gastrointestinal tract: Cellular mechanisms. *In:* AMIDON, G. L., LEE, P. I. & TOPP, E. M. (eds.) *Transport processes in pharmaceutical systems.* New York: Marcel Dekker Inc.

Fleisher, D., Li, C., Zhou, Y., Pao, L. H. & Karim, A. (1999). Drug, meal and formulation interactions influencing drug absorption after oral administration. Clinical implications. *Clin Pharmacokinet,* 36, 233-54.

Gupta, E., Barends, D. M., Yamashita, E., Lentz, K. A., Harmsze, A. M., Shah, V. P., Dressman, J. B. & Lipper, R. A. (2006). Review of global regulations concerning biowaivers for immediate release solid oral dosage forms. *Eur J Pharm Sci,* 29, 315-24.

Haleblian, J. & Mccrone, W. (1969). Pharmaceutical applications of polymorphism. *J Pharm Sci,* 58, 911-29.

Hayduk, W. & Laudie, H. (1974). Prediction of diffusion coefficients for non electrolytes in dilute aqueous solutions. *J Am Inst Chem Eng,* 20, 611-615.

Ho, N. F. & Higuchi, W. I. (1974). Theoretical model studies of intestinal drug absorption. IV. Bile acid transport at premicellar concentrations across diffusion layer-membrane barrier. *J Pharm Sci,* 63, 686-90.

Kasim, N. A., Whitehouse, M., Ramachandran, C., Bermejo, M., Lennernas, H., Hussain, A. S., Junginger, H. E., Stavchansky, S. A., Midha, K. K., Shah, V. P. & Amidon, G. L. (2004). Molecular properties of WHO essential drugs and provisional biopharmaceutical classification. *Mol Pharm,* 1, 85-96.

Lee, V. H. L. & Yang, J. Y. 2001. Oral drug delivery. *In:* HILLERY, A. M., LLOYD, A. W. & SWARBRICK, J. (eds.) *Drug delivery and targeting.* London: Taylor & Francis.

Lennernas, H. (1997). Human jejunal effective permeability and its correlation with preclinical drug absorption models. *J Pharm Pharmacol,* 49, 627-38.

Lennernas, H. & Abrahamsson, B. (2005). The use of biopharmaceutic classification of drugs in drug discovery and development: current status and future extension. *J Pharm Pharmacol,* 57, 273-85.

Levy, G. & Gumtow, R. H. (1963). Effect of Certain Tablet Formulation Factors on Dissolution Rate of the Active Ingredient. Iii. Tablet Lubricants. *J Pharm Sci,* 52, 1139-44.

Lobenberg, R. & Amidon, G. L. (2000). Modern bioavailability, bioequivalence and biopharmaceutics classification system. New scientific approaches to international regulatory standards. *Eur J Pharm Biopharm,* 50, 3-12.

Madan, J., Chawla, G., Arora, V., Malik, R. & Bansal, A. K. (2005). Unbiased membrane permeability parameters for gabapentin using boundary layer approach. *AAPS J,* 7, E224-30.

Nelson, E. (1958). Comparative dissolution rates of weak acids and their sodium salts. *J Am Pharm Assoc Am Pharm Assoc (Baltim),* 47, 297-9.

Papadopoulou, V., Valsami, G., Dokoumetzidis, A. & Macheras, P. (2008). Biopharmaceutics classification systems for new molecular entities (BCS-NMEs) and marketed drugs (BCS-MD): theoretical basis and practical examples. *Int J Pharm,* 361, 70-7.

Salphati, L., Childers, K., Pan, L., Tsutsui, K. & Takahashi, L. (2001). Evaluation of a single-pass intestinal-perfusion method in rat for the prediction of absorption in man. *J Pharm Pharmacol,* 53, 1007-13.

Sinko, P. J. (2006). *Martin's physical pharmacy and pharmaceutical sciences,* Baltimore, Lippincott Williams and Wilkins.

Sinko, P. J., Leesman, G. D. & Amidon, G. L. (1991). Predicting fraction dose absorbed in humans using a macroscopic mass balance approach. *Pharm Res,* 8, 979-88.

Takagi, T., Ramachandran, C., Bermejo, M., Yamashita, S., Yu, L. X. & Amidon, G. L. (2006). A provisional biopharmaceutical classification of the top 200 oral drug products in the United States, Great Britain, Spain, and Japan. *Mol Pharm,* 3, 631-43.

Valizadeh, H., Nokhodchi, A., Qarakhani, N., Zakeri-Milani, P., Azarmi, S., Hassanzadeh, D. & Lobenberg, R. (2004). Physicochemical characterization of solid dispersions of indomethacin with PEG 6000, Myrj 52, lactose, sorbitol, dextrin, and Eudragit E100. *Drug Dev Ind Pharm,* 30, 303-17.

Valizadeh, H., Tajerzadeh, H. & Zakeri-Milani, P. (2008). Comparison of mass transfer models for determination of the intestinal permeability. *Daru,* 16, 203-210.

Valizadeh, H., Zakeri-Milani, P., Barzegar-Jalali, M., Mohammadi, G., Danesh-Bahreini, M. A., Adibkia, K. & Nokhodchi, A. (2007). Preparation and characterization of solid dispersions of piroxicam with hydrophilic carriers. *Drug Dev Ind Pharm,* 33, 45-56.

Varma, M. V., Khandavilli, S., Ashokraj, Y., Jain, A., Dhanikula, A., Sood, A., Thomas, N. S., Pillai, O., Sharma, P., Gandhi, R., Agrawal, S., Nair, V. & Panchagnula, R. (2004). Biopharmaceutic classification system: a scientific framework for pharmacokinetic optimization in drug research. *Curr Drug Metab,* 5, 375-88.

Washington, N., Washington, C. & Wilson, C. G. (2000). *Physiological pharmaceutics, barriers to drug absorption,* London, Taylor & Francis.

Winne, D. (1978). Dependence of intestinal absorption in vivo on the unstirred layer. *Naunyn Schmiedebergs Arch Pharmacol,* 304, 175-81.

Winne, D. (1979). Rat jejunum perfused in situ: effect of perfusion rate and intraluminal radius on absorption rate and effective unstirred layer thickness. *Naunyn Schmiedebergs Arch Pharmacol,* 307, 265-74.

Wood, J., Syarto, J. & Letterman, H. (1965). Improved holder for intrinsic dissolution rate studies. *J Pharm Sci,* 54, 1068.

Wu, C. Y. & Benet, L. Z. (2005). Predicting drug disposition via application of BCS: transport/absorption/ elimination interplay and development of a biopharmaceutics drug disposition classification system. *Pharm Res,* 22, 11-23.

Yu, L. X. (1999). An integrated model for determining causes of poor oral drug absorption. *Pharm Res,* 16, 1883-7.

Yu, L. X., Amidon, G. L., Polli, J. E., Zhao, H., Mehta, M. U., Conner, D. P., Shah, V. P., Lesko, L. J., Chen, M. L., Lee, V. H. & Hussain, A. S. (2002). Biopharmaceutics classification system: the scientific basis for biowaiver extensions. *Pharm Res,* 19, 921-5.

Yu, L. X., Carlin, A. S., Amidon, G. L. & Hussain, A. S. (2004). Feasibility studies of utilizing disk intrinsic dissolution rate to classify drugs. *Int J Pharm,* 270, 221-7.

Zakeri-Milani, P., Barzegar-Jalali, M., Azimi, M. & Valizadeh, H. (2009a). Biopharmaceutical classification of drugs using intrinsic dissolution rate (IDR) and rat intestinal permeability. *Eur J Pharm Biopharm,* 73, 102-6.

Zakeri-Milani, P., Valizadeh, H., Tajerzadeh, H., Azarmi, Y., Islambolchilar, Z., Barzegar, S. & Barzegar-Jalali, M. (2007). Predicting human intestinal permeability using single-pass intestinal perfusion in rat. *J Pharm Pharm Sci,* 10, 368-79.

Zakeri-Milani, P., Valizadeh, H., Tajerzadeh, H. & Islambulchilar, Z. (2009b). The utility of rat jejunal permeability for biopharmaceutics classification system. *Drug Dev Ind Pharm,* 35, 1496-502.

27

Heat and Mass Transfer in Packed Bed Drying of Shrinking Particles

Manoel Marcelo do Prado[1] and Dermeval José Mazzini Sartori[2]
[1]*Federal University of Sergipe, Department of Chemical Engineering*
[2]*Federal University of São Carlos, Department of Chemical Engineering*
Brazil

1. Introduction

Drying is the most widespread heat and mass transport process with applications in several engineering areas and to a wide range of industrial and agricultural products, including grains and seeds. Wet granular materials are usually dried by forced convection using a hot air flow through a packed bed. Besides the low capital cost and low maintenance cost, packed bed dryers have some advantages in relation to moving bed dryers. For instance, the equipment is of simple operation, requiring no additional energy expense to move the solid particles throughout the bed, and minimizing the mechanical damages to the material. Moreover, investigations into packed bed dryers become increasingly important to obtain information on fluid-particle interactions, because this type of dryer provides the base for better understanding the simultaneous phenomena of heat and mass transfer which occur inside each particle in the bed, and the transfer phenomenon between solid and fluid phases of the packed bed, this being a mixture of dried granulated material and air.

The studies on transport phenomena that occur during drying of different particulate materials are not only of technological interest because of the numerous industrial applications, but also of scientific interest because of the material complexity. Within this context, beds of shrinking particles offer important challenges for the analysis of simultaneous heat and mass transfer during drying. A type of shrinking particle includes basically two separated regions: a gel-based coating, which has a highly deformable porous structure, and a wet core, which consists of liquid and solid. Gel-coated particles undergo significant shrinkage during moisture removal, which leads to changes in both the size and shape of the solid, modifying the structural properties of the particulate bed, thus affecting the fluid-particle interaction. The complexity increases as the extent of shrinkage is also process dependent. That is the result of the moisture gradient in the product, which, in turn, induces stresses and, thus, mechanical deformation (Eichler *et al.*, 1997).

A scientific understanding of heat and mass transfer in drying of deformable porous media and the role of shrinkage phenomenon is required for a more rational design and optimization of drying operating conditions. In this sense, mathematical modelling is very important. A large variety of models has been developed to describe the heat and mass transfer inside deep bed dryers. Comprehensive reviews of these models and simulation methods are available in the literature (Brooker *et al.*, 1992; Cenkowski *et al.*, 1993).

The mathematical models traditionally applied to describe packed bed drying of solid particles are based on principles of mass and energy conservation for the gaseous and solid phases in a controlled volume in conjunction with constitutive equations for thermodynamics equilibrium and heat and mass transfer between the phases. Unfortunately, one of the most common oversights in the so-called two-phase models is to neglect the shrinkage phenomenon as well as the changes induced in structural properties of the particulate bed. However, it should be pointed out that the drying of deformable particles is a problem of significant difference from other drying processes involving rigid particulate material, for which the aforementioned assumptions are valid. As a result, efforts have been directed to characterize physical changes in different materials during drying (Koç *et al.*, 2008; Mayor & Sereno, 2004) as well as to adequate models of packed bed dryers to simulate heat and mass transfer in a deformable porous medium (Prado & Sartori, 2008; Bialobrzewski *et al.*, 2008; Mihoubi & Bellagi, 2008; Kowalski *et al.*, 2007), so that particle-fluid interactions can be understood and drying optimized. Such investigations are important for the development of more efficient dryers, once they could lead to reductions in energy consumption when it comes to air flow requirements for materials that present higher bulk porosity upon shrinkage.

One of the most important parameters required in a drying model is the mass transfer coefficient, which expresses the inverse of a resistance to water transfer from the solid to a drying medium (Zanoelo *et al.*, 2007). Although numerous investigations have been performed for measuring and correlating this parameter in packed beds (Geankoplis, 1993; Krokida *et al.* 2002), the changes in shape, dimension and solid structure of deformable particles yield to a particular system where the available coefficients of mass transfer are not suitable to reproduce mass loss during drying of the material. The shrinkage phenomenon affects the length of the diffusion path in dried material, which influences the moisture diffusion coefficient of the material and, as result, influences the drying rate (Bialobrzewski *et al.*, 2008). As a consequence, any attempt to simulate and optimise the operation of drying of shrinking particles requires an experimental investigation that aims to obtain the mass transfer coefficient involved in this operation. Knowledge of the shrinkage mechanism, and of the influence of the process variables on shrinkage, improves the understanding of drying kinetics (Hashemi *et al.*, 2009; Bialobrzewski *et al.*, 2008).

The objective of this chapter is to provide comprehensive information on theoretical-experimental analysis of coupled heat and mass transfer in packed bed drying of shrinking particles. The modelling of the physical problem is first presented and, then the factors that influence its simulation are discussed. The focus is on the shrinkage phenomenon and its effects on the heat and mass transport coefficients. Finally, the validity of the model to predict the bed displacement and the evolution of moisture content and temperature in fluid and solid phases throughout the granular bed during convective drying is confirmed by using sets of data available for the drying of two deformable porous media, a packing of mucilaginous seeds and a packing of seeds having artificial gel coating.

2. Mathematical modelling

The physical problem under consideration is illustrated in Figure 1, in which a drying fluid flowing upward percolates a packed bed of gel-coated seeds. The detail of a volume element extracted from the packed bed shows that part of this element is composed of solid particulate material, whereas the remaining void space is occupied by the fluid phase.

Interactions between the solid and fluid phases by heat and mass transfer occur simultaneously during drying.

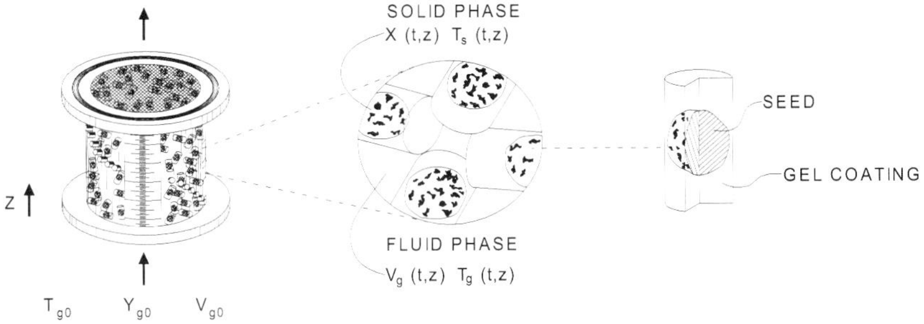

Fig. 1. Sketch of the drying problem

Mass balance in the solid:

$$\frac{\partial X}{\partial t} = f \tag{1}$$

Mass balance in the fluid:

$$\rho_g \left(\frac{v_g}{S_b} \cdot \frac{\partial Y_g}{\partial \xi} + \varepsilon \frac{\partial Y_g}{\partial t} \right) = -\rho_b \cdot \frac{\partial X}{\partial t} \tag{2}$$

Energy balance in the solid:

$$\rho_b \cdot (Cp_S + X \cdot Cp_w) \frac{\partial T_S}{\partial t} = ha_v (T_g - T_S) + \rho_b \cdot \left[L_p + (Cp_v - Cp_w) \cdot T_s \right] \cdot \frac{\partial X}{\partial t} \tag{3}$$

Energy balance in the fluid:

$$\frac{v_g \cdot \rho_g}{S_b} \cdot (Cp_g + Y_g Cp_v) \frac{\partial T_g}{\partial \xi} + \rho_g \cdot \varepsilon \cdot (Cp_g + Y_g Cp_v) \frac{\partial T_g}{\partial t} = \left[ha_v - \rho_b \cdot Cp_v \cdot \frac{\partial X}{\partial t} \right] \cdot (T_s - T_g) \tag{4}$$

Initial and boundary conditions:

$$t = 0 \text{ and } z = 0 \begin{cases} X = X_0 ; T_s = T_{s0} \\ Y_g = Y_{g0;} T_g = T_{g0} \end{cases} \text{ for t=0 and } z \neq 0 \begin{cases} X = X_0 ; T_s = T_{s0} \\ Y_g = Y_{sat}(T_{s0}) ; T_g = T_{s0} \end{cases} \tag{5}$$

$$z = 0 \ \forall \ t, \quad Y_g = Y_{g0} \text{ and } T_g = T_{g0} \tag{6}$$

Table 1. Model equations (Partial differential equations system)

For a complete description of the process, a two-phase model taking into account the effects of bed shrinkage and moisture content on the physical properties is applied. Other assumptions adopted include: the air flow is one-dimensional with uniform distributions of velocity, humidity and temperature in the cross section of the bed; heat losses through dryer walls are negligible; the fluid-solid heat transfer in the packed bed is predominantly convective; and heat and mass transport is one-dimensional. The resulting system of coupled differential equations describing the mass and energy conservation balances for the solid phase and the fluid phase are summarized in the Table 1.

The model equations were written in Lagrangian formulation to incorporate the movement of bed contraction. The dimensionless moving coordinate system, ξ, is related to the spatial coordinate, **z**, by the following equation (Ratti & Mujumdar, 1995):

$$dz = \frac{\rho_{b_0}}{\rho_b} d\xi = \frac{V_b}{V_{b0}} d\xi = S_b \, d\xi \tag{7}$$

The solution of the mathematical model requires knowledge of the shinkage and interphase transport parameters, which are analytically examined in the following sections.

3. Shrinkage

3.1 General aspects

Shrinkage is a physical phenomenon which tends to occur simultaneously with heat and mass transfer in the drying process. In biological materials it occurs when the viscoelastic matrix contracts into the space previously occupied by the water removed from the cells (Aguilera, 2003). Shrinkage is macroscopically characterized by a reduction in volume or area due to evaporation of moisture contained in the solid. According to Mayor & Sereno (2004), heating and water loss cause stresses in the cellular structure of the material and lead to changes in shape and decrease in dimensions.

In deformable porous media, shrinkage or bed displacement can occur as a result of these changes in particles size and shape, and also due to gradients of moisture content generated inside the packed bed.

This phenomenon cannot be ignored for modelling of the drying process, especially for beds of particles with high initial moisture content. Besides the high moisture content, some particles have natural or artificial coating with gel characteristics. Due to the highly deformable polymeric structure of gels, these particles suffer changes in their original dimensions and shape during moisture removal, modifying the thickness and physical properties of the packed bed, as well as the heat and mass exchange area, thus affecting the drying process. However, the influence of shrinkage on heat and mass transfer in the drying of packed beds needs to be better clarified.

Recent experimental and theoretical studies have demonstrated the importance of considering shrinkage for a more realistic analysis of drying phenomena, but most of these studies are concentrated on single particles of food such as fruit and vegetables (Chemkhi *et al.*, 2005; Wang & Brennan, 1995). There are few works reported in the literature on shrinkage of particulate food in bulk (Ratti, 1994) and even less on bulk shrinkage of seeds (Lang *et al.*, 1993).

The quantification of the shrinkage phenomenon as well as of the changes induced in structural parameters of the granular bed is essential for accurate simulation and interpretation of heat and mass transfer phenomena during drying.

3.2 Experimental measurement

Volume changes in individual particles have been evaluated by picnometric techniques and geometric measurements using micrometers. Image analysis has also been used to measure the shrinkage of different materials. These techniques are briefly reviewed in the following paragraphs.

The dimensions method has been commonly used (Burmester & Eggers, 2010; Batista *et al.*, 2007; Dissa *et al.*; 2008) to quantify the changes in apparent volume of samples, by averaging a number of dimensions measurements with micrometers or paquimeters, assuming known geometries. Surface area changes can also be evaluated from geometric measurements and by calculating the approximate surface area of the equivalent sphere, i.e., a sphere that has the same volume as the real particle (Suzuki *et al.*, 1976; Ratti, 1994; Ochoa *et al.*, 2007). This method can only be applied if the initial solid geometry was maintained during drying. When the shrinkage is accompanied by deformations, the use of volumetric displacement methods (buoyant forces) is recommended to determine the volume changes of material (Suzuki *et al.*, 1976; Lozano *et al.*, 1983; Ratti, 1994; Arnosti Jr. *et al.*, 2000; Zielinska & Markowski, 2010; Ochoa *et al.*, 2007).

Volumetric displacement methods have been used by many researchers to determine the apparent volume of different solid materials. Among these, the liquid pycnometric technique is the simplest, and it involves the immersion of the sample in a container of known liquid volume (generally water, hexane or toluene). The apparent volume of sample is determined by measuring the liquid volume displacement. In order to avoid absorption of liquid by the sample, thus ensuring measurement of its apparent volume, an impermeable coating is applied to the solid surface (Aviara *et al.*, 1999).

The use of gas stereopycnometer has been widely spread to determine the true solid volume, which excludes open pore volume (Prado, 2004; Zogzas *et al.*, 1994). However, the measured volume includes a portion of pore spaces inside the solid that are inaccessible to the gas (Chang, 1988).

Image analysis has become one of the most common techniques for shrinkage evaluation. It is a non-destructive method that allows continuous measurement of important parameters during drying such as area, perimeter, major and minor axis length, Feret diameter, elongation and roundness (Prado & Sartori, 2008; Yadollahinia & Jahangiri, 2009; Ramos *et al.*, 2010).

The shrinkage of granular beds has been determined from vertical bed displacement, which can be measured with a linear potentiometer (Lang *et al.*, 1993).

3.3 Experimental and theoretical approaches

Different approaches for shrinkage are found in the literature. The theoretical one involves mechanical laws, in which material stresses and deformations during drying are taken into account. In the experimental one, studies aim to quantify the dependence of materials volume as a function of their moisture content.

The theoretical approach was used for some researchers (Shishido *et al.*, 1986; Towner, 1987) in soil and polymers drying. However, its application to foodstuffs is extremely complex, due to the cellular and multiphase nature of the system and to the necessity of knowing the structure and mechanical and viscoelastic properties of each system phase as well as their variations with moisture content and temperature (Crapiste *et al.*, 1988; Ratti, 1994). Therefore, the most widely used approach for studying food shrinkage is the experimental.

In this approach, the shrinkage coefficient (S_b) is obtained from experimental data according to the ratio between the sample volume at a drying time t and the initial volume of the sample. Changes in S_b with respect to the moisture content constitute the shrinkage curve.

Empirical and theoretical models have been developed to describe the shrinkage behavior of individual particles during drying. In both models, shrinkage has been correlated linearly and non-linearly to moisture content.

In theoretical models, the prediction of changes in volume is based on mass and volume conservation laws assuming, in most cases, additivity of the volumes of the different phases in the system (Mayor & Sereno, 2004).

Suzuki *et al.* (1976) measured the shrinkage of root vegetables that occurred during drying. They attempted to correlate the experimental data with three postulated models; uniform drying, core drying and semicore drying. In the uniform drying model, shrinkage is assumed to be equal to the volume of the water lost by evaporation during all stages of drying. This model results in two equations. The first requires equilibrium moisture content and bulk density, while the second requires the initial moisture.

Based on the laws of mixtures, Sokhansanj & Lang (1996) developed the following two equations to describe the contraction of solid volume, excluding the interparticle volume, and the reduction in grain apparent volumes, respectively, as functions of moisture content:

$$\frac{V_s}{V_{s_0}} = \left[\frac{1-X_0}{1+(\eta-1)\cdot X_0}\right]\cdot\left[\frac{1+(\eta-1)\cdot X}{1-X}\right] \tag{8}$$

and

$$\frac{V_b}{V_{b_0}} = \frac{\left[1-(X_0-X)\right]\cdot\left[1+(\eta-1)\cdot X\right]\cdot(1-\varepsilon_0)}{\left[1+(\eta-1)\cdot X_0\right]\cdot(1-\varepsilon)} \tag{9}$$

where, V_s and V_{s0} are the volumes of solid grain matrix, V_b and V_{b0} are the volumes of particles corresponding to wet basis moisture contents at a certain time (X) and initial time (X_0), respectively, and η is the relative specific mass of the dry solid which is considered constant for each grain.

For the solid matrix, Equation 8 is based on the hypothesis of the additivity of volumes of water and solids, assuming that the volume of air in pores is negligible. In other words, the shrinkage is considered to be equal to the volume of the evaporated water at each period of time during drying. This model has been used for grain kernels, fruit, vegetables and gel-like products by several authors (Zogzas *et al.*, 1994, Krokida & Maroulis, 1997; Arrieche & Sartori, 2004). The model developed for individual particles, Equation (9), taking into account variations of the porosity during the drying process, is particularly interesting, as it can be extended to describe shrinkage behavior of packed beds.

Although there are theoretical models providing physical interpretation for the shrinkage phenomenon, most researchers (Zielinska & Markowski, 2010; Ramos *et al.*, 2010; Ochoa *et al.*, 2007; Prado *et al.*, 2006; Arnosti Jr. *et al.*, 2000; Ratti, 1994; Lozano *et al.*, 1983) have used empirical equations to represent the changes in volume and surface area as functions of moisture content. This is because empirical models allow the prediction of shrinkage without requiring the knowledge of physical properties, such as porosity which is not usually available.

3.4 Experimental shrinkage behavior of individual particles and porous media

To obtain an in-depth understanding of porous bed shrinkage during drying, first, it is necessary to begin with a complete characterization of shrinkage of individual particles, not only in terms of reduced dimensional change in volume, area or thickness, but also of the changes induced in porosity, true and apparent densities, which are clearly dependent upon variations in moisture content. Then, the behavior of deformable porous media is evaluated and related to the characteristics of their individual particles. In order to examine this relationship, experimental data available for thin and deep beds of mucilaginous and gel-coated seeds were chosen to be presented to fulfill the necessities of the present section. Besides the arrangement of particles in the dryer, as single or deep beds, other factors such as volume of water removed and drying conditions affect the magnitude of shrinkage. These factors are also examined in the following paragraphs.

3.4.1 Effect of packing on shrinkage

Information on the shrinkage of papaya seeds both in bulk and as individual particles is presented in Figure 2. The magnitude of shrinkage was significantly affected by the packing of particles. More specifically, the shrinkage coefficient during drying of deep bed of papaya seeds was significantly lower than that of individual seeds arranged in a single layer through which air flows in forced convection. Besides the reduction in particle size, the contraction of the bed porous also appears to be affected by the changes observed in particle shape. Such changes cause a rearrangement of particles within the bed, defining a void space between the particles and, thus leading to a stable packing conformation.

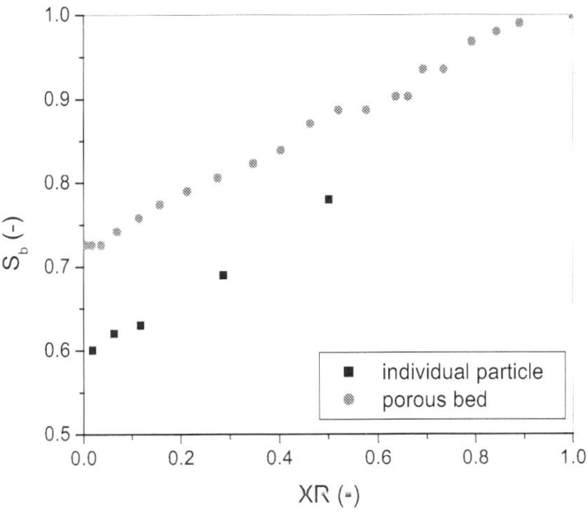

Fig. 2. Volume shrinkage coefficient (S_b) as a function of dimensionless moisture content (XR), for mucilaginous seeds in bulk and as individual particles

3.4.2 Effect of volume of removed water

In the case of deformable media, there is a balance during moisture removal between volume change, density and porosity. In order to know whether the loss in moisture content is converted into volume reduction or into porosity increase, the magnitude of particle shrinkage is usually plotted against normalized volume of water removed (Figure 3). The volume contraction of papaya seeds both in deep bed and as individual particles was lower than the corresponding amount of water removed during drying. According to May & Perré (2002), this behavior reveals that shrinkage of particles and granular bed is accompanied by an increase in porosity, as a result of the reduction in system volume as well as of simultaneous incorporation of gaseous volume.

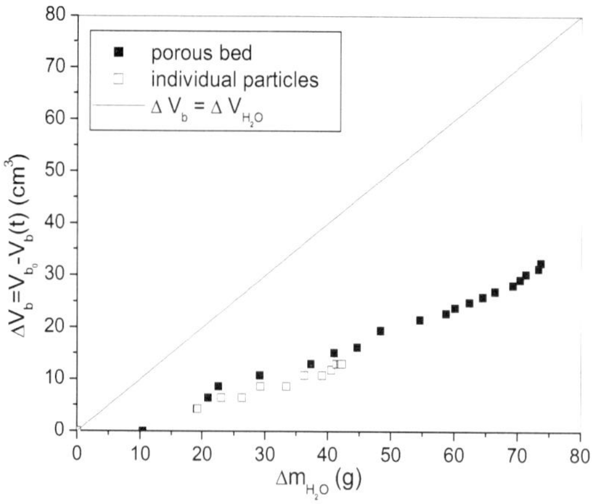

Fig. 3. Relationship between volume reduction and water loss during drying of mucilaginous seeds in bulk and as individual particles, T_{g0} = 50°C, v_{g0}=0.5 m/s

3.4.3 Effect of drying conditions

Volume changes of individual particles and granular deep beds affected by different drying conditions like temperature, air velocity and relative humidity have been investigated by several researchers (Ratti, 1994; Lang *et al.*, 1993). In order to examine this subject, experimental shrinkage data of gel coated millet in bulk and as individual particles are presented in Figures 4 and 5, respectively, as functions of moisture content at different drying conditions.

Shrinkage of individual particles was significantly affected by the operating conditions. This influence is closely related to the rate at which the material is dried, so that the magnitude of volume contraction was higher at lower drying rates. At 30°C and 0.5 m/s, corresponding to low drying rate conditions, external resistance controls the mass transfer, the moisture profiles of the particles are uniform and stresses are minimum (Ratti, 1994). Hence, the particle shrinks uniformly. On contrary, at rapid drying rate conditions, associated with the elevated operating conditions, 50°C and 1.5 m/s, internal resistances control the mass transport and the migration of internal moisture is not uniform, so that it does not compensate the evaporation rate from the particle surface. This induces the formation of a rigid crust in the gel coating, that fixes the volume, hindering subsequent shrinkage.

While for individual particles the volume contraction appears to decrease with an increase in drying rate, no effect of drying conditions on bed shrinkage was found. The deviation within the experimental data obtained under different drying conditions was smaller than the measurement uncertainties. This assures that the bed shrinkage can be correlated with only the bed average moisture content in the experimental range studied.

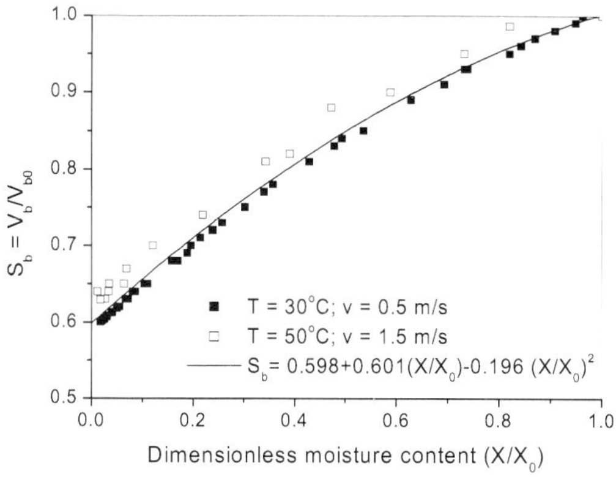

Fig. 4. Bulk shrinkage ratio (S_b) as function of dimensionless bed-averaged moisture content

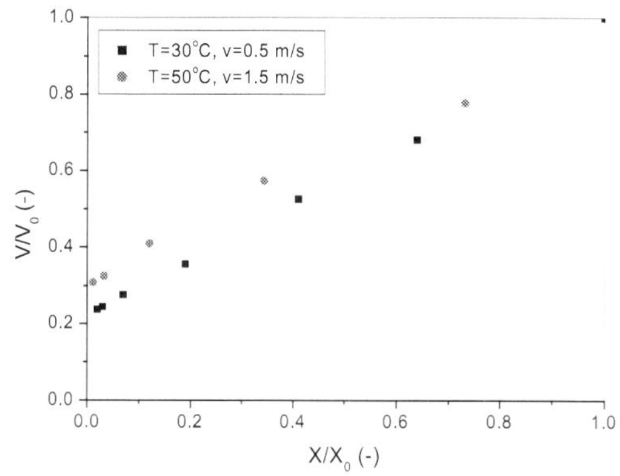

Fig. 5. Shrinkage ratio (S_b) as a function of dimensionless moisture content for artificially coated seeds

3.5 Changes in structural properties during drying

Physical properties of beds composed of very wet particles are clearly dependent upon reduction of both moisture content and volume. Any attempt to raise the shrinkage phenomenon to a higher level of understanding must address the structural properties of the material such as density, porosity and specific area. Moreover, porosity, specific area and bulk density are some of those physical parameters that are required to build drying models; they are particularly relevant in case of porous beds and as such plays an important role in the modelling and understanding of packed bed drying.. Several authors (Ratti, 1994; Wang & Brennan, 1995; Koç *et al.*, 2008) have investigated the changes in structural properties during drying. In what follows, a discussion is directed to changes in physical properties during drying of single and deep bed of skrinking particles. Typical experimental data on changes in structural properties of mucilaginous and gel coated seeds during drying in bulk or as individual particles are presented in Figure 6.

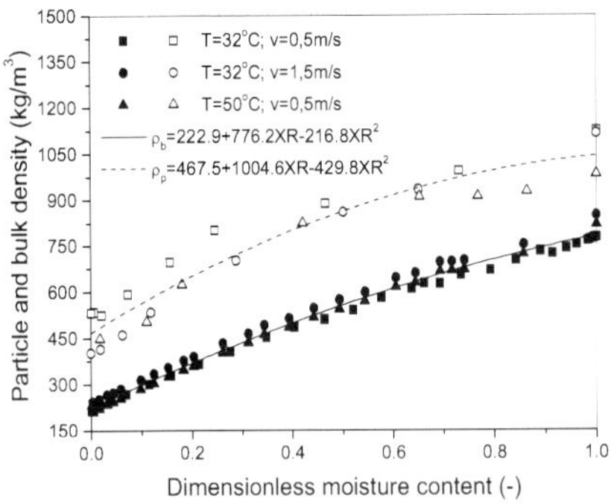

(a) Particle and bulk density versus dimensionless moisture content, for mucilaginous seeds

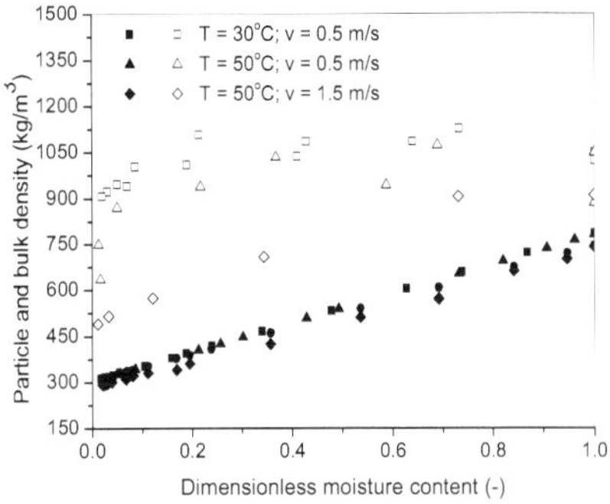

(b) Particle and bulk density versus dimensionless moisture content, for gel coated seeds

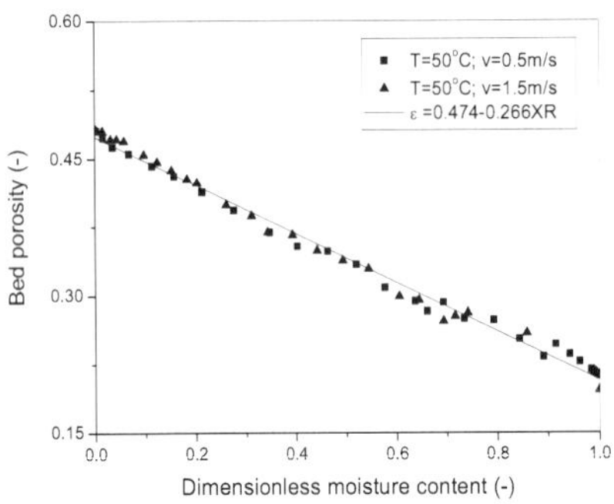

(c) Bed porosity versus dimensionless moisture content, for mucilaginous seeds

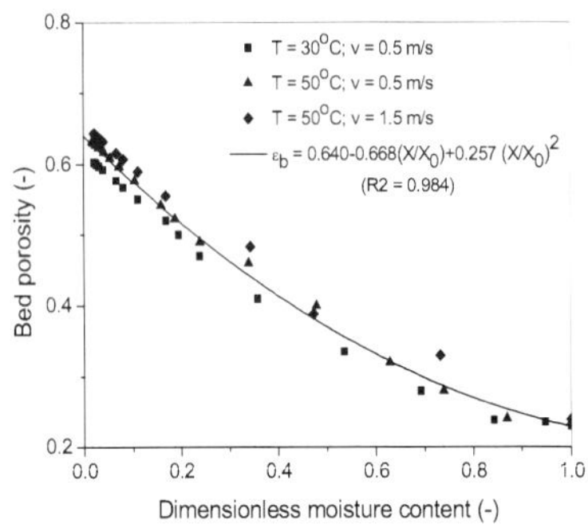

(d) bed porosity versus dimensionless moisture content, for gel coated seeds

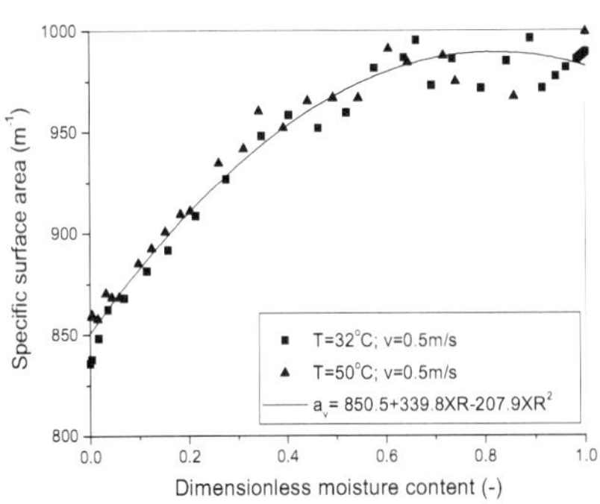

(e) Specific surface area versus dimensionless moisture content for the packed bed

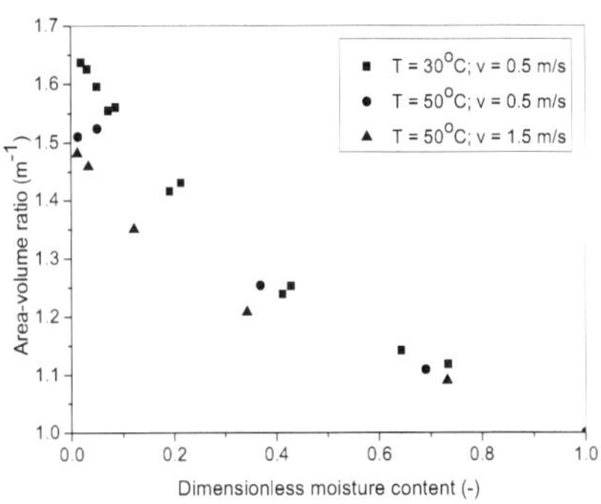

(f) area to volume ratio versus dimensionles moisture content for individual particles

Fig. 6. Structural properties during packed bed drying of mucilaginous and gel coated seeds

For granular beds, three densities have to be distinguished: true, apparent and bulk density. The true density of a solid is the ratio of mass to the volume of the solid matrix, excluding all pores. The apparent density describes the ratio of mass to the outer volume, including an inner volume as a possible pore volume. The term particle density is also used. The bulk density is the density of a packed bed, including all intra and inter particle pores. These structural properties are determined from measurements of mass, true volume, apparent volume and bulk or bed volume. Mass is determined from weighing of samples, whereas the distinct volumes are measured using the methods described in section 4.2. The bed porosity is calculated from the bulk density and the apparent density of the particle.

In Fig. 6 (a) the apparent and bulk densities are shown as functions of dimensionless moisture content of mucilaginous seeds under different drying conditions. The density of packed beds of mucilaginous seeds was found to vary from 777 to 218 kg/m^3, while the particle density varied from 1124 to 400 kg/m^3 as drying proceeded. The decrease in both densities may be attributed to higher weight decrease in both individual particle and packed bed in comparison to their volume contraction on moisture removal.

Apparent density behavior of individual particles and its dependency on temperature were found to be dependent on type of coating (natural or artificial). The distinct behavior of apparent density of mucilaginous and gel coated seeds may be ascribed to the distinct characteristics of reduction in mass and volume of each particle during drying. Contrary to the behavior presented by mucilaginous seeds, the apparent density of gel coated seeds, displayed in Figure 6 (b), was almost constant during most of the drying due to reductions of both particle volume and amount of evaporated water in the same proportions. In the latter drying stages, as the particles undergo negligible volume contraction, the density tends to decrease with moisture loss.

It can also be seen from Figures 6(a) and 6 (b) that drying conditions affected only the apparent density of gel coated seeds. Drying at 50°C and 1.5 m/s, which resulted in the higher drying rate, provided coated seeds with the lower apparent density values. This suggests that there is a higher pore formation within the gel-based coating structure during rapid drying rate conditions. High drying rates induce the formation a stiff outer layer in the early drying stages, fixing particle volume, thus contributing for replacement of evaporated water by air.

The difference between the apparent and bulk densities indicates an increase in bed porosity. Porosity of thick bed of mucilaginous seeds ranged from 0.20, for wet porous beds, to 0.50, for dried porous beds (Fig. 6c). However, beds composed of gel coated seeds presented a higher increase in porosity, with this attaining a porosity of about 0.65 (Fig. 6d). The low porosity of the wet packed beds can be explained in terms of the agglomerating tendency of the particles at high moisture contents. In addition, highly deformable and smooth seed coat facilitates contact among particles within the packed bed, leading to a higher compaction of the porous media and, consequently, to a reduction of porosity.

The increase in bed porosity during drying is firstly due to deformation of external coating that modifies seed shape and size, resulting in larger inter-particle air voids inside the packed bed. Secondly, packed bed and particle shrinkage behaviors are not equivalent, so that the space taken by the evaporated water is air-filled. The variation of about 150% in porosity is extremely high in comparison with other seed beds (Deshpand *et al.*, 1993).

The relationships between the calculated specific surface area of the porous bed as well as of the area volume ratio of the particle with dimensionless moisture content are shown in Figures 6 (e) and 6 (f).

Fig. 6 (f) presents the experimental particle surface area per unit of particle volume made dimensionless using the fresh product value as a function of the dimensionless moisture, for gel-coated seeds. Shrinkage resulted in a significant increase (about 60%) of the area to volume ratio for heat and mass transfer, faciliting the moisture transport, thus increasing the drying rate. This parameter was found to be dependent on drying conditions. The increase in the specific area of the coated particle is reduced as the air drying potential is increased and the shrinkage is limited. On the other hand, the specific surface area of porous bed decreased by about 15 % during drying, Fig. 6 (a). This reduction can be attributed to the significant increase in bed porosity in comparison with the variation in seed shape and size.

The results concerning deep beds indicate that changes in both density, porosity and specific area of the bed are independent of operating conditions in the range tested, and are only related only to the average moisture content of the partially dried bed.

Figure 2 showed a significant contraction of the volume of packed bed composed of mucilaginous particles, of approximately 30%, while Figures 6 (c) and (d) revealed an increase of about 150% in porosity during deep bed drying. Such results corroborate that shrinkage and physical properties such as porosity, bulk density and specific area are important transient parameters in modelling of packed bed drying.

Equations describing the evolution of shrinkage and physical properties as a function of moisture content are implemented in mathematical modeling so as to obtain more realistic results on heat and mass transfer characteristics in drying of deformable media.

4. Drying kinetics

Among the several factors involved in a drying model, a proper prediction of the drying rate in in a volume element of the dryer is required. According to Brooker et al. (1992), the choice of the so-called thin layer equations strongly affects the validity of simulation results for thick-layer bed drying. Several studies related to evaluation of different drying rate equations have been reported in the literature. However, most of works does not deal with the effects of the shrinkage on the drying kinetics although this phenomenon could strongly affect the water diffusion during drying. The influence of shrinkage on drying rates and water diffusion is hence examined in the following paragraphs. To this, experimental kinetic behaviour data of papaya seeds in thin-layer bed are presented. Comparison of the mass transport parameters without and with the shrinkage of gel coated seeds is also presented.

4.1 Experimental determination

Drying kinetics is usually determined by measuring the product moisture content as a function of time, known as the drying curve, for constant air conditions, which are usually obtained from thin layer drying studies Knowledge of the drying kinetics provides useful information on the mechanism of moisture transport, the influence of operating conditions on the drying behaviour as well as for selection of optimal drying conditions for grain quality control. This approach is also widely used for the determination under different drying conditions of the mass transfer parameters, which are required in deep bed drying models (Abalone et al., 2006; Gely and Giner, 2007; Sacilik, 2007; Vega-Gálvez et al, 2008; Saravacos & Maroulis, 2001; Xia and Sun, 2002; Babalis and Belessiotis, 2004).

4.2 Influence of shrinkage on drying behaviour and mass transfer parameters

The theory of drying usually states that the drying behaviour of high moisture materials involves a constant-rate stage followed by one or two falling-rate periods. However, the presence of constant drying rate periods has been rarely reported in food and grain drying studies. A possible explanation for this is that the changes in the mass exchange area during drying are generally not considered (May and Perré, 2002). In order to obtain a better understanding on the subject the discussion is focussed on thin layer drying of shrinking particles, which are characterized by significant area and volume changes upon moisture removal. Experimental drying curves of mucilaginous seeds, obtained under thin-layer drying conditions are presented and examined.

Typical results of water flux density as a function of time with and without consideration of shrinkage are shown in Figure 7.

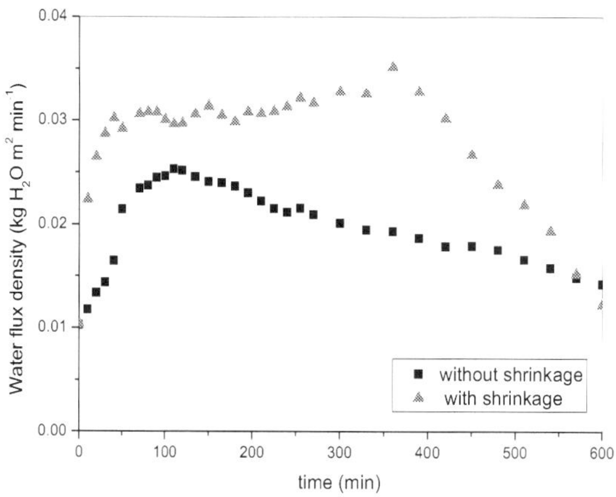

Fig. 7. Water flux density versus moisture content

Because of the initial water content of mucilaginous seeds as well as of their deformable coating structure, the course of drying is accompanied by both shrinking surface area and volume. This results in a large area to volume ratio for heat and mass transfer, faciliting the moisture transport, thus leading to higher values of water flux density. On the other hand, If the exchange surface area reduction is neglected the moisture flux density may decrease even before the completion of the first stage of drying, although free water remains available on the surface (Pabis, 1999). It is evident from Figure 7 that on considering constant mass transfer area the constant-flux period was clearly reduced, providing a critical moisture content which does not reflect the reality.

4.3 Thin-layer drying models

Thin-layer equations are often used for description of drying kinetics for various types of porous materials. Application of these models to experimental data has two purposes: the first is the estimation of mass transfer parameters as the effective diffusion coefficient, mass transfer coefficient, and the second is to provide a proper prediction of drying rates in a volume element of deep bed, in order to be used in dryer simulation program (Brooker et al., 1992.; Gastón et al., 2004).

The thin layer drying models can be classified into three main categories, namely, the theoretical, the semi-theoretical and the empirical ones (Babalis, 2006). Theoretical models are based on energy and mass conservation equations for non-isothermal process or on diffusion equation (Fick's second law) for isothermal drying. Semi-theoretical models are analogous to Newton's law of cooling or derived directly from the general solution of Fick's Law by simplification. They have, thus, some theoretical background. The major differences between the two aforementioned groups is that the theoretical models suggest that the moisture transport is controlled mainly by internal resistance mechanisms, while the other two consider only external resistance. The empirical models are derived from statistical relations and they directly correlate moisture content with time, having no physical fundamental and, therefore, are unable to identify the prevailing mass transfer mechanism. These types of models are valid in the specific operational ranges for which they are developed. In most works on grain drying, semi-empirical thin-layer equations have been used to describe drying kinetics. These equations are useful for quick drying time estimations.

4.3.1 Moisture diffusivity as affected by particle shrinkage

Diffusion in solids during drying is a complex process that may involve molecular diffusion, capillary flow, Knudsen flow, hydrodynamic flow, surface diffusion and all other factors which affect drying characteristics. Since it is difficult to separate individual mechanism, combination of all these phenomena into one, an effective or apparent diffusivity, D_{ef}, (a lumped value) can be defined from Fick's second law (Crank, 1975).

Reliable values of effective diffusivity are required to accurately interpreting the mass transfer phenomenon during falling-rate period. For shrinking particles the determination of D_{ef} should take into account the reduction in the distance required for the movement of water molecules. However, few works have been carried out on the effects of the shrinkage phenomenon on D_{ef}.

Well-founded theoretical models are required for an in-depth interpretation of mass transfer in drying of individual solid particles or thin-layer drying. However, the estimated parameters are usually affected by the model hypothesis: geometry, boundary conditions, constant or variable physical and transport properties, isothermal or non-isothermal process (Gely & Giner, 2007; Gastón *et al.*, 2004).

Gáston *et al.* (2002) investigated the effect of geometry representation of wheat in the estimation of the effective diffusion coefficient of water from drying kinetics data. Simplified representations of grain geometry as spherical led to a 15% overestimation of effective moisture diffusivity compared to the value obtained for the more realistic ellipsoidal geometry. Gely & Giner (2007) provided comparison between the effective water diffusivity in soybean estimated from drying data using isothermal and non-isothermal models. Results obtained by these authors indicated the an isothermal model was sufficiently accurate to describe thin layer drying of soybeans.

The only way to solve coupled heat and mass transfer model or diffusive model with variable effective moisture diffusivity or for more realistic geometries is by numerical methods of finite differences or finite elements. However, an analytical solution of the diffusive model taking into account moisture-dependent shrinkage and a constant average water diffusivity is available in the literature (Souraki & Mowla, 2008; Arévalo-Pinedo & Murr, 2006)

Experimental drying curves of mucilaginous seeds were then fitted to the diffusional model of Fick's law for sphere with and without consideration of shrinkage to determine effective moisture diffusivities. The calculated values of D_{ef} are presented and discussed. First, the adopted approach is described.

The differential equation based on Fick's second law for the diffusion of water during drying is

$$\frac{\partial(\rho_d X)}{\partial t} = \nabla\cdot\left(D_{ef}\nabla(\rho_d X)\right) \tag{10}$$

where D_{ef} is the effective diffusivity, ρ_d is the local concentration of dry solid (kg dry solid per volume of the moist material) that varies with moisture content, because of shrinkage and X is the local moisture content (dry basis).

Assuming one-dimensional moisture transfer, neglected shrinkage and considering the effective diffusivity to be independent of the moisture content, uniform initial moisture distribution, symmetrical radial diffusion and equilibrium conditions at the gas-solid interface, the solution of the diffusion model, in spherical geometry, if only the first term is considered, can be approximated to the form (López et al., 1998):

$$XR = \frac{X - X_e}{X_0 - X_e} = \frac{6}{\pi^2}\exp\left(-\frac{\pi^2 D_{ef}}{R^2}t\right) \tag{11}$$

Where XR is the dimensionless moisture ratio, X_e is the equilibrium moisture content at the operating condition, R is the sphere radius.

In convective drying of solids, this solution is valid only for the falling rate period when the drying process is controlled by internal moisture diffusion for slice moisture content below the critical value. Therefore, the diffusivity must be identified from Eq. (11) by setting the initial moisture content to the critical value X_{cr} and by setting the drying time to zero when the mean moisture content of the sample reaches that critical moisture content.

In order to include the shrinkage effects, substituting the density of dry solid ($\rho_d = m_d/V$), Eq. (11) for constant mass of dry solid could be expressed as (Hashemi et al., 2009; Souraki and Moula, 2008; Arévalo-Pinedo, 2006):

$$\frac{\partial(X/V)}{\partial t} = \nabla\cdot\left(D_{ef}\nabla(X/V)\right) \tag{12}$$

where V is the sample volume. Substituting $Y = X/V$, the following equation is obtained:

$$\frac{\partial Y}{\partial t} = \nabla\cdot\left(D_{ef}^{*}\nabla Y\right) \tag{13}$$

with the following initial and boundary conditions:

$$t = 0,\ Y = X_0/V_0$$
$$t > 0,\ z = 0,\ \frac{\partial Y}{\partial z} = 0$$
$$t > 0,\ z = L,\ Y = X_{eq}/V_{eq}$$

A solution similar to Eq. 11 is obtained:

$$YR = \frac{Y - Y_e}{Y_0 - Y_e} = \frac{6}{\pi^2}\exp\left(-\frac{\pi^2 D_{ef}^{*}}{R^2}t\right) \tag{14}$$

where D_{ef}^* is the effective diffusivity considering the shrinkage, R is the time averaged radius during drying:

The diffusional models without and with shrinkage were used to estimate the effective diffusion coefficient of mucilaginous seeds. It was verified that the values of diffusivity calculated without consideration of the shrinkage were lower than those obtained taking into account the phenomenon. For example, at 50°C the effective diffusivity of papaya seeds without considering the shrinkage (D_{ef} = 7.4 x 10^{-9} m^2/min) was found to be 30% higher than that estimated taking into account the phenomenon (D_{ef}^* = 5.6 x 10^{-9} m^2/min). Neglecting shrinkage of individual particles during thin-layer drying leads, therefore, to a overestimation of the mass transfer by diffusion. This finding agrees with results of previous reports on other products obtained by Souraki & Mowla (2008) and Arévalo-Pinedo and Murr (2006).

4.3.2 Constitutive equation for deep bed models

The equation which gives the evolution of moisture content in a volume element of the bed with time, also known as thin layer equation, strongly affects the predicted results of deep bed drying models (Brooker, 1992). For simulation purpose, models that are fast to run on the computer (do not demand long computing times) are required.

When the constant rate period is considered, it is almost invariably assumed to be an externally-controlled stage, dependent only on air conditions and product geometry and not influenced by product characteristics (Giner, 2009). The mass transfer rate for the evaporated moisture from material surface to the drying air is calculated using heat and mass transfer analogy where the Nusselt and Prandtl number of heat transfer correlation is replaced by Sherwood and Schmidt number, respectively.

With the purpose of describe drying kinetics, when water transport in the solid is the controlling mechanism, many authors have used diffusive models (Gely & Giner, 2007; Ruiz-López & García-Alvarado, 2007; Wang & Brennan, 1995).

When thin-layer drying is a non-isothermal process an energy conservation equation is coupled to diffusion equation. Non-isothermal diffusive models and isothermal models with variable properties do not have an analytical solution; so they must be solved by means of complicated numerical methods. The numerical effort of theoretical models may not compensate the advantages of simplified models for most of the common applications (Ruiz-López & García-Alvarado, 2007). Therefore, simplified models still remain popular in obtaining values for D_{ef}.

Thus, as thin-layer drying behavior of mucilaginous seeds was characterized by occurring in constant and falling rate periods, a two-stage mathematical model (constant rate period and falling rate period) for thin layer drying was developed to be incorporated in the deep bed model, in order to obtain a better prediction of drying rate at each period Prado & Sartori (2008).

The drying rate equation for the constant rate period was described as:

$$\frac{dX}{dt} = -k_c = 1.3 \times 10^{-9} \cdot T_g^{\ 4.112} \cdot v_g^{\ 0.219} \tag{15}$$

valid for 0.5 < v_g < 1.5 m/s 30 < T_g< 50 °C.

For the decreasing rate period, a thin-layer equation similar to Newton's law for convective heat transfer is used, with the driving force or transfer potential defined in terms of free moisture, so that:

$$\frac{dX}{dt} = -K \cdot \left(X - X_{eq}\right) \tag{16}$$

Lewis equation was chosen because combined effects of the transport phenomena and physical changes as the shrinkage are included in the drying constant (Babalis *et al.*, 2006), which is the most suitable parameter for the preliminary design and optimization of the drying process (Sander, 2007).
The relationship between drying constant K and temperature was expressed as:

$$K = 0.011 \exp\left(-201.8/T_g\right) \tag{17}$$

The critical moisture content was found to be independent on drying conditions, in the tested range of air temperature and velocity, equal to (0.99 ± 0.04) d.b.

5. Heat transfer

The convective heat transfer coefficient (h) is one of the most critical parameters in air drying simulation, since the temperature difference between the air and solid varies with this coefficient (Akpinar, 2004). Reliable values of h are, thus, needed to obtain accurate predictions of temperature during drying. The use of empirical equations for predicting h is a common practice in drying, since the heat transfer coefficient depends theoretically on the geometry of the solid, physical properties of the fluid and characteristics of the physical system under consideration, regardless of the product being processed (Ratti & Crapiste, 1995). In spite of the large number of existent equations for estimating h in a fixed bed, the validity of these, for the case of the thick-layer bed drying of shrinking particles has still not been completely established. Table 2 shows the three main correlations found in the literature to predict h in packed beds.

Correlation		Range of validity	Reference
$Nu_p = \left(0.5 Re_p^{1/2} + 0.2 Re_p^{2/3}\right) Pr^{2/3}$	(18)	$20<Re_p< 80000$ $\varepsilon<0.78$	Whitaker (1972)
$h = 3.26\ Cp\ G_g\ Re^{-0.65}\ Pr^{2/3}$	(19)	$20<Re<1000$	Sokhansanj (1987)
$h = \left(\frac{Cp_g\ G_g}{\varepsilon}\right)\left(\frac{2.876}{Re} + \frac{0.302}{Re^{0.35}}\right) Pr^{2/3}$	(20)	$10<Re<10000$	Geankoplis (1993)

Table 2. Empirical equations for predicting the fluid-solid heat transfer coefficient for packed bed dryers

Where, $Re = \frac{\rho_g\ v_g\ dp}{\mu}$, $Re_p = \left(\frac{\rho_g\ v_g\ dp}{\mu}\right)(1-\varepsilon)$, $Nu = \frac{h\ dp}{K_g}$, $Nu_p = \frac{h \cdot dp \cdot \varphi \cdot \varepsilon}{K_g \cdot (1-\varepsilon)}$ and $Pr = \frac{Cp_g\ \mu}{K_g}$.

6. Simultaneous heat and mass transfer in drying of deformable porous media

Drying of deformable porous media as a deep bed of shrinking particles is a complex process due to the strong coupling between the shrinkage and heat and mas transfer phenomena. The degree of shrinkage and the changes in structural properties with the moisture removal influence the heat and mass transport within the porous bed of solid particles. The complexity increases as the extent of bed shrinkage is also dependent on the process as well as on the particle size and shape that compose it.

Theoretical and experimental studies are required for a better understanding of the dynamic drying behavior of these deformable porous systems.

Mathematical modelling and computer simulation are integral parts of the drying phenomena analysis. They are of significance in understanding what happens to the solid particles temperature and moisture content inside the porous bed and in examining the effects of operating conditions on the process without the necessity of extensive time-consuming experiments. Thus, they have proved to be very useful tools for designing new and for optimizing the existing drying systems.

However, experimental studies are very important in any drying research for the physical comprehension of the process. They are essential to determine the physical behavior of the deformable porous system, as a bed composed of shrinking particles, as well as for the credibility and validity of the simulations using theoretical or empirical models.

This section presents the results from a study on the simultaneous heat and mass transfer during deep bed drying of shrinking particles. Two model porous media, composed of particles naturally or artificially coated with a gel layer with highly deformable characteristics, were chosen in order to analyze the influence of the bed shrinkage on the heat and mass transfer during deep bed drying.

First, the numerical method to solve the model equations presented in section 2 is described. The equations implemented in the model to take into account the shrinkage and physical properties as functions of moisture content are also presented. Second, the experimental set up and methodology used to characterize the deep bed drying through the determination of the temperature and moisture distributions of the solid along the dryer are presented. In what follows, the results obtained for the two porous media are presented and compared to the simulated results, in order to verify the numerical solution of the model. A parametric analysis is also conducted to evaluate the effect of different correlations for predicting the heat transfer coefficient in packed beds on the temperature predictions. Lately, simulations with and without consideration of shrinkage and variable physical properties are presented and the question of to what extent heat and mass transfer characteristics are affected by the shrinkage phenomenon is discussed.

6.1 Numerical solution of the model

The numerical solution of the model equations, presented in section 2, provides predictions of the following four drying state variables: solid moisture (X), solid temperature (T_s), fluid temperature (T_g) and air humidity (Y_g) as functions of time (t) and bed height (z). The equations were solved numerically using the finite-difference method. From the discretization of spatial differential terms, the initial set of partial differential equations was transformed into a set of ordinary differential equations. The resulting vector of 4 (N+1) temporal derivatives was solved using the DASSL package (Petzold, 1989), which is based

on the integration method of backwards differential formulation. A computer program in FORTRAN was developed to solve the set of difference equations.
The equations for physical and transport parameters used in model solution are reported in Table 3.

Sorption Properties		
- Equilibrium isotherms	$X_{eq} = \left[\dfrac{-\exp\left(-1.77\times10^{-2}\cdot T+4.25\right)}{\ln(RH)}\right]^{1/1.90}$	(21)
- Moisture desorption heat		
	$L_p = \left(2500.8+2.39\cdot T_g\right)\cdot\left[1+3.2359\cdot\exp\left(-33.6404\cdot X\right)\right]$	(22)
Drying Kinetics ($0.5 < v_g < 1.5$ m/s $30 < T_g < 50$ °C)		
- constant rate period	$\dfrac{dX}{dt} = -k_c = 1.3\times10^{-9}\,T_g^{\,4.112}\,v_g^{\,0.219}$	(15)
- falling rate period	$K = 0.011\exp\left(-201.8/T_g\right)$	(17)
Heat transfer coefficient Equations (18), (19) or (20) from Table 2		
Bulk density	$\rho_b = 240.8+546.1\cdot XR$	(23)
Bed porosity	$\varepsilon = 0.474-0.266\cdot XR$	(24)
Specific area	$a_v = 850.5+339.8\cdot XR-207.9\cdot X^2$	(25)

Table 3. Physical and transport parameters used in model solution

6.2 Experimental study on deep bed drying

The validity of the model which takes into account the shrinkage of the bed and variable physical properties during drying is verified regarding the packed bed drying of particles coated by natural and artificial polymeric structures.
Deep-bed drying experiments were carried out in a typical packed bed dryer (Prado & Sartori, 2008). The experiments were conducted at air temperatures ranging from 30 to 50°C and air velocities from 0.5 to 1.5 m/s, defined by a 2^3 factorial design. These operating conditions satisfy the validity range of the constitutive equations used in the model.
The instrumentation for the drying tests included the measurement of the following variables: temperature of solid particles with time and along the dryer, moisture content of the material with time and along the dryer, and thickness of the bed with time.
Although a two-phase model is more realistic for considering interaction between solid and fluid phases by heat and mass transfer, describing each phase with a conservation equation, it is not simple to use. In addition to the complexity of it solution, there is an additional

difficulty, with regards to its experimental validation, more precisely with the measurement of moisture content and temperature within both the solid phase and the drying air phase.
Techniques for measuring solid moisture and temperature are usually adopted and drying tests are conducted to validate the simulation results of these variables.
To avoid one of the major problems during experimentation on the fixed bed, associated with determination of solid moisture content distribution by continuously taking seed samples from each layer of the deep bed, which can modify the porous structure, possibly causing preferential channels, a stratification method was used. To this, a measuring cell with a height of 0.05 m was constructed with subdivisions of 0.01 m to allow periodical bed fragmentation and measurement of the local moisture by the oven method at $(105 \pm 3)^{\circ}C$ for 24 hours. Afterwards the measuring cell was refilled with a nearly equal mass of seeds and reinstalled in its dryer position. By adjusting the intervals of bed fragmentation appropriately, a moisture distribution history was produced for the packed bed. Although it is a method that requires a large number of experiments and the use of a packing technique to assure the homogeneity and reproducibility of the refilled beds (Zotin, 1985), stratification provides experimental guarantees for model validation.
Temperature distributions were measured using T-type thermocouples located at different heights along the bed.
The overall error in temperature measurements is $0.25^{\circ}C$. In the measurements of airflow, air humidity and solid moisture, the errors are, respectively, equal to 4%, 4% and 1%.
The shrinkage of the packed beds during drying was determined from measurement of its height at three angular positions. From the weighing and vertical displacement of the packed porous bed with time, the parameter of shrinkage (S_b) was obtained as a function of bed-averaged moisture content.

6.3 Experimental verification of the model

The main aim of this section is provide information on the simulation and validation of the drying model by comparison with experimental data of temperature and moisture content distributions of material along the shrinking porous bed and with time. Previously, it is presented a parametric analysis involving the heat transfer coefficient.

6.3.1 Sensitivity analysis of the model

Due to the limited number of reports dealing with external heat transfer in through-flow drying of beds consisting of particles with a high moisture content and susceptible to shrinkage, three correlations found in the literature to predict h in packed beds, Equations (18) to (20) (Table 2), were tested in drying simulation in order to obtain the best reproduction of the experimental data.
Figure 8 shows typical experimental and simulated temperature profiles throughout the packed bed with time, employing in the drying model different empirical equations for predicting the convective heat transfer coefficient (Table 2). Different predictions were obtained, showing the significant effect of h on the numerical solutions. These results are counter to findings for the modeling of thick-layer bed drying of other grains and seeds, specifically rigid particles (Calçada, 1994). In these findings a low sensitivity of the two-phase model to h is generally reported.
When the correlation of Sokhansanj (1987) is used, both the solid and fluid temperatures increase rapidly towards the drying temperature set. However, when the correlations given by Whitaker (1972) and Geankoplis (1993) are applied, the increase in temperature is

gradual and thermal equilibrium between the fluid and solid phases is not reached, so there is a temperature difference between them.

Based on the differences in model predictions, the effects of shrinkage on the estimation of h during drying can be discussed. It should be noted that the Sokhansanj equation is based on the physical properties of air and the diameter of the particle. Thus, it is capable of taking into account only the deformation of individual particles during drying, which produces turbulence at the boundary layer, increasing the fluid-solid convective transport and resulting in an overpredicted rate of heat transfer within the bed (Ratti & Crapiste, 1995). Experiments show that, in the drying of shrinkable porous media, application of correlations capable of incorporating the effects of changes in structural properties, such as Whitaker (1972) and Geankoplis (1993) equations, gives better prediction of the temperature profile. Of these two equations, the Geankoplis equation was chosen, based on a mean relative deviation (MRD) of less than 5%, to be included as an auxiliary equation in drying simulation.

From Figure 8 it can also be verified that during the process of heat transfer, from the increase in saturation temperature up to a temperature approaching equilibrium, the predicted values for the solid phase were closest to the experimental data. This corroborates the interpretation adopted that the temperature measured with the unprotected thermocouple is the seed temperature.

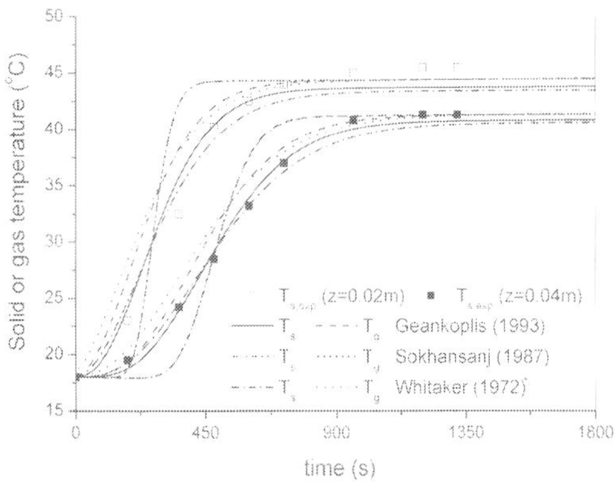

Fig. 8. Dynamic evolution of experimental and simulated temperature profiles obtained from different correlations for h. Drying conditions: v_g = 1.0 m/s, T_{g0} = 50°C, Y_{g0} = 0.01kg/kg, T_{s0} = 18°C and X_0 = 3.9 d.b.

6.3.2 Solid temperature and moisture profiles

In Figures 9 and 10 are presented at different drying times typical experimental and simulated results of moisture and temperature throughout the bed composed of mucilaginous seeds. The mean relative deviations are less that 7% and the maximum absolute error is less than 12% for all the data tested. These results demonstrate the capability of the model to simulate moisture content and temperature profiles during thick-layer bed drying of mucilaginous seeds. From Figure 9 it can also be verified that he model is capable of predicting the simultaneous reduction in the bed depth taking place during packed bed drying.

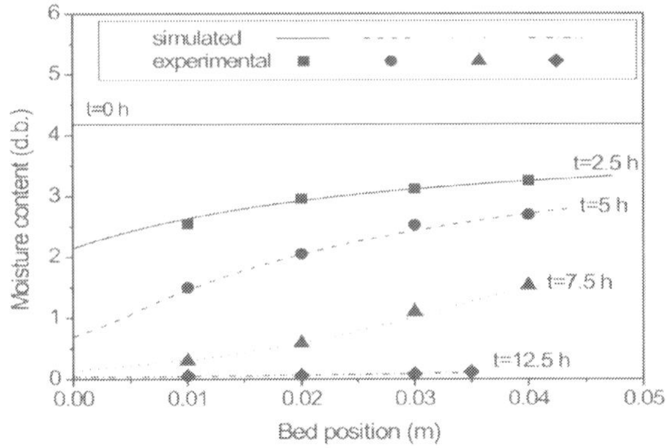

Fig. 9. Experimental and simulated results of moisture of seeds with mucilage throughout the bed, Tgo =50ºC, Ygo =0.01 kg/kg and Vgo =1.0 m/s, Ts =20ºC and Xo=4.1 d.b.

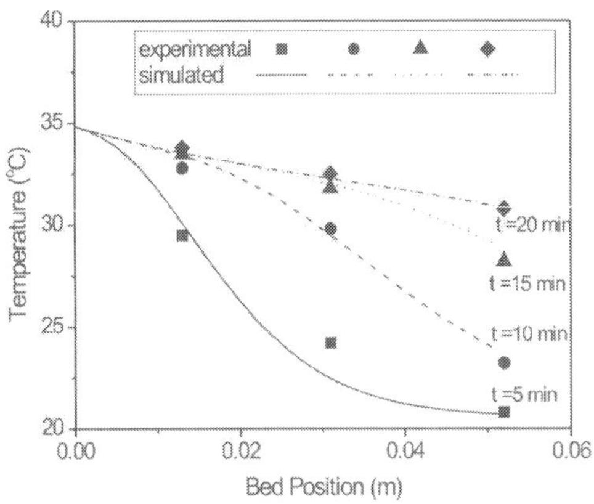

Fig. 10. Experimental and simulated results of temperature of seeds with mucilage along the bed. Tg_0 = 35ºC, Y_{g0} =0.0051 kg/kg and v_{g0} = 0.8 m/s. T_s =20.8ºC and X_0 =3.5 d.b.

6.3.3 Temporal profiles of solid temperature and moisture simulated with and without consideration of shrinkage and variable physical properties

In order to emphasize the importance of shrinkage for a better interpretation of heat and mass transfer in packed bed drying, Figures 11 and 12 show, respectively, the moisture content and temperature profiles of solid with time simulated with and without incorporating bed shrinkage and variable physical properties in the model. It can be verified that there is a significant difference between the sets of data. The model that does not take into account variable physical properties and shrinkage tends to describe a slower drying

process than that accompanied by bed contraction, predicting higher values of moisture content and lower values of temperature at all times. From a practical point of view this would result in higher energy costs and undesirable losses of product quality.
These results suggest that the assumptions of the modelling are essential to simulate adequately solid temperature and moisture content during drying, which have to be perfectly controlled at all times in order to keep the losses in quality to a minimum.

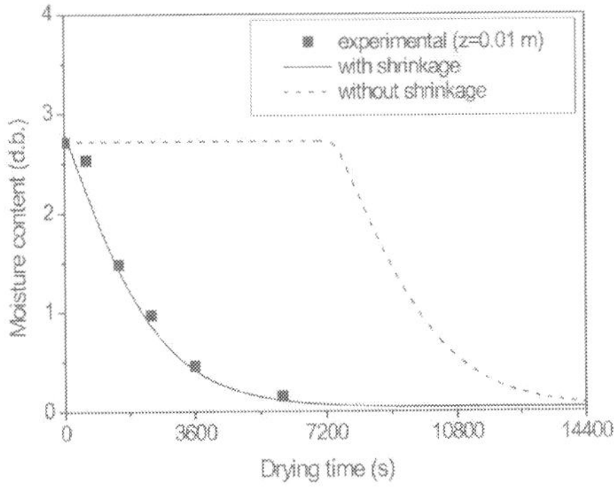

Fig. 11. Moisture content simulated with and without shrinkage. T_{g0}=50°C; v_{g0}=0.5 m/s, Y_{g0} = 0.099 db, Ts_0 = 24°C and X_0 = 2.7 db

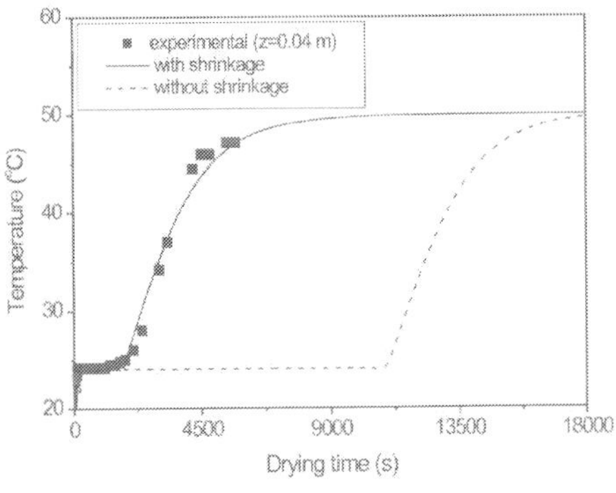

Fig. 12. Seed temperature simulated with and without shrinkage. T_{g0}=50°C; v_{g0}=0.5 m/s, Y_{g0} = 0.099 db, Ts_0 = 24°C and X_0 = 2.7 d.b.

7. Final remarks

This chapter presented a theoretical-experimental analysis of coupled heat and mass transfer in packed bed drying of shrinking particles. Results from two case studies dealing with beds of particles coated by natural and artificial gel structure have demonstrated the importance of considering shrinkage in the mathematical modelling for a more realistic description of

drying phenomena. Bed contraction and variation in properties such as bulk density and porosity cannot be ignored from the point of view of the process dynamics.
Parametric studies showed that the effect of h on the numerical solutions is significant. The best reproduction of the experimental data is obtained when h is calculated using the empirical equation of Geankoplis (1993), which has terms that allow including the effects of changes in structural properties of the packed bed to be taken into account.
In the drying model presented for coated particles, differences in the mass transfer coefficients in the core and external gel layer are not taken into account. This is a limitation of the model that needs to be examined. Moreover, as shrinkage characteristics are directly related to quality attributes, such as density, porosity, sorption characteristics, crust and cracks, the tendency in research is the development of seeds with artificial coating, presenting the best combination of shrinkage and drying characteristics to yield products with higher resistance to deformation and minimal mechanical damages.

8. Nomenclature

a_v	specific surface area, [m^{-1}]
Cp	specific heat, [J/kg °C]
D_{ef}	effective mass diffusivity, [m^2/min]
dp	particle diameter, [m]
G_g	air mass flow rate, [kg m^{-2} s^{-1}]
h	heat transfer coefficient, [J/m^{2}s °C]
k	drying constant, [s^{-1}]
K_g	thermal conductivity [J/s m °C]
L_p	latent heat of vaporization, [J/kg]
N	number of discretized cells
Pr	Prandtl number, [-]
Re	Reynolds number, [-]
RH	relative humidity, [-]
S_b	shrinkage parameter, [-]
t	time, [s]
T	temperature, [°C]
v_g	air velocity, [m/s]
V	volume, [m^3]
X	solid moisture, d.b.,[kg/kg]
XR	dimensionless moisture, [-]
Y_g	air humidity, d.b., [kg/kg]
z	spatial coordinate, [m]

Greek Symbols

ε	porosity, [-]
ϕ	sphericity, [-]
ξ	dimensionless moving coordinate, [-]
ρ	density, [kgm^{-3}]

Abbreviation

d.b.	dry basis
w.b.	wet basis

Subscripts

0	initial
b	bulk
exp	experimental
eq	equilibrium
g	gaseous, fluid
p	particle
s	solid
sat	saturation
v	vapor
w	liquid water

9. Acknowledgements

The authors acknowledge the financial support received from the Foundation for Support of Research of the State of São Paulo, FAPESP, National Council for Research, CNPq, Research and Project Financer, PRONEX/CNPq, and, Organization to the Improvement of Higher Learning Personnel, CAPES.

10. References

Aguilera, J. M. (2003). Drying and dried products under the microscope. *Food Science and Technology International*, Vol. 9, pp. 137–143, ISSN: 1082-0132.

Akpinar, E. K. (2004). Experimental Determination of Convective Heat Transfer Coefficient of Some Agricultural Products in Forced Convective Drying. *International Communications in Heat and Mass Transfer*, Vol. 31, No. 4, pp. 585-595, ISSN 0735-1933.

Arévalo-Pinedo, A., Murr, F. E.X. Kinetics of vacuum drying of pumpkin (*Cucurbita maxima*): Modeling with shrinkage. *Journal of Food Engineering*, Vol. 76 (4), pp. 562-567, ISSN: 0260-8774.

Arnosti Jr. S.; Freire, J. T. & Sartori, D. J. M. (2000). Analysis of shrinkage phenomenon in *Brachiaria brizantha* seeds. *Drying Technology*, Vol. 18, No. 6, pp. 1339-1348, ISSN 0737-3937.

Arrieche, L. S. & Sartori, D. J. M. (2004). Dependence analysis of the shrinkage and shape evolution of a gel system with the forced convection drying periods. *Proceedings of the 14th International Drying Symposium*, São Paulo. DRYING 2004, Vol. A , pp. 152-160, ISSN 0101-2061

Aviara, N.A.; Gwandzang, M.I. & Haque, M.A. (1999). Physical properties of guna seeds. *J. Agric. Eng. Res.*, Vol. 73, pp. 105-111, ISSN 0021-8634.

Babalis, S. J.; Papanicolaou, E.; Kyriakis, N. & Belessiotis, V. G. (2006). Evaluation of thin-layer drying models for describing drying kinetics of figs (*Ficus carica*). *Journal of Food Engineering*, Vo. 75, No. 2, pp. 205-214, ISSN: 0260-8774.

Batista, L. M.; Rosa, Cezar A. & Pinto, L. A. A. (2007). Diffusive model with variable effective diffusivity considering shrinkage in thin layer drying of chitosan. *Journal of Food Engineering*, Vol. 81, pp. 127-132, ISSN: 0260-8774.

Bialobrzewski, I.; Zielínska, M.; Mujumdar, A. S. & Markowski, M. (2008). Heat and mass transfer during drying of a bed of shrinking particles - Simulation for carrot cubes dried in a spout-fluidized-bed drier. *Int. J. Heat Mass Transfer*, Vol. 51, pp. 4704–4716, ISSN 0017+9-9310.

Brooker, D.B.; Bakker-Arkema, F.W. & Hall, C.W. (1992). *Drying and Storage of Grains and Oilseeds*, Van Nostran Reinhold Publishing, ISBN 1-85166-349-5, New York.

Burmester, K. & Eggers, R. (2010). Heat and mass transfer during the coffee drying process. *Journal of Food Engineering*, Vol. 99 , pp. 430–436, ISSN: 0260-8774.

Calçada, L. A. (1994). Modeling and Simulation of Fixed Bed Dryers. M.Sc. Thesis, Federal University of Rio de Janeiro. (in Portuguese).

Cenkowski, S.; Jayas, D.S. & Pabis, S. (1993). Deep-bed grain drying – A review of particular theories. *Drying Technology*, Vol. 11, No. 7, pp. 1553-1581, ISSN 0737-3937

Chang, S. (1988). Measuring density and porosity of grain kernels using a gas pycnometer. *Cereal Chemistry*, Vol. 65, No. 1 , pp. 13-15, ISSN: 0009-0352.

Chemkhi, S.; Zagrouba, F. & Bellagi, A. (2005). Modelling and Simulation of Drying Phenomena with Rheological Behaviour. *Brazilian Journal of Chemical Engineering*, Vol. 22, No. 2 , pp. 153-163, ISSN 0104-6632

Crapiste, G. H.; Whitaker, S. & Rotstein, E. (1988). Drying of cellular material. *Chemical Engineering Science*, Vol. 43, No. 11, pp. 2919-2928, ISSN: 0009-2509.

Deshpande, S. D.; Bal, S. & Ojha, T. P. (1993). Physical Properties of Soybean. *Journal of Agricultural Engineering Research*, Vol. 56, pp. 89-98, ISSN: 1537-5110.

Dissa, A.; Desmorieux, H; Bathiebo J. & Koulidiati, J. (2008). Convective drying characteristics of Amelie mango (*Manguifera Indica* L. cv. 'Amelie') with correction for shrinkage. *Journal of Food Engineering*, Vol. 88, pp. 429–437, ISSN 0260-8774.

Eichler, S.; Ramon, O. & Ladyzhinski, I. (1997). Collapse processes in shrinkage of hydrophilic gels during dehydration. *Food Research International*, Vol. 30, No. 9, pp. 719–726, ISSN 0963-9969.

Gastón, A.; Abalone R.; Giner, S. A. & Bruce, D. M. (2004). Effect of modelling assumptions on the effective water diffusivity in wheat. *Biosystems Engineering*, Vol. 88, No. 2, pp. 175–185, ISSN 1492-9058.

Gastón; A. L.; Abalone, R. M.; Giner, S. A. (2002). Wheat drying kinetics. Diffusivities for sphere and ellipsoid by finite elements. *Journal of Food Engineering*, Vol. 52, No. 4, pp. 313-322, ISSN 0260-8774.

Geankoplis, C. J. (1993). *Transport Processes and Unit Operations*. Prentice-Hall Inc., ISBN 0-13-930439-8, New Jersey.

Gely, M. C; & Giner, S. A. (2007) Diffusion coefficient relationships during drying of soya bean cultivars. *Biosystems Eng*. Vol. 96, No. 2, pp. 213-222.

Giner, S. A.; & Gely, M. C. (2005) Sorptional parameters of sunflower seeds of use in drying and storage stability studies. *Biosys. Engng*, Vol.92, No.2, pp. 217-227, ISSN1537-5110.

Giner, S. A. (2009). Influence of Internal and External Resistances to Mass Transfer on the constant drying rate period in high-moisture foods. *Biosystems Engineering*, Vol. 102, No. 1, pp. 90-94, ISSN 1537-5110.

Hashemi, G.; Mowla, D. & Kazemeini, M. (2009). Moisture diffusivity and shrinkage of broad beans during bulk drying in an inert medium fluidized bed dryer assisted by dielectric heating. *J.Food Eng.*, Vol. 92, pp. 331-338, ISSN 0260-8774.

Koç, B.; Eren, I. & Ertekin, F.K. (2008). Modelling bulk density, porosity and shrinkage of quince during drying: The effect of drying method. *J. Food Eng.*, Vol. 85, No. 3, pp. 340-349, ISSN 0260-8774

Kowalski, S. J.; Musielak, G. & Banaszak, J. (2007). Experimental validation of the heat and mass transfer model for convective drying. *Drying Technology*, Vol. 25, No. 1, pp. 107-121, ISSN 0737-3937

Krokida, M.K. & Z.B. Maroulis. (1997). Effect of drying method on shrinkage and porosity. *Drying Technology*, Vol. 15 , No. 10 , pp. 2441–2458, ISSN 0737-3937.

Krokida M. K.; Maroulis Z. B. & Marinos-Kouris D. (2002). Heat and mass transfer coefficients in drying. Compilation of literature data. *Drying Technology*, Vol. 20, No. 1, pp. 1–18, ISSN 0737-3937

Lang, W.; Sokhansanj, S. & Sosulski, F.W. (1993). Comparative drying experiments with instantaneous shrinkage measurements for wheat and canola. *Canadian Agricultural Engineering*, Vol. 35, No. 2 , pp. 127-132, ISSN 0045-432X.

Lozano, J.E.; Rotstein, E. & Urbicain, M.J. (1983). Shrinkage, porosity and bulk density of foodstuffs at changing moisture contents. *Journal of Food Science*, Vol. 48, pp. 1497-1502, ISSN 0022-1147.

May, B. K. & Perré, P. (2002). The importance of considering exchange surface area reduction to exhibit a constant drying flux period in foodstuffs. *Journal of Food Engineering*, Vol. 54, pp.271-282, ISSN 0260-8774.

Mayor, L. & Sereno, A.M. (2004). Modelling shrinkage during convective drying of food materials: a review. *J. Food Eng.*, Vol. 61, No. 3, pp. 373-386, ISSN 0260-8774

Mihoubi, D. & Bellagi, A. (2008). Two dimensional heat and mass transfer during drying of deformable media. Applied Mathematical Modelling, Vol. 32, pp.303-314, ISSN 0307-904X

Ochoa, M. R.; Kesseler A.G.; Pirone, B.N. & Marquez C.A. (2007). Analysis of shrinkage phenomenon of whole sweet cherry fruits (*Prunus avium*) during convective dehydration with very simple models. *Journal of Food Engineering*, Vol. 79, No. 2 , pp. 657-661, ISSN 0260-8774.

Pabis, S. (1999). The Initial Phase of Convection Drying of Vegetables and Mushrooms and the Effect of Shrinkage. *Journal of Agricultural Engineering Research*, Vol. 72, No. 2, pp. 187-195, ISSN 0021-8634.

Petzold, L.R. (1989). *DASSL: A Differential Algebraic System Solver, Computer and Mathematical Research Division*, Lawrence Livermore National Laboratory, ISBN=8576500493, Livermore, CA.

Prado, M. M. (2004). Drying of seeds with mucilage in packed bed. UFSCar, São Carlos-SP, 162p., *PhD thesis*, University of São Carlos, São Carlos, SP, Brazil.

Prado, M. M.; Ferreira, M. M. & Sartori, D. J. M. (2006). Drying of seeds and gels. *Drying Technology*, Vol. 24 , No. 3 , pp. 281-292, ISSN 0737-3937.

Prado, M. M. & Sartori, D. J. M. (2008). Simultaneous heat and mass transfer in packed bed drying of seeds having a mucilage coating. *Brazilian J. of Chem. Eng.*, Vol. 25, No. 1, pp. 39-50, ISSN 0104-6632.

Ramos, I. N.; Miranda, J. M. R; Brandão, T. R. S. & Silva, C. L. M. (2010). Estimation of water diffusivity parameters on grape dynamic drying. *Journal of Food Engineering*, Vol. 97, No. 3, pp. 519-525, ISSN 0260-8774

Ratti, C. (1994). Shrinkage during drying of foodstuffs. *J. of Food Engineering*, Vol. 23, pp. 91-105, ISSN 0260-8774.

Ratti, C. & Capriste, G. H. (1995) Determination of heat transfer coefficients during drying of foodstuffs, *Journal of Food Process Engineering*, Vol. 18, pp. 41-53, ISSN 0145-8876.

Ratti, C. & Mujumdar, A.S. (1995). Simulation of packed bed drying of foodstuffs with airflow reversal. *J. of Food Engineering*, Vol. 26, No. 3, pp. 259-271, ISSN 0260-8774.

Ruiz-López, I. I. & García-Alvarado, M. A. (2007). Analytical solution for food-drying kinetics considering shrinkage and variable diffusivity. Journal of Food Engineering, Vol. 79 (1), pp. 208–216, ISSN 0260-8774.

Sacilik, K. (2007). Effect of drying methods on thin-layer drying characteristics of hull-less seed pumkin (*Curcubita pepo* L.). *J. Food Engng.*, Vol.79, No.1, pp. 23-30, ISSN 0260-8774.

Sander, A. (2007). Thin-layer drying of porous materials: Selection of the appropriate mathematical model and relationships between thin-layer models parameters. *Chemical Engineering and Processing: Process Intensification*, Vol. 46, No. 12, pp. 1324-1331, ISSN 0255-2701.

Shishido, I.; Suzuki, M. & Ohtani, S. (1986). On the drying mechanism of shrinkage materials. *Proceedings of the III World Congress of Chemical Enginnering*, Tokyo, pp. 21-25, ISBN-13: 978-0852954942

Sokhansanj, S. (1987) Improved heat and mass transfer models to predict quality, *Drying Technology*, Vol. 5, No. 4, pp. 511-525, ISSN 0737-3937.

Sokhansanj, S., & Lang, W. (1996). Prediction of kernel and bulk volume of wheat and canola during adsorption and desorption. *Journal of Agricultural Engineering Research*, Vol. 63, pp. 129-136, ISSN 0021-8634.

Souraki, B.A. & Mowla, D. (2008) Simulation of drying behaviour of a small spherical foodstuff in a microwave assisted fluidized bed of inert particles, *Food Research International*, Vol. 41, No. 3, pp. 255–265, ISSN: 0963-9969.

Suzuki, K.; Kubota, K.; Hasegawa, T. & Hosaka, H. (1976). Shrinkage in dehidration of root vegetables. *Journal of Food Science*, Vol. 41 , pp. 1189-1193, ISSN 0022-1147.

Towner, G.D. (1987). The tensile stress generated in clay through drying. Journal of Agricultural Engineering Research., Vol. 37, pp. 279-289, ISSN 0021-8634.

Yadollahinia A. & Jahangiri M. (2009). Shrinkage of potato slice during drying. Journal of Food Engineering,.Vol. 94 , pp. 52–58, ISSN 0260-8774.

Zanoelo , E.F.; di Celso, G. M. & Kaskantzis, G. (2007). Drying kinetics of mate leaves in a packed bed dryer. Biosystems Engineering, Vol. 96, No. 4, pp. 487-494, ISSN 1537-5110

Zielinska M. & Markowski, M. (2010). Drying behavior of carrots dried in a spout-fluidized bed dryer. Drying Technology, Vol. 25 , No. 1, pp. 261-270, ISSN 0737-3937.

Zogzas, M.K.; Maroulis, Z.B. & Marinos-Kouris, D. (1994). Densities, shrinkage and porosity of some vegetables during air drying. Drying Technology, Vol. 12, No. 7, pp. 1653-1666, ISSN 0737-3937.

Wang, N. & Brennan, J. G. (1995). A mathematical model of simultaneous heat and moisture transfer during drying of potato. Journal of. Food Engineering, Vol. 24, pp. 47-60, ISSN 0260-8774.

Whitaker, S. (1972), Forced convection heat transfer for flow in past flat plates, single cylinders, and for flow in packed bed and tube bundles, *AIChE Journal,* Vol. 8, pp. 361-371, ISSN 0001-1541.

Zotin, F. M. Z. (1985). Wall Effect in the Packed Column. M.Sc. Thesis, Federal University of São Carlos. (in Portuguese).

Impact of Mass Transfer on Modelling and Simulation of Reactive Distillation Columns

Zuzana Švandová, Jozef Markoš and Ľudovít Jelemenský
Institute of Chemical and Environmental Engineering, Slovak University of Technology, Radlinského 9, 812 37, Bratislava, Slovakia

1. Introduction

1.1 Reactive distillation

In chemical process industries, chemical reaction and purification of the desired products by distillation are usually carried out sequentially. In many cases, the performance of this classic chemical process structure can be significantly improved by an integration of reaction and distillation in a single multifunctional process unit. This integration concept is called 'reactive distillation' (RD); when heterogeneous catalysts are applied, the term 'catalytic distillation' is often used. As to the advantages of this integration, chemical equilibrium limitations can be overcome, higher selectivity achieved, by-product formation reduced, heat of the reaction can be used for distillation in-situ, hot spots and run-away effect can be avoided, and azeotropic or closely boiling mixtures can be separated more easily than in a non-RD process. Some of these advantages are realised using a reaction to improve the separation; others are realised using separation to improve the reaction (Sundmacher & Kienle, 2002). Technological advantages as well as financial benefit resulting from this integration are important. Simplification or elimination of the separation system can lead to significant capital savings, increased conversion and total efficiency, which then result in reduced operating costs (Taylor & Krishna, 2000).

1.2 Complexity of RD

The design and operation issue of RD systems are considerably more complex than those involved in either conventional reactors or conventional distillation columns. The introduction of an in-situ separation function within the reaction zone leads to complex interactions between the vapour-liquid equilibrium, vapour-liquid mass transfer, intra-catalyst diffusion (for heterogeneously catalysed processes), chemical kinetics and equilibrium. Such interactions, along with strong nonlinearities introduced by coupling of diffusion and chemical kinetics of counter-current contacting, have been proved to lead to the phenomenon of multiple steady states and complex dynamics, which has been verified in experimental laboratory and pilot plant units (Taylor & Krishna, 2000). Mathematical model of reactive distillation consists of sub-models for mass transfer, reaction and hydrodynamics whose complexity and rigour vary within a broad range (Taylor & Krishna, 2000; Noeres et al., 2003). For example, mass transfer between the gas/vapour and the liquid phase can be described on basis of the most rigorous rate-based approach, using the

Maxwell-Stefan diffusion equations, or it can be accounted for by a simple equilibrium stage model assuming thermodynamic equilibrium between both phases. Homogeneously catalysed reactive distillation, with a liquid catalyst acting as a mixture component, and auto-catalysed reactive distillation present essentially a combination of transport phenomena and reactions taking place in a liquid film (Sláva et al., 2008; Sláva et al., 2009). With heterogeneous systems, it is generally necessary to consider also the particular processes around and inside the solid catalyst particle (Kotora et al., 2009). Modelling of hydrodynamics in multiphase gas/vapour - liquid contactors includes an appropriate description of axial dispersion, liquid hold-up and pressure drop. The correlations providing such descriptions have been published in numerous papers and are collected in several reviews and textbooks. The most suitable approach to reactive distillation modelling depends not only on the model quality and program convergence but also on the quality of model parameters. It is obvious that the choice of the right modelling approach must be harmonised with the availability of the model parameters necessary for the selected model. Optimal complexity of the model for reactive separations depends on one hand on the model accuracy but on the other hand also on the availability of model parameters and the efficiency of simulation methods (Górak, 2006). In this chapter, we focused our attention on vapour-liquid mass transfer influence on the prediction of RD column behaviour neglecting the liquid-solid and intraparticle mass transfer. It means that the bulk phase with solid catalyst was assumed to be homogeneous.

1.3 Mathematical models of a reactive distillation column

Complex behaviour caused by the vapour-liquid interaction, heat effects, thermodynamic and hydrodynamic regimes called for the necessity of models able to describe all these interactions. Starting with the well know McCabe-Thiele graphical method for binary distillation, the approximate shortcut method for multicomponent mixtures according to the Smith-Brinkley or the Fenske-Underwood-Gilliland method, equation tearing procedures using the tridiagonal matrix algorithm or the inside-out method, etc. (Perry et al., 1997) have been used in the history of distillation and reactive distillation modelling. But only with the starting development of computer art, could the researchers start to use standard practices used in chemical engineering calculations without any restrictions in respect to the equations complexity. At the present time, different depth approaches such as the equilibrium (EQ) stage model, EQ stage model with stage efficiencies, nonequilibrium (NEQ) stage model, NEQ cell model and the CFD model can be found in literature on RD column design. Simultaneously, there are several possible versions of the NEQ model formulations with reference to the description of the vapour-liquid equilibrium, reaction equilibrium and kinetics (homogenous/heterogeneous reaction, pseudo-homogenous approach), mass transfer (effective diffusivity method, Maxwell-Stefan approach) and hydrodynamics (completely mixed vapour and liquid, plug-flow vapour, eddy diffusion model for the liquid phase, etc.).

1.3.1 Equilibrium stage model

The main idea is in assuming that the vapour and liquid streams leaving an equilibrium stage are in complete equilibrium with each other and the thermodynamic relations can be used to determine the equilibrium stage temperature and relate the concentrations in the equilibrium streams at a given pressure (Perry et al., 1997). Schematic diagram of an equilibrium stage is shown in Fig. 1. Vapour from the stage below and liquid from the stage above are brought into contact on the stage together with other fresh or recycled feeds. The vapour and liquid streams leaving the stage are assumed to be in equilibrium with each other.

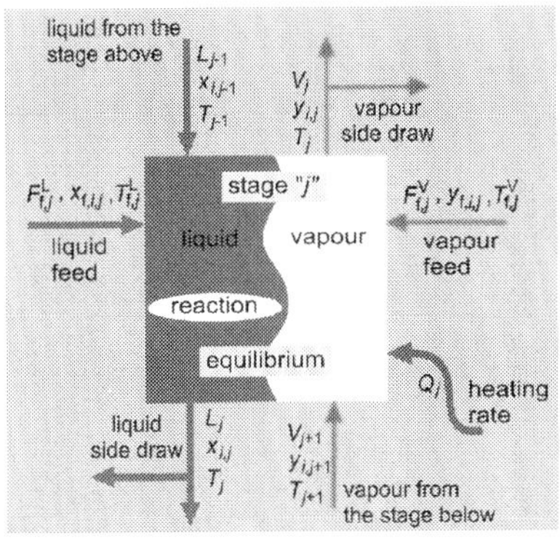

Fig. 1. Equilibrium stage

MESH equations of the equilibrium model for the *j*-th stage

- M equations are the material balance equations;

the total material balance

$$\mathbf{M}_j \equiv \frac{dU_j}{dt} = \sum_{f=1}^{N_F} F_{f,j} + \sum_{r=1}^{N_R}\left(\dot{\xi}_{r,j}\sum_{i=1}^{N_c} \nu_{r,i}\right) + L_{j-1} + V_{j-1} - \left(1+r_j^{\mathrm{V}}\right)V_j - \left(1+r_j^{\mathrm{L}}\right)L_j$$

material balance for component *i*

$$\mathbf{M}_{i,j} \equiv \frac{d\left(U_j x_{i,j}\right)}{dt} = \sum_{f=1}^{N_F} F_{f,j}x_{f,i,j} + \sum_{r=1}^{N_R}\left(\dot{\xi}_{r,j}\nu_{r,i}\right) + L_{j-1}x_{i,j-1} + V_{j-1}y_{i,j-1} - \left(1+r_j^{\mathrm{L}}\right)L_j x_{i,j}$$
$$-\left(1+r_j^{\mathrm{V}}\right)V_j y_{i,j}$$

- E equations are the phase equilibrium relations

$$\mathbf{E}_j \equiv K_{i,j}x_{i,j} - y_{i,j} = 0$$

- S equations are the summation equations in each phase

$$\mathbf{S}_j^{\mathrm{L}} \equiv 1 - \sum_{i=1}^{N_c} x_{i,j} = 0,\ \mathbf{S}_j^{\mathrm{V}} \equiv 1 - \sum_{i=1}^{N_c} y_{i,j} = 0$$

- Heat balance*

$$\mathbf{H}_j \equiv \frac{d(U_j H_j^{\mathrm{L}})}{dt} = \sum_{f=1}^{N_F} F_{f,j}H_{f,j} + \sum_{r=1}^{N_R}\dot{\xi}_{r,j}\left(-\Delta_r H_{r,j}\right) + L_{j-1}H_{j-1}^{\mathrm{L}} + V_{j-1}H_{j-1}^{\mathrm{V}}$$
$$-\left(1+r_j^{\mathrm{L}}\right)L_j H_j^{\mathrm{L}} - \left(1+r_j^{\mathrm{V}}\right)V_j H_j^{\mathrm{V}} + Q_j$$

- Initial conditions, for $t=0$

$$x_{i,j} = x_{i,j}^0,\ y_{i,j} = y_{i,j}^0,\ T_j = T_j^0,\ V_j = V_j^0,\ L_j = L_j^0$$

* *reference state: pure component in the liquid phase at 273.15 K*

Table 1. Specific equations of the equilibrium stage model

A complete reactive distillation column is considered to be a sequence of such stages. Equations describing the equilibrium stages are known as **MESH** equations, **MESH** being an acronym referring to the different types of equations: **M**aterial balances, vapour–liquid **E**quilibrium equations, mole fraction **S**ummations and enthalpy (**H**) balances (Taylor & Krishna, 1993; Kooijman & Taylor, 2000; Taylor & Krishna, 2000). A summary of specific equations is given in Table 1.

1.3.2 Nonequilibrium stage model

An NEQ model for RD follows the philosophy of rate-based models for conventional distillation (Krishnamurthy & Taylor, 1985a; Krishnamurthy & Taylor, 1985b; Taylor & Krishna, 1993; Taylor et al., 1994; Kooijman & Taylor, 2000). In contrary to the EQ model, the NEQ model does not assume thermodynamic equilibrium on the whole stage, but only at the vapour-liquid interface. Mass transfer resistances are located in films near the vapour-liquid and liquid-solid (for heterogeneously catalysed processes) interfaces. The description of the interphase mass transfer, in either fluid phase, is almost invariably based on the film theory and rigorous Maxwell-Stefan theory for the interphase heat and mass transfer rates calculation. Schematic diagram of the nonequilibrium concept is shown in Fig. 2. Vapour and liquid feed streams are contacted on the stage and allowed to exchange mass and energy across their common interface represented in the diagram by a vertical wavy line. A complete reactive distillation column is considered a sequence of these stages. In a nonequilibrium model, separate balance equations are written for each phase on each stage. Conservation equations for each phase are linked by material balances around the interface.

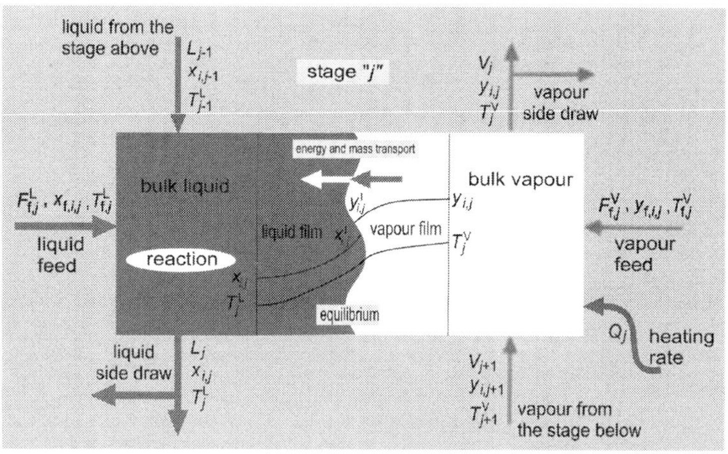

Fig. 2. Nonequilibrium stage

Equilibrium relations are used to relate the composition on each side of the phase interface. The interface composition and temperature must, therefore, be determined as a part of a nonequilibrium column simulation. Equations describing a nonequilibrium stage are **M**aterial balances, **E**nergy balances, **R**ate equations, **S**ummation equations, **H**ydraulic equation, and e**Q**uilibrium relations, i.e. **MERSHQ** equations. A summary of them is given in Table 2.

MERSHQ equations of the nonequilibrium model

- Total **(M)**aterial balances for each phase

$$\mathbf{M}_j^{V} \equiv \frac{dU_j^{V}}{dt} = \sum_{f=1}^{N_j^{V}} F_{f,j}^{V} + V_{j-1} - \left(1 + r_j^{V}\right)V_j - \mathcal{N}_t^{V}$$

$$\mathbf{M}_j^{L} \equiv \frac{dU_j^{L}}{dt} = \sum_{f=1}^{N_j^{L}} F_{f,j}^{L} + \sum_{r=1}^{N_R}\left(\dot{\xi}_{r,j}\sum_{i=1}^{N_c} \nu_{r,i}\right) + L_{j-1} - (1 + r_j^{L})L_j + \mathcal{N}_t^{L}$$

material balances for component i for each phase

$$\mathbf{M}_{i,j}^{V} \equiv \frac{d\left(U_j^{V} y_{i,j}\right)}{dt} = \sum_{f=1}^{N_j^{V}} F_{f,j}^{V} y_{f,i,j} + V_{j-1} y_{i,j-1} - (1 - r_j^{V})V_j y_{i,j} - \mathcal{N}_{i,j}^{V}$$

$$\mathbf{M}_{i,j}^{L} \equiv \frac{d\left(U_j^{L} x_{i,j}\right)}{dt} = \sum_{f=1}^{N_j^{L}} F_{f,j}^{L} x_{f,i,j} + \sum_{r=1}^{N_R}\left(\dot{\xi}_{r,j}\nu_{r,i}\right) + L_{j-1} x_{i,j-1} - (1 + r_j^{L})L_j x_{i,j} + \mathcal{N}_{i,j}^{L}$$

- **(E)**nergy balance equations for each phase*

$$\mathbf{E}_j^{V} \equiv \frac{d(U_j^{V} H_j^{V})}{dt} = \sum_{f=1}^{N_j^{V}} F_{f,j}^{V} H_{f,j}^{V} + V_{j-1} H_{j-1}^{V} - \left(1 + r_j^{V}\right)V_j H_j^{V} + Q_j^{V} - \mathcal{E}^{V}$$

$$\mathbf{E}_j^{L} \equiv \frac{d(U_j^{L} H_j^{L})}{dt} = \sum_{f=1}^{N_j^{L}} F_{f,j}^{L} H_{f,j}^{L} + \sum_{r=1}^{N_R}\dot{\xi}_{r,j}\left(-\Delta_r H_{r,j}\right) + L_{j-1} H_{j-1}^{L}$$

$$-\left(1 + r_j^{L}\right)L_j H_j^{L} + Q_j^{L} + \mathcal{E}^{L}$$

(E)nergy balance around the interface: $\mathbf{E}_j^{I} \equiv \mathcal{E}_j^{V} - \mathcal{E}_j^{L} = 0$

- **(R)**ate equations: $\mathcal{N}_{i,j} - \mathcal{N}_{i,j}^{L} = 0$, $\mathcal{N}_{i,j} - \mathcal{N}_{i,j}^{V} = 0$
- Mole fractions must **(S)**um to unity in each phase:

Bulk liquid: $\mathbf{S}_j^{L} \equiv \sum_{i=1}^{N_c} x_{i,j} - 1 = 0$ bulk vapour: $\mathbf{S}_j^{V} \equiv \sum_{i=1}^{N_c} y_{i,j} - 1 = 0$

liquid film: $\mathbf{S}_j^{LI} \equiv \sum_{i=1}^{N_c} x_{i,j}^{I} - 1 = 0$ vapour film: $\mathbf{S}_j^{VI} \equiv \sum_{i=1}^{N_c} y_{i,j}^{I} - 1 = 0$

(H)ydraulic equations: $\mathbf{H}_j \equiv P_j - P_{j-1} - (\Delta P_{j-1}) = 0$

- Phase e**(Q)**uilibrium is assumed to exist only at the interface: $\mathbf{Q}_{i,j}^{I} \equiv K_{i,j} x_{i,j}^{I} - y_{i,j}^{I} = 0$
- **Initial conditions**, for $t = 0$: $x_{i,j} = x_{i,j}^{0}$, $y_{i,j} = y_{i,j}^{0}$, $x_{i,j}^{I} = \left(x_{i,j}^{I}\right)^{0}$, $y_{i,j}^{I} = \left(y_{i,j}^{I}\right)^{0}$,

$V_j = V_j^{0}$, $L_j = L_j^{0}$, $T_j^{L} = \left(T_j^{L}\right)^{0}$, $T_j^{V} = \left(T_j^{V}\right)^{0}$, $T_j^{I} = \left(T_j^{I}\right)^{0}$, $\mathcal{N}_{i,j} = \mathcal{N}_{i,j}^{0}$, $P_j = \left(P_j\right)^{0}$

* *reference state: pure component in the liquid phase at 273.15 K*

Table 2. Specific equations of the nonequilibrium stage model

Building an NEQ model of reactive separation is not as straightforward as it is with the EQ stage model in which a term accounting for a reaction is added to the liquid-phase material balances (Taylor & Krishna, 2000). A nonequilibrium simulation needs more detailed specifications (in comparison with an equilibrium model). For illustration, we need to know whether the reaction is heterogeneous or homogeneous, also the type and layout of the column internals, mass transfer coefficient model, flow model for both phases, pressure drop model, physical properties models, etc., are important. For the estimation of mass and heat transport properties, the nonequilibrium model requires an evaluation of many more physical properties, such as viscosities, diffusivities, thermal conductivities, surface tension, which are not needed in an equilibrium model. In order to determine the mass transfer coefficients, interfacial area and pressure drop, column type and internals layout must be known.

1.3.2.1 Mass and heat transfer in NEQ model

For the mass transfer on a stage, the well known two-film model which assumes that, for each phase, the transport resistance is concentrated in a thin film adjacent to the phase interface is often adopted. The thickness of these hypothetical films is in the range of 0.001-0.1 mm for the liquid phase and in the range of 0.1-1 mm for the gas phase (Taylor & Krishna, 1993). The most fundamentally sound way to model mass transfer in multicomponent systems is to use the Maxwell-Stefan theory (see Taylor & Krishna (1993) Krishna & Wesselingh (1997)). Another approach to modelling mass transfer in multicomponent systems is enabled by the Fick's law (Sláva et al., 2008; Kotora et al., 2009). Complexity of the Maxwell-Stefan equations and the generalised Fick's law have led to the use of simpler constitutive relations, e.g. effective diffusivity. Extensive and detailed description of multicomponent mass transfer was provided by Taylor & Krishna (1993). The nonequilibrium model uses two sets of mass transfer **(R)**ate equations (see Table 2). At the V-L interface, there is continuity of molar rates:

$$\mathcal{N}_{i,j}^{V} - \mathcal{N}_{i,j}^{L} = 0 \tag{1}$$

Molar rates in each phase are computed from diffusive and convective contributions (Kooijman & Taylor, 2000; Taylor & Krishna, 2000):

$$\mathcal{N}_{i,j}^{L} = a_j^{I} J_{i,j}^{L} + x_{i,j} \mathcal{N}_{t,j}^{L} \tag{2}$$

$$\mathcal{N}_{i,j}^{V} = a_j^{I} J_{i,j}^{V} + y_{i,j} \mathcal{N}_{t,j}^{V} \tag{3}$$

Molar diffusion fluxes $J_{i,j}^{L}$ and $J_{i,j}^{V}$ (in the matrix form) are given as follows:

$$\left(J_j^{L}\right) = c_{t,j}^{L} \left[k_j^{L}\right] \left(x_j^{I} - x_j\right) \tag{4}$$

$$\left(J_j^{V}\right) = c_{t,j}^{V} \left[k_j^{V}\right] \left(y_j - y_j^{I}\right) \tag{5}$$

Note that there are only $(N_I\text{-}1) \times (N_I\text{-}1)$ elements in the $[k_j]$ matrices (Taylor & Krishna, 1993). Matrices of multicomponent mass transfer coefficients assuming ideal behaviour of the vapour phase can be calculated from:

$$\left[k_j^V\right] = \left[R^V\right]^{-1} \tag{6}$$

$$\left[k_j^L\right] = \left[R^L\right]^{-1}\left[\Gamma\right] \tag{7}$$

where $[\Gamma]$ is the matrix of thermodynamic correction factors portraying nonideal behaviour of the liquid phase and $[R^P]$ is the matrix of mass transfer resistances for phase P calculated from:

$$R_{i,i}^P = \frac{z_i^P}{\kappa_{i,N_t}^P} + \sum_{\substack{k=1 \\ k \neq i}}^{N_t} \frac{z_k^P}{\kappa_{i,k}^P} \tag{8}$$

$$R_{i,j(i \neq j)}^P = -z_i^P \left(\frac{1}{\kappa_{i,j}^P} - \frac{1}{\kappa_{i,N_t}^P} \right) \tag{9}$$

where z^P is the mole fraction for phase P and $\kappa_{i,j}^P$ is the mass transfer coefficient for the binary i-j pair in phase P.

A necessary complementary equation is the **(E)**nergy balance for the interface. It presents the continuity of the energy fluxes across the V-L interface (Table 2). Heat transfer rates for the two phases, $\mathcal{E}_j^V, \mathcal{E}_j^L$ are defined as:

$$\mathcal{E}_j^L = h_j^L a_j^I \left(T_j^I - T_j^L\right) + \sum_{i=1}^{N_t} \mathcal{N}_{i,j} H_{i,j}^L \tag{10}$$

$$\mathcal{E}_j^V = h_j^V a_j^I \frac{\varepsilon_j^V}{\exp \varepsilon_j^V - 1} \left(T_j^V - T_j^I\right) + \sum_{i=1}^{N_t} \mathcal{N}_{i,j} H_{i,j}^V \tag{11}$$

where h_j^L and h_j^V are the liquid and the vapour heat transfer coefficients, $H_{i,j}^L$ and $H_{i,j}^V$ are the liquid and the vapour partial molar enthalpies of component i, and ε_j^V is the heat transfer rate factor (Krishnamurthy & Taylor, 1985b) defined as:

$$\varepsilon_j^V = \sum_{i=1}^{N_t} \mathcal{N}_{i,j} C_{pi,j}^V / \left(h_j^V a_j^I\right) \tag{12}$$

To calculate the vapour phase heat transfer coefficients, the Chilton-Colburn analogy between mass and heat transfer may be used (Krishnamurthy & Taylor, 1985b; Kooijman & Taylor, 2000):

$$h_j^V = c_{t,j}^V \kappa_{av,j}^V C_{pm,j}^V \left(Le_j^V\right)^{2/3} \tag{13}$$

For the calculation of the liquid phase heat transfer coefficients, a penetration model may be used (Kooijman & Taylor, 2000):

$$h_j^{\mathrm{L}} = c_{t,j}^{\mathrm{L}} \kappa_{\mathrm{av},j}^{\mathrm{L}} C_{pm,j}^{\mathrm{L}} \left(Le_j^{\mathrm{L}} \right)^{1/2} \tag{14}$$

where $\kappa_{\mathrm{av},j}^{\mathrm{L}}$ and $\kappa_{\mathrm{av},j}^{V}$ are the average values of binary mass transfer coefficients for each phase, Le_j^{L} and Le_j^{V} are the Lewis numbers for each phase, $C_{\mathrm{pm},j}^{V}$ and $C_{\mathrm{pm},j}^{\mathrm{L}}$ are the heat capacities of vapour and liquid mixtures.

Bubble-Cap tray	Sieve tray	Valve tray	Dumped packing	Structured packing
AICHE (1958) Hughmark (1971)	AICHE (1958) Chan-Fair (1984) Zuiderweg (1982) Chen-Chuang (1993) Harris (1965) Bubble-Jet	AICHE (1958)	Onda-Takeuchi-Okumoto (1968) Bravo-Fair (1982) Biller-Schultes (1992) ...	Bravo-Rocha-Fair (1985) Bravo-Rocha-Fair (1992) Billet-Schultes (1992) ...

Table 3. Available mass transfer coefficient correlations per internal type suggested by (Kooijman & Taylor, 2000)

Crucial parameters in a nonequilibrium model are the mass transfer coefficients. The choice of mass transfer coefficient correlations can influence the results of a simulation (Kooijman & Taylor, 2000). In Eqs. (8), (9), (13) and (14), the flux mass transfer coefficients (κ) were used which may be estimated using empirical correlations with the Maxwell-Stefan diffusivity of the appropriate *i-j* pair ($Đ_{i,j}$) replacing the binary Fick's diffusivity ($D_{i,j}$) (Taylor & Krishna, 1993). Table 3 provides a summary of the available mass transfer coefficient correlations for tray and packed columns suggested by Kooijman & Taylor (2000). The binary mass transfer coefficients obtained from these correlations are functions of the tray design and layout, or of the packing type and size, as well as of the operational parameters and physical properties including the binary pair diffusion coefficients. Table 4 introduces four chosen empirical correlations for the estimation of the number of transfer units or relations for the estimation of binary mass transfer coefficients for a sieve tray. Binary mass transfer coefficients can be computed from the number of transfer units (**N**) (Taylor & Krishna, 1993; Kooijman & Taylor, 2000) as follows:

$$\mathbf{N}^{\mathrm{V}} = k^{\mathrm{V}} a^{\mathrm{V}} t^{\mathrm{V}} = k^{\mathrm{V}} a h_f / u_s \tag{15}$$

$$\mathbf{N}^{\mathrm{L}} = k^{\mathrm{L}} a^{\mathrm{L}} t^{\mathrm{L}} = k^{\mathrm{L}} a h_f Z / \left(Q^{\mathrm{L}} / W \right) \tag{16}$$

where Z is the liquid flow path length (m), W is the weir length (m), Q^{L} is the volumetric liquid flow rate ($m^3\ s^{-1}$), h_f is the froth height (m), a^{V} and a^{L} are the interfacial area per unit volume of vapour and liquid ($m^2\ m^{-3}$), respectively, a is the interfacial area per unit volume of froth ($m^2\ m^{-3}$). Parameter ah_f ($m^2\ m^{-2}$) can be estimated from the Zuiderweg's (1982) method (see Table 4), where it is dependent on the regime of operation (spray regime or mixed froth-emulsion flow regime), and u_s is the superficial vapour velocity ($m\ s^{-1}$) based on the bubbling area of the tray, A_b:

$$u_s = \frac{Q^V}{A_b} = \frac{V}{c_t^V A_b} \tag{17}$$

where Q^V is the volumetric vapour flow rate (m^3 s^{-1}), V is the vapour flow rate (mol s^{-1}) and c_t^V is the molar concentration of the vapour mixture (mol m^{-3}).

Method	*Other relations*	*Number of transfer units* (**N**) *[-]*
AICHE (1958)	$F_s = u_s\left(\rho_t^V\right)^{0.5}$ $Sc^V = \mu^V / \left(\rho_t^V D^V\right)$ $t^L = h_L Z W / Q^L$	$\mathbf{N}^L = 19700\left(D^L\right)^{0.5}\left(0.4F_s + 0.17\right)t^L$ $\mathbf{N}^V = \left(0.776 + 4.57h_w - 0.238F_s + \frac{104.8Q^L}{W}\right)\sqrt{Sc^V}$
Chan-Fair (1984)	$t^V = \left(1-\alpha_e\right)h_L / \left(\alpha_e u_s\right)$ $F_f = u_s / u_{sf}$	$\mathbf{N}^V = \left(103000 - 8670F_f\right)F_f\left(D^V\right)^{0.5} t^V h_L^{-0.5}$ $\mathbf{N}^L$ – the AICHE correlation can be used
Chen-Chuang (1993)	$t^V = h_L / u_s$ $F_s = u_s\left(\rho_t^V\right)^{0.5}$ $t^L = t^V \rho_t^L / \rho_t^V$ $\beta = A_h / A_b$	$\mathbf{N}^V = 11\frac{1}{\left(\mu^L\right)^{0.1}\beta^{0.14}}\left(\frac{\rho_t^L F_s^2}{\sigma^2}\right)^{1/3}\sqrt{D^V t^V}$ $\mathbf{N}^L = 14\frac{1}{\left(\mu^L\right)^{0.1}\beta^{0.14}}\left(\frac{\rho_t^L F_s^2}{\sigma^2}\right)^{1/3}\left(\frac{M^V V}{M^L L}\right)\sqrt{D^L t^L}$
		Mass transfer coefficients (k) [m s^{-1}]
Zuiderweg (1982)		$k^V = 0.13 / \rho_t^V - 0.065 / (\rho_t^V)^2 \quad (1 < \rho_t^V < 80\ \text{kg/m}^3)$ $k^L = 0.26 \times 10^{-5} (\mu^L)^{-0.25}$ or $k^L = 0.024 (D^L)^{0.25}$
	Other relations	*Interfacial area per unit bubbling area (ah_f) [m^2 m^{-2}]*
Zuiderweg (1982)	$\text{FP} = \frac{M^L}{M^V}\left(\frac{\rho_t^V}{\rho_t^L}\right)^{0.5}$ $b = W / A_b$ $h_L = 0.6h_w^{0.5}\left(\frac{p\text{FP}}{b}\right)^{0.25}$ $\beta = A_h / A_b$	if $\text{FP} > 3.0bh_L$ then froth-emulsion regime *Spray regime:* $ah_f = \frac{40}{\beta^{0.3}}\left(\frac{u_s^2 \rho_t^V h_L FP}{\sigma}\right)^{0.37}$ *Froth-emulsion regime:* $ah_f = \frac{43}{\beta^{0.3}}\left(\frac{u_s^2 \rho_t^V h_L FP}{\sigma}\right)^{0.53}$

Table 4. Empirical correlation for estimation of the number of transfer units, binary mass transfer coefficients and interfacial area per unit bubbling area

2. Utilisation of mathematical modelling

Mathematical modelling endeavours to provide a true projection of equipment behaviour not only for existing plants but also for designed technologies and equipments when experimental data are insufficient. Steady state modelling provides information about selectivity, conversion, and production in dependence on the column (reactor) configuration and operational mode. Moreover, it gives information about physical properties of the output streams, temperature and concentration profiles in the entire column. This information is needed for the design and optimisation of the process. In the process design phase, dynamic simulation is needed to determine dynamic response to process disturbances. Using dynamic simulations, the start-up and shut-down of current processes can also be optimised. In recent years, mathematical modelling has often been used for the purposes of safety analysis (Molnár et al., 2005).

European direction 96/82/EC on control of major industrial accidents (so called Seveso II) requires a detailed safety analysis not only for existing industrial units, but also for units (technologies, equipments) which are designed. A useful tool for such an analysis is mathematical modelling of equipment linked with a methodology used for safety analysis, like HAZOP or others (Checklists method, What if? Analysis, FMECA (Failure Mode, Effect and Criticality Analysis)). A safety analysis includes an analysis of the multiplicity of steady states and their stability, identification of states leading to hazardous reactor behaviour, i.e. crossing of the practical stability limit, study of safe operating conditions, investigation of conditions and trajectories which can shift the reactor/column from one steady state to another one and the determination of a safe start-up and shut down procedure. Mohl et al. (1999) showed that the multiplicity found for a reactive distillation column has the same cause as that found for a continuous stirred tank reactor (CSTR). The existence of multiple steady states can result in both technological problems and dangerous situations. However, consequences of undesirable situations for a CSTR (other type of reactors) and a reactive distillation column may be completely different. For example, increased production of heat in a CSTR can result in the run-away effect, but in a reactive distillation column, the heat of reaction is used in-situ for evaporation and can (but does not have to) cause technological problems.

HAZOP is one of the best and most rigorous techniques for the identification of hazard and operability problems in a chemical plant (Kletz, 1999). The HAZOP procedure formally examines all equipment step by step as well as deviations from its normal operation conditions and considers what failures can occur. The HAZOP report includes all deviations, their causes, consequences in the equipment performance, analysis of these consequences, implemented protection (active and/or passive), and the resulting suggestions. It can be successfully applied not only for existing plants but also for designed technologies and equipment. On the other hand, two essential drawbacks of the HAZOP study exist. The primary drawback of the HAZOP approach is related to the possibility that hazards and operability problems may be overlooked. HAZOP requires many hours of work of a team of experienced engineers, and it is not easy to record the engineering reasoning, basic input information or the results. Therefore is the application of the HAZOP technique to a detailed chemical plant design a complex and time consuming task. Both these drawbacks can be reduced by integrating a model approach into HAZOP. Usually does the HAZOP analysis not consider the duration and amplitude of the deviations generated during the reactor operation. However, what does the deviation 'less flow' exactly mean; 90 % or 20 % of the usual operational value? Does the deviation occur as an immediate (step) decrease of the flow lasting for 10 min or more, or is it only an impulse? Is this decrease continuous at some rate? Answers to these questions may be obtained by an appropriate mathematical model (Švandová et al., 2005a; Švandová et al., 2005b; Labovský et al., 2007a; Labovský et al., 2007b). In such a model approach, the extent of the deviations may be easily incorporated and possible consequences can be investigated. The basis of this methodology is the integration of mathematical modelling into the HAZOP study. This integration is useful for the identification of consequences of the generated deviations, and the suggestion of corrective actions. Moreover, the integration of mathematical modelling into the HAZOP study may potentially lead to the detection of some unexpected deviations. Other advantages of this approach are:

1. significantly reduced time and effort required in HAZOP;
2. more smooth and detailed HAZOP study;
3. minimisation of the influence of human factors.

3. Impact of mathematical model selection on the prediction of steady state and dynamic behaviour of a reactive distillation column

As a model system, the MTBE reaction system was chosen:

$$CH_3OH + (CH_3)_2 C = CH_2 \Leftrightarrow (CH_3)_3 COCH_3 \tag{18}$$

where iso-butene (IB) reacts with methanol (MeOH) to form MTBE in a reversible, exothermic reaction. The reaction is catalysed by a strong ion-exchange resin. The reaction rate equation and its parameters were given by Rehfinger and Hoffmann (1990). Possible side-reactions were ignored. Reaction rates were calculated assuming a pseudo-homogenous model. The reaction is usually carried out in the presence of inert components.

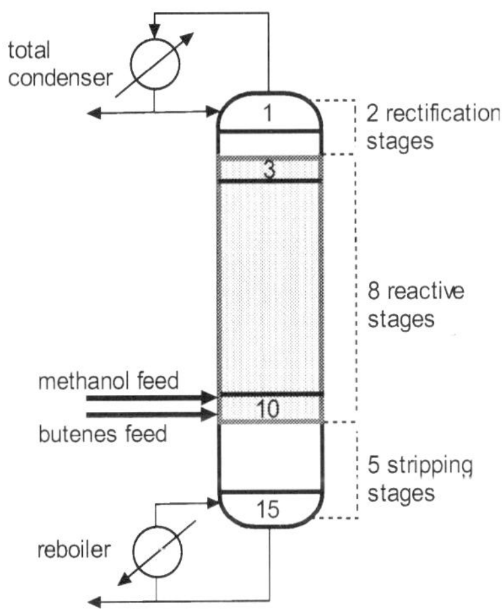

Fig. 3. Reactive distillation column

These inert components result from upstream processing, where isobutene is produced. In our case study, 1-butene was used as an inert. The vapour-liquid equilibrium was calculated using the UNIQUAC model with the binary interaction parameters reported by Rehfinger and Hoffmann (1990) (all binary interactions between MeOH, IB, MTBE) and HYSYS 2.1 (all binary interactions between 1-butene and the other components). The gas phase was assumed to be ideal. Physico-chemical properties of all pure components were taken from the HYSYS 2.1 database. The column configuration chosen for the simulations is the one described by Jacobs and Krishna (1993) (see Fig. 3).The column consisted of a total condenser, 15 sieve trays (2 rectifying stages, 8 reactive stages and 5 stripping stages), and a partial reboiler. On each of the eight stages in the reactive zone, 1000 kg of the catalyst were charged in the form of "envelopes" placed along the flow path length. Details of such a construction are available in a patent (Jones Jr., 1985). The column pressure was 1110 kPa and the column had two feed streams: a methanol feed and a mixed butenes feed, both fed to stage 10. At a standard operating point, the molar flow rates of methanol and the mixture of butenes were 775.8 and 1900 kmol h^{-1}, respectively. The mixed butenes feed consisted of 35.58% isobutene and 64.42% 1-butene. The reflux ratio was set to 7 and the bottom flow rate to 675 kmol h^{-1}. Detailed specifications of the sieve trays are given in Table 5.

Tray specifications			*Tray specifications*		
Type of tray	Sieve	-	Number of liquid flow passes	5	-
Column diameter	6	m	Fractional active area	0.76	-
Total tray area	28.27	m^2	Fractional hole area	0.1	-
Tray spacing	0.61	m	Fractional downcomer area	0.12	-
Hole diameter	4.5	mm	Liquid flow path length	0.97	m
Weir height	50	mm	Total weir length	22	m

Table 5. Specifications of reactive distillation column tray

3.1 Equilibrium vs. nonequilibrium stage model

The objective of this part is to compare the prediction of the equilibrium and nonequilibrium models during the safety analysis of a reactive distillation column focusing on the identification of hazardous situations or operability problems. The safety and operability analyses are based on the application of the HAZOP procedure integrated with a mathematical model with the aim to determine the column response to deviations from normal operation conditions or during a nonstandard procedure, like the start-up of the reactive distillation column. Mathematical models used for this study were described in detail in the previous part of this chapter. In the EQ model, the stage efficiency of 60% was assumed. This value closely corresponded to the calculations of the NEQ model using the AICHE calculation method for sieve tray mass transfer (Baur et al., 2000). Identification of the multiple steady states locus in two parameter planes (remaining operation parameters were kept constant) was performed using the algorithm CONT (Kubíček, 1976; Marek & Schreiber, 1991). The result of these simulations is depicted in Fig. 4, where the methanol feed flow rate and the butenes feed flow rate were chosen as the parameters of interest. The bifurcation diagram contains information about the limit points and the multiplicity intervals. White area in this diagram indicates that for the combination of both investigated parameters only one steady state is possible, meanwhile, the filled area represents the three possible steady states for the actual methanol and butenes feed flow rates. The boundary lines of the filled area present the limit points in the investigated parametric plane.

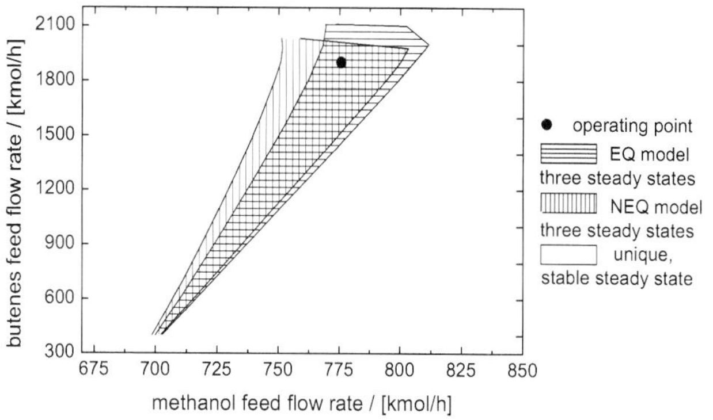

Fig. 4. Bifurcation diagram in the parametric plane methanol feed flow rate and butenes feed flow rate (reprinted from Švandová et al. (2009))

Fig. 4 indicates that the multiplicity zones predicted by EQ and NEQ models do not overlap in the whole range of the investigated parameters; however, for the operational methanol feed flow rate (775.8 kmol h^{-1}) and the operational butenes feed flow rate (1900 kmol h^{-1}), the steady-state multiplicity was predicted assuming both models (black circle in Fig. 4). Fig. 4 provides information about multiplicity intervals; however, it does not give information about conversion in the possible steady states. This information is provided by the solution diagrams (Figs. 5A and B). In the solution diagrams, either the methanol feed flow rate (with constant butenes feed flow rate set to the value of 1900 kmol h^{-1}) or the butenes feed flow rate (with constant methanol feed flow rate set to the value of 775.8 kmol h^{-1}) were considered as continuation parameters. The solution diagrams (Fig. 5A), where the methanol feed flow rate was used as a continuation parameter, indicate three steady states at the operating value of the methanol feed flow rate (775.8 kmol h^{-1}) for both models. These curves of the iso-butene conversion are continuous. The bifurcation diagram (Fig. 5B.), with the butenes feed flow rate used as a continuation parameter, indicates the same three steady states for both models used at the operational value of the butenes feed flow rate (1900 kmol h^{-1}) as those in Fig. 5A. However, very interesting results of these continuations are continuous curves of the iso-butene conversion with isolas located above these curves for both investigated models (Fig. 5B). From both solution diagrams (Figs. 5A and B) follows that the iso-butene conversion in the upper steady states are nearly identical for both used models. The conversion of iso-butene is relatively different in the lower steady state. Analysis of the methanol feed flow rate solution diagram (Fig. 5A) implies that a decrease in the methanol flow rate does not cause rapid decrease of the iso-butene conversion. A different situation occurs when the methanol flow rate is changed to a higher value. An increase in the methanol flow rate can lead to a rapid decrease of the iso-butene conversion. Similar situations can be expected for a rapid decrease or increase of the butenes feed flow rate. As it can be seen, solution diagrams obtained by the EQ and NEQ models are not identical in the whole range of the investigated parameters, what can results in different predictions of the reactive distillation column behaviour during dynamic simulations following the HAZOP procedure. As it was already mentioned, the HAZOP analysis is quite complex and time-consuming. Therefore, only one deviation is analysed and discussed in the text below.

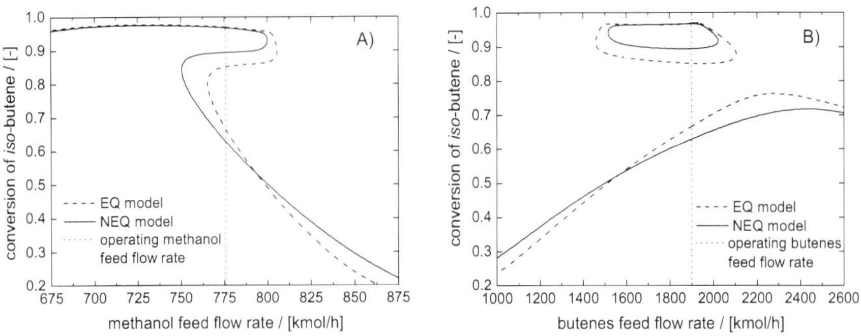

Fig. 5. Solution diagrams obtained by the EQ and NEQ models. Continuation parameter: A) methanol feed flow rate, B) butenes feed flow rate (reprinted from Švandová et al. (2009))

Let us summarise some very important information for this simple example of the HAZOP procedure:

Studied equipment: *reactive distillation column*
Chosen node for HAZOP analysis: *feed stream of butenes*
Investigated parameter: *feed flow rate of butenes*
Designed steady state operation value of this parameter: *1900 kmol* h^{-1}
Used guideword: *higher*
Created deviation: *higher butenes feed flow rate*

Fig. 6A–C shows consequences predicted by the EQ and NEQ models for the deviation "higher butenes feed flow rate". At the time of 1 h, the butenes feed flow rate suddenly increased from the operational value of 1900 to 1995 kmol h^{-1} (+5%; Fig. 6A), to the value of 2090 kmol h^{-1} (+10%; Fig. 6B) and finally to 2185 kmol h^{-1} (+15%; Fig. 6C). In all three cases, time duration of the deviation was 10 h and the reactive distillation column was forced to find a new steady state corresponding to the new value of the butenes feed flow rate. At the time of 11 h, the butenes feed flow rate returned back to the operational value (1900 kmol h^{-1}). Fig. 6A shows a situation when the butenes feed flow rate was changed by about 5%. Both, EQ and NEQ, models predicted that after the butenes flow rate returns to the operational value, the reactive distillation column will return to the original steady state characterised by high conversion of iso-butene. Different situation is depicted in Fig. 6B, the butenes feed flow rate was changed by about 10%. After returning the butenes flow rate to the operational value, the reactive distillation column stabilised in a lower stable steady state considering the NEQ model. However, when the EQ model was used to simulate the reactive distillation column, the conversion of iso-butene returned to the original steady state after the perturbation. These results indicate a notable disagreement in the prediction considering the EQ and the NEQ model. Fig. 6C shows the situation when the butenes feed flow rate was changed by about 15%. Both models predicted that after the butenes flow rate returns to the operating value, the reactive distillation column stabilises in a lower stable steady state. From Figs. 6A–C follows that the magnitude of the deviations has a significant influence on the deviation consequences; however, the consequences prediction can be different considering the EQ or the NEQ model (Fig. 6B). In many cases, duration of the failure may dramatically affect the response of the device although this aspect is not usually investigated during a "typical" HAZOP procedure. Thus, for the next simulation, the same deviation, higher butenes feed flow rate, was taken into account.

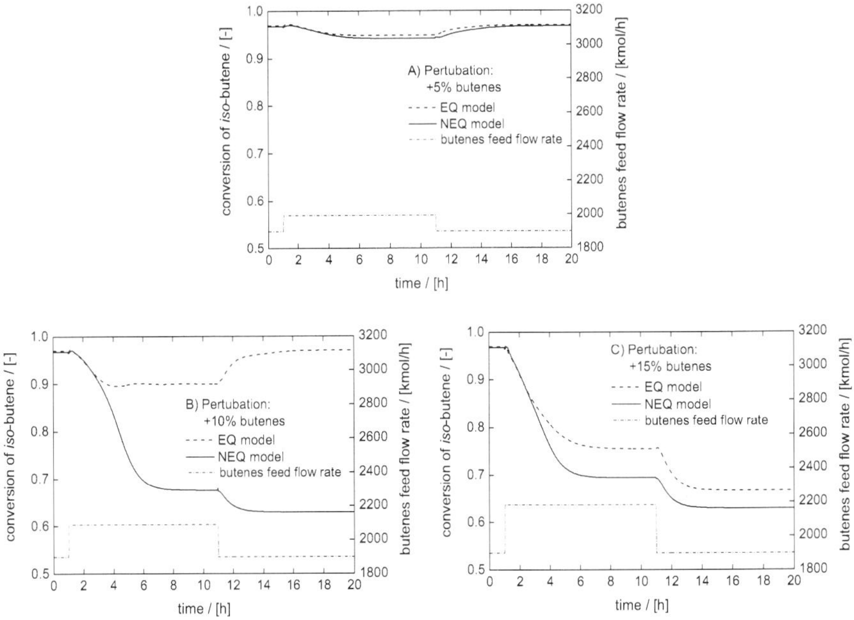

Fig. 6. Conversion changes predicted by the EQ and NEQ models caused by change of the butenes feed flow rate. Deviation extension: (A) +5%, (B) +10% and (C) +15% (reprinted from Švandová et al. (2009))

Fig. 7 presents dynamic simulations of an RD column, starting from the operating steady states characterised by high-conversion of iso-butene. At the time equal to 1 h, a 15% increase of the butenes feed flow rate was simulated (the same deviation as in Fig. 6C). Dynamic responses to the perturbation durations varying from 0.5 to 5 h were investigated using the EQ (Fig. 7A) and NEQ (Fig. 7B) models. A system described by the EQ model switches to the lower conversion steady state only after a perturbation longer than 3 h (Fig. 7A). However, a system described by the NEQ model switches to the lower steady state earlier, at the perturbation duration of 2 h and more (Fig. 7B). From Figs. 7A and B it is clear that the time durations of the feed flow rate disturbances have an extensive impact on the transitions between parallel steady states; however, the prediction of column behaviour is different assuming the EQ and the NEQ model. A more complex comparison of the predictions of the equilibrium and nonequilibrium models during a safety analysis of a reactive distillation column focusing on the identification of hazardous situations or operability problems is provided in the article by Švandová et al. (2009). From the presented analysis follows an important conclusion: the proportion and localisation of the zones of multiple steady states in the solution diagrams predicted by the EQ and NEQ models can be partly different. For this reason, the EQ and NEQ models show different response to the HAZOP deviation "higher butenes feed flow rate". Time duration of the feed flow rate disturbances has an extensive impact on the transitions between parallel steady states predicted by the EQ and NEQ models. It was shown that a reliable prediction of the reactive distillation column behaviour is influenced by the complexity of the mathematical model used for its description. The EQ model is simpler, requiring a lower number of model parameters. On the other hand, the assumption of equilibrium between the vapour and liquid streams leaving the reactor can be difficult to meet, especially if some perturbations of the process parameters occur. The NEQ model takes the interphase mass and heat transfer resistances into account. On the other hand, it is important to point out that prediction of the column behaviour is strongly dependent on the quality of the NEQ model parameters depending on the equipment design; this topic will be analysed in the next section.

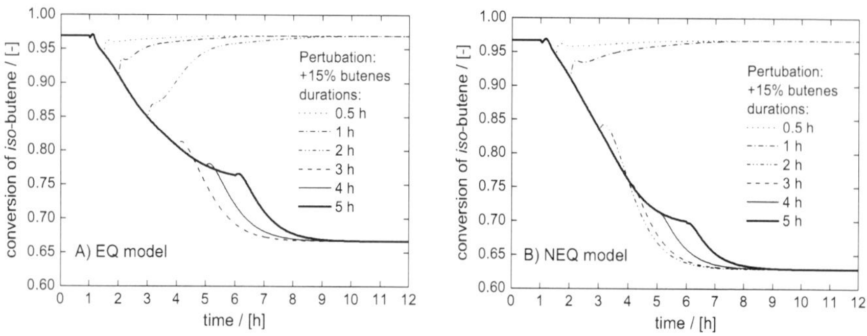

Fig. 7. Dynamic response obtained using (A) EQ model and (B) NEQ model to a 15% increase in the butenes feed flow rate 1 h after the simulation start. The perturbation period varied from 0.5 to 5 h (reprinted from Švandová et al. (2009))

3.2 Impact of mass transfer coefficient correlations on prediction of reactive distillation column behaviour

Model number	Mass transfer coefficient correlation	Style of line in graphs
1	AICHE (1958)	dashed line (■)
2	Chan-Fair (1984)	thick solid line (●)
3	Chen-Chuang (1993)	dash-dotted line (▲)
4	Zuiderweg (1982)	thin solid line (♦)

Table 6. Mass transfer coefficient correlations chosen for investigation

To compare the reactive distillation column behaviour predicted by different correlations for mass transfer coefficients, the four correlations summarised in Table 6 were chosen. The equations are available in the previous section (see Table 4). To estimate the diffusion coefficients in gas mixtures and in dilute liquid mixtures, empirical correlations of Fuller-Schettler-Giddings (1966) and Wilke-Chang (1955), respectively, were used (for details see Reid (1977), Taylor & Krishna (1993), Perry et. al. (1997)). The Maxwell-Stefan diffusion coefficient is defined for each binary pair in the multicomponent liquid mixture using the diffusion coefficients in dilute liquid. The mixing rule used was taken from Wesselingh and

Krishna (1990). In the solution diagrams (Figs. 8a and b), either the methanol feed flow rate or the butenes feed flow rate were the continuation parameters. The conversion of isobutene was examined. The solution diagrams (Fig. 8a), where the methanol feed flow rate was used as the continuation parameter, indicate that multiple steady states are predicted by Models 1 and 2. Using parameters calculated by Models 3 and 4, no multiplicity occurred in the whole range of the investigated parameters. When the isobutene feed flow rate was used as the continuation parameter, multiple steady states were found for three of the four investigated models (Fig. 8b).

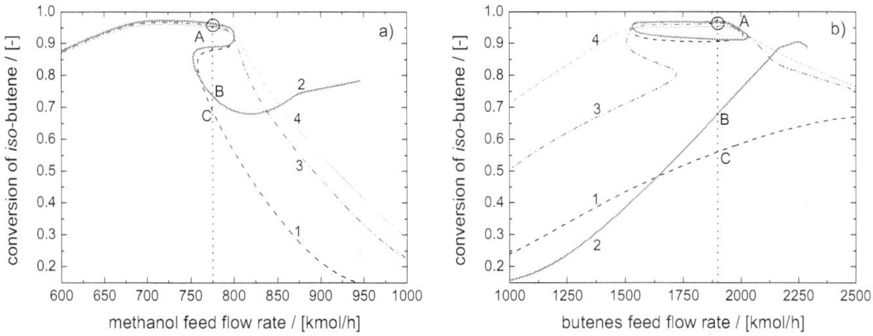

Fig. 8. a) Conversion of isobutene vs. methanol feed flow rate solution diagrams. b) Conversion of isobutene vs. butenes feed flow rate solution diagrams (dashed line—Model 1, thick solid line—Model 2, dash-dotted line—Model 3, thin solid line—Model 4, vertical dotted line—operating feed flow rate, vertical short dotted line—Model 2 out of the validity range) (reprinted from Švandová et al. (2008))

Only when using parameters calculated by Model 4, no multiplicity occurred in the whole range of the investigated parameters. However, a very interesting result is that Models 1 and 2 predict continuous curves of the isobutene conversion with isolas located above these curves. Model 3 predicts a typical 'S' profile of the isobutene conversion and, at the same time, the "window" in which these multiplicities occur is significantly reduced (see Fig. 8b). In general, the AICHE method (Model 1) and the Chan-Fair method (Model 2) behave in essentially the same way, except for the strong dependence on the fraction of flooding in the Chan- Fair method (Model 2). The critics of the Chan-Fair (Model 2) correlation mention that the quadratic dependence on the fractional approach to flooding limits this correlation to the range of fractions of flooding where the quadratic term is positive (the fraction of flooding must lie between 0 and 1.2) (Kooijman & Taylor, 1995). We have encountered situations of the Chan-Fair (Model 2) correlations providing negative mass transfer coefficient because the fraction of flooding was outside this range. It is clear that negative mass transfer coefficients are physically meaningless and at the same time, the program that implements our nonequilibrium model stopped converging to a solution. This situation is represented in the solution diagrams (Fig. 8) by a short vertical dotted line (methanol feed flow rate of ≈948 kmol h^{-1} in Fig. 8a. and butenes feed flow rate of ≈2280 kmol h^{-1} in Fig. 8b). From the solution diagrams (Figs. 8a and b) follows that for the given operating feed flow rate of methanol (775.8 kmol h^{-1}, see dotted line in Fig. 8a) and operating feed flow rate of butenes (1900 kmol h^{-1}, see dotted line in Fig. 8b) predict Models 1 and 2 three steady states (two stable, one unstable); however, Models 3 and 4 predict only one steady state. The steady states obtained by Models 3 and 4 are nearly equal to the upper steady states given by Models 1 and 2. The presence of multiple steady states strongly influences the reactive distillation column behaviour during its start-up as well as during any disturbances of the input parameters. For illustration, disturbances of the butenes feed flow rate were studied (Fig. 9a) starting from the operational steady states characterised by high conversion of isobutene. At 1 h, a very fast increase of the butenes feed flow rate, over 2100 kmol h^{-1}, was simulated. The original flow rate of butenes was reached 10 h later. The duration of the disturbance was so long to show that for all investigated methods, a new steady state corresponding to the butenes feed flow rate of 2100 kmol h^{-1} was reached.

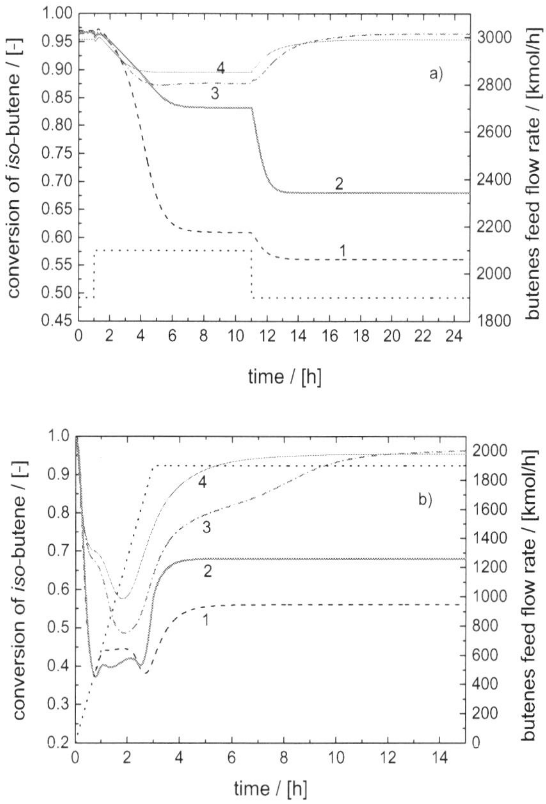

Fig. 9. a) Conversion of isobutene changes after step change of the butenes flow rate, b) Columns start-up considering a gradual increase of the butenes flow rate (dashed line—Model 1, thick solid line—Model 2, dash-dotted line—Model 3, thin solid line—Model 4, dotted line— butenes feed flow rate)

After returning the butenes flow rate to the operational value, the system described with the NEQ model using Models 1 and 2 was stabilised in the lower stable steady state. But providing the NEQ model with Models 3 or 4, the system turned back to the original steady state (see Fig. 9a). From Fig. 9a follows that for situations predicted by Models 1 and 2, the restart of the RD column is needed to switch the conversion to higher steady states. Also, the presence of multiple steady states reduces the operability and controllability of the reactive distillation column during its start-up. This is confirmed by Fig. 9b which represents the column start-up considering a gradual increase of the butenes feed flow rate. Before the start-up procedure, the column was filled with pure methanol. Applying this start-up procedure, after reaching the operational feed flow rate of butenes, Models 1 and 2 predicted column stabilisation in the steady state characterised by low conversion of isobutene. Models 3 and 4 predict only one possible steady state for the operating feed flow rate of butenes (see Fig. 8b), hence, after the start-up procedure, this steady state is reached (Fig. 9b). From Fig. 8 follows that all used models predict the high conversion steady state in good agreement with each other.

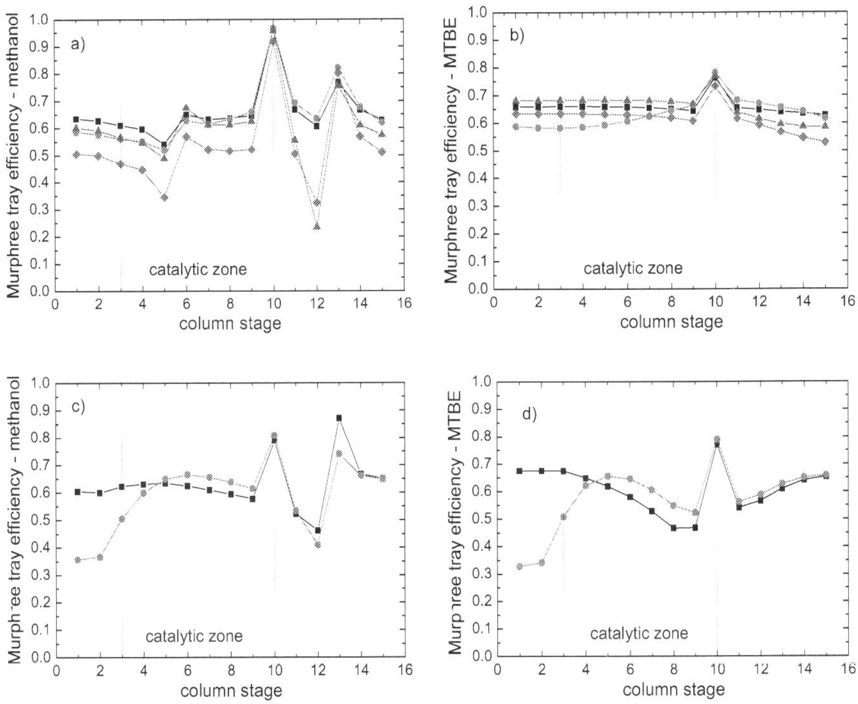

Fig. 10. a) Murphree tray efficiency in upper steady state for methanol, b) Murphree tray efficiency in upper steady state for MTBE, c) Murphree tray efficiency in lower steady state for methanol, d) Murphree tray efficiency in lower steady state for MTBE (■ - Model 1, ● - Model 2, ▲ - Model 3, ♦ - Model 4) (reprinted from Švandová et al. (2008))

This good tendency can be observed also in Figs. 10a-b which show the Murphree tray efficiency for methanol and MTBE, respectively, in the upper (high conversion) steady state.

These pictures indicate, that Model 4 tends to predict lower Murphree tray efficiency compared to other methods, however, this does not markedly affect the conversion of isobutene and the purity of MTBE in the reboiler. For the given operating conditions, the low conversion steady state was predicted only by Models 1 and 2. The difference between the low conversion steady states predicted by Models 1 and 2 is more relevant. Figs. 10c-d show the Murphree tray efficiency for methanol and MTBE, respectively, in the low conversion steady state. These pictures show that in the lower part of the reactive zone (stages 5-10) Model 2 predicts higher Murphree tray efficiency. Simultaneously, Model 2 predicts a radical decrease of the Murphree tray efficiency in the upper part of the column (stages 1-4). This behaviour predicted by Model 2 is caused by the quadratic dependence on the fractional approach to flooding, because in this part of the column is the fraction of flooding close to unity and thus close to flooding. As it was shown, different predictions of the reactive distillation column behaviour during dynamic change of the parameters depend on whether or not multiple steady state are expected for the required values of operational parameters. As it was already mentioned Models 1 and 2 predict multiple steady states for the operating value of butenes feed flow rate; however, Models 3 and 4 predict only one steady state. Fig. 11 shows isobutene conversion dependence on the butenes feed flow rate calculated using Model 1 (Fig. 11a) and Model 3 (Fig. 11b). In Fig. 11a, the AICHE method was used to calculate the mass transfer coefficients (solid line in Fig. 11a). Additional curves in Fig. 11a were calculated under the assumption that all mass transfer coefficients were higher (dashed line in Fig. 11a) or lower (dash-dotted line in Fig. 11a) by 10 %. Fig. 11b shows a similar investigation, however, to calculate the mass transfer coefficients, the Chen-Chuang method (solid line in Fig. 11b) was used. Dashed line and dash-dotted line in Fig. 11b were calculated with the assumption of a 10% increase and a 10 % decrease, respectively, of the mass transfer coefficients compared with those calculated using the Chen-Chuang method.

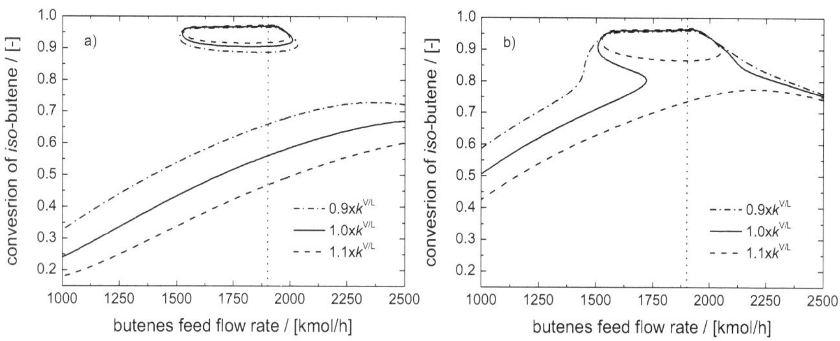

Fig. 11. Conversion of isobutene vs. butenes feed flow rate solution diagrams calculated using a) Model 1, b) Model 3 (solid lines - mass transfer coefficients predicted using correlations (Table 4), dash-dotted lines - mass transfer coefficients lower by 10% than coefficients predicted by correlations, dashed lines - mass transfer coefficient higher by 10% than coefficients predicted by correlations)

From Fig. 11a follows that a 10 % increase of the mass transfer coefficient value calculated using the AICHE method caused a decrease of the conversion in steady states located on the lower branch and a decrease of the mass transfer coefficient caused an increase of the conversion in steady states located on the lower branch; however the number of steady states and quality of higher steady states located on isolas did not change. An interesting result is depicted in Fig. 11b. If the Chen-Chuang method is used to calculate the mass transfer coefficients, only one steady state is predicted for the operational feed flow rate of butenes (1900 kmol h^{-1}). Multiple steady states are predicted only for a short interval of butenes feed flow rate (approximately 1500- 1750 kmol h^{-1}). However, a 10 % increase of the mass transfer coefficients above the value calculated using the Chen- Chuang method (dashed line in Fig. 11b) caused that multiple steady states appeared for the operational feed flow rate of butenes and the shape of the calculated curves were significantly similar to those calculated using the AICHE method. On the other hand, a 10% decrease of the mass transfer coefficients below the value calculated using the Chen- Chuang method (dash-dotted line in Fig. 11a) caused that multiple steady states almost completely disappeared.

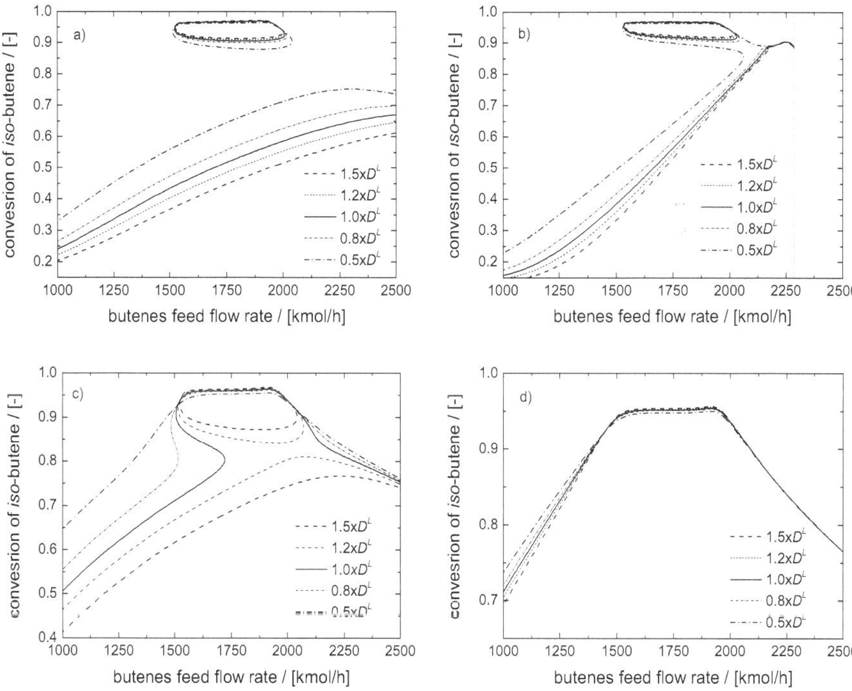

Fig. 12. Conversion of isobutene vs. butenes feed flow rate solution diagrams calculated using different liquid phase diffusion coefficient (±20%, ±50%) for all investigated models: a) Model 1, b) Model 2, c) Model 3 , d) Model 4

From this follows that a 10 % change of the value of mass transfer coefficients may even affect the number of the predicted steady states and consequently the whole prediction of the reactive distillation column behaviour during dynamic change of parameters. Investigations presented in Fig. 11 were made under the assumption that all binary mass transfer coefficients (as well as the liquid in the gas phase) are by 10 % higher or lower than those calculated using empirical correlations (AICHE, Chen-Chuang). This is a very rough assumption which implies a potential uncertainty of the input parameters (diffusivity in the liquid or vapour phase, surface tension, viscosity, density, etc.) needed for the calculation of the mass transfer coefficients according to the correlations. It is important to note that each input parameter needed for the mass transfer coefficient calculation may influence the general NEQ model steady state prediction relatively significantly. Fig. 12 shows isobutene conversion dependence on the butenes feed flow rate calculated using a) Model 1, b) Model 2, c) Model 3 , d) Model 4, whereby several different values of the diffusion coefficients in the liquid phase were used in each model. To calculate the diffusion coefficients in a dilute liquid mixture, the Wilke-Chang (1955) correlation was used, which corresponds to the solid lines in Fig. 12a-d. To show the effect of of the diffusion coefficient uncertainty on the NEQ models steady state prediction, a 20 % and 50 % increase as well as decrease of the calculated diffusion coefficients was assumed. From Fig. 12 follows that the effect of the liquid phase diffusion coefficients on the steady states prediction using different models for mass transfer coefficient prediction is significantly different. The most distinguishable influence can be noticed using Model 3 (i.e., the Chen- Chuang method, see Fig. 12c) where the decrease of the diffusion coefficients led to notable reduction of the multiple steady state zone and the course of the curves was similar to that predicted by Method 4 (i.e., the Zuiderweg method, see Fig. 12d). On the other hand, the increase of the diffusion coefficients led to isola closure and creation of a multiplicity zone similar to that predicted by Method 1 (i.e., the AICHE method, see Fig. 12a) and Method 2 (i.e., the Chan-Fair method, see Fig. 12b). The effect of diffusion coefficients variation is very similar for Method 1 and Method 2 whereas the same equation was used for the number of transfer units in the liquid phase. Method 4 (i.e., the Zuiderweg method, see Fig. 12d) shows the smallest dependence on the diffusion coefficients change.

4. Conclusion

A reliable prediction of the reactive distillation column behaviour is influenced by the complexity of the mathematical model which is used for its description. For reactive distillation column modelling, equilibrium and nonequilibrium models are available in literature. The EQ model is simpler, requiring a lower number of the model parameters; on the other hand, the assumption of equilibrium between the vapour and liquid streams leaving the reactor can be difficult to meet, especially if some perturbations of the process parameters occur. The NEQ model takes the interphase mass and heat transfer resistances into account. Moreover, the quality of a nonequilibrium model differs in dependence of the description of the vapour–liquid equlibria, reaction equilibria and kinetics (homogenous, heterogeneous reaction, pseudo-homogenous approach), mass transfer (effective diffusivity method, Maxwell - Stefan approach) and hydrodynamics (completely mixed vapour and liquid, plug-flow vapour, eddy diffusion model for the liquid phase, etc.). It is obvious that different model approaches lead more or less to different predictions of the reactive distillation column behaviour. As it was shown, different correlations used for the prediction of the mass transfer coefficient estimation lead to significant differences in the prediction of the reactive distillation column behaviour. At the present time, considerable progress has been made regarding the reactive distillation column hardware aspects (tray design and layout, packing type and size). If mathematical modelling is to be a useful tool for optimisation, design, scale-up and safety analysis of a reactive distillation column, the correlations applied in model parameter predictions have to be carefully chosen and employed for concrete column hardware. A problem could arise if, for a novel column hardware, such correlations are still not available in literature, e.g. the correlation and model quality progress are not equivalent to the hardware progress of the reactive distillation column.

As it is possible to see from Figs. 8a and b, for given operational conditions and a "good" initial guess of the calculated column variables (V and L concentrations and temperature profiles, etc.), the NEQ model given by a system of non-linear algebraic equations converged practically to the same steady state with high conversion of isobutene (point A in Fig. 8) with all assumed correlations. If a "wrong" initial guess was chosen, the NEQ model can provide different results according to the applied correlation: point A for Models 3 and 4 with high conversion of isobutene, point B for Model 2 and point C for Model 1. Therefore, the analysis of multiple steady-states existence has to be done as the first step of a safety analysis. If we assume the operational steady state of a column given by point A, and start to generate HAZOP deviations of operational parameters, by dynamic simulation, we can obtain different predictions of the column behaviour for each correlation, see Fig. 9a. Also, dynamic simulation of the column start-up procedure from the same initial conditions (for NEQ model equations) results in different steady states depending on chosen correlation, see Fig. 9b.

Our point of view is that of an engineer who has to do a safety analysis of a reactive distillation column using the mathematical model of such a device. Collecting literature information, he can discover that there are a lot of papers dealing with mathematical modelling. As was mentioned above, Taylor and Krishna (Taylor & Krishna, 2000) cite over one hundred papers dealing with mathematical modelling of RD of different complexicity. And there is a problem: which model is the best and how to obtain parameters for the chosen model. There are no general guidelines in literature. Using correlations suggested by authorities, an engineer can get into troubles. If different models predict different multiple steady states in a reactive distillation column for the same column configuration and the same operational conditions, they also predict different dynamic behaviour and provide different answers to the deviations generated by HAZOP. Consequently, it can lead to different definitions of the operator's strategy under normal and abnormal conditions and in training of operational staff.

5. Acknowledgement

This work was supported by the Slovak Research and Development Agency under the contract No. APVV-0355-07.

6. Nomenclature

A_b	bubbling area of a tray, m^2 (Table 4)
A_h	hole area of a sieve tray, m^2 (Table 4)
A	interfacial area per unit volume of froth, m^2 m^{-3} (Eqs. (15),(16))
a^I	net interfacial area, m^2

b	weir length per unit of bubbling area, m^{-1} (Table 4)
C_p	heat capacity, J mol^{-1} K^{-1}
c	molar concentration, mol m^{-3}
$\mathcal{E}$	energy transfer rate, J s^{-1}
D	Fick's diffusivity, m^2 s^{-1}
$Đ$	Maxwell-Stefan diffusivity, m^2 s^{-1}
F	feed stream, mol s^{-1}
F_f	fractional approach to flooding (Table 4)
FP	flow parameter (Table 4)
F_s	superficial F factor, $kg^{0.5}$ $m^{-0.5}$ s^{-1}(Table 4)
H	molar enthalpy, J mol^{-1}
$\Delta_r H$	reaction enthalpy, J mol^{-1}
h	heat transfer coefficient, J s^{-1} m^{-2} K^{-1}
h_L	clear liquid height, m (Table 4)
h_w	exit weir height, m (Table 4)
J	molar diffusion flux relative to the molar average velocity, mol m^{-2} s^{-1}
K_i	vapour-liquid equilibrium constant for component i
$[k]$	matrix of multicomponent mass transfer coefficients, m s^{-1}
L	liquid flow rate, mol s^{-1}
Le	Lewis number ($\lambda\rho^{-1}C_p^{-1}D^{-1}$)
M	mass flow rates, kg s^{-1} (Table 4)
N	number of transfer units
N_F	number of feed streams
N_I	number of components
N_R	number of reactions
$\mathcal{N}$	transfer rate, mol s^{-1}
n	number of stages
P	pressure of the system, Pa
PF	Pointing correction
ΔP	pressure drop, Pa
p	hole pitch, m (Table 4)
Q	heating rate, J s^{-1}
Q^L	volumetric liquid flow rate, m^3 s^{-1}(Table 4)
Q^V	volumetric vapour flow rate, m^3 s^{-1}(Eq.(17))
$[R]$	matrix of mass transfer resistances, s m^{-1}
r	ratio of side stream flow to interstage flow
Sc^V	Schmidt number for the vapour phase (Table 4)
T	temperature, K
t	time, s
t	residence time, s (Table 4)

U	molar hold-up, mol
u_s	superficial vapour velocity, m s^{-1}
u_{sf}	superficial vapour velocity at flooding, m s^{-1}
V	vapour flow rate, mol s^{-1}
W	weir length, m (Table 4)
x	mole fraction in the liquid phase
y	mole fraction in the vapour phase
Z	the liquid flow path length, m (Table 4)
z^P	mole fraction for phase P

Greek letters

β	fractional free area (Table 4)
$[\Gamma]$	matrix of thermodynamic factors
ε	heat transfer rate factor
κ	binary mass transfer coefficient, m s^{-1}
λ	thermal conductivity, W m^{-1} K^{-1}
μ	viscosity of vapour and liquid phase, Pa s
ν	stoichiometric coefficient
$\dot{\xi}$	reaction rate, mol s^{-1}
ρ	vapour and liquid phase density, kg m^{-3} (Table 4)
σ	surface tension, N m^{-1}

Superscripts

o	initial conditions
I	referring to the interface
L	referring to the liquid phase
V	referring to the vapour phase

Subscripts

av	averaged value
f	feed stream index
i	component index
j	stage index
m	mixture property
r	reaction index
t	referring to the total mixture

7. References

AICHE (1958). *AICHE Bubble Tray Design Manual*, AIChE, New York

Baur, R., Higler, A. P., Taylor, R. & Krishna, R. (2000). Comparison of equilibrium stage and nonequilibrium stage models for reactive distillation, *Chemical Engineering Journal*, 76(1), 33-47; ISSN: 1385-8947.

Fuller, E. N., Schettler, P. D. & Gibbings, J. C. (1966). A new method for prediction of binary gas-phase diffusion coefficints, *Industrial & Enegineering Chemistry,* 58(5), 18-27; ISSN: 0888-5885.

Górak, A. (2006). Modelling reactive distillation. *Proceedings of 33rd International Conference of Slovak Society of Chemical Engineering,* ISBN:80-227-2409-2, Tatranské Matliare, Slovakia, May 2006, Slovak University of Technology, Bratislava, SK, in Publishing House of STU.

Chan, H. & Fair, J. R. (1984). Prediction of point efficiencies on sieve trays. 2. Multicomponent systems, *Industrial & Engineering Chemistry Process Design Development,* 23(4), 820-827; ISSN: 0196-4305.

Chen, G. X. & Chuang, K. T. (1993). Prediction of point efficiency for sieve trays in distillation, *Industrial & Engineering Chemistry Research,* 32, 701-708; ISSN: 0888-5885.

Jacobs, R. & Krishna, R. (1993). Multiple Solutions in Reactive Distillation for Methyl tert-Butyl Ether Synthesis, *Industrial & Engineering Chemistry Research,* 32(8), 1706-1709; ISSN: 0888-5885.

Jones Jr., E. M. (1985). Contact structure for use in catalytic distillation, US Patent 4536373.

Kletz, T. (1999). *HAZOP and HAZAN,* Institution of Chemical Engineers, ISBN: 978-0852954218

Kooijman, H. A. & Taylor, R. (1995). Modelling mass transfer in multicomponent distillation, *The Chemical Engineering Journal,* 57(2), 177-188; ISSN: 1385-8947.

Kooijman, H. A. & Taylor, R. (2000). *The ChemSep book,* Libri Books on Demand, ISBN: 3-8311-1068-9, Norderstedt

Kotora, M., Švandová, Z. & Markoš, J. (2009). A three-phase nonequilibrium model for catalytic distillation, *Chemical Papers,* 63(2), 197-204; ISSN: 0336-6352.

Krishna, R. & Wesselingh, J. A. (1997). The Maxwell-Stefan approach to mass transfer, *Chemical Engineering Science,* 52(6), 861-911; ISSN: 0009-2509.

Krishnamurthy, R. & Taylor, R. (1985a). A nonequilibrium stage model of multicomponent separation processes I-Model description and method of solution, *AIChE Journal,* 31(3), 449-456; ISSN:1547-5905

Krishnamurthy, R. & Taylor, R. (1985b). A nonequilibrium stage model of multicomponent separation processes II-Comparison with experiment, *AIChE Journal,* 31(3), 456-465; ISSN:1547-5905

Kubíček, M. (1976). Algorithm 502. Dependence of solution of nonlinear systems on a parameter [C5], *Transaction on Mathematical Software,* 2(1), 98-107; ISSN 0098-3500.

Labovský, J., Švandová, Z., Markoš, J. & Jelemenský (2007a). Mathematical model of a chemical reactor-Useful tool for its safety analysis and design, *Chemical Engineering Science,* 62, 4915- 4919; ISSN: 0009-2509.

Labovský, J., Švandová, Z., Markoš, J. & Jelemenský (2007b). Model-based HAZOP study of a real MTBE plant, *Journal of Loss Prevention in the Process Industries,* 20(3), 230-237; ISSN: 0950-4230

Marek, M. & Schreiber, I. (1991). *Chaotic Behaviour of Deterministic Dissipative Systems,* Academia Praha, ISBN: 80-200-0186-7, Praha

Mohl, K.-D., Kienle, A., Gilles, E.-D., Rapmund, P., Sundmacher, K. & Hoffmann, U. (1999). Steady-state multiplicities in reactive distillation columns for the production of fuel ethers MTBE and TAME: theoretical analysis and experimental verification, *Chemical Engineering Science,* 54(*8*), 1029-1043; ISSN: 0009-2509.

Molnár, A., Markoš, J. & Jelemenský, L. (2005). Some considerations for safety analysis of chemical reactors, *Trans IChemE, Part A: Chemical Engineering Research and Design,* 83(*A2*), 167-176; ISSN: 0263-8762.

Noeres, C., Kenig, E. Y. & Gorak, A. (2003). Modelling of reactive separation processes: reactive absorption and reactive distillation, *Chemical Engineering and Processing,* 42(*3*), 157-178; ISSN: 0255-2701.

Perry, R. H., Green, D. W. & Maloney, J. O. (1997). *Perry's Chemical Engineers' Handbook,* McGraw-Hill, ISBN 0-07-049841-5, New York

Rehfinger, A. & Hoffmann, U. (1990). Kinetics of methyl tertiary butyl ether liquid phase synthesis catalyzed by ion exchange resin--I. Intrinsic rate expression in liquid phase activities, *Chemical Engineering Science,* 45(*6*), 1605-1617; ISSN: 0009-2509.

Reid, R. C., Prausnitz, J. M., Sherwood, T. K. (1977). *The Properties of Gases and Liquids,* McGraw-Hill, ISBN: 0-07-051790-8, New York

Sláva, J., Jelemenský, L. & Markos, J. (2009). Numerical algorithm for modeling of reactive separation column with fast chemical reaction, *Chemical Engineering Journal,* 150(*1*), 252-260; ISSN: 1385-8947.

Sláva, J., Svandová, Z. & Markos, J. (2008). Modelling of reactive separations including fast chemical reactions in CSTR, *Chemical Engineering Journal,* 139(*3*), 517-522; ISSN: 1385-8947.

Sundmacher, K. & Kienle, A. (2002). *Reactive Distillation, Status a Future Directions,* Wiley-VCH Verlag GmbH & Co. KGaA, ISBN: 3-527-60052-3, Weinheim

Švandová, Z., Jelemenský, Markoš, J. & Molnár, A. (2005b). Steady state analysis and dynamical simulation as a complement in the HAZOP study of chemical reactors, *Trans IChemE, Part B: Process Safety and Environmental Protection,* 83(*B5*), 463 - 471; ISSN: 0957-5820.

Švandová, Z., Labovský, J., Markoš, J. & Jelemenský, Ľ. (2009). Impact of mathematical model selection on prediction of steady state and dynamic behaviour of a reactive distillation column, *Computers & Chemical Engineering,* 33, 788-793; ISSN: 0098-1354.

Švandová, Z., Markoš, J. & Jelemenský (2005a). HAZOP analysis of CSTR with utilization of mathematical modeling, *Chemical Papers,* 59(*6b*), 464-468; ISSN: 0336-6352.

Švandová, Z., Markoš, J. & Jelemenský, Ľ. (2008). Impact of mass transfer coefficient correlations on prediction of reactive distillation column behaviour, *Chemical Engineering Journal,* 140(*1-3*), 381 390; ISSN: 1385-8947.

Taylor, R., Kooijman, H. A. & Hung, J.-S. (1994). A second generation nonequilibrium model for computer simulation of multicomponent separation processes, *Computers & Chemical Engineering,* 18(*3*), 205-217; ISSN: 0098-1354.

Taylor, R. & Krishna, R. (1993). *Multicomponent Mass Transfer,* John Wiley & Sons, Inc., ISBN: 0-471-57417-1, New York

Taylor, R. & Krishna, R. (2000). Modelling reactive distillation, *Chemical Engineering Science,* 55(*22*), 5183-5229; ISSN: 0009-2509.

Wesselingh, J. A., Krishna, R. (1990). *Mass Transfer,* Ellis Horwood Ltd, ISBN:978-0135530252, Chichester, England

Wilke, C. R. & Chang, P. (1955). Correlations of Diffusion Coefficients in Dilute Solutions, *AiChE Journal,* 1(2), 264-270; ISSN:1547-5905

Zuiderweg, F. J. (1982). Sieve trays : A view on the state of the art, *Chemical Engineering Science,* 37(10), 1441-1464; ISSN: 0009-2509.

29

Mass Transfer through Catalytic Membrane Layer

Nagy Endre
University of Pannonia, Research Institute of Chemical and Process Engineering
Hungary

1. Introduction

The catalytic membrane reactor as a promising novel technology is widely recommended for carrying out heterogeneous reactions. A number of reactions have been investigated by means of this process, such as dehydrogenation of alkanes to alkenes, partial oxidation reactions using inorganic or organic peroxides, as well as partial hydrogenations, hydration, etc. As catalytic membrane reactors for these reactions, intrinsically catalytic membranes can be used (e.g. zeolite or metallic membranes) or membranes that have been made catalytic by dispersion or impregnation of catalytically active particles such as metallic complexes, metallic clusters or activated carbon, zeolite particles, etc. throughout dense polymeric- or inorganic membrane layers (Markano & Tsotsis, 2002). In the majority of the above experiments, the reactants are separated from each other by the catalytic membrane layer. In this case the reactants are absorbed into the catalytic membrane matrix and then transported by diffusion (and in special cases by convection) from the membrane interface into catalyst particles where they react. Mass transport limitation can be experienced with this method, which can also reduce selectivity. The application of a sweep gas on the permeate side dilutes the permeating component, thus increasing the chemical reaction gradient and the driving force for permeation (e.g. see Westermann and Melin, 2009). At the present time, the use of a flow-through catalytic membrane layer is recommended more frequently for catalytic reactions (Westermann and Melin, 2009). If the reactant mixture is forced to flow through the pores of a membrane which has been impregnated with catalyst, the intensive contact allows for high catalytic activity with negligible diffusive mass transport resistance. By means of convective flow the desired concentration level of reactants can be maintained and side reactions can often be avoided (see review by Julbe et al., 2001). When describing catalytic processes in a membrane reactor, therefore, the effect of convective flow should also be taken into account. Yamada et al., (1988) reported isomerization of 1-butene as the first application of a catalytic membrane as a flow-through reactor. This method has been used for a number of gas-phase and liquid-phase catalytic reactions such as VOC decomposition (Saracco & Specchia, 1995), photocatalytic oxidation (Maira et al., 2003), partial oxidation (Kobayashi et al., 2003), partial hydrogenation (Lange et a., 1998; Vincent & Gonzales, 2002; Schmidt et al., 2005) and hydrogenation of nitrate in water (Ilinitch et al., 2000).

From a chemical engineering point of view, it is important to predict the mass transfer rate of the reactant entering the membrane layer from the upstream phase, and also to predict the downstream mass transfer rate on the permeate side of the catalytic membrane as a function of the physico-chemical parameters. The outlet mass transfer rate should generally be avoided. The mathematical description of the mass transport enables the reader to choose the operating conditions in order to minimize the outlet mass transfer rate. If this transfer (permeation) rate is known as a function of the reaction rate constant, it can be substituted into the boundary conditions of the full-scale differential mass balance equations for the upstream and/or the downstream phases. Such kind of mass transfer equations can not be found in the literature, yet. For their description, two types of membrane reactors should generally be distinguished, namely intrinsically catalytic membrane and membrane layer with dispersed catalyst particle, either nanometer size or micrometer size catalyst particles. Basically, in order to describe the mass transfer rate, a heterogeneous model can be used for larger particles and/or a pseudo-homogeneous one for very fine catalyst particles (Nagy, 2007). Both approaches, namely the heterogeneous model for larger catalyst particles and the homogeneous one for submicron particles, will be applied for mass transfer through a catalytic membrane layer. Mathematical equations have been developed to describe the simultaneous effect of diffusive flow and convective flow and this paper analyzes mass transport and concentration distribution by applying the model developed.

Membrane bioreactor (MBR) technology is advancing rapidly around the world both in research and commercial applications (Strathman et al., 2006; Yang and Cicek, 2006; Giorno and Drioli, 2000; Marcano and Tsotsis, 2002). Integrating the properties of membranes with biological catalyst such as cells or enzymes forms the basis of an important new technology called membrane bioreactor. Membrane layer is especially useful for immobilizing whole cells (bacteria, yeast, mammalian and plant cells) (Brotherton and Chau, 1990; Sheldon and Small, 2005), bioactive molecules such as enzymes (Rios et al., 2007; Charcosset, 2006; Frazeres and Cabral, 2001) to produce wide variety of chemicals and substances. The main advantages of the membrane, especially the hollow fiber, bioreactor are the large specific surface area (internal and external surface of the membrane) for cell adhesion or enzyme immobilization; the ability to grow cells to high density; the possibility for simultaneous reaction and separation; relatively short diffusion path in the membrane layer; the presence of convective velocity through the membrane if it is necessary in order to avoid the nutrient limitation (Belfort, 1989; Piret and Cooney, 1991; Sardonini and DiBiasio, 1992). This work analyzes the mass transport through biocatalytic membrane layer, either live cells or enzymes, inoculated into the shell and immobilized within the membrane matrix or in a thin layer at the membrane matrix matrix-shell interface. Cells are either grown within the fibers with medium flow outside or across the fibers while wastes and desired products are removed or grown in the extracapillary space with medium flow through the fibers and supplied with oxygen and nutrients (Fig. 12 illustrates this situation). The performance of a hollow-fiber or sheet bioreactor is primarily determined by the momentum and mass transport rate (Calabro et al., 2002; Godongwana et al., 2007) of the key nutrients through the bio-catalytic membrane layer. Thus, the operating conditions (trans-membrane pressure, feed velocity), the physical properties of membrane (porosity, wall thickness, lumen radius, matrix structure, etc.) can considerably influence the performance of a bioreactor, the effectiveness of the reaction. The introduction of convective transport is crucial in overcoming diffusive mass transport limitation of nutrients (Nakajima and Cardoso, 1989) especially of the sparingly soluble oxygen. Several investigators modeled the mass transport through this biocatalyst layer, through enzyme membrane layer (Ferreira et al., 2001; Long et al., 2003; Belfort, 1989; Hossain and Do, 1989; Calabro et al., 2002; Waterland et al., 1975;

Salzman et al., 1999; Carvalho et al, 2000) or cell culture membrane layer (Melo and Oliveira, 2001; Brotherton and Chau, 1990, 1996; Piret and Cooney, 1991; Sardonini and Dibiasio, 1992; Lu et al., 2001; Schonberg and Belfort, 1987). These studies analyze both the mass transport through the membrane and the bulk phase concentration change. Against these detailed studies, there are not known mass transfer equations which define the mass transfer rate through a biocatalytic membrane layer, in closed forms as a function of the transport parameters as membrane Peclet number, reaction rate modulus as well as the Peclet number of the concentration boundary layer. These equations could then be replaced in the full-scale mass transfer models in order to predict the concentration distribution in the bulk liquid phase.

When someone knows the mass transfer rate through the membrane, these rate equations now can be put into the full-scale mass balance equation as boundary value to describe the concentration distribution on the lumen side, feed side or on the shell side, permeate side. The full-scale description of flow in crossflow filtration tubular membrane or in flat sheet membrane is also very often the object of investigations (Damak et al., 2004). A fluid dynamic description of free flows is usually easy to perform, and in a great majority of examples, the well known Navier-Stokes equations can be used to coupling Darcy's law and the Navier-Stokes equations (Mondor & Moresoli, 1999; Damak et al., 2004). A steady-state, laminar, incompressible, viscous and isothermal flow in a cylindrical tube with a permeable wall is considered. The Navier-Stokes equation and Darcy's law describe the transfer in the tube and in the porous wall, respectively.

2. Mass transfer through membrane reactor

Six membrane reactor concepts can be considered related to the catalysts location in the membrane modules (Seidel-Morgenstern, 2010). Topics of this paper are the concept when the catalyst particles are dispersed in the membrane matrix (the membrane serves an active contactor) or the membrane layer is intrinsically catalytic. This concept is illustrated in Fig. 1. The reactants are fed into the reactor from different sides and react within the membrane.

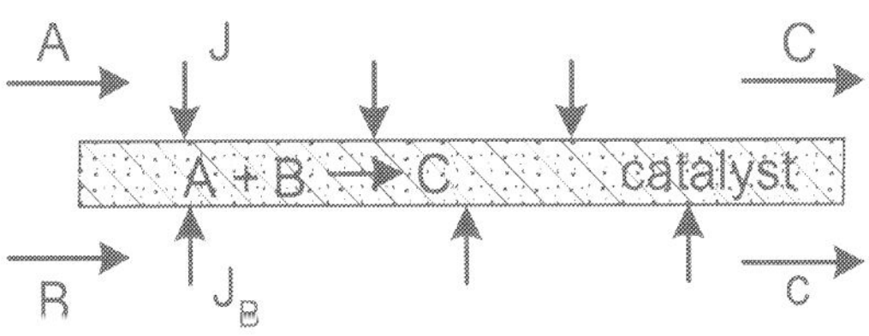

Fig. 1. Schematic illustration of catalytic membrane reactor

Before one can analyze the mass transport in the lumen or shell side of a capillary or on the two sides of a flat membrane, the outlet or inlet mass transfer rate at the membrane interface should be determined. A schematic diagram of the physical model and coordinate system is given in Fig. 2. The mass transfer rate depends strongly on the membrane properties, on the catalyst activity and the mass transfer resistance between the flowing fluid phase and membrane layer. This mass transfer rate should then be taken into account in the mass balance equation for the flowing fluid (liquid or gas) phase, on both sides of membrane reactor. This will be discussed in section 6.

The mass transport through a catalytic membrane layer can be diffusive (there is no transmembrane pressure difference between the two sides of the membrane layer) or diffusive+ convective transport. These two modes of flow will be discussed separately due to its different mathematical treatments in order to get the transfer rate.

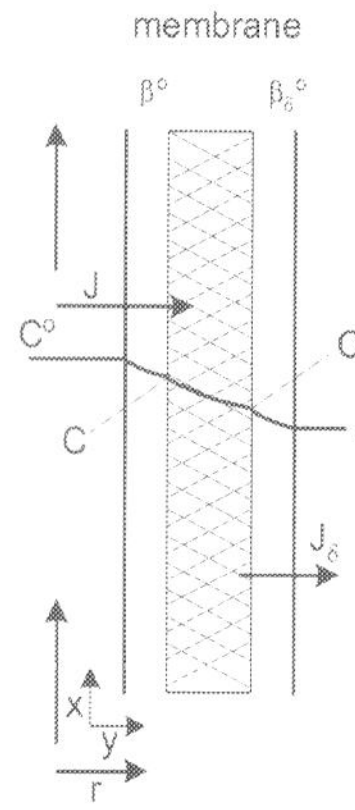

Fig. 2. Illustration of the mass transfer through a membrane reactor

The other important classification of the reactors that, as it was mentioned, the membrane reactor can intrinsically catalytic or it is made catalytic by dispersed catalyst particles distributed uniformly in the membrane matrix. In this latter case two types of mathematical model can be used (Nagy, 2007), namely pseudo-homogeneous or heterogeneous models, depending on the catalyst particle size. It was shown by Nagy (2007) if the size of catalyst particles less than a micron, the simpler homogeneous model can be recommended, in other wise, the heterogeneous model should be applied.
The differential mass balance equation can generally be given by the following equation for the catalytic membrane layer with various geometries, perpendicular to the membrane interface, applying cylindrical coordinate (Ferreira et al., 2001):

$$\frac{\partial}{\partial r}\left(D_m \frac{\partial c}{\partial r}\right)+\frac{(p+1)}{r}\frac{\partial(D_m c)}{\partial r}-\frac{\partial(\upsilon c)}{dr}-Q=\frac{\partial c}{\partial t} \tag{1a}$$

where p denotes a geometrical factor with values of 0 for cylindrical coordinate and -1 for rectangular membranes. The membrane concentration, C is given here in a unit of measure of gmol/m^3. This can be easily obtained by means of the usually applied in the e.g. g/g unit of measure with the equation of C=wρ/M, where w concentration in kg/kg, ρ - membrane density, kg/m^3, M-molar weight, kg/mol. The most often recommended mass balance equation (Marcano & Tsotsis, 2002), in dimensionless form, for membrane reactor is as ($R=r/R_o$; $C=c/c^o$):

$$\frac{\upsilon R_o^2}{D_m}\frac{\partial C}{\partial X}=\frac{1}{R}\frac{\partial}{\partial R}\left(\frac{\partial C}{\partial R}\right)-Q^* \tag{1b}$$

where Q^* reaction term given in dimensionless form. The boundary conditions are as:

$$C=1 \quad \text{at} \quad X=0, \text{ for all } R \tag{2a}$$

$$\frac{\partial C}{\partial R}=0 \quad \text{at} \quad R=0, \text{ for all } X \tag{2b}$$

$$\upsilon C - D\frac{dC}{\partial R} = J \quad \text{at} \quad R=1, \text{ for all } X \tag{2c}$$

The value of the mass transfer rate through the membrane, J will be shown in the next sections under different conditions. From eq. 1, the mass balance equation is easy to get for flat sheet membrane.

2.1 Diffusive mass transport with intrinsic catalytic layer or with fine catalytic particles

In both cases the membrane matrix is regarded as a continuous phase for the mass transport. Assumptions, made for expression of the differential mass balance equation to the catalytic membrane layer, are:

- Reaction occurs at every position within the catalyst layer;
- Mass transport through the catalyst layer occurs by diffusion;
- The partitioning of the components (substrate, product) is taken into account (thus, $CH_m=C^*_m$ where C^*_m denotes membrane concentration on the feed interface; see Fig. 2);
- The mass transport parameters (diffusion coefficient, partitioning coefficient) are constant;
- The effect of the external mass transfer resistance should also be taken into account;
- The mass transport is steady-state and one-dimensional;

In case of dispersed catalyst particles they are uniformly distributed and they are very fine particles with size less than 1 μm, i.e. they are nanometer sized particles. It is assumed that catalyst particles are placed in every differential volume element of the membrane reactor. The reactant firstly enters in the membrane layer and from that it enters into the catalyst particles where the reaction of particles is porous as e.g. active carbon, zeolite (Vital et al., 2001) occurs or it enters onto the particle interface and reacts [particle is nonporous as e.g. metal cluster, (Vancelecom & Jacobs, 2000)]. Consequently, the mass transfer rate into the catalyst particles has to be defined first. In this case, the whole amount of the reactant transported in or on the catalyst particle will be reacted. Then this term should be placed into the mass balance equation of the catalytic membrane layer as a source term. Thus, the differential mass balance equation for intrinsic membrane and membrane with dispersed nanosized particles differ only by their source term. The cylindrical effect can only be significant when the thickness of a capillary membrane can be compared to the internal radius of the capillary tube as it was shown by Nagy (2006). On the other hand, the application of cylindrical coordinate hinders the analytical solution for first or zero-order reactions as well. Thus, the basic equations will be shown here for plane interface and in the section 5 an analytical approach will be presented for cylindrical tube as well.

2.1.1 Mass transfer accompanied by first-order reaction

Herewith first the reaction source term will be defined indifferent cases, namely in cases of intrinsically catalytic membrane and membrane with dispersed catalytic particles and the solution of the differential mass balance equation under different boundary conditions.

2.1.1.1 Reaction terms

Intrinsically catalytic membrane; this is well known in literature ($\Phi = \sqrt{k_1 c_1^o}$):

$$Q = k_1 c_1^o C \equiv \Phi^2 C \tag{3}$$

Catalyst with dispersed particles, reaction takes place inside of the porous particles; For catalytic membrane with dispersed nanometer size particles, the mass transfer rate into the spherical catalyst particle has to be defined. The internal specific mass transfer rate in spherical particles, for steady-state conditions and when the mass transport accompanied by first-order chemical reaction can be given as follows (Nagy & Moser, 1995):

$$j = \beta_p C_p^* \tag{4}$$

where

$$\beta_p = \frac{D_p}{R_p}\left(\frac{Ha_p}{\tanh\left(Ha_p\right)} - 1\right) \tag{5a}$$

and

$$Ha_p = \sqrt{\frac{k_1 R_p^2}{D_p}}$$

The external mass transfer resistance, through the catalyst particle depends on the diffusion boundary layer thickness, δ_p. The value of δ_p could be estimated from the distance of particles from each other (Nagy & Moser, 1995). Namely, its value is limited by the neighboring particles, thus, the value of β_p will be slightly higher than that follows from the well known equation of $2 = \beta_p^o d_p / D_m$, where the value of δ_p is supposed to be infinite. Thus, one can obtain (Nagy et al., 1989):

$$\beta_p^o = \frac{2D_m}{d_p} + \frac{D_m}{\delta_p} \tag{5b}$$

where

$$\delta_p = \frac{h - d_p}{2}$$

From eqs 4 and 5 one can obtain for the mass transfer rate with the overall mass transfer resistance:

$$j = \beta_{tot} c^o C = c^o \frac{C}{\dfrac{1}{\beta_p^o} + \dfrac{1}{H\beta_p}} \tag{6}$$

Accordingly, the Φ value in eq. 3 can be expressed as follows (Nagy et al., 1989):

$$\Phi = \sqrt{\frac{\omega \delta_m^2}{1-\varepsilon}\beta_{tot}} \tag{7}$$

Reaction occurs on the interface of the catalytic particles (Nagy, 2007). It often might occur that the chemical reaction takes place on the interface of the particles, e.g, in cases of metallic clusters, the diffusion inside the dense particles is negligibly. Assuming the Henry's sorption isotherm of the reacting component onto the spherical catalytic surface ($CH_f = q_f$), applying $DdC/dr = k_f H_f C$ boundary condition at the catalyst's interface, at $r=R_p$, the Φ reaction modulus can be given according to eq. (7) with the following β_{sum} value:

$$\beta_\delta = \frac{1}{\dfrac{1}{\beta_p^o} + \dfrac{1}{k_f H_f}} \tag{8}$$

where k_f is the interface reaction rate constant. The above model is obviously a simplified one.

2.1.1.2 Mass transfer rates

The differential mass balance equation for the reactant entering the catalytic membrane layer is as follows in dimensionless form:

$$\frac{d^2C}{dY^2} - \Phi^2 C = 0 \tag{9}$$

Solution of eq. 9 is well known:

$$C = Te^{\Phi Y} + Se^{\Phi Y} \tag{10}$$

For the sake of generalization, in the boundary conditions you should take into account the external mass transfer resistance on both sides of the membrane, though it should be noted that the role of the β_δ^o will be gradually diminish with the increase of the reaction rate. At the end of this subsection the limiting cases will also be briefly given. Thus:

Y=0
$$\beta^o(1-C) = -\frac{D_m}{\delta_m}\left.\frac{dC}{dY}\right|_{Y=0} \tag{11}$$

Y=1
$$\beta_\delta^o\left(C_\delta - C_\delta^o\right) = -\frac{D_m}{\delta_m}\left.\frac{dC}{dY}\right|_{Y=1} \tag{12}$$

The mass transfer rate on the upstream side of the membrane can be given as follows (Nagy, 2007):

$$J = \beta_m H_m c^o\left(1 - TC_\delta^o\right) \tag{13}$$

with

$$\beta_m = \beta_m^o \Phi \frac{\left(1 + \frac{\beta_m^o \Phi H_m}{\beta_\delta^o} \tanh \Phi \right)}{\left(1 + \frac{\left[\beta_m^o \Phi\right]^2 H_m^2}{\beta^o \beta_\delta^o}\right) \tanh \Phi + \beta_m^o \Phi \left(\frac{H_m}{\beta^o} + \frac{H_m}{\beta_\delta^o}\right)} \tag{14}$$

and

$$T = \frac{1}{\cosh \Phi \left(1 + \frac{\beta_m^o \Phi H_m \tanh \Phi}{\beta_\delta^o}\right)} \tag{15}$$

with

$$\beta^o = \frac{D_L}{\delta_L}; \quad \beta_m^o = \frac{D_m}{\delta_m}$$

Similarly, the mass transfer rate for the downstream side of the membrane, at Y=1:

$$J_\delta = \beta_\delta H_m c^o \left(1 - \cosh \Phi \left\langle \tanh \Phi + \frac{\beta_m^o \Phi H_m}{\beta^o} \right\rangle C_\delta^o \right) \tag{14}$$

with

$$\beta_\delta = \frac{\beta_m^o \Phi}{\cosh \Phi} \frac{1}{\tanh \Phi \left(1 + \frac{\left(\beta_m^o \Phi\right)^2}{\beta^o \beta_\delta^o} H_m^2 \right) + \left(\frac{H_m}{\beta^o} + \frac{H_m}{\beta_\delta^o}\right) \beta_m^o \Phi} \tag{15}$$

Limiting cases; The transfer rate without external mass transfer resistances, namely when $\beta^o \to \infty$ and $\beta_\delta^o \to \infty$, can easily be obtained from eq. 13 as limiting case as:

$$J = \frac{\beta_m^o \Phi c^o}{\tanh \Phi} \left(1 - \frac{C_\delta^o}{\cosh \Phi}\right) \tag{16}$$

Eq. 16 is a well known mass transfer equation for liquid mass transfer accompanied by first-order reaction. The mass transfer can similarly be obtained rate for the case when the outlet concentration is zero, and, $\beta_\delta^o \to \infty$:

$$J = \beta_{tot}^o c^o \tag{17}$$

where

$$\beta_{tot}^o = \frac{1}{\frac{\tanh \Phi}{H_m \Phi \beta_m^0} + \frac{1}{\beta^0}} \tag{18}$$

To avoid the outlet flow of reactant is an important requirement for the membrane reactors. For it the operating conditions should be chosen rightly.

2.1.2 Mass transfer accompanied by zero-order reaction

In this case the reaction rate is independent of the concentration of reactant in the membrane layer. The differential mass balance equation can be given as:

$$\frac{d^2C}{dY^2} = \Phi^2 \tag{19}$$

The value of Φ can be given for intrinsically catalytic membrane as:

$$\Phi = \sqrt{\frac{k_0 \delta_m^2}{D_m c^o}} \tag{20}$$

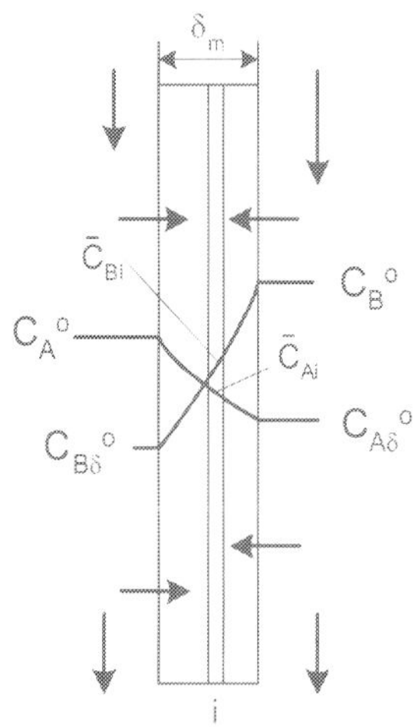

Fig. 3. Illustration of the concentrations for second-order reaction

The case of dispersed catalyst particles in the membrane layer is not discussed here because it unimportance for membrane reactor. For the solution of the eq. 19 let us use the following boundary conditions:

at Y=0 $$\beta^o(1-C) = -\frac{D_m}{\delta_m}\frac{dC}{dY}\bigg|_{Y=0} \tag{21}$$

at Y=1 $$C = HC_\delta^o \tag{22}$$

The mass transfer resistance on the outlet side has not importance in that case because the concentration rapidly decreases down to zero, thus does exist outlet mass transfer in a narrow reaction rate regime, only. After solution, the concentration distribution can be given as:

$$C = \frac{J}{\beta_m^o} Y + \frac{\Phi^2}{2} + B \tag{23}$$

where

$$B = \frac{C_\delta^o - \Phi^2 / 2 + \beta^o / \beta_m^o}{1 + \beta^o / \left(\beta_m^o H\right)} \tag{24}$$

The mass transfer rate can be given as:

$$J = \beta_m^o c^o \left(\frac{H - HC_\delta^o + \Phi^2 / 2}{H + \beta^o / \beta_m^o} \right) \tag{25}$$

From eq. 25, the well known expression of mass transfer rate without chemical reaction can easily be obtained.

2.1.3. Mass transfer accompanied by second-order reaction

It is assumed that the reagents (component A and B) are fed on the both sides of the membrane reactor and they are diffusing through the membrane layer counter-currently (Fig. 3). The reaction term can be given for intrinsically catalytic membrane as follows:

$$Q = k_2 c_A^o c_B^o C_A C_B \tag{26}$$

Substituting the reaction term into eq. (1) for e.g. the A component and plane interface as well as steady-state condition (D_{mA} is constant) one can get:

$$D_{mA} \frac{d^2 C_A}{dy^2} - k_2 c_A^o c_B^o C_A C_B = 0 \tag{27}$$

This equation can be solved either by numerical method or an analytical approach can be developed. Such an analytical approach is given in details in Appendix. The essential of this method that the membrane layer is divided into N very thin sub-layer and the concentration of one of the two components is considered to be constant in this sub-layer (see Fig. 3 and Fig. 13). Thus, one can get a second-order differential equation with linear source term that can be solved analytically. In dimensionless form it is for the ith sub-layer as:

$$\frac{d^2 C_A}{dY^2} - \Phi_{Ai}^2 C_A = 0 \quad \text{for} \quad Y_{i-1} \le Y \le Y_i \tag{28}$$

where

$$\Phi_{Ai} = \sqrt{\frac{k_2 \delta_m^2 c_A^o c_B^o \overline{C}_B}{D_{mA}}}$$

where $\bar{C}_B$ denotes the average concentration of B component in the ith sub-layer. Solution of eq. 28 is well known (see eq. 10). The general solution for every sub-layer has two parameters that should be determined by the suitable boundary conditions (see Appendix):

at Y=0 $$C=1 \qquad (29)$$

at $Y_{i-1} \le Y \le Y_i$ $$\left.\frac{dC_A}{dY}\right|_{Y_i^-} = \left.\frac{dC_A}{dY}\right|_{Y_i^-} \quad \text{with } i=1,2,\ldots,N \qquad (30)$$

at $Y_{i-1} \le Y \le Y_i$ $$C_{A|Y_i^-} = C_{A|Y_i^+} \quad \text{with } i=1,2,\ldots,N \qquad (31)$$

at y=1 $$C_A = C^o_{A\delta} \qquad (32)$$

After solution of the N differential equation with 2N parameters to be determined the T_1 and S_1 parameters for the first sub-layer can be obtained as (ΔY is the thickness of the sub-layers) :

$$T_1 = -\frac{1}{2\xi_N^O \cosh(\Phi_{A1}\Delta Y)}\left(\xi_N^T - \frac{C^o_{A\delta}}{\prod_{i=2}^{N}\cosh(\Phi_{Ai}\Delta Y)}\right) \qquad (33)$$

and

$$S_1 = \frac{1}{2\xi_N^O \cosh(\Phi_{A1}\Delta Y)}\left(\xi_N^S - \frac{C^o_{A\delta}}{\prod_{i=2}^{N}\cosh(\Phi_{Ai}\Delta Y)}\right) \qquad (34)$$

Knowing the T_1 and S_1 the other parameters, namely T_i and S_i (i=2,3,…,N) can be easily be calculated by means of the internal boundary conditions given by eqs. 30 and 31 from starting from T_2 and S_2 up to T_N and S_N.
After differentiating eq. 10 and applying it for the first sub-layer, the mass transfer rate of component A can be expressed as:

$$J = \frac{D_m \Phi_{A1} c^o_A}{\delta_m}\frac{\xi_N^S - \xi_N^T}{2\xi_N^O \cosh(\Phi_{A1}\Delta Y)}\left(1 - \frac{C^o_{A\delta}}{\left(\xi_N^S - \xi_N^T\right)\prod_{j=2}^{N}\cosh\left(\Phi_{Aj}\Delta Y\right)}\right) \qquad (35)$$

where

$$\xi_i^j = \xi_{i-1}^j + \kappa_{i-1}^j \frac{\tanh(\Phi_{Ai}\Delta Y)}{z_i} \quad \text{for } i=2,3,\ldots,N \text{ and } j=S,T,O \qquad (36)$$

and

$$\kappa_i^j = \xi_{i-1}^j \tanh\left(\Phi_{Ai}\Delta Y\right) + \frac{\kappa_{i-1}^j}{z_i} \quad \text{for i=2,3,...,N and j=S,T,O} \tag{37}$$

The starting values of ξ_1^j and κ_1^j are as follows:

$$\xi_1^T = e^{-\Phi_{A1}\Delta Y} \qquad \xi_1^S = e^{\Phi_{A1}\Delta Y} \qquad \xi_1^O = \tanh\left(\Phi_{A1}\Delta Y\right)$$

and

$$\kappa_1^T = -e^{-\Phi_{A1}\Delta Y} \qquad \kappa_1^S = e^{\Phi_{A1}\Delta Y} \qquad \kappa_1^O = 1$$

Obviously, in order to get the inlet mass transfer rate of component A, the concentration distribution of component B is needed. Thus, for prediction of the J value the concentration of component B has to be known. It is easy to learn that trial-error method should be used to get alternately the component concentrations. Steps of calculation of concentration of both components can be as follows:

1. Starting concentration distribution, e.g. for component B should be given and one calculates the concentration distribution of component A;
2. The indices of sub-layer of A component have to be changed adjusted them to that of B started from the permeate side of membrane, i.e. at Y=1, thus, i subscript of A_i should be replaced by N+1-i;
3. Now applying the previously calculated averaged A_i ($\bar{A}_i$), one can predict the concentration distribution of component B, using eqs. 33 to 37, adapted them to component B;
4. These three steps should be repeated until concentrations do not change anymore;

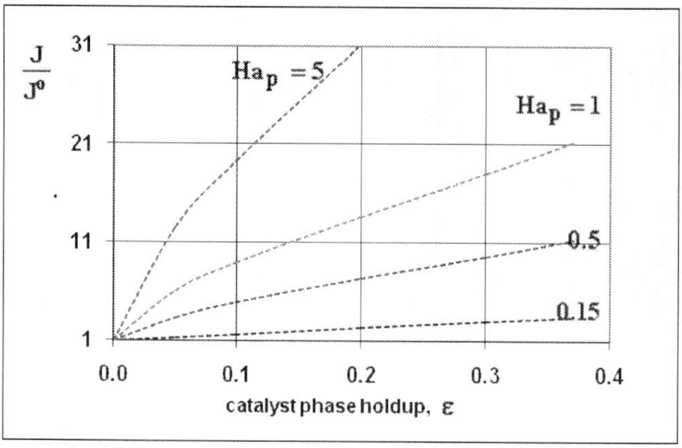

Fig. 4. The mass transfer rate a s a function of the catalyst phase holdup obtained by the pseudo-homogeneous model (H_m=H=1; D_m=1 x 10^{-10} m^2/s; $C_\delta^o = 0$; $\beta^o = \beta_\delta^o \to \infty$; d_p=2 μm; δ_m=30 μm)

2.1.4 Analysis of the mass transport

The detailed discussion of the mass transport through a membrane reactor is not a target of this paper. Applying the equations for mass transfer rate or for concentration distribution presented the reactor performance is easy to calculate. Only a typical figure will be shown in this section. Fig. 4 illustrates the effect of the catalyst holdup and the reaction modulus on the mass transfer rate. Detailed analysis is given in Nagy's paper (2007). Similar results can be obtained by zero-order reaction though its effect is somewhat stronger because its independency of concentration (Nagy, 2007). The concentration of the reactor rapidly decreases down to zero even at rather low reaction rate coefficient. Thus the role of the convective velocity should have got careful attention.
Normally, 3-5 recalculations of concentrations are enough to get the correct results.

2.2 Diffuive+convective mass transport with intrinsic catalytic layer or with fine catalytic particles

Convective mass transport can take place if transmembrane pressure difference exists between the two membrane sides. Recently it was proved in the literature (Ilinitch, 2000, Nagy, 2007) that the presence of convective flow can improve the efficiency of the membrane reactor. Thus, the study of the mass transport in presence of convective mass flow can be important in order to predict the reaction process. On the otherwise, the use of convective flow is rather rare, because the aim is mostly to minimize the outlet rate of the reactant on the permeate side. The source terms of this case are the same as it was showed in subsection 2.1.

2.2.1 Mass transport accompanied by first-order reaction

The differential mass balance equation for the polymeric or macroporous ceramic catalytic membrane layer, for steady-state, taking both diffusive and convective flow into account, can be given as:

$$\frac{d^2C}{dY^2} - Pe_m \frac{dC}{dY} - \Phi^2 C = 0 \tag{38}$$

where

$$Pe_m = \frac{\upsilon \delta_m}{D_m}; \quad \Phi = \sqrt{\frac{\omega \delta_m^2}{D_m(1-\varepsilon)} \beta_{tot}}$$

where υ denotes the convective velocity, D_m is the diffusion coefficient of the membrane, and δ_m is the membrane thickness.

$$\tilde{C} = Ce^{-Pe_m Y/2} \tag{39}$$

Introducing a new variable, $\tilde{C}$ (eq. 39) the following differential equation is obtained from eq. 38):

$$\frac{d^2\tilde{C}}{dY^2} - \Theta^2 \tilde{C} = 0 \tag{40}$$

where

$$\Theta = \sqrt{\frac{Pe_m^2}{4} + \Phi^2}$$

The general solution of eq. 40 is well known, so the concentration distribution in the catalytic membrane layer can be given as follows:

$$C = Te^{\lambda Y} + Se^{\tilde{\lambda} Y} \qquad (41)$$

with

$$\tilde{\lambda} = \frac{Pe_m}{2} - \Theta \qquad \lambda = \frac{Pe_m}{2} + \Theta$$

The inlet and the outlet mass transfer rate can easily be expressed by means of eq. (41). The overall inlet mass transfer rate, namely the sum of the diffusive and convective mass transfer rates, is given by:

$$J = \upsilon C\big|_{Y=0} - \frac{D_m}{\delta_m}\frac{dC}{dY}\bigg|_{Y=0} = \beta_m^o\left(\lambda T + \tilde{\lambda} S\right) \qquad (42)$$

The outlet mass transfer rate is obtained in a similar way to eq. (42) for X=1:

$$J_\delta = \beta_m^o\left(\tilde{\lambda} Te^{\lambda} + \lambda Se^{\tilde{\lambda}}\right) \qquad (43)$$

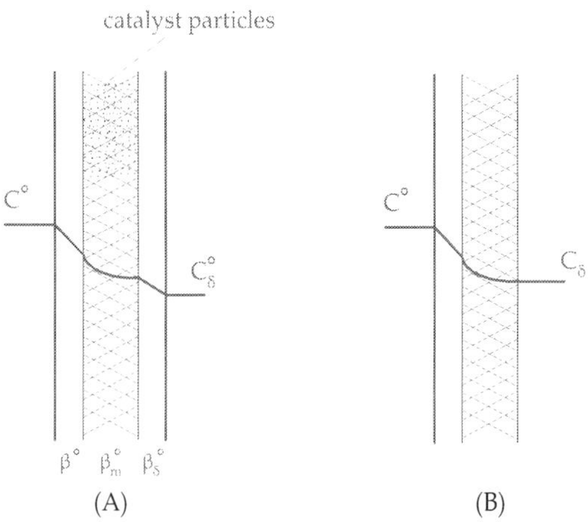

Fig. 5. Illustration of the concentration distribution for models A and B.

The value of parameters T and S can be determined from the boundary conditions. For the sake of generality, two models, namely model A and model B, will be distinguished according to Figure 5 (for details see Nagy, 2010). The essential difference between the models is that, in case of model A, there is a sweeping phase that can remove the transported component from the downstream side providing the low concentration of the reacted component in the outlet phase and due to it, high diffusive mass transfer rate. There is no sweep phase in case of model B, thus the outlet phase is moving from the membrane due to the lower pressure on the permeate side.

Model A. In this case, due to the effect of the sweeping phase, the external mass transfer resistance on both sides of the membrane should be taken into account in the boundary conditions, though the role of β_δ^o is gradually diminished as the catalytic reaction rate increases. The concentration distribution in the catalytic membrane when applying a sweep phase on the two sides of the membrane, as well is illustrated in Fig. 5a. On the upper part of the catalytic membrane layer, in Fig. 5a, the fine catalyst particles are illustrated with black dots. It is assumed that these particles are homogeneously distributed in the membrane matrix. Due to sweeping phase, the concentration of the bulk phase on the permeate side may be lower than that on the membrane interface. The boundary conditions can be given for that case as:

$$\upsilon C + \beta^o\left(C^o - C\right) = J \text{ at Y=0} \tag{44}$$

$$\upsilon C_\delta + \beta_\delta^o\left(C_\delta - C_\delta^o\right) = J_\delta \text{ at Y=1} \tag{45}$$

Boundary conditions given by eqs. (44) and (45) are only valid in two phase flows. Where C_δ^o denotes the concentration on the downstream side, β^o and β_δ^o are mass transfer coefficients in the continuous phase, β_m^o the membrane mass transfer coefficient ($\beta_m^o = D_m / \delta_m$), H_m denotes the distribution coefficient between the continuous phase and the membrane phase. The solution of the algebraic equations obtained, applying eqs. 42 to 45, can be received by means of known mathematical manipulations. Thus, the values of T and S obtained are as follows:

$$T = -\frac{\beta^o\varphi_2 C^o - \beta_\delta^o\varphi_4 C_\delta^o}{\varphi_2\varphi_3 - \varphi_1\varphi_4}\frac{1}{\beta_m^o} \tag{46}$$

and

$$S = \frac{\beta^o\varphi_1 C^o + \beta_\delta^o\varphi_3 C_\delta^o}{\varphi_2\varphi_3 - \varphi_1\varphi_4}\frac{1}{\beta_m^o} \tag{47}$$

where

$$\varphi_1 = \left(\frac{Pe_m}{H_m} + \frac{\beta_\delta^o}{\beta_m^o H_m} - \tilde{\lambda}\right)e^{\lambda}; \quad \varphi_2 = \left(\frac{Pe_m}{H_m} + \frac{\beta_\delta^o}{\beta_m^o H_m} - \lambda\right)e^{\tilde{\lambda}};$$

$$\varphi_3 = \frac{Pe_m}{H_m} - \frac{\beta^o}{\beta_m^o H_m} - \tilde{\lambda}; \quad \varphi_4 = \frac{Pe_m}{H_m} - \frac{\beta^o}{\beta_m^o H_m} - \lambda;$$

An important limiting case should also be mentioned, namely the case when the external diffusive mass transfer resistances on both sides of membrane can be neglected, i.e. when $\beta^o \to \infty$ and $\beta_\delta^o \to \infty$. For that case the concentration distribution and the inlet mass transfer rate can be expressed by eqs. 48 and 49, respectively.

$$C = \frac{e^{Pe_m(Y-1)/2}}{\sinh\Theta} H_m \left\{ C^o e^{Pe_m/2} \sinh\left[\Theta(1-Y)\right] + C_\delta^o \sinh(\Theta Y) \right\} \tag{48}$$

$$J = \beta_m \left(C^o - \frac{\Theta}{e^{Pe_m/2}\left[Pe_m \sinh\Theta / 2 + \Theta\cosh\Theta\right]} C_\delta^o \right) \tag{49}$$

with

$$\beta_m = \frac{\beta_m^o H_m \left(Pe_m \tanh\Theta / 2 + \Theta\right)}{\tanh\Theta}$$

An important limiting case when the outlet concentration is zero, i.e. $C_\delta^o = 0$, accordingly the mass transfer rate is as ($\beta_\delta^o \to \infty$):

$$J = \frac{2\beta^o \Theta c^o}{\frac{\beta^o}{\beta_m^o}(Pe_m - 1)\left(e^{-2\Theta} - 1\right) - \left(\tilde{\lambda} e^{-2\Theta} - \lambda\right)} \tag{50}$$

Model B. For the convective flow catalytic membrane reactor operating in another mode, for instance in dead-end mode as in Figure 4b, the boundary condition on the permeate side of the membrane should be changed. In this case the concentration of the permeate phase does not change during its transport from the membrane interface. If there is no sweeping phase on the downstream side then the correct boundary conditions will be as:

$$\upsilon C + \beta^o \left(C^o - C\right) = J \text{ at Y=0} \tag{51}$$

$$\upsilon C_\delta = J_\delta \text{ at Y=1} \tag{52}$$

After solution one can get as:

$$T = -\frac{\beta_1^o \varphi_2 C^o}{\varphi_2\varphi_3 - \varphi_1\varphi_4} \frac{1}{\beta_m^o} \tag{53}$$

$$S = \frac{\beta^o \varphi_1 C^o}{\varphi_2\varphi_3 - \varphi_1\varphi_4} \frac{1}{\beta_m^o} \tag{54}$$

where

$$\varphi_1 = \left(\frac{Pe_m}{H_m} - \tilde{\lambda}\right) e^{\lambda} \qquad \varphi_2 = \left(\frac{Pe_m}{H_m} - \lambda\right) e^{\tilde{\lambda}}$$

The values of φ_3 and φ_4 are the same as they are given after eq. 47.

2.2.2 Mass transport accompanied by zero-order reaction

The effect of the zero-order reaction will be discussed here for intrinsically catalytic membrane layer, only. This reaction has no important role in the case of membrane reactor.

The differential mass balance equation to be solved is as:

$$D_m \frac{d^2C}{dy^2} - \upsilon \frac{dC}{dy} - k_0 = 0 \tag{55}$$

Similarly to eqs. 19, the differential mass balance equation for the catalytic membrane can be given as:

$$\frac{d^2C}{dY^2} - Pe_m \frac{dC}{dY} = \Phi^2 \tag{56}$$

where

$$\Phi = \sqrt{\frac{k_0 \delta_m^2}{D_m c^o}}$$

Look at first the solution with the following boundary conditions:

$$Y=0 \text{ then } \quad C=1 \tag{57a}$$

$$Y=1 \text{ then } \quad C=C_\delta^o \tag{57b}$$

The general solution of Eq. (56) is as:

$$C = T_m e^{Pe_m Y} - \frac{\Phi^2}{Pe_m} Y + Q_m \tag{58}$$

Applying the boundary conditions [Eqs. (57a) and (57b] one can get:

$$C = \frac{e^{Pe_m Y/2}}{\sinh(Pe_m/2)} \left\{ C_b \left(\sinh\left[\frac{Pe_m}{2}(1-Y) \right] + S \right) + e^{-Pe_m/2} \sinh\left(\frac{Pe_m Y}{2} \right) C_p \right\} \tag{59}$$

The mass transfer rate can be given as:

$$J = \beta_m c^o \left(1 - T C_\delta^o\right) \tag{60}$$

where

$$\beta_m = \beta_m^o c^o \left(1 + \frac{\Phi^2}{Pe_m^2}\right); \quad T = \frac{e^{-Pe_m}}{1 + \Phi^2 / Pe_m^2}$$

The outlet mass transfer rate should also be given:

$$J_\delta = \beta_m^o \alpha_\delta c^o \left(1 - C_\delta^o \frac{e^{-Pe_m}}{\alpha_\delta}\right) \tag{61}$$

where

$$\alpha_\delta = 1 - \frac{\Phi^2}{Pe_m^2} e^{-Pe_m/2}\left(1 - e^{Pe_m}\left[1 - Pe_m\right]\right) \equiv 1 - \frac{\Phi^2}{Pe_m^2}\left(Pe_m - 2e^{-Pe_m/2}\sinh\left(\frac{Pe_m}{2}\right)\right) \quad (62)$$

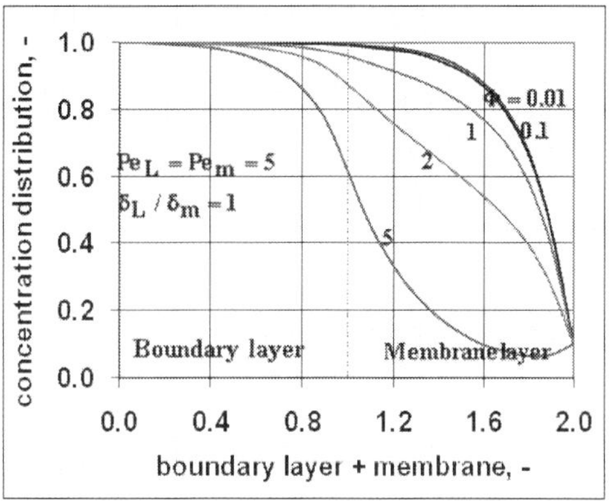

Fig. 6. The effect of first-order chemical reaction on the concentration distribution in the two layers at different values of reaction modulus ($C_\delta^o = 0.1$; $\beta_L^o = \beta_m^o = 1x10^{-4} m/s$)

Now look at the solution with mass transfer resistance in the feed phase, thus, the boundary condition for Y=0 will be as:

$$\upsilon C + \beta^o\left(C^o - C\right) = J \quad (63)$$

The physical mass transfer rate through the boundary layer can be expressed, in case of convective + diffusive flows as follows (Nagy & Kulcsar, 2009):

$$J^o = \beta^o c^o\left(1 - e^{-Pe_L}C\right) \quad (64)$$

with

$$\beta^o = \frac{D_L}{\delta_L} Pe_L \frac{e^{Pe_L}}{e^{Pe_L - 1}} \quad (65)$$

where

$$Pe_L = \frac{\upsilon \delta_L}{D_L}$$

Applying the well known two-layer theory the mass transfer rate can be given by means of the overall mass transfer resistance applying eqs. (60) and (64), the two mass transfer rate are equal to each other, as follows

$$J = \beta_{tot} c^o\left(1 - Te^{-Pe_L} C_\delta^o / H_m\right) \quad (66)$$

with

$$\beta_{tot} = \frac{1}{\frac{1}{\beta^o} + \frac{H_m e^{-Pe_L}}{\beta_m}} \tag{67}$$

The values of β^o and β_m are defined in eqs. 65 and 60.

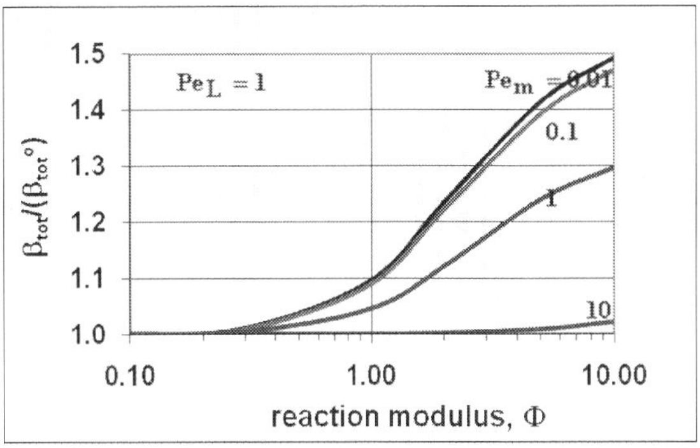

Fig. 7. The overall mass transfer coefficient as a function of first-order reaction rate constant at different values of Pe_m number $\beta_L^o = \beta_m^o = 1x10^{-4} m/s$

2.2.3 Analysis of the mass transport

The simultaneous effect of the convective flow and the reaction rate is less intensively investigated (Nagy, 2009a, 2009b, Nagy & Kulcsár, 2009; Nagy & Borbély, 2009). The importance of convective velocity can essentially improve the efficiency of the membrane reactor (Ilinitch, 2000; Nagy, 2010). Let us show the concentration distribution in case of first-order reaction (Fig. 6). The Φ reaction modulus can significantly lower the concentration in the phases. At larger modulus the curve change from concave one to convex one through an a inflexion point. The dimensionless outlet concentration was chosen to be 0.1. The concentration can decrease below this value indicating that $C_\delta^o = 0.1$ can maintain only when there is inlet flow of reactant on the permeate side. Obviously, this is not a real case. The effect of the first-order chemical reaction on the mass transfer coefficient is illustrated at different, increasing values of the Pe_m and with external resistance (Fig. 7). Obviously, the mass transfer rate will have limiting value due to the increasing effect of the external mass transfer resistance when the reaction rate increases. Description of the reduction of aqueous nitrates applying mono- and bimetallic, palladium-copper catalysts impregnated in γ-Al_2O_3 support layers is discussed by Nagy (2010) applying the mass transfer rate developed.

2.3 Mass transfer with microsized, dispersed, catalyst particles: applying the heterogeneous model

Mostly, depending also on the membrane thickness, when particles are falling into the micrometer size regime, the internal mass transport mechanism, inside of catalyst particles, has to be taken into account. A simple physical model could be applied for the description the process in this case, as it is schematically illustrated in Fig. 8. The gas (or liquid) reactant enters first the catalytic membrane layer and then diffuses to the first catalytic particle, perpendicular to the membrane interface. The chemical reaction, namely a first-order, irreversible chemical reaction, takes place in the catalyst particles only. It is assumed that the concentration of the organic reactant should be much higher in a hydrophobic polymer membrane than that of the reactant investigated, e.g. peroxides, oxygen or hydrogen, etc. Then the unreacted reactant diffuses through the first, catalytic particle to its other side and enters again the polymer membrane matrix and so on (the route of this mass transfer process is illustrated by an arrow denoted by J in Fig. 8). This diffusion path exists only for the heterogeneous part of the membrane interface (which is the projection of the cubic catalyst particle onto the membrane interface). There can be a portion of membrane interface, that is the so called homogeneous part of the interface, where the diffusing reactant does not cross any catalyst particle (this mass stream is denoted by J^o in Fig. 8). This also affects the resultant mass transfer rate. The assumed cubic (Yawalkar et al., 2001, Mehra, 1999, Nagy, 2002) catalyst particles are supposed to be uniformly distributed in the polymer membrane matrix.

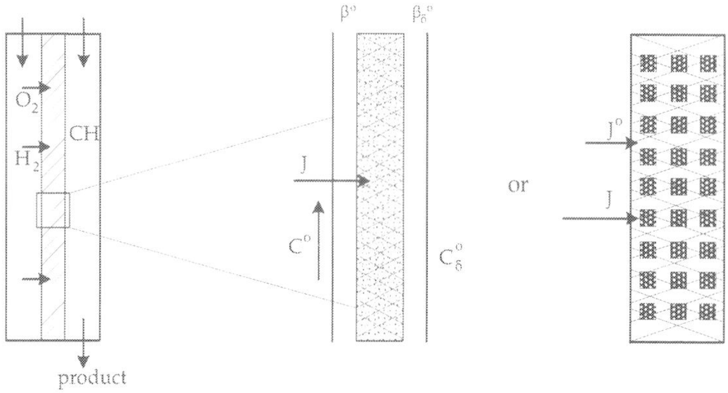

Fig. 8. Membrane reactor with dispersed catalyst particles (for heterogeneous model the spherical particles are modeled as cubic ones.

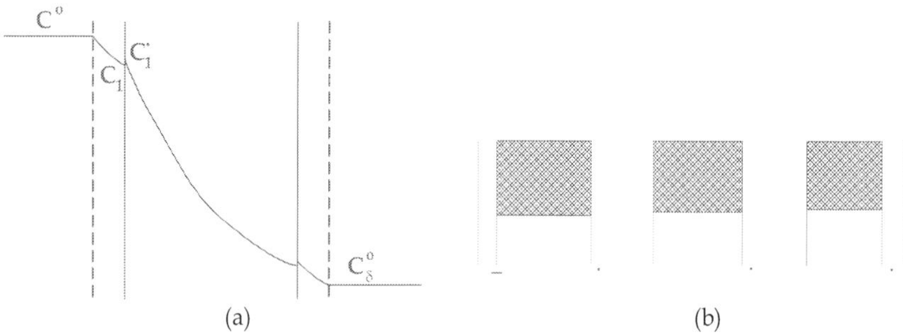

Fig. 9. Concentration distribution in the membrane reactor (a) and a particle line with notations (b)

For the description of this transport process, the catalyst membrane layer should be divided into 2N+1 sub-layers, perpendicular to the membrane interface. Namely, N sub-layers for catalyst particles located perpendicular to the membrane interface, N+1 sub-layers for the polymer membrane matrix between particles (ΔY) and between the first particle (Y_1) and the last particle ($1-Y_N^*$) and the membrane interfaces (Fig. 9b). In order to get a mathematical expression for the mass transfer rates, differential mass balance equation should be given for each sub-layer. Thus, one can obtain a differential equation system containing 2N+1 second order differential equations. This equation system with suitable boundary conditions can be solved analytically which is also demonstrated in this paper. The number of particles, N, and the distance between them, ΔY, can be calculated from the particle size, d_p and the catalyst phase holdup, ε (Nagy, 2007). The distance of the first particle from the membrane interface, Y_1 ($Y_1=y_1/\delta_m$) can be regulated by the preparation method of the catalytic membrane layer. The differential mass balance equations for the sections of the polymer membrane phase and for that of the catalyst particles can be given, in dimensionless form, as follows, respectively:

$$\frac{D_m}{\delta_m^2}\frac{d^2C}{dY^2}=0 \qquad 0\le Y\le Y_1\,,\; Y_i+\frac{d}{\delta}\le Y\le Y_{i+1}\,,\; Y_i\le Y\le 1 \tag{68}$$

$$\frac{d^2C_p}{dY^2}-\frac{k_1\delta_m^2}{D_p}C_p=0 \qquad Y_i\le Y\le Y_i+\frac{d}{\delta} \tag{69}$$

The solutions of the above differential equations, for the ith sections, are well known, namely:

$$C=T_iY+S_i \qquad 1\le i\le N+1 \tag{70}$$

$$C_p=E_i\exp(\Phi Y)+F_i\exp(-\Phi Y) \qquad 1\le i\le N \tag{71}$$

with

$$\Phi=\sqrt{\frac{k_1\delta_m^2}{D_p}}$$

Thus, one can obtain 2N+1 algebraic equations with twice as many parameters, T_i, S_i (i=1, 2, 3, …, N+1) as well as E_i and F_i (i=1, 2, 3, …, N), which are to be determined. Their values can be determined by means of suitable boundary conditions at the external interfaces of the membrane, at Y=0 and Y=1 as well as at the internal interfaces of every segment in the membrane matrix, at Y_i and Y_i^* with i=1, 2, 3, …, N. The effect of the external mass transfer resistances should be taken into account:

At Y=0 then
$$\beta^o\left(C^o-C\right)\equiv\beta^o\left(C^o-\frac{S_1}{H_m}\right)=-\frac{D_m}{\delta_m}\frac{dC}{dY}\equiv-\beta_m^oT_1 \tag{72}$$

At Y=1 then
$$\beta_\delta^o\left(C_\delta-C_\delta^o\right)\equiv\beta_\delta^o\left(\frac{T_{N+1}+S_{N+1}}{H_m}-C_\delta^o\right)=-\frac{D_m}{\delta_m}\frac{dC}{dY}\equiv-\beta_m^oT_{N+1} \tag{73}$$

The boundary conditions for the internal interfaces of the sub-layers are also well known (Fig. 9b):

at $Y = Y_i$ $$(T_i Y_i + S_i)H = E_i e^{\Phi Y_i} + F_i e^{-\Phi Y_i} \tag{74}$$

at $Y = Y_i$ $$D_m A_i = D_p \Phi\left(E_i e^{\Phi Y_i} - F_i e^{-\Phi Y_i}\right) \tag{75}$$

as well as for the other side of the catalyst particles, namely at $Y = Y_i^*$:

at $Y = Y_i + \frac{d}{\delta} = Y_i^*$ $$\left(T_{i+1} Y_i^* + S_{i+1}\right)H = E_i e^{\Phi Y_i^*} + F_i e^{-\Phi Y_i^*} \tag{76}$$

at $Y = Y_i + \frac{d}{\delta} = Y_i^*$ $$D_m T_{i+1} = D_p \Phi\left(E_i e^{\Phi Y_i^*} - F_i e^{-\Phi Y_i^*}\right) \tag{77}$$

Eqs 74 and 76 express that there is equilibrium on the sub-layer interfaces, while eqs 73 and 77 involve that there is no accumulation or source at the internal interfaces. Thus, an algebraic equation system with 2(N+1) equations can be obtained that can be solved analytically with a traditional method using the Cramer rules. The solution is briefly discussed in Nagy's paper (2007).
As a result of this solution, the mass transfer rate on the upstream side of the membrane interface, related to its heterogeneous part (which is the projection of the cubic catalyst particle onto the membrane interface), can be given as follows:

$$J = \beta_m H_m c^o \left(1 - T C_\delta^o\right) \tag{78}$$

where

$$\beta_m = \beta_m^o H \frac{U_{n(N+1)} + H H_m \dfrac{\beta_m^o}{\beta_\delta^o}}{U_{N+1} + H H_m \dfrac{\beta_m^o}{\beta_\delta^o}} \prod_{i=1}^{N} \left(\frac{U_{ni}}{U_i} \frac{\alpha_{ni}}{\alpha_i} \right) \tag{79}$$

as well as

$$T = \frac{1}{\left(U_{n(N+1)} + H H_m \dfrac{\beta_m^o}{\beta_\delta^o} \right) \prod_{i=1}^{N} \left(U_{ni} \alpha_{ni} \right)} \tag{80}$$

with

$$U_{n(N+1)} = \left[H\left(1 - Y_N^*\right) + \frac{\xi_{nN}}{\alpha_{nN}} \right] \tag{81}$$

$$U_{(N+1)} = \left[H\left(1 - Y_N^*\right) + \frac{\xi_N}{\alpha_N} \right] \tag{82}$$

as well as

$$HC = C_p; \quad Y_N^* = Y_1 + (N-1)\Delta Y + N \frac{d}{\delta_m}; \quad H_m C_j = C_j^* \qquad \text{with j = 1,2}$$

The values of U_{ni}, α_{ni}, and ξ_{ni} as well as U_i, α_i, and ξ_i should be calculated from sub-layer to sub-layer, that is from 1 to N (α_{ni}, ξ_{ni}, α_i, ξ_i) or N+1 (U_{ni}, U_i), from equations given in Table 1.

for i=2 to N

$$\alpha_{ni} = \frac{1}{U_{ni}} + \kappa \tanh(Ha_d); \qquad \xi_{ni} = \frac{\tanh(Ha_d)}{U_{ni}\kappa} + 1 \; ; U_{ni} = H\Delta Y + \frac{\xi_{ni}}{\alpha_{ni}}$$

for i=1 to N

$$\alpha_i = \frac{1}{U_i} + \kappa \tanh(Ha_d); \; \xi_i = \frac{\tanh(Ha_d)}{U_i\kappa} + 1; \; U_i = H\Delta Y + \frac{\xi_i}{\alpha_i}; \qquad \Delta Y = Y_i - Y_{i-1}^*$$

as well as

$$U_{n1} = 1 \; ; \; \alpha_{n1} = \kappa \tanh(Ha_d); \qquad \xi_{n1} = 1$$

$$U_1 = \left(HY_1 + HH_m \frac{\beta_m^o}{\beta^o} \right) \qquad Ha_d = \sqrt{\frac{k_1 d^2}{D_p}} \qquad \text{with } \kappa = \frac{\lambda D_d}{D}$$

Table 1. The list of variables that should be applied for the calculation of the mass transfer rates.

It may also be important to know the portion of the reactant that reacts in the catalytic membrane layer during its diffusion, or, there is an unreacted portion of the diffused reactant that passes on the downstream side of the membrane into the continuous phase. This outlet mass transfer rate, for the heterogeneous part of the membrane interface, at X=1 can be given as follows:

$$J_\delta = \beta_{out} H_m c^o \left(1 - \frac{C_\delta^o}{\prod_{i=1}^{N} \cosh(Ha_d) - \frac{\xi_i}{\alpha_i} \sinh(Ha_d)} \right) \tag{83}$$

with

$$\beta_\delta - \beta_m^o H \frac{\prod_{i=1}^{N} \left(\cosh(Ha_d) - \frac{\xi_i}{\alpha_i} \sinh(Ha_d) \right)}{U_{N+1} + HH_m \frac{\beta_m^o}{\beta_\delta^o}} \tag{84}$$

The physical mass transfer rate for the heterogeneous part of the interface is as follows:

$$J^o = \beta_{tot}^o H_m c^o \left(1 - C_\delta^o \right) \tag{85}$$

The physical mass transfer coefficient, with external mass transfer resistances, for the portion of the membrane interface where there are particles in the diffusion path taking into account the effect of the catalyst particles, as well, can be given by the following equation:

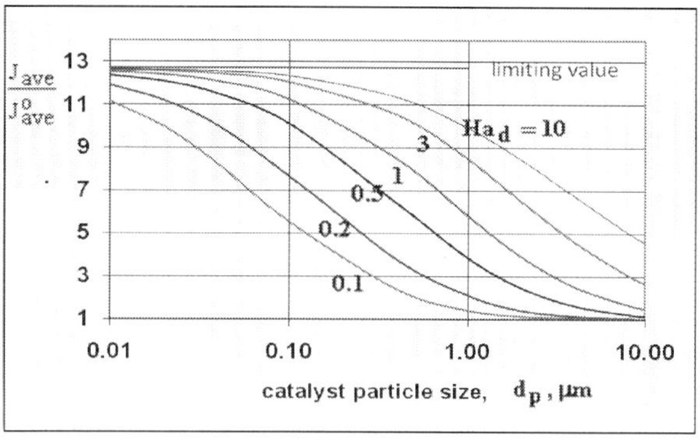

Fig. 10. Effect of the catalyst particle size on the mass transfer rate related to the total membrane interface, as a function of the reaction modulus (H_m=H=1; D_m=1 x 10^{-10} m^2/s; $C^o_\delta = 0$; $\beta^o = \beta^o_\delta \to \infty$; Y_1=1 µm; δ_m=30 µm; ε=0.1)

$$\beta^o_{tot} = \frac{1}{\dfrac{H_m}{\beta^o + \beta^o_\delta} + \dfrac{1}{\beta^o_m} + \dfrac{Nd}{D_p}\left(\dfrac{1}{H} - \dfrac{D_p}{D_m}\right)} \tag{86}$$

Depending on the value of the D_p diffusion coefficient, H solubility coefficient (C_p=CH) as well as the number of particles perpendicular to the interface, N, the value of the physical mass transfer coefficient of the membrane with catalytic particles, β^o_{tot} might be completely different from that of the membrane layer without catalyst particles, β^o_m ($\beta^o_m = D_m / \delta_m$). For example, in the case of a polymer membrane filled with zeolite particles as catalyst, the value of D_p can be lower by about 4 orders of magnitude than that in the polymer matrix (Jawalkar et al., 2001). The specific mass transfer rate related to the total catalytic membrane interface (Nagy, 2007) can be given as:

$$J_{ave} = KJ\varepsilon^{2/3} + J^o\left(1 - K\varepsilon^{2/3}\right) \tag{87}$$

or that for the outlet mass transfer rate:

$$J_{ave\delta} = KJ_\delta\varepsilon^{2/3} + J^o\left(1 - K\varepsilon^{2/3}\right) \tag{88}$$

The value of the mass transfer rate can be easily obtained for the homogeneous part of the interface, J^o, namely:

$$J^o = \beta^o_m H_m c^o\left(1 - C^o_\delta\right) \tag{89}$$

In order to calculate the enhancement during the mass transfer accompanied by chemical reaction, the physical mass transfer rate related to the total membrane interface should also be defined:

$$J^o_{ave} = KJ^o_{tot}\varepsilon^{2/3} + J^o\left(1 - K\varepsilon^{2/3}\right) \tag{90}$$

The value of the factor K can be obtained from the distribution of catalyst particles in the polymer membrane matrix. Its value, depending on the particles distribution in the membrane matrix, should be K=1 or K=1.8715 (for details see Nagy, 2007) The effect of the catalyst particle size is illustrated in Fig. 10 at different reaction rate applying the heterogeneous model. The mass transfer rate is very sensitive to the particle size, namely its value strongly decreases with the increase of the size. With increasing size decreases the number of particle because distance between them increases. Accordingly the particle size should be decreasing as low as possible. Detailed analysis of this model and its comparison to the homogeneous one is given in Nagy's paper (2007).

3. Mass transport through biocatalytic membrane layer

The principle of the mass transport of substrates/nutrients into the immobilized enzyme/cells, through a solid, porous layer (membrane, biofilm) or through a gel layer of enzyme/cells is the same. The structure, the thickness of this mass transport layer can be very different, thus, the mass transport parameters, namely diffusion coefficient, convective velocity, the bio-reaction rate constant, their dependency on the concentration and/or space coordinate is characteristic of the porous layer and of the nature of the biocatalysts. Several investigators modeled the mass transport through this biocatalyst layer, through enzyme membrane layer or cell culture membrane layer (Schonberg & Belfort, 1990, Kelsey et al., 1990, Piret & Cooney, 1991, Ferreira et al., 2001) Recently Nagy (2009a, 2009b) studied the mass transfer rate into a biocatalytic membrane layer with constant mass transport parameters. He defined the mass transfer rates for both side of the membrane surface. The rate equations are expressed as product of the mass transfer coefficient and driving force as it is traditionally applied for the diffusion systems, e.g. in gas-liquid systems. Applying these inlet mass transfer rate, the concentration profile of the two layers, namely that of the boundary layer and biocatalytic membrane layer can be calculated. These will be demonstrated in the case of first- and zero-order reactions as well as in the case of the general Monod kinetics. Assumptions, made for expression of the differential mass balance equation to the biocatalytic membrane layer, are:

- Reaction occurs at every position within the biocatalyst layer;
- Reaction has one rate-limiting substrate/nutrient;
- Mass transport through the biocatalyst layer occurs by diffusion and convection;
- The partitioning of the components (substrate, product) is negligible (Thus, $C^*=C^*_m$);
- The mass transport parameters (diffusion coefficient, convective velocity, bioreaction rate coefficient) are constant or varying;
- The effect of the concentration boundary layer should also be taken into account;

For the sake of simplification, let us regard a steady-state reaction as well as let us use the Cartesian co-ordinate, thus, the differential mass balance equation, applying the Michaelis-Menten kinetics, can be given as follows (y is here the transverse space coordinate, perpendicular to the membrane interface):

$$\left\{\frac{d}{dy}\left(D_m\frac{dC}{dy}\right)-\frac{d(\upsilon C)}{dy}\right\}-\frac{\upsilon_{\max}C}{K_M+C}=0 \tag{91}$$

In general case, as it was mentioned, the diffusion coefficient and/or convective velocity can depend on the space coordinate, thus $D_m=D_m(y)$, υ (y), [or on the concentration, $D_m=D_m(C)$ or both of them, $D_m=D_m(C, y)$]. In the boundary conditions the external mass transfer resistance should also be taken into account. The membrane layer with biomass colony on it and the concentration boundary layer are illustrated, with important notations and concentration profiles, in Fig. 12.
The boundary conditions of eq. 91 are the same, in general case, as it given in eqs. 44 and 45. That is, the fluid phase mass transfer resistances of both sides of the membrane are taken into account. Eq. 91 can be solved numerically or by analytical approach, only. In the following sub-section we give an analytical approach, where the mass transfer rate is expressed in explicit, closed form, to its solution. The general solution of eq. 91 is discussed by Nagy (2008, 2009a). Here a solution without mass transfer resistance will only be shown.

3.1 Approaching analytical solution of the mass transport with variable parameters and/or with Monod kinetics for single membrane layer

Let us first to solve the differential mass balance eq. 91 with the boundary conditions 57a. and 57b. where the diffusive mass transfer resistances are negligible. The solution with boundary conditions given by eqs. 44 and 45 is much more complicated (not shown here). This solution was given by Nagy (2008) for diffusional mass transport through membrane reactor and by Nagy & Borbély, (2009) for diffusive+convective mass transport with variable parameters. In essentials, this solution methodology serves the mass transfer rate and the concentration distribution in closed, explicit mathematical expression. The method can be applied for Cartesian coordinate and cylindrical coordinate as will be shown. For the solution of the eq. (91), the biocatalytic membrane should be divided N sub-layer (Fig. 13), in the direction of the mass transport, that is perpendicular to the membrane interface, with thickness of $\Delta\delta$ ($\Delta\delta=\delta/N$) and with constant transport parameters in every sub-layer [for details of a general solution see (Nagy, 2009a)]. Thus, for the nth sub-layer of the membrane layer, using dimensionless quantities, it can be obtained (for better understanding m subscript for D is not written here):

$$D_n \frac{d^2C_n}{dy^2} - \upsilon \frac{dC_n}{dy} - k_n C_n = 0 \quad y_{n-1} < y < y_n \tag{92}$$

where the value of k_n can be obtained according e.g. to the Michaelis-Menten kinetics as follows:

$$k_n = \frac{v_{max}}{K_M + \bar{C}_n} \tag{93}$$

where $\bar{C}_n$ denotes the average value of C in the nth membrane sub-layer.
In dimensionless form one can get the following equation:

$$\frac{d^2C_n}{dY^2} - Pe_n \frac{dC_n}{dY} - \Phi_n^2 C_n = 0 \, , \tag{94}$$

where

$$\Phi_n = \sqrt{\delta_m^2 k_n / D_n} \, ; \quad Pe_n = \frac{\upsilon \delta_m}{D_n}$$

The solution of eq. 92 can be easily obtained by well known mathematical methods, as it is geiven in sub-section 2.2.1, as it follows:

$$C = T_n e^{(\lambda_n Y)} + S_n e^{(\tilde{\lambda}_n Y)} \quad Y_{n-1} < Y < Y_n \tag{95}$$

with

$$\lambda_n = \frac{Pe_n}{2} + \Theta_n \qquad \tilde{\lambda}_n = \frac{Pe_n}{2} - \Theta_n \qquad \Theta = \sqrt{\frac{Pe_n^2}{4} + \Phi_n^2}$$

T_n and P_n parameters of Eq. (95) can be determined by means of the boundary conditions for the nth sub-layer (with $1 \le n \le N$). The boundary conditions at the internal interfaces of the sub-layers ($1 \le n \le N$-1; Y_m=nΔY; ΔY=1/N) can be obtained from the following two equations [Eqs. (96a) and (96b)]:

$$-\frac{dC_n}{dY} + Pe_n C_n = \frac{D_{n+1}}{D_n}\left(-\frac{dC_{n+1}}{dY} + Pe_{n+1}C_{n+1}\right) \quad \text{at } Y=Y_n \tag{96a}$$

$$C_n\big|_{Y=Y_n^-} = C_{n+1}\big|_{Y=Y_{n-1}^-} \quad \text{at } Y=Y_n \tag{96b}$$

After solution of the algebraic equation system containing 2N equations, the mass transfer rate on the upstream side of the membrane can be given, for that case, as follows:

$$J = \frac{D_1 c^o}{\delta_m}\left(\tilde{\lambda}_1 T_1 + \lambda_1 P_1\right) \tag{97}$$

$$T_1 = \frac{c^o \varphi_N^T}{\varphi_N^O}\left(1 - \frac{C_\delta^o}{\varphi_N^T\left(\prod_{j=2}^{N} \varphi_{j-1}^T D_{mi}\tilde{\lambda}_i - \chi_{j-1}^T\right)}\right) \tag{98a}$$

and

$$S_1 = -\frac{c^o \varphi_N^S}{\varphi_N^O}\left(1 - \frac{C_\delta^o}{\varphi_N^S\left(\prod_{j=2}^{N} \varphi_{j-1}^S D_{mi}\tilde{\lambda}_i - \chi_{j-1}^S\right)}\right) \tag{98b}$$

with

$$\chi_i^j = \frac{\left(\lambda_i e^{\tilde{\lambda}_i \Delta Y} - \tilde{\lambda}_i e^{\lambda_i \Delta Y}\right)}{\lambda_i - \tilde{\lambda}_i} - \frac{D_{mi}\tilde{\lambda}_i e^{\lambda_i \Delta Y}}{D_{mi}\tilde{\lambda}_i - \chi_{i-1}^j / \varphi_{i-1}^j} \quad \text{with i=2,3,…,N and j=T,S,O} \tag{99}$$

and

$$\varphi_i^j = \frac{e^{\tilde{\lambda}_i \Delta Y} - e^{\lambda_i \Delta Y}}{D_{mi}\left(\lambda_i - \tilde{\lambda}_i\right)} - \frac{e^{\lambda_i \Delta Y}}{D_{mi}\tilde{\lambda}_i - \chi_{i-1}^j / \varphi_{i-1}^j} \quad \text{with i=2,3,...,N and j=T,S,O} \tag{100}$$

The starting values of χ_i^T, χ_i^S, χ_i^O and φ_i^T, φ_i^S, φ_i^O are as follows:

$$\chi_1^T = D_{m1}\lambda_1 e^{\tilde{\lambda}_1 \Delta Y}\ ;\ \chi_1^S = D_{m1}\tilde{\lambda}_1 e^{\lambda_1 \Delta Y} \quad \chi_1^O = D_{m1}\left(\lambda_1 e^{\tilde{\lambda}_1 \Delta Y} - \tilde{\lambda}_1 e^{\lambda_1 \Delta Y}\right)$$

as well as

$$\varphi_1^T = e^{\tilde{\lambda}_1 \Delta Y}\ ;\ \varphi_1^S = e^{\lambda_1 \Delta Y} \quad \varphi_1^O = e^{\tilde{\lambda}_1 \Delta Y} - e^{\lambda_1 \Delta Y}$$

If one want to calculate the concentration distribution he should start with values of T_1 and S_1 and applying the known boundary conditions, the values of T_i and S_i (i=2,3,...,N) can easily be calculated. The internal boundary conditions could be as follows:

$$\left(\upsilon C - D_i \frac{\partial C}{\partial Y}\right)\Bigg|_{Y=Y_i^-} = \left(\upsilon C - D_{i+1} \frac{\partial C}{\partial Y}\right)\Bigg|_{Y=Y_i^-} \tag{101}$$

$$C\Big|_{Y=Y_i^-} = C\Big|_{Y=Y_i^-} \tag{102}$$

The outlet mass transfer rate can be similarly given (not shown here). This value should be as less as possible to avoid the loss of the substrate during the process. In the next section some typical figures will be shown to illustrate the effect of the parameters on the mass transport through the catalytic membrane layer.

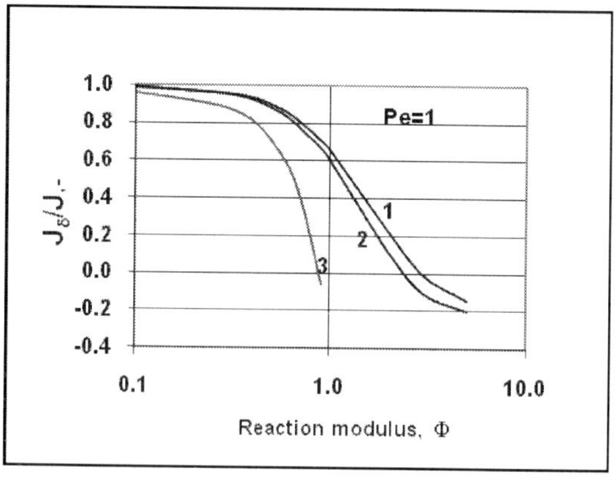

Fig. 11. The relative value of outlet mass transfer rate as a function of reaction modulus applying the Michaelis-Menten kinetics (line 2) and its limiting cases, namely first-order (line 1) and zero-order kinetics. (D_m=5.4 x 10^{-10} m^2/s; $C_\delta^o = 0$; $\beta^o = \beta_\delta^o \to \infty$; K_M/c^o=1; δ_m=100 μm)

In the case of microorganism the substrate concentration should be kept above its critical level. To hold it is important to apply convective mass flow, as well. A typical figure (Fig. 11) illustrates how the reaction rate influences the outlet mass transfer rate related to the inlet one calculated it in limiting cases and with the Michaelis-Menten kinetics (line 2). Fig. 11 illustrates clearly that the approach of the limiting case can cause essential error in prediction of the reactor efficiency.

4. Membrane reactor with biofilm on it

There is an increasing interest in the membrane-aerated biofilm reactor as a technology that can enhance the application of biofilms in wastewater treatment. In this reactor the biofilm is naturally immobilized on an substrate permeable membrane and counter- or (regarding the conventional biofilms where both the dissolved oxygen and substrates diffuse in the same direction) co-diffusion of oxygen and nutrients (organic component, ammonia, etc.) can take place. Several investigators have reported performance advantages of membrane-aerated biofilm reactors for wastewater treatment (Aryal et al., 2009; Monthlagh et al., 2006) oxidation of organic components (Casey et al., 2000; Gross et al., 2007), nitrification (Rittman & Manem, 1992; Wang et al., 2009) etc. The structure of the biofilm can be homogeneous or heterogeneous depending on the substrate concentration (Piciorenau et al., 2001) because the growth rate of microorganisms depends strongly on the substrate concentration. Taking into account that the sum of the reacted amount of substrate in the biofilm, J_r and the inlet mass transfer rate into the membrane layer (layer m) is equal to that of inlet transfer rate of the biofilm layer, ($J=J_r+J_m$, where J_m denotes the mass transfer rate that enters the membrane layer), the J value can be expressed for two reactive layers, similarly to the two-film theory as (Fig. 12):

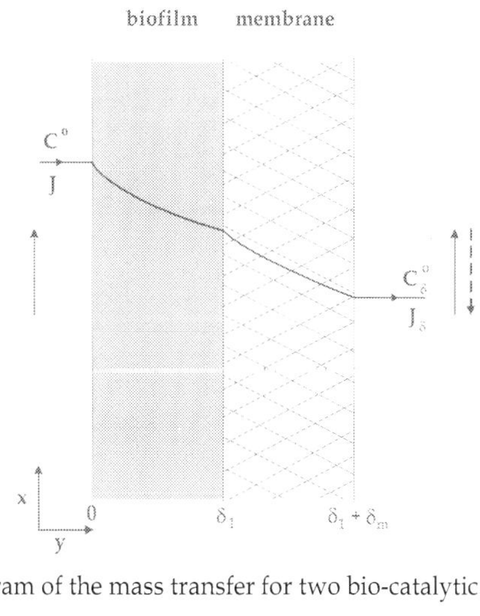

Fig. 12. Schematic diagram of the mass transfer for two bio-catalytic layers as well as the important notations

$$J = \beta_{tot} c^o \left(1 - L C^o_\delta\right) \tag{103}$$

where

$$\beta_{tot} = \frac{1/A + \beta_r/(\beta_M - \beta E)}{1/(\beta_m - \beta E) + 1/(\beta A)} \tag{104}$$

$$\beta = \frac{D_m}{\delta_m}\frac{(Pe_m/2)\tanh\Theta + \Theta}{\tanh\Theta} \tag{105}$$

$$A = \frac{\Theta e^{-Pe/2}}{\cosh\Theta\langle(Pe/2)\tanh\Theta + \Theta\rangle} \tag{106}$$

$$L = \frac{\beta_m A}{(1/A + \beta_r/\langle\beta_m - \beta_r E\rangle)(\beta_m - \beta_r E)} \tag{107}$$

where

$$\beta_r = \frac{D\Phi^2}{\delta\tanh\Theta}\frac{(Pe/2)\tanh\Theta + \Theta\left(e^{-Pe/2}/\cosh\Theta - 1\right)}{(Pe/2)^2 - \Theta^2} \tag{108}$$

and

$$E = \frac{Pe_m\sinh\Theta + \Theta\left(e^{-Pe_m/2} - \cosh\Theta\right)}{\Theta\left(e^{Pe_m/2} - Pe_m e^{-Pe_m/2}/2\right)} \tag{109}$$

This equation does not involve the effect of the external mass transfer resistance.

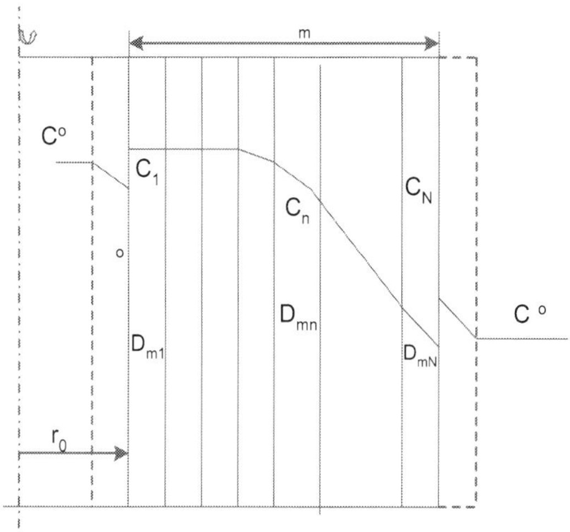

Fig. 13. Division of the membrane layer for prediction of the mass transfer rate for cylindrical and/or variable parameters.

5. Approaching solution in case of cylindrical coordinate

Applying the Michaelis-Menten equation, the mass balance equation with cylindrical coordinate is as:

$$D_m\left(\frac{dc^2}{dr^2}+\frac{1}{r}\frac{dc}{dr}\right)-\upsilon\frac{dc}{dr}-\upsilon_{\max}\frac{c}{K_M+c}=0 \qquad (110)$$

The differential mass balance equation for state-state conditions, for cylindrical coordinate and for the nth sub-layer (Fig. 13) is as:

$$D_{mn}\left(\frac{dc^2}{dr^2}+\frac{1}{r_n}\frac{dc}{dr}\right)-\upsilon\frac{dc}{dr}-\upsilon_{\max}\frac{c}{K_M+\bar{c}_n}=0 \qquad (111)$$

Rearranging eq. 111, the following dimensionless equation is obtained to be solved:

$$\frac{dC^2}{dR^2}-(Pe_n-1)\frac{dC}{dR}-k_nC=0 \qquad (112)$$

where

$$k_n=\frac{\upsilon_{\max}}{K_M+\bar{C}_n} \qquad Pe_n=\frac{\upsilon R_o}{D_{mn}}$$

The solution of eq. 112 is the same as eq. 92:

$$C=T_ne^{(\lambda_nR)}+P_ne^{(\tilde{\lambda}_nR)} \qquad R_{n-1}< R < R_n \qquad (113)$$

with

$$\lambda_n=\frac{(Pe_n-1)+\sqrt{(1-Pe_n)^2+4\Phi_n^2}}{2} \qquad \tilde{\lambda}_n=\frac{(Pe_n-1)-\sqrt{(1-Pe_n)^2+4\Phi_n^2}}{2}$$

as well as

$$\Phi=\sqrt{\frac{k_nR_o^2}{D_n}}$$

where R_o is the radius of the cylindrical membrane, $R=1+n(R_o+\delta_m)/R_o$, $R=r/R_o$.The determination of T_n, S_n parameters is the same as in the case of Cartesian coordinate.

6. Full-scale description of the process in hollow fiber membrane

Mostly used membrane configuration is capillary module, though the industry often applies tubular, spiral-wound and plate-and frame modules (Baker, 2004). Here the mass balance or continuity equations are given in cylindrical coordinate. These can easily be rewritten to Cartesian one, as well (Cebeci, 2005). For cross-flow processes mass balance equation should be given for both the lumen and the shell sides of the membrane. The feed stream in the tube or outside of the membrane, is modeled by the Navier-Stokes equations. The simplified mass and momentum equations, i.e., Navier-Stokes equations expressed in cylindrical coordinates with axisymmetry assumption, are as (Damak et al., 2004):

$$\frac{1}{r}\frac{\partial}{\partial r}(r\upsilon)+\frac{\partial u}{\partial x}=0 \tag{114}$$

$$\rho\left(\upsilon\frac{\partial u}{\partial r}+u\frac{\partial u}{\partial x}\right)=-\frac{\partial P}{\partial x}+\mu\left\{\frac{\partial}{\partial r}\left[\frac{1}{r}\frac{\partial(ru)}{\partial r}\right]+\frac{\partial^2 u}{\partial x^2}\right\} \tag{115}$$

$$\rho\left(\upsilon\frac{\partial \upsilon}{\partial r}+u\frac{\partial \upsilon}{\partial x}\right)=-\frac{\partial P}{\partial x}+\mu\left\{\frac{\partial}{\partial r}\left[\frac{1}{r}\frac{\partial(r\upsilon)}{\partial r}\right]+\frac{\partial^2 \upsilon}{\partial x^2}\right\} \tag{116}$$

as well as for solute concentration in case of constant density

$$u\frac{\partial C}{\partial z}+\upsilon\frac{\partial C}{\partial r}=\frac{\partial}{\partial r}\left(D\frac{\partial C}{\partial r}\right)+\frac{1}{r}\frac{\partial(DC)}{\partial r}+\frac{\partial}{\partial x}\left(D\frac{\partial C}{\partial x}\right) \tag{117}$$

Tranfer equations in the porous wall; The momentum equation, i.e., Darcy's law, is as follows:

$$\upsilon=-\frac{\kappa}{\mu}\frac{\partial P}{\partial r} \tag{118}$$

$$u=-\frac{\kappa}{\mu}\frac{\partial P}{\partial x} \tag{119}$$

where κ is the dimensionless Darcy's law permeability coefficient.
Boundary conditions; At the inlet, a fully developed laminar profile can be considered, i.e., Poiseuille flow which leads to:

$$u=2u_o\left[1-\left(\frac{r}{R}\right)^2\right] \text{ and } \upsilon=0 \text{ at } x=0;\ 0\le r\le R \tag{120}$$

u_o represents the axial velocity at the inlet. At the exit, a fully developed profile is assumed as follows:

$$\frac{\partial u}{\partial x}=0 \text{ and } \upsilon=0 \text{ at } x=L;\ 0\le r\le R \tag{121}$$

At the axis of symmetry there are no momentum fluxes crossing the boundary:

$$\frac{\partial u}{\partial x}=0 \text{ and } \upsilon=0 \text{ at } r=0;\ 0\le x\le L \tag{122}$$

At the permeable wall:

$$\upsilon=\frac{\kappa}{\mu}\frac{\Delta P}{\delta} \text{ and } u=0 \text{ at } r=R;\ 0\le z\le L \tag{123}$$

The most hollow fiber configurations, the ratio R/L are very small, less than 1 x 10^{-3}, thus, the inertial terms can be neglected (Mondor & Moresoli, 1999). Finally, because the velocity gradients are smaller in the axial direction than in radial direction, the axial stress terms can be neglected in the momentum equation. Thus, the simplified form the momentum and the continuity equations are, respectively, given as:

$$\frac{1}{r}\frac{\partial}{\partial r}\left(r\frac{\partial u}{\partial r}\right)=\frac{dP}{dx} \tag{124}$$

and

$$\frac{\partial u}{\partial x}+\frac{1}{r}\frac{\partial (rv)}{\partial r}=0 \tag{125}$$

The solute balance equation with constant diffusion coefficient:

$$u\frac{\partial C}{\partial x}+v\frac{\partial C}{\partial r}=D\left(\frac{\partial^2 C}{\partial r^2}+\frac{1}{r}\frac{\partial C}{\partial r}+\frac{\partial^2 C}{\partial x^2}\right) \tag{126}$$

The J mass transfer rate presented should be inserted into the boundary condition given by eq. 2c of mass balance eq. (1a) or eq. (126), thus this differential equation can be solved. In the case of membrane reactor or bioreactor, the axial pressure gradient within the membrane is often negligible compared to the radial pressure gradient, thus the first term in eq. 125 can often be neglected. When there is no change of volume of the fluid phase because the low convective permeation rate or the case of dilute fluid phase, the mass balance equation given by eq. 126 should be taken into account during the mass transport calculation (Piret & Cooney, 1991).

7. Conclusion

Mass transfer rate and, in some cases, the concentration distribution inside a membrane reactor were defined. Exact solutions of the mass transfer rate were given, taking into account the external mass transfer resistance on the both sides of the catalytic membrane layer. The membrane is either intrinsically catalytic or catalytic particles are dispersed in the membrane matrix. For this letter case, both pseudo-homogeneous model (for nanometer sized particles) and heterogeneous one (for microsized catalyst particles) have been presented. An analytical approaching solution was developed for cylindrical coordinate and/or variable mass transport parameters, as e.g. diffusion coefficient, chemical reaction rate constant. The mass transfer rates obtained then should be inserted as a boundary condition into differential mass balance equations in order to describe the full-scale mass balance equation given for capillary or plate-and-frame modules.

8. Appendix

The differential mass balance equations for the reactants in the membrane layer assuming that $Q=k_2c_Ac_B$, for component A and B, respectively:

$$D_A\frac{d^2C_A}{dy^2}-k_2c_Ac_B=0 \tag{A1}$$

$$D_B \frac{d^2C_B}{dy^2} - k_2 c_A c_B = 0 \tag{A2}$$

Let us apply the following boundary conditions:

y=0 then $$c_A = c_A^o \quad c_B = c_{B\delta} \tag{A3}$$

and

y=δ_m then $$c_A = c_{A\delta} \quad c_B = c_B^o \tag{A4}$$

Dividing the membrane layer into N very thin, sub-layers, the following approach can be applied regarding the concentrations: the mass balance equation is given one of the reactants while its average value, e.g. $(c_{Ai-1}+c_{Ai})/2$ is considered for the other component in this equation. Thus, one can write for e.g. components A the following differential equation, in dimensionless form, for the ith sub-layer:

$$\frac{d^2C_A}{dY^2} - \Phi_{Ai} C_A = 0 \quad Y_{i-1} \le Y \le Y_i \tag{A5}$$

with

$$\Phi_{Ai} = \sqrt{\frac{k_2 \delta_m^2 c_A^o c_B^o \bar{C}_{Bi}}{D_{mA}}} \tag{A6}$$

The mass balance equation can similarly be given for component with the following Φ_{Bi} value:

$$\Phi_{Bi} = \sqrt{\frac{k_2 \delta_m^2 c_A^o c_B^o \bar{C}_{Ai}}{D_{mB}}} \tag{A7}$$

The general solution of eq. A5 for the ith sub-section is as follows:

$$C_A = T_i e^{\Phi_{Ai} Y} + S_i e^{-\Phi_{Ai} Y} \quad Y_{i-1} \le Y \le Y_i \tag{A8}$$

This equation should be given for every sub-layer, thus, one can get N mass balance equation for component A with two parameters, namely T_i and S_i in them. The values of T_i and S_i with i=1,2,…,N can be determined by the following boundary conditions:

at Y=0 $$C=1 \tag{A9}$$

at $Y_{i-1} \le Y \le Y_i$ $$D_{mi-1} \frac{dC_A}{dY} = D_{mi} \frac{dC_A}{dY} \quad \text{with i=1,2,…,N} \tag{A10}$$

at $Y_{i-1} \le Y \le Y_i$ $$C_{Ai-1} = C_{Ai} \quad \text{with i=1,2,…,N} \tag{A11}$$

at Y=1 $$C_A = C_{A\delta} \tag{A12}$$

It is worth to mention that the method presented makes possible to calculate the mass transport when the diffusion coefficient of the reactant is variable. They can depend on the space coordinate and/or on the concentration. In this case a constant diffusion coefficient had to be given for every sub-layer. This is taken into account in eq. A10, where D_{mi} should not be equal to D_{mi-1}. Then the variable diffusion coefficient should be involved in the values of Φ_{Ai} and Φ_{Bi}.

According to eqs. A9 to A12, one can obtain 2N algebraic equations. This equation system can analytically be solved. Thus, the parameters can be given by means of the mass transport parameters, namely diffusion coefficient, reaction rate constant, etc. details on this method can be found in Nagy's papers (Nagy, 2008, 2010).

After solution of the N differential equation with 2N parameters to be determined the T_1 and S_1 parameters for the first sub-layer can be obtained as (ΔY is the thickness of sub-layers):

$$T_1 = -\frac{1}{2\xi_N^O \cosh(\Phi_{A1}\Delta Y)}\left(\xi_N^T - \frac{C_{A\delta}^o}{\prod_{i=2}^{N}\cosh(\Phi_{Ai}\Delta Y)}\right) \quad (A13)$$

and

$$S_1 = \frac{1}{2\xi_N^O \cosh(\Phi_{A1}\Delta Y)}\left(\xi_N^S - \frac{C_{A\delta}^o}{\prod_{i=2}^{N}\cosh(\Phi_{Ai}\Delta Y)}\right) \quad (A14)$$

Knowing the T_1 and S_1 the other parameters, namely T_i and S_i (i=2,3,…,N) can be easily be calculated by means of the internal boundary conditions given by eqs. A10 and A11, from starting from T_2 and S_2 up to T_N and S_N. Thus, one can get the following equations for prediction of the T_i and S_i from T_{i-1} and S_{i-1}:

$$T_i e^{\Phi_i Y_i} + S_i e^{-\Phi_i Y_i} = \Gamma_{i-1} \quad (A15)$$

$$D_{mi}\Phi_i\left(T_i e^{\Phi_i Y_i} - S_i e^{-\Phi_i Y_i}\right) = \Xi_{i-1} \quad (A16)$$

with

$$\Gamma_{i-1} = T_{i-1}e^{\Phi_{i-1}Y_i} + S_{i-1}e^{-\Phi_{i-1}Y_i} \quad (A17)$$

$$D_{mi-1}\Phi_{i-1}\left(T_{i-1}e^{\Phi_{i-1}Y_i} - S_{i-1}e^{-\Phi_{i-1}Y_i}\right) = \Xi_{i-1} \quad (A18)$$

Now knowing the T_i and S_i (with i=1,2,…,N) parameters, the concentration distribution can be calculated easily through the membrane, i.e. its value for every sub-layer.

Notations

c = concentration in the membrane, [$= w\rho / \left(Mc^o\right)$], mol/m³

C = dimensionless concentration in the membrane, ($= c / c^o$),-

c^o = bulk phase concentration, mol/m³
C = concentration at the membrane interface, mol/m³
d_p = particle size, m
d = $d_p\sqrt[3]{6/\pi}/\delta$
D = diffusion coefficient, m²/s
h = distance between cubic particles (Nagy, 2007), m
H = solubility coefficient of reactant between polymer matrix and catalyst particle, -
H_m = solubility constant of reactant between the continuous phase and the polymer membrane matrix,-
Ha_d = Hatta-number of the cubic particles in the heterogeneous model, $\left(=\sqrt{k_1 R_p^2 / D_p}\right)$
Ha_p = Hatta-number of catalyst particles (Ha_d =2.324 Ha_p), $\left(=\sqrt{k_1 R^2 / D_p}\right)$
j = mass transfer rate to catalyst particle, mol/(m²s)
J^o = physical mass transfer rate, mol/(m²s)
J = mass transfer rate in presence of chemical reaction, mol/(m²s)
J_m^o = physical mass transfer rate related to the homogeneous membrane interface, mol/(m²s)
J_δ = outlet mass transfer rate, mol/(m²s)
k = reaction rate constant, 1/s
L = length of capillary, m
M = molecular weight of reactant, g/mol
N = number of particle perpendicular to the membrane interface
P = pressure, Pa
r = radius of the spherical catalyst particles, m
R = dimensiomles radius, (r/R_o)
R_o = capillary radius, m
t = time, s
u = convective velocity in axial direction, m/s
u_o = inlet velocity, m/s
x = axial space coordinate, m
X = dimensionless space coordinate (=x/L)
y = space coordinate through the membrane, m
Y = dimensionless space coordinate $(=y/\delta_m)$
y_1, $\overline{Y}_1$ = distance of first particles from the interface ($Y_1=y_1/\delta_m$, $\overline{Y}_1 = \overline{y}_1/\delta_m$), m
ΔY = distance between particles in the membrane $(\Delta Y=\Delta y/\delta_m)$, m
X_i = distance of the ith particle from the interface, -
Y_i^* = Y_i+d
w = concentration of reactant in the membrane, kg/kg

Greek letters

β^o = physical mass transfer coefficient of fluid phase, m/s
β_m^o = mass transfer coefficient of the polymer membrane layer $(=D_m/\delta_m)$, m/s
β_m = mass transfer coefficient with chemical reaction, m/s
β_{tot}^o = physical mass transfer coefficient with overall resistance, m/s
β_p^o = external mass transfer coefficient around particles $(=2D/d_p + D/\delta_p)$, m/s

β_δ = mass transfer coefficient in the outlet rates, m/s
δ_m = thickness of the membrane layer, m
δ_p = diffusion boundary layer around particles, (=[h-d_p]/2), m
ρ = average density of the membrane, kg/m^3
ε = catalyst phase holdup
ω = specific interface of catalyst particles, m^2/m^3
ω = specific interface of catalyst particles in the membrane, ($=6\varepsilon / d_p$),m^2/m^3

Subscripts

A = reactant A
ave = average
B = reactant B
i = integer parameter
m = polymer membrane
L = fluid phase
p = catalyst particle
δ = permeate side of membrane
1 first-order
0 zero-order

9. References

Aryal R., Lebegue, J., Vigneswaran S., Kandasamy, J., Grasmick, A. (2009) Identification and characterization of biofilm formed on membrane bio-reactor, *Sep. Purif. Technol.*, 67 86-94

Baker, R.W. (2004) *Membrane technology and application*, Wiley & Sons, England

Belfort,G. (1989), Membranes and Bioreactors: A Technical Challenge in Biotechnology, *Biotechnology and Bioengineering*, 33, 1047-1066.

Breure, B.; Peters, E.A.J.F. & Kerkhof, P.J.A.M. (2008). Separation of azeotropic mixtures of alcohols and water with FricDiff, *Sep. Purif. Technology*, 62, 350-363.

Brotherton, J.D. & Chau, P.C. (1990). Modeling analysis of an intercalated-spiral alternate-dead-ended hollow fiber bioreactor for mammalian cell cultures, *Biotechnology and Bioengineering*, 35, 375-394.

Brotherton, J.D. & Chau, P.C. (1996) Biotechnology Progress, 112, 575-590

Cabral, J.M.S. & Tramper, J. (1994) Bioreactor Design, In *Applied Biocatalysis.* (Ed. By J.M.S. Cabral, D. Best, L. Boross, J. Tramper, Harwood Academic Publishers, Switzerland, pp. 330 370.

Calabro, V., Curcio, S., & Iorio, G. (2002), A theoretical analysis of transport phenomena in a hollow fiber membrane bioreactor with immobilized biocatalyst, *J. Membrane Sci.*, 206, 217-241.

Carvalho CML, Aires-Barros M.R. & Cabral JMS. (2000) A continuous membrane bioreactor for ester synthesis in organic media: II Modeling of MBR continuous operation, *Biotechnology and Bioengineering*, 72, 136-143

Casey, E., Glennon, B., Hamer, G. (2000) Biofilm development in a membrane-areated biofilm reactor: Effect of flow velocity on performance, *Biotechnol. Bioeng.*, 67 (4) 476-486.

Cebeci, T., Shao, J.P., Kafyeke, F. & Laurendeau, L. (2005) *Computational fluid dynamics for engineers*, Horizons Publishing Inc. and Springer, Heidelberg.

Charcosset,C. (2006), Membrane processes in biotechnology: An overview, *Biotechnology Advances*, 24, 482-492.

Champagnie, A.M., Tsotsis. T.T., Minet, R.G. & Webster, I.A. (1990) *Chem. Engng. Sci.*, 45, 2423-2429.

Damak, K.; Ayadi, A.; Zeghmati, B. & Schmitz, P. (2004). A new Navier-Stokes and Darcy's law combined model for fluid flow in crossflow filtration tubular membranes, *Desalination*, 161, 67-77.

Ferreira, B.S., Fernandes, P. & Cabral J.M.S. (2001) Design and modeling of immobilized biocatalytic reactors, in *Multiphase bioreactor design* (Ed. by J.M.S. Cabral, M. Mota, J. Tramper), Taylor and Francis, London, UK., pp. 85-180.

Frazeres, D.M.F. & Cabral, J.M.S. (2001) Enzymatic membrane reactors, In: J.M.S. Cabral, M. Mota, J. Tramper (Eds.), *Multiphase bioreactor design*, Taylor Francis., pp. 135-184.

Giorno L. & Drioli E. (2000) Biocatalytic membrane reactors: applications and perspectives, *Trends in biotechnology* 18, 339-349.

Godongwana, B.; Sheldon, M.S. & Solomons, D.M. (2007). Momentum transfer inside a vertically orientated capillary membrane bioreactor, *J. Membr. Sci.*, 303, 86-99.

Gross, R., Hauer, B., Otto, K., Schmid, A. (2007) Microbial biofilms: New catalysis for maximizing productivity of long-term biotransformation, *Biotechnol. Bioeng.*, 98 (6) 1123-1134..

Habulin,M. & Knez,Z. (1991), Enzymatic synthesis of n-butyl oleate in a hollow fiber membrane reactor, *J. Membrane Sci.*, 61, 315-324.

Hossain MM. & Do DD. (1989) General Theory of Determining Intraparticle Active Immobilized Enzyme Distribution and Rate Parameters, *Biotechnology and Bioengineering*, 33, 963-975.

Ilinitch O.M., Cuperus, F.P., Nosova, L.V. & Gribov, E.N., (2000) Catalytic membrane in reduction of aqueous nitrates: operational principles and catalytic performance, *Catalysis Today*, 56, 137-145.

Yawalkar, A.A., Pangarkar, V.G. & Baron, G.V. (2001) Alkene epoxidation with peroxide in a catalytic membrane reactor: a theoretical study, *J. Membrane Science*, 182, 129-137.

Julbe, A.; Farusseng, D. & Guizard, C. (2001) Porous ceramic membranes for catalytic reactors-overview and new ideas, *J. Membrane Sci.*, 181, 3-20.

Kelsey, L.J., Pillarella, M.R. & Zydney,A.L. (1990), Theoretical analysis of convective flow profiles in a hollow-fiber membrane bioreactor, *Chem.Eng.Sci.*, 45 (11), 3211-3220.

Lu SG, Imai T, Ukita M, Sekine M, Higuchi T. & Fukagawa M. (2001) A model for membrane bioreactor process based on the concept of formation and degradation of soluble microbial products, *Wat. Res.*,35, 2038-2048.

Long, W.S.; Bhatia S. & Kamaruddin, A. (2003). Modeling and simulation of enzymatic membrane reactor for kinetic resolution of ibuprofen ester, *J. Membr. Sci.*, 219, 69-88.

Marcano, J.G.S. & Tsotsis, T.T. (2002). *Catalytic membranes and membrane reactiones*, Wiley-VCH, Veinheim.

Maira, A.J., Lau, W.N.; Lee, C.Y.; Chan, C.K. & Yeung, K.L. (2003) Performance of a membrane-catalyst for phocatalyc oxidation of volatile organic compounds, *Chem. Eng. Sci.*, 58, 959-962.

Marror,B., Barnos-Martinez,A., Moulin,P. & Roche,N. (2004), Industrial Wastewater Treatment in a Membrane Bioreactor: A Review, *Environmental Prograss*, 23, 59-68.

Mehra, A. (1999) Heterogeneous modeling of gas absorption in emulsion, *Ind. Eng. Chem. Res.*, 38, 2460-2468.

Melo, L.F. & Oliveira, R. (2001). Biofilm reactors, In: J.M.S. Cabral, M. Mota, J. Tramper (Eds.), *Multiphase bioreactor design,* Taylor Francis., pp. 271-308.

Mondor, M. & Moresoli C. (1999). Theoretical analysis of the influence of the axial variation of the transmembrane pressure in cross-flow filtration of rigid spheres, *J. Membr. Sci.*, 152, 71-87.

Mothlagh, A.R., Voller, V.R., Semmens, M.J. (2006) Advective flow through membrane-areated biofilms, Modeling results, *J. Membr. Sci.*, 273 143-151.

Mulder, M.H.V. Polarization phenomena and membrane fouling. In Noble and Stern (Ed.) *Membrane separation technology, principles and applications*, Elsevier, Oxford, 1995

Nagy E. (1995) Three-phase mass transfer: One-dimensional heterogeneous model, *Chem. Engng. Sci.*, 50, 827-836.

Nagy, E. (2002) Three-phase oxygen absorption and its effect on fermentation, *Adv. Biochem. Eng./Biotechnol.*, 75, 51-81.

Nagy E. (2009a) Mathematical Modeling of Biochemical Membrane Reactors, in *Membrane Operations, Innovative Separations and Transformations,* Ed. by E. Drioli and L. Giorno, WILEY-VCH Velag, Weinheim, pp. 309-334

Nagy, E. (2009b) Basic equations of mass transfer through biocatalytic membrane layer, *Asia-Pacific J. of Chem. Eng.*, 4, 270-278.

Nagy E., Kulcsár E. (2009) Mass transport through biocatalytic membrane reactors, *Desalination* 245, 422-436.

Nagy E., Borbély G. (2009) Mass transport through anisotropic membrane layer, *Desalination,* 240, 54-63.

Nagy E. (2007) Mass transfer through a dense, polymeric, catalytic membrane layer with dispersed catalyst, *Ind. Eng. Chem. Res.*, 46, 2295-2306.

Nagy, E. (2006) Binary, coupled mass transfer with variable diffusivity through cylindrical membrane, *J. Membrane Science,* 274, 159-168

Nagy E. & Moser, A. (1995) Three-phase mass transfer: Improved pseudo-homogeneous model, *AIChE J., 41*, 23-34.

Nagy, E. (2008) Mass transport with varying diffusion- and solubility coefficient through a catalytic membrane layer, *Chem. Eng. Res. Design*, 86, 723-730

Nagy, E., Blickle, T. & Ujhidy, A. (1989) Spherical effect on mass transfer between fine solid particles and liquid accompanied by chemical reaction, *Chem. Engng. Sci., 44*, 198-201.

Nagy E. (2010) Mass transport through a convection flow catalytic membrane layer with dispersed nanometer-sized catalyst, *Ind. Eng. Chem Res.*, 49, 1057-1062.

Nakajima, M. & Cardoso, J.P. (1989) Forced-flow bioreactor for sucrose inversion using ceramic membrane activated by silanization, *Biotechnol. Bioeng.*, 33, 856-861.

Picioreanu, C., van Loosdrecht, M.C.M., Heijnen, J.J. (2001) Two-dimensional model of biofilm detachment caused by internal stress from liquid flow, *Biotechnol. Bioeng.*, 72 (2) 205-218.

Piret, J.M. & Cooney, C.L. (1991). Model of oxygen transport limitations in hollow fiber bioreactors, *Biotechnology and Bioengineering*, 37, 80-92.

Rios,G.M., Belleville,M-P. & Paolucci-Jeanjean,D. (2007), Membrane engineering in biotechnology: quo vamus? *Trends in Biotechnology*, 25, 242-246.

Rittman, B.E., Manem J.A. (1992) Development and experimental evaluation of a steady-state, multispecies biofil model, *Biotechnol. Bioeng.*, 39 914-922.

Rosenberger,S., Krüger,U., Witzig,R., Manz,W., Szewzyk,U. & Kraume,M. (2002), Performance of a bioreactor with submerged membranes for aerobic treatment of municipal waste water, *Water Research*, 36, 413-420.

Salzman,G., Tadmor,R., Guzy,S., Sideman,S., & Lotan,N. (1999), Hollow fiber enzymic reactors for a two substrate process: analytical modeling and numerical simulations, Chem.Eng.Proc., 38, 289-299.

Saracco, G.& Specchia, V.; (1995) Catalytic ceramic filters for flue gas cleaning. 2: catalytic performance and modeling thereof. *Ind. Eng. Chem. Res.*, 34, 1480-1487.

Sardonini, C.A. & Dibiasio, D. (1992), An Investigation of the Diffusion-Limited Growth of Animal Cells Around Single Hollow Fibers, *Biotechnology and Bioengineering*, 40, 1233-1242.

Schonberg JA. & Belfort G. (1987) Enhanced nutrient transport in hollow fiber perfusion bioreactors: a theoretical analysis, *Biotechnology Progress*, 3, 81-89.

Schmidt, A.; Haidar, R. & Schomacker, R. (2005) Selectivity of partial hydrogenation reactions performed in a pore-through-flow catalyst membrane reactor, *Catal. Today*, 104, 305-312.

Seidel-Morgenstern, A. (2010), *Membrane reactors*, Wiley-WCH, Weinheim

Sheldon,M.S. & Small, H.J. (2005), Immobilisation and biofilm development of *Phanerochaete chrysosporium* on polysulphone and ceramic membranes, *J.Membrane Sci.*, 263, 30-37.

Strathmann H, Giorno L, & Drioli E. (2006) *An introduction to membrane science and technology*, Institute on Membrane Technology, Italy

Yamada, M.; Fugii, K.; Haru, H. & Itabashi, K.; (1988) Preparation and catalytic properties of special alumina membrane formed by anodic oxidation of aluminum. *Tech. Rep. The Light Metal Educational Foundation*, Inc. Research Group for Functionalizing of Aluminum and its Surface Films

Yang, W., Cicek, N.& Ilg, J. (2006) State-of-the-art of membrane bioreactors: Worldwide research and commercial applications in North America, *J. Membrane Sci.*, 270, 201-211

Vancelecom, I.F.J. & Jacobs, P.A. (2000). Dense organic catalytic membrane for fine chemical synthesis, *Catalysis Today*, 56, 147-157.

Vincent, M.J.& Gonzales, R.D. (2002) Selective hydrogenation of acetylene through a short contact time reactor, *AIChE J.*, 48, 1257-1263.

Vital, J., Ramos, A.M., Silva, I.F., Valenete, H. & Castanheiro, J.E. (2001) Hydration of α-pinene over zeolites and activated carbons dispersed in polymeric membranes, *Cataysis Today*, 67, 217-223.

Wang, R., Terada, A., Lackner, S., Smets, B.F., Henze, M., Xia, S., Zhao, J. (2009) Nitritation performance and biofilm development of co- and counter-diffusion biofilm reactors: Modeling and experimental comparison, *Water Research*, 43, 2699-2709.

Waterland, L.R., Robertson, C.R. & Michaelis, A.S. (1975) Enzymatic catalysis using asymmetric hollow fiber membranes, *Chem. Eng. Commun.*, 2: 37-47.

Waterland, L.R., Robertson, C.R. & Michaelis, A.S. (1974). A theoretical model for enzymatic catalysis using asymmetric hollow fiber membranes, *AIChE Journal*, 20, 50-59

Westermann T., Melin T. (2009) Flow-through catalytic membrane reactors- Principles and applications, *Chem. Eng. and Processing*, 48, 17-28.

30

Mass Transfer in Bioreactors

Ma. del Carmen Chávez[1], Linda V. González[2], Mayra Ruiz[3], Ma. de la Luz X. Negrete[4], Oscar Martín Hernández[5] and Eleazar M. Escamilla[6]

[1]Facultad de Ingeniería Química, Universidad Michoacana de San Nicolás de Hidalgo, Francisco J. Mújica s/n, Col. Felicitas del Río, 58060, Morelia, Michoacán.

[2]Centro de Investigación y Desarrollo Tecnológico en Electroquímica, Parque Tecnológico Querétaro Sanfandila, 76703 Sanfandila, Pedro Escobedo, Qro.,

[3]Facultad de Ingeniería Química, Benemérita Universidad Autónoma de Puebla. 4 sur 104 centro histórico C.P. 72000, Puebla.,

[4]Departamento de Ingeniería Ambiental, Instituto Tecnológico de Celaya, Ave. Tecnológico y Antonio García Cubas S/N, Celaya, Gto., C.P. 38010,

[5]Universidad Autónoma de Sinaloa. Facultad de Ciencias Químico Biológicas. Ciudad Universitaria, C.p. 80090, Culiacán, Sinaloa.

[6]Instituto Tecnológico de Celaya, Departamento de Ingeniería Química, Ave. Tecnológico y Antonio García Cubas S/N, Celaya, Gto., C.P. 38010,Sinaloa. México

1. Introduction

The study of transport in biological systems is complicated for two reasons: 1. because each system is different, we cannot generalize it and 2. Because always take place in more than one phase. If we talk about microorganism, there is a range of them with physicochemical and biological characteristics very different, and certain microorganisms can be filamentous and can grow branched or dispersed, in some the viscosity and density increases with time. In some times their maximum growth rate is achieved in two hours while others in 15 days. Some are affected by the light, others agitation rate, others require air for developing others not. If we talk about production of plants by tissue culture systems have become more complex, that the transport properties are affected by agitation rate, type of agitation, the growth of tissues. To design the bioreactors of these biological systems requires knowledge of the nature of what is to be produced, the dynamics of transport, rheology, to decide what type of reactor we can used. Biological fluids such reactors behave as highly non-Newtonian systems and as such require special treatment. This paper will discuss three types of reactors: air-lift, packed column and fluidized bed and stirred tank, where case studies are applied to biological systems. 1. Production of Gibberellic acid and Bikaverin 2. Biodegradation of azodyes in textile industry and 3. Gibberellins Production. It is intended that in these three cases brought to appreciate as engineering parameters are evaluated where they involve the transport mass balances and the type of bioreactor and feature you in l fluid. On the other hand show a combination of experimental results and simulations with mathematical models developed to strengthen the knowledge of chemical engineering applied to biological systems.

2. Case I. Hydrodynamics, mass transfer and rheological studies of gibberellic acid production in an airlift bioreactor

2.1 Introduction

Gibberellic acid is an endogenous hormone in higher plants, belonging to the group of gibberellins, and also a product of the secondary metabolism in certain fungi. Approximately 126 gibberellins have been characterized (Tudzynski 1999; Shukla et al. 2003) but only a few are commercially available. Gibberellic acid is the most important and its effects on higher plants are: marked stem elongation, reversal of dwarfism, promotion of fruit setting, breaking of dormancy, acceleration of seed fermentation, among others (Bru¨ ckner and Blechschmidt 1991; Tudzynski 1999). Currently, gibberellic acid is microbiologically produced in a submerged culture (SmF) fashion but another fermentation techniques such as solid sate fermentation or with immobilized mycelium are also reported (Heinrich and Rehm 1981; Jones and Pharis 1987; Kumar and Lonsane 1987, 1988; Nava Saucedo et al. 1989; Escamilla et al. 2000; Gelmi et al. 2000, 2002). Nevertheless stirred tank bioreactors with or without a fed-batch scheme have been the most employed in gibberellic acid production. Other geometries and type of bioreactors have also been reported. Only Chavez (2005) has described gibberellic acid production employing an airlift bioreactor. Airlift bioreactors are pneumatically agitated and circulation takes place in a defined cyclic pattern through a loop, which divides the reactor into two zones: a flow-upward and a flow-downward zone. The gas-sparged zone or the riser has higher gas holdup than the relatively gas-free zone, the downcomer, where the flow is downward (Gouveia et al. 2003). Practical application of airlift bioreactors depends on the ability to achieve the required rates of momentum; heat and mass transfer at acceptable capital and operating costs. The technical and economic feasibility of using airlift devices has been conclusively established for a number of processes and these bioreactors find increasing use in aerobic fermentations, in treatment of wastewater and other similar operations. The simplicity of their design and construction, better defined flow patterns, low power input, low shear fields, good mixing and extended aseptic operation, made possible by the absence of stirrer shafts, seals and bearings, are important advantages of airlift bioreactors in fermentation applications (Chisti 1989).

Even though gibberellic acid has been produced on an industrial scale since the last century, hydrodynamics, mass transfer and rheological studies are sparse. Flow regime, bubble size distribution, and coalescence characteristics, gas holdup, interfacial mass transfer coefficients, gas-liquid interfacial area, dispersion coefficients and heat transfer coefficients are important design parameters for airlift bioreactors. A thorough knowledge of these interdependent parameters is also necessary for a proper scale-up of these bioreactors (Shah et al. 1982). Besides hydrodynamics and mass transfer studies, rheological studies are important since in many chemical process industries, the design and performance of operations involving fluid handling like mixing, heat transfer, chemical reactions and fermentations are dependent on the rheological properties of the processed media (Brito-De la Fuente et al. 1998). Mycelial fermentation broths present challenging problems in the design and operation of bioreactors since the system tends to have highly non-Newtonian flow behaviour and this has a very significant effect on mixing and mass transfer within the bioreactor.

The main objective of this work was to study hydrodynamic, mass transfer and rheological aspects of gibberellic acid production by *Gibberella fujikuroi* in an airlift reactor.

2.2 Materials and methods

Microorganism and inoculum preparation *Gibberella fujikuroi* (Sawada) strain CDBB H-984 maintained on potato dextrose agar slants at 4_C and sub-cultured every 2 months was used in the present work (Culture collection of the Department of Biotechnology and Bioengineering, CINVESTAV-IPN, Mexico). Fully developed mycelia materials from a slant were removed by adding an isotonic solution (0.9% NaCl). The removed mycelium was used to inoculate 300 ml of fresh culture medium contained in an Erlenmeyer flask. The flask was placed in a radial shaker (200 rev min−1) for 38 h at 29 ± 1_C. Subsequent to this time; the contents of the flask were used to inoculate the culture medium contained in the airlift bioreactor. The culture medium employed for the inoculum preparation is reported by Barrow et al. (1960).

Batch culture in the airlift bioreactor

An airlift bioreactor (Applikon, Netherlands, working volume, 3.5 l) was employed in the present work. It consists of two concentric tubes of 4.0 and 5.0 cm of internal diameter with a settler. The air enters the bioreactor through the inner tube. A jacket filled with water allowing temperature control surrounds the bioreactor. It is also equipped with sensors of pH and dissolved oxygen to control these variables. Moreover it allows feed or retiring material from the bioreactor employing peristaltic pumps. Typical culture medium contained glucose (50 g l^{-1}), NH_4Cl (0.75 g l^{-1}) or NH_4NO_3 (1.08 g l^{-1}), KH_2PO_4 (5 g l^{-1}), $MgSO_4$. 7 H_2O (1 g l^{-1}) and trace elements (2 ml l^{-1}). A stock solution of the trace elements used contained (g l^{-1}) 1.0 Fe SO_4. 7 H_2O, 0.15 $CuSO_4$. 5 H_2O, 1.0 $ZnSO_4$. 7 H_2O, 0.1 $MnSO_4$. 7 H_2O, 0.1 $NaMoO_4$, 3.0 EDTA (Na_2 salt) 1 l of distilled water, and hydrochloric acid sufficient to clarify the solution (Barrow et al. 1960). During the fermentation period, the pH was controlled to 3.0, temperature to 29°C and aeration rate to 1.6 vvm. These conditions promoted gibberellic acid production with the studied strain but they are not optimized values. About 30 ml subsamples were withdrawn from the bioreactor at different times and were used to perform rheological studies. Biomass concentration was quantified by the dry weight method.

2.3 Hydrodynamics and mass transfer studies

Gas holdup was determined in the actual culture medium using an inverted U-tube manometer as described by Chisti (1989). Liquid velocities in the riser were determined measuring the time required for the liquid to travel through the riser by means of a pulse of concentrated sulphuric acid using phenolphthalein as an indicator; the same was done for the downcomer. The mixing time was calculated as the time required obtaining a pH variation within 5% of the final pH value. For doing this, pH variation was followed after injection of a pulse of a concentrated solution of ammonium hydroxide. The volumetric mass transfer coefficient was determined employing the gassing-out method as described elsewhere (Quintero 1981).

2.4 Rheological studies

Rheological studies of fermentation broth were performed in a rotational rheometer (Haake, Model CV20N) equipped with a helical impeller to perform torque measurements. This type of geometry is appropriate when dealing with complex fluids and the measurement methodology is reported by Brito-de la Fuente et al. (1998). Rheological results, like hydrodynamics and mass transfer, are given as the average of two replicates for each sample. All the experiments were carried out in triplicate and the results that are presented are an average.

2.5 Results and discussion

Gas holdup

The importance of gas holdup is multifold. The gas holdup determines the residence time of the gas in the liquid and, in combination with the bubble size, influences the gas–liquid interfacial area available for mass transfer. The gas holdup impacts upon the bioreactor design because the total design volume of the bioreactor for any range of operating conditions depends on the maximum gas holdup that must be accommodated (Chisti 1989). Figure 1 shows the gas holdup (ε) variation with superficial gas velocity in the riser (v_{gr}). Experimental data were fitted to a correlation of the type of Eq. 1.

$$F = A\, v_{gr}^{B} \tag{1}$$

Where F could be the gas holdup (ε), the liquid velocity in the riser (v_{lr}), liquid velocity in the downcomer (v_{ld}) or the volumetric mass transfer coefficient (k_La). This type of correlation has been applied by many investigators (Shah et al. 1982; Godbole et al. 1984; Chisti 1989; Gravilescu and Tudose 1998; Abashar et al. 1998) and was derived empirically. Chisti (1989) presented an analysis for Newtonian and non-Newtonian fluids where shows the theoretical basis of Eq. 1 (for the gas holdup case). He found that parameters A and B were dependent on the flow regime and on the flow behaviour index of the fluid. Moreover, parameter A is dependent on the consistency index of the fluid, on the fluid densities and on the gravitational field. Equation 2 was obtained from fitting experimental data.

$$\varepsilon = 0.7980\, v_{gr}^{1.0303} \tag{2}$$

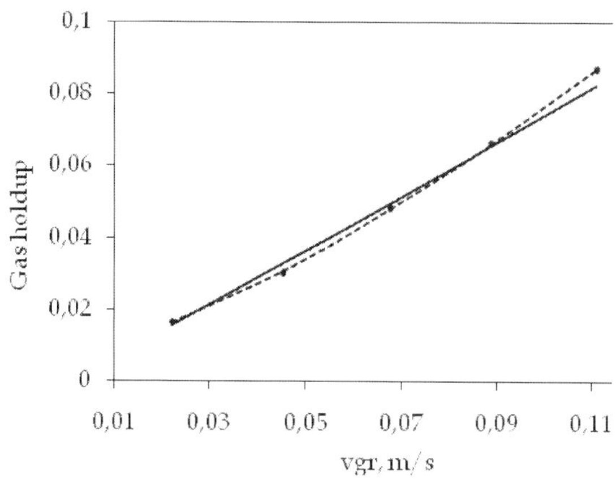

Fig. 1. Gas holdup variation with superficial gas velocity in the riser.
• Experimental data --- Equation 2 --- Equation 12

An increase in superficial gas velocity in the riser implies an increase in the quantity of gas present in the riser, that is, an increase of gas fraction in the riser (Chisti 1989; Gravilescu and Tudose 1998). Chisti (1989) reports a correlation that calculates the value of B in Eq. 1 (for the gas holdup case). The value obtained employing this correlation is 1.2537.

Gravilescu and Tudose (1998) present a similar correlation, which predicts a value of 0.8434 for B. The B value obtained in the present work is between the B values obtained from these correlations that employ the flow behaviour index obtained from rheological studies. Shah et al. (1982) reported that B values in Eq. 1 oscillate between 0.7 and 1.2.

Liquid velocity

The liquid circulation in airlift bioreactors originates from the difference in bulk densities of the fluids in the riser and the downcomer. The liquid velocity, while itself controlled by the gas holdups in the riser and the downcomer, in turn affects these gas holdups by either enhancing or reducing the velocity of bubble rise. In addition, liquid velocity affects turbulence, the fluidreactor wall heat transfer coefficients, the gas-liquid mass transfer and the shear forces to which the microorganism are exposed. Figure 2 shows liquid velocities variation in the riser and the downcomer as a function of superficial gas velocity in the riser. Liquid velocities in the riser (v_{lr}) and in the downcomer (v_{ld}) were fitted to correlations of the type of Eq. 1 and Eqs. 3 and 4 were obtained.

$$v_{lr} = 1.3335\ v_{gr}^{0.3503} \tag{3}$$

$$v_{ld} = 0.8716\ v_{gr}^{0.2970} \tag{4}$$

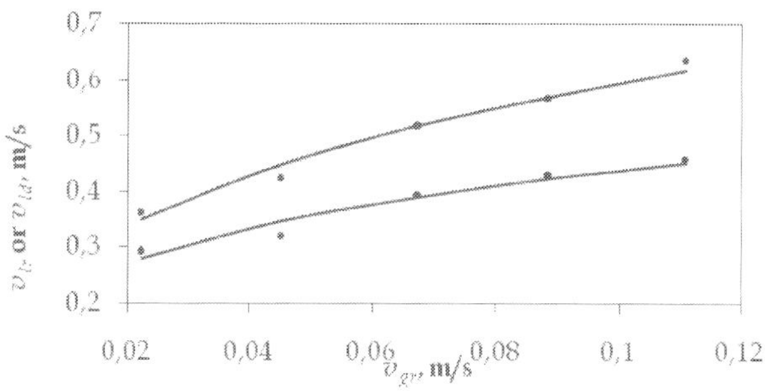

Fig. 2. Liquid velocities as a function of superficial gas velocity in the riser.
• Experimental data --- Equation 3 or 4

The B value in Eq. 1 must be close to 0.3333 as was reported by Freitas and Teixeira (1998) for the liquid velocity in the riser, Kawase (1989) theoretically derived this value. The B value obtained in the present work is closer to 0.3333. Freitas and Teixeira (1998) also showed that the B values for the liquid velocity in the downcomer were lower than the B value for the liquid velocity in the riser, which agrees with the results obtained in this work. Liquid velocities in the riser and in the downcomer increase with an increase in gas velocity in the riser due to an increase in the density difference of the fluids in the riser and the downcomer.

Mixing time

Mixing in airlift bioreactors may be considered to have two contributing components: back mixing due to recirculation and axial dispersion in the riser and
downcomer due to turbulence and differential velocities of the gas and liquid phases (Choi et al. 1996).

Mixing time is used as a basis for comparing various reactors as well as a parameter for scaling up (Gravilescu and Tudose 1999). Figure 3 shows the mixing time variation with the superficial gas velocity in the riser. Once again, the mixing time variation was fitted to a correlation of the type of Eq. 1 and Eq. 5 was obtained.

$$t_m = 5.0684\, v_{gr}^{-0.3628} \tag{5}$$

Choi et al. (1996) reported a B value in Eq. 5 of -0.36 while Freitas and Teixeira (1998) reported a B value equal to -0.417. The B value obtained in this work is similar to the value reported by Choi et al. (1996). The mixing time decreases with an increase in superficial gas velocity in the riser since the fluid moves more often to the degassing zone where most of the mixing phenomenon takes place, due to the ring vortices formed above the draught tube (Freitas and Teixeira 1998).

Volumetric mass transfer coefficient

One of the major reasons that oxygen transfer can play an important role in many biological processes is certainly the limited oxygen capacity of the fermentation broth due to the low solubility of oxygen. The volumetric mass transfer coefficient (k_La) is the parameter that characterizes gas-liquid oxygen transfer in bioreactors. One of the commonest employed scale-up criteria is constant k_La. The influences of various design (i.e., bioreactor type and geometry), system (i.e., fluid properties) and operation (i.e., liquid and gas velocities) variables on k_La must be evaluated so that design and operation are carried out to optimize k_La (Chisti, 1989).

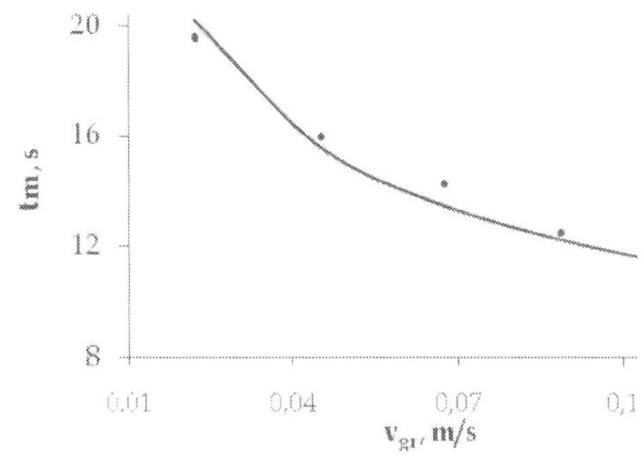

Fig. 3. Mixing time as a function of superficial gas velocity in the riser.

The value of the volumetric mass transfer coefficient determined for a microbial system can differ substantially from those obtained for the oxygen absorption in water or in simple aqueous solutions, i.e., in static systems with an invariable composition of the liquid media along the time. Hence k_La should be determined in bioreactors which involve the actual media and microbial population (Tobajas and García-Calvo, 2000). Figure 4 shows the volumetric mass transfer coefficient variation with the superficial gas velocity in the riser. Experimental data shown in Figure 4 were fitted to a correlation of the type of Equation 1 and Equation 6 was obtained.

$$k_L a = 0.4337\ v_{gr}^{1.2398} \tag{6}$$

Barboza *et al.*, (2000) report a B value in Equation 6 equal to 1.33 and Schügerl *et al.*, (1977) report a value of 1.58. The value of 1.2398, obtained in this work, is close to these last values.

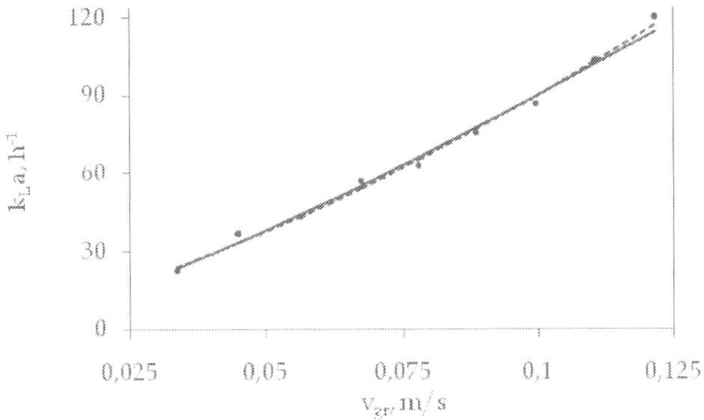

Fig. 4. Effect of the superficial gas velocity in the riser on $k_L a$.

Volumetric mass transfer coefficient ($k_L a$) increases with an increase in superficial gas velocity in the riser due to an increase in gas holdup which increases the available area for oxygen transfer. Moreover an increase in the superficial gas velocity in the riser increases the liquid velocity which decreases the thickness of the gas-liquid boundary layer decreasing the mass transfer resistance. Figure 5 shows the evolution of $k_L a$ through fermentation course employing two different nitrogen sources. The $k_L a$ decreases in the first hours of fermentation and reaches a minimum value at about 24 hours. After this time the $k_L a$ starts to increase and after 48 hours of fermentation it reaches a more or less constant value which remains till the end of fermentation process. This behaviour is similar irrespective of the nitrogen source and will be discussed with the rheological results evidence.

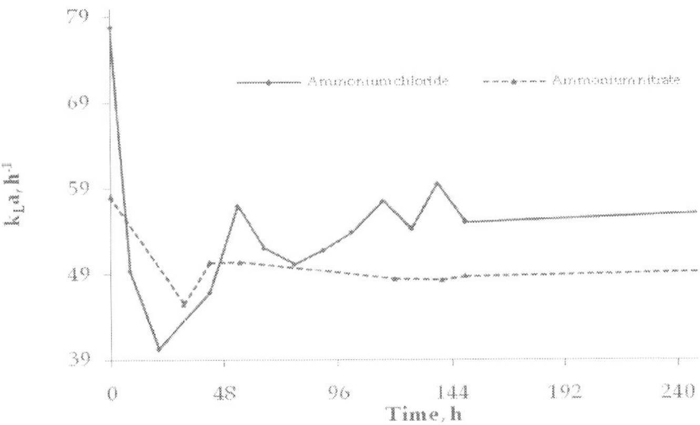

Fig. 5. $k_L a$ through fermentation time in the airlift bioreactor.

Figure 6 shows the relation between gas holdup and k_La. McManamey and Wase (1986) point out that the volumetric mass transfer coefficient is dependent on gas holdup in pneumatically agitated systems. The later was experimentally determined in bubble columns by Akita and Yoshida (1973) and Prokop *et al.*, (1983). Shah *et al.*, (1982) mention that this was expectable since both the volumetric mass transfer coefficient and the gas holdup present similar correlations with the superficial gas velocity. McManamey and Wase (1996) proposed a correlation similar to Equation 1 to relate volumetric mass transfer coefficient with gas holdup. Equation 7 presents the obtained result.

$$k_L a = 0.2883\ \varepsilon^{0.9562} \tag{7}$$

Akita and Yoshida (1973) and Prokop *et al.* (1983) found that the exponent in Equation 7 oscillates between 0.8 and 1.1.

$$\ln k_L a = \ln\left(6\frac{k_L}{d_B}\right) + \ln\frac{\varepsilon}{(1-\varepsilon)} \tag{8}$$

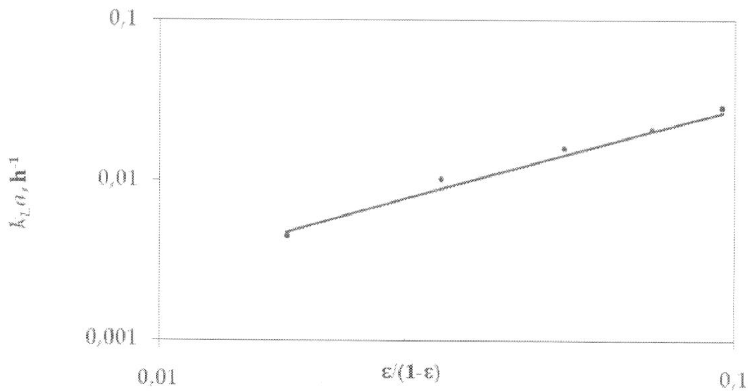

Fig. 6. k_La vs. gas holdup in the airlift bioreactor, unit slope.

It is well known (Chisti, 1989) that logarithmic scale plots of k_La vs. $\varepsilon/(1-\varepsilon)$ for any particular data set should have a unit slope according to Equation 8. Where k_L is the mass transfer coefficient and d_B is the bubble diameter. Even though the later is a generally known fact, few investigators determined these slopes for their data to ascertain the validity of their experimental results. Figure 6 shows this analysis for the experimental data of the present work obtaining a slope of 1.034. Chisti (1989) shows the same analysis for two different data set and obtained slopes of 1.020 and 1.056.

A rearrangement of Equation 8 leads to Equation 9 which results are shown in Figure 7. As is showed in the Figure 7 the gas superficial velocity practically did not affect the k_L/d_B values, therefore it can be taken as a value average and constant to slant the superficial velocity changes.

$$\frac{k_L}{d_B} = \frac{k_L a(1-\varepsilon)}{6\varepsilon} \tag{9}$$

The average value of k_L/d_B obtained in the present work is 0.050 s^{-1}. Chisti (1989) performed a similar analysis for 97 data points obtained from several different reactors and found an average value of 0.053 s^{-1}. The foregoing observations have important scale-up implications. In large industrial fermenters the k_La determination is not only difficult, but there is uncertainty as to whether the measured results reflect the real k_La or not. The gas holdup measurements on these reactors are relatively easy to carry out, however. Thus, Equation 9 can help to estimate k_La in these reactors once gas holdup measurements have been made (Chisti, 1989).

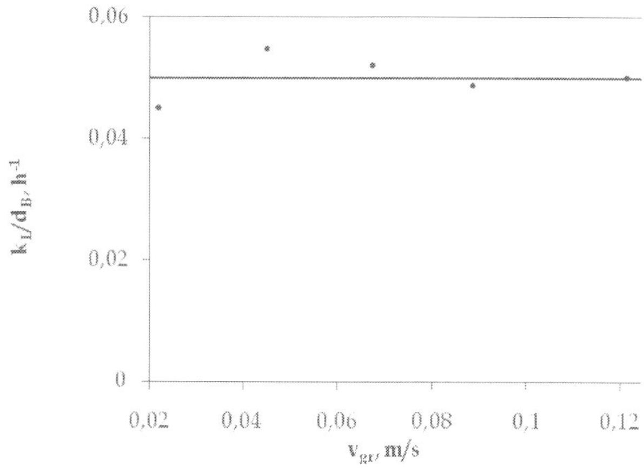

Fig. 7. The k_L/d_B ratio as a function of superficial gas velocity.

2.6 Rheology

Rheological parameters such as the flow index (n) and the consistency index (K) depend on such factors as the concentration of solids in the broth, the morphology (length, diameter, degree of branching, shape) of the particles, the growth conditions (flexibility of cell wall and particle), the microbial species and the osmotic pressure of the suspending liquid, among others possible factors. For the case of mycelial cultures, as the biomass concentration increases the broth becomes more viscous and non-Newtonian; leading to substantial decreases in oxygen transfer rates. This effect is often important since for many aerobic processes involving viscous non-Newtonian broths oxygen supply is the limiting factor determining bioreactor productivity (Moo-Young *et al.*, 1987). Apparent viscosity is a widely used design parameter which correlates mass transfer and hydrodynamic parameters for viscous non-Newtonian systems (Al-Masry and Dukkan, 1998).

It is worth to mention that the present work uses impeller viscometry for performing rheological studies avoiding the use of other geometries, i.e., concentric tubes or cone and plate, overcoming associated problems with these geometries such sedimentation, solids compacting and jamming between measuring surfaces or pellet destruction (Metz *et al.*, 1979). Impeller viscometry was used to obtain torque data at different velocities of the impeller, these data were transformed to shear stress (τ) and shear rate (γ) data and typical results are shown in Figure 8. As can be seen in Figure 8, the experimental data follow a straight line and can be represented by the Ostwald-de Waele model (Equation 10).

$$\tau = K\gamma^n \tag{10}$$

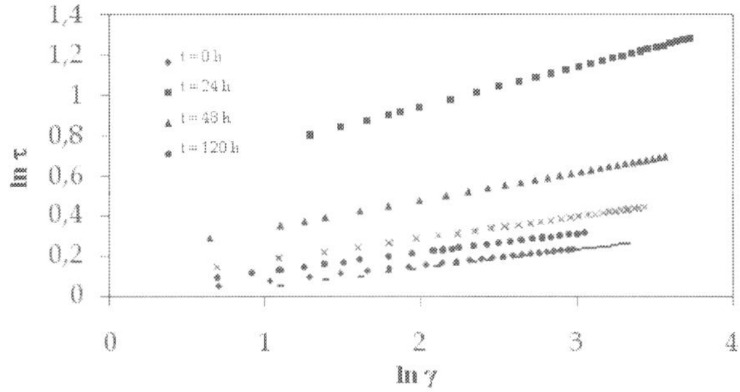

Fig. 8. Typical rheogram employing impeller viscometry

Rheograms obtained from fermentations employing different nitrogen source show a pseudo plastic behaviour for the culture medium during the fermentation period since the exponent, n, in Equation 10 is always lower than unity. Figure 9 shows the results of consistency and flow indexes for the different fermentations, employing ammonium chloride or ammonium nitrate as nitrogen source, where similar results were obtained.

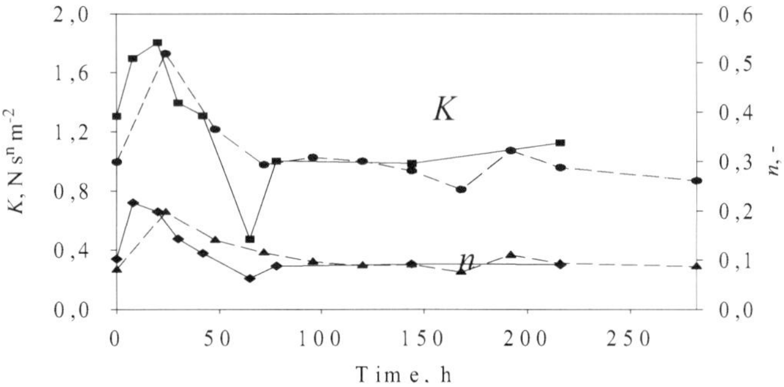

Fig. 9. K and n through fermentation time in the airlift bioreactor • K for ammonium nitrate ▲ n for ammonium nitrate ■ K for ammonium chloride ◆ n for ammonium chloride.

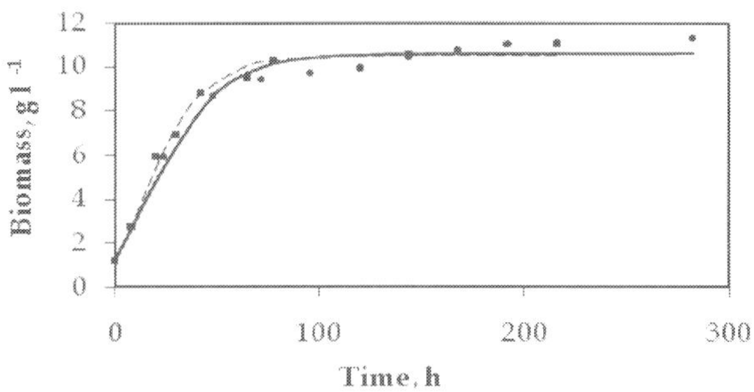

Fig. 10. Growth kinetics employing ammonium chloride (■) or ammonium nitrate (•) as nitrogen source.

Figure 10 shows the growth kinetics of *Gibberella fujikuroi* obtained during different fermentations. As can be seen in Figure 10, the growth kinetics is similar irrespective of the employed nitrogen source. Experimental data where fitted to two-parameter Gompertz model proposed by Chavez-Parga et al., (2005).As can be seen in Figure 10, there is no lag phase and exponential growth of mycelia starts immediately and ceases during the first 24 hours of fermentation. The later causes the medium viscosity to increase (K and n increase in Figure 9) which causes a k_La decrease in Figure 5. After 24 hours of fermentation, the formation of pellets by the fungus starts to occur reflected in a decrease of medium viscosity (K and n start to decrease in Figure 9) and hence an increase in k_La value in Figure 5. After 72 hours of fermentation the medium viscosity was practically unchanged (K and n remain constant in Figure 9) because the stationary growth phase is reached by the fungus reflected in practically constant values of medium viscosity and k_La. Also, after 72 hours of fermentation, the pellet formation process by the fungus stops.

Figure 11 shows the correlation between consistency and flow indexes with biomass concentration. Experimental data were fitted to Equations 11 and 12 proposed in the present work. Optimized values for constants in Equations 11 and 12 are summarized in Table 1.

$$K = \frac{c_1}{\left(1+\frac{c_2}{x}\right)+\left(\frac{x}{c_3}\right)^2} \tag{11}$$

$$n = \frac{c_1}{\left(1+\frac{c_2}{x}\right)+\left(\frac{x}{c_3}\right)^2} \tag{12}$$

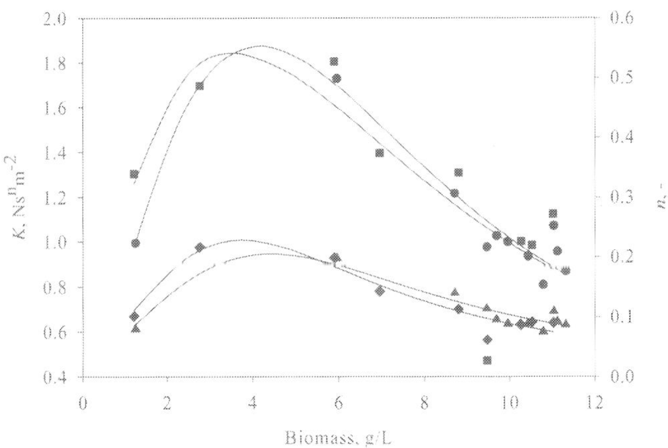

Fig. 11. K and n as a function of biomass concentration in the airlift bioreactor.
• K for ammonium nitrate ▲ n for ammonium nitrate ■ K for ammonium chloride
◆ n for ammonium chloride.

Consistency index			
Nitrogen source	c_1	c_2	c_3
Ammonium nitrate	6.31	6.55	4.69
Ammonium chloride	3.43	2.05	6.67
Flow index			
Nitrogen source	c_1	c_2	c_3
Ammonium nitrate	7.64	106.49	1.27
Ammonium chloride	7.63	80.91	1.14

Table 1. Optimized values found for constants of Equations 11 and 12.

With the aid of rheological studies is possible to use correlations of the type of Equation 13 to relate gas holdup and volumetric mass transfer coefficient with fermentation medium viscosity (Godbole *et al.*, 1984; Halard *et al.*, 1989; Al-Masry and Dukkan, 1998; Barboza *et al.*, 2000) to obtain Equations 14 and 15.

$$F = Av_{gr}^{B}\mu_{app}^{C} \tag{13}$$

$$k_L a = 0.0036\, v_{gr}^{0.3775}\, \mu_{app}^{-0.5488} \tag{14}$$

$$\varepsilon = 0.0072\, v_{gr}^{0.2381}\, \mu_{app}^{-0.5703} \tag{15}$$

Figures 1 and 4 show experimental data fitting for gas holdup and k_La, respectively. As it was expectable, Equations 14 and 15 present a better fit to experimental data than that obtained with the aid of Equations 2 and 3 due to the existence of an extra adjustable parameter.

2.7 Conclusions

In the present work preliminary hydrodynamics, mass transfer and rheological studies of gibberellic acid production in an airlift bioreactor were achieved and basic correlations between gas holdup, liquid velocity in the riser, and liquid velocity in the downcomer, mixing time and volumetric mass transfer coefficient with superficial gas velocity in the riser were obtained. Adjustable parameters calculated for each variable were compared with literature reported values and a good agreement was obtained. Gassing out method was successfully applied in determining volumetric mass transfer through fermentation time employing two different nitrogen sources. Irrespective of the nitrogen source the volumetric mass transfer behaviour was similar and it was explained in terms of the fungus growth and changes in its morphology which affect the culture medium rheology. Pellet formation by the fungus was used to explain the increase of k_La or the decrease of medium viscosity. In both fermentations, k_La decreases as exponential growth of the fungus occurs and reaches an asymptotic value once the stationary growth phase is reached. A helical impeller was employed successfully for rheological studies, avoiding problems of settling, jamming or pellet destruction, finding that the culture medium behaves as a pseudoplastic fluid. Rheological measurements were used to correlate gas holdup and k_La with apparent culture medium viscosity. Once again, for both fermentations, apparent viscosity increases as exponential growth of the fungus occurs and reaches an asymptotic value once the stationary growth phase is reached. A satisfactory validation of experimental data for gas holdup and volumetric mass transfer coefficient was performed which allows to employ these data in scale-up strategies.

3. Case 2. Dynamic transport and reaction model for the removal of azo dye in a UAFB reactor

3.1 Introduction

Azo dye degradation from textile effluents has been the objective of research for some years due to the pollution problem they generate. For the removal of these compounds different processes have been applied: physicochemical, advanced oxidation, and biological. However there is a continuous search for an efficient, low cost and low environmental impact process to eliminate this problem. In particular, Reactive dyes are highly water soluble due to the sulphonated groups in their molecule so it cannot be reduced under the ordinary wastewater treatment processes (Beydilli, 2005). Anaerobic bioreactors have an important role in the treatment process of hazardous wastes, besides they can treat higher organic loads than aerobic reactors. Fixed bed reactors can be immerse, usually upflow, or trickle bed, downflow, the main characteristic is that the biomass is forming a biofilm covering a material that works as a support or carrier for the growth and maintenance of the microorganisms; in this way, the reactor efficiency is improved because the substrate-biomass contact is increased (effective surface area), and the process is more stable. The use of a carrier in the reactor is to improve the mechanical properties of the biomass and cell retention; in addition, the carrier may participate in the degradation process (Van der Zee, et al. 2003). A biofilm usually do not grow in a homogeneous way on the support, but rather forms clusters on the surface; the way in which a biofilm is grown and their internal structure is formed depends on the superficial velocity of the flow through the reactor, it is also affected by the mass transfer velocity and microorganism activity (Beyenal, 2002). The degree of biomass buildup affects the hydrodynamic behavior of the reactor. In this work, an Upflow Anaerobic Fixed Bed (UAFB) bioreactor with activated carbon (AC) as the carrier was used to remove azo dye from the effluent. It has been proved that AC possess good properties for biofilm growth and to remove diverse pollutants (Fan, et al. 1987; Fan, et al. 1990; Herzberg and Dosoretz, 2003; McCarty and Meyer,2005), moreover, AC could accelerate azo dye degradation due to its redox mediator function through the chemical groups on its surface (Van der Zee, et al. 2003) Di Iaconi et al (2005) proposed a mechanism for biofilm growth: 1) formation of a thin film covering the support by the microorganisms, 2) increment of the biofilm thickness, 3) the break of the added biofilm clusters and release of particles (biomass due to the excess of growth) and 4) small pellet formation by detached particles. In UAFB reactors it is common to have the bioparticles (carrier plus biofilm), some free cells and biomass pellets as a function of the superficial velocity on the reactor; the water flowing through the bioreactor can carry out the drag of small biomass pellets. The mass transport through this bioparticles occurs on three stages: diffusion of the dye molecule from the solution to the biofilm, diffusion through the biofilm, adsorption-diffusion through the carbon surface and reaction. One disadvantage of using upflow fixed bed reactors is that the liquid flow is non-ideal and dispersion, backmixing and bypassing flow are considerable (Iliuta, et al. 1996), therefore it is important to carry out the hydraulic characterization of the reactor through tracer test, although it is common to consider plug flow to model the reactor. The reasons of modelling a reactor of this kind are to estimate all the important parameters in its function, to optimize the efficiency and to predict its behaviour, besides its future scale-up. However, scaling a reactor from laboratory models is often difficult, since some factors which are negligible when modelling small reactors have to be included in real reactor models, such as the transport between static and dynamic zones. Therefore, the main objective of this paper is to propose a dynamic mathematical model for an UAFB bioreactor with AC as carrier, to attach microorganisms and enhance biodegradation, in the removal of the azo dye reactive red 272 (Fig. 1).

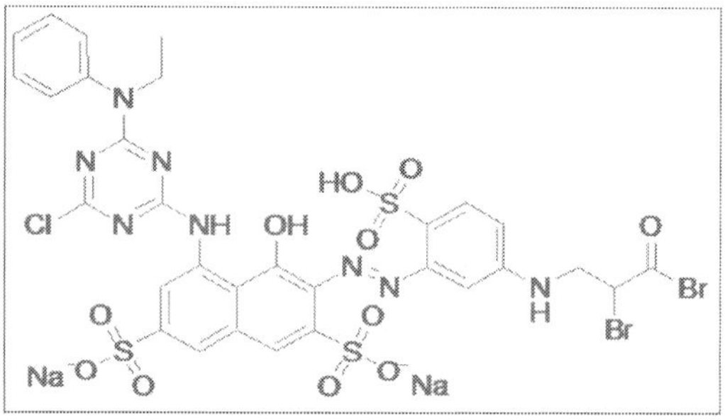

Fig. 1. Reactive red 272

The presented mathematical model includes all the transport phenomena: convection, dispersion, diffusion and mass transfer from one phase to another, along the reactor and through the bioparticle, as well as the reaction of dye reduction. The balance equations are coupled and solve together as a system. We try to include in the model all the possible phenomena that take place in the reactor in order to describe it and obtain enough information about it.

3.2 Materials and methods

3.2.1 Reactor assembling.

It was started up to work an anaerobic upflow reactor of the kind of the UAFB, made from Pyrex glass, with a fixed bed of AC of 42% of its operation volume, equivalent to 1.244 L and 541.17 g of AC. The reactor is outlined in Figure 2 and its characteristics are shown in Table 1.

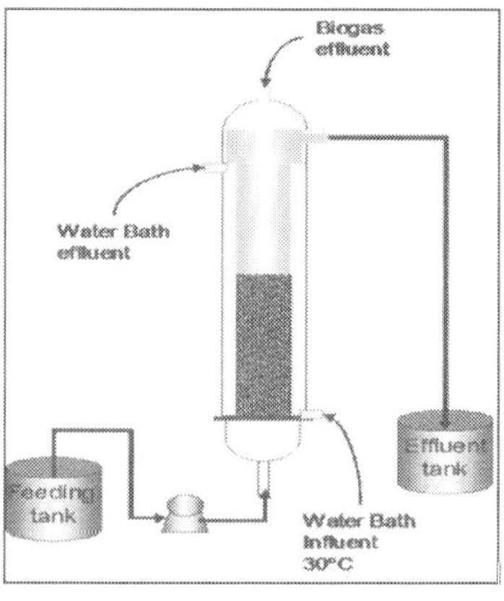

Fig. 2. Upflow Anaerobic Fixed Bed (UAFB) Reactor.

Work volume, L	3.3
Inside diameter, cm	6
Inside diameter of the settle, cm	9.5
Total longitude, cm	105.5
Initial and steady state porosity of the bed	0.53, 0.19
Fixed bed volume, L	1.244
Fixed bed longitude, cm	48
Superficial velocity (average), cm/min	0.52
Volumetric flow (average), mL/min	18
RT_m (average), min	206.25

Table 1. UAFB Reactor Characteristics

At the beginning, there was an adsorption stage to saturate the AC in the reactor with dye and do not attribute the removal efficiency to simply adsorption on to AC. Afterwards, the rector was inoculated by recirculating water with 10% v/v of adapted sludge for a period of 15 days; this was a consortium of microorganisms adapted to azo dye reduction using textile wastewater enriched with reactive red 272. In this stage, 11.586 mg biomass/g AC was adsorbed, forming a biofilm on the AC surface. The reactor was operated using synthetic wastewater, containing different azo dye concentration, from 100 to 500 mg/L, and 1 g/L of dextrose and yeast extract as carbon and nitrogen source to the microorganism.

3.2.2 Residence time distribution.

A lithium chloride solution was used as a tracer in order to determine the hydraulic characteristics of the reactor and to obtain the residence time distribution. Smith and Elliot (1996) used LiCl as a tracer and recommend a concentration of 5 mg Li^+/L to avoid toxicity problems. In this case it was applied a one minute pulse of a 2000 mg/L LiCl solution. Dextrose and yeast extract with a concentration of 1 g/L were used as the substrates during the test. Samples were taken in the reactor effluent every 30 min during approximately 3 times the half residence time (RT_m), in this case, during 10 hr. Lithium concentration was analyzed in an atomic adsorption spectrophotometer (Perkin Elmer model 2280; USA).
The hydraulic residence time (HRT) was calculated as:

$$HRT = \frac{\int_0^{\infty} t(C - C_0)dt}{\int_0^{\infty} (C - C_0)dt} \quad (1)$$

Were C is the tracer concentration at a time t and C_0 is the tracer concentration at $t=0$. The parameters and non-dimensional numbers necessary to describe the reactor as well as the axial dispersion and mass transfer coefficients were calculated according to the next equations.
Dispersion number (d) and Péclet (Pe). These numbers indicate the dispersion grade in the reactor. A Pe above 1 indicates that convection is the leading factor in the mass transport, and if it is lesser than 1, the leading factor is the dispersion. The numbers are calculated as (Levenspiel, 2004):

$$d = \frac{1}{2}\frac{\sigma_{\Delta C}^2}{RT_m{}^2} = \frac{D}{uL} \quad (2)$$

$$Pe = \frac{uL}{D} = \frac{1}{d} \tag{3}$$

Where u is the superficial velocity in the reactor, L is the longitude and D is the axial dispersion coefficient.

Dispersion coefficient (D). It can be calculated by the dispersion number or by other correlations as the presented through the Reynolds number.

$$D = d\ u\ L = 1.01\nu\ N_{\text{Re}}{}^{0.875} \tag{4}$$

Here, ν is the cinematic viscosity of the water in the reactor (Levenspiel, 2004).

Sherwood number (Sh) and mass transfer coefficient (k_m). It was calculated by the Frössling correlation (Fogler, 1999), which is applied to the mass transfer or flux around a spherical

$$Sh = 2 + 0.6 N_{\text{Re}}{}^{1/2} Sc^{1/3} \tag{5}$$

particle. Supposing this the following equation was used:

And the mass transfer coefficient of the dye was estimated by the equation:

Where d_p is the average particle diameter of the carbon particles and biomass in the bed. For this analysis it was taken the d_p values of the carbon particles at the beginning of the study, 1.03 mm.

$$k_m = \frac{D_{ef} Sh}{d_p} \tag{6}$$

3.2.3 Kinetic model

The applied kinetic model to represent the dye biodegradation (reduction) was derived according to experimental observations, after fitting kinetic data at dye concentrations from 100 to 500 mg/l. The model expresses a change in the reaction order since it was noticed that the reaction in the system is a function of dye concentration and occur in two stages: first order, the dye is adsorbed by the bioparticle and reduced, and second, the enzymatic reactions take place to degrade the dye to certain extent. This is shown by Equation 7, here: C_{A0} and C_A are the initial and every moment dye concentration, k_1 and k_2 are 1st and 2nd order specific reaction rate, (h^{-1}, L/mg·h). The deduction is explained in another paper (in revision).

$$r_A = -\frac{dC_A}{dt} = k_1 C_A - k_2 C_A (C_{A_0} - C_A) \tag{7}$$

3.2.4 Model dimensionless numbers

From the dimensionless analysis of the model, the dimensionless numbers that explain the transport process in the reactor were obtained. These were: Biot's number (Bi), that relates mass transfer with diffusivity, Fourier's number (Fo), that relates the diffusivity in the reaction area in the reaction time, Wagner's module (Φ^2), by means of which it is obtained the Thiele's number (Φ) that indicates if diffusion modifies the reaction rate; from this number, the Effectiveness factor (η) is calculated, which relates the real reaction rate with the reaction rate without diffusion resistance, in other words, it expresses the influence of the diffusion on the reaction rate. Thiele number is calculated by Equation 8 (according the proposed kinetic model there is a Thiele number for the first order term and other for the second order term). The Effectiveness factor for the reduction rate of the dye by volume unit of bioparticle was calculated using Equation 9, according to the definition of volume average (Escamilla-Silva et al, 2001) and using the proposed kinetic model expressed in Equation 7. Here, $\overline{R_A}$ is the average reaction rate in the biofilm and $R_A\big|_{\xi=1}$ is the reaction rate in the bioparticle surface in the liquid boundary; Fo_b is the characteristic Fourier number for the biofilm defined in Equation 19, in the next section.

$$\Phi_1 = \delta\sqrt{\frac{k_1}{D_{eb}}}\ ;\ \Phi_2 = \delta\sqrt{\frac{k_2 C_{A0}}{D_{eb}}} \tag{8}$$

$$\eta = \frac{4\pi\int_0^1 R_A\xi^2 d\xi}{\frac{4}{3}\pi R_A\big|_{\xi=1}} = \frac{3\int_0^1 R_A\xi^2 d\xi}{R_A\big|_{\xi=1}} = \frac{3\int_0^1\left[\Phi_1^2 Fo_b\omega_b - \Phi_2^2 Fo_b\omega_b\left(\omega_L-\omega_b\right)\right]\xi^2 d\xi}{\left[\Phi_1^2 Fo_b\omega_b - \Phi_2^2 Fo_b\omega_b\left(\omega_L-\omega_b\right)\right]\Big|_{\xi=1}} = \frac{\overline{R_A}}{R_A\big|_{\xi=1}} \tag{9}$$

3.3 Results and discussion.

The UAFB reactor efficiently removes the reactive red dye, from 91.35% to 98.64% and up to 56% of DQO, at inflow concentration from 100 to 500 mg/L and at a RT_m from 3 to 5 hours. Higher removal rates can be obtained at higher residence times. The difference between colour and COD removal is because the first step in the biodegradation of the dye takes place when the azo bond is broken, and this results in the lost of colour of the solution. There are aromatics amines and other organic compounds in the water as products of dye reduction, which can be degraded to a certain extent in to other low molecular weight molecules, as carboxylic acids. The results described in this section are in regard to an analysis of the transport and reaction phenomena inside the reactor and to obtain predictions about its performance. The balance equations are proposed according to theoretical principles. Some of the parameters used were calculated according to experimental and real results and others in base to references. The model can be used and applied to similar problems, but it will need a parameter fit. Because of this, the real removal rate of the reactor is higher than the predicted for the model, at the highest dye concentration used (400-500 mg/L).

3.4 Residence time distribution.

The parameters and non-dimensional numbers that describe the transport in the reactor fixed bed are shown in Table 1. The superficial velocity was calculated as $u_L = Q/\varepsilon_L \pi R_i^2$ and the porosity of the bed ε_L was 0.19 after equilibrium was reached. The hydraulic behaviour of the reactor was approximated to a plug flow with axial dispersion. Figure 3 show HRT distribution curves; it was observed that when Q was increased, the dispersion was reduced and the reactor was closer to ideal plug flow behaviour. This is a hydrodynamic effect, but for packed beds it is attributed to the particle size of the packing material. This result can be attributed to the fine particles formed with time operation in the inter-particle space in the reactor, because it reduces the bed porosity and as a result the by-pass fluxes. Kulkarni et al (Kulkarni, 2005) established that the fine particles formed in packed bed reactors reduce the by-pass flux because there is a better spreading of the water flow, and therefore the dispersion is reduced. During the residence time distribution tests there was biogas production due to the digestion of dextrose in the synthetic wastewater, however, the biogas production with or without dye in the water is not enough to consider the reactor as a mixed tank. Besides, the rise of biogas through the bed is very slow, this generates by-pass flow due to the biogas bubbles trapped, and when the superficial velocity in the reactor is increased, the bubbles are pushed and can flow better and the by-pass flux is reduced; thus the reactor became closer to a plug flow. The HRT was from 1.6 to 1.8 times the RT_m at the less volumetric flow in the reactor, and from 1.1 to 1.3 times at high volumetric flow. The residence time distribution achieved for all the tests was fitted by a statistical distribution of extreme value (Fisher-Tippet) shown by Equation 10; this is a frequency distribution function for slant peaks.

$$P(t) = y_0 + a * Exp\left[-Exp(-f1) - f1 * s + 1\right]$$
$$f1 = \frac{t - t_C}{w} \tag{10}$$

Where $P(t)$ is the normalized tracer concentration, y_0 represents the distribution displacement (time 0 tracer concentration), a is the amplitude of the distribution and w the wide of the peak. These calculated parameters, the error and correlation coefficient (p<5). Figure 3 shows the residence time distribution in the reactor for each test and the fit by the Extreme model. Iliuta et al (1996), states that a long tail in the residence time distribution could be caused because of axial dispersion in the dynamic flow as well as the mass transfer between the dynamic and static zones. Also, when the reactor has porous particles in the packed bed, as in this case, the alteration in the distribution can be attributed to internal diffusion and reversible adsorption of the tracer, however, this is negligible, but it happens for the dye.

The reactive red 272 dye molecule is adsorbed reversible and superficially on to biomass, also is adsorbed onto AC surface, but poorly, and has a limited adsorption capacity due to its molecule size and chemical groups as naphthalene disulphonic acid, a choloro-triazine and phenyl-amine (see Fig. 1); also AC has limited active sites for adsorption. There is an equilibrium between adsorption, reaction and desorption. In this case it is considered a constant saturation of the bioparticle, so the adsorption dynamics will be negligible.

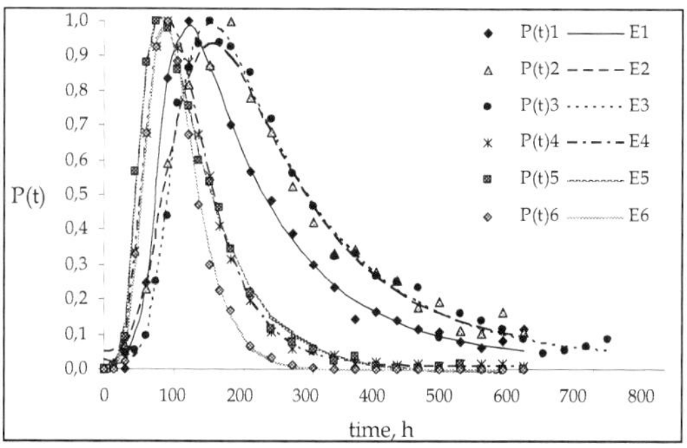

Fig. 3. HRT distribution; tracer pulse applied to the bioreactor.

Parameter	HRT1	HRT2	HRT3	HRT4	HRT5	HRT6
RT_m (min)	57.264	66.203	66.203	42.412	42.412	42.412
u_L (cm/s)	0.0735	0.0631	0.0631	0.0985	0.0985	0.0985
N_{ReL}	54.962	47.468	47.468	74.096	74.096	74.096
Bo_L	11029.2	9461.54	9461.54	14769.2	14769.2	14769.2
d_L	2.864	2.255	2.407	2.332	1.208	0.413
D_L (cm²/s)	10.026	6.828	7.287	11.022	5.709	1.953
Pe_L	0.349	0.443	0.415	0.429	0.828	2.420
Sh_L	45.403	42.246	42.246	52.283	52.283	52.283
k_L (cm/s)	0.0176	0.0164	0.0164	0.0203	0.0203	0.0203
y_0	0.0270	0.0470	0.0251	7.6×10^{-3}	9.3×10^{-3}	5.5×10^{-4}
a	0.5697	0.6629	0.6175	0.8944	0.6959	1.0029
w	22.033	46.312	33.212	29.893	19.756	32.259
SE[a]	0.0255	0.0443	0.0325	0.0182	0.0226	0.0162
$R^{2\,b}$	0.9945	0.9823	0.9920	0.9977	0.9966	0.9980

Table 2. Hidrodinamic, mass transfer and and fitted extreme model Parameters for the UAFB reactor (L = fixed bed).

3.5 Dynamic model: Transport and reaction in the fixed bed

After the calculus of HRT and all the necessary parameters and dimensionless numbers, a mathematical model was deduced and executed to represent the transport and reaction of the dye in the water upflow through the reactor (the clarifier zone is not considered here due to complexity of the process).

The dynamic model is based in the next assumptions:

1. Radial dispersion is negligible.
2. The reactor is divided in two zones: the fixed bed and the clarifier (not considered) and there is mass transfer between them.
3. The dispersion coefficient is constant in each zone.
4. There is mass transfer between the water flowing through the bed and the bioparticles.
5. The superficial velocity trough the bed is constant and calculated as $u_L = Q/\varepsilon_L \pi R_i^2$.
6. The dye can be reversible adsorbed into bioparticles.
7. The dye is diffused, adsorbed and reacts in the biofilm and there is superficial diffusion in the activated carbon core macropores because the molecule size (21 Å) would limit the micropore diffusion (pore average diameter 23.3 Å).
8. The particle is spherical and with a uniform and porous biofilm.
9. The biomass on the activated carbon core grow up to certain thickness, and afterwards the biomass excess is detached and form small granules, then the biomass attached on the activated carbon came back to their normal or equilibrium thickness. Thus, there is a dynamic in the biofilm thickness but will be taken as an average value supposing equilibrium between grow and detachment. The model includes the balance in the bioparticle divided in two zones: zone I represents the AC particle (core) and zone II the microorganism biofilm surrounding the AC core. The reaction term is included in two balance equations, in the zone II of the bioparticle and in the liquid balance; this because is an extracellular process and there are free cells or small biomass granules in the flowing liquid. Figure 4 shows a graphical scheme of the model; here C_{AL} is the liquid concentration of the dye, C_{AP} is the dye concentration in the AC core, and C_{Ab} the biofilm concentration.

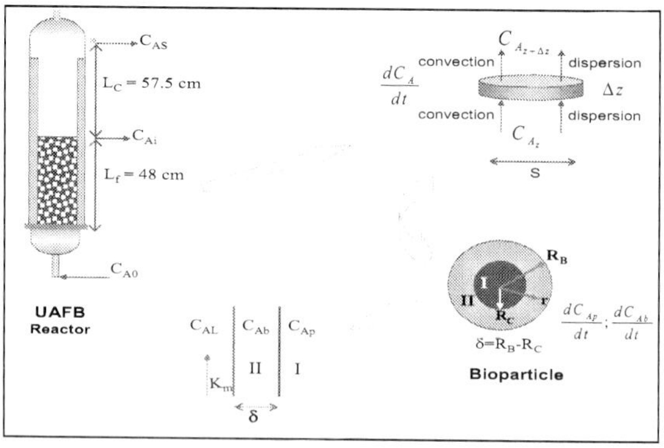

Fig. 4. Transport and reaction model in the UAFB, Reactor scheme.

The governing equations are:

***a)* Balance for the liquid flow in the fixed bed.**

$$\frac{\partial C_{AL}}{\partial t} = D_L \frac{\partial^2 C_{AL}}{\partial Z^2} - u_L \frac{\partial C_{AL}}{\partial Z} - K_m a_{sb}(C_{AL} - C_{Ab}\big|_{r=R_B}) - k_1 C_{AL} + k_2 C_{AL}(C_{A0} - C_{AL}) \quad (11)$$

This equation describes the dye convection, dispersion and mass transfer from the liquid phase to the biofilm surface and the reaction. The initial and boundary conditions are:

$$\begin{aligned} t = 0 \quad & C_{AL} = C_{A0} \\ Z = 0 \quad & C_{AL} = C_{A0} \\ Z = L_f \quad & \frac{\partial C_{AL}}{\partial Z} = 0 \end{aligned} \quad (12)$$

***b)* Balance for the bioparticle.**

For the AC core:

$$\frac{\partial C_{Ap}}{\partial t} = D_{ep}\left(\frac{2}{r}\frac{\partial C_{Ap}}{\partial r} + \frac{\partial^2 C_{Ap}}{\partial r^2}\right) \quad (13)$$

This equation describes the dye convection and diffusion. The initial and boundary conditions that expresses field equality in the interface are:

$$\begin{aligned} t = 0 \quad & C_{Ap} = 0 \\ r = 0 \quad & \frac{\partial C_{Ap}}{\partial r} = 0 \\ r = R_C \quad & C_{Ap} = C_{Ab} \end{aligned} \quad (14)$$

For the biofilm (diffusion and reaction):

$$\frac{\partial C_{Ab}}{\partial t} = D_{eb}\left(\frac{2}{r}\frac{\partial C_{Ab}}{\partial r} + \frac{\partial^2 C_{Ab}}{\partial r^2}\right) - k_1 C_{AL} + k_2 C_{AL}(C_{A0} - C_{AL}) \quad (15)$$

This equation describes the dye convection, diffusion and reaction. The initial and boundary conditions that expresses flux equality in the interface and mass transfer from the bioparticle to the liquid phase are:

$$t = 0 \quad C_{Ab} = 0$$

$$r = R_C \quad -D_{ep}\frac{\partial C_{Ap}}{\partial r} = -D_{eb}\frac{\partial C_{Ab}}{\partial r} \quad (16)$$

$$r = R_B \quad -D_{eb}\frac{\partial C_{Ab}}{\partial r} = K_m\left(C_{Ab} - C_{AL}\right)$$

c) **Dimensionless model.** The dimensionless governing equations are shown next.

For the liquid flow in the fixed bed:

$$\frac{\partial \omega_L}{\partial \tau} = d_L\frac{\partial^2 \omega_L}{\partial \zeta^2} - \frac{\partial \omega_L}{\partial \zeta} - \beta_m(\omega_L - \omega_b) - \Phi_1^2 Fo_b \omega_L + \Phi_2^2 Fo_b \omega_L\left(1 - \omega_L\right) \quad (17)$$

$$\tau = 0 \quad \omega_L = 1$$

$$\zeta = 0 \quad \omega_L = 1 \quad (18)$$

$$\zeta = 1 \quad \frac{\partial \omega_L}{\partial \zeta} = 0$$

Using the dimensionless numbers:

$$\omega_L = \frac{C_{AL}}{C_{A0}} \; ; \; \zeta = \frac{Z}{L_f} \; ; \; \tau = \frac{t}{t_{mL}} = \frac{tu_L}{L_f} \; ; \; d_L = \frac{D_L}{u_L L_f} = \frac{1}{Pe_L} \; ; \; \beta_m = \frac{K_m a_{sb} L_f}{u_L}$$

$$\Phi_1^2 = \frac{\delta^2 k_1}{D_{eb}} \; ; \; \Phi_2^2 = \frac{\delta^2 k_2 C_{A0}}{D_{eb}} \; ; \; Fo_b = \frac{D_{eb} L_f}{\delta^2 u_L} \quad (19)$$

In order to become dimensionless the two zones of the bioparticle, it was proposed a parallel model; the AC core and the biofilm are then expressed as a single particle with a radius from 0 to 1 (see Fig. 5):

For the AC core:

$$\frac{\partial \omega_p}{\partial \tau} = Fo_p\left(\frac{2}{\xi}\frac{\partial \omega_p}{\partial \xi} + \frac{\partial^2 \omega_p}{\partial \xi^2}\right) \quad (20)$$

$$\tau = 0 \quad \omega_p = 1$$

$$\xi = 0 \qquad (21)$$

$$\xi = 1 \quad \omega_p = \omega_b$$

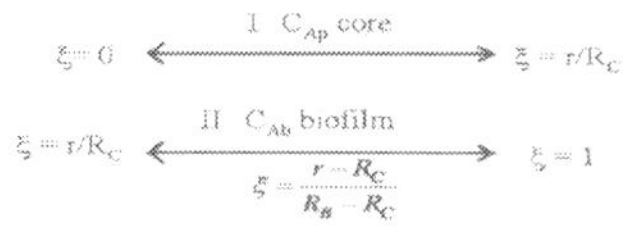

Fig. 5. Parallel dimensionless model.

Using the dimensionless numbers:

$$\omega_p = \frac{C_{Ap}}{C_{A0}} \;;\; Fo_p = \frac{D_{ep}L_f}{R_C^2 u_L} \;;\; \tau = \frac{t}{t_{mL}} = \frac{tu_L}{L_f} \qquad (22)$$

For the biofilm:

$$\frac{\partial \omega_b}{\partial \tau} = Fo_b\left[\frac{\partial^2 \omega_b}{\partial \xi^2} + \left(\frac{2}{\xi+\beta}\right)\frac{\partial \omega_b}{\partial \xi}\right] - \Phi_1^2 Fo_b \omega_b + \Phi_2^2 Fo_b \omega_b\left(\omega_L - \omega_b\right) \qquad (23)$$

$$\tau = 0 \quad \omega_b = 0$$

$$\xi = 0 \quad \frac{\partial \omega_b}{\partial \xi} = \alpha\,\beta \frac{\partial \omega_p}{\partial \xi} \qquad (24)$$

$$\xi = 1 \quad \frac{\partial \omega_b}{\partial \xi} + Bi\;\omega_b = Bi\;\omega_L$$

Using the dimensionless numbers:

$$\omega_b = \frac{C_{Ab}}{C_{A0}} \;;\; \tau = \frac{t}{t_{mL}} = \frac{tu_L}{L_f} \;;\; \Phi_1^2 = \frac{\delta^2 k_1}{D_{eb}} \;;\; \Phi_2^2 = \frac{\delta^2 k_2 C_{A0}}{D_{eb}} \;;\; Fo_b = \frac{D_{eb}L_f}{\delta^2 u_L}$$

$$Bi = \frac{K_m R_B}{D_{eb}} \;;\; \alpha = \frac{D_{eb}}{D_{ep}} \;;\; \beta = \frac{R_C}{R_B - R_C} = \frac{R_C}{\delta} \qquad (25)$$

The reactor model then consist of three differential parabolic equations; to solve the model, all the parameters are calculated and is used the finite differences with a 5th order Runge-Kutta-Fehlberg method, programmed in Fortran language.

3.6 Model solution.

The data that we had previous to the model solving are: from the AC properties: ρ_p = 0.435 g/cm^3 (density), ε_p = 0.1392 (porosity), R_C = 0.0515 cm (average particle radius), and about the reactor we have: L_L = 48 cm (longitude), R_i = 3 cm (radius), ε_L = 0.19 (bed porosity). The dispersion coefficient (D_L), the dispersion number (d_L), the Peclet number (Pe), the superficial velocity (u_L) and the half residence time (RT_m), were estimated in the residence

time distribution analysis, as was explained in the methodology section. The diffusion coefficient in the AC core was calculated by the method explained in Hines and Hines and Maddox (1987), supposing Knudsen diffusivity, but the value was changed in two orders of magnitud in order to fit the model (Fan et al. 1990; Lee et al.2006), comparing the values reported for other dyes in references. The diffusivity in the biofilm was assumed to be 100 times larger than the diffusivity in the AC core, based in values reported for other dyes of the same kind (Fan et al. 1987; Chen et al. 2003). The kinetic constants were estimated from experimental data and fitted to model.
The mass transfer surface of the bioparticles was calculated by the Equation 20 proposed by Iliuta and Larachi (2005)

$$a_{sb} = \frac{6}{d_b} = \frac{6}{d_p + 2\gamma} \tag{26}$$

The parameters used to solve the mathematical model are shown in Table 4; in this case, the Thiele number of second order is applied for an initial dye concentration of 250 mg/L, for 400 mg/L was 1.43 and for 500 mg/L was 1.60. This indicates that when the dye concentration is augmented in the reactor influent, the mass transfer effects are increased, specifically the resistance to the diffusion, and therefore, it decreases dye removal. Figure 6 shows the concentration profile along the reactor to different dye concentration at the reactor inflow, and Figure 7 shows the concentration profile within the bioparticle (AC core plus biofilm) for an initial concentration of 250 mg/L.
It can be seen in the Figure 7 that the particles close to the reactor influent (ζ = 0.045) contain higher dye concentration than the ones in the effluent extreme, whose concentration is close to the dye effluent concentration. The concentration profile in the bioparticle changes with time as the bioparticle is saturated and reaches equilibrium, while a curve indicative of the reaction in the biofilm is observed. When the bioparticle is close to the effluent, the profile is becoming flat, and the concentration is approximately uniform within the bioparticle. This is because at the effluent zone there is the smallest dye concentration in the reactor height, and the kinetics depends on this.

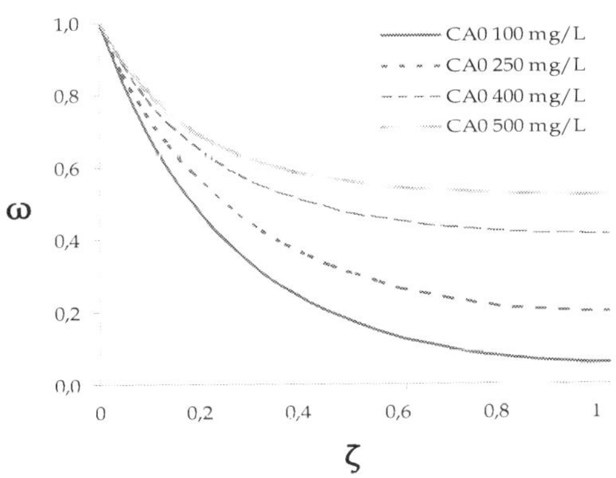

Fig. 6. Predicted concentration profile along the reactor

D_L (cm²/min)	423.4	$\Phi_1^{2\ a}$	1.061
u (cm/min)	3.786	$\Phi_2^{2\ b}$	1.276
K_m (cm/min)	0.984	$\Phi_1^{\ c}$	1.030
D_{ep} (cm²/min)	2.58×10^{-5}	$\Phi_2^{\ d}$	1.130
D_{eb} (cm²/min)	2.58×10^{-3}	β	2
k_1 (min⁻¹)	3.046	α	100
k_2 (L/mg min)	1.47×10^{-2}	Bi	34.27
a_{sb} (cm²/cm³)	36.81	B_m	459.22
F_{op}	0.091	d_L	2.33
F_{ob}	36.404		

Table 4. Parameters used in model solution.

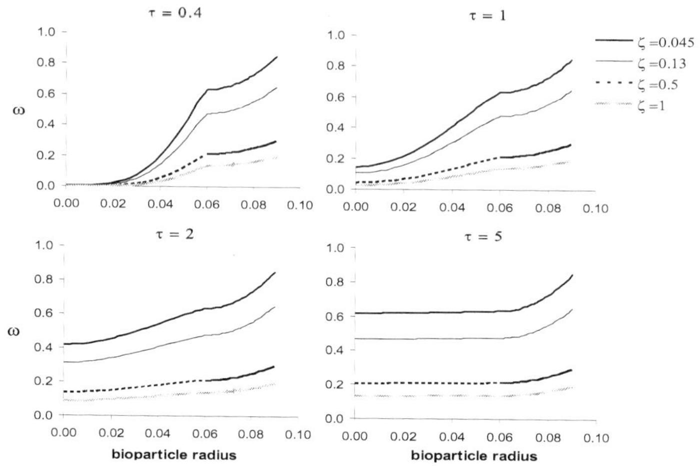

Fig. 7. Predicted concentration profile in the bioparticle at different τ and ζ in the bed. Radius in cm.

The RT_m in the reactor does not have much influence on the removal of the dye according the predicted results of the model. Figure 8 shows the concentration profile along the reactor at different conditions of RT_m (See Tab. 3), and using a dye inflow concentration of 250 mg/L; it can be seen that there is not a noticeable effect in the concentration profile as RT_m is incremented. The concentration profile in the bioparticle is affected in a different way by the RT_m; in Figure 9 we can see that as RT_m is increased, the bioparticles are saturated faster, this is because the contact time for adsorption and for reaction is augmented and the mass transfer between the liquid phase and the bioparticles is improved.

As a result, the RT_m in the reactor affects only the mass transport rate in the reactor but not the biodegradation reaction and thus does not have an influence on the concentration profile along the reactor and consequently, in the removal efficiency. Therefore, analyzing the

profiles expressed in Figure 6, the main factor that affects the removal efficiency is the inflow dye concentration.
Similar results were found by other authors. Spigno et al (2004) presented a mathematical model for the steady state degradation of phenol along a biofilter reactor and obtained a concentration profile for phenol reduction at two different concentrations, and they found the same profile for both conditions displaced by a concentration gradient, more reduction at the lesser concentration. Mammarella and Rubiolo (2006) could predict the concentration profile for lactose hydrolysis in an immobilized enzyme packed bed reactor under different operation conditions, and they obtained an asymptotic profile for lactose conversion along the reactor height and obtained a much higher conversion at the lesser volumetric flow (100 mL/h); on the contrary, in this work the volumetric flow does not have much influence. These authors did not include the dynamic of the pollutant inside the bioparticle, as is shown in this work.

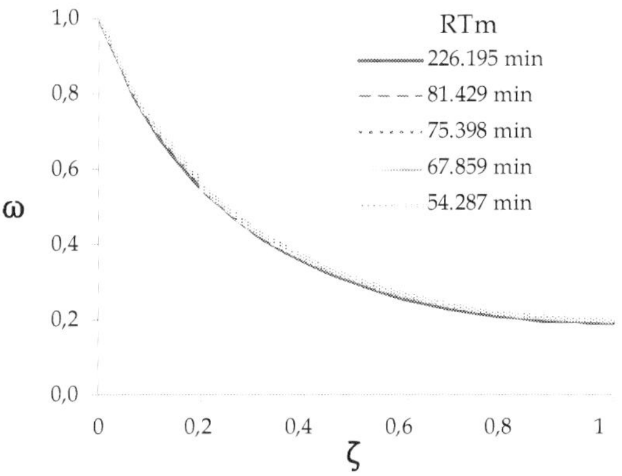

Fig. 8. Predicted concentration profile along the reactor a different RT_m

RT_m (min)	226.195	81.429	75.398	67.859	54.287
Q (cm^3/min)	6.000	16.67	18.00	20.00	25.00
u (cm/min)	1.117	3.102	3.351	3.723	4.654
K_m (cm/min)	0.695	0.913	0.939	0.978	1.083
d_L	7.899	2.843	2.633	2.370	1.896
β_m	1099.8	519.7	495.1	464.4	411.5
F_{op}	0.308	0.111	0.103	0.093	0.074
F_{ob}	123.4	44.42	41.13	37.02	29.62
Bi	24.21	31.78	32.70	34.07	37.75

Table 5. Parameters used in model solution at different RT_m

Leitão and Rodrigues (1996) presented the influence of the biofilm thickness on the removal of a substrate when the support carrier material is an adsorbent and when it is not, and they obtained the concentration profile within the bioparticle but in the absence of reaction; the profiles show the saturation of the bioparticle as time increases, and the concentration augments as the biofilm thickness increase, but it was not observed the reaction zone in the biofilm as in the results presented in this work. Leitão and Rodrigues (1998) proposed an intraparticle model for biofilms on to a carrier material including the convective flow inside the particle, and they obtained the concentration profile and biofilm thickness regarding time. They conclude that bioreactors have to be operated under conditions such that allow the liquid movement to occur in the void space of the biofilm, in order to improve mass transfer and as a result the efficiency of the process.

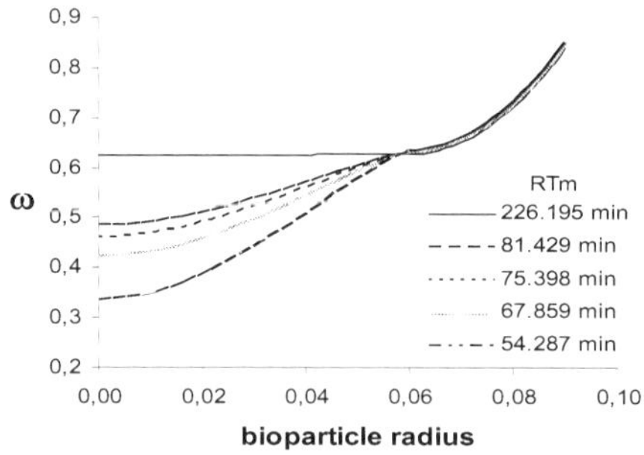

Fig. 9. Predicted concentration profile in the bioparticle at different RT_m.
Radius in cm.τ = 2 and ζ = 0.045 (close to the influent) in the bed.

In biofilms the hydrodynamics and kinetics of the system is related to the fact that most biofilm reactions are diffusion limited, therefore the shape of the concentration profiles will be determined by diffusivity and convection (Lewandowski, 1994). We suppose that the biofilm is a continuous phase, but it is not, biofilms are formed by clusters, void spaces and channels, so the convection in the system is important, and the measured diffusion coefficients are always approximations. The effect of convection in the bioparticle is shown in Figure 9; mass transfer in the biofilm is fast but in the core the RT_m affects the mass transfer and as a result the concentration profile.

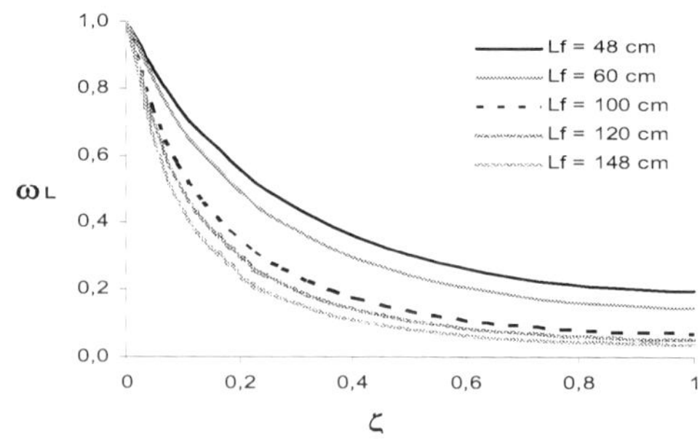

Fig. 10. Predicted concentration profile along the reactor, increasing the reactor bed height (L_f). C_{A0} = 250 mg/L at the inlet.

The effect of the bed height is shown in Figure 10, with a constant inlet dye concentration of 250 mg/L. Here we can see that the dimensionless dye concentration is reduced as the longitude of the bed height (L_f) is increased, this means higher removal efficiency; a higher L_f implies more bioparticles and time for degradation reaction, however the most of the reduction of the dye is carried out in the lower third of the bed longitude.

3.7 Effectiveness factor.

The effectiveness factor (η) was calculated at different azo dye concentration at the reactor influent, from 100 to 500 mg/L, and at different RT_m from 54.3 to 226.2 min. Each calculation was carried out taking the predicted dimensionless concentration value in the biofilm and in the liquid, and solving the model for a $\tau = 2$. The results explain that the average reaction rate in the biofilm and at the biofilm surface is changed regarding the height of the reactor, and in consequence the effectiveness factor; furthermore, the effectiveness factor depends mainly on the azo dye concentration. The change of the reaction rate regarding the reactor height is explained because there is a higher amount of active biomass at the inlet, at the bottom of the reactor. In this zone, dye and substrate concentration (dextrose and yeast extract) is always higher and in consequence there is always a major reaction activity; therefore, in this zone takes place the most of the degradation of the dye, as it was also noticed in Figure 6. The calculated effectiveness factor decreased from 0.78 to 0.47 when dye concentration was increased from 100 to 500 mg/l, proving that, when dye concentration is increased the diffusion of dye through biofilm has a major influence on the reaction rate, what is reflected in a reduction of the efficiency of removal. On the contrary, changes in RT_m in the reactor at a constant inflow dye concentration did not influence the value of the effectiveness factor; this did vary from 0.74 to 0.73 when RT_m was decreased from 226.2 to 54.3 min. These values indicate that there is a slight effect of dye diffusion on the reaction rate when RT_m is changed.

3.8 Conclusions

The proposed mathematical model for a UAFB bioreactor was able to predict the concentration profiles along the reactor and within the bioparticle (carbon core and biofilm) for the biodegradation of a reactive red azo dye, using a kinetic model with a change in reaction order. The profiles at different inflow dye concentration showed an asymptotic curve with a major activity of reaction in the lower zone of the reactor, and this concentration fall is reduced as the inflow dye concentration is augmented. Nevertheless, the RT_m does not have much influence on the concentration profile along the reactor. In addition, the model predicts a larger removal rate as the longitude of the bed is increased for the same inlet dye concentration.

The profiles within the bioparticle illustrate the saturation of the particle and reflect the zone of reaction in the biofilm; it can be seen the differences in the concentration values with regard to the reaction zone along the reactor. The saturation rate of the bioparticles change with the RT_m, at a larger time, the mass transfer is improved and the bioparticles are saturated faster, without affecting the reaction. The calculation of the effectiveness factor showed that the rate of reaction is changed in regard to the position at the height of the reactor and depends of the dye diffusion when the concentration is increased. By means of the presented dynamic model it can be predicted the dye and COD removal rate in a UAFB reactor, specifying its characteristics, dye inflow concentration, and residence time.

3.9 Nomenclature

a_{sb}	Specific surface of the particles, cm^2/cm^3
C, C_A	Dye or tracer concentration, mg/L
C_0,C_{A0}	Initial tracer, dye concentration, mg/L
C_m	Average concentration, mg/L
d	Dispersion number
D	Axial dispersion coefficient, cm^2/s
D_e	Effective diffusivity coefficient, cm^2/s
K_m	Mass transfer coefficient, cm/s
k_1	1st order specific reaction rate, h^{-1}
k_2	2nd order specific reaction rate, L/mg·h
L_f	Reactor bed height, cm
Q	Volumetric flow
r_A	Reaction rate, mg/L·h
R_B	Bioparticle radius, cm
R_C	AC core radius, cm
R_i	Reactor internal radius, cm
S	Crosswise area to the flux, cm^2
HRT	Hydraulic residence time
t	Time
t_m, RT_m	Half residence time
V_P	Pore volume
u	Superficial velocity
ρ	Density, g/cm^3
ε	Porosity
δ	Biofilm thickness
ζ	Dimensionless longitude
ξ	Dimensionless particle radius
ω	Dimensionless concentration
τ	Dimensionless time
Fo	Characteristic Fourier number
Bi	Biot number
Φ^2	Wagner number
Φ	Thiele number
βm	Dimensionless parameter
α	Dimensionless parameter
β	Dimensionless parameter

Subindex

b	biofilm
L	fixed bed
p	AC particle
1	For the first order term
2	For the second order term

4. Case III. A method to evaluate the isothermal effectiveness factor for dynamic of oxygen in mycelial pellets in submerged cultures

4.1 Introduction

Although a lot of research has been done into modelling microbial processes, the applicability of these concepts to problems specific for bioreactor design and optimization of process conditions is limited. This is partly due to the tendency to separate the two essential factors of bioreactor modelling, i.e. physical transport processes and microbial kinetics. The deficiencies of these models become especially evident in industrial production processes where O_2 supply is likely to become the limiting factor, e.g. production of gibberellic acid and other organic acids (Takamatsu et al. 1981; Qian et al. 1994; Lu et al. 1995; Hollmann,et al.1995). Important physical transport processes related to O_2 supply have not always received sufficient attention (el-Enshasy et al. 1999; Yamane and Shimizu, 1984; Wang and Stephanopoulos, 1984; Parulekar and Lim 1985; Sharon et al. 1999; Stephanopoulos and Tsiveriotis, 1989). Attempts to obtain optimal trajectories for various control variables in different bioreactors, in particular at different scale of operation, are likely to cause unrealistic conditions far from the expected optimum. Another reason to combine the intrinsic O_2 kinetics for growth and product formation with mass transfer is related to O_2 gradients on production scale. Recent attempts to account for these effects in stirrer bioreactors use different structured models and include the O_2 supply in viscous mycelial fermentation broths (Reuss et al. 1986; Nielsen and Villadsen, 1994; Sunil and Subhash, 1996; Cui et al. 1998; Goosen, 1999; Fan et al.1996). It is important when evaluating the potential of a fermentation process using mycelial pellets to have a mathematical model that can predict the O_2 consumption, fungal growth rates and substrate consumption.

In this study, we present a model that combines transport phenomena and kinetic mechanisms for O_2 consumption and applied it to a process in which gibberellin was produced in a submerged fermentation with mycelia of *Gibberella fujikuroi*. The effectiveness factors found in this work are compared to those reported in literature for other fungi.

4.2 Mathematical modelling

The mathematical model was developed assuming a bioreactor in which the continuous phase was well mixed and experimentally verified. O_2 concentration in the bioreactor was measured so the model only considered O_2 transport in the pellets. The method of volume-averaging was used to simulate bulk transport in a two-phase system (Carbonell and Whitaker, 1984; Ochoa, 1986):

$$\frac{\partial C}{\partial t} + \nabla.(\mathbf{v}C) = \mathbf{D} : \nabla\nabla C + k_o \rho \tag{1}$$

with C the dissolved O_2 concentration (kg-moles O_2 m^{-3}), t time (h), k_o the specific O_2 uptake rate per unit dry mycelial weight (kg-moles O_2 kg^{-1} of dry cell h^{-1}) and ρ the pellet suspension density (kg m^{-3}).

The use of average transport equations imposes constraints on the length and physicochemical parameters of the system (Whitaker, 1991). The following assumptions were made to model the O_2 diffusion and reaction in the pellets:

1. The pellet is an effective system, isotropic with constant thermodynamic properties. Effective diffusivity of O_2 in porous medium can be used and the scalar D_{eff} or the effective diffusivity coefficient of dissolved O_2 in mycelial pellet ($m^2 h^{-1}$) replaces the tensor **D** of eq 1. Effective diffusivity contains the convective effects generated inside the pellets.
   1.03 mm.
2. D_{eff} is only a function of void fraction as proposed by different authors (e.g. Aris, 1975).
3. Mycelial pellets are spherical, and the O_2 transport occurs only in the radial direction.
4. An analogous Michaelis-Menten model was used to describe the O_2 consumption rate:

$$k_o = \frac{(k_o)_{max} C}{K_m + C} \tag{2}$$

where $(k_o)_{max}$ is the maximum specific O_2 uptake rate per unit dry mycelial weight (kg-moles O_2 kg^{-1} of dry cell h^{-1}) and K_m the apparent Michaelis constant for mycelia (kg-moles m^{-3}) and the transport equation (eq 1) can be reduced to :

$$\frac{\partial C}{\partial t} = D_{eff}\left[\frac{1}{r^2}\frac{\partial}{\partial r}\left(r^2\frac{\partial C}{\partial r}\right)\right] - k_o\,\rho \tag{3}$$

with r the radial distance from the centre of the mycelial pellet (m). Boundary conditions for eq 3 were:

$$1 \quad @ \quad r = R \quad -D_{eff}\frac{\partial C}{\partial r} = k_P(C_S - C_L) \quad t>0 \tag{4a}$$

$$2 \quad @ \quad r=0 \quad \frac{\partial C}{\partial r} = 0 \quad t>0 \tag{4b}$$

with R the radius of the mycelial pellet (m), k_P the mass-transfer coefficient for the liquid film around cells or pellets (m h^{-1}), C_L the concentration of dissolved O_2 in bulk of liquid (kg-moles O_2 m^{-3}) and C_S the concentration of dissolved O_2 at the liquid-pellet interface (kg-moles O_2 m^{-3}).
Initial conditions were:

$$t = 0 \quad C = C_o \quad 0 \le r \le R \tag{5}$$

with C_O the initial concentration of dissolved O_2 (kg mole O_2 m^{-3}).
The dissolved O_2 concentration was monitored during the fermentation and is expressed as:

$$C_L = C_o f(t).$$

Using the dimensionless variables:

$$u = \frac{C}{C_o} \quad \xi = \frac{r}{R} \quad \tau = \frac{D_{eff}\,t}{R^2} \tag{6}$$

and introducing them into eqs 3 to 5 gives the dimensionless boundary value problem:

$$\frac{\partial u}{\partial \tau} = \frac{1}{\xi^2}\frac{\partial}{\partial \xi}\left(\xi^2 \frac{\partial u}{\partial \xi}\right) - \phi^2 \frac{u}{\beta + u} \tag{7}$$

with ϕ the Thiele modulus and subjected to the following :

$$\text{at} \quad \xi = 1 \qquad -\frac{\partial u}{\partial \xi} = N_{Sh}(u - u_L) \tag{8a}$$

$$\xi = 0 \qquad \frac{\partial u}{\partial \xi} = 0 \tag{8b}$$

When $\tau = 0 \quad u = 1 \quad 0 \le \xi \le 1$ (9)

Where $\phi^2 = \dfrac{R^2 (k_O)_{max} \rho}{D_{eff}\, Co_{\varepsilon\varphi\varphi}} \qquad N_{Sh} = \dfrac{k_p R}{D_{eff}} \qquad \beta = \dfrac{K_m}{Co}$ (10)

with N_{Sh} the Sherwood number and u_L the dimensionless O_2 concentration when the external mass transfer resistance was not neglected.
From the mass balance in the pellet we obtain:

$$\frac{\partial u}{\partial \tau} = \frac{1}{\xi^2}\frac{\partial}{\partial \xi}\left(\xi^2 \frac{\partial u}{\partial \xi}\right) - \Re \tag{11}$$

with the reaction rate ($\Re$)defined by:

$$\Re = \phi^2 \frac{u}{\beta + u} \tag{12}$$

Whitaker (Whitaker, 1984; 1991) defines the volume-averaging function ($\bar{\Psi}$) as:

$$\bar{\Psi} = \frac{4\pi}{\frac{4}{3}\pi r_p^3}\int_0^{r_p} \Psi\, r^2 dr = 3\int_0^1 \Psi\, \xi^2 d\xi \tag{13}$$

with r_p the radius of one pellet (m) and Ψ. The mean O_2 concentration ($\bar{u}$) in the pellet was:

$$\bar{u} = 3\int_0^1 u\, \xi^2 d\xi \tag{14}$$

Substituting eq 14 in eq 11 gives:

$$\frac{\partial \bar{u}}{\partial \tau} = 3\int_{\xi=0}^{\xi=1} \frac{\partial}{\partial \xi}\left(\xi^2 \frac{\partial u}{\partial \xi}\right) - \bar{\Re}$$

$$\frac{\partial \bar{u}}{\partial \tau} = 3\xi^2 \frac{\partial u}{\partial \xi}\bigg|_{\xi=0}^{\xi=1} - \bar{\Re} = 3\frac{\partial u}{\partial \xi}\bigg|_{\xi=1} - \bar{\Re} = \frac{\partial \bar{u}}{\partial \tau} \tag{15}$$

and then the mean reaction rate ($\bar{\Re}$) is defined by :

$$\bar{\Re} = 3\int_0^1 \Re\,\xi^2\,d\xi \tag{16}$$

The effectiveness factor for O_2 consumption rate per unit of mycelial pellet (η) is defined as:

$$\eta = \frac{4\pi\int_0^1 \Re\xi^2 d\xi}{\frac{4}{3}\pi\Re\big|_{u_L}} = \frac{4\pi\int_0^{r_p} \Re R^2 dr}{\frac{4}{3}\pi R^3\Re\big|_{uL}} \tag{17}$$

And

$$\eta = \frac{1}{\phi^2}\left(\frac{\beta + u_L}{u_L}\right)\bar{\Re} \tag{18}$$

Using eq 15 and eq 18 gives:

$$\eta = \frac{1}{\phi^2}\left(\frac{\beta + u_L}{u_L}\right)\left(3\frac{\partial u}{\partial \xi}\Big|_{\xi=1} - \frac{\partial \bar{u}}{\partial \tau}\right) \tag{19}$$

By solving eq 7 to 15, $\partial u/\partial\xi$ can be determined and used to calculate the effectiveness factor with eq 19. Eq 7 was discretized in radial direction with 13 orthogonal collocation points, using Legendre polynomials (Finlayson, 1980). The set of ordinary differential equations generated was solved with the Runge-Kutta-Fehlberg method with an adaptive control of each step.

Micro-organism. *Gibberella fujikuroi* (Sawada) strain CDBB H-984 conserved in potato glucose slants at 4°C and sub-cultured every two months was used in the experiment (Culture collection of the Department of Biotechnology and Bioengineering, CINVESTAV-IPN, Mexico).

Culture medium. The culture medium contained 100 g of glucose l^{-1}, 3 g of NH_4Cl l^{-1}, 5 g of KH_2PO_4 l^{-1}, 1.5 g of $MgSO_4$ l^{-1}, 2 g of rice flour l^{-1}.

Culture conditions and equipment. The fungus was cultured on potato dextrose agar (PDA) slants at 29°C for seven days. A 1000 ml Erlenmeyer flask containing 500 ml of medium was inoculated with spores and mycelium taken from the slants and incubated on a rotary shaker at 180 rpm and 29°C for 36-38 h. A 30 dm^3 turbine-agitated fermenter (Chemap A.G., Zurich) containing 20 dm^3 of sterilized culture medium (pH 5) was inoculated with 5 % v/v of this culture. The aeration rate was 1 volume air volume^{-1} of medium min^{-1} (vvm), the temperature and agitation-speed were automatically controlled at 29°C and 700 rpm, respectively. pH and dissolved O_2, measured with three polarographic electrodes (Ingold, USA) installed at different depths in the culture medium, were monitored each h for 7 days. Every two h, 60 ml of medium was sampled and analysed for biomass, density (ρ) and diameter of the wet and dry pellet, reductive sugars, NH_4^+-N and gibberellic acid concentrations.

Pellets characterisation. The pellets in the sub-sample of the medium were filtered, washed twice with distilled H_2O and dried to constant weight at 90°C in a vacuum-oven. The fermented broth was centrifuged in conical graduate tubes at 3000 rpm for 20 min and density, volume and weight of the wet pellets were determined while their diameter was measured with a microscope (Leica, MSD) on a calibrated micrometer grid. The pellets were vacuum dried at 90°C for 16 h and their dry weight measured.

Volumetric mass-transfer coefficient (k_La). The gas flow rate was measured with a Brooks Mass controller 5851E while O_2 and CO_2 were monitored at the in and outlet with a paramagnetic O_2 analyser (Sybron 540A) and infrared CO_2 analyser (Sybron, Anatek PSA 402). The volumetric mass transfer coefficient obtained at maximum pellet concentration (k_La) (h^{-1}) was derived from the O_2 mass balance in the bioreactor (Sano et al. 1974).
O_2 uptake rate. Different concentrations of dissolved O_2 in the bioreactor were obtained by changing the compositions of the inlet air while keeping agitation speed and volumetric gas flow rate constant. The rate of O_2 uptake was determined by measuring the O_2 concentrations at the *in* and *outlet* and, as such, kinetics of O_2 were obtained without disturbing the system, i.e. power supply and gas hold-up (Wang and Fewkes, 1977).
Mixing time. The model assumed perfect mixing and two methods were used to verify this. First, the bioreactor with agitation speed of 700 rpm, a temperature of 29°C and an airflow of 1 vvm was filled with 0.1 M NaOH and phenolphthalein as a tracer. Samples were taken every 10 to 15 s at four different depths in the bioreactor (A, B, C, D), and analysed for absorbance at 550 nm (Figure 1). Second, a culture of *G. fujikuroi* in its maximum growth phase to which dextran blue was added as a tracer, was sampled every 10-15 sec at four different depths in the bioreactor and analysed for absorbance at 617.1 nm. Dextran blue was used as it is not affected by pH or by oxide-reduction processes, which take place during fermentation.
The distribution ages were determined by fitting the normalized equation (Levenspiel, 1999):

$$\int_0^\infty A dt = \int_0^\infty \frac{A}{Q} = dt \tag{20}$$

To the dynamics of the tracer with $Q = \int_0^\infty A dt$ the area under the curve of absorbance. A is absorbance of the tracer and t is time. The mixing grade was determined by:

$$m = \left(\frac{A - A_\infty}{A_\infty - A_0} \right) 100 \tag{21}$$

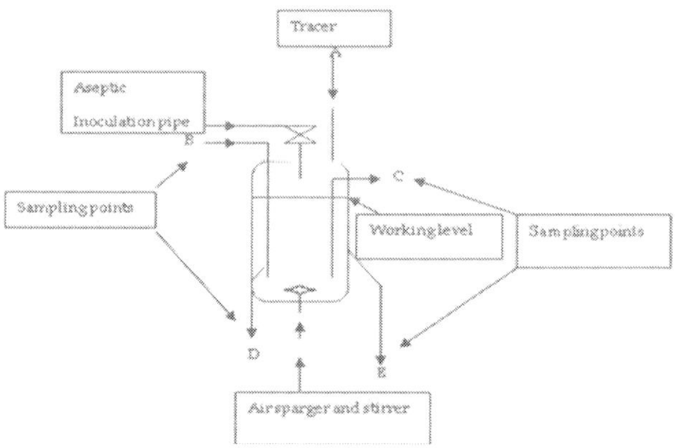

Fig. 1. Diagram of the bioreactor

4.3 Parameter estimation

k_La determinations. The O_2 transfer rate (OTR) was derived from:

$$OTR = \frac{7.32\times10^5}{V_L}\left[\frac{Q_i\,P_i\,y_i}{T_i} - \frac{Q_o\,P_o\,y_o}{T_o}\right] \qquad (22)$$

where 7.32×10^5 is a conversion factor (60 min h^{-1}) [mole (22.4 dm^3)$^{-1}$(standard conditions of Temperature and Pressure)] (273° K atm^{-1}), Q_i and Q_o is the volumetric air flow rate at the air in and outlet (dm^3 min^{-1}), P_i and P_o is the total pressure at the bioreactor air in and outlet (atm absolute), T_i and T_o is the temperature of the gases at the in and outlet (°K), V_L is the volume of the broth contained in the vessel in dm^3, and y_i and y_o is the mole fraction of O_2 at the in and outlet (Wang et al. 1979).

The experimental values of k_La obtained from the *G. fujikuroi* culture were used to determine the volume fraction (θ_p) of the pellet using the empirical equation (Van Suijdam, 1982):

$$\frac{k_La}{(k_La)_0} = 0.5\left[1-\tanh\left(15\theta_p - 7.5\right)\right] \qquad (23)$$

with $(k_La)_o$ the initial volumetric mass transfer coefficient (h^{-1}).

The liquid to pellet mass-transfer coefficient (k_pa_p) was calculated using the Sano, Yamaguchi and Adachi correlation (Sano et al. 1974). This correlation is based on Kolmoghorov's theory of local isotropic turbulence and is independent of the geometry of the equipment or the method energy input used. The Sherwood number N_{Sh} is:

$$N_{Sh} = 2.0 + 0.4\left(N_{Re}\right)^{1/4}\left(N_{Sc}\right)^{1/3} \qquad (24)$$

where N_{Re} is the Reynolds number and N_{Sc} the Schmidt number.

N_{Sh} is given by:

$$N_{Sh} = \frac{k_p d_p}{D_{eff}} \qquad (25)$$

with k_p is defined by eq (4a) and d_p is the diameter of the pellet (m). N_{Re} is defined as:

$$N_{Re} = \frac{\overline{\epsilon}\, d_p^4}{\nu^3} \qquad (26)$$

where $\overline{\epsilon}$ is the mean of local energy dissipation per unit mass of suspension (W kg^{-1}) and ν is the kinematics viscosity of the suspending medium (9.18×10^{-6} m^2 s^{-1}). N_{Sc} is equal to νD_L^{-1} and approximately 3991 with D_L the molecular diffusion coefficient of dissolved O_2 in H_2O (m^2 h^{-1}).

$\overline{\epsilon}$ in the impeller jet stream can be given as a function of the distance from the impeller shaft (r_{is}), the stirrer speed (N), and the stirrer diameter (D_R) (Van Suijdam and 1981, Metz):

$$\overline{\epsilon} = \frac{0.86 N^3 D_R^6}{r_{is}^4} \qquad (27)$$

$\overline{\epsilon}$ obtained was 140 W kg^{-1}; acceptable for inter-medium viscosity in the region of the impeller as the mycelial pellet suspensions showed Newtonian characteristics. The specific surface area of the these pellets (a_p) was estimated using

$$a_p = \frac{6\theta_p}{d_p} \tag{28}$$

The value for the liquid-solid mass transfer coefficient was estimated using eqs 25 to 30 with

$$\frac{dCs}{dt} = k_P a_P (C_L - C_S) - \overline{k}_o \tag{29}$$

with $\overline{k}_o$ the mean O_2 consumption rate per unit of mycelial pellet (kg-moles of O_2 kg^{-1} of dry cell h^{-1}). Experimental radii, pellet density, maximal O_2 uptake rate and the effective diffusivity coefficient (D_{eff}) were used to calculate the Thiele modulus (eq 10).

O_2 uptake. The O_2 uptake rate was derived from the measured inlet gas flow rate ($\dot{V}_G^\alpha$), volume of the broth contained in the vessel (V_L), and gas compositions at the in and outlet using the gas balance taking into account the differences in inner and outlet gas flow rates:

$$k_o = \frac{\dot{V}_G^\alpha}{V_L}\left[Y_{O_2}^\alpha - Y_{O_2}^\omega\left(\frac{1 - Y_{O_2}^\alpha - Y_{CO_2}^\alpha}{1 - Y_{O_2}^\omega - Y_{CO_2}^\omega}\right)\right] \tag{30}$$

where Y_{O_2} and Y_{CO_2} are the volume fractions of O_2 and carbon dioxide in gas (α = inlet, ω = outlet).

Effective diffusivity estimation. Miura (Miura 1976) assumed that the effective diffusion coefficient is proportional to the void fraction within the pellet

$D_{eff} = D_L\ \varepsilon$ (31)

with D_L being 9×10^{-6} m^2 h^{-1} at 29°C (Perry, 1997). Although eq 31 implies only the rectilinear paths inside the particles, similar results have been obtained with other empirical equations that consider tortuosity (Riley et al.. 1995; Riley et al. 1996) or intra-particle convection (Sharonet al. 1999).

Void fraction (ε) was defined as:

$$\varepsilon = 1 - \frac{\rho_v}{\rho_c} \tag{31}$$

where ρ_c is the density of the dry pellet (kg m^{-3}) and ρ_v is the density of the wet pellet (kg m^{-3}). Both were experimentally determined.

The intrapellet Peclet number (Pe_{in}):

$$Pe_{in} \cong \left(\frac{\chi}{\varepsilon}\right) Pe_{out} \tag{32}$$

was calculated to estimate the contribution of intrapellet convection (Parulekar and Lim,1985). The extra-Peclet number Pe_{out} is defined by:

$$Pe_{out} \cong \left(\frac{N_{Sh}}{0.6245} \right)^3 \tag{33}$$

where the dimensionless number χ is defined as:

$$\chi = \frac{\kappa}{d_p^2} \tag{34}$$

where κ is the hydraulic permeability of the pellet (m^2) and estimated through Johnson's equation (Johnson and Kamm, 1987):

$$\frac{\kappa}{r_p^2} = 0.31(\theta_P)^{-1.17} \tag{35}$$

Numerical method. To fit the experimental oxygen uptake values with the non-linear ξ with parameters ϕ (involving $(k_o)_{max}$) and β (involving K_m), a least square algorithm coupled with the discretization of eq 7 via orthogonal collocation using Legendre polynomials and Runge-Kutta-Fehlberg methods was used (Jiménez-Islas et al. 1999). The set of non-linear equations derived in the minimization process, are solved with the Newton-Raphson method with LU factorisation. The optimization sequence is shown in Figure 2.

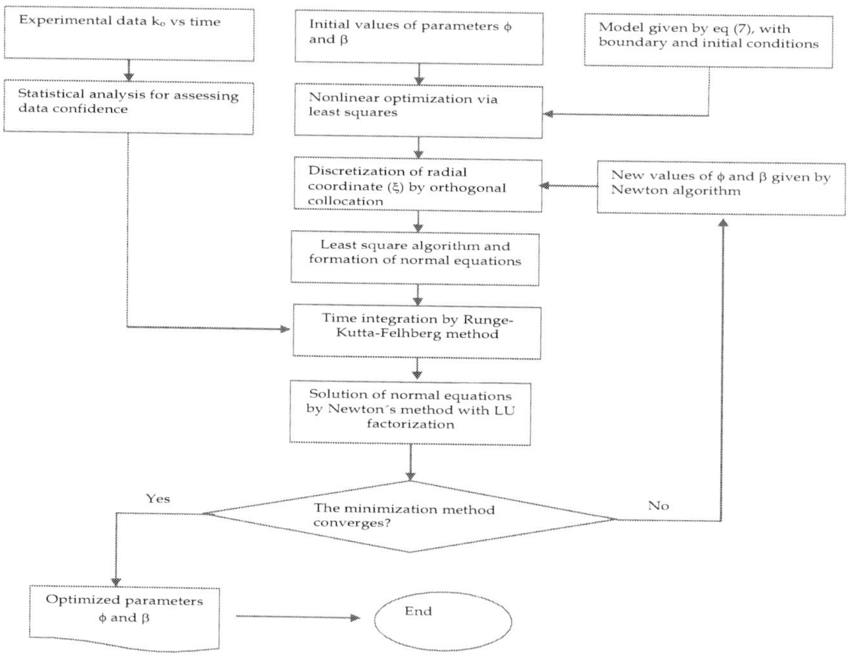

Fig. 2. Flow diagram for the optimization of the parameters ϕ and β (eq. 7).

4.4 Results and discussion

The bioreactor was well mixed (Figure 3). *G. fujikuroi* grew in dispersed mycelia (10%) or in the form of pellets (90%) within 38 h of culturing. The mean size of the pellets increased from 39 to 60 h and remained constant thereafter (Table I). The density of the pellets increased and gave a maximum after 82 h whereupon it decreased. O_2 uptake rates were simulated using eq 7 with a program specifically written for this purpose and the parameters were varied to fit the experimental data (Figure 4). These results included the resistance effects in the Michaelis-Menten equation (eq 3) not optimised before in this way. The estimated values for $(k_o)_{max}$ were $1.80\times10^{-4}\pm 3.05\times10^{-6}$ kg mole kg^{-1} dry cell h^{-1} and for K_m $2.49\times10^{-5} \pm 2.28\times10^{-6}$ kg-moles m^{-3} (Table II). These values are similar to those reported for *Aspergillus niger* (Miura et al.1975) and *Aspergillus orizae* (Kim et al. 1983) but lower than

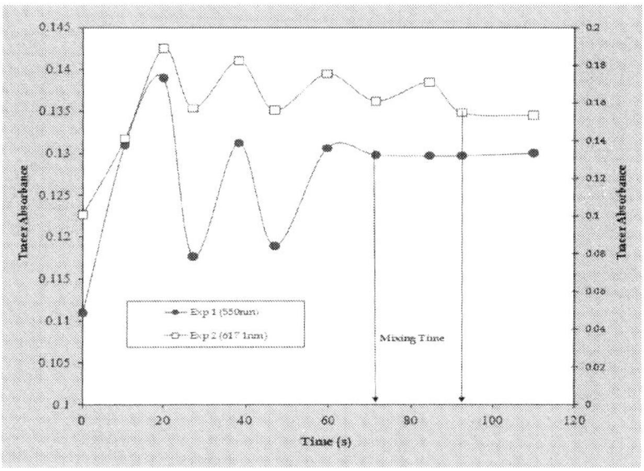

Fig. 3. Tracer absorbance of phenolphthalein measured at 550 nm (•) and dextran blue measured at 617.1 nm (□) used to verify the mixing behaviour in the bioreactor.

Fermentation time (h)	Size of the pellet ($\times10^3$ m)	Pellet density (kg m^{-3})
39	1.50 (0.30)†	16.0 (0.90)
45	1.67 (0.40)	17.5 (0.75)
60	1.90 (0.32)	18.0 (0.50)
66	1.91 (0.28)	18.8 (0.23)
82	2.05 (0.51)	20.7 (0.45)
95	1.92 (0.40)	20.3 (0.30)
108	1.92 (0.30)	19.5 (0.45)
132	1.92 (0.29)	18.4 (0.23)
† values between parenthesis are standard deviations of five replicates		

Table I Size and density of *Gibberella fujikuroi* pellets during fermentation.

those obtained for *Penicillium chrysogenum* (Aiba, S.; Kobayashi,1975; Kobayashi et al. 1973). Differences between simulated and experimental data were less than 6 % and differences can be due to:

1. O_2 transfer rate in the mycelial pellet increases with agitation (Miura and Miyamoto 1977),
2. mycelial density is not uniform (Miura, 1976),
3. respiratory activity is not uniform in radial direction within the pellet (Wittler et al. 1986),
4. and internal convection (Sharon 1999).

The importance of each of these factors has not been assessed separately but they are indistinguishable in a model using D_{eff} and a homogeneous pellet. A summary of experimental and estimated parameters of O_2 diffusion in a bioreactor with *G. fujikuroi* (eq 7 to 34) is given in Table II. D_{eff} was derived from eqs 30 and 31 and is comparable to values reported in literature for other fungi. θ_p values below 30 % did not affect k_La values but they decreased when θ_p values were between 40 % and 60 % (Figure 5). The calculated θ_p value for pellets of *G. fujikuroi* was 39.8 % and allowed calculation of κ (eq 35) and Pe_{in} (eq 32). Pe_{in} for *G. fujikuroi* was 1.38 and κ was 8.22×10^{-7} m^2 (Table II). Stephanopoulos and Tsiveriotis (Sharon et al 1999) stated that the O_2 flow through the pellet does not affect the external mass transfer when Pe_{in} was close to 1 as found in this study. A constant D_{eff} can thus be assumed in our model. O_2 concentration derived from numerical solutions of eq 7 indicated that ϕ = 1 gave an overall reaction rate of O_2 lower than the diffusion rate.

Parameter	value	Dimension	Remarks
d_p	2.090×10^{-3}	m	Experimental data
D_{eff}	4.15×10^{-6}	$m^2\ h^{-1}$	estimated from eq 31
$\overline{\epsilon}$	139.96	$W\ kg^{-1}$	estimated from eq 27
K_m	2.49×10^{-5} (700 rpm)	kg mol O_2 m^{-3}	fitted from experimental data of Figure 4
$(k_o)_{max}$	1.80×10^{-4} (700 rpm)	kg mol O_2 h^{-1} kg^{-1} dry pellet	fitted from experimental data of Figure 4
k_La	91.93	h^{-1}	Experimental data
$(k_La)_o$	188.92	h^{-1}	Experimental data
N_{Re}	2.36×10^{6}	dimensionless	estimated from eq 26
κ	8.22×10^{-7}	m^2	estimated from eq 36
Pe_{out}	5.856	dimensionless	estimated from eq 34
Pe_{in}	1.38	dimensionless	estimated from eq 33
R	0.95×10^{-3}	m	Experimental data
N_{Sh}	250.6	dimensionless	estimated from eq 25
ϕ	1.12 to 2.4	dimensionless	estimated from eq 10
θ_p	0.398	dimensionless	estimated from Figure 5
ν	9.18×10^{-6}	$m^2\ s^{-1}$	Experimental data
ρ	18.65	$kg\ m^{-3}$	Experimental data

Table II Summary of experimental and estimated parameters used for the solution of eqs 7 to 26.

The O_2 concentration did not change substantially in the pellet when ϕ > 5 and the O_2 uptake was limited by diffusion and by mycelial activity. O_2 is then mostly consumed in the external core of the pellet.

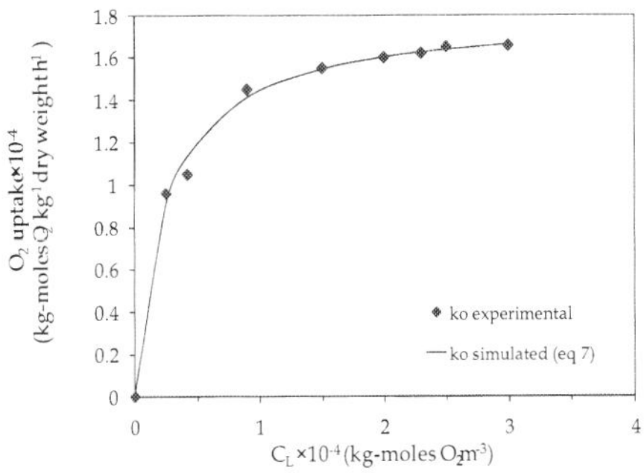

Fig. 4. Measured (♦) and simulated (—) O_2 uptake (kg-moles kg^{-1} dry weight h^{-1}) by *Gibberella fujikuroi* in function of the O_2 dissolved in bulk liquid (kg-moles m^{-3}).

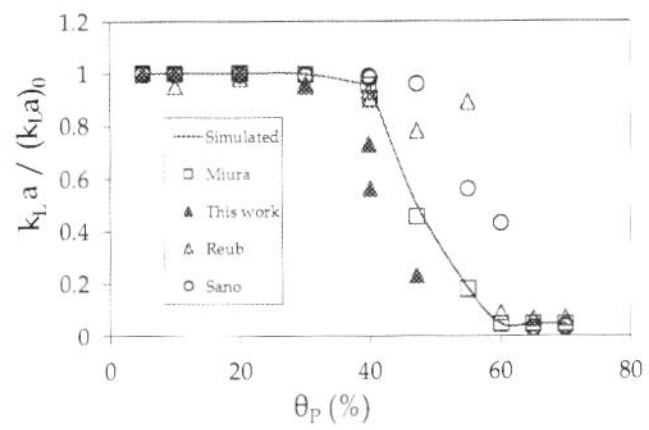

Fig. 5. Simulation of relationship between the dimensionless gas-liquid mass-transfer coefficient k_La $(k_La)_o^{-1}$ and the volume fraction of *Gibberella fujikuroi* pellets.

Experimental values for ϕ in fermentation with *G. fujikuroi* varied between 1.125 to 2.4 (Figure 6). The transport within the pellet depends on both diffusion and kinetics of the O_2 reaction. The mycelial activity in the inner zone of the pellet was reduced by O_2 limitation. Our model predicted that for $\phi<1.875$, η was close to 1 (Figure 7), consistent with other model predictions (Miura, 1976). Under these conditions, the respiratory activity is not limited by O_2 transport. For $\phi>1.875$, η is inversely proportional to ϕ. The estimated ϕ for *G. fujikuroi* indicated a small limitation of O_2 diffusion into the pellet. The large agitation rates and the small size of the pellet formed could explain this.

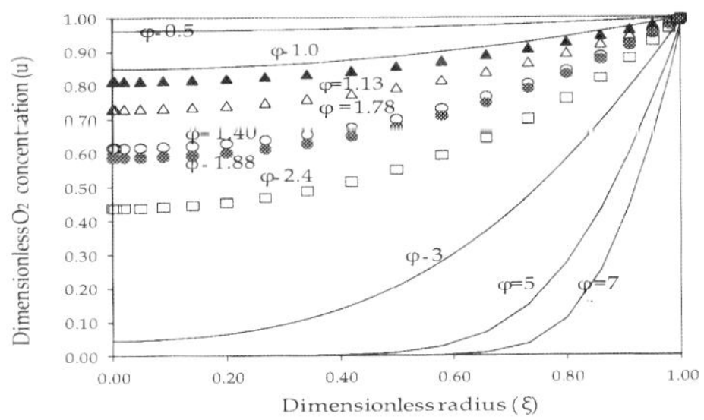

Fig. 6. O_2 concentration (u) in the pellet in function of the dimensionless radius (ξ) and the Thiele modulus (ϕ) with theoretical values of 0.5, 1.0, 3, 5 and 7 and other values calculated from experimental data 1.13 (▲), 1.40 (Δ), 1.78 (o), 1.88 (•), and 2.4 (□).

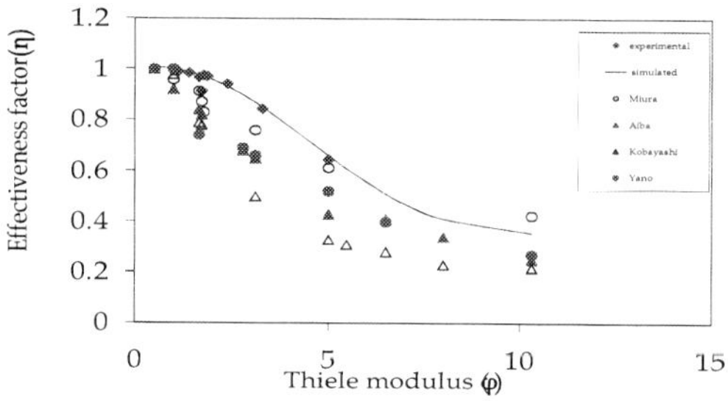

Fig. 7. Effect of Thiele Modulus (ϕ) on the effectiveness factor for mycelial pellets (η) as measured (♦) and simulated (—) in this experiment for *Gibberella fujikuroi* and as reported by Aiba et al. *(30)* (▲), Kobayashi et al. *(31)* (Δ), Miura et al. *(28)* (o) and Yano *(38)* (•).

Data from different authors were recalculated and expressed for ϕ in function of η (Figure 7). The effectiveness model was used to simulate those data and only η obtained with the data reported by Miura (Miura 1976) were comparable with those η values found in this experiment. A possible explanation is that Miura (Miura 1976) used a Michaelis-Menten type kinetic to calculate K_m and $(k_o)_{max}$ while the other authors used a zero and first-order kinetic resulting in values that were unrealistically large. η was not limited by transport for $\tau < 0.2$ (45.7 h) (Figure 8). After that, limitation of O_2 diffusion into the pellet started and a minimum for η was found for τ 0.8 (183.1 h). η remained constant thereafter (Wittler 1986).

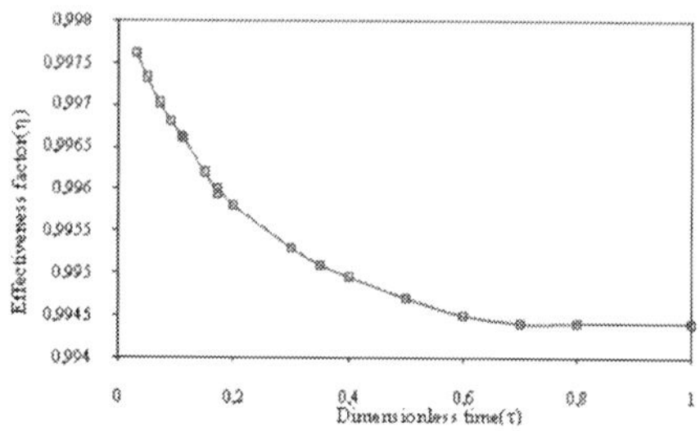

Fig. 8. Typical effectiveness factor through the dimensionless time for a representative experiment (pellet size ≥2 mm, air flow rate 1 vvm, 700 rpm, 29 °C, Thiele modulus 2.8).

4.5 Conclusions

Limitations in models simulating O_2 transfer into mycelial pellets with different strains of fungi have been reported, e.g. Sunil and Subhash, 1996; Miura, 1976; Aiba and Kobayashi, 197; Metz and Kossen, 1977; Chiam and Harris, 1981; Reuss et al. 1982; Nienow 1990. Explanations for these shortcomings can be related to unrealistically large values for D_{eff}, K_m and $(k_o)_{max}$ Experimental data of O_2 diffusion into pellets of *G. fujikuroi* were simulated satisfactorily. The O_2 reaction rate in pellets of 1.7-2.0 mm was only marginally inhibited by diffusion constraints under the conditions tested. Pe_{in} was small enough to justify a constant effective diffusivity and an isotropic pellet system with constant thermodynamic characteristics. O_2 transfer into the mycelial pellet can become the limiting factor in submerged fermentation of fungi when pellets larger than 2 mm are formed in the bioreactor. Eqs 7 and 19 allows to identify conditions critical for fermentations and to derive values for process parameters.

4.6 Nomenclature

a_P = specific surface area of pellets (m^2)
C = concentration of dissolved O_2 (kg-moles O_2 m^{-3})
C_O = initial concentration of dissolved O_2 (kg-moles O_2 m^{-3})
C_L = concentration of dissolved O_2 in bulk of liquid (kg-moles O_2 m^{-3})
C_S = concentration of dissolved O_2 at liquid-pellet interface (kg-moles O_2 m^{-3})
d_P = diameter of the pellet (m)
D_L = molecular diffusion coefficient of dissolved O_2 in H_2O (m^2 h^{-1})
D_{eff} = effective diffusivity coefficient of dissolved O_2 in mycelial pellet (m^2 h^{-1})
D_R = stirrer diameter (m)
k_O = specific O_2 uptake rate per unit dry mycelial weight (kg-moles of O_2 kg^{-1} of dry cell h^{-1})
$\overline{k_O}$ = mean O_2 consumption rate per unit of mycelial pellet (kg-moles of O_2 kg^{-1} of dry cell h^{-1})
$(k_O)_{max}$ = maximum specific O_2 consumption rate per unit dry mycelial weight (kg-moles of O_2 kg^{-1} of dry cell h^{-1})
k_P = mass-transfer coefficient for the liquid film around cells or pellets defined by eq 4a (m h^{-1})
$k_P a_P$ = liquid to pellet mass-transfer coefficient (m^2 h^{-1})
k_La = volumetric mass transfer coefficient obtained at maximum pellet concentration (h^{-1})
$(k_La)_0$ = initial volumetric mass transfer coefficient (h^{-1})
K_m = apparent Michaelis constant for mycelia (kg-moles m^{-3})
N = stirrer speed (rpm)
N_{Re} = Reynolds number
N_{Sc} = Schmidt number
N_{sh} = Sherwood number
Pe_{in} = intraparticle Peclet number $\left(\frac{\chi}{(\varepsilon)}\right)Pe_{out}$
Pe_{out} = extra-Peclet number $\left(\frac{N_{Sh}}{0.6245}\right)^3$
P_i P_o = the total pressure at the bioreactor air in and outlet (atm absolute),
Q_i Q_o = the volumetric air flow rate at the air in and outlet (dm^3 min^{-1})
r = radial distance from centre of mycelial pellet (m)
R = radius of mycelial pellet (m)
r_{is} = radius from the impeller shaft (m)
r_p = radius of one pellet (m)
t = time (h)
T_i T_o = the temperature of the gases at the in and outlet (°K)
u = dimensionless concentration of O_2 defined in eq 6
$\bar{u}$ = dimensionless mean concentration of O_2 defined in eq 14
u_L = dimensionless O_2 concentration when the external mass transfer resistance was not neglected defined in eq 8a
$\dot{V}_G^{\alpha}$ = gas flow rate (m^3 h^{-1})

V_L = volume of broth contained in the vessel (dm^3)
y_i y_o is the mole fraction of O_2 at the in and outlet
$Y^{\alpha}_{O_2}$ $Y^{\omega}_{O_2}$ = are the volume fractions of O_2 in gas (α = inlet, ω = outlet)

$Y^{\alpha}_{CO_2}$ $Y^{\omega}_{CO_2}$ = are the volume fractions of CO_2 in gas (α = inlet, ω = outlet)

Greek

β = constant defined in eq 10 (dimensionless)
ε = void fraction (dimensionless)
$\bar{\epsilon}$ = mean local energy dissipation per unit mass ($W\ kg^{-1}$)
θ_p = volume fraction of pellets (dimensionless)
ξ = ratio of radial distance to radius of the pellet (dimensionless)
κ = effective hydraulic permeability of the pellet (m^2)
η = Effectiveness factor for O_2 consumption rate per unit mycelial pellet (dimensionless)
ρ_c = density of the dried pellet ($kg\ m^{-3}$)
ρ_v = density of wet pellet ($kg\ m^{-3}$)
ρ = pellet suspension density ($kg\ m^{-3}$)
τ = dimensionless time defined in eq 6
ϕ = Thiele modulus (dimensionless)
ν = kinematics viscosity of the suspending medium ($m^2\ s^{-1}$)
χ = dimensionless parameter defined in eq 35
$\bar{\Re}$ = mean reaction rate defined in eq 16
$\Re$ = reaction rate defined by eq 12
Ψ = volume function
$\bar{\Psi}$ = volume averaging function defined in eq 13

umerical values for process parameters.

5. Reference

5.1 References cited in Case I

Abashar, M. E., Narsingh, U., Rouillard, A. E. and Judd, R. (1998). Hydrodynamic flow regimes, gas holdup, and liquid circulation in *airlift* reactors. *Ind. Eng. Chem. Res.* 37: 1251-1259.

Akita, K. and Yoshida, F. (1973). Gas holdup and volumetric mass transfer coefficient in bubble columns. Effects of liquid properties. *Ind. Eng. Chem. Process Des. Develop.* 12: 76-80.

Al-Masry, W. A.. and Dukkan, A. R. (1998). Hydrodynamics and mass transfer studies in a pilot-plant *airlift* reactor: non-Newtonian systems. *Ind. Eng. Chem. Res.* 37: 41-48.

Barboza, M., Zaiat, M. and Hokka, C.O. (2000). General relationship for volumetric oxygen transfer coefficient (k_La) prediction in tower bioreactors utilizing immobilized cells. *Bioprocess Eng.* 22: 181-184.

Barrow, A., Jefferys, E. G. and Nixon, I. S. (1960). Process for the production of gibberellic acid. ICI Patent. GB 838,032.

Brito-De la Fuente, E., Nava, J. A., López, L. M., Medina, L., Ascanio, G. and Tanguy, P. A. (1998). Process viscometry of complex fluids and suspensions with helical ribbon agitators. *Can. J. Chem. Eng.* 76: 689-695.

Brückner, B. and Blechschmidt, D. (1991). The Gibberellin Fermentation. *Crit. Rev. Biotech.* 11, 163-192.

Chavez Parga, M. C. (2005). "Producción de ácido giberélico en un biorreactor *airlift*". Ph. D. Thesis. Instituto Tecnológico de Celaya. Celaya, Gto., México.

Chisti, M. Y. (1989). *Airlift bioreactor.* London-New York: Elsevier Appl. Science.

Choi, K. H., Chisti, Y. and Moo-Young, M. (1996). Comparative evaluation of hydrodynamic and gas-liquid mass transfer characteristics in bubble column and airlift slurry reactors. *Biochem. Eng. J.* 62:223-229.

Escamilla-Silva, E. M., Dendooven, L., Magaña, I. P., Parra-Saldivar, R. and De la Torre, M. (2000). Optimization of Gibberellic acid production by immobilized *Gibberella fujikuroi* mycelium in fluidized bioreactors. *J. Biotechnol.* 76:147-155.

Freitas, C. and Teixeira, J. A. (1998). Hydrodynamic studies in an *airlift* reactor with an enlarged degassing zone. *Bioprocess Eng.*18: 267-279.

Gelmi, C., Pérez-Correa, R., González, M. and Agosin, E. (2000). Solid substrate cultivation of *Gibberella fujikuroi* on an inert support. *Process Biochem.* 35:1227-1233.

Gelmi, C., Pérez-Correa, R. and Agosin, E. (2002). Modelling *Gibberella fujikuroi* growth and GA_3 production in solid-state fermentation. *Process Biochem.* 37:1033-1040.

Godbole, S. P., Schumpe, A., Shah, T. and Carr, N. L. (1984). Hydrodynamics and mass transfer in non-Newtonian solutions in a bubble column. *AIChE J.* 30: 213-220.

Gouveia, E. R., Hokka, C. O. and Badino-Jr, A. C. (2003). The effects of geometry and operational conditions on gas holdup, liquid circulation and mass transfer in an airlift reactor. *Braz. J. Chem. Eng.* 20:363-374.

Gravilescu, M. and Tudose, R. Z. (1998). Hydrodynamics of non-Newtonian liquids in external-loop *airlift* bioreactor. Part I. Study of the gas holdup. *Bioprocess. Eng.* 18:17-26.

Gravilescu, M. and Tudose, R. Z. (1999). Modelling mixing parameters in concentric-tube airlift bioreactors. Part I. Mixing time. *Bioprocess Eng.* 20:423-428.

Halard, B., Kawase, Y. and Moo-Young, M. (1989). Mass transfer in a pilot plant scale *airlift* column with non-Newtonian fluids. *Ind. Eng. Chem. Res.* 28: 243-245.

Heinrich, M. and Rehm, H. J. (1981). Growth of *Fusarium moniliforme* on n-alkanes: comparison of an immobilization method with conventional processes. *Eur. J. Appl. Microbiol. Biotechnol.* 11:239.

Jones, A. and Pharis, R. P. (1987). Production of gibberellins and bikaverin by cells of *Gibberella fujikuroi* immobilized in carrageenan. *J. Ferment. Technol.* 65:717-722.

Kawase, Y. (1989). Liquid circulation in external-loop airlift bioreactors. *Biotechnol. Bioeng.* 35:540-546.

Kumar, P. K. P. and Lonsane, B. K. (1987) Gibberellic acid by solid state fermentation: consistent and improved yields. *Biotechnol. Bioeng.* 30:267-271.

Kumar, P. K. P. and Lonsane, B. K. (1988) Immobilized growing cells of *Gibberella fujikuroi* P-3 for production of gibberellic acid and pigment in batch and semi-continuous cultures. *Appl. Microbiol. Biotechnol.* 28:537-542.

Metz, B., Kossen, N. W. F. and van Suijdam, J. C. (1979). The rheology of mould suspensions. *Adv. Biochem. Eng.* 11:103-156.

McManamey, W. J. and Wase, D. A. J. (1986). Relationship between the volumetric mass transfer coefficient and gas holdup in *airlift* fermentors. *Biotechnol. Bioeng.* 28:1446-1448.

Moo-Young, M., Halard, B., Allen, D. G., Burrell, R. and Kawase, Y. (1987). Oxygen transfer to mycelial fermentation broths in an airlift fermentor. *Biotechnol. Bioeng.* 30:746-753.

Nava Saucedo, J. E., Barbotin, J. N. and Thomas, D. (1989). Continuous production of gibberellic acid in a fixed-bed reactor by immobilized myclia of *Gibberella fujikuroi* in calcium alginate beads. *Appl. Microbiol. Biotechnol.* 30:226-233.

Prokop, A., Janík, P., Sobotka, M. and Krumphanzi, V. (1983). Hydrodynamics, mass transfer, and yeast culture performance of a column bioreactor with ejector. *Biotechnol. Bioeng.* 25: 114-1160.

Quintero, R. R. (1981). *Ingeniería bioquímica, Teoría y aplicaciones.* Ed. Alambra. México.

Schügerl, K., Lücke, J. and Oels, U. (1977). Bubble column bioreactors. *Adv. Biochem. Eng.* 7:1-81.

Shah, Y. T., Kelkar, B. G., Godbole, S. P. and Deckwer, W. D. (1982). Design parameters estimations for bubble column reactors. *AIChE J.* 28:353-379.

Shukla, R., Srivastava, A. K. and Chand, S. (2003). Bioprocess strategies and recovery processes in gibberellic acid fermentation. *Biotechnol. Bioprocess Eng.* 8:269-278.

Tobajas, M. and García-Calvo, E. (2000). Comparison of experimental methods for determination of the volumetric mass transfer coefficient in fermentation processes. *Heat and mass transfer* 36: 201-207.

Tudzynski, B. (1999). Biosynthesis of gibberellins in *Gibberella fujikuroi*: biomolecular aspects. *Appl. Microbiol. Biotechnol.* 52:298-310.

5.2 References cited in Case II

Beydilli M I, Pavlostasthis S G. (2005). Decolorization kinetics of the azo dye Reactive Red 2 under methanogenic conditions: effect of long-term culture acclimation. Biodegradation. 16: 135–146.

Van der Zee F P, Bisschops I A E, Lettinga G. (2003). Activated carbon as an electron acceptor and redox mediator during the anaerobic biotransformation of azo dyes. Environ Sci Technol. 37: 402-408.

Beyenal H, Lewandowski Z. (2002). Internal and external mass transfer in biofilms grown at various flow velocities. Biotechnol Prog. 18: 55-61.

Fan L S, Fujie K, Long T R, Tang W T. (1987). Characteristics of draft tube gas-liquid-solid fluidized-bed bioreactor with immobilized living cells for phenol degradation. Biotechnol Bioeng. 30: 498-504.

Fan L-S, Leyva-Ramos R, Wisecarver K D, Zehner B J. (1990). Diffusion of phenol through a biofilm grown on activated carbon particles in a draft-tube three-phase fluidized-bed biorreactor. Biotechnol Bioeng. 35: 279-286.

Herzberg M, Dosoretz C G, Tarre S, Green M. (2003). Patchy biofilm coverage can explain the potential advantage of BGAC reactors. Environ Sci Technol. 37: 4274-4280.

McCarty P L, Meyer T E. (2005) Numerical model for biological fluidized-bed reactor treatment of perchlorate-contaminated groundwater. Environ Sci Technol. 39: 850-858.

Di Iaconi C, Ramadori R, Lopez A, Pasión R. (2005). Hydraulic shear stress calculation in a sequencing batch biofilm reactor with granular biomass. Environ Sci Techno. 139: 889-894.

Iliuta I, Thyrion F C, Muntean O, Giot M. (1996). Residence time distribution of the liquid in gas-liquid cocurrent upflow fixed-bed reactors. Chem Eng Sci. 51(20): 4579-4593.

Smith L C, Elliot D J, James A. (1996). Mixing in upflow anaerobic filters and its influence on performance and scale-up. Water Res. 30(12): 3061-3073.

Levenspiel O. Chemical reaction engineering (Spanish translation). 3rd Ed. México: Limusa Wiley; 2004. p. 668.

Fogler H S. Elements of chemical reaction engineering, 3rd Ed. New Jersey, USA: Prentice Hall PTR; 1999. p. 967.

Escamilla-Silva E M, Gutierrez G F, Dendooven L, Jimenez-Islas H, Ochoa-Tapia J A. (2001). A Method to Evaluate the Isothermal Effectiveness Factor for Dynamic Oxygen into Mycelial Pellets in Submerged Cultures. Biotechnol Prog.17(1): 95-103.

Kulkarni R R, Wood J, Winterbottom J M, Stitt E H. (2005). Effect of fines and porous catalyst on hydrodynamics of trickle bed reactors. Ind Eng Chem Res; 44(25): 9497-9501.

Hines A L, Maddox R N. Mass transfer, fundamentals and applications (spanish translation). México: Prentice Hall Hispanoamericana; 1987. p. 600.

Lee J-W, Choi S-P, Thiruvenkatachari R, Shim W-G, Moon H. (2006). Evaluation of the performance of adsorption and coagulation processes for the maximum removal of reactive dyes. Dyes and Pigments. 69: 196-203.

Chen K-C, Wu J-Y, Yang W-B, Hwang S-C J. Evaluation of Effective Diffusion Coefficient and Intrinsic Kinetic Parameters on Azo Dye Biodegradation Using PVA-Immobilized Cell Beads. Biotechnol Bioeng 2003; 83(7): 821-832.

Iliuta I, Larachi F. (2005). Modeling simultaneous biological clogging and physical plugging in trickle-bed bioreactors for wastewater treatment. Chem Eng Sci. 60: 1477 - 1489.

Spigno G, Zilli M, Nicolella C. (2004). Mathematical modelling and simulation of phenol degradation in biofilters. Biochem Eng J. 19: 267–275.

Mammarella E J, Rubiolo A C. (2006). Predicting the packed-bed reactor performance with immobilized microbial lactase. Process Biochem. 41: 1627–1636.

Leitão A, Rodriguez A. (1996). Modeling of biodegradation/adsorption combined processes in fixed-bed biofilm reactors: effects of the intraparticle convective flow. Chem Eng Sci. 51(20): 4595 4604.

Leitão A, Rodriguez A. Dynamic behavior of a fixed-bed biofilm reactor: analysis of the role of the intraparticle convective flow under biofilm growth. Biochem Eng J 1998; 2: 1-9.

Lewandowski Z, Stoodley P, Altobelli S, Fukushima E. (1994). Hydrodynamics and kinetics in Biofilm systems - Recent advances and new problems. Wat Sci Tech. 29(10-11): 223-229.

5.3 References cited in Case III

Takamatsu, T.; Shioya, S.; Furuya, T. (1981). Mathematical Model for Gluconic Acid Fermentation by *Aspergillus niger*. *J. Chem. Tech. Biotechnol.* *31*, 697-704.

Qian, X.M.; du Preez, J.C.; Kilian, S. G. (1994). Factors Affecting Gibberellic Acid Production by *Fusarium moniliforme* in Solid-State Cultivation on Starch. *World J. Microbiol. Biotechnol. 10*, 93-99.

Lu, Z.X.; Xie, Z.C.; Kumakura, M. (1995). Production of Gibberellic Acid in *Gibberella fujikuroi* Adhered onto Polymeric Fibrous Carriers. *Process Biochem. 30*, 661-665.

Hollmann, D.; Switalski, J.; Geipel, S.; Onken, U. (1995). Extractive Fermentation of Gibberellic Acid by *Gibberella fujikuroi. J. Ferment. And Bioeng.79*, 594-600.

el-Enshasy, H., Hellmuth, K., Rinas U. (1999). Fungal morphology in submerged cultures and its relation to glucose oxidase excretion by recombinant *Aspergillus niger*. Appl. Biochem. Biotechnol., *81*, 1-11.

Yamane, T.; Shimizu, S. (1984). Fed-batch techniques in microbial processes. In: Fiechter A. Ed.: *Adv. Biochem. Eng./Biotechnol., 30*, 147-194.

Wang, N. S.; Stephanopoulos, G. (1984). Computer Applications for Fermentations Processes. *CRC Critical Reviews in Biotechnology. 2*, 1-103.

Parulekar, S. J.; Lim, H.C. (1985). Modelling Optimization and Control of Semi-Batch Bioreactors. In: Fiechter A.(De.): *Adv. Biochem. Eng./Biotechnol. 32*, 207-258.

Sharon, C., Nakazato, M., Ogawa H.I., Kato Y. (1999). Bioreactor operated production of lipase: castor oil hydrolysis using partially-purified lipase. Indian. J. Exp. Biol. *37*, 481-486.

Stephanopoulos, G.; Tsiveriotis, K. (1989). The Effect of Intraparticle Convection on Nutrient Transport in Porous Biological Pellets. *Chem. Eng. Sci. 44*, 2031-2039.

Reuss, M.; Fröhlich, S.; Kramer, B.; Messerschmidt, K.; Pommerening, G. (1986). Coupling of Microbial Kinetics and oxygen transfer for Analysis and Optimization of Gluconic Acid Production with *Aspergillus niger. 1*,79-91.

Nielsen, J.; Villadsen, J. In *Bioreaction Engineering Principles*. Plenum Press: New York 1994.

Sunil, N.; Subhash, Ch. (1996). Mass Transfer and Biochemical Reaction in Immobilised Cell Packed Bed Reactors: Correlation of Experimental with Theory. *J. Chem. Tech. Biotechnol. 66*, 286-292.

Cui, Y.Q., van der Lans, R.G., Luyben, K.C. (1998). Effects of dissolved oxygen tension and mechanical forces on fungal morphology in submerged fermentation. Biotechnol. Bioeng. 57, 409-410

Goosen, M.F. (1999). Physico-chemical and mass transfer considerations in micro-encapsulation. Ann. N. Y. Acad. Sci. *875*, 84-104.

Fan, D., Shang, L., Yu, J. (1996). Research on fermentation scale-up based on the OUR obtained from a shake flask. Chin. J. Biotechnol. *12*, 177-184.

Carbonell, R. G.; Whitaker S. Heat and mass transfer in porous media. In *Fundamentals of Transport in Porous Media*; Bear, J.; Corapcioglu, M. Y. Eds.; Martinus Nijhoff: Brussels, 1984; pp 121-198.

Ochoa, J.A.; Strove P.; Whitaker, S. (1986). Diffusion and Reaction in Cellular Media. *Chem. Eng. Sci. 41*, 2999-3013.

Whitaker, S. (1991). Improved Constraints for the Principle of Local Equilibrium. *I & E C Research. 29*, 983-997.

Aris, R. The Mathematical Theory of Diffusion and Reaction in Permeable Catalysts. Vol.1, The theory of the Steady State; Vol.2, Questions of Uniqueness, Stability and Transient Behaviour; Clarendon Press: Oxford, 1975.

Whitaker, S. Flow in Porous Media I: A Theoretical Derivation of Darcy's Law. *Transport in Porous Media*. 1986, *1*, 3-25.

Finlayson, B. A. *Nonlinear Analysis in Chemical Engineering*. McGraw-Hill Book Co.: USA, 1980.

Sano, Y.; Yamaguchi, N.; Adachi, T. (1974). Mass Transfer Coefficients for Suspended Particles in Agitated Vessels and Bubble Columns. *J. Chem. Eng. Jpn*. *7*, 255-261.

Wang, D. I. C.; Fewkes, R. C. Mass Transfer Studies in Fermentation Broths. *J. Dev. Ind. Microbiol*. 1977, *18*, 39-44.

Levenspiel O. *Chemical Reaction Engineering*. Wiley International, Second Edition, 1972.

Wang, D. I. C.; Cooney, C. L.; Demain, A. L.; Dunnill., P.; Humphrey, A. E.; Lilly, M. D. Fermentation and Enzyme Technology. Wiley, N.Y. 1979.

Van Suijdam, J. C.; Hols, H.; Kosen, N. W. F. (1982). Unstructured Model for Growth of Mycelial Pellets in Submerged Cultures. Biotechnol. Bioeng. 24, 177-191.

Van Suijdam, J. C.; Metz, B. (1981). Influence of Engineering Variables upon the Morphology of Filamentous Molds. *Biotechnol. Bioeng*. *23*, 111-148.

Miura, Y. (1976). Transfer of Oxygen and Scale-Up in Submerged Aerobic Fermentation. *Adv. Biochem. Eng*. *4*, 1-40.

Perry, H. R.; Green, Don W.; Maloney, J.O. *Perry's Chemical Engineers Handbook*; Mc Graw-Hill: New York, 1997.

Riley, M. R.; Muzzio F. J.; Buettner H. M.; Reyes, S. C. (1995). A Diffusion in Heterogeneous Media: Applications to Immobilized Cell Systems. *AIChE J*. *41*, 691-700.

Riley, M. R.; Muzzio, F. J.; Buettner, H. M.; Reyes S. C. (1996). A Simple Correlation for Predicting Effective Diffusivities in Immobilized Cell Systems. *Biotechnol. Bioeng*. *49*, 223-227.

Johnson, M. R.; Kamm, C. R.; Ethier, T. P. (1987). Scaling Laws and the Effects of Concentration Polarization on the Permeability of Hyaluronic Acid. PhysicoChem. Hydrody. 9, 427.

Jiménez-Islas, H., López-Isunza, F., Ochoa-Tapia, J.A. (1999). Natural convection in a cylindrical porous cavity with internal heat source: a numerical study with Brinkman-extended Darcy model. *Int. J. Heat Mass Transfer*. *42*, 4185-4195.

Miura, Y.; Miyamoto, K.; Kanamori, T.; Ohira, N. (1975). Oxygen Transfer within Fungal Pellets. *J. Chem. Eng. Jpn*. *8*, 300-304.

Kim. J. H., Lebeault, J. M., Reuss, M. (1983). Comparative Study of Rheological Properties of Mycelial Broth in Filamentous and Pelleted Forms. Eur. J. App. Microbiol. Biotechnol. *18*, 11-23.

Aiba, S.; Kobayashi, K. (1971). Comments on Oxygen Transfer Within a Mold Pellet. *Biotechnol. Bioeng*. *15*, 27-29.

Kobayashi, T.; Van Dedem, G.; Moo-Young, M. (1973) Oxygen Transfer Into Mycelial Pellets. *Biotechnol. Bioeng*. *15*, 27-32.

Miura, Y.; Miyamoto, K. (1977). Oxygen Transfer within Fungal Pellets. *Biotechnol. Bioeng*. *19*, 1407-1409.

Wittler, R.; Baumgartl, H.; Lübbers, D.W.; Schürgerl, K.(1986). Investigations of Oxygen Transfer into *Penicillium chrysogenum* Pellets by Microprobe Measurement. *Biotechnol. Bioeng*. *28*, 1024-1036.

Metz, B.; Kossen, N. W. F. (1977). The Growth of Molds in the Form of the Pellets A Literature Review. *Biotechnol. Bioeng*. *19*, 781-799.

Chiam, H. F.; Harris I. J. (1981). Microelectrode Studies of Oxygen Transfer in Trickling Filter Slimes. *Biotechnol. Bioeng. 23,* 781-792.

Reuss, M.; Bajpai, R. K.; Berke, W. (1982). Effective Oxygen-Consumption Rates in Fermentation Broths with Filamentous Organisms. *J. Chem. Tech. Biotechnol. 32,* 81-91.

Nienow, A. W. 1990. Agitation for Mycelial Fermentations. *Trends Biotechnol. 8,* 224-233.

Yano, T., Kodama, T., Yamada, K. (1961). Fundamental Studies on the Aerobic Fermentation. *Agr. Biol. Chem.* 25(7), 580-584.

Huang, M.Y.; Bungay, H.R. (1973).Microprobe Measurements of Oxygen Concentration in Mycelial Pellets. *Biotechnol. Bioeng. 15,* 1193-1201.

Ngian, K. F.; Lin S. H. (1976). Diffusion Coefficient of Oxygen in Microbial Aggregates. *Biotechnol. Bioeng. 18,* 1623-1627.

Fan, L.S.; Leyva Ramos, R.; Wisecarver, K. D.; Zehener, B. J. (1990). Diffusion of Phenol through a Biofilm Grown on Activated Carbon Particles in a Draft-Tube Three-Phase Fluidized-Bed Bioreactor. *Biotechnol. Bioeng. 39,* 279-286.

Ross, L. W.; Updegraff, D. M. (1971). Kinetics of Diffusion-Coupled Fermentation Processes: Conversion of Cellulose to Protein. *Biotechnol. Bioeng. 13,* 99-111.

Analytical Solutions of Mass Transfer around a Prolate or an Oblate Spheroid Immersed in a Packed Bed

J.M.P.Q. Delgado[1] and M. Vázquez da Silva[2]
[1]*LFC – Building Physics Laboratory, Civil Engineering Department Faculty of Engineering, University of Porto*
[2]*Departamento de Ciências, Instituto Superior de Ciências da Saúde – Norte, CESPU. Rua Central de Gandra, n° 1317, 4585-116 Gandra PRD Portugal*

1. Introduction

There are several situations of practical interest, both in nature and in man made processes, in which fluid flows through a bed of inert particles, packed around a large solid mass, which is soluble or reacts with the flowing fluid.

In order to predict the rate of mass transfer between the solid and the flowing fluid it is necessary to understand in detail the mechanics of the flow and the associated processes of diffusion and convection.

For many years, intense research on diffusion phenomena in porous bodies using the diffusion model has been applied to different materials (Delgado, 2007 and Delgado and Vázquez da Silva, 2009). Fundamental solutions of the diffusion problems for spheres, cylinders, plates and parallelepipeds have been provided by Crank (1992) and Gebhart (1993), for example. However, in many situations the shape of the particles immersed in a fluid or porous media is not perfectly spherical, and may be classified as prolate or oblate spheroids. Numerical and analytical solutions of the diffusion equation for prolate spheroids have been reported by Coutelieris et al. (2004), Lima et al. (2002), Coutelieris et al. (1995), etc., and for oblate spheroids by Carmo and Lima (2008), Coutelieris et al. (1995), etc.

Fluid flow along buried spheroidal surfaces is an important model situation (e.g. Clift et al., 1978) and in the present work it is investigated analytically. The treatment of irregular shapes could only be done by numerical methods; therefore it was decided to take the prolate and the oblate spheroid as a model of non spherical particle and study the problem of diffusion around it, for two limiting cases: pure diffusion regime and high values of Peclet number.

The main objective of this work is to provide an analytical solution to the considered problem, as it can be very useful in situations such as the analytical models of continuous injection of solute at a point source, in a uniform stream, to estimate the distance from the "contaminant source" beyond which the levels of contaminant are expected to fall below some safe limit, etc.

2. Analytical solutions

In many practical situations it is often required to consider operations in which there are physico-chemical interactions between a solid particle and the fluid flowing around it. In the treatment of these operations it is common practice to assume the soluble particle to be spherical, because the treatment of irregular shapes could only be done by numerical methods. Spheroidal particles can be either prolate or oblate, and using a simple transformation, one obtains the results for an oblate spheroid from the prolate results. If we consider the situation of a prolate spheroid of major and minor axes c and a, respectively, and an oblate spheroid with the major and minor axes, respectively, a and c, the surface of the spheroid ($\theta = \theta_0$) is described by

$$\frac{x^2 + y^2}{a^2} + \frac{z^2}{c^2} = 1 \tag{1}$$

Since $r^2 = x^2 + y^2$, Eq. (1) can be written as

$$z = c\sqrt{1 - (r/a)^2} \tag{2}$$

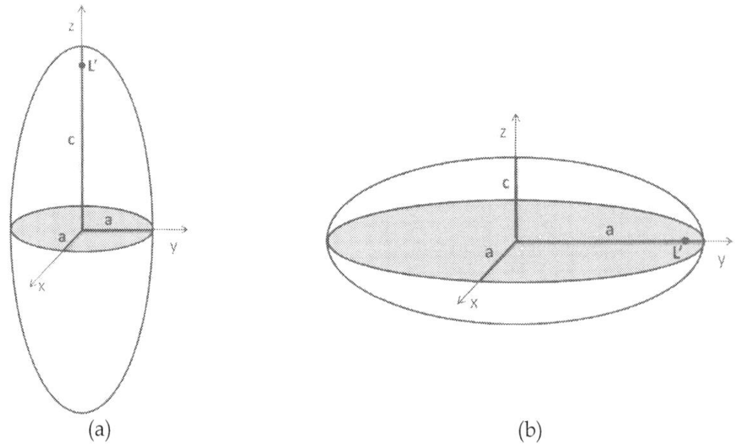

Fig. 1. The coordinates system of: (a)-prolate spheroid and (b)-oblate spheroid.

The surface area S and volume V of a prolate/oblate spheroid are given by

$$S = 2\pi a^2 \left(1 + \frac{c/a}{\sqrt{1 - a^2/c^2}} \sin^{-1}\left(\sqrt{1 - a^2/c^2}\right)\right) \quad \text{Prolate} \tag{3a}$$

$$S = 2\pi a^2 \left(1 + \frac{(c/a)^2}{2\sqrt{1 - c^2/a^2}} \ln\left(\frac{1 + \sqrt{1 - c^2/a^2}}{1 - \sqrt{1 - c^2/a^2}}\right)\right) \quad \text{Oblate} \tag{3b}$$

$$V = \frac{4}{3}\pi a^2 c \tag{4}$$

where $e = \sqrt{1 - a^2 / c^2}$ is the eccentricity for a prolate spheroid and $e = \sqrt{1 - c^2 / a^2}$ for an oblate spheroid; where $e = 0$ corresponds to a sphere. Figure 1 describes the prolate/oblate spheroidal coordinate system. The dimensional Cartesian coordinates (x, y, z) are related to the prolate spheroidal ones (θ, η, β) through the equations (see Moon and Spencer, 1971)

$$x = L'\sinh\theta\sin\eta\cos\beta \tag{5a}$$

$$y = L'\sinh\theta\sin\eta\sin\beta \tag{5b}$$

$$z = L'\cosh\theta\cos\eta \tag{5c}$$

and for the case of an oblate spheroid by

$$x = L'\cosh\theta\sin\eta\cos\beta \tag{6a}$$

$$y = L'\cosh\theta\sin\eta\sin\beta \tag{6b}$$

$$z = L'\sinh\theta\cos\eta \tag{6c}$$

where L' is the focal distance ($L' = \sqrt{c^2 - a^2}$, for a prolate and $L' = \sqrt{a^2 - c^2}$, for an oblate spheroid) and the coordinates range are: $0 \le \theta < \infty$, $0 \le \eta \le \pi$ and $0 \le \beta \le 2\pi$.

2.1 Mass transfer around a prolate spheroid

2.1.1 Pure diffusional regime

The spheroid of slightly soluble solid is assumed to be buried in a packed bed, of "infinite extent", the interstices of the bed being filled with a stagnant fluid that is assumed to be free of solute, at a large distance from the spheroid.
In steady state, a mass balance on the solute, without chemical reaction, leads to

$$\frac{\partial}{\partial\theta}\left(\sinh\theta\frac{\partial C}{\partial\theta}\right) = 0 \tag{7}$$

along coordinate θ. The boundary conditions are

$$C = C^* \quad \theta = \theta_0 \tag{8a}$$

$$C \to C_\infty \quad \theta \to \infty \tag{8b}$$

and the solution is given by

$$\frac{C - C_\infty}{C^* - C_\infty} = \frac{\ln|\tan(\theta/2)|}{\ln|\tanh(\theta/2)|} \tag{9}$$

The mass transfer rate is given by the following expression

$$n = \int_0^{\pi}\int_0^{2\pi} - D_m \left(\frac{\partial C}{\partial \theta}\right)_{\theta_0} dS \tag{10}$$

with the elemental area, dS, of the prolate surface ($\theta = \theta_0$) given by

$$dS = \left(L'\sqrt{\sinh^2 \theta_0 + \sin^2 \eta} \right) d\eta \times (L' \sinh \theta_0 \sin \eta) d\beta \tag{11}$$

From integration of Eq. (10), the total mass transfer rate from the active prolate spheroid is obtained as

$$n = kS(C^* - C_\infty) = \frac{4\pi L' D_m}{\ln\left|\tanh(\theta_0 / 2)\right|}(C^* - C_\infty) \tag{12}$$

After rearranging Eq. (11), with S given by Eq. (3) and using the useful mathematical relations $\sinh \theta_0 = a / L'$ and $\cosh \theta_0 = c / L' = 1/e$, it is possible to obtain the Sherwood number, Sh, for mass transfer by pure diffusion,

$$\mathrm{Sh} = \frac{k2a}{D_m} = \frac{4}{\left[\sqrt{1/e^2 - 1} + 1/e^2 \sin^{-1}(e)\right] \ln\left|\frac{\sqrt{1-e^2}}{1+e}\right|} \tag{13}$$

and using the equivalent sphere diameter (*i.e.* a sphere with the same volume), $d_{eq} = 2(a^2 c)^{1/3}$, the previous expression results in

$$\mathrm{Sh} = \frac{kd_{eq}}{D_m} = \frac{4(1-e^2)^{-1/6}}{\left[\sqrt{1/e^2 - 1} + 1/e^2 \sin^{-1}(e)\right] \ln\left|\frac{\sqrt{1-e^2}}{1+e}\right|} \tag{14}$$

For the special case of $e \approx 0$, the well known result of $\mathrm{Sh} = 2,$, corresponding to pure diffusion around a sphere in an unbounded fluid, is obtained.

2.1.2 High Peclet numbers

The theory is based on the assumption that the inert particles in the bed are packed with uniform voidage, ε , and that the gas flow may be approximated everywhere by Darcy's law, $\mathbf{u} = -K\,\mathbf{grad}\,p$. Furthermore, if the fluid is treated as incompressible, mass conservation leads to $\mathrm{div}\,\mathbf{u} = 0$, Laplace's equation is obtained $\nabla^2 \phi = 0$. This result is well known to hydrologists (see Scheidegger, 1974) and shows that incompressible Darcy flow through a packed bed obeys to the laws of potential flow. Darcy's law is strictly valid only for laminar flow through the packing, but according to Bear (1988) it is still a good approximation for values of the Reynolds number (based on superficial velocity) up to ~ 10, which for beds with $\varepsilon \sim 0.4$ is equivalent to $\mathrm{Re} \sim 25$, the upper limit for the validity of this analysis.

When a solid prolate spheroid is immersed in a packed bed of significantly smaller particles, through which fluid flows with uniform interstitial velocity u_0, far from the spheroid, the solution of Laplace's equation, in terms of spheroidal coordinates (θ, η, β), is (see Alassar and Badr, 1997)

$$\phi = -u_0 L' \cos\eta \left[\cosh\theta - \frac{\cosh\theta \coth^{-1}(\cosh\theta) - 1}{\coth^{-1}(\cosh\theta_0) - \dfrac{\cosh\theta_0}{\sinh^2\theta_0}} \right] \tag{15}$$

and the corresponding stream function is given by

$$\psi = u_0 \frac{L^2}{4} \sinh\theta \left[\sinh\theta - \cos(2\eta)\sinh\theta - (1-\cos(2\eta)) \frac{\sinh\theta \coth^{-1}(\cosh\theta) - \coth\theta}{\coth^{-1}(\cosh\theta_0) - \dfrac{\cosh\theta_0}{\sinh^2\theta_0}} \right] \tag{16}$$

The stream and potential functions are related to the dimensionless velocity components (u_θ, u_η) by the equations (see Batchelor, 1997)

$$u_\theta = \frac{1}{L'\sqrt{\sinh^2\theta + \sin^2\eta}} \frac{\partial\phi}{\partial\theta} = \frac{-1}{L'^2\sqrt{\sinh^2\theta + \sin^2\eta}\, \sinh\theta \sin\eta} \frac{\partial\psi}{\partial\eta} \tag{17}$$

$$u_\eta = \frac{1}{L'\sqrt{\sinh^2\theta + \sin^2\eta}} \frac{\partial\phi}{\partial\eta} = \frac{1}{L'^2\sqrt{\sinh^2\theta + \sin^2\eta}\, \sinh\theta \sin\eta} \frac{\partial\psi}{\partial\theta} \tag{18}$$

resulting in the following velocity components

$$u_\theta = \frac{-u_0 \cos\eta}{\sqrt{\sinh^2\theta + \sin^2\eta}} \left[\sinh\theta - \frac{\sinh\theta \coth^{-1}(\cosh\theta) - \coth(\theta)}{\coth^{-1}(\cosh\theta_0) - \dfrac{\cosh\theta_0}{\sinh^2\theta_0}} \right] \tag{19}$$

$$u_\eta = \frac{u_0 \sin\eta}{\sqrt{\sinh^2\theta + \sin^2\eta}} \left[\cosh\theta - \frac{\cosh\theta \coth^{-1}(\cosh\theta) - 1}{\coth^{-1}(\cosh\theta_0) - \dfrac{\cosh\theta_0}{\sinh^2\theta_0}} \right] \tag{20}$$

The tangential velocity at the surface of the prolate spheroid ($\theta = \theta_0$) can be found through

$$u_{\eta 0} = \frac{1}{L'\sqrt{\sinh^2\theta + \sin^2\eta}} \left. \frac{\partial\phi}{\partial\eta} \right|_{\theta=\theta 0} \tag{21}$$

and the resulting expression is

$$u_{\eta 0} = \frac{u_0 \sin\eta}{\left(1/e^2 - 1 + \sin^2\eta\right)^{0.5} \left[1/e - (1/e^2 - 1)\tanh^{-1} e\right]} \tag{22}$$

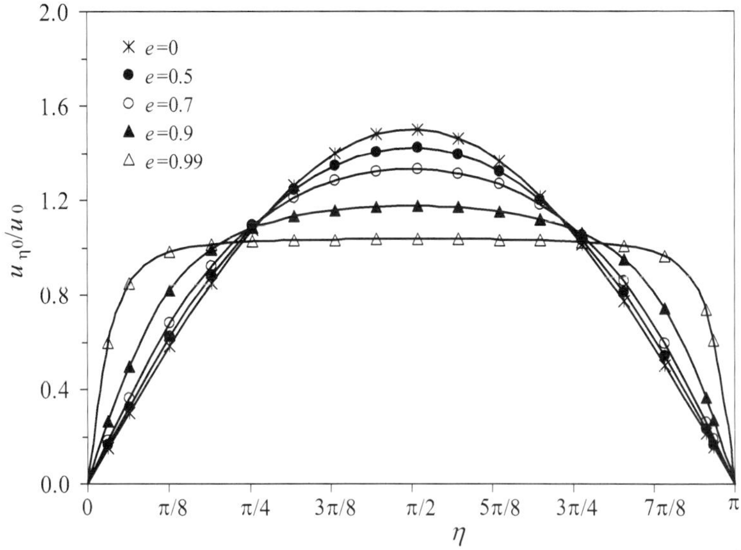

Fig. 2. The dimensionless tangential surface velocity, $u_{\eta 0} / u_0$, of a prolate spheroid as a function of η, for different values of the eccentricity, e.

Figure 2 shows the dimensionless tangential surface velocity, $u_{\eta 0} / u_0$, of a prolate spheroid as a function of η, for different values of the eccentricity, e. Note that, for the case of a sphere, $e \approx 0$, the well-known result of $u_{\eta 0} = 1.5 u_0 \sin \eta$ is obtained, for potential flow over the surface of the sphere. On the other hand, for a slender prolate, $e \to 1$, as expected $u_{\eta 0} / u_0 \to 1$.

A convenient way of expressing the differential mass balance on the solute is to take a control volume along a stream tube, between two nearby potential surfaces. The resulting expression, for convection with molecular diffusion, is (see Coelho and Guedes de Carvalho, 1988)

$$\frac{\partial C}{\partial \phi} = \frac{\partial}{\partial \phi}\left(D'_{\mathrm{m}} \frac{\partial C}{\partial \phi} \right) + \frac{\partial}{\partial \psi}\left(D'_{\mathrm{m}}\, \omega^2 \frac{\partial C}{\partial \psi} \right) \tag{23}$$

For high values of the Peclet number the concentration boundary layer will be thin and the first term on the right hand side of Eq. (23) may be neglected (see Guedes de Carvalho *et al.*, 2004). After some algebraic manipulation and a suitable change of variables, it is then possible to obtain

$$\frac{\partial C}{\partial \xi} = \frac{\partial^2 C}{\partial \psi^2} \tag{24}$$

where ξ is defined by

$$\xi = \int_0^{\eta} L'^3 \sinh^2 \theta_0 \sin^2 \eta' (\sinh^2 \theta_0 + \sin^2 \eta')^{0.5}\, u_\eta\, D'_{\mathrm{m}}\, \mathrm{d}\eta' \tag{25}$$

with $\omega = L' \sinh\theta_0 \sin\eta$. The boundary conditions for Eq. (24), in our problem, are

$$C = C_0 \quad \xi = 0 \quad \psi > 0 \tag{26a}$$

$$C = C^* \quad \xi > 0 \quad \psi = 0 \tag{26b}$$

$$C \to C_0 \quad \xi > 0 \quad \psi \to \infty \tag{26c}$$

and the corresponding solution is

$$\frac{C - C_0}{C^* - C_0} = 1 - erf\left(\frac{\psi}{2\sqrt{\xi}}\right) \tag{27}$$

The value of ξ varies over the surface of the spheroid. Now, for potential flow, u_η is given by Eq. (22) over the surface of the spheroid ($\theta = \theta_0$) and the integral in Eq. (25) is

$$\xi = u_0 D'_{\mathrm{m}} \frac{c^3 e^3 (1 - e^2)}{e - (1 - e^2)\tanh^{-1} e}\left(\frac{2}{3} - \cos\eta + \frac{1}{3}\cos^3\eta\right) \tag{28}$$

The flux of solute at any point on the surface of the spheroid is

$$N = -D'_{\mathrm{m}}\,\varepsilon\left(\frac{\partial C}{\partial b}\right) = -\frac{D'_{\mathrm{m}}\,\varepsilon}{L'\sqrt{\sinh^2\theta_0 + \sin^2\eta}}\left(\frac{\partial C}{\partial\theta}\right)_{\theta=\theta 0}$$

$$= -D'_{\mathrm{m}}\,\varepsilon\, u_\eta L' \sinh\theta_0 \sin\eta \left(\frac{\partial C}{\partial\psi}\right)_{\psi=0} \tag{29}$$

and from Eq. (29) it may be shown that $\left(\partial C/\partial\psi\right)_{\psi=0} = \left(-1/\sqrt{(\pi\xi)}\right)\left(C^* - C_0\right)$. The rate of dissolution of the spheroid in the region $0 < \eta < \eta_1$ will then be

$$n(\eta_1) = \int_0^{\eta 1} N 2\pi L'^2 \sqrt{\sinh^2\theta_0 + \sin^2\eta}\,\sinh\theta_0 \sin\eta\, d\eta = \int_0^{\xi} 2\pi\varepsilon\left(C^* - C_0\right)\left(1/\sqrt{\pi z}\right) dz$$
$$= 4\sqrt{\pi}\;\varepsilon\;\xi^{1/2}\left(C^* - C_0\right) \tag{30}$$

with $\xi(\eta_1)$ given by Eq. (28). In particular, the total rate of dissolution of the spheroid, n_T, may be obtained taking $\eta_1 = \pi$. By definition, the average mass transfer coefficient, k, is

$$k = n_T / \left[S\left(C^* - C_0\right)\right] \tag{31}$$

the resulting expression for k (from Eqs. (30) and (31)) is

$$k = \frac{4\varepsilon\sqrt{\pi}}{2\pi a^2\left(1 + \dfrac{\sqrt{1/(1-e^2)}}{e}\sin^{-1}(e)\right)}\left(\frac{4}{3}\frac{c^3(1-e^2)e^3 u_0 D'_{\mathrm{m}}}{e - (1-e^2)\tanh^{-1} e}\right)^{1/2} \tag{32}$$

It is convenient to express the rate of dissolution in terms of the Sherwood number, $\mathrm{Sh}' = k d_{\mathrm{eq}} / D'_{\mathrm{m}}$, with $d_{\mathrm{eq}} = 2(a^2 c)^{1/3}$, and the expression obtained, after some algebraic manipulation is

$$\frac{\mathrm{Sh}'}{\varepsilon} = \sqrt{\frac{4}{\pi}\mathrm{Pe}'}\left(\frac{2}{3}\frac{e^3}{e-(1-e^2)\tanh^{-1} e}\right)^{1/2} \frac{2}{(1-e^2)^{1/3} + \dfrac{(1-e^2)^{-1/6}}{e}\sin^{-1} e} \tag{33}$$

where $\mathrm{Pe}' = u_0 d_{\mathrm{eq}} / D'_{\mathrm{m}}$ is the Peclet number.
For the special case of a sphere, $e \approx 0$, the result of $\mathrm{Sh}'/\varepsilon = \left[4\,\mathrm{Pe}' / \pi\right]^{1/2}$ is obtained, which corresponds to the asymptotic behaviour for thin concentration layer (high values of Peclet number) when dispersion is constant and tend to D'_{m} over the surface of the sphere (see Guedes de Carvalho and Alves, 1999).
For moderate values of Peclet number, Eq. (23), without any simplification, only should be solved with an indispensable numerical analyse.
Nevertheless, it is important to bear in mind the results obtained by Guedes de Carvalho and Alves (1999), for mass transfer around a single sphere buried in a granular bed of inert particles (a limiting case, with $e \approx 0$). In this work, the authors showed that the values of Sh'/ε calculated from the expression obtained by the quadratic mean of the two asymptotes, $\mathrm{Sh}'/\varepsilon = \left[4 + 4\,\mathrm{Pe}' / \pi\right]^{1/2}$, differ at most by 10% from the corresponding numerical solution obtained,

$$\frac{\mathrm{Sh}'}{\varepsilon} = \left[4 + \frac{4}{5}(\mathrm{Pe}')^{2/3} + \frac{4}{\pi}\mathrm{Pe}'\right]^{1/2} \tag{34}$$

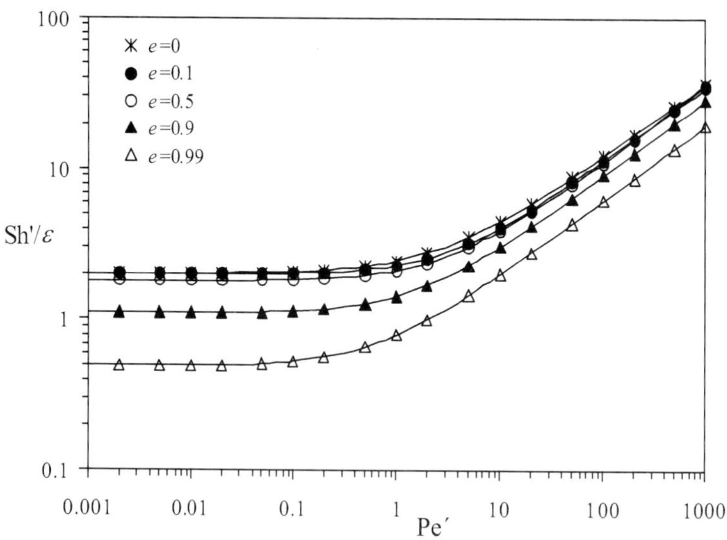

Fig. 3. Dependence of Sh'/ε on Pe', for different values of the eccentricity, e.

In our study, the expression obtained using the quadratic mean of the Sh'/ε values when $\mathrm{Pe}' \to 0$, Eq. (14), and the asymptote for convection with molecular diffusion across a thin boundary layer, Eq. (33), is

$$\frac{\mathrm{Sh}'}{\varepsilon} = \left(\left[\frac{16(1-e^2)^{-1/3}}{\left[\left(\frac{1}{e^2} - 1 \right)^{1/2} + \frac{1}{e^2} \sin^{-1} e \right]^2 \ln \left| \frac{\sqrt{1-e^2}}{1+e} \right|} \right]^2 + \left[\frac{\frac{4}{\pi} \mathrm{Pe}' \frac{8}{3} \frac{e^3}{e-(1-e^2)\tanh^{-1} e}}{\left[(1-e^2)^{1/3} + \frac{(1-e^2)^{-1/6}}{e} \sin^{-1} e \right]^2} \right] \right)^{1/2} \tag{35}$$

and it is expected that Eq. (35) does not differ by more than 10% from the exact solution obtained numerically.

Figure 3 shows the dependence of Sh'/ε on Pe', for different values of the eccentricity, e. Analyzing the Figure, is possible to conclude that the total quantity of material transferred from a prolate spheroid is smaller than that of a soluble sphere (*i.e.* values of Sh'/ε decreasing with eccentricity increase).

2.2 Mass transfer around an oblate spheroid

2.2.1 High peclet numbers

For high values of Peclet numbers, the theory is similar to the case of a prolate spheroid. When a solid oblate spheroid is immersed in a packed bed of significantly smaller particles, through which fluid flows with uniform interstitial velocity u_0, far from the spheroid, the solution of Laplace's equation and the corresponding stream function, in terms of spheroidal coordinates (θ, η, β), are (see Alassar and Badr, 1997)

$$\phi = -u_0 L' \cos\eta \left[\sinh\theta - \frac{\sinh\theta \cot^{-1}(\sinh\theta) - 1}{\cot^{-1}(\sinh\theta_0) - \frac{\sinh\theta_0}{\cosh^2\theta_0}} \right] \tag{36}$$

$$\psi = u_0 \frac{L^2}{4} \cosh\theta \left[\cosh\theta - \cos(2\eta)\cosh\theta - (1-\cos(2\eta)) \frac{\cosh\theta \cot^{-1}(\sinh\theta) - \tanh\theta}{\cot^{-1}(\sinh\theta_0) - \frac{\sinh\theta_0}{\cosh^2\theta_0}} \right] \tag{37}$$

The stream and potential functions are related to the dimensionless velocity components (u_θ, u_η) by

$$u_\theta = \frac{1}{L'\sqrt{\cosh^2\theta - \sin^2\eta}} \frac{\partial\phi}{\partial\theta} = \frac{-1}{L'^2 \sqrt{\cosh^2\theta - \sin^2\eta}\, \cosh\theta \sin\eta} \frac{\partial\psi}{\partial\eta} \tag{38}$$

$$u_\eta = \frac{1}{L'\sqrt{\cosh^2\theta - \sin^2\eta}} \frac{\partial\phi}{\partial\eta} = \frac{1}{L'^2 \sqrt{\cosh^2\theta - \sin^2\eta}\, \cosh\theta \sin\eta} \frac{\partial\psi}{\partial\theta} \tag{39}$$

resulting the following velocity components

$$u_\theta = \frac{-u_0 \cos\eta}{\sqrt{\cosh^2\theta - \sin^2\eta}} \left[\cosh\theta - \frac{\cosh\theta \cot^{-1}(\sinh\theta) - \tanh(\theta)}{\cot^{-1}(\sinh\theta_0) - \dfrac{\sinh\theta_0}{\cosh^2\theta_0}} \right] \tag{40}$$

$$u_\eta = -\frac{u_0 \sin\eta}{\sqrt{\cosh^2\theta - \sin^2\eta}} \left[\sinh\theta - \frac{\sinh\theta \cot^{-1}(\sinh\theta) - 1}{\cot^{-1}(\sinh\theta_0) - \dfrac{\sinh\theta_0}{\cosh^2\theta_0}} \right] \tag{41}$$

The tangential velocity at the surface of the oblate spheroid ($\theta = \theta_0$) can be found through

$$u_{\eta 0} = \frac{1}{L'\sqrt{\cosh^2\theta - \sin^2\eta}} \left.\frac{\partial \phi}{\partial \eta}\right|_{\theta=\theta 0} = \frac{u_0 \sin\eta}{\left(\dfrac{1}{e^2} - \sin^2\eta\right)^{0.5} \left[-\sqrt{\dfrac{1-e^2}{e^2}} + \dfrac{1}{e^2}\cot^{-1}\left(\sqrt{\dfrac{1-e^2}{e^2}}\right)\right]} \tag{42}$$

Figure 4 shows the dimensionless tangential surface velocity, $u_{\eta 0} / u_0$, of an oblate spheroid as a function of η, for different values of the eccentricity, e. Note that, for the case of a sphere, $e \approx 0$, the well-known result of $u_{\eta 0} = 1.5 u_0 \sin\eta$ is obtained, for potential flow over the surface of the sphere. On the other hand, for a slender oblate, $e \to 1$, no simple expression for the velocity profile can be given, since it must be described by a series with many terms.

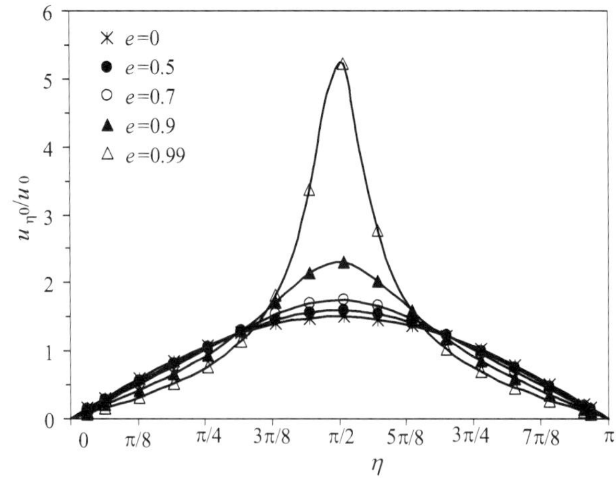

Fig. 4. The dimensionless tangential surface velocity, $u_{\eta 0} / u_0$, as a function of η, for different values of the eccentricity, e.

A convenient way of expressing the differential mass balance on the solute is to take a control volume along a stream tube, between two nearby potential surfaces. The resulting expression, for convection with molecular diffusion, is given by Eq. (23).
For high values of the Peclet number the concentration boundary layer will be thin and the first term on the right hand side of Eq. (23) may be neglected. After some algebraic manipulation and a suitable change of variables, it is then possible to obtain

$$\frac{\partial C}{\partial \xi} = \frac{\partial^2 C}{\partial \psi^2} \tag{43}$$

where ξ is defined by

$$\xi = \int_0^{\eta} L'^3 \cosh^2\theta_0 \sin^2\eta' (\cosh^2\theta_0 - \sin^2\eta')^{1/2} u_\eta D'_m \, d\eta' \tag{44}$$

with $\omega = L'\cosh\theta_0 \sin\eta$. The boundary conditions for Eq. (43), in our problem, are given by Eqs. (26a) to (26c), and the corresponding solution is

$$\frac{C - C_0}{C^* - C_0} = 1 - erf\left(\frac{\psi}{2\sqrt{\xi}}\right) \tag{45}$$

The value of ξ varies over the surface of the spheroid. Now, for potential flow, u_η is given by Eq. (42) over the surface of the spheroid ($\theta = \theta_0$) and the integral in Eq. (44) is

$$\xi = u_0 D'_m \frac{e^3 c^3 / (1-e^2)}{-e + e^3 + \sqrt{1-e^2}\cot^{-1}(\sqrt{1/e^2 - 1})}\left(\frac{2}{3} - \cos\eta + \frac{1}{3}\cos^3\eta\right) \tag{46}$$

The flux of solute at any point on the surface of the spheroid is

$$N = -D'_m \varepsilon \left(\frac{\partial C}{\partial b}\right) = -\frac{D'_m \varepsilon}{L'\sqrt{\cosh^2\theta_0 - \sin^2\eta}} = -D'_m \varepsilon u_\eta L'\cosh\theta_0 \sin\eta \left(\frac{\partial C}{\partial \psi}\right)_{\psi=0} \left(\frac{\partial C}{\partial \theta}\right)_{\theta=\theta 0} \tag{47}$$

and from Eq. (47) it may be shown that $(\partial C/\partial \psi)_{\psi=0} = \left(-1/\sqrt{(\pi\xi)}\right)\left(C^* - C_0\right)$. The rate of dissolution of the spheroid in the region $0 < \eta < \eta_1$ will then be

$$n(\eta_1) = \int_0^{\eta 1} N 2\pi L'^2 \sqrt{\cosh^2\theta_0 - \sin^2\eta} \cosh\theta_0 \sin\eta \, d\eta = 4\sqrt{\pi}\,\varepsilon\,\xi^{1/2}(C^* \; C_0) \tag{48}$$

with $\xi(\eta_1)$ given by Eq. (46). In particular, the total rate of dissolution of the spheroid, n_T, may be obtained taking $\eta_1 = \pi$, and the resulting expression for k is

$$k = \frac{4\varepsilon\sqrt{\pi}}{2\pi a^2\left(1 + \frac{1-e^2}{2e}\ln\left(\frac{1+e}{1-e}\right)\right)}\left(\frac{2}{3}\frac{e^3 u_0 D'_m c^3/(1-e^2)}{-e + e^3 + \sqrt{1-e^2}\cot^{-1}(\sqrt{1/e^2 - 1})}\right)^{1/2} \tag{49}$$

It is convenient to express the rate of dissolution in terms of the Sherwood number, $\mathrm{Sh}' = kd_{\mathrm{eq}} / D'_{\mathrm{m}}$, with $d_{\mathrm{eq}} = 2(a^2c)^{1/3}$, and the expression obtained, after some algebraic manipulation, is

$$\frac{\mathrm{Sh}'}{\varepsilon} = \sqrt{\frac{4}{\pi}\mathrm{Pe}'}\left(\frac{4}{3}\frac{e^3}{-e+e^3+\sqrt{1-e^2}\cot^{-1}\left(\sqrt{1/e^2-1}\right)}\right)^{1/2}\frac{(1-e^2)^{1/3}}{1+\dfrac{1-e^2}{2e}\ln\left(\dfrac{1+e}{1-e}\right)} \tag{50}$$

were $\mathrm{Pe}' = u_0 d_{\mathrm{eq}} / D'_{\mathrm{m}}$ is the Peclet number.

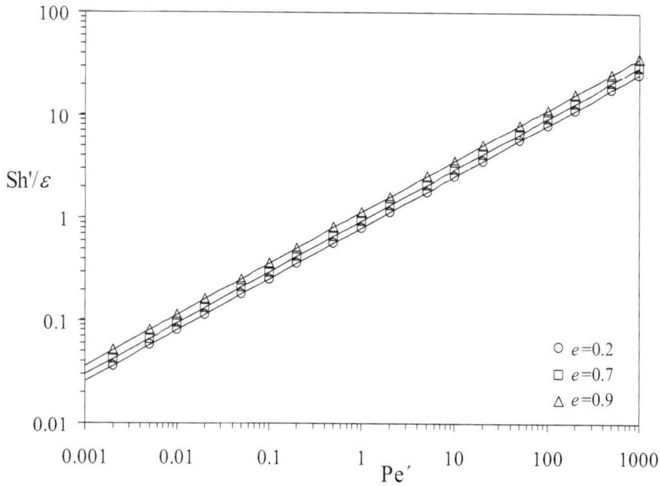

Fig. 5. Dependence of Sh'/ε on Pe', for different values of the eccentricity, e.

For the special case of a sphere, $e \approx 0$, the result of $\mathrm{Sh}'/\varepsilon = [4\,\mathrm{Pe}'/\pi]^{1/2}$ is obtained, which corresponds to the asymptotic behaviour for thin concentration layer (high values of Peclet number) when dispersion is constant and tends to D'_{m} over the surface of the sphere.
Figure 5 shows the dependence of Sh'/ε on Pe', for different values of the eccentricity, e. The total quantity of material transferred from an oblate spheroid is greater than that of a soluble sphere which is explained by the increasing of the Sh'/ε values with eccentricity.

3. Concentration profiles

3.1 Prolate spheroid buried in a packed bed

The analytical solution for a continuous point source has also been derived by Wexler (1992), solving the three-dimensional solute-transport equation from a point source. The solution is given by

$$c = \frac{n_{\mathrm{T}}}{4\varepsilon\pi(x^2+y^2)^{1/2}D'_{\mathrm{m}}}\exp\left[\frac{u_0\left[x-(x^2+y^2)^{1/2}\right]}{2D'_{\mathrm{m}}}\right] \tag{51}$$

with $D_T \cong D_L \cong D'_m$ and a good estimate for n_T is required. The point source is located at the point $(x,y)=(0,0)$. In our case, the problem of mass transfer around a soluble prolate spheroid immersed in a granular bed of inerts through which fluid flow with uniform interstitial velocity, the mass flux rate is expressed as,

$$n_T = \varepsilon D'_m \sqrt{\frac{4}{\pi}\mathrm{Pe'}_{ps}} \left(\frac{2}{3}\frac{e^3}{e-(1-e^2)\tanh^{-1} e} \right)^{1/2} \frac{2\pi a}{(1-e^2)^{1/6}} (C^*-C_0) \tag{52}$$

with Sherwood number given by Eq. (35). Making use of the dimensionless variables, Eq. (51) results, after re-arrangement, in

$$\frac{C-C_0}{C^*-C_0} = \frac{\frac{1}{2}\left[\frac{4}{\pi}\mathrm{Pe'}_{ps}\right]^{1/2} \left(\frac{2}{3}\frac{e^3}{e-(1-e^2)\tanh^{-1} e} \right)^{1/2}}{4\left[(x/d_{eq})^2+(y/d_{eq})^2\right]^{1/2}} \times \\ \times \exp\left\{ \frac{\mathrm{Pe'}_{ps}}{2}\left[x/d_{eq} - \left[(x/d_{eq})^2+(y/d_{eq})^2\right]^{1/2} \right] \right\} \tag{53}$$

3.2 Oblate spheroid buried in a packed bed

If a soluble oblate spheroid, buried in a packed bed, is exposed to uniform fluid flow with uniform interstitial velocity u_0, it will then release solute at a rate n, given by

$$n_T = \varepsilon D'_m \sqrt{\frac{4}{\pi}\mathrm{Pe'}} \left(\frac{4}{3}\frac{e^3}{-e+e^3+\sqrt{1-e^2}\cot^{-1}(\sqrt{1/e^2-1})} \right)^{1/2} \frac{2\pi a^2(1-e^2)^{1/3}}{d_{eq}} (C^*-C_0) \tag{54}$$

with Sherwood number, Sh', given by Eq. (50). Making use of the dimensionless variables, Eq. (51) results, after re-arrangement, in

$$\frac{C-C_0}{C^*-C_0} = \frac{\frac{1}{2}\left[\frac{4}{\pi}\mathrm{Pe'}\right]^{1/2} \left(\frac{4}{3}\frac{e^3}{-e+e^3+\sqrt{1-e^2}\cot^{-1}(\sqrt{1/e^2-1})} \right)^{1/2}}{4\left[(x/d_{eq})^2+(y/d_{eq})^2\right]^{1/2}} \times \\ \times \exp\left\{ \frac{\mathrm{Pe'}}{2}\left[x/d_{eq} - \left[(x/d_{eq})^2+(y/d_{eq})^2\right]^{1/2} \right] \right\} \tag{55}$$

For the special case of a sphere, $e \approx 0$, the result given by Eq. (56) is obtained. It is important to bear in mind that this result was obtained by Guedes de Carvalho *et al.* (2004) for mass transfer around a single sphere buried in a granular bed of inert particles,

$$\frac{C-C_0}{C^*-C_0} = \frac{\left[\frac{4}{\pi}\mathrm{Pe}\right]^{1/2}}{4\left[(x/d_1)^2+(y/d_1)^2\right]^{1/2}} \exp\left\{ \frac{\mathrm{Pe'}}{2}\left[\frac{x}{d_1} - \left[\left(\frac{x}{d_1}\right)^2+\left(\frac{y}{d_1}\right)^2\right]^{1/2} \right] \right\} \tag{56}$$

where $\mathrm{Pe}' = u_0 d_1 / D'_m$ is the Peclet number and d_1 is the diameter of the sphere ($d_1 = d_{eq}$). Figure 6 show the concentration contour plots obtained, taking $\mathrm{Pe}' = 1000$ and $e = 0.5$ as an example, for low values of the dimensionless concentration. As the value of C decreases, the distance of the contour surfaces to the solid soluble particle increases and the solution for the "continuous point source" approach to the "exact" solution, possible to obtain numerically (*i.e.* if a correct value of n_T is used, true coincidence is observed).

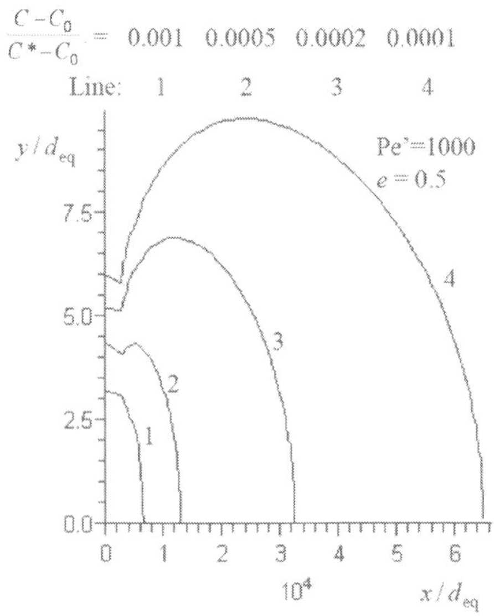

Fig. 6. Dimensionless concentration contour plots obtained with the solution for the "continuous point source", at long distances from the spheroid.

4. Conclusions

The problem of mass transfer around a spheroid buried in a granular bed (be it packed or incipiently fluidised) lends itself to a simple full theoretical analysis, under an appropriate set of conditions. If Darcy flow is considered in the packing, the differential equation describing mass transfer may be obtained analytically considering two asymptotes: one for $\mathrm{Pe} \to 0$ and the other for convection with molecular diffusion across a thin boundary layer, being the results described by Eq. (35), for the case of a prolate spheroid and Eq. (50), for the case of an oblate spheroid.

Results of the analytical solutions were also used to predict the solute migration from an active prolate or oblate spheroid buried in a packed bed of inert particles, through which fluid flows with uniform velocity. The concentration contour surfaces were obtained using an analytical solution of continuous injection of solute at a point source in a uniform stream and the proposed correlations for the mass transfer rate developed.

5. Nomenclature

a, c	Semi-axis of the spheroid
C	Solute concentration
C_0	Bulk concentration of solute
C^*	Saturation concentration of solute
d_{eq}	Equivalent diameter
D_m	Molecular diffusion coefficient
D'_m	Effective molecular diffusion coefficient ($= D_m / \tau$)
e	Eccentricity
K	Permeability in Darcy's law
k	Average mass transfer coefficient
L'	Focus distance
n	Mass transfer rate
n_T	Total mass transfer rate
N	Local flux of solute
p	Pressure
Pe'	Peclet number ($= u_0 d_{eq} / D'_m$)
S	Surface area
Sh'	Sherwood number ($= k d_{eq} / D'_m$)
u	Interstitial velocity (vector)
u_0	Absolute value of interstitial velocity far from the active spheroid
u_η, u_θ	Components of fluid interstitial velocity
V	Volume
x, y, z	Cartesian coordinates

5.1 Greek letters

β	Spheroidal coordinate
ε	Bed voidage
ϕ	Potential function (defined in Eq. (15) and Eq. (36))
θ	Spheroidal coordinate
η	Spheroidal coordinate
τ	Tortuosity
ϖ	Cylindrical radial coordinate, distance to the axis ($= L' \sinh\theta_0 \sin\eta$)
ξ	Variable (defined in Eq. (25) and Eq. (44))
ψ	Stream function (defined in Eq. (16) and Eq. (37))

6. References

Alassar, R.S. & Badr, H.M. (1997). Analytical Solution of Oscillating Inviscid Flow over Oblate Spheroids with Spheres and Flat Disks as Special Cases. *Ocean Engineering*, Vol. 24, No. 3, pp. 217-225.

Batchelor, G.K. (1997). *An Introduction to Fluid Dynamics*, University-Press, Cambridge.

Bear, J. (1988). *Dynamics of Fluids in Porous Media*, Elsevier, New York.

Clift, R.; Grace, J.R. & Weber, M.E. (1978). *Bubbles, Drops and Particles*, Academic Press, New York.

Carmo, J.E.F. & Lima, A.G.B. (2008). Mass Transfer inside Oblate Spheroidal Solids: Modelling and Simulation. *Brazilian Journal of Chemical Engineering*, Vol. 25, No. 1, pp. 19–26.

Coelho, M.A.N. & Guedes de Carvalho, J.R.F. (1988). Transverse dispersion in Granular Beds: Part II - Mass-Transfer from Large Spheres Immersed in Fixed or Fluidized-Beds of Small Inert Particles. *Chemical Engineering Research & Design*, Vol. 66, No. 2, pp. 178-189.

Coutelieris, F.A.; Burganos, V.N. & Payatakes, A.C., (2004). Model of Adsorption-Reaction-Desorption in a Swarm of Spheroidal Particles. *AIChE Journal*, Vol. 50, No. 4, pp. 779-785.

Coutelieris, F.A.; Burganos, V.N. & Payatakes, A.C. (1995). Convective Diffusion and Adsorption in a Swarm of Spheroidal Particles, *AICHE Journal*, Vol. 41, No. 5, pp. 1122-1134.

Crank, J. (1992). *The Mathematics of Diffusion*, Oxford Science Publications, New York.

Delgado, J.M.P.Q. (2007). Mass Transfer around a Spheroid Buried in Granular Beds of Small Inert Particles and Exposed to Fluid Flow. *Chemical Engineering and Technology*, Vol. 30, No. 6, pp. 797–801.

Delgado, J.M.P.Q. & Vázquez da Silva, M. (2009). Mass Transfer and Concentration Contours Between an Oblate Spheroid Buried in Granular Beds and a Flowing Fluid. *Chemical Engineering Research & Design*, Vol. 87, No. 12, pp. 1667-1671.

Gebhart, B. (1993). *Heat Conduction and Mass Diffusion*, McGraw-Hill Inc, New York.

Guedes de Carvalho, J.R.F.; Delgado, J.M.P.Q. & Alves, M.A. (2004). Mass Transfer between Flowing Fluid and Sphere Buried in Packed Bed of Inerts. *AIChE Journal*, Vol. 50, No. 1, pp. 65-74.

Guedes de Carvalho, J.R.F. & Alves, M.A. (1999). Mass Transfer and Dispersion around Active Sphere Buried in a Packed Bed, *AIChE Journal*, Vol. 45, no. 12, pp. 2495-2502.

Lima, A.G.B.; Nebra, S.A. & Queiroz, M.R. (2002). Simultaneous Moisture Transport and Shrinkage during Drying of Solids with Ellipsoidal Configuration. *Chemical Engineering Journal*, Vol. 86, No. 1-2, pp. 85-93.

Moon, P. & Spencer, D.E. (1971). *Field Theory Handbook*, Second edition, Springer, Berlin.

Scheidegger, A.E. (1974). *The Physics of Flow through Porous Media*, Third edition, University of Toronto Press, Toronto.

Wexler, E.J. (1992). *Analytical Solutions for One-, Two-, and Three-Dimensional Solute Transport in Ground-Water Systems with Uniform Flow*, U.S. Geological Survey Techniques of Water-Resources Investigations, Book 3, Chap. B7.